NEUROSCIENCE OF NICOTINE

NEUROSCIENCE OF NICOTINE

MECHANISMS AND TREATMENT

Edited By

VICTOR R. PREEDY
Faculty of Life Science and Medicine
King's College London
London, United Kingdom

ELSEVIER

ACADEMIC PRESS
An imprint of Elsevier

Academic Press is an imprint of Elsevier
125 London Wall, London EC2Y 5AS, United Kingdom
525 B Street, Suite 1650, San Diego, CA 92101, United States
50 Hampshire Street, 5th Floor, Cambridge, MA 02139, United States
The Boulevard, Langford Lane, Kidlington, Oxford OX5 1GB, United Kingdom

Library of Congress Cataloging-in-Publication Data
A catalog record for this book is available from the Library of Congress

British Library Cataloguing-in-Publication Data
A catalogue record for this book is available from the British Library

ISBN 978-0-12-813035-3

For information on all Academic Press publications
visit our website at https://www.elsevier.com/books-and-journals

Publisher: Nikki Levy
Acquisition Editor: Joslyn Chaiprasert
Editorial Project Manager: Timothy Bennett
Production Project Manager: Anusha Sambamoorthy
Cover Designer: Mark Rogers

Typeset by SPi Global, India

Contents

16. Functional Magnetic Resonance Imaging of Acute Nicotine Effects

CHRISTIANE M. THIEL

17. Nicotine Dependence in Schizophrenia: Contributions of Nicotinic Acetylcholine Receptors

ROBERT D. COLE, AND VINAY PARIKH

18. Attentional Bias and Smoking

DAVID J. DROBES, JASON A. OLIVER, JOHN B. CORREA, AND DAVID E. EVANS

19. Effects of Nicotine on Inhibitory Control in Humans

ULRICH ETTINGER, AND VEENA KUMARI

20. Nicotine, Corticotropin-Releasing Factor, and Anxiety-Like Behavior

ADRIAAN W. BRUIJNZEEL

21. 6-Hydroxy-L-Nicotine and Memory Impairment

LUCIAN HRITCU, AND MARIUS MIHASAN

22. Cotinine and Memory: Remembering to Forget

VALENTINA ECHEVERRIA, AND ROSS ZEITLIN

23. Nicotine in Aberrant Learning and Corticostriatal Plasticity

JESSICA L. KORANDA, AND JEFF A. BEELER

Contributors

Jessica L. Ables Department of Psychiatry, The Icahn School of Medicine at Mount Sinai, New York, NY, United States

Yael Abreu-Villaça Department of Physiological Sciences, Institute of Biology, State University of Rio de Janeiro, Rio de Janeiro, Brazil

Armando Alberola-Die Division of Physiology, Department of Physiology, Genetics and Microbiology, Universidad de Alicante (Spain), Alicante, Spain

Tursun Alkam Japanese Drug Organization of Appropriate Use and Research, Nagoya, Japan; Department of Basic Medical Sciences, College of Osteopathic Medicine of the Pacific, Western University of Health Sciences, Pomona, CA, United States

M. Inês G.S. Almeida School of Chemistry, The University of Melbourne, Parkville, Melbourne, VIC, Australia

Goel Ankit Department of Psychiatry, Lady Hardinge Medical College, New Delhi, India

Beatriz Antolin-Fontes Faculty of Medicine, Imperial College London, Hammersmith Hospital Campus, London, United Kingdom

Insa Backhaus Department of Public Health and Infectious Diseases, Sapienza University of Rome, Piazzale Aldo Moro, Rome, Italy

Deniz Bagdas Department of Pharmacology & Toxicology, Virginia Commonwealth University, Richmond, VA, United States

Zsolt Bagosi Department of Pathophysiology, Faculty of Medicine, University of Szeged, Szeged, Hungary

Dzejla Bajrektarevic Department of Neuroscience, SISSA, Trieste, Italy

Luisa Barreiros Department of Chemical Sciences, Faculty of Pharmacy, University of Porto, Porto, Portugal

Elizabeth S. Barrie Institute for Genomic Medicine, Nationwide Children's Hospital, Columbus, OH, United States

Jeff A. Beeler Department of Psychology, Queens College and the Graduate Center, CUNY, Flushing, NY, United States

Ramon O. Bernabeu Departamento de Fisiología e Instituto de Fisiología y Biofísica (IFIBIO-Houssay, UBA-CONICET), Universidad de Buenos Aires (UBA), Paraguay 2155, Buenos Aires, Argentina

Russell W. Brown Department of Biomedical Sciences, East Tennessee State University, James H. Quillen College of Medicine, Johnson City, TN, United States

Adriaan W. Bruijnzeel Department of Psychiatry, University of Florida, Gainesville, FL, United States

Beata Budzyńska Department of Histology and Embryology with Experimental Cytology Unit, Medical University, Lublin, Poland

Barbara Budzyńska Department of Pharmacology and Pharmacodynamics, Medical University of Lublin, Lublin, Poland

S. Caille Aquitaine Institute for Cognitive and Integrative Neuroscience, University of Bordeaux, CNRS UMR5287, Bordeaux, France

Vince D. Calhoun The Mind Research Network, Albuquerque, NM, United States

Yik Lung Chan Respiratory Cellular and Molecular Biology, Woolcock Institute of Medical Research, The University of Sydney, Glebe, NSW, Australia

Gary C.K. Chan Centre for Youth Substance Abuse Research, University of Queensland, Brisbane, QLD, Australia

Adriano José Maia Chaves Filho Department of Physiology and Pharmacology, Drug Research and Development Centre, Faculty of Medicine, Federal University of Ceara, Fortaleza, Brazil

Hui Chen School of Life Sciences, Faculty of Science, University of Technology Sydney, Sydney, NSW, Australia

Chidera C. Chukwueke Department of Pharmacology and Toxicology, University of Toronto, Toronto, ON, Canada

Patrycja Chylińska-Wrzos Department of Histology and Embryology with Experimental Cytology Unit, Medical University, Lublin, Poland

Kelly J. Clemens School of Psychology, University of New South Wales, Sydney, NSW, Australia

Raúl Cobo Division of Physiology, Department of Physiology, Genetics and Microbiology, Universidad de Alicante (Spain), Alicante, Spain

Robert D. Cole Department of Psychology and Neuroscience Program, Temple University, Philadelphia, PA, United States

John B. Correa Tobacco Research and Intervention Program, Department of Health Outcomes and Behavior, Moffitt Cancer Center, Tampa, FL, United States

Silvia Corsini Neuroscience Paris Seine—Institute of Biology Paris Seine, CNRS, UMR 8246—Inserm U1130, Université Pierre et Marie Curie (UPMC), Sorbonne Universités, Paris, France

Fiammetta Cosci Department of Health Sciences, University of Florence, Florence, Italy

Luca Cucullo Department of Pharmaceutical Sciences, TTUHSC, School of Pharmacy, Amarillo, TX, United States

Antonio Gomes de Castro-Neto Study Group on Alcohol and other Drugs, Research Group on Biomedical Nanotechnology, Department of Pharmaceutical Sciences, Federal University of Pernambuco, Cidade Universitária, Recife, Pernambuco, Brazil

David Freitas de Lucena Department of Physiology and Pharmacology, Drug Research and Development Centre, Faculty of Medicine, Federal University of Ceara, Fortaleza, Brazil

Pollyanna Fausta Pimentel de Medeiros Study Group on Alcohol and other Drugs, Research Group on Biomedical Nanotechnology, Department of Pharmaceutical Sciences, Federal University of Pernambuco, Cidade Universitária, Recife, Pernambuco, Brazil

Philip DeCicca Department of Economics, Ball State University, Muncie, IN; NBER, Cambridge, MA, United States

Armani P. Del Franco Department of Neuroscience, University of Minnesota, Minneapolis, MN, United States

Manuel Delgado-Vélez Department of Biology, University of Puerto Rico and the Clinical Bioreagent Center, Molecular Sciences Research Center, San Juan, Puerto Rico

Kataria Dinesh Department of Psychiatry, Lady Hardinge Medical College, New Delhi, India

David J. Drobes Tobacco Research and Intervention Program, Department of Health Outcomes and Behavior, Moffitt Cancer Center, Tampa, FL, United States

Valentina Echeverria Research & Development Service, Bay Pines VA Healthcare System, Bay Pines, FL, United States; Fac. Cs de la Salud, Universidad San Sebastián, Lientur 1457, Concepción, Chile

Branden Eggan Department of Liberal Arts and Science, Maria College, Albany, NY, United States

Ulrich Ettinger Department of Psychology, University of Bonn, Bonn, Germany

David E. Evans Tobacco Research and Intervention Program, Department of Health Outcomes and Behavior, Moffitt Cancer Center, Tampa, FL, United States

Maria Paula Faillace Departamento de Fisiología e Instituto de Fisiología y Biofísica (IFIBIO-Houssay, UBA-CONICET), Universidad de Buenos Aires (UBA), Paraguay 2155, Buenos Aires, Argentina

Fabrizio Ferretti School of Social Sciences, Department of Communication and Economics (DCE), University of Modena and Reggio Emilia (UNIMORE), Reggio Emilia, Italy

Lorra Garey Department of Psychology, University of Houston, Houston, TX, United States

W. Drew Gill Department of Biomedical Sciences, East Tennessee State University, James H. Quillen College of Medicine, Johnson City, TN, United States

Cassandra D. Gipson Department of Psychology, Arizona State University, Tempe, AZ, United States

Stanley D. Glick Department of Neuroscience and Experimental Therapeutics, Albany Medical College, Albany, New York, United States

Julianna G. Goenaga Department of Psychology, Arizona State University, Tempe, AZ, United States

Patrícia Xavier Lima Gomes Department of Physiology and Pharmacology, Drug Research and Development Centre, Faculty of Medicine, Federal University of Ceara, Fortaleza, Brazil

Britta Hahn University of Maryland School of Medicine, Maryland Psychiatric Research Center, Baltimore, MD, United States

Meghan Harding Family Medicine Physician Candidate, UCONN School of Medicine, Hartford, CT, United States

Brandon J. Henderson Department of Biomedical Sciences, Marshall University, Joan C. Edwards School of Medicine, Huntington, WV, United States

David C. Hodgins Department of Psychology, University of Calgary, Calgary, AB, Canada

Lucian Hritcu Department of Biology, Alexandru Ioan Cuza University of Iasi, Iasi, Romania

Yun Hu Department of Biological Systems Engineering, Virginia Tech, Blacksburg, VA, United States

Ines Ibañez-Tallon Laboratory of Molecular Biology, The Rockefeller University, New York, NY, United States

M. Imad Damaj Department of Pharmacology & Toxicology, Virginia Commonwealth University, Richmond, VA, United States

Isabel Ivorra Division of Physiology, Department of Physiology, Genetics and Microbiology, Universidad de Alicante (Spain), Alicante, Spain

Sari Izenwasser Department of Psychiatry & Behavioral Sciences, University of Miami Miller School of Medicine, Miami, FL, United States

Doris Clark Jackson Brigham Young University, Physiology and Developmental Biology, Provo, UT, United States

Hrvoje Jakovac Department of Physiology, Immunology and Pathophysiology, School of Medicine, University of Rijeka, Rijeka, Croatia

Barbara Jodłowska-Jędrych Department of Histology and Embryology with Experimental Cytology Unit, Medical University, Lublin, Poland

Do-Un Jung Department of Psychiatry, Busan Paik Hospital, Inje University College of Medicine, Busan, Republic of Korea

Adrian B. Kelly School of Psychology and Counselling, Institute for Health and Biomedical Innovation, Queensland University of Technology, Brisbane, QLD, Australia

Shaun Yon-Seng Khoo Center for Studies in Behavioral Neurobiology, Department of Psychology, Concordia University, Montreal, QC, Canada

Sungroul Kim Department of Environmental Health Sciences, SoonChunHyang University, Asan, South Korea

Sung-Jin Kim Department of Psychiatry, Busan Paik Hospital, Inje University College of Medicine, Busan, Republic of Korea

Ari P. Kirshenbaum Department of Psychology, Neuroscience Program, Saint Michael's College, Colchester, VT, United States

Kristi A. Kohlmeier Department of Drug Design and Pharmacology, Faculty of Health and Medical Sciences, University of Copenhagen, Copenhagen, Denmark

Spas D. Kolev School of Chemistry, The University of Melbourne, Parkville, Melbourne, VIC, Australia

Jessica L. Koranda Department of Neurobiology, University of Chicago, Chicago, IL, United States

Veena Kumari Centre for Cognitive Neuroscience, Division of Psychology, College of Health and Life Sciences, Brunel University London, Uxbridge, United Kingdom

Giuseppe La Torre Department of Public Health and Infectious Diseases, Sapienza University of Rome, Piazzale Aldo Moro, Rome, Italy

José A. Lasalde-Dominicci Department of Biology, University of Puerto Rico and the Clinical Bioreagent Center, Molecular Sciences Research Center, San Juan, Puerto Rico

S. Lauren Kyte Department of Pharmacology & Toxicology, Virginia Commonwealth University, Richmond, VA, United States

Bernard Le Foll Translational Addiction Research Laboratory, Centre for Addiction and Mental Health, Toronto, ON, Canada

Sung-Ha Lee Center for Pharmacogenomics, College of Medicine, The Ohio State University, Columbus, OH, United States

Edward D. Levin Department of Psychiatry and Behavioral Sciences, Duke University Medical Center, Durham, NC, United States

Aldo Liccardi Department of Public Health and Infectious Diseases, Sapienza University of Rome, Piazzale Aldo Moro, Rome, Italy

Taylor Liles Department of Pharmaceutical Sciences, TTUHSC, School of Pharmacy, Amarillo, TX, United States

Marta Lis-Sochocka Department of Histology and Embryology with Experimental Cytology Unit, Medical University, Lublin, Poland

Yudan Liu Department of Neuroendocrine Pharmacology, School of Pharmacy, China Medical University, Shenyang, Liaoning, China

Danielle Macedo Department of Physiology and Pharmacology, Drug Research and Development Centre, Faculty of Medicine, Federal University of Ceara, Fortaleza, Brazil

Alex Christian Manhães Department of Physiological Sciences, Institute of Biology, State University of Rio de Janeiro, Rio de Janeiro, Brazil

Alice Mannocci Department of Public Health and Infectious Diseases, Sapienza University of Rome, Piazzale Aldo Moro, Rome, Italy

John J. Maurer Pharmacology Graduate Group, Perelman School of Medicine; Department of Biobehavioral Health Sciences, School of Nursing, University of Pennsylvania, Philadelphia, PA, United States

Sarah McCallum Department of Neuroscience and Experimental Therapeutics, Albany Medical College, Albany, NY, United States

Daniel S. McGrath Department of Psychology, University of Calgary, Calgary, AB, Canada

Gavan P. McNally School of Psychology, University of New South Wales, Sydney, NSW, Australia

Agnieszka Michalak Department of Pharmacology and Pharmacodynamics, Medical University of Lublin, Lublin, Poland

Marius Mihasan Department of Biology, Alexandru Ioan Cuza University of Iasi, Iasi, Romania

Long Chiau Ming Faculty of Pharmacy, Quest International University Perak, Perak, Malaysia; Unit for Medication Outcomes Research and Education, Pharmacy, University of Tasmania, Hobart, Tasmania, Australia

Andrés Morales Division of Physiology, Department of Physiology, Genetics and Microbiology, Universidad de Alicante (Spain), Alicante, Spain

Toshitaka Nabeshima Japanese Drug Organization of Appropriate Use and Research, Nagoya; Advanced Diagnostic System Research Laboratory, Graduate School of Health Sciences, Fujita Health University, Toyoake; Aino University, Ibaraki, Japan

Sergej Nadalin Department of Biology and Medical Genetics, School of Medicine, University of Rijeka, Rijeka, Croatia

Mark D. Namba Department of Psychology, Arizona State University, Tempe, AZ, United States

Erik Nesson Department of Economics, Ball State University, Muncie, IN, United States

Andrea Nistri Department of Neuroscience, SISSA, Trieste, Italy

Brian G. Oliver School of Life Sciences, Faculty of Science, University of Technology Sydney, Sydney; Respiratory Cellular and Molecular Biology, Woolcock Institute of Medical Research, The University of Sydney, Glebe, NSW, Australia

Jason A. Oliver Center for Addiction Science and Technology, Department of Psychiatry and Behavioral Sciences, Duke University School of Medicine, Durham, NC, United States

Oné R. Pagán Department of Biology, West Chester University, West Chester, PA, United States

Vinay Parikh Department of Psychology and Neuroscience Program, Temple University, Philadelphia, PA, United States

Carol A. Pollock Renal Group, Department of Medicine, Kolling Institute, Royal North Shore Hospital, St Leonards, NSW, Australia

Gregory L. Powell Department of Psychology, Arizona State University, Tempe, AZ, United States

Harry Prapavessis Faculty of Health Sciences, School of Kinesiology, The University of Western Ontario, London, ON, Canada

Shikha Prasad Department of Neurology, Northwestern University Feinberg School of Medicine, Chicago, IL, United States

Victor R. Preedy Diabetes and Nutritional Sciences Research Division, Faculty of Life Science and Medicine, King's College London, London, United Kingdom

Kukreti Prerna Department of Psychiatry, Lady Hardinge Medical College, New Delhi, India

Rajkumar Rajendram Department of Medicine, King Abdulaziz Medical City, Ministry of National Guard Health Affairs, Riyadh, Saudi Arabia; Diabetes and Nutritional Sciences Research Division, Faculty of Life Science and Medicine, King's College London, London, United Kingdom

Rossana Carla Rameh-de-Albuquerque Study Group on Alcohol and other Drugs, Federal Institute of Education, Science and Technology of Pernambuco, Cidade Universitária, Recife, Pernambuco, Brazil

Amir H. Rezvani Department of Psychiatry and Behavioral Sciences, Duke University Medical Center, Durham, NC, United States

Anderson Ribeiro-Carvalho Department of Sciences, Faculty of Teacher Training, State University of Rio de Janeiro, São Gonçalo, Brazil

Linda Richter Director of Policy Research and Analysis, Center on Addiction, New York, NY, United States

Emma V. Ritchie Department of Psychology, University of Calgary, Calgary, AB, Canada

Scott Rollo Faculty of Health Sciences, School of Kinesiology, The University of Western Ontario, London, ON, Canada

Sonia Saad Renal Group, Department of Medicine, Kolling Institute, Royal North Shore Hospital, St Leonards, NSW, Australia

Wolfgang Sadee Center for Pharmacogenomics, Department of Cancer Biology and Genetics, College of Medicine, The Ohio State University, Columbus, OH, United States

Lia Lira Olivier Sanders Department of Physiology and Pharmacology, Drug Research and Development Centre, Faculty of Medicine, Federal University of Ceara, Fortaleza, Brazil

Beate Saegesser Santos Study Group on Alcohol and other Drugs, Research Group on Biomedical Nanotechnology, Department of Pharmaceutical Sciences, Federal University of Pernambuco, Cidade Universitária, Recife, Pernambuco, Brazil

Michael A. Sayette Department of Psychology, University of Pittsburgh, Pittsburgh, PA, United States

Kyle Saylor Department of Biological Systems Engineering, Virginia Tech, Blacksburg, VA, United States

Heath D. Schmidt Department of Biobehavioral Health Sciences, School of Nursing; Department of Psychiatry, Perelman School of Medicine, University of Pennsylvania, Philadelphia, PA, United States

Marcela A. Segundo Department of Chemical Sciences, Faculty of Pharmacy, University of Porto, Porto, Portugal

M. Sibel Gurun Department of Pharmacology, Faculty of Medicine, Uludag University, Bursa, Turkey

Ryan M. Smith Division of Pharmaceutics and Translational Therapeutics, Department of Pharmaceutical Sciences and Experimental Therapeutics, College of Pharmacy, University of Iowa, Iowa City, IA, United States

Emily A. Stockings National Drug and Alcohol Research Centre (NDARC), UNSW Sydney, Randwick, NSW, Australia

Tiwari Sucheta Department of Psychiatry, Lady Hardinge Medical College, New Delhi, India

Sterling N. Sudweeks Brigham Young University, Physiology and Developmental Biology, Provo, UT, United States

Wuyou Sui Faculty of Health Sciences, School of Kinesiology, The University of Western Ontario, London, ON, Canada

Chee Fai Sui Department of Pharmacy, Hospital Tapah, Perak, Malaysia

Taraneh Taghavi Departments of Pharmacology and Toxicology, and Psychiatry, University of Toronto, Toronto, ON, Canada

S. Tannous Aquitaine Institute for Cognitive and Integrative Neuroscience, University of Bordeaux, CNRS UMR5287, Bordeaux, France

Christiane M. Thiel Biological Psychology Lab, Department of Psychology, Department for Medicine and Health Sciences, Carl von Ossietzky University Oldenburg, Oldenburg, Germany

Rebekah Thomas School of Psychology, University of Queensland, Brisbane, QLD, Australia

Wisam Toma Department of Pharmacology & Toxicology, Virginia Commonwealth University, Richmond, VA, United States

Maria Tortora MRC London Institute of Medical Science, Imperial College of London, London, United Kingdom

Rachel F. Tyndale Campbell Family Mental Health Research Institute and Addictions Division; Department of Psychiatry, Campbell Family Mental Health Research Institute and Addictions Division, University of Toronto, Centre for Addiction and Mental Health, Toronto, ON, Canada

Roberta Uchôa Department of Social Work, Federal University of Pernambuco, Economistas Avenue, Cidade Universitária, Recife, Pernambuco, Brazil

Victor M. Vergara The Mind Research Network, Albuquerque, NM, United States

Ewelina Wawryk-Gawda Department of Histology and Embryology with Experimental Cytology Unit, Medical University, Lublin, Poland

Stephen J. Wilson Department of Psychology, The Pennsylvania State University, University Park, PA, United States

Paul Zammit Department of Geriatrics, Karen Grech Hospital, Pieta, Malta

Ross Zeitlin Research & Development Service, Bay Pines VA Healthcare System, Bay Pines, FL, United States

Chenming Zhang Department of Biological Systems Engineering, Virginia Tech, Blacksburg, VA, United States

Zongmin Zhao Department of Biological Systems Engineering, Virginia Tech, Blacksburg, VA, United States

Michael J. Zvolensky Department of Psychology, University of Houston, Houston, TX, United States

Preface

Tobacco usage has historical origins in the Americas, and tobacco was cultivated on a commercial scale about 400 years ago. Its usage is now widespread on a worldwide basis. Tobacco use is synonymous with nicotine addiction or smoking. It is estimated that, globally, 1 billion individuals are classified as tobacco users and about 7 million deaths occur annually because of its use. About 40 million adults in the United States use tobacco, and three-quarters of these are classified as daily smokers. About half a million Americans die from the effects of smoking each year.

Nicotine alters a variety of neurobiological processes: from molecular and cellular biology to functional change in the brain. A detailed understanding of nicotine and its wide-ranging effects is necessary to lay the foundations for treatment regimens. However, "quitting" is exceedingly difficult due to the multitude of withdrawal symptoms that smokers experience. Nicotine withdrawal is apparent at the behavioral level, but this too has a molecular and cellular basis. On the other hand, some reports suggest that nicotine may enhance cognition. In fact, the scientific material relating to nicotine is vast, interlinked between different disciplines and found in different resources. Thus, finding the information that leads to greater understanding of the neuroscience of nicotine in a single source has hitherto been difficult. This is addressed in *The Neuroscience of Nicotine: Mechanisms and Treatments*.

The book is divided into seven parts as follows:

(1) Introductory chapters: setting the scene
(2) Neurobiology
(3) Psychology, behavior, craving and withdrawal
(4) Pharmacology, neuroactives, and molecular and cellular biology
(5) Nicotine and other addictions
(6) Biomarkers and screening
(7) Treatments, strategies, and resources

It has been difficult to ascribe particular chapters to different parts of the book as many chapters can be placed into two or more sections. Nevertheless, the navigation of the chapters, areas, and key aspects related to nicotine is aided by the excellent index at the end of the book.

The Neuroscience of Nicotine: Mechanisms and Treatments transcends both the multiple disciplinary and intellectual divides as each chapter has the following:

- **A set of Key Facts**
- **A Mini-Dictionary of Terms**
- **A set of Summary Points**

The Neuroscience of Nicotine: Mechanisms and Treatments is designed for research and teaching purposes. It is suitable for neurologists, health scientists, public health workers, doctors, pharmacologists, and research scientists. The audience also includes federal, state, and local tobacco research and service program directors. It may also be of interest to physicians leading efforts for treatment and prevention of nicotine misuse. It is valuable as a personal reference book and also for academic libraries that cover the domains of neurology, health sciences, or addictions. Contributions are from leading national and international experts including those from world-renowned institutions. It is suitable for undergraduates, postgraduates, lecturers, and academic professors.

Victor R. Preedy
King's College London, London, United Kingdom

Editorial Advisors

Dr. Vinood Patel. PhD, FRSC
Reader, Department of Biomedical Sciences, University of Westminister, London, United Kingdom

Dr. Rajkumar Rajendram AKC BSc (Hons) MBBS (Dist) EDIC FRCP (Edin)
Consultant, Department of Internal Medicine, King Abdulaziz Medical City, Riyadh, Saudi Arabia

Chairman, Medication Utilisation & Process Evaluation Subcommittee, Medication Safety Program, Ministry of the National Guard Health Affairs, King Abdulaziz Medical City, Riyadh, Saudi Arabia

Lecturer, Department of Nutrition, Faculty of Life Sciences and Medicine, King's College, London, United Kingdom

CHAPTER

1

Understanding Tobacco Use in Different Countries

Fabrizio Ferretti

School of Social Sciences, Department of Communication and Economics (DCE), University of Modena and Reggio Emilia (UNIMORE), Reggio Emilia, Italy

Abbreviations

CDC	Centers for Disease Control and Prevention, United States
NIH	National Institutes of Health, United States
OECD	Organisation for Economic Co-operation and Development
WBG	The World Bank Group
WHO	World Health Organization
GBD	Global Burden of Diseases
IARC	International Agency for Research on Cancer

1.1 INTRODUCTION

There is a good variety of tobacco-based products that allow people around the world to consume tobacco in many different forms, according to local preferences and customs. For instance, tobacco products may be combusted (e.g., cigarettes and bidis), heated (e.g., water pipes and hookah), or even taken orally or nasally (e.g., snuff, betel quid, and chewing tobacco). Each population has its own cultural traditions—for example, bidis in India or hookah and snuff in the Middle East and South Asian countries, respectively (Hammond, 2009, p. 3).

However, along with industrialization, urbanization, and globalization, the consumption of manufactured cigarettes has grown sharply and spread across virtually all countries over the last century. As a result, cigarettes nowadays have become the predominant form of tobacco use worldwide, accounting for about 92% of total tobacco product sales globally. This is why in tobacco epidemiology, looking at current public health challenges, the terms "tobacco use" and "cigarette smoking" are often used as synonyms (NIH, 2016, chap. 2).

1.2 BASIC CONCEPTS OF TOBACCO EPIDEMIOLOGY

In each population, the magnitude and pattern of tobacco consumption result from the interplay between various individual and collective influences (Warner & MacKay, 2006). An adapted epidemiological "triangle model" of agent, host, vector, and environment, as depicted in Fig. 1.1, provides a useful framework to describe and conceptualize these complex relationships (Penn State, 2016; Slade, 1993, chap. 1). Tobacco—in the form, for instance, of cigarette smoking—acts as the agent (i.e., the "necessary" factor that is required for a disease to occur, although it may not inevitably lead to disease). All habitual smokers are hosts who, at least potentially, due to cigarette consumption, may develop one or more tobacco-related diseases (that usually result in disability and death). The tobacco companies play the role of vectors (i.e., anything that transports and disseminates the agent to susceptible individuals) by producing cigarettes and promoting their use within the population to expand the size of the market (i.e., the number of smokers and the number of cigarettes consumed per smoker). Finally, hosts and vectors operate and interact in a social environment determined by the interplay of a wide range of psychological, cultural, legal, and economic factors (Giovino, 2002).

Given the outstanding role of tobacco as a major risk factor for several chronic noncommunicable diseases (such as cardiovascular and respiratory diseases and many types of cancer), measuring the distribution and intensity of tobacco use across the world is crucial for (1) better understanding the determinants of smoking behavior, (2) developing effective public health programs,

https://doi.org/10.1016/B978-0-12-813035-3.00001-0

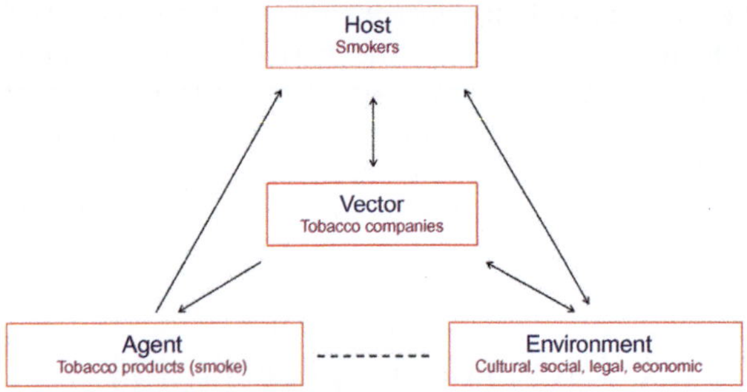

FIG. 1.1 Interactions between agent, hosts, vectors, and environment in tobacco epidemic. *From: Penn State. (2016). Epidemiologic Triad. Department of Statistics. Pennsylvania State University. Available at: https://onlinecourses.science.psu.edu/stat507/node/25. Originally adapted from Egger, G., Swinburn, B., & Rossner, S. (2003). Dusting off the epidemiological triad: could it work with obesity? Obesity Reviews 4(2), 115–119.*

and (3) monitoring countries' progress (GBD, 2015). Overall, to study the pattern of tobacco use and to simulate the impact of tobacco control policies in specific populations, epidemiologists rely today on aggregate (i.e., compartmental) and individual (i.e., agent-based) comprehensive and sophisticated models (CDC, 2014, chap. 15). However, the "bread and butter" of monitoring tobacco use at country level and over time still consists of measuring (or better estimating) two key health policy variables—the prevalence of smoking and the intensity of smoking—by using either a direct and an indirect approach or both (IARC, 2008, chap. 3).

The prevalence of smoking in a given population is usually assessed using a direct approach—that is, by asking a sample of representative subjects their smoking status and behavior (Bonnie, Stratton, & Wallace, 2007). These surveys provide information about the identity of smokers (e.g., age, gender, ethnicity, educational attainment, and income levels) and about each smoker's habits and attitudes. Specifically, according to self-reported information, respondents are classified, as shown in Fig. 1.2, into three main categories: never, current, and former smokers (CDC, 2014, chap. 15). The number of people in each group is a stock variable (i.e., a quantity measured at a given point in time). Monitoring of the population under study over time gives the number of smoking initiations, cessations, and relapses. These are flow variables

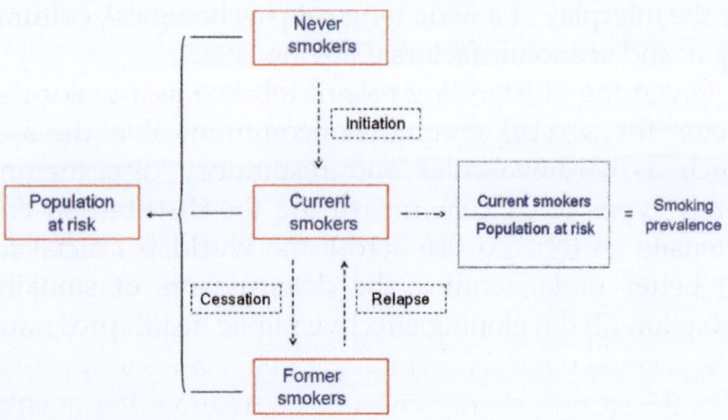

FIG. 1.2 Stock and flow variables in tobacco epidemiology.

(i.e., quantities measured over a period of time) that represent the amount of change in each stock during a given time lapse (for instance, a year).

Data on stock and flow variables are more meaningful if converted into rates by dividing the number of cases in a given category (e.g., the number of smokers) by the corresponding number of people in the population at risk (the sum of never, current, and former smokers), where both the numerator and denominator may refer to the entire sample under study (usually composed of the population aged 15 and over) or to a specific subset, disaggregated by sex, gender, age, and so on (Bonita, Beaglehole, & Kjellstrom, 2006, chap. 2). Advanced statistical techniques are usually applied to gather results from different surveys of a given population to obtain country-level estimates of the prevalence of smoking. Within this framework, the prevalence rate of smoking among the general population provides information on the proportion of current tobacco users in a given country. Finally, these national simple (crude) rates are age-standardized to allow fair comparisons across countries with populations of different age structures (WHO, 2015).

Tobacco use surveys also register self-reported data on the number of cigarettes consumed per day by each smoker. These figures, combined with prevalence data, yield estimates of the total number of cigarettes consumed in the population under study over a given period of time (usually a year). Total cigarette consumption is a basic indicator of the size of the tobacco market in a given economy. Dividing aggregate consumption by the country's number of smokers gives the average number of cigarettes consumed by each smoker. This is a useful country measure of the intensity of smoking, typically expressed as mean daily consumption—that is, the number of cigarettes consumed by the "average smoker" per day (Guindon & Boisclair, 2003). Total and average cigarette consumption, however, are also assessed at country level by using an indirect approach. Indirect estimates are mainly based on national commodity balance-sheet statistics. In fact, the algebraic sum of cigarette production, net exports (i.e., import minus export), and change in inventories provides a measure of apparent tobacco consumption, being the quantity of cigarettes theoretically available for domestic consumption. In developed countries, these estimates are often integrated with data from tobacco tax revenue statistics. The ratio between the total annual apparent consumption and the population aged 15 or older measures the country's cigarette consumption per capita (WHO, 2017).

Overall, both the direct and the indirect approach have strengths and weaknesses. The indirect approach is a relatively easy way to generate basic statistics that give an overview of tobacco consumption in different countries, whereas the direct approach, although more challenging, provides information on the identity (e.g., sex, age, and gender) of smokers and distinguishes between the prevalence and the intensity of smoking (i.e., between the number of smokers and the consumption per smoker) by allowing in-depth epidemiological investigations into the pattern and evolution of tobacco use within and between countries (Lopez, Neil, & Tapani, 1994).

1.3 THE DEMAND FOR TOBACCO PRODUCTS: A BRIEF INTRODUCTION

In a market-oriented economy, cigarettes and other tobacco-related products, like any manufactured product, are typically available to current and potential consumers for a price. Price, however, is only one of the many variables affecting consumers' willingness to buy. Consumers' incomes and tastes, the prices of related goods (complements and substitutes, such as alcohol and electronic cigarettes), and tobacco control policies are among the main factors that influence how many cigarettes (or tobacco-related products) consumers buy in a given market and period of time (Gallus, Schiaffino, LaVecchia, Townsend, & Fernandez, 2006). With so many interrelated factors influencing the demand, some abstraction is needed to understand the determinants of tobacco consumption in different countries. In economics, this problem is mainly approached by first analyzing the effect of price changes on quantity demanded—holding constant all other factors that may influence purchases—and then by examining the impact on consumption of changes in any variables other than price (for instance, disposable income or mandatory health warnings), considered one at a time (Mankiw, 2012, chap. 4).

The result of this "thought experiment" is a fundamental tool in economic analysis: the demand-curve model. Broadly speaking, a demand curve describes the relationship between the quantity of a good that consumers are willing to buy and the good's price, holding constant everything else that influences how much of the good consumers want to buy. This relationship can be written as an equation, such as $Q=D(P; Z)$, where Q and P denote quantity and price, respectively; Z is a vector (i.e., a list of variables) used to account for all factors other than price that affect the demand; and $D(\,)$ is the function expressing how the variables in parentheses determine the quantity demanded (Allen, Doherty, Weigelt, & Mansfield, 2005, chap. 3). The relationship can also be graphically depicted, as in Fig. 1.3A, which shows a simplified demand-curve model for cigarettes.

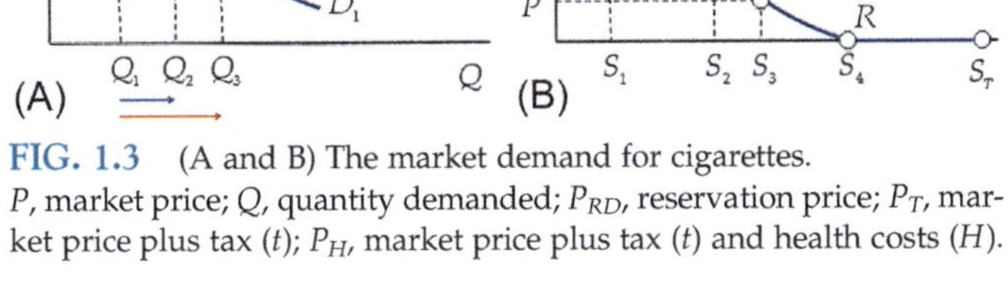

FIG. 1.3 (A and B) The market demand for cigarettes. P, market price; Q, quantity demanded; P_{RD}, reservation price; P_T, market price plus tax (t); P_H, market price plus tax (t) and health costs (H).

Specifically, in Fig. 1.3A, the horizontal axis measures Q, the quantity of cigarettes demanded in the market under study during a specified period, while the vertical axis measures P, the country's average price of cigarettes. The blue curve labeled D_1 shows the quantity demanded at different prices, under the assumption that there is no change in any other factors—listed in Z—that might also influence Q. For instance, if the price is P_1, then the quantity demanded will be equal to Q_1. By dividing this total quantity of cigarettes consumed in the market under study by the corresponding population at risk (i.e., the population aged 15 or older) or by the estimated number of smokers, we obtain the two main country-level indicators of the intensity of smoking described in the previous section—that is, the average cigarette consumption per capita or per smoker, respectively.

1.3.1 Price Elasticity and Changes in Quantity Demanded

Numerous studies have established that the demand for cigarettes is generally downward sloping—that is, it follows the law of demand (Perucic, 2015). In other words, a lower price tends to encourage current smokers—who are already buying cigarettes—to consume larger quantities, and it may also allow potential smokers, who were previously unable to afford cigarettes, to start smoking. This means that, other things being equal, a lower price (for instance, P_2, in Fig. 1.3A) leads to an increase in either the intensity of smoking or the prevalence of smoking or both (consumption increases from Q_1 to Q_2). The shape and position of the demand curve reflect the degree to which changes in cigarette price affect the number of smokers and the number of cigarettes consumed per smoker. To measure the responsiveness of the quantity demanded to changes in price, economists usually compute the price elasticity of demand, being the ratio between the percentage change in quantity demanded and the percentage change in price (Allen et al., 2005, chap. 3).

For example, if the total number of cigarettes demanded, other things being equal, decreases by 6% in response to an increase in the average market price of 8%, then the price elasticity of demand is $(-6\%)/(8\%)=-0.75$. The demand for cigarettes is typically relatively inelastic—that is, a price increase causes a less than proportionate fall in the quantity demanded. According to recent estimates, the average price elasticity of cigarettes ranges from -0.3 to -0.5 so that a price rise of 10% reduces the quantity demanded, on average, by 4% (Chaloupka & Warner, 2000).

1.3.2 Changes in Demand and Income Elasticity

A change in price, other things being equal, leads to a change in the quantity demanded—that is, a movement along a given demand curve (for instance, from point F to point E along the D_1 curve in Fig. 1.3A). Conversely, a change in any factor other than price causes a change in the demand—that is, a shift of the entire demand curve (for instance, a rightward shift in the demand from D_1 to D_2, as shown in Fig. 1.3A). The latter is a change in the market conditions. The new demand curve D_2, depicted in red, implies a greater quantity of cigarettes demanded by consumers at any given price.

The distinction between a movement along the demand curve (due to a change in price) and a shift of the demand curve (due to a change in any other variable, such as income and taste) is also crucial for a better understanding of the impact of tobacco control policies on cigarette consumption. A tax on cigarettes, for example, by raising the market price, discourages cigarette consumption but affects only the quantity demanded, whereas better education about smoking's adverse health effects reduces consumption by structurally decreasing the demand for cigarettes (this is shown in Fig. 1.3A by the leftward shift of the demand curve from D_2 to D_1). A strategy of price reduction by tobacco companies may completely offset the effects of a tax on cigarettes but not those of an education program, because, if successful, the program will lower the quantity demanded at any given price.

Besides taste, regulation, and the price of related goods, populations' smoking habits and behaviors are mainly affected by the level and distribution of consumers' income. As with changes in any factor other than price, changes in disposable income affect consumption by shifting the demand curve. However, an increasing income per capita might shift the demand curve right or left (i.e., from D_1 to D_2 or vice versa, in Fig. 1.3A) and therefore increase or decrease the consumption of cigarettes.

The income elasticity of demand—computed as the ratio of the percentage change in quantity demanded to the percentage change in income—is a measure of how and to what extent the quantity consumed responds to a change in consumers' income, other factors being equal. This coefficient is important in determining differences in cigarette consumption between countries. If the income elasticity is positive, cigarettes are a normal good (i.e., a good for which the quantity demanded rises when income rises). In such a context, economic development tends to increase the number of smokers and the consumption per smoker, exacerbating the incidence of tobacco health-related problems. Otherwise, negative income elasticity indicates that cigarettes are an inferior good (i.e., a good whose quantity demanded falls as income rises). In the majority of developed countries, over the last decades, cigarettes have generally moved from being a normal good (with an income elasticity less than 1—around 0.5) to being an inferior good. In many developing countries, however, manufactured cigarettes, as opposed to the traditional forms of tobacco consumption, still have a positive and large income elasticity (Wilkins, Yurekli, & Hu, 2007, chap. 3).

1.3.3 Using the Demand Model to Understand Smoking Prevalence

This basic demand model can be further developed to take into account other specific features of tobacco consumption (Phillips, 2016). To this aim, in Fig. 1.3B, the x-axis measures the number of actual and potential smokers, S, and the y-axis measures the reservation price, P_{RD} (i.e., the highest price that each consumer will accept to pay). Along the downward-sloping demand curve (the D curve, depicted in blue), consumers are placed in descending order according to their willingness to pay. In other words, the height of each point on the D curve indicates the reservation price of a given consumer. This demand curve for cigarettes intersects the x-axis at point R, beyond which it becomes horizontal. The difference between S_T (the population at risk) and S_4 gives the number of consumers whose reservation price is zero (i.e., those individuals who would not choose to smoke even if cigarettes were available for free). Conversely, the behavior of the remainder of consumers (those to the left of S_4) is the function of the "total cost of smoking": the sum of cigarettes' free-market price, taxes, and smoking health-related costs.

If there were no information (or education) about the health risks of smoking and the government did not undertake any tobacco control policies, the total cost of smoking would simply be P, the cigarettes' free-market price. In such a society, all consumers whose willingness to pay is greater than (or equal to) P would choose to smoke (S_3 in Fig. 1.3B). The ratio between the number of smokers and the population at risk (in this case, S_3/S_T) indicates the rate of prevalence of smoking (as described in the previous section). An increase in the price of cigarettes (up to $P_T = P + t$) due to an excise tax (t) leads to a lower prevalence rate by decreasing the number of smokers to S_2. However, the sum of the free-market price and taxes represents only a small fraction of the total cost of smoking. A sharp reduction in smoking prevalence is mainly the result of a widespread awareness of the serious health effects of tobacco use. This is shown in Fig. 1.3B by the decline in the number of smokers until S_1, when consumers take into account the full cost of smoking, $P_H = P + t + H$ (where H denotes the health-related costs).

A positive and significant prevalence rate of smoking (measured here by the ratio S_1/S_T), despite the well-known harmful effects of cigarette consumption, highlights one of the main features of tobacco use: addictiveness. From an economic standpoint, the addiction to tobacco is quite similar to other drug addictions, being based on the same three key dimensions: tolerance, reinforcement, and withdrawal. Specifically, tolerance implies a gradual adaptation—that is, current consumption tends to become less satisfying as cumulative past consumption increases. Reinforcement reflects the key role of smoking habits in current smokers' behavior, whereas withdrawal means some degree of irreversibility of tobacco consumption (i.e., quitting smoking is normally not easy or costless). As a result, consumption in the current period becomes a function of past consumption choices.

A way to capture addiction in a basic model is by writing the equation of the demand curve as $Q=D(P; Q^*, Z)$, where Q^* measures the cumulative sum of cigarettes consumed in previous periods. The concept of addiction may help in explaining some specific properties of the demand for cigarettes, such as (1) the irreducible high level of long-standing smokers' willingness to pay; (2) the greater responsiveness of long-term demand to changes in price and income (i.e., both price and income elasticity tend to increase the longer the time available to consumers to modify their health-related habits is); and (3) the rise of various alternative tobacco products, like electronic cigarettes (IARC, 2011, chap. 4).

Finally, given the high tangible (medical expenditure) and intangible (the loss of quality of life and loss of life) costs that tobacco use imposes on individuals and societies, even a number of smokers equal to S_1 are likely to result in a market failure (i.e., an inefficiency condition that occurs when the individual pursuit of self-interest leads to bad results for society as a whole). Tobacco markets are indeed characterized by two main sources of inefficiency: (1) There is an information failure about both the health risks of smoking and the addictive nature of tobacco consumption, and (2) smoking imposes costs on nonsmokers and on health systems. There is thus an economic rationale for tobacco control measures (besides taxes) that aim to decrease not only the quantity demanded but also and foremost the demand for cigarettes (i.e., measures that shift leftward the D curve), such as a smoking ban in public places, mandatory health warnings on packages, and the prohibition of cigarette advertising (Nguyen, Rosenqvist, & Pekurinen, 2012).

1.4 AN OVERVIEW OF TOBACCO USE IN DIFFERENT COUNTRIES

According to recent comprehensive estimations and projections (Ng et al., 2014; NIH, 2016, chap. 2), at the time of writing, there are about 1.11 billion smokers (aged 15 or older) worldwide. Around half of them live in the Western Pacific and South East Asia regions, especially in three countries—China, India, and Indonesia—that together account for nearly 45% of the global number of smokers. Gender plays a key role in determining smoking habits. About 85% (938 million) of all smokers are males who live in developing countries. Conversely, female smokers are 175 million, and around 46% are inhabitants of high-income OECD countries.

The total number of smokers has remained quite stable during the last decade, but it is projected to rise to 1.15 billion in 2025. This increase is mainly due to the population growth expected to occur in developing countries. The prevalence of smoking, in contrast, is declining in the vast majority of the world's regions, except for the male population in the Eastern Mediterranean and African WHO regions (where the prevalence rates are around 36% and 26% and are predicted to reach 45% and 34%, respectively, over the next 10 years). Globally, the prevalence of smoking was around 26% 17 years ago; this has declined gradually to about 20% in 2015. The Russian Federation, China, and several European countries (Fig. 1.4) still have prevalence rates well above the world average. For example, in Eastern Europe and Russia, nearly one-third of the adult population (both sexes combined) currently smoke (Ng et al., 2014; NIH, 2016, chap. 2).

An estimation of the number of cigarettes consumed indicates that about 5.8 trillion cigarettes were smoked worldwide in 2014. The first six countries for the total number of cigarettes consumed (i.e., China, Russian Federation, India, United States, Indonesia, and Japan) account for around 60% of the total world consumption. There has been an increase in the global number of cigarettes consumed since 2000, mainly due to the increasing consumption registered in several Western Pacific and Eastern Mediterranean countries. The intensity of smoking—measured by the annual per capita cigarette consumption among adults (people aged 15 and older)—has globally declined during the last 10 years, from about 1200 to 1000 cigarettes per person and per year. This basic indicator of tobacco addictiveness is especially declining in OECD countries, where the average consumption was substantially higher than the world average in 2000 (around 2200 cigarettes per capita) and has fallen sharply to about 1400 in 2015. Mean annual consumption is over 2500 cigarettes per capita in some Balkan and Eastern European countries (Fig. 1.5), such as Croatia (2771), Belarus (2896), and the Russian Federation (2838). Furthermore, consumption per capita is around 2000 cigarettes not only in small Mediterranean countries (Greece, Lebanon, and Cyprus) but also in large population countries—for example, Saudi Arabia, South Korea, and especially China (Eriksen, Mackay, Schluger, Gomeshtapeh, & Drope, 2016; Ferretti, 2015; Ng et al., 2014; NIH, 2016, chap. 2).

Finally, one key driver of cigarette consumption is affordability—usually measured as the ratio between the price of a pack of cigarettes and the daily average

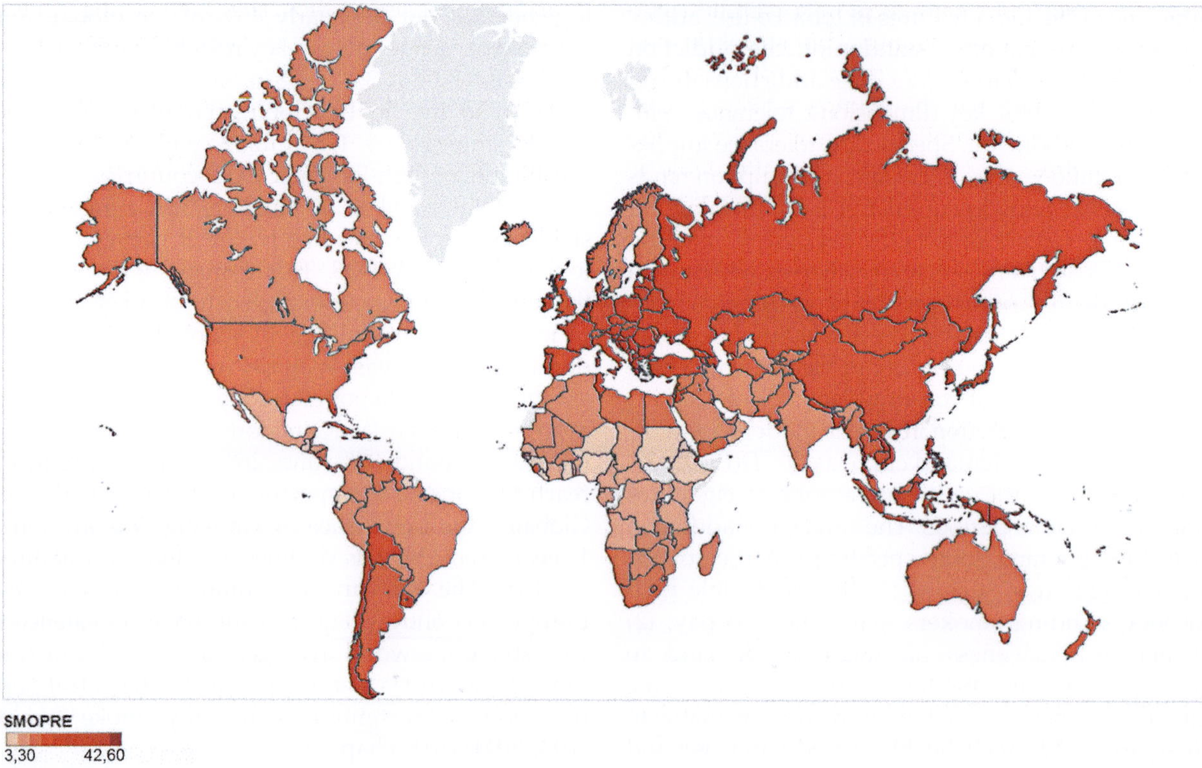

FIG. 1.4 Smoking prevalence (age-standardized rate, both sex), 2015. *SMOPRE*, smoking prevalence. *Source: Author's calculation on Ng et al. (2014).*

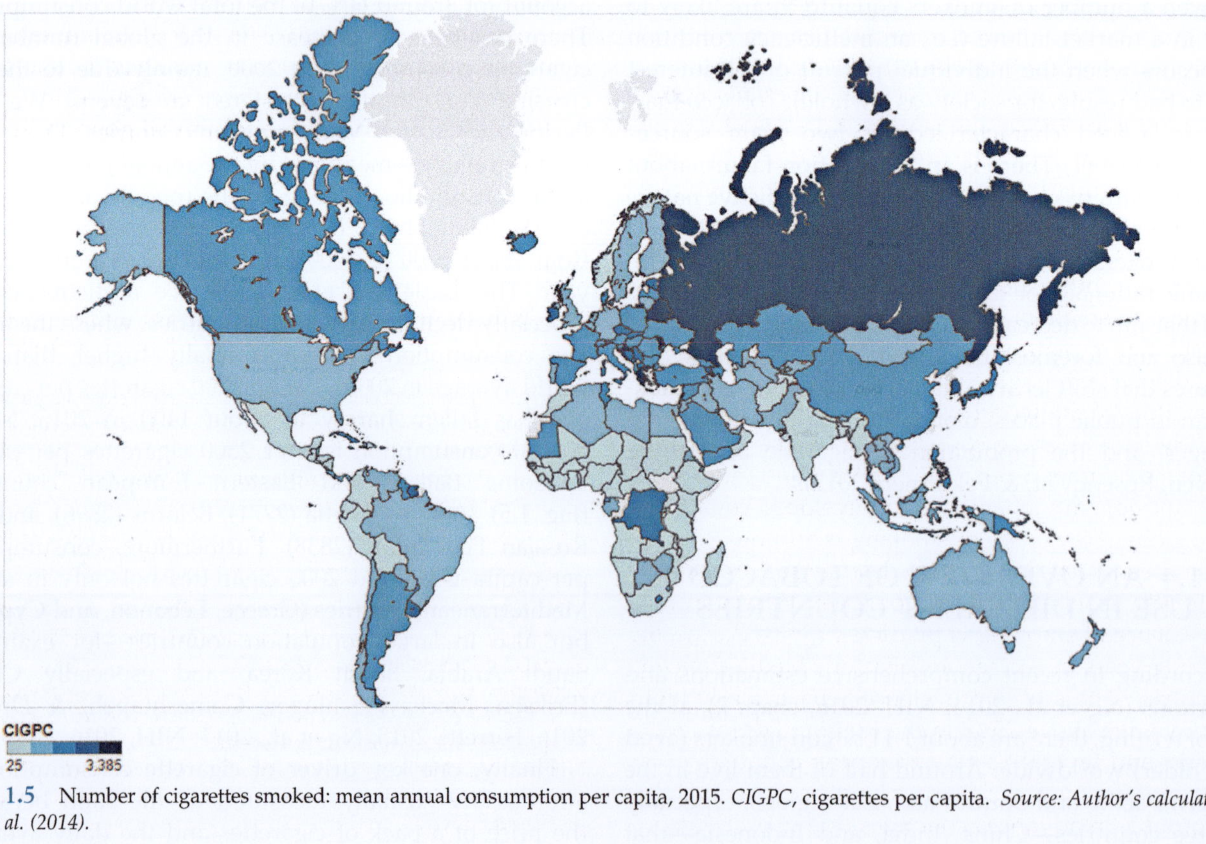

FIG. 1.5 Number of cigarettes smoked: mean annual consumption per capita, 2015. *CIGPC*, cigarettes per capita. *Source: Author's calculation on Ng et al. (2014).*

disposable income (both expressed in a common currency to account for cross-country differences in cost of living)—that captures the amount of resources (money or labor time) that the country's average consumer has to give up to smoking. Recent evidence suggests that cigarettes are becoming more affordable in developing countries than in developed ones, mainly as a result of the rapid increase in income per capita in emerging economies and the growing burden of excise tax on the price of cigarettes in almost all high-income countries (Blecher & van Walbeek, 2008).

MINI-DICTIONARY OF TERMS

Cigarette affordability A measure of the quantity of resources (in terms of money or time) required to buy a pack of cigarettes.
Demand curve for cigarettes The relationship between the quantity that consumers demand and the price of cigarettes, holding constant the other factors that influence purchase such as consumers' income and tastes, the price of related goods, and government regulation.
Normal (inferior) good A good for which quantity demanded increases (decreases) when income rises.
Price elasticity of cigarettes The percentage change in quantity demanded resulting from a 1% change in the price of cigarettes (other things being equal).
Smoking intensity The ratio between the total number of cigarettes consumed and the number of smokers in a given country and year.
Smoking prevalence The ratio between the number of smokers and the population at risk in a given country and year.

Key Facts on Tobacco Use

- Tobacco may be consumed in different forms. Nowadays, however, manufactured cigarettes account for more than 90% of the total world sales of tobacco products.
- There are around 1.11 billion smokers worldwide. About half of them live in the Western Pacific and South East Asia regions and especially in three countries: China, India, and Indonesia. The total number of smokers is increasing slowly due to the population growth in developing countries.
- Gender matters in global smoking habits. About four-fifths of the world smokers are males who live mostly in developing countries. Conversely, female smokers number around 175 million, and more than 40% of them are inhabitants of developed countries.
- The global prevalence of smoking is around 21%, and it has been declining over the last decade, especially in highly developed OECD countries. However, the prevalence of smoking is predicted to rise during the coming years in African and Eastern Mediterranean countries. The rate of prevalence is particularly high (on average above 25%) in the Russian Federation, China, and several Eastern European countries.
- During recent years, cigarettes have become much less affordable in advance economies due to excise tax and generally more affordable in emerging economies, where smokers' income is rising more than the average price of cigarettes.
- The intensity of smoking has declined steadily since 2000. A major reduction in cigarette consumption per capita has been registered in OECD countries (from 2240 cigarettes per person in 2000 to 1450 cigarettes per person in 2013). However, mean annual consumption is above 2500 cigarettes per capita in some Balkan and Eastern European countries such as Croatia (2771), Belarus (2896), and the Russian Federation (2838).

Summary Points

- This chapter is a brief guide to some essential concepts of tobacco economics and epidemiology.
- A "triangle model" of agent (smoke), hosts (smokers), vectors (tobacco companies), and social environment is used to describe the interaction between smoking determinants in a given population.
- There are two fundamental measures of tobacco use in a given country: smoking prevalence and smoking intensity.
- Changes in the price of cigarettes—due, for instance, to excise tax—lead to changes in the quantity demanded (i.e., movements along a given demand curve), whereas changes in consumers' income and tastes or in tobacco regulation policies lead to changes in demand (i.e., shifts of the entire demand curve).
- The demand for cigarettes tends to be relatively inelastic to changes in price, especially in the short run. Cigarettes often behave as an inferior good in developed countries and as a normal good in developing ones.

References

Allen, W. B., Doherty, N., Weigelt, K., & Mansfield, E. (2005). *Managerial economics. Theory, applications, and cases* (6th ed.). New York: W.W. Norton & Company.
Blecher, E., & van Walbeek, C. (2008). *An analysis of cigarette affordability* (pp. 4–14). Paris: International Union Against Tuberculosis and Lung Disease.
Bonita, R., Beaglehole, R., & Kjellstrom, T. (2006). *Basic epidemiology*. Geneva: WHO.
Bonnie, R. J., Stratton, K., & Wallace, R. B. (Eds.), (2007). *Ending the tobacco problem: A blueprint for the nation*. Washington, DC: The National Academies Press.
CDC. (2014). *The health consequences of smoking—50 years of progress: A report of the surgeon general*. Atlanta, GA: Centers for Disease Control and Prevention [Appendix 15.1].
Chaloupka, F., & Warner, K. (2000). The economics of smoking. In A. J. Culyer & J. P. Newhouse (Eds.), *Handbook of health economics* (pp. 1539–1627). Amsterdam: Elsevier.
Eriksen, M., Mackay, J., Schluger, N., Gomeshtapeh, F. I., & Drope, J. (2016). *The tobacco atlas* (4th ed.). Atlanta, GA: American Cancer Society.

Ferretti, F. (2015). Unhealthy behaviours: an international comparison. *PLoS ONE*, *10*(10), e0141834.

Gallus, S., Schiaffino, A., LaVecchia, C., Townsend, J., & Fernandez, E. (2006). Price and cigarette consumption in Europe. *Tobacco Control*, *15*(2), 114–119.

GBD Tobacco Collaborators. (2015). Smoking prevalence and attributable disease burden in 195 countries and territories, 1990–2015: a systematic analysis from the Global Burden of Disease Study 2015. *Lancet*, *389*(10082), 1885–1906.

Giovino, G. A. (2002). Epidemiology of tobacco use in the United States. *Oncogene*, *21*(48), 7326–7340.

Guindon, G. E., & Boisclair, D. (2003). *Past, current and future trends in tobacco use*. Discussion paper, February, no. 3 Washington DC: WB.

Hammond, K. S. (2009). Global patterns of nicotine and tobacco consumption. *Nicotine Psychopharmacology* (pp. 3–28). In J. E. Henningfield, E. D. London, & S. Pogure (Eds.). *Handbook of Experimental Pharmacology* (pp. 2–28). Berlin, Heidelberg: Springer-Verlag.

IARC. (2008). *Methods for evaluating tobacco control policies*. Lyon: International Agency for Research on Cancer and WHO.

IARC. (2011). *Effectiveness of tax and price policies for tobacco control*. Lyon: International Agency for Research on Cancer and WHO.

Lopez, A. D., Neil, E. C., & Tapani, P. (1994). A descriptive model of the cigarette epidemic in developed countries. *Tobacco Control*, *3*(3), 242–247.

Mankiw, N. G. (2012). *Principles of microeconomics* (6th ed.). Mason, OH: South-Western Cengage Learning.

Ng, M., Freeman, M. K., Fleming, T. D., Robinson, M., Dwyer-Lindgren, L., Thomson, B., et al. (2014). Smoking prevalence and cigarette consumption in 187 countries, 1980–2012. *Journal of the American Medical Association*, *311*(2), 183–192.

Nguyen, L., Rosenqvist, G., & Pekurinen, M. (2012). *Demand for tobacco in Europe*. Report no. 6 Helsinki: National Institute for Health and Welfare.

NIH. (2016). *The economics of tobacco and tobacco control. National Cancer Institute, tobacco control monograph, series no 21*. Bethesda, MD: NIH and WHO.

Penn State. (2016). *Epidemiologic triad*. Department of Statistics, Pennsylvania State University. Available at: https://onlinecourses.science.psu.edu/stat507/node/25.

Perucic, A. M. (2015). The demand for cigarettes and other tobacco products. In *Tobacco control economics tobacco free initiative*. Geneva: WHO.

Phillips, C. V. (2016). *Understanding the basic economics of tobacco harm reduction*. Discussion paper no. 72 London: The Institute of Economic Affairs.

Slade, J. (1993). Nicotine delivery systems. In C. T. Orleans & J. Slade (Eds.), *Nicotine addiction: Principles and management*. Oxford: Oxford University Press.

Warner, K. E., & MacKay, J. (2006). The global tobacco disease pandemic: nature, causes, and cures. *Global Public Health*, *1*(1), 65–86.

WHO. (2015). *Global report on trends in prevalence of tobacco smoking*. Geneva: WHO.

WHO. (2017). *Report on the global tobacco epidemic: Monitoring tobacco use and prevention policies*. Geneva: WHO [Appendix X].

Wilkins, N., Yurekli, A., & Hu, T. (2007). Economic analysis of tobacco demand. In A. Yurekli & J. de Beyer (Eds.), *Economics of tobacco toolkit*. Washington, DC: WB.

CHAPTER

2

Maternal Smoking and Fetal Brain Outcome: Mechanisms and Possible Solutions

Hui Chen, Yik Lung Chan†, Brian G. Oliver*,†, Carol A. Pollock‡, Sonia Saad‡*

*School of Life Sciences, Faculty of Science, University of Technology Sydney, Sydney, NSW, Australia
†Respiratory Cellular and Molecular Biology, Woolcock Institute of Medical Research, The University of Sydney, Glebe, NSW, Australia
‡Renal Group, Department of Medicine, Kolling Institute, Royal North Shore Hospital, St Leonards, NSW, Australia

Abbreviations

Drp	dynamin-related protein
Fis	fission protein
HI	hypoxic-ischemic
LC	microtubule-associated protein light chain
MnSOD	manganese superoxide dismutase
NO	nitric oxide
Opa-1	optic atrophy 1 protein
OXPHOS	oxidative phosphorylation
Pink	PTEN-induced putative kinase
ROS	reactive oxygen species
SE	cigarette smoke exposure
TLR	Toll-like receptor
TOM	translocase of mitochondrial outer membrane

2.1 INTRODUCTION

There has been a rapid advancement in the understanding of fetal programming of diseases in adulthood. Changes in certain gene expression at birth can persist until adulthood, which significantly increase the susceptibility to certain diseases. This highlights the critical role of the ideal intrauterine environment to optimize fetal health outcomes. Maternal smoking can disturb the stability of the intrauterine environment, leading to brain inflammatory response and oxidative stress in the offspring. This can result in neonatal hypoxic-ischemic (HI) injury and adulthood cognitive change, such as depression and anxiety.

2.2 MATERNAL SMOKING AND BRAIN DEVELOPMENT

Despite the general education on the risks of smoking during pregnancy, it is estimated that approximately 20%–45% of women still smoke during pregnancy in some countries, while the rate is even higher in certain indigenous communities. Additionally, ~82% of the world's population is not protected from secondhand smoking, including pregnant women.

Maternal smoking/cigarette smoke exposure (SE) is a major contributor to intrauterine growth restriction, low birth weight, perinatal morbidity and mortality, and long-term consequences in the offspring, including behavioral problems (Chen & Morris, 2007). The vasoconstriction effect of nicotine reduces placental blood flow resulting in intrauterine shortage of nutrients and oxygen that restricts fetal growth and subsequently permanently changes the physiological functions.

Although the brain receives priority nutrition delivery, smoking during pregnancy is closely linked to small brain weight, frontal lobe, and cerebellar volumes (Fig. 2.1). Maternal nicotine administration alone does not seem to change brain size in the offspring (Grove et al., 2001), whereas maternal SE reduced brain size at birth (Chan, Saad, Pollock, et al., 2016). Clearly, other chemicals in cigarette smoke play a critical role in fetal brain underdevelopment. Due to the complex nature of more than 5000 chemicals in tobacco smoke, it is unlikely that one single component will cause all pathologies.

https://doi.org/10.1016/B978-0-12-813035-3.00002-2

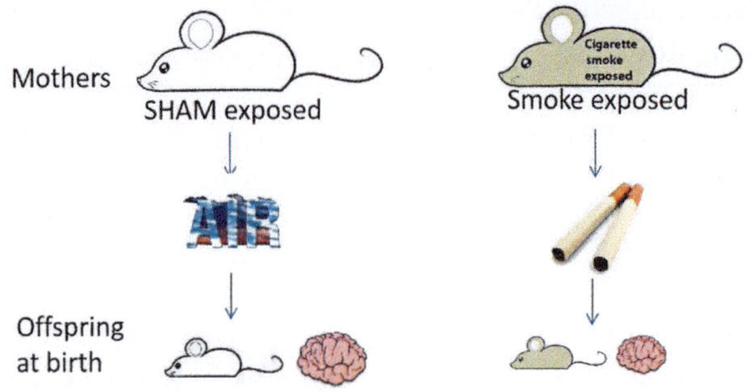

FIG. 2.1 Birth outcome of maternal smoking.
Maternal smoking reduces the birth weight and brain size in the newborn offspring.

2.3 MATERNAL SMOKING AND NEUROCOGNITIVE OUTCOME

Maternal smoking causes long-lasting adverse effects on the structural and functional development of the fetal brain leading to cognitive disorders (Bublitz & Stroud, 2012). Small brain volume significantly correlates with lower intelligent quotient (Haier, Jung, Yeo, Head, & Alkire, 2004), while verbal ability positively correlates with cerebral volume (Witelson, Beresh, & Kigar, 2006). In humans, 13–16-year-old offspring from the smoking mothers had lower verbal and visual memory abilities than those from nonsmokers (Fried, Watkinson, & Gray, 2003). Heavy smoking (>20 cigarettes/day) during pregnancy can increase the risk of internalizing behaviors such as fear and anxiety in young children (Moylan et al., 2015). In addition, maternal smoking increases the risk of attention deficit hyperactivity disorder in a dose-dependent manner (Altink et al., 2009). Hence, smoking during pregnancy is a significant public health issue.

However, some confounding factors, such as socioeconomic status of the parents, alcohol consumption, and paternal smoking, can result in inconsistent findings in humans studies (Moylan et al., 2015). Therefore, animal models have the advantage of removing these confounding factors to determine the impact of maternal smoking alone. There are a limited number of studies on the direct impact of SE, whereas most studies adopted nicotine, limiting the data interpretation. This chapter will focus on the animal models using direct SE.

2.4 MATERNAL SMOKING AND HI ENCEPHALOPATHY

Oxygen deprivation before and around birth can result in HI brain damage in newborns (Johnston & Hoon Jr., 2006). Nicotine reduces blood flow to the placenta. Smoking also increases carboxyhemoglobin levels that can reduce the oxygen-carrying capacity of both fetal and maternal red blood cells. Thus, maternal smoking has been shown to cause hypoxia in the fetus in animal study (Socol, Manning, Murata, & Druzin, 1982). HI itself can cause cerebral palsy and associated disabilities in children (Johnston & Hoon Jr., 2006), whereas smoking ≥10 cigarettes/day during pregnancy has been shown to increase the risk of cerebral palsy (Streja et al., 2013).

During HI encephalopathy, blood oxygen saturation and blood flow are decreased, interrupting normal fetal brain development (Li, Gonzalez, & Zhang, 2012). Microglia responds rapidly to hypoxia and accumulates in injured tissues, where excessive amounts of inflammatory cytokines such as TNF-α and IL-1β along with reactive oxygen species (ROS) are produced, leading to inflammation and oxidative stress. In mice, increased brain inflammation and oxidative stress are already present in the offspring from the SE mothers even without injury (Chan, Saad, Pollock, et al., 2016). As such, more cell death occurs when such offspring suffer from HI encephalopathy (Chan et al., 2017). Cerebral cortex, hippocampus, and subventricular regions are the most vulnerable to HI damage. Infarct size is increased in male pups with HI encephalopathy, but not in the females (Li, Xiao, et al., 2012), indicating a gender difference with males more seriously affected.

2.5 POTENTIAL MECHANISMS

2.5.1 Brain Inflammatory Response

ROS produced by burning tobacco are not removed by the cigarette filters, leading to the activation of inflammatory pathways in various myeloid and lymphoid cells (Qiu et al., 2017). ROS can also activate macrophages, which further produce more ROS (Rahman & Adcock, 2006). Prolonged systemic inflammation in pregnant smokers also affects the offspring. In adult male offspring of SE mothers, brain pro-inflammatory cytokine IL-6, IL-1α receptor, and Toll-like receptor (TLR) 4 expression are increased (Fig. 2.2) (Chan, Saad, Pollock, et al., 2016). The activation of TLRs stimulates the production of IL-1β and IL-6 in monocytes (Fig. 2.2), which in turn enhances TLR expression via a positive feedback loop. The female offspring had similar changes persistent from weaning to adulthood (Chan, Saad, Al-Odat, et al., 2016).

Neuroinflammation plays a crucial role in the development of neurodegeneration. Increased levels of TLR4 and IL-1 can both lead to an elevation of β-amyloid, linking to the development of Alzheimer's disease. Increased brain IL-6 level is also associated with increased anxiety, autism-like behavior, and the progression of neurodegenerative diseases (Wei et al., 2012). Indeed, increased severity of schizophrenia or autism has been found in

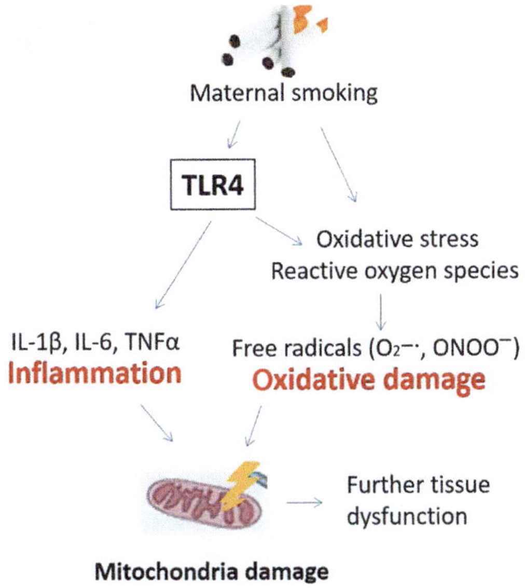

FIG. 2.2 Mechanism of maternal-smoking-induced brain disorder in the offspring.
Maternal smoking increases brain inflammation and oxidative stress in the offspring's brain, while both lead to mitochondrial damage and result in neurological dysfunction.

offspring of smoking mothers who have high blood level of inflammatory cytokines (Ashwood et al., 2011; Potvin et al., 2008).

2.5.2 Brain Oxidative Stress

2.5.2.1 ROS

When the cellular production of oxidative molecules overwhelms endogenous antioxidant defense systems, oxidative stress occurs. Brain tissue is especially susceptible to ROS damage since it is a major organ to metabolize oxygen (20% of the body consumption). The increase in ROS has been linked to the increase in permeability of mitochondrial membrane and eventually cell death (Popa-Wagner, Mitran, Sivanesan, Chang, & Buga, 2013).

Long-term SE itself can increase oxidative stress and cellular damage in the mother's brain (Chan, Saad, Pollock, et al., 2016). Breast milk is rich in antioxidants, which can temporarily protect the newborn. However, once the pups gradually wean from the breastfeeding, male offspring start to display increased brain oxidative stress persisting until adulthood, with significant cellular damage in the adult brain (Chan, Saad, Pollock, et al., 2016). Certain toxic chemicals in the cigarette smoke may induce oxidative stress in both mothers and offspring, as maternal antioxidant supplementation can reverse such effects (Chan et al., 2017). Interestingly, female offspring seem to be protected from such adverse impact of maternal smoking (Chan, Saad, Al-Odat, et al., 2016). The potential mechanisms will be discussed in Section 2.6.

2.5.2.2 Antioxidant Defense System

There is a complex antioxidant defense system to scavenge excess ROS. This is especially important in the brain as the neurons are vulnerable to oxidative stress. The most crucial antioxidant in the brain is manganese superoxide dismutase (MnSOD), which is present at a higher concentration in the mitochondria than the other intracellular components. Mitochondrial oxidative phosphorylation (OXPHOS) complexes I and III produce ROS during normal energy metabolism; therefore, mitochondrial MnSOD is important for removing excessive ROS. Interestingly, there is an MnSOD surge during late gestation and newborn periods, which is subsequently reduced as mice reach postnatal day 4 (Khan & Black, 2003). Several studies have shown that MnSOD is crucial for neuroprotection, which prevents neuronal apoptosis and reduces ischemic brain injury through preventing mitochondrial dysfunction (Keller et al., 1998).

MnSOD levels are increased in peripheral tissues such as esophagus and lung, in order to scavenge overproduced ROS by long-term smoking. To date, only two studies reported brain MnSOD change in response to maternal SE (Chan, Saad, Al-Odat, et al., 2016; Chan, Saad, Pollock, et al., 2016). In the brains from both mothers and adult male offspring, MnSOD levels are reduced (Chan, Saad, Pollock, et al., 2016). An exhaustion of MnSOD can cause mitochondrial and DNA damage due to ROS overproduction in the long term (Chan et al., 2017).

2.5.3 Mitochondrial Function and Integrity

Mitochondria are the major site for ATP production. In the brain, there is a high density of mitochondria in neurons due to their high metabolic and energy requirement. Mitochondrial dysfunction has been discovered in a number of neurological disorders such as amyotrophic lateral sclerosis and Alzheimer's disease, suggesting that healthy mitochondria are critical to maintain nervous health (Jiang, Wang, Perry, Zhu, & Wang, 2015).

2.5.3.1 Mitochondrial Membrane Functional Units

ATP is produced through OXPHOS at mitochondrial cristae facilitated by OXPHOS complexes I–V (Fig. 2.3). Complexes I and II act as the first and second entry points of electrons in the respiratory chain, respectively. Complex III facilitates electron transfer to complex IV, which is a crucial regulator for ATP production. Protons from complexes I, III, and IV drive the conversion of ADP to ATP. Brain levels of all five OXPHOS complexes are increased by long-term SE, suggesting increased demand for energy supply (Chan, Saad, Pollock, et al., 2016). Indeed, brain ATPase activities are reduced by SE (Vani, Anbarasi, & Shyamaladevi, 2015). Although brain

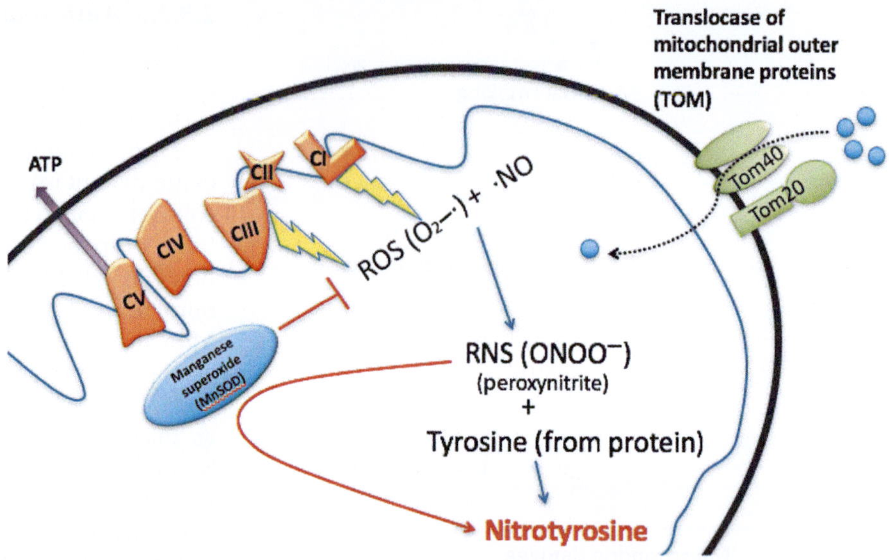

FIG. 2.3 Mitochondrial energy metabolic units.
Reactive oxygen species (ROS) is generated by oxidative phosphorylation (OXPHOS) complexes (CI–CV). The proteins enter mitochondria through Tom20 and Tom40. ROS combines with nitric oxide (NO) to form peroxynitrite (RNS), which interacts with nitroxylate to form nitrotyrosine. MnSOD suppresses this process.

complex I and V levels are reduced at weaning, all complexes I–V are increased in adulthood by maternal SE, similar as their mothers, suggesting mitochondrial function may be inheritable from the mother (Chan, Saad, Al-Odat, et al., 2016; Chan, Saad, Pollock, et al., 2016).

During the OXPHOS process, 90% of ROS are generated as a by-product in complexes I and III (Fig. 2.3). ROS can form peroxynitrate with nitric oxide (NO), which further causes protein tyrosine nitration to form 3-nitrotyrosine to damage mitochondria (Beal, 1998). MnSOD competes with NO to react with superoxide that prevents the generation of peroxynitrate and 3-nitrotyrosine. Nitration itself can also inactivate MnSOD (Surmeli, Litterman, Miller, & Groves, 2010). When MnSOD is reduced and nitrotyrosine levels are high, mitochondria are less protected from oxidative stress, such as during smoking and maternal SE (Chan, Saad, Pollock, et al., 2016).

The translocase of mitochondrial outer membrane (TOM) protein complex is the main entry portal for most mitochondrial protein precursors synthesized in the cytoplasm. Tom40 forms ion channels in lipid bilayers during transportation (Rapaport, Neupert, & Lill, 1997). TOM20 (a peripheral subunit of the TOM40 complex) recognizes and imports the protein precursors, by facilitating protein insertion at the outer mitochondrial membrane. TOM20 can be degraded under oxidative stress. While brain TOM20 is unchanged by SE, in the male offspring, its level is reduced in at weaning but increased in the adulthood by maternal smoking (Chan, Saad, Pollock, et al., 2016).

2.5.3.2 *Mitochondrial Integrity*

Mitochondrial structure is highly dynamic and maintained through "mitophagy." Phagy means to eat; autophagy means "self-eating," which is to degrade cellular constituents to maintain intercellular homeostasis. "Mitophagy" is the removal of mitochondria by autophagy.

During autophagy, microtubule-associated protein light chain (LC) forms autophagosome to engulf intracellular components. The conversion of LC3A/B-I to LC3A/B-II is used as an indicator of autophagic activity, and LC3A/B-II level correlates with autophagosome formation. Maternal SE decreases brain LC3A/B-II levels in both weaning and adult male offspring but increases LC3A/B-II levels in the female offspring at the same ages (Chan et al., 2017). This suggests reduced autophagy capacity in the male versus female offspring.

Mitophagy is facilitated by fission and fusion (steps 1 and 2 in Fig. 2.4, respectively). Fission separates damaged mitochondrial portion from the healthy fragment, while fusion combines two healthy fragments to form a new mitochondrion. These two processes are balanced to maintain the overall morphology of the mitochondria. High fusion-to-fission ratio leads to less mitochondria, with an elongated and more interconnected shape; low fusion-to-fission ratio leads to small spheres and short rods of mitochondria, often referred as "fragmented mitochondria." Failure to trigger mitophagy in the brain can lead to neurodegenerative diseases (Cheung & Ip, 2009).

2.5.3.2.1 FISSION MACHINERY

Dynamin-related protein (Drp)-1 presents at sites of mitochondrial division to separate the damaged mitochondrial fragment (Fig. 2.4, step 1). Fission protein (Fis)-1 anchored at the outer membrane of mitochondrion, which serves as a platform to adapt Drp-1. Following the segregation, PTEN-induced putative kinase (Pink)-1 accumulates on the outer membrane of damaged mitochondrion leading to the recruitment of Parkin (Fig. 2.4, step 3).

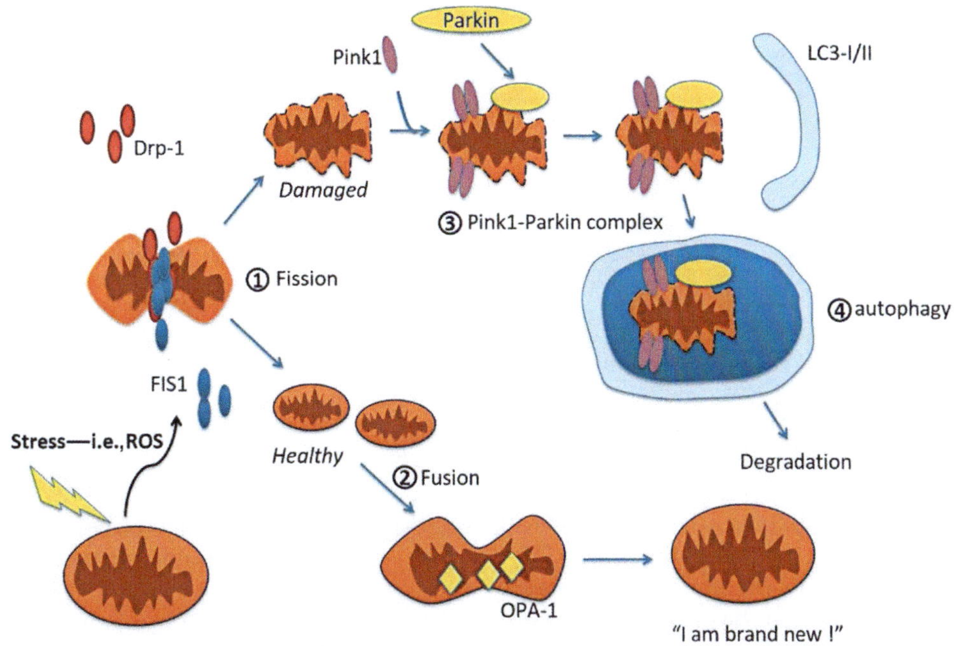

FIG. 2.4 Mitophagy and autophagy machinery.
Mitophagy and autophagy. Damaged mitochondrial fragments are separated from the healthy part facilitated by dynamin-related protein (Drp)-1 and fission protein (Fis)-1. Damaged mitochondria attract PTEN-induced putative kinase (Pink)-1 and Parkin. This complex is then engulfed by microtubule-associated protein light chain (LC3) A/B-I/II to form autophagosome for degradation. The healthy part of a mitochondrion can bind to the healthy part of another mitochondrion through optic atrophy 1 protein (Opa)-1.

Mitochondrial Drp-1 is increased at postnatal day 1 but reduced in adult male offspring by maternal smoking (Chan et al., 2017). Mitochondrial Fis-1 is not increased until postnatal day 20 (weaning age), which is also reduced in the adult male offspring, suggesting reduced fission capacity by maternal smoking. This may be related to reduced brain mitochondrial density due to increased neural apoptosis. On the contrary, Drp-1 is reduced at postnatal day 1 but increased in adult female offspring (Chan et al., 2017). Parkin is also reduced in female offspring at postnatal day 1 and 20, but Pink-1 is somewhat increased in adult female offspring. Such increase is associated with increased mitochondrial MnSOD and reduced apoptotic marker levels. This suggests that increased fission activity can prevent the brain from increased apoptosis induced by maternal smoking in the female offspring.

2.5.3.2.2 FUSION MACHINERY

Optic atrophy 1 protein (Opa-1) regulates fusion process (Fig. 2.4, step 2). Opa-1 knockout mice die at embryonic day 9; thus, Opa-1 is essential for embryonic development (Rahn, Stackley, & Chan, 2013). Mitochondrial fusion appears to protect cells from apoptosis and prolongation of lifespan, although the mechanism is unknown.

In response to maternal smoking, brain Opa-1 is significantly reduced in the male offspring at adulthood with increased brain apoptosis, suggesting less healthy mitochondrial fragments are available for recycling (Chan et al., 2017). A reduction of brain mitochondrial fusion was observed in Alzheimer's disease (Zhang et al., 2016). Although it has been well studied that smoking itself is closely linked to Alzheimer's disease and dementia (Anstey, von Sanden, Salim, & O'Kearney, 2007), such risks in the offspring are yet to be investigated in humans. On the other hand, Opa-1 level is increased in the female offspring at postnatal day 1 but unchanged in the adulthood (Chan et al., 2017). This suggests that unknown mechanism promotes mitochondrial fusion machinery to prevent excess brain apoptosis in the female offspring. Similar observation in mitochondrial protection has been found in young women (Azarashvili, Stricker, & Reiser, 2010).

2.6 GENDER DIFFERENCE IN THE RESPONSE TO MATERNAL SMOKING

Although the brain structures in men and women are similar, they have different susceptibility to specific neural diseases. Males are more vulnerable to mental illness, such as autism and attention deficit and hyperactivity disorders (Davies, 2014). When females suffer from these brain disorders, they start at an older age than men (Zagni, Simoni, & Colombo, 2016).

The male offspring are also more vulnerable to maternal-smoking-induced inflammatory response,

TABLE 2.1 Impact of Maternal SE on Brain Mitophagy in the Offspring

Gender	Male offspring	Female offspring
Mitophagy—fission	↓	↑
Mitophagy—fusion	↓	↑
Autophagy	No change	↑

Maternal smoking reduced mitochondrial fission and fusion activities in the male offspring's brain with no impact on autophagy activity; however, it increased fission, fusion, and autophagy.

TABLE 2.2 Impact of Maternal L-Carnitine Supplementation on Mitophagy in the Offspring

Gender	Male offspring	Female offspring
Mitophagy—fission	↑	No change
Mitophagy—fusion	↑	No change
Autophagy	No change	↓

Maternal L-carnitine supplementation during pregnancy can improve brain mitochondrial fission and fusion activities in the male offspring from the smoking mothers, while it reduces autophagy in the female offspring.

oxidative stress, mitochondrial injury, and brain apoptosis compared to female offspring. This gender difference is speculated to be driven by estrogen, which has been considered neuroprotective and antiinflammatory and thus protects the female's brain (Brann, Dhandapani, Wakade, Mahesh, & Khan, 2007). Another role of estrogen is to act as an antioxidant to prevent lipid peroxidation, protein oxidation, and DNA damage (Escalante, Mora, & Bolaños, 2017). Estrogen has also been shown to maintain mitochondrial membrane potential during mitochondrial toxin exposure (Wang, Green, & Simpkins, 2001).

In the male and female brains, glial cells react to the environmental insults differently. The astrocytes are the most abundant glial cells, which support nutrition homeostasis and neural transmission of electric impulses. The astrocytes obtained from the males express higher levels of IL-1β mRNA that can result in worse outcomes following neuronal injury (Santos-Galindo, Acaz-Fonseca, Bellini, & Garcia-Segura, 2011). The astrocytes from the females are more resistant to stressors, such as oxidant-induced cell death, than those from the males (Liu, Oyarzabal, Yang, Murphy, & Hurn, 2008). Lipopolysaccharide found in cigarette smoke (Hasday, Bascom, Costa, Fitzgerald, & Dubin, 1999) can increase IL-6, TNF-α, and IL1β mRNA expression in the astrocytes from the males compared to those from the females (Santos-Galindo et al., 2011). Similarly, maternal smoking SE increased brain IL-6 in the adult male offspring, but not the females (Chan, Saad, Al-Odat, et al., 2016; Chan, Saad, Pollock, et al., 2016). The gender difference in mitophagy response to maternal smoking has been described in Section 2.5.3 and summarized in Table 2.1.

2.7 L-CARNITINE AS A THERAPEUTIC STRATEGY

L-Carnitine is an endogenous natural quaternary ammonium compound found in all mammalian species. It is a vital component for mitochondrial fatty acid oxidation (Gülçin, 2006). L-Carnitine acts as an energy carrier in the mitochondrial inner membrane to control acetyl-CoA supply and support OXPHOS complex activities (Virmani & Binienda, 2004). L-Carnitine also acts as a ROS scavenger that protects MnSOD from oxidative damage, suggesting that it may be useful for improving mitochondrial function (Gülçin, 2006).

L-Carnitine is also neuroprotective. Pretreatment with L-carnitine before mitochondrial toxin exposure can increase the activities of endogenous ROS scavengers to protect against oxidative stress (Virmani & Binienda, 2004). In SE mice mothers, L-carnitine supplementation during gestation and lactation can increase brain MnSOD and TOM20 levels in newborn male offspring, leading to a marked improvement in mitophagy markers in adulthood (summarized in Table 2.2) (Chan et al., 2017). Brain apoptosis and cellular DNA damage are also reduced, suggesting sustained neural protection of L-carnitine against maternal smoking. Similar improvement in brain mitophagy markers has also been observed in the female offspring's brain. Thus, L-carnitine supplementation might be a good candidate to mitigate oxidative-stress-induced mitochondrial dysfunction in the offspring of smokers.

2.8 CONCLUSION

Maternal smoking during gestation and lactation can induce significant inflammatory response and oxidative stress in male offspring's brain, which impairs mitochondrial integrity leading to cell death. Female offspring seems to be protected from such impact by maternal smoking. Maternal supplementation of L-carnitine has shown promising neural protective effect in the offspring of the smoking mothers.

MINI-DICTIONARY OF TERMS

Acetyl-CoA A produce from glycolysis, which is a metabolic intermediate that can convert into carbohydrate, protein, and fat.

Apoptosis A programmed cell death in response to stress, such as environmental pollution exposure and smoking.

Autophagy Means "self-eating," which is a strategy for cells to remove damaged proteins by self-digesting.

Encephalopathy Brain disease or brain disorders.

Hypoxic-ischemic A condition due to reduced oxygen and blood supply, commonly occurs when an artery is blocked.

Mitochondrion A cellular "powerhouse," where the energy substance ATP is produced from energy substrate acetyl-CoA and oxygen.

Mitophagy Damaged mitochondrial self-eating or self-renewal process to maintain the healthy mitochondrial population in the body.

Oxidative phosphorylation When the electrons produced through the citric acid cycle are deposited in the electron transport chain in the inner mitochondrial membrane, they are captured by enzymes to generate ATP.

Oxidative stress The production of free radicals, mainly during the process of ATP synthesis, is over the capacity of endogenous antioxidant to clear them from the cell.

Reactive oxygen species Chemically reactive chemical species containing additional oxygen molecule that can oxidize other cellular components.

Key Facts of Maternal Smoking

- Offspring from smoking mother have brain underdevelopment resulting in impaired learning and memory functions.
- Offspring from smoking mother are more likely to have oxygen-shortage- and blood-shortage-induced brain damage.
- Male offspring from smoking mother have more brain cell death than the female offspring.
- Brain cellular powerhouse mitochondrial function is damaged by maternal smoking.
- Female offspring's brain is more protected from maternal smoking than the male offspring.

Summary Points

- Maternal smoking delays brain development in the offspring.
- Maternal smoking impairs cognitive function in the offspring.
- Maternal smoking is linked to high risk of hypoxic-ischemic encephalopathy.
- Maternal smoking increases brain apoptosis in the male offspring.
- Maternal smoking increases oxidative stress and impairs brain mitochondrial function in the male offspring.
- Female offspring's brain is more protected from the detrimental impact of maternal smoking than the male offspring.

References

Altink, M. E., Slaats-Willemse, D. I. E., Rommelse, N. N. J., Buschgens, C. J. M., Fliers, E. A., Arias-Vásquez, A., et al. (2009). Effects of maternal and paternal smoking on attentional control in children with and without ADHD. *European Child & Adolescent Psychiatry*, *18*(8), 465–475.

Anstey, K. J., von Sanden, C., Salim, A., & O'Kearney, R. (2007). Smoking as a risk factor for dementia and cognitive decline: a meta-analysis of prospective studies. *American Journal of Epidemiology*, *166*(4), 367–378.

Ashwood, P., Krakowiak, P., Hertz-Picciotto, I., Hansen, R., Pessah, I., & Van de Water, J. (2011). Elevated plasma cytokines in autism spectrum disorders provide evidence of immune dysfunction and are associated with impaired behavioral outcome. *Brain, Behavior, and Immunity*, *25*(1), 40–45.

Azarashvili, T., Stricker, R., & Reiser, G. (2010). The mitochondria permeability transition pore complex in the brain with interacting proteins - promising targets for protection in neurodegenerative diseases. *Biological Chemistry*, *391*(6), 619–629.

Beal, M. F. (1998). Mitochondrial dysfunction in neurodegenerative diseases. *Biochimica et Biophysica Acta (BBA): Bioenergetics*, *1366*(1–2), 211–223.

Brann, D. W., Dhandapani, K., Wakade, C., Mahesh, V. B., & Khan, M. M. (2007). Neurotrophic and neuroprotective actions of estrogen: basic mechanisms and clinical implications. *Steroids*, *72*(5), 381–405.

Bublitz, M. H., & Stroud, L. R. (2012). Maternal smoking during pregnancy and offspring brain structure and function: review and agenda for future research. *Nicotine & Tobacco Research*, *14*(4), 388–397.

Chan, Y. L., Saad, S., Al-Odat, I., Oliver, B. G., Pollock, C., Jones, N. M., et al. (2017). Maternal L-carnitine supplementation improves brain health in offspring from cigarette smoke exposed mothers. *Frontiers in Molecular Neuroscience*, *10*(33).

Chan, Y. L., Saad, S., Al-Odat, I., Zaky, A. A., Oliver, B., Pollock, C., et al. (2016). Impact of maternal cigarette smoke exposure on brain and kidney health outcomes in female offspring. *Clinical and Experimental Pharmacology & Physiology*, *43*(12), 1168–1176.

Chan, Y. L., Saad, S., Pollock, C., Oliver, B., Al-Odat, I., Zaky, A. A., et al. (2016). Impact of maternal cigarette smoke exposure on brain inflammation and oxidative stress in male mice offspring. *Scientific Reports*, *6*, 25881.

Chen, H., & Morris, M. J. (2007). Maternal smoking—a contributor to the obesity epidemic? *Obesity Research & Clinical Practice*, *1*, 155–163.

Cheung, Z. H., & Ip, N. Y. (2009). The emerging role of autophagy in Parkinson's disease. *Molecular Brain*, *2*, 29.

Davies, W. (2014). Sex differences in attention deficit hyperactivity disorder: candidate genetic and endocrine mechanisms. *Frontiers in Neuroendocrinology*, *35*(3), 331–346.

Escalante, C. G., Mora, S. Q., & Bolaños, L. N. (2017). Hormone replacement therapy reduces lipid oxidation directly at the arterial wall: a possible link to estrogens' cardioprotective effect through atherosclerosis prevention. *Journal of Mid-Life Health*, *8*(1), 11–16.

Fried, P. A., Watkinson, B., & Gray, R. (2003). Differential effects on cognitive functioning in 13- to 16-year-olds prenatally exposed to cigarettes and marihuana. *Neurotoxicology and Teratology*, *25*(4), 427–436.

Grove, K. L., Sekhon, H. S., Brogan, R. S., Keller, J. A., Smith, M. S., & Spindel, E. R. (2001). Chronic maternal nicotine exposure alters neuronal systems in the arcuate nucleus that regulate feeding behavior in the newborn rhesus macaque. *The Journal of Clinical Endocrinology & Metabolism*, *86*(11), 5420–5426. PMID:11701716.

Gülçin, İ. (2006). Antioxidant and antiradical activities of l-carnitine. *Life Sciences*, *78*(8), 803–811.

Haier, R. J., Jung, R. E., Yeo, R. A., Head, K., & Alkire, M. T. (2004). Structural brain variation and general intelligence. *NeuroImage*, *23*(1), 425–433.

Hasday, J. D., Bascom, R., Costa, J. J., Fitzgerald, T., & Dubin, W. (1999). Bacterial endotoxin is an active component of cigarette smoke. *Chest*, *115*(3), 829–835.

Jiang, Z., Wang, W., Perry, G., Zhu, X., & Wang, X. (2015). Mitochondrial dynamic abnormalities in amyotrophic lateral sclerosis. *Translational Neurodegeneration*, *4*, 14.

Johnston, M. V., & Hoon, A. H., Jr. (2006). Cerebral palsy. *Neuromolecular Medicine*, *8*(4), 435–450.

Keller, J. N., Kindy, M. S., Holtsberg, F. W., St. Clair, D. K., Yen, H.-C., Germeyer, A., et al. (1998). Mitochondrial manganese superoxide dismutase prevents neural apoptosis and reduces ischemic brain injury: suppression of peroxynitrite production, lipid peroxidation, and mitochondrial dysfunction. *The Journal of Neuroscience*, *18*(2), 687–697.

Khan, J. Y., & Black, S. M. (2003). Developmental changes in murine brain antioxidant enzymes. *Pediatric Research*, *54*(1), 77–82.

Li, Y., Gonzalez, P., & Zhang, L. (2012). Fetal stress and programming of hypoxic/ischemic-sensitive phenotype in the neonatal brain: mechanisms and possible interventions. *Progress in Neurobiology*, *98*(2), 145–165.

Li, Y., Xiao, D., Dasgupta, C., Xiong, F., Tong, W., Yang, S., et al. (2012). Perinatal nicotine exposure increases vulnerability of hypoxic-ischemic brain injury in neonatal rats: role of angiotensin II receptors. *Stroke 43*(9), 2483–2490.

Liu, M., Oyarzabal, E. A., Yang, R., Murphy, S. J., & Hurn, P. D. (2008). A novel method for assessing sex-specific and genotype-specific response to injury in astrocyte culture. *Journal of Neuroscience Methods*, *171*(2), 214–217.

Moylan, S., Gustavson, K., Øverland, S., Karevold, E. B., Jacka, F. N., Pasco, J. A., et al. (2015). The impact of maternal smoking during pregnancy on depressive and anxiety behaviors in children: the Norwegian mother and child cohort study. *BMC Medicine*, *13*(1), 24.

Popa-Wagner, A., Mitran, S., Sivanesan, S., Chang, E., & Buga, A. M. (2013). ROS and brain diseases: the good, the bad, and the ugly. *Oxidative Medicine and Cellular Longevity*, *2013*, 963520.

Potvin, S., Stip, E., Sepehry, A. A., Gendron, A., Bah, R., & Kouassi, E. (2008). Inflammatory cytokine alterations in schizophrenia: a systematic quantitative review. *Biological Psychiatry*, *63*(8), 801–808.

Qiu, F., Liang, C.-L., Liu, H., Zeng, Y.-Q., Hou, S., Huang, S., et al. (2017). Impacts of cigarette smoking on immune responsiveness: up and down or upside down? *Oncotarget*, *8*(1), 268–284.

Rahman, I., & Adcock, I. M. (2006). Oxidative stress and redox regulation of lung inflammation in COPD. *The European Respiratory Journal*, *28*(1), 219–242.

Rahn, J. J., Stackley, K. D., & Chan, S. S. (2013). Opa1 is required for proper mitochondrial metabolism in early development. *PLoS ONE*, *8*(3)e59218.

Rapaport, D., Neupert, W., & Lill, R. (1997). Mitochondrial protein import. Tom40 plays a major role in targeting and translocation of preproteins by forming a specific binding site for the presequence. *The Journal of Biological Chemistry*, *272*(30), 18725–18731.

Santos-Galindo, M., Acaz-Fonseca, E., Bellini, M. J., & Garcia-Segura, L. M. (2011). Sex differences in the inflammatory response of primary astrocytes to lipopolysaccharide. *Biology of Sex Differences*, *2*, 7.

Socol, M. L., Manning, F. A., Murata, Y., & Druzin, M. L. (1982). Maternal smoking causes fetal hypoxia: experimental evidence. *American Journal of Obstetrics and Gynecology*, *142*(2), 214–218.

Streja, E., Miller, J. E., Bech, B. H., Greene, N., Pedersen, L. H., Yeargin-Allsopp, M., et al. (2013). Congenital cerebral palsy and prenatal exposure to self-reported maternal infections, fever, or smoking. *American Journal of Obstetrics and Gynecology*, *209*(4), 332.e1–332.e10.

Surmeli, N. B., Litterman, N. K., Miller, A. F., & Groves, J. T. (2010). Peroxynitrite mediates active site tyrosine nitration in manganese superoxide dismutase. Evidence of a role for the carbonate radical anion. *Journal of the American Chemical Society*, *132*(48), 17174–17185.

Vani, G., Anbarasi, K., & Shyamaladevi, C. S. (2015). Bacoside A: role in cigarette smoking induced changes in brain. *Evidence-based Complementary and Alternative Medicine: eCAM*, *2015*, 286137.

Virmani, A., & Binienda, Z. (2004). Role of carnitine esters in brain neuropathology. *Molecular Aspects of Medicine*, *25*(5–6), 533–549.

Wang, J., Green, P. S., & Simpkins, J. W. (2001). Estradiol protects against ATP depletion, mitochondrial membrane potential decline and the generation of reactive oxygen species induced by 3-nitroproprionic acid in SK-N-SH human neuroblastoma cells. *Journal of Neurochemistry*, *77*(3), 804–811.

Wei, H., Chadman, K. K., McCloskey, D. P., Sheikh, A. M., Malik, M., Brown, W. T., et al. (2012). Brain IL-6 elevation causes neuronal circuitry imbalances and mediates autism-like behaviors. *Biochimica et Biophysica Acta (BBA): Molecular Basis of Disease*, *1822*(6), 831–842.

Witelson, S. F., Beresh, H., & Kigar, D. L. (2006). Intelligence and brain size in 100 postmortem brains: sex, lateralization and age factors. *Brain*, *129*(2), 386–398.

Zagni, E., Simoni, L., & Colombo, D. (2016). Sex and gender differences in central nervous system-related disorders. *Neuroscience Journal*, *2016*, 2827090.

Zhang, L., Trushin, S., Christensen, T. A., Bachmeier, B. V., Gateno, B., Schroeder, A., et al. (2016). Altered brain energetics induces mitochondrial fission arrest in Alzheimer's disease. *Scientific Reports*, *6*, 18725.

CHAPTER

3

Nicotine Effects in Adolescents

Sari Izenwasser

Department of Psychiatry & Behavioral Sciences, University of Miami Miller School of Medicine, Miami, FL, United States

Abbreviations

5HT	serotonin
AD	adult
Adol	adolescent
CB recs	cannabinoid receptors
CPP	conditioned place preference
DA	dopamine
DAT	dopamine transporter
nAChR	nicotinic acetylcholine receptor
NC	no change
Nic	nicotine
S-A	self-administration
sens	sensitization
SERT	serotonin transporter
STR	striatum

For many years, preclinical studies on nicotine focused solely on adults, primarily on male adults. In the past 15–20 years, more research has been done examining the effects of nicotine in adolescents. To date, there still are more studies in males; however, a number of studies in males and females have shown that there are different effects across sex and across age, when compared to data in adults. This chapter will focus on the effects of nicotine initiated during adolescence in both males and females, where possible. In addition, an attempt will be made to include studies where both adolescents and adults are included, to try to determine whether or not adolescence is unique. As mentioned above, however, many studies still report data on only males and often do not include a direct comparison with adults. Thus, it often is necessary to compare results across labs. Due to space and reference limitations, this cannot be an exhaustive review of the literature; however, an attempt will be made to include studies that show the breadth of the field.

The average age of initiation of nicotine use occurs during adolescence with 90% of users first trying smoking before 18 years of age (Centers for Disease Control and Prevention, 2017). Although cigarette smoking declined in middle and high school students between 2011 and 2016, use of electronic cigarettes and hookahs increased. Approximately 10% of high school students and 3% of middle school students report having used two or more tobacco products within the past 30 days, and lifetime use of multiple tobacco products by high school students was just over 30%. In general, rates of use in high school and middle school students are higher in males than in females, although slightly more high school females than males report having used a hookah.

3.1 NICOTINE REWARD IN ADOLESCENT AND ADULT MALES AND FEMALES

Preclinical laboratory studies often show that adolescents often exhibit unique responses to nicotine compared to adults (see Table 3.1). In general, adolescents tend to be more sensitive than adults to the effects of nicotine, although males and females often respond differently from one another. Most studies show that nicotine stimulates locomotor activity and that sensitization to this effect does not occur in adolescent males (Collins & Izenwasser, 2004; Collins, Montano, & Izenwasser, 2004; Cruz, Delucia, & Planeta, 2005; Faraday, Elliott, & Grunberg, 2001; Schochet, Kelley, & Landry, 2004). In contrast, a number of studies show that sensitization does occur in adult males (Bracken, Chambers, Berg, Rodd, & McBride, 2011; Collins & Izenwasser, 2004; Collins, Montano, et al., 2004; Cruz et al., 2005; Schochet et al., 2004) and in adult and adolescent females (Collins & Izenwasser, 2004). The sensitization in male adolescents occurs during the course of daily nicotine administration (Collins & Izenwasser, 2004; Collins, Montano, et al., 2004) and remains evident if the animals are challenged 3 days after the end of the nicotine treatment period (Cruz et al., 2005). One potential

Neuroscience of Nicotine
https://doi.org/10.1016/B978-0-12-813035-3.00003-4

TABLE 3.1 Behavioral Effects of Nicotine During Adolescence vs Adulthood in Males and Females

Nic 7 days; inj PND 28, 60; test 1 or 30 days later	Locomotor activity	Adol: males no sens, females sens; AD: males and females sens	Collins and Izenwasser (2004) and Collins, Montano, et al. (2004)
Nic 7 days; inj PND 28–34, 90–96; test 3 days later	Locomotor activity	Male Adol: no sens; male AD: sens	Cruz et al. (2005)
Nic 10 days; PND 28, 70; males only	Locomotor activity, cue conditioning	Adol: no sens, cue conditioning; AD: sens and cue conditioning	Schochet et al. (2004)
Nic PND 31–42; test 7 days later; males only	Locomotor activity	Sens as adults to loco and stereotypy	Bracken et al. (2011)
Nic continuous 21 days; PND 30, 60	Locomotor activity	Adol, AD: males, females NC during Tx	Faraday et al. (2001)
Nic continuous 12 days; PND 25, 55; males only	Locomotor activity	Adol, AD: sens during Tx	Faraday et al. (2003)
Nic daily; PND 49, 69; females only	Locomotor activity in figure-8 maze	Female Adol: no effect; AD: ↓ activity	Levin et al. (2003)
Nic continuous PND 30–47	Grooming	Adol: male no effect, female decreased; AD: no effect	Trauth, Seidler, and Slotkin (2000)
Nic acute	Social anxiety	Nic produced anxiety in male and female Adol; more potent in females	Cheeta, Irvine, Tucci, Sandhu, and File (2001), Elliott, Faraday, Phillips, and Grunberg (2004), and Slawecki, Gilder, Roth, and Ehlers (2003)
Nic 5 days; PND 25–30, 55–60	Elevated plus maze	Adol: males ↑ anxiety, females ↓ anxiety; AD: males/females ↓ anxiety	Cheeta et al. (2001), Elliott et al. (2004), and Slawecki et al. (2003)
Nic 6 days; PND 31–36	Open field test as AD	Males ↑ anxiety (less time in center)	Cheeta et al. (2001), Elliott et al. (2004), and Slawecki et al. (2003)
Nic PND 34–43, 60–69; test 5 weeks later	Nic S-A	Male increased after Tx as Adol not AD	Adriani et al. (2003)
Nic S-A start PND 32, 64	Nic S-A	Adol females: 2× AD intake; Adol males: 3× AD intake	Levin et al. (2003, 2007)
Nic PND 28, 38, 90	Nic CPP	Males CPP only on PND 28	Belluzzi et al. (2004)
Nic PND 28, 58; test PND 40, 70	Nic CPP	CPP only in Adol, not in AD	Vastola, Douglas, Varlinskaya, and Spear (2002)
Nic CPP PND 34–38, 66–70	Nic CPP	CPP Adol more sensitive than AD. Nic more potent in male in Adol/AD	Lenoir et al. (2015)
Nic or cigarette smoke PND 28–41, 63–76; test 28 days later	Nic CPP, S-A	↑ CPP after Nic or smoke, ↑ Nic S-A after smoke only	de la Pena et al. (2015)

All studies used rats unless noted. Abbreviations: *AD*, adult; *Adol*, adolescent; *CPP*, conditioned place preference; *NC*, no change; *Nic*, nicotine; *S-A*, self-administration; *sens*, sensitization.

contributor to the lack of sensitization in adolescent males is that the acute effects of nicotine on activity are much greater in this group than in adult males or in adult or adolescent females. With initial responses to nicotine so high, it might be difficult for a significant increase in activity to occur. Continuous administration of nicotine via osmotic minipump also produced sensitization to the locomotor effects of nicotine; however, this occurred in both adolescent and adult males (Faraday, Elliott, Phillips, & Grunberg, 2003).

When first exposure to nicotine is during adolescence, both male and female rats self-administer significantly larger amounts of nicotine than adults. Female adolescent rats self-administer about twice as much nicotine as adult females (Levin, Rezvani, Montoya, Rose, & Swartzwelder, 2003), and male adolescents self-administer three times as much as adult males (Levin et al., 2007). These increased rates of administration persist into adulthood once they have been established during adolescence. After repeated administration of nicotine to male rats during adolescence, but not adulthood, nicotine self-administration (S-A) later in adults is increased significantly (Adriani et al., 2003).

Nicotine conditioned place preference (CPP) is evident in adolescent and adult males and females. At both ages, male rats exhibited a significant CPP at lower doses than females, suggesting that nicotine is more potent as a reward in males than in females (Lenoir et al., 2015). This is interesting in that the opposite is generally observed for other psychostimulant drugs such as cocaine and amphetamine (e.g., Dow-Edwards, 2010; Festa et al., 2004; Hu & Becker, 2003; Lynch & Carroll, 1999; Russo et al., 2003; Zakharova, Wade, & Izenwasser, 2009). Overall, adolescents were more sensitive than adults, although the way in which this was manifest differed across sex. In males, the dose-response curve was shifted to the left in adolescents compared to adults. In contrast, the curve was shifted upward in adolescent females compared to adults, suggesting greater efficacy of reward in adolescents than adults (Lenoir et al., 2015). An earlier study in males showed that nicotine CPP was evident only early in adolescence, but not later in adolescence or during adulthood (Belluzzi, Lee, Oliff, & Leslie, 2004). One month after repeated administration of nicotine during either adolescence or adulthood, nicotine CPP was increased in male rats (de la Pena et al., 2015). Exposure to cigarette smoke, however, only led to increased nicotine CPP when the exposure occurred during adolescence. These findings suggest that early adolescence may be a period of particular vulnerability to the effects of nicotine.

3.2 NICOTINE DURING ADOLESCENCE INCREASES EFFECTS OF SUBSEQUENT DRUG ADMINISTRATION

Many studies show that adolescent males are more susceptible to cross sensitization with other drugs (e.g., an increased response to another drug after exposure to nicotine during adolescence), promoting the idea that nicotine may be a "gateway" drug that leads to the use of other substances of abuse (see Table 3.2). Repeated exposure to nicotine during adolescence led to sensitization to the locomotor stimulant effects of cocaine (Collins & Izenwasser, 2004) and amphetamine (Collins, Montano, et al., 2004) in male rats, and these effects persisted into adulthood (Collins & Izenwasser, 2004; Collins, Montano, et al., 2004). This cross sensitization was not evident at either time period in female or adult rats.

Adolescent nicotine exposure increased cocaine conditioned place preference (CPP) in rats when tested later during adulthood (McMillen, Davis, Williams, & Soderstrom, 2005). Studies in mice, however, show it was decreased (Kelley & Middaugh, 1999; Kelley & Rowan, 2004). It is not clear what contributes to this species difference. In contrast, Pomfrey et al. (Pomfrey, Bostwick, Wetzell, & Riley, 2015) reported that nicotine during adolescence had no effect on cocaine CPP or self-administration. Although multiple doses of cocaine were tested in this study, the data are presented in an unusual manner whereby all of the doses are collapsed into a single bar on a graph. Thus, it is difficult to determine whether or not there was a dose-response or any curve shift or to draw any conclusions as to why this lab found results that differ from other labs.

Nicotine self-administration during adolescence led to increased ethanol iv self-administration in adolescent

TABLE 3.2 Effects of Nicotine During Adolescence vs Adulthood on Other Drugs in Males and Females

Nic 7 days PND 28, 60; test 1 or 30 days later	Locomotor activity—cocaine	Adol: male sens to cocaine, female NC; AD: male/female NC	Collins and Izenwasser (2004)
Nic 7 days PND 28, 60; test 1 or 30 days later	Locomotor activity—amphetamine	Adol: male sens to amphetamine, female NC; AD: male/female NC	Collins, Montano, et al. (2004)
Nic PND 35–50; S-A starts PND 51; males only	Methamphetamine S-A	↑ Methamphetamine S-A after Adol pretreatment	Pipkin et al. (2014)
Nic PND 42–60	Fentanyl S-A	↑ Males ↑ S-A; females NC	Klein (2001)
Nic PND 25–60 (mice); tested 12 or 28 days later	Cocaine CPP	Males ↓ CPP	Kelley and Middaugh (1999) and Kelley and Rowan (2004)
Nic PND 35–44; test PND 80	Cocaine CPP	Males ↑ CPP	McMillen et al. (2005)
Nic PND 28–42; test PND 66; males only	Cocaine CPP, S-A	NC	
Cigarette smoke PND 35–66; males only; mice	Ethanol drinking	↑ Ethanol	Burns and Proctor (2013)
NIC daily PND 28–34, 60–66; test 30 days later; males only	Cocaine S-A	↑ Cocaine S-A after Adol pretreatment only	Reed and Izenwasser (2017)
Nic S-A starts PND 32	Ethanol IV S-A	Adol males: ↑ Ethanol; females/adults: NC	Larraga et al. (2017)
Nic PND 45–59, Inj every 3 days; test as adults	Ethanol drinking	Sens when Nic was paired with operant chamber	Zipori et al. (2017)

↑ means that an increase was observed, ↓ means a decrease was observed. All studies used rats unless noted. Abbreviations: *AD*, adult; *Adol*, adolescent; *CPP*, conditioned place preference; *NC*, no change; *Nic*, nicotine; *S-A*, self-administration; *sens*, sensitization.

males but not females or adults of either sex (Larraga, Belluzzi, & Leslie, 2017). Similarly, exposure to cigarette smoke for 3 weeks during adolescence into young adulthood led to greater oral ethanol consumption in male mice (Burns & Proctor, 2013). Similarly, in male rats, nicotine injections during adolescence resulted in increased methamphetamine self-administration (Pipkin et al., 2014). Repeated nicotine also increased fentanyl self-administration in males with no change in females (Klein, 2001). Male adolescent rats exposed to nicotine also increased cocaine self-administration, whereas nicotine in adult males had no significant effect on cocaine self-administration (Reed & Izenwasser, 2017).

3.3 NICOTINE EFFECTS ON NEUROCHEMISTRY IN ADOLESCENCE

As for the behavioral studies, numerous studies have been done examining multiple receptors and brain pathways after nicotine administration in adolescent and adult rats (see Table 3.3). Studies from our lab showed that daily injections of nicotine increased nicotinic acetylcholine receptor (nAChR) densities in the caudate-putamen and nucleus accumbens of adult, but not adolescent, male rats (Collins, Wade, Ledon, & Izenwasser, 2004). There were no changes in dopamine transporter, dopamine D_1 or D_2 receptor, or serotonin transporter densities in adult rats pretreated with nicotine. In adolescent rats, dopamine transporter densities were increased, and serotonin transporter densities were decreased following repeated nicotine administration. These data are consistent with the behavioral data showing that sensitization to the locomotor stimulant effects of cocaine and amphetamine, but not nicotine, occurs subsequent to nicotine administration in adolescents, while the opposite is true in adults (Collins & Izenwasser, 2004; Collins, Montano, et al., 2004). In addition to the increase in nAChR density, there also is evidence of increased gene expression of several nicotine receptor subunits (e.g., α5, α6, and β2) (Adriani et al., 2003) following repeated injections of nicotine. Following continuous infusion of nicotine for 14 days, there were significant increases in α4 and β2 receptors, but no change in α5 receptors in male and female adolescents and adults (Hoegberg, Lomazzo, Lee, & Perry, 2015). Thus, there may be differences in the effects of repeated administration versus continuous infusion of nicotine. Other studies using different regimens of nicotine administration have shown changes in nAChRs in multiple brain regions. For example, continuous administration of nicotine via minipump led to increased nAChR densities in the hippocampus, cerebral cortex, and midbrain (Abreu-Villaca et al., 2003; Trauth, Seidler, McCook, & Slotkin, 1999). Continuous administration of nicotine also led to increases in serotonin transporter binding in female rats,

TABLE 3.3 Neurochemical Effects of Treatment With Nicotine During Adolescence vs Adulthood in Males and Females

Nic from PND 28–34, 60–66 daily inj	Adol: ↑ DAT, ↓ SERT, NC nAChR; AD: ↑ nAChR, NC DAT, SERT	Collins and Izenwasser (2002)
Nic from PND 34–43, 60–69 daily inj Male only	Adol: ↑ α5, α6, β2; AD: NC	Adriani et al. (2003)
Nic continuous, PND 28–42, 70–84	Adol/AD: males/females ↑ α4, β2; NC α5	Hoegberg et al. (2015)
Nic 2 weeks, PND 30 Adol only	Initial ↑ DA turnover in STR in males, PND 50–60 ↓ DA turnover in males and females	Trauth, Seidler, Ali, and Slotkin (2001)
Nic from PND 30–47 or 30–37 continuous	Adol, AD: ↑ nAChR during Tx; some persistence 30 days later in males, not females	Abreu-Villaca et al. (2003), Abreu-Villaca, Seidler, Tate, Cousins, and Slotkin (2004), and Trauth et al. (1999)
Nic continuous, PND 30–47.5, Adol only	Males/females: ↑ SERT PND 75; ↑ $5HT_2$ PND 45, 60	Xu et al. (2001)
Nic continuous, PND 30–47 then 90–107; test PND 180	Males: ↑ 5HT turnover; females: much smaller and delayed effect	Slotkin et al. (2014)
Nic continuous, PND 30–47 then 90–107; test PND 180	Long term ↑ 5HT recs in males, ↓ in females	Slotkin and Seidler (2009)
Nic PND 30–47.5, Adol only. Males and females analyzed together	Adol: ↓ M2 muscarinic, ↑ basal and FSK-stim AC	Chow et al. (2000)
Nic PND 30, 70. Male only	Baseline arc and c-fos: Adol > AD; >↑ in arc in PFC in Adol than AD after nic	Schochet, Kelley, and Landry (2005)
Nic 4 days PND 31, 41, 56. Males and females analyzed together	Nic acutely did not ↑ DA in Adol; tolerance to ↑ DA in AD	Badanich and Kirstein (2004)
Nic from PND 34–43 daily inj. Test 1 or 30 days later	1 day: NC. 30 days: males/females ↓ CB recs in STR, ↑ in hippo, ↓ Mu opioid in STR and hippo	Marco et al. (2007)
Nic 7 days, inj PND 28–34, 90–96. Test 3 days later. Male only	AD: tolerance to elevation in corticosterone; Adol: NC	Cruz et al. (2005)

↑ means that an increase was observed, ↓ means a decrease was observed. Abbreviations: *5-HT*, serotonin; *AD*, adult; *Adol*, adolescent; *CB recs*, cannabinoid receptors; *DA*, dopamine; *DAT*, dopamine transporter; *nAChR*, nicotinic acetylcholine receptor; *NC*, no change; *Nic*, nicotine; *SERT*, serotonin transporter; *STR*, striatum.

much more quickly than in male rats. In females, the increases were observed during adolescence, and they were not seen in males until later when they were adults (Xu, Seidler, Ali, Slikker, & Slotkin, 2001). In contrast, serotonin 5HT2 binding was increased in male, but not female rats (Xu, Seidler, Cousins, Slikker, & Slotkin, 2002).

Numerous studies have been done by Slotkin and colleagues examining the interaction of nicotine with the brain serotonergic system (e.g., Slotkin et al., 2007; Slotkin, Card, & Seidler, 2014; Slotkin & Seidler, 2007, 2009). They have shown that nicotine during adolescence has a great impact on serotonin that persists long into adulthood and that there are age and sex differences. For example, exposure to nicotine during adolescence greatly increases the acute effect of nicotine on serotonin turnover in adults, leading to a significant depletion of serotonin (Slotkin et al., 2014). Furthermore, females exposed to nicotine in adolescence blunted the spike in serotonin turnover seen upon withdrawal from nicotine in adults, and this was not observed in males (Slotkin et al., 2014).

Although much focus has been on nAChRs, dopamine, and serotonin, nicotine also has effects on other receptor systems, for example, decreases in M2 muscarinic receptors (Chow, Seidler, McCook, & Slotkin, 2000), cannabinoid receptors (Marco et al., 2006), and mu-opioid receptors (Marco et al., 2006) after repeated administration in adolescence.

3.4 CONCLUSIONS

In conclusion, preclinical studies of the behavioral effects of nicotine suggest that adolescent males are particularly sensitive to the effects of this drug. Nicotine is more potent as a reward and a behavioral stimulant in adolescent males than in adolescent females or adults, and the long-term effects are more pronounced in male adolescents than in the other groups. Exposure to nicotine during adolescence increases the rewarding and reinforcing properties of other drugs of abuse including stimulants (cocaine, amphetamine, and methamphetamine), opioids (fentanyl), and ethanol in adolescent males (see Table 3.3). Thus, these findings suggest that adolescents may be particularly vulnerable to the effects of nicotine and to the risk of using other drugs of abuse subsequent to exposure to nicotine and further that adolescent males may exhibit the greatest vulnerability. These studies also show that it is important to study different developmental stages and both sexes before drawing conclusions about the effects of nicotine and other drugs. The results of most studies show that there are both acute differences and persistent adaptations to repeated nicotine administration that differ across age and sex.

MINI-DICTIONARY OF TERMS

Conditioned place preference Method used to study drug reward.
Self-administration Method used to study drug reinforcement where the subject administers drugs by performing an operant task.
Sensitization Increased effect of a drug after exposure to some stimulus (e.g., another drug).

Key Facts of Adolescent Nicotine Effects

- Most smokers initiate cigarette use before age 18; thus, it is important to study this developmental period.
- Drug self-administration is a model used in the laboratory whereby the subject administers drug to itself by performing a task, such as a lever press. This mimics the human condition whereby people take their own drugs.
- Conditioned place preference is a laboratory model used to determine whether or not a drug is considered to be rewarding.
- Sensitization indicates that a drug has an even greater effect after repeated exposure than it did upon first exposure.

Summary Points

- The average age of initiation of nicotine use occurs during adolescence with 90% of users first trying smoking before 18 years of age.
- Preclinical laboratory studies often show that adolescents often exhibit unique responses to nicotine compared to adults.
- Adolescent males are particularly sensitive to the effects of nicotine, and they differ from adults and females.
- Nicotine during adolescence often increases subsequent administration of other drugs of abuse.
- It is important to study different developmental stages and both sexes before drawing conclusions about the effects of nicotine and other drugs.

References

Abreu-Villaca, Y., Seidler, F. J., Qiao, D., Tate, C. A., Cousins, M. M., Thillai, I., et al. (2003). Short-term adolescent nicotine exposure has immediate and persistent effects on cholinergic systems: critical periods, patterns of exposure, dose thresholds. *Neuropsychopharmacology*, *28*, 1935–1949.

Abreu-Villaca, Y., Seidler, F. J., Tate, C. A., Cousins, M. M., & Slotkin, T. A. (2004). Prenatal nicotine exposure alters the response to nicotine administration in adolescence: effects on cholinergic systems during exposure and withdrawal. *Neuropsychopharmacology*, *29*, 879–890.

Adriani, W., Spijker, S., Deroche-Gammonet, V., Laviola, G., Le Moal, M., Smit, A. B., et al. (2003). Evidence for enhanced neurobehavioral vulnerability to nicotine during periadolescence in rats. *The Journal of Neuroscience*, *23*, 4712–4716.

Badanich, K. A., & Kirstein, C. L. (2004). Nicotine administration significantly alters accumbal dopamine in the adult but not in the adolescent rat. *Annals of the New York Academy of Sciences*, *1021*, 410–417.

Belluzzi, J. D., Lee, A. G., Oliff, H. S., & Leslie, F. M. (2004). Age-dependent effects of nicotine on locomotor activity and conditioned place preference in rats. *Psychopharmacology*, *174*, 389–395.

Bracken, A. L., Chambers, R. A., Berg, S. A., Rodd, Z. A., & McBride, W. J. (2011). Nicotine exposure during adolescence enhances behavioral sensitivity to nicotine during adulthood in Wistar rats. *Pharmacology, Biochemistry, and Behavior*, *99*, 87–93.

Burns, B. E., & Proctor, W. R. (2013). Cigarette smoke exposure greatly increases alcohol consumption in adolescent C57BL/6 mice. *Alcoholism, Clinical and Experimental Research*, *37*(Suppl. 1), E364–E372.

Centers for Disease Control and Prevention. (2017). Tobacco use among middle and high school students—United States, 2011–2016. *Morbidity and Mortality Weekly Report*,.

Cheeta, S., Irvine, E. E., Tucci, S., Sandhu, J., & File, S. E. (2001). In adolescence, female rats are more sensitive to the anxiolytic effect of nicotine than are male rats. *Neuropsychopharmacology*, *25*, 601–607.

Chow, F. A., Seidler, F. J., McCook, E. C., & Slotkin, T. A. (2000). Adolescent nicotine exposure alters cardiac autonomic responsiveness: b-adrenergic and m2-muscarinic receptors and their linkage to adenylyl cyclase. *Brain Research*, *878*, 119–126.

Collins, S. L., & Izenwasser, S. (2002). Cocaine differentially alters behavior and neurochemistry in periadolescent versus adult rats. *Developmental Brain Research*, *138*, 27–34.

Collins, S. L., & Izenwasser, S. (2004). Chronic nicotine differentially alters cocaine-induced locomotor activity in adolescent vs. adult male and female rats. *Neuropharmacology*, *46*, 349–362.

Collins, S. L., Montano, R., & Izenwasser, S. (2004). Nicotine treatment produces persistent increases in amphetamine-stimulated locomotor activity in periadolescent male but not female or adult male rats. *Developmental Brain Research*, *153*, 175–187.

Collins, S. L., Wade, D., Ledon, J., & Izenwasser, S. (2004). Neurochemical alterations produced by daily nicotine exposure in periadolescent vs. adult male rats. *European Journal of Pharmacology*, *502*, 75–85.

Cruz, F. C., Delucia, R., & Planeta, C. S. (2005). Differential behavioral and neuroendocrine effects of repeated nicotine in adolescent and adult rats. *Pharmacology, Biochemistry, and Behavior*, *80*, 411–417.

de la Pena, J. B., Ahsan, H. M., Tampus, R., Botanas, C. J., de la Pena, I. J., Kim, H. J., et al. (2015). Cigarette smoke exposure during adolescence enhances sensitivity to the rewarding effects of nicotine in adulthood, even after a long period of abstinence. *Neuropharmacology*, *99*, 9–14.

Dow-Edwards, D. (2010). Sex differences in the effects of cocaine abuse across the life span. *Physiology & Behavior*, *100*, 208–215.

Elliott, B. M., Faraday, M. M., Phillips, J. M., & Grunberg, N. E. (2004). Effects of nicotine on elevated plus maze and locomotor activity in male and female adolescent and adult rats. *Pharmacology, Biochemistry, and Behavior*, *77*, 21–28.

Faraday, M. M., Elliott, B. M., & Grunberg, N. E. (2001). Adult vs. adolescent rats differ in biobehavioral responses to chronic nicotine administration. *Pharmacology, Biochemistry, and Behavior*, *70*, 475–489.

Faraday, M. M., Elliott, B. M., Phillips, J. M., & Grunberg, N. E. (2003). Adolescent and adult male rats differ in sensitivity to nicotine's activity effects. *Pharmacology, Biochemistry, and Behavior*, *74*, 917–931.

Festa, E. D., Russo, S. J., Gazi, F. M., Niyomchai, T., Kemen, L. M., Lin, S.-N., et al. (2004). Sex differences in cocaine-induced behavioral responses, pharmacokinetics, and monoamine levels. *Neuropharmacology*, *46*, 672–687.

Hoegberg, B. G., Lomazzo, E., Lee, N. H., & Perry, D. C. (2015). Regulation of alpha4beta2alpha5 nicotinic acetylcholinergic receptors in rat cerebral cortex in early and late adolescence: sex differences in response to chronic nicotine. *Neuropharmacology*, *99*, 347–355.

Hu, M., & Becker, J. B. (2003). Effects of sex and estrogen on behavioral sensitization to cocaine in rats. *The Journal of Neuroscience*, *23*, 693–699.

Kelley, B. M., & Middaugh, L. D. (1999). Periadolescent nicotine exposure reduces cocaine reward in adult mice. *Journal of Addictive Diseases*, *18*, 27–39.

Kelley, B. M., & Rowan, J. D. (2004). Long-term, low-level adolescent nicotine exposure produces dose-dependent changes in cocaine sensitivity and reward in mice. *International Journal of Developmental Neuroscience*, *22*, 339–348.

Klein, L. C. (2001). Effects of adolescent nicotine exposure on opioid consumption and neuroendocrine responses in adult male and female rats. *Experimental and Clinical Psychopharmacology*, *9*, 251–261.

Larraga, A., Belluzzi, J. D., & Leslie, F. M. (2017). Nicotine increases alcohol intake in adolescent male rats. *Frontiers in Behavioral Neuroscience*, *11*, 25.

Lenoir, M., Starosciak, A. K., Ledon, J., Booth, C., Zakharova, E., Wade, D., et al. (2015). Sex differences in conditioned nicotine reward are age-specific. *Pharmacology, Biochemistry, and Behavior*, *132*, 56–62.

Levin, E. D., Lawrence, S. S., Petro, A., Horton, K., Rezvani, A. H., Seidler, F. J., et al. (2007). Adolescent vs. adult-onset nicotine self-administration in male rats: duration of effect and differential nicotinic receptor correlates. *Neurotoxicology and Teratology*, *29*, 458–465.

Levin, E. D., Rezvani, A. H., Montoya, D., Rose, J. E., & Swartzwelder, H. S. (2003). Adolescent-onset nicotine self-administration modeled in female rats. *Psychopharmacology*, *169*, 141–149.

Lynch, W., & Carroll, M. (1999). Sex differences in the acquisition of intravenously self-administered cocaine and heroin in rats. *Pyschopharmacology*, *144*, 77–82.

Marco, E. M., Granstrem, O., Moreno, E., Llorente, R., Adriani, W., Laviola, G., et al. (2007). Subchronic nicotine exposure in adolescence induces long-term effects on hippocampal and striatal cannabinoid-CB1 and mu-opioid receptors in rats. *European Journal of Pharmacology*, *557*, 37–43.

Marco, E. M., Llorente, R., Moreno, E., Biscaia, J. M., Guaza, C., & Viveros, M. P. (2006). Adolescent exposure to nicotine modifies acute functional responses to cannabinoid agonists in rats. *Behavioural Brain Research*, *172*, 46–53.

McMillen, B. A., Davis, B. J., Williams, H. L., & Soderstrom, K. (2005). Periadolescent nicotine exposure causes heterologous sensitization to cocaine reinforcement. *European Journal of Pharmacology*, *509*, 161–164.

Pipkin, J. A., Kaplan, G. J., Plant, C. P., Eaton, S. E., Gil, S. M., Zavala, A. R., et al. (2014). Nicotine exposure beginning in adolescence enhances the acquisition of methamphetamine self-administration, but not methamphetamine-primed reinstatement in male rats. *Drug and Alcohol Dependence*, *142*, 341–344.

Pomfrey, R. L., Bostwick, T. A., Wetzell, B. B., & Riley, A. L. (2015). Adolescent nicotine exposure fails to impact cocaine reward, aversion and self-administration in adult male rats. *Pharmacology, Biochemistry, and Behavior*, *137*, 30–37.

Reed, S. C., & Izenwasser, S. (2017). Nicotine produces long-term increases in cocaine reinforcement in adolescent but not adult rats. *Brain Research*, *1654*, 165–170.

Russo, S. J., Jenab, S., Fabian, S. J., Festa, E. D., Kemen, L. M., & Quinones-Jenab, V. (2003). Sex differences in the conditioned rewarding effects of cocaine. *Brain Research*, *970*, 214–220.

Schochet, T. L., Kelley, A. E., & Landry, C. F. (2004). Differential behavioral effects of nicotine exposure in adolescent and adult rats. *Psychopharmacology*, *175*, 265–273.

Schochet, T. L., Kelley, A. E., & Landry, C. F. (2005). Differential expression of arc mRNA and other plasticity-related genes induced by nicotine in adolescent rat forebrain. *Neuroscience*, *135*, 285–297.

Slawecki, C. J., Gilder, A., Roth, J., & Ehlers, C. L. (2003). Increased anxiety-like behavior in adult rats exposed to nicotine as adolescents. *Pharmacology, Biochemistry, and Behavior, 75*, 355–361.

Slotkin, T. A., Card, J., & Seidler, F. J. (2014). Nicotine administration in adolescence reprograms the subsequent response to nicotine treatment and withdrawal in adulthood: sex-selective effects on cerebrocortical serotonergic function. *Brain Research Bulletin, 102*, 1–8.

Slotkin, T. A., MacKillop, E. A., Rudder, C. L., Ryde, I. T., Tate, C. A., & Seidler, F. J. (2007). Permanent, sex-selective effects of prenatal or adolescent nicotine exposure, separately or sequentially, in rat brain regions: indices of cholinergic and serotonergic synaptic function, cell signaling, and neural cell number and size at 6 months of age. *Neuropsychopharmacology, 32*, 1082–1097.

Slotkin, T. A., & Seidler, F. J. (2007). A unique role for striatal serotonergic systems in the withdrawal from adolescent nicotine administration. *Neurotoxicology and Teratology, 29*, 10–16.

Slotkin, T. A., & Seidler, F. J. (2009). Nicotine exposure in adolescence alters the response of serotonin systems to nicotine administered subsequently in adulthood. *Developmental Neuroscience, 31*, 58–70.

Trauth, J. A., Seidler, F. J., Ali, S. F., & Slotkin, T. A. (2001). Adolescent nicotine exposure produces immediate and long-term changes in CNS noradrenergic and dopaminergic function. *Brain Research, 892*, 269–280.

Trauth, J. A., Seidler, F. J., McCook, E. C., & Slotkin, T. A. (1999). Adolescent nicotine exposure causes persistent upregulation of nicotinic cholinergic receptors in rat brain regions. *Brain Research, 851*, 9–19.

Trauth, J. A., Seidler, F. J., & Slotkin, T. A. (2000). Persistent and delayed behavioral changes after nicotine treatment in adolescent rats. *Brain Research, 880*, 167–172.

Vastola, B. J., Douglas, L. A., Varlinskaya, E. I., & Spear, L. P. (2002). Nicotine-induced conditioned place preference in adolescent and adult rats. *Physiology & Behavior, 77*.

Xu, Z., Seidler, F. J., Ali, S. F., Slikker, W. J., & Slotkin, T. A. (2001). Fetal and adolescent nicotine administration: effects on CNS serotonergic systems. *Brain Research, 914*, 166–178.

Xu, Z., Seidler, F. J., Cousins, M. M., Slikker, W. J., & Slotkin, T. A. (2002). Adolescent nicotine administration alters serotonin receptors and cell signaling mediated through adenylyl cyclase. *Brain Research, 951*, 280–292.

Zakharova, E., Wade, D., & Izenwasser, S. (2009). Sensitivity to cocaine conditioned reward depends on sex and age. *Pharmacology, Biochemistry, and Behavior, 92*, 131–134.

Zipori, D., Sadot-Sogrin, Y., Goltseker, K., Even-Chen, O., Rahamim, N., Shaham, O., et al. (2017). Re-exposure to nicotine-associated context from adolescence enhances alcohol intake in adulthood. *Scientific Reports, 7*, 2479.

CHAPTER

4

The Impact of Traditional Cigarettes and E-Cigarettes on the Brain

Ewelina Wawryk-Gawda, Marta Lis-Sochocka, Patrycja Chylińska-Wrzos, Beata Budzyńska, Barbara Jodłowska-Jędrych

Department of Histology and Embryology with Experimental Cytology Unit, Medical University, Lublin, Poland

Abbreviations

Bcl-2	B-cell lymphoma 2
CREB/BDNF	cyclic adenosine monophosphate (cAMP) response element binding protein/brain-derived neurotrophic factor
fALFF	fractional amplitude of low-frequency fluctuation
fMRI	functional magnetic resonance imaging
FTND	Fagerström Test for Nicotine Dependence
GABA	γ-aminobutyric acid
GLT-1	glutamate transporter-1
IL-1β	interleukin 1β
nAChR	nicotinic acetylcholine receptor
NGF	nerve growth factor
RSFC	resting-state functional connectivity
TNF-α	tumor necrosis factor-alpha

4.1 INTRODUCTION

Nicotine is present in many products, and one of the oldest nicotine-containing products is conventional cigarettes (Fig. 4.1). Nowadays, new nicotine devices such as electronic cigarettes (e-cigarettes) are becoming increasingly popular (Oh & Kacker, 2014). The original intent of electronic cigarettes (Fig. 4.2) being introduced on the market was to help smokers stop smoking tobacco and, hence, to treat nicotine addiction (Kaisar, Prasad, Liles, & Cucullo, 2016). However, more and more data point to the fact that some people, especially adolescents, use the e-cigarettes as a first means of obtaining their fix of nicotine, thus starting a tobacco addiction. The use of e-cigarettes is among young people "modern" and is perceived as healthier than traditional smoking (Cooper, Harell, Pérez, Delk, & Perry, 2016; Kaisar et al., 2016; Lauterstein et al., 2016; Lee, Grana, & Glantz, 2014; Makadia, Roper, Adrews, & Tingen, 2017). The difference between smoking traditional cigarettes and the use of e-cigarettes is significant (Table 4.1), but in both cases, nicotine leads to the emergence of adverse effects on the body (Geiss, Bianchi, Barahona, & Barrero-Moreno, 2015). Electronic cigarettes do not burn tobacco to deliver flavor. Instead, they contain a liquid-based flavorant (e-liquid) that is thermally vaporized by an electric element. This liquid is a mixture of water, glycerin, and/or propylene glycol; sometimes, formaldehyde and other mixtures contain psychoactive substances. The liquid also contains nicotine and a variety of flavors (Bertholon et al., 2013). Although aerosol is formed as a result of the heating of e-liquid, it does not contain carcinogenic combustion products, and the amount of phenolics and carbonyls is not detectable. Still, both active and passive "smoking" of e-cigarettes bring about adverse effects on the human body (Long, 2014; Tayyarah & Long, 2014). Contrary to earlier opinions, the use of e-cigarettes also leads to nicotine addiction and even raises the risk of later addiction to traditional cigarettes (Arane & Goldman, 2016).

4.2 STRUCTURAL CHANGES OF BRAIN AMONG THE NICOTINE USERS

The impact of smoking of traditional cigarettes on the brain has been the object of many studies, yet literature on the impact of e-cigarettes on the central nervous system (CNS) is relatively scanty. Previously published data show that the changes observed in people using nicotine in the form of electronic cigarettes affect the same area in which changes have been observed in people who use traditional cigarettes: mainly the frontal cortex, insula, thalamus, and striatum (Fig. 4.3). These changes are

https://doi.org/10.1016/B978-0-12-813035-3.00004-6

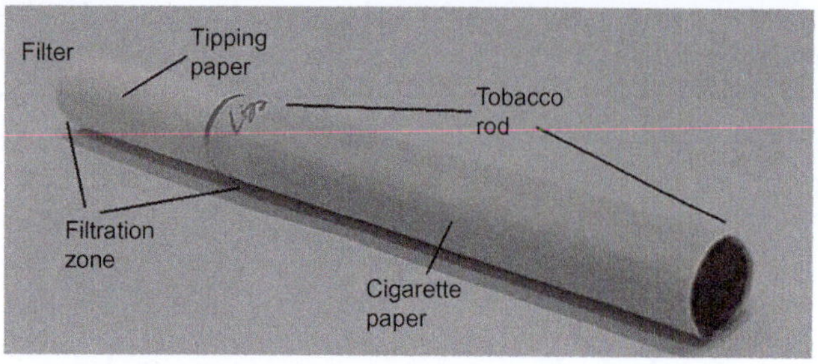

FIG. 4.1 Conventional cigarette.
The picture shows the conventional cigarette with the filter.

probably sex-dependent (Janes et al., 2017). Such studies also showed structural and functional changes in the brain caused by the use of nicotine (Dumais et al., 2017). Imaging such as that gained from magnetic resonance imaging (MRI) and functional magnetic resonance imaging (fMRI) allows for the observation of structural and functional changes that come about as a result of

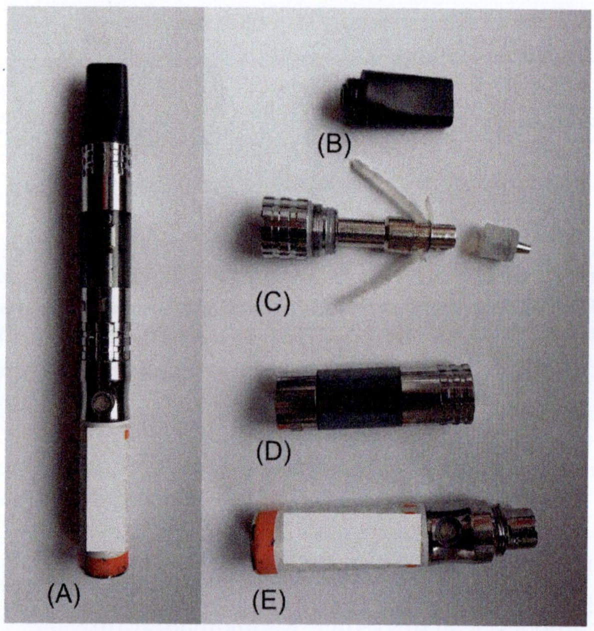

FIG. 4.2 Electronic cigarette.
The picture shows the electronic cigarette (A). The mouthpiece (B) resembling a tobacco cigarette's filter, an atomizer (C) that is a heating element and vaporizes the liquid in the mouthpiece and generates the mist with each puff. A cartridge (D) with liquid. The body of the device: battery component (E) that houses a rechargeable battery to power the atomizer. The body of the device also houses an electronic airflow sensor to automatically activate the heating element upon inhalation and to light up a LED indicator to signal the activation of the device with each puff. When the user pushes a button, it activates a pressure sensor by inhaling the heating element and then atomizes the liquid solu tion. The e-liquid reaches a temperature of roughly 100–250°C within a chamber to create an aerosolized vapor (Polosa et al., 2011).

TABLE 4.1 The Comparison of Traditional Cigarettes and Electronic Cigarettes

Feature	Electronic cigarettes	Conventional cigarettes
Exhaled	Aerosol, vapor	Smoke
Main ingredients of exhaled air	Water, 73, 3–75, 7%; glycerin, 24, 2–26, 7% (Long, 2014) Water, 15%; glycerin, 73% (Tayyarah & Long, 2014)	Water + glycerin, 83 ± 21%, the remaining exhaled aerosol mass for cigarette samples are attributed to particulates from combustion processes known to comprise more than 70% of mainstream conventional cigarette smoke (Long, 2014) Water, 20%; glycerin, 2%; favors and combustion products, 41% (Tayyarah & Long, 2014)
Nicotine in user environment	8–33 μg/puff (Tayyarah & Long, 2014)	194–232 μg (Tayyarah & Long, 2014).
Carbon monoxide in user environment	Not detectable	20 mg/cig (Long, 2014) 34% (Tayyarah & Long, 2014)
Phenolics[a] in user environment	Not significantly different than the amounts observed in exhaled breaths	66 μg/session (Long, 2014) 22–32 μg/puff (Tayyarah & Long, 2014)
Carbonyls[b] in user environment		24 μg/session (Long, 2014) 211–251 μg/puff (Tayyarah & Long, 2014)
Submicron particles (SMPs) in user environment	• Bigger than those of cigarettes • Occurs only during the "smoking" period • 48% of the deposited SMPs resulted so small as to be able to reach the alveolar region of passively exposed subjects (Bertholon et al., 2013; Protano et al., 2016)	• Deposition in the respiratory tract higher than those released by e-cigarette • 52%–53% of the deposited SMPs resulted so small as to be able to reach the alveolar region of passively exposed subjects

[a] *Phenolics: Hydroquinone, resorcinol, catechol, phenol, m/p-cresol, and o-cresol.*
[b] *Carbonyls: Formaldehyde, acetaldehyde, acrolein propionaldehyde, crotonaldehyde, MEK, and butyraldehyde.*

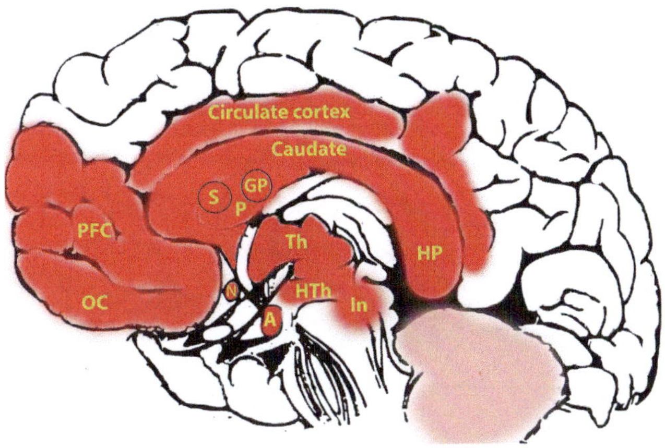

FIG. 4.3 Areas of the brain affected by conventional cigarette smoking and by electronic cigarette vaping.
The data concerning the impact of nicotine on the brain show that using of electronic cigarette and tobacco smoking affects specific area of the brain. These are the nucleus accumbens (NAc), amygdala (A), caudate, cingulate cortex, striatum (S), insula (In), frontal cortex, prefrontal cortex (PFC), orbitofrontal cortex (OC), thalamus (Th), hypothalamus (HTh), hippocampus (HP), putamen (P), and globus pallidus (GP).

addiction to tobacco (nicotine) and for the changes resulting from this addictive substance withdrawal. The results of such tests show that when compared to nonsmokers, cigarette smoking reduces the volume of the bilateral thalamus. In addition, the left thalamus volume in young smokers has been correlated with the severity of nicotine addiction, as assessed using the Fagerström Test for Nicotine Dependence (FTND) (Yu et al., 2018). What is more, in the study of Li et al. (2015), the authors observed significant cortical thinning in the frontal cortex, left insula, left middle temporal gyrus, right inferior parietal lobule, and right parahippocampus in young adult smokers. The authors stress that these brain structure changes may be the cause of an imbalance between the cognitive control and reward drive behaviors that are associated with nicotine addiction and relapse (Li et al., 2015).

4.3 FUNCTIONAL CHANGES OF THE BRAIN AMONG CONVENTIONAL CIGARETTES SMOKERS

Resting-state functional MRI (fMRI) scans are used in the researches on nicotine dependence and improve the understanding of the neuroplastic changes that occur during the development of nicotine dependence (Shen et al., 2016). Using these methods, researchers have shown that smokers, when compared with nonsmokers, had significantly lower resting-state functional connectivity (RSFC) in the reward circuit between the orbitofrontal cortex, superior frontal gyrus, temporal lobe, and insula (Zhou et al., 2017). Moreover, Yuan et al. (2016) have revealed the existence of pathological changes in the structure and function of the brain in smokers. In this research, tested smokers were also shown to have an increased volume of the right caudate and reduced frontostriatal circuits RSFC between the caudate, dorsolateral prefrontal cortex, and orbitofrontal cortex. Furthermore, Yuan et al. (2016) saw significant positive correlation between right caudate volume and the results of the Questionnaire on Smoking Urges (QSU) that assesses craving state. Wang et al. (2017) reveal reduced resting-state functional connectivity of the thalamus with the right dorsolateral prefrontal cortex, the anterior cingulate cortex, the insula, and the caudate, in smokers. They indicate as well that the thalamus also plays a role in nicotine addiction. Shen et al. (2016) reveal that, compared with low-dependent smokers, high-dependent smokers show greater RSFC between the right amygdala, the left nucleus accumbens, and between the bilateral hippocampus (Shen et al., 2016). In the research of Zhou et al. (2017), the RSFC between the orbitofrontal cortex, temporal lobe, inferior parietal cortex, occipital lobe, and insula was positively correlated with FTND. FMRI also allows for the evaluation of fractional amplitude of low-frequency fluctuation (fALFF) in the different regions of the brain. Utilizing this technique, Chu et al. (2014) observed an increase of fALFF in the left middle occipital gyrus, left limbic lobe, and left cerebellum posterior lobe but a decrease of the same in the right middle frontal gyrus, right superior temporal gyrus, right extra nuclear, left postcentral gyrus, and left cerebellum anterior lobe in smokers (as compared with nonsmokers) and suggested that these changes were associated with the cumulative amount of nicotine intake and the severity of nicotine dependence.

4.4 FUNCTIONAL CHANGES OF THE BRAIN AMONG ELECTRONIC CIGARETTES USERS

For now, data about electronic cigarette impact on the brain are limited; however, the fMRI study of Wall et al. (2017) shows brain activation in a network of cortical regions (the motor cortex, insula, cingulum, and amygdala) and subcortical regions (the putamen, thalamus, globus pallidus, and cerebellum) in e-cigarettes users, while concomitant relative deactivations were seen in the ventral striatum and orbitofrontal cortex. Hobkirk et al. (2018) also reported that electronic cigarette users showed decreased RSFC of two clusters in the right frontal pole and frontal medial cortexes with an attentional control salience network and a decreased RSFC of five clusters in the left thalamus and insula, in addition to a brain stem with a reward network encompassing the striatum. Moreover, Hobkirk et al. (2018) noted that after habit cessation, withdrawal symptoms are observed.

4.5 NICOTINE WITHDRAWAL SYMPTOMS

Nicotine withdrawal symptoms are negative feelings that occur after cessation of nicotine consumption. Among the symptoms are irritability, frustration, anxiety, depressed mood, difficulty concentrating, increased appetite, insomnia, and restlessness (McLaughlin, Dani, & De Biasi, 2015). Tests with fMRI have shown that the nicotine withdrawal syndrome is associated with increased connectivity between the anterior cingulate cortex and the precuneus, insula, orbital frontal gyrus, superior frontal gyrus, posterior cingulate cortex, superior temporal, and inferior temporal lobe (Huang et al., 2014). These data suggest that smoking triggers structural and functional neural adaptations in the brain that support withdrawal-induced craving (Samochowiec, Rogoziński, Hajduk, Skrzypińska, & Arentowicz, 2001). The withdrawal syndrome symptoms are the consequence of changes caused by nicotine input in the activation of certain neurotransmitter receptors (Fig. 4.6). Nicotine, through the activation of the N-cholinergic receptors (nAChRs, Fig. 4.4), stimulates the secretions of the neurotransmitters of the dopaminergic and glutamatergic neurons (Ma, Liu, Neumann, & Gao, 2017).

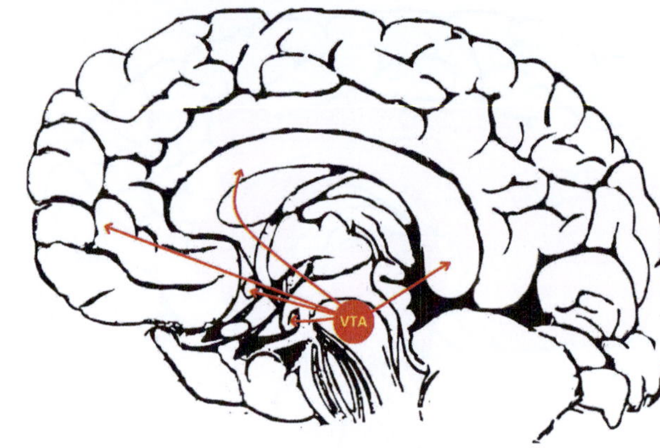

FIG. 4.5 Activation of mesolimbic system by nicotine. Nicotine activates nAChR within the ventral tegmental area (VTA). Dopaminergic projections transport the signal from the VTA to the frontal cortex, nucleus accumbens (NAc), striatum, hippocampus, and amygdala. The dopaminergic projection target areas are involved in reinforcement (the ventral striatum), learning and declarative memory (the hippocampus), emotional memory (the amygdala), habit formation (the dorsal striatum), and executive functions and working memory (the prefrontal cortex and orbitofrontal cortex). Chronic nicotine exposure indirectly affects motivational systems by altering synthesis and the release of opioid peptides. The addiction process produces cellular adaptations and alterations in brain neurotransmission that participate in the nicotine withdrawal syndrome.

4.6 NICOTINE AND REWARD CIRCUIT

Nicotine activates the mesolimbic system by way of increasing dopamine release in the central nervous system (Fig. 4.5), and stimulating the reward system, as well as enhancing the activity of locomotor endurance and preserving appetitive behaviors focused on drug seeking and self-administration (Berrendero, Robledo, Trigo, Martin-Garcia, & Maldonado, 2010; Kostowski, 2001; Seo, Kim, Yu, & Kang, 2016; Suchanecka, 2013). In addition, the components of tobacco smoke block monoamine oxidases A and B (Mao-A and Mao-B), additionally activating the dopaminergic system (Kostowski, 2001). Tobacco smoking addiction is a complex process (Fig. 4.6), mainly associated with the addictive effects of nicotine, but the desire to reach for traditional cigarettes is also enhanced by the taste and aroma and activity (the rituals) that is associated with smoking (Kostowski, 2001).

The N-cholinergic receptor that is affected by tobacco addiction is a pentameric structure (Fig. 4.4) built with two types of subunits—alpha (α) and beta (β). There are six known subunits of the alpha and three beta forms of these subunits. The differing effects of nicotine on the individual nAChR subunits bring about varying effects upon the nervous system. Baldassarri et al. (2018) saw

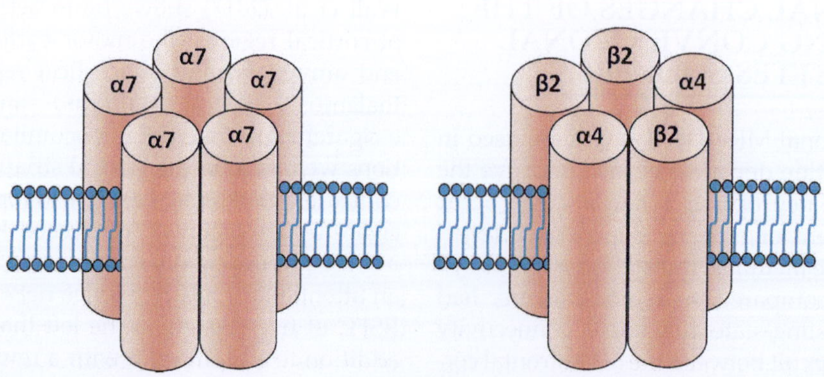

FIG. 4.4 The nicotinic cholinergic receptor. The N-cholinergic receptor is a pentameric structure built with two types of subunits—alpha (α) and beta (β). The propensity of these receptors to desensitize depends largely on the subunit composition of the nAChR assembly (homooligomeric or heterooligomeric).

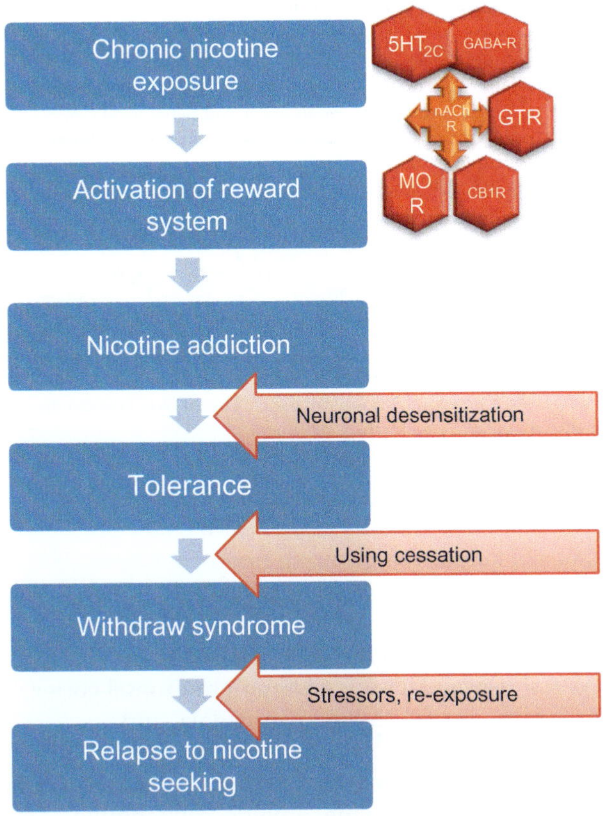

FIG. 4.6 The effects of nicotine on the central nervous system. Nicotine activates nAChR on neurons in mesolimbic structures that increase secretion of dopamine in the reward system. Another neurotransmitters involve in the rewarding system and increase dopamine release after nicotine intake are serotonin (through 5-HT_{2c} receptor subtype), glutamate (glutamatergic receptor, GTR), opioid peptides (by μ-opioid receptor, MOR), cannabinoids (by cannabinoid receptor type 1, CB1R) and GABA. Chronic nicotine exposure induces reinforcing effects and development of cigarette craving (nicotine addiction). Long-time activation of reward system by nicotine leads to neuronal desensitization and upregulation of receptors that is the cause of tolerance development. Nicotine use cessation causes the behavioral and somatic symptoms of withdrawal syndrome. Exposure to stressors or nicotine in environment in the abstinent time triggers nicotine craving and relapse to nicotine seeking (Berrendero et al., 2010).

that the average β2-nAChR occupancy was higher after the vaporization of e-cigarettes containing 36 mg/mL (84 ± 3%) nicotine as compared to a level of 8 mg/mL (64 ± 17%), while the average β2-nAChR occupancy after tobacco cigarette smoking was 68 ± 18% (Baldassarri et al., 2018). In another study by Alasmari et al. (2017), the inhalation of e-cigarette vapor for 6 months increased α7-nAChR expression in the frontal cortex and striatum, but not in the hippocampus. In addition, chronic e-cigarette exposure reduced glutamate transporter-1 (GLT-1) expression only in striatum but induced downregulation of cystine/glutamate antiporter (xCT) in the frontal cortex, the striatum, and the hippocampus. In the study of Hernandez and Terry Jr. (2005), the authors showed that male Wistar rats exposed to relatively low dosage of nicotine (0.35 mg/kg every 12 h) for 14 days demonstrated improved memory performance when compared with controls. Hernandez and Terry Jr. (2005) hold that probably, this effect was dependent on nicotine-induced changes in the expression of specific acetylcholine receptor subtypes—key cholinergic proteins that are regulated by nerve growth factor (NGF). The argument that nicotine increases the expression of nAChRs in different learning- and memory-related brain regions, hence possibly enhancing the short-term memory function, is used for suggesting a nicotinic agonist-based therapy for memory disorders such as Alzheimer's disease (Xue et al., 2015).

4.7 IMPACT OF NICOTINE ON DEVELOPING BRAIN

On the other hand, some of the data resulting from the analysis of the impact of nicotine on the developing brain suggest that the drug leads to neurodegenerative disease and that using chronic nicotine has a negative impact on the learning and cognitive function (Connor & Gould, 2017; Suter, Mastrobattista, Sachs, & Aagaard, 2015). In addition, the data suggest that secondhand smoke and exposure to nicotine during prenatal life may impact upon brain development (Peterson & Hecht, 2017; Protano, Manigrasso, Avino, & Vitali, 2017; Smith et al., 2015). Still, more research is necessary, as the effects of secondhand cigarette smoking and e-cigarette vaping on the developing fetus and young child are described in only a small number of studies (Protano, Manigrasso, Avino, Sernia, & Vitali, 2016; Suter et al., 2015; Makadia et al., 2017). However, based on animal studies, nicotine leads to serotonin, dopamine, and norepinephrine disorders in the forebrain and midbrain (Suter et al., 2015). Moreover, according to Lauterstein et al. (2016), the offspring of mice exposed to aerosol with e-cigarettes exhibited gene alterations in the frontal cortex that can cause cognitive and behavioral deficits later in life. Indeed, Lavezzi et al. (2017) suggest that alterations in cerebellar cholinergic transmission in certain areas of the fetus' brain of smoking mothers may be the cause of sudden unexplained fetal and infant death, as they had observed a negative or low expression of α7-nAChRs in the granular layers of the cerebellar cortex in 66% of the sudden unexplained perinatal deaths, as compared to the 11% of the (nonsmoking) controls. Furthermore, as seen by Mohamed, Loy, Lim, Al Mamun, & Jan Mohamed (2017), prenatal nicotine concentration levels were negatively associated with communication and fine motor skills, while postnatal nicotine concentration levels were inversely associated with fine motor and problem-solving skills.

4.8 THE NICOTINE-INDUCED OXIDATIVE STRESS AND NEURONAL APOPTOSIS

The described neurodenegeration and alterations in the brain development may be the consequence of the nicotine-induced increase of oxidative stress and neuronal apoptosis. In the study of Motaghinejad, Motevalian, Fatima, Faraji, and Mozaffari (2017), the authors observed that nicotine treatment increased lipid peroxidation and the levels of IL-1β, TNF-α, and proapoptotic protein Bax while reducing the neuroprotective CREB/BDNF pathway and Bcl-2 levels in the hippocampus. Nicotine also reduced the activity of superoxide dismutase, glutathione peroxidase, and glutathione reductase in the hippocampus. The toxic effects of nicotine may be increased by flavors and other substances added to tobacco or to liquid smoke. Menthol is a commonly used flavorant in tobacco and e-cigarettes and could contribute to nicotine sensitivity (Fait et al. 2017; Thompson et al., 2017). Accidental ingestion of e-cigarette liquid can lead to death as a result of the toxic action of nicotine on the brain. Indeed, the described cases of accidental ingestion of e-cigarette liquid to child and adult exist, showing that ingestion leads to respiratory failure or death as a consequence of severe anoxic brain injury and acute infarcts and brain edema (Chen, Bright, Trivedi, & Valento, 2015; Noble, Longstreet, Hendrickson, & Gerona, 2017; Seo et al., 2016).

4.9 CONCLUSION

The data presented demonstrate the highly damaging action upon users, those close by, and the unborn of traditional cigarettes and e-cigarettes. Hence, it is most important to disseminate knowledge regarding the actual effects of the different products containing nicotine and to raise awareness among the public that not only tobacco smoking but also vaping or other nicotine delivery devices are harmful to both users and bystanders. Thus, activities aimed at promoting health and encouraging healthy living should place emphasis on helping to stop nicotine consumption.

MINI-DICTIONARY OF TERMS

Electronic cigarettes Nicotine delivery battery-powered devices that look like cigarettes and deliver vapor (aerosol-containing nicotine). The vapor is produced by the heating of liquids localized in the e-cigarette containers. This liquid contains varied concentrations of nicotine, glycerin, propylene glycol, flavorants, and sometimes several psychoactive substances.

Nicotine dependence An addiction syndrome with a catalog of behavioral and physiological symptoms produced by repeated nicotine consumption.

Reward system The neural system in the brain that is part of the limbic system, which is related to motivation and behavioral control. This system is activated by satisfying "needs" such as drinking, eating, and reproduction and also by the administering of psychoactive addictive substances such as alcohol, nicotine, amphetamine, and other psychoactive drugs.

Tobacco combustion products Tobacco products (tobacco and cellulose, as well as other items) that when lit results in smoke that contains solid ultrafine particles of differing diameter and free radicals and tumorigenic gases (nitrogen oxide, carbon monoxide, and others).

Traditional (conventional) cigarettes Compositions of tobacco and different additive substances rolled in thin paper that is then smoldered and the smoke inhaled by users. The smoke, beyond nicotine, contains several combustion products.

Key Facts of the Reward Circuit

- Reward circuit is a group of brain structures responsible for motivation and behavioral control.
- The reward system includes structures such as the ventral tegmental area, striatum, prefrontal cortex, thalamus, globus pallidus, amygdala, and hippocampus.
- The neurotransmitters involved in reward system are dopamine, acetylcholine, opioids, cannabinoids, serotonin, GABA, and glutamatergic acid.
- Every subsequent satisfying the need gives a "positive reinforcement," excitation, pleasure, and incentive motivation.
- The intrinsic rewards (palatable food and sex) are inherently pleasurable and serve to increase the likelihood of survival and reproduction.
- The extrinsic rewards (money) are conditioned rewards that are attractive and motivate behavior.
- Psychoactive drugs and risky behaviors activate reward circuit and lead to addiction.

Summary Points

- This chapter focuses on the impact on the brain of users and secondarily exposed people of tobacco smoking and electronic cigarette liquid vaping.
- Conventional cigarettes are a long-standing legally available consumed product.
- Electronic cigarettes (e-cigarettes) are new nicotine delivery devices that partially replace traditional cigarettes.
- Vaping (the use of electronic cigarettes) is the act of inhaling vapor that is produced by heating liquids containing nicotine, propylene glycol, glycerin, flavorants, and other substances. In vaping, no combustion occurs.

- Both conventional cigarette smoking and electronic cigarette vaping induce nicotine dependence and withdrawal syndromes.
- Consumption of nicotine by these two devices leads to structural and functional changes of the specific areas of the brain related to the reward system.
- Consumption of nicotine by pregnant women leads to neurodegeneration and alterations of the brain development in their offspring, when such individuals are exposed to smoke or vapor products prenatally or during the early life stages.

References

Alasmari, F., Crotty, A. L. E., Nelson, J. A., Schiefer, I. T., Breen, E., Drummond, C. A., et al. (2017). Effects of chronic inhalation of electronic cigarettes containing nicotine on glial glutamate transporters and α-7 nicotinic acetylcholine receptor in female CD-1 mice. *Progress in Neuro-Psychopharmacology & Biological Psychiatry*, *77*, 1–8.

Arane, K., & Goldman, R. D. (2016). Electronic cigarettes and adolescents. *Canadian Family Physician Médecin de famille canadien*, *62*(11), 897–898.

Baldassarri, S. R., Hillmer, A. T., Anderson, J. M., Jatlow, P., Nabulsi, N., Labaree, D., et al. (2018). Use of electronic cigarettes leads to significant beta 2-nicotinic acetylcholine receptor occupancy: evidence from a PET imaging study. *Nicotine & Tobacco Research*, *20*(4), 425–433. https://doi.org/10.1093/ntr/ntx091.

Berrendero, F., Robledo, P., Trigo, J. M., Martín-García, E., & Maldonado, R. (2010). Neurobiological mechanisms involved in nicotine dependence and reward: participation of the endogenous opioid system. *Neuroscience and Biobehavioral Reviews*, *35*(2), 220–231.

Bertholon, J. F., Becquemin, M. H., Roy, M., Roy, F., Ledur, D., Annesi Maesano, I., et al. (2013). Comparison of the aerosol produced by electronic cigarettes with conventional cigarettes and the shisha. *Revue des Maladies Respiratoires*, *30*(9), 752–757.

Chen, B. C., Bright, S. B., Trivedi, A. R., & Valento, M. (2015). Death following intentional ingestion of e-liquid. *Clinical Toxicology*, *53*(9), 914–916.

Chu, S., Xiao, D., Wang, S., Peng, P., Xie, T., He, Y., et al. (2014). Spontaneous brain activity in chronic smokers revealed by fractional amplitude of low frequency fluctuation analysis: a resting state functional magnetic resonance imaging study. *Chinese Medical Journal*, *127*(8), 1504–1509.

Connor, D. A., & Gould, T. J. (2017). Chronic fluoxetine ameliorates adolescent chronic nicotine exposure-induced long-term adult deficits in trace conditioning. *Neuropharmacology*, *2*(125), 272–283.

Cooper, M., Harrell, M. B., Pérez, A., Delk, J., & Perry, C. L. (2016). Flavorings and perceived harm and addictiveness of E-cigarettes among youth. *Tobacco Regulatory Science*, *2*(3), 278–289.

Dumais, K. M., Franklin, T. R., Jagannathan, K., Hager, N., Gawrysiak, M., Betts, J., et al. (2017). Multi-site exploration of sex differences in brain reactivity to smoking cues: consensus across sites and methodologies. *Drug and Alcohol Dependence*, *1*(178), 469–476.

Fait, B. W., Thompson, D. C., Mose, T. N., Jatlow, P., Jordt, S. E., Picciotto, M. R., et al. (2017). Menthol disrupts nicotine's psychostimulant properties in an age and sex-dependent manner in C57BL/6J mice. *Behavioural Brain Research*, *22*(334), 72–77.

Geiss, O., Bianchi, I., Barahona, F., & Barrero-Moreno, J. (2015). Characterisation of mainstream and passive vapours emitted by selected electronic cigarettes. *International Journal of Hygiene and Environmental Health*, *218*(1), 169–180.

Hernandez, C. M., & Terry, A. V., Jr. (2005). Repeated nicotine exposure in rats: effects on memory function, cholinergic markers and nerve growth factor. *Neuroscience*, *130*(4), 997–1012.

Hobkirk, A. L., Nichols, T. T., Foulds, J., Yingst, J. M., Veldheer, S., Hrabovsky, S., et al. (2018). Changes in resting state functional brain connectivity and withdrawal symptoms are associated with acute electronic cigarette use. *Brain Research Bulletin*, *138*, 56–63. https://doi.org/10.1016/j.brainresbull.2017.05.010. Epub 2017 May 17.

Huang, W., King, J. A., Ursprung, W. W., Zheng, S., Zhang, N., Kennedy, D. N., et al. (2014). The development and expression of physical nicotine dependence corresponds to structural and functional alterations in the anterior cingulate-precuneus pathway. *Brain and Behavior*, *4*(3), 408–417.

Janes, A. C., Gilman, J. M., Radoman, M., Pachas, G., Fava, M., & Evins, A. E. (2017). Revisiting the role of the insula and smoking cue-reactivity in relapse: a replication and extension of neuroimaging findings. *Drug and Alcohol Dependence*, *12*(179), 8–12.

Kaisar, M. A., Prasad, S., Liles, T., & Cucullo, L. (2016). A decade of e-cigarettes: limited research & unresolved safety concerns. *Toxicology*, *15*(365), 67–75.

Kostowski, W. (2001). Współczesna farmakoterapia uzależnienia od nikotyny. *Alkoholizm i Narkomania*, *14*(1), 129–136.

Lauterstein, D. E., Tijerina, P. B., Corbett, K., Akgol Oksuz, B., Shen, S. S., Gordon, T., et al. (2016). Frontal cortex transcriptome analysis of mice exposed to electronic cigarettes during early life stages. *International Journal of Environmental Research and Public Health*, *13*(4), 417.

Lavezzi, A. M., Ferrero, S., Roncati, L., Piscioli, F., Matturri, L., & Pusiol, T. (2017). Nicotinic receptor abnormalities in the cerebellar cortex of sudden unexplained fetal and infant death victims-possible correlation with maternal smoking. *American Society for Neurochemistry Neuro*, *9*(4), 1–10.

Lee, S., Grana, R. A., & Glantz, S. A. (2014). Electronic cigarette use among Korean adolescents: a cross-sectional study of market penetration, dual use, and relationship to quit attempts and former smoking. *The Journal of Adolescent Health*, *54*(6), 684–690.

Li, Y., Yuan, K., Cai, C., Feng, D., Yin, J., Bi, Y., et al. (2015). Reduced frontal cortical thickness and increased caudate volume within fronto-striatal circuits in young adult smokers. *Drug and Alcohol Dependence*, *1*(151), 211–219.

Long, G. A. (2014). Comparison of select analytes in exhaled aerosol from E-cigarettes with exhaled smoke from a conventional cigarette and exhaled breaths. *International Journal of Environmental Research and Public Health*, *11*, 11177–11191.

Ma, C., Liu, Y., Neumann, S., & Gao, X. (2017). Nicotine from cigarette smoking and diet and Parkinson disease: a review. *Translational Neurodegeneration*, *2*(6), 18.

Makadia, L. D., Roper, P. J., Andrews, J. O., & Tingen, M. S. (2017). Tobacco use and smoke exposure in children: new trends, harm, and strategies to improve health outcomes. *Current Allergy and Asthma Reports*, *17*(8), 55.

McLaughlin, I., Dani, J. A., & De Biasi, M. (2015). Nicotine withdrawal. *Current Topics in Behavioral Neurosciences*, *24*, 99–123.

Mohamed, N. N., Loy, S. L., Lim, P. Y., Al Mamun, A., & Jan Mohamed, H. J. (2017). Early life secondhand smoke exposure assessed by hair nicotine biomarker may reduce children's neurodevelopment at 2 years of age. *Science of the Total Environment*, *10* (610–611), 147–153.

Motaghinejad, M., Motevalian, M., Fatima, S., Faraji, F., & Mozaffari, S. (2017). The neuroprotective effect of curcumin against nicotine-induced neurotoxicity is mediated by CREB-BDNF signaling pathway. *Neurochemical Research*, *42*(10), 2921–2932. https://doi.org/10.1007/s11064-017-2323-8.

Noble, M. J., Longstreet, B., Hendrickson, R. G., & Gerona, R. (2017). Unintentional pediatric ingestion of electronic cigarette nicotine refill liquid necessitating intubation. *Annals of Emergency Medicine*, *69*(1), 94–97.

Oh, A. Y., & Kacker, A. (2014). Do electronic cigarettes impart a lower potential disease burden than conventional tobacco cigarettes? Review on E-cigarette vapor versus tobacco smoke. *Laryngoscope*, *124*(12), 2702–2706.

Peterson, L. A., & Hecht, S. S. (2017). Tobacco, e-cigarettes, and child health. *Current Opinion in Pediatrics*, *29*(2), 225–230.

Polosa, R., Caponnetto, P., Morjaria, J. B., Papale, G., Campagna, D., & Russo, C. (2011). Effect of an electronic nicotine delivery device (e-Cigarette) on smoking reduction and cessation: a prospective 6-month pilot study. *BMC Public Health*, *11*, 786.

Protano, C., Manigrasso, M., Avino, P., Sernia, S., & Vitali, M. (2016). Second-hand smoke exposure generated by new electronic devices (IQOS® and e-cigs) and traditional cigarettes: submicron particle behaviour in human respiratory system. *Annali di Igiene: Medicina Preventiva e di Comunita*, *28*(2), 109–112.

Protano, C., Manigrasso, M., Avino, P., & Vitali, M. (2017). Second-hand smoke generated by combustion and electronic smoking devices used in real scenarios: ultrafine particle pollution and age-related dose assessment. *Environment International*, *107*, 190–195.

Samochowiec, J., Rogoziński, D., Hajduk, A., Skrzypińska, A., & Arentowicz, G. (2001). Diagnostyka, mechanizm uzależnienia i metody leczenia uzależnienia od nikotyny. *Alkoholizm i Narkomania*, *14*(3), 323–340.

Seo, A. D., Kim, D. C., Yu, H. J., & Kang, M. J. (2016). Accidental ingestion of E-cigarette liquid nicotine in a 15-month-old child: an infant mortality case of nicotine intoxication. *Korean Journal of Pediatrics*, *59*(12), 490–493.

Shen, Z., Huang, P., Qian, W., Wang, C., Yu, H., Yang, Y., et al. (2016). Severity of dependence modulates smokers' functional connectivity in the reward circuit: a preliminary study. *Psychopharmacology*, *233*(11), 2129–2137.

Smith, D., Aherrera, A., Lopez, A., Neptune, E., Winickoff, J. P., Klein, J. D., et al. (2015). Adult behavior in male mice exposed to E-cigarette nicotine vapors during late prenatal and early postnatal life. *PLoS ONE*, *10*(9). e0137953.

Suchanecka, A. (2013). Rola dopaminy w procesach motywacyjnych i powstawaniu uzależnień. *Annales Academiae Medicae Stetinensis Roczniki Pomorksiej Akademii Medycznej w Szczecinie Neurokongwi Nistyka w patologii i zdrowiu*, 158–161.

Suter, M. A., Mastrobattista, J., Sachs, M., & Aagaard, K. (2015). Is there evidence for potential harm of electronic cigarette use in pregnancy? *Birth Defects Research Part A*, *103*(3), 186–195.

Tayyarah, R., & Long, G. A. (2014). Comparison of select analytes in aerosol from e-cigarettes with smoke from conventional cigarettes and with ambient air. *Regulatory Toxicology and Pharmacology*, *70* (3), 704–710.

Thompson, M. F., Poirier, G. L., Dávila-García, M. I., Huang, W., Tam, K., Robidoux, M., et al. (2017). Menthol enhances nicotine-induced locomotor sensitization and in vivo functional connectivity in adolescence. *Journal of Psychopharmacology*, *1*. 332–343.

Wall, M. B., Mentink, A., Lyons, G., Kowalczyk, O. S., Demetriou, L., & Newbould, R. D. (2017). Investigating the neural correlates of smoking: feasibility and results of combining electronic cigarettes with fMRI. *Scientific Reports*, *7*(1), 11352.

Wang, C., Bai, J., Wang, C., von Deneen, K. M., Yuan, K., & Cheng, J. (2017). Altered thalamo-cortical resting state functional connectivity in smokers. *Neuroscience Letters*, *13*(653), 120–125.

Xue, M., Zhu, L., Zhang, J., Qiu, J., Du, G., Qiao, Z., et al. (2015). Low dose nicotine attenuates Aβ neurotoxicity through activation early growth response gene 1 pathway. *PLoS ONE*, *3*, e0120267.

Yu, D., Yuan, K., Cheng, J., Guan, Y., Li, Y., Bi, Y., et al. (2018). Reduced thalamus volume may reflect nicotine severity in young male smokers. *Nicotine and Tobacco Research*, *20*(4), 434–439. https://doi.org/10.1093/ntr/ntx146.

Yuan, K., Yu, D., Bi, Y., Li, Y., Guan, Y., Liu, J., et al. (2016). The implication of frontostriatal circuits in young smokers: a resting-state study. *Human Brain Mapping*, *37*(6), 2013–2026.

Zhou, S., Xiao, D., Peng, P., Wang, S. K., Liu, Z., Qin, H. Y., et al. (2017). Effect of smoking on resting-state functional connectivity in smokers: an fMRI study. *Respirology*, *22*(6), 1118–1124.

CHAPTER

5

Reduction of Nicotine in Tobacco and Impact

Yael Abreu-Villaça, Alex Christian Manhães*, Anderson Ribeiro-Carvalho†*

*Department of Physiological Sciences, Institute of Biology, State University of Rio de Janeiro, Rio de Janeiro, Brazil
†Department of Sciences, Faculty of Teacher Training, State University of Rio de Janeiro, São Gonçalo, Brazil

Abbreviations

DAergic dopaminergic
MAO monoamine oxidase
NAcc nucleus accumbens
nAChR nicotinic cholinergic receptor
VTA ventral tegmental area

5.1 STRATEGIES TO PREVENT ADDICTION AND REDUCE TOBACCO USE

In an effort to reduce the burden impinged on the individual and on society by the need to deal with the health problems associated with tobacco consumption, several strategies aimed at fostering quitting attempts and continued abstinence have been developed since the first clear indications were uncovered that smoking constitutes a health hazard (West, 2017). These strategies can be divided into two main lines of action: The first one involves preventing addiction, and the second one is meant to reduce or altogether stop tobacco consumption. Smokers, for instance, may decide to quit upon realizing that they are indeed suffering from the deleterious effects of their addiction. In trying to do so, they have several options at their disposal, with varying degrees of success (West, 2017). They may stop smoking cold turkey, a method that is fraught with the consequences of severe withdrawal syndrome, or they may try to ease into abstinence by progressively reducing their consumption, selecting brands with lower contents of nicotine prior to quitting, and resorting to pharmacological treatments and/or nicotine replacement therapy. Psychological support is also sometimes used as a coadjutant during the quitting attempt. Given the socioeconomic impact of the illnesses associated with smoking, public policies intended to prevent or reduce smoking have also been implemented (Hoffman & Tan, 2015). These include the restriction of advertisement in public venues; establishment of a minimal age for legal purchase; obligation of health risk notices in cigarette packing; prohibition of indoor consumption in public spaces; and, more recently, implementation of regulatory legislation requiring that cigarette nicotine content is reduced to levels that do not result in addiction (World Health Organization, 2015). Regarding this last strategy, for example, the Family Smoking Prevention and Tobacco Control Act in the United States gave the Food and Drug Administration authority to greatly reduce the nicotine content of cigarettes if doing so would improve public health (United States Congress, 2009). This strategy is based on the fact that nicotine is the primary addictive constituent in cigarettes (United States Department of Health and Human Services, 1988), and therefore, the premise is that its reduction will prevent the development of addiction in those willing to experiment with smoking and suppress nicotine-seeking behaviors in smokers (Donny et al., 2012; Hatsukami, Benowitz, Donny, Henningfield, & Zeller, 2013).

If forcibly reducing nicotine content in cigarettes is to be considered a viable strategy in addressing the health issues associated with smoking, the consumption of very-low-nicotine cigarettes must eventually lead to either one, preferably both, of the following outcomes: (1) a reduction in the percentage of individuals that progress from experimentation to continued use and (2) an

Neuroscience of Nicotine
https://doi.org/10.1016/B978-0-12-813035-3.00005-8

increase in the percentage of smokers that either reduce consumption or successfully quit smoking. Put simply, while it is true that the total amount of consumed nicotine might be reduced by severely limiting its content in cigarettes, it is also true that the contents of the other 4000 plus substances that have been identified in tobacco smoke, including several known carcinogens (Smith, Perfetti, Garg, & Hansch, 2003), are not considerably affected (Denlinger-Apte, Joel, Strasser, & Donny, 2016). Although the notion that serious health consequences may result from smoking tobacco-containing products has become widespread over the years, even among smokers, evidence has been found indicating that incorrect perceptions regarding the deleterious effects of smoking very-low-nicotine cigarettes are present. Below, following a discussion of these public perceptions, we present results from studies, both in humans and animal models, that address the issue of whether reducing nicotine content in cigarettes is indeed an advantageous strategy regarding the management of smoking and its health consequences.

5.2 PUBLIC PERCEPTION OF LOW-NICOTINE CIGARETTE HEALTH RISKS

A troublesome aspect of reducing nicotine content in cigarettes is that it may have effects that are not in line with the desired ones, either keeping consumption of cigarettes by current smokers in levels compatible with those observed prior to the adoption of low-nicotine cigarettes or actually driving consumption up (Mercincavage et al., 2017). Even worst is the possibility that nonsmokers, particularly susceptible adolescents, and those that had successfully quit start smoking due to a misguided view that nicotine is the major source of illnesses associated with the habit (Ambrose et al., 2014). While many smokers acknowledge that nicotine causes addiction, they may not realize that the majority of the diseases resulting from smoking are actually caused by the other components present in tobacco smoke. In fact, approximately 75% of US adults either incorrectly believe that nicotine causes cancer or are unsure of whether this causal link exists (O'Brien, Nguyen, Persoskie, & Hoffman, 2017). That being the case, the use of low-nicotine cigarettes may be wrongfully considered less harmful, and as a consequence, a smoker, instead of quitting altogether, may choose a low-nicotine brand believing that its use will be beneficial to his/her health while avoiding the discomfort associated with withdrawal. In fact, Denlinger-Apte et al. (2016) and O'Brien et al. (2017) showed that subjects consider low-nicotine cigarettes as less harmful to their health overall when compared to an average nicotine cigarette and that this effect holds true for specific smoking-related diseases.

5.3 IS IT ADVANTAGEOUS TO REDUCE NICOTINE CONTENT IN TOBACCO PRODUCTS? EVIDENCE FROM EPIDEMIOLOGICAL AND EXPERIMENTAL STUDIES IN HUMANS

Contrasting with the idea that a less unfavorable perception toward low-nicotine cigarettes may eventually drive consumption up is evidence that these cigarettes produce a desirable set of outcomes, including increased desire to quit smoking, reduced exposure to nicotine, reduced smoking, and reduced dependence (Box 5.1). For instance, Denlinger-Apte et al. (2016) demonstrated that while their subjects rated a very-low-nicotine cigarette as having less desirable subjective effects than an average nicotine cigarette, they also predicted that they would have greater interest in quitting smoking in the future if only the very-low-nicotine cigarette was available. More to the point, the consumption of low-nicotine cigarettes results in a reduced exposure to this drug, which may be accompanied by reductions in dependence and craving (Donny et al., 2015; Higgins et al., 2017). These effects were observed in the absence of significant indications of compensatory consumption. There are special situations in which using low-nicotine cigarettes might prove particularly useful as a stopgap measure prior to a determined attempt on quitting. For instance, smokers that want to reduce their exposure to the harmful consequences of tobacco smoke but do not want to deal with effects of nicotine withdrawal and find the act of smoking psychologically very rewarding is one such case (Benowitz, Donny, & Hatsukami, 2017). If provided with the possibility of using noncombusted nicotine alternatives, such as e-cigarettes, along with the

BOX 5.1

FAVORABLE ASPECTS ASSOCIATED WITH LOW-NICOTINE CIGARETTE CONSUMPTION

- Reduced exposure to nicotine
- Reduced potential for progressing from experimentation to dependency
- Reduced dependency
- Reduced craving
- Suppression of nicotine-seeking behaviors
- Reduced smoking
- Reduced withdrawal symptoms
- Higher success rates in quitting attempts

use of very-low-nicotine cigarettes, their smoking topography is altered in a way that results in reduced exposure to tobacco smoke (Hatsukami et al., 2017).

The favorable outcomes observed in some studies regarding the use of low-nicotine cigarettes would seem to suggest that widespread adoption of regulation requiring severe reductions in nicotine content is justified. However, several unfavorable aspects were also identified (Box 5.2). One of them is compensatory consumption (Strasser, Lerman, Sanborn, Pickworth, & Feldman, 2007; Zacny & Stitzer, 1988). This compensation is present when the total number of cigarettes that are smoked per day increases (Mercincavage et al., 2017) and/or when alterations in smoking topography, such as increases in the volume, frequency, and number of puffs, take place (Scherer & Lee, 2014). For instance, adolescents will show a partial compensation by increasing the number of puffs that they take when using denicotinized cigarettes relative to high-nicotine-yield ones (Kassel et al., 2007). The extent of compensation in adults varies considerably among studies, particularly when brand-switching strategies are involved (Scherer & Lee, 2014). The aspect of the smoking topography most significantly changed in adults is the puff volume, and the compensatory increase in this parameter was found to be more pronounced when the differences in nicotine yield between brands were narrower. There are indications that smoking low-nicotine cigarettes results in effects that might discourage continued use in a situation in which a reduction in craving is desired. Short-term weight gain is one such example. Rupprecht et al. (2017) demonstrated that both men and women might gain more than 1 kg over the course of a 6-week period of continued use of a very-low-nicotine cigarette. Besides the esthetic impact of the body mass gain, obesity is also a health concern, with its own set of issues. Another concern is associated with reduction in cognitive performance. Deficits in working memory, associated with alterations in the subjacent brain function, have also been reported (McClernon et al., 2016).

Although our knowledge regarding the potential benefits and drawbacks associated with smoking low-nicotine cigarettes has been increasing steadily, much is still unknown, including the effects of gradual versus abrupt changes in nicotine content and effects in vulnerable populations, particularly adolescents (Donny et al., 2014). In order to help bridge this gap in information, animal models have been systematically used to address potential issues associated with the exposure to tobacco products that have reduced concentrations of nicotine (Fig. 5.1).

BOX 5.2

UNFAVORABLE ASPECTS ASSOCIATED WITH LOW-NICOTINE CIGARETTE CONSUMPTION

- Incorrect perception that low-nicotine cigarettes are less harmful than average nicotine ones
- Facilitation of experimentation
- Compensatory smoking
- Altered smoking topography (e.g., increases in volume, frequency, and number of puffs)
- Continued or increased exposure to toxic nonnicotine cigarette compounds
- Altered mood
- Reduced cognitive performance
- Body mass gain

5.4 IS IT ADVANTAGEOUS TO REDUCE NICOTINE CONTENT IN TOBACCO PRODUCTS? MOST EVIDENCE FROM ANIMAL MODELS OF EXPOSURE INDICATES OTHERWISE

Smokers are exposed to several other substances present in tobacco smoke besides nicotine, which may act

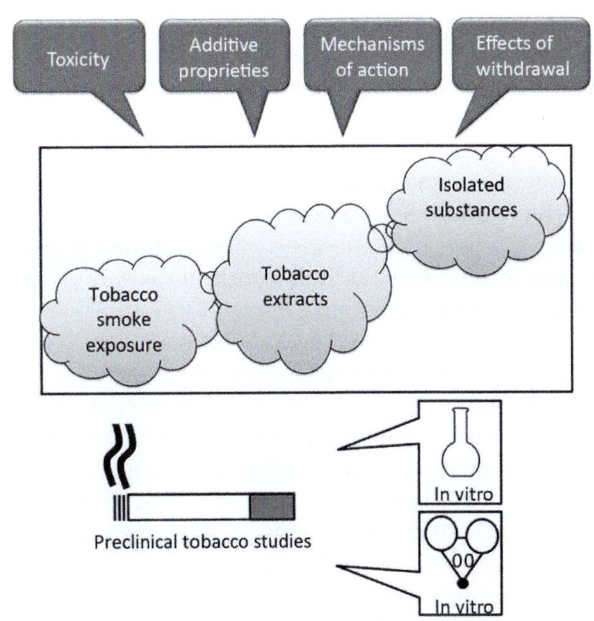

FIG. 5.1 Models of preclinical tobacco exposure.
The effects of smoking are best modeled by exposing subjects to the many compounds found in tobacco, therefore allowing for the full range of interactions that are observed in humans to be present. Research cigarettes and tobacco extracts with varying levels of nicotine have proved particularly useful in this regard. However, the study of the effects of specific compounds other than nicotine has also provided valuable information. Evidence shows that tobacco addiction can no longer be understood as an addiction to nicotine only.

independently from or interfere with nicotine effects. In this regard, preclinical models of tobacco smoke exposure are relevant approaches to study the human condition of smoking. Most data generated by the use of animal models and in vitro studies indicate that the reduction of nicotine content in cigarettes is not accompanied by obvious reductions in harm (Deutsch, 2009). For example, in vitro studies show that, when compared to regular cigarettes, differences in the chemical composition of mainstream smoke generated from cigarettes containing low levels of nicotine do not reduce its mutagenicity and cytotoxicity proprieties (Coffa, Coggins, Werley, Oldham, & Fariss, 2016). In the brain, evidence shows that very-low-nicotine and nicotine-free cigarettes can be more damaging to the blood-brain barrier endothelium than regular ones, increasing the risk for cerebrovascular and central nervous system disorders (Naik et al., 2014). Given the deleterious effects that have been identified so far, the use of low-nicotine or denicotinized cigarettes as a possible harm reduction strategy would be justified only if these cigarettes are less addictive than the regular ones, therefore allowing for a significant reduction in consumption.

The development of cigarette addiction is impacted not only by its nicotine content but also by the contents of nonnicotine compounds, which may either facilitate or hamper nicotine actions. As a result, exposure to nicotine alone does not accurately reflect the effects of smoking and being exposed to the myriad compounds present in tobacco. For example, rats preexposed to tobacco smoke during adolescence are more susceptible to the effects of a later (at adulthood) exposure to nicotine than rats preexposed only to nicotine (de la Pena et al., 2014). The impact of varying nicotine levels in tobacco has also been investigated. Smoke generated from cigarettes containing very low nicotine levels elicits significant short- and long-term effects on brain function that are distinct from those identified in mice exposed to tobacco smoke containing high nicotine levels (Abreu-Villaca et al., 2010, 2015, 2016). Even large reductions of nicotine yield do not spare the brain. Exposure of adolescent mice to cigarette smoke containing very low nicotine levels results in decreased novelty-seeking behavior, an effect that persists well after the end of exposure (Abreu-Villaca et al., 2015). Furthermore, the interruption of exposure leads to increased anxiety levels, a particularly disconcerting fact since anxiety is considered a relevant factor to smoking relapse (Abreu-Villaca et al., 2015). Being an acetylcholine analog, nicotine has the cholinergic system as its primary target. It is therefore interesting to note that, despite a tenfold difference in nicotine content between the very-low-nicotine cigarette and the conventional one, cholinergic effects of exposure, such as nicotinic cholinergic receptor (nAChR) upregulation in the cerebral cortex, were present in mice exposed to both types of cigarette (Abreu-Villaca et al., 2016). In addition, mice exposed to very-low-nicotine cigarette smoke exhibit downregulation of the high-affinity choline transporter in the cerebral cortex long after the end of the period of exposure. This transporter is a cholinergic marker highly regulated by neural impulse activity, which suggests that very low levels of nicotine, when combined to other compounds of tobacco smoke, are still capable of promoting cholinergic synaptic impairment during withdrawal (Abreu-Villaca et al., 2016). Altogether, these data show that the cholinergic system remains vulnerable even to the low levels of nicotine exposure that would be expected from smoking cigarettes in which the content of this drug is considerably reduced.

While animal models are fundamental for the study of potential neurochemical and behavioral effects and underlying mechanisms of tobacco smoke exposure, these models do not address the rewarding effects associated with the consumption of tobacco-containing products. In this regard, tobacco extracts have proved to be particularly useful. Rodents can be trained to self-administer tobacco extracts containing distinct levels of nicotine, and their patterns of self-administration can be compared to those obtained with equivalent levels of nicotine. Interestingly, tobacco extracts are more powerful than nicotine alone in both the acquisition and maintenance of self-administration and more potent in stress-induced reinstatement (Costello et al., 2014). In addition, nicotine alone and tobacco extracts have similar affinities for different nAChR types, but extracts are more potent in the ability to induce nAChR upregulation (Ambrose et al., 2007). Furthermore, tobacco extracts, but not nicotine alone, increase the dopamine transporter function (Danielson, Putt, Truman, & Kivell, 2014), evoke a more robust increase in dopamine extracellular levels than nicotine in the striatum (Khalki, Navailles, Piron, & De Deurwaerdere, 2013), and are more potent than nicotine in promoting inhibition of serotonergic activity in raphe neurons (Touiki, Rat, Molimard, Chait, & de Beaurepaire, 2007).

Preclinical studies have already identified a number of neuroactive nonnicotine tobacco constituents that may contribute to tobacco addiction (Fig. 5.2). Monoamine oxidase (MAO) inhibitors, acetaldehyde, and minor tobacco alkaloids mimic or enhance the behavioral and neuropharmacological effects of nicotine in animal models. MAO catalyzes the degradation of monoamines, including dopamine, while MAO inhibition potentiates the reinforcing effects of nicotine, mainly at low doses (Hogg, 2016). The acetaldehyde generated during tobacco pyrolysis also seems to increase nicotine self-administration (Belluzzi, Wang, & Leslie, 2005). Furthermore, tobacco has several other alkaloids similar to nicotine, such as nornicotine, cotinine, and anatabine, that

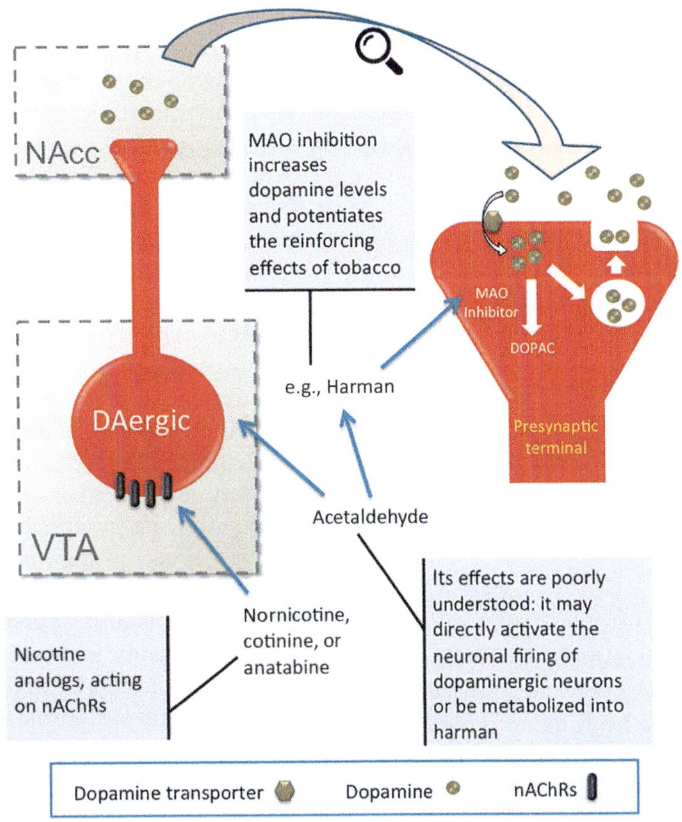

FIG. 5.2 Effects of active nonnicotine compounds of tobacco smoke on the mesocorticolimbic system.
Here, we show a schematic representation of midbrain ventral tegmental area (VTA) dopaminergic (DAergic) projections to the nucleus accumbens (NAcc). This pathway plays important roles in mediating rewarding effects of drugs of abuse, including tobacco smoke. Minor tobacco alkaloids (e.g., nornicotine, cotinine, and anatabine) present in tobacco products interact directly with nicotinic cholinergic receptors (nAChRs) located in DAergic neurons. Monoamine oxidase (MAO) inhibitors (e.g., harman) increase dopamine levels and potentiate the reinforcing effects of tobacco. Acetaldehyde, generated during tobacco pyrolysis, activates the neuronal firing of DAergic neurons or is metabolized into harman.

participate in the reinforcing effects of tobacco (Bardo, Green, Crooks, & Dwoskin, 1999). Finally, there are also substances that reinforce sensory stimulation and substances that minimize the irritation caused by tobacco smoke inhalation. For example, menthol, a frequently used cigarette additive that increases the palatability of tobacco products (Fan et al., 2016), alters the effects of nicotine in the brain by increasing nicotine-induced desensitization and upregulation of nAChR and, consequently, dopamine neuron excitability and locomotor sensitization (Henderson et al., 2017; Thompson et al., 2017).

Altogether, these data suggest that reduced levels of nicotine neither protect the central nervous system from the toxic effects of tobacco nor are less addictive than tobacco with regular nicotine content. Considering that there is a growing body of evidence that nonnicotine compounds impact tobacco addictive properties, tobacco addiction can no longer be understood as an addiction to nicotine. Compounds that have synergistic actions with nicotine or have chemosensory effects could promote smoking initiation, discourage cessation, and facilitate relapse and should be regulated. This assumption is crucial when it is considered that tobacco manufacturers have developed a range of substances to increase acceptance and sales of low-nicotine cigarettes (Alpert, Agaku, & Connolly, 2016; FDA, 1995).

5.5 CONCLUSIONS

The reduction of nicotine content in tobacco products is based on the premise that this strategy prevents the development of addiction, suppresses nicotine-seeking behaviors, and reduces deleterious health effects. We showed that there are epidemiological and experimental studies carried out both in humans and animal models evidencing that the use of low-nicotine cigarettes decreases the level of exposure to this drug, favoring a reduction in dependency. However, there is also evidence that these cigarettes may not suppress nicotine-seeking behaviors and may drive consumption up as a compensatory strategy and that toxic effects are very much present. Continued research is needed if the controversy surrounding this issue is to be definitively settled.

MINI-DICTIONARY OF TERMS

Desensitization Phenomenon that occurs when a receptor decreases its response to an agonist molecule.

Locomotor sensitization Phenomenon that occurs when subsequent exposures to a drug evoke a greater locomotor hyperactivity than seen initially.

Monoamine oxidase (MAO) inhibitors Substances that inhibit the activity of the MAO enzyme family. Enzymes of this family catalyze the oxidation of monoamines.

Nicotinic cholinergic receptor (nAChR) upregulation Increased number of nAChRs promoted by cholinergic agonists such as nicotine.

Self-administration Method for modeling various aspects of drug addiction using animal subjects. In general, animals are trained to press a lever that delivers (typically via an intravenous catheter) a dose of a drug.

Key Facts of Nicotine Reduction in Tobacco Research

- The knowledge that nicotine is a major addictive component of tobacco smoke has led to the development of cigarettes containing reduced levels of this alkaloid.
- These include cigarettes made from tobacco blends that contain low nicotine levels or in which nicotine (intentionally) and other constituents (unintentionally) were partially or completely removed.
- These cigarettes have been used by the research community to identify the roles of nicotine and nonnicotine constituents of tobacco in health.
- There is no perfect placebo cigarette, as none of the available options are identical to regular ones in composition (disregarding nicotine) and sensory characteristics.
- Preclinical studies, by using low-nicotine cigarettes or tobacco smoke extracts, contribute to the current state of knowledge on the roles of nicotine.
- Animal models allow the investigation of reward and neurotoxic effects of varying levels of nicotine in tobacco smoke or tobacco extracts and further unveil the role of other substances present in tobacco.
- Complementary data from clinical and preclinical studies evaluating toxicity and additive potential of tobacco brands are useful to governmental agencies to establish tobacco control legislation.

Summary Points

- This chapter focuses on the advantages and disadvantages of lowering nicotine content in tobacco.
- Nicotine is the primary addictive component in tobacco products.
- A reduction in nicotine content may prevent addiction mechanisms from taking place and/or may suppress nicotine-seeking behaviors in smokers.
- Therefore, a low content of nicotine in cigarettes is being considered as a possible strategy to prevent or reduce smoking.
- Very-low-nicotine cigarettes reduce exposure to nicotine, increase desire to quit smoking, and reduce dependence and craving.
- There is also contrasting evidence that reducing nicotine content may not be an advantageous strategy regarding the management of smoking habit and its health consequences.
- Smokers may adjust their smoking inhalation patterns to compensate for low nicotine content.
- The content of toxic components of tobacco smoke other than nicotine is not reduced in "light" and denicotinized cigarettes.
- Substances other than nicotine present in tobacco smoke may act independently from or interfere with nicotine effects in the brain and contribute to tobacco toxicity and/or addiction.
- There are substances present in cigarettes that may reinforce sensory stimulation and/or minimize excessive irritation of smoke.
- Findings suggest that nicotine is not the sole factor responsible for tobacco addictive properties.

References

Abreu-Villaca, Y., Correa-Santos, M., Dutra-Tavares, A. C., Paes-Branco, D., Nunes-Freitas, A., Manhaes, A. C., et al. (2016). A ten fold reduction of nicotine yield in tobacco smoke does not spare the central cholinergic system in adolescent mice. *International Journal of Developmental Neuroscience*, *52*, 93–103. https://doi.org/10.1016/j.ijdevneu.2016.06.002.

Abreu-Villaca, Y., Filgueiras, C. C., Correa-Santos, M., Cavina, C. C., Naiff, V. F., Krahe, T. E., et al. (2015). Tobacco smoke containing high or low levels of nicotine during adolescence: effects on novelty-seeking and anxiety-like behaviors in mice. *Psychopharmacology*, *232*(10), 1693–1703. https://doi.org/10.1007/s00213-014-3801-1.

Abreu-Villaca, Y., Filgueiras, C. C., Guthierrez, M., Medeiros, A. H., Mattos, M. A., Pereira Mdos, S., et al. (2010). Exposure to tobacco smoke containing either high or low levels of nicotine during adolescence: differential effects on choline uptake in the cerebral cortex and hippocampus. *Nicotine & Tobacco Research*, *12*(7), 776–780. https://doi.org/10.1093/ntr/ntq075.

Alpert, H. R., Agaku, I. T., & Connolly, G. N. (2016). A study of pyrazines in cigarettes and how additives might be used to enhance tobacco addiction. *Tobacco Control*, *25*(4), 444–450. https://doi.org/10.1136/tobaccocontrol-2014-051943.

Ambrose, V., Miller, J. H., Dickson, S. J., Hampton, S., Truman, P., Lea, R. A., et al. (2007). Tobacco particulate matter is more potent than nicotine at upregulating nicotinic receptors on SH-SY5Y cells. *Nicotine & Tobacco Research*, *9*(8), 793–799. https://doi.org/10.1080/14622200701485117.

Ambrose, B. K., Rostron, B. L., Johnson, S. E., Portnoy, D. B., Apelberg, B. J., Kaufman, A. R., et al. (2014). Perceptions of the relative harm of cigarettes and e-cigarettes among U.S. youth. *American Journal of Preventive Medicine*, *47*(2 Suppl. 1), S53–S60. https://doi.org/10.1016/j.amepre.2014.04.016.

Bardo, M. T., Green, T. A., Crooks, P. A., & Dwoskin, L. P. (1999). Nornicotine is self-administered intravenously by rats. *Psychopharmacology*

(Berl), *146*(3), 290–296. Retrieved from: http://www.ncbi.nlm.nih.gov/pubmed/10541729.

Belluzzi, J. D., Wang, R., & Leslie, F. M. (2005). Acetaldehyde enhances acquisition of nicotine self-administration in adolescent rats. *Neuropsychopharmacology*, *30*(4), 705–712. https://doi.org/10.1038/sj.npp.1300586.

Benowitz, N. L, Donny, E. C., & Hatsukami, D. K. (2017). Reduced nicotine content cigarettes, e-cigarettes and the cigarette end game. *Addiction*, *112*(1), 6–7. https://doi.org/10.1111/add.13534.

Coffa, B. G., Coggins, C. R. E., Werley, M. S., Oldham, M. J., & Fariss, M. W. (2016). Chemical, physical, and in vitro characterization of research cigarettes containing denicotinized tobacco. *Regulatory Toxicology and Pharmacology*, *79*, 64–73. https://doi.org/10.1016/j.yrtph.2016.05.016.

Costello, M. R., Reynaga, D. D., Mojica, C. Y., Zaveri, N. T., Belluzzi, J. D., & Leslie, F. M. (2014). Comparison of the reinforcing properties of nicotine and cigarette smoke extract in rats. *Neuropsychopharmacology*, *39*(8), 1843–1851. https://doi.org/10.1038/npp.2014.31.

Danielson, K., Putt, F., Truman, P., & Kivell, B. M. (2014). The effects of nicotine and tobacco particulate matter on dopamine uptake in the rat brain. *Synapse*, *68*(2), 45–60. https://doi.org/10.1002/syn.21715.

de la Pena, J. B., Ahsan, H. M., Botanas, C. J., Sohn, A., Yu, G. Y., & Cheong, J. H. (2014). Adolescent nicotine or cigarette smoke exposure changes subsequent response to nicotine conditioned place preference and self-administration. *Behavioural Brain Research*, *272*, 156–164. https://doi.org/10.1016/j.bbr.2014.06.044.

Denlinger-Apte, R. L., Joel, D. L., Strasser, A. A., & Donny, E. C. (2016). Low nicotine content descriptors reduce perceived health risks and positive cigarette ratings in participants using very low nicotine content cigarettes. *Nicotine & Tobacco Research*, *19*(10), 1149–1154. https://doi.org/10.1093/ntr/ntw320.

Deutsch, M. E. (2009). Less-toxic cigarette use may backfire. *Science*, *325* (5943), 944. https://doi.org/10.1126/science.325_944a.

Donny, E. C., Denlinger, R. L., Tidey, J. W., Koopmeiners, J. S., Benowitz, N. L., Vandrey, R. G., et al. (2015). Randomized trial of reduced-nicotine standards for cigarettes. *The New England Journal of Medicine*, *373*(14), 1340–1349. https://doi.org/10.1056/NEJMsa1502403.

Donny, E. C., Hatsukami, D. K., Benowitz, N. L., Sved, A. F., Tidey, J. W., & Cassidy, R. N. (2014). Reduced nicotine product standards for combustible tobacco: building an empirical basis for effective regulation. *Preventive Medicine*, *68*, 17–22. doi: 10.1056/NEJMc1513886 https://doi.org/10.1016/j.ypmed.2014.06.020.

Donny, E. C., Taylor, T. G., LeSage, M. G., Levin, M., Buffalari, D. M., Joel, D., et al. (2012). Impact of tobacco regulation on animal research: new perspectives and opportunities. *Nicotine & Tobacco Research*, *14*(11), 1319–1938. https://doi.org/10.1093/ntr/nts162.

Fan, L., Balakrishna, S., Jabba, S. V., Bonner, P. E., Taylor, S. R., Picciotto, M. R., et al. (2016). Menthol decreases oral nicotine aversion in C57BL/6 mice through a TRPM8-dependent mechanism. *Tobacco Control*, *25*(Suppl. 2), ii50–ii54. https://doi.org/10.1136/tobaccocontrol-2016-053209.

FDA. (1995). Regulations restricting the sale and distribution of cigarettes and smokeless tobacco products to protect children and adolescents; proposed rule analysis regarding FDA's jurisdiction over nicotine-containing cigarettes and smokeless tobacco products; notice. *Federal Register*, *60*, 41314–41792.

Hatsukami, D. K., Benowitz, N. L., Donny, E., Henningfield, J., & Zeller, M. (2013). Nicotine reduction: strategic research plan. *Nicotine & Tobacco Research*, *15*(6), 1003–1013. https://doi.org/10.1093/ntr/nts214.

Hatsukami, D. K., Luo, X., Dick, L., Kangkum, M., Allen, S. S., Murphy, S. E., et al. (2017). Reduced nicotine content cigarettes and use of alternative nicotine products: exploratory trial. *Addiction*, *112*(1), 156–167. https://doi.org/10.1111/add.13603.

Henderson, B. J., Wall, T. R., Henley, B. M., Kim, C. H., McKinney, S., & Lester, H. A. (2017). Menthol enhances nicotine reward-related behavior by potentiating nicotine-induced changes in nAChR function, nAChR upregulation, and DA neuron excitability. *Neuropsychopharmacology*. https://doi.org/10.1038/npp.2017.72.

Higgins, S. T., Heil, S. H., Sigmon, S. C., Tidey, J. W., Gaalema, D. E., Stitzer, M. L., et al. (2017). Response to varying the nicotine content of cigarettes in vulnerable populations: an initial experimental examination of acute effects. *Psychopharmacology*, *234*(1), 89–98. https://doi.org/10.1007/s00213-016-4438-z.

Hoffman, S. J., & Tan, C. (2015). Overview of systematic reviews on the health-related effects of government tobacco control policies. *BMC Public Health*, *15*, 744. https://doi.org/10.1186/s12889-015-2041-6.

Hogg, R. C. (2016). Contribution of monoamine oxidase inhibition to tobacco dependence: a review of the evidence. *Nicotine & Tobacco Research*, *18*(5), 509–523. https://doi.org/10.1093/ntr/ntv245.

Kassel, J. D., Greenstein, J. E., Evatt, D. P., Wardle, M. C., Yates, M. C., Veilleux, J. C., et al. (2007). Smoking topography in response to denicotinized and high-yield nicotine cigarettes in adolescent smokers. *The Journal of Adolescent Health*, *40*(1), 54–60. https://doi.org/10.1016/j.jadohealth.2006.08.006.

Khalki, H., Navailles, S., Piron, C. L., & De Deurwaerdere, P. (2013). A tobacco extract containing alkaloids induces distinct effects compared to pure nicotine on dopamine release in the rat. *Neuroscience Letters*, *544*, 85–88. https://doi.org/10.1016/j.neulet.2013.03.047.

McClernon, F. J., Froeliger, B., Rose, J. E., Kozink, R. V., Addicott, M. A., Sweitzer, M. M., et al. (2016). The effects of nicotine and non-nicotine smoking factors on working memory and associated brain function. *Addiction Biology*, *21*(4), 954–961. https://doi.org/10.1111/adb.12253.

Mercincavage, M., Saddleson, M. L., Gup, E., Halstead, A., Mays, D., & Strasser, A. A. (2017). Reduced nicotine content cigarette advertising: how false beliefs and subjective ratings affect smoking behavior. *Drug and Alcohol Dependence*, *173*, 99–106. https://doi.org/10.1016/j.drugalcdep.2016.12.022.

Naik, P., Fofaria, N., Prasad, S., Sajja, R. K., Weksler, B., Couraud, P. O., et al. (2014). Oxidative and pro-inflammatory impact of regular and denicotinized cigarettes on blood brain barrier endothelial cells: is smoking reduced or nicotine-free products really safe? *BMC Neuroscience*, *23*, 15–51. https://doi.org/10.1186/1471-2202-15-51.

O'Brien, E. K., Nguyen, A. B., Persoskie, A., & Hoffman, A. C. (2017). U. S. adults' addiction and harm beliefs about nicotine and low nicotine cigarettes. *Preventive Medicine*, *96*, 94–100. https://doi.org/10.1016/j.ypmed.2016.12.048.

Rupprecht, L. E., Koopmeiners, J. S., Dermody, S. S., Oliver, J. A., al'Absi, M., Benowitz, N. L., et al. (2017). Reducing nicotine exposure results in weight gain in smokers randomised to very low nicotine content cigarettes. *Tobacco Control*, *26*(e1), e43–e48. https://doi.org/10.1136/tobaccocontrol-2016-053301.

Scherer, G., & Lee, P. N. (2014). Smoking behaviour and compensation: a review of the literature with meta-analysis. *Regulatory Toxicology and Pharmacology*, *70*(3), 615–628. https://doi.org/10.1016/j.yrtph.2014.09.008.

Smith, C. J., Perfetti, T. A., Garg, R., & Hansch, C. (2003). IARC carcinogens reported in cigarette mainstream smoke and their calculated log P values. *Food and Chemical Toxicology*, *41*(6), 807–817. https://doi.org/10.1016/S0278-6915(03)00021-8.

Strasser, A. A., Lerman, C., Sanborn, P. M., Pickworth, W. B., & Feldman, E. A. (2007). New lower nicotine cigarettes can produce compensatory smoking and increased carbon monoxide exposure. *Drug and Alcohol Dependence*, *86*(2–3), 294–300. https://doi.org/10.1016/j.drugalcdep.2006.06.017.

Thompson, M. F., Poirier, G. L., Davila-Garcia, M. I., Huang, W., Tam, K., Robidoux, M., et al. (2017). Menthol enhances nicotine-induced locomotor sensitization and in vivo functional connectivity in adolescence. *Journal of Psychopharmacology*. https://doi.org/10.1177/0269881117719265.

Touiki, K., Rat, P., Molimard, R., Chait, A., & de Beaurepaire, R. (2007). Effects of tobacco and cigarette smoke extracts on serotonergic raphe neurons in the rat. *NeuroReport*, *18*(9), 925–929. https://doi.org/10.1097/WNR.0b013e32811d6d21.

United States Congress. (2009). *Family smoking prevention and tobacco control act*. US Government Printing Office.

United States Department of Health and Human Services. (1988). *The health consequences of smoking: nicotine addiction. a report of the surgeon general*. US Department of Health and Human Services. DHHS Publication No. (CDC) 88-8406: Public Health Service, Centers for Disease Control, Center for Chronic Disease Prevention and Health Promotion, Office of Smoking and Health.

West, R. (2017). Tobacco smoking: Health impact, prevalence, correlates and interventions. *Psychology and Health*, *32*(8), 1018–1036. https://doi.org/10.1080/08870446.2017.1325890.

World Health Organization. (2015). *Advisory note: Global nicotine reduction strategy: WHO study group on tobacco product regulation*. Geneva: WHO Press.

Zacny, J. P., & Stitzer, M. L. (1988). Cigarette brand-switching: effects on smoke exposure and smoking behavior. *The Journal of Pharmacology and Experimental Therapeutics*, *246*(2), 619–627.

CHAPTER

6

Prenatal Nicotine Exposure and Neuronal Progenitor Cells

Tursun Alkam[*,†], *Toshitaka Nabeshima*[*,‡,§]

*Japanese Drug Organization of Appropriate Use and Research, Nagoya, Japan
†Department of Basic Medical Sciences, College of Osteopathic Medicine of the Pacific, Western University of Health Sciences, Pomona, CA, United States
‡Advanced Diagnostic System Research Laboratory, Graduate School of Health Sciences, Fujita Health University, Toyoake, Japan
§Aino University, Ibaraki, Japan

Abbreviations

ACh	acetylcholine
CNS	central nervous system
DG	the dentate gyrus
E	embryonic day
G	gestational day
IPC	intermediate progenitor cell
mRNA	messenger RNA
nAChRs	nicotinic acetylcholine receptors
NMDAR	*N*-methyl-D-aspartate receptor
NPCs	neuronal progenitor cells
NSCs	neuronal stem cells
P	postnatal day
PNE	prenatal nicotine exposure
RGCs	radial glial cells
RT-PCR	reverse transcription-polymerase chain reaction
SVZ	subventricular zone

6.1 INTRODUCTION

Offspring of women who smoke during pregnancy suffer from cognitive and emotional abnormalities including inattentiveness, learning disability, impaired decision-making, impulsive behaviors, and restlessness (Blood-Siegfried & Rende, 2010; Ernst, Moolchan, & Robinson, 2001; Hamosh, Simon, & Hamosh, 1979). Nicotine, obtained through tobacco smoking or use of nicotine products for tobacco cessation therapy during pregnancy, is thought to be a disrupter of the development of the fetal brain by interfering with the functions of nAChRs (Dempsey & Benowitz, 2001; Slotkin, 1998). Animal studies on the toxicity of prenatal nicotine exposure (PNE) (Fig. 6.1) confirmed nicotine-induced emotional and cognitive behavioral abnormalities in adult offspring (Alkam, Kim, Mamiya, et al., 2013; Alkam, Kim, Hiramatsu, et al., 2013; Alkam et al., 2017; Berner et al., 2008; Britton, Vann, & Robinson, 2007; Gold, Keller, & Perry, 2009; Hall et al., 2016; Oliff & Gallardo, 1999). These behavioral abnormalities are associated with deficits in neurotransmitter production and signaling in the adult brain (Alkam, Kim, Mamiya, et al., 2013; Alkam et al., 2017; Huang, Liu, Griffith, & Winzer-Serhan, 2007; Matta et al., 2007; Slotkin et al., 2005, 2007; Vaglenova et al., 2008; Vaglenova, Birru, Pandiella, & Breese, 2004; Zhu et al., 2012; Zhu, Lee, Spencer, Biederman, & Bhide, 2014), while the disruption of neuronal progenitor cell proliferation and impairments in neurogenesis of the developing brain are also implicated (Abrous et al., 2002; Cohen et al., 2015; He, Wang, Wang, Shen, & Yin, 2013; Scerri, Stewart, Breen, & Balfour, 2006; Wei et al., 2012).

6.2 NEURONAL PROGENITOR CELLS IN DEVELOPING BRAIN

The mammalian central nervous system (CNS) is one of the earliest systems to appear during the embryogenesis. The CNS is derived from the outermost tissue layer, the ectoderm, of the embryo. The ectoderm gives rise to neuroectoderm that forms the neural plate. The neural plate then shapes the neural tube with a single layer

Neuroscience of Nicotine
https://doi.org/10.1016/B978-0-12-813035-3.00006-X

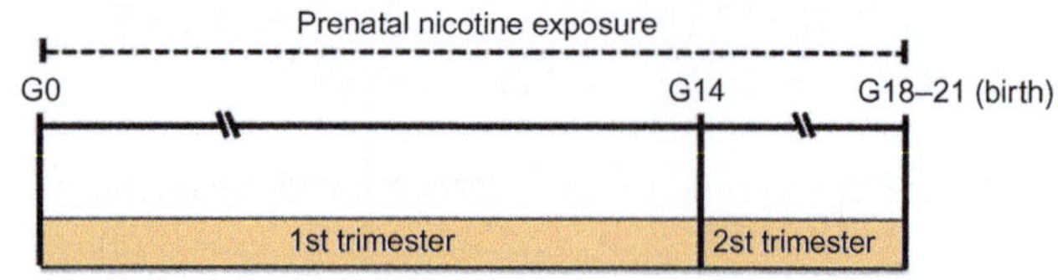

FIG. 6.1 Prenatal nicotine exposure period in the mouse.
The whole gestational period of a rodent is 18–21 days and consists of the first and second trimesters of human. *Modified from Dwyer et al. (2008).*

of neuroepithelial cells as the neuroepithelium. These neuroepithelial cells, function as primary neuronal stem cells (NSCs), give rise to differentiated neural cell types early in embryonic development, producing neurons initially at the early stage but gradually switching over to produce different types of glial cells (Aaku-Saraste, Oback, Hellwig, & Huttner, 1997; Kintner, 2002). The NSCs are defined as multipotent proliferative cells with a seemingly unlimited capacity for self-renewal with two mitotic pathways, the symmetrical division and the asymmetrical division. In the symmetrical division, both of the daughter cells become multipotent NSCs. In the asymmetrical division, one daughter cell becomes a multipotent neuronal stem cell, while the other becomes a radial glial cell that can renew itself and produce an intermediate progenitor cell (IPC) that moves toward neuronal and glial differentiation (Gotz & Huttner, 2005; Potten & Loeffler, 1990; Seaberg & van der Kooy, 2003; Weiss et al., 1996) (Fig. 6.2).

The NSCs begin to acquire features associated with glial cells, when cortical neurogenesis begins, on approximately around embryonic day 9 (E9) and E10 in the mouse (Kriegstein & Alvarez-Buylla, 2009). The asymmetrical cell division of a radial glial cell results in self-renewal of itself and the birth of a neuron or a glial cell (astrocyte or oligodendrocyte) depending on the differentiation (Fishell & Kriegstein, 2003). The proliferation versus differentiation of these cells and the type of cellular division affect the rate of neurogenesis (Gotz & Huttner, 2005). In the rat embryogenesis on E15–16, one radial glial cell generates one neuron and one radial glial cell per cell cycle (Noctor, Flint, Weissman, Dammerman, & Kriegstein, 2001). Thus, by both self-renewal of themselves and the production of radial glial cells (RGCs), NSCs generate a large number of differentiated cells including neurons and glial cells in the developing brain (Seaberg & van der Kooy, 2003; Weiss et al., 1996). Interestingly, while isolated RGCs from the neocortex of mouse and rat on

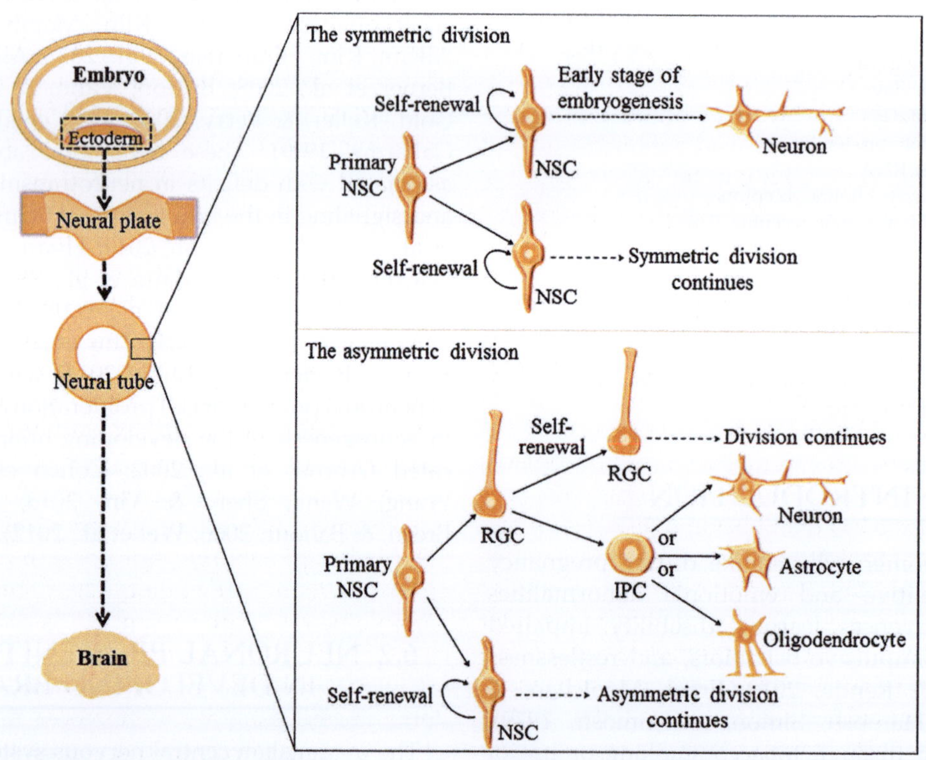

FIG. 6.2 Development of the central nervous system.
The primary neuronal stem cell in neural tube is a multipotent proliferative cell with a seemingly unlimited capacity for self-renewal with two mitotic pathways, the symmetrical division and the asymmetrical division. In the symmetrical division, both of the daughter cells become multipotent neuronal stem cells while producing neurons at the early stage. In the asymmetrical division, one daughter cell becomes a multipotent neuronal stem cell, while the other becomes a radial glial cell that can renew itself and produce an intermediate progenitor cell that moves toward neuronal or glial (astrocyte or oligodendrocyte) differentiation. *IPC*, intermediate progenitor cell; *NSC*, neuronal stem cell; *RGC*, radial glial cell.

E14, when cortical neurogenesis is at its peak, mainly generate neurons in vitro, they mainly generate astrocytes if isolated from the cortex on E18 when cortical neurogenesis is nearly complete (Malatesta, Hartfuss, & Gotz, 2000). Besides being the source of neuronal generation, RGCs also play a major guidance role for several generations of their neuronal progeny during brain histogenesis and degenerate or become astrocytes when neuronal migration is completed in the cerebral cortex and radial glial fibers are no longer required as guides for neuronal migration (Chanas-Sacre, Rogister, Moonen, & Leprince, 2000; Mission, Takahashi, & Caviness Jr., 1991; Noctor et al., 2001; Rakic, 1972).

Obviously, during the development of the brain, neurons not only are directly generated by NSCs but also originate from transit-amplifying intermediate progenitor cells, the RGCs (Kriegstein & Alvarez-Buylla, 2009). Therefore, both NSCs and RGCs during the embryonic and fetal neurogenesis are regarded as neuronal progenitor cells (NPCs) (Fig. 6.2). All NPCs can be identified by nestin, an intermediate filament protein, while nestin-expressing RGCs can be distinguished from NSCs by glial fibrillary acidic protein (GFAP), which is also expressed by astrocytes (Noctor et al., 2001).

Coordinated proliferation and differentiation of NPCs is the base for the production of appropriate numbers of neurons and glia during neuronal development in order to establish normal brain functions (Resende, Alves, Britto, & Ulrich, 2008). During embryonic and early postnatal neurogenesis, the transition of multipotent NSCs to RGCs is accompanied by a series of cellular structural and functional changes including the appearance of "glial" features of the radial glial cells and the appearance of new signaling molecules and proteins to support their newly assigned functions (Kriegstein & Gotz, 2003; Urban & Guillemot, 2014). The coordinated action of multiple signals acting on embryonic NPCs gives rise to a vast diversity of neuronal and glial cells that begin to shape the brain (Urban & Guillemot, 2014). With the help of these signaling molecules and proteins, newly differentiated neurons also migrate from their birthplaces to their final destinations to generate the primary anatomical regions of the brain and begin to establish a lifelong functional communication route with fellow neurons by their axons, dendrites, and spines with the help of synapses. The development of the brain continues, after birth, to early childhood to be fully completed. Henceforward, except the continuation of embryonic neurogenesis within the subventricular zone (SVZ) and the dentate gyrus (DG) of the hippocampus, the completed brain has no chance to regenerate its neuronal population during adulthood. Therefore, any changes in neurogenesis by any abnormal signaling pathways during the embryonic or prenatal development period of CNS compromise the structures and functions of the brain for whole life.

6.3 NICOTINIC ACETYLCHOLINE RECEPTORS IN NPCs

During embryonic or early postnatal development of CNS, various intrinsic factors and extracellular signaling pathways drive the proliferation and differentiation of neuronal stem cells (Urban & Guillemot, 2014). During the early development, activation of critical transcription factors at different time points of the CNS formation by patterning genes that are under the control of diverse intrinsic signaling pathways is required for the proliferation and differentiation of NPCs (Kintner, 2002). Once embryonic neurogenesis is initiated, neurotransmitter signaling also has an impact on several aspects of neurogenesis including proliferation, migration, and differentiation in various locations in the CNS (Berg, Belnoue, Song, & Simon, 2013).

One of the neurotransmitters that involved both in embryonic and adult neurogenesis is acetylcholine (ACh), which exerts its functions via nicotinic acetylcholine receptors (nAChRs) (Itou, Nochi, Kuribayashi, Saito, & Hisatsune, 2011). Functional subtypes of nAChR in NPCs play an important role in the mechanism underlying the embryonic neurogenesis and in early brain development.

The proteins of α4 and α7 subunits of nAChRs are detected in neuronal progenitors of the early embryonic mouse cerebral cortex as early as E10, and patch-clamp electrophysiological measurements indicate that nicotine and ACh evoke sizable inward currents characteristic of nicotinic receptor-evoked cytosolic Ca^{2+} signals (Atluri et al., 2001), while messenger RNA (mRNA) of β2 subunit is detected in the cerebral cortex as early as on E12 (Zoli, Le Novere, Hill Jr., & Changeux, 1995). Other subunits of nAChRs are also detected in undifferentiated nestin-positive but GFAP-negative NPCs that isolated from the mouse or rat neocortex at later fetal days. The reverse transcription-polymerase chain reaction (RT-PCR) analysis of embryonic mouse NPCs that isolated from the neocortex on E15.5 reveals mRNA expression of α3, α4, α5, α7, α9, β2, and β4 subunits of nAChRs, but not of α2, α6, and β3 subunits. In the same study, mRNA expression of α2, α3, α4, α5, α7, α9, β2, and β4 subunits, but not of α6 and β3 subunits, of nAChRs in progenitor cells prepared from the embryonic rat neocortex on E18 is detected (Takarada et al., 2012). The temporal and regional expression of individual subunits and their various combinations can form diverse subtypes of neuronal nAChRs. Dynamic stimulation of emerging functional subtypes of nAChRs regulates the proliferation of NPCs, neurogenesis, neuronal migration and neuronal maturation and plasticity, and neuronal survival as well as gliogenesis (Abreu-Villaca, Filgueiras, & Manhaes, 2011; Asrican, Paez-Gonzalez, Erb, & Kuo, 2016).

6.4 IMPACT OF NICOTINE ON PRENATAL NEUROGENESIS

The activity of nAChRs is also regulated by other exogenous chemicals such as nicotine and results in the disruption of fetal brain development (Navarro et al., 1989). Studies on the effects of nicotine almost consistently reported the decrease of neurogenesis in the developing brain and cognitive deficits in adult life (Abrous et al., 2002; Cohen et al., 2015; He et al., 2013; Scerri et al., 2006; Wei et al., 2012), suggesting the most immediate impact of nicotine on the fetal brain is the disruption of the ongoing neurogenesis.

In in vivo studies, neurogenesis is studied by examining the proliferation of NPCs and neuronal differentiation of newborn cells (Table 6.1). The proliferating cells can be labeled with injected 5-bromo-2′-deoxyuridine (BrdU), an analog of thymidine. The BrdU-containing newborn cells can be traced by anti-BrdU antibodies in immunohistochemistry, while neuronal differentiation of these cells can be identified by colabeling BrdU with a neuronal marker in a single neuron. Rat offspring who are exposed to nicotine in utero via osmotic minipumps, between gestational day 3 (G3) and birth (postnatal day 0, P0), and then through breast milk until P21 fail to show significant changes in cellular proliferation on P15 in the DG of the hippocampus. In the same study, the BrdU-labeled (on P15) newborn cells both in nicotine and control groups are equally differentiated into neurons in the DG of the hippocampus when examined on P41, suggesting no changes in neurogenesis (Mahar et al., 2012). Exposure to nicotine via osmotic minipumps during G7–P0 does not change the numbers of neurons in CA1, CA3, or DG of the hippocampus on P14 compared to that of control offspring (Wang & Gondre-Lewis, 2013). Nicotine exposure via drinking water during G6–P21 does not change proliferating cell indexes in the subgranular zone of the hippocampal DG while transiently suppresses the late-stage differentiation of NPCs in rat offspring on P21, but this suppression disappears on P77 (Ohishi et al., 2014). These findings, however, do not reflect the direct effects of nicotine on prenatal neurogenesis, because the newborn cells in these studies are not labeled during PNE before birth but labeled or examined during the perinatal stage when the direct stimulatory effect of nicotine is weak or absent. The lack of influences of PNE on hippocampal neurogenesis in these studies may also be due to the continuous contribution of active neurogenesis in the hippocampus on the testing days that are not under direct influences of nicotine. The hippocampal neurogenesis is still prominent during early postnatal days (Altman & Bayer, 1990), and this robust neurogenesis may make it difficult to evaluate the effects of PNE on fetal neurogenesis if the newborn cells are not labeled before birth during a certain time window. Further, the birth after PNE would result in nicotine withdrawal in offspring during immediate postnatal days and may change the sensitivity and functions of nAChRs (Slotkin, 1998). These functional changes may complicate the observation of the effects of nicotine on NPCs during the prenatal period and make the observed findings less meaningful, if the effects of PNE on neurogenesis are studied later on postnatal days. Therefore, the effects of PNE on NPCs and consequent neurogenesis should be investigated during the prenatal period or before birth, to provide supporting explanations for the behavioral dysfunctions in adult offspring. When the proliferating cells during a series time windows of G14, G15, and G16 in mice are labeled with BrdU, PNE during G14–P0 are found to impair neurogenesis by suppressing the proliferation of NPCs in ventricular zone and SVZ (Aoyama et al., 2016). These impairments result in fewer glutamatergic neurons in the medial prefrontal cortex on P70–P84 in offspring. The antineurogenic effects of PNE are blocked by pretreatment with a α7 nAChR antagonist methyllycaconitine, but not with a α4β2 nAChR antagonist dihydro-β-erythroidine (Aoyama et al., 2016). These results suggest that α7-nAChR-mediated decrease in

TABLE 6.1 Impact of Prenatal Nicotine Exposure on Neurogenesis

Time exposure and the route	Time of BrdU labeling	Time of examination	Brain regions	Impact on neurogenesis	References
G3–P0, via osmotic minipumps and until P21 through breast milk	P15	P41	The hippocampus	No changes	Mahar et al. (2012)
G7–P0, via osmotic minipumps	No labeling	P14	The hippocampus	No changes in the number of neurons	Wang and Gondre-Lewis (2013)
G6–P21, via drinking water	No labeling	P21 and P77	The hippocampus	Transiently suppresses the late-stage neuronal differentiation on P21, but this suppression disappears on P77	Ohishi et al. (2014)
G14–P0, via drinking water	G14, G15, G16	P70–P84	The medial prefrontal cortex	Decreases the number of glutamatergic neurons	Aoyama et al. (2016)

the proliferation of neuronal progenitors during prenatal period induced by PNE is critical to behavioral dysfunctions of offspring observed during adolescent and adult life (Alkam, Kim, Mamiya, et al., 2013; Alkam, Kim, Hiramatsu, et al., 2013; Alkam et al., 2017; Aoyama et al., 2016).

In an in vitro study, chronically exposed nicotine for 12 days significantly inhibits the proliferation of undifferentiated NPCs that isolated from the mouse or rat neocortex on E15.5 or on E18; the NPCs then subsequently differentiated into neurons (Takarada et al., 2012). These effects of nicotine on proliferation and neuronal differentiation are significantly prevented by the antagonists of α4β2 subtype of nAChRs, dihydro-β-erythroidine and 4-(5-ethoxy-3-pyridinyl)-*N*-methyl-(3*E*)-3-buten-1-amine, but not by an antagonist of α7 subtype of nAChRs methyllycaconitine (Takarada et al., 2012). While the subtypes of nAChRs involved nicotine-inhibited proliferation of NPCs differ on these in vivo and in vitro studies, these findings confirm antiproliferation effects of PNE.

6.5 THE RELEVANT FUNCTIONS OF INVOLVED nAChRs

Regional and temporal coordinative expressions of distinct subtypes of nAChRs during critical prenatal and early postnatal periods of neuronal development involve mechanisms associated with neurogenesis. The expression of subunits of nAChRs is strongly regulated during brain development, and these receptors can upregulate their expression when chronically exposed to nicotine (Abreu-Villaca et al., 2011). Both α7 subtype and α4β2 subtype are the most abundant classes of nAChRs found in the brain regions that implicated in cognitive functions (Corradi & Bouzat, 2016; Mazzaferro, Bermudez, & Sine, 2017; Pandya & Yakel, 2011). While α4β2 subtype of nAChRs upregulates strongly when exposed to nicotine and displays high-affinity nicotine binding site, some nAChR subtypes may even be downregulated (Albuquerque, Pereira, Alkondon, & Rogers, 2009). Differences in relative subunit association and final receptor assembly have been proposed to explain these apparently conflicting results (Albuquerque et al., 2009).

The diversity of nAChR assembly from different subunits contributes to functional differences of nAChRs such as Ca^{2+}, Na^{+}, and K^{+} ion permeability and desensitization (Albuquerque et al., 2009). Receptors composed of α7 subunits are known to desensitize rapidly and have a high Ca^{2+} permeability ratio that exceeds that of the glutamate *N*-methyl-D-aspartate receptor (NMDAR) and most other nAChRs (Albuquerque et al., 2009). Following the stimulation of nAChRs by nicotine, extracellular Ca^{2+} enters cells and induces Ca^{2+} release from intracellular stores. Calcium signals are pivotal in shaping nAChR-mediated effects (Abreu-Villaca et al., 2011). It is well accepted that calcium signaling regulates many fundamental steps of the cell cycle, including the transitions among the various phases of the cell cycle, the transcription of immediate early genes, and the regulating events that control proliferation, neuronal differentiation, and migration (Toth, Shum, & Prakriya, 2016). Chronic exposure to nicotine significantly inhibits the proliferation of undifferentiated NPCs that are defective of NMDAR subunit 1, suggesting activation of nAChRs would likely result in negative regulation of self-replication through a Ca^{2+} signaling mechanism independent of NMDAR signal in undifferentiated NPCs (Takarada et al., 2012). Interestingly, activation of NMDAR in NPCs isolated from the developing rat neocortex inhibits proliferation and leads subsequent differentiation toward a neuronal lineage (Nakamichi, Takarada, & Yoneda, 2009; Yoneyama et al., 2008).

These findings suggest calcium signaling may mediate the effects of prenatally exposed nicotine on NPCs (Fig. 6.3), while neurogenesis could be also under the coordinative control by other signal inputs in the developing brain.

6.6 FUTURE DIRECTIONS

Currently available literature about the inhibitory effects of PNE on the proliferation and differentiation of NPCs suggests the involvement of α7 and α4β2 subtypes of nAChRs through a Ca^{2+} signaling mechanism during brain development in rodents. However, developing conditional knockout and knock-in model animals for individual nAChR subtypes will crucially help to identify how these and other subtypes of nAChRs are involved in the mediation of the inhibitory effects of PNE on neurogenesis in the developing brain.

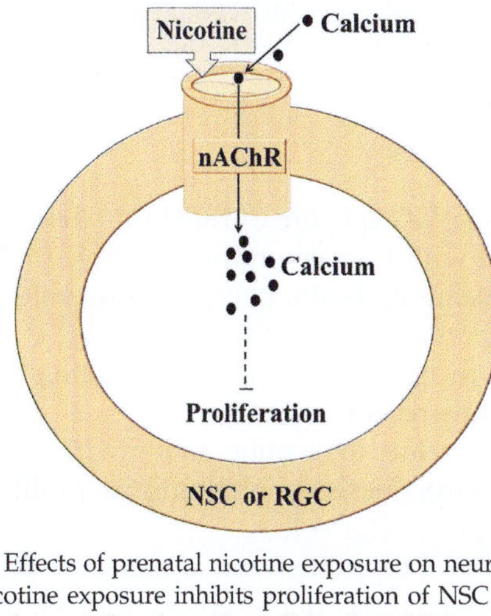

FIG. 6.3 Effects of prenatal nicotine exposure on neurogenesis. Prenatal nicotine exposure inhibits proliferation of NSC and RGC via increasing calcium influx through nAChRs. Both NSC and RGC during the embryonic and fetal neurogenesis are neuronal progenitor cells. *NSC*, neuronal stem cell; *RGC*, radial glial cell.

MINI-DICTIONARY OF TERMS

Cell proliferation The process of increasing the number of cells by cell divisions.

Cellular differentiation The process by which an intermediate progenitor cell becomes a neuron, an astrocyte, or an oligodendrocyte during the development of the central nervous system.

Cellular self-renewal The process by which a stem cell itself divides into two stem cells that, at least one of them, have an identical property to the original stem cell.

Neurogenesis The process of producing new neurons from neuronal stem cells or neuronal progenitor cells in the embryonic and early postnatal brain.

Key Facts About Prenatal Tobacco Exposure

- Tobacco smoking during pregnancy can disrupt the normal development of the fetal brain and affects the optimum function of the brain in whole life.
- Prenatal exposure through passive smoking can also disrupt the normal development of the fetal brain.
- Sixteen subunits of nicotinic acetylcholine receptor have been identified in mammals.
- Neuronal nicotinic acetylcholine receptors in mammals have 11 subunits that can regulate emotion-, addiction-, and cognition-related neurotransmitter signaling pathways in the brain.
- More than 80% of the world's smokers live in low-income and middle-income countries. Tobacco use in pregnant women in these countries, except Turkey, is low.
- Higher prevalence of tobacco smoking in pregnant women have been reported in high-income countries including the United States and the United Kingdom from population-based studies, while smoking during pregnancy has been reported decreasing in Japan.
- Nicotine replacement therapy for tobacco cessation during pregnancy is controversial and not better than placebo treatment.

Summary Points

- Tobacco smoke exposure during brain development is associated with later emerging cognitive problems.
- In the developing brain, neuronal progenitor cells include neuronal stem cells and radial glial cells.
- Neuronal progenitor cells express nicotinic acetylcholine receptors.
- Prenatal exposure to nicotine, an ingredient of tobacco, impairs neurogenesis by inhibiting the proliferation of neuronal progenitor cells.
- Inhibition of the proliferation of neuronal progenitor cells by prenatal nicotine exposure may be mediated by calcium signaling through nicotinic acetylcholine receptors.

References

Aaku-Saraste, E., Oback, B., Hellwig, A., & Huttner, W. B. (1997). Neuroepithelial cells downregulate their plasma membrane polarity prior to neural tube closure and neurogenesis. *Mechanisms of Development*, *69*(1–2), 71–81.

Abreu-Villaca, Y., Filgueiras, C. C., & Manhaes, A. C. (2011). Developmental aspects of the cholinergic system. *Behavioural Brain Research*, *221*(2), 367–378.

Abrous, D. N., Adriani, W., Montaron, M. F., Aurousseau, C., Rougon, G., Le Moal, M., et al. (2002). Nicotine self-administration impairs hippocampal plasticity. *The Journal of Neuroscience*, *22*(9), 3656–3662.

Albuquerque, E. X., Pereira, E. F., Alkondon, M., & Rogers, S. W. (2009). Mammalian nicotinic acetylcholine receptors: from structure to function. *Physiological Reviews*, *89*(1), 73–120.

Alkam, T., Kim, H. C., Hiramatsu, M., Mamiya, T., Aoyama, Y., Nitta, A., et al. (2013). Evaluation of emotional behaviors in young offspring of C57BL/6J mice after gestational and/or perinatal exposure to nicotine in six different time-windows. *Behavioural Brain Research*, *239*, 80–89.

Alkam, T., Kim, H. C., Mamiya, T., Yamada, K., Hiramatsu, M., & Nabeshima, T. (2013). Evaluation of cognitive behaviors in young offspring of C57BL/6J mice after gestational nicotine exposure during different time-windows. *Psychopharmacology*, *230*(3), 451–463.

Alkam, T., Mamiya, T., Kimura, N., Yoshida, A., Kihara, D., Tsunoda, Y., et al. (2017). Prenatal nicotine exposure decreases the release of dopamine in the medial frontal cortex and induces atomoxetine-responsive neurobehavioral deficits in mice. *Psychopharmacology*, *234*(12), 1853–1869.

Altman, J., & Bayer, S. A. (1990). Migration and distribution of two populations of hippocampal granule cell precursors during the perinatal and postnatal periods. *The Journal of Comparative Neurology*, *301* (3), 365–381.

Aoyama, Y., Toriumi, K., Mouri, A., Hattori, T., Ueda, E., Shimato, A., et al. (2016). Prenatal nicotine exposure impairs the proliferation of neuronal progenitors, leading to fewer glutamatergic neurons in the medial prefrontal cortex. *Neuropsychopharmacology*, *41*(2), 578–589.

Asrican, B., Paez-Gonzalez, P., Erb, J., & Kuo, C. T. (2016). Cholinergic circuit control of postnatal neurogenesis. *Neurogenesis (Austin)*, *3*(1).

Atluri, P., Fleck, M. W., Shen, Q., Mah, S. J., Stadfelt, D., Barnes, W., et al. (2001). Functional nicotinic acetylcholine receptor expression in stem and progenitor cells of the early embryonic mouse cerebral cortex. *Developmental Biology*, *240*(1), 143–156.

Berg, D. A., Belnoue, L., Song, H., & Simon, A. (2013). Neurotransmitter-mediated control of neurogenesis in the adult vertebrate brain. *Development*, *140*(12), 2548–2561.

Berner, J., Ringstedt, T., Brodin, E., Hokfelt, T., Lagercrantz, H., & Wickstrom, R. (2008). Prenatal exposure to nicotine affects substance p and preprotachykinin-A mRNA levels in newborn rat. *Pediatric Research*, *64*(6), 621–624.

Blood-Siegfried, J., & Rende, E. K. (2010). The long-term effects of prenatal nicotine exposure on neurologic development. *Journal of Midwifery & Women's Health*, *55*(2), 143–152.

Britton, A. F., Vann, R. E., & Robinson, S. E. (2007). Perinatal nicotine exposure eliminates peak in nicotinic acetylcholine receptor response in adolescent rats. *The Journal of Pharmacology and Experimental Therapeutics*, *320*(2), 871–876.

Chanas-Sacre, G., Rogister, B., Moonen, G., & Leprince, P. (2000). Radial glia phenotype: origin, regulation, and transdifferentiation. *Journal of Neuroscience Research*, *61*(4), 357–363.

Cohen, A., Soleiman, M. T., Talia, R., Koob, G. F., George, O., & Mandyam, C. D. (2015). Extended access nicotine self-administration with periodic deprivation increases immature neurons in the hippocampus. *Psychopharmacology*, *232*(2), 453–463.

Corradi, J., & Bouzat, C. (2016). Understanding the bases of function and modulation of alpha7 nicotinic receptors: implications for drug discovery. *Molecular Pharmacology*, *90*(3), 288–299.

Dempsey, D. A., & Benowitz, N. L. (2001). Risks and benefits of nicotine to aid smoking cessation in pregnancy. *Drug Safety*, *24*(4), 277–322.

Dwyer, J. B., Broide, R. S., & Leslie, F. M. (2008). Nicotine and brain development. *Birth Defects Research. Part C, Embryo Today*, *84*(1), 30–44.

Ernst, M., Moolchan, E. T., & Robinson, M. L. (2001). Behavioral and neural consequences of prenatal exposure to nicotine. *Journal of the American Academy of Child and Adolescent Psychiatry*, *40*(6), 630–641.

Fishell, G., & Kriegstein, A. R. (2003). Neurons from radial glia: the consequences of asymmetric inheritance. *Current Opinion in Neurobiology*, *13*(1), 34–41.

Gold, A. B., Keller, A. B., & Perry, D. C. (2009). Prenatal exposure of rats to nicotine causes persistent alterations of nicotinic cholinergic receptors. *Brain Research*, *1250*, 88–100.

Gotz, M., & Huttner, W. B. (2005). The cell biology of neurogenesis. *Nature Reviews: Molecular Cell Biology*, *6*(10), 777–788.

Hall, B. J., Cauley, M., Burke, D. A., Kiany, A., Slotkin, T. A., & Levin, E. D. (2016). Cognitive and behavioral impairments evoked by low-level exposure to tobacco smoke components: comparison with nicotine alone. *Toxicological Sciences*, *151*(2), 236–244.

Hamosh, M., Simon, M. R., & Hamosh, P. (1979). Effect of nicotine on the development of fetal and suckling rats. *Biology of the Neonate*, *35*(5–6), 290–297.

He, N., Wang, Z., Wang, Y., Shen, H., & Yin, M. (2013). ZY-1, a novel nicotinic analog, promotes proliferation and migration of adult hippocampal neural stem/progenitor cells. *Cellular and Molecular Neurobiology*, *33*(8), 1149–1157.

Huang, L. Z., Liu, X., Griffith, W. H., & Winzer-Serhan, U. H. (2007). Chronic neonatal nicotine increases anxiety but does not impair cognition in adult rats. *Behavioral Neuroscience*, *121*(6), 1342–1352.

Itou, Y., Nochi, R., Kuribayashi, H., Saito, Y., & Hisatsune, T. (2011). Cholinergic activation of hippocampal neural stem cells in aged dentate gyrus. *Hippocampus*, *21*(4), 446–459.

Kintner, C. (2002). Neurogenesis in embryos and in adult neural stem cells. *The Journal of Neuroscience*, *22*(3), 639–643.

Kriegstein, A., & Alvarez-Buylla, A. (2009). The glial nature of embryonic and adult neural stem cells. *Annual Review of Neuroscience*, *32*, 149–184.

Kriegstein, A. R., & Gotz, M. (2003). Radial glia diversity: a matter of cell fate. *Glia*, *43*(1), 37–43.

Mahar, I., Bagot, R. C., Davoli, M. A., Miksys, S., Tyndale, R. F., Walker, C. D., et al. (2012). Developmental hippocampal neuroplasticity in a model of nicotine replacement therapy during pregnancy and breastfeeding. *PLoS ONE*, *7*(5), e37219.

Malatesta, P., Hartfuss, E., & Gotz, M. (2000). Isolation of radial glial cells by fluorescent-activated cell sorting reveals a neuronal lineage. *Development*, *127*(24), 5253–5263.

Matta, S. G., Balfour, D. J., Benowitz, N. L., Boyd, R. T., Buccafusco, J. J., Caggiula, A. R., et al. (2007). Guidelines on nicotine dose selection for in vivo research. *Psychopharmacology*, *190*(3), 269–319.

Mazzaferro, S., Bermudez, I., & Sine, S. M. (2017). Alpha4beta2 nicotinic acetylcholine receptors: relationships between subunit stoichiometry and function at the single channel level. *The Journal of Biological Chemistry*, *292*(7), 2729–2740.

Mission, J. P., Takahashi, T., & Caviness, V. S., Jr. (1991). Ontogeny of radial and other astroglial cells in murine cerebral cortex. *Glia*, *4*(2), 138–148.

Nakamichi, N., Takarada, T., & Yoneda, Y. (2009). Neurogenesis mediated by gamma-aminobutyric acid and glutamate signaling. *Journal of Pharmacological Sciences*, *110*(2), 133–149.

Navarro, H. A., Seidler, F. J., Eylers, J. P., Baker, F. E., Dobbins, S. S., Lappi, S. E., et al. (1989). Effects of prenatal nicotine exposure on development of central and peripheral cholinergic neurotransmitter systems. Evidence for cholinergic trophic influences in developing brain. *The Journal of Pharmacology and Experimental Therapeutics*, *251*(3), 894–900.

Noctor, S. C., Flint, A. C., Weissman, T. A., Dammerman, R. S., & Kriegstein, A. R. (2001). Neurons derived from radial glial cells establish radial units in neocortex. *Nature*, *409*(6821), 714–720.

Ohishi, T., Wang, L., Akane, H., Shiraki, A., Itahashi, M., Mitsumori, K., et al. (2014). Transient suppression of late-stage neuronal progenitor cell differentiation in the hippocampal dentate gyrus of rat offspring after maternal exposure to nicotine. *Archives of Toxicology*, *88*(2), 443–454.

Oliff, H. S., & Gallardo, K. A. (1999). The effect of nicotine on developing brain catecholamine systems. *Frontiers in Bioscience*, *4*, D883–D897.

Pandya, A., & Yakel, J. L. (2011). Allosteric modulators of the alpha4-beta2 subtype of neuronal nicotinic acetylcholine receptors. *Biochemical Pharmacology*, *82*(8), 952–958.

Potten, C. S., & Loeffler, M. (1990). Stem cells: attributes, cycles, spirals, pitfalls and uncertainties. Lessons for and from the crypt. *Development*, *110*(4), 1001–1020.

Rakic, P. (1972). Mode of cell migration to the superficial layers of fetal monkey neocortex. *The Journal of Comparative Neurology*, *145*(1), 61–83.

Resende, R. R., Alves, A. S., Britto, L. R., & Ulrich, H. (2008). Role of acetylcholine receptors in proliferation and differentiation of P19 embryonal carcinoma cells. *Experimental Cell Research*, *314*(7), 1429–1443.

Scerri, C., Stewart, C. A., Breen, K. C., & Balfour, D. J. (2006). The effects of chronic nicotine on spatial learning and bromodeoxyuridine incorporation into the dentate gyrus of the rat. *Psychopharmacology*, *184*(3–4), 540–546.

Seaberg, R. M., & van der Kooy, D. (2003). Stem and progenitor cells: the premature desertion of rigorous definitions. *Trends in Neurosciences*, *26*(3), 125–131.

Slotkin, T. A. (1998). Fetal nicotine or cocaine exposure: which one is worse? *The Journal of Pharmacology and Experimental Therapeutics*, *285*(3), 931–945.

Slotkin, T. A., MacKillop, E. A., Rudder, C. L., Ryde, I. T., Tate, C. A., & Seidler, F. J. (2007). Permanent, sex-selective effects of prenatal or adolescent nicotine exposure, separately or sequentially, in rat brain regions: indices of cholinergic and serotonergic synaptic function, cell signaling, and neural cell number and size at 6 months of age. *Neuropsychopharmacology*, *32*(5), 1082–1097.

Slotkin, T. A., Seidler, F. J., Qiao, D., Aldridge, J. E., Tate, C. A., Cousins, M. M., et al. (2005). Effects of prenatal nicotine exposure on primate brain development and attempted amelioration with supplemental choline or vitamin C: neurotransmitter receptors, cell signaling and cell development biomarkers in fetal brain regions of rhesus monkeys. *Neuropsychopharmacology*, *30*(1), 129–144.

Takarada, T., Nakamichi, N., Kitajima, S., Fukumori, R., Nakazato, R., Le, N. Q., et al. (2012). Promoted neuronal differentiation after activation of alpha4/beta2 nicotinic acetylcholine receptors in undifferentiated neural progenitors. *PLoS ONE*, *7*(10), e46177.

Toth, A. B., Shum, A. K., & Prakriya, M. (2016). Regulation of neurogenesis by calcium signaling. *Cell Calcium*, *59*(2–3), 124–134.

Urban, N., & Guillemot, F. (2014). Neurogenesis in the embryonic and adult brain: same regulators, different roles. *Frontiers in Cellular Neuroscience*, *8*, 396.

Vaglenova, J., Birru, S., Pandiella, N. M., & Breese, C. R. (2004). An assessment of the long-term developmental and behavioral teratogenicity of prenatal nicotine exposure. *Behavioural Brain Research*, *150*(1–2), 159–170.

Vaglenova, J., Parameshwaran, K., Suppiramaniam, V., Breese, C. R., Pandiella, N., & Birru, S. (2008). Long-lasting teratogenic effects of nicotine on cognition: gender specificity and role of AMPA receptor function. *Neurobiology of Learning and Memory*, *90*(3), 527–536.

Wang, H., & Gondre-Lewis, M. C. (2013). Prenatal nicotine and maternal deprivation stress de-regulate the development of CA1, CA3, and dentate gyrus neurons in hippocampus of infant rats. *PLoS ONE, 8*(6), e65517.

Wei, Z., Belal, C., Tu, W., Chigurupati, S., Ameli, N. J., Lu, Y., et al. (2012). Chronic nicotine administration impairs activation of cyclic AMP-response element binding protein and survival of newborn cells in the dentate gyrus. *Stem Cells and Development, 21*(3), 411–422.

Weiss, S., Reynolds, B. A., Vescovi, A. L., Morshead, C., Craig, C. G., & van der Kooy, D. (1996). Is there a neural stem cell in the mammalian forebrain? *Trends in Neurosciences, 19*(9), 387–393.

Yoneyama, M., Nakamichi, N., Fukui, M., Kitayama, T., Georgiev, D. D., Makanga, J. O., et al. (2008). Promotion of neuronal differentiation through activation of N-methyl-D-aspartate receptors transiently expressed by undifferentiated neural progenitor cells in fetal rat neocortex. *Journal of Neuroscience Research, 86*(11), 2392–2402.

Zhu, J., Lee, K. P., Spencer, T. J., Biederman, J., & Bhide, P. G. (2014). Transgenerational transmission of hyperactivity in a mouse model of ADHD. *The Journal of Neuroscience, 34*(8), 2768–2773.

Zhu, J., Zhang, X., Xu, Y., Spencer, T. J., Biederman, J., & Bhide, P. G. (2012). Prenatal nicotine exposure mouse model showing hyperactivity, reduced cingulate cortex volume, reduced dopamine turnover, and responsiveness to oral methylphenidate treatment. *The Journal of Neuroscience, 32*(27), 9410–9418.

Zoli, M., Le Novere, N., Hill, J. A., Jr., & Changeux, J. P. (1995). Developmental regulation of nicotinic ACh receptor subunit mRNAs in the rat central and peripheral nervous systems. *The Journal of Neuroscience, 15*(3 Pt 1), 1912–1939.

CHAPTER

7

Synaptically Located Nicotinic Acetylcholine Receptor Subunits in Neurons Involved in Dependency to Nicotine

Kristi A. Kohlmeier

Department of Drug Design and Pharmacology, Faculty of Health and Medical Sciences, University of Copenhagen, Copenhagen, Denmark

7.1 INTRODUCTION

Nicotine does not occur naturally in the mammalian body. However, recognition of the ability of nicotine to strongly activate one of the receptors for the endogenous signaling transmitter, acetylcholine (ACh), resulted in the naming of this ACh receptor as the nicotinic acetylcholine receptor (nAChR). Nicotine is considered the prototypic agonist of the mammalian nAChR, and in fact, this receptor is activated by nicotine in some ways more efficaciously than it is by its natural agonist, ACh (Whiteaker, Sharples, & Wonnacott, 1998). nAChRs are found throughout the human body, and while activated by endogenous ACh, they can also be activated by exogenously mediated exposure to nicotine via traditional, combustible cigarettes and vapor from e-cigarettes, as well as via abstinence aids such as transdermal nicotine patches and nicotine-containing gum. The psychopharmacological actions of nicotine are due to the activation of nAChRs in the brain, and therefore, examination of how nAChRs function at synapses within the central nervous system is relevant to considerations of neurobiological effects elicited by nicotine-containing products, which, in addition to underlying the potentially positive actions of cognitive enhancement and increases in arousal, also induce the negative action of engendering psychological dependency.

7.2 nAChRs STRUCTURE AND BINDING

Synaptically located nAChRs are pentameric, ligand-gated membrane-bound protein structures, containing an extracellular disulfide bond at the N-terminal domain that creates a cysteine loop characteristic of other "cys-loop" ligand-gated receptors, such as the chloride permeable $GABA_A$ receptor and the glycine receptor, to which nAChRs are related. nAChRs can form as either homo- or heteropentameric structures, with the five subunits creating a central, ion-permeable pore (Albuquerque, Pereira, Alkondon, & Rogers, 2009). To date, 16 homologous mammalian subunits of the nAChR have been identified, which are designated as being either α or β: α2–10 and β2–10 (Fig. 7.1). The five subunits coassemble, with heteromeric receptors consisting of two or three α subunits and two or three β subunits, resulting in a receptor complex which exhibits a high affinity for agonist. Lower-affinity receptors are usually homomeric, with five α7 subunits being the most common. Homomeric nAChRs of the α8 and 9 subunits exist; however, these subunits can also participate in heteromers. Interestingly, while long thought to only exist as part of a homomeric structure within the human brain, recent data indicate that the α7 subunit can also be expressed with other subunits in functional nAChR clusters (Moretti et al., 2014). Coexpression with β subunits is an apparent requirement for functionality of nAChRs containing α2–6, and association with the α9 subunit is a prerequisite for the activity of nAChRs containing the α10 subunit (Elgoyhen et al., 2001).

The binding site for orthosteric ligands, which includes ACh and exogenously applied nicotine, is positioned at the interface between subunits, and multiple allosteric binding sites have been identified as well, depending on the stoichiometry of the subunits (Fig. 7.2). Upon

Neuroscience of Nicotine
https://doi.org/10.1016/B978-0-12-813035-3.00007-1

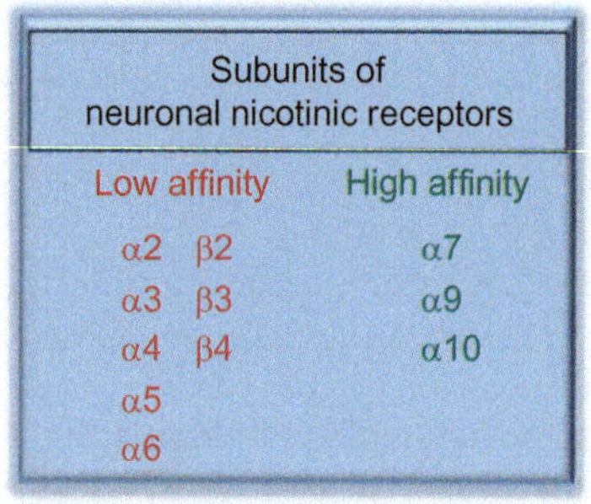

FIG. 7.1 Neuronal nicotinic receptors.
Neuronal nicotinic receptors are a family of homo- and heteropentameric receptors composed of subunits that determine the affinity of the receptor.

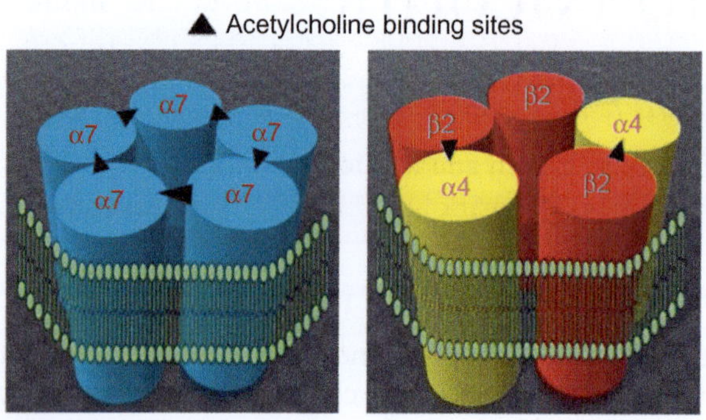

FIG. 7.2 Structure of homo- or heteropentameric neuronal nicotinic receptors (nAChRs).
The binding sites for acetylcholine are at subunit interfaces (*black arrowheads*), and the composition of subunits dictates ion permeability.

binding of agonist, nAChRs undergo a change in conformation that allows conduction of sodium and potassium ions, which can be altered by binding of agents at allosteric sites (Albuquerque et al., 2009). Calcium permeability can be conferred by inclusion within the pentameric structure of subunits, which allow passage to this cation via charged residues along the pore and polar residues located in the outer region of the pore, such as is present in homomeric α9-containing nAChRs (Dani, 1986). In general, nAChRs are considered to rapidly desensitize, and prolonged exposure to nicotine is likely to result in significant desensitization of the nAChR, leaving it closed and in a nonresponsive state (Dani, Radcliffe, & Pidoplichko, 2000).

However, the rate of onset and recovery from this refractory period is highly dependent on subunit composition (McGehee & Role, 1995). Further, multiple time constants appear to be involved in the longevity of desensitization. Dependence to nicotine is characterized by constant levels of the drug in the blood, and the constant exposure engendered by habitual users is likely to shift the desensitization kinetic of nAChRs to the long-lived state (Dani & Heinemann, 1996). Therefore, in addition to conferring ion permeability, the constellation of subunits defines the pharmacokinetic properties of the binding of agonists and modulators, and a further complexity to the kinetic profile is conferred by the dynamic temporal pattern of nicotine exposure.

Although expression systems suggest a plethora of subunit combinations, in native systems, only a few constellations have been definitively identified, with the high-affinity α4β2 and the lower-affinity homomeric α7 being considered the most abundant and, without doubt, the best well characterized. The α4β2 can be further distinguished from the α7 by its more rapid rate of desensitization to low levels of agonist, its more relatively slow kinetics and differential pharmacological sensitivity. In α- and β-containing nAChRs, the positioning of the subunits relative to each other can result in differences in sensitivity to ACh, with the (α4)2(β2)3 (or Type I) constellation exhibiting a high affinity to ACh and the (α4)3(β2)2 (or Type II) relationship resulting in a lower-affinity profile to the naturally present agonist (Miwa, Freedman, & Lester, 2011; Moroni & Bermudez, 2006). Other regulatory factors including calcium, protein kinases, and intracellular and extracellular peptides can also alter nAChR kinetic functioning (Miwa et al., 2011; Parri & Dineley, 2010).

One very curious feature of α4β2 nAChRs is that chronic exposure to agonist leads to upregulation of numbers of synaptically located, membrane-bound α4β2-containing receptors, which is a phenomenon not shown with other cys-loop receptors (Fenster, Whitworth, Sheffield, Quick, & Lester, 1999). Chronic exposure to nicotine-containing products was found to upregulate high-affinity labeled nicotine binding sites in the human brain; however, the mechanism underlying this finding is unknown (Marks, Burch, & Collins, 1983). Later studies in isolated systems showed a relationship between desensitization and upregulation (Fenster et al., 1999). One implication of this work is that chronic exposure to the agonist nicotine, such as experienced by habitual users of nicotine-containing products, is likely to have fostered increases in nAChRs in synapses in the brain that likely play a role in processes underlying addiction, tolerance, and withdrawal (Fenster et al., 1999)

7.3 LOCATION

7.3.1 Presynaptic

Within the human brain, nAChRs exhibit a strong presence on presynaptic sites (Fig. 7.3), where they can be activated endogenously by the release of ACh in the synapse either from their location within the presynaptic membrane or at extrasynaptic sites, where slower acting volume transmission or spillover of ACh could play a role (Bennett, Arroyo, Berns, & Hestrin, 2012). However, the biological significance of this effect in processes governed

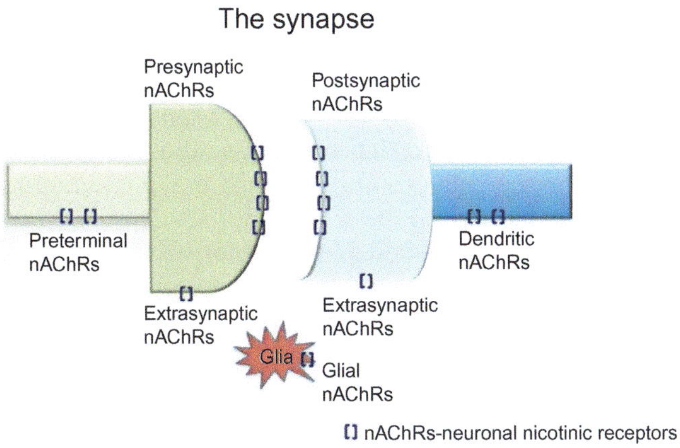

FIG. 7.3 Neuronal nicotinic receptors are located at several different positions.
nAChRs can be found on neurons at presynaptic, postsynaptic, and extrasynaptic locations.

by ACh may not be as relevant as that of physically mediated, rapid synaptic signaling (Sarter, Parikh, & Howe, 2009). Regardless of the actions of ACh, nicotine applied exogenously would be expected to activate presynaptic nAChRs, located both within the presynaptic membrane or extrasynaptically, which can result in alterations in cellular excitability. If the presynaptic nAChR is permeable to calcium, increases in the terminal of this ion could lead to the release of the signature neurotransmitter (Alkondon, Pereira, Eisenberg, & Albuquerque, 1999). Increases in transmitter could also be generated even in the absence of calcium permeability of the nAChR pore, as it would be expected that activation of the nAChR would lead to depolarization of the terminal that would lead indirectly to activation of voltage-gated calcium channels, with the result of the release of neurotransmitter. The phenotype of neurons upon which these presynaptic nAChRs are located determines if their activation leads to the release of excitatory modulators or transmitters such as glutamate or inhibitory transmitters such as GABA onto the postsynaptic cell. In this way, although cellular actions following agonist binding to presynaptic nAChRs are almost entirely excitatory, in fact, if located on inhibitory cells, their activation could result in a net inhibition of the network via downstream inhibitory actions on the postsynaptic cell.

7.3.2 Postsynaptic

While most commonly thought of as located presynaptically, many studies have identified nAChRs located on postsynaptic membranes (Fig. 7.3). In addition to their presence on dendrites and axons, calcium-permeable α7 homomeric receptors have been localized on somata where they mediate calcium-dependent events including activation of intracellular messengers that could result in stimulation of gene expression that can play a role in synaptic plasticity (Lagostena, Trocme-Thibierge, Morain, & Cherubini, 2008). Homomeric α7-containing receptors exhibit a rapid rate of desensitization and a lower affinity for ACh than that exhibited by some of the other nAChR constellations. However, recognition of the coupling of this nAChR constellation to membrane trafficking events has led to the suggestion that this receptor, despite its rapid desensitization, could participate in cellular events along more extended timelines than previously believed (Liu, Tearle, Nai, & Berg, 2005), and elucidation of associations between this receptor and other intracellular trafficking proteins could suggest other functions in the postsynaptic cell. The α4β2 constellation of nAChRs are also found postsynaptically. However, the high- and low-affinity configurations appear to exhibit differential anatomical positioning on the cell. While low-affinity receptor (Type I) is predominantly present on axons, the high-affinity receptor (Type II) is present on dendrites and somata (Alkondon et al., 1999).

7.4 THE nAChR IN MESOLIMBIC CIRCUITRY

Although nAChRs are found pre and postsynaptically throughout the brain, their location within the mesolimbic circuitry is believed to play a large role in the dependency-inducing properties of nicotine. The mesolimbic pathway is composed of dopamine (DA)-containing projections from the ventral tegmental area (VTA) to the nucleus accumbens (nAc). The nAc sends excitatory projections to the prefrontal cortex (PFC). Activity within the VTA-nAc-PFC pathway has been shown to be involved in decisions involving motivated choices and emotion. While multiple circuits across the brain are involved (Koob & Volkow, 2016), dysfunctions or alterations in the functioning of the mesolimbic pathway due to age, genetics, or changes wrought by external behaviors are implicated in several psychologically and cognitively based disorders, including drug-seeking behavior (Volkow, Koob, & McLellan, 2016).

Although the mechanisms of action of drugs of abuse are diverse, one feature shared by most dependency-inducing drugs is the ability to stimulate large phasic rises in DA within the shell of the nAc, which receives DA-containing afferents projecting from the VTA (Grace, 1991; Schultz, 2007a, 2007b; Zhang et al., 2009). Basal firing of DA VTA cells results in tonic DA release; however, stimuli are incentivized by large rises in DA mediated by this phasic firing, and it is phasic rises in DA that are believed to be involved in the initiation of processes underlying drug dependency (Grace, 1991). Nicotine has been shown to stimulate rises in DA in mesolimbic circuitry via a variety of pre and postsynaptically mediated mechanisms resulting from actions at nAChRs. Smoking a cigarette leads to brain level rises in nicotine up to 100 nM (Henningfield, Stapleton, Benowitz, Grayson, & London, 1993), which is a concentration sufficient to activate and lead to substantial desensitization of nAChRs (Mansvelder, Keath, & McGehee, 2002; Mansvelder & McGehee, 2002).

Quantification of nAChR subunit mRNA within the VTA has revealed the presence of high levels of α4, α6, α7, β2, and β3 transcripts across multiple VTA cell types, including the DA and GABA cells present in this nucleus (Champtiaux et al., 2002; Klink, de Kerchove d'Exaerde, Zoli, & Changeux, 2001; Wooltorton, Pidoplichko, Broide, & Dani, 2003). Therefore, nicotine is expected to excite VTA cells at kinetically distinct nAChR subtypes (Dani et al., 2000; Fisher, Pidoplichko, & Dani, 1998; Klink et al., 2001; Pidoplichko, DeBiasi, Williams, & Dani, 1997; Wooltorton et al., 2003). Within the VTA, high-affinity α4β2 nAChRs are present postsynaptically on DA and GABA-containing neurons. GABAergic neurons within the VTA have been shown to send projections to DA VTA cells (Omelchenko & Sesack, 2009). Nicotine acting at nAChRs on the DA and GABAergic VTA neurons would excite these cells directly (Mansvelder et al., 2002; Mansvelder & McGehee, 2000). The α7-containing nAChR is present on presynaptic glutamatergic terminals deriving from the cortex and subcortical areas (Omelchenko & Sesack, 2007), and its excitation by nicotine would result in enhanced excitatory transmission directed to postsynaptic VTA cells (Dani et al., 2000; Mansvelder et al., 2002; Mansvelder & McGehee, 2000). Activation of the postsynaptic α4β2 nAChRs would lead to cationic flow, which, in combination with enhanced glutamatergic tone acting at postsynaptic NMDA and AMPA receptors, would induce depolarization of the cell sufficient to elicit action potentials, which are believed to underlie phasic rises in DA in the shell of the nAc (Zhang et al., 2009). Interestingly, a concurrent decrease in tonic DA release would ensure an enhancement of the signal-to-noise ratio sourcing from DA afferents to the shell of the nAc (Zhang et al., 2009).

Excitation of the GABAergic nAChRs was found to source not only from nicotine exposure but also from endogenous ACh afferents (Mansvelder & McGehee, 2000). However, desensitization of α4β2 nAChRs on GABAergic VTA cells is believed to result in reductions in inhibitory drive directed to DA VTA cells (Dani, 2015; Mansvelder & McGehee, 2002). Evidence calling to question this widely accepted model has been provided, which suggests that coordinated activity of GABAergic nAChRs may also be required for behaviorally relevant DA cellular functioning (Tolu et al., 2013). Despite rapid desensitization of α4β2, excitation of DA VTA cells persists due to the development of long-term synaptic potentiation (LTP) shown to be elicited within synapses of the DA VTA cells by nicotine exposure (Mansvelder & McGehee, 2000). In addition to requiring functional α7-containing receptors, mechanisms underlying nicotine-induced LTP of DA VTA cells depend, in part, on postsynaptic depolarization and subsequent removal of the magnesium block of the postsynaptic NMDA receptor.

In addition to α4β2- and α7-containing nAChRs, the α6 subunit has been shown to play a role in cellular mechanisms underlying persistent DA VTA cellular excitability; however, the necessity of this subunit for nicotine responses has been questioned. These subunits have been shown to be expressed postsynaptically on DA neurons projecting to the nAc shell, their activation is long lasting and produced by low concentrations of nicotine (Zhao-Shea et al., 2011), and data suggest inclusion of the α6 subunit in a α4β2 complex (Grady, Wageman, Patzlaff, & Marks, 2012; Zhao-Shea et al., 2011). This complex, which to date exhibits the highest sensitivity to nicotine of any native nAChR identified (Grady et al., 2007; Salminen et al., 2007), is believed to play a significant role in sustained DA VTA neuronal excitability in the face of desensitization and reduced nicotine concentrations following intake of the drug (Gotti et al., 2010; Zhao-Shea et al., 2011). However, whether the α4 or α6 subunit in a complex is critical for nicotine-induced enhancement of firing of DA VTA cells remains a debated issue (Exley et al., 2011). The α6 subunit was also shown to be involved in enhancing AMPA receptor functioning and altering the AMPA/NMDA ratio on DA VTA cells, suggesting a role in LTP induction (Berry, Engle, McIntosh, & Drenan, 2015; Engle, McIntosh, & Drenan, 2015; Engle, Shih, McIntosh, & Drenan, 2013). When taken together, these studies suggest that the α6 subunit could play a role in nAChR-mediated cellular activation within the mesolimbic pathways; however, elucidation of the exact mechanisms underlying this role remains.

7.5 ENDOGENOUS ACh AND nAChRs

Although nAChRs on DA VTA neurons respond quite well to exogenous application of nicotine, they are naturally activated by ACh (Mameli-Engvall et al., 2006), and this endogenous ACh plays a large role in naturally motivated behaviors (Forster & Blaha, 2000; Forster, Falcon, Miller, Heruc, & Blaha, 2002; Jerlhag, Janson, Waters, & Engel, 2012; Lammel et al., 2012; Schmidt, Famous, & Pierce, 2009; Shinohara, Kihara, Ide, Minami, & Kaneda, 2014; Steidl & Veverka, 2015). One model postulates that the endogenous actions of ACh are in some ways usurped by nicotine, as the constant presence of nicotine leads to desensitization of the nAChRs at which ACh acts, thereby reducing the impact of normal cholinergic signaling (Dani, Kosten, & Benowitz, 2014).

However, drug-stimulated rises in ACh could also play a role in incentivizing drug stimuli, leading to continued use. Local ACh input sources from VTA interneurons, which if stimulated, results in excitation of DA VTA cells sufficient to release DA in the nAc, and it is likely that nicotine activates nAChRs on these cells (Cachope et al., 2012). Cholinergic input to the VTA also sources from ACh-containing projections from two pontine nuclei: the lateral dorsal tegmentum (LDT) and the pedunculopontine tegmentum (Omelchenko & Sesack, 2005, 2006) (Fig. 7.4). The majority of the VTA cholinergic input from the LDT forms anatomically defined, excitatory synapses preferentially on DA VTA cells that project to the nAc (Omelchenko & Sesack, 2005, 2006). Interestingly, integrity of this projection has been found to be critical for firing patterns of DA neurons underlying phasic bursting, which is the firing pattern believed to be required for behaviorally relevant, large effluxes of DA in the nAc (Chen & Lodge, 2013; Grace, 1991; Grace, Floresco, Goto, & Lodge, 2007; Lodge & Grace, 2006; Maskos, 2008). Glutamate projections are also sent from the LDT to the VTA (Omelchenko & Sesack, 2007), and this excitatory neurotransmitter is likely involved in switching tonic firing of DA VTA cells to burst firing (Lammel et al., 2012). When taken together, recent studies firmly place the LDT as an important regulator of DA VTA cellular and behavioral functioning (Chen & Lodge, 2013; Forster et al., 2002; Forster & Blaha, 2000; Lammel et al., 2012; Steidl & Veverka, 2015; Xiao et al., 2016).

Nicotine can result in activation of DA VTA cells directly through actions on nAChRs located in the VTA, but enhancement of DA activity can also arise from the activation of nAChRs located on cells sending excitatory projections to the VTA. nAChRs are present in the LDT, and they can be functionally activated by nicotine sufficient to induce cell firing (Christensen, Ishibashi, Nielsen, Leonard, & Kohlmeier, 2014). Cholinergic and putative glutamatergic cells of the LDT could be activated by nicotine sufficiently to stimulate the output of these cells to their target structures, including those within the VTA (Christensen et al., 2014). Postsynaptic nAChRs containing α7 and β2 subunits have been implicated in nicotinic activation of LDT cells (Kohlmeier, 2013). Therefore, nicotine actions in enhancing DA release in the nAc could include activation of nAChRs in cholinergic and glutamatergic pontine neuronal groups, which would be expected to contribute to shifting firing of DA VTA cells from basal levels to the behaviorally relevant, phasic pattern characteristic of neurally encoding reward (for review, see Kohlmeier, 2013).

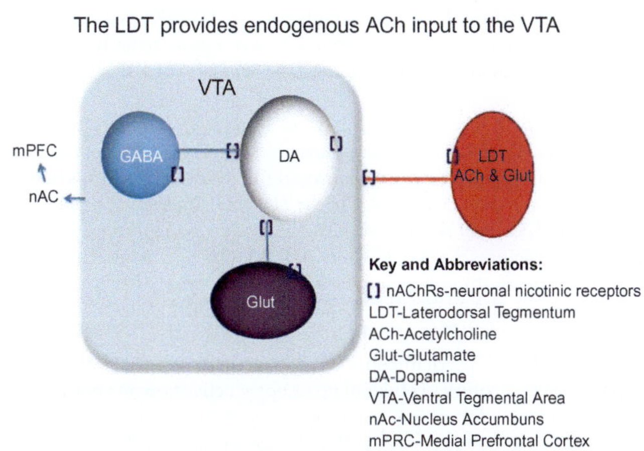

FIG. 7.4 Mesolimbic neuronal circuitry.
Acetylcholine from the laterodorsal tegmentum (LDT) activates nAChRs located on GABAergic, glutamatergic, and dopamine neurons of the ventral tegmental area (VTA).

7.6 CONCLUSIONS

The psychobiological actions of nicotine are due to its actions at neuronal nAChRs. These receptors are located throughout the brain; however, their activation by nicotine within the mesoaccumbal circuitry critical in motivated behavior is believed to underlie the addicting properties of this drug. Activation of nAChRs located on neuronal regions, which project to the VTA and control behaviorally relevant firing activity of cells in this circuit, is also likely involved. nAChRs are composed of multiple subunits; however, the α4β2, α7, and perhaps the α6 subunits appear to play a crucial role in the cellular actions underlying the rewarding properties of nicotine. nAChRs containing these subunits are located on multiple neuronal phenotypes within the mesoaccumbal circuitry, and they can exert their actions via pre or postsynaptic locations on these cells. The nAChRs are one of the most prominent and promising targets for pharmacologically based nicotine abstinence programs. However, although much information has been gained regarding how nicotine works at these receptors in the mesoaccumbal pathway, more remains to be learned at the synaptic level about the basis of nicotine addiction, if we are to efficaciously pharmacologically target these receptors in strategies of nicotine cessation.

Acknowledgment

The author thanks A. Sabina Kristensen for the assistance in the making of figures.

Mini-Dictionary of Terms

nAChR The endogenous neurotransmitter acetylcholine (ACh) is an agonist at two types of receptor, an ionotropic receptor known as the nicotinic acetylcholine receptor (nAChR) and a metabotropic receptor known as the muscarinic receptor.

VTA The ventral tegmental area (VTA) is a midbrain nucleus composed primarily of dopamine (DA) and GABA-containing cells. Release of DA from the VTA signals saliency of the triggering stimulus.

nAc The nucleus accumbens (nAc), which is composed of spiny GABA-containing cells, is the principle target of DA-containing VTA cells, and its activation by large rises in DA is involved in signaling saliency.
LDT The laterodorsal tegmental nucleus is a pontine nucleus composed of acetylcholine and GABAergic and glutamatergic neurons that send projections to the VTA. Although it has been recognized to play a role in arousal for decades, the LDT has recently been added to circuit models of the brain's reward pathway.
Postsynaptic In a tripartite synapse, the localization of receptors on dendritic or somatic compartments designates the receptor as being located postsynaptically or at the receiving end of the synapse.
Presynaptic In a tripartite synapse, the location of receptors on the axonal terminal of the cell designates the receptor as being presynaptic, where it is often involved in the release of neuroactive chemicals from the terminal.

Key Facts of nAChRs

- nAChRs are stimulated endogenously by acetylcholine (ACh).
- nAChRs are stimulated by exogenous application of nicotine.
- nAChRs are composed of five subunits of the α and/or β types.
- Two major subunit constellations of the nAChR have been implicated in drug addiction, the $\alpha4\beta2$ constellation and a homomeric constellation composed of five $\alpha7$ subunits.
- Differences in the localization of nACh receptors on the pre and postsynaptic terminal and the composition of subunits confer differential activation of GABAergic, glutamatergic, and dopamine neurons in addiction-related circuitry.

Key Facts of Drug Addiction Circuitry

- Release of DA from the VTA from afferents directed to the nAc is a key component of signaling saliency to stimuli such as drugs of abuse.
- Activation of nAChRs by nicotine alters their responsivity to endogenous ACh.
- ACh is endogenously provided to the VTA by afferents from the LDT, and via this connection, the LDT alters the excitability of DA VTA cells.
- nAChRs are present in the LDT, and exogenous nicotine activates these receptors in addition to those in the VTA. Therefore, exposure of the brain to nicotine would alter the endogenous release of ACh from LDT afferents located in the VTA.

Summary Points

- Psychobiological actions of nicotine are due to the activation of nAChRs.
- The subunit composition of the nAChR confers the kinetics of the response to activation by agonist.
- The most common constellation of subunits in the brain is the heteromeric $\alpha4\beta2$ combination and the homomeric $\alpha7$ organization.
- nAChRs can be located pre or postsynaptically and outside the synapse.
- Addictive properties of nicotine are conferred by the activation of pre and postsynaptically located nAChRs in mesoaccumbal circuitry.
- Nicotine induces large rises in dopamine via the activation of nAChRs located on presynaptic GABAergic and glutamatergic terminals and nAChRs located postsynaptically on dopamine-containing VTA cells.
- Cholinergic afferents sourcing from the LDT also are influenced by nicotine, leading to changes in the release of the endogenous agonist, ACh, at the nAChR in the VTA.

References

Albuquerque, E. X., Pereira, E. F., Alkondon, M., & Rogers, S. W. (2009). Mammalian nicotinic acetylcholine receptors: from structure to function. *Physiological Reviews, 89*, 73–120.

Alkondon, M., Pereira, E. F., Eisenberg, H. M., & Albuquerque, E. X. (1999). Choline and selective antagonists identify two subtypes of nicotinic acetylcholine receptors that modulate GABA release from CA1 interneurons in rat hippocampal slices. *The Journal of Neuroscience, 19*, 2693–2705.

Bennett, C., Arroyo, S., Berns, D., & Hestrin, S. (2012). Mechanisms generating dual-component nicotinic EPSCs in cortical interneurons. *The Journal of Neuroscience, 32*, 17287–17296.

Berry, J. N., Engle, S. E., McIntosh, J. M., & Drenan, R. M. (2015). alpha6-Containing nicotinic acetylcholine receptors in midbrain dopamine neurons are poised to govern dopamine-mediated behaviors and synaptic plasticity. *Neuroscience, 304*, 161–175.

Cachope, R., Mateo, Y., Mathur, B. N., Irving, J., Wang, H. L., Morales, M., et al. (2012). Selective activation of cholinergic interneurons enhances accumbal phasic dopamine release: setting the tone for reward processing. *Cell Reports, 2*, 33–41.

Champtiaux, N., Han, Z. Y., Bessis, A., Rossi, F. M., Zoli, M., Marubio, L., et al. (2002). Distribution and pharmacology of alpha 6-containing nicotinic acetylcholine receptors analyzed with mutant mice. *The Journal of Neuroscience, 22*, 1208–1217.

Chen, L., & Lodge, D. J. (2013). The lateral mesopontine tegmentum regulates both tonic and phasic activity of VTA dopamine neurons. *Journal of Neurophysiology, 110*, 2287–2294.

Christensen, M. H., Ishibashi, M., Nielsen, M. L., Leonard, C. S., & Kohlmeier, K. A. (2014). Age-related changes in nicotine response of cholinergic and non-cholinergic laterodorsal tegmental neurons: implications for the heightened adolescent susceptibility to nicotine addiction. *Neuropharmacology, 85*, 263–283.

Dani, J. A. (1986). Ion-channel entrances influence permeation. Net charge, size, shape, and binding considerations. *Biophysical Journal, 49*, 607–618.

Dani, J. A. (2015). Neuronal nicotinic acetylcholine receptor structure and function and response to nicotine. *International Review of Neurobiology, 124*, 3–19.

Dani, J. A., & Heinemann, S. (1996). Molecular and cellular aspects of nicotine abuse. *Neuron, 16*, 905–908.

Dani, J. A., Kosten, T. R., & Benowitz, N. L. (2014). The pharmacology of nicotine and tobacco. In: Ries R.K., Fiellin D.A., Miller S.C., & Saitz R. (Eds.), *The ASAM principles of addiction medicine* (pp. 201–216). Philadelphia, PA: Lippincott Williams & Wilkins, Wolters Kluwer. (Chapter 12).

Dani, J. A., Radcliffe, K. A., & Pidoplichko, V. I. (2000). Variations in desensitization of nicotinic acetylcholine receptors from hippocampus and midbrain dopamine areas. *European Journal of Pharmacology, 393*, 31–38.

Elgoyhen, A. B., Vetter, D. E., Katz, E., Rothlin, C. V., Heinemann, S. F., & Boulter, J. (2001). alpha10: a determinant of nicotinic cholinergic receptor function in mammalian vestibular and cochlear mechanosensory hair cells. *Proceedings of the National Academy of Sciences of the United States of America, 98*, 3501–3506.

Engle, S. E., McIntosh, J. M., & Drenan, R. M. (2015). Nicotine and ethanol cooperate to enhance ventral tegmental area AMPA receptor function via alpha6-containing nicotinic receptors. *Neuropharmacology, 91*, 13–22.

Engle, S. E., Shih, P. Y., McIntosh, J. M., & Drenan, R. M. (2013). alpha4alpha6beta2* nicotinic acetylcholine receptor activation on ventral tegmental area dopamine neurons is sufficient to stimulate a depolarizing conductance and enhance surface AMPA receptor function. *Molecular Pharmacology, 84*, 393–406.

Exley, R., Maubourguet, N., David, V., Eddine, R., Evrard, A., Pons, S., et al. (2011). Distinct contributions of nicotinic acetylcholine receptor subunit alpha4 and subunit alpha6 to the reinforcing effects of nicotine. *Proceedings of the National Academy of Sciences of the United States of America, 108*, 7577–7582.

Fenster, C. P., Whitworth, T. L., Sheffield, E. B., Quick, M. W., & Lester, R. A. (1999). Upregulation of surface alpha4beta2 nicotinic receptors is initiated by receptor desensitization after chronic exposure to nicotine. *The Journal of Neuroscience, 19*, 4804–4814.

Fisher, J. L., Pidoplichko, V. I., & Dani, J. A. (1998). Nicotine modifies the activity of ventral tegmental area dopaminergic neurons and hippocampal GABAergic neurons. *Journal of Physiology, Paris, 92*, 209–213.

Forster, G. L., & Blaha, C. D. (2000). Laterodorsal tegmental stimulation elicits dopamine efflux in the rat nucleus accumbens by activation of acetylcholine and glutamate receptors in the ventral tegmental area. *The European Journal of Neuroscience, 12*, 3596–3604.

Forster, G. L., Falcon, A. J., Miller, A. D., Heruc, G. A., & Blaha, C. D. (2002). Effects of laterodorsal tegmentum excitotoxic lesions on behavioral and dopamine responses evoked by morphine and d-amphetamine. *Neuroscience, 114*, 817–823.

Gotti, C., Guiducci, S., Tedesco, V., Corbioli, S., Zanetti, L., Moretti, M., et al. (2010). Nicotinic acetylcholine receptors in the mesolimbic pathway: primary role of ventral tegmental area alpha6beta2* receptors in mediating systemic nicotine effects on dopamine release, locomotion, and reinforcement. *The Journal of Neuroscience, 30*, 5311–5325.

Grace, A. A. (1991). Phasic versus tonic dopamine release and the modulation of dopamine system responsivity: a hypothesis for the etiology of schizophrenia. *Neuroscience, 41*, 1–24.

Grace, A. A., Floresco, S. B., Goto, Y., & Lodge, D. J. (2007). Regulation of firing of dopaminergic neurons and control of goal-directed behaviors. *Trends in Neurosciences, 30*, 220–227.

Grady, S. R., Salminen, O., Laverty, D. C., Whiteaker, P., McIntosh, J. M., Collins, A. C., et al. (2007). The subtypes of nicotinic acetylcholine receptors on dopaminergic terminals of mouse striatum. *Biochemical Pharmacology, 74*, 1235–1246.

Grady, S. R., Wageman, C. R., Patzlaff, N. E., & Marks, M. J. (2012). Low concentrations of nicotine differentially desensitize nicotinic acetylcholine receptors that include alpha5 or alpha6 subunits and that mediate synaptosomal neurotransmitter release. *Neuropharmacology, 62*, 1935–1943.

Henningfield, J. E., Stapleton, J. M., Benowitz, N. L., Grayson, R. F., & London, E. D. (1993). Higher levels of nicotine in arterial than in venous blood after cigarette smoking. *Drug and Alcohol Dependence, 33*, 23–29.

Jerlhag, E., Janson, A. C., Waters, S., & Engel, J. A. (2012). Concomitant release of ventral tegmental acetylcholine and accumbal dopamine by ghrelin in rats. *PLoS One, 7*, e49557.

Klink, R., de Kerchove d'Exaerde, A., Zoli, M., & Changeux, J. P. (2001). Molecular and physiological diversity of nicotinic acetylcholine receptors in the midbrain dopaminergic nuclei. *The Journal of Neuroscience, 21*, 1452–1463.

Kohlmeier, K. A. (2013). Off the beaten path: drug addiction and the pontine laterodorsal tegmentum. *ISRN Neuroscience, 2013*, 604847.

Koob, G. F., & Volkow, N. D. (2016). Neurobiology of addiction: a neurocircuitry analysis. *Lancet Psychiatry, 3*, 760–773.

Lagostena, L., Trocme-Thibierge, C., Morain, P., & Cherubini, E. (2008). The partial alpha7 nicotine acetylcholine receptor agonist S 24795 enhances long-term potentiation at CA3-CA1 synapses in the adult mouse hippocampus. *Neuropharmacology, 54*, 676–685.

Lammel, S., Lim, B. K., Ran, C., Huang, K. W., Betley, M. J., Tye, K. M., et al. (2012). Input-specific control of reward and aversion in the ventral tegmental area. *Nature, 491*, 212–217.

Liu, Z., Tearle, A. W., Nai, Q., & Berg, D. K. (2005). Rapid activity-driven SNARE-dependent trafficking of nicotinic receptors on somatic spines. *The Journal of Neuroscience, 25*, 1159–1168.

Lodge, D. J., & Grace, A. A. (2006). The laterodorsal tegmentum is essential for burst firing of ventral tegmental area dopamine neurons. *Proceedings of the National Academy of Sciences of the United States of America, 103*, 5167–5172.

Mameli-Engvall, M., Evrard, A., Pons, S., Maskos, U., Svensson, T. H., Changeux, J. P., et al. (2006). Hierarchical control of dopamine neuron-firing patterns by nicotinic receptors. *Neuron, 50*, 911–921.

Mansvelder, H. D., Keath, J. R., & McGehee, D. S. (2002). Synaptic mechanisms underlie nicotine-induced excitability of brain reward areas. *Neuron, 33*, 905–919.

Mansvelder, H. D., & McGehee, D. S. (2000). Long-term potentiation of excitatory inputs to brain reward areas by nicotine. *Neuron, 27*, 349–357.

Mansvelder, H. D., & McGehee, D. S. (2002). Cellular and synaptic mechanisms of nicotine addiction. *Journal of Neurobiology, 53*, 606–617.

Marks, M. J., Burch, J. B., & Collins, A. C. (1983). Effects of chronic nicotine infusion on tolerance development and nicotinic receptors. *The Journal of Pharmacology and Experimental Therapeutics, 226*, 817–825.

Maskos, U. (2008). The cholinergic mesopontine tegmentum is a relatively neglected nicotinic master modulator of the dopaminergic system: relevance to drugs of abuse and pathology. *British Journal of Pharmacology, 153*(Suppl. 1), S438–S445.

McGehee, D. S., & Role, L. W. (1995). Physiological diversity of nicotinic acetylcholine receptors expressed by vertebrate neurons. *Annual Review of Physiology, 57*, 521–546.

Miwa, J. M., Freedman, R., & Lester, H. A. (2011). Neural systems governed by nicotinic acetylcholine receptors: emerging hypotheses. *Neuron, 70*, 20–33.

Moretti, M., Zoli, M., George, A. A., Lukas, R. J., Pistillo, F., Maskos, U., et al. (2014). The novel alpha7beta2-nicotinic acetylcholine receptor subtype is expressed in mouse and human basal forebrain: biochemical and pharmacological characterization. *Molecular Pharmacology, 86*, 306–317.

Moroni, M., & Bermudez, I. (2006). Stoichiometry and pharmacology of two human alpha4beta2 nicotinic receptor types. *Journal of Molecular Neuroscience, 30*, 95–96.

Omelchenko, N., & Sesack, S. R. (2005). Laterodorsal tegmental projections to identified cell populations in the rat ventral tegmental area. *The Journal of Comparative Neurology, 483*, 217–235.

Omelchenko, N., & Sesack, S. R. (2006). Cholinergic axons in the rat ventral tegmental area synapse preferentially onto mesoaccumbens dopamine neurons. *The Journal of Comparative Neurology, 494*, 863–875.

Omelchenko, N., & Sesack, S. R. (2007). Glutamate synaptic inputs to ventral tegmental area neurons in the rat derive primarily from subcortical sources. *Neuroscience, 146*, 1259–1274.

Omelchenko, N., & Sesack, S. R. (2009). Ultrastructural analysis of local collaterals of rat ventral tegmental area neurons: GABA phenotype and synapses onto dopamine and GABA cells. *Synapse, 63*, 895–906.

Parri, R. H., & Dineley, T. K. (2010). Nicotinic acetylcholine receptor interaction with beta-amyloid: molecular, cellular, and physiological consequences. *Current Alzheimer Research, 7*, 27–39.

Pidoplichko, V. I., DeBiasi, M., Williams, J. T., & Dani, J. A. (1997). Nicotine activates and desensitizes midbrain dopamine neurons. *Nature, 390*, 401–404.

Salminen, O., Drapeau, J. A., McIntosh, J. M., Collins, A. C., Marks, M. J., & Grady, S. R. (2007). Pharmacology of alpha-conotoxin MII-sensitive subtypes of nicotinic acetylcholine receptors isolated by breeding of null mutant mice. *Molecular Pharmacology, 71*, 1563–1571.

Sarter, M., Parikh, V., & Howe, W. M. (2009). Phasic acetylcholine release and the volume transmission hypothesis: time to move on. *Nature Reviews Neuroscience, 10*, 383–390.

Schmidt, H. D., Famous, K. R., & Pierce, R. C. (2009). The limbic circuitry underlying cocaine seeking encompasses the PPTg/LDT. *The European Journal of Neuroscience, 30*, 1358–1369.

Schultz, W. (2007a). Behavioral dopamine signals. *Trends in Neurosciences, 30*, 203–210.

Schultz, W. (2007b). Multiple dopamine functions at different time courses. *Annual Review of Neuroscience, 30*, 259–288.

Shinohara, F., Kihara, Y., Ide, S., Minami, M., & Kaneda, K. (2014). Critical role of cholinergic transmission from the laterodorsal tegmental nucleus to the ventral tegmental area in cocaine-induced place preference. *Neuropharmacology, 79*, 573–579.

Steidl, S., & Veverka, K. (2015). Optogenetic excitation of LDTg axons in the VTA reinforces operant responding in rats. *Brain Research, 1614*, 86–93.

Tolu, S., Eddine, R., Marti, F., David, V., Graupner, M., Pons, S., et al. (2013). Co-activation of VTA DA and GABA neurons mediates nicotine reinforcement. *Molecular Psychiatry, 18*, 382–393.

Volkow, N. D., Koob, G. F., & McLellan, A. T. (2016). Neurobiologic advances from the brain disease model of addiction. *The New England Journal of Medicine, 374*, 363–371.

Whiteaker, P., Sharples, C. G., & Wonnacott, S. (1998). Agonist-induced up-regulation of alpha4beta2 nicotinic acetylcholine receptors in M10 cells: pharmacological and spatial definition. *Molecular Pharmacology, 53*, 950–962.

Wooltorton, J. R., Pidoplichko, V. I., Broide, R. S., & Dani, J. A. (2003). Differential desensitization and distribution of nicotinic acetylcholine receptor subtypes in midbrain dopamine areas. *The Journal of Neuroscience, 23*, 3176–3185.

Xiao, C., Cho, J. R., Zhou, C., Treweek, J. B., Chan, K., McKinney, S. L., et al. (2016). Cholinergic mesopontine signals govern locomotion and reward through dissociable midbrain pathways. *Neuron, 90*, 333–347.

Zhang, T., Zhang, L., Liang, Y., Siapas, A. G., Zhou, F. M., & Dani, J. A. (2009). Dopamine signaling differences in the nucleus accumbens and dorsal striatum exploited by nicotine. *The Journal of Neuroscience, 29*, 4035–4043.

Zhao-Shea, R., Liu, L., Soll, L. G., Improgo, M. R., Meyers, E. E., McIntosh, J. M., et al. (2011). Nicotine-mediated activation of dopaminergic neurons in distinct regions of the ventral tegmental area. *Neuropsychopharmacology, 36*, 1021–1032.

CHAPTER

8

Cotinine as a Possible Allosteric Modulator of Nicotine Effects in Various Models

Oné R. Pagán

Department of Biology, West Chester University, West Chester, PA, United States

Abbreviations

CYP2A6 cytochrome P450 2A6
LGIC ligand-gated ion channels
nAChR nicotinic acetylcholine receptor
PMR the principle of microscopic reversibility

8.1 INTRODUCTION

Nicotine's most active isomer (the (*S*)-isomer of 3-(1-methyl-2-pyrrolidinyl) pyridine) is widely recognized as one of the most addictive compounds known to humans and at the very least is consistently ranked as one of the "top five" substances in this regard (Stolerman & Jarvis, 1995). The historical and scientific importance of nicotine use and abuse, its modes of administration, etc. are reviewed elsewhere in this volume. The main natural source of this compound is the tobacco plant, *Nicotiana tabacum*, as well as related *Nicotiana* species (Firn, 2011). However, a surprising variety of related plants (mainly but not limited to other members of the *Solanaceae* family) possess detectable levels of nicotine, although in much lower quantities compared with tobacco. These other plant species include edible plants such as tomatoes, eggplant, and cauliflower (Domino, Hornbach, & Demana, 1993). Similarly to many plant secondary metabolites, nicotine's role in nature seems to be to act as an insecticide (Soloway, 1976). Purposeful crop spraying with nicotine and the selection of tobacco plants with higher nicotine production are well-established agricultural practices (Jackson, Johnson, & Stephenson, 2002; Schmeltz, 1971) with the purpose of combating agricultural pests. An example of a direct investigational support for this idea includes a research report in which nicotine-producing pathways of the *Nicotiana attenuata* plants were silenced, reducing their nicotine production by more than 95%. These modified plants were significantly less resistant to insect predation compared with their wild-type counterparts (Steppuhn, Gase, Krock, Halitschke, & Baldwin, 2004).

8.2 NICOTINIC ACETYLCHOLINE RECEPTORS

Nicotine defines a whole class of macromolecules collectively known as nicotinic acetylcholine receptors (nAChRs), which belong to the general family of ligand-gated ion channels (LGICs). Interestingly, apart from the nAChR, nicotine seems to interact with proteasome-ubiquitin complexes in several models (Caldeira, Salazar, Curcio, Canzoniero, & Duarte, 2014; Chapman, 2009; Kane, Konu, Ma, & Li, 2004; Massaly, Francès, & Moulédous, 2014; Rezvani, Teng, Shim, & De Biasi, 2007). Investigation on these alternate nicotine targets could lead to the development of novel pharmacotherapies. Nonetheless, to date, the best understood molecular targets of nicotine are still the nAChRs.

To date, the best understood LGIC is the muscle-type nAChR. This receptor occupies a unique historical status among ion channels. It was the first to be purified, the first to have its primary structure determined, the first to be functionally reconstituted in artificial lipid bilayers, and the first from which single-channel recordings were obtained (reviewed in Hucho, Tsetlin, & Machold, 1996; Karlin & Akabas, 1995). nAChRs are also one of the prototypic models to describe allosteric receptors (Bertrand & Changeaux, 1995). There is little doubt that

Neuroscience of Nicotine
https://doi.org/10.1016/B978-0-12-813035-3.00008-3

this receptor will be the first LGIC to be understood at the atomic level (Chen, 2010; Giastas, Zouridakis, & Tzartos, 2018; Gu, Zhong, & Wei, 2011; Tsetlin & Hucho, 2009). There are two broadly defined types of nAChRs: the muscle-type receptor, largely responsible for physiological events leading to striated muscle contraction, and the much more heterogeneous neuronal type, predominantly found in nervous tissue, which is involved in a variety of nervous system functions (Dani, 2015; Hogg & Bertrand, 2004; Hurst, Rollema, & Bertrand, 2013).

For the last four decades or so, neuronal nAChRs have been intensively studied as possible targets to treat an assorted list of neuropsychiatric conditions (reviewed in Freedman, 2014; Pogocki, Ruman, Danilczuk, Celuch, & Wałajtys-Rode, 2007; Rollema, Bertrand, & Hurst, 2015; Terry, Callahan, Hall, & Webster, 2011; Terry, Callahan, & Hernandez, 2015). Alas, despite much progress on the understanding on the specific structural, biochemical, and physiological characteristics of each neuronal nAChR subtype known to date, the promise of developing useful pharmacotherapies has been modest at best, with the exception of smoking cessation therapies (a rather insightful review of the general state of affairs in the field was recently published by Bertrand & Terry, 2018).

All nAChRs are pentameric proteins forming an ion channel. The muscle-type nAChR is a heteromeric complex, formed by α, β, γ, and δ subunits, displaying a stoichiometry of $\alpha_2\beta\gamma\delta$. In the case of the muscle-type nAChR, upon development, the γ subunit is replaced by an epsilon (ε) subunit. By convention, the muscle-type α and β subunits are considered the first of their class (i.e., α1 and β1). In contrast, neuronal nAChRs are found in heteromeric combinations consisting of two α subunits (α2–α10) and three β subunits (β2–β4); alternatively, homomeric nAChRs consist of five identical α subunits. For example, the best understood homomeric nicotinic receptors are the ones composed of five α7 subunits. A variety of mutations on several types of nAChRs are associated with various neuromuscular and neuronal channelopathies (reviewed in Wu & Lukas, 2011; Zoli, Pistillo, & Gotti, 2015).

The nAChRs are discussed in more detail elsewhere in this volume. However, the main point that I want to state here is that neuronal nAChRs display a rich variety of subtypes in vertebrate nervous systems by virtue of the multiplicity of possible combinations of the 11 neuronal subunits α2–α10/β2–β4. Each different combination is preferentially expressed in specific neuronal structures and tissues, endowing each receptor type with subtly different physiological properties and regulatory domains. This point is especially important to the discussion of the interactions of nicotine and its main metabolite, cotinine.

8.3 COTININE

In humans, nicotine is metabolized into about 25 metabolites (reviewed in Benowitz, Hukkanen, & Jacob, 2009). Most of this metabolism is carried out by a liver enzyme called cytochrome P450 2A6 (CYP2A6), which catalyzes the conversion of about 80% of absorbed nicotine into cotinine (1-methyl-5-(pyridin-3-yl)pyrrolidin-2-one; Grizzell & Echeverría, 2015). Thus, cotinine is widely considered the main nicotine metabolite in humans. In turn, cotinine is further metabolized by CYP2A6 into a series of compounds, albeit at a lower rate than nicotine. Structurally, cotinine and nicotine are quite similar, the only difference between them being a carbonyl group (Fig. 8.1). Despite this close structural similarity between nicotine and cotinine, there are significant differences in the effects of either compound on the nervous systems (Grizzell & Echeverría, 2015). This fact is likely a reflection of their respective interactions with neuronal nAChRs and is of particular importance for the ideas in this chapter.

Pharmacokinetically, cotinine displays a longer half-life than nicotine (De Schepper, Van Hecken, Daenens, & Van Rossum, 1987), giving credence to the idea that most of the longer term (as opposed to the immediate) cholinergic effects attributed to nicotine are in fact due to cotinine (Grizzell & Echeverría, 2015). Additionally, cotinine seems to display many of the beneficial effects of nicotine while lacking most of its negative effects, including its addictive properties (Hatsukami, Grillo, Pentel, Oncken, & Bliss, 1997; Moran, 2012; reviewed in Grizzell & Echeverría, 2015). The initial impetus for the study of cotinine was as an indicator of tobacco consumption (Benowitz et al., 2009; Moran, 2012). More recently, cotinine has caught the interest of researchers pursuing novel pharmacotherapies. Cotinine is being investigated as an antiinflammatory agent (Echeverría, Grizzell, & Barreto, 2016) and as a possible treatment for Parkinson's disease (Barreto, Yarkov, Avila-Rodriguez, Aliev, & Echeverría, 2015), Alzheimer's disease (Echeverría & Zeitlin, 2012), and posttraumatic stress disorder (Barreto, Iarkov, & Moran, 2014; Barreto et al., 2015; Mendoza et al., 2018). Interestingly, cotinine

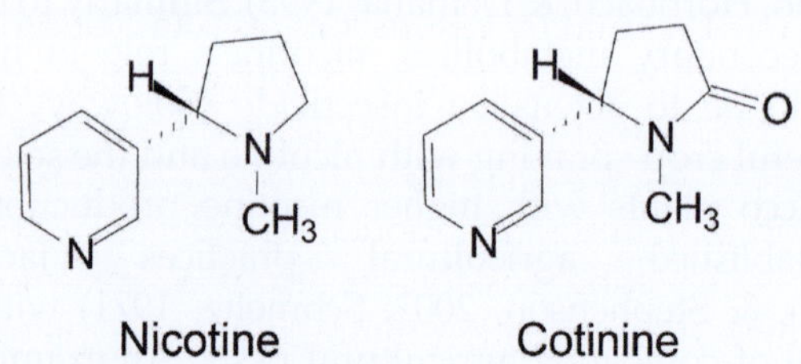

FIG. 8.1 Compounds reviewed in this work. Nicotine and cotinine, as indicated.

has also been shown to display antidepressant effects (Grizzell et al., 2014).

Even though cotinine generally seems to display physiological effects similar to those induced by nicotine (i.e., nicotinic agonist; Grizzell & Echeverría, 2015; Moran, 2012), there are instances when, depending on the model system and the specific type of nicotinic receptor studied, cotinine and nicotine display opposite physiological effects (Grizzell & Echeverría, 2015) and are mutually antagonistic, especially in radioligand binding studies (Vainio & Tuominen, 2001; Vainio, Törnquist, & Tuominen, 2000; Vainio, Viluksela, & Tuominen, 1998a,b).

8.4 NICOTINE/COTININE AND PLANARIANS

Until recently, cotinine has been studied exclusively in vivo through vertebrate models or ex vivo using cell lines derived from vertebrates. However, cotinine has also been detected in certain invertebrates as a result of nicotine metabolism. These invertebrates include a species of blowfly (*Calliphora vomitoria*, Magni et al., 2016), tobacco budworm (*Heliothis virescens*, Orth, Head, & Mierkowski, 2007), and honeybees (du Rand, Pirk, Nicolson, & Apostolides, 2017). At the time of this writing, the physiological and behavioral effects of cotinine have not been systematically investigated in any invertebrate model. Data from my laboratory indicate that cotinine acts as a nicotine antagonist in a particular invertebrate system: planarians. Planarians are well-established model organisms in regeneration and developmental biology. Since the 1970s, they have been increasingly studied in a pharmacological and neurobiological context, particularly the behavioral effects of abused drugs and their mechanism of action (reviewed in Pagán, 2017). We recently reported that cotinine prevents the induction of three different nicotine-induced behaviors in the planarian *Girardia tigrina* in a concentration-dependent manner (Bach et al., 2016). The studied behaviors were motility decrease, seizure-like movements, and withdrawal-like behaviors. Additionally, we are in the process of further exploring the antagonistic effect of cotinine in a specific behavior of planarians. Preliminary results suggest that cotinine acts as a noncompetitive inhibitor of nicotine in this organism (Pagán et al., unpublished data). To our knowledge, our results represent the first study of nicotine/cotinine interactions in an invertebrate model, opening the door to a series of interesting possibilities, from ecotoxicology to agriculture. It is currently unknown whether planarians metabolize nicotine into cotinine. Several studies have established the presence of cholinergic systems in planarians. Acetylcholine and nicotine induce behavioral effects in planarians (Buttarelli, Pellicano, & Pontieri, 2008; Buttarelli, Pontieri, Margotta, & Palladini, 2000; Pagán et al., 2009, 2013; Pagán, Montgomery, Deats, Bach, & Baker, 2015; Rawls et al., 2011). Indirect evidence for candidate nAChR examples in planarians has been found using in situ hybridization (Cebrià et al., 2002), microarray technology (Nakazawa et al., 2003), and expression sequence tags (Mineta et al., 2003). Also, using the planarian *Schmidtea mediterranea* database (http://smedgd.neuro.utah.edu/; Robb, Gotting, Ross, & Sánchez Alvarado, 2015), we have found several nicotinic receptor candidate sequences. These putative planarian nicotine binding sites have not been directly identified. Thus, to date, the specific molecular targets for nicotine in planarians are unknown. Nonetheless, it is reasonable to expect that these flatworms will display a similar degree of diversity on their nAChRs. Our work about nicotine and cotinine in planarians (Bach et al., 2016) shows that cotinine by itself does not induce any evident behavioral effects yet cotinine antagonizes nicotine-induced behaviors in this class of organisms. Based on that, I hypothesize that regulatory sites exist in at least a subtype of putative planarian nAChRs that when bound by cotinine inhibit nicotine's behavioral effects.

Taken together, the heterogeneity of in vivo and ex vivo pharmacological interactions between nicotine and cotinine in a variety of models, including planarians, suggests that similarly to nicotine, cotinine interacts with multiple subtypes of neuronal nAChRs. Since some of cotinine's effects seem to mimic nicotine's (i.e., cholinergic agonist), the simplest interpretation is that in these cases, cotinine binds to the same or at least overlapping binding sites as nicotine does. On the other hand, the cases where cotinine behaves as a nicotine antagonist suggest a more complex interaction. A possible interpretation is an allosteric regulatory model.

8.5 THE PRINCIPLE OF MICROSCOPIC REVERSIBILITY

The principle of microscopic reversibility (PMR) was proposed by Gilbert N. Lewis on thermodynamic grounds in 1925; originally, Lewis called it "*The Law of Entire Equilibrium*." He articulated it as "*Corresponding to every individual process there is a reverse process, and in a state of equilibrium the average rate of every process is equal to the average rate of its reverse process.*" (Lewis, 1925). The same year when Lewis' paper was published, Richard C. Tolman tried to refute Lewis' idea, and oddly, in the process, he ended up renaming it as "The principle of microscopic reversibility" (Tolman, 1925). Despite Tolman's challenge, experimental evidence of the PMR began to be obtained, and ironically, Tolman's naming of the concept became permanently associated with the model ever since. The PMR has been demonstrated to apply not only at equilibrium conditions but also in systems that are far from equilibrium, finding its way to model enzymatic reactions and eventually ligand-receptor systems using transient kinetic techniques (Astumian, 2012; Blackmond, 2009a; Hess, 2003). In this context, the PMR essentially states that the free energy required for a ligand to bind to its receptor target is independent of whether the compound is bound to the active or inactive form of the receptor. In other words, regardless of the binding or enzymatic pathway taken, there is constant energy cost to the synthesis of a product or the activation of a receptor system (Kenakin, 1997). One of the earliest applications of this principle was proposed by Katz and Thesleff (1957) to describe the desensitization to acetylcholine in the motor end-plate model. Further examination of the principle suggested a mechanism of activation and inactivation for the nAChR (Fig. 8.2; reviewed in Hess, 2003). Eventually, this was the basis to hypothesize the existence of regulatory domains in the nAChR. It was predicted that compounds binding to a regulatory site specific for the inactive conformation of the receptor with higher affinity than to an alternate site specific to the active state of the receptor would shift the equilibrium toward the inactive conformation of the receptor. Conversely, compounds binding preferentially to the active receptor regulatory site(s) would not inhibit the receptor; furthermore, such a ligand would displace inhibitors from the receptor. This was the theoretical basis for predicting the existence of previously unknown activators and inhibitors of this receptor and, by extension, to other related receptor systems. Evidence supporting these predictions was provided using transient kinetic techniques to examine the mechanism of inhibition of the nAChR by cocaine and the glutamatergic drug (+) dizocilpine (Hess et al., 2000), eventually identifying compounds alleviating such inhibition (Hess et al., 2003; Sivaprakasam, Pagán, & Hess, 2010). Interestingly, ecgonine methyl ester, a major cocaine metabolite, and close analogue, 3-acetoxy ecgonine methyl ester (Ambre, Ruo, Smith, Backes, & Smith, 1982, Fig. 8.3) were found to be alleviators of cocaine inhibition of the muscle-type nAChR (Chen, Banerjee, & Hess, 2004) and of a neuronal type nAChR (Krivoshein & Hess, 2004). Additionally, similar mechanisms were demonstrated by a close structural analogue of the nAChR, the gamma-aminobutyric receptor type A, and an inhibitory receptor (GABA-A, Ramakrishnan & Hess, 2004, 2005, 2010).

The idea of a metabolite (ecgonine methyl ester) acting as an inhibitor of its parent compound (cocaine) is consistent with the idea of cotinine acting as a nicotine antagonist at least in some nAChR subtypes. Circumstantial evidence for this idea is the heterogeneous nature of the physiological effects of cotinine in various systems (reviewed in Grizzell & Echeverría, 2015) and our own studies using planarians. The evolutionary significance of the PMR is seldom explicitly alluded to in the biochemical and pharmacological literature, with some notable exceptions, particularly in light of the polymeric nature of many enzymes (and receptors, e.g., please see Blackmond, 2009b; Burbaum, Raines, Albery, & Knowles, 1989; Ricard, 1978). From this perspective, the evolutionary explanation hints at maximizing catalytic efficiency and regulatory power (Ricard, 1978). Applying these principles to the discovery of new regulatory mechanisms modulating receptor function is an exciting area of research with the potential of obtaining new pharmacotherapies for a wide variety of pathological conditions. Further studies of the nAChR and related proteins, coupled to novel applications using established model organisms like planarians, can provide significant contributions to the understanding of how cotinine serves as a regulatory ligand, which might throw light on general mechanisms of neurotransmitter allosteric regulation.

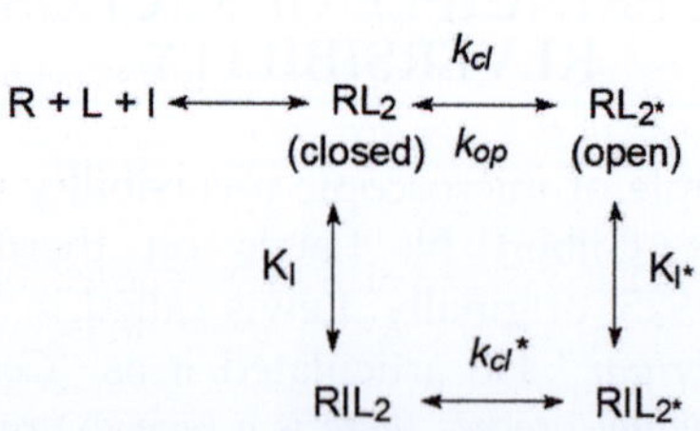

FIG. 8.2 Proposed inhibition mechanism for the nicotinic acetylcholine receptor.
I, allosteric ligand (inhibitor or activator); K_I and K_I* are the observed dissociation constants of the allosteric ligand for the closed- and open-channel form, respectively.; k_{op} and k_{cl}, rate constants for channel opening and closing, respectively; k_{op}* and k_{cl}*, rate constants for channel opening and closing, respectively, when the allosteric ligand bound to the open-channel form; *L*, ligand; *R*, receptor; *RL*, *RIL*$_2$, and *RL*$_2$, closed-channel conformations; *RL*$_2$* and *RIL*$_2$*, open-channel conformations. *After Hess et al. (2000).*

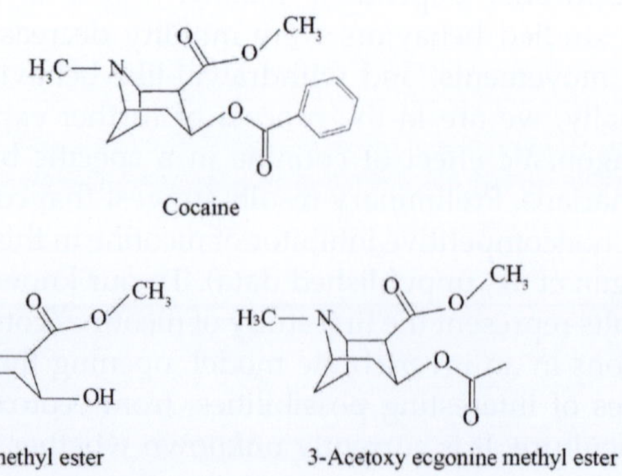

FIG. 8.3 Examples of allosteric inhibitors of cocaine (see text). Cocaine; ecgonine methyl ester, a major cocaine metabolite; and close analogue, 3-acetoxy ecgonine methyl ester, as indicated.

Acknowledgments

I wish to dedicate this chapter to the memory of my PhD advisor, Prof. George P. Hess, of Cornell University, from whom I learned about the principle of microscopic reversibility; he was a friend and a mentor. I am very grateful for the financial support from the Department of Biology, the Office of Sponsored Research, College of Sciences and Mathematics, West Chester University, and the National Institutes of Health (NIH; R03DA026518). I declare no conflict of interests.

Mini-Dictionary of Terms

Cotinine Major nicotine metabolite in humans (1-methyl-5-(pyridin-3-yl)pyrrolidin-2-one).

Nicotinic acetylcholine receptor The best-known ligand-gated ion channel to date. Defined by its capacity of being activated by nicotine.

Planarians Invertebrate organisms belonging to the phylum Platyhelminthes. These organisms are an emerging model in biochemical and behavioral pharmacology.

Principle of microscopic reversibility Describes the bioenergetic equivalence of dissimilar activation pathways of an enzyme or allosteric receptor.

Key Facts

- Cotinine is being investigated as a pharmacological agent against Parkinson's and Alzheimer's diseases, posttraumatic stress disorder, and depression, among others.
- Cotinine acts as either an agonist or an antagonist of nicotine depending on the specific model.
- The nicotinic acetylcholine receptor is the best-known protein receptor to date.
- Planarians are an increasingly popular model in neuropharmacology.

Summary Points

- Cotinine and nicotine bind to the same protein target: the nicotinic acetylcholine receptor (nAChR).
- The nicotine/nicotine-specific interaction mechanisms are not completely understood.
- When cotinine acts as an agonist of the nAChR, the most likely explanation is that nicotine and cotinine bind to the same site(s) on this receptor.
- When cotinine acts as an antagonist of the nAChR, the most likely explanation is that the nicotine and cotinine binding sites display an allosteric interaction.
- The principle of microscopic reversibility can provide a suitable mechanistic model to elucidate cotinine/nicotine interactions.
- Planarians are a suitable in vivo model to study cotinine/nicotine interactions.

References

Ambre, J. J., Ruo, T. I., Smith, G. L., Backes, D., & Smith, C. M. (1982). Ecgonine methyl ester, a major metabolite of cocaine. *Journal of Analytical Toxicology*, *6*(1), 26–29.

Astumian, R. D. (2012). Microscopic reversibility as the organizing principle of molecular machines. *Nature Nanotechnology*, *7*(11), 684–688.

Bach, D. J., Tenaglia, M., Baker, D. L., Deats, S., Montgomery, E., & Pagán, O. R. (2016). Cotinine antagonizes the behavioral effects of nicotine exposure in the planarian *Girardia tigrina*. *Neuroscience Letters*, *632*, 204–208.

Barreto, G. E., Iarkov, A., & Moran, V. E. (2014). Beneficial effects of nicotine, cotinine and its metabolites as potential agents for Parkinson's disease. *Frontiers in Aging Neuroscience*, *6*, 340.

Barreto, G. E., Yarkov, A., Avila-Rodriguez, M., Aliev, G., & Echeverría, V. (2015). Nicotine-derived compounds as therapeutic tools against post-traumatic stress disorder. *Current Pharmaceutical Design*, *21*(25), 3589–3595.

Benowitz, N. L., Hukkanen, J., & Jacob, P. (2009). Nicotine chemistry, metabolism, kinetics and biomarkers. *Handbook of Experimental Pharmacology*, *192*, 29–60.

Bertrand, D., & Changeaux, J.-P. (1995). Nicotinic receptor: an allosteric protein specialized for intercellular communication. *Seminars in the Neurosciences*, *7*, 75–90.

Bertrand, D., & Terry, A. V., Jr. (2018). The wonderland of neuronal nicotinic acetylcholine receptors. *Biochemical Pharmacology*, *151*, 214–225.

Blackmond, D. G. (2009a). Challenging the concept of "recycling" as a mechanism for the evolution of homochirality in chemical reactions. *Chirality*, *21*(3), 359–362.

Blackmond, D. G. (2009b). "If pigs could fly" chemistry: a tutorial on the principle of microscopic reversibility. *Angewandte Chemie (International Ed. in English)*, *48*(15), 2648–2654.

Burbaum, J. J., Raines, R. T., Albery, W. J., & Knowles, J. R. (1989). Evolutionary optimization of the catalytic effectiveness of an enzyme. *Biochemistry*, *28*(24), 9293–9305.

Buttarelli, F. R., Pellicano, C., & Pontieri, F. E. (2008). Neuropharmacology and behavior in planarians: translations to mammals. *Comparative Biochemistry and Physiology, Part C: Toxicology & Pharmacology*, *147*(4), 399–408.

Buttarelli, F. R., Pontieri, F. E., Margotta, V., & Palladini, G. (2000). Acetylcholine/dopamine interaction in planaria. *Comparative Biochemistry and Physiology, Part C: Toxicology & Pharmacology*, *125*(2), 225–231.

Caldeira, M. V., Salazar, I. L., Curcio, M., Canzoniero, L. M., & Duarte, C. B. (2014). Role of the ubiquitin-proteasome system in brain ischemia: friend or foe? *Progress in Neurobiology*, *112*, 50–69.

Cebrià, F., Kudome, T., Nakazawa, M., Mineta, K., Ikeo, K., Gojobori, T., et al. (2002). The expression of neural-specific genes reveals the structural and molecular complexity of the planarian central nervous system. *Mechanisms of Development*, *116*(1–2), 199–204.

Chapman, M. A. (2009). Does smoking reduce the risk of Parkinson's disease through stimulation of the ubiquitin-proteasome system? *Medical Hypotheses*, *73*(6), 887–891.

Chen, L. (2010). In pursuit of the high-resolution structure of nicotinic acetylcholine receptors. *The Journal of Physiology*, *588*(Pt 4), 557–564.

Chen, Y., Banerjee, A., & Hess, G. P. (2004). Mechanism-based discovery of small molecules that prevent noncompetitive inhibition by cocaine and MK-801 mediated by two different sites on the nicotinic acetylcholine receptor. *Biochemistry*, *43*(31), 10149–10156.

Dani, J. A. (2015). Neuronal nicotinic acetylcholine receptor structure and function and response to nicotine. *International Review of Neurobiology*, *124*, 3–19.

De Schepper, P. J., Van Hecken, A., Daenens, P., & Van Rossum, J. M. (1987). Kinetics of cotinine after oral and intravenous administration to man. *European Journal of Clinical Pharmacology*, *31*(5), 583–588.

Domino, E. F., Hornbach, E., & Demana, T. (1993). The nicotine content of common vegetables. *The New England Journal of Medicine*, *329*(6), 437.

du Rand, E. E., Pirk, C. W. W., Nicolson, S. W., & Apostolides, Z. (2017). The metabolic fate of nectar nicotine in worker honey bees. *Journal of Insect Physiology*, *98*, 14–22.

Echeverría, V., Grizzell, J. A., & Barreto, G. E. (2016). Neuroinflammation: a therapeutic target of cotinine for the treatment of psychiatric disorders? *Current Pharmaceutical Design, 22*(10), 1324–1333.

Echeverría, V., & Zeitlin, R. (2012). Cotinine: a potential new therapeutic agent against Alzheimer's disease. *CNS Neuroscience & Therapeutics, 18*(7), 517–523.

Firn, R. (2011). *Nature's chemicals: The natural products that shaped our world.* New York: Oxford University Press.

Freedman, R. (2014). α7-nicotinic acetylcholine receptor agonists for cognitive enhancement in schizophrenia. *Annual Review of Medicine, 65,* 245–261.

Giastas, P., Zouridakis, M., & Tzartos, S. J. (2018). Understanding structure-function relationships of the human neuronal acetylcholine receptor: insights from the first crystal structures of neuronal subunits. *British Journal of Pharmacology, 175*(11), 1880–1891.

Grizzell, J. A., & Echeverría, V. (2015). New insights into the mechanisms of action of cotinine and its distinctive effects from nicotine. *Neurochemical Research, 40*(10), 2032–2046.

Grizzell, J. A., Mullins, M., Iarkov, A., Rohani, A., Charry, L. C., & Echeverría, V. (2014). Cotinine reduces depressive-like behavior and hippocampal vascular endothelial growth factor downregulation after forced swim stress in mice. *Behavioral Neuroscience, 128*(6), 713–721.

Gu, R. X., Zhong, Y. Q., & Wei, D. Q. (2011). Structural basis of agonist selectivity for different nAChR subtypes: insights from crystal structures, mutation experiments and molecular simulations. *Current Pharmaceutical Design, 17*(17), 1652–1662.

Hatsukami, D. K., Grillo, M., Pentel, P. R., Oncken, C., & Bliss, R. (1997). Safety of cotinine in humans: physiologic, subjective, and cognitive effects. *Pharmacology, Biochemistry, and Behavior, 57*(4), 643–650.

Hess, G. P. (2003). Rapid chemical reaction techniques developed for use in investigations of membrane-bound proteins (neurotransmitter receptors). *Biophysical Chemistry, 100*(1–3), 493–506.

Hess, G. P., Gameiro, A. M., Schoenfeld, R. C., Chen, Y., Ulrich, H., Nye, J. A., et al. (2003). Reversing the action of noncompetitive inhibitors (MK-801 and cocaine) on a protein (nicotinic acetylcholine receptor)-mediated reaction. *Biochemistry, 42*(20), 6106–6114.

Hess, G. P., Ulrich, H., Breitinger, H. G., Niu, L., Gameiro, A. M., Grewer, C., et al. (2000). Mechanism-based discovery of ligands that counteract inhibition of the nicotinic acetylcholine receptor by cocaine and MK-801. *Proceedings of the National Academy of Sciences of the United States of America, 97*(25), 13895–13900.

Hogg, R. C., & Bertrand, D. (2004). Nicotinic acetylcholine receptors as drug targets. *Current Drug Targets. CNS and Neurological Disorders, 3*(2), 123–130.

Hucho, F., Tsetlin, V. I., & Machold, J. (1996). The emerging three-dimensional structure of a receptor. The nicotinic acetylcholine receptor. *European Journal of Biochemistry, 239*(3), 539–557.

Hurst, R., Rollema, H., & Bertrand, D. (2013). Nicotinic acetylcholine receptors: from basic science to therapeutics. *Pharmacology & Therapeutics, 137*(1), 22–54.

Jackson, D. M., Johnson, A. W., & Stephenson, M. G. (2002). Survival and development of *Heliothis virescens* (Lepidoptera: Noctuidae) larvae on isogenic tobacco lines with different levels of alkaloids. *Journal of Economic Entomology, 95*(6), 1294–1302.

Kane, J. K., Konu, O., Ma, J. Z., & Li, M. D. (2004). Nicotine coregulates multiple pathways involved in protein modification/degradation in rat brain. *Brain Research Molecular Brain Research. 132*(2), 181–191.

Karlin, A., & Akabas, M. H. (1995). Toward a structural basis for the function of nicotinic acetylcholine receptors and their cousins. *Neuron, 15*(6), 1231–1244.

Katz, B., & Thesleff, S. (1957). A study of the desensitization produced by acetylcholine at the motor end-plate. *The Journal of Physiology, 138*(1), 63–80.

Kenakin, T. (1997). *Molecular pharmacology: A short course* (1st ed.). Wiley-Blackwell.

Krivoshein, A. V., & Hess, G. P. (2004). Mechanism-based approach to the successful prevention of cocaine inhibition of the neuronal (alpha 3 beta 4) nicotinic acetylcholine receptor. *Biochemistry, 43*(2), 481–489.

Lewis, G. N. (1925). A new principle of equilibrium. *Proceedings of the National Academy of Sciences of the United States of America, 11*(3), 179–183.

Magni, P. A., Pazzi, M., Vincenti, M., Alladio, E., Brandimarte, M., & Dadour, I. R. (2016). Development and validation of a GC-MS method for nicotine detection in *Calliphora vomitoria* (L.) (Diptera: Calliphoridae). *Forensic Science International, 261,* 53–60.

Massaly, N., Francès, B., & Moulédous, L. (2014). Roles of the ubiquitin proteasome system in the effects of drugs of abuse. *Frontiers in Molecular Neuroscience, 7,* 99.

Mendoza, C., Barreto, G. E., Iarkov, A., Tarasov, V. V., Aliev, G., & Echeverría, V. (2018). Cotinine: a therapy for memory extinction in post-traumatic stress disorder. *Molecular Neurobiology.* https://doi.org/10.1007/s12035-018-0869-3.

Mineta, K., Nakazawa, M., Cebria, F., Ikeo, K., Agata, K., & Gojobori, T. (2003). Origin and evolutionary process of the CNS elucidated by comparative genomics analysis of planarian ESTs. *Proceedings of the National Academy of Sciences of the United States of America, 100*(13), 7666–7671.

Moran, V. E. (2012). Cotinine: beyond that expected, more than a biomarker of tobacco consumption. *Frontiers in Pharmacology, 3,* 173.

Nakazawa, M., Cebrià, F., Mineta, K., Ikeo, K., Agata, K., & Gojobori, T. (2003). Search for the evolutionary origin of a brain: planarian brain characterized by microarray. *Molecular Biology and Evolution, 20*(5), 784–791.

Orth, R. G., Head, G., & Mierkowski, M. (2007). Determining larval host plant use by a polyphagous lepidopteran through analysis of adult moths for plant secondary metabolites. *Journal of Chemical Ecology, 33*(6), 1131–1148.

Pagán, O. R. (2017). Planaria: an animal model that integrates development, regeneration and pharmacology. *The International Journal of Developmental Biology, 61*(8–9), 519–529.

Pagán, O. R., Deats, S., Baker, D., Montgomery, E., Wilk, G., Tenaglia, M., et al. (2013). Planarians require an intact brain to behaviorally react to cocaine, but not to react to nicotine. *Neuroscience, 246,* 265–270.

Pagán, O. R., Montgomery, E., Deats, S., Bach, D., & Baker, D. (2015). Evidence of nicotine-induced, curare-insensitive, behavior in planarians. *Neurochemical Research, 40*(10), 2087–2090.

Pagán, O. R., Rowlands, A. L., Fattore, A. L., Coudron, T., Urban, K. R., Bidja, A. H., et al. (2009). A cembranoid from tobacco prevents the expression of nicotine-induced withdrawal behavior in planarian worms. *European Journal of Pharmacology, 615*(1–3), 118–124.

Pogocki, D., Ruman, T., Danilczuk, M., Celuch, M., & Wałajtys-Rode, E. (2007). Application of nicotine enantiomers, derivatives and analogues in therapy of neurodegenerative disorders. *European Journal of Pharmacology, 563*(1–3), 18–39.

Ramakrishnan, L., & Hess, G. P. (2004). On the mechanism of a mutated and abnormally functioning gamma-aminobutyric acid (A) receptor linked to epilepsy. *Biochemistry, 43*(23), 7534–7540.

Ramakrishnan, L., & Hess, G. P. (2005). Picrotoxin inhibition mechanism of a gamma-aminobutyric acid A receptor investigated by a laser-pulse photolysis technique. *Biochemistry, 44*(23), 8523–8532.

Ramakrishnan, L., & Hess, G. P. (2010). Mechanism of potentiation of a dysfunctional epilepsy-linked mutated GABA(A) receptor by a neurosteroid (3alpha, 21-dihydroxy-5alpha-pregnan-20-one): Transient kinetic investigations. *Biochemistry, 49*(36), 7892–7901. https://doi.org/10.1021/bi901241g.

Rawls, S. M., Patil, T., Tallarida, C. S., Baron, S., Kim, M., Song, K., et al. (2011). Nicotine behavioral pharmacology: clues from planarians. *Drug and Alcohol Dependence*, *118*(2–3), 274–279.

Rezvani, K., Teng, Y., Shim, D., & De Biasi, M. (2007). Nicotine regulates multiple synaptic proteins by inhibiting proteasomal activity. *The Journal of Neuroscience*, *27*(39), 10508–10519.

Ricard, J. (1978). Generalized microscopic reversibility, kinetic co-operativity of enzymes and evolution. *The Biochemical Journal*, *175*(3), 779–791.

Robb, S. M., Gotting, K., Ross, E., & Sánchez Alvarado, A. (2015). SmedGD 2.0: the *Schmidtea mediterranea* genome database. *Genesis*, *53*(8), 535–546.

Rollema, H., Bertrand, D., & Hurst, R. S. (2015). Nicotinic agonists and antagonists. In I. Stolerman, & L. Price (Eds.), *Encyclopedia of psychopharmacology*. Berlin, Heidelberg: Springer.

Schmeltz, I. (1971). Nicotine and other tobacco alkaloids. In M. Jacobson, & D. G. Crosby (Eds.), *Naturally occurring insecticides* (pp. 99–136). New York, NY: Marcel Dekker.

Sivaprakasam, K., Pagán, O. R., & Hess, G. P. (2010). Minimal RNA aptamer sequences that can inhibit or alleviate noncompetitive inhibition of the muscle-type nicotinic acetylcholine receptor. *The Journal of Membrane Biology*, *233*(1–3), 1–12.

Soloway, S. B. (1976). Naturally occurring insecticides. *Environmental Health Perspectives*, *14*, 109–117.

Steppuhn, A., Gase, K., Krock, B., Halitschke, R., & Baldwin, I. T. (2004). Nicotine's defensive function in nature. *PLoS Biology*, *2*(8), E217.

Stolerman, I. P., & Jarvis, M. J. (1995). The scientific case that nicotine is addictive. *Psychopharmacology*, *117*(1), 2–10 (discussion 14–20).

Terry, A. V., Callahan, P. M., Hall, B., & Webster, S. J. (2011). Alzheimer's disease and age-related memory decline (preclinical). *Pharmacology, Biochemistry, and Behavior*, *99*(2), 190–210.

Terry, A. V., Callahan, P. M., & Hernandez, C. M. (2015). Nicotinic ligands as multifunctional agents for the treatment of neuropsychiatric disorders. *Biochemical Pharmacology*, *97*(4), 388–398.

Tolman, R. C. (1925). The principle of microscopic reversibility. *Proceedings of the National Academy of Sciences of the United States of America*, *11*(7), 436–439.

Tsetlin, V., & Hucho, F. (2009). Nicotinic acetylcholine receptors at atomic resolution. *Current Opinion in Pharmacology*, *9*(3), 306–310.

Vainio, P. J., Törnquist, K., & Tuominen, R. K. (2000). Cotinine and nicotine inhibit each other's calcium responses in bovine chromaffin cells. *Toxicology and Applied Pharmacology*, *163*(2), 183–187.

Vainio, P. J., & Tuominen, R. K. (2001). Cotinine binding to nicotinic acetylcholine receptors in bovine chromaffin cell and rat brain membranes. *Nicotine & Tobacco Research*, *3*(2), 177–182.

Vainio, P. J., Viluksela, M., & Tuominen, R. K. (1998a). Inhibition of nicotinic responses by cotinine in bovine adrenal chromaffin cells. *Pharmacology & Toxicology*, *83*(5), 188–193.

Vainio, P. J., Viluksela, M., & Tuominen, R. K. (1998b). Nicotine-like effects of cotinine on protein kinase C activity and noradrenaline release in bovine adrenal chromaffin cells. *Journal of Autonomic Pharmacology*, *18*(4), 245–250.

Wu, J., & Lukas, R. J. (2011). Naturally-expressed nicotinic acetylcholine receptor subtypes. *Biochemical Pharmacology*, *82*(8), 800–807.

Zoli, M., Pistillo, F., & Gotti, C. (2015). Diversity of native nicotinic receptor subtypes in mammalian brain. *Neuropharmacology*, *96*(Pt B), 302–311.

CHAPTER

9

Nicotine, Neural Plasticity, and Nicotine's Therapeutic Potential

Russell W. Brown, W. Drew Gill

Department of Biomedical Sciences, East Tennessee State University, James H. Quillen College of Medicine, Johnson City, TN, United States

Abbreviations

AD	Alzheimer's disease
BDNF	brain-derived neurotrophic factor
CREB	cyclic AMP-responsive element-binding protein
FGF	fibroblast growth factor
GDNF	glial-cell- line-derived neurotrophic factor
nAChR	nicotinic receptor
NGF	nerve growth factor
PD	Parkinson's disease

9.1 INTRODUCTION

The focus of this review is to explore findings analyzing the effects of nicotine on neural plasticity. This review is aimed at analyzing two ultimate consequences of nicotine on neural plasticity: (1) the role of increases of neurotrophic factors related to the addictive consequences of nicotine and (2) the possible role of changes in neural plasticity proteins toward the neuroprotective impact produced by nicotine and antineuroinflammatory effects. An additional area of analysis will be the effects of nicotine on behavior to support the consequences of nicotine on neural plasticity. The majority of the review will focus on the preclinical literature, although the translational impact of this work is emphasized.

Nicotine (1-methyl-2-[3-pyridyl]pyrrolidine) is an alkaloid and the primary ingredient in tobacco that produces physiological and psychological effects that leads to nicotine being absorbed into the bloodstream and ultimately underlies the addictive features of tobacco use. In 1988, the U.S. Surgeon General proposed that nicotine be included with alcohol, opiates, amphetamine, and cocaine in the category of addictive or dependency-producing drugs (USPHS, 1988). Notably, tobacco use is also particularly widespread among individuals that are diagnosed with different mental disorders, and tobacco abuse is particularly highly comorbid among the mentally ill, especially individuals diagnosed with schizophrenia (Prochaska, Das, & Young-Wolff, 2017).

9.2 NICOTINE AS A NEUROPROTECTANT

Nicotine is a drug that has properties that are potentially advantageous in cases of insults to the brain. Nicotine serves as an agonist to nicotinic receptors (nAChRs), and these receptors are found on the terminals of most of the small molecular neurotransmitters in the brain. Nicotine has been shown to increase the activity of all of these neurotransmitters (McGehee & Role, 1996). The fact that nicotine can enhance neuronal activity likely plays a role in its increase of neurotrophic factor proteins, many of which are activity-dependent (Mitre, Mariga, & Chao, 2017). Nicotine's interaction with neurotrophic factor proteins has led investigators to speculate that these increases may underlie nicotine's ability to enhance cognitive performance, reduce cognitive deficits in models of brain insults, and may also be the mechanism underlying the lesser incidence of Parkinson's disease (PD) in the tobacco smoking population (Mudo, Belluardo, & Fuxe, 2007). In other words, nicotine, which has been associated with lung cancer due to tobacco smoking, may have therapeutic potential when administered alone, and there has been enough evidence to suggest over the past 30 years that this may be the case. Although nicotine is not likely a therapeutic drug in and of itself, there have been some very interesting findings over these past decades that may change the way in which we approach nicotine and its neuroprotective potential.

Neuroscience of Nicotine
https://doi.org/10.1016/B978-0-12-813035-3.00009-5

9.3 NICOTINE AND NEUROTROPHIC FACTORS

There have been many studies analyzing the role of neurotrophic factors in neuroprotection, in recovery of function after brain injury, stroke, or in neurodegenerative diseases such as Alzheimer's disease (AD) and PD. Neurotrophic factors are a family of biomolecules that support growth, survival, differentiation, and maintenance of both developing and mature neurons (Koskela et al., 2017). Additionally, due to their activity-dependent release, neurotrophic factors are ideally suited to modify the strength of neuronal connections by enhancing synaptic activity through direct actions at presynaptic terminals (Tyler, Perrett, & Pozzo-Miller, 2002). Neurotrophins bind to the tyrosine kinase (Trk) receptor subtypes, which produces retrograde signaling from the terminal to the cell body, resulting in changes of overall cellular function (Friedman & Greene, 1999). Although increases of neurotrophic factors are generally viewed as a positive aspect of the effects of a particular drug when it comes to neuroprotection, it can also be the mechanism underlying its addictive properties (Pickens et al., 2011). Several addictive psychostimulants, including amphetamine, cocaine, and methamphetamine, have been shown to increase neurotrophic factors in brain areas that mediate addiction (Angelucci et al., 2009).

9.4 NICOTINE, nAChRs, AND RELATIONSHIP TO NEUROTROPHIC FACTORS

Nicotine is an acetylcholinergic agonist at the nicotinic receptor (nAChR) and readily crosses the blood-brain barrier. In the brain, the homomeric α7 and heteromeric α4/β2 are the largest number of nAChRs (Dineley, Anshul, & Yakel, 2015). There are multiple nAChRs that have been isolated and identified, and these receptors have an alpha (α2–α7, α9, and α10) and beta (β2–β4) subunits, although not all receptors contain both subunits. These subunits that constitute the receptor are assembled around a central hydrophilic pore that mediates the flow of the cations such as potassium (K^+), sodium (Na^+), and calcium (Ca^{++}) all of which play a vital role in neuronal transmission. In addition, these subunits assemble in different combinations to generate a variety of nAChR subtypes with distinct electrophysiological properties and brain localization (Brown et al., 2017). However, in a study from our laboratory, we analyzed GDNF at 24 h, and there is evidence that nicotine's effects on GDNF may be time-locked, because a different study reported that nicotine increased GDNF 1 h after drug administration (Takarada et al., 2012). In addition, nicotine and/or nicotinic agonists have also been shown to increase receptor expression of many of these neurotrophic factors including both Trk A, which binds NGF, and Trk B, which binds BDNF.

9.5 NICOTINE, BDNF, AND ADDICTION

Regarding addiction, most research interest has been in investigating BDNF's role underlying the rewarding aspects of nicotine (Perna & Brown, 2013). This focus on BDNF is likely at least in part due to not only its ubiquitous location across brain areas, but also its important role in other psychiatric disorders in which smoking is a common comorbidity, including schizophrenia and major depressive disorder (Jamal, Van der Does, & Penninx, 2015). There appears to be a strong relationship between nicotine and BDNF in brain areas that mediate drug reward that undoubtedly have an impact on addictive behavior (Naha, Gandhi, Gautam, & Prakash, 2017). Several studies have shown that nicotine increases BDNF expression in different brain areas up to 72 h after drug administration in normal rats (Kenny, File, & Rattray, 2000). However, most of these studies have focused on the hippocampus and prefrontal cortex, and there has been little research to analyze BDNF in brain areas that mediate drug addiction outside of work from our laboratory. Maggio et al. (1997) were the first to show that nicotine enhanced BDNF in the dorsal striatum, but this study was directed more at the potential neuroprotective role of nicotine in PD versus addiction. We have shown in a series of studies that neonatal quinpirole (dopamine D_2-like agonist) treatment to rats, which results in increases of dopamine D_2 receptor sensitivity throughout the animal's lifetime, enhances the BDNF response to nicotine in the nucleus accumbens and hippocampus when brain tissue was assayed 24 h after the last nicotine treatment (Brown et al., 2012). Therefore, it is well established that nicotine increases BDNF across a number of brain areas.

There is much less information regarding nicotine and most other neurotrophic factors relative to their role in addiction. However, there have been several studies to analyze several downstream regulators of neurotrophic factors on the MAPK/ERK pathway, including cyclic AMP response element binding protein (CREB). CREB is a cellular transcription factor which binds to certain DNA sequences called cAMP response elements (CRE) which will increase or decrease transcription of downstream genes. CREB regulates several genes, including both BDNF and tyrosine hydroxylase (the precursor to norepinephrine and dopamine biosynthesis) which may be related to the neural plasticity consequences of nicotine. Brunzell, Mineur, Neve, and Picciotto (2009) demonstrated that nicotine increased CREB activity in the nucleus accumbens of mice that demonstrated nicotine conditioned place preference

(CPP). Importantly, experimentally induced decreased expression of accumbal CREB via a viral vector was sufficient to block the expression of nicotine CPP. A different study replicated this effect and demonstrated blockade of nAChRs using the nAChR general antagonist mecamylamine that blocked both nicotine CPP and CREB expression (Pascual, Pastor, & Bernabeu, 2009). In addition, nicotine has been shown to increase CREB-ser133 in the prefrontal cortex after behavioral sensitization (Gomez, Midde, Mactutus, Booze, & Zhu, 2012). Essentially, the increase in CREB produced by nicotine in brain areas known to be important in addiction, such as dopamine terminal areas in the nucleus accumbens and prefrontal cortex, could be a mechanism underlying increases of BDNF or even changes in dopamine levels. Regardless, other neurotrophic factors may be playing a role in the effects of nicotine in addiction, and this avenue of research needs to be pursued more thoroughly to better understand these effects.

9.6 FUNCTIONAL IMPROVEMENT IN ANIMAL MODELS: PARKINSON'S DISEASE

Because of nicotine's effects on neurotrophic factors and synaptogenesis, there has been a large research focus on the effects of nicotine on neuroprotection. Certainly, the majority of the work in functional improvement by nicotine has been in PD. There has been a considerable amount of epidemiological evidence over the past 20 years that have consistently shown that PD is less prevalent among smokers than among that of individuals that have never smoked. Dr. Maryka Quik and colleagues have shown that nicotine and/or nAChR agonists protect against nigrostriatal damage, findings that may help to explain the well-established decline in PD incidence with tobacco use (Quik, 2004). Based on these observations, there have been several studies to analyze whether nicotine may be a promising drug for preventing risk of PD or possibly slow disease progression. The rationale for this approach has been the close relationship between nAChRs and the dopaminergic system in the striatum. It is well known that nAChRs are localized on dopaminergic neuron terminals in both the nucleus accumbens and striatum. Nicotine acts as an agonist at these nAChRs and increases dopamine release, which ultimately not only underlies addiction to the drug but also increases overall neuronal activity in these regions. Nicotine has been shown to protect against 1-methyl-4-phenyl-1,2,3,6-tetrahydropyridine (MPTP)-induced striatal damage and improve motor function in animal models of PD (Janson, Fuxe, & Goldstein, 1992). In addition, studies have shown that nicotine reduces L-dopa-induced abnormal involuntary movements, a debilitating complication of L-dopa therapy for PD (Quik, Bordia, Zhang, & Perez, 2015). These combined observations suggest that nAChR stimulation may represent a useful treatment strategy for PD and alleviate behavioral deficits associated with the disease.

Providing further support for nicotine as a therapeutic target for PD is that this effect was also dose-dependent. Tanner et al. (2002) discovered that within twin pairs, the risk of PD was inversely correlated with the amount of cigarette smoking. However, some authors argue that the link between the environmental factor "smoking" and PD may be in part due to an epiphenomenon caused by specific personality traits in individuals with susceptibility to PD (Evans et al., 2006). On the other hand, further evidence supporting that nicotine may be neuroprotective against PD is that nicotine upregulates antiapoptotic proteins that have been shown to prevent or slow down neurodegeneration (Dasgupta et al., 2006) and to increase cytochrome P450 enzymes that can protect against neurotoxicity (Miksys & Tyndale, 2006). There appears to be sufficient evidence to support that nicotine may be neuroprotective toward PD, although there are limitations to these discoveries, especially involving α7 nAChR agonists.

9.7 FUNCTIONAL IMPROVEMENT IN ANIMAL MODELS: ALZHEIMER'S DISEASE

Naturally, with nicotine's agonist action at acetylcholinergic nAChRs, there has been a fair amount of research focus analyzing on whether nicotine may produce similar neuroprotection toward AD, where cholinergic agonists (acetylcholinesterase inhibitors) are a common treatment option. AD is characterized by several neuropathologic hallmarks consisting of an accumulation of amyloid β peptide (Aβ), intracellular deposits of tau protein, neuronal loss, and prominent synaptic loss (Spires-Jones, Attems, & Thal, 2017). Binding studies performed with the use of [3H]-nicotine and [3H]-ACh showed a significant reduction in nicotine and ACh binding sites in the cerebral cortex of patients suffering from AD, demonstrating a decrease of both nAChR and muscarinic acetylcholine receptor (mAChR) populations (see Lombardo & Maskos, 2015). Considering the large literature of the role of ACh in cognition, the degeneration of acetylcholinergic neurons in AD led to the formulation of the "cholinergic hypothesis" of geriatric disorders (Bartus, Dean, Beer, & Lippa, 1982; Davies & Maloney, 1976) according to which the reduction in cholinergic innervation is responsible for the cognitive decline observed in AD patients.

A series of studies have revealed that α7 nAChRs colocalize with Amyloid β peptide (Wang et al., 2000), and this was not shown with the α4 nAChR subunit. The interaction of the nAChR and Aβ interaction initiates molecular pathways associated with neuroprotection, synaptic plasticity, and cognition (Dineley et al., 2015).

For example, Akt phosphorylation mediates the downstream activation of an antiapoptotic pathway, which is also activated by nicotine treatment (Maldifassi et al., 2014). In in vitro hippocampal slice preparations, subsequent to incubation with concentrations of $A\beta_{1-42}$ monomers and oligomers, an increase of hippocampal long-term potentiation (LTP) was observed. LTP is the cellular phenomenon that has been implicated in long-term memory storage (see Kumar, 2011). It was found that LTP and the activation of intracellular pathways are mediated by the activation of α7 nAChRs (Puzzo et al., 2008). Functionally, an in vivo Aβ infusion into the dorsal hippocampus in mice was able to enhance performance in hippocampus-dependent tasks such as the Morris water maze and contextual fear conditioning. Interestingly, the Aβ-nAChR interaction has also been shown to inhibit survival pathways, and one study has shown that Aβ inhibits the neuroprotective effect of α7 nAChR activation (Liu et al., 2007). Clearly, this is a very complex interaction that deserves further investigation as a possible treatment target in AD.

9.8 NICOTINE NEUROPROTECTION AGAINST COGNITIVE IMPAIRMENTS DUE TO BRAIN INSULT

Nicotine has been shown to increase neurotrophic factors in brain areas that mediate cognitive impairment due to traumatic brain injury (TBI) or other neurodegenerative diseases, especially the hippocampal formation. Based on those effects, there has been a research focus toward analyzing nAChR agonists without addictive properties aimed at neuroprotection (Jurado-Coronel et al., 2016). It is important to establish that if nicotine has been shown to increase neurotrophic factors that may relate to neuroprotection, then it must be established that nicotine also alleviates behavioral deficits associated with brain insults. In normal control animals, nicotine has been shown to enhance cognitive performance in rodents that had not been administered any brain lesions and/or trauma. Dr. Ed Levin and colleagues at Duke University have demonstrated over a number of studies that nicotine's action at cholinergic and glutamatergic synapses appears to mediate its enhancing effects on cognition (Levin, Bradley, Addy, & Sigurani, 2002). Therefore, it has long been established that nicotine, when administered alone, can enhance cognitive performance that leads to this drug having the potential to lessen cognitive impairment in TBI.

Although the mechanism underlying these enhancing effects has not completely been delineated, it does appear that the α7 nAChR is important. Activation of α7 nAChRs can improve cognitive performance in aged rats (Arendash, Sanberg, & Sengstock, 1995), whereas blockade of those receptors impairs performance (Levin et al., 2002). Consistent with these animal studies, recent data from a clinical study suggest that the α7 nAChR partial agonist ABT-126 positively influenced memory and attention in individuals diagnosed with schizophrenia that demonstrated cognitive impairment (Haig, Bain, Robieson, Baker, & Othman, 2016). Preclinically, chronic intermittent nicotine administration reduced cognitive impairment caused by either lesions to the fornix or TBI in rats tested on the Morris water maze, which is a spatial memory task primarily mediated by the function of the hippocampal formation (Brown, Gonzalez, Whishaw, & Kolb, 2001; Verbois, Hopkins, Scheff, & Pauly, 2003). Verbois, Scheff, and Pauly (2003) also reported that nicotine attenuated decreases of α7 nAChRs in the hippocampus of animals, which was produced by TBI. Although there has not been much recent research attention on the role of nAChRs in TBI, this is an area of therapeutic potential, especially considering the antiinflammatory properties of nicotine.

9.9 NICOTINE AND NEUROINFLAMMATION

Neuroinflammation is common among a number of disorders, including not only AD and PD but also mental disorders such as schizophrenia, major depressive disorder, and bipolar disorder. The study of the role of α7 nAChRs in the cascade involved in microglia/macrophage cell activation has been facilitated by the fact that only α7 nAChRs are expressed in both monocytes and macrophages (Wang et al., 2003). There is strong evidence that α7 nAChR activation by nicotine and nAChR agonists decreases the abnormal activation of microglia, which is so prevalent in neuroinflammation. This has led to the conclusion that nicotine's suppressive effects on microglia activation must mediate some of the neuroprotective actions of nicotine (Morioka et al., 2015).

It has been known for some time that α7 nAChR activation plays an important role in mediating antiinflammation and activating antiapoptotic pathways. In addition, increases of acetylcholinergic activity lead to attenuation of the release of several proinflammatory cytokines including TNF, IL-1β, and IL-6 (Marrero & Bencherif, 2009). The presence of cholinergic antiinflammatory pathways mediated by the α7 nAChR in the brain offers new therapeutic avenues for neurological disorders that are characterized by neuroinflammation. Further, there has been a deficit in α7 nAChRs found in the hippocampal formation of patients diagnosed with AD and schizophrenia (Olincy & Freedman, 2012), leading to focusing on α7 nAChR as a therapeutic target. However, the possible downfall of this target is that α7 nAChRs become rapidly desensitized by its agonists and, therefore, unresponsive to binding. This characteristic of α7 nAChRs limits the benefits of drugs directed to increase acetylcholine levels such as the aforementioned acetylcholinesterase inhibitors and agonists of the α7 nAChR (Williams, Wang, & Papke, 2011).

On the other hand, it is known that not only α7 nAChRs are activated by conformational changes induced by ligands such as competitive antagonists and full and partial agonists that bind the receptor at the primary (orthosteric) binding site but also α7 nAChRs are positively or negatively regulated by allosteric modulators that bind to allosteric sites (Echeverria, Yarkov, & Aliev, 2016). Through binding at the allosteric binding site, an allosteric modulator would change the conformation of the receptor that would enhance the representation of receptors in functional states, permitting higher rates of spontaneous openings in the absence of an orthosteric agonist. In addition, modulators of the allosteric binding site of α7 nAChRs will increase the efficacy of agonists to induce cation currents, such as calcium and sodium, throughout the activated channel through inhibition of the desensitized states of α7 nAChRs. Positive allosteric modulators (PAMs) may switch the receptor's conformation from nonactivatable to activatable states and may have the ability to circumvent the problem of the fast desensitizing state of α7 nAChRs.

9.10 CONCLUDING REMARKS

Nicotine and nAChR agonists essentially constitute a double-edged sword toward addiction and neuroprotection. However, it has been established that nicotine and nAChR agonists increase neurotrophic factors, which lead to synaptogenesis and neuroprotection. There appears to be a lot of attention paid to the α7 nAChR as a specific target, because of its antiinflammatory properties and potential as enhancing cognition while avoiding the addictive properties of nicotine. Because of the unique desensitization properties of the α7 nAChR, drug discovery may be pushed toward the PAMs, as suggested by Echeverria and colleagues, which may lead to nAChRs as therapeutic targets.

Summary Points

- Nicotine and nicotinic receptor (nAChR) agonists enhance neurotrophic factors across the brain.
- Changes in synaptogenesis and synaptic maintenance underlie the addictive and neuroprotective characteristics of nicotine.
- The strongest evidence for nicotine's neuroprotective properties has been in the Parkinson's disease research field.
- Agonists at the α7 nAChR appear to have the strongest therapeutic potential.
- Nicotine has antiinflammatory properties that should be kept in mind when analyzing its potential for neuroprotection.

References

Angelucci, F., Ricci, V., Spalletta, G., Caltagirone, C., Mathé, A. A., & Bria, P. (2009). Effects of psychostimulants on neurotrophins implications for psychostimulant-induced neurotoxicity. *International Reviews in Neurobiology*, *88*, 1–24.

Arendash, G. W., Sanberg, P. R., & Sengstock, G. J. (1995). Nicotine enhances the learning and memory of aged rats. *Pharmacology Biochemistry & Behavior*, *52*, 517–523.

Bartus, R. T., Dean, R. L., Beer, B., & Lippa, A. S. (1982). The cholinergic hypothesis of geriatric memory dysfunction. *Science*, *217*, 408–414.

Brown, R. W., Gonzalez, C. L., Whishaw, I. Q., & Kolb, B. (2001). Nicotine improvement of Morris water task performance after fimbria-fornix lesion is blocked by mecamylamine. *Behavioural Brain Research*, *119*, 185–192.

Brown, R. W., Kirby, S. L., Denton, A. R., Dose, J. M., Cummins, E. D., Gill, W. D., et al. (2017). An analysis of the rewarding and aversive associative properties of nicotine in the neonatal quinpirole model: effects on glial cell line-derived neurotrophic factor (GDNF). *Schizophrenia Research*, *17*, 30161–30165.

Brown, R. W., Maple, A. M., Perna, M. K., Sheppard, A. B., Cope, Z. A., & Kostrzewa, R. M. (2012). Schizophrenia and substance abuse comorbidity: nicotine addiction and the neonatal quinpirole model. *Developmental Neuroscience*, *34*, 140–151.

Brunzell, D. H., Mineur, Y. S., Neve, R. L., & Picciotto, M. R. (2009). Nucleus accumbens CREB activity is necessary for nicotine conditioned place preference. *Neuropsychopharmacology*, *34*, 1993–2001.

Dasgupta, P., Kinkade, R., Joshi, B., Decook, C., Haura, E., & Chellappan, S. (2006). Nicotine inhibits apoptosis induced by chemotherapeutic drugs by up-regulating XIAP and survivin. *Proceeding of the National Academy of Science of the United States of America*, *103*, 6332–6337.

Davies, P., & Maloney, A. J. (1976). Selective loss of central cholinergic neurons in Alzheimer's disease. *Lancet*, *2*, 1403.

Dineley, K. T., Anshul, A. P., & Yakel, J. L. (2015). Nicotinic ACh receptors as therapeutic targets in CNS disorders. *Trends in Pharmacological Sciences*, *36*(2), 96–108.

Echeverria, V., Yarkov, A., & Aliev, G. (2016). Positive modulators of the α7 nicotinic receptor against neuroinflammation and cognitive impairment in Alzheimer's disease. *Progress in Neurobiology*, *144*, 142–157.

Evans, A. H., Lawrence, A. D., Potts, J., MacGregor, L., Katzenschlager, R., Shaw, K., et al. (2006). Relationship between impulsive sensation seeking traits, smoking, alcohol and caffeine intake, and Parkinson's disease. *Journal of Neurology, Neurosurgery and Psychiatry*, *77*, 317–321.

Friedman, W. J., & Greene, L. A. (1999). Neurotrophin signaling via Trks and p75. *Experimental Cell Research*, *253*, 131–142.

Gomez, A. M., Midde, N. M., Mactutus, C. F., Booze, R. M., & Zhu, J. (2012). Environmental enrichment alters nicotine-mediated locomotor sensitization and phosphorylation of DARPP-32 and CREB in rat prefrontal cortex. *PLoS One*, *7*, e44149.

Haig, G. M., Bain, E. E., Robieson, W. Z., Baker, J. D., & Othman, A. A. (2016). A randomized trial to assess the efficacy and safety of ABT-126, a selective α7 nicotinic acetylcholine receptor agonist, in the treatment of cognitive impairment in schizophrenia. *American Journal of Psychiatry*, *173*, 827–835.

Jamal, M., Van der Does, W., & Penninx, B. W. (2015). Effect of variation in BDNF Val(66)Met polymorphism, smoking, and nicotine dependence on symptom severity of depressive and anxiety disorders. *Drug Alcohol Dependence, 1*(148), 150–157.

Janson, A. M., Fuxe, K., & Goldstein, M. (1992). Differential effects of acute and chronic nicotine treatment on MPTP-(1-methyl-4-phenyl-1,2,3,6-tetrahydropyridine) induced degeneration of nigrostriatal dopamine neurons in the black mouse. *Clinical Investigations, 70*, 232–238.

Jurado-Coronel, J. C., Avila-Rodriguez, M., Capani, F., Gonzalez, J., Moran, V. E., & Barreto, G. E. (2016). Targeting the nicotinic acetylcholine receptors (nAChRs) in astrocytes as a potential therapeutic target in Parkinson's disease. *Current Pharmaceutical Design, 22*, 1305–1311.

Kenny, P. J., File, S. E., & Rattray, M. (2000). Acute nicotine decreases, and chronic nicotine increases the expression of brain-derived neurotrophic factor mRNA in rat hippocampus. *Brain Research Molecular Brain Research, 85*, 234–238.

Koskela, M., Bäck, S., Võikar, V., Richie, C. T., Domanskyi, A., Harvey, B. K., et al. (2017). Update of neurotrophic factors in neurobiology of addiction and future directions. *Neurobiology, 97*(Pt B), 189–200.

Kumar, A. (2011). Long-term potentiation at CA3–CA1 hippocampal synapses with special emphasis on aging, disease, and stress. *Frontiers in Aging Neuroscience, 3*, 7–11.

Levin, E. D., Bradley, A., Addy, N., & Sigurani, N. (2002). Hippocampal alpha 7 and alpha 4 beta 2 nicotinic receptors and working memory. *Neuroscience, 109*, 757–765.

Liu, Q., Zhang, J., Zhu, H., Qin, C., Chen, Q., & Zhao, B. (2007). Dissecting the signaling pathway of nicotine-mediated neuroprotection in a mouse Alzheimer disease model. *FASEB Journal, 21*, 61–73.

Lombardo, S., & Maskos, U. (2015). Role of the nicotinic acetylcholine receptor in Alzheimer's disease pathology and treatment. *Neuropharmacology, 96*(Pt B), 255–262.

Maggio, R., Riva, M., Vaglini, F., Fornai, F., Racagni, G., & Corsini, G. U. (1997). Striatal increase of neurotrophic factors as a mechanism of nicotine protection in experimental parkinsonism. *Journal of Neural Transmission, 104*, 1113–1123.

Maldifassi, M. C., Atienza, G., Arnalich, F., López-Collazo, E., Cedillo, J. L., Martín-Sánchez, C., et al. (2014). A new IRAK-M-mediated mechanism implicated in the anti-inflammatory effect of nicotine via α7 nicotinic receptors in human macrophages. *PLoS One, 9*(9), e108397.

Marrero, M. B., & Bencherif, M. (2009). Convergence of alpha 7 nicotinic acetylcholine receptor-activated pathways for anti-apoptosis and anti-inflammation: central role for JAK2 activation of STAT3 and NF-kappaB. *Brain Research, 1256*, 1–7.

McGehee, D. S., & Role, L. W. (1996). Neurobiology: memories of nicotine. *Nature, 383*, 670–671.

Miksys, S., & Tyndale, R. F. (2006). Nicotine induces brain CYP enzymes: relevance to Parkinson's disease. *Journal of Neural Transmission, 70*, 177–180.

Mitre, M., Mariga, A., & Chao, M. V. (2017). Neurotrophin signalling: novel insights into mechanisms and pathophysiology. *Clinical Science, 131*, 13–23.

Morioka, N., Harano, S., Tokuhara, M., Idenoshita, Y., Zhang, F. F., Hisaoka-Nakashima, K., et al. (2015). Stimulation of α7 nicotinic acetylcholine receptor regulates glutamate transporter GLAST via basic fibroblast growth factor production in cultured cortical microglia. *Brain Research, 1625*, 111–120.

Mudo, G., Belluardo, N., & Fuxe, K. (2007). Nicotinic receptor agonists as neuroprotective/neurotrophic drugs. Progress in molecular mechanisms. *Journal of Neural Transmission, 114*, 135–147.

Naha, N., Gandhi, D. N., Gautam, A. K., & Prakash, J. R. (2017). Nicotine and cigarette smoke modulate Nrf2-BDNF-dopaminergic signal and neurobehavioral disorders in adult rat cerebral cortex. *Human Experimental Toxicology 37*, 540–566.

Olincy, A., & Freedman, R. (2012). Nicotinic mechanisms in the treatment of psychotic disorders: a focus on the α7 nicotinic receptor. *Handbook of Experimental Pharmacology, 213*, 211–232.

Pascual, M. M., Pastor, V., & Bernabeu, R. O. (2009). Nicotine-conditioned place preference induced CREB phosphorylation and Fos expression in the adult rat brain. *Psychopharmacology, 207*, 57–71.

Perna, M. K., & Brown, R. W. (2013). Adolescent nicotine sensitization and effects of nicotine on accumbal dopamine release in a rodent model of increased dopamine D2 receptor sensitivity. *Behavioural Brain Research, 242*, 102–109.

Pickens, C. L., Airavaara, M., Theberge, F., Fanous, S., Hope, B. T., & Shaham, Y. (2011). Neurobiology of the incubation of drug craving. *Trends in Neuroscience, 34*, 411–420.

Prochaska, J. J., Das, S., & Young-Wolff, K. C. (2017). Smoking, mental Illness, and public Health. *Annual Review of Public Health, 38*, 165–185.

Puzzo, D., Privitera, L., Leznik, E., Fà, M., Staniszewski, A., Palmeri, A., et al. (2008). Picomolar amyloid-β positively modulates synaptic plasticity and memory in hippocampus. *Journal of Neuroscience, 28*, 14537–14545.

Quik, M. (2004). Smoking, nicotine and Parkinson's disease. *Trends in Neuroscience, 27*, 561–568.

Quik, M., Bordia, T., Zhang, D., & Perez, X. A. (2015). Nicotine and nicotinic receptor drugs: potential for Parkinson's disease and drug-induced movement disorders. *International Review of Neurobiology, 124*, 247–271.

Spires-Jones, T. L., Attems, J., & Thal, D. R. (2017). Interactions of pathological proteins in neurodegenerative diseases. *Acta Neuropathology, 134*, 187–205.

Takarada, T., Nakamichi, N., Kawagoe, H., Ogura, M., Fukumori, R., Nakazato, R., et al. (2012). Possible neuroprotective property of nicotinic acetylcholine receptors in association with predominant upregulation of glial cell line-derived neurotrophic factor in astrocytes. *Journal of Neuroscience Research, 90*, 2074–2085.

Tanner, C. M., Goldman, S. M., Aston, D. A., Ottman, R., Ellenberg, J., Mayeux, R., et al. (2002). Smoking and Parkinson's disease in twins. *Neurology, 58*, 581–588.

Tyler, W. J., Perrett, S. P., & Pozzo-Miller, L. D. (2002). The role of neurotrophins in neurotransmitter release. *Neuroscientist, 8*, 524–531.

United States Department of Health and Human Services. (1988). *The Health Consequences of Smoking: A Report of the Surgeon General* (pp. 1–643). United States Department of Health and Human Services.

Verbois, S. L., Hopkins, D. M., Scheff, S. W., & Pauly, J. R. (2003). Chronic intermittent nicotine administration attenuates traumatic brain injury-induced cognitive dysfunction. *Neuroscience, 119*, 1199–1208.

Verbois, S. L., Scheff, S. W., & Pauly, J. R. (2003). Chronic nicotine treatment attenuates alpha 7 nicotinic receptor deficits following traumatic brain injury. *Neuropharmacology, 44*, 224–233.

Wang, H.-Y., Lee, D. H. S., D'Andrea, M. R., Peterson, P. A., Shank, R. P., & Reitz, A. B. (2000). β-Amyloid1–42 binds to α7 nicotinic acetylcholine receptor with high affinity implications for Alzheimer's disease pathology. *Journal of Biological Chemistry, 275*, 5626–5632.

Wang, H., Yu, M., Ochani, M., Amella, C. A., Tanovic, M., Susarla, S., et al. (2003). Nicotinic acetylcholine receptor alpha7 subunit is an essential regulator of inflammation. *Nature, 421*, 384–388.

Williams, D. K., Wang, J., & Papke, R. L. (2011). Positive allosteric modulators as an approach to nicotinic acetylcholine receptor-targeted therapeutics: advantages and limitations. *Biochemical Pharmacology, 82*, 915–930.

CHAPTER

10

Habenular Synapses and Nicotine

Jessica L. Ables, Beatriz Antolin-Fontes†, Ines Ibañez-Tallon‡*

*Department of Psychiatry, The Icahn School of Medicine at Mount Sinai, New York, NY, United States
†Faculty of Medicine, Imperial College London, Hammersmith Hospital Campus, London, United Kingdom
‡Laboratory of Molecular Biology, The Rockefeller University, New York, NY, United States

Abbreviations

3V third ventricle
4V fourth ventricle
ACh acetylcholine
AP action potential
BAC bacterial artificial chromosome
ChAT choline acetyltransferase
cKO conditional knockout
CPA conditioned place aversion
DR dorsal raphe
DTg dorsal tegmental nuclei
EC entorhinal cortex
EPSP excitatory postsynaptic potential
FR fasciculus retroflexus
GWAS genome-wide association studies
Hb-IPN habenulo-interpeduncular
HC hippocampus
HCN hyperpolarization-activated cyclic nucleotide-gated
Hyp hypothalamus
IL-18 interleukin 18
IPA interpeduncular nucleus apical
IPC interpeduncular nucleus central
IPDL interpeduncular nucleus dorsolateral
IPDM interpeduncular nucleus dorsomedial
IPI interpeduncular nucleus intermediate
IPL interpeduncular nucleus lateral
IPN interpeduncular nucleus
IPR interpeduncular nucleus rostral
KCC2 K^+/Cl^- cotransporter 2
LC locus coeruleus
LDTg laterodorsal tegmental nuclei
LHb lateral habenula
LV lateral ventricle
MHb medial habenula
MHbD medial habenula dorsal
MHbS medial habenula superior
MHbV medial habenula ventral
MHbVc medial habenula ventrocentral
MHbVl medial habenula ventrolateral
MHbVm medial habenula ventromedial
MnR median raphe
MS medial septum
nAChR nicotinic acetylcholine receptor
NDB nucelus of diagonal band
NI nucleus incertus
OPRM μ-opioid receptor
PAG periaqueductal gray
sEPSC spontaneous excitatory postsynaptic current
SFi septofimbrial nucleus
SNP single nucleotide polymorphism
SP substance P
SV synaptic vesicle
Thal thalamus
TRAP translational ribosomal affinity purification
TS triangular septum
VACHT vesicular acetylcholine transporter
VGLUT vesicular glutamate transporter
VTA ventral tegmental area

10.1 INTRODUCTION

Human genome-wide association studies (GWAS) have established that genetic factors are involved in the development of drug addiction. In 2008, two studies reported that variants in the gene cluster *CHRNA3-B4-A5* encoding α3β4α5 nicotinic receptors are associated with lung cancer disease susceptibility (Hung et al., 2008) and with nicotine dependence lung cancer and peripheral disease (Thorgeirsson et al., 2008). This was a remarkable finding because although α4β2 and α7 nAChRs are the major nicotinic receptor subtypes present in the brain and periphery, GWAS did not link these genes to nicotine abuse. Rather, they indicated that α3β4α5 nAChR subtypes, which are highly enriched in the MHb-IPN, are critical for acquisition of nicotine dependence and difficulties in smoking cessation. Many groups replicated these GWAS across different human populations (Bierut, 2010; Bierut et al., 2008; Liu et al., 2010), and recent studies in

Neuroscience of Nicotine
https://doi.org/10.1016/B978-0-12-813035-3.00010-1

rodents revealed the molecular mechanisms (Fowler, Lu, Johnson, Marks, & Kenny, 2011; Frahm et al., 2011). These studies renewed the interest in this ancient brain structure and have led to major efforts to understand this circuit and its significance for addiction (reviewed in Antolin-Fontes, Ables, Gorlich, & Ibanez-Tallon, 2015). In this chapter, we will discuss the current understanding of the MHb and its role in nicotine dependence.

10.2 THE HABENULA: A HIGHLY CONSERVED CIRCUIT

The habenula is a paired structure in the epithalamus, highly conserved across vertebrates, and serves to connect more recently evolved structures involved in executive function in the neocortex and limbic forebrain with ancient areas that process sleep, pain, and reward in the midbrain and hindbrain (reviewed in Antolin-Fontes et al., 2015; Boulos, Darcq, & Kieffer, 2017). While all vertebrates have habenular nuclei, there are distinctions between mammals and lower vertebrates in terms of target structures and organization. The habenula of fish, reptiles, and birds is asymmetrical with respect to left and right sides, while in mammals, the habenula appears largely symmetrical. In fish and amphibians, the habenula is organized into dorsal and ventral nuclei, which both project to the IPN via the fasciculus retroflexus (FR) and together are homologous to the MHb of mammals, while it is unclear if a homologue of the mammalian lateral habenula (LHb) exists in fish. In amphibians, the ventral habenula appears equivalent to the LHb of higher vertebrates (Aizawa, Amo, & Okamoto, 2011). Reptiles, birds, and mammals demonstrate both an LHb that projects around the IPN to more rostral structures, including the ventral tegmental area (VTA), rostromedial tegmentum (RMTg), and median and dorsal raphe nuclei, and an MHb that projects to the IPN, which in turn projects to the raphe, laterodorsal tegmentum (LDTg), and nucleus incertus. Both the LHb and MHb receive input from the septum, the prefrontal cortex, and the hypothalamus, via the median forebrain bundle, though the inputs to each of the nuclei differ (for a thorough review on the connectivity, see Boulos et al., 2017). Thus, the habenular nuclei are strategically situated to regulate dopaminergic, serotoninergic, and adrenergic tone in response to input from higher structures and integrate cortical decision-making with reward and external stimuli (Fig. 10.1).

Indeed, several animal studies have demonstrated a role for the habenula in mental illness and drug addiction. Studies in animal models demonstrated that the FR degenerated after chronic administration of nicotine and cocaine and that the degeneration was specific to either the MHb or LHb neurons, respectively, suggesting that the habenula may be a "weak link" of the brain that is especially susceptible to drugs of abuse (Ciani, Severi, Bartesaghi, & Contestabile, 2005). On a more functional level, both the LHb and MHb have been found to play a role in the expression of withdrawal, consistent with their activation during aversive stimuli. Withdrawal from cocaine or from nicotine increases activity in the LHb or MHb, respectively (Ciani et al., 2005). MHb-dependent activation of the IPN is necessary for behavioral expression of nicotine withdrawal, and conversely, inhibiting IPN activity attenuates nicotine withdrawal (Zhao-Shea, Liu, Pang, Gardner, & Tapper, 2013). In the following sections, we will discuss in more detail the role of the MHb-IPN pathway in nicotine addiction, due in large part to its high expression of a diverse number of nAChRs.

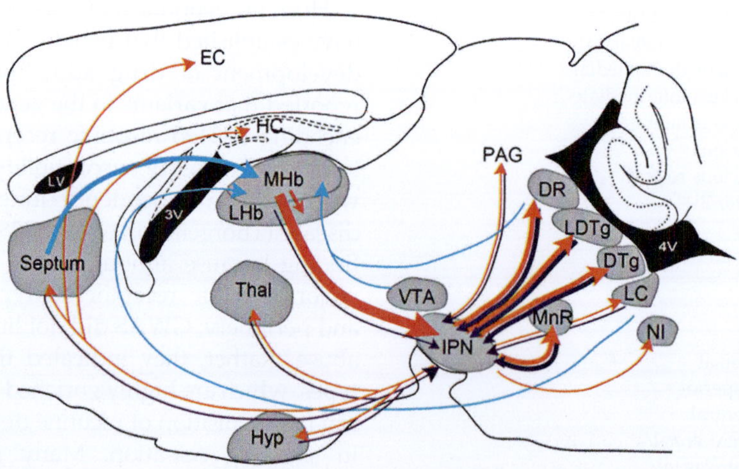

FIG. 10.1 MHb-IPN connectivity.
Schematic representation of a sagittal mouse brain showing the MHb afferents (*blue*), MHb efferents (*red*), IPN afferents (*purple*), and IPN efferents (*orange*). The thickness of the arrows reflects the strength of the connection. *Image taken from Antolin-Fontes et al. (2015),* Neuropharmacology, *with permission from the publishers.*

10.3 NICOTINIC ACETYLCHOLINE RECEPTOR DIVERSITY IN THE MHB-IPN

The MHb-IPN circuit expresses the highest density of nAChRs in the mammalian brain (Le Novere, Zoli, & Changeux, 1996; Marks et al., 1992; Yeh et al., 2001). All known neuronal nAChR subunits, except α9 and α10, are found in the MHb-IPN. The α3 and β4 nAChR subunits are strongly enriched throughout the ventral MHb and IPN (Gorlich et al., 2013; Quick, Ceballos, Kasten, McIntosh, & Lester, 1999; Salas, Pieri, Fung, Dani, & De Biasi, 2003; Sheffield, Quick, & Lester, 2000; Shih et al., 2014), whereas α4, α6, β2, and β3 subunits are selectively found in certain subnuclei of the ventral MHb (Shih et al., 2014). Specifically, α4 subunits are localized in the lateral portion of the vMHb (Fonck et al., 2009; Nashmi et al., 2007; Shih et al., 2014), α6 subunits are mainly found in the inferior portion of vMHb, and β2 and β3 subunits are located in the central and inferior vMHb (Shih et al., 2014). The auxiliary α5 subunit not only assembles with the α3β4* combination in the MHb and IPN but also can be incorporated in α4β2* complexes and is strongly expressed in the IPN (Fig. 10.2).

In agreement with human GWAS, cumulative evidence from animal models points to α3, β4, and α5 nAChRs expressed in the MHb-IPN as key regulators of nicotine intake and nicotine withdrawal (Fowler et al., 2011; Frahm et al., 2011; Salas, Sturm, Boulter, & De Biasi, 2009). A critical role for β4 in nicotine aversion was uncovered using transgenic mice that overexpress the β4 nAChR subunit in endogenous sites. These mice, called Tabac mice, displayed enhanced sensitivity to nicotine and reduced consumption (Frahm et al., 2011). β4 expression was shown to be rate-limiting and competed with α5 to form pentameric α3β4α5 nAChRs (Frahm et al., 2011; Slimak et al., 2014). Thus, overexpression of β4 led to increased number of α3β4* receptors and enhanced α3β4* currents in vitro and in vivo (Frahm et al., 2011). The ability of β4 to increase nicotine-evoked currents depended on a unique single residue, S435, located in the intracellular vestibule of the receptor (Frahm et al., 2011) (Fig. 10.3). Sequence alignments identified other SNP mappings to the intracellular vestibule, one of them being the most common polymorphism linked to high risk of nicotine dependence in humans, rs16969968 in *CHRNA5* (Bierut et al., 2008). Functional

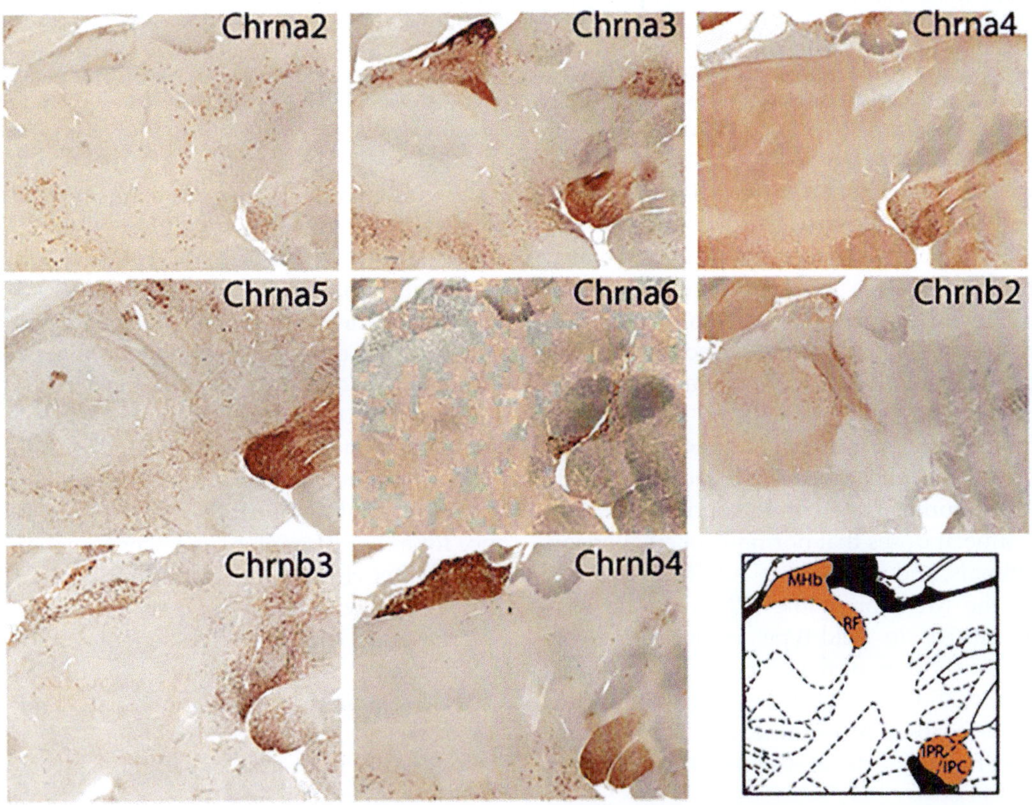

FIG. 10.2 nAChR subunits expressed in the MHb-IPN.
Sagittal sections of the habenula and IPN in BAC transgenic mice (Gensat.org) expressing eGFP under the corresponding gene regulatory regions of the indicated nAChR genes. Diagram indicates the MHb and IPN regions. *From the links http://www.gensat.org/imagenavigator.jsp?imageID=94262; http://www.gensat.org/imagenavigator.jsp?imageID=93930; http://www.gensat.org/imagenavigator.jsp?imageID=72032; http://www.gensat.org/imagenavigator.jsp?imageID=97051; http://www.gensat.org/imagenavigator.jsp?imageID=7868; http://www.gensat.org/imagenavigator.jsp?imageID=54061; http://www.gensat.org/imagenavigator.jsp?imageID=92495; http://www.gensat.org/imagenavigator.jsp?imageID=94924. Used with permission from GENSAT.*

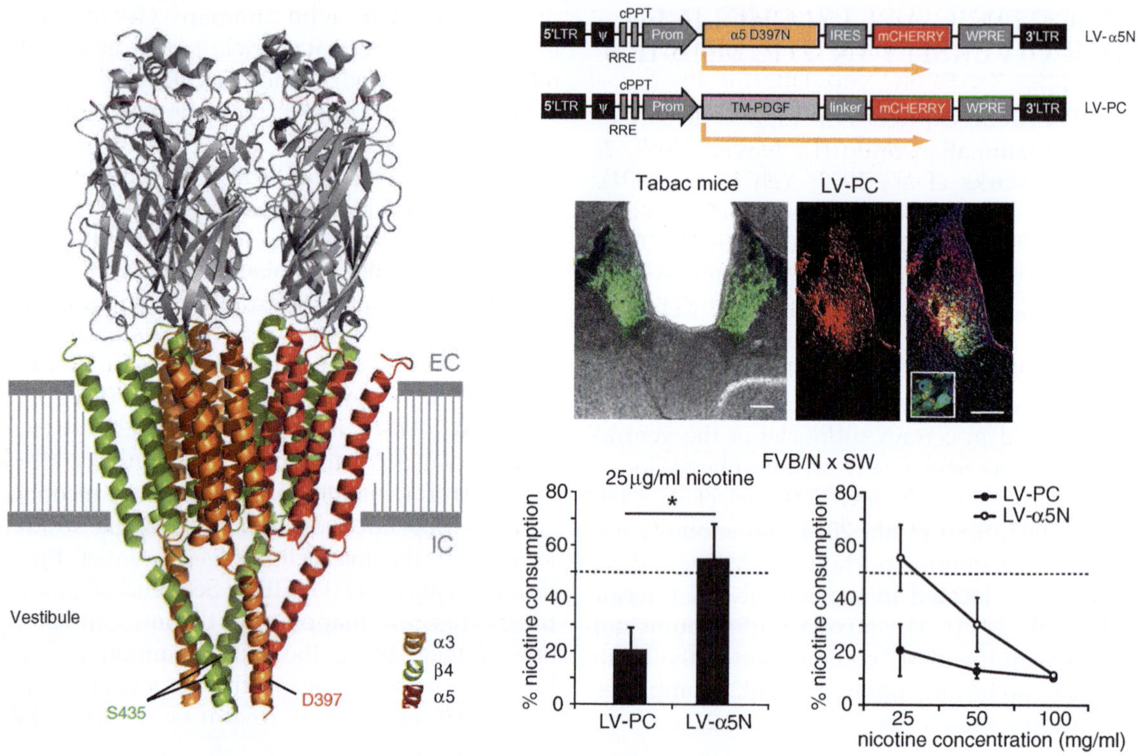

FIG. 10.3 Nicotine aversion in Tabac mice is reversed by the expression of the α5 D397N variant in the MHb. (*Left*) Model of the $\alpha3_2\beta4_2\alpha5_1$ nAChR showing the transmembrane and intracellular domains of α3 (*orange*), α5 (*red*), and β4 (*green*) subunits and the specific residues S435 in β4 and D397 in α5, which are the most frequent variant associated to heavy smoking. (*Right*) The lentiviral (LV) constructs used for injections in the MHb of Tabac mice. Colocalization of eGFP fluorescence of Tabac mice and mCherry fluorescence from either LV control (LV-PC) or LV expressing the a5D397N genetic variant (LV-α5N) in injected mice. Scale bars, 100μm. Tabac mice injected with LV-α5N no longer display nicotine aversion compared to Tabac mice injected with LV-PC. *Modified from Frahm et al. (2011), Neuron, with permission from the publishers.*

analysis of this SNP in the α5 subunit, which results in the amino acid substitution aspartic acid to asparagine (D398N), showed significantly reduced nicotine-evoked currents. Behaviorally, this nonsynonymous variant had profound effects: the nicotine aversion observed in Tabac mice overexpressing β4 was reversed upon viral-mediated expression of the α5 D398N variant in the MHb (Frahm et al., 2011) (Fig. 10.3).

The role of the α5 nAChR subunit in nicotine aversion was supported by the phenotype observed in animals lacking the α5 subunit. These mice continue to self-administer nicotine at doses that normally elicit aversion in wild-type animals, and when the α5 nAChR subunit is reexpressed in the MHb or in the IPN, nicotine self-administration returns to wild-type levels. Moreover, viral-mediated knockdown of the α5 nAChR subunit in the MHb-IPN tract did not alter the reward-enhancing properties of lower doses of nicotine, but significantly reduced the aversive effects of higher doses in rats (Fowler et al., 2011).

α3β4* or α3β4α5 nAChRs have a lower affinity for nicotine than α4β2 receptors and are likely less desensitized at nicotine levels found in smokers than α4β2 nAChRs since they retain their sensitivity to fluctuating nicotine levels in smokers (Rose et al., 2007). This, together with the role of α3, β4, and α5 subunits in nicotine aversion, resulted in a model in which nicotine reward and nicotine aversion are mediated by different circuits (Fowler & Kenny, 2014): Low doses of nicotine would activate the high-affinity α4α6β2β3* nAChR subtype (Grady et al., 2007) in the mesoaccumbens pathway leading to reward, whereas high doses of nicotine would activate the lower affinity α3β4* or α5* nAChRs in the MHb-IPN tract leading to aversion (Fowler & Kenny, 2014).

In addition to nicotine aversion, both β4 and α5 nAChRs located in the MHb-IPN pathway contribute to the physical symptoms of nicotine withdrawal. Null mice for either β4 or α5 have fewer somatic signs of withdrawal after chronic nicotine administration and attenuated withdrawal-induced hyperalgesia (Jackson, Martin, Changeux, & Damaj, 2008; Salas et al., 2009; Salas, Pieri, & De Biasi, 2004). Manifestations of nicotine withdrawal are also altered in mice null for α2, β2, and β6 subunits (Jackson et al., 2008; Lotfipour et al., 2013), which are also present in the MHb-IPN axis (Shih et al., 2014).

10.4 ELECTROPHYSIOLOGY OF HABENULAR NEURONS

The anatomical and electrophysiological properties of the habenula are quite unique. Within the MHb, there are two distinct populations: a cholinergic population in the ventral MHb that projects to the rostral (IPR), intermediate (IPI), and central (IPC) subnuclei of the IPN and a peptidergic (substance P positive) population in the dorsal MHb that projects to the lateral part of the IPN (IPL) (Fig. 10.4). Both types of projections release glutamate, but the cholinergic population also coreleases acetylcholine, as demonstrated by a recent landmark study using channelrhodopsin-2 transgenic mice to study synaptic transmission upon blue-light stimulation. Brief stimulation of MHb efferents elicits glutamatergic but not cholinergic responses, while tetanic stimulation is required to generate slow inward currents mediated by nAChRs, establishing that habenular neurons use two modes of transmission: wired transmission for glutamate and volume transmission for ACh (Ren et al., 2011).

Consistent with this corelease, the vesicular transporters for ACh (VAChT) and glutamate (VGLUT1/2) have been found in the same MHb terminals in the IPN by confocal microscopy (Fig. 10.4), immunoprecipitation analysis (Frahm et al., 2015; Ren et al., 2011), and electron microscopy (Frahm et al., 2015). Remarkably, within the same habenular terminal, these two transporters are also found together in the same synaptic vesicles (Frahm et al., 2015) (Fig. 10.5).

The corelease of two neurotransmitters is not uncommon in the nervous system (for reviews, see El Mestikawy, Wallen-Mackenzie, Fortin, Descarries, & Trudeau, 2011; Hnasko & Edwards, 2012) and allows for precise local control of excitatory-inhibitory balance at individual synapses (Shabel, Proulx, Piriz, & Malinow, 2014). Local elimination of ACh in MHb neurons in ChAT-cKO mice decreased the amplitude of glutamatergic miniature excitatory postsynaptic potentials (mEPSCs) in IPN neurons, because in the absence

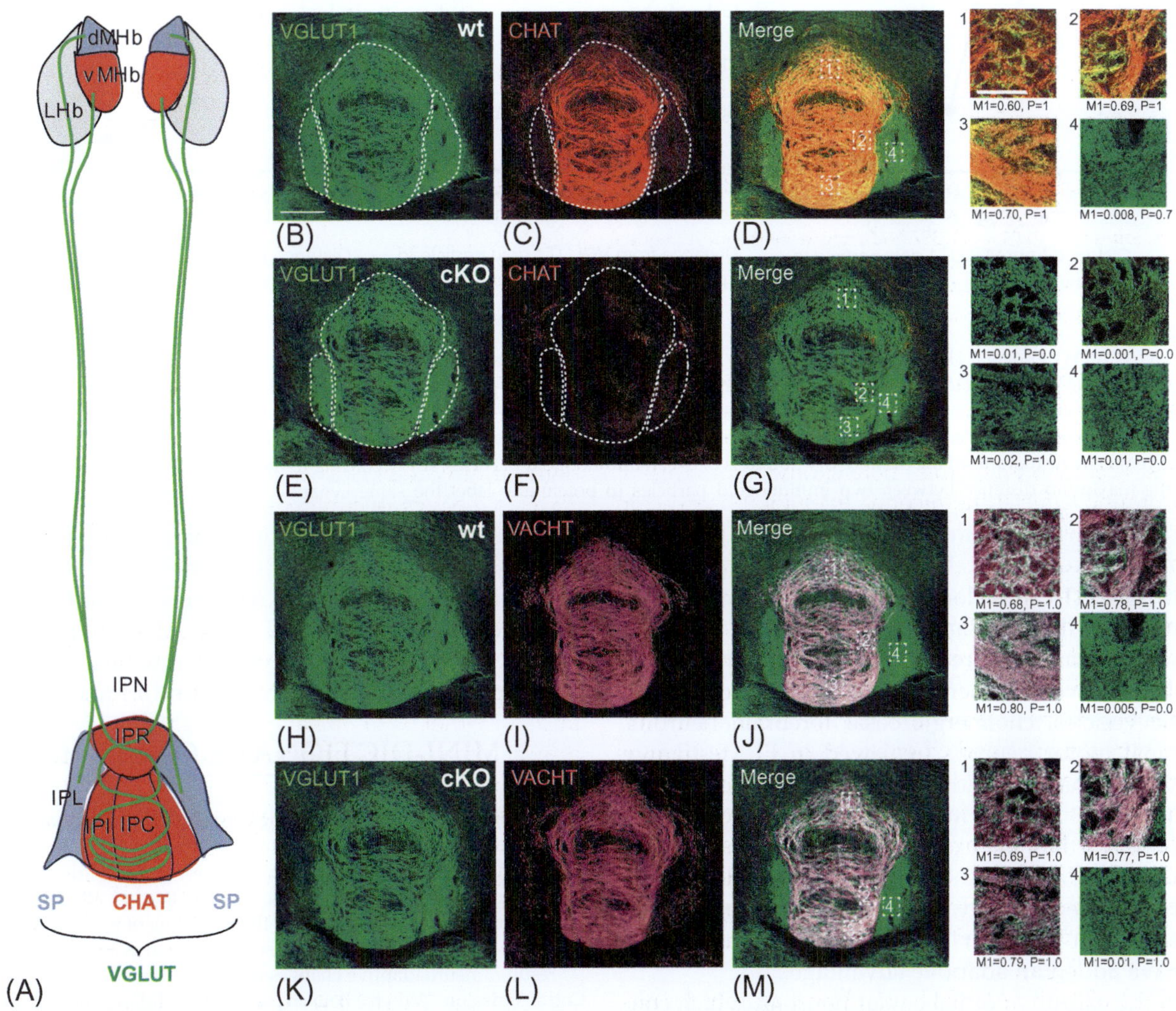

FIG. 10.4 Distribution of cholinergic and glutamatergic axonal terminals in IPN subnuclei. (A) Schematic representation of the segregated projections of MHb neurons to IPN subnuclei.
Peptidergic (substance P, SP) neurons in the dorsal MHb (*blue*) project to the lateral IPN (IPL, *blue*). Cholinergic (ChAT positive) neurons in the ventral MHb (vMHb, *red*) project to the rostral (IPR), intermediate (IPI), and central (IPC) subnuclei of the IPN (*red*). Both types of projections corelease glutamate (VGLUT, *green*). (B–M) Double immunostaining and Manders' colocalization coefficient analyses of glutamatergic and cholinergic markers in different subnuclei of the IPN. *Figure taken from Frahm et al. (2015),* eLife, *with permission from the publishers.*

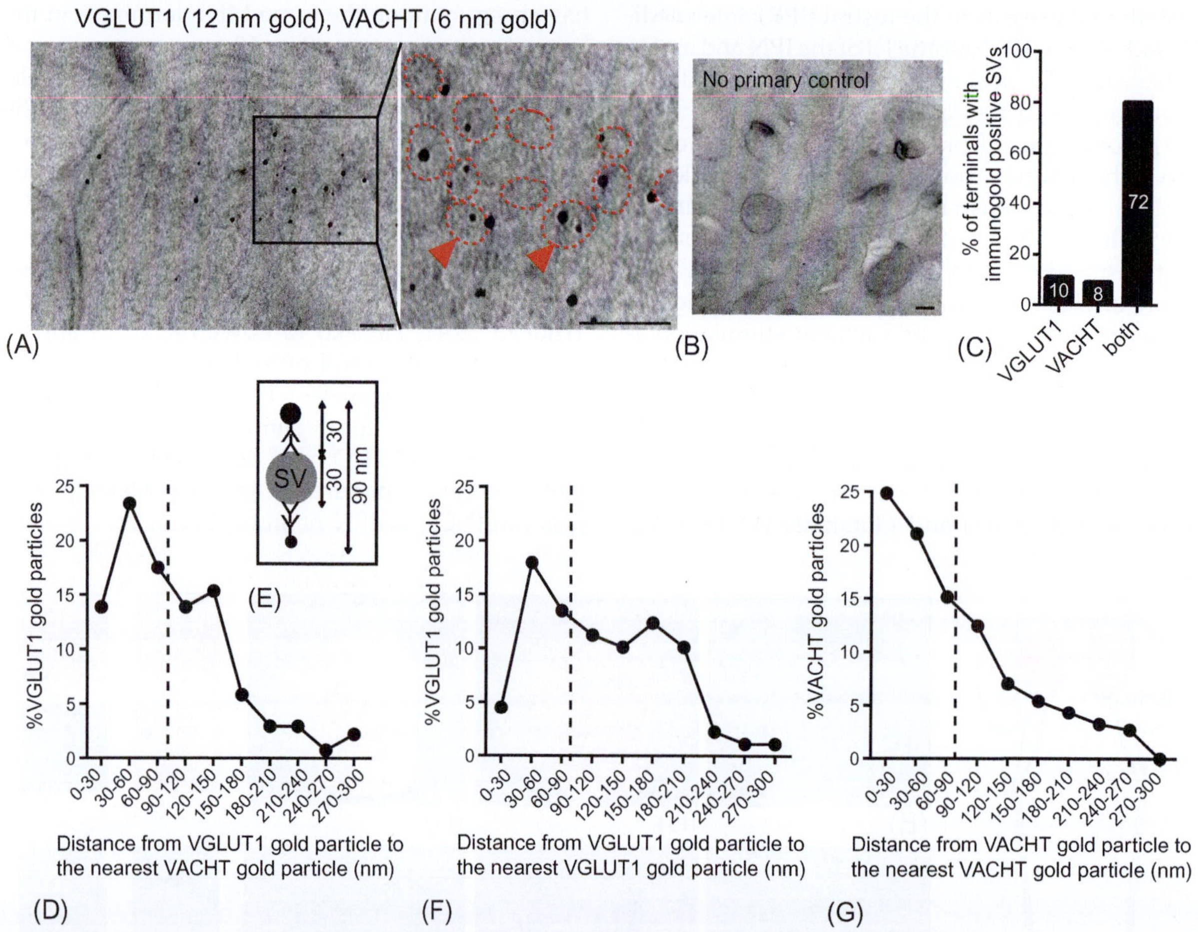

FIG. 10.5 VACHT and VGLUT1 colocalize in the same synaptic vesicles in axonal terminals in the central IPN. (A,B) Colocalization of the vesicular transporters for glutamate (VGLUT, 12nm gold particles) and ACh (VACHT, 6nm gold particles) in the same habenular synaptic vesicles (SV, outlined in *red*) by immunogold electron microscopy. Arrowheads indicate synaptic vesicles that contain both transporters. Scale bar, 100nm. (C) Percentage of axonal terminals that contain SV positive only for VGLUT1 or VACHT or both vesicular transporters. (D, F, G) Frequency histograms of the distance distributions between immunogold particles. (E) According to the size of an SV and conjugated antibodies, the maximum distance between two immunogold particles to potentially label the same synaptic vesicle is 90nm. *Figure taken from Frahm et al. (2015),* eLife, *with permission from the publishers.*

of ACh, VACHT can no longer promote glutamate coentry through VGLUT1/2 and consequently the vesicular content of glutamate is reduced. Vesicular synergy has been also observed in other synapses, for instance, striatal synaptic vesicles, cholinergic basal forebrain neurons, and spinal motor neurons (reviewed in El Mestikawy et al., 2011; Hnasko & Edwards, 2012). Cotransmission of ACh and glutamate might therefore be considered the rule rather than the exception raising a fundamental question about the role of cotransmitter release and neurotransmitter synergy at cholinergic synapses and suggesting that vesicular synergy of these transmitters must have added an adaptive advantage.

With the exception of habenular neurons, which contain a high concentration of postsynaptic nAChR in the soma (Frahm et al., 2011; Gorlich et al., 2013), most nAChRs present in the brain are presynaptic to facilitate glutamate release (Girod & Role, 2001; McGehee, Heath, Gelber, Devay, & Role, 1995). This presynaptic facilitation also occurs in the Hb-IPN since MHb neurons also express particularly high levels of nAChRs along their axonal projections to the IPN (Tables 10.1 and 10.2).

MINI-DICTIONARY OF TERMS

Addiction A pattern of drug-taking behavior in humans characterized by the lack of control over intake or compulsive use, escalating use, development of tolerance, and use despite adverse consequences. Addiction is often associated with guilt after consuming the substance in question rather than a feeling of reward. Addiction results from repeated use of a substance and is thought to be the result of neuroadaptive changes in brain circuitry.

Cotransmission When a neuron releases two different neurotransmitters upon stimulation. In most cases, cotransmission requires the coexistence of two distinct vesicular neurotransmitter transporters in the same terminal. If both transporters are in the same synaptic vesicle, they can cooperate to fill the vesicle with the two neurotransmitters, a process termed vesicular synergy.

GWAS Genome-wide association studies search for an association between genetic variants (single nucleotide polymorphisms) found in the human population to a specific trait or disease.

Habenula-interpeduncular circuit Conserved brain structure in the epithalamus of vertebrates distinguished by the high density of neurons in the habenula that send axonal projections via the fasciculus retroflexus to the interpeduncular nucleus. Habenula derives from the Latin name *habena* referring to "whip" or "rein" because of the conspicuous long projections of the fasciculus retroflexus. Interpeduncular nucleus refers to the fact that it is between the cerebral peduncles.

Presynaptic facilitation Facilitation of neurotransmitter release by receptors located at the presynapse. nAChRs in the brain are mostly presynaptic, and their activation by volumetric transmission of ACh enhances calcium influx in the presynapse that in turn potentiates neurotransmitter release.

Withdrawal The experience of physical and/or psychological symptoms upon abrupt discontinuation of a chronically used substance. Withdrawal is not synonymous with addiction and may occur in the absence of addiction. Withdrawal is a result of neuroadaptive changes that occur with chronic use that are maladaptive in the absence of the substance.

TABLE 10.1 Key Facts of Habenula-IPN circuit

- The habenula-IPN circuit is conserved among all vertebrate species.
- The primary function of the habenula is to link the forebrain with the hindbrain to regulate dopamine, serotonin, and adrenaline in the brain.
- The main output track of the habenula is called the fasciculus retroflexus.
- The habenula can be divided into lateral and medial nuclei. Lateral nuclei project to the VTA, raphe, and other midbrain nuclei, while the main target of the medial habenula is the interpeduncular nucleus.
- Fibers from the lateral nuclei are segregated to the outer portion of the fasciculus retroflexus, while fibers from the medial habenula form the core.
- The medial habenula consists of peptidergic, cholinergic, and glutamatergic neurons, while the lateral habenula consists of glutamatergic and GABAergic neurons.

This table lists the key facts of the habenula-IPN circuit including its function in general and its anatomy.

TABLE 10.2 Key Facts of the nAChRs in the MHb-IPN

- The MHb, together with the IPN, expresses the highest density of nAChRs in the mammalian brain.
- nAChRs are highly enriched in three compartments of the MHb-IPN tract: the soma of MHb neurons, presynaptic terminals of MHb neurons in the IPN, and postsynaptic IPN neurons.
- The predominant nAChR subunits expressed in the MHb and IPN are α3, β4, and α5.
- α5 and β4 nAChR subunits mediate the aversive response to nicotine intake and nicotine withdrawal.
- Presynaptic nAChRs facilitate glutamate release from MHb terminals.

This table lists the key facts of the nAChRs of the MHb-IPN tract including the main subunits expressed in MHb and IPN neurons, their localization, and their role in nicotine aversion.

Summary Points

- The habenula-interpeduncular circuit plays a role in the expression of aversion to high doses of nicotine and the negative experience of nicotine withdrawal.
- The habenula-interpeduncular circuit expresses a wide repertoire of nicotinic acetylcholine receptor subunits, including subunits that have been linked with nicotine addiction.
- Chronic nicotine exposure, withdrawal from nicotine, and reexposure to nicotine alter the rate of baseline firing of the habenula.
- Acetylcholine is coreleased with glutamate by habenular cholinergic neurons.
- Vesicular synergy between glutamate and acetylcholine is observed in cholinergic MHb terminals.
- ACh release by MHb terminals is necessary for presynaptic nAChR-mediated facilitation.

References

Aizawa, H., Amo, R., & Okamoto, H. (2011). Phylogeny and ontogeny of the habenular structure. *Frontiers in Neuroscience*, *5*, 138.

Antolin-Fontes, B., Ables, J. L., Gorlich, A., & Ibanez-Tallon, I. (2015). The habenulo-interpeduncular pathway in nicotine aversion and withdrawal. *Neuropharmacology*, *96*(Pt B), 213–222.

Bierut, L. J. (2010). Convergence of genetic findings for nicotine dependence and smoking related diseases with chromosome 15q24-25. *Trends in Pharmacological Sciences*, *31*(1), 46–51.

Bierut, L. J., Stitzel, J. A., Wang, J. C., Hinrichs, A. L., Grucza, R. A., Xuei, X., et al. (2008). Variants in nicotinic receptors and risk for nicotine dependence. *The American Journal of Psychiatry*, *165*(9), 1163–1171.

Boulos, L. J., Darcq, E., & Kieffer, B. L. (2017). Translating the Habenula-from rodents to humans. *Biological Psychiatry*, *81*(4), 296–305.

Ciani, E., Severi, S., Bartesaghi, R., & Contestabile, A. (2005). Neurochemical correlates of nicotine neurotoxicity on rat habenulo-interpeduncular cholinergic neurons. *Neurotoxicology*, *26*(3), 467–474.

El Mestikawy, S., Wallen-Mackenzie, A., Fortin, G. M., Descarries, L., & Trudeau, L. E. (2011). From glutamate co-release to vesicular synergy: vesicular glutamate transporters. *Nature Reviews Neuroscience*, *12*(4), 204–216.

Fonck, C., Nashmi, R., Salas, R., Zhou, C., Huang, Q., De Biasi, M., et al. (2009). Demonstration of functional alpha4-containing nicotinic receptors in the medial habenula. *Neuropharmacology*, *56*(1), 247–253.

Fowler, C. D., & Kenny, P. J. (2014). Nicotine aversion: neurobiological mechanisms and relevance to tobacco dependence vulnerability. *Neuropharmacology*, *76*(Pt B), 533–544.

Fowler, C. D., Lu, Q., Johnson, P. M., Marks, M. J., & Kenny, P. J. (2011). Habenular alpha5 nicotinic receptor subunit signalling controls nicotine intake. *Nature*, *471*(7340), 597–601.

Frahm, S., Antolin-Fontes, B., Gorlich, A., Zander, J. F., Ahnert-Hilger, G., & Ibanez-Tallon, I. (2015). An essential role of acetylcholine-glutamate synergy at habenular synapses in nicotine dependence. *eLife*, *4*, e11396.

Frahm, S., Slimak, M. A., Ferrarese, L., Santos-Torres, J., Antolin-Fontes, B., Auer, S., et al. (2011). Aversion to nicotine is regulated by the balanced activity of beta4 and alpha5 nicotinic receptor subunits in the medial habenula. *Neuron*, *70*(3), 522–535.

Girod, R., & Role, L. W. (2001). Long-lasting enhancement of glutamatergic synaptic transmission by acetylcholine contrasts with response adaptation after exposure to low-level nicotine. *The Journal of Neuroscience, 21*(14), 5182–5190.

Gorlich, A., Antolin-Fontes, B., Ables, J. L., Frahm, S., Slimak, M. A., Dougherty, J. D., et al. (2013). Reexposure to nicotine during withdrawal increases the pacemaking activity of cholinergic habenular neurons. *Proceedings of the National Academy of Sciences of the United States of America, 110*(42), 17077–17082.

Grady, S. R., Salminen, O., Laverty, D. C., Whiteaker, P., McIntosh, J. M., Collins, A. C., et al. (2007). The subtypes of nicotinic acetylcholine receptors on dopaminergic terminals of mouse striatum. *Biochemical Pharmacology, 74*(8), 1235–1246.

Hnasko, T. S., & Edwards, R. H. (2012). Neurotransmitter corelease: mechanism and physiological role. *Annual Review of Physiology, 74*, 225–243.

Hung, R. J., McKay, J. D., Gaborieau, V., Boffetta, P., Hashibe, M., Zaridze, D., et al. (2008). A susceptibility locus for lung cancer maps to nicotinic acetylcholine receptor subunit genes on 15q25. *Nature, 452*(7187), 633–637.

Jackson, K. J., Martin, B. R., Changeux, J. P., & Damaj, M. I. (2008). Differential role of nicotinic acetylcholine receptor subunits in physical and affective nicotine withdrawal signs. *The Journal of Pharmacology and Experimental Therapeutics, 325*(1), 302–312.

Le Novere, N., Zoli, M., & Changeux, J. P. (1996). Neuronal nicotinic receptor alpha 6 subunit mRNA is selectively concentrated in catecholaminergic nuclei of the rat brain. *The European Journal of Neuroscience, 8*(11), 2428–2439.

Liu, J. Z., Tozzi, F., Waterworth, D. M., Pillai, S. G., Muglia, P., Middleton, L., et al. (2010). Meta-analysis and imputation refines the association of 15q25 with smoking quantity. *Nature Genetics, 42*(5), 436–440.

Lotfipour, S., Byun, J. S., Leach, P., Fowler, C. D., Murphy, N. P., Kenny, P. J., et al. (2013). Targeted deletion of the mouse alpha2 nicotinic acetylcholine receptor subunit gene (Chrna2) potentiates nicotine-modulated behaviors. *The Journal of Neuroscience, 33*(18), 7728–7741.

Marks, M. J., Pauly, J. R., Gross, S. D., Deneris, E. S., Hermans-Borgmeyer, I., Heinemann, S. F., et al. (1992). Nicotine binding and nicotinic receptor subunit RNA after chronic nicotine treatment. *The Journal of Neuroscience, 12*(7), 2765–2784.

McGehee, D. S., Heath, M. J., Gelber, S., Devay, P., & Role, L. W. (1995). Nicotine enhancement of fast excitatory synaptic transmission in CNS by presynaptic receptors. *Science, 269*(5231), 1692–1696.

Nashmi, R., Xiao, C., Deshpande, P., McKinney, S., Grady, S. R., Whiteaker, P., et al. (2007). Chronic nicotine cell specifically upregulates functional alpha 4* nicotinic receptors: basis for both tolerance in midbrain and enhanced long-term potentiation in perforant path. *The Journal of Neuroscience, 27*(31), 8202–8218.

Quick, M. W., Ceballos, R. M., Kasten, M., McIntosh, J. M., & Lester, R. A. (1999). Alpha3beta4 subunit-containing nicotinic receptors dominate function in rat medial habenula neurons. *Neuropharmacology, 38*(6), 769–783.

Ren, J., Qin, C., Hu, F., Tan, J., Qiu, L., Zhao, S., et al. (2011). Habenula "cholinergic" neurons co-release glutamate and acetylcholine and activate postsynaptic neurons via distinct transmission modes. *Neuron, 69*(3), 445–452.

Rose, J. E., Behm, F. M., Salley, A. N., Bates, J. E., Coleman, R. E., Hawk, T. C., et al. (2007). Regional brain activity correlates of nicotine dependence. *Neuropsychopharmacology, 32*(12), 2441–2452.

Salas, R., Pieri, F., & De Biasi, M. (2004). Decreased signs of nicotine withdrawal in mice null for the beta4 nicotinic acetylcholine receptor subunit. *The Journal of Neuroscience, 24*(45), 10035–10039.

Salas, R., Pieri, F., Fung, B., Dani, J. A., & De Biasi, M. (2003). Altered anxiety-related responses in mutant mice lacking the beta4 subunit of the nicotinic receptor. *The Journal of Neuroscience, 23*(15), 6255–6263.

Salas, R., Sturm, R., Boulter, J., & De Biasi, M. (2009). Nicotinic receptors in the habenulo-interpeduncular system are necessary for nicotine withdrawal in mice. *The Journal of Neuroscience, 29*(10), 3014–3018.

Shabel, S. J., Proulx, C. D., Piriz, J., & Malinow, R. (2014). Mood regulation. GABA/glutamate co-release controls habenula output and is modified by antidepressant treatment. *Science, 345*(6203), 1494–1498.

Sheffield, E. B., Quick, M. W., & Lester, R. A. (2000). Nicotinic acetylcholine receptor subunit mRNA expression and channel function in medial habenula neurons. *Neuropharmacology, 39*(13), 2591–2603.

Shih, P. Y., Engle, S. E., Oh, G., Deshpande, P., Puskar, N. L., Lester, H. A., et al. (2014). Differential expression and function of nicotinic acetylcholine receptors in subdivisions of medial habenula. *The Journal of Neuroscience, 34*(29), 9789–9802.

Slimak, M. A., Ables, J. L., Frahm, S., Antolin-Fontes, B., Santos-Torres, J., Moretti, M., et al. (2014). Habenular expression of rare missense variants of the beta4 nicotinic receptor subunit alters nicotine consumption. *Frontiers in Human Neuroscience, 8*, 12.

Thorgeirsson, T. E., Geller, F., Sulem, P., Rafnar, T., Wiste, A., Magnusson, K. P., et al. (2008). A variant associated with nicotine dependence, lung cancer and peripheral arterial disease. *Nature, 452*(7187), 638–642.

Yeh, J. J., Yasuda, R. P., Davila-Garcia, M. I., Xiao, Y., Ebert, S., Gupta, T., et al. (2001). Neuronal nicotinic acetylcholine receptor alpha3 subunit protein in rat brain and sympathetic ganglion measured using a subunit-specific antibody: regional and ontogenic expression. *Journal of Neurochemistry, 77*(1), 336–346.

Zhao-Shea, R., Liu, L., Pang, X., Gardner, P. D., & Tapper, A. R. (2013). Activation of GABAergic neurons in the interpeduncular nucleus triggers physical nicotine withdrawal symptoms. *Current Biology, 23*(23), 2327–2335.

CHAPTER

11

Nicotine Neuroprotection of Brain Neurons: The Other Side of Nicotine Addiction

Dzejla Bajrektarevic, Silvia Corsini†, Andrea Nistri*, Maria Tortora‡,a*

*Department of Neuroscience, SISSA, Trieste, Italy

†Neuroscience Paris Seine—Institute of Biology Paris Seine, CNRS, UMR 8246—Inserm U1130, Université Pierre et Marie Curie (UPMC), Sorbonne Universités, Paris, France

‡MRC London Institute of Medical Science, Imperial College of London, London, United Kingdom

Abbreviations

ACh	acetylcholine
AD	Alzheimer's disease
AIF	apoptotic inducing factor
Akt	protein kinase B
ALS	amyotrophic lateral sclerosis
BDNF	brain-derived neurotrophic factor
CaKMII	Ca^{2+}/calmodulin-dependent protein kinase II
CREB	cyclic AMP-responsive element-binding protein
cyt-c	cytochrome *c*
ERK1/2	extracellular regulated protein kinase 1/2
FGF-2	fibroblast growth factor-2
HMs	hypoglossal motoneurons
Hsp70	heat-shock protein 70
MAPK	mitogen-activated protein kinase
MPTP	mitochondrial permeability transition pore
nAChRs	nicotinic acetylcholine receptors
NFTs	neurofibrillary tangles
NGF	nerve growth factor
PI3K	phosphoinositide 3-kinase
PLC	phospholipase C
ROS	reactive oxygen species
SP	senile plaques
TBOA	DL-threo-β-benzyloxyaspartate
TrkA	tyrosine kinase A

11.1 INTRODUCTION

Nicotinic acetylcholine receptors (nAChRs; mainly α4β2 and α7 subtypes; Chapters 2 and 4) are important targets for pharmaceutical research and development related to neurodegenerative diseases in which the loss of high-affinity nicotine binding sites is significant (Newhouse, Potter, & Singh, 2004; Picciotto & Zoli, 2008). Currently, acetylcholinesterase inhibitors are the only cholinergic drugs for clinical use in cognitive decline (Pepeu & Giovannini, 2009) because they are expected to raise the concentration of acetylcholine (ACh) and its action on nAChRs (Anand, Gill, & Mahdi, 2014; Geldenhuys & Darvesh, 2015). As indicated in Chapter 1, the field of tobacco and nicotine studies has highlighted their link to major neurological disorders. Hence, these data have provided the impetus for a top-down approach to understand whether nicotine can change the disease course through specific molecular or cellular mechanisms best explored with preclinical models.

11.2 EPIDEMIOLOGICAL STUDIES OF NEURODEGENERATIVE DISEASES: NICOTINE AND SMOKING

In the 16th century, Jean Nicot brought to France the tobacco plant from which nicotine takes its name and proposed its medical use. Despite the well-established action of nicotine as a potent cholinergic agonist to activate and desensitize nAChRs (Giniatullin, Nistri, & Yakel, 2005), sustained administration of nicotine is expected to induce more complex effects, including large upregulation of nAChR expression (Parker et al., 2004). Since smoking is a very effective way to administer nicotine that reaches

[a] The authors' names are in alphabetic order because they all provided equal contribution to this chapter.

Neuroscience of Nicotine
https://doi.org/10.1016/B978-0-12-813035-3.00011-3

the brain twice as fast as via the intravenous route (Tyas, 1996), early epidemiological studies showed a negative correlation between smoking and the incidence of PD as detailed in Chapter 2. The issue of nicotine effects on Alzheimer's disease (AD) remains a complex subject not fully resolved. As illustrated in Table 11.1, meta-analyses examining risk factors for AD have reported an inverse association between smoking and AD. Despite early studies suggesting that the density of nicotinic binding sites may be higher in smokers, thus protecting them against AD, many investigations have methodological limitations such as case heterogeneity, small sample size, and controls not assessed to be AD-free (Tyas, 1996). Subsequent prospective studies have not shown lower incidence of AD in smokers or even increased AD risk (Table 11.1). Nonetheless, the validity of these prospective studies is also limited by the length of follow-up, indirect diagnosis, the variable interval between stopping smoking and disease onset, and the amount of smoking. Notwithstanding these unresolved issues, work on clinical neuroprotection by nicotine (and related agonists) has been carried out with the goal of enhancing the activity of nAChRs in AD and neurodegeneration in general.

TABLE 11.1 Proposed Role of Neuronal nAChRs in Alzheimer's Disease (AD)

Clinical results	Study protocols
Lower incidence of AD in smokers	Meta-analysis of case-control studies[a], review of epidemiological evidence[b]
Higher incidence of AD in smokers	Systematic review and meta-analysis[c], prospective population-based cohort study[d], observational study[e]
Clinical studies of nicotine administration	Acute subcutaneous nicotine[f,g], intravenous nicotine[h], nicotine patches[i–k]
Clinical studies of nicotine agonist administration	Agonist for the α4β2 receptor[l], agonist for the α7 receptor[m]
Smoking influence on AD neuropathology	Autopsy findings[n]

[a] *Graves et al., Int J Epidemiol, 1991; 2, S48–S57.*
[b] *Lee, Neuroepidemiology, 1994; 4, 131–144.*
[c] *Peters et al., BMC Geriatr., 2008; 23, 8–36.*
[d] *Reitz et al., Neurology, 2007; 10, 998–1005.*
[e] *Almeida et al., Am J Geriatr Psychiatry, 2008; 16, 92–98.*
[f] *Jones et al., Psychopharmacology, 1992; 108, 485–494.*
[g] *Sahakian et al., Br J Psychiatry, 1989; 154, 797–800.*
[h] *Newhouse et al., Psychopharmacology, 1988; 95, 171–217.*
[i] *Newhouse et al., Neurology, 2012; 10(78), 91–101.*
[j] *White and Levin, Psychopharmacology, 1999; 143, 158–165.*
[k] *Wilson et al., Pharmacol Biochem Behav, 1995; 51, 509–514.*
[l] *Dunbar et al., Psychopharmacology, 2007; 191, 919–929.*
[m] *Haydar et al., Bioorg Med Chem, 2009; 17, 5247–5258.*
[n] *Urlich et al., Acta Neuropathol, 1997; 94, 450–454.*

11.3 CLINICAL STUDIES OF NEURODEGENERATIVE DISEASES: THE INFLUENCE OF TOBACCO VS NICOTINE

Although several reports have demonstrated the activation of human nAChRs by nicotine itself (Table 11.1), pathology studies of nicotine effect on AD have not provided unequivocal conclusions. Thus, chronic smoking directly affects neuronal function with negative implications for long-term neuronal survival because of decreased gray matter density in several brain regions (Table 11.1), whereas another *postmortem* study has shown that smoking can decrease the number of senile plaques (SP) and neurofibrillary tangles (NFTs) in women only (Table 11.1). Conversely, Swan and Lessov-Schlaggar (2007) have suggested smoking to be associated with accelerated cognitive decline in the areas of executive functions and verbal memory and increased risk of dementia, despite the observation that nicotine may improve vigilance, memory, and attention. Because of the presence of toxic compounds in tobacco, understanding the action of nicotine on AD or other neurodegenerative diseases should be based on data originating from the administration of nicotine itself.

Prospective studies of subcutaneous or transdermal administration of nicotine and nicotine agonists (Table 11.1) as treatment for AD seem to support potential cognitive benefits. Other small-scale trials have indicated that subcutaneous nicotine (Table 11.1) improves visual attention, information processing, perception, and vigor. In agreement with this notion, chronic transdermal nicotine (Table 11.1) is reported to significantly improve attention with minimal side effects. Transdermal nicotine (Table 11.1) can be safely administered to non-smoking subjects with mild cognitive impairment over 6 months with improved attention, memory, and mental processing.

Large trials of nAChR agonists have also been performed (Table 11.1). For example, ispronicline (AZD3480; formerly TC-1734) has high affinity and selectivity for the α4β2 receptor, while SEN12333/WAY-317538 has a high affinity for the α7 receptor (Table 11.1). Because AZD3480 and SEN12333/WAY-317538 apparently improve cognitive performance in humans, they might be candidates for AD treatment and other cognitive disorders (Table 11.1).

Notwithstanding the current data on the role of nicotine and nAChR agonists in AD, it is difficult to treat chronic diseases like AD when substantial neuronal damage starts long before clinical manifestations. Because the primary cause for SP and NFTs remains unclear, damage limitation rather than arrest of disease progression is usually the target and is time-limited. In principle, it would be more useful to deal with the early stage of the disease when the chance for a disease-modifying approach may be more realistic.

A different scenario may be offered in the case of another neurodegenerative disease, namely, amyotrophic lateral sclerosis (ALS; Lou Gehrig's disease) with progressive motoneuron degeneration and muscle weakness (Rothstein, Martin, & Kuncl, 1992). Large-scale studies have indicated a positive association between smoking and ALS (Alonso, Logroscino, & Hernán, 2010; de Jong et al., 2012), possibly due to accompanying respiratory and cardiovascular deficits that tobacco is known to induce. Nevertheless, because in a large number of patients, the earliest pathophysiology of ALS may be due to impaired transport of the excitatory transmitter glutamate (Rothstein et al., 1992), it is possible to examine the effect of nicotine on the primary cellular processes that are prodromal to neurodegeneration. To this end and on the assumption that the modulation of nAChRs is important to stave off the pathological deterioration, it is helpful to understand the cellular effects of nicotine on simple neurodegenerative models in vitro and in vivo with translation value for clinical application.

11.4 CELLULAR EFFECTS OF NICOTINE ON IN VITRO MODELS

nAChRs are expressed at synapses (Chapter 2) and on presynaptic boutons where they regulate the secretion of other neurotransmitters (Alkondon & Albuquerque, 2004) and modulate spike discharge (Mulle, Vidal, Benoit, & Changeux, 1991). The action of nicotine, therefore, varies among brain areas because it depends on the local nAChR expression. Nicotine, at nanomolar concentrations similar to those in the plasma after cigarette smoking, enhances GABAergic, glutamatergic, and cholinergic transmission by acting on presynaptic nAChRs to increase presynaptic Ca^{2+} influx as exemplified in the rat hippocampus (Gray, Rajan, Radcliff, Yakehiro, & Dani, 1996), mouse amygdala (Barazangi & Role, 2001), and chick habenula (McGehee, Heath, Gelber, Devay, & Role, 1995). In chick brain slices, nicotine modulates GABAergic synaptic transmission in a dose-dependent manner as low concentrations (50–100nM) potentiate electrically evoked GABAergic currents, whereas higher doses (0.5–1.0μM) have either no effect or diminish evoked GABA transmission (Zhu & Chiappinelli, 1999).

In vitro studies support the notion of a neuroprotective role of nicotine administration whose effects on nAChRs are time- and dose-dependent (Table 11.2). Regarding nicotine neuroprotection against glutamate excitotoxicity, Dajas-Bailador, Lima, and Wonnacott (2000) have reported that the activation of α7 receptors protects

TABLE 11.2 Nicotine Effects on in In Vitro Preparations

Animal model or cell type	Preparation	Neuroprotective effects of nicotine
Rat	Primary culture of cortical neurons	Improved cell viability[a–c]
Rat	Hypoglossal nucleus slice	Reduction of ROS and mitochondrial energy dysfunction via nAChR activation[d,e]
Rat	Hypoglossal nucleus slice	Reduction of Connexin 36 activity, increased Hsp70 expression[f]
Mouse	Isolated mitochondria from the mouse liver	Regulation of MPTP-related signaling through PI_3K/Akt activation[g]
Rat and α7 nAChR knockout mice	Hippocampal slices	Reduction of LDH release taken as parameter of neuronal damage in an in vitro model of ischemia. Failure to contrast LDH release from α7 nAChR knockout mouse cells[h]
Rat	Brain mitochondria	Reduction of ROS not mediated by cholinergic receptors[i]
Mouse	Ventral midbrain cultures	Reduction of the downstream effector of UPR signaling pathway (CHOP) and its activator (eIF2α). Impairment of nuclear migration of two stress factors, XBP1 and ATF6[j]
Mouse	Organotypic hippocampal cultures	In the presence of a neuroinflammatory stimulus (LPS, 1ng/ml) and mitochondrial dysfunction (using the mitochondrial complex III blocker antimycin A), nicotine reduces toxicity by the activation of PI3K/Akt signaling pathway[k]
Cell line	PC12 cells	Prevention of cell death induced by nerve growth factor deprivation[l]
Rat	Primary neuronal cultures	Reduction of toxicity by coincubating the cultures with nicotine (0.5μM, 7days), glutamate for 24h (20μM), and Aβ1–40 (1nM) and Aβ1–42 (100pM). Nicotine protects through the activation of PI3K, the

Continued

TABLE 11.2 Nicotine Effects on in In Vitro Preparations—cont'd

Animal model or cell type	Preparation	Neuroprotective effects of nicotine
		consequent Akt phosphorylation, and the increase of the prosurvival protein Bcl-2[m]
Rat	Mesencephalic primary cultures	Coadministration of nicotine (0.1–100 μM) together with rotenone (100 nM) for 24 h contrasts rotenone toxicity through the activation of the PI3K-Akt pathway[n]
Cell line	Human SH-SY5Y neuroblastoma cells	Pretreatment with nicotine (0.01–100 μM) for 12 h followed by $A\beta_{25-35}$ exposure attenuates the activation of caspase-3 and increased Bcl-2, Bcl-xL, and Mcl-1 expression in a time-dependent manner[o]
Mouse	Primary cortical microglial cultures, primary hippocampal and cortical neuronal cultures	Maintenance by nicotine of microglia in an inactive state. Nicotine preserves neurons in the presence of LPS-treated microglia[p]
Mouse	Primary astrocyte cultures	Reduction of astrocyte activation and suppression of the increase in phosphorylated Erk and p38 in the presence of mitochondrial or inflammatory stressors[q]

[a] *Akaike et al., Brain Res, 1994; 644, 181–187.*
[b] *Semba et al., Brain Res, 1996; 735, 335–338.*
[c] *Kaneko et al., Brain Res, 1997; 765, 135–140.*
[d] *Corsini et al., J Physiol, 2016; 594, 6777–6798.*
[e] *Tortora et al., Neurosci Lett, 2017; 639, 43–48.*
[f] *Corsini et al., Cell Death Dis, 2017; 8, e2881.*
[g] *Gergalova et al., Int J Biochem Cell Biol, 2014; 49, 26–31*
[h] *Egea et al., Neuroscience, 2007; 145, 866–872.*
[i] *Cormier et al., Brain Res, 2001; 900, 72–79.*
[j] *Srinivasan et al., J Neurosci, 2016; 36, 65–79.*
[k] *Navarro et al., Biochem Pharmacol, 2015; 97, 473–481.*
[l] *Yamashita et al., Neurosci Lett, 1996; 213, 145–147.*
[m] *Kihara et al., J Biol Chem, 2001; 276, 13541–13546.*
[n] *Takeuchi et al., J Neurosci Res, 2009; 87, 576–585.*
[o] *Xue et al., Int J Mol Med, 2014; 33, 925–933.*
[p] *Noda and Kobayashi, J Physiol Sci, 2017; 67, 235–245.*
[q] *Liu et al., J Neuroinflam, 2017; 141, 473–474.*

hippocampal neurons against NMDA-induced cell death. In line with this finding, nicotine strongly neuroprotects brain stem hypoglossal motoneurons (HMs) preventing their loss after exposure to the glutamate transport blocker DL-threo-β-benzyloxyaspartate (TBOA; Table 11.2). By using TBOA to mimic the early pathophysiological phase of ALS (Urushitani et al., 1998), gradual buildup of extracellular glutamate develops with subsequent triggering of network bursting and delayed death (Sharifullina & Nistri, 2006). Nicotine (1–10 μM) inhibits TBOA-evoked bursting activity in HMs, decreases excitatory neurotransmission, and enhances synaptic inhibition (Table 11.2). Likely, nicotine neuroprotection is due to increased Ca^{2+} influx through nAChRs (Fig. 11.1), especially α7 subtypes, which are rather permeable to this cation (Shen & Yakel, 2009). The suggestion that α7 receptors play a dominant role in mediating neuroprotection is confirmed by failed neuroprotection in α7 nAChR knockout mice (Table 11.2). In accordance with this observation, the selective α7 agonist PNU282987 protects against toxicity by restoring the mitochondrial membrane potential, reducing superoxide anion levels, and decreasing the production of reactive oxygen species (ROS) in organotypic hippocampal cultures (Table 11.2).

11.5 NEUROPROTECTIVE EFFECT OF NICOTINE ON MOUSE, RAT, OR MONKEY MODELS IN VIVO

The molecular and cellular mechanisms activated by nicotine to elicit neuroprotection have also been validated with in vivo experiments. Beneficial effects of nicotine (injected subcutaneously or intravenously) have been reported in a range of animal models in vivo as indicated in Table 11.3.

Pharmacological strategies (Table 11.3; Bitner et al., 2010; Li, Arias, Jonnala, Mruthinti, & Buccafusco, 2005) and experiments with knockout mice (Table 11.3) have confirmed that α7 and α4β2 receptors mediate neuroprotection, whose relative contribution depends on the animal model, age, and brain area. For example, the neuroprotective action of the α7-containing receptors has been observed in the hippocampus and entorhinal cortex, but not in the cerebral cortex or hypothalamus of a rat AD model (Li et al., 2005). Selective α7 nAChR agonists can also enhance cognitive performance in AD models in accordance with the improved performance detected in a variety of memory testing protocols (Bitner et al., 2010).

Using pharmacological (α4β2 or α7 nAChR agonists) and molecular ($\beta2^{-/-}$ or $\alpha7^{-/-}$ knockout mice) tools, it has been observed that α4β2 nAChRs are involved in the neuroprotection of neonatal murine excitotoxic brain injury, while α7 receptors, which are neuroprotective in adult animals, are not at this early age (Table 11.3). The differential role of α7 nAChRs in neonatal and adult models may be due to the transient overexpression of this highly Ca^{2+}-permeable receptor type within the cortex during the postnatal period (Table 11.3).

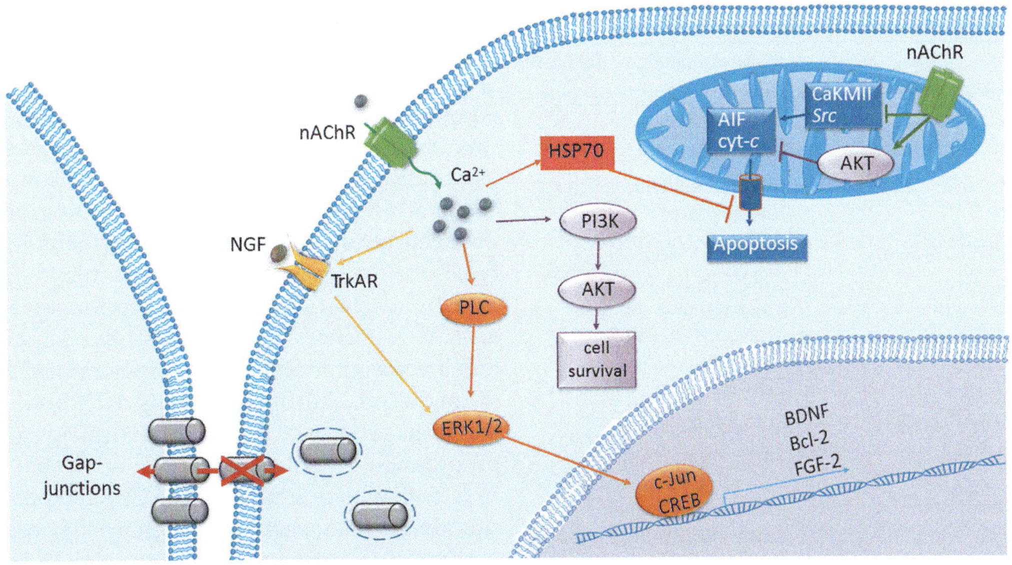

FIG. 11.1 Idealized diagram to account for neuroprotective effects of nicotine on in vitro and in vivo models. Nicotine induces the opening of nicotinic acetylcholine receptors (nAChRs) located on the cell and mitochondrial membranes. The resulting Ca^{2+} influx activates a complex set of pathways promoting cell survival and blocking apoptosis. Mitochondrial nAChRs activate the phosphoinositide 3-kinase/protein kinase B (PI3K/Akt) pathway, preventing apoptotic inducing factor (AIF) and cytochrome *c* (cyt-*c*) release, and inhibit Ca^{2+}/calmodulin-dependent protein kinase II (CaKMII) and tyrosine-protein kinase Src-related pathways. Lethal AIF release/translocation is also limited by increased level of heat-shock protein 70 (Hsp70). The extracellular regulated protein kinase 1/2 (ERK1/2) pathway is activated directly, through phospholipase C (PLC), or indirectly, via increased expression of tyrosine kinase A (TrkA) receptors bound by extracellular nerve growth factor (NGF). In the cell nucleus, ERK1/2 activates transcription factors such as c-Jun and cyclic AMP-responsive element-binding protein (CREB), to upregulate genes related to cell survival, like brain-derived neurotrophic factor (BDNF), B-cell lymphoma 2 (Bcl-2), and fibroblast growth factor-2 (FGF-2). Furthermore, nicotine blocks gap-junction communication and favors connexin 36 redistribution from the membrane to the cytosol to uncouple neurons during collective excitation. S. Corsini, unpublished.

11.6 NICOTINE AFFECTS INTRACELLULAR SIGNALING

The downstream effectors following receptor activation by nicotine remain incompletely understood and might involve survival and apoptotic signaling cascades (Buckingham, Jones, Brown, & Sattelle, 2009) as schematized in Fig. 11.1. In brain stem motoneurons, nicotine leads to the redistribution of connexin 36 gap-junction proteins from the membrane to the cytosol, thus decoupling HMs and preventing the spreading of death stimuli from injured to uninjured motoneurons (Table 11.2). Furthermore, during excitotoxic stress, nicotine enhances the expression of the protective factor heat-shock protein 70 (Hsp70) to prevent nuclear migration of the apoptotic inducing factor (AIF) causative of cell death via DNA damage. In the presence of β-amyloid-induced neurotoxicity, nicotine via α7 nAChR activation triggers the phosphoinositide 3-kinase (PI3K) pathway and decreases AIF release/translocation (Yu, Mechawar, Krantic, & Quirion, 2011). The PI3K-protein kinase B (Akt) and the mitogen-activated protein kinase (MAPK) pathways play important roles in promoting cell survival by inhibiting crucial steps of apoptosis signaling (Datta, Brunet, & Greenberger, 1999; Erhardt, Schremser, & Cooper, 1999). Nicotine (via α7 receptors) significantly reduces toxicity elicited by glutamate and β-amyloid in primary neuronal cultures through the modulation of PI3K activity (Table 11.2). Importantly, nicotine also induces a significant enhancement of antiapoptotic proteins such as extracellular regulated protein kinase 1/2 (ERK1/2), p38, and c-Jun MAPK (Table 11.2). In a model of oxidative stress caused by the mitochondrial complex I inhibitor rotenone, α7 and α4β2 nAChR signaling (Table 11.2) through the activation of PI3K-Akt pathway contrasts the aggregation of unfolded proteins and activation of apoptotic pathways.

The neuroprotective MAPK/ERK pathways (Hetman & Gozdz, 2004) activated by nicotine play putative roles in in vitro and in vivo models (Tables 11.2 and 11.3; Bitner et al., 2010; Brunzell, Russell, & Picciotto, 2003; Jonnala, Terry Jr., & Buccafusco, 2002; Li et al., 2005; Toborek et al., 2007) via phospholipase C (PLC) activation (Mai, May, Gao, Jin, & Deng, 2003). Chronic nicotine administration could modulate ERK signaling via Ca^{2+} influx and release from internal stores or mediation of neurotrophic factors like the brain-derived neurotrophic factor (BDNF; Brunzell et al., 2003). Neuroprotection due to the activation of the ERK pathway has been observed in cultured spinal cord neurons, considered to be models of spinal cord trauma (Toborek et al., 2007) and AD (Table 11.2; Bitner et al., 2010), and it appears to be dependent on the upregulation of antiapoptotic protein Bcl-2 (Tables 11.2 and 11.3; Heusch & Maneckjee, 1998).

TABLE 11.3 Nicotine Effects on In Vivo Animal Models

Experimental condition	Animal model	Neuroprotective effects of nicotine
Parkinson's disease	C57BL/6J mouse	nAChR activation can protect dopaminergic neurons against degeneration[a]
Alzheimer's disease	C57BL/6J mouse	Nicotine protects against Aβ25–35-induced neurotoxicity via the upregulation of the antiapoptotic protein (Bcl-2)[b]
Intracerebral hemorrhage	C57BL/6J mouse	Activation of α7 nAChRs increases the number of surviving neurons and decreases activated microglia/macrophages[c]
Excitotoxicity	Mouse genetic ablation of nAChR subunits	Nicotine decreases cortical lesions in newborn mice via the activation of the α4β2 nAChRs[d]
Inflammation	B6 (H-2^b), nAChR $\alpha7^{-/-}$, $FoxP3^{GFP}$, and $CX_3CR1^{+/GFP}$ mice	α7 nAChRs contribute only partially to nicotinic effects on brain inflammation. Non-α7-nAChRs are involved in the antiinflammatory nicotine properties[e]
Aging	Sprague-Dawley rats	Nicotine upregulates FGF-2 mRNA in glial and neuronal cells[f]
Impairment of learning and memory	Sprague-Dawley rats	Nicotine prevents performance impairment in memory tests[g]

[a] *Takeuchi et al., J Neurosci Res, 2009; 87, 576–585.*
[b] *Xue et al., Int J Mol Med, 2014; 33, 925–933.*
[c] *Hijioka et al., Neuroscience, 2012; 222, 10–19.*
[d] *Laudenbach et al., FASEB J, 2002; 16, 423–425.*
[e] *Hao et al., Exp Neurol, 2011; 227, 110–119.*
[f] *Belluardo et al., Neurobiol Aging, 2004; 25, 1333–1342.*
[g] *Hiramatsu et al., J Neural Transm, 2002; 109, 361–375.*

Furthermore, nicotine increases the expression of the tyrosine kinase A (TrkA) receptor (Jonnala et al., 2002; Li et al., 2005) that is a member of a tyrosine kinase family affecting neuronal plasticity, differentiation, and survival via the activation of several signal cascades. In particular, TrkA is a membrane-bound receptor that, once activated by a neurotrophin-like nerve growth factor (NGF), phosphorylates itself and certain proteins connected with the MAPK/ERK pathway. The neuroprotective effect of nicotine associated with TrkA has been mainly reported in AD models (Jonnala et al., 2002; Li et al., 2005).

Loss of basal forebrain cholinergic neurons is a hallmark of human AD: in the rat nucleus basalis, it is related to the loss of α7 nAChRs and apparently coincides with a decrease in TrkA-positive neurons (Li et al., 2005). Thus, the origin of the cholinergic deficit may be attributed to toxic effects of the Aβ peptide, $A\beta_{1-42}$, which is a high-affinity α7 nAChR blocker (Li & Buccafusco, 2004) and decreases TrkA receptor expression with subsequent cholinergic neuronal death (Li et al., 2005). To counteract this phenomenon, it is noteworthy that nicotine elicits ERK phosphorylation essential for activating the cyclic AMP-responsive element-binding protein (CREB; Brunzell et al., 2003; Pandey, Roy, Xu, & Mittal, 2001; Fig. 11.1) that plays an important role in cognitive function by regulating the transcription of BDNF (Bitner et al., 2010). Since AD pathology has been associated with the downregulation of the phosphorylated form of CREB (Bitner et al., 2010), enhancing CREB expression with nicotinic agonists is a potential strategy to contrast AD progression.

Nicotine also acts on α7 nAChRs expressed by mitochondrial outer membranes that, via ion-independent mechanisms, activate the intramitochondrial PI3K/Akt pathway and inhibit the tyrosine-protein kinase *Src* and Ca^{2+}/calmodulin-dependent protein kinase II (CaKMII; Table 11.2; Fig. 11.1). These effects prevent the onset of preapoptotic events like the mitochondrial permeability transition pore (MPTP) formation and the consequent cytochrome *c* (cyt-*c*) release (Table 11.2). In line with these observations, the toxic effects by the βA peptide ($A\beta_{25-35}$) via apoptosis, caspase-3 activation, and release of cyt-*c* in human SH-SY5Y neuroblastoma cells are blocked by nicotine (Table 11.2).

Beyond promoting cell survival, nAChR activation depresses the formation of ROS and mitochondrial energy dysfunction, two key downstream effectors of glutamate-mediated excitotoxicity (Table 11.2). Intrinsic nAChR activity is insufficient to contrast these two processes (Table 11.2). On rat brain mitochondria, the antioxidant properties of nicotine are not receptor-mediated: in fact, nicotine can interact directly with the complex I of the respiratory chain by inhibiting NADPH binding to it, thus inducing a decrease in superoxide anion generation via a nAChR-independent process (Table 11.2). Linert et al. (1999) had actually proposed the beneficial/protective effects of nicotine in PD and AD to be, at least partly, due to antioxidant mechanisms. Table 11.2 lists a few additional mechanisms likely to be involved in nicotine-mediated neuroprotection.

Several other genes related to nicotine neuroprotection are involved in expression regulation, early growth response, transcription factors, and homeobox activity (Belluardo et al., 2005). These genes under nicotinic receptors control may play important roles in coupling receptor activation to long-term neuronal responses including the neuroprotective-neurotrophic actions of nicotine (Belluardo et al., 2005). Among them, it is important to consider the role of the neurotrophic fibroblast growth factor-2 (FGF-2), which has been implicated in neuroprotection and neuronal survival in several in vitro and in vivo insult models (Table 11.3).

In recent years, the role of nAChRs in glial cells, especially microglia and astrocytes, has been investigated. Microglia comprises the immune cells of the CNS, which are rapidly activated with changes in their morphology under pathological conditions such as ischemia, trauma, and stroke (Kettenmann, Hanish, Noda, & Verkhratsky, 2011). Nicotine prevents the morphological changes of microglia cells stimulated by inflammatory processes and neuronal damage (Table 11.2). Furthermore, as glial cells are active components of complex brain networks, it would be interesting to understand if nicotine protection of microglia might impact on neurons. In addition, inhibition of astrocytes has been proposed as a novel strategy for the treatment of neurodegenerative disease (Table 11.2) because pretreatment with nicotine suppresses, via α7 nAChRs, astrocyte activation by chemical stressors. These issues clearly require future investigation.

MINI-DICTIONARY OF TERMS

Apoptosis Programmed cell death process occurring in multicellular organisms.

Excitotoxicity Pathological process induced by high extracellular glutamate causing excessive excitatory receptor activity. It can damage or kill nerve cells.

In vitro Studies performed with organisms, cells, or molecules in laboratory conditions outside their biological context.

In vivo Studies carried out on whole, living organisms.

Oxidative stress Imbalance in the cell energy metabolism due to mismatch between production and removal of intracellular reactive oxygen species.

Presynaptic bouton Specialized area at the terminal end of the axon containing neurotransmitter to be released into the synaptic cleft.

Key Facts of Nicotine Neuroprotection

- nAChRs are important targets for pharmaceutical research and development.
- Nicotine administered via smoking reaches the brain twice as fast as via the intravenous route.
- Epidemiological studies reported negative correlations between smoking and PD or AD incidence.
- Nicotine may improve vigilance, memory, attention, and mental processing.
- Nicotinic neuronal effects should be balanced against damage to vascular and respiratory systems.

Summary Points

- Notwithstanding the serious health risks inherent in tobacco smoking, nicotine can protect brain neurons from cell death.
- Neuronal death can be caused by either excitotoxicity induced by excessive extracellular glutamate or toxic peptides like β-amyloid fragments.
- This effect of nicotine is mediated by the activation of neuronal nicotinic receptors expressed by cell and mitochondrial membranes.
- Activation of nicotinic receptors triggers multiple intracellular pathways to prevent DNA damage.
- Certain synthetic agonists of nicotinic receptors can mimic the neuroprotective effects of nicotine.
- The neuroprotective role of nicotinic receptors outlines a future pharmacological strategy to arrest neurodegeneration at an early stage.

References

Alkondon, M., & Albuquerque, E. X. (2004). The nicotinic acetylcholine receptor subtypes and their function in the hippocampus and cerebral cortex. *Progress in Brain Research*, *145*, 109–120.

Alonso, A., Logroscino, G., & Hernán, M. A. (2010). Smoking and the risk of amyotrophic lateral sclerosis: a systematic review and meta-analysis. *Journal of Neurology Neurosurgery, and Psychiatry*, *81*, 1249–1252.

Anand, R., Gill, K. D., & Mahdi, A. A. (2014). Therapeutics of Alzheimer's disease: past, present and future. *Neuropharmacology*, *76*, 27–50.

Barazangi, N., & Role, L. (2001). Nicotine-induced enhancement of glutamatergic and GABAergic synaptic transmission in the mouse amygdala. *Journal of Neurophysiology*, *86*, 463–474.

Belluardo, N., Olsson, P. A., Mudò, G., Sommer, W. H., Amato, G., & Fuxe, K. (2005). Transcription factor gene expression profiling after acute intermittent nicotine treatment in the rat cerebral cortex. *Neuroscience*, *133*, 787–796.

Bitner, R. S., Bunnelle, W. H., Decker, M. W., Drescher, K. U., Kohlhaas, K. L., Markosyan, S., et al. (2010). In vivo pharmacological characterization of a novel selective α7 neuronal nicotinic acetylcholine receptor agonist ABT-107: preclinical considerations in Alzheimer's disease. *The Journal of Pharmacology and Experimental Therapeutics*, *334*, 875–886.

Brunzell, D. H., Russell, D. S., & Picciotto, M. R. (2003). In vivo nicotine treatment regulates mesocorticolimbic CREB and ERK signaling in C57Bl/6J mice. *Journal of Neurochemistry*, *84*, 1431–1441.

Buckingham, S., Jones, A., Brown, L. A., & Sattelle, B. D. (2009). Nicotinic acetylcholine receptor signalling: roles in Alzheimer's disease and amyloid neuroprotection. *Pharmacological Reviews*, *61*, 39–61.

Dajas-Bailador, F. A., Lima, P. A., & Wonnacott, S. (2000). The α7 nicotinic acetylcholine receptor subtype mediates nicotine protection against NMDA excitotoxicity in primary hippocampal cultures through a Ca^{2+} dependent mechanism. *Neuropharmacology*, *39*, 2799–2807.

Datta, S. R., Brunet, A., & Greenberger, M. E. (1999). Cellular survival: a play in three Akts. *Genes and Development*, *13*, 2905–2927.

de Jong, S. W., Huisman, M. H., Sutedja, N. A., van der Kooi, A. J., de Visser, M., Schelhaas, H. J., et al. (2012). Smoking, alcohol consumption, and the risk of amyotrophic lateral sclerosis: a population-based study. *American Journal of Epidemiology*, *176*, 233–239.

Erhardt, P., Schremser, E. J., & Cooper, G. M. (1999). B-Raf inhibits programmed cell death downstream of cytochrome c release from mitochondria by activating the MEK/Erk pathway. *Molecular and Cellular Biology*, *19*, 5308–5315.

Geldenhuys, W. J., & Darvesh, A. S. (2015). Pharmacotherapy of Alzheimer's disease: current and future trends. *Expert Review of Neurotherapeutics*, *15*, 3–5.

Giniatullin, R., Nistri, A., & Yakel, J. L. (2005). Desensitization of nicotinic ACh receptors: shaping cholinergic signaling. *Trends in Neurosciences*, *28*, 371–378.

Gray, R., Rajan, A., Radcliff, K., Yakehiro, M., & Dani, J. (1996). Hippocampal synaptic transmission enhanced by low concentrations of nicotine. *Nature*, *383*, 713–716.

Hetman, M., & Gozdz, A. (2004). Role of extracellular signal regulated kinases 1 and 2 in neuronal survival. *European Journal of Biochemistry, 271*, 2050–2055.

Heusch, W. L., & Maneckjee, R. (1998). Signalling pathways involved in nicotine regulation of apoptosis of human lung cancer cells. *Carcinogenesis, 19*, 551–556.

Jonnala, R. R., Terry, A. V., Jr., & Buccafusco, J. J. (2002). Nicotine increases the expression of high affinity nerve growth factor receptors in both in vitro and in vivo. *Life Sciences, 70*, 1543–1554.

Kettenmann, H., Hanish, U. K., Noda, M., & Verkhratsky, A. (2011). Physiology of microglia. *American Physiological Society, 91*, 461–553.

Li, X. D., Arias, E., Jonnala, R. R., Mruthinti, S., & Buccafusco, J. J. (2005). Effect of amyloid peptides on the increase in TrkA receptor expression induced by nicotine in vitro and in vivo. *Journal of Molecular Neuroscience, 27*, 325–336.

Li, X. D., & Buccafusco, J. J. (2004). Role of α7 nicotinic acetylcholine receptors in the pressor response to intracerebroventricular injection of choline: blockade by amyloid peptide Aβ1-42. *Journal of Pharmacology and Experimental Therapeutics, 309*, 1206–1212.

Linert, W., Bridge, M. H., Huber, M., Bjugstad, K. B., Grossman, S., & Arendash, G. W. (1999). In vitro and in vivo studies investigating possible antioxidant actions of nicotine: relevance to Parkinson's and Alzheimer's diseases. *Biochimica et Biophysica Acta, 1454*, 143–152.

Mai, H., May, W. S., Gao, F., Jin, Z., & Deng, X. (2003). A functional role for nicotine in Bcl2 phosphorylation and suppression of apoptosis. *The Journal of Biological Chemistry, 278*, 1886–1891.

McGehee, D. S., Heath, M. J., Gelber, S., Devay, P., & Role, L. W. (1995). Nicotine enhancement of fast excitatory synaptic transmission in CNS by presynaptic receptors. *Science, 269*, 1692–1696.

Mulle, C., Vidal, C., Benoit, P., & Changeux, J. P. (1991). Existence of different subtypes of nicotinic acetylcholine receptors in the rat habenulo-interpeduncular system. *The Journal of Neuroscience, 11*, 2588–2597.

Newhouse, P. A., Potter, A., & Singh, A. (2004). Effects of nicotinic stimulation on cognitive performance. *Current Opinion in Pharmacology, 1*, 36–46.

Pandey, U. C., Roy, A., Xu, T., & Mittal, N. (2001). Effects of protracted nicotine exposure and withdrawal on the expression and phosphorylation of the CREB gene transcription factor in rat brain. *Journal of Neurochemistry, 77*, 943–952.

Parker, S. L., Fu, Y., McAllen, K., Luo, J., McIntosh, J. M., & Lindstrom, J. M. (2004). Sharp BM Up-regulation of brain nicotinic acetylcholine receptors in the rat during long-term self administration of nicotine: disproportionate increase of the α6 subunit. *Molecular Pharmacology, 65*, 611–622.

Pepeu, G., & Giovannini, M. G. (2009). Cholinesterase inhibitors and beyond. *Current Alzheimer Research, 2*, 86–96.

Picciotto, M. R., & Zoli, M. (2008). Neuroprotection via nAChRs: the role of nAChRs in neurodegenerative disorders such as Alzheimer's and Parkinson's disease. *Frontiers in Bioscience, 13*, 492–504.

Rothstein, J. D., Martin, L. J., & Kuncl, R. W. (1992). Decreased glutamate transport by the brain and spinal cord in amyotrophic lateral sclerosis. *The New England Journal of Medicine, 326*, 1464–1468.

Sharifullina, E., & Nistri, A. (2006). Glutamate uptake block triggers deadly rhythmic bursting of neonatal rat hypoglossal motoneurons. *The Journal of Physiology, 572*, 407–423.

Shen, J. X., & Yakel, J. L. (2009). Nicotinic acetylcholine receptor-mediated calcium signaling in the nervous system. *Acta Pharmacologica Sinica, 30*, 673–680.

Swan, G. E., & Lessov-Schlaggar, C. N. (2007). The effects of tobacco smoke and nicotine on cognition and the brain. *Neuropsychology review, 17*, 259–273.

Toborek, M., Son, K. W., Pudelko, A., King-Pospisil, K., Wylegala, E., & Malecki, A. (2007). ERK 1/2 signaling pathway is involved in nicotine-mediated neuroprotection in spinal cord neurons. *Journal of Cellular Biochemistry, 100*, 279–292.

Tyas, S. L. (1996). Are tobacco and alcohol use related to Alzheimer's disease? A critical assessment of the evidence and its implications. *Addiction Biology, 1*, 237–254.

Urushitani, M., Shimohama, S., Kihara, T., Sawada, H., Akaike, A., Ibi, M., et al. (1998). Mechanism of selective motor neuronal death after exposure of spinal cord to glutamate: involvement of glutamate-induced nitric oxide in motor neuron toxicity and nonmotor neuron protection. *Annals of Neurology, 44*, 796–807.

Yu, W., Mechawar, N., Krantic, S., & Quirion, R. (2011). α7 Nicotinic receptor activation reduces β-amyloid-induced apoptosis by inhibiting caspase-independent death through phosphatidylinositol 3-kinase signaling. *Journal of Neurochemistry, 119*, 848–858.

Zhu, P. J., & Chiappinelli, V. A. (1999). Nicotine modulates evoked GABAergic transmission in the brain. *Journal of Neurophysiology, 82*, 3041–3045.

CHAPTER

12

Linking Nicotine, Menthol, and Brain Changes

Brandon J. Henderson

Department of Biomedical Sciences, Marshall University, Joan C. Edwards School of Medicine, Huntington, WV, United States

Abbreviations

GABA γ-aminobutyric acid
nAChRs nicotinic acetylcholine receptors
SNc substantia nigra pars compacta
SNr substantia nigra pars reticulata
VTA ventral tegmental area

12.1 INTRODUCTION

Menthol is the most popular tobacco flavorant worldwide. In America and European countries, the smoking rate of menthol cigarettes is ~30% of the smoking population (Caraballo & Asman, 2011). Elsewhere, the rates of smoking menthol cigarettes vary with the highest being the Philippines (60%) (WHO, 2016). African Americans exhibit an incredibly high rate of smoking menthol cigarettes: 75% of adults and 90% of youth smokers (D'Silva, Boyle, Lien, Rode, & Okuyemi, 2012). The general youth smoking population in America also prefers menthol cigarettes over nonflavored cigarettes (Villanti et al., 2016). Menthol not only is found in high doses (3–20 mg/cigarette) in menthol-labeled cigarettes but also is present in 98% of nonmenthol cigarettes at low doses (2–70 μg/cigarette) (Ai et al., 2015).

Menthol is a sweet-scented monoterpenoid derived from peppermint oil. It produces cooling sensations topically through its agonist activity on transient receptor potential melastatin-8 (TrpM8) (Journigan & Zaveri, 2013). Research into menthol's actions is currently expanding due to increased public health concerns about its role in reduced smoking cessation rates, cardiovascular effects, and incidents of smoking-related cancer (discussed further below). Here, we discuss menthol, nicotine, and the changes in the brain that result from their combined actions.

12.2 BRIEF HISTORY OF MENTHOL CIGARETTES

The first menthol cigarette was likely the "Spud," which was marketed by Axton-Fisher in 1926 (Proctor, 2012). Soon after (1933), the tobacco company Brown and Williamson launched their Kool brand of menthol cigarettes. Since the 1950s, there have been several arguments whether menthol cigarettes were disproportionately marketed to African Americans. Discussing this would take away from the focus of this chapter, but a thorough review can be found elsewhere (Lochlann Jain, 2003).

The most notable public health concerns involve menthol's role in nicotine addiction. When we examine smokers of all ages, smokers of menthol cigarettes quit at a significantly reduced rate when compared to smokers of nonmenthol cigarettes (Delnevo, Gundersen, Hrywna, Echeverria, & Steinberg, 2011; D'Silva et al., 2012). This of course increases the risk of smoking-related negative outcomes: cancers of the lung, esophagus, oral cavity, and larynx and proposed higher incidence in heart disease and hypertension. The concern has become so great that the U.S. Food and Drug Administration (FDA) and World Health Organization (WHO) conducted independent investigations and concluded that menthol does increase smoking initiation and causes more intense addiction (FDA, 2012; WHO, 2016).

Young smokers of menthol cigarettes are twice as likely to transition to lifelong smoking compared to youths smoking nonmenthol cigarettes (D'Silva et al., 2012). As e-cigarettes grow in popularity, smoking rates of flavored products (including menthol) are increasing, especially among youth smokers (CDC, 2016; Villanti et al., 2016). This troubles many public health experts, as they fear this may contribute to a new generation of nicotine addicts.

Neuroscience of Nicotine
https://doi.org/10.1016/B978-0-12-813035-3.00012-5

12.3 MENTHOL AND NICOTINE: EARLY CLINICAL FINDINGS

How could menthol reduce the quit rates of smokers? Some studies suggest that menthol decreases airway irritation and allows smokers to inhale more nicotine. Clinical reports show that the puff duration, puff volume, and number of puffs are significantly increased with menthol cigarettes (Ahijevych & Parsley, 1999). There is also evidence that menthol decreases the metabolism of nicotine, allowing elevated concentrations of nicotine in the plasma (Alsharari et al., 2015; Benowitz, Herrera, & Jacob 3rd, 2004). Despite this, there are counterarguments as other reports show that plasma levels of nicotine fail to increase significantly with menthol (Ashley, Dixon, Sisodiya, & Prasad, 2012).

The most dramatic finding regarding the smoking of menthol and nonmenthol cigarettes is the effect on nicotinic acetylcholine receptor (nAChR) upregulation. The upregulation of β2-containing (β2*) nAChRs is a well-characterized feature of chronic nicotine exposure. This occurs in human smokers (Jasinska, Zorick, Brody, & Stein, 2013), rodents (Henderson et al., 2017), and cell lines (Srinivasan et al., 2011). This is believed to be a biomarker for nicotine addiction as this event plays an important role in altering the dopamine reward pathway (Faure, Tolu, Valverde, & Naude, 2014; Nashmi et al., 2007). In the context of humans that smoke menthol or nonmenthol cigarettes, Brody et al. (2013) found that smokers of menthol cigarettes exhibit more β2* nAChR upregulation in several brain regions compared to nonmenthol smokers. These brain areas included the prefrontal cortex, corpus callosum, hypothalamus, cerebellum, and brain stem. Many of these regions play an important role in the dopamine reward pathway (Fig. 12.1), and this finding was a key step toward understanding why smokers of menthol cigarettes quit at lower rates.

12.4 MENTHOL'S ACTIONS ON CYS-LOOP RECEPTORS

The molecular actions of menthol have been studied on several classes of ion channels (for a complete review, see Oz, El Nebrisi, Yang, Howarth, and Al Kury (2017)) including the Cys-loop family of receptors. Many of these studies use fairly high doses (>50 μM) of menthol (Table 12.1), while a few use lower doses (<1 μM) that are believed to be smoking-relevant (discussed below). The Cys-loop ligand-gated ion channel family includes nAChRs, γ-aminobutyric acid (GABA), glycine,

TABLE 12.1 Acute Actions of Menthol on Cys-loop Receptors

		IC_{50} or EC_{50}	Stereoselective?
$GABA_A$	Positive allosteric modulator[a,b]	~60 μM, (+)-menthol ~160 μM, (−)-menthol	Yes
Glycine	Positive allosteric modulator[a]	75 μM	No
5-HT_3	Noncompetitive antagonist[c]	163 μM	No
α3β4 nAChR	Noncompetitive antagonist[d]	69–100 μM	Unknown
α7 nAChR	Noncompetitive antagonist[e]	33 μM	No
α4β2 nAChR	Negative allosteric modulator[f]	111 μM	Unknown

[a] *Hall et al. (2004).*
[b] *Corvalan, Zygadlo, and Garcia (2009).*
[c] *Ashoor et al. (2013a).*
[d] *Ton et al. (2015).*
[e] *Ashoor et al. (2013b).*
[f] *Hans, Wilhelm, and Swandulla (2012).*

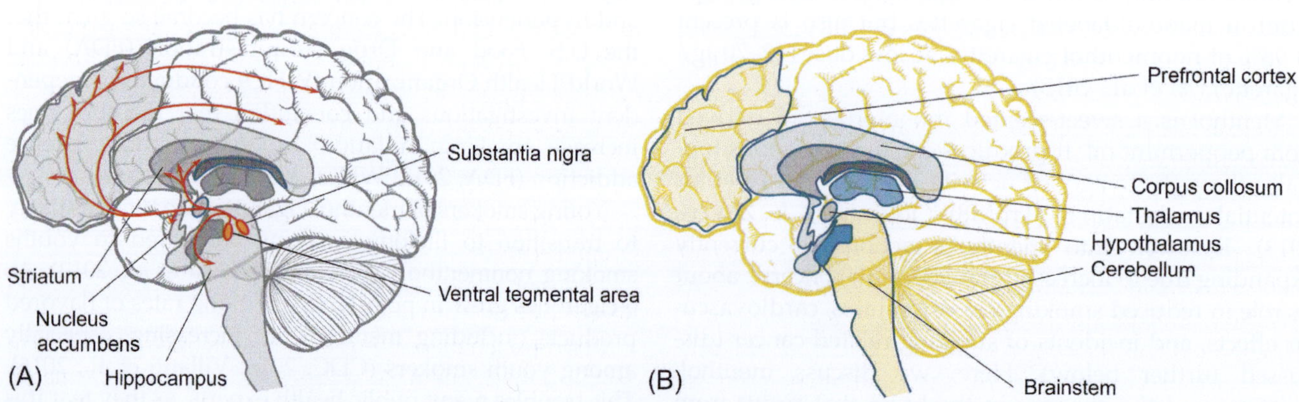

FIG. 12.1 Human smokers exhibit an upregulation of brain β2* nAChRs in several regions that play a role in addictive behavior. (A) Schematic of dopamine neuron projections (*red*) that originate from the substantia nigra and ventral tegmental area. (B) Schematic of brain regions that exhibit β2* nAChR upregulation in the brains of smokers (see Jasinska et al., 2013).

5-hydroxytryptophan 3 (5-HT_3), and zinc-activated ion channel receptors. Each of these receptors is composed of a pentameric protein assembly and have a well-characterized loop formed by highly conserved amino acids between two cysteine (Cys) residues that form a disulfide bond (Sine & Engel, 2006). Menthol was discovered to be a negative allosteric modulator of α4β2 nAChRs and a noncompetitive antagonist of α7 and α3β4 nAChRs, with IC_{50} values of ~111, ~35, and ~70 μM, respectively (Ashoor et al., 2013b; Hans et al., 2012; Ton et al., 2015) (Table 12.1). Additionally, menthol has been shown to be a noncompetitive antagonist of 5-HT_3 receptors (IC_{50}, 163 μM), a positive allosteric modulator of $GABA_A$ receptors (EC_{50}, 60–160 μM), and a positive allosteric modulator of glycine receptors (EC_{50}, 50 μM) (Ashoor et al., 2013a; Hall et al., 2004; Lau, Karim, Goodchild, Vaughan, & Drew, 2014). While many investigations examined (−)-menthol only, one identified a significant difference in the pharmacological actions of (+)-menthol and (−)-menthol (Hall et al., 2004). Here, both (+)-menthol and (−)-menthol acted as potentiators of $GABA_A$ currents, but (+)-menthol was significantly more potent than (−)-menthol. Thus, there may be more stereospecific interactions of menthol on Cys-loop receptors that are yet to be discovered. The concentrations used in these investigations probably exceed those found *in vivo* as the smoking-relevant concentration of menthol is likely <8 μM (Benowitz et al., 2004). These findings using high μM concentrations of menthol may not explain the observations of enhanced β2* nAChR upregulation in menthol smokers (Brody et al., 2013), but several implications can be made from these observations. GABA neurons play an important role in the nicotine reward pathway; thus, menthol's ability to potentiate GABA signaling could exert significant changes in the brain that contribute to nicotine addiction (discussed further below).

Given menthol's ability to pharmacologically manipulate Cys-loop receptors, there has been great interest in identifying its binding site. In studies that compared menthol to the actions of benzodiazepines and the anesthetic propofol, it was revealed that menthol and propofol may share the same binding site on $GABA_A$ receptors (Watt et al., 2008). Given the conserved chemical features between menthol and propofol, it is not unbelievable that menthol binds $GABA_A$ receptors in a manner similar to propofol. Computational methods have been employed to identify an allosteric binding site on nAChRs (Ashoor et al., 2013b). Here, the muscle-type nAChR was used as a model due to its similarity to α7 nAChRs and because, at the time, it was the only nAChR that had a solved, full-length crystal structure. Here, the binding of menthol was found to be in the transmembrane region, similar to the findings of Watt et al. (2008). Despite this, the field is still far from identifying a menthol-binding site on nAChRs.

12.5 MENTHOL ENHANCES NICOTINE REWARD BY ALTERING MIDBRAIN DOPAMINE NEURONS

To understand how menthol acts with nicotine to enhance nAChR upregulation (as observed by Brody *et al.*) and possibly reduce smoking quit rates, several investigations were completed to examine menthol's ability to alter nicotine's actions in the brain. From rodent models, we know that nAChR subunits are important for nicotine reward and reinforcement: Individual deletions of α4, α6, or β2 nAChR subunits prevent self-administration of nicotine in mice, and selective reexpression of these deleted subunits in the VTA (not SNc or SNr) reinstates self-administration behavior (Pons et al., 2008) (see Fig. 12.2 for a schematic of neuronal inputs to VTA dopamine neurons). Selective activation of α6β2* or α4β2* nAChRs by smoking-relevant concentrations of nicotine stimulates depolarization and elevates firing frequency of VTA DA neurons (Engle, Shih, McIntosh, & Drenan, 2013; Liu, Zhao-Shea, McIntosh, Gardner, & Tapper, 2012). Together, these results clearly show that VTA α4β2*, α6β2*, and α4α6β2* nAChRs are the primary targets of nicotine and mediate its rewarding and reinforcing properties.

12.5.1 Menthol Enhances Nicotine-Induced Upregulation of nAChRs

Nicotine's effect on the brain has been well characterized. Nicotine upregulates nAChRs, alters the firing frequency and excitability of dopamine and GABA neurons, and enhances dopamine release in the nucleus accumbens (see Faure et al. (2014) for a complete review). Recently, an investigation into menthol's effect on these well-characterized nicotine-induced changes in the brain was completed (Henderson et al., 2017). Experiments using mice revealed that menthol plus nicotine upregulates α4β2* and α4α6β2* nAChRs (not α6β2* nAChRs) in the VTA and SNr significantly more than nicotine alone (Henderson et al., 2017) (see Fig. 12.3). Another study using mice also reported that menthol enhances nicotine's upregulation of β2* nAChR subunits in the prefrontal cortex (Alsharari et al., 2015). These studies provided a nAChR subtype-specific, region-specific, and cell-specific correlate to the assays using human brain imaging (Brody et al., 2013).

Menthol Enhances Nicotine Reward and Reinforcement

Conditioned place preference assays have been used to examine how menthol alters nicotine reward-related behavior (Henderson et al., 2017). Conditioned place preference is an experimental method based on Pavlovian training, used to study drug reward-related behavior.

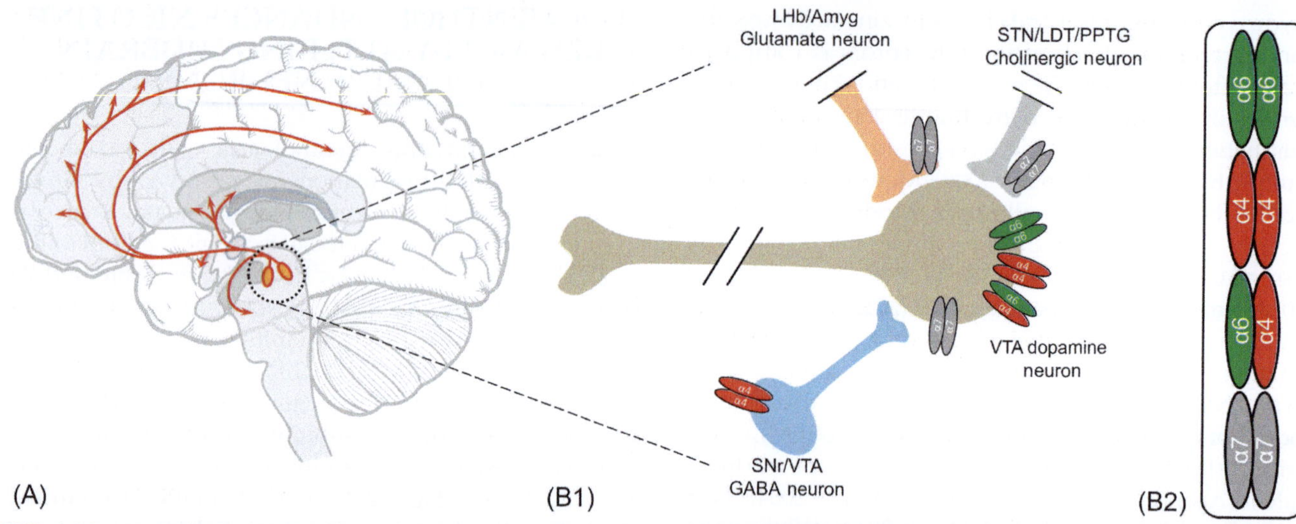

FIG. 12.2 (A) Schematic of dopamine reward pathway with the midbrain (VTA and SNr) highlighted for panel B_1. (B_1) Schematic of neurons contributing to nicotine reward in the midbrain and the nAChR subtypes involved. (B_2) Key for nAChR subtypes in B_1. Additional abbreviations: *Amyg*, amygdala; *LDT*, lateral dorsal tegmentum; *LHB*, lateral habenula; *PPTG*, pedunculopontine tegmental nucleus; *STN*, subthalamic nucleus.

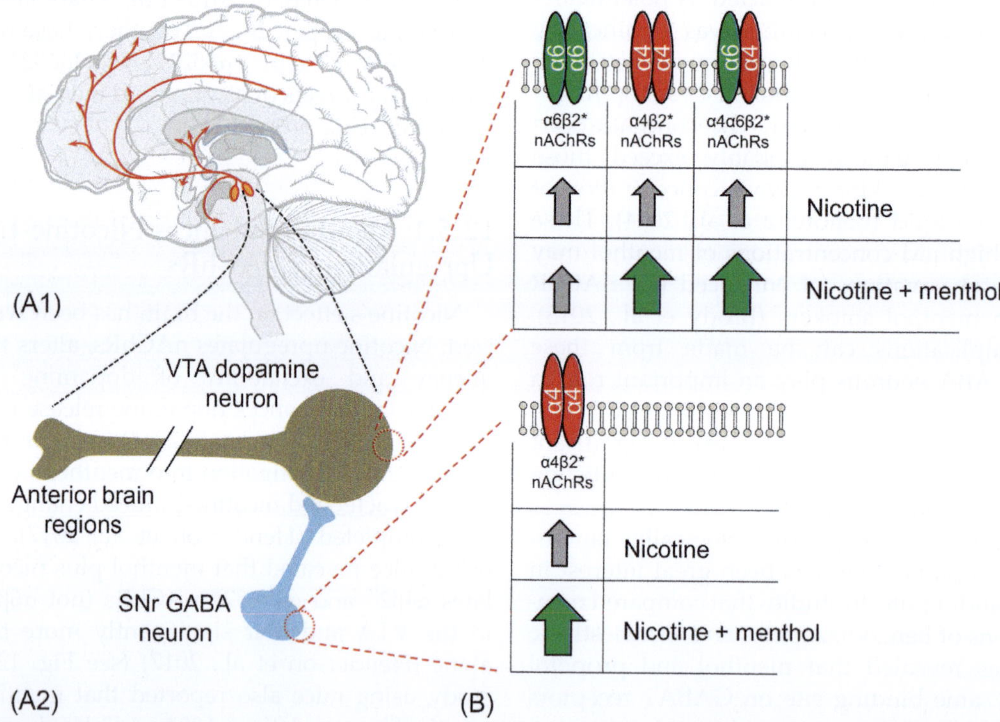

FIG. 12.3 α4β2*, α6β2*, and α4α6β2* nAChRs on VTA dopamine neurons are upregulated by nicotine and nicotine plus menthol. (A_{1-2}) Schematic of a VTA dopamine neuron projecting from the midbrain to anterior regions of the brain (the nucleus accumbens, prefrontal cortex, and striatum). (B) Schematic of nAChRs on the cell surface that are upregulated by nicotine or nicotine plus menthol. Sizes of arrows indicate the degree of upregulation.

This method has frequently been applied to the study of nicotine reward-related behavior (Henderson et al., 2017; Tapper et al., 2004). In this assay, menthol was added to nicotine at a dose perceived to be smoking-relevant and produced a significant increase in nicotine reward-related behavior (Henderson et al., 2017).

Rat self-administration assays were also used to examine menthol's effect on nicotine reward and reinforcement. Like conditioned place preference, these experiments are used to examine drug reward, but unlike conditioned place preference, drug delivery is contingent to rodents triggering a mechanism. In rat nicotine intravenous self-administration assays, menthol was also reported to enhance nicotine intake (Biswas et al., 2016; Wang, Wang, & Chen, 2014). It is assumed that menthol itself is not rewarding as these studies also revealed that menthol at many doses failed to produce reward in mice using the conditioned place preference assay (Henderson et al., 2017). Together, these experiments clearly document that menthol enhances nicotine reward and reinforcement. More importantly, this suggests that the lowered quit rate observed with smokers of menthol cigarettes could be due to the fact that menthol makes nicotine more rewarding (more addictive). Of course, there are other features that likely underlie the reduced cessation rate as it was recently revealed that menthol also enhances nicotine withdrawal (Alsharari et al., 2015).

Although reports of menthol enhancing nicotine-induced upregulation of β2* nAChRs were made for several brain regions (Alsharari et al., 2015) (Henderson et al., 2017), these studies were not done with concentrations of menthol or nicotine that produce changes in reward-related behavior. When upregulation assays were repeated using doses of nicotine and menthol sufficient to alter reward-related behavior (Henderson et al., 2017), there was only an enhancement in the nicotine-induced upregulation of α4α6β2* nAChRs in the VTA (see Fig. 12.3). Given that α4α6β2* nAChRs are the most sensitive to nicotine and selectively expressed on dopamine neurons in the midbrain, this suggests that the selective upregulation of these nAChRs would dramatically alter dopamine neuron excitability and signaling (discussed further below). This marks α4α6β2* nAChRs as a primary contributor to menthol's ability to enhance nicotine reward and reinforcement.

12.5.2 Menthol Enhances Nicotine-Induced Changes in Dopamine Neuron Excitability

Midbrain dopamine neurons exhibit spontaneous activity of two distinct types: one characterized by slow, regular spikes (tonic) and one characterized by bursts of activity (phasic). Nicotine, through its ability to upregulate nAChRs in a cell-specific manner (Nashmi et al., 2007), alters both the tonic and phasic activity of dopamine neurons (see Faure et al. (2014) for a full review). GABA neurons are critical for this observed nicotine-induced change in dopamine neuron firing. GABA neurons from the VTA and SNr provide inhibitory input to the VTA dopamine neurons that project to anterior brain regions. Following chronic exposure to nicotine, α4* nAChRs are upregulated on VTA and SNr GABAergic neurons (Nashmi et al., 2007). The upregulated α4* nAChRs on GABA neurons provide enhanced inhibitory signaling and decrease tonic dopamine neuron firing. However, acute nicotine exposure enhances GABAergic activity transiently, followed by a prolonged depression in activity due to the upregulated α4* nAChRs' desensitization. This triggers increased dopamine neuron burst firing (Mansvelder, Keath, & McGehee, 2002) (summarized in Fig. 12.4). This likely originates from α4α6β2* nAChRs, as the nicotine-induced

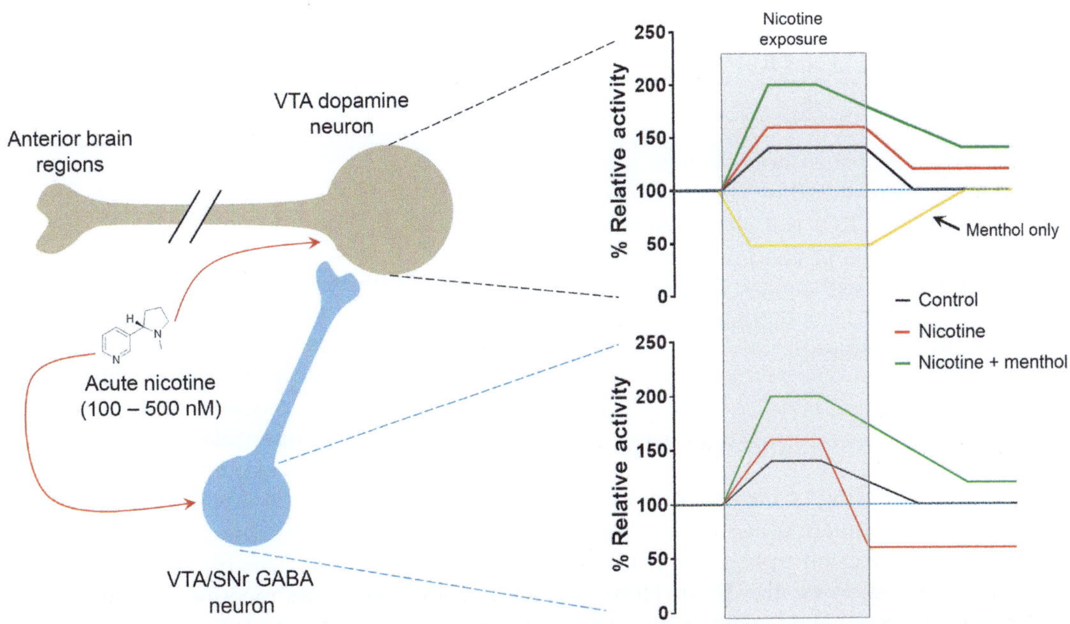

FIG. 12.4 Nicotine is well characterized to enhance both dopamine and GABA neuron firing transiently (denoted by transparent gray shading). It is also known that following chronic exposure to nicotine (>7 days), GABA neurons rapidly desensitize and exhibit reduced firing frequencies (for up to 1 h). This results in a disinhibition of target dopamine neurons. This contributes to an increase in dopamine neuron excitability. Chronic treatment with menthol and nicotine prevents GABA neurons from desensitization yet dopamine neurons exhibit an enhanced increase in excitability. See Henderson et al. (2017).

enhancement in dopamine neuron firing is dependent upon nAChRs that contain both α4 and α6 subunits (Engle et al., 2013; Liu et al., 2012).

Menthol, combined with nicotine, was also found to alter dopamine neuron firing (Fig. 12.4). Menthol plus nicotine produces increases in nicotine-stimulated dopamine neuron firing frequency that is significantly greater than nicotine alone. Once again, this is likely due to the enhanced upregulation of α4α6β2* nAChRs (see Fig. 12.3A). Despite menthol enhancement of nicotine's actions on dopamine neurons, menthol was found to reverse the actions of nicotine on GABA neuron firing. As previously mentioned, long-term exposure to nicotine causes GABA neurons to exhibit a sustained depression in activity when they are acutely exposed to nicotine (Fig. 12.4). In the case of menthol plus nicotine, GABA neuron firing frequency increases transiently but slowly returns to a state resembling baseline firing frequency. Thus, no signs of sustained depression in activity were observed. It has been shown that menthol (alone) reduces α4β2* nAChR desensitization (Henderson et al., 2016). This may explain how GABA neurons treated with exposed to long-term menthol and nicotine may fail to desensitize during nicotine applications since they have only α4 (non-α6)β2* nAChRs. To summarize, when we consider the characterized effects of nicotine on dopamine neuron firing, menthol enhances these changes in dopamine neuron firing that ultimately leads to enhanced excitability.

12.6 MENTHOL BY ITSELF ALTERS nAChRs ON MIDBRAIN DOPAMINE NEURONS

Much of our knowledge of menthol's actions comes from studies that examined menthol combined with nicotine. From these studies, it is not clear if menthol produces an effect on its own and is additive to that of nicotine or if menthol by itself causes distinct brain changes. To that end, several investigations have been completed to address this question.

12.6.1 Menthol Alone Upregulates nAChRs

To gain a better understanding of how menthol produces enhanced nAChR upregulation in smokers, menthol was examined for its effect on upregulation in the absence of nicotine (Henderson et al., 2016). Here, it was found that menthol upregulated α4β2 and α6β2β3 nAChRs that were transiently transfected into neuro-2a cells. More importantly, this was achieved using a concentration of menthol (0.5 μM) that was much lower than the concentrations necessary for allosteric antagonism or enhancement of Cys-loop receptors (see Table 12.1). In mice, upregulation of nAChRs was observed using menthol concentrations <2.5 μM (Henderson et al., 2016). Menthol also exhibited cell-specific upregulation of midbrain α4* nAChRs that differs from nicotine-induced upregulation. Nicotine robustly upregulates α4* nAChRs on SNr and VTA GABA neurons (Nashmi et al., 2007), but menthol upregulates α4* nAChRs on SNc and VTA DA neurons and does not upregulate SNr α4* nAChRs (Henderson et al., 2016).

There also was a difference in the upregulated nAChRs when you compare menthol plus nicotine to menthol alone: nicotine plus menthol upregulates high-sensitivity nAChRs, while the menthol alone upregulates only low-sensitivity nAChRs (Fig. 12.5). This was observed only with in vitro preparations as the studies using mice could not provide information in regard to nAChR stoichiometry.

12.6.2 Menthol-Alone Abolishes Nicotine Reward-Related Behavior

Interestingly, menthol exposure prior to nicotine prevents nicotine reward-related behavior in a conditioned place preference assay using mice (Henderson et al., 2016). In this assay, mice were treated with menthol alone for 10 days prior to the training for conditioned place preference. Therefore, the prevention of nicotine reward-related behavior likely resulted from menthol's ability to upregulate low-affinity nAChRs. The result is that nicotine failed to activate midbrain dopamine and GABA neurons robustly due to the presence of low-affinity and not high-affinity nAChRs being available.

12.7 SUMMARY

Menthol, nicotine, and their combination cause several changes in the brain. Although menthol has been shown to act on several Cys-loop receptors, the long-term effects have only been characterized on nAChRs. Alone or together, nicotine and menthol upregulate brain nAChRs. The upregulation of highly nicotine sensitive nAChRs cause significant changes in dopamine reward signaling (summarized in Table 12.2). Most notably, these neurons exhibit increased excitability in response to nicotine. This likely is the reason why menthol, combined with nicotine, causes greater positive reward when compared to nicotine alone. Together, these changes highlight the fact that the reduced cessation rates for smokers of menthol cigarettes can be attributed, at least in part, to menthol-induced enhancement in nicotine reward.

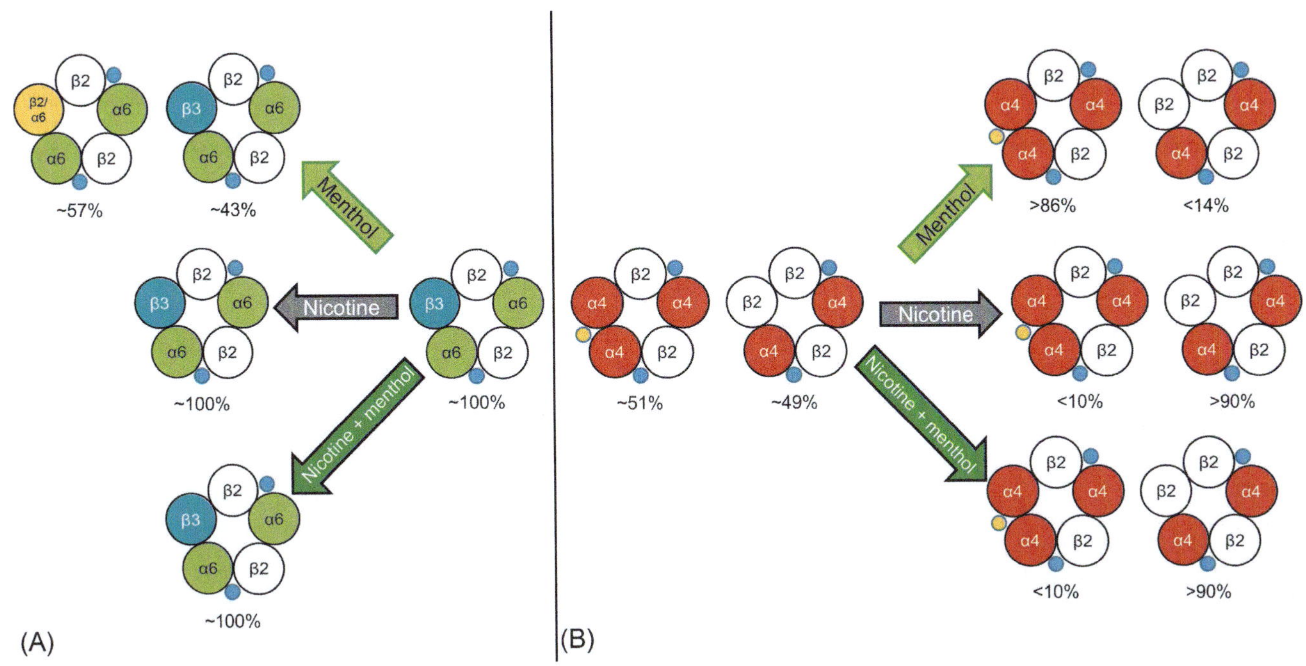

FIG. 12.5 Menthol, nicotine, and menthol plus nicotine alter (A) α6* and (B) α4* nAChR stoichiometry.
This diagram represents findings from cultured neuro-2a cells transiently transfected with α6, β2, and β3 nAChR subunits or α4 and β2 nAChR subunits (Henderson et al., 2017, 2016). Menthol alone stabilizes low-sensitivity nAChRs (α6β2(non-β3) and $\alpha4_{(3)}\beta2_{(2)}$). Nicotine and nicotine plus menthol, stabilize high-sensitivity nAChRs (α6β2β3 and $\alpha4_{(2)}\beta2_{(3)}$). Blue dots denote the agonist binding site for nicotine and acetylcholine. The yellow dots denote the noncanonical binding site that has been recently characterized for ACh.

TABLE 12.2 Comparison of Brain Changes Caused by Nicotine, Nicotine Plus Menthol, and Menthol

	Nicotine	Nicotine +menthol	Menthol
Upregulation			
α4* nAChR upregulation	VTA[a], SNc[a], SNr[a,d], and striatum[b]	VTA[a], SNc[a], SNr[a], striatum[b], and prefrontal cortex[b]	VTA[c], SNc[c], striatum[c], and prefrontal cortex[c]
α6* nAChR upregulation	VTA[a] and SNc[a]	VTA[a]	VTA[c] and SNc[c]
α4α6* nAChR upregulation	VTA[a] and SNc[a]	VTA[a] and SNc[a]	*Not yet examined*
Neuron baseline firing			
Dopamine neurons	Decreased firing[a,d]	Decreased firing[a]	Decreased firing[c]
GABA neurons	Increased firing[a]	Increased firing[a]	*Not yet examined*
Neuron excitability			
Dopamine neurons	Increased excitability[a,d]	Increased excitability[a]	Decreased excitability[c]
GABA neurons	Decreased excitability[a]	Increased excitability[a]	*Not yet examined*

[a] *Henderson et al. (2017).*
[b] *Alsharari et al. (2015).*
[c] *Henderson et al. (2016).*
[d] *Nashmi et al. (2007).*

MINI-DICTIONARY OF TERMS

Conditioned place preference An assay of drug reward-related behavior that is noncontingent to an instrumental/animal reaction.

Drug reward and reinforcement The study of addiction with animal models is complex. For this reason, assays are optimized to measure individual components of addiction: reward, reinforcement, reinstatement, withdrawal, etc.

Electrophysiology The study of electric properties of cells and neurons.

Nicotinic acetylcholine receptors (nAChRs) Cholinergic receptors that are sensitive to nicotine. These receptors are present in the peripheral and central nervous system. Acetylcholine (ACh) is the endogenous neurotransmitter of both nAChRs and muscarinic ACh receptors.

Nucleus accumbens A region of the ventral striatum where the terminals from mesolimbic dopamine neurons terminate onto GABAergic medium spiny neurons.

Prefrontal cortex The region of the cerebral cortex that covers anterior brain regions and is implicated in many aspects of cognitive behavior.

Self-administration An assay of drug reinforcement-related behavior that is contingent to an action (choice) of the animal.

Substantia nigra pars reticulata Midbrain region mostly consisted of inhibitory GABAergic neurons.

Upregulation Upregulation is a complex process, but this term is most widely used to designate an increase in the number of a particular protein. This can occur with many proteins, but this is a hallmark feature of nicotine: nicotine increases the number (upregulates) of nAChRs.

Ventral tegmental area Brain region close to the midline of the ventral midbrain and is the origin of dopamine projections of the mesolimbic and nigrostriatal pathways. This region is critical for reward circuitry.

Key Facts of Menthol

- Menthol is the only legal flavorant allowed in nonelectronic cigarettes in America.
- Menthol is present even in ~98% of the nonmenthol cigarettes sold.
- Menthol is present at $\geq$3 mg/cigarette in menthol-labeled cigarettes.
- In nonmenthol cigarettes, menthol is present in trace amounts (<0.1 mg).
- In addition to its use as a tobacco flavorant, menthol is used as a local anesthetic, decongestant, antipruritic, and flavorant for mint and peppermint flavored products (gum, candy, etc.).

Summary Points

- Both nicotine and menthol upregulate nAChRs on midbrain dopamine and GABA neurons.
- When nicotine is combined with menthol, the upregulation that occurs is significantly greater than either alone.
- Individually, nicotine increases phasic dopamine neuron firing, while menthol decreases this pattern of firing.
- Combined menthol and nicotine enhance phasic dopamine neuron firing to a level that is greater than nicotine alone.
- Ultimately, these changes increase dopamine neuron excitability and enhance nicotine reward.

References

Ahijevych, K., & Parsley, L. A. (1999). Smoke constituent exposure and stage of change in black and white women cigarette smokers. *Addictive Behaviors, 24*, 115–120.

Ai, J., Taylor, K. M., Lisko, J. G., Tran, H., Watson, C. H., & Holman, M. R. (2015). Menthol content in US marketed cigarettes. *Nicotine & Tobacco Research*. https://doi.org/10.1093/ntr/ntv162.

Alsharari, S. D., King, J. R., Nordman, J. C., Muldoon, P. P., Jackson, A., Zhu, A. Z. X., et al. (2015). Effects of menthol on nicotine pharmacokinetic, pharmacology and dependence in mice. *PLoS One, 10*, e0137070.

Ashley, M., Dixon, M., Sisodiya, A., & Prasad, K. (2012). Lack of effect of menthol level and type on smokers' estimated mouth level exposures to tar and nicotine and perceived sensory characteristics of cigarette smoke. *Regulatory toxicology and pharmacology: RTP, 63*, 381–390.

Ashoor, A., Nordman, J. C., Veltri, D., Yang, K. H., Shuba, Y., Al Kury, L., et al. (2013a). Menthol inhibits 5-HT3 receptor-mediated currents. *The Journal of Pharmacology and Experimental Therapeutics, 347*, 398–409.

Ashoor, A., Nordman, J. C., Veltri, D., Yang, K.-H. S., Al Kury, L., Shuba, Y., et al. (2013b). Menthol binding and inhibition of α7-nicotinic acetylcholine receptors. *PLoS One, 8*, e67674.

Benowitz, N. L., Herrera, B., & Jacob, P., 3rd (2004). Mentholated cigarette smoking inhibits nicotine metabolism. *The Journal of Pharmacology and Experimental Therapeutics, 310*, 1208–1215.

Biswas, L., Harrison, E., Gong, Y., Avusula, R., Lee, J., Zhang, M., et al. (2016). Enhancing effect of menthol on nicotine self-administration in rats. *Psychopharmacology, 233*, 3417–3427.

Brody, A., Mukhin, A., La Charite, J., Ta, K., Farahi, J., Sugar, C., et al. (2013). Up-regulation of nicotinic acetylcholine receptors in menthol cigarette smokers. *The International Journal of Neuropsychopharmacology, 16*, 957–966.

Caraballo, R. S., & Asman, K. (2011). Epidemiology of menthol cigarette use in the United States. *Tobacco Induced Diseases, 9*(Suppl 1), S1.

CDC. (2016). Tobacco use among middle and high school students—United States, 2011–2015. *Morbidity and Mortality Weekly Report, 65*, 361–367.

Corvalan, N. A., Zygadlo, J. A., & Garcia, D. A. (2009). Stereo-selective activity of menthol on GABA(A) receptor. *Chirality, 21*, 525–530.

Delnevo, C. D., Gundersen, D. A., Hrywna, M., Echeverria, S. E., & Steinberg, M. B. (2011). Smoking-cessation prevalence among U.S. smokers of menthol versus non-menthol cigarettes. *American Journal of Preventive Medicine, 41*, 357–365.

D'Silva, J., Boyle, R. G., Lien, R., Rode, P., & Okuyemi, K. S. (2012). Cessation outcomes among treatment-seeking menthol and nonmenthol smokers. *American Journal of Preventive Medicine, 43*, S242–S248.

Engle, S. E., Shih, P. Y., McIntosh, J. M., & Drenan, R. M. (2013). alpha4 alpha6 beta2* nicotinic acetylcholine receptor activation on ventral tegmental area dopamine neurons is sufficient to stimulate a depolarizing conductance and enhance surface AMPA receptor function. *Molecular Pharmacology, 84*, 393–406.

Faure, P., Tolu, S., Valverde, S., & Naude, J. (2014). Role of nicotinic acetylcholine receptors in regulating dopamine neuron activity. *Neuroscience, 282*, 86–100.

FDA. (2012). *Priliminary scientific evaluation of the possible public health effects of menthol versus nonmenthol cigarettes.* (2012). https://www.fda.gov/downloads/ucm361598.pdf.

Hall, A. C., Turcotte, C. M., Betts, B. A., Yeung, W. Y., Agyeman, A. S., & Burk, L. A. (2004). Modulation of human GABAA and glycine receptor currents by menthol and related monoterpenoids. *European Journal of Pharmacology, 506*, 9–16.

Hans, M., Wilhelm, M., & Swandulla, D. (2012). Menthol suppresses nicotinic acetylcholine receptor functioning in sensory neurons via allosteric modulation. *Chemical Senses, 37*, 463–469.

Henderson, B. J., Wall, T., Henley, B. M., Kim, C. H., Nichols, W. A., Moaddel, R., et al. (2016). Menthol alone upregulates midbrain nAChRs, alters nAChR sybtype stoichiometry, alters dopamine neuron firing frequency, and prevents nicotine reward. *The Journal of Neuroscience, 36*, 2957–2974.

Henderson, B. J., Wall, T. R., Henley, B. M., Kim, C. H., McKinney, S., & Lester, H. A. (2017). Menthol enhances nicotine reward-related behavior by potentiating nicotine-induced changes in nAChR function, nAChR upregulation, and DA neuron excitability. *Neuropsychopharmacology*. https://doi.org/10.1038/npp.2017.72.

Jasinska, A. J., Zorick, T., Brody, A. L., & Stein, E. A. (2013). Dual role of nicotine in addiction and cognition: a review of neuroimaging studies in humans. *Neuropharmacology, 84*, 111–122.

Journigan, V. B., & Zaveri, N. T. (2013). TRPM8 ion channel ligands for new therapeutic applications and as probes to study menthol pharmacology. *Life Sciences, 92*, 425–437.

Lau, B. K., Karim, S., Goodchild, A. K., Vaughan, C. W., & Drew, G. M. (2014).Menthol enhances phasic and tonic GABAA receptor-mediated currents in midbrain periaqueductal grey neurons. *British Journal of Pharmacology, 171*, 2803–2813.

Liu, L., Zhao-Shea, R., McIntosh, J. M., Gardner, P. D., & Tapper, A. R. (2012).Nicotine persistently activates ventral tegmental area dopaminergic neurons via nicotinic acetylcholine receptors containing alpha4 and alpha6 subunits. *Molecular Pharmacology, 81*, 541–548.

Lochlann Jain, S. S. (2003). "Come up to the Kool taste": African American upward mobility and the semiotics of smoking menthols. *Public Culture, 15*, 295–322.

Mansvelder, H. D., Keath, J. R., & McGehee, D. S. (2002). Synaptic mechanisms underlie nicotine-induced excitability of brain reward areas. *Neuron, 33*, 905–919.

Nashmi, R., Xiao, C., Deshpande, P., McKinney, S., Grady, S. R., Whiteaker, P., et al. (2007). Chronic nicotine cell specifically upregulates functional α4* nicotinic receptors: basis for both tolerance in midbrain and enhanced long-term potentiation in perforant path. *The Journal of Neuroscience*, *27*, 8202–8218.

Oz, M., El Nebrisi, E. G., Yang, K. S., Howarth, F. C., & Al Kury, L. T. (2017). Cellular and molecular targets of menthol actions. *Frontiers in Pharmacology*, *8*, 472.

Pons, S., Fattore, L., Cossu, G., Tolu, S., Porcu, E., McIntosh, J. M., et al. (2008). Crucial role of α4 and α6 nicotinic acetylcholine receptor subunits from ventral tegmental area in systemic nicotine self-administration. *The Journal of Neuroscience*, *28*, 12318–12327.

Proctor, R. N. (2012). *Golden holocaust: Origins of the cigarette catastrophe and the case for abolition*. University of California Press.

Sine, S. M., & Engel, A. G. (2006). Recent advances in Cys-loop receptor structure and function. *Nature*, *440*, 448–455.

Srinivasan, R., Pantoja, R., Moss, F. J., Mackey, E. D. W., Son, C., Miwa, J., et al. (2011). Nicotine upregulates α4β2 nicotinic receptors and ER exit sites via stoichiometry-dependent chaperoning. *The Journal of General Physiology*, *137*, 59–79.

Tapper, A. R., McKinney, S. L., Nashmi, R., Schwarz, J., Deshpande, P., Labarca, C., et al. (2004). Nicotine activation of α4* receptors: sufficient for reward, tolerance and sensitization. *Science*, *306*, 1029–1032.

Ton, H. T., Smart, A. E., Aguilar, B. L., Olson, T. T., Kellar, K. J., & Ahern, G. P. (2015). Menthol enhances the desensitization of human alpha3 beta4 nicotinic acetylcholine receptors. *Molecular Pharmacology*, *88*, 256–264.

Villanti, A. C., Mowery, P. D., Delnevo, C. D., Niaura, R. S., Abrams, D. B., & Giovino, G. A. (2016). Changes in the prevalence and correlates of menthol cigarette use in the USA, 2004-2014. *Tobacco Control*, *25*, ii14–ii20.

Wang, T., Wang, B., & Chen, H. (2014). Menthol facilitates the intravenous self-administration of nicotine in rats. *Frontiers in Behavioral Neuroscience*, *8*, 437.

Watt, E. E., Betts, B. A., Kotey, F. O., Humbert, D. J., Griffith, T. N., Kelly, E. W., et al. (2008). Menthol shares general anesthetic activity and sites of action on the GABA(A) receptor with the intravenous agent, propofol. *European Journal of Pharmacology*, *590*, 120–126.

WHO (2016) Banning menthol in tobacco products. http://apps.who.int/iris/bitstream/10665/205928/1/9789241510332_eng.pdf.

CHAPTER

13

Cigarette Smoking and Nicotine: Effects on Multiple Sclerosis

Insa Backhaus, Alice Mannocci, Giuseppe La Torre, Aldo Liccardi

Department of Public Health and Infectious Diseases, Sapienza University of Rome, Piazzale Aldo Moro, Rome, Italy

Abbreviations

Anti-EBNA	serum antibody titers to the Epstein-Barr virus nuclear antigens
CI	confidence interval
CNS	central nervous system
EBV	Epstein-Barr virus
EU	European Union
HLA	human leukocyte antigen
MS	multiple sclerosis
OR	odds ratio

13.1 INTRODUCTION

Cigarette smoking is a leading cause of preventable diseases worldwide. Annually, 700,000 premature deaths are attributable to smoking in the EU alone. While cigarette smoking is known to be a main cause of respiratory diseases, cardiovascular diseases, and autoimmune diseases, growing data suggest that nicotine, a cigarette component known for its addictive properties, perhaps decreases the incidence and the severity of diseases such as Parkinson's disease (Quik, Bordia, & O'Leary, 2007) and Multiple Sclerosis (MS).

MS is a complex progressive disease of the central nervous system (CNS), an inflammatory demyelinating condition, and one of the most common neurological disorders and causes of nontraumatic disability in young adults. Estimates of the number of people affected by MS throughout the world range between 1.1 and 2.5 million. The onset of MS is typically during early adulthood, between the ages of 20 and 50 years, the peak of productivity. MS is therefore a pressing public health issue. The manifestations of MS vary from a harmless illness to a rapidly evolving and disabling disease requiring intense lifestyle adjustments. The course of MS can be relapsing-remitting or progressive (Hauser & Goodin, 2012).

Autoimmune disorders such as MS occur when the immune system attacks the body's own healthy tissue by mistake (Longo, Longo, & Harrison, 2012). In the case of MS, within the CNS, the immune system attacks myelin, a fatty substance that surrounds and insulates the nerve fibers, known as axons. During this process, these insulating covers of the brain and the spinal cord are damaged (Fig. 13.1).

This process is also known as demyelization. The myelin sheath becomes thinner and eventually breaks down; consequently, the axons cannot function properly anymore. Nerve impulses traveling to and from the brain and spinal cord are interrupted, resulting in a wide range of symptoms such as a tingling sensation, weakness, impaired balance, and blurred vision (Hauser & Goodin, 2012) (Table 13.1).

Despite the fact that almost 200 years has passed since its discovery in 1868, the etiology of MS is not fully known (Gohil, 2015). Researchers believe that genetic predisposition and certain environmental factors play a major role in disease development. Environmental factors, for instance, include infection with Epstein-Barr virus, latitude, and cigarette smoking. Particularly, cigarette smoking has been brought forward negatively and is emerging as a strong risk factor for chronic inflammatory diseases such as MS. A systematic review and meta-analysis of 26 studies revealed that ever-smokers had an increased risk of 55% compared with never-smokers (Zhang et al., 2016). Similar results were shown in another meta-analysis of 34 studies. This meta-analysis claimed that, compared with nonsmokers, the OR estimate was 1.46 (1.33 and1.59) among ever-smokers, 1.57 (1.34 and 1.80) among current smokers, and 1.36 (1.27 and 1.46) among ex-smokers (Poorolajal, Bahrami, Karami, & Hooshmand, 2016). Numerous studies support the argument that cigarette smoking not only is an

Neuroscience of Nicotine
https://doi.org/10.1016/B978-0-12-813035-3.00013-7

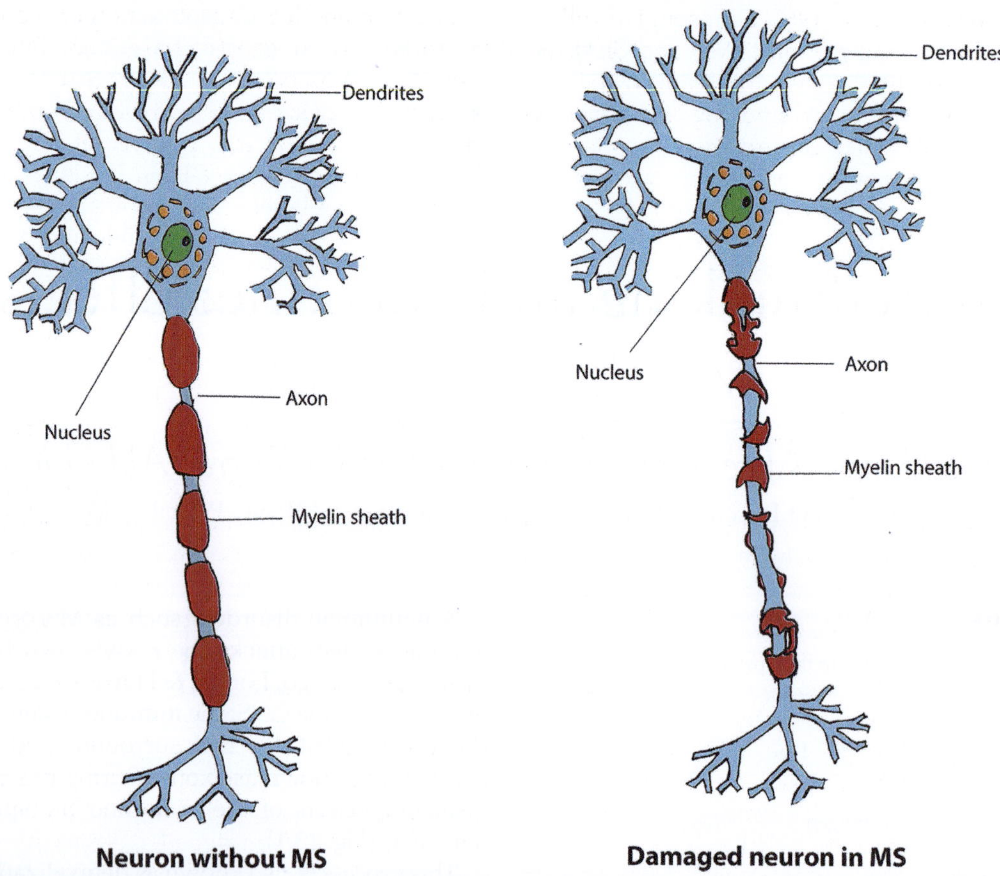

FIG. 13.1 Healthy neuron without MS (left) and damaged neuron with MS (right).
It shows a normal neuron on the left side vs demyelinated neuron found in MS on the right side. Axons are responsible for electric impulses that travel from the cell body of the neuron to the central nervous system and the peripheral nervous system. A fatty protective layer, known as myelin, surrounds the axon and aids in the transmission of the electric impulses along the axon. MS results in the demyelination, loss, or breakdown of myelin and therefore interrupts or distorts the message along the axon to the CNS.

TABLE 13.1 Key Facts of Multiple Sclerosis

- Multiple sclerosis is an autoimmune disease affecting the central nervous system
- The typical onset of Multiple Sclerosis is between the ages 20 and 40 years
- MS prevalence is higher in the north than in the south
- The course can be either relapsing-remitting or progressive
- There are four types:
  - Relapsing-remitting multiple sclerosis
  - Secondary-progressive multiple sclerosis
  - Primary-progressive multiple sclerosis
  - Progressive-relapsing multiple sclerosis
- Researchers believe that genetic predisposition and the environment play a crucial role in disease development
- Symptoms are numbness or a tingling sensation, weakness, impaired balance, blurred vision, and bladder or bowel problems as well as cognitive and emotional changes

This table lists the key facts of multiple sclerosis, including the epidemiology, the course and types of MS, and the risk factors and symptoms patient may experience.

important risk factor for increasing MS risk but also has an impact on the development and progression of MS (Hernán et al., 2005).

This chapter provides a comprehensive review of the findings on the relationship between cigarette smoking, nicotine, and MS. The specific questions addressed in this chapter are as follows:

- Is there an association between cigarette smoking and multiple sclerosis susceptibility?
- Is there an association between nicotine and multiple sclerosis susceptibility?
- What are the possible mechanisms linking smoking and MS?

13.2 CIGARETTE SMOKING AND MS RISK

The history of establishing a possible link between multiple sclerosis and cigarette smoking dates back to the 1960s when observational studies were first published (Antonovsky et al., 1965; Simpson, Newell, & Schapira, 1966). Antonovsky et al. (1965) and Simpson et al. (1966) were one of the first research teams who independently published studies on smoking and MS risk. However, these studies had controversial results and did not lead to any meaningful conclusion. Antonovsky et al. (1965) and his research team questioned 241 MS patients from Israel about whether they had ever smoked prior to disease onset. The results showed that among MS cases, a significantly higher percentage had smoked before the age of onset than compared with the control group (44% vs 36%, $p = 0.02$) (Antonovsky et al., 1965). Simpson and colleagues instead concluded that, among cases and controls from England, there was no overall difference in the proportion of patients and controls who were smokers (Simpson et al., 1966).

Over the past several decades, the topic of a possible relationship between MS and smoking has aroused the interest of many research teams, and the past decade has seen a striking growth in the number of published articles. Table 13.2 gives an overview of observational studies (case-control, cohort, and cross-sectional studies). In the 1990s, two longitudinal studies from the United Kingdom were published. Both studies observed that women who regularly smoked had a higher risk of MS, although these findings were not significant (Thorogood & Hannaford, 1998; Villard-Mackintosh & Vessey, 1993). Nevertheless, it must be noted that both studies primarily focused on the relation between oral contraceptives and MS risk and smoking was assessed only as a secondary factor. Hernán, Oleky, and Ascherio (2001) studied the association between MS incidence and smoking in two cohort studies of US women, the US Nurses' Health Study (121,700 women) and the US Nurses' Health Study II (116,671 women). Results indicate a relative incidence rate of 1.6 (CI: 95% 1.2–2.1) among current smokers, after adjusting for age, latitude, and ancestry. A retrospective case-control study from Canada (2001), for instance, detected a direct and significant association between cigarette smoking and the risk of MS (Ghadirian et al., 2001). The researcher studied cigarette-smoking habits of 200 individuals and 202 age- and sex-matched controls. The results indicated an OR of 1.6 (CI: 95% 1.0–2.4) for ever-smokers compared with never-smokers. For heavy smokers, who smoked 20–40 cigarettes per day, the risk rose almost twofold with an OR of 1.9 (CI: 95% 1.2–3.2) (Ghadirian et al., 2001), giving evidence of a dose-response correlation between the number of cigarettes smoked and the risk of developing MS. Ghadirian et al. (2001) based their analysis on reported cigarette smoking in the year prior to MS diagnosis, adjusted for age, sex, and education. A case-control study from Belgrade confirmed these findings. Pekmezovic et al. (2006) demonstrated that cigarette smoking among MS patients was more common than in the controls (Pekmezovic et al., 2006). Furthermore, the researcher observed a dose-response relationship between the risk of MS and both duration (years) of smoking ($p = 0.027$) and number of cigarettes smoked daily ($p = 0.021$). Despite all that, MS is a complex disease, and it is probable that smoking interacts with other MS risk factors and that such a possible interplay might even increase MS risk. Simon and colleagues (Simon et al., 2010) examined exactly such an interplay.

The researcher studied how smoking interacts with serum antibody titers to the Epstein-Barr virus nuclear antigens (anti-EBNA) and HLA-DR15 (Simon et al., 2010). They conducted individual and pooled analyses among about 440 cases and 860 controls from three MS case-control studies. The results showed that increased MS risk associated with higher anti-EBNA titers was stronger among "ever-smokers" than "never-smokers." However, this was only observed among people who had high anti-EBNA titers. In fact, investigating the combined effects of smoking and anti-EBNA Ab titer, ever-smoker with the highest anti-EBNA titers had a sevenfold increased MS risk (RR: 7.4, 95% CI: 3.6–15.0) (Simon et al., 2010).

A crucial point in determining causation of any environmental factor is the interaction between susceptibility and dose (van der Mei, Simpson, Stankovich, & Taylor, 2011). Though many of the studies reported an increasing MS susceptibility with an increasing numbers of cigarettes smoked (Carlens et al., 2010; Thorogood & Hannaford, 1998), others showed no effect (Antonovsky et al., 1965; Jafari et al., 2009).

Nevertheless, when combining case-control studies, a summary OR of 1.40 (95% CI: 1.29–1.52) among "ever"- vs "never"-smokers and 1.42 (95% CI: 1.26–1.60) among "current smokers" vs "nonsmokers" is shown (Backhaus, Mannocci, Lemmens, & La Torre, 2016). Fig. 13.2 shows a forest plot of a recent meta-analysis conducted by Backhaus et al. (2016).

Similar results were also demonstrated in other meta-analyses (Handel et al., 2011; Hawkes, 2007; Poorolajal et al., 2016; Zhang et al., 2016). Poorolajal et al. (2016), for instance, revealed that the estimated OR of MS was 1.46 (95% CI: 1.33–1.59) among ever-smokers and 1.57 (95% CI: 1.34–1.80) among current smokers. Zhang et al. (2016) who divided their analysis into conservative and nonconservative analyses propose that smoking is a risk factor for MS. Whereas the conservative model included only studies where smoking behavior was described prior to disease onset, the nonconservative model included all studies irrespective of whether smoking behavior occurred before onset or simultaneously. Both analyses revealed that smoking is a risk factor for MS with an OR of 1.55 (95% CI:

TABLE 13.2 Characteristics of Observational Studies Assessing Smoking and MS Risk

Lead author (year)	Country	Study design	Diagnostic criteria	Number of cases	Number of controls	RR/OR	Confidence interval	Female/male ratio cases
Antonovsky et al. (1965)	Israel	Case-control	No stated	241	964	1.4	1.05–1.86	1.2
Alonso, Cook, Maghzi, and Divani (2011)	Iran	Case-control	McDonald criteria	394	394	1.7	0.9–3.07	3.7
Al-Afasy et al. (2013)	Kuwait	Case-control	Neurologist diagnosis	101	202	1.7	0.9–3.4	1.2
Briggs et al. (2014)	The United States/ Sweden	Case-control	Neurologist diagnosis	1506	1070	1.97	1.29–3.00	NA
Carlens et al. (2010)	Sweden	Cohort	Medical record diagnosis	277,777 (total cohort)		1.9	1.4–2.6	All males
Engdahl et al. (2014)	Sweden	Case-control	McDonald criteria	446	487	1.57	1.16–2.12	2.5
Ghadirian, Dadgostar, Azani, and Maisonneuve (2001)	Canada	Case-control	No stated	200	202	1.6	1–2.4	2.2
Healy et al. (2009)	The United States	Cross-sectional	McDonald criteria	1465		2.41	1.09–5.34	NA
Hedström, Bäärnhielm, Olsson, and Alfredsson (2009)	Sweden	Case-control	McDonald criteria	902	1855	1.5	1.3–1.8	2.6
Hedström et al. (2011)	Sweden	Case-control	McDonald criteria	843	1209	1.6	1.3–2.0	NA
Hedström, Hillert, Olsson, and Alfredsson (2013)	Sweden	Case-control	McDonald criteria	7883	9264	1.5	1.4–1.6	2.6
Hernán et al. (2001)	The United States	Cohort	Poser criteria	238,371 (total cohort)		1.6	1.2–2.1	All females
Hernán et al. (2005)	The United Kingdom	Case-control	Poser criteria	201	1913	1.3	1–1.7	2.4
Jafari, Hoppenbrouwers, Hop, Breteler, and Hintzen (2009)	The Netherlands	Case-control	McDonald criteria	136	204	1.09	0.68–1.73	1.7
Mansouri et al. (2014)	Iran	Case-control	Poser and McDonald criteria	1403	883	1.93	1.31–2.73,	3.11
Maghzi et al. (2011)	Iran	Case-control	McDonald criteria	516	1090	2.67	1.70–4.21	3.7
Kotzamani et al. (2012)	Crete	Case-control	Clinical MRI	657	593	1.9 (f: 1.8, 2.0)		1.6
Pekmezovic et al. (2006)	Serbia	Case-control	Poser criteria	210	210	2.4	1.3–4.2	NA
O'Gorman, Bukhari, Todd, Freeman, and Broadley (2014)	Australia	Case-control	McDonald criteria	560	480	1.9	1.5–2.5	4.5
Rodríguez Regal et al. (2009)	Spain	Case-control	McDonald criteria	138	138	2.18	1.3–3.6	2.2

TABLE 13.2 Characteristics of Observational Studies Assessing Smoking and MS Risk—cont'd

Lead author (year)	Country	Study design	Diagnostic criteria	Number of cases	Number of controls	RR/OR	Confidence interval	Female/ male ratio cases
Roudbari, Ansar, and Yousefzad (2013)	Iran	Cross-sectional	Poser criteria, old Schumacher and, McDonald criteria	514		Hazard ratio 2.25	95% CI: 1.3–3.88	1.4
Ramagopalan et al. (2013)	Canada	Case-Control	Not stated	3157	756	1.32	1.10–1.60	3.3
Salzer et al. (2013)	Sweden	Case-control	ICD	192	384	1.5	0.98–2.3	NA
da Silva et al. (2009)	Brazil	Case-control	Poser criteria	81	81	7.6	2.1–28.2	N/A
Simon et al. (2010)	The United States	Case-control	Poser criteria and imaging	442	865	1.5	1–2.4	2.1
Sundström, Nyström, and Hallmans (2008)	Sweden	Case-control	Poser criteria	109	218	1.4	0.78–2.6	1.3
Simpson et al. (1966)	Australia	Case-control	Not stated	584	No data	1.1	0.9–1.3	1.5
Thorogood and Hannaford (1998)	The United States	Cohort	GP diagnosis	No data		1.2	0.8–1.8	Females
Villard-Mackintosh and Vessey (1993)	The United States	Cohort	Diagnosed by specialist	No data		1.5	0.94–2.53	Females
Zorzon et al. (2003)	The United States	Case-control	Not stated	140	131	1.9 OR (current smokers)	1.1–3.2	1.8

It shows the characteristics of observational studies conducted between 1965 and 2014. Most studies were conducted in Europe (e.g., Sweden).

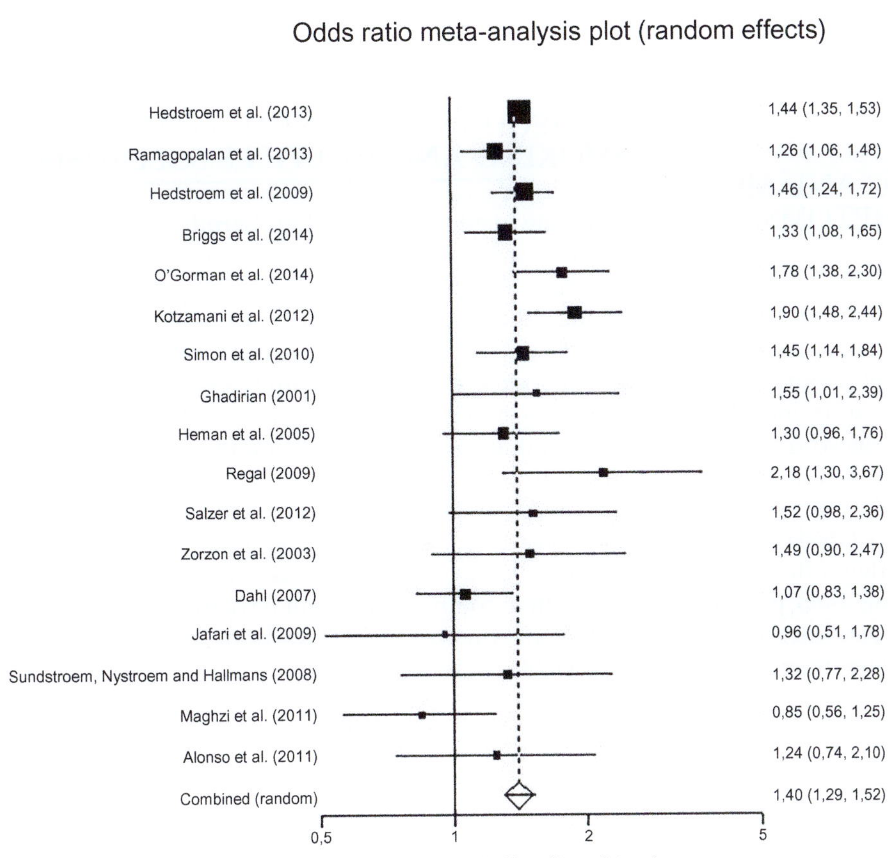

FIG. 13.2 Forest plot of the risk of MS for "ever"- vs "never"-smokers. The forest plot shows the pooled analysis for the comparison of ever- vs never-smokers. Each horizontal line shows the 95% CI for the estimate for each study. The open diamond represents the overall summary estimate. *The figure was extracted from Backhaus et al. (2016).*

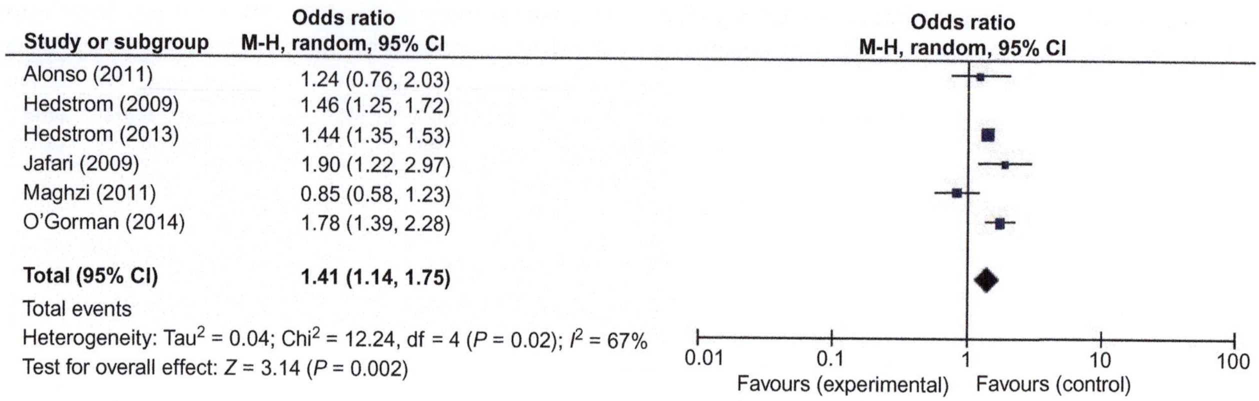

FIG. 13.3 Forest plot of the risk of MS for "ever"- vs "never"-smokers for studies using McDonald criteria.
It shows the pooled analysis for the comparison of ever- vs never-smokers for studies using the McDonald criteria. Each horizontal line shows the 95% CI for the estimate for each study. The open diamond represents the overall summary estimate.

TABLE 13.3 Characteristics of Studies Assessing Swedish Snuff, Nicotine, and MS Risk

Lead author (year)	Country	Study design	Diagnostic criteria	Number of cases	Number of control	RR/OR	Lower 95% and upper CI
Carlens et al. (2010)	Sweden	Cohort	Not stated	277,777 total cohort		1	0.8–1.4
Hedström et al. (2009)	Sweden	Case-control	McDonald criteria	902	1855	0.3	0.1–0.8
Hedström et al. (2013)	Sweden	Case-control	McDonald criteria	7883	9437	0.83	0.75–0.92

The studies in this table show that snuff takers have a decreased risk of MS development compared with nonsnuff takers.

1.48–1.62) for the conservative and an OR of 1.57 (95% CI: 1.50–1.64) for the nonconservative analysis. In addition, when performing a pooled analysis of studies using the McDonald criteria to ascertain MS cases, comparable results can be seen (OR: 1.41, 95% CI: 1.14–1.75) (Fig. 13.3).

13.3 CIGARETTE SMOKING AND MS PROGRESSION AND SYMPTOMS WORSENING

The association between cigarette smoking and disease progression is still uncertain. Whereas some studies suggest an increased risk, others do not. Hernán et al. (2005) conducted a nested case-control study including MS 201 patients and 1913 controls. Patients who smoked had a threefold higher risk of a relapsing-remitting clinical form transforming into a secondary-progressive form. Courville, Maschmeyer, and Delay (1964) demonstrated symptoms worsening immediately after smoking. A cross-sectional survey, including 1465 patients from the United States, reported that at baseline, current smokers had significantly worse disease than never-smokers in terms of Expanded Disability Status Scale score and Multiple Sclerosis Severity Score (Healy et al., 2009). Current smokers were also more likely to have primary-progressive MS compared with never-smokers. Compared with previously conducted studies, the study included a considerably larger sample size and had more statistical power.

13.4 MECHANISMS LINKING CIGARETTE SMOKING AND MULTIPLE SCLEROSIS

Smoking in general has various harmful effects on the immune system causing relative immune deficiency and higher rates of infection. However, exactly how cigarette smoking influences MS risk has not been fully understood. Several mechanisms have been proposed including immunomodulatory and direct effects of smoking on the blood-brain barrier, neurotoxic effects of cigarette smoke, and effects on the respiratory tract (Pugliatti et al., 2008).

Cigarette smoke contains over 6000 potentially toxic components such as nicotine, carbon monoxide, and hydrogen cyanide (Arnson, Shoenfeld, & Amital, 2010). Some of these components might have direct toxic effects on the CNS (Costenbader & Karlson, 2006). One main and probably most influential component that has gained attention is hydrogen cyanide. In the late 1960s, Levine (1967) discovered that some components of tobacco smoke such as hydrogen cyanide and its main metabolite thiocyanate might have a toxic effect on the CNS including the corpus callosum. Interestingly, not only is the corpus callosum most often abnormal in most MS cases, but also it is most vulnerable to cyanide damage (Evangelou et al., 2000; Ozturk et al., 2010). Inhaled cigarette smoke is a main source of hydrogen cyanide. This would explain the idea that cigarette smoke is the main contributor to increased MS risk. Another potential explanation is that cigarette smoke has been linked to an increasing susceptibility of respiratory diseases, which in turn is another risk factor associated with raised MS risk (Egesten et al., 2008; Jafari & Hintzen, 2011). This is in line with the studies from Sweden suggesting that cigarette smoking, but not nicotine, is responsible for the observed effect. Components of tobacco smoke mediated through the airway may cause irritations in the lungs (Hedström et al., 2009).

Experimental studies using animal models have shown that components of cigarettes affect parts of the immune system, including innate immunity, B and T cells, and natural killer cells (Fusby et al., 2010; Handel et al., 2011). Lastly, cigarette smoking has been linked to other autoimmune diseases (e.g., Graves' disease and Crohn's disease) that allow for analogies to MS to be made (Costenbader & Karlson, 2006).

13.5 NICOTINE AND MULTIPLE SCLEROSIS

A substantial body of evidence suggests that smoking cigarettes increases the risk of developing MS and may even speed up disease progression. Whether that also applies to smokeless tobacco and nicotine, the ingredient in tobacco, accountable for its addictive natures, remains unknown. Unlike cigarette smoke, nicotine is not yet considered carcinogenic and might even have therapeutic and protective effects on inflammatory diseases (Piao et al., 2009). In comparison with cigarette smoking, literature about nicotine and its effect on MS is lacking. Table 13.3 gives an overview of the studies focusing on Swedish snuff, nicotine, and MS risk. Two studies from Sweden explored that nicotine could have a protective effect in the etiology of MS (Hedström et al., 2009; Hedström et al., 2013). In a case-control study, the research team studied the snuff habit of around 7880 MS patients and 9440 controls. The results suggest that people who use snuff have a decreased risk (17%) of developing MS in comparison with people who have never used snuff (Hedström et al., 2013). In addition, the research group demonstrated evidence of an inverse relationship between increased cumulative dose of snuff use and the risk of developing the disease (Hedström et al., 2013), thus indicating that the more snuff is used, the lower the risk of MS. Like cigarette smoking, the mechanism behind this association is not fully understood yet. Aside from nicotine, chewable tobacco contains numerous different substances that could result in the observed effect (Hedström et al., 2013). A possible explanation could be the effect nicotine has on myelin antigens. A research group from Iran examined the role of nicotine in the experimental autoimmune encephalomyelitis model. They demonstrated, using animals models, that mice who received nicotine as treatment presented less inflammation in histopathologic evaluation alongside myelin sheath protection (Naddafi, Reza Haidari, Azizi, Sedaghat, & Mirshafiey, 2013).

13.6 CONCLUSION

Unlike cigarette smoke, nicotine is not yet considered a carcinogenic and might even have therapeutic potential as a neuroprotective and antiinflammatory agent. It must be realized that MS is a very complex disease and risk factors are most likely influenced by the interaction with other environmental and genetic factors. Furthermore, observational studies have some limitations that must be acknowledged. Retrospective and questionnaire-based studies are prone to recall bias (Hassan, 2005). The problem of recall bias is especially important for studies concerning multiple sclerosis since recalling information depends completely on memory that can be limited among MS cases. Additional studies are needed to explain how cigarette smoking and nicotine affect the immune system and their role in MS development and progression. Nevertheless, in light of the results, people and patients would benefit from smoking cessation. Physicians, health-care professionals, and policy makers should carefully consider this association, especially because cigarette smoking remains a main modifiable risk factor.

MINI-DICTIONARY OF TERMS

Autoimmune disease When body tissues are attacked by its own immune system. This can lead to damage of the attacked body tissue.

B cell Also known as B lymphocytes. Responsible for producing antibodies.

Causation The relationship between cause and effect. Also referred to as causality.

Corpus callosum Connects the left and right hemisphere and is responsible for their communication.

McDonald criteria Diagnostic criteria for multiple sclerosis. These criteria were introduced in 2001 and revised in 2005 and focus on clinical, radiological, and laboratory evidence of lesions at different times and in different areas.

Multiple sclerosis An autoimmune inflammatory demyelinating disease affecting the central nervous system.

Odds ratio A measure of association between an exposure and an outcome. The odds ratio is the odds that an outcome will occur given a specific exposure, compared with the odds of the outcome when the exposure is not present.

Prevalence The proportion of a population who have or who had a particular characteristic in a given time period.

Snuff/moist snuff Smokeless tobacco that is usually inhaled or snuffed through the nasal cavity. It is also known as dipping tobacco. It is a finely ground and moistened tobacco product.

T cell Also known as T lymphocyte and is important for cell-mediated immunity.

Summary Points

- This chapter focuses on the relationship between cigarette smoking, nicotine, and multiple sclerosis.
- Data from observational studies suggest that cigarette smoking is an important factor in multiple sclerosis development.
- Nicotine on the other hand might have a protective effect.
- A research group from Sweden demonstrated that people who use snuff have a 17% lower risk of developing multiple sclerosis compared with individuals who have not used snuff.
- More research is needed to fully understand the mechanism behind the association.

References

Al-Afasy, H. H., Al-Obaidan, M. A., Al-Ansari, Y. A., Al-Yatama, S. A., Al-Rukaibi, M. S., Makki, N. I., et al. (2013). Risk factors for multiple sclerosis in Kuwait: a population-based case-control study. *Neuroepidemiology, 40*(1), 30–35. (2013). https://doi.org/10.1159/000341240.

Alonso, A., Cook, S. D., Maghzi, A.-H., & Divani, A. A. (2011). A case-control study of risk factors for multiple sclerosis in Iran. *Multiple Sclerosis (Houndmills, Basingstoke, England), 17*(5), 550–555. (2011). https://doi.org/10.1177/1352458510397685.

Antonovsky, A., Leibowitz, U., Smith, H. A., Medalie, J. M., Balogh, M., Kats, R., et al. (1965). Epidemiologic study of multiple sclerosis in Israel. I. An overall review of methods and findings. *Archives of Neurology, 13*, 183–193.

Arnson, Y., Shoenfeld, Y., & Amital, H. (2010). Effects of tobacco smoke on immunity, inflammation and autoimmunity. *Journal of Autoimmunity, 34*(3), J258–J265. (2010). https://doi.org/10.1016/j.jaut.2009.12.003.

Backhaus, I., Mannocci, A., Lemmens, P. H. H. M., & La Torre, G. (2016). Smoking as a risk factor for developing multiple sclerosis: a meta-analysis of observational studies. *La Clinica Terapeutica, 167*(3), 82–92.

Briggs, F. B. S., Acuna, B., Shen, L., Ramsay, P., Quach, H., Bernstein, A., et al. (2014). Smoking and risk of multiple sclerosis: evidence of modification by NAT1 variants. *Epidemiology (Cambridge, Mass), 25*(4), 605–614. (2014). https://doi.org/10.1097/EDE.0000000000000089.

Carlens, C., Hergens, M.-P., Grunewald, J., Ekbom, A., Eklund, A., Höglund, C. O., et al. (2010). Smoking, use of moist snuff, and risk of chronic inflammatory diseases. *American Journal of Respiratory and Critical Care Medicine, 181*(11), 1217–1222. (2010). https://doi.org/10.1164/rccm.200909-1338OC.

Costenbader, K. H., & Karlson, E. W. (2006). Cigarette smoking and autoimmune disease: what can we learn from epidemiology? *Lupus, 15*(11), 737–745. (2006). https://doi.org/10.1177/0961203306069344.

Courville, C. B., Maschmeyer, J. E., & Delay, C. P. (1964). Effect of smoking on the acute exacerbations of multiple sclerosis. *Bulletin of the Los Angeles Neurological Society, 29*, 1–6.

da Silva, K. R. P., Alvarenga, R. M. P., Fernandez, Y., Fernandez, O., Alvarenga, H., & Thuler, L. C. S. (2009). Potential risk factors for multiple sclerosis in Rio de Janeiro: a case-control study. *Arquivos de Neuro-Psiquiatria, 67*(2A), 229–234.

Egesten, A., Brandt, L., Olsson, T., Granath, F., Inghammar, M., Löfdahl, C.-G., et al. (2008). Increased prevalence of multiple sclerosis among COPD patients and their first-degree relatives: a population-based study. *Lung, 186*(3), 173–178. (2008). https://doi.org/10.1007/s00408-008-9081-y.

Engdahl, E., Gustafsson, R., Ramanujam, R., Sundqvist, E., Olsson, T., Hillert, J., et al. (2014). HLA-A(*)02, gender and tobacco smoking, but not multiple sclerosis, affects the IgG antibody response against human herpesvirus 6. *Human Immunology, 75*(6), 524–530. (2014). https://doi.org/10.1016/j.humimm.2014.03.001.

Evangelou, N., Konz, D., Esiri, M. M., Smith, S., Palace, J., & Matthews, P. M. (2000). Regional axonal loss in the corpus callosum correlates with cerebral white matter lesion volume and distribution in multiple sclerosis. *Brain: A Journal of Neurology, 123*(Pt 9), 1845–1849.

Fusby, J. S., Kassmeier, M. D., Palmer, V. L., Perry, G. A., Anderson, D. K., Hackfort, B. T., et al. (2010). Cigarette smoke-induced effects on bone marrow B-cell subsets and CD4+:CD8+ T-cell ratios are reversed by smoking cessation: influence of bone mass on immune cell response to and recovery from smoke exposure. *Inhalation Toxicology, 22*(9), 785–796. (2010). https://doi.org/10.3109/08958378.2010.483258.

Ghadirian, P., Dadgostar, B., Azani, R., & Maisonneuve, P. (2001). A case-control study of the association between socio-demographic, lifestyle and medical history factors and multiple sclerosis. *Canadian Journal of Public Health, 92*(4), 281–285.

Gohil, K. (2015). Multiple sclerosis: progress, but no cure. *Pharmacy and Therapeutics, 40*(9), 604–605.

Handel, A. E., Williamson, A. J., Disanto, G., Dobson, R., Giovannoni, G., & Ramagopalan, S. V. (2011). Smoking and multiple sclerosis: an updated meta-analysis. *PLoS One, 6*(1)e16149(2011). https://doi.org/10.1371/journal.pone.0016149.

Hassan, E. (2005). Recall Bias can be a threat to retrospective and prospective research designs. *The Internet Journal of Epidemiology, 3*(2) Retrieved from (2005). http://ispub.com/IJE/3/2/13060.

Hauser, S. L., & Goodin, D. S. (2012). Chapter 380. Multiple sclerosis and other demyelinating diseases. In D. L. Longo, A. S. Fauci, D. L. Kasper, S. L. Hauser, J. L. Jameson, & J. Loscalzo (Eds.), Harrison's Principles of Internal Medicine (18th ed.). New York, NY: The McGraw-Hill Companies. Retrieved from accessmedicine.mhmedical.com/content.aspx?aid=9147780

Hawkes, C. H. (2007). Smoking is a risk factor for multiple sclerosis: a metanalysis. *Multiple Sclerosis (Houndmills, Basingstoke, England), 13*(5), 610–615. (2007). https://doi.org/10.1177/1352458506073501.

Healy, B. C., Ali, E., Guttmann, C. R. G., Chitnis, T., Glanz, B. I., Buckle, G., et al. (2009). Smoking and disease progression in multiple sclerosis. *Archives of Neurology, 66*(7), 858–864. (2009). https://doi.org/10.1001/archneurol.2009.122.

Hedström, A. K., Bäärnhielm, M., Olsson, T., & Alfredsson, L. (2009). Tobacco smoking, but not Swedish snuff use, increases the risk of multiple sclerosis. *Neurology, 73*(9), 696–701. (2009). https://doi.org/10.1212/WNL.0b013e3181b59c40.

Hedström, A. K., Hillert, J., Olsson, T., & Alfredsson, L. (2013). Nicotine might have a protective effect in the etiology of multiple sclerosis. *Multiple Sclerosis (Houndmills, Basingstoke, England), 19*(8), 1009–1013. (2013). https://doi.org/10.1177/1352458512471879.

Hedström, A. K., Sundqvist, E., Bäärnhielm, M., Nordin, N., Hillert, J., Kockum, I., et al. (2011). Smoking and two human leukocyte antigen genes interact to increase the risk for multiple sclerosis. *Brain: A Journal of Neurology, 134*(Pt 3), 653–664. (2011). https://doi.org/10.1093/brain/awq371.

Hernán, M. A., Jick, S. S., Logroscino, G., Olek, M. J., Ascherio, A., & Jick, H. (2005). Cigarette smoking and the progression of multiple sclerosis. *Brain: A Journal of Neurology*, *128*(Pt 6), 1461–1465. (2005). https://doi.org/10.1093/brain/awh471.

Hernán, M. A., Oleky, M. J., & Ascherio, A. (2001). Cigarette smoking and incidence of multiple sclerosis. *American Journal of Epidemiology*, *154*(1), 69–74. https://doi.org/10.1093/aje/154.1.69.

Jafari, N., & Hintzen, R. Q. (2011). The association between cigarette smoking and multiple sclerosis. *Journal of the Neurological Sciences*, *311*(1–2), 78–85. (2011). https://doi.org/10.1016/j.jns.2011.09.008.

Jafari, N., Hoppenbrouwers, I. A., Hop, W. C. J., Breteler, M. M. B., & Hintzen, R. Q. (2009). Cigarette smoking and risk of MS in multiplex families. *Multiple Sclerosis (Houndmills, Basingstoke, England)*, *15*(11), 1363–1367. (2009). https://doi.org/10.1177/1352458509345907.

Kotzamani, D., Panou, T., Mastorodemos, V., Tzagournissakis, M., Nikolakaki, H., Spanaki, C., et al. (2012). Rising incidence of multiple sclerosis in females associated with urbanization. *Neurology*, *78*(22), 1728–1735. (2012). https://doi.org/10.1212/WNL.0b013e31825830a9.

Levine, S. (1967). Experimental cyanide encephalopathy: Gradients of susceptibility in the corpus callosum. *Journal of Neuropathology and Experimental Neurology*, *26*(2), 214–222.

Longo, H., Longo, D. L., & Harrison, T. R. (2012). In D. L. Longo (Ed.), *Harrison's principles of internal medicine*. (18th ed.). New York: McGraw-Hill.

Maghzi, A.-H., Etemadifar, M., Heshmat-Ghahdarijani, K., Moradi, V., Nonahal, S., Ghorbani, A., et al. (2011). Cigarette smoking and the risk of multiple sclerosis: a sibling case-control study in Isfahan, Iran. *Neuroepidemiology*, *37*(3–4), 238–242. (2011). https://doi.org/10.1159/000332765.

Mansouri, B., Asadollahi, S., Heidari, K., Fakhri, M., Assarzadegan, F., Nazari, M., et al. (2014). Risk factors for increased multiple sclerosis susceptibility in the Iranian population. *Journal of Clinical Neuroscience: Official Journal of the Neurosurgical Society of Australasia*, *21*(12), 2207–2211. (2014). https://doi.org/10.1016/j.jocn.2014.04.020.

Naddafi, F., Reza Haidari, M., Azizi, G., Sedaghat, R., & Mirshafiey, A. (2013). Novel therapeutic approach by nicotine in experimental model of multiple sclerosis. *Innovations in Clinical Neuroscience*, *10*(4), 20–25.

O'Gorman, C., Bukhari, W., Todd, A., Freeman, S., & Broadley, S. A. (2014). Smoking increases the risk of multiple sclerosis in Queensland, Australia. *Journal of Clinical Neuroscience: Official Journal of the Neurosurgical Society of Australasia*, *21*(10), 1730–1733. https://doi.org/10.1016/j.jocn.2014.01.009.

Ozturk, A., Smith, S. A., Gordon-Lipkin, E. M., Harrison, D. M., Shiee, N., Pham, D. L., et al. (2010). MRI of the Corpus callosum in multiple sclerosis: association with disability. *Multiple Sclerosis (Houndmills, Basingstoke, England)*, *16*(2), 166. (2010). https://doi.org/10.1177/1352458509353649.

Pekmezovic, T., Drulovic, J., Milenkovic, M., Jarebinski, M., Stojsavljevic, N., Mesaros, S., et al. (2006). Lifestyle factors and multiple sclerosis: a case-control study in Belgrade. *Neuroepidemiology*, *27*(4), 212–216. (2006). https://doi.org/10.1159/000096853.

Piao, W.-H., Campagnolo, D., Dayao, C., Lukas, R. J., Wu, J., & Shi, F.-D. (2009). Nicotine and inflammatory neurological disorders. *Acta Pharmacologica Sinica*, *30*(6), 715–722. (2009). https://doi.org/10.1038/aps.2009.67.

Poorolajal, J., Bahrami, M., Karami, M., & Hooshmand, E. (2016). Effect of smoking on multiple sclerosis: a meta-analysis. *Journal of Public Health*, *2010*, fdw030(2016). https://doi.org/10.1093/pubmed/fdw030.

Pugliatti, M., Harbo, H. F., Holmøy, T., Kampman, M. T., Myhr, K.-M., Riise, T., et al. (2008). Environmental risk factors in multiple sclerosis. *Acta Neurologica Scandinavica. Supplementum*, *188*, 34–40. (2008). https://doi.org/10.1111/j.1600-0404.2008.01029.x.

Quik, M., Bordia, T., & O'Leary, K. (2007). Nicotinic receptors as CNS targets for Parkinson's disease. *Biochemical Pharmacology*, *74*(8), 1224–1234. (2007). https://doi.org/10.1016/j.bcp.2007.06.015.

Ramagopalan, S. V., Lee, J. D., Yee, I. M., Guimond, C., Traboulsee, A. L., Ebers, G. C., et al. (2013). Association of smoking with risk of multiple sclerosis: A population-based study. *Journal of Neurology*, *260* (7), 1778–1781. https://doi.org/10.1007/s00415-013-6873-7.

Rodríguez Regal, A., del Campo Amigo, M., Paz-Esquete, J., Martínez Feijoo, A., Cebrián, E., Suárez Gil, P., et al. (2009). A case-control study of the influence of the smoking behaviour in multiple sclerosis. *Neurologia (Barcelona, Spain)*, *24*(3), 177–180.

Roudbari, S. A., Ansar, M. M., & Yousefzad, A. (2013). Smoking as a risk factor for development of secondary progressive multiple sclerosis: a study in IRAN, Guilan. *Journal of the Neurological Sciences*, *330*(1–2), 52–55. (2013). https://doi.org/10.1016/j.jns.2013.04.003.

Russo, C., Morabito, F., Luise, F., Piromalli, A., Battaglia, L., Vinci, A., et al. (2008). Hyperhomocysteinemia is associated with cognitive impairment in multiple sclerosis. *Journal of Neurology*, *255*(1), 64–69. (2008). https://doi.org/10.1007/s00415-007-0668-7.

Salzer, J., Hallmans, G., Nyström, M., Stenlund, H., Wadell, G., & Sundström, P. (2013). Smoking as a risk factor for multiple sclerosis. *Multiple Sclerosis (Houndmills, Basingstoke, England)*, *19*(8), 1022–1027. (2013). https://doi.org/10.1177/1352458512470862.

Simon, K. C., van der Mei, I. A. F., Munger, K. L., Ponsonby, A., Dickinson, J., Dwyer, T., et al. (2010). Combined effects of smoking, anti-EBNA antibodies, and HLA-DRB1*1501 on multiple sclerosis risk. *Neurology*, *74*(17), 1365–1371. (2010). https://doi.org/10.1212/WNL.0b013e3181dad57e.

Simpson, C. A., Newell, D. J., & Schapira, K. (1966). Smoking and multiple sclerosis. *Neurology*, *16*(10), 1041–1043. passim.

Sundström, P., Nyström, L., & Hallmans, G. (2008). Smoke exposure increases the risk for multiple sclerosis. *European Journal of Neurology*, *15*(6), 579–583. https://doi.org/10.1111/j.1468-1331.2008.02122.x.

Thorogood, M., & Hannaford, P. C. (1998). The influence of oral contraceptives on the risk of multiple sclerosis. *British Journal of Obstetrics and Gynaecology*, *105*(12), 1296–1299.

van der Mei, I. A. F., Simpson, S., Stankovich, J., & Taylor, B. V. (2011). Individual and joint action of environmental factors and risk of MS. *Neurologic Clinics*, *29*(2), 233–255. (2011). https://doi.org/10.1016/j.ncl.2010.12.007.

Villard-Mackintosh, L., & Vessey, M. P. (1993). Oral contraceptives and reproductive factors in multiple sclerosis incidence. *Contraception*, *47*(2), 161–168.

Zhang, P., Wang, R., Li, Z., Wang, Y., Gao, C., Lv, X., et al. (2016). The risk of smoking on multiple sclerosis: a meta-analysis based on 20,626 cases from case-control and cohort studies. *PeerJ*, *4*, e1797 (2016). https://doi.org/10.7717/peerj.1797.

Zorzon, M., Zivadinov, R., Nasuelli, D., Dolfini, P., Bosco, A., Bratina, A., et al. (2003). Risk factors of multiple sclerosis: a case-control study. *Neurological Sciences: Official Journal of the Italian Neurological Society and of the Italian Society of Clinical Neurophysiology*, *24*(4), 242–247. (2003). https://doi.org/10.1007/s10072-003-0147-6.

CHAPTER

14

Tobacco and Positron-Emission Tomography (PET) of the Dopaminergic System: A Review of Human Studies

Chidera C. Chukwueke, Bernard Le Foll†*

*Department of Pharmacology and Toxicology, University of Toronto, Toronto, ON, Canada

†Translational Addiction Research Laboratory, Centre for Addiction and Mental Health, Toronto, ON, Canada

Abbreviations

BP_{ND}	binding potential (nondisplaceable)
D1/D2/D3	dopamine receptor subtype 1/2/3
DA	dopamine
Denic	denicotinized cigarette
MDD	major depressive disorder
Nic	nicotinized cigarette
PET	positron-emission tomography

14.1 TOBACCO SMOKING AND THE DOPAMINERGIC SYSTEM

As tobacco smoking is the leading cause of preventable mortality in the United States (Surgeon General, 2014), there is a significant need to better understand the mechanisms that maintain smoking in order to improve treatment. Nicotine, the psychoactive ingredient in tobacco, is thought to exert its rewarding and reinforcing effects through modulation of dopamine (DA) neurotransmission (Di Chiara, 2000). The advent of positron-emission tomography (PET) currently makes it possible to assess DA activity in vivo in humans in response to nicotine use.

14.1.1 Positron-Emission Tomography

With earlier iterations of this technology dating back to the 1950s, PET imaging has evolved to be capable of producing high-resolution images used to assess neurochemical brain activity (Portnow, Vaillancourt, & Okun, 2013). PET scanners reconstruct images of spatial density by detecting photons or gamma rays produced from the use of positron-emitting radiotracer. By employing radiotracers that bind to receptors of interest, PET imaging allows the assessment of regional receptor density and neurotransmitter fluctuation. This is done by measuring the radiotracer binding potential (BP), which has been defined as the ratio of B_{max} (receptor density) to K_D (radiotracer equilibrium dissociation constant) (Mintun, Raichle, Kilbourn, Wooten, & Welch, 1984). As radiotracer binding affinity is the inverse of K_D, BP can be seen as the product of receptor density and radiotracer affinity. In PET studies, BP_{ND} refers to the ratio of specifically bound radiotracer to nondisplaceable radiotracer in tissue at equilibrium. Used frequently in reference tissue methods, BP_{ND} compares the concentration of radiotracer in receptor-rich with that in receptor-free regions.

In PET studies investigating tobacco smoker's brain activity, BP_{ND} provides an indirect measure of neurotransmission and receptor availability. By examining the difference between radiotracer BP_{ND} before and after an acute challenge (e.g., smoking a cigarette), changes in DA neurotransmitter concentration are inferred as a result of increased challenge-induced competition between radiotracer and endogenous neurotransmitter. Therefore, in comparison to a baseline BP_{ND} value, observed decreases in BP_{ND} following an acute challenge are thought to reflect an increase in DA concentration (while increases in BP_{ND} likewise reflect neurotransmitter decrease). On the other hand, receptor availability can be observed by comparing baseline BP_{ND} between smokers and nonsmokers. This design infers regional receptor availability under stable baseline conditions, whereby differences between

Neuroscience of Nicotine
https://doi.org/10.1016/B978-0-12-813035-3.00014-9

groups may reflect neuroadaptive changes in response to chronic tobacco use. In this chapter, we will review studies that use PET to assess the effects of tobacco smoking or nicotine on DA activity in humans.

14.1.2 Dopamine

DA has been implicated in drug addiction (Volkow, Wang, Fowler, & Tomasi, 2012), therefore necessitating the evaluation of its role in tobacco dependence. Like most substances of abuse, tobacco smoking increases extracellular DA concentration, particularly in the striatum (Brody, Mandelkern, et al., 2009). Drug-elicited DA concentration is also generally associated with the drug's rewarding effects (Volkow et al., 2012). This review focuses on nicotine PET studies analyzing DA synthesis, neurotransmission, and DA receptor (e.g., subtypes D1, D2, and D3) availability inferred by using radiotracers that have distinct regional BP_{ND} and receptor affinities. DA activity in the striatum is indirectly measured using radiotracers [^{11}C]-raclopride and [^{18}F]-fallypride for D2 activity and [^{11}C]-SCH 23390 for D1 activity. Furthermore, this chapter reviews extrastriatal D3 activity measured with [^{11}C]-(+)-PHNO, cortical D2 activity measured with [^{11}C]-FLB-457, and DA synthesis measured with [^{18}F]-fluorodopa.

Though the studies reviewed here utilize various radiotracers, some methodological commonalities exist. First, some studies assess DA release by examining the difference between radiotracer BP_{ND} before and after acute nicotine administration. As explained above, DA release is indirectly measured by inferring that challenge-induced changes in radiotracer BP_{ND} proportionally reflect challenge-induced changes in DA concentration. Second, some studies compare baseline BP_{ND} between chronic and nonsmokers to assess DA receptor levels in response to long-term nicotine use. However, it is important to note that, with this design, it remains difficult to ascertain whether observed differences in BP_{ND} reflect preexisting neural configurations or whether they are consequences of nicotine use.

14.2 PET STUDIES IN HUMAN SMOKERS

14.2.1 [^{11}C]-Raclopride

DA release. [^{11}C]-raclopride, a D2/D3 receptor antagonist with modest preference for striatal D2 over D3 (Lammertsma & Hume, 1996), has been a primary radiotracer used in PET studies investigating nicotine-related D2 activity. Studies using this tracer (Table 14.1) have demonstrated DA release in the ventral striatum of smokers who did, compared to those who did not, smoke a regular cigarette during an outdoor midscan break (Brody et al., 2004). A double-blind, randomized, placebo-controlled study confirmed DA release following nicotine gum consumption in the ventral striatum of smokers, but not nonsmokers (Takahashi et al., 2008). More recent reports exhibit smoking-induced DA release in the dorsal striatum (Weinstein et al., 2016). These studies provide multimodal evidence in support of nicotine-induced DA release. Sex differences have also been reported in studies using [^{11}C]-raclopride (Cosgrove et al., 2014) and with different radiotracers (Brown et al., 2012; Okita, Petersen, et al., 2016). With [^{11}C]-raclopride, male smokers displayed DA release in the ventral striatum, while females displayed faster smoking responses in the dorsal striatum following smoking a regular cigarette (Cosgrove et al., 2014). This report adds differential sex-specific regional and temporal DA responses to nicotine.

By contrast, some studies have reported no DA release following either regular cigarette smoking in a scanner or intranasal nicotine administration (Barrett, Boileau, Okker, Pihl, & Dagher, 2004; Montgomery, Lingford-Hughes, Egerton, Nutt, & Grasby, 2007). Though using the same radiotracer, these results may be attributable to discrepant methodology. Studies supporting nicotine-induced DA release typically show this in the ventral striatum (Brody et al., 2004; Takahashi et al., 2008) and more recently in the dorsal striatum (Weinstein et al., 2016), though this may be exemplified more ventrally in males, while females display time-course sensitive DA responses in the dorsal region (Cosgrove et al., 2014). However, nicotine administration inconsistently induces apparent DA release (Barrett et al., 2004; Montgomery et al., 2007). This inconsistency in the literature might suggest that nicotine-induced DA release is sensitive to the route of nicotine administration (e.g., smoking vs nasal spray), the smoking methods (e.g., outdoors vs in a scanner), or other factors (e.g., withdrawal in smokers and multiple chemicals in tobacco) that exist in these studies. It should be noted that because of small sample sizes and the lack of systematic control of some factors in most of these studies, it is difficult to conclude on the influence of those factors at this point.

Nicotinized vs denicotinized cigarettes. To elucidate the specific contribution of nicotine on DA release, studies have employed a nicotinized (nic; containing regular amounts of nicotine) versus denicotinized (denic; containing significantly reduced amounts of nicotine, i.e., almost no nicotine) cigarette paradigm. This paradigm is beneficial in that smoking a denic cigarette allows us to examine the effects of smoking cues (i.e., the act of smoking minus nicotine delivery) on DA activity, if a proper control condition is tested.

Studies employing this paradigm have produced relatively consistent results. One study showed that smokers

TABLE 14.1 [^{11}C]-Raclopride Studies

Authors	Subjects	Objective	Intervention	Results/conclusion
Wiers et al. (2017)	Eight current smokers (7.4 cigarettes per day), 10 ex-smokers, and 18 nonsmokers	Evaluate the effect of smoking status on baseline D2 receptor availability and DA release	Retrospectively analyzed data of subjects who underwent one PET scan after either placebo or methylphenidate injection	Smokers displayed significantly lower receptor levels, while ex-smokers did not differ from nonsmokers. No significant different in DA release between current smokers and nonsmokers after methylphenidate injection
Weinstein et al. (2016)	10 Smokers smoking 21 cigarettes/day on average	Assessed the association between pretreatment cigarette use and smoking-induced DA release using [^{11}C]-raclopride	One PET scan on the seventh week of bupropion treatment where subjects smoked during a midscan break in a different room	Smoking-induced DA release in caudate and putamen
Cosgrove et al. (2014)	16 Smokers (eight males, eight females)	Examined the differential effects of smoking between males and females using [^{11}C]-raclopride	One PET scan where subjects smoked one or two cigarettes while in the scanner	Found significant right ventral striatum activation in males and none in females. Females displayed faster responses to smoking in the dorsal putamen
Albrecht et al. (2013)	81 Total subjects. 34 nontreatment-seeking alcoholics. 21 social drinking smokers. 26 social drinking nonsmokers	Conducted a retrospective study using [^{11}C]-raclopride PET data to examine the differential contribution of alcohol and tobacco in chronic comorbid use in striatal DA concentration	N/A	Found lower binding in smoking groups compared to the nonsmoking group in the dorsal striatum and primarily in the dorsal putamen. No differences were found in the ventral striatum
Domino et al. (2013)	17 Males smoking a mean of 18.5 cigarettes/day	Investigated the effects of plasma nicotine concentration on mood and DA release by comparing nicotinized with denicotinized cigarettes and using [^{11}C]-raclopride	Two PET scans, the first after smoking denicotinized cigarettes and the second after smoking nicotinized cigarettes	Denicotinized cigarettes released DA in the right caudate and putamen, while nicotinized cigarettes released DA in the whole striatum
Domino et al. (2012)	20 Male smokers smoking 15–40 cigarettes/day	Examined the modulating role of the OPRM1 A118G polymorphism on DA release following smoking a nicotinized vs a denicotinized cigarette using [^{11}C]-raclopride	Two PET scans. First, after smoking two denicotinized cigarettes, then again after smoking two nicotinized cigarettes	Those with the G allele displayed greater nicotine-induced DA release in the right caudate and ventral pallidum when compared to individuals with homozygous AA alleles
Brody, Mandelkern, et al. (2009)	62 Dependent smokers smoking 24.2 cigarettes/day	Compared smoking-induced DA release using [^{11}C]-raclopride in nicotinized with that in denicotinized cigarettes	One PET scan during which they smoked either a nicotinized or a denicotinized cigarette	Both cigarette types relieved craving; however, only nicotinized cigarettes increased DA release in the ventral striatum and improved mood
Brody, Olmstead, et al. (2009)	56 Smokers smoking 24.7 cigarettes/day with and without MDD	Used [^{11}C]-raclopride to examine smoking-induced DA release between smokers with and without depressive episodes	One PET scan during which subjects smoked a regular cigarette	In comparison to MDD−, MDD+ subjects showed heightened smoking-induced DA release. Greater smoking-induced DA release was associated with greater prescan depression and anxiety levels
Busto et al. (2009)	17 MDD+/− smokers, smoking an average of 16.65 cigarettes/day, compared to 21 MDD+/− nonsmokers	Compared the effects of an acute amphetamine challenge across healthy controls and depressed subjects with or without tobacco dependence using [^{11}C]-raclopride	Two PET scans, a baseline scan, then a scan following oral administration of amphetamine	Smokers, but not MDD+ subjects, showed lower binding at baseline. MDD+ smokers displayed the greatest amphetamine-induced decreases in binding

Continued

TABLE 14.1 [^{11}C]-Raclopride Studies—cont'd

Authors	Subjects	Objective	Intervention	Results/conclusion
Takahashi et al. (2008)	Six smokers smoking over 15 cigarettes/day compared to six nonsmokers	A double-blind, randomized, placebo-controlled study that investigated the differences in [^{11}C]-raclopride binding between smokers and nonsmokers using nicotine gum	Two PET scans. Both scans were done, while the subject chewed either a nicotine or placebo gum throughout the scan	Nicotine gum elicited significantly lower binding when compared to placebo in the striatal region of smokers but not nonsmokers. Striatal binding correlated with degree of nicotine dependence
Scott et al. (2007)	Six smokers smoking a mean of 16.7 cigarettes/day compared to six age and sex matched nonsmokers	Examined the effects of nicotinized vs denicotinized cigarette smoking on D2 and μ-opioid receptors using [^{11}C]-raclopride and [^{11}C]-carfentanil	One PET scan. During the scan, subjects were sequentially injected with both radiotracers in a counterbalanced manner. Smokers smoked the denicotinized cigarette during the first half of the scan and smoked the nicotinized cigarette during the second half	Smokers did not display a significant difference in D2 binding compared to nonsmokers. Nicotinized cigarettes resulted in lower D2/D3 binding compared to nicotinized cigarettes in the left ventral basal ganglia
Montgomery et al. (2007)	10 Smokers smoking a mean of 10.9 cigarettes/day	Determined the effect of nicotine administration via nasal spray on DA release by using [^{11}C]-raclopride	One PET scan with nicotine nasal spray administration midscan	No significant difference in binding between pre- and postnicotine administration. Found several associations between nicotine-induced changes in mood and dorsal striatal binding
Brody et al. (2006)	45 Smokers smoking 23.9 cigarette/day	Examined the genetic variability involved in smoking-induced DA release using [^{11}C]-raclopride	One PET scan under either a smoking condition or a nonsmoking condition	Found associations between specific genetic polymorphisms and greater smoking-induced DA release
Barrett et al. (2004)	10 Male smokers smoking an average of 18 cigarettes/day	Assessed DA activity in response to acute cigarettes smoking using [^{11}C]-raclopride, as well as associations to nicotine craving and the hedonic effects of smoking	Two PET scans under a smoking condition, where subjects smoked cigarettes repeatedly while in the scanner, and a nonsmoking condition where they did not smoke	No significant difference in DA binding between smoking and the nonsmoking condition; however, the hedonic effects of smoking were significantly correlated with DA binding in the dorsal striatum
Brody et al. (2004)	20 Smokers smoking at least 15 cigarettes/day	Assessed the effect of nicotine administration via regular cigarette on DA release using [^{11}C]-raclopride	One PET scan where half the subjects underwent a smoking condition while the other half underwent a control condition	Smoking condition resulted in significant reductions in binding in the ventral striatum that was associated with reductions in self-reported craving

This table outlines the subjects, objectives, and conclusions of nicotine PET studies that use the radiotracer [^{11}C]-raclopride. *DA*, dopamine; *MDD+/−*, major depressive disorder positive or negative; *PET*, positron-emission tomography.

($N=6$) exhibited DA release in the ventral striatum after smoking a nic cigarette compared to a denic cigarette (Scott et al., 2007). These findings were replicated in another study with a much larger sample ($N=62$; Brody, Mandelkern, et al., 2009). Interestingly, another study assessing DA release in response to nic versus denic smoking confirmed nic-induced DA release, though in the whole striatum (i.e., ventral and dorsal), while also showing a denic-induced DA release in the right dorsal striatum (Domino et al., 2013).

Collectively, these studies provide further evidence to support smoking-induced DA release. Potentially, while nicotine itself may contribute to smoking-induced DA release in the ventral striatum, DA release in the dorsal striatum may be attributable to smoking cues. The drug addiction literature supports this view as ventral striatal

DA release has been linked to drugs rewarding and reinforcing effects while dorsal striatal DA release has been implicated in habit learning and exposure to drug-associated cues (Volkow, Fowler, Wang, Baler, & Telang, 2009; Volkow et al., 2012).

Subjective correlates. Subjective measures of craving, mood regulation, and nicotine dependence severity have been reported in relation to nicotine-induced DA activity. Smoking-induced mood changes and hedonic effects have been reported to correlate with DA release in the dorsal striatum (Barrett et al., 2004; Montgomery et al., 2007). On the other hand, studies correlate ventral striatal DA release to the reduction of subjective craving scores and mood improvements (Brody et al., 2004; Brody, Mandelkern, et al., 2009; Takahashi et al., 2008). Also, subjective measures of nicotine dependence severity are reported to correlate with the ventral striatal DA release (Scott et al., 2007; Takahashi et al., 2008) and baseline dorsal striatum DA receptor availability (Montgomery et al., 2007), making the regional distinctions less clear. While more research is needed to distinguish regional contribution in nicotine dependence, these results do support the idea that both regions play an important role in behavioral components of addiction (Volkow et al., 2012).

Nicotine and depression. Tobacco dependence may occur in the context of comorbid major depressive disorder (MDD). In studies using [^{11}C]-raclopride, smokers with past depressive episodes displayed greater smoking-induced DA release compared to those without (Brody, Olmstead, et al., 2009). On the other hand, smokers with MDD showed blunted amphetamine challenge-induced DA response compared to nonsmokers with or without MDD (Busto et al., 2009). Clear differences in the study design may account for the discrepant results. However, it is plausible that in smokers with comorbid mental health issues, the administration of the substance of dependence (e.g., nicotine for smokers) stimulates an overactive DA response while a less familiar substance (e.g., amphetamine for smokers) produces a weakened response as has been previously reported (Parsey et al., 2001). Altogether, it is clear that comorbid existence of MDD and nicotine dependence dysregulates the DA system more than any one disorder alone.

Genetic variation. Studies examining genetic variability in smokers using [^{11}C]-raclopride have shown an association to various genetic polymorphisms (Brody et al., 2006; Domino et al., 2012). According to these studies, carriers of the minor G allele of the μ-opioid receptor and genes associated with low resting DA tone (nine-repeat allele of the dopamine transporter, Val/Val genotype of the catechol-*O*-methyltransferase enzyme, and seven-repeat allele of the D4 receptor) show greater magnitudes of smoking-induced DA release. These studies recruited a majority male population, thus warranting further research. Though far from conclusive, genetic variability may be able to account for a significant portion of individual differences in smoking-induced DA release.

Receptor availability. Studies assessing receptor availability changes in tobacco dependence using [^{11}C]-raclopride have produced mixed results. While earlier studies show no difference in baseline receptor availability in smokers versus nonsmokers (Scott et al., 2007; Takahashi et al., 2008), more recent studies challenge these findings. Studies show that smokers with comorbid alcohol use and depressive issues have presented lower D2 receptor availability compared to nonsmokers (Albrecht, Kareken, & Yoder, 2013; Busto et al., 2009). This finding has recently been supported as Wiers et al. (2017) found lower D2 receptor availability in current smokers compared to nonsmokers using a retrospective study design. In this study, ex-smokers did not differ from nonsmokers in D2 receptor availability, suggesting that periods of abstinence may normalize receptor levels. While these reports are inconsistent, addiction-induced DA receptor downregulation is well documented (Volkow et al., 2009) and is seen in nicotine PET studies using different radiotracers (Dagher et al., 2001; Fehr et al., 2008; Yasuno et al., 2007), therefore supporting the narrative that tobacco dependence does indeed reduce DA receptor availability with the possibility of restoration upon abstinence. Interestingly, a recent study negatively correlated daily cigarette use with DA release in the dorsal striatum, suggesting neuroadaptive tolerance to smoking (Weinstein et al., 2016); however, these findings are preliminary.

14.2.2 [^{18}F]-Fallypride

Nicotine PET studies have utilized [^{18}F]-fallypride (Table 14.2), a high-affinity D2 radioligand (Riccardi et al., 2008, 2005), to assess D2/3 receptor availability in striatal and extrastriatal areas. [^{18}F]-Fallypride research shows lower D2 receptor availability in the dorsal striatum of smokers but not nonsmokers (Fehr et al., 2008). However, sex differences complicates this finding as studies reports lower dorsal striatum D2 availability in male smokers (Brown et al., 2012) in combination with higher midbrain D2 availability, positively correlated with dorsal striatal D2 availability, in female smokers (Okita, Petersen, et al., 2016). This indicates an underactive DA system in nicotine dependence where males experience more DA receptor downregulation than females.

[^{18}F]-Fallypride studies have confirmed associations between the DA system and craving, cigarette use, and dependence severity. Research positively correlates craving with ventral striatal receptor availability but negatively correlates craving with the anterior cingulate and inferior temporal cortex (Fehr et al., 2008). Furthermore, one study shows lower D2 availability in the ventral striatum correlated with higher cigarette use and dependence severity (Okita, Mandelkern, & London, 2016),

TABLE 14.2 [^{18}F]-Fallypride Studies

Authors	Subjects	Objective	Intervention	Results/conclusion
Okita, Mandelkern, et al. (2016)	20 Nondependent smokers	Correlated [^{18}F]-fallypride binding to both daily cigarette smoking and nicotine dependence	One PET scan	Lower binding in the ventral striatum correlated with higher cigarette use. Higher dependence scores correlated with lower binding in the whole striatum (ventral and dorsal striatum). This suggests a downregulation of DA receptors as cigarette use escalates and dependence develops
Okita, Petersen, et al. (2016)	18 Smokers smoking 5–20 cigarettes/day compared to 19 nonsmokers	Assessed sex differences in midbrain DA receptor availability using [^{18}F]-fallypride	One PET scan	Female smokers exhibited higher midbrain binding than male smokers. Midbrain binding was positively correlated with striatal binding and negatively correlated with dependence scores
Brown et al. (2012)	19 Light smokers smoking 4–14 cigarettes/day compared to 18 nonsmokers	Investigated sex differences in light smokers vs nonsmokers using [^{18}F]-fallypride	One PET scan	Male smokers showed significantly lower D2/D3 receptor availability in the caudate and putamen when compared to females and nonsmokers. Females did not differ from nonsmokers in receptor availability
Fehr et al. (2008)	17 Smokers smoking a mean of 19.5 cigarettes/day compared to 21 nonsmokers	Used [^{18}F]-fallypride to compare D2/D3 receptor availability in striatal and extrastriatal regions between smokers and nonsmokers	Two PET scans, first following regular cigarette consumption and, secondly, following 24-hour abstinence	Smokers displayed lower D2/D3 receptor availability in the putamen, but not in extrastriatal regions, following both regular cigarette consumption and abstinence. This study found correlations between binding in several brain regions and behavioral measures

This table outlines the subjects, objectives, and conclusions of nicotine PET studies that use the radiotracer [^{18}F]-fallypride. *DA*, dopamine; *PET*, positron-emission tomography.

while another shows dependence scores negatively correlated with dorsal striatum D2 availability (Okita, Petersen, et al., 2016). This suggests that craving may be differentially mediated by both striatal and extrastriatal areas, and in accordance with [^{11}C]-raclopride research, level of dependence may be related to DA receptor availability in the both striatal regions.

14.2.3 [^{11}C]-FLB-457

A single study has examined smoking-induced cortical DA release using [^{11}C]-FLB-457 (Table 14.3). [^{11}C]-FLB-457 is a high-affinity D2/3 receptor ligand that shows greater sensitivity to DA changes in cortical areas with low DA receptor density (Narendran et al., 2009). The sole study examining cortical DA transmission shows that smoking a cigarette releases DA primarily in the cingulate cortex (Wing, Payer, Houle, George, & Boileau, 2015). In this study, craving was not correlated with cortical DA release; thus, the role of cortical DA transmission in nicotine craving remains unclear. However, this does add to the literature that smoking a cigarette results in not only striatal DA release but also cortical DA release.

14.2.4 [^{11}C]-(+)-PHNO

[^{11}C]-(+)-PHNO is a D2/3 agonist reported to be more sensitive than [^{11}C]-raclopride to acute DA fluctuations (Shotbolt et al., 2012) and is a D3 preferring agonist in extrastriatal areas such as the substantia nigra (Tziortzi et al., 2011). To determine the specific involvement of

TABLE 14.3 The [^{11}C]-FLB-457 Study

Authors	Subjects	Objective	Intervention	Results/ conclusion
Wing et al. (2015)	10 Smokers smoking an average of 16 cigarettes/day	Examined DA release by tobacco smoking in extrastriatal and cortical brain regions using [^{11}C]-FLB-457	Two PET scans. First scan following 16-hour abstinence, the second scan following smoking to satiation	Found significant binding reductions in primarily the cingulate gyrus. Greater puff volume correlated with greater binding reductions

This table outlines the subjects, objectives, and conclusions of a nicotine PET study that used the radiotracer [^{11}C]-FLB-457. *DA*, dopamine; *PET*, positron-emission tomography.

D3 in nicotine dependence, studies using [^{11}C]-(+)-PHNO (Table 14.4) suggest that smoking a regular cigarette elevates DA in the limbic (i.e., ventral) striatum and the D3 rich ventral pallidum (Fig. 14.1) (Le Foll et al., 2014). Motivation to smoke, craving, and withdrawal symptoms have been correlated with DA release in the limbic striatum (Le Foll et al., 2014). However, presentation of smoking-related cues alone elicits no detectable DA response in a pilot study (Chiuccariello et al., 2013). At this point, these studies support striatal involvement and indicate the involvement of D3-rich areas (ventral pallidum) in response to smoking but not smoking-related environmental cues.

14.2.5 [11C]-SCH 23390

Research using [^{11}C]-SCH 23390, a D1 receptor antagonist, to measure D1 receptor availability (Table 14.5) has demonstrated significantly lower D1 receptor availability in the ventral striatum of smokers compared to nonsmokers (Dagher et al., 2001). These findings were replicated by Yasuno et al. (2007), showing lower ventral striatal D1 receptor availability in smokers that increased following 6 months of abstinence. Yasuno et al. (2007) also negatively correlated ventral striatum D1 receptors with cue-induced craving scores, suggesting an association between lower D1 receptor density and larger craving. These studies suggest that while chronic smoking induces the downregulation of D1 receptors, which is associated with larger craving, this downregulation may be reversible upon abstinence as seen with D2 receptors (Wiers et al., 2017).

14.2.6 [^{18}F]-Fluorodopa

Studies investigating DA synthesis capacity using [^{18}F]-fluorodopa in smokers have produced mixed results (Table 14.6). One study reports significantly higher DA activity in the dorsal striatum of male smokers than in nonsmokers (Salokangas et al., 2000), while Bloomfield et al. (2014) found no difference in the whole striatum related to smoking status. These inconsistent results

TABLE 14.4 [^{11}C]-(+)-PHNO Studies

Authors	Subjects	Objective	Intervention	Results/conclusion
Le Foll et al. (2014)	10 Smokers smoking 13.2 cigarettes/day	Investigated smoking-induced DA release in the striatum, ventral pallidum, and substantia nigra using [^{11}C]-(+)-PHNO	Two PET scans under a smoking and abstinent condition	The smoking condition resulted in significant reductions in binding in the limbic striatum and the ventral pallidum when compared to the abstinent condition. There were also correlations between PHNO binding and smoking behaviors
Chiuccariello et al. (2013)	18 Smokers smoking 18.2 cigarettes/day	Investigated DA release by presentation of smoking cue vs neutral cue using [^{11}C]-(+)-PHNO	One PET scan with smoking cue vs neutral cue (visual and tactile components)	No significant difference in DA levels in any brain region between smoking and neutral cue presentation

This table outlines the subjects, objectives, and conclusions of nicotine PET studies that use the radiotracer [^{11}C]-(+)-PHNO. *DA*, dopamine; *PET*, positron-emission tomography.

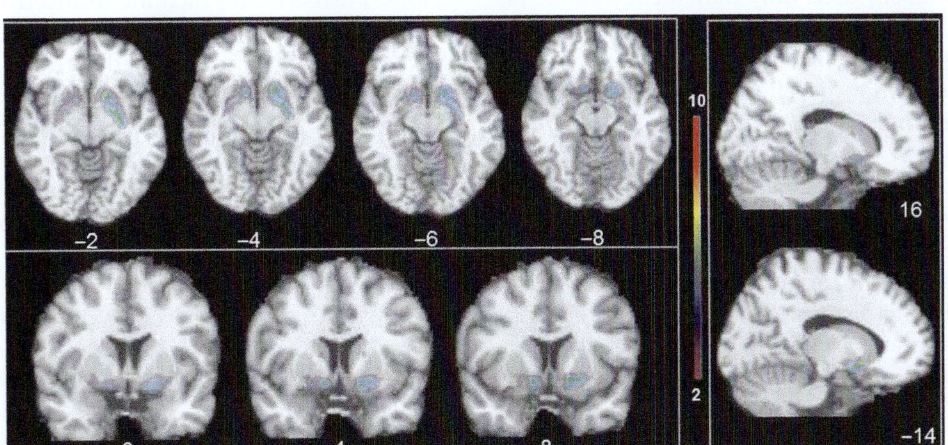

FIG. 14.1 [^{11}C]-(+)-PHNO binding in response to cigarette smoking.
T-statistical overlaid average T1 MRI showing clusters of significantly reduced [^{11}C]-(+)-PHNO BPND after smoking a cigarette. Greatest decreases in [^{11}C]-(+)-PHNO BPND cluster in the ventral part of the striatum and in the area that corresponds to the ventral pallidum. For visualization purposes, the image is threshold at a $P<.05$ uncorrected. *Adapted from Le Foll et al. (2014).*

TABLE 14.5 [^{11}C]-SCH23390 Studies

Authors	Subjects	Objective	Intervention	Results/conclusion
Yasuno et al. (2007)	18 Smokers smoking at least 15 cigarettes/day vs 12 nonsmokers	Used [^{11}C]-SCH23390 studies to investigate binding levels in D1 receptors in smokers vs nonsmokers during regular cigarette consumption and after months of abstinence	Subjects underwent one PET scan at baseline and subjects who achieved abstinence underwent subsequent PET scans	Smokers showed significantly lower D1 receptor binding than nonsmokers. D1 receptor binding in the bilateral nucleus accumbens increased after 3 and 6 months of abstinence. Finally, there was a significant negative correlation in the nucleus accumbens between smoking-induced brain activation, craving scores, and D1 receptor binding
Dagher et al. (2001)	11 Smokers smoking a mean of 19.4 cigarettes/day compared to 18 nonsmokers	Examined D1 receptor density in smokers vs nonsmokers using [^{11}C]-SCH23390 studies	Two PET scans, one following overnight abstinence and the other during regular smoking rates	Smokers displayed significantly less D1 receptor binding than nonsmokers in the whole striatum; however, there was no difference in D1 receptor binding between overnight abstinence and regular smoking

This table outlines the subjects, objectives, and conclusions of nicotine PET studies that use the radiotracer [^{11}C]-SCH23390. *DA*, dopamine; *PET*, positron-emission tomography.

TABLE 14.6 [^{18}F]-Fluorodopa

Authors	Subjects	Objective	Intervention	Results/conclusion
Bloomfield et al. (2014)	15 Smokers (average of 8.1 cigarettes/day) matched against 15 nonsmokers	Compared DA synthesis in smokers with nonsmokers using [^{18}F]-fluorodopa	One PET scan; smokers were abstinent for 3 hours prior to scanning	Found no associations in any striatal region between DA synthesis and smoker status, daily cigarette use or sex
Salokangas et al. (2000)	Nine smokers smoking 19.8 cigarettes/day compared to 10 nonsmokers	Examined whether dopamine synthesis in the basal ganglia differs between smokers and nonsmokers using [^{18}F]-fluorodopa	One PET scan	Smokers showed increased DA synthesis in the putamen and the caudate, especially in the right caudate and left putamen

This table outlines the subjects, objectives, and conclusions of nicotine PET studies that use the radiotracer [^{18}F]-fluorodopa. *DA*, dopamine; *PET*, positron-emission tomography.

may be a result of discrepant sample populations. Salokangas et al. (2000) used only male heavy smokers, while Bloomfield et al. (2014) recruited both male and female moderate smokers. Therefore, it is plausible that dopamine synthesis may be altered in heavy smokers but not moderate smokers.

14.3 CONCLUSION

Using various radiotracers, PET studies have elucidated much about the nicotine-DA relationship. Firstly, nicotine administration seems to elicit DA release in striatal (both ventral and dorsal), cortical (primarily the cingulate cortex), and D3-rich (the ventral pallidum) areas. Though nicotine itself may be a major contributor to DA release, smoking-related cues may also play an important role. Nicotine-induced DA release has not been consistently reported, however, and may be sensitive to route of nicotine administration, smoking method, and other factors. Furthermore, nicotine-induced DA release may differ between sexes, across various genetic polymorphisms, and within comorbid disorders.

Secondly, chronic tobacco use results in a downregulation of D1 and D2 receptors, though this may be reversible for upon abstinence. As with DA release, downregulation of DA receptors may also be experienced differently between sexes. DA synthesis may also be compromised as nicotine dependence develops, though this may affect heavier users more severely.

Finally, DA activity is regionally correlated with several smoking behaviors. Craving seems related to ventral striatal DA activity and receptor (D1 and D2) availability, though possibly also related to cortical regions. Mood changes and the pleasurable effects of smoking may be related to the dorsal striatum, while dependence severity may be related to activity in both striatal regions. These behaviors and tobacco dependence as a whole are effectively combatted with pharmacotherapy such as varenicline and bupropion. Interestingly, bupropion (Brody et al., 2010) and varenicline (Di Ciano et al., 2016) appear to affect DA transmission differently.

As tobacco dependence continues to be a considerable clinical issue, more research is needed to understand the complex tobacco-DA relationship. Overall, very limited PET studies have been conducted in this area, and we know very little on DA synthesis and how receptors evolve over time. There is clearly a need for more and larger studies. It would be important to delineate the impact of smoking on DA transmission in different brain areas and to explore the role of individual factors into the variability of the signal that is measured. Better understanding how subjects respond to drugs used clinically would also provide a better understanding of the mechanism of action of those drugs in human smokers.

MINI-DICTIONARY OF TERMS

Acute challenge The administration of a substance in order to elicit a measurable change on a defined parameter.

Binding potential A measure of a radiolabeled ligand's binding ability. This measure is typically a combination of receptor availability and ligand affinity.

Denicotinized cigarette A cigarette that contains a significantly reduced amount of nicotine, typically for research purposes.

Dorsal striatum A brain region that is part of the subcortical basal ganglia. This area typically consists of the caudate nucleus and the putamen.

Induced dopamine release A measure of the change in dopamine concentration elicited by a substance or action. For example, smoking-induced dopamine release is the release of dopamine following smoking.

Limbic striatum Similar to the ventral striatum and sometimes used interchangeably by authors. This brain area refers to the subdivisions of the striatum that receives neuronal projections from limbic structures (e.g., amygdala and hippocampus).

Nicotinized cigarette A cigarette that contains regular concentrations of nicotine. This cigarette is similar to commercially bought cigarettes but may also be produced specifically for research purposes.

Receptor availability The density or amount of receptors, in a defined brain region, to which ligand binding is possible.

Smoking-related cue This refers to stimuli associated with smoking. For example, pictures of a cigarette or lighter or the smell of cigarette smoke.

Ventral striatum A brain region that is part of the subcortical basal ganglia. This area typically includes the nucleus accumbens.

Key Facts of Radiotracers in PET Scans

- Radiotracers used in PET scans are positron-emitting chemical ligands that bind to receptors of interest.
- There are many radiotracers with different affinities to several receptors.
- They allow the indirect measure of neurotransmitter concentration by competing with endogenous neurotransmitter for receptor binding.
- Radiotracers allow to get a measure of receptor density by the estimation of BP_{ND} when the measures are done at rest.

Summary Points

- This chapter reviews human dopamine (DA) activity in nicotine dependence using positron-emission tomography (PET).
- PET employs different radiotracers to assess DA neurotransmission in different brain regions and the availability of different receptor subtypes in vivo.
- Research shows that tobacco administration increase DA levels in the striatum, D3-rich extrastriatal areas, and cortical areas.
- There have been some inconsistent reports on the ability of tobacco to elevate DA levels.
- Sex and genetic variability may influence those effects.
- Chronic tobacco use downregulates DA receptors, and this may be reversible after a period of abstinence.
- There are several tobacco-related phenotypes that are associated with DA changes: withdrawal, craving, mood changes, and dependence severity.
- Future research should focus on elucidating the mechanistic underpinnings of individual variability as it relates to acute neurotransmission and neuroadaptive changes in response to chronic tobacco use for the improvement of treatment.

References

Albrecht, D. S., Kareken, D. A., & Yoder, K. K. (2013). Effects of smoking on D2/D3 striatal receptor availability in alcoholics and social drinkers. *Brain Imaging and Behavior*, *7*(3), 326–334. https://doi.org/10.1007/s11682-013-9233-4.

Barrett, S. P., Boileau, I., Okker, J., Pihl, R. O., & Dagher, A. (2004). The hedonic response to cigarette smoking is proportional to dopamine release in the human striatum as measured by positron emission tomography and [11C]raclopride. *Synapse*, *54*(2), 65–71. https://doi.org/10.1002/syn.20066.

Bloomfield, M. A., Pepper, F., Egerton, A., Demjaha, A., Tomasi, G., Mouchlianitis, E., et al. (2014). Dopamine function in cigarette smokers: an [18F]-DOPA PET study. *Neuropsychopharmacology*, *39*(10), 2397–2404.

Brody, A. L., London, E. D., Olmstead, R. E., Allen-Martinez, Z., Shulenberger, S., Costello, M. R., et al. (2010). Smoking-induced change in intrasynaptic dopamine concentration: effect of treatment for tobacco dependence. *Psychiatry Research: Neuroimaging*, *183*(3), 218–224. https://doi.org/10.1016/j.pscychresns.2009.06.004.

Brody, A. L., Mandelkern, M. A., Olmstead, R. E., Allen-Martinez, Z., Scheibal, D., Abrams, A. L., et al. (2009). Ventral striatal dopamine release in response to smoking a regular vs a denicotinized cigarette. *Neuropsychopharmacology*, *34*(2), 282–289.

Brody, A. L., Mandelkern, M. A., Olmstead, R. E., Scheibal, D., Hahn, E., Shiraga, S., et al. (2006). Gene variants of brain dopamine pathways and smoking-induced dopamine release in the ventral caudate/nucleus accumbens. *Archives of General Psychiatry*, *63*(7), 808. https://doi.org/10.1001/archpsyc.63.7.808.

Brody, A. L., Olmstead, R. E., Abrams, A. L., Costello, M. R., Khan, A., Kozman, D., et al. (2009). Effect of a history of major depressive disorder on smoking-induced dopamine release. *Biological Psychiatry, 66*(9), 898–901. https://doi.org/10.1016/j.biopsych.2009.06.011.

Brody, A. L., Olmstead, R. E., London, E. D., Farahi, J., Meyer, J. H., Grossman, P., et al. (2004). Smoking-induced ventral striatum dopamine release. *American Journal of Psychiatry, 161*(7), 1211–1218.

Brown, A. K., Mandelkern, M. A., Farahi, J., Robertson, C., Ghahremani, D. G., Sumerel, B., et al. (2012). Sex differences in striatal dopamine D2/D3 receptor availability in smokers and non-smokers. *The International Journal of Neuropsychopharmacology, 15*(07), 989–994. https://doi.org/10.1017/S1461145711001957.

Busto, U. E., Redden, L., Mayberg, H., Kapur, S., Houle, S., & Zawertailo, L. A. (2009). Dopaminergic activity in depressed smokers: a positron emission tomography study. *Synapse, 63*(8), 681–689. https://doi.org/10.1002/syn.20646.

Chiuccariello, L., Boileau, I., Guranda, M., Rusjan, P. M., Wilson, A. A., Zawertailo, L., et al. (2013). Presentation of smoking-associated cues does not elicit dopamine release after one-hour smoking abstinence: a [11C]-(+)-PHNO PET study. *PLoS ONE, 8*(3), e60382. https://doi.org/10.1371/journal.pone.0060382.

Cosgrove, K. P., Wang, S., Kim, S.-J., McGovern, E., Nabulsi, N., Gao, H., et al. (2014). Sex differences in the brain's dopamine signature of cigarette smoking. *Journal of Neuroscience, 34*(50), 16851–16855. https://doi.org/10.1523/JNEUROSCI.3661-14.2014.

Dagher, A., Bleicher, C., Aston, J. A., Gunn, R. N., Clarke, P., & Cumming, P. (2001). Reduced dopamine D1 receptor binding in the ventral striatum of cigarette smokers. *Synapse, 42*(1), 48–53.

Di Chiara, G. (2000). Role of dopamine in the behavioural actions of nicotine related to addiction. *European Journal of Pharmacology, 393*(1), 295–314. https://doi.org/10.1016/S0014-2999(00)00122-9.

Di Ciano, P., Guranda, M., Lagzdins, D., Tyndale, R. F., Gamaleddin, I., Selby, P., et al. (2016). Varenicline-induced elevation of dopamine in smokers: a preliminary [11C]-(+)-PHNO PET study. *Neuropsychopharmacology*. Retrieved from: (2016). http://www.nature.com/npp/journal/vaop/ncurrent/full/npp2015305a.html.

Domino, E. F., Evans, C. L., Ni, L., Guthrie, S. K., Koeppe, R. A., & Zubieta, J.-K. (2012). Tobacco smoking produces greater striatal dopamine release in G-allele carriers with mu opioid receptor A118G polymorphism. *Progress in Neuro-Psychopharmacology and Biological Psychiatry, 38*(2), 236–240. https://doi.org/10.1016/j.pnpbp.2012.04.003.

Domino, E. F., Ni, L., Domino, J. S., Yang, W., Evans, C., Guthrie, S., et al. (2013). Denicotinized versus average nicotine tobacco cigarette smoking differentially releases striatal dopamine. *Nicotine & Tobacco Research, 15*(1), 11–21. https://doi.org/10.1093/ntr/nts029.

Fehr, C., Yakushev, I., Hohmann, N., Buchholz, H.-G., Landvogt, C., Deckers, H., et al. (2008). Association of low striatal dopamine D 2 receptor availability with nicotine dependence similar to that seen with other drugs of abuse. *American Journal of Psychiatry, 165*(4), 507–514.

Lammertsma, A. A., & Hume, S. P. (1996). Simplified reference tissue model for PET receptor studies. *NeuroImage, 4*(3), 153–158.

Le Foll, B., Guranda, M., Wilson, A. A., Houle, S., Rusjan, P. M., Wing, V. C., et al. (2014). Elevation of dopamine induced by cigarette smoking: novel insights from a [11C]-(+)-PHNO PET study in humans. *Neuropsychopharmacology, 39*(2), 415–424.

Mintun, M. A., Raichle, M. E., Kilbourn, M. R., Wooten, G. F., & Welch, M. J. (1984). A quantitative model for the in vivo assessment of drug binding sites with positron emission tomography. *Annals of Neurology, 15*(3), 217–227.

Montgomery, A. J., Lingford-Hughes, A. R., Egerton, A., Nutt, D. J., & Grasby, P. M. (2007). The effect of nicotine on striatal dopamine release in man: a [11C]raclopride PET study. *Synapse, 61*(8), 637–645. https://doi.org/10.1002/syn.20419.

Narendran, R., Frankle, W. G., Mason, N. S., Rabiner, E. A., Gunn, R. N., Searle, G. E., et al. (2009). Positron emission tomography imaging of amphetamine-induced dopamine release in the human cortex: a comparative evaluation of the high affinity dopamine D2/3 radiotracers [11C]FLB 457 and [11C]fallypride. *Synapse, 63*(6), 447–461. https://doi.org/10.1002/syn.20628.

Okita, K., Mandelkern, M. A., & London, E. D. (2016). Cigarette use and striatal dopamine D2/3 receptors: possible role in the link between smoking and nicotine dependence. *International Journal of Neuropsychopharmacology, 19*(11). pyw074.

Okita, K., Petersen, N., Robertson, C. L., Dean, A. C., Mandelkern, M. A., & London, E. D. (2016). Sex differences in midbrain dopamine D2-type receptor availability and association with nicotine dependence. *Neuropsychopharmacology, 41*(12), 2913–2919.

Parsey, R. V., Oquendo, M. A., Zea-Ponce, Y., Rodenhiser, J., Kegeles, L. S., Pratap, M., et al. (2001). Dopamine D(2) receptor availability and amphetamine-induced dopamine release in unipolar depression. *Biological Psychiatry, 50*(5), 313–322.

Portnow, L. H., Vaillancourt, D. E., & Okun, M. S. (2013). The history of cerebral PET scanning from physiology to cutting-edge technology. *Neurology, 80*(10), 952–956.

Riccardi, P., Baldwin, R., Salomon, R., Anderson, S., Ansari, M. S., Li, R., et al. (2008). Estimation of baseline dopamine D2 receptor occupancy in striatum and extrastriatal regions in humans with positron emission tomography with [18F] fallypride. *Biological Psychiatry, 63*(2), 241–244. https://doi.org/10.1016/j.biopsych.2007.03.022.

Riccardi, P., Li, R., Ansari, M. S., Zald, D., Park, S., Dawant, B., et al. (2005). Amphetamine-induced displacement of [18F] fallypride in striatum and extrastriatal regions in humans. *Neuropsychopharmacology, 31*(5), 1016–1026. https://doi.org/10.1038/sj.npp.1300916.

Salokangas, R. K., Vilkman, H., Ilonen, T., Taiminen, T., Bergman, J., Haaparanta, M., et al. (2000). High levels of dopamine activity in the basal ganglia of cigarette smokers. *American Journal of Psychiatry, 157*(4), 632–634.

Scott, D. J., Domino, E. F., Heitzeg, M. M., Koeppe, R. A., Ni, L., Guthrie, S., et al. (2007). Smoking modulation of μ-opioid and dopamine D2 receptor-mediated neurotransmission in humans. *Neuropsychopharmacology, 32*(2), 450–457.

Shotbolt, P., Tziortzi, A. C., Searle, G. E., Colasanti, A., van der Aart, J., Abanades, S., et al. (2012). Within-subject comparison of [11C]-(+)-PHNO and [11C]raclopride sensitivity to acute amphetamine challenge in healthy humans. *Journal of Cerebral Blood Flow & Metabolism, 32*(1), 127–136. https://doi.org/10.1038/jcbfm.2011.115.

Surgeon General (2014). *The health consequences of smoking—50 years of progress: a report of the surgeon general.* US Department of Health and Human Services.

Takahashi, H., Fujimura, Y., Hayashi, M., Takano, H., Kato, M., Okubo, Y., et al. (2008). Enhanced dopamine release by nicotine in cigarette smokers: a double-blind, randomized, placebo-controlled pilot study. *The International Journal of Neuropsychopharmacology. 11*(03). https://doi.org/10.1017/S1461145707008103.

Tziortzi, A. C., Searle, G. E., Tzimopoulou, S., Salinas, C., Beaver, J. D., Jenkinson, M., et al. (2011). Imaging dopamine receptors in humans with [11C]-(+)-PHNO: dissection of D3 signal and anatomy. *NeuroImage, 54*(1), 264–277. https://doi.org/10.1016/j.neuroimage.2010.06.044.

Volkow, N. D., Fowler, J. S., Wang, G. J., Baler, R., & Telang, F. (2009). Imaging dopamine's role in drug abuse and addiction. *Neuropharmacology, 56*, 3–8. https://doi.org/10.1016/j.neuropharm.2008.05.022.

Volkow, N. D., Wang, G.-J., Fowler, J. S., & Tomasi, D. (2012). Addiction circuitry in the human brain. *Annual Review of Pharmacology and Toxicology, 52*(1), 321–336. https://doi.org/10.1146/annurev-pharmtox-010611-134625.

Weinstein, A. M., Freedman, N., Greif, J., Yemini, Z., Mishani, E., London, E., et al. (2016). Negative association of pretreatment cigarette use with smoking-induced striatal dopamine release in smokers receiving bupropion treatment: pre-treatment cigarette use and dopamine release. *The American Journal on Addictions*, *25*(6), 486–492. https://doi.org/10.1111/ajad.12419.

Wiers, C. E., Cabrera, E. A., Tomasi, D., Wong, C. T., Demiral, Ş. B., Kim, S. W., et al. (2017). Striatal dopamine D2/D3 receptor availability varies across smoking status. *Neuropsychopharmacology*, https://doi.org/10.1038/npp.2017.131.

Wing, V. C., Payer, D. E., Houle, S., George, T. P., & Boileau, I. (2015). Measuring cigarette smoking-induced cortical dopamine release: a [11C] FLB-457 PET study. *Neuropsychopharmacology*, *40*(6), 1417–1427.

Yasuno, F., Ota, M., Ando, K., Ando, T., Maeda, J., Ichimiya, T., et al. (2007). Role of ventral striatal dopamine D1 receptor in cigarette craving. *Biological Psychiatry*, *61*(11), 1252–1259. https://doi.org/10.1016/j.biopsych.2006.06.028.

CHAPTER

15

Resting-State Functional Connectivity Imaging and Nicotine Dependence

Victor M. Vergara, Vince D. Calhoun

The Mind Research Network, Albuquerque, NM, United States

Abbreviations

ACC	anterior cingulate cortex
BART	Balloon Analogue Risk Task
BOLD	blood-oxygenation-level-dependent
dlPFC	dorsolateral prefrontal cortex
DMN	default mode network
ECN	executive control network
fMRI	functional magnetic resonance imaging
rsFC	resting-state functional connectivity
SN	salience network

15.1 INTRODUCTION

The neurological effects exerted by nicotine not only are constrained to task-demanding brain activity but also extend to moments of unchallenged rest. Even during this resting condition, the brain remains in ceaseless activity exhibiting behaviors that are important for healthy functioning. Resting-state fluctuations are observed at low frequencies presenting important patterns of coactivations through the brain. Coactivation between separate areas of the brain is regarded as functional connectivity (van den Heuvel & Hulshoff Pol, 2010). Different patterns of whole-brain connectivity have been discovered to spontaneously iterate through time. Disruption of these whole-brain patterns and point-to-point connectivity has been linked to the presence of neurological dysfunctions in subjects with different types of addiction (Sutherland, McHugh, Pariyadath, & Stein, 2012). In particular, nicotine affects functional connectivity on areas commonly associated with nicotine addiction such as the insula and the anterior cingulate cortex. The following compilation presents the most relevant results and conclusions obtained from studying functional connectivity of the resting brain.

15.2 ASSESSING RESTING-STATE FUNCTIONAL CONNECTIVITY

Assessment of resting-state functional connectivity (rsFC) is based on the measurement of dynamic changes in brain activity. Of the several methods utilized for functional connectivity estimation, functional magnetic resonance imaging (fMRI) is one of the most utilized due to its relatively higher temporal resolution and minimal invasive action. In this section, we will describe rsFC in terms of fMRI.

15.2.1 Regions of Interest and Resting-State Networks in fMRI Data

Neuronal activity is assessed from fMRI data thanks to the magnetization properties of the blood. Deoxyhemoglobin in venous blood acts as a blood-oxygenation-level-dependent (BOLD) contrast agent based on paramagnetic differences between oxygenated and deoxygenated blood (Ogawa, Lee, Kay, & Tank, 1990). Instantaneous BOLD measurements provide information about synaptic activity as oxygen metabolism evolves through time (Ekstrom, 2010; Heeger & Ress, 2002). Anatomically separate brain regions show synchronic neuronal activity that can be assessed using the BOLD signal (Biswal, Zerrin Yetkin, Haughton, & Hyde, 1995). Measured coherent activity among different neuronal populations is an indication of their functional connection, which has been shown to be correlated with white matter tracts (Honey et al., 2009).

One important technique for rsFC assessment is to focus on the BOLD signal of a region of interest by dropping a seed on a specific location and estimating coactivation with all other areas of the brain (Biswal et al., 1995). This method is utilized when prior information exists.

Neuroscience of Nicotine
https://doi.org/10.1016/B978-0-12-813035-3.00015-0

The seed is relevant for hypothesis testing. Another important option promotes the identification of functionally and spatially different BOLD signals proceeding from differentiated sources. Recognition of brain sources is commonly achieved using blind source separation methods (the word blind indicates the lack of prior information) among which we can mention dictionary learning (Lee et al., 2016) and independent component analysis (Calhoun & Adali, 2012). Because no prior information or hypothesis is imposed, these methods are utilized in exploratory and data-driven studies.

15.2.2 Static rsFC Using Functional Magnetic Resonance Imaging

The correlation is one of the first methods utilized as an estimation of rsFC (van den Heuvel & Hulshoff Pol, 2010). The simplest approach is to assume that rsFC is constant through the duration of the fMRI scan. The minimum fMRI scanning time in practice is 5 min (Allen et al., 2011), but longer scanning times are better with 10 min being considered high enough in terms of capturing signal power (Murphy, Bodurka, & Bandettini, 2007). The scanning time will determine the amount of temporal data in the correlation estimation, thus having an important effect in the significance of the assessment. Fig. 15.1 displays a typical rsFC correlation matrix and the brain areas involved. These brain areas were recently utilized for a nicotine and alcohol study (Vergara, Liu, Claus, Hutchison, & Calhoun, 2017). Blind source separation techniques such as independent component analysis tend to be more resilient to nuisances, permitting a single pass at preprocessing (Vergara, Mayer, Damaraju, Hutchison, & Calhoun, 2017), while other techniques might require a more elaborated cleaning preprocessing (Power et al., 2014).

15.2.3 Dynamic rsFC Using Functional Magnetic Resonance Imaging

Evidence collected through the literature indicates that functional connectivity does not remain constant during resting-state experiments. Instead, brain connectivity suffers noticeable temporal variations that can be observed within typical scanning lengths (Chang & Glover, 2010). Capturing this dynamicity requires the use of coactivation assessments at relatively short lapses of time. The shorter the lapse, the finer the temporal resolution; however, shorter time spans decrease the number of data samples available for correlation and thus degrade the assessment of connectivity. Functional connectivity dynamics have been recommended to be assessed using time lapses between 40 and 100 s (Leonardi & Van De Ville, 2015; Zalesky & Breakspear, 2015).

Dynamic rsFC studies have identified a finite set of connectivity patterns that iterate through time as whole-brain dynamics evolve (Allen et al., 2014; Sakoğlu et al., 2010). These patterns are known as dynamic states and are highly replicable (Abrol, Chaze, Damaraju, & Calhoun, 2016). Evidence supports the link between states and psychological and emotional mental states (Cribben, Haraldsdottir, Atlas, Wager, & Lindquist, 2012). Fig. 15.1 shows a set of dynamic states found in a study involving nicotine, alcohol, and cannabis (Vergara, Weiland, Hutchison, & Calhoun, 2018). Connectivity analysis can be applied after isolating rsFC on each state in a similar way it is applied to static rsFC. In this case, there are as many analyses as identified states. The most common measurements are the frequency and duration of each state. More frequently occurring states have a stronger influence in the overall rsFC. These characteristics of dynamics states are not available through static rsFC, but can have an impact on the analysis of rsFC with demographic and neuropathic measures.

15.3 RESTING-STATE CONNECTIVITY DYSFUNCTIONS

Abnormal functional connectivity linked to nicotine use has been observed in some important areas of the brain including the dorsolateral prefrontal cortex (dlPFC), the insula, the anterior cingulate cortex (ACC), the sensorimotor areas, the executive control network (ECN), and the default mode network (DMN). These areas belong to networks involved in exteroception, interoception, craving, and general resting-state dynamics. This section makes a summary of resting-state findings in the brain linked to nicotine.

15.3.1 Insula

The insula has an important place in nicotine addiction given its role in sensing the physiological condition of the body, which is a characteristic known as interoception (Craig, 2009). The insula resolves bodily feelings including pain, temperature, itch, touch, hunger, thirst, the need for air, and muscular sensations (Craig, 2003). This interoceptive integration influences feelings and decision-making contributing to nicotine addiction (Naqvi & Bechara, 2009) to the extent that trauma to this area ceases conscious urges resulting in smoking cessation (Naqvi, Rudrauf, Damasio, & Bechara, 2007). These observations indicate that the insula is a critical area for the maintenance of smoking urge rooted in the interoceptive relationship between the brain and the body.

In addition to its action in the interoceptive system, the relationship between the insula and an "exteroceptive"

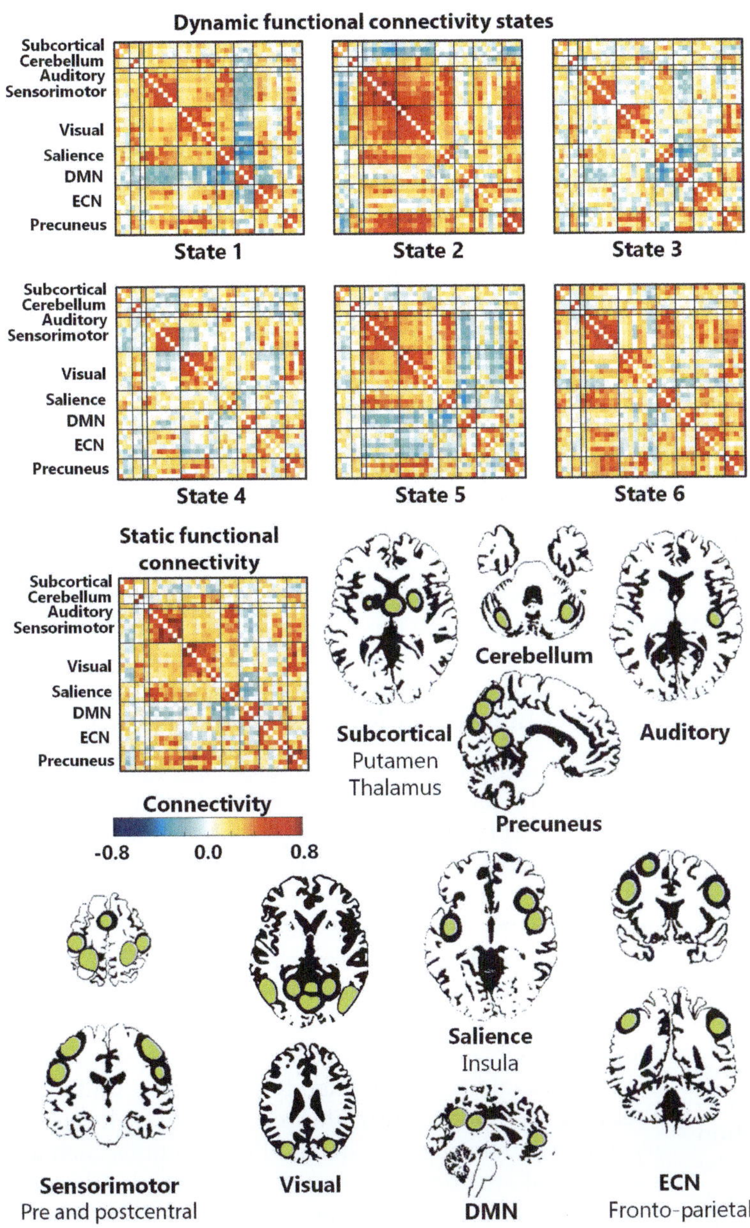

FIG. 15.1 Mean correlation matrices for static and dynamic functional connectivity and brain regions involved. The figure displays the mean correlation matrices for static (Vergara, Liu, et al., 2017) and dynamic (Vergara et al., 2018) functional connectivity. Included are locations of areas representing each of the functional groups: the subcortical, cerebellum, auditory, sensorimotor, visual, salience, default mode network (DMN), executive control network (ECN), and precuneus. The figure was rerendered using data from 534 subjects from a previous publication (Vergara et al., 2018).

system rooted in the somatosensory cortex can be linked to the maintenance of smoking behavior. For example, smokers who relapse show weaker rsFC between the insula and bilateral pre- and postcentral gyri (Addicott, Sweitzer, Froeliger, Rose, & McClernon, 2015). This effect is related to a poor inhibitory control in smokers consistent with observations from go/no-go studies (Perry & Carroll, 2008). The insula and the somatosensory cortex are further implicated in a response inhibition network along with the ACC, prefrontal cortex, basal ganglia, and thalamus (Congdon et al., 2010; Janes et al., 2010). Networks identified in Congdon's study (Congdon et al., 2010) closely match networks from other resting-state studies, providing further evidence that rsFC and demanding task effects linked to nicotine involve similar brain structures.

The effects of nicotine on the insula exhibit a longitudinal abstinence component. In addition to dopamine modulation following nicotine administration (Marshall, Redfern, & Wonnacott, 2002), a negative functional connectivity (measured as a negative correlation) between insula-dlPFC and insula-temporal cortex is observed after nicotine administration in regular smokers, who consumed a cigarette just before the scan, as compared to healthy controls (Stoeckel, Chai, Zhang, Whitfield-Gabrieli, & Evins, 2016). The same study found that smokers had decreased gray matter density in the mentioned brain regions in relation to healthy controls.

Successful abstinence after a period of 24h was predicted by the functional connectivity between dlPFC and insula (Zelle, Gates, Fiez, Sayette, & Wilson, 2017). These functional connectivity effects are a response to the presence of nicotine in the brain that also modulates future self-control when attempting to remain abstinent.

The insula has then a dual function of interoceptive and exteroceptive processing that can be altered by nicotine (Addicott et al., 2015). Changes of insula rsFC also extend to known brain networks such as the ECN (Stoeckel et al., 2016) and the DMN (Wilcox et al., 2017), which is a major influential network during the resting state (Buckner, Andrews-Hanna, & Schacter, 2008). The insula is at the center of a competitive network dynamics as a brain structure offering a salient processing function that toggles between opposing networks (Sutherland et al., 2013). This interplay with other important brain regions sets the insula as an area of high interest in addiction (Naqvi & Bechara, 2009).

15.3.2 Anterior cingulate cortex

The ACC is another brain region where rsFC has been linked to nicotine. An increment of insula-ACC resting-state connectivity was linked to enhanced smoking cue reactivity measured during a task performed after resting data were collected (Janes, Farmer, Peechatka, deB Frederick, & Lukas, 2015). The influence of nicotine produces important changes between the ACC and the insula as their functional relationship might aid in initiating smoking urges. The insula and ACC are central parts of a saliency network that is coactive with the DMN during self-referential brain function (Spreng, Stevens, Chamberlain, Gilmore, & Schacter, 2010). The DMN is less active during tasks that require external attention (Spreng, Mar, & Kim, 2009) and more active during the resting state (Buckner et al., 2008). However, abnormally high activity in DMN during exposure to external smoking cues is modulated by glutamate in the ACC and results in higher reactivity (Janes, Betts, Jensen, & Lukas, 2016). Dysfunctional ACC allows resting-state areas to coactivate with task-related regions to increase smoking cue reactivity and promote smoking behavior.

In addition to a dysfunctional relationship with self-referential areas, the ACC of smokers have also been found to be abnormally connected to areas of cognitive control. Several studies have indicated higher risky behavior in nicotine-dependent subjects (Galvan et al., 2013). Risk behavioral measures using the Balloon Analogue Risk Task (BART) (Lejuez et al., 2002) have been able to differentiate smokers and nonsmokers (Lejuez et al., 2003). The ACC is an important part of risky behavior as observed after analyzing data from BART experiments (Rao, Korczykowski, Pluta, Hoang, & Detre, 2008). Although the relationship between the brain and risky behavior must be measured using task designs, effects observed during tasks are reflected in the resting state. A tripartite relationship between nicotine dependence, risky behavior (measured using BART), and rsFC in a network where the ACC and thalamus are the main nodes has been identified (Wei et al., 2016). The same study determined that brain activation during a task was not related to either the task or nicotine dependence.

The relationship of the ACC with nicotine seems to be explained through its relationships with other brain areas forming a network. Nicotine addiction has also been related to the strength of ACC-striatum connectivity (Hong et al., 2009), indicating that the effects of the ACC reach reward regulation areas. The important nodes in this dysfunctional network are the ACC, insula, thalamus, and ventral striatum. Some studies show evidence that the amygdala, involved in salient emotional processing, can be added to this network (Bi et al., 2017).

15.3.3 The Default Mode Network

A well identified set of brain areas are known to be active during passive periods of rest indicating the existence of a default brain activity (Buckner & Vincent, 2007). The areas involved form a network of brain regions including the cingulate, parietal, and prefrontal cortices and other areas (Buckner et al., 2008). This DMN is a major focus of study in resting-state experiments. DMN functioning has been found to be affected by nicotine consumption. The presence of nicotine in the brain reduces activity within the DMN during the resting state (Tanabe et al., 2011; Weiland, Sabbineni, Calhoun, Welsh, & Hutchison, 2015). This reduction of DMN activity aids in enhancing task-demanding functions such as visuospatial attention (Hahn et al., 2007; Thiel, Zilles, & Fink, 2005) and working memory (Bentley, Husain, & Dolan, 2004). These alterations in DMN functioning are not toll-free. Nicotine produces a counteracting effect of diminishing DMN suppression observed during smoking cue exposure (Janes et al., 2016; Wilcox et al., 2017). A dual role of nicotine is then evident that while it enhances some cognitive abilities, it also promotes addiction (Jasinska, Zorick, Brody, & Stein, 2014). Adding to the negative side of nicotine consumption is the link found with decline in cognitive performance as smokers ages (Kalmijn, 2002).

15.3.4 Executive Control and Salience Networks

The other two networks of high relevance in resting-state studies of addiction are the ECN and the salience network (SN). The central component of the SN is the insula as an area of high relevance in nicotine addiction (Naqvi et al., 2007). Similar to the DMN, resting-state experiments uncovered reduced functional connectivity in areas of the parietal cortex and dlPFC, two of the main areas of the ECN. Detrimental effects related to ECN, cognition, and nicotine are not consistently observed, but they might interplay with abstinence. For example, nicotine withdrawal produces difficulty concentrating and restlessness (Hughes, 2007), contrasting with enhancements of attention and memory after nicotine administration (Bentley et al., 2004; Hahn et al., 2007; Thiel et al., 2005). In addition to cognitive alterations, dysfunctional ECN contributes to addictive behavior as it has greater cue response (Goldstein & Volkow, 2002) and higher activation during inhibitory control (Luijten, O'Connor, Rossiter, Franken, & Hester, 2013; Zhang et al., 2011).

15.3.5 DMN-SN-ECN—Network Model of Nicotine Addiction

The three resting-state networks (DMN, SN, and ECN) maintain a functional balance of interactions. The influence of nicotine in the brain alters this balance in such a way that it promotes addictive behavior. The network model of nicotine addiction (Sutherland et al., 2012) describes the interaction changes among these resting-state networks. Nicotine administration effects described by this model are illustrated in Fig. 15.2. During administration, nicotine shifts brain processing away from the DMN. The rsFC in the ECN and its interaction with the SN receive a boost due to a diminished influence from the DMN. More attention is directed toward extrinsic stimuli resulting in performance improvement of task-based activity. This model is consistent with the cognitive enhancing effects observed in nicotine-naive populations (Heishman, Kleykamp, & Singleton, 2010) and evidence

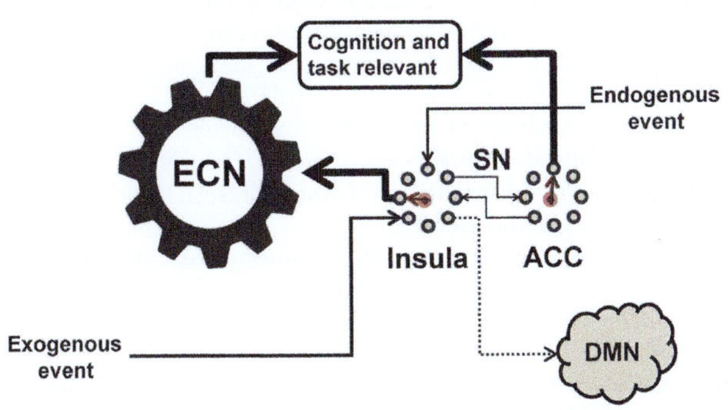

FIG. 15.2 Resting-state interactions corresponding to nicotine administration.
Administration of nicotine suppresses the default mode network (DMN) (Tanabe et al., 2011), allowing for a boost of the executive control network (ECN) consistent with enhancement of attention and memory (Bentley et al., 2004; Hahn et al., 2007).

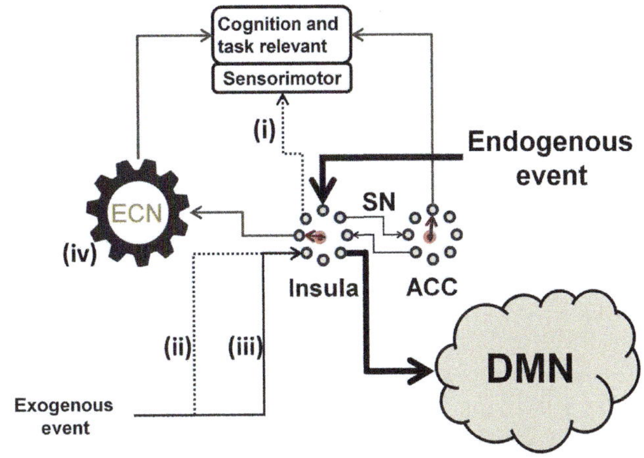

FIG. 15.3 Resting-state interactions corresponding to nicotine abstinence.
Endogenous and exogenous effects are processed differently. During nicotine abstinence, former default mode network (DMN) suppression (due to nicotine administration) is diminished (Janes et al., 2016; Wilcox et al., 2017) along with other effects: (i) weak insula-sensorimotor connectivity (Addicott et al., 2015), (ii) reduced connectivity with sensorial input including occipital areas (Addicott et al., 2015; Vergara, Liu, et al., 2017), (iii) enhanced functioning of visuospatial attention areas (Hahn et al., 2007; Vergara, Liu, et al., 2017), and (iv) reduced resting-state functioning of the executive control network (ECN) (Weiland et al., 2015).

that nicotine suppresses the DMN activity in nonsmokers (Tanabe et al., 2011).

Fig. 15.3 illustrates network interactions for nicotine abstinence. During abstinence, the model (Sutherland et al., 2012) predicts (1) increased rsFC between the insula and DMN related to withdrawal symptoms and impaired task performance, (2) reduced rsFC between the ECN and insula, (3) increased rsFC within the DMN, (4) reduced rsFC within the ECN, and finally (5) the condition that resulted in a dysfunctional relationship between the DMN and ECN that diminishes their negative coupling (Cole et al., 2010).

15.4 TREATMENT IMPLICATIONS OF RESTING-STATE FUNCTIONAL CONNECTIVITY

The emergence of nicotine delivery products, differing from conventional cigarettes, brings new challenges for clinicians (Schick et al., 2017). Nicotine dependence tests developed during the last century focused on cigarette use with an accuracy that has been revisited several times along with recent evidence that suggests a new modification might be advisable (Chabrol, Niezborala, Chastan, Montastruc, & Mullet, 2003). Alternatively, interactions among major brain networks (DMN, ECN, and SN) and nicotine in the brain provide important features for neuroimaging-based addiction biomarking (Fedota & Stein, 2015), resulting in the possibility of identifying nicotine addiction (Pariyadath, Stein, & Ross, 2014). Included in biomarking development is the possibility of predicting and tracking treatment outcomes. Higher rsFC among the putamen, ACC, caudate, and dlPFC (Wilcox et al., 2017) was found to be a predictor of nicotine treatment outcome allowing for the identification of individuals with a higher risk of relapse. In a similar context, rsFC has provided evidence for the existence of at least three subgroups of nicotine users (Ding et al., 2018). These outcomes indicate that rsFC is an important tool for surveying brain nicotine status in further assessments of subject addictive behavior.

MINI-DICTIONARY OF TERMS

Blood-oxygenation-level-dependent Contrast agent based on paramagnetic differences between oxygenated and deoxygenated blood.

Default mode network A network of brain areas with preferentially greater activity during resting moments when compared to a wide range of activity demanding tasks. The key components are medial prefrontal cortex, angular gyrus, and posterior cingulate cortex.

Dynamic resting-state functional connectivity Measurement of synchronic activation at relatively short (less than 100s) time lapses.

Executive control network A network of brain areas of executive and cognitive functions necessary for behavior control and goal directed tasks. The key components are the dorsolateral prefrontal and parietal cortices.

Resting-state functional connectivity Resting-state relationship between brain areas based on synchronic neuronal activations. In fMRI, neuronal activation is measured using the blood-oxygenation-level-dependent signal.

Resting-state Moments of unchallenged rest where the brain is not occupied in demanding tasks.

Static resting-state functional connectivity Measurement of synchronic activation at relatively long (5min) time lapses.

Key Facts of Resting-State Brain Networks

- The activity of the default mode network and task-positive networks fluctuates in an anticorrelated manner.
- Task-positive networks are those required during externally demanding tasks.
- Among the task-positive networks are the executive control, somatosensory, and visual networks.
- Anticorrelation is important for healthy functioning of the brain.

Summary Points

- Resting-state brain activation is assessed through the measurement of a blood-oxygenation-level-dependent contrast.
- Resting-state functional connectivity relates to the synchronic activation of brain areas during rest.
- The three important resting-state networks in nicotine addiction are the default mode network, the executive control network, and the salience network.
- The functional balance of brain networks identified in resting-state studies is affected by nicotine.
- Nicotine administration enhances function of the executive control network.
- Nicotine withdrawal enhances function of the default mode network.
- Dysfunctional interplay among resting-state networks promotes nicotine-seeking behavior.

References

Abrol, A., Chaze, C., Damaraju, E., & Calhoun, V. D. (2016). *The chronnectome: evaluating replicability of dynamic connectivity patterns in 7500 resting fMRI datasets* (pp. 5571–5574)*Conference Proceedings: Annual International Conference of the IEEE Engineering in Medicine and Biology Society, 2016*. https://doi.org/10.1109/EMBC.2016.7591989.

Addicott, M. A., Sweitzer, M. M., Froeliger, B., Rose, J. E., & McClernon, F. J. (2015). Increased functional connectivity in an insula-based network is associated with improved smoking cessation outcomes. *Neuropsychopharmacology*, *40*(11), 2648–2656. https://doi.org/10.1038/npp.2015.114.

Allen, E. A., Damaraju, E., Plis, S. M., Erhardt, E. B., Eichele, T., & Calhoun, V. D. (2014). Tracking whole-brain connectivity dynamics in the resting state. *Cerebral Cortex*, *24*(3), 663–676. https://doi.org/10.1093/cercor/bhs352.

Allen, E. A., Erhardt, E. B., Damaraju, E., Gruner, W., Segall, J. M., Silva, R. F., et al. (2011). A baseline for the multivariate comparison of resting-state networks. *Frontiers in Systems Neuroscience*, *5*, 2. https://doi.org/10.3389/fnsys.2011.00002.

Bentley, P., Husain, M., & Dolan, R. J. (2004). Effects of cholinergic enhancement on visual stimulation, spatial attention, and spatial working memory. *Neuron*, *41*(6), 969–982. https://doi.org/10.1016/s0896-6273(04)00145-x.

Bi, Y., Yuan, K., Guan, Y., Cheng, J., Zhang, Y., Li, Y., et al. (2017). Altered resting state functional connectivity of anterior insula in young smokers. *Brain Imaging and Behavior*, *11*(1), 155–165. https://doi.org/10.1007/s11682-016-9511-z.

Biswal, B., Zerrin Yetkin, F., Haughton, V. M., & Hyde, J. S. (1995). Functional connectivity in the motor cortex of resting human brain using echo-planar MRI. *Magnetic Resonance in Medicine*, *34*(4), 537–541. https://doi.org/10.1002/mrm.1910340409.

Buckner, R. L., Andrews-Hanna, J. R., & Schacter, D. L. (2008). The brain's default network: anatomy, function, and relevance to disease. *Annals of the New York Academy of Sciences*, *1124*(1), 1–38. https://doi.org/10.1196/annals.1440.011.

Buckner, R. L., & Vincent, J. L. (2007). Unrest at rest: default activity and spontaneous network correlations. *NeuroImage*, *37*(4), 1091–1096. discussion 1097–1099 (2007). https://doi.org/10.1016/j.neuroimage.2007.01.010.

Calhoun, V. D., & Adali, T. (2012). Multisubject independent component analysis of fMRI: a decade of intrinsic networks, default mode, and neurodiagnostic discovery. *IEEE Reviews in Biomedical Engineering*, *5*, 60–73. https://doi.org/10.1109/RBME.2012.2211076.

Chabrol, H., Niezborala, M., Chastan, E., Montastruc, J.-L., & Mullet, E. (2003). A study of the psychometric properties of the Fagestrom Test for Nicotine Dependence. *Addictive Behaviors*, *28*(8), 1441–1445. https://doi.org/10.1016/s0306-4603(02)00236-8.

Chang, C., & Glover, G. H. (2010). Time–frequency dynamics of resting-state brain connectivity measured with fMRI. *NeuroImage*, *50*(1), 81–98.

Cole, D. M., Beckmann, C. F., Long, C. J., Matthews, P. M., Durcan, M. J., & Beaver, J. D. (2010). Nicotine replacement in abstinent smokers improves cognitive withdrawal symptoms with modulation of resting brain network dynamics. *Neuroimage*, *52*(2), 590–599.

Congdon, E., Mumford, J. A., Cohen, J. R., Galvan, A., Aron, A. R., Xue, G., et al. (2010). Engagement of large-scale networks is related to individual differences in inhibitory control. *NeuroImage*, *53*(2), 653–663. https://doi.org/10.1016/j.neuroimage.2010.06.062.

Craig, A. (2003). Interoception: the sense of the physiological condition of the body. *Current Opinion in Neurobiology*, *13*(4), 500–505.

Craig, A. D. (2009). How do you feel–now? The anterior insula and human awareness. *Nature Reviews Neuroscience*, *10*(1), 59–70. https://doi.org/10.1038/nrn2555.

Cribben, I., Haraldsdottir, R., Atlas, L. Y., Wager, T. D., & Lindquist, M. A. (2012). Dynamic connectivity regression: determining state-related changes in brain connectivity. *NeuroImage*, *61*(4), 907–920. https://doi.org/10.1016/j.neuroimage.2012.03.070.

Ding, X., Salmeron, B. J., Wang, J., Yang, Y., Stein, E. A., & Ross, T. J. (2018). Evidence of subgroups in smokers as revealed in clinical measures and evaluated by neuroimaging data: a preliminary study. *Addiction Biology*. https://doi.org/10.1111/adb.12620.

Ekstrom, A. (2010). How and when the fMRI BOLD signal relates to underlying neural activity: the danger in dissociation. *Brain Research Reviews*, *62*(2), 233–244. https://doi.org/10.1016/j.brainresrev.2009.12.004.

Fedota, J. R., & Stein, E. A. (2015). Resting-state functional connectivity and nicotine addiction: prospects for biomarker development. *Annals of the New York Academy of Sciences*, *1349*(1), 64–82. https://doi.org/10.1111/nyas.12882.

Galvan, A., Schonberg, T., Mumford, J., Kohno, M., Poldrack, R. A., & London, E. D. (2013). Greater risk sensitivity of dorsolateral prefrontal cortex in young smokers than in nonsmokers. *Psychopharmacology*, *229*(2), 345–355. https://doi.org/10.1007/s00213-013-3113-x.

Goldstein, R. Z., & Volkow, N. D. (2002). Drug addiction and its underlying neurobiological basis: neuroimaging evidence for the involvement of the frontal cortex. *The American Journal of Psychiatry*, *159*(10), 1642–1652. https://doi.org/10.1176/appi.ajp.159.10.1642.

Hahn, B., Ross, T. J., Yang, Y., Kim, I., Huestis, M. A., & Stein, E. A. (2007). Nicotine enhances visuospatial attention by deactivating areas of the resting brain default network. *The Journal of Neuroscience*, *27*(13), 3477–3489.

Heeger, D. J., & Ress, D. (2002). What does fMRI tell us about neuronal activity? *Nature Reviews Neuroscience*, *3*(2), 142–151. https://doi.org/10.1038/nrn730.

Heishman, S. J., Kleykamp, B. A., & Singleton, E. G. (2010). Meta-analysis of the acute effects of nicotine and smoking on human performance. *Psychopharmacology*, *210*(4), 453–469.

Honey, C., Sporns, O., Cammoun, L., Gigandet, X., Thiran, J.-P., Meuli, R., et al. (2009). Predicting human resting-state functional connectivity from structural connectivity. *Proceedings of the National Academy of Sciences*, *106*(6), 2035–2040.

Hong, L. E., Gu, H., Yang, Y., Ross, T. J., Salmeron, B. J., Buchholz, B., et al. (2009). Association of nicotine addiction and nicotine's actions with separate cingulate cortex functional circuits. *Archives of General Psychiatry*, *66*(4), 431–441.

Hughes, J. R. (2007). Effects of abstinence from tobacco: valid symptoms and time course. *Nicotine & Tobacco Research*, *9*(3), 315–327. https://doi.org/10.1080/14622200701188919.

Janes, A. C., Betts, J., Jensen, J. E., & Lukas, S. E. (2016). Dorsal anterior cingulate glutamate is associated with engagement of the default mode network during exposure to smoking cues. *Drug and Alcohol Dependence*, *167*, 75–81. https://doi.org/10.1016/j.drugalcdep.2016.07.021.

Janes, A. C., Farmer, S., Peechatka, A. L., deB Frederick, B., & Lukas, S. E. (2015). Insula-dorsal anterior cingulate cortex coupling is associated with enhanced brain reactivity to smoking cues. *Neuropsychopharmacology*, *40*(7), 1561–1568. https://doi.org/10.1038/npp.2015.9.

Janes, A. C., Pizzagalli, D. A., Richardt, S., deB Frederick, B., Chuzi, S., Pachas, G., et al. (2010). Brain reactivity to smoking cues prior to smoking cessation predicts ability to maintain tobacco abstinence. *Biological Psychiatry*, *67*(8), 722–729. https://doi.org/10.1016/j.biopsych.2009.12.034.

Jasinska, A. J., Zorick, T., Brody, A. L., & Stein, E. A. (2014). Dual role of nicotine in addiction and cognition: a review of neuroimaging studies in humans. *Neuropharmacology*, *84*, 111–122. https://doi.org/10.1016/j.neuropharm.2013.02.015.

Kalmijn, S. (2002). Cigarette smoking and alcohol consumption in relation to cognitive performance in middle age. *American Journal of Epidemiology*, *156*(10), 936–944. https://doi.org/10.1093/aje/kwf135.

Lee, Y. B., Lee, J., Tak, S., Lee, K., Na, D. L., Seo, S. W., et al. (2016). Sparse SPM: group sparse-dictionary learning in SPM framework for resting-state functional connectivity MRI analysis. *NeuroImage*, *125*, 1032–1045. https://doi.org/10.1016/j.neuroimage.2015.10.081.

Lejuez, C. W., Aklin, W. M., Jones, H. A., Richards, J. B., Strong, D. R., Kahler, C. W., et al. (2003). The balloon analogue risk task (BART) differentiates smokers and nonsmokers. *Experimental and Clinical Psychopharmacology*, *11*(1), 26–33. https://doi.org/10.1037/1064-1297.11.1.26.

Lejuez, C. W., Read, J. P., Kahler, C. W., Richards, J. B., Ramsey, S. E., Stuart, G. L., et al. (2002). Evaluation of a behavioral measure of risk taking: the Balloon Analogue Risk Task (BART). *Journal of Experimental Psychology: Applied*, *8*(2), 75–84. https://doi.org/10.1037//1076-898x.8.2.75.

Leonardi, N., & Van De Ville, D. (2015). On spurious and real fluctuations of dynamic functional connectivity during rest. *NeuroImage*, *104*, 430–436. https://doi.org/10.1016/j.neuroimage.2014.09.007.

Luijten, M., O'Connor, D. A., Rossiter, S., Franken, I. H., & Hester, R. (2013). Effects of reward and punishment on brain activations associated with inhibitory control in cigarette smokers. *Addiction*, *108*(11), 1969–1978. https://doi.org/10.1111/add.12276.

Marshall, D. L., Redfern, P. H., & Wonnacott, S. (2002). Presynaptic nicotinic modulation of dopamine release in the three ascending pathways studied by in vivo microdialysis: comparison of naive and chronic nicotine-treated rats. *Journal of Neurochemistry*, *68*(4), 1511–1519. https://doi.org/10.1046/j.1471-4159.1997.68041511.x.

Murphy, K., Bodurka, J., & Bandettini, P. A. (2007). How long to scan? The relationship between fMRI temporal signal to noise ratio and necessary scan duration. *NeuroImage*, *34*(2), 565–574. https://doi.org/10.1016/j.neuroimage.2006.09.032.

Naqvi, N. H., & Bechara, A. (2009). The hidden island of addiction: the insula. *Trends in Neurosciences*, *32*(1), 56–67. https://doi.org/10.1016/j.tins.2008.09.009.

Naqvi, N. H., Rudrauf, D., Damasio, H., & Bechara, A. (2007). Damage to the insula disrupts addiction to cigarette smoking. *Science*, *315* (5811), 531–534. https://doi.org/10.1126/science.1135926.

Ogawa, S., Lee, T. M., Kay, A. R., & Tank, D. W. (1990). Brain magnetic resonance imaging with contrast dependent on blood oxygenation. *Proceedings of the National Academy of Sciences of the United States of America*, *87*(24), 9868–9872.

Pariyadath, V., Stein, E. A., & Ross, T. J. (2014). Machine learning classification of resting state functional connectivity predicts smoking status. *Frontiers in Human Neuroscience*, *8*, 425. https://doi.org/10.3389/fnhum.2014.00425.

Perry, J. L., & Carroll, M. E. (2008). The role of impulsive behavior in drug abuse. *Psychopharmacology, 200*(1), 1–26. https://doi.org/10.1007/s00213-008-1173-0.

Power, J. D., Mitra, A., Laumann, T. O., Snyder, A. Z., Schlaggar, B. L., & Petersen, S. E. (2014). Methods to detect, characterize, and remove motion artifact in resting state fMRI. *NeuroImage, 84*, 320–341.

Rao, H., Korczykowski, M., Pluta, J., Hoang, A., & Detre, J. A. (2008). Neural correlates of voluntary and involuntary risk taking in the human brain: an fMRI study of the Balloon Analog Risk Task (BART). *NeuroImage, 42*(2), 902–910. https://doi.org/10.1016/j.neuroimage.2008.05.046.

Sakoğlu, Ü., Pearlson, G. D., Kiehl, K. A., Wang, Y. M., Michael, A. M., & Calhoun, V. D. (2010). A method for evaluating dynamic functional network connectivity and task-modulation: application to schizophrenia. *Magnetic Resonance Materials in Physics, Biology and Medicine, 23*(5–6), 351–366.

Schick, S. F., Blount, B. C., Jacob, P. R., Saliba, N. A., Bernert, J. T., El Hellani, A., et al. (2017). Biomarkers of exposure to new and emerging tobacco delivery products. *American Journal of Physiology Lung Cellular and Molecular Physiology, 313*(3), L425–L452. https://doi.org/10.1152/ajplung.00343.2016.

Spreng, R. N., Mar, R. A., & Kim, A. S. (2009). The common neural basis of autobiographical memory, prospection, navigation, theory of mind, and the default mode: a quantitative meta-analysis. *Journal of Cognitive Neuroscience, 21*(3), 489–510. https://doi.org/10.1162/jocn.2008.21029.

Spreng, R. N., Stevens, W. D., Chamberlain, J. P., Gilmore, A. W., & Schacter, D. L. (2010). Default network activity, coupled with the frontoparietal control network, supports goal-directed cognition. *NeuroImage, 53*(1), 303–317. https://doi.org/10.1016/j.neuroimage.2010.06.016.

Stoeckel, L. E., Chai, X. J., Zhang, J., Whitfield-Gabrieli, S., & Evins, A. E. (2016). Lower gray matter density and functional connectivity in the anterior insula in smokers compared with never smokers. *Addiction Biology, 21*(4), 972–981. https://doi.org/10.1111/adb.12262.

Sutherland, M. T., Carroll, A. J., Salmeron, B. J., Ross, T. J., Hong, L. E., & Stein, E. A. (2013). Down-regulation of amygdala and insula functional circuits by varenicline and nicotine in abstinent cigarette smokers. *Biological Psychiatry, 74*(7), 538–546. https://doi.org/10.1016/j.biopsych.2013.01.035.

Sutherland, M. T., McHugh, M. J., Pariyadath, V., & Stein, E. A. (2012). Resting state functional connectivity in addiction: lessons learned and a road ahead. *NeuroImage, 62*(4), 2281–2295. https://doi.org/10.1016/j.neuroimage.2012.01.117.

Tanabe, J., Nyberg, E., Martin, L. F., Martin, J., Cordes, D., Kronberg, E., et al. (2011). Nicotine effects on default mode network during resting state. *Psychopharmacology, 216*(2), 287–295.

Thiel, C. M., Zilles, K., & Fink, G. R. (2005). Nicotine modulates reorienting of visuospatial attention and neural activity in human parietal cortex. *Neuropsychopharmacology, 30*(4), 810–820. https://doi.org/10.1038/sj.npp.1300633.

van den Heuvel, M. P., & Hulshoff Pol, H. E. (2010). Exploring the brain network: a review on resting-state fMRI functional connectivity. *European Neuropsychopharmacology, 20*(8), 519–534. https://doi.org/10.1016/j.euroneuro.2010.03.008.

Vergara, V. M., Liu, J., Claus, E. D., Hutchison, K., & Calhoun, V. (2017). Alterations of resting state functional network connectivity in the brain of nicotine and alcohol users. *NeuroImage, 151*, 45–54. https://doi.org/10.1016/j.neuroimage.2016.11.012.

Vergara, V. M., Mayer, A. R., Damaraju, E., Hutchison, K., & Calhoun, V. D. (2017). The effect of preprocessing pipelines in subject classification and detection of abnormal resting state functional network connectivity using group ICA. *Neuroimage, 145*(Pt B), 365–376. https://doi.org/10.1016/j.neuroimage.2016.03.038.

Vergara, V. M., Weiland, B. J., Hutchison, K. E., & Calhoun, V. D. (2018). The impact of combinations of alcohol, nicotine, and cannabis on dynamic brain connectivity. *Neuropsychopharmacology, 43*(4), 877–890. https://doi.org/10.1038/npp.2017.280.

Wei, Z., Yang, N., Liu, Y., Yang, L., Wang, Y., Han, L., et al. (2016). Resting-state functional connectivity between the dorsal anterior cingulate cortex and thalamus is associated with risky decision-making in nicotine addicts. *Scientific Reports, 6*, 21778. https://doi.org/10.1038/srep21778.

Weiland, B. J., Sabbineni, A., Calhoun, V. D., Welsh, R. C., & Hutchison, K. E. (2015). Reduced executive and default network functional connectivity in cigarette smokers. *Human Brain Mapping, 36*(3), 872–882. https://doi.org/10.1002/hbm.22672.

Wilcox, C. E., Calhoun, V. D., Rachakonda, S., Claus, E. D., Littlewood, R. A., Mickey, J., et al. (2017). Functional network connectivity predicts treatment outcome during treatment of nicotine use disorder. *Psychiatry Research, 265*, 45–53. https://doi.org/10.1016/j.pscychresns.2017.04.011.

Zalesky, A., & Breakspear, M. (2015). Towards a statistical test for functional connectivity dynamics. *NeuroImage, 114*, 466–470. https://doi.org/10.1016/j.neuroimage.2015.03.047.

Zelle, S. L., Gates, K. M., Fiez, J. A., Sayette, M. A., & Wilson, S. J. (2017). The first day is always the hardest: functional connectivity during cue exposure and the ability to resist smoking in the initial hours of a quit attempt. *NeuroImage, 151*, 24–32. https://doi.org/10.1016/j.neuroimage.2016.03.015.

Zhang, X., Salmeron, B. J., Ross, T. J., Gu, H., Geng, X., Yang, Y., et al. (2011). Anatomical differences and network characteristics underlying smoking cue reactivity. *NeuroImage, 54*(1), 131–141. https://doi.org/10.1016/j.neuroimage.2010.07.063.

CHAPTER

16

Functional Magnetic Resonance Imaging of Acute Nicotine Effects

Christiane M. Thiel

Biological Psychology Lab, Department of Psychology, Department for Medicine and Health Sciences, Carl von Ossietzky University Oldenburg, Oldenburg, Germany

Abbreviations

BOLD blood-oxygen-level-dependent contrast
fMRI functional magnetic resonance imaging
PET positron-emission tomography

16.1 INTRODUCTION

Neuroimaging techniques such as functional magnetic resonance imaging (fMRI) or positron-emission tomography (PET) allow to investigate the neurobiological mechanisms underlying the effects of nicotine in humans. fMRI, which was introduced in humans in 1991, measures brain activity via the so-called blood-oxygen-level-dependent (BOLD) contrast, which exploits the different magnetic properties of oxygenated and deoxygenated blood (Belliveau et al., 1991). Even though this measure is an indirect measure of neuronal activity that relies on neurovascular coupling, it is strongly correlated with the local field potential of neurons (Logothetis, 2008). Studies that investigate the effects of drugs with this method in the human brain are often subsumed under the term "pharmacological fMRI" or "pharmaco-fMRI." There are, however, different pharmacological fMRI approaches that should be distinguished. Most pharmacological fMRI studies investigate the effects of drug administration in combination with a cognitive, sensory, or motor task. Participants or patients are administered a drug or placebo prior to undergoing a regular fMRI study. This approach enables to investigate which brain regions or networks are responsive to the task and how they are modulated by the drug. For example, one can analyze which brain regions increase or decrease their activity under nicotine in an attention task. Note that the brain regions identified by this approach do not necessarily correspond to those areas with high receptor densities. The regions that show a drug-induced modulation of task-related activity depend to a large extent on the paradigm used. For example, nicotine may affect BOLD activity within the motor cortex if subjects perform a motor task but in prefrontal cortex if a working memory task is used.

Recently, a growing number of fMRI studies have examined the effects of nicotine on so-called resting-state networks. In this approach, the participant or patient receives the drug prior to the fMRI examination. In contrast to task-based fMRI, resting-state studies measure brain activity while the subject is at rest, that is, doing nothing inside the MRI with either open or closed eyes. During such rest, the BOLD signal exhibits low-frequency (<0.1 Hz) spontaneous fluctuations showing strong temporal coherence across distributed brain regions (Buckner & Vincent, 2007; Greicius, 2008; Lu & Stein, 2014). Several resting-state networks have been isolated, which are consistent with task-induced activations and deactivations and with structural connectivity (Beckmann DeLuca, Devlin, & Smith, 2005). Three of those networks, which have often been reported with respect to the action of nicotine, are the default mode network, which includes the medial prefrontal cortex, the posterior cingulate cortex, and the parahippocampal gyrus and is usually deactivated during task performance; the executive control network with nodes in dorsolateral prefrontal and lateral posterior parietal cortices; and the salience network consisting of the dorsal anterior cingulate and frontoinsular cortices.

Another less frequent approach attempts to measure drug-induced changes in BOLD activity, taking pharmacokinetic parameters instead of task-based parameters into account (Bloom et al., 1999). Since nicotine shows a rapid onset and offset of effects, such pharmacokinetic

Neuroscience of Nicotine
https://doi.org/10.1016/B978-0-12-813035-3.00016-2

approach is feasible. Using this approach, Stein et al. (1998) has shown that nicotine induces a dose-dependent increase of BOLD activity in limbic brain regions. The pharmacokinetic approach and further advancements may be of specific interest for drug development (Black, Koller, & Miller, 2013).

In this book chapter, we will summarize major findings of task-based and resting-state fMRI studies using a placebo-controlled acute nicotine application in nonsmokers or smokers. Findings from fMRI studies focusing mainly on reward-related processes and nicotine dependence are reviewed elsewhere and will not be addressed in detail (Fedota & Stein, 2015; Sutherland & Stein, 2018). fMRI studies where nicotine or nicotinic compounds are used to target dysfunctional neuronal circuits in psychiatric disease, such as schizophrenia, are also not addressed (see Smucny & Tregellas, 2017, for review).

16.2 TASK-BASED fMRI STUDIES

Acute nicotine administration in nonsmokers or minimally deprived smokers has beneficial behavioral effects on fine motor abilities, alertness, orienting of attention, short-term episodic memory, and working memory (Heishman, Kleykamp, & Singleton, 2010). Therefore, most task-based fMRI studies aimed to identify the neural correlates of acute nicotine effects with respect to attention and memory. We will focus in the following on these; a broader overview of neural nicotine effects on cognitive function can be found in Newhouse, Potter, Dumas, and Thiel (2011).

One of the first fMRI studies on nicotine effects was performed by Lawrence, Ross, and Stein (2002) and focused on sustained attention. Nondeprived smokers received a 21 mg nicotine or placebo patch 2 hours prior to entering the MRI. Within the MRI, they performed a rapid visual information processing and control task, while brain activity was measured. The major finding was that nicotine increased neural activity during the sustained attention task bilaterally in several attention-related areas, including the parietal and occipital cortices, the thalamus, and the caudate. Behaviorally, nicotine improved task performance, however, only when given on the second session. A subsequent study by the same group investigated orienting of visuospatial attention in a cued target detection task in minimally deprived smokers under a 21 mg nicotine or placebo patch (Hahn et al., 2007). Data analysis focused on BOLD activity during the cue period, when subjects oriented attention to a possible upcoming target. Results revealed an increased deactivation in several brain regions, including the anterior and posterior cingulate cortices, the left angular gyrus, the left anterior middle frontal gyrus, and the bilateral cuneus. These brain regions are part of the so-called default mode network that shows deactivation during task performance and is often anticorrelated with activity increases in task-positive networks (Fox et al., 2005). The increased deactivation may therefore represent an increased neural efficiency. This interpretation was supported by the finding that increased deactivations were associated with performance improvements under nicotine.

The increased deactivation in the default mode network under nicotine was partly reproduced in a similar approach with a slightly different paradigm tapping onto feature-based selective visual attention, divided attention, and simple target detection (Hahn et al., 2009). A second finding was that, independent of task condition, nicotine reduced neural activity in frontal, temporal, thalamic, and visual regions. This result is in contrast to the finding of increased BOLD activity in attention-related brain regions reported above (Lawrence et al., 2002). Because the same subject population and mode of nicotine administration were used, the authors suggest that the directionality of neural nicotine effects may depend on task demands. Increases of neural activity in attention-related brain regions may primarily occur in tasks with fast stimulus presentation that require maximal processing capacity.

The studies described so far investigated the effects of nicotine in smokers. A limitation of studies in smokers is that changes in neural activity under nicotine may reflect a reversal of withdrawal effects, even though the studies reviewed above tried to minimize such effects by reducing withdrawal time to 2–3 hours. A second limitation is that there is evidence for changes in brain structure and function in smokers, such as prefrontal gray matter damage, which may confound the response to nicotine (Fritz et al., 2014; Zhang et al., 2011). Testing for nicotine effects in nonsmokers avoids these confounds.

The neural effects of nicotine in nonsmokers have been investigated in our lab. Several studies focused on orienting and reorienting of visuospatial attention. Subjects chewed a 2 mg nicotine or placebo gum for 30 minutes prior to entering the MRI, where they performed a cued target detection task with validly and invalidly cued targets (Thiel & Fink, 2008; Thiel, Zilles, & Fink, 2005; Vossel, Thiel, & Fink, 2008). Data analysis focused on comparing neural activity in invalidly with that in validly cued trials under nicotine and placebo. The major finding across all those studies was that nicotine reduced neural activity in attention-related brain regions, including the right inferior parietal cortex (the right angular gyrus and right precuneus), the middle temporal gyrus, and the middle frontal gyrus. Behaviorally, nicotine improved reorienting of attention by reducing reaction times in invalidly cued trials, although behavioral effects of nicotine were small, especially in situations with low cue validity.

Reductions of BOLD activity in inferior parietal cortex were also found in a sustained attention task with prospective memory trials in nonsmokers after 0.5 mg of nicotine nasal spray (Rusted, Ruest, & Gray, 2011). Therefore, the effects of nicotine-induced reductions of neural activity in attention-related brain regions seen here in nonsmokers and by Hahn et al. (2009) in minimally deprived smokers seem to be independent of smoking status and the mode of drug application.

The effects of nicotine on auditory attention were investigated by Smucny, Olincy, Eichman, and Tregellas (2015) using a 7 mg nicotine or placebo patch application in nonsmokers. The task was a sustained attention task with fast presentation of auditory stimuli. Two factors were additionally manipulated: task difficulty and distraction. Independent on task condition, nicotine increased BOLD activity in the motor and somatosensory cortices (see however Wylie, Tanabe, Martin, Wongngamnit, and Tregellas (2013) for an absence of nicotine effects in these regions during finger tapping). Depending on task condition, both nicotine-induced increases and decreases in BOLD activity were found in the anterior cingulate cortex, hippocampus, and inferior parietal cortex. In the anterior cingulate cortex and parietal cortex, the largest increase in BOLD response was found in the easiest condition with no distraction, which argues against the suggestion by Hahn et al. (2009) that BOLD signal increases in attention-related brain regions are found under nicotine if stimuli are presented fast and require maximal processing capacity.

Warbrick et al. (2011) investigated the relationship between BOLD signal increases and decreases and behavioral benefit after administration of 1 mg of nicotine via nasal spray. They studied a mixed group of minimally deprived smokers and nonsmokers in a visual oddball task. While nicotine induced an overall increase in neural activity in a variety of brain regions, there was large individual variability with some subjects showing increases in BOLD activity and others showing decreases. Decreases in BOLD activity in several frontal, parietal, and occipital brain regions correlated with decreases in reaction times and reaction time variability, which may suggest that decreased neural activity is beneficial for behavioral performance. However, an open question exists: why do such reductions in neural activity benefit performance? One suggestion, with respect to benefits in selective attention, was tested by Thiel and Fink (2008). The hypothesis was that the reduced attention-related activity may reflect a reduced need for attentional reorienting in invalidly cued trials, since nicotine may have reduced the reliance on top-down information provided by the cue (Yu & Dayan, 2005). However, neural data could not provide any evidence for a reduced reliance on top-down information under nicotine.

Attention-related brain regions largely overlap with brain regions recruited during working memory processes (Mayer et al., 2007). Therefore, one could expect similar nicotine-induced changes in working memory tasks. However, there are only few studies that investigated the effects of nicotine in working memory tasks. Kumari et al. (2003) tested nonsmokers after a 12 μg/kg subcutaneous nicotine or a respective placebo administration using a visual n-back task. Nicotine improved task performance and increased BOLD activity in the anterior cingulate cortex, superior frontal cortex, and superior parietal cortex, which resembles the findings in smokers performing a sustained attention task under nicotine (Lawrence et al., 2002). A newer study that tested however working memory under nicotine in smokers did not yield any evidence for nicotine-induced changes in brain activity (Sutherland, Ross, Shakleya, Huestis, & Stein, 2011). Therefore, it remains open, whether the common neural networks shared by working memory and attention are similarly modulated by nicotine.

Albeit several pharmacological fMRI studies on nicotine are published and some consistent patterns have been described above, there are still many inconsistencies. In order to aggregate the findings of different nicotine neuroimaging studies, Sutherland et al. (2015) performed a coordinate-based metaanalysis on activation foci across all placebo-controlled studies with acute administration of a nicotinic agonist. Even though such an analysis will miss nicotinic modulations that are specific to certain cognitive tasks and show a bias toward those brain regions relevant for the most studied cognitive functions, it nevertheless constitutes a valuable approach for the identification of common mechanisms across tasks. The results of this metaanalysis across smokers and nonsmokers and a variety of tasks show a widespread pattern of both nicotine-induced increases and decreases in BOLD activity (Fig. 16.1). While decreases were mostly observed in regions of the default mode network and the attention-related right posterior parietal cortex, other attention-related brain regions including cortical and subcortical (thalamus) regions mostly showed increases in BOLD activity under nicotine.

16.3 RESTING-STATE fMRI STUDIES

BOLD signal changes during task performance are affected by neural activity triggered by the task and by underlying states of endogenous BOLD signal fluctuations in the low-frequency range (Buckner & Vincent, 2007). Resting-state networks have received increasing interest in understanding large-scale alterations in psychiatric and neurological disorders. Endogenous BOLD signal fluctuations during task and at rest are also an important aspect that should be considered in order to understand the effects of the drug at systems level.

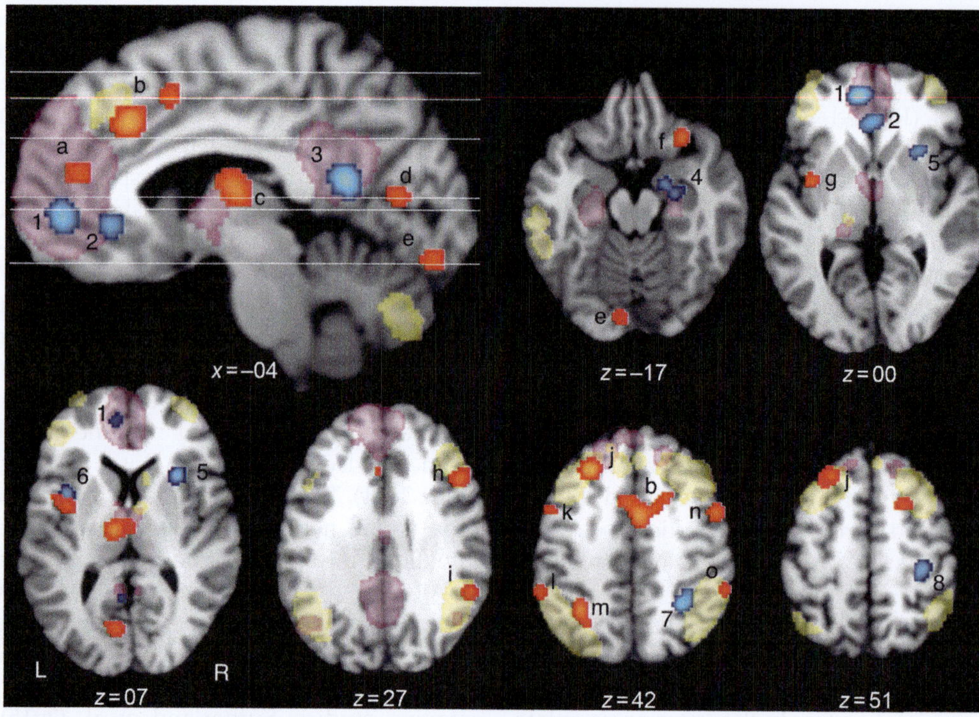

FIG. 16.1 Overview of activations and deactivations after administration of nicotinic acetylcholine receptor agonists. Depicted are estimated activations maps (ALE metaanalysis) of 38 studies involving 796 participants (pooled over smokers and nonsmokers and including a few studies with subjects with schizophrenia). *Red areas* signify regions that show increased neural activity under drug; *blue areas* depict activity decreases. *Purple regions* delineate the default mode network; *yellow regions* delineate the executive control network. Regions with activity decreases: 1+2, anterior cingulate cortex; 3, posterior cingulate cortex; 4, parahippocampal gyrus; 5+6, insula; 7, superior parietal lobe; and 8, precentral gyrus/insula. Regions with activity increases: a, medial frontal gyrus; b, anterior cingulate; c, thalamus; d, cuneus; e, lingual gyrus; f, inferior frontal gyrus; g, insula; h+j, middle frontal gyrus; i, supramarginal gyrus; k+n, precentral gyrus; l+o, inferior parietal lobe; and m, superior parietal lobe. *Reprinted by permission of Elsevier from Sutherland, M. T., Ray, K. L., Riedel, M. C., Yanes, J. A., Stein, E. A., & Laird, A. R. (2015). Neurobiological impact of nicotinic acetylcholine receptor agonists: an activation likelihood estimation meta-analysis of pharmacologic neuroimaging studies. Biological Psychiatry, 78(10), 711–720. Copyright 2015 by the Society of Biological Psychiatry.*

One of the first resting-state fMRI studies on nicotine effects was performed by Tanabe et al. (2011). Nonsmokers received a 7mg nicotine or placebo patch 90minutes prior to entering the MRI. Within the MRI, subjects were instructed to rest with eyes closed. fMRI data analysis used an independent component analysis approach to identify resting-state networks. Nicotine decreased activity in regions of the default mode network, including the precuneus, posterior cingulate cortex, and medial frontal cortex. Increased activity after nicotine was observed in the extrastriate resting-state network. The coupling between the default mode resting-state network and the executive resting-state network was analyzed in deprived smokers (8hours) under placebo and 4mg of nicotine gum (Cole et al., 2010). The results showed a correlation between network coupling and cognitive symptoms of smoking abstinence. A region of interest-based functional connectivity analysis, which was centered on the cingulate cortex, was performed by Hong et al. (2009). Mildly deprived smokers received a 21 or 35mg nicotine patch depending on smoking habits prior to undergoing a resting-state MRI scan. Nicotine increased functional connectivity of different cingulate subdivisions with parietal and frontal cortices (Fig. 16.2). Note however that the latter two studies performed the resting-state scan after a longer period of task-based fMRI, which may impact on resting-state activity (Breckel et al., 2013).

Two further studies on resting-state connectivity under nicotine choose a graph-theory-based analysis approach (Giessing, Thiel, Alexander-Bloch, Patel, & Bullmore, 2013; Wylie, Rojas, Tanabe, Martin, & Tregellas, 2012). In contrast to the resting-state analyses described above, this approach describes the topology of functionally connected brain networks consisting of nodes and edges (Bullmore & Sporns, 2009). Brain regions usually constitute the nodes, while edges relate to the functional interactions between them. With respect to drug effects, topological approaches allow to quantify whether drugs change network topology and facilitate or hinder information processing (Giessing & Thiel, 2012). Using such approach and a 4mg nicotine gum in mildly deprived smokers undergoing an eye-open resting-state scan, Giessing et al. (2013) were able to show that nicotine tended to increase global brain network efficiency, that is, how well information is exchanged across the whole

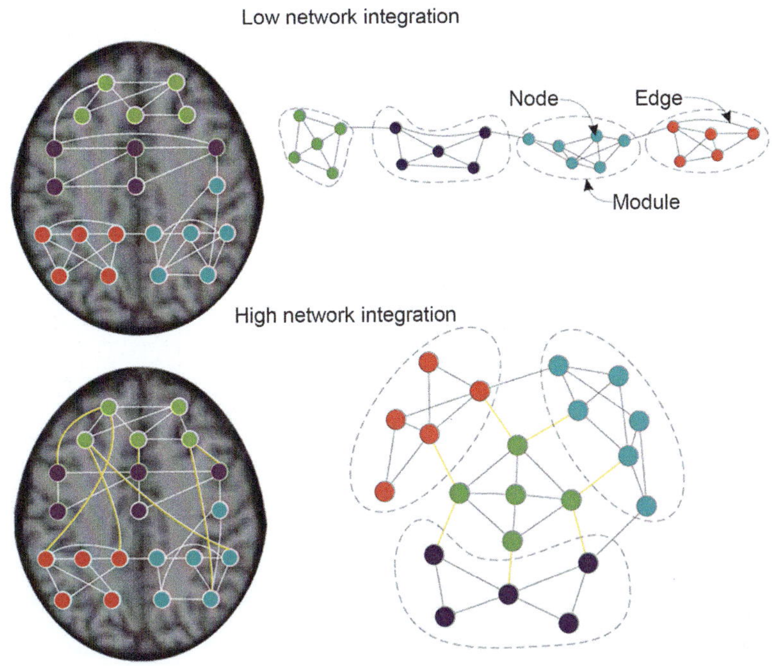

FIG. 16.2 Illustration of increased integration of functional brain graphs after nicotine.
The figure depicts two graphs with different topology, lower and higher network integration. The left part depicts the graphs within the physical space of the brain. An increased network integration as seen under nicotine is characterized by less serial processing and higher capacity for parallel information transfer, that is, increased global brain network efficiency. Modules are connected by shorter path lengths and are less clustered. Further, there were increases in the number of functional long-distance connections *(yellow)* under nicotine. Node: a unique homogenous part of the brain, for example, an anatomically defined brain region; edge: functional connectivity between two nodes; module: smaller subnetwork. *Reprinted from Giessing and Thiel (2012). Frontiers, permission not necessary.*

network and significantly decreased mean clustering, that is, how well information is exchanged across neighbors, in posterior brain regions. Further, the number of functional long-distance connections was increased. The findings suggest that brain networks under nicotine are shifted to a more integrative network configuration. Nonsmokers were studied in the resting-state with eyes closed under 7mg transdermal nicotine by Wylie et al. (2012). This study also showed a trend for increased global brain network efficiency under nicotine, however, together with an increase in local efficiency, which is related to clustering. Note that effects of nicotine may differ depending on whether resting-state activity is investigated under eyes open or eyes closed (Ranzi, Thiel, & Herrmann, 2016), which may be one explanation for the discrepancy.

16.4 OUTLOOK

Even though the availability of neuroimaging methods has enabled us over the last decades to use these techniques for mapping the effects of nicotine on the system level in the human brain, we are far away from drawing a convincing picture on which neural circuits are modulated in which direction by nicotine at rest or during task performance. Apart from different modes of drug application, major problems in comparing across studies are the wide array of different paradigms and/or data analysis techniques used. Further, many early pharmacological fMRI studies used small subject numbers and lenient statistical thresholds, which may be prone to false positives and bias narrative reviews or metaanalyses based on published activation foci. In the neuroimaging community, data sharing has become a major aspect to pool findings across paradigms and subjects, and pharmacological fMRI approaches may benefit from these developments (Poldrack & Gorgolewski, 2014). Since there is less variability in resting-state as compared to task-based studies, a first approach would be an open sharing of raw resting-state nicotine data for larger analyses. Recent approaches provide guidelines for standardized data acquisition and analysis, which is required for this effort (Khalili-Mahani et al., 2017). Pooling across subjects and studies would also allow to investigate the large interindividual variability with respect to the effects of nicotine (Newhouse, Potter, & Singh, 2004).

MINI-DICTIONARY OF TERMS

Attention-related brain regions These are regions that are usually active when subjects perform an attention task in a task-based fMRI approach. They include parietal and frontal cortical regions and subcortical regions, such as the thalamus. If attention is tested in the visual modality, higher visual areas are also active.

BOLD effect The BOLD effect/contrast is the signal obtained in fMRI studies that relates to neural activity. It depends on the relative levels of oxy- and deoxyhemoglobin, which can be picked up by MRI machines to map neural activity in the human brain.

Default mode network This network consists of a set of brain regions, mostly localized in medial parts of the brain, that are active when a person is not involved in a task but at rest or daydreaming.

Graph theory Graph theory is a mathematical framework that allows to model pairwise relations among objects. It has been introduced into the neuroimaging community to model the brain's functional and structural connectivity.

Pharmacological fMRI studies Pharmacological fMRI is not a technique on its own; it is a summary term for all fMRI studies (e.g., task-based and resting-state) that involve the administration of a drug to investigate its effect on neural activity and circuits.

Resting-state fMRI studies In these studies, subjects are asked to do nothing with open or closed eyes, while their brain activity is measured. Data are analyzed with respect to low-frequency spontaneous fluctuations, which show strong temporal coherence across spatially separate brain regions.

Task-based fMRI studies In these studies, subjects perform a motor, sensory, or cognitive task, while their brain activity is measured. Data analysis compares brain activity during task performance and a respective control condition or between different stimuli within a task.

Key Facts on Functional Magnetic Resonance Imaging

- The major technique to track noninvasively neural activity across the human brain.
- Since the 1990s, the number of studies using this technique to identify neural correlates of cognitive functions has massively increased.
- The fMRI signal is an indirect measure of neural activity measuring changes in oxygenated and deoxygenated hemoglobin that occur in brain areas where cells are active.
- To uncover the signal, a variety of data analysis steps are necessary that include signal reconstruction, preprocessing, and statistical analysis approaches.
- In recent years, the number of different statistical analysis and modeling approaches has massively increased.
- The analysis of fMRI data shifted from analyzing changes in local neural activity to changes in functional connectivity in widely distributed brain networks.

Summary Points

- This chapter focuses on acute effects of nicotine in the human brain studied in conjunction with cognitive tasks or under rest to gauge functional connectivity changes.
- Nicotine-induced deactivations in parietal and frontal brain regions were found in several visuospatial selective attention tasks in nonsmokers and in one study in smokers.
- Nicotine-induced activation increases were found in parietal and frontal brain regions in a sustained visual attention and working memory task.
- A metaanalysis of all neuroimaging nicotine studies across tasks in smokers and nonsmokers shows decreases of neural activity under nicotine in the default mode network and posterior parietal cortex and increases in attention-related brain regions.
- Resting-state fMRI studies provide evidence for decreased activity in the default mode network in nonsmokers under nicotine.
- Graph-based analysis approaches to resting-state data suggest increased global network efficiency under nicotine in smokers and nonsmokers.

References

Beckmann, C. F., DeLuca, M., Devlin, J. T., & Smith, S. M. (2005). Investigations into resting-state connectivity using independent component analysis. *Philosophical Transactions of the Royal Society London B Biological Sciences*, *360*(1457), 1001–1013. https://doi.org/10.1098/rstb.2005.1634.

Belliveau, J. W., Kennedy, D. N., Jr., McKinstry, R. C., Buchbinder, B. R., Weisskoff, R. M., Cohen, M. S., et al. (1991). Functional mapping of the human visual cortex by magnetic resonance imaging. *Science*, *254* (5032), 716–719.

Black, K. J., Koller, J. M., & Miller, B. D. (2013). Rapid quantitative pharmacodynamic imaging by a novel method: theory, simulation testing and proof of principle. *PeerJ*, *1*. e117:https://doi.org/10.7717/peerj.117.

Bloom, A. S., Hoffmann, R. G., Fuller, S. A., Pankiewicz, J., Harsch, H. H., & Stein, E. A. (1999). Determination of drug-induced changes in functional MRI signal using a pharmacokinetic model. *Human Brain Mapping*, *8*(4), 235–244.

Breckel, T. P., Thiel, C. M., Bullmore, E. T., Zalesky, A., Patel, A. X., & Giessing, C. (2013). Long-term effects of attentional performance on functional brain network topology. *PLoS ONE*. *8*(9)e74125: https://doi.org/10.1371/journal.pone.0074125.

Buckner, R. L., & Vincent, J. L. (2007). Unrest at rest: default activity and spontaneous network correlations. *NeuroImage*, *37*(4), 1091–1096. discussion 1097–1099(2007). https://doi.org/10.1016/j.neuroimage.2007.01.010.

Bullmore, E., & Sporns, O. (2009). Complex brain networks: graph theoretical analysis of structural and functional systems. *Nature Reviews. Neuroscience*, *10*(3), 186–198.

Cole, D. M., Beckmann, C. F., Long, C. J., Matthews, P. M., Durcan, M. J., & Beaver, J. D. (2010). Nicotine replacement in abstinent smokers improves cognitive withdrawal symptoms with modulation of resting brain network dynamics. *NeuroImage*, *52*(2), 590–599.

Fedota, J. R., & Stein, E. A. (2015). Resting-state functional connectivity and nicotine addiction: prospects for biomarker development. *Annals of the New York Academy of Sciences*, *1349*, 64–82. https://doi.org/10.1111/nyas.12882.

Fox, M. D., Snyder, A. Z., Vincent, J. L., Corbetta, M., Van Essen, D. C., & Raichle, M. E. (2005). The human brain is intrinsically organized into dynamic, anticorrelated functional networks. *Proceedings of the National Academy of Sciences of the United States of America*, *102*(27), 9673–9678. https://doi.org/10.1073/pnas.0504136102.

Fritz, H. C., Wittfeld, K., Schmidt, C. O., Domin, M., Grabe, H. J., Hegenscheid, K., et al. (2014). Current smoking and reduced gray matter volume-a voxel-based morphometry study. *Neuropsychopharmacology*, *39*(11), 2594–2600. https://doi.org/10.1038/npp.2014.112.

Giessing, C., & Thiel, C. M. (2012). Pro-cognitive drug effects modulate functional brain network organization. *Frontiers in Behavioral Neuroscience*, *6*, 53. https://doi.org/10.3389/fnbeh.2012.00053.

Giessing, C., Thiel, C. M., Alexander-Bloch, A. F., Patel, A. X., & Bullmore, E. T. (2013). Human brain functional network changes associated with enhanced and impaired attentional task performance. *The Journal of Neuroscience*, *33*(14), 5903–5914. https://doi.org/10.1523/JNEUROSCI.4854-12.2013.

Greicius, M. (2008). Resting-state functional connectivity in neuropsychiatric disorders. *Current Opinion in Neurology, 21*(4), 424–430.

Hahn, B., Ross, T. J., Wolkenberg, F. A., Shakleya, D. M., Huestis, M. A., & Stein, E. A. (2009). Performance effects of nicotine during selective attention, divided attention, and simple stimulus detection: an fMRI study. *Cerebral Cortex, 19*(9), 1990–2000.

Hahn, B., Ross, T. J., Yang, Y., Kim, I., Huestis, M. A., & Stein, E. A. (2007). Nicotine enhances visuospatial attention by deactivating areas of the resting brain default network. *The Journal of Neuroscience, 27*(13), 3477–3489.

Heishman, S. J., Kleykamp, B. A., & Singleton, E. G. (2010). Meta-analysis of the acute effects of nicotine and smoking on human performance. *Psychopharmacology, 210*(4), 453–469.

Hong, L. E., Gu, H., Yang, Y., Ross, T. J., Salmeron, B. J., Buchholz, B., et al. (2009). Association of nicotine addiction and nicotine's actions with separate cingulate cortex functional circuits. *Archives of General Psychiatry, 66*(4), 431–441.

Khalili-Mahani, N., Rombouts, S. A., van Osch, M. J., Duff, E. P., Carbonell, F., Nickerson, L. D., et al. (2017). Biomarkers, designs, and interpretations of resting-state fMRI in translational pharmacological research: a review of state-of-the-art, challenges, and opportunities for studying brain chemistry. *Human Brain Mapping, 38*(4), 2276–2325. https://doi.org/10.1002/hbm.23516.

Kumari, V., Gray, J. A., ffytche, D. H., Mitterschiffthaler, M. T., Das, M., Zachariah, E., et al. (2003). Cognitive effects of nicotine in humans: an fMRI study. *NeuroImage, 19*(3), 1002–1013.

Lawrence, N. S., Ross, T. J., & Stein, E. A. (2002). Cognitive mechanisms of nicotine on visual attention. *Neuron, 36*(3), 539–548.

Logothetis, N. K. (2008). What we can do and what we cannot do with fMRI. *Nature, 453*(7197), 869–878. https://doi.org/10.1038/nature06976.

Lu, H., & Stein, E. A. (2014). Resting state functional connectivity: its physiological basis and application in neuropharmacology. *Neuropharmacology, 84*, 79–89. https://doi.org/10.1016/j.neuropharm.2013.08.023.

Mayer, J. S., Bittner, R. A., Nikolic, D., Bledowski, C., Goebel, R., & Linden, D. E. (2007). Common neural substrates for visual working memory and attention. *NeuroImage, 36*(2), 441–453. https://doi.org/10.1016/j.neuroimage.2007.03.007.

Newhouse, P. A., Potter, A. S., Dumas, J. A., & Thiel, C. M. (2011). Functional brain imaging of nicotinic effects on higher cognitive processes. *Biochemical Pharmacology, 82*(8), 943–951. https://doi.org/10.1016/j.bcp.2011.06.008.

Newhouse, P. A., Potter, A., & Singh, A. (2004). Effects of nicotinic stimulation on cognitive performance. *Current Opinion in Pharmacology, 4*(1), 36–46.

Poldrack, R. A., & Gorgolewski, K. J. (2014). Making big data open: data sharing in neuroimaging. *Nature Neuroscience, 17*(11), 1510–1517. https://doi.org/10.1038/nn.3818.

Ranzi, P., Thiel, C. M., & Herrmann, C. S. (2016). EEG source reconstruction in male nonsmokers after nicotine administration during the resting state. *Neuropsychobiology, 73*(4), 191–200. https://doi.org/10.1159/000445481.

Rusted, J., Ruest, T., & Gray, M. A. (2011). Acute effects of nicotine administration during prospective memory, an event related fMRI study. *Neuropsychologia, 49*(9), 2362–2368.

Smucny, J., Olincy, A., Eichman, L. S., & Tregellas, J. R. (2015). Neuronal effects of nicotine during auditory selective attention. *Psychopharmacology, 232*(11), 2017–2028. https://doi.org/10.1007/s00213-014-3832-7.

Smucny, J., & Tregellas, J. R. (2017). Targeting neuronal dysfunction in schizophrenia with nicotine: evidence from neurophysiology to neuroimaging. *Journal of Psychopharmacology, 31*(7), 801–811. https://doi.org/10.1177/0269881117705071.

Stein, E. A., Pankiewicz, J., Harsch, H. H., Cho, J. K., Fuller, S. A., Hoffmann, R. G., et al. (1998). Nicotine-induced limbic cortical activation in the human brain: a functional MRI study. *The American Journal of Psychiatry, 155*(8), 1009–1015.

Sutherland, M. T., Ray, K. L., Riedel, M. C., Yanes, J. A., Stein, E. A., & Laird, A. R. (2015). Neurobiological impact of nicotinic acetylcholine receptor agonists: an activation likelihood estimation meta-analysis of pharmacologic neuroimaging studies. *Biological Psychiatry, 78*(10), 711–720. https://doi.org/10.1016/j.biopsych.2014.12.021.

Sutherland, M. T., Ross, T. J., Shakleya, D. M., Huestis, M. A., & Stein, E. A. (2011). Chronic smoking, but not acute nicotine administration, modulates neural correlates of working memory. *Psychopharmacology, 213*(1), 29–42.

Sutherland, M. T., & Stein, E. A. (2018). Functional neurocircuits and neuroimaging biomarkers of tobacco use disorder. *Trends in Molecular Medicine, 24*(2), 129–143. https://doi.org/10.1016/j.molmed.2017.12.002.

Tanabe, J., Nyberg, E., Martin, L. F., Martin, J., Cordes, D., Kronberg, E., et al. (2011). Nicotine effects on default mode network during resting state. *Psychopharmacology.*

Thiel, C. M., & Fink, G. R. (2008). Effects of the cholinergic agonist nicotine on reorienting of visual spatial attention and top-down attentional control. *Neuroscience, 152*(2), 381–390. https://doi.org/10.1016/j.neuroscience.2007.10.061.

Thiel, C. M., Zilles, K., & Fink, G. R. (2005). Nicotine modulates reorienting of visuospatial attention and neural activity in human parietal cortex. *Neuropsychopharmacology, 30*(4), 810–820. https://doi.org/10.1038/sj.npp.1300633.

Vossel, S., Thiel, C. M., & Fink, G. R. (2008). Behavioral and neural effects of nicotine on visuospatial attentional reorienting in non-smoking subjects. *Neuropsychopharmacology, 33*(4), 731–738. https://doi.org/10.1038/sj.npp.1301469.

Warbrick, T., Mobascher, A., Brinkmeyer, J., Musso, F., Stoecker, T., Shah, N. J., et al. (2011). Direction and magnitude of nicotine effects on the fMRI BOLD response are related to nicotine effects on behavioral performance. *Psychopharmacology (Berl), 215*(2), 333–344.

Wylie, K. P., Rojas, D. C., Tanabe, J., Martin, L. F., & Tregellas, J. R. (2012). Nicotine increases brain functional network efficiency. *NeuroImage, 63*(1), 73–80. https://doi.org/10.1016/j.neuroimage.2012.06.079.

Wylie, K. P., Tanabe, J., Martin, L. F., Wongngamnit, N., & Tregellas, J. R. (2013). Nicotine increases cerebellar activity during finger tapping. *PLoS ONE, 8*(12), e84581:https://doi.org/10.1371/journal.pone.0084581.

Yu, A. J., & Dayan, P. (2005). Uncertainty, neuromodulation, and attention. *Neuron, 46*(4), 681–692.

Zhang, X., Salmeron, B. J., Ross, T. J., Geng, X., Yang, Y., & Stein, E. A. (2011). Factors underlying prefrontal and insula structural alterations in smokers. *NeuroImage, 54*(1), 42–48.

CHAPTER

17

Nicotine Dependence in Schizophrenia: Contributions of Nicotinic Acetylcholine Receptors

Robert D. Cole, Vinay Parikh

Department of Psychology and Neuroscience Program, Temple University, Philadelphia, PA, United States

Abbreviations

DA	dopamine
nAChRs	nicotinic acetylcholine receptors
PFC	prefrontal cortex

17.1 INTRODUCTION

Nicotine, the psychoactive ingredient found in tobacco, remains one of the most widely abused drugs in the world despite its devastating health effects. While the incidence of nicotine consumption among the "healthy" population hovers around 20%–35%, the incidence within the population suffering from schizophrenia (a chronic and debilitating mental illness) is reported to be three to four times higher. Smokers with schizophrenia not only consume more cigarettes but also extract more nicotine per cigarette. Additionally, patients with schizophrenia experience greater reinforcing effects of smoking and more severe withdrawal symptoms during abstinence than smokers without schizophrenia. Thus, subjects with schizophrenia are more vulnerable to develop nicotine dependence, and smoking cessation in these individuals remains very challenging.

Nicotine dependence is also found to be associated with symptom severity and poor outcome in patients with schizophrenia. Furthermore, heavy smoking in these individuals makes them highly susceptible to develop various disorders such as cardiovascular diseases, lung cancer, and pulmonary conditions, resulting in premature mortality. Therefore, a fundamental understanding of the neurobiological mechanisms that contribute to comorbid nicotine use in schizophrenia is essential to develop therapeutic strategies to improve smoking cessation, treatment outcome, and life span of patients suffering from schizophrenia. This chapter will focus on the neurological underpinnings, namely, alterations in the function of neuronal nicotinic acetylcholine receptors (nAChRs) that may lead to nicotine dependence in schizophrenia. People diagnosed with schizophrenia display a combination of positive (hallucinations, delusions, and thought impairments), negative (diminished emotional expression, loss of motivation, and psychosocial impairments), and cognitive (disorganized thought, attention and working memory deficits, and executive dysfunction) symptoms. The empirical evidence supporting the link between these symptoms and nAChR dysfunction will be reviewed. Lastly, this chapter will delve into the use of nAChR-targeted therapies as new means to treat the symptomology of schizophrenia and to ameliorate smoking cessation in patients with schizophrenia.

17.2 PHARMACOLOGY OF nAChRs

Nicotine exerts its behavioral and neuromodulatory effects by primarily binding to nAChRs located throughout the central nervous system. nAChRs are a family of ligand-gated ionotropic receptors that mediate fast synaptic transmission by altering cation channel currents. These receptors belong to one of the two classes of cholinergic receptors activated by the endogenous neurotransmitter ACh. The other types of cholinergic receptors stimulated by ACh are muscarinic receptors, which are G-protein-coupled receptors.

Neuronal nAChRs are pentameric structures that are formed from a combination of five membrane-spanning units consisting of nine isoforms of α subunits (α2–α10)

Neuroscience of Nicotine
https://doi.org/10.1016/B978-0-12-813035-3.00017-4

and three isoforms of β subunits (β2–β4) and arranged as either heteromeric or homomeric assemblies (Gotti et al., 2009). Within the mammalian brain, homomeric α7 and heteromeric α4β2 are the most predominant and widely distributed nAChRs. α4β2 nAChRs that possess a higher affinity for nicotine have been shown to desensitize and upregulate under continued nicotine administration (Marks, Grady, & Collins, 1993). On the other hand, α7 nAChRs are characterized by faster receptor kinetics and enhanced Ca^{2+} permeability but possess a lower affinity for nicotine (Castro & Albuquerque, 1995). By virtue of higher Ca^{2+} permeability, α7 nAChRs are also known to facilitate downstream signaling pathways such as extracellular signaling-regulated kinase and cAMP-response element-binding protein (Bitner et al., 2007). nAChRs are localized on both the pre- and postsynaptic sites. Presynaptic nAChRs exist as both auto- and hetero-receptors and are known to modulate the release of multiple neurotransmitters including dopamine (DA) and glutamate either directly or by interacting with other coexisting receptors at the terminals (Parikh, Ji, Decker, & Sarter, 2010; Wonnacott, 1997). Animal studies involving the genetic and pharmacological manipulations of α4β2 and α7 nAChRs have implicated these receptors in specific behavioral processes such as reinforcing and rewarding effects of nicotine (Dani & De Biasi, 2001), mood and anxiety (Picciotto, Lewis, van Schalkwyk, & Mineur, 2015), and cognition (Guillem et al., 2011; Young et al., 2007). Continuing research is now beginning to uncover the roles of other nAChRs containing ternary subunit complexes such as α3β4α5 and α4β2α5 and their contributions to specific neurophysiological and behavioral processes.

17.3 nAChR EXPRESSION IN SCHIZOPHRENIA

Along with the myriad of neurotransmitter dysfunctions associated with schizophrenia, alterations in neuronal nAChR expression have been uncovered through various postmortem imaging and genetic profiling studies. Postmortem evaluations of the brains of patients with schizophrenia show reduced expression of α7 nAChRs in multiple brain regions as compared to the healthy controls (Court et al., 1999; Freedman, Hall, Adler, & Leonard, 1995; Guan, Zhang, Blennow, & Nordberg, 1999). However, the differences in the mRNA and protein levels of α7 nAChRs were not observed between schizophrenic smokers and healthy smokers raising the possibility that aberrant posttranslational modification or trafficking of these receptors in schizophrenia might have been improved by nicotine use (Mexal et al., 2010).

Receptor autoradiographic studies indicated an absence of the upregulation of the high-affinity α4β2 nAChRs in the brains of schizophrenic smokers, which is typically seen in normal smokers (Breese et al., 2000; Durany et al., 2000). In addition, human neuroimaging studies employing positron-emission tomography also show reduced β2* nAChR availability in smokers with schizophrenia as compared to normal smokers (Brasic et al., 2012; D'Souza et al., 2012) but higher upregulation in comparison to nonsmoker schizophrenia patients (Esterlis et al., 2014). These findings suggested that either desensitization or turnover of α4β2 nAChRs is abnormal in schizophrenic subjects.

Accumulating evidence from genetic studies points toward an association between schizophrenia and nAChR subunit genes that may account for alterations in nAChR expression and function. Polymorphism in genes encoding for α7 (CHRNA7) subunit located on chromosome 15 (15q13–14) has been reported in patients with schizophrenia (Freedman et al., 2001). In addition, an overexpression of the chimeric gene α7 nAChRs (CHRFAM7A) that is considered to be a dominant negative regulator of α7 nAChR function was found in the dorsolateral prefrontal cortex (PFC) of schizophrenia subjects (Kunii et al., 2015). A genetic interaction between the α4 and β2 subunits and schizophrenia has also been reported suggesting a possible link between schizophrenia and α4β2 nAChRs (De Luca, Voineskos, Wong, & Kennedy, 2006; Voineskos et al., 2007). Furthermore, there is emerging evidence that variants in the gene cluster CHRNA5/A3/B4 on chromosome 15 that encodes for the α3, α5, and β4 subunits might account for smoking behavior (Lassi et al., 2016). Together, these studies indicate that nAChRs are dysregulated in schizophrenia patients and higher incidence of smoking in these subjects might represent an attempt to restore nAChR function. Table 17.1 provides a summary of clinical studies evaluating nAChR densities in smoking and nonsmoking schizophrenia patients.

17.4 nAChRs AND THE POSITIVE SYMPTOMS OF SCHIZOPHRENIA

Mild, but not high, nicotine consumption has been reported to reduce positive symptom (psychosis) score on the positive and negative syndrome scale in patients with schizophrenia (Aguilar, Gurpegui, Diaz, & de Leon, 2005). Moreover, higher ratios of *CHRFAM7A* to *CHRNA7* are suggested to be involved in both the development of nicotine dependence that precedes clinical diagnosis of schizophrenia and psychotic symptoms themselves (Kunii et al., 2015). Excess *CHRFAM7A* expression disrupts α7 nAChR signaling at both pre- and postsynaptic compartments through a decrease in the cation channel function and therefore might contribute to the positive symptoms. Although the high-affinity

TABLE 17.1 Studies Examining nAChR Density in Smoking and Nonsmoking Populations of Schizophrenia Patients

nAChRs	Technique	Comparison	Brain regions	Effects	Study
α7 nAChRs	Autoradiography	SZ vs normal	The hippocampus, thalamus, and prefrontal cortex	Reduced α7 nAChR density	Freedman et al. (1995), Court et al. (1999), and Guan et al. (1999)
	Fluorescent in situ hybridization	SZ vs normal	The dorsolateral prefrontal cortex	Overexpression of the chimeric gene α7 nAChRs (CHRFAM7A)	Kunii et al. (2015)
	Western blotting	SZ vs normal	The dorsolateral prefrontal cortex	Reduced α7 nAChR protein expression	Mexal et al. (2010)
α4β2 nAChRs	Autoradiography	SZ smokers vs normal smokers	The cortex, striatum, and hippocampus	Reduced α4β2 nAChR density	Breese et al. (2000) and Durany et al. (2000)
	Positron-emission tomography	SZ smokers vs normal smokers	The frontal cortex, parietal cortex, and thalamus	Reduced β2 nAChR availability	Brasic et al. (2012) and D'Souza et al. (2012)
	Positron-emission tomography	SZ smokers vs SZ nonsmokers	The frontal cortex, parietal cortex, and striatum	Higher β2 nAChR upregulation	Esterlis et al. (2014)

SZ, schizophrenia.
Adapted from Parikh et al. (2016). Reprinted with permission from Elsevier.

nAChR α4β2* has not been directly implicated in the development of psychosis in schizophrenia, β2* nAChR mRNA transcription was altered in bipolar disorder patients that experience psychotic symptoms (Severance & Yolken, 2007). On the other hand, there are studies that found either no change in positive symptoms (Patkar et al., 2002) or increases in the intensity of these symptoms with smoking in schizophrenia patients (Ziedonis, Kosten, Glazer, & Frances, 1994). Thus, the evidence for the exact role that nAChRs play in the mediation of positive symptoms of schizophrenia is not conclusive, and more research is needed in this direction.

17.5 NICOTINE AND COGNITION IN SCHIZOPHRENIA: NEUROMODULATION BY nAChRs

Cognitive deficits are considered to be a central feature of schizophrenia. Patients with schizophrenia show reduced sensory gating as compared to normal subjects, and this effect was found to be linked to the polymorphism in the *CHRNA7* gene (Freedman et al., 2001). Deficits in sensory gating contribute to disruptions in higher cognitive processes such as attention and working memory. Animal studies conducted in mice with the homozygous deletion of the α7 nAChR gene reported performance impairments in the tasks of cognition including the five-choice serial-reaction time task and delayed nonmatch to sample task (Fernandes, Hoyle, Dempster, Schalkwyk, & Collier, 2006; Young et al., 2007). Moreover, acute systemic administration of SSR180711, an α7 nAChR partial agonist, improved attention set-shifting in the rat neonatal ventral hippocampus lesion model that reproduces many developmental aspects of schizophrenia (Brooks et al., 2012). Schizophrenic smokers exhibited marked improvement in P50 sensory gating function (Adler, Hoffer, Wiser, & Freedman, 1993). Additionally, performance on the tasks of sustained and selective attention and working memory was superior in schizophrenia patients that smoked cigarettes as compared to the nonsmokers (Segarra et al., 2011). Collectively, these studies indicated that perturbations in α7 nAChR signaling may account for cognitive deficits in schizophrenia and higher nicotine use in patients with schizophrenia may represent an attempt to normalize cognitive deficits by keeping the low-affinity nAChR activated. Although direct clinical evidence supporting the involvement of α4β2 nAChRs in the cognitive symptoms of schizophrenia is lacking, smoking-induced activation of these high-affinity nAChRs in subjects with schizophrenia is hypothesized to improve attention. In this context, animal studies utilizing mice with the genetic deletion of β2 nAChR subunits reported deficits in attentional processing and conflict resolution (Granon, Faure, & Changeux, 2003; Guillem et al., 2011).

DA deficiency and disrupted signaling via D1 receptors in the PFC are linked to cognitive deficits in schizophrenia (Goldman-Rakic, Castner, Svensson, Siever, & Williams, 2004). Both α7- and β2-subunit-containing nAChRs are known to exist on prefrontal dopaminergic terminals and modulate DA release (Livingstone et al., 2009). Moreover, nAChRs are also expressed on the cell bodies of midbrain DA neurons and could potentially influence DA release in the PFC by modulating the firing rate of these neurons. Thus, it is possible that cigarette smoking or nicotine administration may partly relieve cognitive symptoms of schizophrenia by activating both the α7 and α4β2* nAChRs and normalizing DA transmission in the PFC.

There is compelling evidence that nAChRs also regulate synaptic glutamate signaling. Thalamocortical glutamatergic afferents projecting to layer V pyramidal neurons of the PFC are endowed with functional β2*-subunit-containing nAChRs. Additionally, glutamate release and spontaneous excitatory postsynaptic currents in the PFC were abolished following local application of nicotine in mice lacking the β2 nAChRs (Lambe, Picciotto, & Aghajanian, 2003; Parikh et al., 2010). These studies indicated that α4β2 nAChRs regulate thalamocortical glutamatergic transmission that is critical for attentional processing. Andre and colleagues (Andre, Leach, & Gould, 2011) found that deficits in fear conditioning induced by NMDA receptor antagonist MK-801 were ameliorated with nicotine and this effect was mediated by β2*-subunit-containing nAChRs in the hippocampus. This report provided clear evidence that beneficial cognitive effects of nicotine on hippocampus-dependent memory involve the facilitation of postsynaptic NMDA-mediated glutamate signaling via α4β2 nAChRs. α7 nAChR activation has also been shown to be essential for the excitation of postsynaptic NMDA receptors in the prefrontal networks (Yang et al., 2013). Furthermore, aberrations in NMDA glutamate-receptor-mediated signaling have been proposed to be one of the underlying mechanisms contributing to the cognitive symptoms of schizophrenia (Moghaddam & Javitt, 2012). Thus, augmentation of prefrontal glutamate signaling associated with nAChR activation might rectify attentional deficits in patients with schizophrenia.

DA and glutamate neurotransmitter systems work in conjunction with each other to facilitate synaptic plasticity in the PFC. Specifically, D1 receptor activation facilitates NMDA responsiveness via intracellular Ca^{2+} and protein kinase A, and synergistic interactions between these two receptors are critical for normal cortical pyramidal cell excitability and PFC function (Tseng & O'Donnell, 2004). As noted above, pharmacological antagonism of nAChRs disrupts PFC-dependent cognitive functions. Therefore, stimulation of prefrontal nAChRs either by nicotine or specific α4β2/α7 nAChR agonists may normalize DA-glutamate release and restore D1/NMDA receptor signaling and synergism to ameliorate some of the cognitive deficits in patients with schizophrenia (Parikh, Kutlu, & Gould, 2016). A hypothetical model illustrating how nAChR-mediated facilitation of DA-glutamate cross talk in the PFC could fine-tune cognitive functions in schizophrenia is shown in Fig. 17.1.

17.6 CONTRIBUTIONS OF nAChRs TO THE NEGATIVE SYMPTOMS OF SCHIZOPHRENIA

Negative symptoms of schizophrenia including the lack of motivation to pursue goals, anhedonia, social withdrawal, and reduced emotional reactivity to affective stimuli predicted smoking behavior in schizophrenia patients (Patkar et al., 2002). Deficits in reward processing are known to contribute to the manifestation of negative symptoms in schizophrenia. Moreover, α4β2 nAChRs are known to modulate reward-motivated behaviors, and a significant association was found between *CHRNA4* gene polymorphism and heavy smoking among schizophrenia patients (Voineskos et al., 2007). Recent nuclear imaging studies that employed a β2 nAChR agonist radiotracer [123I]5-IA-85380 showed not only a lower α4β2 nAChR availability in smokers with schizophrenia as compared to normal smokers but also a negative correlation between the receptor availability and negative symptoms (D'Souza et al., 2012; Esterlis et al., 2014). It is therefore possible that higher nicotine reinforcement in patients with schizophrenia might be linked to the negative symptoms that occur due to α4β2 nAChR dysfunction.

The motivation to obtain natural reward was not affected in α7 nAChR mutant mice (Young et al., 2007). Moreover, the administration of α7 agonist or genetic deletion of α7 nAChR subunit did not affect nicotine reward and reinforcement (Grottick et al., 2000; Pons et al., 2008). However, the administration of TC-5619 (α7 nAChR full agonist) in a transgenic mouse model that features developmental aspects of schizophrenia improved social behavior in these animals (Hauser et al., 2009). Based on available evidence, it appears that disruption in α7 nAChRs does not contribute to altered reward sensitivity in schizophrenia patients and their association with negative symptoms might be unrelated to reward-processing deficits.

17.7 nAChRs MODULATION AS POSSIBLE TREATMENTS OF SCHIZOPHRENIA

Although nicotine is known to provide symptomatic relief to the patients with schizophrenia, the highly addictive properties and the development of tolerance following prolonged treatment limit its use as a drug candidate. Substantial evidence indicates that α7 nAChRs may serve as a therapeutic target for schizophrenia and several α7 nAChR ligands are already under clinical investigation. Phase II clinical trials involving the treatment of schizophrenia with a partial α7 nAChR agonist GTS-21 showed improvements in P50 sensory gating function, attention, working memory, and negative symptoms in schizophrenia patients (Olincy & Freedman, 2012). Clinical studies with selective α7 nAChR partial agonists such as encenicline and tropisetron, conducted in patients with schizophrenia, reported positive effects on various measures of cognition (Noroozian et al., 2013; Preskorn et al., 2014). Moreover, the α7 nAChR full agonist TC-5619 was also found to exert beneficial effects on working memory performance in smokers with schizophrenia (Lieberman et al., 2013).

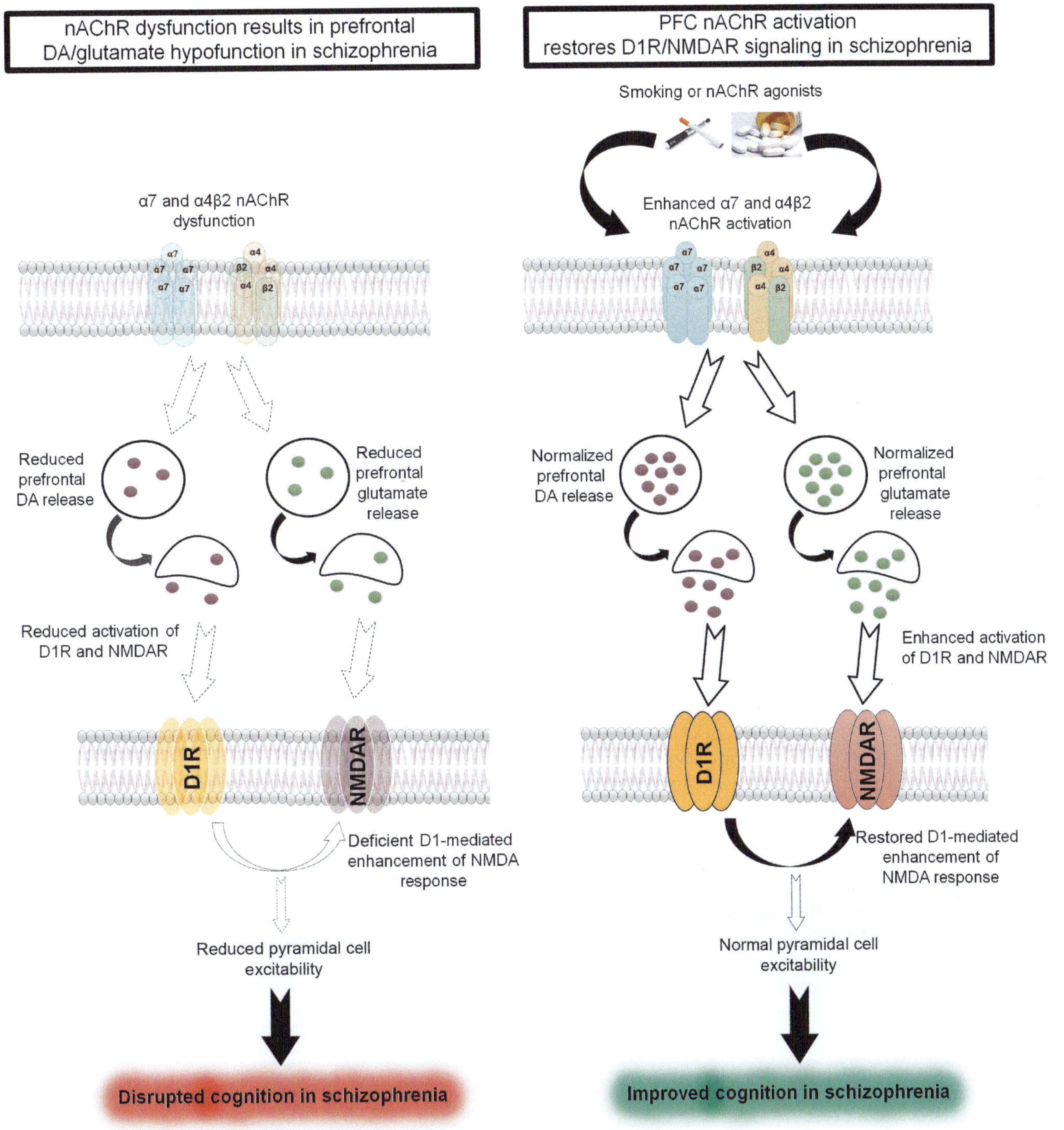

FIG. 17.1 Cartoon depicting the role of prefrontal nAChRs in cognitive deficits of schizophrenia. *(Left panel)* nAChR dysfunction in schizophrenia produces deficient DA and glutamate release in the PFC resulting in diminished D1 and NMDA receptor signaling. Reduced synergy between D1 and NMDA receptor disrupts PFC excitability and cognitive functions. *(Right panel)* Smoking (nicotine) or activation of α4β2 and/or α7 nAChRs increases DA/glutamate release in the PFC and restores synergistic interactions between the DA and glutamate signaling to normalize PFC function and improve cognition in schizophrenia. *Source: Parikh et al. (2016). Reprinted with permission from Elsevier.*

A new and exciting avenue to control cognitive symptoms of schizophrenia is to develop the positive allosteric modulators (PAMs) for α7 nAChRs. PAMs are unique in their action because they preferentially bind to a receptor allosteric site, leaving the endogenous ligand site open. PAMs also potentiate the effects of endogenous ligands while not directly activating receptor signaling. A key advantage of using this approach would be that α7 nAChR modulation may not only amplify the behavioral/cognitive effects produced by α7 nAChR activation but also prevent the tolerance to these effects that may occur due to the desensitization of these receptors. In this regard, galantamine, a cholinesterase inhibitor that is also an α7 nAChR PAM, has shown neurocognitive improvement in subjects with schizophrenia (Buchanan et al., 2008).

Varenicline, an α4β2 nAChR partial agonist, given as an adjunctive therapy with the antipsychotic drugs significantly improved cognitive performance of smokers with schizophrenia on the Wisconsin card sorting test and the digit symbol substitution test (Shim et al., 2012). Another clinical study reported that maintenance therapy with varenicline prolonged tobacco abstinence rates in patients with schizophrenia (Evins et al., 2014). Furthermore, schizophrenia patients with lower baseline symptoms of affective flattening were more likely to achieve smoking abstinence and demonstrated larger increases in reward sensitivity during varenicline treatment (Dutra, Stoeckel, Carlini, Pizzagalli, & Evins, 2012). Based on the existing evidence, it appears that both α7 and α4β2 nAChR ligands may hold promise as potential therapeutic targets in alleviating the negative and cognitive symptoms of schizophrenia. Additionally, targeting α4β2 nAChRs may also be beneficial in reducing smoking relapse in schizophrenia patients. Table 17.2 depicts the results of clinical trials of nAChR modulators in patients with schizophrenia.

Recent genome-wide association studies indicated that polymorphism in the human *CHRNA5* gene that encodes the α5 nAChR subunit increases the risks for both smoking and schizophrenia. Moreover, mice with a single nucleotide polymorphism of the α5 nAChR gene exhibited PFC-dependent behavioral deficits, and chronic nicotine reversed the hypofrontality in these animals (Koukouli et al., 2017). Whether precise targeting of α5 nAChRs improves the cognitive symptoms of schizophrenia that are known to be associated with the PFC dysfunction and ameliorate nicotine dependence in schizophrenia patients remains to be seen.

TABLE 17.2 Clinical Studies Assessing the Effects of nAChR Ligands on Cognitive and Negative Symptoms and Tobacco Cessation in Schizophrenia Patients

nAChR ligand	Specificity	Clinical stage	Neurobehavioral effects
GTS-21	α7 nAChR partial agonist	Phase II clinical trial	Improved P50 sensory gating function, attention, working memory, and negative symptoms
Encenicline	α7 nAChR partial agonist	Phase II clinical trial	Exerted beneficial effects on mismatch negativity, P300 potentials, visual attention, nonverbal working memory, and decision-making
Tropisetron	α7 nAChR partial agonist	Phase II clinical trial	Improved auditory sensory gating and attention
TC-5619	α7 nAChR full agonist	Exploratory trial	Enhanced working memory performance
Galantamine	α7 nAChR PAM	Phase III clinical trial	Improved attention and memory
Varenicline	α4β2 nAChR partial agonist	Phase III clinical trial	Improved cognitive flexibility, visuospatial working memory performance, prolonged tobacco abstinence rates, increased in reward sensitivity, reduced negative symptoms

Adapted from Parikh et al. (2016). Reprinted with permission from Elsevier.

17.8 CONCLUSIONS

The prevalence for smoking in patients with schizophrenia is enormously higher than normal population. There is considerable evidence to support that the drastic increase in nicotine consumption may stem from a dysregulation in nAChRs in the brain that contributes to various symptoms of the schizophrenia pathology presumably by altering neurotransmitter systems including the DA and glutamate and their downstream targets. Therefore, the development of nAChR-subtype-specific agonists represents a viable therapeutic option to attenuate symptoms of schizophrenia. Additionally, nAChR modulators could also be effective in improving smoking cessation to circumvent smoking-related health hazards in smokers with schizophrenia.

MINI-DICTIONARY OF TERMS

Agonist A chemical that binds to a receptor and activates the receptor to produce a biological response.

Cognition The mental process by which knowledge is acquired including perception, attention, memory, and judgment.

Dopamine A catecholamine neurotransmitter involved in reward-motivated behavior and cognition. Addictive drugs increase dopamine release in the brain.

Gene polymorphism Variation that occurs in allele in a DNA sequence.

Glutamate A major excitatory neurotransmitter in the brain. Glutamate is involved in synaptic plasticity and learning and memory.

Heteromeric receptor A receptor that is composed of different types of protein subunits.

Homomeric receptor A receptor that is composed of similar protein subunits.

Nicotine An alkaloid found in the leaves of tobacco plant *Nicotiana tabacum*. Nicotine possesses psychoactive properties and is highly addictive.

Nicotinic acetylcholine receptor Receptor proteins that respond to either neurotransmitter acetylcholine or nicotine. These proteins are linked to ion channels that in the activated state allow passage of cations inside a cell to produce depolarization.

Pentameric protein A protein structure that consists of five protein subunits.

Key Facts of Schizophrenia

- Schizophrenia is a disabling heterogeneous mental disorder that interferes with the ability to socialize normally, think clearly, and distinguish aspects of reality.
- The symptoms of schizophrenia fall under three categories: positive, negative, and cognitive.
- Interactions between the genetic and environmental factors contribute to the risk of developing schizophrenia.
- Imbalance in brain chemicals including the neurotransmitters dopamine and glutamate is linked to the symptoms of schizophrenia.
- There is no cure for schizophrenia, but symptoms can be managed by antipsychotic medications and psychotherapy.

Summary Points

- Those diagnosed with schizophrenia consume nicotine at rates three to four times higher than that of the healthy population.
- Nicotine interacts with nicotinic acetylcholine receptors (nAChRs) to produce diverse behavioral effects.
- The expression of nAChRs is dysregulated in the brains of patients suffering from schizophrenia.
- nAChRs influence the actions of glutamate and dopamine to alter cognitive symptoms of schizophrenia.
- Development of specific nAChR agonists holds promise to manage symptoms of schizophrenia and improve smoking cessation.

References

Adler, L. E., Hoffer, L. D., Wiser, A., & Freedman, R. (1993). Normalization of auditory physiology by cigarette smoking in schizophrenic patients. *The American Journal of Psychiatry*, *150*(12), 1856–1861. https://doi.org/10.1176/ajp.150.12.1856.

Aguilar, M. C., Gurpegui, M., Diaz, F. J., & de Leon, J. (2005). Nicotine dependence and symptoms in schizophrenia: naturalistic study of complex interactions. *The British Journal of Psychiatry*, *186*, 215–221. https://doi.org/10.1192/bjp.186.3.215.

Andre, J. M., Leach, P. T., & Gould, T. J. (2011). Nicotine ameliorates NMDA receptor antagonist-induced deficits in contextual fear conditioning through high-affinity nicotinic acetylcholine receptors in the hippocampus. *Neuropharmacology*, *60*(4), 617–625. https://doi.org/10.1016/j.neuropharm.2010.12.004.

Bitner, R. S., Bunnelle, W. H., Anderson, D. J., Briggs, C. A., Buccafusco, J., Curzon, P., et al. (2007). Broad-spectrum efficacy across cognitive domains by α7 nicotinic acetylcholine receptor agonism correlates with activation of ERK1/2 and CREB phosphorylation pathways. *Journal of Neuroscience*, *27*(39), 10578–10587. https://doi.org/10.1523/JNEUROSCI.2444-07.2007.

Brasic, J. R., Cascella, N., Kumar, A., Zhou, Y., Hilton, J., Raymont, V., et al. (2012). Positron emission tomography experience with 2-[(1)(8)F]fluoro-3-(2(S)-azetidinylmethoxy)pyridine (2-[(1)(8)F]FA) in the living human brain of smokers with paranoid schizophrenia. *Synapse*, *66*(4), 352–368. https://doi.org/10.1002/syn.21520.

Breese, C. R., Lee, M. J., Adams, C. E., Sullivan, B., Logel, J., Gillen, K. M., et al. (2000). Abnormal regulation of high affinity nicotinic receptors in subjects with schizophrenia. *Neuropsychopharmacology*, *23*(4), 351–364. https://doi.org/10.1016/S0893-133X(00)00121-4.

Brooks, J. M., Pershing, M. L., Thomsen, M. S., Mikkelsen, J. D., Sarter, M., & Bruno, J. P. (2012). Transient inactivation of the neonatal ventral hippocampus impairs attentional set-shifting behavior: reversal with an alpha7 nicotinic agonist. *Neuropsychopharmacology*, *37*(11), 2476–2486. https://doi.org/10.1038/npp.2012.106.

Buchanan, R. W., Conley, R. R., Dickinson, D., Ball, M. P., Feldman, S., Gold, J. M., et al. (2008). Galantamine for the treatment of cognitive impairments in people with schizophrenia. *The American Journal of Psychiatry*, *165*(1), 82–89. https://doi.org/10.1176/appi.ajp.2007.07050724.

Castro, N. G., & Albuquerque, E. X. (1995). Alpha-bungarotoxin-sensitive hippocampal nicotinic receptor channel has a high calcium permeability. *Biophysical Journal*, *68*(2), 516–524. https://doi.org/10.1016/S0006-3495(95)80213-4.

Court, J., Spurden, D., Lloyd, S., McKeith, I., Ballard, C., Cairns, N., et al. (1999). Neuronal nicotinic receptors in dementia with Lewy bodies and schizophrenia: alpha-bungarotoxin and nicotine binding in the thalamus. *Journal of Neurochemistry*, *73*(4), 1590–1597.

Dani, J. A., & De Biasi, M. (2001). Cellular mechanisms of nicotine addiction. *Pharmacology, Biochemistry, and Behavior*, *70*(4), 439–446.

De Luca, V., Voineskos, S., Wong, G., & Kennedy, J. L. (2006). Genetic interaction between alpha4 and beta2 subunits of high affinity nicotinic receptor: analysis in schizophrenia. *Experimental Brain Research*, *174*(2), 292–296. https://doi.org/10.1007/s00221-006-0458-y.

D'Souza, D. C., Esterlis, I., Carbuto, M., Krasenics, M., Seibyl, J., Bois, F., et al. (2012). Lower ss2*-nicotinic acetylcholine receptor availability in smokers with schizophrenia. *The American Journal of Psychiatry*, *169*(3), 326–334. https://doi.org/10.1176/appi.ajp.2011.11020189.

Durany, N., Zochling, R., Boissl, K. W., Paulus, W., Ransmayr, G., Tatschner, T., et al. (2000). Human post-mortem striatal alpha4beta2 nicotinic acetylcholine receptor density in schizophrenia and Parkinson's syndrome. *Neuroscience Letters*, *287*(2), 109–112.

Dutra, S. J., Stoeckel, L. E., Carlini, S. V., Pizzagalli, D. A., & Evins, A. E. (2012). Varenicline as a smoking cessation aid in schizophrenia: effects on smoking behavior and reward sensitivity. *Psychopharmacology*, *219*(1), 25–34. https://doi.org/10.1007/s00213-011-2373-6.

Esterlis, I., Ranganathan, M., Bois, F., Pittman, B., Picciotto, M. R., Shearer, L., et al. (2014). In vivo evidence for beta2 nicotinic acetylcholine receptor subunit upregulation in smokers as compared with nonsmokers with schizophrenia. *Biological Psychiatry*, *76*(6), 495–502. https://doi.org/10.1016/j.biopsych.2013.11.001.

Evins, A. E., Cather, C., Pratt, S. A., Pachas, G. N., Hoeppner, S. S., Goff, D. C., et al. (2014). Maintenance treatment with varenicline for smoking cessation in patients with schizophrenia and bipolar disorder: a randomized clinical trial. *JAMA*, *311*(2), 145–154. https://doi.org/10.1001/jama.2013.285113.

Fernandes, C., Hoyle, E., Dempster, E., Schalkwyk, L. C., & Collier, D. A. (2006). Performance deficit of alpha7 nicotinic receptor knockout mice in a delayed matching-to-place task suggests a mild impairment of working/episodic-like memory. *Genes, Brain, and Behavior*, *5*(6), 433–440. https://doi.org/10.1111/j.1601-183X.2005.00176.x.

Freedman, R., Hall, M., Adler, L. E., & Leonard, S. (1995). Evidence in postmortem brain tissue for decreased numbers of hippocampal nicotinic receptors in schizophrenia. *Biological Psychiatry*, *38*(1), 22–33. https://doi.org/10.1016/0006-3223(94)00252-X.

Freedman, R., Leonard, S., Gault, J. M., Hopkins, J., Cloninger, C. R., Kaufmann, C. A., et al. (2001). Linkage disequilibrium for schizophrenia at the chromosome 15q13-14 locus of the alpha7-nicotinic acetylcholine receptor subunit gene (CHRNA7). *American Journal of Medical Genetics*, *105*(1), 20–22.

Goldman-Rakic, P. S., Castner, S. A., Svensson, T. H., Siever, L. J., & Williams, G. V. (2004). Targeting the dopamine D1 receptor in schizophrenia: Insights for cognitive dysfunction. *Psychopharmacology*, *174*(1), 3–16. https://doi.org/10.1007/s00213-004-1793-y.

Gotti, C., Clementi, F., Fornari, A., Gaimarri, A., Guiducci, S., Manfredi, I., et al. (2009). Structural and functional diversity of native brain neuronal nicotinic receptors. *Biochemical Pharmacology*, *78*(7), 703–711. https://doi.org/10.1016/j.bcp.2009.05.024.

Granon, S., Faure, P., & Changeux, J.-P. (2003). Executive and social behaviors under nicotinic receptor regulation. *Proceedings of the National Academy of Sciences*, *100*(16), 9596–9601.

Grottick, A. J., Trube, G., Corrigall, W. A., Huwyler, J., Malherbe, P., Wyler, R., et al. (2000). Evidence that nicotinic alpha(7) receptors are not involved in the hyperlocomotor and rewarding effects of nicotine. *The Journal of Pharmacology and Experimental Therapeutics*, *294*(3), 1112–1119.

Guan, Z. Z., Zhang, X., Blennow, K., & Nordberg, A. (1999). Decreased protein level of nicotinic receptor alpha7 subunit in the frontal cortex from schizophrenic brain. *NeuroReport*, *10*(8), 1779–1782.

Guillem, K., Bloem, B., Poorthuis, R. B., Loos, M., Smit, A. B., Maskos, U., et al. (2011). Nicotinic acetylcholine receptor beta2 subunits in the medial prefrontal cortex control attention. *Science*, *333*(6044), 888–891. https://doi.org/10.1126/science.1207079.

Hauser, T. A., Kucinski, A., Jordan, K. G., Gatto, G. J., Wersinger, S. R., Hesse, R. A., et al. (2009). TC-5619: an alpha7 neuronal nicotinic receptor-selective agonist that demonstrates efficacy in animal models of the positive and negative symptoms and cognitive dysfunction of schizophrenia. *Biochemical Pharmacology*, *78*(7), 803–812. https://doi.org/10.1016/j.bcp.2009.05.030.

Koukouli, F., Rooy, M., Tziotis, D., Sailor, K. A., O'Neill, H. C., Levenga, J., et al. (2017). Nicotine reverses hypofrontality in animal models of addiction and schizophrenia. *Nature Medicine*, *23*(3), 347–354. https://doi.org/10.1038/nm.4274.

Kunii, Y., Zhang, W., Xu, Q., Hyde, T. M., McFadden, W., Shin, J. H., et al. (2015). CHRNA7 and CHRFAM7A mRNAs: Co-localized and their expression levels altered in the postmortem dorsolateral prefrontal cortex in major psychiatric disorders. *The American Journal of Psychiatry*, *172*(11), 1122–1130. https://doi.org/10.1176/appi.ajp.2015.14080978.

Lambe, E. K., Picciotto, M. R., & Aghajanian, G. K. (2003). Nicotine induces glutamate release from thalamocortical terminals in prefrontal cortex. *Neuropsychopharmacology*, *28*(2), 216–225. https://doi.org/10.1038/sj.npp.1300032.

Lassi, G., Taylor, A. E., Timpson, N. J., Kenny, P. J., Mather, R. J., Eisen, T., et al. (2016). The CHRNA5-A3-B4 gene cluster and smoking: From discovery to therapeutics. *Trends in Neurosciences*, *39*(12), 851–861. https://doi.org/10.1016/j.tins.2016.10.005.

Lieberman, J. A., Dunbar, G., Segreti, A. C., Girgis, R. R., Seoane, F., Beaver, J. S., et al. (2013). A randomized exploratory trial of an alpha-7 nicotinic receptor agonist (TC-5619) for cognitive enhancement in schizophrenia. *Neuropsychopharmacology*, *38*(6), 968–975.

Livingstone, P. D., Srinivasan, J., Kew, J. N., Dawson, L. A., Gotti, C., Moretti, M., et al. (2009). alpha7 and non-alpha7 nicotinic acetylcholine receptors modulate dopamine release in vitro and in vivo in the rat prefrontal cortex. *The European Journal of Neuroscience*, *29*(3), 539–550. https://doi.org/10.1111/j.1460-9568.2009.06613.x.

Marks, M. J., Grady, S. R., & Collins, A. C. (1993). Downregulation of nicotinic receptor function after chronic nicotine infusion. *The Journal of Pharmacology and Experimental Therapeutics*, *266*(3), 1268–1276.

Mexal, S., Berger, R., Logel, J., Ross, R. G., Freedman, R., & Leonard, S. (2010). Differential regulation of alpha7 nicotinic receptor gene (CHRNA7) expression in schizophrenic smokers. *Journal of Molecular Neuroscience*, *40*(1–2), 185–195. https://doi.org/10.1007/s12031-009-9233-4.

Moghaddam, B., & Javitt, D. (2012). From revolution to evolution: the glutamate hypothesis of schizophrenia and its implication for treatment. *Neuropsychopharmacology*, *37*(1), 4–15. https://doi.org/10.1038/npp.2011.181.

Noroozian, M., Ghasemi, S., Hosseini, S. M., Modabbernia, A., Khodaie-Ardakani, M. R., Mirshafiee, O., et al. (2013). A placebo-controlled study of tropisetron added to risperidone for the treatment of negative symptoms in chronic and stable schizophrenia. *Psychopharmacology*, *228*(4), 595–602.

Olincy, A., & Freedman, R. (2012). Nicotinic mechanisms in the treatment of psychotic disorders: a focus on the alpha7 nicotinic receptor. *Handbook of Experimental Pharmacology*, (213), 211–232. https://doi.org/10.1007/978-3-642-25758-2_8.

Parikh, V., Ji, J., Decker, M. W., & Sarter, M. (2010). Prefrontal beta2 subunit-containing and alpha7 nicotinic acetylcholine receptors differentially control glutamatergic and cholinergic signaling. *The Journal of Neuroscience*, *30*(9), 3518–3530. https://doi.org/10.1523/JNEUROSCI.5712-09.2010.

Parikh, V., Kutlu, M. G., & Gould, T. J. (2016). nAChR dysfunction as a common substrate for schizophrenia and comorbid nicotine addiction: current trends and perspectives. *Schizophrenia Research*, *171*(1–3), 1–15. https://doi.org/10.1016/j.schres.2016.01.020.

Patkar, A. A., Gopalakrishnan, R., Lundy, A., Leone, F. T., Certa, K. M., & Weinstein, S. P. (2002). Relationship between tobacco smoking and positive and negative symptoms in schizophrenia. *The Journal of Nervous and Mental Disease*, *190*(9), 604–610. https://doi.org/10.1097/01.NMD.0000030525.40247.B9.

Picciotto, M. R., Lewis, A. S., van Schalkwyk, G. I., & Mineur, Y. S. (2015). Mood and anxiety regulation by nicotinic acetylcholine receptors: a potential pathway to modulate aggression and related behavioral states. *Neuropharmacology*, *96*(Pt B), 235–243. https://doi.org/10.1016/j.neuropharm.2014.12.028.

Pons, S., Fattore, L., Cossu, G., Tolu, S., Porcu, E., McIntosh, J. M., et al. (2008). Crucial role of alpha4 and alpha6 nicotinic acetylcholine receptor subunits from ventral tegmental area in systemic nicotine self-administration. *The Journal of Neuroscience*, *28*(47), 12318–12327. https://doi.org/10.1523/JNEUROSCI.3918-08.2008.

Preskorn, S. H., Gawryl, M., Dgetluck, N., Palfreyman, M., Bauer, L. O., & Hilt, D. C. (2014). Normalizing effects of EVP-6124, an alpha-7 nicotinic partial agonist, on event-related potentials and cognition: a proof of concept, randomized trial in patients with schizophrenia. *Journal of Psychiatric Practice*, *20*(1), 12–24. https://doi.org/10.1097/01.pra.0000442935.15833.c5.

Segarra, R., Zabala, A., Eguiluz, J. I., Ojeda, N., Elizagarate, E., Sanchez, P., et al. (2011). Cognitive performance and smoking in first-episode psychosis: the self-medication hypothesis. *European Archives of Psychiatry and Clinical Neuroscience*, *261*(4), 241–250. https://doi.org/10.1007/s00406-010-0146-6.

Severance, E. G., & Yolken, R. H. (2007). Lack of RIC-3 congruence with beta2 subunit-containing nicotinic acetylcholine receptors in bipolar disorder. *Neuroscience*, *148*(2), 454–460. https://doi.org/10.1016/j.neuroscience.2007.06.008.

Shim, J. C., Jung, D. U., Jung, S. S., Seo, Y. S., Cho, D. M., Lee, J. H., et al. (2012). Adjunctive varenicline treatment with antipsychotic medications for cognitive impairments in people with schizophrenia: a randomized double-blind placebo-controlled trial. *Neuropsychopharmacology*, *37*(3), 660–668. https://doi.org/10.1038/npp.2011.238.

Tseng, K. Y., & O'Donnell, P. (2004). Dopamine-glutamate interactions controlling prefrontal cortical pyramidal cell excitability involve multiple signaling mechanisms. *The Journal of Neuroscience*, *24*(22), 5131–5139. https://doi.org/10.1523/JNEUROSCI.1021-04.2004.

Voineskos, S., De Luca, V., Mensah, A., Vincent, J. B., Potapova, N., & Kennedy, J. L. (2007). Association of alpha4beta2 nicotinic receptor and heavy smoking in schizophrenia. *Journal of Psychiatry & Neuroscience*, *32*(6), 412–416.

Wonnacott, S. (1997). Presynaptic nicotinic ACh receptors. *Trends in Neurosciences*, *20*(2), 92–98.

Yang, Y., Paspalas, C. D., Jin, L. E., Picciotto, M. R., Arnsten, A. F., & Wang, M. (2013). Nicotinic alpha7 receptors enhance NMDA cognitive circuits in dorsolateral prefrontal cortex. *Proceedings of the National Academy of Sciences of the United States of America*, *110*(29), 12078–12083. https://doi.org/10.1073/pnas.1307849110.

Young, J. W., Crawford, N., Kelly, J. S., Kerr, L. E., Marston, H. M., Spratt, C., et al. (2007). Impaired attention is central to the cognitive deficits observed in alpha 7 deficient mice. *European Neuropsychopharmacology*, *17*(2), 145–155. https://doi.org/10.1016/j.euroneuro.2006.03.008.

Ziedonis, D. M., Kosten, T. R., Glazer, W. M., & Frances, R. J. (1994). Nicotine dependence and schizophrenia. *Hospital & Community Psychiatry*, *45*(3), 204–206.

CHAPTER

18

Attentional Bias and Smoking

David J. Drobes*, Jason A. Oliver[†], John B. Correa*, David E. Evans*

*Tobacco Research and Intervention Program, Department of Health Outcomes and Behavior, Moffitt Cancer Center, Tampa, FL, United States

[†]Center for Addiction Science and Technology, Department of Psychiatry and Behavioral Sciences, Duke University School of Medicine, Durham, NC, United States

Abbreviations

AB attentional bias
ABR attentional bias retraining
EEG electroencephalogram
RT reaction time

18.1 INTRODUCTION

It has long been observed that stimuli and thoughts related to appetitive or desirable substances or objects disproportionately capture one's attention. A person with a *sweet tooth* is apt to see chocolate cake, a compulsive gambler spots a poker chip on the floor, or a heavily dependent smoker identifies someone across the street smoking a cigarette. These observations are examples of attentional bias (AB), and such occurrences appear to have key importance in the phenomenology and treatment of addictive behavior. To the extent that one's attention is drawn to or held by substance-related stimuli, this may initiate or extend the motivational force underlying addiction-related ideation and behavior. Over the past several decades, both theoretical and empirical work on the topic of AB have drawn heavily from the field of cognitive psychology and in recent years have been extended to the study of smoking and other forms of addictive behavior.

Learning-based theories of addiction assert that drug-related environmental stimuli, through the processes of classical conditioning, become conditioned stimuli that promote drug motivational states. Stemming from these theories, traditional cue reactivity research paradigms present salient drug-paired cues to users, with a typical response profile consisting of subjective craving, physiological arousal, and/or drug-seeking or ingestive behaviors. Elevated responding to drug cues among addicts has increasingly been interpreted from the perspective of cognitive (e.g., information processing) or neural theories of addiction, in which cue saliency is associated with increased attentional capture and/or maintenance at very early stages of information processing and often with little conscious awareness (e.g., Franken, 2003; Robinson & Berridge, 1993; Tiffany, 1990). The processes that subserve attentional biases to drug-related cues are assumed to be fundamental to the development and expression of drug seeking and intake. The present chapter will review paradigms that have been used to study smoking-related AB and will discuss the clinical relevance of this work. Furthermore, we will discuss current challenges to research in this area and future directions for research and application.

18.2 PARADIGMS AND MEASURES

Modified Stroop tasks. In the classic Stroop task (Stroop, 1935), color words are presented in congruent (e.g., the word "blue" in blue font) and incongruent (e.g., the word "red" in blue font) colors. Participants more rapidly identify the font color for congruent words than incongruent words, reflecting cognitive interference from automatic processing of the semantic content of words, despite it being unnecessary for task performance. This task has been modified to study processing biases for salient words, including those related to drug use and abuse. In the smoking area, this is accomplished through the presentation of both smoking and neutral (e.g., tools and household items) words, with the goal of identifying font colors while ignoring word content. It is important to note that this adaptation of the Stroop task may tap distinct processes from the original Stroop, which evokes a more direct conflict between reading and seeing the color.

Neuroscience of Nicotine
https://doi.org/10.1016/B978-0-12-813035-3.00018-6

Findings generally confirm that smokers take longer than nonsmokers to identify the color of smoking-related words relative to neutral words (e.g., Johnsen, Thayer, Laberg, & Asbjornsen, 1997), yet these effects are relatively subtle as compared to interference effects from the original Stroop task. In addition, other addiction Stroop effects appear to be at least partially dependent on design factors. In one study using an alcohol-related Stroop task, interference effects were present only when participants were instructed to suppress alcohol-related thoughts, suggesting that Stroop effects may stem more from efforts to avoid processing semantic content than biases in attentional capture (Klein, 2007). Another study found equivocal interference effects between patients and staff in a substance use clinic, indicating effects may be driven by familiarity rather than addiction, per se (Ryan, 2002). Neither of these two issues has been explicitly examined in smoking research to date.

A metaanalysis of smoking Stroop studies (Cox, Fadardi, & Pothos, 2006) indicated larger interference effects when using a card format (all words presented concurrently on a single card) relative to a single-trial format (i.e., smoking-related and control words presented in either random or pseudorandom order) or when words are presented in blocks (i.e., all words from one category presented successively, followed by a separate block of control category words) as compared to a randomly interspersed order. Similar findings exist for emotion Stroop tasks (Kindt, Bierman, & Brosschot, 1996). If the response is to the entire block of smoking words, as opposed to the initial orienting response to each word, this may call into question the validity of the smoking Stroop paradigm as a pure measure of attentional bias (i.e., early orienting to stimuli). Thus, although the presence of interference effects in response to smoking words among regular smokers appears reliable and is not under scientific debate, the fundamental meaning of these findings remains uncertain.

Visual/dot probe. Another common task employed to assess attentional bias to addiction stimuli is the visual (or dot) probe task. In this task, a smoking-related stimulus and a neutral stimulus are presented side by side. After a set time period, the images disappear, and one is replaced by the "probe" to which participants must respond—for instance, by indicating which direction an arrow is pointing or simply the location of the probe. By contrasting how quickly participants respond to probes that replace smoking stimuli relative to neutral stimuli, inferences can be made about where individuals are directing their attention at the time of stimulus offset. That is, orienting to the smoking stimuli causes a reduced response time to probes replacing the smoking picture, whereas response times to probes replacing the neutral stimulus are protracted. Furthermore, by varying the duration of stimulus presentation, researchers have attempted to parse the subcomponents of attention (i.e., initial capture vs maintenance). A number of studies have used this task to examine whether attentional bias to smoking cues differs as a function of smoking status (e.g., Ehrman et al., 2002). Interestingly, some research has shown greater attentional bias among lighter or less dependent smokers (Hogarth, Mogg, Bradley, Duka, & Dickinson, 2003; Mogg, Field, & Bradley, 2005). However, as with the addiction Stroop, there have been questions raised surrounding interpretation. For example, there is debate about the exact duration of stimulus presentation needed to assess attentional orienting. Some researchers argue that only durations less than 200 ms reflect an orienting response (Field & Cox, 2008), whereas others suggest that up to 500 ms captures orienting (Robbins & Ehrman, 2004).

Eye-tracking. Eye-tracking is a methodology that has been used extensively in neuroscience, psychology, advertising, and marketing research (Duchowski, 2007), and it has demonstrated efficacy in identifying attentional biases associated with various types of psychopathology (e.g., mood and anxiety disorders; Armstrong & Olatunji, 2012). Eye-tracking tasks are primarily computer-based and involve measuring where individuals direct their gaze when they are presented with a visual stimulus on a computer screen. It is possible that eye-tracking tasks may produce more direct measures of attention allocation and consequently more valid estimates of AB than modified Stroop or visual/dot probe tasks, both of which rely on the production of a response that may require cognitive effort beyond simple attention allocation. Many contemporary eye-tracking systems are video-based and utilize high-resolution cameras, infrared or near-infrared light sources, head mounts, and advanced image processing hardware and software. Typical eye-tracking systems can produce several objective measures of visual attention (e.g., orienting and maintenance), such as total fixation time (time spent looking at a stimulus), latency to fixation (time elapsed before gaze was directed toward a stimulus), and area of initial fixation (the type of stimulus where gaze was first directed). Eye-tracking studies have shown associations between AB for smoking-related stimuli and current smoking (Kwak, Na, Kim, Kim, & Lee, 2007), self-reported craving (Kang et al., 2012), and nicotine deprivation (Field, Mogg, & Bradley, 2004).

Neural indices. Neural markers of attentional bias to drug cues may be more sensitive than behavioral markers (e.g., RT), as behavior may reflect the summation of multiple cognitive processes. A metaanalysis (Engelmann et al., 2012) of fMRI studies examining more protracted smoking cue reactivity confirmed a number of affected brain areas (e.g., the precuneus, anterior and posterior cingulate cortices, areas of prefrontal and parietal cortices, dorsal striatum, and insula). While these findings inform our understanding of smoking cue reactivity, more broadly, the temporal resolution of fMRI is insufficient to capture initial orienting or attentional bias to smoking-related stimuli. In contrast, the EEG-based event-related potential (ERP) has much higher temporal resolution and is better suited to mark neural events during the first several hundred milliseconds upon stimulus exposure (Van Veen & Carter, 2002) and therefore can capture the early orienting response to smoking/drug cues (e.g., Oliver, Jentink, Drobes, & Evans, 2016). In particular, the late positive potential (LPP) is enhanced by affective- (e.g., Schupp et al., 2000) and smoking-related (e.g., Littel & Franken, 2011; Oliver et al., 2016; Versace et al., 2010) cues. However, there have been mixed findings in relation to smoking cues, with some studies finding equal LPPs to smoking and control cues among smokers and nonsmokers (Bloom, Potts, Evans, & Drobes, 2013; Warren & McDonough, 1999). Nonetheless, recent findings have shown that LPP amplitude to smoking cues is positively related to smoking behavior (Littel & Franken, 2011; Versace et al., 2011). Other ERP components also seem to be impacted by smoking cues among smokers. For example, the "N300" response to smoking cues was reduced among smokers compared to nonsmokers (Warren & McDonough, 1999), with even greater reductions during smoking withdrawal (McDonough & Warren, 2001). Even earlier processing of smoking cues has been observed with occipital P1 (peaking at 140ms) responses to smoking cues being positively associated with smoking level (Versace et al., 2011). These early ERP components (P1 and N300) may be reflective of unconscious/preconscious activity (Dehaene, 2014), whereas the LPP may reflect conscious attention to stimuli.

Other paradigms. A number of additional paradigms have been developed or adapted for assessing the role of AB in smoking behavior. These include the following: (1) Dual task (e.g., Sayette & Hufford, 1994): smokers are presented with visual smoking or neutral cues but are tasked with responding to stimuli presented in a different modality. (2) Attentional blink (Chanon, Sours, & Boettiger, 2010): participants respond to target stimuli superimposed over smoking and neutral cues presented in a rapid sequence. (3) N-back working memory (Evans, Craig, Oliver, & Drobes, 2011): a series of smoking and neutral words are presented, and participants must respond whether the word matches the one presented one, two, or three trials previously. (4) Visual search (Oliver & Drobes, 2012): smoking and neutral cues are embedded in grids of distracting stimuli. A detailed description of these and other paradigms is beyond the scope of the present review, but this is clearly an active area of research, and these other paradigms continue to provide evidence of how various components of attention may govern smoking behavior.

18.3 CLINICAL RELEVANCE

A robust literature has examined relationships between AB, smoking-related behaviors, and smoking cessation. Studies utilizing the modified Stroop task have found that AB to smoking-related cues may be associated with current smoking (Munafò, Mogg, Roberts, Bradley, & Murphy, 2003), self-reported craving (Mogg & Bradley, 2002), positive outcome expectancies for smoking (Waters et al., 2009), and anticipated opportunity to smoke (Wertz & Sayette, 2001). Visual/dot probe studies have corroborated some of these relationships and identified additional ones (e.g., Ehrman et al., 2002; Vollstädt-Klein et al., 2011; Waters, Shiffman, Bradley, & Mogg, 2003). Finally, eye-tracking studies have produced associations between AB to smoking-related cues and clinically relevant outcomes, including craving, smoking behavior, and nicotine deprivation effects (Field et al., 2004; Kang et al., 2012; Kwak et al., 2007). A metaanalysis supported that measures of AB and craving/urge are correlated, though the strength of this relationship appeared to be modest (Field, Munafò, & Franken, 2009).

Given that, across paradigms, AB to smoking-related cues appears to be associated with clinically relevant smoking behavior, it is reasonable to assume that AB to smoking-related cues may represent a potential barrier to successful smoking cessation. Indeed, one study found that AB to smoking-related words (in a modified Stroop task) predicted relapse at 1week, though not at 1 or 3months post quit (Powell, Dawkins, West, Powell, & Pickering, 2010). Another modified Stroop study found that prolonged RT to the first block of smoking-related words predicted time to relapse, but that RT during subsequent trial blocks did not show this effect (Waters, Shiffman, Sayette, et al., 2003). Although replication of these findings would be valuable going forward, researchers have begun to develop intervention strategies to decrease smokers' AB to smoking-related cues, with the presumption that this would reduce smoking behavior and improve cessation outcomes. Attentional bias retraining (ABR), which had demonstrated efficacy in improving symptoms associated with other types of psychopathology (Hakamata et al., 2010), has emerged as an intriguing new approach in the area of smoking cessation research. Studies evaluating ABR have utilized traditional AB methodologies (primarily visual/dot probe tasks) and attempted to retrain smokers' automatic attentional processes. Smokers are trained either to avoid smoking-related stimuli when presented or associate smoking-related stimuli with negatively valenced descriptors.

Early ABR studies found that smokers could indeed be trained to reduce AB to smoking-related stimuli, at least within time-limited lab settings (Attwood, O'Sullivan,

Leonards, Mackintosh, & Munafò, 2008; Field, Duka, Tyler, & Schoenmakers, 2009; Lopes, Pires, & Bizarro, 2014). However, more recent studies have produced mixed findings for the ability to successfully retrain attention among smokers (e.g., Begh et al., 2015; Elfeddali, de Vries, Bolman, Pronk, & Wiers, 2016), and these training procedures do not appear to readily generalize to smoking cessation outcomes (e.g., Begh et al., 2015; Kerst & Waters, 2014). There have also been inconsistencies in the effects of ABR on cigarette cravings, with some studies finding that ABR effectively reduces cravings (e.g., Attwood et al., 2008; Kerst & Waters, 2014) and others finding no such effects (e.g., Begh et al., 2015; Field, Duka, et al., 2009; Field, Munafò, et al., 2009; McHugh, Murray, Hearon, Calkins, & Otto, 2010). Nonetheless, continued research regarding this potential clinical application of AB is warranted. As suggested by Franken and van de Wetering (2015), the clinical utility of AB (and other neurocognitive measures) should be evaluated in the context of larger, more systematic research efforts that frame these indexes as aspects of diagnosis, treatment selection, treatment outcomes, and moderators of treatment efficacy. AB measures could prove useful in any of these clinical domains, and to enable the effective translation of AB procedures from laboratory to clinic, more emphasis should be placed on refining experimental procedures and applying sound methodology to measure this construct in the natural environment.

18.4 CURRENT CHALLENGES

Multiple psychometric and interpretive challenges are evident in the study of AB. For example, a reanalysis of findings from seven independent studies of drug-related AB found that the visual/dot probe task exhibited internal consistencies far below acceptable levels (Ataya et al., 2012). Test-retest reliability was not possible to assess in these studies but has been low in studies of threat processing using the visual/dot probe task (Schmukle, 2005). Similar concerns have been raised about other cognitive measures and manipulations commonly employed in the AB literature (e.g., overnight smoking abstinence; Rhodes & Hawk, 2016).

Another design issue has been a lack of "positive" or "active" control conditions within the AB literature. Despite several notable exceptions (Drobes, Elibero, & Evans, 2006; Oliver et al., 2016; Versace et al., 2011), the prototypical AB study contrasts attention to smoking cues with attention to emotionally inert "control" stimuli. When other emotional (salient) categories are included, findings suggest traditional interpretation of AB tasks—that substance-related cues acquire incentive motivational properties that cause these cues to "grab" attention—is not necessarily what drives effects (Field & Cox, 2008). Indeed, salient cues (i.e., those perceived as appetitive or aversive) often produce similar levels of AB. For instance, in one study, smokers exhibited an AB for both smoking and pleasant images, suggesting that AB findings may be picking up on an underlying predisposition or "appetitive overdrive" in smokers that may not be a function of smoking, per se (Oliver et al., 2016).

18.5 SUMMARY/FUTURE DIRECTIONS

Multiple paradigms have been developed to assess initial orienting and maintenance of attention to smoking cues. In addition to demonstrations that AB occurs among smokers, studies have shown relationships between AB and clinically relevant smoking-related variables, such as craving, nicotine dependence, and smoking cessation/relapse. A new area of focus within this literature is examining training procedures designed to reduce AB to smoking cues in the context of smoking cessation. ABR research has thus far produced mixed evidence, and more work is needed to determine its potential role as a component of smoking cessation treatment. A number of methodological issues, including low reliability and insufficient replication of associations between attentional bias and smoking behavior, suggest that larger well-designed studies are necessary to better elucidate the nature of AB and how to best measure such processes and develop effective treatment strategies to address their role in smoking addiction.

In addition to AB paradigms for understanding and intervening in smoking behavior at the individual level, these methods are now being applied within the tobacco policy and regulatory environment. For instance, eye-tracking paradigms have shown how graphic warning labels and other features of cigarette packs draw smokers' attention (e.g., Strasser, Tang, Romer, Jepson, & Cappella, 2012). Similar AB methods are being used to evaluate various cigarette-related messaging formats, including print ads (e.g., Lochbuehler et al., 2016) and point-of-sale displays (e.g., Dutra et al., 2018).

MINI-DICTIONARY OF TERMS

Attentional bias retraining This refers to methods for decreasing attentional bias to relevant cues.

Attentional bias Attentional bias refers to elevated attention to stimuli with enhanced saliency or relevance for certain individuals or groups. Key subcomponents of attentional bias include initial orienting and maintenance stages of information processing.

Eye-tracking Eye-tracking involves measuring where individuals direct their gaze when they are presented with a visual stimulus on a computer screen, providing a direct assessment of attention allocation.

Modified Stroop task This task involves naming the color of a word, which is delayed when the word has saliency or relevance that grabs attention.

Visual/dot probe This task involves detecting the presence of a target that replaces one of two stimuli in the visual field. The speed for target detection is faster when it replaces the stimulus that has greater relevance and, therefore, receives greater attention.

Key Facts of Attentional Bias Assessment

- AB as an important phenomenon in addiction was initially highlighted in theories that focused on automatic processing of drug-paired cues, such as Tiffany's (1990) cognitive theory of drug urges and drug use.
- AB assessment captures the implicit (or unconscious) processing of salient cues, which is thought to reflect the incentive salience of drug-paired cues. This may provide a key index of addiction.
- AB assessment requires controlled presentation of salient drug-related cues, with simultaneous measurement of attention to these cues.
- Paradigms for AB assessment among smokers have appeared in the literature since the 1990s, with the first studies using modified versions of the classic Stroop task.
- Many subsequent paradigms have been used to assess AB among smokers, including visual/dot probe tasks, eye-tracking tasks, and measurement of neural processing of salient cues.

Summary Points

- This chapter focuses on attentional bias (AB) for smoking-related cues, which applies models and paradigms from cognitive psychology and neuroscience.
- Attentional bias may underlie the motivation to smoke, for both ongoing smokers and those attempting to quit.
- Common methods for assessing smoking-related AB include modified Stroop tasks, visual/dot probe tasks, eye tracking, and direct neural indexes.
- AB tasks address different subcomponents of attention, and each is subject to methodological and interpretational issues.
- We suggest that addressing methodological limitations will improve the conceptual clarity of AB findings.
- Recent attempts to retrain individuals to reduce attentional bias have not established that these methods are associated with improved smoking cessation outcomes.
- AB paradigms are increasingly being used to address tobacco policy and regulatory issues.

References

Armstrong, T., & Olatunji, B. O. (2012). Eye tracking of attention in the affective disorders: a meta-analytic review and synthesis. *Clinical Psychology Review, 32*, 704–723.

Ataya, A. F., Adams, S., Mullings, E., Cooper, R. M., Attwood, A. S., & Munafò, M. R. (2012). Internal reliability of measures of substance-related cognitive bias. *Drug and Alcohol Dependence, 121*, 148–151.

Attwood, A. S., O'Sullivan, H., Leonards, U., Mackintosh, B., & Munafò, M. R. (2008). Attentional bias training and cue reactivity in cigarette smokers. *Addiction, 103*, 1875–1882.

Begh, R., Munafò, M. R., Shiffman, S., Ferguson, S. G., Nichols, L., Mohammed, M. A., et al. (2015). Lack of attentional retraining effects in cigarette smokers attempting cessation: a proof of concept double-blind randomised controlled trial. *Drug and Alcohol Dependence, 149*, 158–165.

Bloom, E. L., Potts, G. F., Evans, D. E., & Drobes, D. J. (2013). Cue reactivity in smokers: an event-related potential study. *International Journal of Psychophysiology, 90*(2), 258–264.

Chanon, V. W., Sours, C. R., & Boettiger, C. A. (2010). Attentional bias toward cigarette cues in active smokers. *Psychopharmacology, 212*, 309–320.

Cox, W. M., Fadardi, J. S., & Pothos, E. M. (2006). The addiction-stroop test: theoretical considerations and procedural recommendations. *Psychological Bulletin, 132*, 443–476.

Dehaene, S. (2014). *Consciousness and the brain: Deciphering how the brain codes our thoughts.* New York: Penguin.

Drobes, D. J., Elibero, A., & Evans, D. E. (2006). Attentional bias for smoking and affective stimuli: a Stroop task study. *Psychology of Addictive Behaviors, 20*, 490–495.

Duchowski, A. T. (2007). *Eye tracking methodology: Theory and practice* (2nd ed.). London: Springer.

Dutra, L. M., Nonnemaker, J., Guillory, J., Bradfield, B., Taylor, N., & Kim, A. (2018). Smokers' attention to point-of-sale antismoking ads: an eye-tracking study. *Tobacco Regulatory Science, 4*, 631–643.

Ehrman, R. N., Robbins, S. J., Bromwell, M. A., Lankford, M. E., Monterosso, J. R., & O'Brien, C. P. (2002). Comparing attentional bias to smoking cues in current smokers, former smokers, and non-smokers using a dot-probe task. *Drug and Alcohol Dependence, 67*, 185–191.

Elfeddali, I., de Vries, H., Bolman, C., Pronk, T., & Wiers, R. W. (2016). A randomized controlled trial of web-based attentional bias modification to help smokers quit. *Health Psychology, 35*, 870–880.

Engelmann, J. M., Versace, F., Robinson, J. D., Minnix, J. A., Lam, C. Y., Cui, Y., et al. (2012). Neural substrates of smoking cue reactivity: a meta-analysis of fMRI studies. *NeuroImage, 60*, 252–262.

Evans, D. E., Craig, C., Oliver, J. A., & Drobes, D. J. (2011). The smoking N-back: a measure of biased cue processing at varying levels of cognitive load. *Nicotine & Tobacco Research, 13*, 88–93.

Field, M., & Cox, W. M. (2008). Attentional bias in addictive behaviors: a review of its development, causes, and consequences. *Drug and Alcohol Dependence, 97*, 1–20.

Field, M., Duka, T., Tyler, E., & Schoenmakers, T. (2009). Attentional bias modification in tobacco smokers. *Nicotine & Tobacco Research, 11*, 812–822.

Field, M., Mogg, K., & Bradley, B. P. (2004). Eye movements to smoking-related cues: effects of nicotine deprivation. *Psychopharmacology, 173*, 116–123.

Field, M., Munafò, M. R., & Franken, I. H. (2009). A meta-analytic investigation of the relationship between attentional bias and subjective craving in substance abuse. *Psychological Bulletin, 135*, 589–607.

Franken, I. H. (2003). Drug craving and addiction: Integrating psychological and neuropsychopharmacological approaches. *Progress in Neuro-Psychopharmacology & Biological Psychiatry, 27*, 563–579.

Franken, I. H., & van de Wetering, B. J. (2015). Bridging the gap between the neurocognitive lab and the addiction clinic. *Addictive Behaviors, 44*, 108–114.

Hakamata, Y., Lissek, S., Bar-Haim, Y., Britton, J. C., Fox, N. A., Leibenluft, E., et al. (2010). Attention bias modification treatment: a meta-analysis toward the establishment of novel treatment for anxiety. *Biological Psychiatry, 68*, 982–990.

Hogarth, L. C., Mogg, K., Bradley, B. P., Duka, T., & Dickinson, A. (2003). Attentional orienting towards smoking-related stimuli. *Behavioural Pharmacology, 14*, 153–160.

Johnsen, B. H., Thayer, J. F., Laberg, J. C., & Asbjornsen, A. E. (1997). Attentional bias in active smokers, abstinent smokers, and nonsmokers. *Addictive Behaviors, 22*, 813–817.

Kang, O. S., Chang, D. S., Jahng, G. H., Kim, S. Y., Kim, H., Kim, J. W., et al. (2012). Individual differences in smoking-related cue reactivity in smokers: an eye-tracking and fMRI study. *Progress in Neuro-Psychopharmacology and Biological Psychiatry, 38*, 285–293.

Kerst, W. F., & Waters, A. J. (2014). Attentional retraining administered in the field reduces smokers' attentional bias and craving. *Health Psychology, 33*, 1232–1240.

Kindt, M., Bierman, D., & Brosschot, J. F. (1996). Stroop versus Stroop: comparison of a card format and a single-trial format of the standard color-word Stroop task and the emotional Stroop task. *Personality and Individual Differences, 21*, 653–661.

Klein, A. A. (2007). Suppression-induced hyperaccessibility of thoughts in abstinent alcoholics: a preliminary investigation. *Behaviour Research and Therapy, 45*, 169–177.

Kwak, S. M., Na, D. L., Kim, G., Kim, G. S., & Lee, J. H. (2007). Use of eye movement to measure smokers' attentional bias to smoking-related cues. *Cyberpsychology & Behavior, 10*, 299–304.

Littel, M., & Franken, I. H. (2011). Implicit and explicit selective attention to smoking cues in smokers indexed by brain potentials. *Journal of Psychopharmacology, 25*, 503–513.

Lochbuehler, K., Tang, K. Z., Souprountchouk, V., Campetti, D., Cappella, J. N., Kozlowski, L. T., et al. (2016). Using eye-tracking to examine how embedding risk corrective statements improves cigarette risk beliefs: implications for tobacco regulatory policy. *Drug and Alcohol Dependence, 164*, 97–105.

Lopes, F. M., Pires, A. V., & Bizarro, L. (2014). Attentional bias modification in smokers trying to quit: a longitudinal study about the effects of number of sessions. *Journal of Substance Abuse Treatment, 47*, 50–57.

McDonough, B. E., & Warren, C. A. (2001). Effects of 12-h tobacco deprivation on event-related potentials elicited by visual smoking cues. *Psychopharmacology, 154*, 282–291.

McHugh, R. K., Murray, H. W., Hearon, B. A., Calkins, A. W., & Otto, M. W. (2010). Attentional bias and craving in smokers: the impact of a single attentional training session. *Nicotine & Tobacco Research, 12*, 1261–1264.

Mogg, K., & Bradley, B. P. (2002). Selective processing of smoking-related cues in smokers: manipulation of deprivation level and comparison of three measures of processing bias. *Journal of Psychopharmacology, 16*, 385–392.

Mogg, K., Field, M., & Bradley, B. P. (2005). Attentional and approach biases for smoking cues in smokers: an investigation of competing theoretical views of addiction. *Psychopharmacology, 180*, 333–341.

Munafò, M., Mogg, K., Roberts, S., Bradley, B. P., & Murphy, M. (2003). Selective processing of smoking-related cues in current smokers, ex-smokers and never-smokers on the modified Stroop task. *Journal of Psychopharmacology, 17*, 310–316.

Oliver, J. A., & Drobes, D. J. (2012). Visual search and attentional bias for smoking cues: the role of familiarity. *Experimental and Clinical Psychopharmacology, 20*, 489–496.

Oliver, J. A., Jentink, K. G., Drobes, D. J., & Evans, D. E. (2016). Smokers exhibit biased neural processing of smoking and affective images. *Health Psychology, 35*, 866–869.

Powell, J., Dawkins, L., West, R., Powell, J., & Pickering, A. (2010). Relapse to smoking during unaided cessation: clinical, cognitive and motivational predictors. *Psychopharmacology, 212*, 537–549.

Rhodes, J. D., & Hawk, L. W. (2016). Smoke and mirrors: the overnight abstinence paradigm as an index of disrupted cognitive function. *Psychopharmacology, 233*, 1395–1404.

Robbins, S. J., & Ehrman, R. N. (2004). The role of attentional bias in substance abuse. *Behavioral and Cognitive Neuroscience Reviews, 3*, 243–260.

Robinson, T. E., & Berridge, K. C. (1993). The neural basis of drug craving: an incentive-sensitization theory of addiction. *Brain Research, 18*, 247–291.

Ryan, F. (2002). Attentional bias and alcohol dependence: a controlled study using the modified Stroop paradigm. *Addictive Behaviors, 27*, 471–482.

Sayette, M. A., & Hufford, M. R. (1994). Effects of cue exposure and deprivation on cognitive resources in smokers. *Journal of Abnormal Psychology, 103*, 812–818.

Schmukle, S. C. (2005). Unreliability of the dot probe task. *European Journal of Personality, 19*, 595–605.

Schupp, H. T., Cuthbert, B. N., Bradley, M. M., Cacioppo, J. T., Ito, T., & Lang, P. J. (2000). Affective picture processing: the late positive potential is modulated by motivational relevance. *Psychophysiology, 37*, 257–261.

Strasser, A. A., Tang, K. Z., Romer, D., Jepson, C., & Cappella, J. N. (2012). Graphic warning labels in cigarette advertisements: recall and viewing patterns. *American Journal of Preventive Medicine, 43*, 41–47.

Stroop, J. R. (1935). Studies of interference in serial verbal reactions. *Journal of Experimental Psychology, 18*, 643.

Tiffany, S. T. (1990). A cognitive model of drug urges and drug-use behavior: role of automatic and nonautomatic processes. *Psychological Review, 97*, 147–168.

Van Veen, V., & Carter, C. S. (2002). The anterior cingulate as a conflict monitor: fMRI and ERP studies. *Physiology and Behavior, 77*, 477–482.

Versace, F., Minnix, J. A., Robinson, J. D., Lam, C. Y., Brown, V. L., & Cinciripini, P. M. (2011). Brain reactivity to emotional, neutral and cigarette-related stimuli in smokers. *Addiction Biology, 16*, 296–307.

Versace, F., Robinson, J. D., Lam, C. Y., Minnix, J. A., Brown, V. L., Carter, B. L., et al. (2010). Cigarette cues capture smokers' attention: Evidence from event-related potentials. *Psychophysiology, 47*, 435–441.

Vollstädt-Klein, S., Loeber, S., Winter, S., Leménager, T., von der Goltz, C., Dinter, C., et al. (2011). Attention shift towards smoking cues relates to severity of dependence, smoking behavior and breath carbon monoxide. *European Addiction Research, 17*, 217–224.

Warren, C. A., & McDonough, B. E. (1999). Event-related brain potentials as indicators of smoking cue-reactivity. *Clinical Neurophysiology, 110*, 1570–1584.

Waters, A. J., Carter, B. L., Robinson, J. D., Wetter, D. W., Lam, C. Y., Kerst, W., et al. (2009). Attentional bias is associated with incentive-related physiological and subjective measures. *Experimental and Clinical Psychopharmacology, 17*, 247–257.

Waters, A. J., Shiffman, S., Bradley, B. P., & Mogg, K. (2003). Attentional shifts to smoking cues in smokers. *Addiction, 98*, 1409–1417.

Waters, A. J., Shiffman, S., Sayette, M. A., Paty, J. A., Gwaltney, C. J., & Balabanis, M. H. (2003). Attentional bias predicts outcome in smoking cessation. *Health Psychology, 22*, 378–387.

Wertz, J. M., & Sayette, M. A. (2001). Effects of smoking opportunity on attentional bias in smokers. *Psychology of Addictive Behaviors, 15*, 268–271.

CHAPTER

19

Effects of Nicotine on Inhibitory Control in Humans

Ulrich Ettinger, Veena Kumari†*

*Department of Psychology, University of Bonn, Bonn, Germany

†Centre for Cognitive Neuroscience, Division of Psychology, College of Health and Life Sciences, Brunel University London, Uxbridge, United Kingdom

Abbreviations

ADHD attention deficit hyperactivity disorder
ANT attention network test
mg milligram
RT reaction time
SSRT stop signal reaction time

19.1 INTRODUCTION

The ability to inhibit unwanted actions, thoughts, or emotions is an important function serving goal-directed behavior. Inhibiting impulses and inappropriate (re) actions is frequently required in everyday life, and the ability to do so is related to advantageous individual and societal functioning. Accordingly, enhancing the ability to inhibit unwanted behaviors has been in the focus of several streams of research. While some researchers have aimed to improve inhibitory control via training, another widely adopted approach is the application of cognition-enhancing substances such as the nicotinic acetylcholine receptor agonist nicotine.

The study of pharmacological influences on inhibitory control is important for a number of reasons. First, such work has the potential to identify domains of cognition that can be enhanced in healthy humans to improve function in occupational, educational, and other societal contexts. Second, inhibitory impairments are observed in some psychiatric and neurological patient groups. Developing treatments to alleviate these deficits is an unmet clinical need. Third, regarding nicotine, it is important to understand the factors that cause and maintain cigarette smoking. These may include beneficial effects on aspects of cognition. Finally, the experimental study of drug effects on inhibitory control may inform models of cognition and brain function.

This chapter focuses on nicotine effects on inhibitory control. Inhibitory control includes both the inhibition of a motor response and the attentional control of interference from distracting information. The chapter discusses nicotine effects on the most widely studied inhibitory paradigms, namely, the antisaccade, stop-signal, go/no-go, Stroop, and flanker tasks. Given the considerable methodological heterogeneity in this literature, findings from controlled nicotine applications (e.g., patch, tablet, injection, or gum) and cigarette smoking in deprived smokers are discussed. For reasons of space, only human studies are considered.

19.2 EFFECTS OF NICOTINE ON RESPONSE INHIBITION AND INTERFERENCE CONTROL

19.2.1 Antisaccade Tasks

In the antisaccade task (Fig. 19.1), participants are instructed to make a rapid eye movement, a saccade, away from a sudden-onset peripheral target while avoiding a saccade toward it. The primary measure of inhibitory control on this task is the direction error rate, that is, percentage of trials on which the participant first looked toward the peripheral target. Additionally, latencies (RTs) of correct responses are often reported. Prosaccades are typically included as sensorimotor control condition.

Numerous studies have provided evidence of beneficial effects of nicotine on antisaccade task performance.

Neuroscience of Nicotine
https://doi.org/10.1016/B978-0-12-813035-3.00019-8

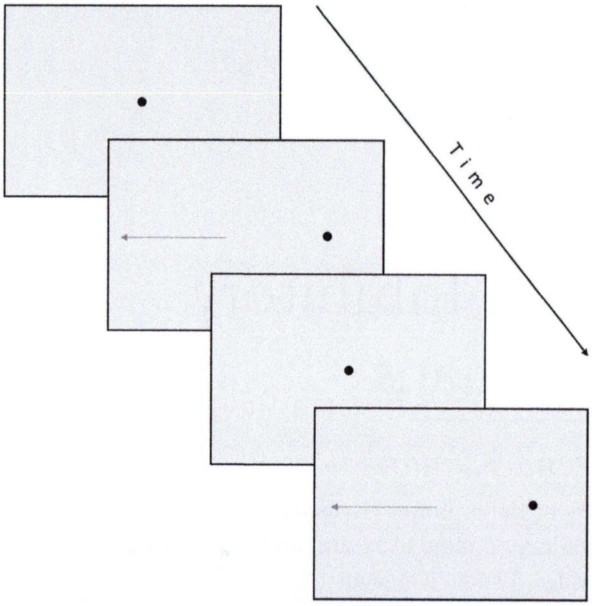

FIG. 19.1 Flowchart of an antisaccade task.
In the antisaccade task, a rapid eye movement has to be made in the opposite direction of a peripheral stimulus (indicated here by a *gray arrow*). Antisaccade trials may be presented blocked, as in this example, or interleaved with prosaccade trials (saccades toward the target).

These effects have been reported with study designs differing considerably with regard to task characteristics (e.g., blocked or mixed antisaccade/prosaccade tasks), method of nicotine application (e.g., cigarette smoking after withdrawal, gum, nasal spray, injection, and patch), clinical characteristics (e.g., schizophrenia patients, schizotypal individuals, and controls), and smoking status (e.g., smoker and nonsmoker).

Reductions in direction errors with nicotine have been observed in most (Dawkins, Powell, Pickering, Powell, & West, 2009; Dawkins, Powell, West, Powell, & Pickering, 2007; Dépatie et al., 2002; Larrison, Briand, & Sereno, 2004; Larrison-Faucher, Matorin, & Sereno, 2004; Petrovsky et al., 2012, 2013; Pettiford et al., 2007; Powell, Dawkins, & Davis, 2002; Rycroft, Hutton, & Rusted, 2006; Schmechtig et al., 2013), but not all, studies (Ettinger et al., 2009, 2017; Rycroft et al., 2007; Thaker, Ellsberry, Moran, Lahti, & Tamminga, 1991; Wachter & Gilbert, 2013).

Nicotine may also reduce antisaccade latencies (Bowling & Donnelly, 2010; Ettinger et al., 2009; Larrison et al., 2004; Larrison-Faucher et al., 2004; Petrovsky et al., 2013; Rycroft et al., 2006, 2007; Wachter & Gilbert, 2013), although not all studies have observed this effect (Dépatie et al., 2002; Ettinger et al., 2017; Petrovsky et al., 2012; Schmechtig et al., 2013; Thaker et al., 1991). It should be noted that not all studies have reported latencies.

While these effects are generally independent of clinical (schizophrenia, schizotypy, and control) and smoking (smoker and nonsmoker) status, there may be effects of baseline dependency with regard to antisaccade performance (however, see Ettinger et al., 2017). For example, a study of schizophrenia patients showed greater improvement in low-performing patients (Larrison-Faucher et al., 2004). Additionally, Petrovsky et al. (2012) reported that nicotine reduced antisaccade errors in low-performing but not in high-performing healthy participants, without affecting antisaccade latencies or prosaccade performance. Baseline dependency was also observed by Wachter and Gilbert (2013), who reported lower antisaccade but not prosaccade latency following a nicotine patch in poor and average (but not high) performing smokers, without any effects on error rates.

To conclude, nicotine has beneficial effects on either direction errors or antisaccade latencies in the majority of studies. Given that most studies report no significant effects on performance on the prosaccade control task (Dawkins et al., 2007; Dépatie et al., 2002; Ettinger et al., 2009; Larrison et al., 2004; Larrison-Faucher et al., 2004; Petrovsky et al., 2012; Pettiford et al., 2007; Schmechtig et al., 2013; Thaker et al., 1991; Wachter & Gilbert, 2013; however, see Ettinger et al., 2017; Petrovsky et al., 2013), these findings may reflect beneficial effects on the top-down control of saccadic eye movements.

19.2.2 Stop-Signal Tasks

The stop-signal task (Fig. 19.2) is a measure of stopping or canceling a motor response. The task has been widely used to model the cognitive and neural mechanisms of

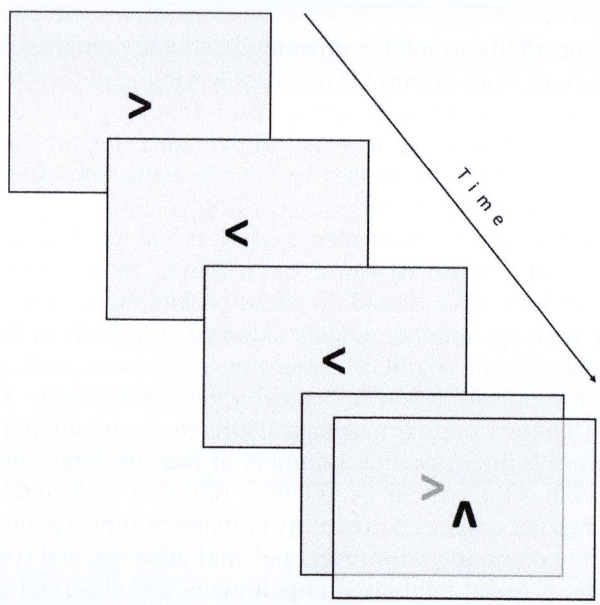

FIG. 19.2 Flowchart of a stop-signal task.
In the stop-signal task, a response in a choice reaction task has to be stopped when a sudden-onset stop signal occurs. Here, the participant has to respond by indicating the direction of the target arrow (left or right) but has to stop this response when shortly after the target a stop signal (here, *an arrow pointing up*) is shown. The stop signal may also be an auditory stimulus.

response inhibition as well as deficits in inhibitory control in psychiatric and neurological patients. The key dependent variable is the SSRT, an indirect measure of the speed of the stopping process, but go RTs and percentages of correct responses are also reported in some studies.

Beneficial effects of nicotine on SSRT have been observed in nonsmoking adolescents (Potter & Newhouse, 2004) and adults (Potter & Newhouse, 2008) with attention deficit hyperactivity disorder (ADHD); in nonsmoking young adults with high levels of impulsivity (Potter, Bucci, & Newhouse, 2012); and, weakly, in healthy nonsmokers (Logemann, Böcker, Deschamps, Kemner, & Kenemans, 2014b). Other studies, however, have failed to observe effects in healthy nonsmokers (Ettinger et al., 2017; Wignall & De Wit, 2011), smokers, (Bekker, Böcker, Van Hunsel, van den Berg, & Kenemans, 2005) or a combined sample of smokers and nonsmokers (Logemann, Böcker, Deschamps, Kemner, & Kenemans, 2014a).

More consistent evidence of beneficial effects of nicotine on SSRT, however, comes from studies of smokers in a state of abstinence. Better performance following smoking as usual compared to abstinence was reported in most studies (Ashare & Hawk, 2012; Charles-Walsh, Furlong, Munro, & Hester, 2014; Tsaur, Strasser, Souprountchouk, Evans, & Ashare, 2015), although one study showed an *increase* in SSRT following smoking in smokers, suggesting disinhibition with nicotine (Austin, Duka, Rusted, & Jackson, 2014).

Overall, it appears that nicotine improves stop-signal task performance in individuals with inhibitory impairments, that is, patients with ADHD and cigarette-withdrawn smokers. The experimentally controlled application of nicotine to healthy smokers and nonsmokers does not, however, appear to reliably enhance performance on this task.

19.2.3 Go/No-Go Tasks

Go/no-go tasks (Fig. 19.3) are widely used to assess the ability to inhibit a dominant motor response. The most common measure of inhibitory control on the task is the (commission) error rate, that is, the percentage of responses on no-go trials, although studies have frequently also reported the RT of responses on go trials.

While there is evidence of beneficial effects of cigarette smoking on no-go errors in smokers (Kozink, Kollins, & McClernon, 2010), most studies appear not to find significant effects of nicotine, including designs using cigarette-deprived smokers (Ettinger et al., 2017; Evans, Park, Maxfield, & Drobes, 2009; Giessing, Thiel, Alexander-Bloch, Patel, & Bullmore, 2013; Harrison, Coppola, & McKee, 2009; Smucny, Olincy, Eichman, & Tregellas, 2015). One study of smokers with ADHD and smoking controls observed that the patients made more commission errors when abstinent compared to a satiated state and more errors than the control group when abstinent (McClernon et al., 2008).

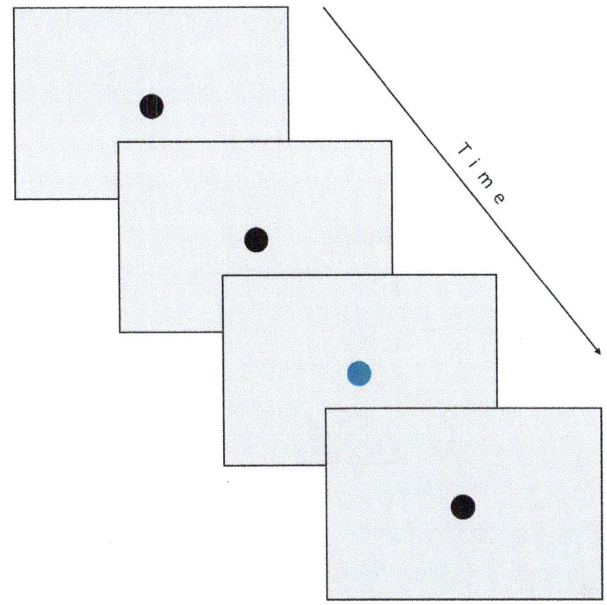

FIG. 19.3 Flowchart of a go/no-go task.
In go/no-go tasks, the participant has to respond to a frequent go stimulus (here, *a black circle*) but withhold the response when a no-go stimulus is shown (here, *a blue circle*). In variants of the task, oddball stimuli are included. These also differ from the go stimulus in one stimulus dimension (e.g., color; an oddball stimulus in this example may thus be a green circle [not shown]) but have to be responded to like go stimuli.

There is also evidence of shorter (Giessing et al., 2013; Harrison et al., 2009; Houlihan, Pritchard, Krieble, Robinson, & Duke, 1996) and less variable (Houlihan et al., 1996) RTs on go trials with nicotine in nicotine-deprived smokers (however, see Ettinger et al., 2017; Kozink et al., 2010; Smucny et al., 2015).

To conclude, the majority of studies have failed to provide evidence of beneficial nicotine effects on response inhibition in go/no-go tasks, and evidence is mixed with regard to effects on go RTs.

19.2.4 Stroop Tasks

In the Stroop task (Fig. 19.4), the prepotent tendency to name a color word has to be inhibited; instead, the nondominant response of naming the print color of the word has to be performed. The Stroop task and its variants are widely applied in cognitive neuroscience, psychopharmacology, and psychopathology research. The key performance measure is the Stroop effect, that is, the difference in RTs or error rates between incongruent and congruent conditions.

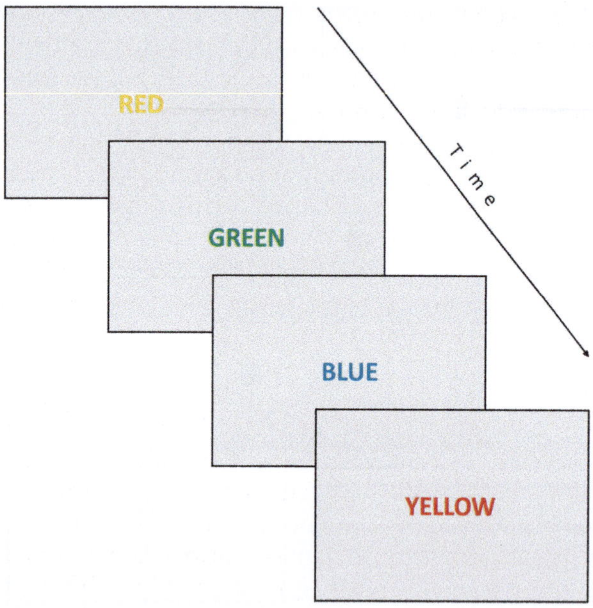

FIG. 19.4 Flowchart of a Stroop task.
In the Stroop task, the participant is required to name the color that the word stimulus is printed in. The flowchart shows a randomized sequence of incongruent and congruent stimuli. Stimuli may also be presented blocked, that is, in separate lists of congruent and incongruent stimuli. Stroop tasks often also include a color bar condition to measure simple color naming time and frequently employ a central fixation cross before stimulus onset (not shown here).

A large number of studies have examined the influence of nicotine on the Stroop effect. Studies have employed both experimentally controlled applications and naturalistic methods involving cigarette smoking in smokers after variable durations of deprivation. Overall, the majority of studies have failed to observe beneficial influences of smoking or controlled applications of nicotine on the size of the Stroop effect (Cook, Gerkovich, Graham, Hoffman, & Peterson, 2003; Della Casa, Höfer, Weiner, & Feldon, 1999; Ettinger et al., 2017; Evins et al., 2005; Foulds et al., 1996; George et al., 2002; Ilan & Polich, 2001; Kos, Hasenfratz, & Bättig, 1997; Levin et al., 1996; Mancuso, Warburton, Mélen, Sherwood, & Tirelli, 1999; Parrott & Craig, 1992; Poltavski & Petros, 2006; Rusted, Caulfield, King, & Goode, 2000; Suter, Buzzi, Woodson, & Bättig, 1983; Tsaur et al., 2015; Wesnes & Revell, 1984; Xu et al., 2007; Xu & Domino, 2000; Zack, Belsito, Scher, Eissenberg, & Corrigall, 2001).

Other studies, however, have reported *general* enhancement of task performance (Atzori, Lemmonds, Kotler, Durcan, & Boyle, 2008; Azizian et al., 2010; Mancuso et al., 1999; Pomerleau, Teuscher, Goeters, & Pomerleau, 1994; Potter & Newhouse, 2008; Rusted et al., 2000; Zack et al., 2001). Yet, other studies have observed selective improvements in the Stroop effect in healthy individuals (Domier et al., 2007; Hasenfratz & Bättig, 1992; Landers, Crews, Boutcher, Skinner, & Gustafsen, 1992; Provost & Woodward, 1991; Wesnes & Warburton, 1978; Wignall & De Wit, 2011) and patients with ADHD (Potter & Newhouse, 2004) and schizophrenia (Barr et al., 2008). There is also evidence that nicotine effects on Stroop performance may depend on *CHRNA5* genotype (Jensen et al., 2015) and smoker status, with beneficial effects observed in smokers but not in nonsmokers (Grundey et al., 2015).

Overall, therefore, nicotine effects on inhibitory control in the Stroop task appear not to be reliable. The reasons for this are not clear, as both positive and negative effects have been obtained across smoker and nonsmoker samples as well as application methods (e.g., cigarettes in smokers vs controlled application of patches).

19.2.5 Flanker Tasks

Flanker tasks (Fig. 19.5) are popular measures of attentional interference. In flanker tasks, a response to a central stimulus is required, whereas peripheral flankers are to be ignored. The flanker effect describes the decrement in performance when peripheral flankers are incongruent with the central stimulus, compared to congruent or neutral conditions. A variant of the flanker task is the attention network test (ANT), which in addition to the conflict

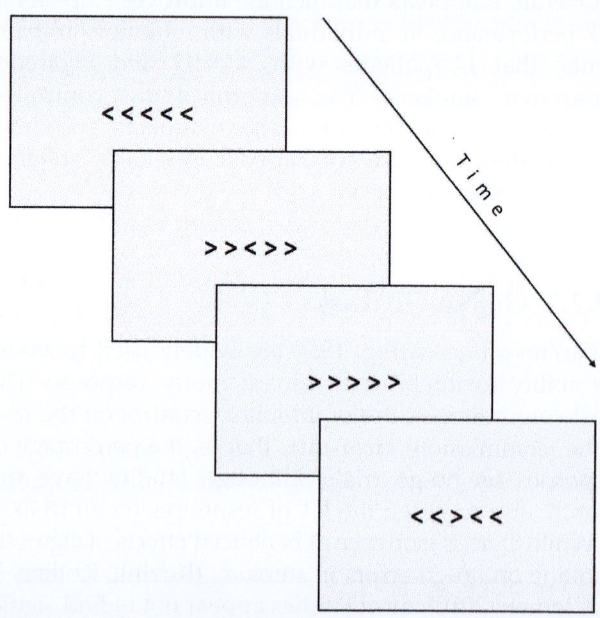

FIG. 19.5 Flowchart of a flanker task.
In this variant of the flanker task, the participant is required to indicate with a button press the direction of the central arrow (right or left) while ignoring the flanking distractors. This flowchart shows a randomized sequence of congruent and incongruent trials. Trials may also include neutral flankers and frequently employ a central fixation cross before stimulus onset (not shown here).

effect due to incongruent flankers taps alerting and orienting aspects of attention.

While one study observed more efficient interference control on the ANT following a 21 mg nicotine patch in smokers with schizophrenia and smoking controls (AhnAllen, Nestor, Shenton, McCarley, & Niznikiewicz, 2008), other studies have failed to observe nicotine effects on flanker interference control in snuff users (Lindgren, Stenberg, & Rosén, 1996), nonsmokers (Ettinger et al., 2017; Kleykamp, Jennings, Blank, & Eissenberg, 2005; Wignall & De Wit, 2011), or samples of smokers and nonsmokers (Myers, Taylor, Salmeron, Waters, & Heishman, 2013). The study by Lindgren and colleagues, however, reported an overall reduction in RT and improvement in correct response rate following oral snuff, suggesting general performance-enhancing effects of nicotine (Lindgren et al., 1996).

Flanker tasks have also been applied to investigate nicotine effects in smokers abstinent from smoking. One study observed that smokers had *generally* (irrespective of flanker congruency) higher RTs after overnight abstinence compared to ad lib smoking (Schlienz, Hawk, & Rosch, 2013). In another study, smokers showed reduced speeded accuracy during abstinence compared to smoking as usual (Schlienz & Hawk, 2017).

Overall, therefore, effects of nicotine administration on interference control on flanker tasks are likely to be small, with most studies failing to report interference-specific effects or any effects at all.

19.3 CONCLUSIONS

Overall, while there is evidence of beneficial effects of nicotine on inhibitory control, the literature is rather mixed. The most consistent effects are observed with the antisaccade task, where most studies report improvements in error rate and/or latency. The majority of studies also report beneficial effects on the SSRT, especially in impaired populations. The weight of evidence points against consistently positive effects on go/no-go, Stroop, and flanker tasks.

These findings may be reconciled with earlier conclusions that nicotine improves more basic functions of attention, such as alertness and orienting, more strongly than the selectivity function of attention (Heishman, Kleykamp, & Singleton, 2010; Provost & Woodward, 1991). It should be noted, however, that improvements in basic, bottom-up attentional processing with nicotine may conceivably result in impaired top-down control (Ettinger et al., 2017; Rycroft et al., 2006), a pattern that is rarely observed. This suggests that nicotine may have positive effects on components of information processing that not only benefit basic attentional functions but also may be drawn upon in top-down control processes.

The reasons for inconsistent effects on go/no-go, Stroop, and flanker tasks are unclear but may involve task factors, participant factors, and drug dose factors as well as their interactions.

Task factors, such as inhibitory load, attentional load, or general task complexity, may contribute to variable nicotine effects across tasks. It is intriguing that effects appear to be more consistent on antisaccade and stop-signal tasks than on other tasks. This issue has not been systematically investigated. Designs that selectively vary relevant task features and inhibitory requirements are needed.

Participant factors were found to play a role on antisaccade, stop-signal, and Stroop tasks in some studies. These factors include baseline performance (antisaccades), clinical conditions and/or level of impulsivity (stop-signal and go/no-go), genotype (Stroop), and smoking status (Stroop). These factors likely play an important role in explaining interindividual differences in nicotine response and interstudy differences in overall results (given differences in participant recruitment and selection between studies).

Finally, dose dependency of nicotine effects on inhibition has been observed in some (Larrison-Faucher et al., 2004; Poltavski, Petros, & Holm, 2012) but not in other studies (Bekker et al., 2005; Foulds et al., 1996; Kleykamp et al., 2005; Meyers et al., 2015; Parrott & Craig, 1992). A related issue is that certain methods of nicotine administration allow a better titration than others (e.g., injection vs smoking) to achieve optimal dosages to meet individual participants' needs. Generally, however, the issue of dose dependency merits further attention. Additionally, dose dependency may interact with participant and task factors.

MINI-DICTIONARY OF TERMS

Antisaccade A saccade made in opposite direction to a sudden-onset, peripheral stimulus.

Commission error A response to a stimulus that should not have been responded to.

Flanker effect The difference in RT between incongruent and congruent conditions in flanker tasks.

Reaction time (RT) The time, typically given in milliseconds, from stimulus appearance until the participant's behavioral response.

Saccade A rapid, conjugate movement of the eyes.

Stop-signal reaction time (SSRT) The latency of the stopping process in the stop-signal task. The SSRT cannot be directly observed but is estimated.

Stroop effect The interference of the meaning of a color word with the response to its print color. Naming the color in which a color word is printed is faster and less error-prone when the print color corresponds to the word (e.g., the word BLUE printed in blue) than when print color and word do not correspond (e.g., the word YELLOW printed in green).

Key Facts of Inhibitory Control

- The term inhibition denotes a heterogeneous construct that comprises a number of different phenomena in neuroscience and psychology.
- In cognitive psychology, inhibitory control is a key executive function that refers to the ability to suppress actions, thoughts, and emotions.
- Having to inhibit unwanted reactions, impulses, or emotions features prominently in everyday life.
- Deficits in inhibitory control are observed in numerous neurological and psychiatric conditions.
- Inhibitory control generally shows good temporal stability and significant heritability.
- The neural networks underlying inhibitory control include the prefrontal cortex and its posterior cortical and subcortical projection targets.

Summary Points

- This chapter reviews the literature on nicotine effects on inhibitory control in humans.
- The included inhibitory control paradigms are the antisaccade, stop-signal, go/no-go, Stroop, and flanker tasks.
- Studies of both controlled applications (e.g., patch, tablet, injection, lozenge, spray, or gum) and cigarette smoking in deprived smokers are included.
- The majority of studies of antisaccade tasks observed beneficial effects.
- Beneficial effects on stop-signal tasks are observed in populations with inhibitory impairment.
- Effects on go/no-go, Stroop, and flanker tasks are less consistent.
- Overall, the literature suggests that nicotine effects on inhibitory control may be subtle, with task, participant, and dosage factors likely playing a role.

References

AhnAllen, C. G., Nestor, P. G., Shenton, M. E., McCarley, R. W., & Niznikiewicz, M. A. (2008). Early nicotine withdrawal and transdermal nicotine effects on neurocognitive performance in schizophrenia. *Schizophrenia Research*, *100*(1), 261–269.

Ashare, R. L., & Hawk, L. W. (2012). Effects of smoking abstinence on impulsive behavior among smokers high and low in ADHD-like symptoms. *Psychopharmacology*, *219*(2), 537–547.

Atzori, G., Lemmonds, C. A., Kotler, M. L., Durcan, M. J., & Boyle, J. (2008). Efficacy of a nicotine (4mg)-containing lozenge on the cognitive impairment of nicotine withdrawal. *Journal of Clinical Psychopharmacology*, *28*(6), 667–674.

Austin, A. J., Duka, T., Rusted, J., & Jackson, A. (2014). Effect of varenicline on aspects of inhibitory control in smokers. *Psychopharmacology*, *231*(18), 3771–3785.

Azizian, A., Nestor, L. J., Payer, D., Monterosso, J. R., Brody, A. L., & London, E. D. (2010). Smoking reduces conflict-related anterior cingulate activity in abstinent cigarette smokers performing a Stroop task. *Neuropsychopharmacology*, *35*(3), 775–782.

Barr, R. S., Culhane, M. A., Jubelt, L. E., Mufti, R. S., Dyer, M. A., Weiss, A. P., et al. (2008). The effects of transdermal nicotine on cognition in nonsmokers with schizophrenia and nonpsychiatric controls. *Neuropsychopharmacology*, *33*(3), 480–490.

Bekker, E. M., Böcker, K. B. E., Van Hunsel, F., van den Berg, M. C., & Kenemans, J. L. (2005). Acute effects of nicotine on attention and response inhibition. *Pharmacology, Biochemistry, and Behavior*, *82*(3), 539–548.

Bowling, A. C., & Donnelly, J. F. (2010). Effect of nicotine on saccadic eye movement latencies in non-smokers. *Human Psychopharmacology*, *25*(5), 410–418.

Charles-Walsh, K., Furlong, L., Munro, D. G., & Hester, R. (2014). Inhibitory control dysfunction in nicotine dependence and the influence of short-term abstinence. *Drug and Alcohol Dependence*, *143*, 81–86.

Cook, M. R., Gerkovich, M. M., Graham, C., Hoffman, S. J., & Peterson, R. C. (2003). Effects of the nicotine patch on performance during the first week of smoking cessation. *Nicotine & Tobacco Research*, *5*(2), 169–180.

Dawkins, L., Powell, J. H., Pickering, A., Powell, J., & West, R. (2009). Patterns of change in withdrawal symptoms, desire to smoke, reward motivation and response inhibition across 3 months of smoking abstinence. *Addiction*, *104*(5), 850–858.

Dawkins, L., Powell, J. H., West, R., Powell, J., & Pickering, A. (2007). A double-blind placebo-controlled experimental study of nicotine: II--effects on response inhibition and executive functioning. *Psychopharmacology*, *190*(4), 457–467.

Della Casa, V., Höfer, I., Weiner, I., & Feldon, J. (1999). Effects of smoking status and schizotypy on latent inhibition. *Journal of Psychopharmacology*, *13*(1), 45–57.

Dépatie, L., O'Driscoll, G. A., Holahan, A.-L. V., Atkinson, V., Thavundayil, J. X., Kin, N. N. Y., et al. (2002). Nicotine and behavioral markers of risk for schizophrenia: a double-blind, placebo-controlled, cross-over study. *Neuropsychopharmacology*, *27*(6), 1056–1070.

Domier, C. P., Monterosso, J. R., Brody, A. L., Simon, S. L., Mendrek, A., Olmstead, R., et al. (2007). Effects of cigarette smoking and abstinence on stroop task performance. *Psychopharmacology*, *195*(1), 1–9.

Ettinger, U., Faiola, E., Kasparbauer, A.-M., Petrovsky, N., Chan, R. C. K., Liepelt, R., et al. (2017). Effects of nicotine on response inhibition and interference control. *Psychopharmacology*, *234*(7), 1093–1111.

Ettinger, U., Williams, S. C. R. R., Patel, D., Michel, T. M., Nwaigwe, A., Caceres, A., et al. (2009). Effects of acute nicotine on brain function in healthy smokers and non-smokers: estimation of inter-individual response heterogeneity. *NeuroImage*, *45*(2), 549–561.

Evans, D. E., Park, J. Y., Maxfield, N., & Drobes, D. J. (2009). Neurocognitive variation in smoking behavior and withdrawal: genetic and affective moderators. *Genes, Brain and Behavior*, *8*(1), 86–96.

Evins, A. E., Deckersbach, T., Cather, C., Freudenreich, O., Culhane, M. A., Henderson, D. C., et al. (2005). Independent effects of tobacco abstinence and bupropion on cognitive function in schizophrenia. *Journal of Clinical Psychiatry*, *66*(9), 1184–1190.

Foulds, J., Stapleton, J., Swettenham, J., Bell, N., McSorley, K., & Russell, M. A. (1996). Cognitive performance effects of subcutaneous nicotine in smokers and never-smokers. *Psychopharmacology*, *127*(1), 31–38.

George, T., Vessicchio, J. C., Termine, A., Sahady, D. M., Head, C. A., Pepper, W. T., et al. (2002). Effects of smoking abstinence on visuospatial working memory function in schizophrenia. *Neuropsychopharmacology*, *26*(1), 75–85.

Giessing, C., Thiel, C. M., Alexander-Bloch, A. F., Patel, A. X., & Bullmore, E. T. (2013). Human brain functional network changes

associated with enhanced and impaired attentional task performance. *Journal of Neuroscience*, *33*(14), 5903–5914.
Grundey, J., Amu, R., Ambrus, G. G., Batsikadze, G., Paulus, W., & Nitsche, M. A. (2015). Double dissociation of working memory and attentional processes in smokers and non-smokers with and without nicotine. *Psychopharmacology*, *232*(14), 2491–2501.
Harrison, E. L. R., Coppola, S., & McKee, S. A. (2009). Nicotine deprivation and trait impulsivity affect smokers' performance on cognitive tasks of inhibition and attention. *Experimental and Clinical Psychopharmacology*, *17*(2), 91–98.
Hasenfratz, M., & Bättig, K. (1992). Action profiles of smoking and caffeine: Stroop effect, EEG, and peripheral physiology. *Pharmacology, Biochemistry, and Behavior*, *42*(1), 155–161.
Heishman, S. J., Kleykamp, B. A., & Singleton, E. G. (2010). Meta-analysis of the acute effects of nicotine and smoking on human performance. *Psychopharmacology*, *210*(4), 453–469.
Houlihan, M. E., Pritchard, W. S., Krieble, K. K., Robinson, J. H., & Duke, D. W. (1996). Effects of cigarette smoking on EEG spectral-band power, dimensional complexity, and nonlinearity during reaction-time task performance. *Psychophysiology*, *33*(6), 740–746.
Ilan, A. B., & Polich, J. (2001). Tobacco smoking and event-related brain potentials in a Stroop task. *International Journal of Psychophysiology*, *40* (2), 109–118.
Jensen, K. P., DeVito, E. E., Herman, A. I., Valentine, G. W., Gelernter, J., & Sofuoglu, M. (2015). A CHRNA5 smoking risk variant decreases the aversive effects of nicotine in humans. *Neuropsychopharmacology*, *40*(12), 2813–2821.
Kleykamp, B. A., Jennings, J. M., Blank, M. D., & Eissenberg, T. (2005). The effects of nicotine on attention and working memory in never-smokers. *Psychology of Addictive Behaviors*, *19*(4), 433–438.
Kos, J., Hasenfratz, M., & Bättig, K. (1997). Effects of a 2-day abstinence from smoking on dietary, cognitive, subjective, and physiologic parameters among younger and older female smokers. *Physiology & Behavior*, *61*(5), 671–678.
Kozink, R. V., Kollins, S. H., & McClernon, F. J. (2010). Smoking withdrawal modulates right inferior frontal cortex but not presupplementary motor area activation during inhibitory control. *Neuropsychopharmacology*, *35*(13), 2600–2606.
Landers, D. M., Crews, D. J., Boutcher, S. H., Skinner, J. S., & Gustafsen, S. (1992). The effects of smokeless tobacco on performance and psychophysiological response. *Medicine and Science in Sports and Exercise*, *24*(8), 895–903.
Larrison, A. L., Briand, K. A., & Sereno, A. B. (2004). Nicotine improves antisaccade task performance without affecting prosaccades. *Human Psychopharmacology*, *19*(6), 409–419.
Larrison-Faucher, A. L., Matorin, A. A., & Sereno, A. B. (2004). Nicotine reduces antisaccade errors in task impaired schizophrenic subjects. *Progress in Neuro-Psychopharmacology & Biological Psychiatry*, *28*(3), 505–516.
Levin, E. D., Conners, C. K., Sparrow, E., Hinton, S. C., Erhardt, D., Meck, W. H., et al. (1996). Nicotine effects on adults with attention-deficit/hyperactivity disorder. *Psychopharmacology*, *123*(1), 55–63.
Lindgren, M., Stenberg, G., & Rosén, I. (1996). Effects of nicotine in visual attention tasks. *Human Psychopharmacology: Clinical and Experimental*, *11*(1), 47–51.
Logemann, H. N. A., Böcker, K. B. E., Deschamps, P. K. H., Kemner, C., & Kenemans, J. L. (2014a). Differences between nicotine-abstinent smokers and non-smokers in terms of visuospatial attention and inhibition before and after single-blind nicotine administration. *Neuroscience*, *277*, 375–382.
Logemann, H. N. A., Böcker, K. B. E., Deschamps, P. K. H., Kemner, C., & Kenemans, J. L. (2014b). The effect of enhancing cholinergic neurotransmission by nicotine on EEG indices of inhibition in the human brain. *Pharmacology Biochemistry and Behavior*, *122*, 89–96.
Mancuso, G., Warburton, D. M., Mélen, M., Sherwood, N., & Tirelli, E. (1999). Selective effects of nicotine on attentional processes. *Psychopharmacology*, *146*(2), 199–204.
McClernon, F. J., Kollins, S. H., Lutz, A. M., Fitzgerald, D. P., Murray, D. W., Redman, C., et al. (2008). Effects of smoking abstinence on adult smokers with and without attention deficit hyperactivity disorder: results of a preliminary study. *Psychopharmacology*, *197*(1), 95–105.
Meyers, K. K., Crane, N. A., O'Day, R., Zubieta, J. K., Giordani, B., Pomerleau, C. S., et al. (2015). Smoking history, and not depression, is related to deficits in detection of happy and sad faces. *Addictive Behaviors*, *41*, 210–217.
Myers, C. S., Taylor, R. C., Salmeron, B. J., Waters, A. J., & Heishman, S. J. (2013). Nicotine enhances alerting, but not executive, attention in smokers and nonsmokers. *Nicotine & Tobacco Research*, *15*(1), 277–281.
Parrott, A. C., & Craig, D. (1992). Cigarette smoking and nicotine gum (0, 2 and 4mg): effects upon four visual attention tasks. *Neuropsychobiology*, *25*(1), 34–43.
Petrovsky, N., Ettinger, U., Quednow, B. B., Landsberg, M. W., Drees, J., Lennertz, L., et al. (2013). Nicotine enhances antisaccade performance in schizophrenia patients and healthy controls. *International Journal of Neuropsychopharmacology*, *16*(7), 1473–1481.
Petrovsky, N., Ettinger, U., Quednow, B. B., Walter, H., Schnell, K., Kessler, H., et al. (2012). Nicotine differentially modulates antisaccade performance in healthy male non-smoking volunteers stratified for low and high accuracy. *Psychopharmacology*, *221*(1), 27–38.
Pettiford, J., Kozink, R. V., Lutz, A. M., Kollins, S. H., Rose, J. E., & McClernon, F. J. (2007). Increases in impulsivity following smoking abstinence are related to baseline nicotine intake and boredom susceptibility. *Addictive Behaviors*, *32*(10), 2351–2357.
Poltavski, D. V., & Petros, T. (2006). Effects of transdermal nicotine on attention in adult non-smokers with and without attentional deficits. *Physiology & Behavior*, *87*(3), 614–624.
Poltavski, D. V., Petros, T. V., & Holm, J. E. (2012). Lower but not higher doses of transdermal nicotine facilitate cognitive performance in smokers on gender non-preferred tasks. *Pharmacology Biochemistry and Behavior*, *102*(3), 423–433.
Pomerleau, C. S., Teuscher, F., Goeters, S., & Pomerleau, O. F. (1994). Effects of nicotine abstinence and menstrual phase on task performance. *Addictive Behaviors*, *19*(4), 357–362.
Potter, A. S., Bucci, D. J., & Newhouse, P. A. (2012). Manipulation of nicotinic acetylcholine receptors differentially affects behavioral inhibition in human subjects with and without disordered baseline impulsivity. *Psychopharmacology*, *220*(2), 331–340.
Potter, A. S., & Newhouse, P. A. (2004). Effects of acute nicotine administration on behavioral inhibition in adolescents with attention-deficit/hyperactivity disorder. *Psychopharmacology*, *176*(2), 182–194.
Potter, A. S., & Newhouse, P. A. (2008). Acute nicotine improves cognitive deficits in young adults with attention-deficit/hyperactivity disorder. *Pharmacology Biochemistry and Behavior*, *88*(4), 407–417.
Powell, J., Dawkins, L., & Davis, R. E. (2002). Smoking, reward responsiveness, and response inhibition: tests of an incentive motivational model. *Biological Psychiatry*, *51*(2), 151–163.
Provost, S. C., & Woodward, R. (1991). Effects of nicotine gum on repeated administration of the stroop test. *Psychopharmacology*, *104* (4), 536–540.
Rusted, J. M., Caulfield, D., King, L., & Goode, A. (2000). Moving out of the laboratory: does nicotine improve everyday attention? *Behavioural Pharmacology*, *11*(7–8), 621–629.
Rycroft, N., Hutton, S. B., Clowry, O., Groomsbridge, C., Sierakowski, A., & Rusted, J. M. (2007). Non-cholinergic modulation of antisaccade performance: a modafinil-nicotine comparison. *Psychopharmacology*, *195*(2), 245–253.
Rycroft, N., Hutton, S. B., & Rusted, J. M. (2006). The antisaccade task as an index of sustained goal activation in working memory: modulation by nicotine. *Psychopharmacology*, *188*(4), 521–529.
Schlienz, N. J., & Hawk, L. W. (2017). Probing the behavioral and neurophysiological effects of acute smoking abstinence on drug and nondrug reinforcement during a cognitive task. *Nicotine & Tobacco Research*, *19*(6), 729–737.

Schlienz, N. J., Hawk, L. W., & Rosch, K. S. (2013). The effects of acute abstinence from smoking and performance-based rewards on performance monitoring. *Psychopharmacology, 229*(4), 701–711.

Schmechtig, A., Lees, J., Grayson, L., Craig, K. J., Dadhiwala, R., Dawson, G. R., et al. (2013). Effects of risperidone, amisulpride and nicotine on eye movement control and their modulation by schizotypy. *Psychopharmacology, 227*(2), 331–345.

Smucny, J., Olincy, A., Eichman, L. S., & Tregellas, J. R. (2015). Neuronal effects of nicotine during auditory selective attention. *Psychopharmacology, 232*(11), 2017–2028.

Suter, T. W., Buzzi, R., Woodson, P. P., & Bättig, K. (1983). Psychophysiological correlates of conflict solving and cigarette smoking. *Activitas Nervosa Superior, 25*(4), 261–272.

Thaker, G. K., Ellsberry, R., Moran, M., Lahti, A., & Tamminga, C. (1991). Tobacco smoking increases square-wave jerks during pursuit eye movements. *Biological Psychiatry, 29*(1), 82–88.

Tsaur, S., Strasser, A. A., Souprountchouk, V., Evans, G. C., & Ashare, R. L. (2015). Time dependency of craving and response inhibition during nicotine abstinence. *Addiction Research and Theory, 23*(3), 205–212.

Wachter, N. J., & Gilbert, D. G. (2013). Nicotine differentially modulates antisaccade eye-gaze away from emotional stimuli in nonsmokers stratified by pre-task baseline performance. *Psychopharmacology, 225*(3), 561–568.

Wesnes, K., & Revell, A. (1984). The separate and combined effects of scopolamine and nicotine on human information processing. *Psychopharmacology, 84*(1), 5–11.

Wesnes, K., & Warburton, D. M. (1978). The effect of cigarette smoking and nicotine tablets upon human attention. In R. E. Thornton (Ed.), *Smoking behaviour: Physiological and psychological influences* (pp. 131–147). Edinburgh: Churchill Livingstone.

Wignall, N. D., & De Wit, H. (2011). Effects of nicotine on attention and inhibitory control in healthy nonsmokers. *Experimental and Clinical Psychopharmacology, 19*(3), 183–191.

Xu, X., & Domino, E. F. (2000). Effects of tobacco smoking on topographic EEG and Stroop test in smoking deprived smokers. *Progress in Neuro-Psychopharmacology & Biological Psychiatry, 24*(4), 535–546.

Xu, J., Mendrek, A., Cohen, M. S., Monterosso, J., Simon, S., Jarvik, M., et al. (2007). Effect of cigarette smoking on prefrontal cortical function in nondeprived smokers performing the Stroop task. *Neuropsychopharmacology, 32*(6), 1421–1428.

Zack, M., Belsito, L., Scher, R., Eissenberg, T., & Corrigall, W. A. (2001). Effects of abstinence and smoking on information processing in adolescent smokers. *Psychopharmacology, 153*(2), 249–257.

CHAPTER

20

Nicotine, Corticotropin-Releasing Factor, and Anxiety-Like Behavior

Adriaan W. Bruijnzeel

Department of Psychiatry, University of Florida, Gainesville, FL, United States

Abbreviations

CeA	central nucleus of the amygdala
CRF	corticotropin-releasing factor
CRF1	corticotropin-releasing factor type 1 receptor
CRF2	corticotropin-releasing factor type 2 receptor
ICSS	intracranial self-stimulation
IPN	interpeduncular nucleus
PVN	hypothalamic paraventricular nucleus
VTA	ventral tegmental area

20.1 INTRODUCTION

Tobacco addiction is characterized by a loss of control over smoking and intense craving and negative affective withdrawal signs upon smoking cessation. Worldwide, there are about 1 billion smokers, and the great majority would like to quit (WHO, 2011). Smoking rates are on the decline in Western societies (WHO, 2015). Fewer people start smoking, and increasing numbers of people are able to quit (Kulik & Glantz, 2016). In the late 1990s, almost 25% of US adults smoked, and in 2014, this was less than 16% (Clarke, Norris, & Schiller, 2017). Smoking is particularly on the decline among highly educated, affluent people. Smoking rates remain high, however, in economically disadvantaged groups and in people with mental illnesses (Bruijnzeel, 2017; de Leon, Diaz, Rogers, Browne, & Dinsmore, 2002; Hughes, Hatsukami, Mitchell, & Dahlgren, 1986). Hughes and colleagues were one of the first to investigate the smoking prevalence in psychiatric patients (Hughes et al., 1986). They reported that almost 90% of people with schizophrenia smoke and 50% of depressed people. Having an anxiety disorder also increases the risk for smoking. A large study with adolescents showed that nonsmokers and nondependent smokers with social fears are at an increased risk for becoming nicotine-dependent later in life (Sonntag, Wittchen, Höfler, Kessler, & Stein, 2000).

Clinical studies indicate that there is an association between smoking, depression, and anxiety disorders (Bruijnzeel, 2012). A study with children from New Zealand showed that there is a strong association between depression and nicotine dependence in adolescents and young adults (Fergusson, Goodwin, & Horwood, 2003). This is in line with another large study in the United States that investigated the association between smoking, depression, and anxiety disorders (Johnson et al., 2000). In this study, adolescents were interviewed when they were 16 years old (1985–86) and again when they were 22 years old. Being depressed at the age of 16 increased the risk for having a nicotine dependency during the same developmental stage. There was no association between having an anxiety disorder during adolescence and tobacco smoking during this period. However, heavy smoking (>20 cigarettes/day) during adolescence increased the risk for agoraphobia, generalized anxiety disorder, and panic disorder in adulthood. This suggests that smoking during adolescence induces brain changes that increase the likelihood of developing an anxiety disorder later in life. Having an anxiety disorder also increases the risk for relapse when trying to quit smoking. Smokers with a high level of anxiety sensitivity are more likely to have a smoking lapse after trying to quit than smokers with a low level of anxiety sensitivity (Zvolensky et al., 2017).

Overall, smoking cessation leads to an increase in anxiety, and this contributes to relapse to smoking (Bruijnzeel, 2012). Smokers are more likely to have an anxiety disorder than nonsmokers, and smokers with an anxiety disorder are less motivated to quit (Johnson et al., 2000; Zvolensky et al., 2007). In this chapter overview, the role of CRF in nicotine-withdrawal-induced anxiety-like behavior will be discussed.

Neuroscience of Nicotine
https://doi.org/10.1016/B978-0-12-813035-3.00020-4

20.2 CORTICOTROPIN-RELEASING FACTOR (CRF)-LIKE PEPTIDES

Corticotropin-releasing factor plays a role in regulating the hypothalamic-pituitary adrenal (HPA) axis, autonomic responses to stress, and behavior. CRF is produced in the hypothalamus and transported to the median eminence. After being released in the portal circulation, it stimulates the release of proopiomelanocortin (POMC)-derived peptides (e.g., ACTH) from the anterior pituitary. ACTH stimulates the release of glucocorticoids from the adrenal cortex. CRF is also expressed in extrahypothalamic brain sites. CRF positive cells have been detected in the central nucleus of the amygdala (CeA), bed nucleus of the stria terminalis (BNST), locus coeruleus, parabrachial nucleus, dorsal vagal complex, and prefrontal cortex (Swanson, Sawchenko, Rivier, & Vale, 1983). The role of CRF in stress responses is supported by studies showing that stressors induce the release of CRF in the hypothalamus and the CeA (Cook, 2004; Hand et al., 2002; Merali, McIntosh, Kent, Michaud, & Anisman, 1998; Merlo Pich et al., 1993).

Urocortin 1 was the second member of the CRF receptor family to be discovered (Vaughan et al., 1995). The distribution of urocortin 1 in the brain is very different compared to that of CRF. Urocortin 1 is expressed in the Edinger-Westphal nucleus and the lateral superior olive. Urocortin 1 projections from the Edinger-Westphal nucleus to the lateral septal nucleus may mediate anxiety-like behavior (Radulovic, Ruhmann, Liepold, & Spiess, 1999). High levels of urocortin 1 have also been detected in the periphery. Urocortin 1 is expressed in the gastrointestinal tract, adrenal glands, and testes (Oki et al., 1998). More recently discovered members of the CRF family are urocortin 2 and urocortin 3 (Hauger et al., 2003; Hsu & Hsueh, 2001; Lewis et al., 2001; Reyes et al., 2001). Urocortin 2 has been detected in the locus coeruleus, hypothalamic paraventricular nucleus (PVN), and arcuate nucleus. High levels of urocortin 2 have also been detected in the gastrointestinal tract, adrenal glands, heart, blood cells, and testes (Hsu & Hsueh, 2001). Urocortin 3 has been detected in the hypothalamus, brain stem, lateral septum, BNST, gastrointestinal tract, adrenal glands, and skin (Hsu & Hsueh, 2001; Lewis et al., 2001).

20.3 CORTICOTROPIN-RELEASING FACTOR RECEPTORS

CRF binding sites have been detected in the anterior pituitary, amygdala, cerebral cortex, and hippocampus (De Souza et al., 1985; De Souza, Perrin, Rivier, Vale, & Kuhar, 1984). A decade after the discovery of the CRF binding sites, two CRF receptors were cloned (Chen, Lewis, Perrin, & Vale, 1993; Lovenberg, Liaw, et al., 1995; Perrin, Donaldson, Chen, Lewis, & Vale, 1993). The CRF type 1 (CRF1) receptor is expressed in the anterior pituitary and regulates HPA-axis activity. The CRF1 receptor is also expressed in the cortical areas, the hippocampus, and the basolateral amygdala (Potter et al., 1994). The CRF type 2 (CRF2) receptor is mainly expressed in subcortical brain areas including the lateral septum, PVN, and ventromedial hypothalamic nucleus (Lovenberg, Chalmers, Liu, & De Souza, 1995). CRF and urocortin 1 bind the CRF1 receptor with a high affinity, and urocortin 1 also binds the CRF2 receptor with high affinity (Vaughan et al., 1995). Urocortin 2 and urocortin 3 have a low affinity for the CRF1 receptor and a high affinity for the CRF2 receptor (Lewis et al., 2001).

20.4 ANIMAL MODELS TO STUDY NICOTINE-WITHDRAWAL-INDUCED ANXIETY-LIKE BEHAVIOR

A wide range of animal models have been developed to study nicotine-withdrawal-induced anxiety-like behavior. In most of these studies, anxiety-like behavior was investigated in rats or mice that received nicotine noncontingently using osmotic minipumps or injections (Epping-Jordan, Watkins, Koob, & Markou, 1998; Malin, 2001). In mouse studies, nicotine is often added to drinking water to induce dependence (Grabus et al., 2005). There is extensive evidence that nicotine dependence can be established by noncontingent nicotine administration. Spontaneous and precipitated somatic signs have been observed in rats and mice after the cessation of nicotine administration using pumps or injections (Epping-Jordan et al., 1998; Malin, 2001). Somatic signs have also been reported in mice after the cessation of oral nicotine intake (Grabus et al., 2005). The intracranial self-stimulation (ICSS) paradigms have been widely used to investigate the anhedonia associated with nicotine withdrawal. In this procedure, electrodes are implanted in the medial forebrain bundle, and rats can self-stimulate their reward system in operant chambers. Cessation of nicotine administration leads to anhedonia, and this is reflected in an elevation in brain reward thresholds. After the cessation of nicotine administration, rats become less sensitive to rewarding electric stimuli, and therefore, higher currents are needed to maintain ICSS behavior. Cessation of noncontingent nicotine administration leads to elevations in brain reward thresholds in rats and mice (i.e., anhedonia) (Epping-Jordan et al., 1998; Johnson, Hollander, & Kenny, 2008). Cessation of nicotine administration has also been shown to increase anxiety-like behavior in rats and mice. Damaj and colleagues showed that cessation of nicotine administration (24 and 48 mg/kg/day for 14 days, minipumps) leads to an increase in anxiety-like behavior in the elevated plus-maze test (Damaj, Kao, & Martin, 2003). A lower dose of nicotine, 6mg/kg/day, did not induce dependence and did not increase anxiety-like behavior. In this study, anxiety-like behavior was investigated 24h after cessation of nicotine administration. On a similar note, rats display increased anxiety-like behavior in the social interaction test 72h after the cessation of chronic nicotine administration (injections for 7 or 14days) (Irvine, Cheeta, & File, 1999). The great majority of studies have investigated anxiety-like behavior in rats and mice after the cessation of noncontingent nicotine self-administration. However, a few studies have also investigated anxiety-like behavior after the self-administration of nicotine. Irvine and colleagues investigated anxiety-like behavior in the social interaction test, 0, 24, and 72h after the cessation of 4weeks of nicotine self-administration (Irvine et al., 2001). The rats display increased anxiety-like behavior immediately after the last self-administration session, but not 24 or 48h later. This suggests that chronic nicotine self-administration increases anxiety-like behavior but that cessation of nicotine self-administration does not increase anxiety-like behavior. More recent research suggests that it depends on the nicotine self-administration paradigm whether or not rats display anxiety-like behavior after the cessation of nicotine self-administration. Cohen and colleagues investigated anxiety-like behavior in rats after the cessation of short- and long-access nicotine self-administration (Cohen et al., 2015). The short-access rats had self-administered nicotine 1h/day and the long-access rats 21h/day. In this study, the rats were allowed to self-administer nicotine for 14weeks, and anxiety-like behavior was investigated 72h after the last self-administration session. Rats displayed anxiety-like behavior after the cessation of long-access nicotine self-administration but not short-access nicotine self-administration. It was also shown that rats with long access displayed somatic withdrawal signs upon the cessation of nicotine self-administration but rats with short access did not. This indicates that nicotine self-administration under the long-access paradigm leads to dependence but self-administration under the short-access paradigm does not. This suggests that only rats that have developed nicotine dependence display anxiety-like behavior upon the cessation of nicotine self-administration.

20.5 CORTICOTROPIN-RELEASING FACTOR, NICOTINE WITHDRAWAL, AND ANXIETY

Some of the first studies with CRF showed that it increases anxiety-like behavior (Dunn & File, 1987; Swerdlow, Geyer, Vale, & Koob, 1986). CRF mediates anxiety-like behavior mainly by activating CRF1 receptors. CRF1 receptors have been found in the brain and outside the blood-brain barrier (e.g., the gastrointestinal tract and anterior pituitary) (Van Pett et al., 2000; Yuan et al., 2012). A recent study showed that the CRF-induced increase in anxiety-like behavior is mediated by CRF1 receptors in the brain and not by CRF1 receptor outside the blood-brain barrier (Tanaka et al., 2017). CRF also contributes to anxiety-like behavior after the cessation of nicotine administration. Cohen and colleagues showed that rats with long access to nicotine (21h) display an increase in anxiety-like behavior 72h after the last self-administration session (Cohen et al., 2015). This increase in anxiety-like behavior was blockade by systemic and intra-CeA administration of the CRF1 receptor antagonist MPZP.

The habenula-interpeduncular system plays a critical role in the regulation of emotional states and drug addiction (Hikosaka, 2010). The medial habenula receives input from the septum and diagonal band of Broca and connects via the fasciculus retroflexus to the interpeduncular nucleus (IPN). The habenula and the IPN play a role in the development of nicotine dependence. Blockade of nAChRs in the habenula and IPN of nicotine-dependent mice leads to somatic withdrawal signs (Salas, Sturm, Boulter, & De, 2009). In addition, mice that express hypersensitive $\alpha6/\alpha4\beta2$ nAChRs in the habenula are more anxious than control mice (Pang et al., 2016). Zhao-Shea and colleagues explored the role of CRF in the habenula-IPN circuit in anxiety-like behavior during nicotine withdrawal (Zhao-Shea et al., 2015). They reported that CRF release in the IPN mediates anxiety-like behavior. Infusion of CRF in the IPN increased anxiety-like behavior and blockade of CRF1 receptors in the IPN-diminished nicotine-withdrawal-induced anxiety-like behavior. Decreasing CRF levels in the ventral tegmental area (VTA) with short hairpin RNAs decreased the nicotine-withdrawal-induced activation of the IPN and anxiety-like behavior. This suggests that the increase in anxiety-like behavior after the cessation of nicotine intake is mediated by the activation of a CRF projection from the VTA to the IPN. CRF receptors in the VTA have also been implicated in the anxiogenic effects of cocaine in rats (Ettenberg et al., 2015). Overall, these findings indicate that CRF transmission in the VTA and IPN contributes to anxiety-like behavior in animals with a history of drug intake.

20.6 ROLE OF CRF2 RECEPTORS IN NICOTINE WITHDRAWAL

Bagosi and colleagues investigated the role of CRF2 receptors in nicotine withdrawal (Bagosi et al., 2016). Mice were treated with nicotine for 7days, and 24h after the last injection, the effect of nicotine withdrawal on anxiety- and depressive-like behavior was investigated (Bagosi et al., 2016). The nicotine-withdrawing mice displayed increased anxiety-like behavior in the elevated plus-maze test and depressive-like behavior in the forced swim test. Central (icv) administration of urocortin 2 or urocortin 3 prevented the nicotine-withdrawal-induced increase in anxiety- and depressive-like behavior. Administration of the CRF2 receptor agonists also diminished the elevations in corticosterone levels in the nicotine-withdrawing mice. This is in line with studies that showed that stimulation of CRF2 receptors diminishes anxiety-like behavior during alcohol withdrawal (Valdez, Sabino, & Koob, 2004). It should be noted that not all studies have found that stimulation of CRF2 receptors diminishes anxiety-like behavior associated with drug withdrawal. One study reported that blockade of CRF2 receptors with antisauvagine-30 decreases anxiety-like behavior during amphetamine withdrawal (Reinbold, Scholl, Oliver, Watt, & Forster, 2014). On a similar note, CRF2 receptor knockout mice do not display an anhedonia-like state during opioid withdrawal (Ingallinesi, Rouibi, Le Moine, Papaleo, & Contarino, 2012). Therefore, conflicting findings with regard to the role of the CRF2 receptor in drug withdrawal have been reported. However, the discussed studies would suggest that stimulating CRF2 receptors might diminish anxiety- and depressive-like behavior associated with nicotine and alcohol withdrawal.

20.7 CRF AND ANHEDONIA ASSOCIATED WITH NICOTINE WITHDRAWAL

My laboratory has extensively explored the role of CRF in the anhedonia associated with nicotine withdrawal. In one of our first experiments, we investigated if blockade of both CRF1 and CRF2 receptors diminishes the elevations in brain thresholds associated with precipitated nicotine withdrawal (Bruijnzeel, Zislis, Wilson, & Gold, 2007). Central (icv) administration of the nonselective CRF1/CRF2 receptor antagonist D-Phe $CRF_{(12–41)}$ diminished mecamylamine-induced elevations in brain reward thresholds. In a follow-up experiment, the role of CRF1 and CRF2 receptors in the anhedonia associated with precipitated nicotine withdrawal was investigated (Bruijnzeel, Prado, & Isaac, 2009). CRF1 receptors were blocked with R278995/CRA0450, and CRF2 receptors were blocked with astressin 2B. Blockade of CRF1, but not CRF2, receptors prevented the elevations in brain reward thresholds associated with precipitated nicotine withdrawal. These findings suggest that CRF1 receptors play a role in the anhedonia associated with smoking cessation. It was then investigated which brain areas mediate the anhedonia associated with nicotine withdrawal (Marcinkiewcz et al., 2009). Administration of the nonselective CRF receptor antagonist D-Phe $CRF_{(12–41)}$ into the CeA and Nacc shell prevented the elevations in brain reward thresholds in the nicotine-dependent rats. In contrast, blockade of CRF1/CRF2 receptors in the BNST was ineffective. To investigate if CRF1 receptors in the CeA play a role in nicotine withdrawal, we conducted a study with the selective CRF1 receptor antagonist R278995/CRA0450 (Bruijnzeel et al., 2012). We found that blockade of CRF1 receptors in the CeA diminishes the anhedonia associated with nicotine withdrawal. Overall, these studies indicate that CRF signaling in the CeA and Nacc shell plays a role in the negative mood state associated with nicotine withdrawal.

20.8 CONCLUDING REMARKS

Clinical studies show that there is a positive association between smoking and anxiety and that high levels of anxiety contribute to relapse. The reviewed animal studies indicate that the cessation of noncontingent nicotine administration and the cessation of long-access nicotine self-administration lead to an increase in anxiety-like behavior. Prolonged exposure to nicotine dysregulates brain CRF systems, and blockade of CRF1 receptors or stimulation of CRF2 receptors diminishes anxiety-like behavior associated with nicotine withdrawal. Therefore, drugs that block CRF1 receptors or stimulate CRF2 receptors may decrease anxiety during a smoking cessation attempt and thereby decrease the risk for relapse.

MINI-DICTIONARY OF TERMS

Addiction Brain disorder that is characterized by loss of control over drug use and withdrawal signs upon cessation of drug use.

Anhedonia Negative mood state that prevents people from experiencing pleasure. It is one of the core symptoms of depression.

Corticotropin-releasing factor (CRF) Neuropeptide that is produced in the brain and plays an important role in stress responses, anxiety, and depression.

Intracranial self-stimulation (ICSS) A method that is used to determine the state of the brain reward system in rodents.

Rat drug self-administration Experimental method in which rats are prepared with intravenous catheters to investigate the positive reinforcing effects of drugs of abuse.

Key Facts of Tobacco Smoking

- Worldwide, there are about 1 billion people who smoke.
- When people stop smoking, they feel anxious and depressed and have strong urges to smoke.
- It takes most people several attempts to be able to quit smoking.

- In the 1930s, scientists started to realize that tobacco smoking has adverse health effects.
- There are several treatment options for smoking cessation, but the relapse rates are still high.
- Smoking rates are on the decline among healthy affluent people, but it is more difficult to quit for people with psychiatric disorders.

Summary Points

- This chapter focuses on the role of CRF in anxiety- and depressive-like behavior associated with nicotine withdrawal.
- Smoking is more common in people with psychiatric disorders.
- Cessation of nicotine administration leads to an increase in anxiety- and depressive-like behavior.
- Increased CRF release contributes to anxiety- and depressive-like behavior during nicotine withdrawal.
- CRF mediates its effects in the brain by binding to the CRF1 and CRF2 receptor.
- Blockade of CRF1 receptors diminishes anxiety- and depressive-like behavior associated with nicotine withdrawal.
- Stimulation of CRF2 receptors diminishes anxiety- and depressive-like behavior associated with nicotine withdrawal.

Acknowledgments

A.B. was supported by the National Institute on Drug Abuse of the National Institutes of Health under Award Number R01DA042530 when working on this book chapter.

References

Bagosi, Z., Palotai, M., Simon, B., Bokor, P., Buzás, A., Balangó, B., et al. (2016). Selective CRF2 receptor agonists ameliorate the anxiety- and depression-like state developed during chronic nicotine treatment and consequent acute withdrawal in mice. *Brain Research, 1652*, 21–29.

Bruijnzeel, A. W. (2012). Tobacco addiction and the dysregulation of brain stress systems. *Neuroscience and Biobehavioral Reviews, 36*, 1418–1441.

Bruijnzeel, A. W. (2017). Reducing the prevalence of smoking: policy measures and focusing on specific populations. *Nicotine & Tobacco Research 19*, 1003–1004.

Bruijnzeel, A. W., Ford, J., Rogers, J. A., Scheick, S., Ji, Y., Bishnoi, M., et al. (2012). Blockade of CRF1 receptors in the central nucleus of the amygdala attenuates the dysphoria associated with nicotine withdrawal in rats. *Pharmacology, Biochemistry, and Behavior, 101*(1), 62–68.

Bruijnzeel, A. W., Prado, M., & Isaac, S. (2009). Corticotropin-releasing factor-1 receptor activation mediates nicotine withdrawal-induced deficit in brain reward function and stress-induced relapse. *Biological Psychiatry, 66*, 110–117.

Bruijnzeel, A. W., Zislis, G., Wilson, C., & Gold, M. S. (2007). Antagonism of CRF receptors prevents the deficit in brain reward function associated with precipitated nicotine withdrawal in rats. *Neuropsychopharmacology, 32*(4), 955–963.

Chen, R., Lewis, K. A., Perrin, M. H., & Vale, W. W. (1993). Expression cloning of a human corticotropin-releasing-factor receptor. *Proceedings of the National Academy of Sciences of the United States of America, 90*(19), 8967–8971.

Clarke, T. C., Norris, T., & Schiller, J. S. (2017). *Early release of selected estimates based on data from the 2016 National Health Interview Survey.* National Center for Health Statistics. Retrieved from: *https://www.cdc.gov/nchs/data/nhis/earlyrelease/earlyrelease201705.pdf.*

Cohen, A., Treweek, J., Edwards, S., Leao, R. M., Schulteis, G., Koob, G. F., et al. (2015). Extended access to nicotine leads to a CRF1 receptor dependent increase in anxiety-like behavior and hyperalgesia in rats. *Addiction Biology, 20*(1), 56–68.

Cook, C. J. (2004). Stress induces CRF release in the paraventricular nucleus, and both CRF and GABA release in the amygdala. *Physiology & Behavior, 82*(4), 751–762.

Damaj, M. I., Kao, W., & Martin, B. R. (2003). Characterization of spontaneous and precipitated nicotine withdrawal in the mouse. *The Journal of Pharmacology and Experimental Therapeutics, 307*(2), 526–534.

de Leon, J., Diaz, F. J., Rogers, T., Browne, D., & Dinsmore, L. (2002). Initiation of daily smoking and nicotine dependence in schizophrenia and mood disorders. *Schizophrenia Research, 56*(1), 47–54.

De Souza, E. B., Insel, T. R., Perrin, M. H., Rivier, J., Vale, W. W., & Kuhar, M. J. (1985). Corticotropin-releasing factor receptors are widely distributed within the rat central nervous system: an autoradiographic study. *The Journal of Neuroscience, 5*(12), 3189–3203.

De Souza, E. B., Perrin, M. H., Rivier, J., Vale, W. W., & Kuhar, M. J. (1984). Corticotropin-releasing factor receptors in rat pituitary gland: autoradiographic localization. *Brain Research, 296*(1), 202–207.

Dunn, A. J., & File, S. E. (1987). Corticotropin-releasing factor has an anxiogenic action in the social interaction test. *Hormones and Behavior, 21*(2), 193–202.

Epping-Jordan, M. P., Watkins, S. S., Koob, G. F., & Markou, A. (1998). Dramatic decreases in brain reward function during nicotine withdrawal. *Nature, 393*(6680), 76–79.

Ettenberg, A., Cotten, S. W., Brito, M. A., Klein, A. K., Ohana, T. A., Margolin, B., et al. (2015). CRF antagonism within the ventral tegmental area but not the extended amygdala attenuates the anxiogenic effects of cocaine in rats. *Pharmacology, Biochemistry, and Behavior, 138*, 148–155.

Fergusson, D. M., Goodwin, R. D., & Horwood, L. J. (2003). Major depression and cigarette smoking: results of a 21-year longitudinal study. *Psychological Medicine, 33*(8), 1357–1367.

Grabus, S. D., Martin, B. R., Batman, A. M., Tyndale, R. F., Sellers, E., & Damaj, M. I. (2005). Nicotine physical dependence and tolerance in the mouse following chronic oral administration. *Psychopharmacology, 178*(2–3), 183–192.

Hand, G. A., Hewitt, C. B., Fulk, L. J., Stock, H. S., Carson, J. A., Davis, J. M., et al. (2002). Differential release of corticotropin-releasing hormone (CRH) in the amygdala during different types of stressors. *Brain Research, 949*(1–2), 122–130.

Hauger, R. L., Grigoriadis, D. E., Dallman, M. F., Plotsky, P. M., Vale, W. W., & Dautzenberg, F. M. (2003). International Union of Pharmacology. XXXVI. Current status of the nomenclature for receptors for corticotropin-releasing factor and their ligands. *Pharmacological Reviews, 55*(1), 21–26.

Hikosaka, O. (2010). The habenula: from stress evasion to value-based decision-making. *Nature Reviews Neuroscience, 11*(7), 503–513.

Hsu, S. Y., & Hsueh, A. J. (2001). Human stresscopin and stresscopin-related peptide are selective ligands for the type 2 corticotropin-releasing hormone receptor. *Nature Medicine, 7*(5), 605–611.

Hughes, J. R., Hatsukami, D. K., Mitchell, J. E., & Dahlgren, L. A. (1986). Prevalence of smoking among psychiatric outpatients. *The American Journal of Psychiatry, 143*(8), 993–997.

Ingallinesi, M., Rouibi, K., Le Moine, C., Papaleo, F., & Contarino, A. (2012). CRF2 receptor-deficiency eliminates opiate withdrawal distress without impairing stress coping. *Molecular Psychiatry*, *17*(12), 1283.

Irvine, E., Bagnalasta, M., Marcon, C., Motta, C., Tessari, M., File, S., et al. (2001). Nicotine self-administration and withdrawal: modulation of anxiety in the social interaction test in rats. *Psychopharmacology*, *153*(3), 315–320.

Irvine, E. E., Cheeta, S., & File, S. E. (1999). Time-course of changes in the social interaction test of anxiety following acute and chronic administration of nicotine. *Behavioural Pharmacology*, *10*(6–7), 691–697.

Johnson, J. G., Cohen, P., Pine, D. S., Klein, D. F., Kasen, S., & Brook, J. S. (2000). Association between cigarette smoking and anxiety disorders during adolescence and early adulthood. *JAMA*, *284*(18), 2348–2351.

Johnson, P. M., Hollander, J. A., & Kenny, P. J. (2008). Decreased brain reward function during nicotine withdrawal in C57BL6 mice: evidence from intracranial self-stimulation (ICSS) studies. *Pharmacology, Biochemistry, and Behavior*, *90*(3), 409–415.

Kulik, M. C., & Glantz, S. A. (2016). The smoking population in the USA and EU is softening not hardening. *Tobacco Control*, *25*(4), 470–475.

Lewis, K., Li, C., Perrin, M. H., Blount, A., Kunitake, K., Donaldson, C., et al. (2001). Identification of urocortin III, an additional member of the corticotropin-releasing factor (CRF) family with high affinity for the CRF2 receptor. *Proceedings of the National Academy of Sciences of the United States of America*, *98*(13), 7570–7575.

Lovenberg, T. W., Chalmers, D. T., Liu, C., & De Souza, E. B. (1995). CRF2 alpha and CRF2 beta receptor mRNAs are differentially distributed between the rat central nervous system and peripheral tissues. *Endocrinology*, *136*(9), 4139–4142.

Lovenberg, T. W., Liaw, C. W., Grigoriadis, D. E., Clevenger, W., Chalmers, D. T., De Souza, E. B., et al. (1995). Cloning and characterization of a functionally distinct corticotropin-releasing factor receptor subtype from rat brain. *Proceedings of the National Academy of Sciences of the United States of America*, *92*(3), 836–840.

Malin, D. H. (2001). Nicotine dependence: studies with a laboratory model. *Pharmacology, Biochemistry, and Behavior*, *70*(4), 551–559.

Marcinkiewcz, C. A., Prado, M. M., Isaac, S. K., Marshall, A., Rylkova, D., & Bruijnzeel, A. W. (2009). Corticotropin-releasing factor within the central nucleus of the amygdala and the nucleus accumbens shell mediates the negative affective state of nicotine withdrawal in rats. *Neuropsychopharmacology*, *34*(7), 1743–1752.

Merali, Z., McIntosh, J., Kent, P., Michaud, D., & Anisman, H. (1998). Aversive and appetitive events evoke the release of corticotropin-releasing hormone and bombesin-like peptides at the central nucleus of the amygdala. *The Journal of Neuroscience*, *18*(12), 4758–4766.

Merlo Pich, E., Koob, G. F., Heilig, M., Menzaghi, F., Vale, W., & Weiss, F. (1993). Corticotropin-releasing factor release from the mediobasal hypothalamus of the rat as measured by microdialysis. *Neuroscience*, *55*(3), 695–707.

Oki, Y., Iwabuchi, M., Masuzawa, M., Watanabe, F., Ozawa, M., Iino, K., et al. (1998). Distribution and concentration of urocortin, and effect of adrenalectomy on its content in rat hypothalamus. *Life Sciences*, *62* (9), 807–812.

Pang, X., Liu, L., Ngolab, J., Zhao-Shea, R., McIntosh, J. M., Gardner, P. D., et al. (2016). Habenula cholinergic neurons regulate anxiety during nicotine withdrawal via nicotinic acetylcholine receptors. *Neuropharmacology*, *107*, 294–304.

Perrin, M. H., Donaldson, C. J., Chen, R., Lewis, K. A., & Vale, W. W. (1993). Cloning and functional expression of a rat brain corticotropin releasing factor (CRF) receptor. *Endocrinology*, *133*(6), 3058–3061.

Potter, E., Sutton, S., Donaldson, C., Chen, R., Perrin, M., Lewis, K., et al. (1994). Distribution of corticotropin-releasing factor receptor mRNA expression in the rat brain and pituitary. *Proceedings of the National Academy of Sciences of the United States of America*, *91*(19), 8777–8781.

Radulovic, J., Ruhmann, A., Liepold, T., & Spiess, J. (1999). Modulation of learning and anxiety by corticotropin-releasing factor (CRF) and stress: differential roles of CRF receptors 1 and 2. *The Journal of Neuroscience*, *19*(12), 5016–5025.

Reinbold, E. D., Scholl, J. L., Oliver, K. M., Watt, M. J., & Forster, G. L. (2014). Central CRF 2 receptor antagonism reduces anxiety states during amphetamine withdrawal. *Neuroscience Research*, *89*, 37–43.

Reyes, T. M., Lewis, K., Perrin, M. H., Kunitake, K. S., Vaughan, J., Arias, C. A., et al. (2001). Urocortin II: a member of the corticotropin-releasing factor (CRF) neuropeptide family that is selectively bound by type 2 CRF receptors. *Proceedings of the National Academy of Sciences of the United States of America*, *98*(5), 2843–2848.

Salas, R., Sturm, R., Boulter, J., & De, B. M. (2009). Nicotinic receptors in the habenulo-interpeduncular system are necessary for nicotine withdrawal in mice. *The Journal of Neuroscience*, *29*(10), 3014–3018.

Sonntag, H., Wittchen, H., Höfler, M., Kessler, R., & Stein, M. (2000). Are social fears and DSM-IV social anxiety disorder associated with smoking and nicotine dependence in adolescents and young adults? *European Psychiatry*, *15*(1), 67–74.

Swanson, L. W., Sawchenko, P. E., Rivier, J., & Vale, W. W. (1983). Organization of ovine corticotropin-releasing factor immunoreactive cells and fibers in the rat brain: an immunohistochemical study. *Neuroendocrinology*, *36*, 165–186.

Swerdlow, N. R., Geyer, M. A., Vale, W. W., & Koob, G. F. (1986). Corticotropin-releasing factor potentiates acoustic startle in rats: blockade by chlordiazepoxide. *Psychopharmacology*, *88*(2), 147–152.

Tanaka, M., Tomimatsu, Y., Sakimura, K., Ootani, Y., Sako, Y., Kojima, T., et al. (2017). Characterization of Crf1 receptor antagonists with differential peripheral vs central actions in Crf challenge in rats. *Peptides*, *95*, 40–50.

Valdez, G. R., Sabino, V., & Koob, G. F. (2004). Increased anxiety-like behavior and ethanol self-administration in dependent rats: reversal via corticotropin-releasing factor-2 receptor activation. *Alcoholism, Clinical and Experimental Research*, *28*(6), 865–872.

Van Pett, K., Viau, V., Bittencourt, J. C., Chan, R. K., Li, H. Y., Arias, C., et al. (2000). Distribution of mRNAs encoding CRF receptors in brain and pituitary of rat and mouse. *The Journal of Comparative Neurology*, *428*(2), 191–212.

Vaughan, J., Donaldson, C., Bittencourt, J., Perrin, M. H., Lewis, K., Sutton, S., et al. (1995). Urocortin, a mammalian neuropeptide related to fish urotensin I and to corticotropin-releasing factor. *Nature*, *378*(6554), 287–292.

WHO (2011). *Systematic review of the link between tobacco and poverty. Retrieved from: http://www.who.int/tobacco/publications/syst_rev_tobacco_poverty/en/index.html.*

WHO (2015). *WHO report on the global tobacco epidemic, 2015: raising taxes on tobacco. Retrieved from: http://www.who.int/tobacco/global_report/2015/en/.*

Yuan, P.-Q., Wu, S. V., Elliott, J., Anton, P. A., Chatzaki, E., Million, M., et al. (2012). Expression of corticotropin releasing factor receptor type 1 (CRF 1) in the human gastrointestinal tract and upregulation in the colonic mucosa in patients with ulcerative colitis. *Peptides*, *38* (1), 62–69.

Zhao-Shea, R., DeGroot, S. R., Liu, L., Vallaster, M., Pang, X., Su, Q., et al. (2015). Increased CRF signalling in a ventral tegmental area-interpeduncular nucleus-medial habenula circuit induces anxiety during nicotine withdrawal. *Nature Communications*, *6*, 6770.

Zvolensky, M. J., Rodríguez-Cano, R., Paulus, D. J., Kotov, R., Bromet, E., Gonzalez, A., et al. (2017). Respiratory problems and anxiety sensitivity in smoking lapse among treatment seeking smokers. *Addictive Behaviors*, *75*, 25–29.

Zvolensky, M. J., Vujanovic, A. A., Miller, M. O., Bernstein, A., Yartz, A. R., Gregor, K. L., et al. (2007). Incremental validity of anxiety sensitivity in terms of motivation to quit, reasons for quitting, and barriers to quitting among community-recruited daily smokers. *Nicotine & Tobacco Research*, *9*(9), 965–975.

CHAPTER

21

6-Hydroxy-L-Nicotine and Memory Impairment

Lucian Hritcu, Marius Mihasan

Department of Biology, Alexandru Ioan Cuza University of Iasi, Iasi, Romania

Abbreviations

2,3,6-THP	2,3,6-trihydroxypyridine
2,6-DHP	2,6-dihydroxypyridine
2,6-DHPON	2,6-dihydroxypseudooxynicotine
6HDNO	6-hydroxy-D-nicotine oxidase
6HLN	6-hydroxy-L-nicotine
6HLNO	6-hydroxy-L-nicotine oxidase
6-HMM	6-hydroxy-methylmyosmine
6-HPON	6-hydroxy-pseudooxynicotine
AD	Alzheimer's disease
CHL	chlorisondamine
DHPH	2,6-dihydroxypyridine-3-hydroxylase
DHPONH	2,6-dihydroxypseudooxynicotine hydrolase
KDH	ketone dehydrogenase
MGABA	γ-*N*-methylaminobutyrate
nAChBP	nicotinic receptor binding domain homolog
nAChR	nicotinic acetylcholine receptors
NB	nicotine blue
NDH	nicotine dehydrogenase

21.1 INTRODUCTION

Alzheimer's disease (AD) is a progressive neurodegenerative disorder with a gradual decline in cognition. It is considered to be the most common pathological cause of dementia (Sase, Yamamoto, Kawashima, Tan, & Sawa, 2017). It has been reported that in 2016, about 2 million people who have AD are at the age of 85 or older, accounting for 37% of all people with the AD (Hebert, Weuve, Scherr, & Evans, 2013). By 2050, as many as 7 million people at the age of 85 and older may have AD, accounting for half (51%) of all people at the age of 65 and older with the AD (Hebert et al., 2013).

It has been reported that nicotinic receptors are therapeutic targets in the nervous system (Taly, Corringer, Guedin, Lestage, & Changeux, 2009). Also, the involvement of the nicotinic acetylcholine receptor (nAChR) subtypes α7 and α4β2 in AD pathology was evidenced (Parri, Hernandez, & Dineley, 2011). The AD is characterized by substantial decreasing in both nicotinic and muscarinic acetylcholine receptor expression in the basal forebrain, cerebral cortex, and hippocampus (Kása, Rakonczay, & Gulya, 1997). Consequently, in the cerebral cortex, a significant loss of α4β2 nAChR subtypes (Aubert et al., 1992) was evidenced, while in the hippocampus, a loss of α7 nAChRs was reported being correlated with the progressive alteration of the memory function (Nordberg, 2001).

Some epidemiological studies have shown data suggesting an inverse relationship between tobacco consumption and the development of AD (Echeverria Moran, 2013). The putative beneficial effect of tobacco has been mainly attributed to nicotine, which has been reported to improve cognitive abilities and reduce plaques in a mouse model of the AD (Nordberg et al., 2002). Numerous studies suggested that acute nicotine administration significantly improved the performance of patients with AD in cognitive tasks (Taly et al., 2009). Since nAChRs play an important role in attention and learning and memory, the positive effects of nicotine on memory have been mostly credited to the activation of these receptors (Sabbagh, Lukas, Sparks, & Reid, 2002). Moreover, the postmortem levels of the beta-amyloid peptide that considered the neurotoxic agents in the AD brains were significantly decreased in the brains of smoking AD patients compared to nonsmokers with the disease (Boopathi & Kolandaivel, 2014). The high potency of nicotine as a cognition-enhancing agent and AD therapeutic strategy (Murray & Abeles, 2002) can be explained not only by its ability to bind and modulate nAChRs but also through its antioxidant effects at low concentrations (Newman et al., 2002) and through its ability to interact with beta-amyloid peptide.

However, nicotine has not demonstrated in clinical studies to be a useful treatment for the AD (López-Arrieta & Sanz, 2001). Because of its short halftime (about 2 h), its proved negative effects on various other organs such as the lungs (Hecht, Hochalter, Villalta, &

https://doi.org/10.1016/B978-0-12-813035-3.00021-6

Murphy, 2000), its linkage to cigarettes, and the negative publicity associated with smoking (Buccafusco, 2004), nicotine did not impose itself as a feasible therapeutic agent for AD.

Numerous studies have focused primarily on the effects of nicotine and nicotine metabolites found in mammalian cells, because of their potential contribution to the neuropharmacological effects resulting from nicotine exposure (Pogocki et al., 2007; Riveles, Huang, & Quik, 2008). In mammalian cells, nicotine is extensively metabolized in the liver to six primary metabolites (nicotine glucuronide, nicotine N-oxide, nornicotine, isomethonium ion, cotinine, and 2-hydroxynicotine). The predominant pathway during the first-pass metabolism yields cotinine (70%–80% of nicotine is metabolized to cotinine in humans) that may have some relevance in the diverse neurobiological effects of smoking as a ligand of nAChRs (Dome, Lazary, Kalapos, & Rihmer, 2010). Furthermore, there are several other indications that cotinine is a pharmacologically active substance. It has neuroprotective and/or cell-protective effects in the cell-culture model, attention-enhancing effects, antipsychotic-like effects (in prepulse inhibition), dopamine-releasing effects in striatal tissue samples, etc. (O'Leary, Parameswaran, McIntosh, & Quik, 2008).

All of these results indicate that the impact of nicotine and nicotine metabolites on brain functions is not negligible. Thus, it has become increasingly important to obtain more information on the biological effects of nicotine metabolites derived from various sources, in order to possibly manage neurological abnormalities in some neurodegenerative afflictions.

21.2 IMPLICATION TO TREATMENTS

There are many treatments available for subjects who have an addiction to nicotine/tobacco products such as nicotine replacement therapy (the patch and the nicotine gum), nicotine sprays and inhalers, and medication (varenicline, Chantix, and bupropion, Wellbutrin). Additionally, there are mentioned different psychological and behavioral treatments such as hypnotherapy, cognitive-behavioral therapy, and neurolinguistic programming that help people to fight against tobacco addiction.

21.3 NICOTINE DERIVATIVES IN BACTERIA

The crystal structure determination of the nicotinic receptor binding domain homolog (nAChBP) with bound nicotine (Celie et al., 2004) followed by the structure of α7 nAChRs (Mowrey et al., 2013) has spurred the interest of several academic and pharmaceutical laboratories in the possibility that new nicotinic drugs could be designed. From this point of view, nicotine derivatives are ideal candidates, offering a wide array of possibilities. The difficulty resides, firstly, in the identification of molecules that have the beneficial effects of nicotine but elude its side effects (Pogocki et al., 2007) and, secondly, in providing simple and reliable methods for production and isolation of the identified compounds.

In this context, the nicotine-degrading bacteria with their ability to use this alkaloid as the sole carbon source offer a wide range of nicotine derivatives with unexplored biotechnological potential. A multitude of reports indicate that bacteria including *Pseudomonas* (Liu et al., 2014; Tang et al., 2011), *Bacillus* and *Pusillimonas* (Ma et al., 2015) strains, *Ochrobactrum* sp. SJY1 (Yu, Tang, Zhu, Li, & Xu, 2015), *Agrobacterium tumefaciens* S33 (Wang, Liu, & Xu, 2009), *Shinella* sp. HZN7 (Ma et al., 2014), and *Achromobacter nicotinophagum* (Hylin, 1958) can degrade nicotine by using three related pathways: the pyridine pathway (Brandsch, 2006), the pyrrolidine pathway (Tang et al., 2013), and a variant of pyridine and pyrrolidine pathways named the VPP pathway (Wang, Huang, Xie, & Xu, 2012). The intermediates and enzymes involved in each pathway have been extensively presented in two reviews as described by Brandsch (2006) and Gurusamy and Natarajan (2013), respectively.

21.4 NICOTINE CATABOLISM IN *ARTHROBACTER NICOTINOVORANS*

By far, the most extensively studied is the nicotine metabolism of *A. nicotinovorans* pAO1. This bacterium makes use of the pyridine pathway to degrade nicotine. Briefly, nicotine degradation starts with the hydroxylation of the pyridine ring to 6-hydroxy-L-nicotine (6HLN) by nicotine dehydrogenase (NDH) (Andreesen & Fetzner, 2002). Next, the pyrrolidine ring is oxidized by 6-hydroxy-L-nicotine oxidase (6HLNO) (Schenk & Decker, 1999), and 6-hydroxy-methylmyosmine (6-HMM) is formed. The compound reacts spontaneously with water and turns to 6-hydroxy-pseudooxynicotine (6-HPON; *N*-methylamino-propyl-(6-hydroxypyridyl-3)-ketone). 6HLNO has a high stereoselectivity and acts only on 6-hydroxy-L-nicotine. As small amounts of D-nicotine can be produced by the tobacco plant, there is a 6-hydroxy-D-nicotine oxidase (6HDNO) made by *A. nicotinovorans* pAO1 that can transform 6-hydroxy-L-nicotine to 6-hydroxy-methylmyosmine and, consecutively, to 6-HPON (Decker & Bleeg, 1965). A second hydroxylation that takes place at C2 of 6-HPON yields 2,6-dihydroxypseudooxynicotine (2,6-DHPON). This step is encoded by a multimeric ketone dehydrogenase (KDH). A key step in nicotine catabolism is catalyzed by

2,6-dihydroxypseudooxynicotine hydrolase (DHPONH) (Sachelaru, Schiltz, Igloi, & Brandsch, 2005) and consists of the cleavage of 2,6-DHPON into two main metabolites: 2,6-dihydroxypyridine (2,6-DHP) and γ-*N*-methylaminobutyrate (MGABA).

2,6-DHP is further hydroxylated by a FAD-dependent monooxygenase named 2,6-dihydroxypyridine-3-hydroxylase (DHPH) yielding 2,3,6-trihydroxypyridine (2,3,6-THP). In the presence of oxygen, 2,3,6-THP suffers an apparently spontaneous dimerization. The dimerization takes place inside the cell, and the product of the reaction is 4,4′,5,5′-tetrahydroxy-3,3′-diazadiphenoquinone-(2,2′), which is then excreted and accumulates in the medium forming characteristic nicotine blue (NB) color. Even though the metabolic pathway in *A.nicotinovorans* was described in detail by the group of Prof. Roderich Brandsch, Albert Ludwigs University, Freiburg, Germany, until now, there are no biotechnological applications of the identified metabolic products.

21.5 6HLN IS ABLE TO INTERACT WITH nAChRs

We were interested to see if any of the nicotine intermediates found in the *A. nicotinovorans* catabolic pathway have any affinity for nAChRs, and we used in silico experiments to evaluate their binding potential (Mihasan, Capatina, Neagu, Stefan, & Hritcu, 2013). For this, the 3-D structures of all of the intermediates found in the nicotine degradation pathway of *A.nicotinovorans* were downloaded from PubChem (Kim et al., 2016) and converted into a format suitable for docking using FROG v.1.01—free online drug conformation generation (Leite et al., 2007). All steps required for directed docking (molecular surface generation, sphere box, and energy grid generation) were achieved using ADT 1.5.4 (Morris et al., 2009). In silico docking was performed using Autodock 4 (Morris et al., 2009) using the default parameters and the acetylcholine binding protein (AChBP, PDB ID: 1UW6) (Celie et al., 2004) as a receptor. The receptor was kept rigid, and all ligands were flexible. All bonds within the ligand molecules were allowed to rotate. The targeted region for docking was a cube of approx 150 Å centered on Tyr143 from the subunit A of AChBP.

First, L-nicotine was docked at the binding site. The computationally determined orientation of L-nicotine in the binding site of AChBP was compared with the experimentally observed orientation described by Celie et al. (2004). A very good fit between the computationally obtained pose of L-nicotine with the experimentally determined one (root-mean-square deviation, RMSD, of 0.2 Å for 12 superimposed atoms of the ligand) was obtained, indicating the reliability of the docking method employed.

The resulted theoretical binding pose of 6HLN was very similar to the one experimentally observed for nicotine with an RMSD of 0.19 Å for 12 atoms of the ligand. Not only this compound would fit into the binding pocket, but also the increased interaction energy calculated by AutoDockTools for 6HLN is due to an extra H bond between the hydroxyl group of the 6HLN and the Tyr185 residue. Moreover, the antioxidant activity of 6HLN was evaluated using the quantum chemical quantitative structure-activity relationship (QSAR) equation proposed by Rastija and Medić-Šarić (2009). We showed that the antioxidant activity of 6HLN calculated as $IC_{50} = 43.06$ is improved compared to the $IC_{50} = 54.54$ for nicotine.

21.6 6HLN PRODUCED USING AN *ARTHROBACTER*-BASED BIOTECHNOLOGY

As a result of the promising data produced by this computational pilot study, one might easily see the high biotechnological potential of 6HLN. Currently, the compound can be chemically synthesized using a diazotization reaction on 6-aminonicotine and is available from different suppliers (Santa Cruz Biotechnology or Carbosynth) at a fairly expensive price. *Arthrobacter nicotinovorans* cells could nevertheless provide an alternative way to produce this compound using waste and by-products from the tobacco industry.

HPLC analysis of *A. nicotinovorans* growth medium showed a small accumulation of 6HLN during the growth of the bacteria in the presence of nicotine (Boiangiu, Guzun, & Mihasan, 2014). Two previously published articles indicate that 6HLN can be found in not only the nicotine growth media of *A. nicotinovorans* (Hochstein & Rittenberg, 1959) but also a possible export mechanism involving a facilitated transport system (Ganas, Igloi, & Brandsch, 2009). The accumulation is only temporary, and as soon as the amount of nicotine in the medium is depleted, the 6HLN levels drop (Fig. 21.1). Apparently, a metabolic imbalance is involved.

NDH, the trimeric enzyme containing a molybdenum pyranopterin dinucleotide cofactor, FAD, and two iron-sulfur clusters (Andreesen & Fetzner, 2002) responsible for converting L-nicotine to 6HNLO, has roughly the same affinity (K_m of 0.037 mM) for its substrate but a higher reaction rate (specific activity of 29.2 μmol/min/mg) (Hochstein & Dalton, 1967) compared to downstream enzyme 6-hydroxy-L-nicotine oxidase (6HLNO). The dimeric enzyme 6HLNO contains one FAD per subunit (Schenk & Decker, 1999) and has a K_m of 0.02 mM and a much lower specific activity—4.73 μmol/min/mg (Decker, Dai, Möhler, & Brühmüller, 1972; Kachalova et al., 2010). Basically, as long as nicotine is still in the

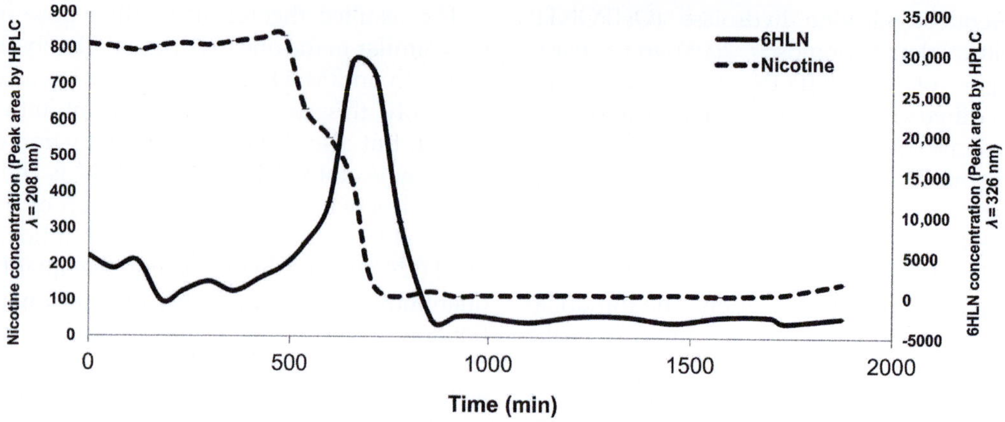

FIG. 21.1 Dynamics of nicotine and 6HLN in the growth medium of *A. nicotinovorans*.
A temporary accumulation of 6HLN can be observed (*continuous line*). As soon as nicotine is depleted from the medium (*dashed line*), 6HLN levels drop.

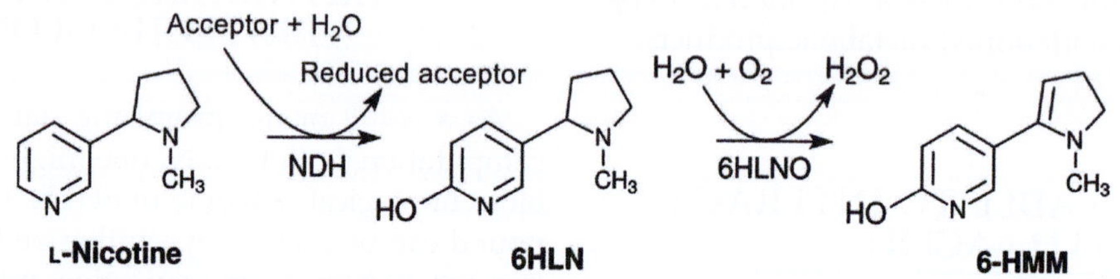

FIG. 21.2 Upstream and downstream reactions of 6HLN in the nicotine catabolic pathway of *A. nicotinovorans*.
NDH, nicotine dehydrogenase; *6HLN*, 6-hydroxy-L-nicotine; *6HLNO*, 6-hydroxy-L-nicotine oxidase; *6-HMM*, 6-hydroxy-methylmyosmine.

medium, NDH converts it to 6HLN at a rather fast pace, 6HLNO cannot keep up, and 6HLN accumulates in the medium. As soon as nicotine is depleted, 6HLNO can quickly recover, and 6HLN is consumed (Fig. 21.2).

The observed metabolic imbalance could be used to produce 6HLN. In order to increase yield, a genetically engineered strain of *A. nicotinovorans* was created. The *ndh* genes were cloned in the custom pART2 expression vector (Andrei & Mihasan, 2013; Sandu, Chiribau, Sachelaru, & Brandsch, 2005) that allows the overexpression of the NDH in a nicotine-dependent manner (Fig. 21.3). Currently, our lab uses low-density *A. nicotinovorans* pART2ndh cell cultures cultivated in flasks under constant agitation at 28°C in order to produce 6HLN. By overexpressing NDH and chemically inhibiting 6HLNO using 0.05 mM $ZnSO_4$, we are able to commonly produce approx 50–60 mg of 6HLN per 100 mL medium.

6HLN produced has memory-enhancing effects.

Next, 6HLN was tested in vivo on male naive Wistar rats and proved to have surprisingly positive effects on memory processes and oxidative brain status.

Hritcu, Stefan, Brandsch, and Mihasan (2013) continue to demonstrate that in vivo chronic treatment of 6HLN (0.3 mg/kg, seven consecutive days) significantly increased spontaneous alternations in Y-maze task and working memory in radial-arm-maze task, suggesting effects on short-term memory, without affecting long-term memory, explored by reference memory in radial-arm-maze task. Also, 6HLN significantly improved locomotor activity, as evidenced by the number of arm entries within Y-maze test (Table 21.1). In addition, 6HLN increased antioxidant enzyme activity (SOD and GPX) and decreased the production of lipid peroxidation (MDA) in the rat temporal lobe homogenates, suggesting antioxidant effects (Table 21.2). We concluded that positive effects of 6HLN on spatial memory may occur by antioxidant actions.

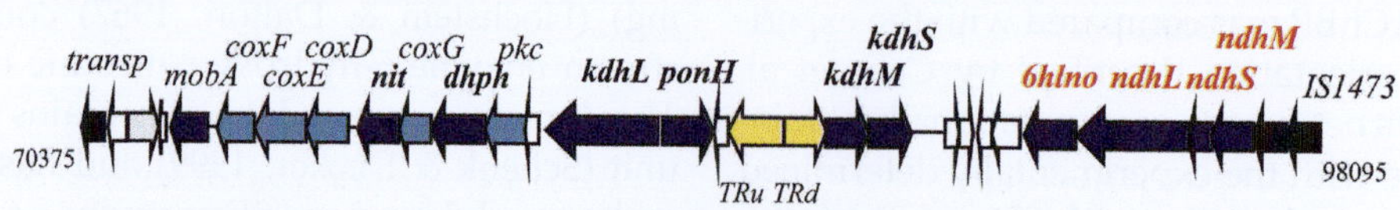

FIG. 21.3 Location of the *ndh* and *6hlno* genes (*red*) in the *nic*-gene cluster from the pAO1 megaplasmid of *A. nicotinovorans*
Dark blue, genes experimentally shown to be involved in nicotine metabolism; light blue, genes putatively involved in nicotine metabolism; white, genes with no experimental or putative function; black, transposons and insertion elements; yellow, putative transcription factors.

TABLE 21.1 Effects of 6HLN in Experimental Models of Memory

Nicotine derivative	Administration	Animal species	Behavioral test	Observed effects	Reference
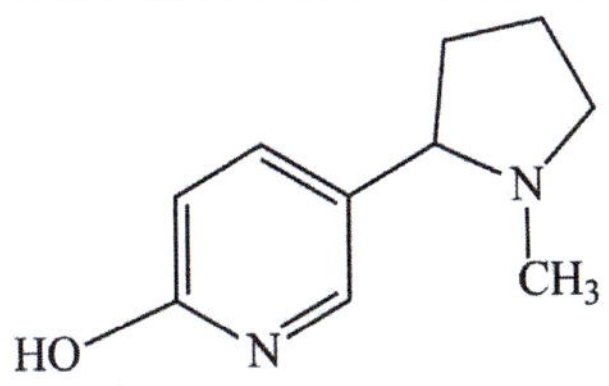 6-hydroxy-L-nicotine	Intraperitoneal in naive rats 0.3 mg/kg	Rats	Y-maze	Improved short-term memory and locomotor activity	Hritcu et al. (2013)
	Intraperitoneal in naive rats 0.3 mg/kg	Rats	Radial-arm-maze	Enhanced working memory, without affecting reference memory	Hritcu et al. (2013)
	Intraperitoneal in scopolamine (0.7 mg/kg)-treated rats 0.3 mg/kg	Rats	Y-maze	Sustained short-term memory and locomotor activity	Hritcu, Stefan, Brandsch, and Mihasan (2015)
	Intraperitoneal in scopolamine (0.7 mg/kg)-treated rats 0.3 mg/kg	Rats	Radial-arm-maze	Enhanced working memory and reference memory	Hritcu et al. (2015)
	Intraperitoneal in chlorisondamine (10 mg/kg)-treated rats 0.3 mg/kg	Rats	Y-maze	Enhanced spontaneous alternation	Hritcu et al. (2017)
	Intraperitoneal in chlorisondamine (10 mg/kg)-treated rats 0.3 mg/kg	Rats	Radial-arm-maze	Improved working memory and reference memory	Hritcu et al. (2017)

Hritcu et al. (2015) demonstrated that 6HLN ameliorated scopolamine-induced spatial memory impairment (Table 21.1). Decreased activities of superoxide dismutase, glutathione peroxidase, and catalase along with a decrease of total content of reduced glutathione were observed in the rat hippocampal homogenates of scopolamine-treated animals as compared with control. Production of malondialdehyde (lipid peroxidation) significantly increased in the rat hippocampal homogenates of scopolamine-treated animals as compared with control, as a consequence of impaired antioxidant enzyme activities. Additionally, in scopolamine-treated rats, 6HLN significantly improved memory formation and decreased oxidative stress (Table 21.2), suggesting memory-enhancing and antioxidant effects. Therefore, our results suggest that the administration of 6HLN ameliorates scopolamine-induced spatial memory impairment by attenuation of the oxidative stress in the rat hippocampus.

Recently, Hritcu et al. (2017) reported that 6HLN administration attenuated the cognitive deficits (Table 21.1) and recovered the antioxidant capacity (Table 21.2) in the rat hippocampus of the chlorisondamine (CHL) rat model. Taken together, administration of 6HLN as compared to nicotine improved significantly spatial memory in the Y-maze, and radial-arm-maze

TABLE 21.2 Effects of 6HLN on Oxidative Stress

Nicotine derivative	Administration	Animal species	Biochemical markers	Observed effects	References
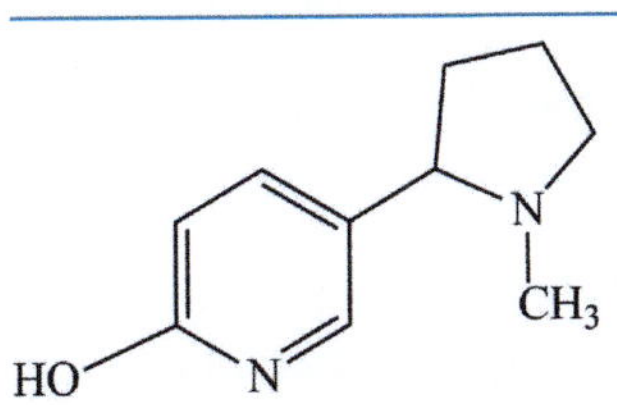 6-hydroxy-L-nicotine	Intraperitoneal in naive rats 0.3 mg/kg	Rats	SOD, GPX, MDA	Increased the activity of SOD and GPX and decreased MDA level	Hritcu et al. (2013)
	Intraperitoneal in scopolamine (0.7 mg/kg)-treated rats 0.3 mg/kg	Rats	SOD, GPX, CAT, GSH, MDA	Increased the activity of SOD, GPX, CAT, and GSH and decreased MDA level	Hritcu et al. (2015)
	Intraperitoneal in chlorisondamine (10 mg/kg)-treated rats 0.3 mg/kg	Rats	SOD, GPX, CAT, GSH, MDA	Increased the activity of SOD, GPX, CAT, and GSH and decreased MDA level	Hritcu et al. (2017)

SOD, superoxide dismutase; *GPX*, glutathione peroxidase; *CAT*, catalase; *GSH*, glutathione; *MDA*, malondialdehyde.

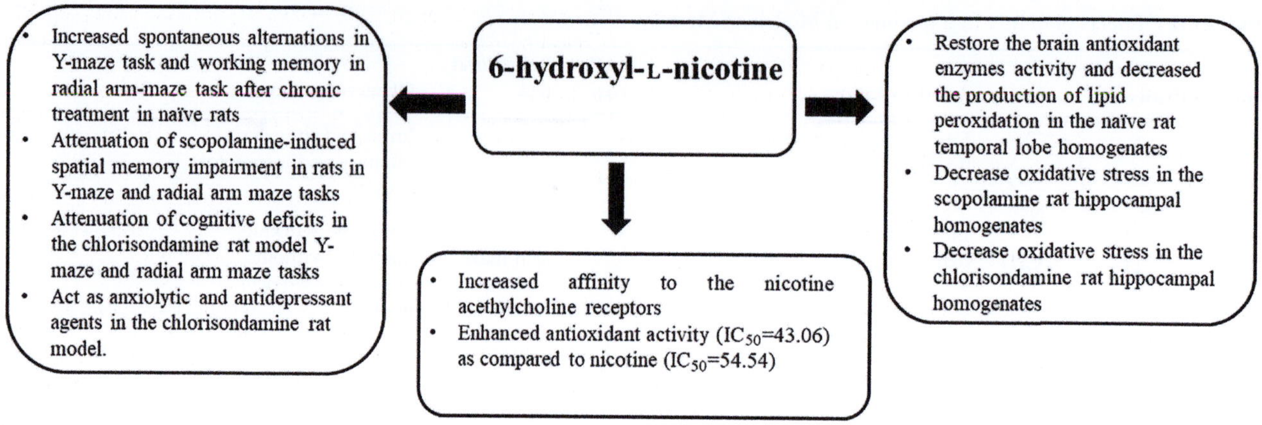

FIG. 21.4 Possible mechanism of action of 6HLN with memory-enhancing and antioxidant activity.

tasks decreased lipid peroxidation and enhanced antioxidant status in the hippocampus of the CHL-treated rats. This observed effect could suggest possible neuroprotective effects of both nicotine and 6HLN. However, the effects of 6HLN are significantly increased as compared with nicotine in the CHL-treated rats. From the observation of this study, we can suggest that 6HLN may have a beneficial role in the treatment and/or management of memory disorders such as AD and indicate the important significance of the nicotinic acetylcholine system in brain dysfunction. Additionally, we also reported (Ionita et al., 2017) that both nicotine and 6HLN act as anxiolytic and antidepressant agents in the chlorisondamine rat model. The observed effects could be mediated by the nicotinic acetylcholine receptors.

In summary, our data contribute with additional support concerning the role of nicotine and nicotine metabolites in memory processes and brain oxidative status. Moreover, 6HLN should be considered for its beneficial pharmacological properties and for its improved safety profile relative to nicotine. Finally, we summarize in Fig. 21.4 the mechanisms by which 6HLN acts as memory-enhancing and antioxidant agent.

MINI-DICTIONARY OF TERMS

Arthrobacter nicotinovorans Gram-positive actinobacteria isolated from soil that has the ability to degrade nicotine. It was previously described as *A. oxydans*.

Brain oxidative stress An imbalance between oxidants and antioxidants, characterized by decreasing ability of the brain to counteract the free-radical production.

In silico method A method based on computer programs and algorithms.

pAO1 megaplasmid A large bacterial plasmid isolated from *A. nicotinovorans* and harbors the genes responsible for bacterial degradation of nicotine.

Radial-arm-maze test A behavioral test used to evaluate working memory and reference memory in rodents.

Y-maze test A behavioral test used to evaluate spatial short-term memory in rodents.

Key Facts of Memory impairment

- Alzheimer's disease (AD) is a progressive neurodegenerative disorder mainly characterized by the loss of learning and memory.
- Loss of cholinergic neurons and a significant fall in acetylcholine level were evidenced during the AD, leading to memory impairment.
- Increasing of brain oxidative stress as a consequence of decreasing of the antioxidant enzyme activity may occur in the AD.
- The search for new therapeutic options is of high interest and growth.

Summary Points

- This chapter focuses on 6-hydroxy-L-nicotine (6HLN), which is a nicotine metabolite.
- 6HLN is produced by nicotine degradation within *A. nicotinovorans* pAO1 strain.
- In silico and in vivo behavioral approaches were used to assess the target-ligand interactions and the effects on memory processes and brain oxidative stress of 6HLN using animal models of cognitive impairment.
- 6HLN is a potent agonist of the nicotine acetylcholine receptors and poses memory-enhancing and antioxidant profile.
- However, we argue that 6HLN could represent a valuable therapeutic agent to improve memory impairment in the AD.

Acknowledgment

This work was supported by PED-PN-III-P2-2.1-PED-2016-0177.

References

Andreesen, J. R., & Fetzner, S. (2002). The molybdenum-containing hydroxylases of nicotinate, isonicotinate, and nicotine. *Metal Ions in Biological Systems*, *39*, 405–430.

Andrei, A., & Mihasan, M. (2013). Molecular gene cloning of nicotine-dehidrogenase from the pAO1 megaplasmid of *Arthrobacter nicotinovorans*. *Analele Stiintifice ale Universitatii "Alexandru Ioan Cuza" din Iasi Sec. II a. Genetica si Biologie Moleculara*, *14*, 15–19.

Aubert, I., Araujo, D. M., Cécyre, D., Robitaille, Y., Gauthier, S., & Quirion, R. (1992). Comparative alterations of nicotinic and muscarinic binding sites in Alzheimer's and Parkinson's diseases. *Journal of Neurochemistry*, *58*(2), 529–541. https://doi.org/10.1111/j.1471-4159.1992.tb09752.x.

Boiangiu, R., Guzun, D., & Mihasan, M. (2014). Time dependent accumulation of nicotine derivatives in the culture medium of *Arthrobacter nicotinovorans* pAO1. *Analele Stiintifice ale Universitatii "Alexandru Ioan Cuza" din Iasi Sec. II a. Genetica si Biologie Moleculara*, *15*, 19–25.

Boopathi, S., & Kolandaivel, P. (2014). Targeted studies on the interaction of nicotine and morin molecules with amyloid β-protein. *Journal of Molecular Modeling*, *20*(3), 1–15. https://doi.org/10.1007/s00894-014-2109-8.

Brandsch, R. (2006). Microbiology and biochemistry of nicotine degradation. *Applied Microbiology and Biotechnology*, *69*, 493–498.

Buccafusco, J. (2004). Neuronal nicotinic receptor subtypes: defining therapeutic targets. *Molecular Interventions*, *4*, 285–295.

Celie, P. H. N., van Rossum-Fikkert, S. E., van Dijk, W. J., Brejc, K., Smit, A. B., & Sixma, T. K. (2004). Nicotine and carbamylcholine binding to nicotinic acetylcholine receptors as studied in AChBP crystal structures. Neuron, 41(6), 907-914. Doi:https://doi.org/10.1016/S0896-6273(04)00115-1

Decker, K., & Bleeg, H. (1965). Induction and purification of stereospecific nicotine oxidizing enzymes from *Arthrobacter oxidans*. *Biochimica et Biophysica Acta (BBA): Enzymology and Biological Oxidation*, *105*, 313–324.

Decker, K., Dai, V., Möhler, H., & Brühmüller, M. (1972). D- and L-6-hydroxynicotine oxidase, enantiozymes of *Arthrobacter oxidans*. *Zeitschrift fur Naturforschung. Teil B. Anorganische Chemie,organische Chemie, Biochemie, Biophysik, Biologie*, *27*, 1072–1073.

Dome, P., Lazary, J., Kalapos, M. P., & Rihmer, Z. (2010). Smoking, nicotine and neuropsychiatric disorders. *Neuroscience & Biobehavioral Reviews*, *34*(3), 295–342. https://doi.org/10.1016/j.neubiorev.2009.07.013.

Echeverria Moran, V. (2013). Brain effects of nicotine and derived compounds. *Frontiers in Pharmacology*, *4*, 60. https://doi.org/10.3389/fphar.2013.00060.

Ganas, P., Igloi, G. L., & Brandsch, R. (2009). The megaplasmid pAO1 of *Arthrobacter nicotinovorans* and nicotine catabolism. In E. Schwartz (Ed.), *Microbial Megaplasmids* (pp. 271–282). Berlin, Heidelberg: Springer Berlin Heidelberg.

Gurusamy, R., & Natarajan, S. (2013). Current status on biochemistry and molecular biology of microbial degradation of nicotine. *The Scientific World Journal*, *2013*, 125385. https://doi.org/10.1155/2013/125385.

Hebert, L. E., Weuve, J., Scherr, P. A., & Evans, D. A. (2013). Alzheimer disease in the United States (2010–2050) estimated using the 2010 census. *Neurology*, *80*(19), 1778–1783. https://doi.org/10.1212/WNL.0b013e31828726f5.

Hecht, S. S., Hochalter, J. B., Villalta, P. W., & Murphy, S. E. (2000). 2′-Hydroxylation of nicotine by cytochrome P450 2A6 and human liver microsomes: Formation of a lung carcinogen precursor. *Proceedings of the National Academy of Sciences of the United States of America*, *97*(23), 12493–12497. https://doi.org/10.1073/pnas.220207697.

Hochstein, L. I., & Dalton, B. P. (1967). The purification and properties of nicotine oxidase. *Biochimica et Biophysica Acta (BBA): Enzymology*, *139* (1), 56–68. https://doi.org/10.1016/0005-2744(67)90113-1.

Hochstein, L. I., & Rittenberg, S. C. (1959). The bacterial oxidation of nicotine: I. Nicotine oxidation by cell-free preparations. *Journal of Biological Chemistry*, *234*(1), 151–155.

Hritcu, L., Ionita, R., Motei, D. E., Babii, C., Stefan, M., & Mihasan, M. (2017). Nicotine versus 6-hydroxy-L-nicotine against chlorisondamine induced memory impairment and oxidative stress in the rat hippocampus. *Biomedicine & Pharmacotherapy*, *86*, 102–108. (2017). https://doi.org/10.1016/j.biopha.2016.12.008.

Hritcu, L., Stefan, M., Brandsch, R., & Mihasan, M. (2013). 6-hydroxy-l-nicotine from *Arthrobacter nicotinovorans* sustain spatial memory formation by decreasing brain oxidative stress in rats. *Journal of Physiology and Biochemistry*, *69*(1), 25–34. https://doi.org/10.1007/s13105-012-0184-9.

Hritcu, L., Stefan, M., Brandsch, R., & Mihasan, M. (2015). Enhanced behavioral response by decreasing brain oxidative stress to 6-hydroxy-L-nicotine in Alzheimer's disease rat model. *Neuroscience Letters*, *591*, 41–47. (2015). https://doi.org/10.1016/j.neulet.2015.02.014.

Hylin, J. W. (1958). Microbial degradation of nicotine I. *Achromobacter nicotinophagumn*. sp.: morphology and physiology. *Journal of Bacteriology*, *76*(1), 36–40.

Kachalova, G. S., Bourenkov, G. P., Mengesdorf, T., Schenk, S., Maun, H. R., Burghammer, M., et al. (2010). Crystal structure analysis of free and substrate-bound 6-hydroxy-L-nicotine oxidase from *Arthrobacter nicotinovorans*. *Journal of Molecular Biology*, *396*(3), 785–799. https://doi.org/10.1016/j.jmb.2009.12.009.

Kása, P., Rakonczay, Z., & Gulya, K. (1997). The cholinergic system in Alzheimer's disease. *Progress in Neurobiology*, *52*(6), 511–535. https://doi.org/10.1016/S0301-0082(97)00028-2.

Kim, S., Thiessen, P. A., Bolton, E. E., Chen, J., Fu, G., Gindulyte, A., et al. (2016). PubChem substance and compound databases. *Nucleic Acids Research*, *44*(Database issue), D1202–D1213. https://doi.org/10.1093/nar/gkv951.

Leite, T. B., Gomes, D., Miteva, M. A., Chomilier, J., Villoutreix, B. O., & Tufféry, P. (2007). Frog: a FRee online druG 3D conformation generator. *Nucleic Acids Research*, *35*(Web Server issue), W568–W572. https://doi.org/10.1093/nar/gkm289.

Liu, Y., Wang, L., Huang, K., Wang, W., Nie, X., Jiang, Y., et al. (2014). Physiological and biochemical characterization of a novel nicotine-degrading bacterium *Pseudomonas geniculata* N1. *PLoS One*. *9*(1) e84399. https://doi.org/10.1371/journal.pone.0084399.

López-Arrieta, J., & Sanz, F. J. S. (2001). Nicotine for Alzheimer's disease. *Cochrane Database of Systematic Reviews*. *2*, https://doi.org/10.1002/14651858.CD001749.

Ma, Y., Wei, Y., Qiu, J., Wen, R., Hong, J., & Liu, W. (2014). Isolation, transposon mutagenesis, and characterization of the novel nicotine-degrading strain *Shinella* sp. HZN7. *Applied Microbiology and Biotechnology*, *98*(6), 2625–2636. https://doi.org/10.1007/s00253-013-5207-0.

Ma, Y., Wen, R., Qiu, J., Hong, J., Liu, M., & Zhang, D. (2015). Biodegradation of nicotine by a novel strain *Pusillimonas*. *Research in Microbiology*, *166*(2), 67–71. https://doi.org/10.1016/j.resmic.2014.12.009.

Mihasan, M., Capatina, L., Neagu, E., Stefan, M., & Hritcu, L. (2013). *In-silico* identification of 6-hydroxy-L-nicotine as a novel neuroprotective drug. *Romanian Biotechnological Letters*, *18*, 8333–8340.

Morris, G. M., Huey, R., Lindstrom, W., Sanner, M. F., Belew, R. K., Goodsell, D. S., et al. (2009). AutoDock4 and AutoDockTools4: automated docking with selective receptor flexibility. *Journal of Computational Chemistry*, *30*(16), 2785–2791. https://doi.org/10.1002/jcc.21256.

Mowrey, D. D., Liu, Q., Bondarenko, V., Chen, Q., Seyoum, E., Xu, Y., et al. (2013). Insights into distinct modulation of α7 and α7β2 nicotinic acetylcholine receptors by the volatile anesthetic isoflurane. *Journal of Biological Chemistry*, *288*(50), 35793–35800. https://doi.org/10.1074/jbc.M113.508333.

Murray, K. N., & Abeles, N. (2002). Nicotine's effect on neural and cognitive functioning in an aging population. *Aging & Mental Health, 6* (2), 129–138.

Newman, M. B., Arendash, G. W., Shytle, R. D., Bickford, P. C., Tighe, T., & Sanberg, P. R. (2002). Nicotine's oxidative and antioxidant properties in CNS. Life Sciences, 71(24), 2807-2820. Doi: https://doi.org/10.1016/S0024-3205(02)02135-5

Nordberg, A. (2001). Nicotinic receptor abnormalities of Alzheimer's disease: Therapeutic implications. *Biological Psychiatry, 49*, 200–210.

Nordberg, A., Hellström-Lindahl, E., Lee, M., Johnson, M., Mousavi, M., Hall, R., et al. (2002). Chronic nicotine treatment reduces β-amyloidosis in the brain of a mouse model of Alzheimer's disease (APPsw). *Journal of Neurochemistry, 81*(3), 655–658. https://doi.org/10.1046/j.1471-4159.2002.00874.x.

O'Leary, K., Parameswaran, N., McIntosh, J. M., & Quik, M. (2008). Cotinine selectively activates a subpopulation of alpha3/alpha6-beta2 nicotinic receptors in monkey striatum. *Journal of Pharmacology and Experimental Therapeutics, 325*, 646–654.

Parri, H. R., Hernandez, C. M., & Dineley, K. T. (2011). Research update: alpha7 nicotinic acetylcholine receptor mechanisms in Alzheimer's disease. *Biochemical Pharmacology, 82*(8), 931–942. (2011). https://doi.org/10.1016/j.bcp.2011.06.039.

Pogocki, D., Ruman, T., Danilczuk, M., Danilczuk, M., Celuch, M., & Wałajtys-Rode, E. (2007). Application of nicotine enantiomers, derivatives and analogues in therapy of neurodegenerative disorders. *European Journal of Pharmacology, 563*, 18–39.

Rastija, V., & Medić-Šarić, M. (2009). QSAR study of antioxidant activity of wine polyphenols. *European Journal of Medicinal Chemistry, 44*(1), 400–408. https://doi.org/10.1016/j.ejmech.2008.03.001.

Riveles, K., Huang, L. Z., & Quik, M. (2008). Cigarette smoke, nicotine and cotinine protect against 6-hydroxydopamine-induced toxicity in SH-SY5Y cells. *Neurotoxicology, 29*, 421–427.

Sabbagh, M., Lukas, R., Sparks, D., & Reid, R. (2002). The nicotinic acetylcholine receptor, smoking, and Alzheimer's disease. *Journal of Alzheimer's disease: JAD, 4*, 317–325.

Sachelaru, P., Schiltz, E., Igloi, G. L., & Brandsch, R. (2005). An α/β-fold C—C bond hydrolase is involved in a central step of nicotine catabolism by *Arthrobacter nicotinovorans*. *Journal of Bacteriology, 187*(24), 8516–8519. https://doi.org/10.1128/JB.187.24.8516-8519.2005.

Sandu, C., Chiribau, C.-B., Sachelaru, P., & Brandsch, R. (2005). Plasmids for nicotine-dependent and -independent gene expression in *Arthrobacter nicotinovorans* and other arthrobacter species. *Applied and Environmental Microbiology, 71*(12), 8920–8924. https://doi.org/10.1128/AEM.71.12.8920-8924.2005.

Sase, S., Yamamoto, H., Kawashima, E., Tan, X., & Sawa, Y. (2017). Discrimination between patients with Alzheimer disease and healthy subjects using layer analysis of cerebral blood flow and xenon solubility coefficient in xenon-enhanced computed tomography. *Journal of Computer Assisted Tomography, 41*(3), 477–483. https://doi.org/10.1097/rct.0000000000000525.

Schenk, S., & Decker, K. (1999). Horizontal gene transfer involved in the convergent evolution of the plasmid-encoded enantioselective 6-hydroxynicotine oxidases. *Jounal of Molecular Evolution, 48*, 178–186.

Taly, A., Corringer, P.-J., Guedin, D., Lestage, P., & Changeux, J.-P. (2009). Nicotinic receptors: allosteric transitions and therapeutic targets in the nervous system. *Nature Reviews Drug Discovery, 8*, 733. https://doi.org/10.1038/nrd2927.

Tang, H., Wang, L., Wang, W., Yu, H., Zhang, K., Yao, Y., et al. (2013). Systematic unraveling of the unsolved pathway of nicotine degradation in *Pseudomonas*. *PLoS Genetics. 9*(10), e1003923. https://doi.org/10.1371/journal.pgen.1003923.

Tang, H., Yao, Y., Zhang, D., Meng, X., Wang, L., Yu, H., et al. (2011). A novel NADH-dependent and FAD-containing hydroxylase is crucial for nicotine degradation by *Pseudomonas putida*. *The Journal of Biological Chemistry, 286*(45), 39179–39187. https://doi.org/10.1074/jbc.M111.283929.

Wang, S., Huang, H., Xie, K., & Xu, P. (2012). Identification of nicotine biotransformation intermediates by *Agrobacterium tumefaciens* strain S33 suggests a novel nicotine degradation pathway. *Applied Microbiology and Biotechnology, 95*(6), 1567–1578. https://doi.org/10.1007/s00253-012-4007-2.

Wang, S. N., Liu, Z., & Xu, P. (2009). Biodegradation of nicotine by a newly isolated *Agrobacterium* sp. strain S33. *Journal of Applied Microbiology, 107*(3), 838–847. https://doi.org/10.1111/j.1365-2672.2009.04259.x.

Yu, H., Tang, H., Zhu, X., Li, Y., & Xu, P. (2015). Molecular mechanism of nicotine degradation by a newly isolated strain, *Ochrobactrum* sp. strain SJY1. *Applied and Environmental Microbiology, 81*(1), 272–281. https://doi.org/10.1128/AEM.02265-14.

CHAPTER

22

Cotinine and Memory: Remembering to Forget

*Valentina Echeverria**,†, *Ross Zeitlin**

*Research & Development Service, Bay Pines VA Healthcare System, Bay Pines, FL, United States
†Fac. Cs de la Salud, Universidad San Sebastián, Lientur 1457, Concepción, Chile

Abbreviations

AD	Alzheimer's disease
ASDs	autism spectrum disorders
Aβ	amyloid-β peptide
BDNF	brain-derived neurotrophic factor
CNS	central nervous system
FXS	fragile X syndrome
GFAP	glial fibrillary acidic protein
GSK3	glycogen synthase kinase 3
IL	interleukin
KO	knockout
mGluR5	metabotropic glutamate receptor 5
mRNA	messenger ribonucleic acid
nAChRs	nicotinic acetylcholine receptors
NFAT	nuclear factor of activated T cells
NFκB	the nuclear factor kappa B
NT-3	neurotrophin
PD	Parkinson's disease
PTSD	posttraumatic stress disorder
SCHZ	schizophrenia
STAT	the signal transducers and activators of transcription factors
Tg	transgenic
VEGFR	vascular endothelial growth factor receptor

22.1 INTRODUCTION

Numerous studies have highlighted the high incidence of tobacco consumption among patients with PTSD, schizophrenia (SCHZ), and major depression disorder. It has been thought that the hope of finding relief from the undesirable symptoms associated with these conditions leads to smoking in people with these conditions (Echeverria, Grizzell, & Barreto, 2016; Moran, 2012). Cotinine, the main metabolite of nicotine, has beneficial effects over brain function under normal and pathological conditions (Grizzell & Echeverria, 2015; Moran, 2012; Zevin, Jacob, Geppetti, & Benowitz, 2000). Safety studies permitted to confirm the safety of cotinine and its differential effects from nicotine regarding addiction, toxicity, and vascular function in mammals (Benowitz, Kuyt, Jacob 3rd, Jones, & Osman, 1983; Benowitz & Sharp, 1989; Hatsukami et al., 1998; Hatsukami et al., 1998; Hatsukami, Grillo, Pentel, Oncken, & Bliss, 1997; Keenan, Hatsukami, Pentel, Thompson, & Grillo, 1994; Kuo, Dryjski, & Bjornsson, 1989). The positive effects of cotinine in promoting higher cognitive functions is broadly documented in animal models of different pathological conditions and aging (Buccafusco & Terry Jr., 2003; Echeverria, Grizzell, & Barreto, 2016; Echeverria & Zeitlin, 2012; Grizzell & Echeverria, 2015) (Fig. 22.1). A previous study, more than two decades ago, revealed that cotinine improved performance accuracy on delayed matching-to-sample task by macaque monkeys (Buccafusco & Terry Jr., 2003). However, likely due to the availability of more effective agonists of the nicotinic acetylcholine receptors (nAChRs), its development as a therapeutic drug was discouraged. Considering that smoking was associated with lower incidence of Parkinson's disease (PD) and Alzheimer's disease (AD) and other psychiatric disorders affecting memory such as PTSD and major depression, the effect of cotinine on memory abilities has been increasingly investigated (Table 22.1).

22.2 COTININE AMELIORATES MEMORY IN AD MICE

Our memories define us, and its malleability affects any aspect of our lives. As stated by Gabriel García Marquez, "Life is not what one lived, but what one remembers and how one remembers it in order to recount it" (García Márquez, 2003).

The first evidence of cotinine effects against AD pathology came from toxicity studies trying to prove that it would exacerbate the neurotoxic effects of amyloid-β

Neuroscience of Nicotine
https://doi.org/10.1016/B978-0-12-813035-3.00022-8

TABLE 22.1 Preclinical studies showing beneficial effects of cotinine on memory

Animal Model/ Condition	Year	Signaling Factors Involved	References
Monkey/aging	2005	–	Terry AV Jr. et al.
Monkey/SCHZ	2009	–	Buccafusco JJ et al.
Tg6799 mice/AD	2011	Akt, GSK3	Echeverria V et al.
Male Wistar rats/SCHZ	2012	–	Terry AV Jr. et al.
Mice/PTSD	2014	ERK	Zeitlin R et al.
Mice/depression	2014	GSK3, synaptophysin	Grizzell JA et al.
Mice/depression	2014	VEGF, Erb	Grizzell JA et al.
Male young rats/PTSD	2014	α7, α4β2 nAChRs, ERK	de Aguiar RB
DBA/2 mice/sensory gating	2014	α7, α4β2 nAChRs	Wildeboer-Andrud KM et al.
Tg6799 mice/AD	2014	PSD95	Patel et al.
Male Sprague-Dawley rats/donepezil treatment	2015	α7, α4β2 nAChRs	Terry AV Jr. et al.
Wistar female rat/ chemotherapy	2016	–	Iarkov A et al.
FXM (–/–) Mice/autism	2017	Akt, GSK3	Pardo M et al.
Tg6799 mice/AD	2017	Tau, CREB	Grizzell JA et al.
C57BL/6 Mice/ depression	2017	GFAP	Perez-Urrutia et al.

peptide (Aβ). Unexpectedly, these studies revealed that cotinine diminished the toxicity of Aβ by inhibiting its aggregation (Burgess, Zeitlin, Gamble-George, & Echeverria Moran, 2008; Echeverria et al., 2011). Similar results were reported by other group later (Gao, La, Chapman, Bertrand, & Terry, 2012) (Fig. 22.2). These studies were followed by investigations using the transgenic (Tg) 6799 AD mice that revealed that cotinine was effective in diminishing Aβ aggregation and tau hyperphosphorylation, at various stages of the disease. These effects were accompanied by an improvement in working memory as tested in several memory tests including the circular platform, radial arm water maze, interference, and novel object recognition tests (Echeverria et al., 2011; Echeverria & Zeitlin, 2012; Grizzell, Patel, Barreto, & Echeverria, 2017). Cotinine not only helped with memory but also diminished depressive-like behavior in the Tg6799 mice (Patel et al., 2014).

22.3 COTININE PREVENTS MEMORY LOSS IN MOUSE MODELS OF POSTTRAUMATIC STRESS DISORDER

PTSD is induced by experiencing unescapable terror because of suffering or witnessing violent situations such as a car accident, war combat, natural disasters, rape, or murder (Young, 2017). Actual medications for PTSD have negative side effects, and their positive effects are restricted to a portion of patients (Thomas & Stein, 2017). Furthermore, they usually have short-term effectivity, with a high percentage of patients having a reinstatement of the disease. Recent clinical studies have reported encouraging results with prazosin to alleviate nightmares in PTSD patients (Breen, Blankley, & Fine, 2017; Singh, Hughes, Mehta, Erwin, & Parsaik, 2016).

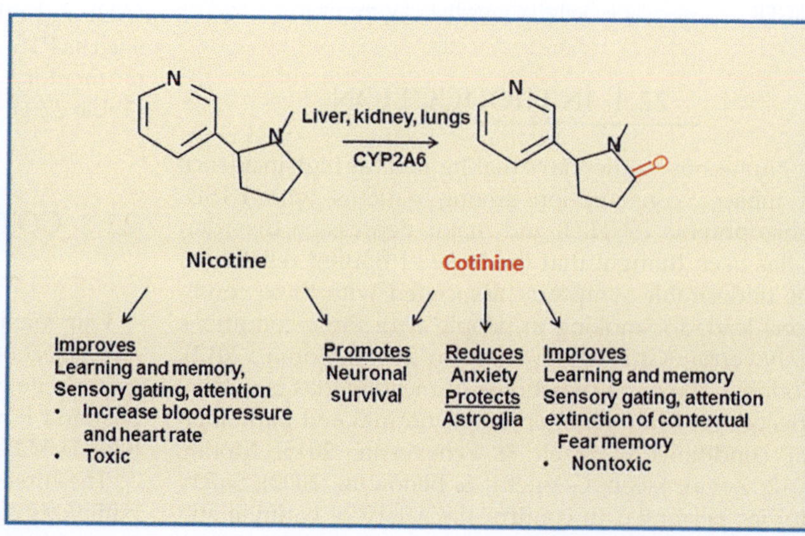

FIG. 22.1 Cotinine has distinctive properties to nicotine.
Cotinine the main derivative of nicotine is metabolized by the cytochrome (cyt) P450. This transformation provides cotinine new beneficial effects.

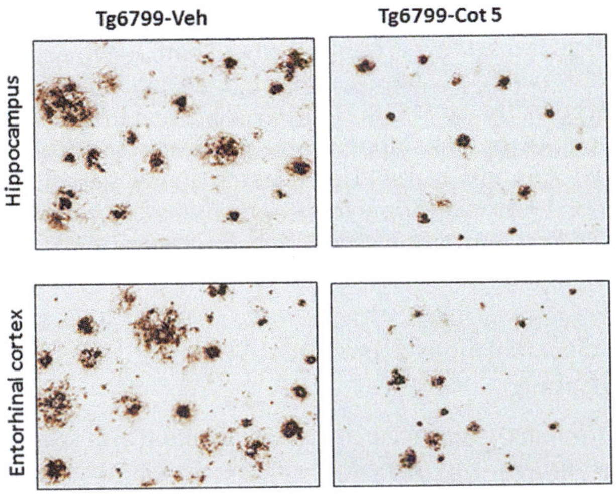

FIG. 22.2 Cotinine inhibits amyloid burden in the hippocampus and cortex of the Tg6799 AD mice.
The figure depicts the immunohistochemical analysis of Aβ plaques in the brain of transgenic (Tg) 6799 mice treated with vehicle (Veh) or cotinine 5 mg/kg (Cot 5). The differences were significant between groups with a significant decrease in the cotinine-treated Tg mice. *Taken from Patel et al. (2014).*

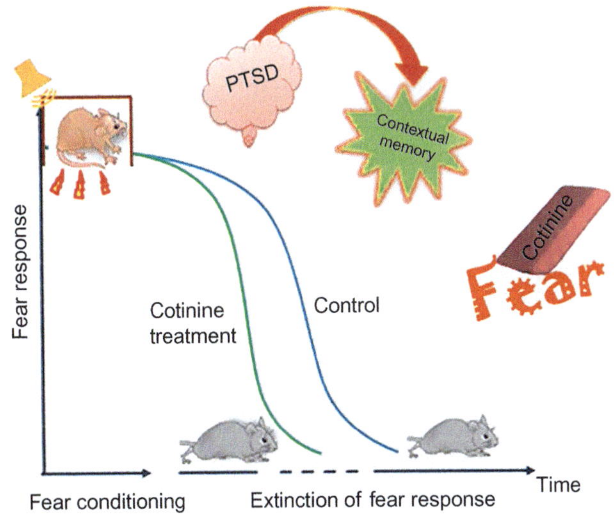

FIG. 22.3 Cotinine enhances the extinction of contextual fear memory.
Diagram representing the facilitation of fear extinction in mice. Note that nicotine does not help promoting contextual fear extinction.

The search of new therapies, with less side effects and inducing a permanent extinction of fear, has currently been investigated.

The high incidence of tobacco consumption among PTSD patients (Kelly, Jensen, & Sofuoglu, 2015) and its effects on memory motivated the investigation of cotinine on PTSD. Cotinine showed antidepressant, anxiolytic, and memory-enhancing effects in the animal models of PTSD induced by forced swim, restraint stress, and fear conditioning. Also, cotinine enhanced contextual fear extinction in conditioned rodents (de Aguiar, Parfitt, Jaboinski, & Barros, 2013; Echeverria, Barreto, Avila-Rodriguez, Tarasov, & Aliev, 2017; Grizzell, Iarkov, Holmes, Mori, & Echeverria, 2014; Patel et al., 2014; Perez-Urrutia et al., 2017; Zeitlin et al., 2012) (Fig. 22.3).

22.4 COTININE RESTORE ASTROCYTES NUMBER AND FUNCTION AFFECTED BY CHRONIC STRESS

Recent studies have indicated that astroglia is severely affected by fear conditioning and restraint stress paradigms in rodents (Perez-Urrutia et al., 2017; Saur et al., 2016). Astrocytes are considered versatile caretakers in the central nervous system (CNS) (Moraga-Amaro, Jerez-Baraona, Simon, & Stehberg, 2014). Astrocytes subdivide the gray matter of the CNS into domains encompassing blood vessels, neurons, and synapses. They are indispensable for proper brain development; modulate neuronal signaling; and play central roles maintaining ions, neurotransmitters, water, and energy homeostasis, as well as can exert the function of neural progenitor cells. In rats, fear conditioning decreased the density of GFAP+ astrocytes and changed astrocyte morphology, diminishing their total number of primary processes, and arborization complexity in the CA1 region of the hippocampus (Saur et al., 2016). Stress decreases the expression of the astroglial marker GFAP (Czeh, Simon, Schmelting, Hiemke, & Fuchs, 2006; Gomez-Galan, De Bundel, Van Eeckhaut, Smolders, & Lindskog, 2013; Perez-Urrutia et al., 2017). Also, changes in GFAP expression has been found in the blood-brain barrier of animals subjected to restraint stress (Santha et al., 2015). Recent studies have shown that cotinine increases synaptic density in the hippocampus of restrained mice (Grizzell & Echeverria, 2015; Grizzell, Iarkov, et al., 2014). Amazingly, posttreatment with intranasal cotinine induced a restoration of astrocytes in the hippocampus and frontal cortex of restrained mice (Perez-Urrutia et al., 2017). This is a relevant finding since both immobilization stress and fear conditioning in rodents induce cognitive impairment and a deficit in astrocytes in brain regions that are involved in memory. Therefore, memory impairment, depression, and astrocyte deficit seem to be connected as astrocytes play a significant role in supporting neuronal plasticity, energy balance, and brain recovery from injury. These functions are exerted by the release of energy compounds and neurotrophic factors. Based on this evidence, it is reasonable to say that restoration of astrocyte function is a cellular mechanism of action of cotinine in the brain.

22.5 SIGNALING PATHWAYS ACTIVATED BY COTININE IN RODENT MODELS OF DEPRESSION AND PTSD

22.5.1 Cotinine Has Antiinflammatory Effects by Modulating the nAChRs

Neuroinflammation involves a chain of events affecting neurons, endothelial cells, and glial cells (Echeverria, Grizzell, & Barreto, 2016; Echeverria, Yarkov, & Aliev, 2016; Mendoza, Barreto, Avila-Rodriguez, & Echeverria, 2016; Wager-Smith & Markou, 2011) that occur in PTSD, AD, and depression. As discussed, glial cells have beneficial effects; however, their chronic activation impairs brain function by inducing neuroinflammation and cell death (Cabezas et al., 2017).

Biochemical studies of downstream effectors activated by cotinine and pilot electrophysiological studies using *Xenopus laevis* oocytes expressing the human α7 nAChR (communicated by Dr. Zeitlin, September 28, 2009) permitted to hypothesize that cotinine was a PAM of the α7 nAChR (Echeverria & Zeitlin, 2012) (Fig. 22.4). Later, cotinine's pharmacodynamics was confirmed and showed that cotinine, as expected for a PAM, potentiated the effect of the acetylcholinesterase inhibitors such as donepezil in rats (Terry Jr., Callahan, & Bertrand, 2015). Since α7 nAChR has anti-inflammatory effects and inhibits the release of inflammatory factors from immune cells (Han et al., 2014), cotinine by activating this receptor may have similar effects in the brain.

On the other hand, it has been shown that the facilitation of fear extinction by cotinine in rats requires the activation of α7 and α4β2 nAChRs (de Aguiar et al., 2013). This dependence can rely on the fact that the cholinergic system impacts synaptic plasticity in both the prefrontal cortex and hippocampus, enhancing the inhibition of the amygdala by the PFC and the consolidation of inhibitory memories by the hippocampus (Liberzon & Sripada, 2008). An additional study in mice also found that cotinine affected sensory gating abilities by modulating the nAChRs (Wildeboer-Andrud, Zheng, Choo, & Stevens, 2014).

22.5.2 Cotinine Stimulates Akt and ERK Signaling

Cotinine's beneficial effects on cognition also seem to depend on the enhancement of VEGF expression (Echeverria et al., 2017) and the stimulation of the downstream effector the protein kinase B (Akt) (Echeverria, Grizzell, & Barreto, 2016). This kinase regulates protein synthesis, neurogenesis, neural survival, and neuronal plasticity in the brain and is downregulated in the brain by many psychiatric and neurological disorders. Akt is activated by the nAChRs, the tyrosine receptor kinase B, the brain-derived neurotrophic factor (BDNF), neurotrophin(NT)-3 growth factor receptors (Wang et al., 2017), and the vascular endothelial growth factor receptor (VEGFR) that promote brain plasticity, neuronal survival neurogenesis, and angiogenesis (Fournier, Lee, Banasr, Elsayed, & Duman, 2012; Wu et al., 2008). Since cotinine increases the expression of VEGF, it may stimulate the VEGFRs in the brain (Conklin, Zhao, Zhong, & Chen, 2002; Echeverria et al., 2017; Grizzell et al., 2014). In line with this discovery, cotinine stimulated the Akt-GSK3 pathway and promoted neuronal survival and

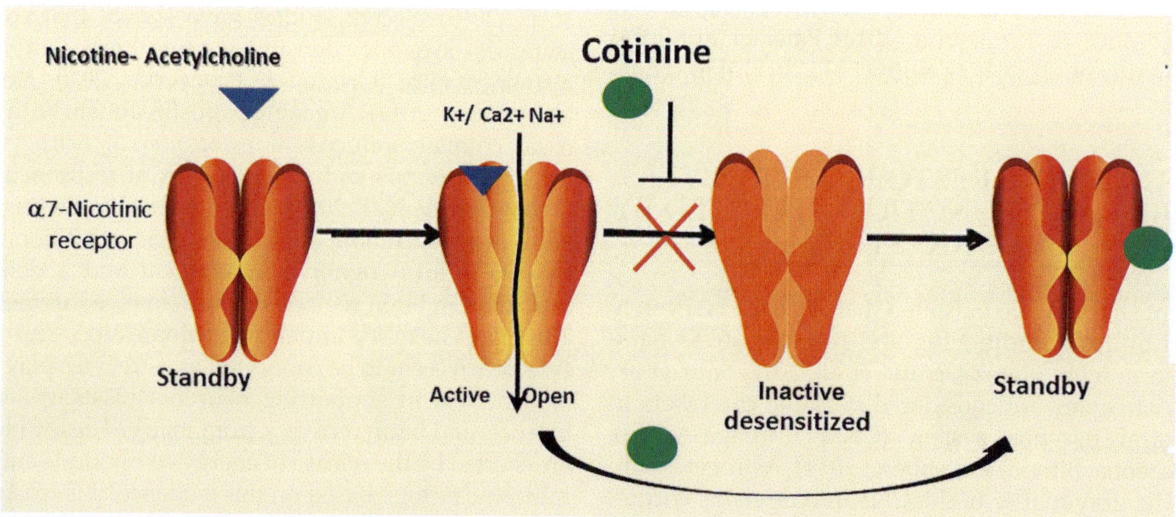

FIG. 22.4 Cotinine positive modulates the α7 nAChRs.
Schematic representation of the positive allosteric modulation of the α7 nicotinic receptor by cotinine that would increase the number of receptors available in the membrane to be activated by acetylcholine (ACh).

the expression of synaptophysin, in the brain of mice (Grizzell, Iarkov, et al., 2014). Furthermore, Akt activation by cotinine was accompanied by an increase in postsynaptic density protein 95 (PSD95) levels in the brain of Tg6799 mice. Also, cotinine inhibited GSK3β by phosphorylation at serine 9. The inhibition of GSK3 by Akt can explain the decrease of apoptosis and tau phosphorylation. GSK3 is also a key modulator of cytokine expression and T-cell proliferation, differentiation, and survival by controlling the expression of the nuclear factor kappa B (NFκβ), nuclear factor of activated T cells (NFAT), and the signal transducers and activators of transcription (STAT) factors (Fig. 22.5). Neuroinflammation is involved in the progression of several psychiatric diseases including AD. For this reason, GSK3 inhibitors such as cotinine and lithium can reduce neuroinflammation and reduce tau pathology and neuronal cell death (Gates & Dunnett, 2001). The added effects of cotinine inhibiting GSK3 activation, tau phosphorylation, and Aβ aggregation merit its clinical investigation as a therapy against AD.

The activation of another protein kinases ERK1/2 in the hippocampus plays a critical role in promoting contextual fear extinction and neuroplasticity (Fischer et al., 2007; Guedea et al., 2011; Herry, Trifilieff, Micheau, Luthi, & Mons, 2006; Matsuda et al., 2010). In agreement with this role, cotinine has shown to increase ERK activation by phosphorylation while promoting fear extinction in rodents (de Aguiar et al., 2013; Zeitlin et al., 2012) (Fig. 22.5).

22.6 COTININE RESTORE WORKING MEMORY ABILITIES AFTER CHEMOTHERAPY IN FEMALE RATS

Chemotherapy has profound effects on cognitive abilities including attention deficits, diminished working memory, and slower thinking, a syndrome referred as "chemo brain"(Simo, Rifa-Ros, Rodriguez-Fornells, & Bruna, 2013). Using a rat model of chemotherapy-induced cognitive impairment, it has been shown that cotinine enhanced reference memory and decreased depressive-like behavior in rats treated with chemotherapy (Iarkov, Appunn, & Echeverria, 2016). This study suggests that cotinine may help cancer patients to recover from the "chemo brain."

22.7 COTININE RESTORES MEMORY IN ANIMAL MODELS OF AUTISM SPECTRUM DISORDER AND SCHIZOPHRENIA

Autism spectrum disorders (ASDs) affect millions of individuals worldwide ("Prevalence of autism spectrum disorders—Autism and Developmental Disabilities Monitoring Network, 14 sites, United States, 2008, 2012). ASDs involve erroneous synaptic development and function (Spooren, Lindemann, Ghosh, & Santarelli, 2012). Fragile X syndrome (FXS) is the leading heritable cause of intellectual disability and the most prevalent known cause of autism (Garber,

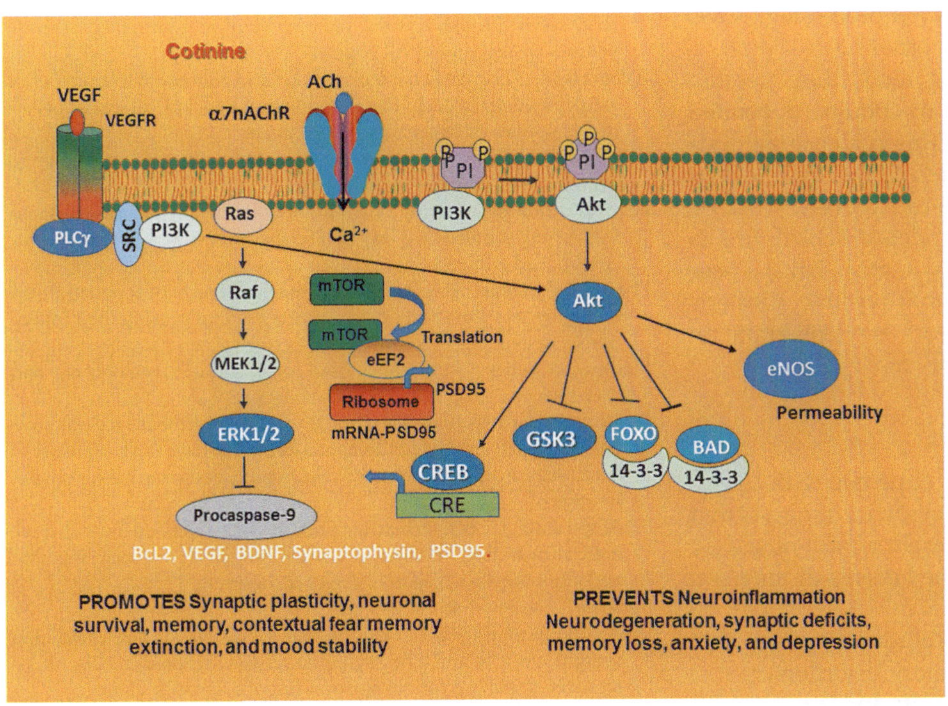

FIG. 22.5 Cotinine stimulate signaling pathways that preserve healthy brain functions.
Cotinine by activating the nAChR and VEGF signaling may stimulate the expression of survival factors such as Akt and inhibits apoptotic factors such as GSK3β that promotes tau hyperphosphorylation and neuronal cell death and the expression of synaptic proteins.

Visootsak, & Warren, 2008). FXS results from a loss of fragile X mental retardation protein (FMRP) expression. Children with FXS exhibit cognitive and behavioral deficits including social withdrawal, attention deficits, hyperactivity, and stereotyped behaviors (Chudley et al., 1983). The *FMR1* knockout (KO) mouse (FXS mice) is one of the best characterized animal models for ASD (Moy, Nadler, Magnuson, & Crawley, 2006). These mice exhibit autistic-like core symptoms including increased errors in learning paradigms. FXS mice show reduced synaptic plasticity and spine density, as well as a hyperactivation of the metabotropic glutamate receptor 5 (mGluR5) in the brain. The mGluR5 stimulates GSK3β signaling. The FMRP-mediated GSK3β dysregulation in FXS mice is considered a key event in ASD pathology, because hyperactivation of GSK3β leads to neuroinflammation and negatively affects neurogenesis, cognitive abilities (King et al., 2013), and brain development (Martel et al., 2013). Thus, GSK3β signaling may be a novel therapeutic target for disease-modifying interventions for AD, FXS, and related ASDs (Hooper, Killick, & Lovestone, 2008). Previous studies have shown that stimulation of the cholinergic system with nicotine had positive effects on autism symptoms such as repetitive behavior and social withdraw. Interestingly, we previously found that cotinine inhibited GSK3β in the brain of mice. Using the $Fmr1^{-/-}$ knockout and control littermate mice, it has been shown that cotinine increased the inhibitory serine phosphorylation of GSK3β by Akt, in the hippocampus of the $Fmr1^{-/-}$ mice (Portis, Giunta, Obregon, & Tan, 2012). Treatment with cotinine improved cognitive abilities in spatial processing, novel object recognition, and temporal ordering in $Fmr1^{-/-}$ mice. Furthermore, using genetic tools, they demonstrated that the beneficial effects of cotinine depended on the inhibition of GSK3 by Akt. This evidence suggests that cotinine can be a promising therapy for ASD. Several studies have shown that cotinine can prevent working memory loss in rat and monkey models of schizophrenia (Buccafusco & Terry Jr., 2009; Terry Jr. et al., 2012). These studies are in line with the idea that positive allosteric modulators of the nicotinic receptors may constitute an effective therapy for memory impairment in SCHZ.

Key Facts

- Tobacco consumption is higher in people with psychiatric disorders, and new evidence suggests that tobacco consumption may represent a form of self-medication directed to obtain the benefits of the main derivative of nicotine, cotinine.
- Current evidence suggests that cotinine may represent a powerful therapy against memory disorders characterized by cholinergic deficits and/or neuroinflammation.
- Cotinine has antiinflammatory, mnemonic, antidepressant, and anxiolytic effects in mammals under normal and pathological conditions.
- Numerous clinical studies have shown the safety and nonaddictive properties of cotinine in humans at doses equivalent to the ones with therapeutic effects in mammals.
- Clinical studies are required to confirm the effectiveness of cotinine to improve memory in humans.

Summary Points

- Cotinine prevents cognitive impairment by a mechanism involving the activation of ERK and the Akt/GSK3 pathway in the frontal cortex and hippocampus.
- Cotinine stimulates synaptogenesis as reflected by the stimulation of synaptophysin and PSD95 in the hippocampus and frontal cortex of PTSD and AD mice.
- Cotinine stimulates the expression of the VEGF factor in stressed mice.
- Cotinine enhances contextual fear memory extinction by a mechanism involving the activation of the nAChR signaling and its downstream effector ERK1/2.
- Cotinine recovers astrocyte number and morphology affected by chronic stress.

MINI-DICTIONARY OF TERMS

Alzheimer's disease Progressive neurodegenerative disease and main cause of dementia in the elderly characterized by synaptic loss, neuronal cell death, astrocytosis, tau pathology, and formation of senile plaques of aggregated Aβ peptide.

Astrocytosis Change in the morphology and increase in the number of astrocytes that may lead to neuroinflammation and the release of oxidative species of oxygen and nitric oxide.

Extinction Diminution of fear response after repetitive exposure to cues related to the traumatic event.

Fear conditioning Pairing of environmental cues with a noxious and frightening stimulus in a way that further exposure to the cues elicit a conditioned fear response.

Pharmacodynamic It is the study of the mechanism of action of drugs including receptors and other proteins binding causing their inhibition or activation.

Pharmacokinetics Absorbance and distribution, metabolism, and excretion of a drug in the body.

Posttraumatic stress disorder (PTSD) It is a conditneurodegenerative disease and main cause of dementia ion that arouse from exposure to a traumatic event that involved unescapable danger to life for oneself or others.

Schizophrenia It is psychiatric condition characterized by cognitive deficits including impaired prepulse inhibition, working memory, and symptoms such as delusions and delirium.

Synapse Neuronal structures that represent the neurochemical connection among neurons and involve a presynaptic site containing synaptic vesicles and a postsynaptic site having ion channels, receptors, and signaling factors surrounding the synaptic cleft.

Synaptogenesis It refers to the appearance of new synapses in neurons.

References

Benowitz, N. L., Kuyt, F., Jacob, P., 3rd, Jones, R. T., & Osman, A. L. (1983). Cotinine disposition and effects. *Clinical Pharmacology and Therapeutics*, *34*(5), 604–611.

Benowitz, N. L., & Sharp, D. S. (1989). Inverse relation between serum cotinine concentration and blood pressure in cigarette smokers. *Circulation*, *80*(5), 1309–1312.

Breen, A., Blankley, K., & Fine, J. (2017). The efficacy of prazosin for the treatment of posttraumatic stress disorder nightmares in U.S. military veterans. *Journal of the American Association of Nurse Practitioners*, *29*(2), 65–69. https://doi.org/10.1002/2327-6924.12432.

Buccafusco, J. J., & Terry, A. V., Jr. (2003). The potential role of cotinine in the cognitive and neuroprotective actions of nicotine. *Life Sciences*, *72* (26), 2931–2942.

Buccafusco, J. J., & Terry, A. V., Jr. (2009). A reversible model of the cognitive impairment associated with schizophrenia in monkeys: potential therapeutic effects of two nicotinic acetylcholine receptor agonists. *Biochemical Pharmacology*, *78*(7), 852–862.

Burgess, S., Zeitlin, R. S., Gamble-George, J., & Echeverria Moran, V. (2008). P2-319: Cotinine is neuroprotective against beta-amyloid toxicity. *Alzheimer's and Dementia*, *4*(4). https://doi.org/10.1016/j.jalz.2008.05.1396.

Cabezas, R., Vega-Vela, N. E., Gonzalez-Sanmiguel, J., Gonzalez, J., Esquinas, P., Echeverria, V., et al. (2017). PDGF-BB preserves mitochondrial morphology, attenuates ROS production, and upregulates neuroglobin in an astrocytic model under rotenone insult. *Molecular Neurobiology*. https://doi.org/10.1007/s12035-017-0567-6.

Chudley, A. E., Knoll, J., Gerrard, J. W., Shepel, L., McGahey, E., & Anderson, J. (1983). Fragile (X) X-linked mental retardation I: relationship between age and intelligence and the frequency of expression of fragil (X)(q28). *American Journal of Medical Genetics*, *14*(4), 699–712. https://doi.org/10.1002/ajmg.1320140412.

Conklin, B. S., Zhao, W., Zhong, D. S., & Chen, C. (2002). Nicotine and cotinine up-regulate vascular endothelial growth factor expression in endothelial cells. *The American Journal of Pathology*, *160*(2), 413–418. https://doi.org/10.1016/S0002-9440(10)64859-6.

Czeh, B., Simon, M., Schmelting, B., Hiemke, C., & Fuchs, E. (2006). Astroglial plasticity in the hippocampus is affected by chronic psychosocial stress and concomitant fluoxetine treatment. *Neuropsychopharmacology*, *31*(8), 1616–1626. https://doi.org/10.1038/sj.npp.1300982.

de Aguiar, R. B., Parfitt, G. M., Jaboinski, J., & Barros, D. M. (2013). Neuroactive effects of cotinine on the hippocampus: behavioral and biochemical parameters. *Neuropharmacology*, *71*, 292–298.

Echeverria, V., Barreto, G. E., Avila-Rodriguez, M., Tarasov, V. V., & Aliev, G. (2017). Is VEGF a key target of cotinine and other potential therapies against Alzheimer disease? *Current Alzheimer Research*. https://doi.org/10.2174/1567205014666170329113007.

Echeverria, V., Grizzell, J. A., & Barreto, G. E. (2016). Neuroinflammation: a therapeutic target of cotinine for the treatment of psychiatric disorders? *Current Pharmaceutical Design*, *22*(10), 1324–1333.

Echeverria, V., Yarkov, A., & Aliev, G. (2016). Positive modulators of the alpha7 nicotinic receptor against neuroinflammation and cognitive impairment in Alzheimer's disease. *Progress in Neurobiology*, *144*, 142–157. https://doi.org/10.1016/j.pneurobio.2016.01.002.

Echeverria, V., & Zeitlin, R. (2012). Cotinine: a potential new therapeutic agent against Alzheimer's disease. *CNS Neuroscience & Therapeutics*, *18*(7), 517–523. https://doi.org/10.1111/j.1755-5949.2012.00317.x.

Echeverria, V., Zeitlin, R., Burgess, S., Patel, S., Barman, A., Thakur, G., et al. (2011). Cotinine reduces amyloid-beta aggregation and improves memory in Alzheimer's disease mice. *Journal of Alzheimer's Disease*, *24*(4), 817–835. https://doi.org/10.3233/JAD-2011-102136.

Fischer, A., Radulovic, M., Schrick, C., Sananbenesi, F., Godovac-Zimmermann, J., & Radulovic, J. (2007). Hippocampal Mek/Erk signaling mediates extinction of contextual freezing behavior. *Neurobiology of Learning and Memory*, *87*(1), 149–158.

Fournier, N. M., Lee, B., Banasr, M., Elsayed, M., & Duman, R. S. (2012). Vascular endothelial growth factor regulates adult hippocampal cell proliferation through MEK/ERK- and PI3K/Akt-dependent signaling. *Neuropharmacology*, *63*(4), 642–652. https://doi.org/10.1016/j.neuropharm.2012.04.033.

J. Gao, B. La, J. M. Chapman, D. Bertrand, A. V. Terry; (2012). Neuroprotective effects of the nicotine metabolite, cotinine, and several structural analogs of cotinine paper presented at the Society for Neuroscience, New Orleans, LA.

Garber, K. B., Visootsak, J., & Warren, S. T. (2008). Fragile X syndrome. *European Journal of Human Genetics*, *16*(6), 666–672.

Márquez G. G., *Living to Tell the Tale* (2003) (Edith Grossman, translator), New York, 2004, Vintage International, pp 1–493. ISBN 1–4000–3454.

Gates, M. A., & Dunnett, S. B. (2001). The influence of astrocytes on the development, regeneration and reconstruction of the nigrostriatal dopamine system. *Restorative Neurology and Neuroscience*, *19*(1–2), 67–83.

Gomez-Galan, M., De Bundel, D., Van Eeckhaut, A., Smolders, I., & Lindskog, M. (2013). Dysfunctional astrocytic regulation of glutamate transmission in a rat model of depression. *Molecular Psychiatry*, *18*(5), 582–594.

Grizzell, J. A., & Echeverria, V. (2015). New insights into the mechanisms of action of cotinine and its distinctive effects from nicotine. *Neurochemical Research*, *40*(10), 2032–2046. https://doi.org/10.1007/s11064-014-1359-2.

Grizzell, J. A., Iarkov, A., Holmes, R., Mori, T., & Echeverria, V. (2014). Cotinine reduces depressive-like behavior, working memory deficits, and synaptic loss associated with chronic stress in mice. *Behavioural Brain Research*, *268*, 55–65. https://doi.org/10.1016/j.bbr.2014.03.047.

Grizzell, J. A., Mullins, M., Iarkov, A., Rohani, A., Charry, L. C., & Echeverria, V. (2014). Cotinine reduces depressive-like behavior and hippocampal vascular endothelial growth factor downregulation after forced swim stress in mice. *Behavioral Neuroscience*, *128* (6), 713–721. https://doi.org/10.1037/bne0000021.

Grizzell, J. A., Patel, S., Barreto, G. E., & Echeverria, V. (2017). Cotinine improves visual recognition memory and decreases cortical Tau phosphorylation in the Tg6799 mice. *Progress in Neuro-Psychopharmacology & Biological Psychiatry*, *78*, 75–81. https://doi.org/10.1016/j.pnpbp.2017.05.010.

Guedea, A. L., Schrick, C., Guzman, Y. F., Leaderbrand, K., Jovasevic, V., Corcoran, K. A., et al. (2011). ERK-associated changes of AP-1 proteins during fear extinction. *Molecular and Cellular Neurosciences*, *47*(2), 137–144. https://doi.org/10.1016/j.mcn.2011.03.009.

Han, Z., Li, L., Wang, L., Degos, V., Maze, M., & Su, H. (2014). Alpha-7 nicotinic acetylcholine receptor agonist treatment reduces neuroinflammation, oxidative stress, and brain injury in mice with ischemic stroke and bone fracture. *Journal of Neurochemistry*, *131*(4), 498–508.

Hatsukami, D., Lexau, B., Nelson, D., Pentel, P. R., Sofuoglu, M., & Goldman, A. (1998). Effects of cotinine on cigarette self-administration. *Psychopharmacology*, *138*(2), 184–189.

Hatsukami, D., Pentel, P. R., Jensen, J., Nelson, D., Allen, S. S., Goldman, A., et al. (1998). Cotinine: effects with and without nicotine. *Psychopharmacology*, *135*(2), 141–150.

Hatsukami, D. K., Grillo, M., Pentel, P. R., Oncken, C., & Bliss, R. (1997). Safety of cotinine in humans: physiologic, subjective, and cognitive effects. *Pharmacology, Biochemistry, and Behavior*, *57*(4), 643–650.

Herry, C., Trifilieff, P., Micheau, J., Luthi, A., & Mons, N. (2006). Extinction of auditory fear conditioning requires MAPK/ERK activation in the basolateral amygdala. *The European Journal of Neuroscience*, *24*(1), 261–269.

Hooper, C., Killick, R., & Lovestone, S. (2008). The GSK3 hypothesis of Alzheimer's disease. *Journal of Neurochemistry*, *104*(6), 1433–1439.

Iarkov, A., Appunn, D., & Echeverria, V. (2016). Post-treatment with cotinine improved memory and decreased depressive-like behavior after chemotherapy in rats. *Cancer Chemotherapy and Pharmacology*, *78*(5), 1033–1039. https://doi.org/10.1007/s00280-016-3161-0.

Keenan, R. M., Hatsukami, D. K., Pentel, P. R., Thompson, T. N., & Grillo, M. A. (1994). Pharmacodynamic effects of cotinine in abstinent cigarette smokers. *Clinical Pharmacology and Therapeutics, 55* (5), 581–590.

Kelly, M. M., Jensen, K. P., & Sofuoglu, M. (2015). Co-occurring tobacco use and posttraumatic stress disorder: Smoking cessation treatment implications. *The American Journal on Addictions, 24*(8), 695–704. https://doi.org/10.1111/ajad.12304.

King, M. K., Pardo, M., Cheng, Y., Downey, K., Jope, R. S., & Beurel, E. (2013). Glycogen synthase kinase-3 inhibitors: rescuers of cognitive impairments. *Pharmacology & Therapeutics*. https://doi.org/10.1016/j.pharmthera.2013.07.010.

Kuo, B. S., Dryjski, M., & Bjornsson, T. D. (1989). Influence of nicotine and cotinine on the expression of plasminogen activator activity in bovine aortic endothelial cells. *Thrombosis and Haemostasis, 61*(1), 70–76.

Liberzon, I., & Sripada, C. S. (2008). The functional neuroanatomy of PTSD: a critical review. *Progress in Brain Research, 167*, 151–169. https://doi.org/10.1016/S0079-6123(07)67011-3.

Martel, C., Allouche, M., Esposti, D. D., Fanelli, E., Boursier, C., Henry, C., et al. (2013). Glycogen synthase kinase 3-mediated voltage-dependent anion channel phosphorylation controls outer mitochondrial membrane permeability during lipid accumulation. *Hepatology, 57*(1), 93–102. https://doi.org/10.1002/hep.25967.

Matsuda, S., Matsuzawa, D., Nakazawa, K., Sutoh, C., Ohtsuka, H., Ishii, D., et al. (2010). d-serine enhances extinction of auditory cued fear conditioning via ERK1/2 phosphorylation in mice. *Progress in Neuro-Psychopharmacology & Biological Psychiatry, 34*(6), 895–902.

Mendoza, C., Barreto, G. E., Avila-Rodriguez, M., & Echeverria, V. (2016).Role of neuroinflammation and sex hormones in war-related PTSD. *Molecular and Cellular Endocrinology, 434*, 266–277. https://doi.org/10.1016/j.mce.2016.05.016.

Moraga-Amaro, R., Jerez-Baraona, J. M., Simon, F., & Stehberg, J. (2014). Role of astrocytes in memory and psychiatric disorders. *Journal of Physiology, Paris, 108*(4–6), 240–251.

Moran, V. E. (2012). Cotinine: beyond that expected, more than a biomarker of tobacco consumption. *Frontiers in Pharmacology, 3*, 173.

Moy, S. S., Nadler, J. J., Magnuson, T. R., & Crawley, J. N. (2006). Mouse models of autism spectrum disorders: the challenge for behavioral genetics. *American Journal of Medical Genetics Part C, Seminars in Medical Genetics, 142C*(1), 40–51.

Patel, S., Grizzell, J. A., Holmes, R., Zeitlin, R., Solomon, R., Sutton, T. L., et al. (2014). Cotinine halts the advance of Alzheimer's disease-like pathology and associated depressive-like behavior in Tg6799 mice. *Frontiers in Aging Neuroscience, 6*, 162. https://doi.org/10.3389/fnagi.2014.00162.

Perez-Urrutia, N., Mendoza, C., Alvarez-Ricartes, N., Oliveros-Matus, P., Echeverria, F., Grizzell, J. A., et al. (2017). Intranasal cotinine improves memory, and reduces depressive-like behavior, and GFAP+ cells loss induced by restraint stress in mice. *Experimental Neurology, 295*, 211–221. https://doi.org/10.1016/j.expneurol.2017.06.016.

Portis, S., Giunta, B., Obregon, D., & Tan, J. (2012). The role of glycogen synthase kinase-3 signaling in neurodevelopment and fragile X syndrome. *International Journal of Physiology, Pathophysiology, and Pharmacology, 4*(3), 140–148.

Prevalence of autism spectrum disorders—Autism and Developmental Disabilities Monitoring Network, 14 sites, United States, 2008 (2012). *MMWR Surveillance Summaries, 61*(3), 1–19.

Santha, P., Veszelka, S., Hoyk, Z., Meszaros, M., Walter, F. R., Toth, A. E., et al. (2015). Restraint stress-induced morphological changes at the blood-brain barrier in adult rats. *Frontiers in Molecular Neuroscience, 8*, 88. https://doi.org/10.3389/fnmol.2015.00088.

Saur, L., Baptista, P. P., Bagatini, P. B., Neves, L. T., de Oliveira, R. M., Vaz, S. P., et al. (2016). Experimental post-traumatic stress disorder decreases astrocyte density and changes astrocytic polarity in the CA1 Hippocampus of male rats. *Neurochemical Research, 41*(4), 892–904. https://doi.org/10.1007/s11064-015-1770-3.

Simo, M., Rifa-Ros, X., Rodriguez-Fornells, A., & Bruna, J. (2013). Chemobrain: a systematic review of structural and functional neuroimaging studies. *Neuroscience and Biobehavioral Reviews*. https://doi.org/10.1016/j.neubiorev.2013.04.015.

Singh, B., Hughes, A. J., Mehta, G., Erwin, P. J., & Parsaik, A. K. (2016). Efficacy of prazosin in posttraumatic stress disorder: a systematic review and meta-analysis. *Prime Care Companion to CNS Disorders. 18*(4)https://doi.org/10.4088/PCC.16r01943.

Spooren, W., Lindemann, L., Ghosh, A., & Santarelli, L. (2012). Synapse dysfunction in autism: a molecular medicine approach to drug discovery in neurodevelopmental disorders. *Trends in Pharmacological Sciences*,.

Terry, A. V., Jr., Buccafusco, J. J., Schade, R. F., Vandenhuerk, L., Callahan, P. M., Beck, W. D., et al. (2012). The nicotine metabolite, cotinine, attenuates glutamate (NMDA) antagonist-related effects on the performance of the five choice serial reaction time task (5C-SRTT) in rats. *Biochemical Pharmacology, 83*(7), 941–951.

Terry, A. V., Jr., Callahan, P. M., & Bertrand, D. (2015). R-(+) and S-(−) isomers of cotinine augment cholinergic responses in vitro and in vivo. *The Journal of Pharmacology and Experimental Therapeutics, 352*(2), 405–418. https://doi.org/10.1124/jpet.114.219881.

Thomas, E., & Stein, D. J. (2017). Novel pharmacological treatment strategies for posttraumatic stress disorder. *Expert Review of Clinical Pharmacology, 10*(2), 167–177. https://doi.org/10.1080/17512433.2017.1260001.

Wager-Smith, K., & Markou, A. (2011). Depression: a repair response to stress-induced neuronal microdamage that can grade into a chronic neuroinflammatory condition? *Neuroscience and Biobehavioral Reviews, 35*(3), 742–764.

Wang, Z. G., Li, H., Huang, Y., Li, R., Wang, X. F., Yu, L. X., et al. (2017). Nerve growth factor-induced Akt/mTOR activation protects the ischemic heart via restoring autophagic flux and attenuating ubiquitinated protein accumulation. *Oncotarget, 8*(3), 5400–5413. https://doi.org/10.18632/oncotarget.14284.

Wildeboer-Andrud, K. M., Zheng, L., Choo, K. S., & Stevens, K. E. (2014). Cotinine impacts sensory processing in DBA/2 mice through changes in the conditioning amplitude. *Pharmacology, Biochemistry, and Behavior, 117*, 144–150. https://doi.org/10.1016/j.pbb.2013.12.005.

Wu, H., Lu, D., Jiang, H., Xiong, Y., Qu, C., Li, B., et al. (2008). Simvastatin-mediated upregulation of VEGF and BDNF, activation of the PI3K/Akt pathway, and increase of neurogenesis are associated with therapeutic improvement after traumatic brain injury. *Journal of Neurotrauma, 25*(2), 130–139.

Young, G. (2017). PTSD in court II: Risk factors, endophenotypes, and biological underpinnings in PTSD. *International Journal of Law and Psychiatry, 51*, 1–21. https://doi.org/10.1016/j.ijlp.2017.02.002.

Zeitlin, R., Patel, S., Solomon, R., Tran, J., Weeber, E. J., & Echeverria, V. (2012). Cotinine enhances the extinction of contextual fear memory and reduces anxiety after fear conditioning. *Behavioural Brain Research, 228*(2), 284–293. https://doi.org/10.1016/j.bbr.2011.11.023.

Zevin, S., Jacob, P., Geppetti, P., & Benowitz, N. L. (2000). Clinical pharmacology of oral cotinine. *Drug and Alcohol Dependence, 60*(1), 13–18.

CHAPTER

23

Nicotine in Aberrant Learning and Corticostriatal Plasticity

Jessica L. Koranda, Jeff A. Beeler†*

*Department of Neurobiology, University of Chicago, Chicago, IL, United States

†Department of Psychology, Queens College and the Graduate Center, CUNY, Flushing, NY, United States

Abbreviations

AMPAR	AMPA glutamate receptor
cNIC	chronic nicotine
HFS	high-frequency stimulation
L-DOPA	levodopa
LDR	long-duration response
LTD	long-term depression
LTP	long-term potentiation
MSN	medium spiny neuron
NMDAR	NMDA glutamate receptor
NR2B	2B subunit of NMDA glutamate receptor
PD	Parkinson's disease

23.1 INTRODUCTION

The basal ganglia, via cortico-basal ganglia-cortical reentrant loops, profoundly affect cortical activity. Alterations in dopamine function alter this circuit, and such alterations are implicated in generating key features of Parkinson's disease (PD) and behavioral compulsions associated with drug addiction. Dopamine signaling can alter corticostriatal transmission in response to ongoing stimuli in the environment; however, dopamine also plays a critical role in regulating synaptic plasticity at corticostriatal synapses (Augustin, Beeler, McGehee, & Zhuang, 2014; Kreitzer & Malenka, 2008; Lovinger, 2010), allowing experience-dependent modification—learning—to occur in cortico-basal ganglia loops, modifying *future* behavior. Chronic nicotine exposure, such as occurs in smoking, induces a multitude of neural and physiological adaptations with myriad functional and behavioral consequences. In this chapter, we focus on changes in the regulation of corticostriatal plasticity induced by chronic nicotine in the context of PD and addiction.

23.2 PARKINSON'S DISEASE, NICOTINE, AND NEUROPROTECTION

Parkinson's disease, characterized by motor impairments, arises from the progressive degeneration of dopamine (DA) neurons in the substantia nigra compacta (SNc). Dopamine replacement therapy with L-DOPA remains the mainstay of treatment. Effective at ameliorating motor symptoms early in disease progression, L-DOPA treatment does not prevent degeneration and becomes less effective in later stages, underscoring the need for novel therapies, particularly neuroprotective strategies.

Epidemiological studies have identified an inverse relationship between smoking and PD incidence. This decreased risk translates across nicotine products, implicating nicotine rather than another component in cigarette smoke (Chen et al., 2010; Van Der Mark et al., 2014). Decreased PD risk correlates more strongly with duration (years smoked) rather than amount (heavy vs light smokers). PD risk gradually normalizes upon smoking cessation, suggesting that putative nicotine-induced neuroadaptations forestall a progressive neurodegenerative process.

How nicotine decreases PD risk remains unclear. While studies performed in cell culture suggest nicotine may protect against cellular insults such as oxidative stress and excitatory stimuli (e.g., Jin, Gao, Flagg, & Deng, 2004), studies in animals are more contradictory. In studies that report protection, chronic nicotine dose-dependently preserves biochemical markers of striatal dopamine activity, often without preserving the number of tyrosine hydroxylase immunoreactive fibers or cells in the SNc (Huang, Lin, et al., 2009). Nicotine treatment does not improve dopaminergic integrity in animals with

Neuroscience of Nicotine
https://doi.org/10.1016/B978-0-12-813035-3.00023-X

preexisting nigrostriatal damage (Huang, Lin, et al., 2009), suggesting that nicotine may mitigate the functional consequences of denervation.

Despite apparent neuroprotective properties of nicotine, clinical studies examining nicotine in treating PD have yielded inconclusive results. The discrepancy may arise from differences in length of treatment. Where nicotinic agents improved motor symptoms, nicotine treatment lasted from 2 weeks to 1 year (Hanagasi, Lees, Johnson, Singleton, & Emre, 2007; Kelton, Kahn, Conrath, & Newhouse, 2000), while in studies reporting no change or worsening in symptoms (Vieregge, Sieberer, Jacobs, Hagenah, & Vieregge, 2001), nicotine treatment was acute, restricted to a few days. Kelton et al. (2000) found that improved motor symptoms persisted up to 1 month following discontinuation of nicotine treatment. This persistence of effects following nicotine discontinuation, together with the observed inverse correlation between the number of years smoked and the degree of reduced PD incidence, suggests that *chronic* nicotine (cNIC) induces progressive neuroadaptations that arise and decline gradually.

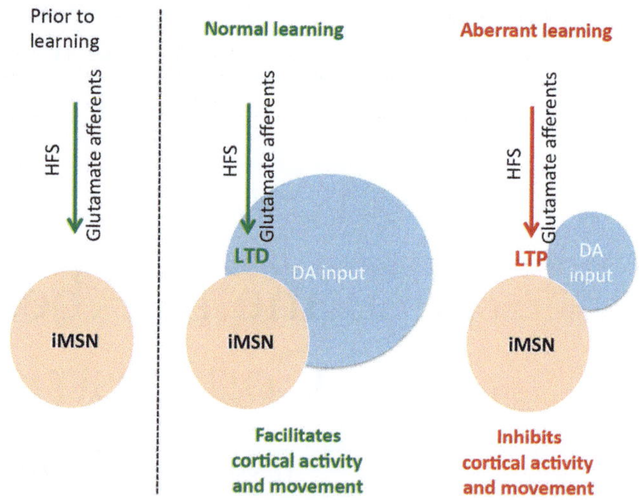

FIG. 23.1 Aberrant motor learning hypothesis. High afferent activity combined with dopaminergic stimulation induces depression at corticostriatal synapses in indirect-pathway MSNs, promoting movement and selection of appropriate behaviors. Under dopamine blockade or denervation, HFS of cortical afferents induces potentiation, inhibiting movement and yielding maladaptive aberrant motor learning.

23.3 THE ABERRANT MOTOR LEARNING HYPOTHESIS

Although dopamine denervation alters corticostriatal plasticity, how this contributes to PD is poorly understood. In indirect-pathway medium spiny neurons (MSNs), dopamine depletion or blockade impairs long-term depression (LTD) and can induce aberrant long-term potentiation (LTP) under conditions (high-frequency stimulation) that normally induce LTD (Calabresi et al., 1997; Kreitzer & Malenka, 2007; Shen, Flajolet, Greengard, & Surmeier, 2008). Corticostriatal LTD in the indirect pathway facilitates movement by selectively reducing basal ganglia inhibition of cortical activity (Kreitzer & Malenka, 2008). Inappropriate LTP where normally LTD would arise would have the opposite effect: synaptically encoding inhibition of movement rather than facilitation, a pathophysiology of synaptic plasticity we term aberrant learning (Fig. 23.1; Beeler, 2011; Beeler et al., 2010, 2012; Beeler, Petzinger, & Jakowec, 2013; Zhuang, Mazzoni, & Kang, 2013). Inappropriate inhibition arising from aberrant learning is *targeted* specifically to cortical afferent activity that *should be facilitated*: in essence, turning plasticity against movement and hastening functional deterioration by unraveling a lifetime of learned motor skills.

To experimentally dissociate dopamine's effect on synaptic plasticity and learning (affecting future behavior) from its effects on synaptic transmission (modulating current behavior), we employ a *reversible* blockade of dopamine in a multiphase variation of the accelerating rotarod, a mouse behavioral task assessing motor performance and learning (Fig. 23.2). In phase 1, mice are exposed to 5–7 days of training under administration of a cocktail of dopamine D1 and D2 antagonists (control and saline). After a break of minimally 72 h, mice are provided additional drug-free training sessions in phase 2, the "recovery phase." Impaired performance during the drug-free phase 2 reflects learning that occurred during the training sessions under the influence of dopamine antagonists.

Treatment with dopamine antagonists impairs performance during phase 1, as expected. However, mice treated with antagonists continue to show impaired performance during the drug-free recovery phase 2. These mice exhibit very slow recovery, taking over a week to obtain asymptotic performance, normally obtained by naive mice within 2–3 days of training (Fig. 23.2A). Impaired performance during phase 2, recovery, is not a residual pharmacological effect as mice administered the same regimen of dopamine antagonists but placed in their homecage, without rotarod training, show no subsequent impairment (Fig. 23.2A). Thus, subsequent impaired performance following dopamine antagonism is experience-dependent, indicating a learning process. We interpret these results as an aberrant learning that occurs during initial training under dopamine antagonism. Once in place, prior aberrant learning interferes with appropriate relearning, and performance improves more slowly than initial, naive acquisition. In testing D1 and D2 receptor antagonists individually, D1 blockade, though it impairs performance, has no effect on subsequent recovery, while D2 blockade alone

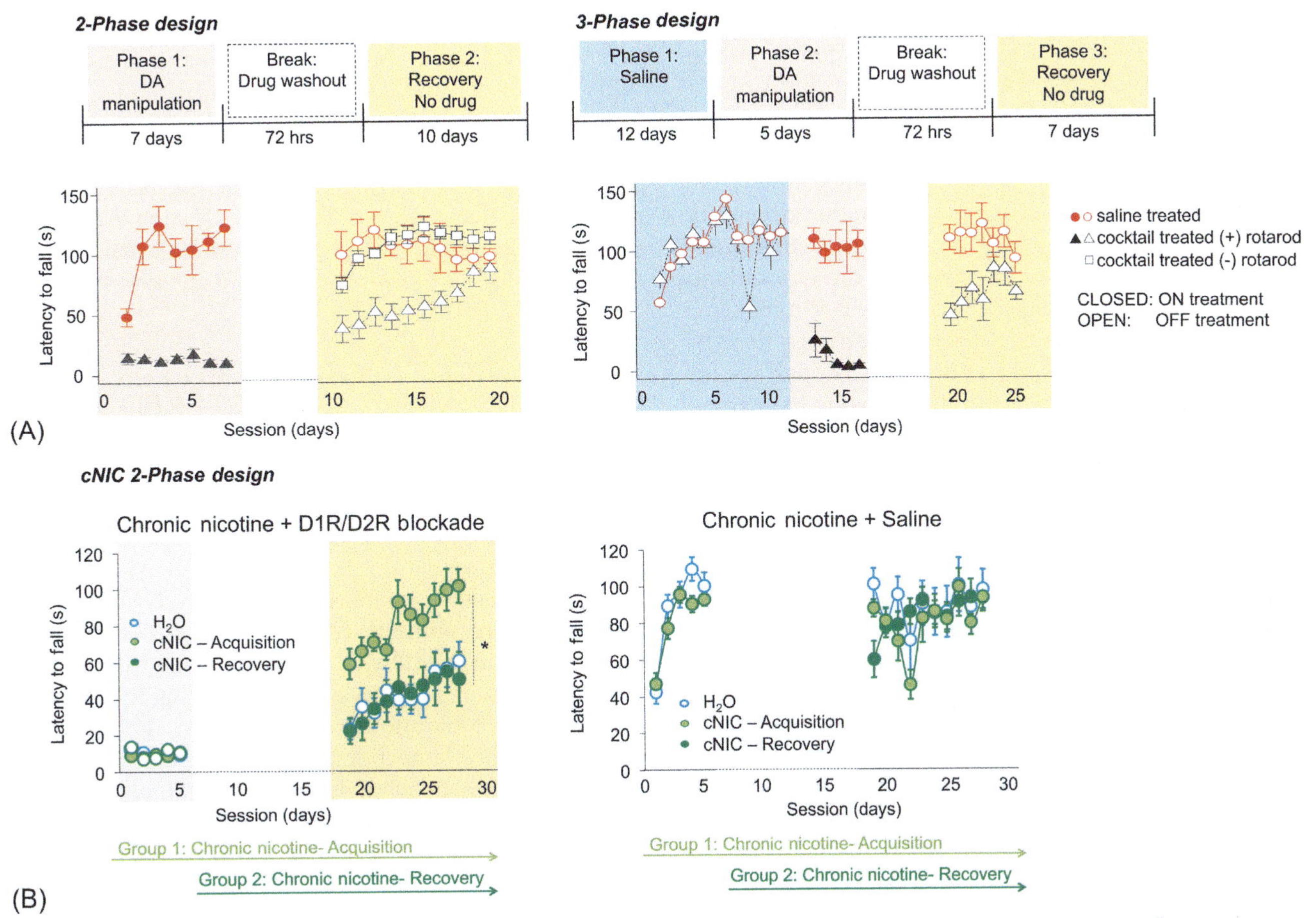

FIG. 23.2 cNIC mitigates aberrant learning.
(A) D1 (SCH23390, 0.10mg/kg) and D2 (eticlopride, 0.16mg/kg) dopamine antagonists used to induce dopamine blockade in a two- (*left*) or three-phase (*right*) design. After rotarod training under dopamine blockade, experience-dependent aberrant learning is expressed as increased time to asymptotic performance during recovery. (B) cNIC (100μg/mL in drinking water) prior to but not after rotarod training under dopamine blockade diminishes aberrant motor learning (*left*). cNIC does not alter learning under normal dopamine (*right*). Data are presented as mean latency to fall ± SEM. *Figure adapted with permission from Koranda et al. (2016), Journal of Neuroscience.*

does, suggesting that aberrant learning is specific to the indirect pathway. A three-phase version of this task tests aberrant learning in an established motor skill: mice are first trained to asymptotic performance, and then during phase 2, mice are trained with dopamine antagonists, followed by a subsequent, drug-free recovery phase 3 (Fig. 23.2A).

A similar unraveling of established motor learning is observed in mice lacking the transcription factor Pitx3. These mice have selective loss of nigrostriatal dopamine neurons, resulting in 90% reduced dorsal striatal dopamine (Beeler, Cao, Kheirbek, & Zhuang, 2009). Pitx3 mice are unable to learn the rotarod task unless treated with L-DOPA, which restores normal motor learning. Upon discontinuation of L-DOPA treatment, subsequent performance is not immediately impaired, but declines gradually. This is not a residual effect of L-DOPA as the gap between discontinuation can be extended to 2 weeks with the same results; the decline in performance upon L-DOPA discontinuation is experience-dependent. These results led us to propose that the poorly understood but clinically critical long-duration response (LDR) to L-DOPA may arise from the correction of dysregulated corticostriatal plasticity and aberrant learning (Beeler, 2011; Beeler et al., 2010, 2012; Zhuang et al., 2013), explaining both its gradual onset during initial treatment and its gradual decline following treatment discontinuation. We hypothesize that targeting aberrant learning may represent a neuroprotective strategy in PD: by protecting established skills and a lifetime of learning, the rate of functional deterioration may be slowed.

23.4 NICOTINE MITIGATES ABERRANT MOTOR LEARNING INDUCED BY DOPAMINE DEFICIENCY

We hypothesized that cNIC induces neuroadaptations that reduce aberrant learning associated with the loss of dopamine, potentially contributing to its putative protective effect against PD (Koranda et al., 2016). Mice were administered cNIC in their drinking water for 2 weeks and then trained on the rotarod with dopamine antagonists as described above. cNIC dose-dependently reduces aberrant motor learning induced by dopamine antagonists (Fig. 23.2B) or by a D2R antagonist alone. In contrast, cNIC instituted following aberrant learning does not facilitate recovery (Fig. 23.2B), suggesting cNIC diminishes aberrant learning arising under dopamine blockade but does not facilitate subsequent "relearning." Moreover, acute injections did not mitigate aberrant learning and worsened recovery if administered during phase 2 (Koranda et al., 2016), suggesting the benefit arises from cNIC-induced neuroadaptations developing over time in response to chronic exposure to nicotine. This cNIC-induced neuroadaptation likely involves chronic desensitization and/or decreased function of β2 subunit containing nicotinic acetylcholine receptors as mice lacking β2 subunits, including selective deletion from dopamine cells, also exhibit reduced aberrant learning (Koranda et al., 2016).

These results suggest that cNIC protects against aberrant learning arising from the loss of dopamine signaling. We propose that this induces an "LDR-like" therapeutic effect that is neuroprotective. Over the course of progressive denervation, this effect may mitigate a gradual unraveling of established synaptic connections that support behavior and, in turn, slow the functional decline associated with progressive dopamine denervation. The longer this putative protective effect is in place—the longer an individual smokes—the greater the accrued benefit of slowing the process. As cNIC actually reduces dopamine function (Koranda et al., 2014; Perez, Ly, McIntosh, & Quik, 2012), neuroprotective adaptations likely arise downstream of dopamine signaling. Our work suggests cNIC counteracts the dysregulation of corticostriatal plasticity and aberrant learning associated with dopamine denervation.

23.5 NICOTINE ADDICTION AND ABERRANT LEARNING

Quitting smoking is widely construed as exceptionally difficult, with poor long-term success rates. The difficulty arises from persistent cravings that lead to relapse. Such cravings can continue for extended periods, sometimes years. While one might expect cravings to diminish over time, evidence suggests that cravings actually increase in the months following cessation, a phenomenon termed incubation (Li, Caprioli, & Marchant, 2015). These cravings are attributed to drug-reinforced associative learning during periods of drug use that link countless stimuli, activities, and states (e.g., stress) to the compulsion to use. What is not clear is why this learning is so remarkably resistant to extinction; that is, at the level of synaptic physiology, it is unknown what would make synaptic connections representing particular learning more resistant to change than other "normal" synapses. One possibility is that chronic nicotine and other drugs of abuse induce neuroadaptations that alter mechanisms of synaptic plasticity, including the *regulation* of synaptic plasticity. That is, nicotine may induce aberrant synaptic plasticity: not *impaired* per se, but *altered* such that acquiring new learning and revising established learning operate differently than prior to chronic drug exposure.

23.6 NICOTINE-INDUCED SILENT SYNAPSES AND ALTERED CORTICOSTRIATAL PLASTICITY

Repeated cocaine and opiates can induce silent synapses in striatal medium spiny cells (Graziane et al., 2016; Huang, Parameswaran, Bordia, Michael McIntosh, & Quik, 2009). Silent synapses contain only NMDA glutamate receptors without AMPARs. As a consequence, Mg^{++} block in NMDARs prevents fast glutamatergic transmission at resting potentials, rendering these synapses "silent." Ubiquitous during early development, where synaptic connections are overexpressed and then selected or pruned, silent synapses are largely absent in adult animals (Hanse, Seth, & Riebe, 2013), with some exceptions. Notably, silent synapses on MSNs persist in adulthood, comprising ~5%–15% of excitatory synapses in normal, healthy mice (Brown et al., 2011; Huang, Parameswaran, et al., 2009; Koya et al., 2012; Whitaker et al., 2015; Xia, Meyers, & Beeler, 2017). Though the functional role of these residual silent synapses in adult MSNs remains unknown, the discovery that cocaine and opioids can increase their prevalence substantially (35%–55%) has led to the "rejuvenation hypothesis" that suggests these silent synapses function as a substrate for circuit remodeling that is "hijacked" by drugs of abuse (Dong & Nestler, 2014), much like the idea that drugs of abuse "hijack the reward system."

In a recent study (Xia et al., 2017), we demonstrated that cNIC also increases silent synapses in MSNs (Fig. 23.3A). Cocaine and opioids differentially increase silent synapses in direct- and indirect-pathway MSNs, respectively, by different mechanisms (Graziane et al., 2016). Cocaine upregulates the NR2B NMDA subunit increasing silent synapses via de novo generation of new silent synapses. Opioids increase silent synapses via removal of AMPARs from established synapses. We have demonstrated that cNIC, like opioids, increases silent synapses in the *indirect* pathway but via NR2B-mediated mechanisms, like cocaine, though we have not assessed its effects on the direct pathway. Elegant work has examined the link between silent synapses and drug-associated behavior (Brown et al., 2011; Koya et al., 2012; Lee et al., 2013; Ma et al., 2014; Whitaker

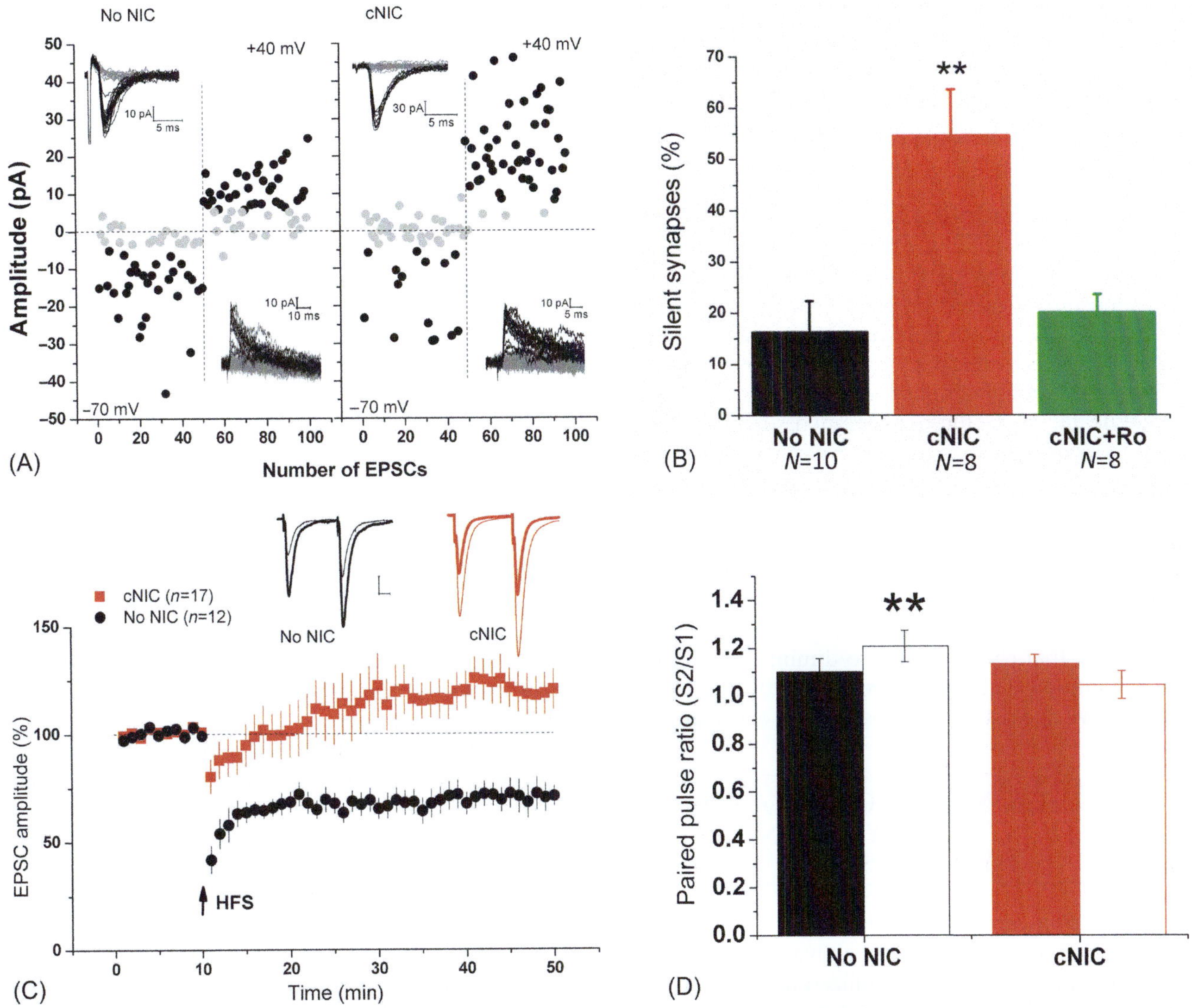

FIG. 23.3 cNIC increases silent synapses and inverts corticostriatal plasticity in indirect-pathway MSNs.
(A) Minimal stimulation assay assessing prevalence of silent synapses in indirect-pathway MSNs comparing cNIC-treated with control mice. A decrease in failures (*gray circles*) at a holding potential of +40 mV compared to −70 mV reflects the presence of silent synapses as the Mg^{++} block of NMDARs is relieved at +40 mV, allowing NMDA-only synapses to contribute. Insets: traces from all cells, black for successes and gray for failures. (B) cNIC increases silent synapses (*red bar*), which is dependent upon NR2B subunits as the increased success rate is blocked with the NR2B selective antagonist. (C) Patch-clamp recordings of indirect-pathway MSNs from mice treated with chronic nicotine (*red*) or controls (*black*). After HFS (*arrow*), control mice exhibit synaptic depression, while cNIC-treated mice show potentiation. (D) Paired-pulse ratio before and after HFS. Control mice show the expected paired-pulse facilitation, indicating presynaptic depression, while cNIC-treated mice do not, consistent with postsynaptic locus of potentiation. Data are presented as mean ± SEM. *Figure adapted with permission from Xia et al. (2017), Neuropsychopharmacology.*

et al., 2015), demonstrating in particular that silent synapses are associated with incubation (Lee et al., 2013; Ma et al., 2014). However, this relationship is complex because upon drug discontinuation, resilencing activated silent synapses in some regions diminishes incubation but, in other regions increases it (Ma et al., 2014). These data suggest that silent synapses per se are not mediating particular behavioral effects, but rather reflect alterations in synaptic plasticity where the behavioral consequences are region-dependent.

How is corticostriatal plasticity altered by cNIC? We tested high-frequency stimulation (HFS) in indirect MSNs in mice that received cNIC (100 μg/ml) in their drinking water. Normally, HFS induces LTD at indirect-pathway excitatory synapses, critical for selectively reducing basal ganglia inhibition on cortical activity to facilitate behavior. In mice treated with cNIC, HFS induces LTP instead, inverting the normal direction of plasticity (Fig. 23.3B). This altered corticostriatal plasticity is accompanied by an overall increase in

NR2B-enriched NMDAR current, concomitant with dramatically increased prevalence of silent synapses. Thus, exposure to cNIC induces neuroadaptations that alter the regulation of corticostriatal synaptic plasticity in the indirect pathway.

23.7 THE CHRONIC NICOTINE PUZZLE

cNIC underlies two very different phenomena, one positive and one negative: on one hand, cNIC provides some form of neuroprotection that reduces the PD risk, though how remains unclear. At the same time, cNIC is addictive and, through smoking, a leading cause of preventable death. Both these phenomena may involve nicotine-induced neuroadaptations in corticostriatal plasticity. If so, we might expect that cNIC diminishes LTP and promotes LTD in indirect-pathway MSNs. However, we observe precisely the opposite: cNIC itself appears to invert corticostriatal plasticity in the indirect pathway, hypothesized to be the cause of the aberrant learning that cNIC demonstrably reduces. This only compounds the prior paradox that cNIC, known to diminish dopamine function (Koranda et al., 2014; Perez et al., 2012), is somehow protective against the progressive loss of dopamine.

23.8 CHRONIC NICOTINE AND SYNAPTIC STABILITY: AN INITIAL HYPOTHESIS

Silent synapses can easily be unsilenced—activated—by the insertion of AMPARs. Once activated, these unsilenced synapses can either be stabilized and retained as normal, functional synapses or resilenced and possibly removed (see Hanse et al., 2013 for review). The prevalence of silent synapses in a cell arises as a dynamic balance between the generation of new silent synapses and their activation, selection and stabilization as functional synapses, and/or their removal upon resilencing. An increase in silent synapse prevalence, then, can arise from either an increased rate of generation, a decreased rate of utilization (selection/stabilization), or decreased removal (Fig. 23.4).

In our patch-clamp slice recordings of mice treated with cNIC, we observe inverted corticostriatal plasticity in the indirect-pathway MSNs (Fig. 23.3B), where these cells express LTP rather than LTD following HFS. However, "long-term potentiation" is defined within the temporal scale of slice recordings (45–50 min); whether this potentiation last for days, weeks, or years in living animals is unknown. Assuming that cNIC from smoking induces similar changes, this would argue *against* the hypothesis that this LTP is stable and long lasting. Potentiation of high-frequency cortical afferents to the indirect pathway consistently over years would dramatically increase inhibitory tone on cortical activity—afferent activity normally facilitated by LTD. That is, smoking would induce aberrant learning where one might expect the emergence of Parkinsonian symptoms. Smokers not only do not develop PD-like symptoms but also have a reduced risk of PD. An alternative explanation is that the HFS-LTP we observe in cNIC-treated mice is not stable but transient (Fig. 23.4). In this case, the increased prevalence of silent synapses observed in cNIC (and perhaps other drugs of abuse) may arise from a *failure to appropriately utilize this substrate for plasticity* through selection and stabilization, perhaps coupled with a failure to *remove* silent synapses upon unsilencing, allowing their accumulation. As a consequence, the cells would sustain a large pool of silent synapses, resulting in a pronounced LTP when activated, i.e., unsilenced, but this plasticity would be unstable, being easily resilenced with those activated synapses returning again to the silent synapse pool. The net result of this would be the appearance of an increased substrate for plasticity and circuit remodeling—increased silent synapses—but *reduced functionality*. In short, we hypothesize that increased silent synapses reflect impairment in the utilization of silent synapses.

This is not to suggest that cNIC induces overt learning impairments, as it does not. Mice exposed to cNIC exhibit normal motor learning (Koranda et al., 2016). Rather, we suggest cNIC increases the stringency or threshold regulating when new learning and circuit remodeling is permitted, effectively protecting established, prior learning and diminishing, though not eliminating, new learning. One of the puzzles in the literature linking silent synapses to incubation is how a substrate meant to facilitate *new* learning appears to lie dormant and encodes *prior* learning (i.e., drug craving) upon unsilencing (e.g., Lee et al., 2013). One possibility is that in addition to a failure to utilize silent synapses, there is a concomitant impairment in the ability to *remove* these synapses. That is, while normally there is a process by which an unsilenced synapse that is *not* selected may be removed, effectively pruning unneeded connections, cNIC (and other drugs of abuse) may prevent this removal, creating a network of "ghost" connections that upon unsilencing *and stabilization*—a capacity potentially restored upon abstinence—encode prior connectivity that should have been pruned: prior connectivity associated with drug-related stimuli. This same instability of activated silent synapses may facilitate one aspect of cNIC neuroprotection in PD. If dopamine denervation induces maladaptive synaptic plasticity, i.e., aberrant learning, then impaired selection and stabilization of new (aberrant) learning may be beneficial and slow functional degradation of established learning and skills that may otherwise occur with progressive denervation.

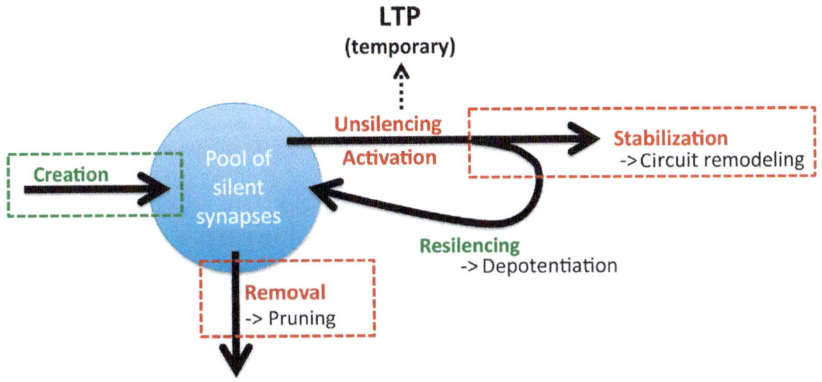

FIG. 23.4 Schematic of cycling processes determining prevalence/pool of silent synapses.
Mechanisms that increase silent synapse prevalence are notated in green, including (i) creation through insertion of NMDARs into a new site or via removal of AMPARs from an existing synapse and (ii) the resilencing of an activated (unsilenced) silent synapse via removal of recently inserted AMPARs. Mechanisms that decrease the silent synapse pool are notated in red, including (i) unsilencing via AMPAR insertion (activation) and *subsequent stabilization* or (ii) the removal of a silent synapse altogether. When unsilenced, these synapses contribute to synaptic potentiation. Activated silent synapses can be stabilized and subsequently be modified, contributing to circuit remodeling, or be resilenced, resulting in depotentiation. The removal of a silent synapse reflects a pruning and selection process. Points in the life cycle of silent synapses at which chronic nicotine may alter the pool of available synapses, indicated with dotted boxes, include altering the rate of creation of new silent synapses, altering the regulation by which activated silent synapses are selected and stabilized vs resilenced, and modulating the removal and pruning of silent synapses.

This hypothesis is necessarily partial and speculative as so little is known about the role of silent synapses in adults. Understanding how nicotine and other drugs of abuse alter this adult substrate of silent synapses may provide insight into how drugs of abuse alter synaptic plasticity and give rise to abnormally resilient associative learning and peculiar phenomena such as incubation. Such work may also provide insight into the apparent cumulative protective effect of cNIC against PD, which remains poorly understood. Work on regulation of silent synapses under cNIC and other drugs of abuse may yield insights into a fundamental aspect of neural plasticity within the striatum, where continued low prevalence of silent synapses may, like neurogenesis in the hippocampus, reflect a developmental mechanism retained in adulthood serving an important role in mediating neural plasticity and adaptive behavior.

MINI-DICTIONARY OF TERMS

Cortical-basal ganglia reentrant loops The cortex projects densely to the striatum, chief input to the basal ganglia, which project back to the cortex, constituting reentrant loops that allow basal ganglia modulation of cortical activity.

Direct/indirect pathways Principle pathways in the basal ganglia. The direct pathway, via disinhibition, facilitates cortical activity and movement, while the indirect pathway inhibits cortical activity.

Incubation The phenomenon where craving for a drug increases over time following abstinence.

Long-duration response (LDR) Treatment with levodopa (L-DOPA) induces gradual improvement in symptoms that does not correspond to L-DOPA in the bloodstream, observed as a lessening of symptoms during "troughs" in drug levels between doses and sustained improvement following L-DOPA discontinuation, despite the absence of the drug.

Long-term potentiation/depression (LTP/LTD) An increase/decrease in synaptic efficacy increasing/reducing a neuron's response to the same stimulus in the future.

Medium spiny neuron (MSN) Projection neurons of the striatum composed of direct- and indirect-pathway MSNs, distinguished by protein expression, including expression of dopamine D1 and D2 receptors, respectively.

Key Facts of Silent Synapses

- Silent synapses contain NMDA but not AMPA receptors and are unresponsive to glutamate at resting membrane potentials.
- Widely expressed throughout the brain during early development, silent synapses are pruned in an experience-dependent way, associated with critical periods in development.
- Discovery of silent synapses in the hippocampus was an important step in establishing the postsynaptic locus of long-term potentiation in the 1990s.
- Largely absent in adults, silent synapses persist in some neuronal populations, including within the striatum.
- Drugs of abuse increase expression of silent synapses in striatal medium spiny neurons and have been linked to incubation of craving.

Summary Points

- This chapter focuses on chronic nicotine and corticostriatal plasticity.
- Loss of dopamine signaling induces abnormal corticostriatal plasticity, giving rise to aberrant learning that may contribute to functional deterioration in Parkinson's disease.
- Smoking reduces risk for Parkinson's disease, though how is unknown. Chronic nicotine can mitigate aberrant learning associated with dopamine blockade, suggesting a possible mechanism for nicotine neuroprotection.
- Chronic nicotine upregulates silent synapses and inverts the direction of corticostriatal plasticity in indirect-pathway medium spiny neurons.
- We hypothesize that increased silent synapses induced by chronic nicotine reflects impaired utilization of this substrate for circuit modification, reducing plasticity. In addition, this reduces behavioral flexibility by protecting established drug-reinforced learning while yielding positive effects in Parkinson's disease by reducing aberrant learning.

References

Augustin, S. M., Beeler, J. A., McGehee, D. S., & Zhuang, X. (2014). Cyclic AMP and afferent activity govern bidirectional synaptic plasticity in Striatopallidal neurons. *The Journal of Neuroscience, 34*, 6692–6699.

Beeler, J. A. (2011). Preservation of function in Parkinson's disease: what's learning got to do with it? *Brain Research, 1423*, 96–113.

Beeler, J. A., Cao, Z. F. H., Kheirbek, M. A., Ding, Y., Koranda, J., Murakami, M., et al. (2010). Dopamine-dependent motor learning insight into Levodopa's long-duration response. *Annals of Neurology, 67*, 639–647.

Beeler, J. A., Cao, Z. F. H., Kheirbek, M. A., & Zhuang, X. (2009). Loss of cocaine locomotor response in Pitx3-deficient mice lacking a nigrostriatal pathway. *Neuropsychopharmacology, 34*, 1149–1161.

Beeler, J. A., Frank, M. J., McDaid, J., Alexander, E., Turkson, S., Sol Bernandez, M., et al. (2012). A role for dopamine-mediated learning in the pathophysiology and treatment of Parkinson's disease. *Cell Reports, 2*, 1747–1761.

Beeler, J. A., Petzinger, G., & Jakowec, M. W. (2013). The enemy within: propagation of aberrant corticostriatal learning to cortical function in Parkinson's disease. *Frontiers in Neurology, 4*, 134.

Brown, T. E., Lee, B. R., Mu, P., Ferguson, D., Dietz, D., Ohnishi, Y. N., et al. (2011). A silent synapse-based mechanism for cocaine-induced locomotor sensitization. *The Journal of Neuroscience, 31*, 8163–8174.

Calabresi, P., Saiardi, A., Pisani, A., Baik, J., Centonze, D., Mercuri, N. B., et al. (1997). Abnormal synaptic plasticity in the striatum of mice lacking dopamine D2 receptors. *The Journal of Neuroscience, 17*, 4536–4544.

Chen, H., Huang, X., Guo, X., Mailman, R. B., Park, Y., Kamel, F., et al. (2010). Smoking duration, intensity, and risk of Parkinson disease. *Neurology, 74*, 878–884.

Dong, Y., & Nestler, E. J. (2014). The neural rejuvenation hypothesis of cocaine addiction. *Trends in Pharmacological Sciences, 35*, 374–383.

Graziane, N. M., Sun, S., Wright, W. J., Jang, D., Liu, Z., Huang, Y. H., et al. (2016). Opposing mechanisms mediate morphine- and cocaine-induced generation of silent synapses. *Nature Neuroscience, 19*, 915–925.

Hanagasi, H. A., Lees, A., Johnson, J. O., Singleton, A., & Emre, M. (2007). Smoking-responsive juvenile-onset Parkinsonism. *Movement Disorders: Official Journal of the Movement Disorder Society, 22*, 115–119.

Hanse, E., Seth, H., & Riebe, I. (2013). AMPA-silent synapses in brain development and pathology. *Nature Reviews. Neuroscience, 14*, 839–850.

Huang, L. Z., Parameswaran, N., Bordia, T., Michael McIntosh, J., & Quik, M. (2009). Nicotine is neuroprotective when administered before but not after nigrostriatal damage in rats and monkeys. *Journal of Neurochemistry, 109*, 826–837.

Huang, Y. H., Lin, Y., Mu, P., Lee, B. R., Brown, T. E., Wayman, G., et al. (2009). In vivo cocaine experience generates silent synapses. *Neuron, 63*, 40–47.

Jin, Z., Gao, F., Flagg, T., & Deng, X. (2004). Nicotine induces multi-site phosphorylation of bad in association with suppression of apoptosis. *The Journal of Biological Chemistry, 279*, 23837–23844.

Kelton, M. C., Kahn, H. J., Conrath, C. L., & Newhouse, P. A. (2000). The effects of nicotine on Parkinson's disease. *Brain and Cognition, 43*, 274–282.

Koranda, J. L., Cone, J. J., McGehee, D. S., Roitman, M. F., Beeler, J. A., & Zhuang, X. (2014). Nicotinic receptors regulate the dynamic range of dopamine release in vivo. *Journal of Neurophysiology, 111*, 103–111.

Koranda, J. L., Krok, A. C., Xu, J., Contractor, A., McGehee, D. S., Beeler, J. A., et al. (2016). Chronic nicotine mitigates aberrant inhibitory motor learning induced by motor experience under dopamine deficiency. *The Journal of Neuroscience, 36*, 5228–5240.

Koya, E., Cruz, F. C., Ator, R., Golden, S. A., Hoffman, A. F., Lupica, C. R. et al. (2012). Silent synapses in selectively activated nucleus accumbens neurons following cocaine sensitization. *Nature Neuroscience, 15*, 1556–1562.

Kreitzer, A. C., & Malenka, R. C. (2007). Endocannabinoid-mediated rescue of striatal LTD and motor deficits in Parkinson's disease models. *Nature, 445*, 643–647.

Kreitzer, A. C., & Malenka, R. C. (2008). Striatal plasticity and basal ganglia circuit function. *Neuron, 60*, 543–554.

Lee, B. R., Ma, Y.-Y., Huang, Y. H., Wang, X., Otaka, M., Ishikawa, M., et al. (2013). Maturation of silent synapses in amygdala-accumbens projection contributes to incubation of cocaine craving. *Nature Neuroscience, 16*, 1644–1651.

Li, X., Caprioli, D., & Marchant, N. J. (2015). Recent updates on incubation of drug craving: a mini-review: updates: incubation of craving. *Addiction Biology, 20*, 872–876.

Lovinger, D. M. (2010). Neurotransmitter roles in synaptic modulation, plasticity and learning in the dorsal striatum. *Neuropharmacology, 58*, 951–961.

Ma, Y.-Y., Lee, B. R., Wang, X., Guo, C., Liu, L., Cui, R., et al. (2014). Bidirectional modulation of incubation of cocaine craving by silent synapse-based remodeling of prefrontal cortex to accumbens projections. *Neuron, 83*, 1453–1467.

Perez, X. A., Ly, J., McIntosh, J. M., & Quik, M. (2012). Long-term nicotine exposure depresses dopamine release in nonhuman primate nucleus accumbens. *The Journal of Pharmacology and Experimental Therapeutics, 342*, 335–344.

Shen, W., Flajolet, M., Greengard, P., & Surmeier, D. J. (2008). Dichotomous dopaminergic control of striatal synaptic plasticity. *Science, 321*, 848–851.

Van Der Mark, M., Nijssen, P. C., Vlaanderen, J., Huss, A., Mulleners, W. M., Sas, A. M., et al. (2014). A case-control study of the protective effect of alcohol, coffee, and cigarette consumption on Parkinson disease risk: time-since-cessation modifies the effect of tobacco smoking. *PLoS One, 9*. e95297.

Vieregge, A., Sieberer, M., Jacobs, H., Hagenah, J. M., & Vieregge, P. (2001). Transdermal nicotine in PD: a randomized, double-blind, placebo-controlled study. *Neurology*, *57*, 1032–1035.

Whitaker, L. R., Carneiro de Oliveira, P. E., McPherson, K. B., Fallon, R. V., Planeta, C. S., Bonci, A., et al. (2015). Associative learning drives the formation of silent synapses in neuronal ensembles of the nucleus accumbens. *Biological Psychiatry*, *80*, 246–256.

Xia, J., Meyers, A. M., & Beeler, J. A. (2017). Chronic nicotine alters corticostriatal plasticity in the Striatopallidal pathway mediated by NR2B-containing silent synapses. *Neuropsychopharmacology*. https://doi.org/10.1038/npp.2017.87 (Epub ahead of print).

Zhuang, X., Mazzoni, P., & Kang, U. J. (2013). The role of neuroplasticity in dopaminergic therapy for Parkinson disease. *Nature Reviews. Neurology*, *9*, 248–256.

CHAPTER

24

Prenatal Nicotine Exposure and Impact on the Behaviors of Offspring

Tursun Alkam[*,†], *Toshitaka Nabeshima*[*,‡,§]

[*]Japanese Drug Organization of Appropriate Use and Research, Nagoya, Japan
[†]Department of Basic Medical Sciences, College of Osteopathic Medicine of the Pacific, Western University of Health Sciences, Pomona, CA, United States
[‡]Advanced Diagnostic System Research Laboratory, Graduate School of Health Sciences, Fujita Health University, Toyoake, Japan
[§]Aino University, Ibaraki, Japan

Abbreviations

5-CSRT task	5-choice serial-reaction time task
5-HT	serotonin
5-HT-ergic	serotonergic
ACh	acetylcholine
ADHD	attention deficit hyperactivity disorder
CA test	cliff-avoidance test
DA	dopamine
DAergic	dopaminergic
EPM test	elevated plus-maze test
LDB test	light and dark box test
LDT	laterodorsal tegmental
MBB test	marble-burying behavior test
mPFC	medial prefrontal cortex
NA	noradrenaline.
nAChRs	nicotinic acetylcholine receptors
NAergic	noradrenergic
NRT	nicotine replacement therapy
NSF test	novelty-suppressed feeding test
OBA test	object-based attention test
OF test	open-field test
PNE	prenatal nicotine exposure
PPI test	prepulse inhibition test
SN test	social novelty test
SS task	stop-signal task
TBFC test	two-bottle free-choice test
Y-M test	Y-maze test

24.1 INTRODUCTION

Tobacco smoking during pregnancy has significant adverse impact on the overall quality of offspring. Epidemiological studies and clinical investigations report that children born from women who smoke during pregnancy suffer from neurobehavioral impairments, such as persistent anxiety, learning disabilities, impaired decision-making, impulsive behaviors, restlessness and attention deficit hyperactivity disorder (ADHD), and addiction (Fig. 24.1) (Abbott & Winzer-Serhan, 2012; Clifford, Lang, & Chen, 2012; De Genna, Goldschmidt, Day, & Cornelius, 2016; Fitzpatrick, Barnett, & Pagani, 2014; Garcia-Rill, Buchanan, McKeon, Skinner, & Wallace, 2007; Thapar et al., 2003). Among the myriad ingredients in tobacco smoke, nicotine diffuses into vessels in the lung of the smoking mother and moves quickly with the blood circulation to her brain, where it binds and stimulates nicotinic acetylcholine receptors (nAChRs). Stimulation of nAChRs in neurons causes the influx of sodium or calcium that triggers the ultimate increase of intracellular calcium in neurons and causes the release of neurotransmitters that are related to feelings of pleasure and dependence on nicotine, which is the cause of tobacco addiction. Although pregnant women know that tobacco smoke contains many toxic substances that are harmful for the fetus, they cannot easily quit smoking due to the addiction to nicotine. To decrease the cravings for nicotine in tobacco and to help quit smoking, nicotine replacement therapy (NRT), in conjunction with cognitive-behavioral therapy, is applied to smoking cessation during pregnancy (Osadchy, Kazmin, & Koren, 2009). However, either through smoking tobacco or through using NRT, nicotine gets into the blood of the mother, easily penetrates blood-placenta barrier,

https://doi.org/10.1016/B978-0-12-813035-3.00024-1

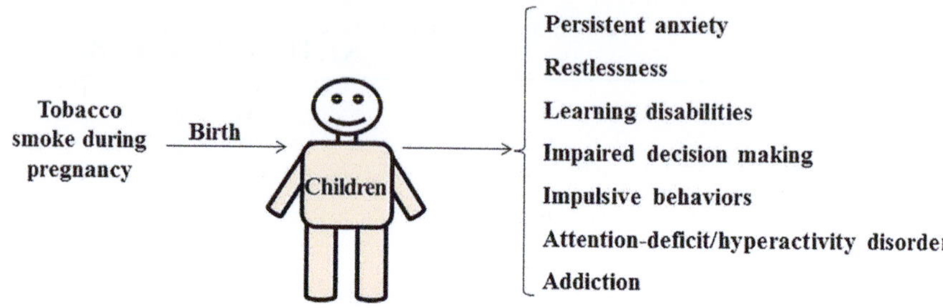

FIG. 24.1 Behaviors of children born from mothers who smoked tobacco during pregnancy.

interacts with nAChRs in developing fetal brain, interferes with normal functional development of the brain by affecting neurogenesis and synaptogenesis, and causes long-lasting abnormalities in neurotransmitter signaling (Dempsey & Benowitz, 2001; Slotkin, 1998).

24.2 PRENATAL NICOTINE EXPOSURE

Reports from animal models indicate that prenatal nicotine exposure (PNE) compromises the development of critical neural pathways in the developing brain and leads to emotional and cognitive-behavioral abnormalities that are noted in the offspring born from women who smoke during pregnancy. The majority of these studies are done in rodents (mice and rats) whose embryonic development shares many similarities to that of humans, while there are some important differences. The whole gestational period of a rodent is developmentally equivalent to only the first and second trimesters of human gestation (Fig. 24.2) (Cross, Linker, & Leslie, 2017; Dwyer et al., 2008). Many critical cerebral developmental events, such as neurogenesis and early synaptogenesis, occur during the second trimester following the appearance of nAChRs in the end of the first trimester (Cross et al., 2017; Dwyer et al., 2008). The first 2 weeks of postnatal development of the mouse is equivalent to the third trimester of human development, while the brain reaches structurally mature stage in the first week (Cross et al., 2017; Dwyer et al., 2008). The nAChRs appear as early as on gestational day 10 in the mouse cerebral cortex (Atluri et al., 2001) and at the end of first trimester in the human brain (Cross et al., 2017; Dwyer et al., 2008). The nAChRs play a critical role in many aspects of brain development, depending on their regional expression and the pharmacology of their subunit composition, by regulating neurogenesis and synaptogenesis. The dynamic roles of nAChRs in neurodevelopment are consistently driven by acetylcholine (ACh), the endogenous neurotransmitter. Nicotine exposure during fetal neurodevelopment either exaggerates the role of ACh or interferes with the regulatory role of nAChRs and affects the timing and intensity of critical neurotransmitter signals that emerge as necessary to ensure the proper development of numerous neural circuits and pathways (Pauly & Slotkin, 2008).

When animal models of PNE are evaluated, factors to consider are the route of administration, the dose of nicotine, and the length of exposure (Slotkin, 1998). Animal studies of PNE applied maternal multitime injections during the light phase of the day or subcutaneously implanted osmotic minipump during the whole 24 h of the day or administered nicotine through the drinking water to mimic the pharmacokinetics of smoking with time intervals (Alkam et al., 2013; Alkam et al., 2013; Slotkin, 1998). Some PNE studies in rodents prefer additional postnatal nicotine exposure through maternal lactation during the early postnatal days to mimic the nicotine exposure in all three trimesters in human pregnancy. Regardless of inconsistency of the choices of nicotine dose, the route of administration, and the onset and the length of exposure in various animal models of PNE, all of the reports support the notion that PNE induces behavioral abnormalities in offspring.

24.3 IMPACT ON BEHAVIORS OF OFFSPRING

Available studies related to the impact of PNE on offspring behavior provide systematic understanding for behavioral neurotoxicity of PNE (Fig. 24.3). Behavioral

FIG. 24.2 Prenatal period in humans and rodents. *Modified from Dwyer, Broide, and Leslie (2008).*

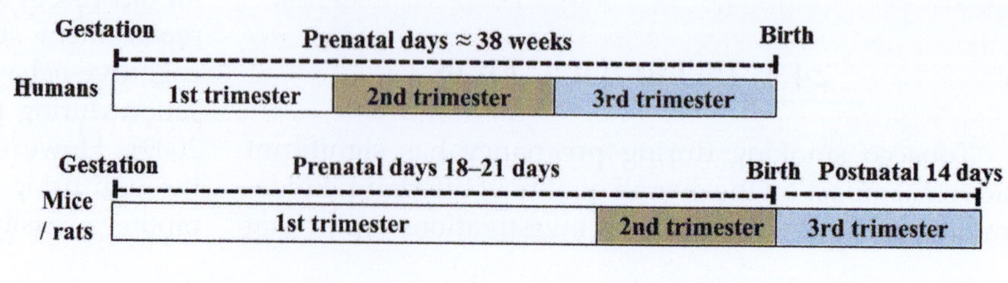

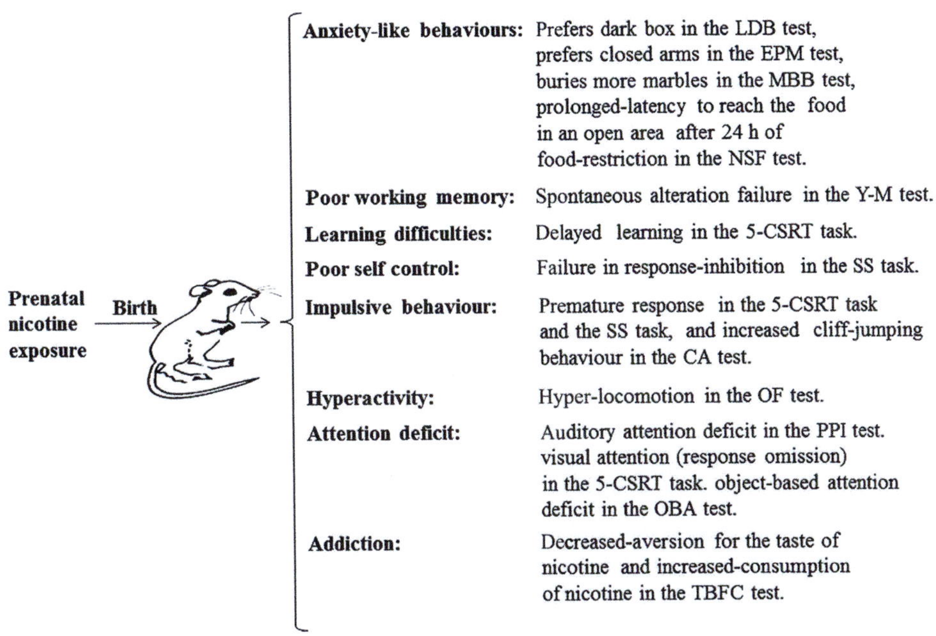

FIG. 24.3 Prenatal-nicotine-exposure-induced behavioral changes in rodent offspring.
5-CSRT task, 5-choice serial-reaction time task; *CA test*, cliff-avoidance test; *EPM test*, elevated plus-maze test; *LDB test*, light and dark box test; *MBB test*, marble-burying behavior test; *NSF test*, novelty-suppressed feeding test; *OBA test*, object-based attention test; *PPI test*, prepulse inhibition test; *SN test*, social novelty test; *SS task*, stop-signal task; *Y-M test*, Y-maze test; *TBFC test*, two-bottle free-choice test.

impact of PNE during the different time windows of fetal brain development could be different due to the availability of either the functional nAChRs in particular brain region or nicotine itself at different concentrations.

24.4 EMOTIONAL BEHAVIOR

Emotional behaviors are investigated using the light and dark box test (LDB test), the novelty-suppressed feeding test (NSF test), the marble-burying behavior test (MBB test), the social novelty test (SN test), the elevated plus-maze test (EPM test), the cliff-avoidance test (CA test), the 5-choice serial-reaction time task (5-CSRT task), and the stop-signal task (SS task) (explained in Mini-Dictionary of Terms).

In the LDB test, mice with PNE spend more time in the dark box that seems safer, an indicator of anxiety-like behavior, than in the brightly lightened box that seems unsafe (Alkam, Kim, Hiramatsu, et al., 2013). Maternal nicotine exposure during late gestation and lactation likewise increases anxiety-like behavior of rat offspring in the LDB test (Lee, Chung, & Noh, 2016). In MBB test, PNE mice burry more marbles (unfamiliar shining objects) than male counterparts in control group (Alkam, Kim, Hiramatsu, et al., 2013; Aoyama et al., 2016). In the EPM test, mice with PNE spend significantly more hiding time exploring the closed arms and have more entries into the closed arms, both of which indicate anxiety-like behavior, than do in the open arms (Alkam, Kim, Hiramatsu, et al., 2013; Hall et al., 2016; Parameshwaran et al., 2012). In SN test, PNE mice prefer to stay with the immobilized familiar mouse than with the immobilized unfamiliar mouse while showing an increased preference for staying alone in empty chamber (Alkam, Kim, Hiramatsu, et al., 2013). In the same study, the latency to initiate feeding in open arena is prolonged in 24 h food-restricted hungry mice with PNE in the NSF test. In the CA test, PNE mice jump off the platform significantly earlier, which is a sign of impulsive behavior, than control counterparts (Alkam et al., 2017; Zhu et al., 2017). PNE makes rats impulsive, which is indicated by more premature responses to visual stimuli presented in ordered and predictable sequences in both a 5-CSRT task and an SS task (Bryden et al., 2016; Schneider et al., 2011).

Rats with PNE show increased numbers of premature responses, which are responses that occur after the initiation of the trial but prior to visual stimuli presented in an ordered and predictable sequences, and omission errors; more variable response times; and lower accuracy with evidence of delayed learning in the 5-CSRT task and suggest a direct effect of PNE on important aspects of attention, inhibitory control, or learning later in life (Schneider et al., 2011).

PNE leads to significant deficits in choline acetyltransferase, an enzyme that is present at the presynaptic ends of axons and catalyzes the synthesis of excitatory neurotransmitter ACh, in the cerebral cortex in offspring (Slotkin et al., 2007). PNE elicits damage to serotonin (5-HT) projections in the cerebral cortex (Xu, Seidler, Ali, Slikker Jr., & Slotkin, 2001) and decreases the turnover of 5-HT while increasing 5-HT transporter density

(Muneoka et al., 1997; Muneoka et al., 2001). 5-HT mediates both excitatory and inhibitory neurotransmission through its various receptors and serves as the neurobiological basis of emotional regulations. PNE also reduces 5-HT_{1A} receptor binding, while 5-HT2 receptor binding showed significant overall increases in the cerebral cortex in offspring (Slotkin et al., 2007). Microdialysis studies indicated that postsynaptic stimulation of 5-HT_{1A} receptors could increase noradrenaline (NA) release and turnover in a variety of brain regions of the awake rat including the hippocampus and the frontal cortex (Done & Sharp, 1994; Hajos-Korcsok, McQuade, & Sharp, 1999; Suzuki et al., 1995). It is possible that the reduction in the level of 5-HT_{1A} receptor may result in the decreased turnover of NA. Interestingly, PNE reduces the content and turnover rate of NA in the frontal cortex (Alkam, Kim, Mamiya, et al., 2013; Ribary & Lichtensteiger, 1989). PNE disrupts the firing of medial prefrontal cortex (mPFC) neurons that carry signals related to response direction and conflict monitoring (Bryden et al., 2016). These decreased signal transductions may be responsible for the behavioral changes in offspring with PNE.

24.5 COGNITIVE BEHAVIOR

Cognitive behaviors are evaluated by the Y-maze test (Y-M test), the object-based attention test (OBA test), the prepulse inhibition test (PPI test), the open-field test (OF test), the 5-choice serial-reaction time task (5-CSRT task), and the stop-signal task (SS task) (explained in Mini-Dictionary of Terms).

PNE decreases spontaneous alternation behaviors, a measure of short-term working memory, in the Y-M test in mice (Alkam, Kim, Mamiya, et al., 2013; Parameshwaran et al., 2012). In the OBA test, mice with PNE show decreased familiarity with a "familiar object" that is presented to them (to get familiar) shortly before the test, suggesting that PNE mice are not attentive to the details (Alkam, Kim, Mamiya, et al., 2013). When the preattentive process of sensorimotor gating is investigated by PPI test, sensorimotor processing is impaired in mice and rats with PNE (Alkam et al., 2017; Alkam, Kim, Mamiya, et al., 2013; Lacy, Mactutus, & Harrod, 2011). PNE increases the spontaneous locomotor activity of mice and rats in the OF test (Schneider, Bizarro, Asherson, & Stolerman, 2012; Zhu et al., 2012). PNE makes rats less self-restraining on "STOP" trials during which rats have to stop an already initiated response in an SS task (Bryden et al., 2016). Rats with PNE show increased numbers of premature responses, which are responses that occur after the initiation of the trial but prior to visual stimuli presented in ordered and predictable sequences, and omission errors; more variable response times; and lower accuracy with evidence of delayed learning in the 5-CSRT task and suggest a direct effect of PNE on important aspects of attention, inhibitory control, or learning later in life (Schneider et al., 2011). These behavioral abnormalities induced by PNE mimic ADHD-like behaviors.

Although the exact etiology of ADHD is still unknown, the defects in both dopaminergic (DAergic) and NAergic transmission are thought to be responsible for the symptoms of ADHD (Halperin et al., 1997; Shaywitz, Cohen, & Bowers Jr., 1977; Shekim et al., 1987; Shekim, Javaid, Davis, & Bylund, 1983). PNE decreases the contents of dopamine (DA) and its release and turnover rate in the PFC of mice offspring (Alkam et al., 2017; Alkam, Kim, Mamiya, et al., 2013; Zhu et al., 2012; Zhu, Lee, Spencer, Biederman, & Bhide, 2014). PNE also disrupts both DAergic and NAergic transmissions as well as dendritic morphology and synaptic connectivity throughout the cortex (Muhammad et al., 2012; Mychasiuk, Muhammad, Gibb, & Kolb, 2013; Seidler, Levin, Lappi, & Slotkin, 1992; Zhu et al., 2012). PNE reduces the number of tyrosine hydroxylase-positive varicosities, in which the synthesis, release, and reuptake of NA and DA take place, in the mPFC (Alkam et al., 2017). PNE-induced ADHD-like behavioral deficits are attenuated by acute treatment with drugs that increase synaptic DA and NA (Alkam et al., 2017; Zhu et al., 2017). Taken together, these findings support the notion that PNE induces behavioral abnormalities in mouse offspring by disrupting the NAergic and DAergic neurotransmitter systems.

24.6 ADDICTIVE BEHAVIOR

PNE is a risk factor for developing an addiction to nicotine at a later stage in life (Christensen, Nielsen, & Kohlmeier, 2015). PNE increases preference for nicotine odor in early postnatal and adolescent rats (Mantella, Kent, & Youngentob, 2013). PNE induces decreased aversion for nicotine solution and increases nicotine consumption of adolescent rats in TBFC test (Schneider et al., 2012). Rats with PNE also exhibit significantly increased consumption of nicotine and other substances of abuse with no changes in chow and water intake or body weight (Chang, Karatayev, & Leibowitz, 2013). PNE induces changes in cholinergic laterodorsal tegmental (LDT) nucleus, which is importantly involved in responding to motivational stimuli and in the development of drug-addiction-associated behaviors (Christensen et al., 2015; McNair & Kohlmeier, 2015). PNE alters glutamate signaling within LDT, and PNE increases calcium concentration in LDT cells in the brain of young mice with PNE and larger amplitude excitatory postsynaptic currents (Christensen et al., 2015; McNair &

Kohlmeier, 2015). These reports support that PNE induces craving responses to nicotine and possible addiction later in life.

24.7 CONCLUSION

The behavioral findings from PNE studies in animal models are consistent with the symptoms observed in the children who are born from mothers who smoke during their pregnancy. Currently available reports provide sufficient evidence to support the adverse effects of PNE on the emotional, cognitive, and addictive behaviors of offspring (Fig. 24.3) and to explain the underlying mechanisms (Fig. 24.4). These findings urge serious considerations for nicotine exposure via any means of exposure including NRT during pregnancy.

MINI-DICTIONARY OF TERMS

5-choice serial-reaction time task This task measures attention performance in rodents using visual discrimination and impulse control performance in a response to a visual stimulus in an operant-based task. An inattentive rodent will show omission of responses to short duration of visual stimuli. An impulsive rodent will show premature responses before visual stimuli are given.

Cliff-avoidance test This test assesses the eagerness of cliff-jumping behaviors, with no regard to danger of being hurt, of rodents. A rodent with impulsive-like behavior jumps off the cliff immediately.

Elevated plus-maze test This test measures the time and number of entries of a rodent when given the choice of spending in open unprotected arms or in enclosed protected arms that are elevated from the floor. A rodent with anxiety generally tends to avoid the open areas, favoring enclosed "safe" spaces.

Light-dark box test This test takes advantage of the natural tendency of rodents to explore novel environments. The key measure for assessing anxiety-related behavior in this test is reflected in time spent in the light (unprotected) box over the dark (protected) box. A rodent with anxiety prefers the dark (protected) compartment.

Novelty-suppressed feeding test This test measures a rodent's avoidance from novel open area, in which they might be more vulnerable to predation. A food-restricted hungry rodent with anxiety generally displays longer latency (due to the considerably longer investigation for safety confirmation) to approach the food in the novel open area.

Object-based attention test This test takes advantage of the natural tendency of mice to explore novel objects. In this task, a mouse is exposed to the five objects for 3 min (training session); then, after a short interval, the mouse is exposed to two objects that include a "familiar object" that is presented in training session and a novel object. A mouse with attention deficit will explore both novel object and "familiar object" as novel objects.

Open-field test In this test, a rodent is placed in the open-field box or cage with photobeam motion sensors placed parallel to the floor. Each move of the animal breaks a photobeam. A hyperactive animal breaks more photobeams during the test.

Prepulse inhibition test This test measures the attenuation in startle response to a strong stimulus (pulse) that follows a weak nonstartling stimulus (prepulse) by a very short interval (e.g., 30–500 ms). It is thought that the prepulse activates a preattentional gating mechanism that inhibits the startle response. A rodent with attention deficits shows impaired prepulse inhibition.

Stop-signal task This task measures the ability of the test subject to inhibit its response (to a stimulus) that has already been initiated. Response inhibition is a hallmark of executive control. A subject with impaired goal-directed behavior shows no immediate response inhibition to a learned stimulus.

The marble-burying test This test is a useful method to examine neophobia, anxiety, and obsessive-compulsive behavior in mice. A mouse with any of these emotional disorders tends to bury more marbles.

Two-bottle free-choice test In this test, a rodent has access to two drinking bottles in the cage. One bottle contains nicotine solution (sweetened with n% sucrose), and the other bottle contained n% sucrose solution. The bottle positions are alternated every day.

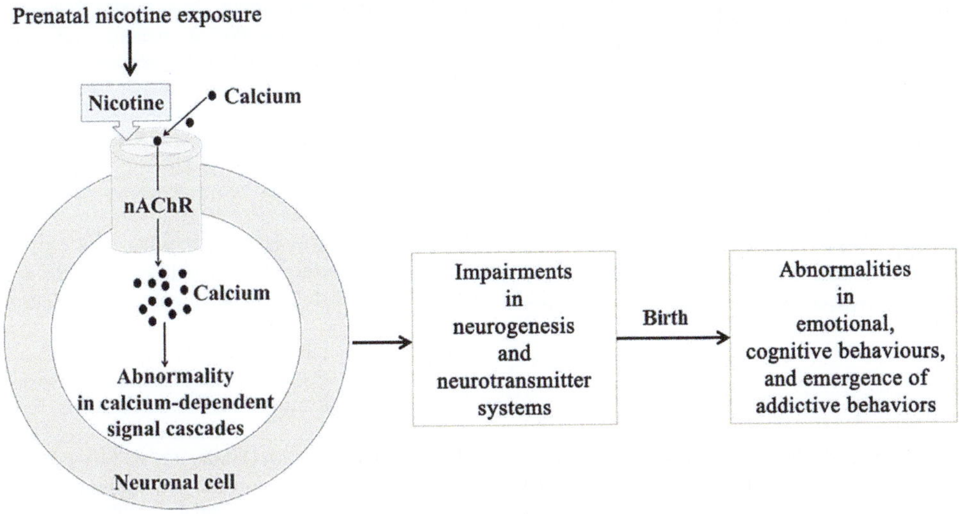

FIG. 24.4 Possible mechanisms of prenatal-nicotine-exposure-induced behavioral changes.
The neurotransmitter systems that are impaired by prenatal nicotine exposure, as described in this chapter, include acetylcholinergic, dopaminergic, serotonergic, and noradrenergic neurotransmitter systems. For the details of impairment of neurogenesis induced by prenatal nicotine exposure, please see Chapter "Prenatal Nicotine Exposure and Neuronal Progenitor Cells."

The volume of each solution is measured daily to access the preference of the animal toward any solution. An addictive rodent shows preference to nicotine solution and drinks more.

Y-maze test This test takes advantage of the natural tendency of rodents to explore novel spaces alternatively and evaluates spontaneous alternation behavior, an indicator of spatial working memory in a rodent. A rodent with poor working memory tends to repeat the "visit" to an arm (of Y-maze) that it just exited, instead of alternatingly "visit" the other arm of the maze.

Key Facts of Nicotine Replacement Therapy during Pregnancy

- To decrease the cravings for tobacco and to help quit smoking in general population, nicotine replacement therapy (NRT) has been developed.
- NRT, in conjunction with cognitive-behavioral therapy, is thought to be the most effective strategy to achieve smoking cessation during pregnancy, although it is equal to PNE in terms of nicotine exposure.
- Although NRT in pregnancy increases smoking cessation rates, both NRT and placebo are found to have similar efficacy.
- Preclinical studies using animal models provide sufficient data about the unsafety of PNE, which is the equivalent of NRT during human pregnancy.
- Placebo-randomized controlled clinical trials are needed to investigate whether or not NRT is safe in human pregnancy for smoking cessation.

Summary Points

- The harmful effects of tobacco smoking during pregnancy on the brain development are becoming an increasingly important public health issue.
- Children born from mothers who smoked during pregnancy suffer from neurobehavioral abnormalities.
- Nicotine is the main compound in tobacco that makes people smoke tobacco.
- Nicotine replacement treatments are being applied as an aid to smoking cessation during pregnancy.
- Prenatal nicotine exposure is harmful for fetal brain functional development and causes behavioral abnormalities in offspring.
- Prenatal nicotine exposure induces long-lasting abnormality in neurotransmitter system in offspring.

References

Abbott, L. C., & Winzer-Serhan, U. H. (2012). Smoking during pregnancy: lessons learned from epidemiological studies and experimental studies using animal models. *Critical Reviews in Toxicology*, *42*(4), 279–303.

Alkam, T., Kim, H. C., Hiramatsu, M., Mamiya, T., Aoyama, Y., Nitta, A., et al. (2013). Evaluation of emotional behaviors in young offspring of C57BL/6J mice after gestational and/or perinatal exposure to nicotine in six different time-windows. *Behavioural Brain Research*, *239*, 80–89.

Alkam, T., Kim, H. C., Mamiya, T., Yamada, K., Hiramatsu, M., & Nabeshima, T. (2013). Evaluation of cognitive behaviors in young offspring of C57BL/6J mice after gestational nicotine exposure during different time-windows. *Psychopharmacology*, *230*(3), 451–463.

Alkam, T., Mamiya, T., Kimura, N., Yoshida, A., Kihara, D., Tsunoda, Y., et al. (2017). Prenatal nicotine exposure decreases the release of dopamine in the medial frontal cortex and induces atomoxetine-responsive neurobehavioral deficits in mice. *Psychopharmacology*, *234*(12), 1853–1869.

Aoyama, Y., Toriumi, K., Mouri, A., Hattori, T., Ueda, E., Shimato, A., et al. (2016). Prenatal nicotine exposure impairs the proliferation of neuronal progenitors, leading to fewer glutamatergic neurons in the medial prefrontal cortex. *Neuropsychopharmacology*, *41*(2), 578–589.

Atluri, P., Fleck, M. W., Shen, Q., Mah, S. J., Stadfelt, D., Barnes, W., et al. (2001). Functional nicotinic acetylcholine receptor expression in stem and progenitor cells of the early embryonic mouse cerebral cortex. *Developmental Biology*, *240*(1), 143–156.

Bryden, D. W., Burton, A. C., Barnett, B. R., Cohen, V. J., Hearn, T. N., Jones, E. A., et al. (2016). Prenatal nicotine exposure impairs executive control signals in medial prefrontal cortex. *Neuropsychopharmacology*, *41*(3), 716–725.

Chang, G. Q., Karatayev, O., & Leibowitz, S. F. (2013). Prenatal exposure to nicotine stimulates neurogenesis of orexigenic peptide-expressing neurons in hypothalamus and amygdala. *The Journal of Neuroscience*, *33*(34), 13600–13611.

Christensen, M. H., Nielsen, M. L., & Kohlmeier, K. A. (2015). Electrophysiological changes in laterodorsal tegmental neurons associated with prenatal nicotine exposure: implications for heightened susceptibility to addict to drugs of abuse. *Journal of Developmental Origins of Health and Disease*, *6*(3), 182–200.

Clifford, A., Lang, L., & Chen, R. (2012). Effects of maternal cigarette smoking during pregnancy on cognitive parameters of children and young adults: a literature review. *Neurotoxicology and Teratology*, *34*(6), 560–570.

Cross, S. J., Linker, K. E., & Leslie, F. M. (2017). Sex-dependent effects of nicotine on the developing brain. *Journal of Neuroscience Research*, *95*(1–2), 422–436.

De Genna, N. M., Goldschmidt, L., Day, N. L., & Cornelius, M. D. (2016). Prenatal and postnatal maternal trajectories of cigarette use predict adolescent cigarette use. *Nicotine & Tobacco Research*, *18*(5), 988–992.

Dempsey, D. A., & Benowitz, N. L. (2001). Risks and benefits of nicotine to aid smoking cessation in pregnancy. *Drug Safety*, *24*(4), 277–322.

Done, C. J., & Sharp, T. (1994). Biochemical evidence for the regulation of central noradrenergic activity by 5-HT1A and 5-HT2 receptors: microdialysis studies in the awake and anaesthetized rat. *Neuropharmacology*, *33*(3–4), 411–421.

Dwyer, J. B., Broide, R. S., & Leslie, F. M. (2008). Nicotine and brain development. *Birth Defects Research. Part C, Embryo Today*, *84*(1), 30–44.

Fitzpatrick, C., Barnett, T. A., & Pagani, L. S. (2014). Parental bad habits breed bad behaviors in youth: exposure to gestational smoke and child impulsivity. *International Journal of Psychophysiology*, *93*(1), 17–21.

Garcia-Rill, E., Buchanan, R., McKeon, K., Skinner, R. D., & Wallace, T. (2007). Smoking during pregnancy: postnatal effects on arousal and attentional brain systems. *Neurotoxicology*, *28*(5), 915–923.

Hajos-Korcsok, E., McQuade, R., & Sharp, T. (1999). Influence of 5-HT1A receptors on central noradrenergic activity: microdialysis studies using (+/−)-MDL 73005EF and its enantiomers. *Neuropharmacology*, *38*(2), 299–306.

Hall, B. J., Cauley, M., Burke, D. A., Kiany, A., Slotkin, T. A., & Levin, E. D. (2016). Cognitive and behavioral impairments evoked by low-level exposure to tobacco smoke components: comparison with nicotine alone. *Toxicological Sciences*, *151*(2), 236–244.

Halperin, J. M., Newcorn, J. H., Koda, V. H., Pick, L., McKay, K. E., & Knott, P. (1997). Noradrenergic mechanisms in ADHD children with and without reading disabilities: a replication and extension. *Journal of the American Academy of Child and Adolescent Psychiatry*, *36*(12), 1688–1697.

Lacy, R. T., Mactutus, C. F., & Harrod, S. B. (2011). Prenatal IV nicotine exposure produces a sex difference in sensorimotor gating of the auditory startle reflex in adult rats. *International Journal of Developmental Neuroscience*, *29*(2), 153–161.

Lee, H., Chung, S., & Noh, J. (2016). Maternal nicotine exposure during late gestation and lactation increases anxiety-like and impulsive decision-making behavior in adolescent offspring of rat. *Toxicology Research*, *32*(4), 275–280.

Mantella, N. M., Kent, P. F., & Youngentob, S. L. (2013). Fetal nicotine exposure increases preference for nicotine odor in early postnatal and adolescent, but not adult, rats. *PLoS One*, *8*(12), e84989.

McNair, L. F., & Kohlmeier, K. A. (2015). Prenatal nicotine is associated with reduced AMPA and NMDA receptor-mediated rises in calcium within the laterodorsal tegmentum: a pontine nucleus involved in addiction processes. *Journal of Developmental Origins of Health and Disease*, *6*(3), 225–241.

Muhammad, A., Mychasiuk, R., Nakahashi, A., Hossain, S. R., Gibb, R., & Kolb, B. (2012). Prenatal nicotine exposure alters neuroanatomical organization of the developing brain. *Synapse*, *66*(11), 950–954.

Muneoka, K., Ogawa, T., Kamei, K., Mimura, Y., Kato, H., & Takigawa, M. (2001). Nicotine exposure during pregnancy is a factor which influences serotonin transporter density in the rat brain. *European Journal of Pharmacology*, *411*(3), 279–282.

Muneoka, K., Ogawa, T., Kamei, K., Muraoka, S., Tomiyoshi, R., Mimura, Y., et al. (1997). Prenatal nicotine exposure affects the development of the central serotonergic system as well as the dopaminergic system in rat offspring: involvement of route of drug administrations. *Brain Research Developmental Brain Research*, *102* (1), 117–126.

Mychasiuk, R., Muhammad, A., Gibb, R., & Kolb, B. (2013). Long-term alterations to dendritic morphology and spine density associated with prenatal exposure to nicotine. *Brain Research*, *1499*, 53–60.

Osadchy, A., Kazmin, A., & Koren, G. (2009). Nicotine replacement therapy during pregnancy: recommended or not recommended? *Journal of Obstetrics and Gynaecology Canada*, *31*(8), 744–747.

Parameshwaran, K., Buabeid, M. A., Karuppagounder, S. S., Uthayathas, S., Thiruchelvam, K., Shonesy, B., et al. (2012). Developmental nicotine exposure induced alterations in behavior and glutamate receptor function in hippocampus. *Cellular and Molecular Life Sciences*, *69*(5), 829–841.

Pauly, J. R., & Slotkin, T. A. (2008). Maternal tobacco smoking, nicotine replacement and neurobehavioural development. *Acta Paediatrica*, *97* (10), 1331–1337.

Ribary, U., & Lichtensteiger, W. (1989). Effects of acute and chronic prenatal nicotine treatment on central catecholamine systems of male and female rat fetuses and offspring. *The Journal of Pharmacology and Experimental Therapeutics*, *248*(2), 786–792.

Schneider, T., Bizarro, L., Asherson, P. J., & Stolerman, I. P. (2012). Hyperactivity, increased nicotine consumption and impaired performance in the five-choice serial reaction time task in adolescent rats prenatally exposed to nicotine. *Psychopharmacology*, *223*(4), 401–415.

Schneider, T., Ilott, N., Brolese, G., Bizarro, L., Asherson, P. J., & Stolerman, I. P. (2011). Prenatal exposure to nicotine impairs performance of the 5-choice serial reaction time task in adult rats. *Neuropsychopharmacology*, *36*(5), 1114–1125.

Seidler, F. J., Levin, E. D., Lappi, S. E., & Slotkin, T. A. (1992). Fetal nicotine exposure ablates the ability of postnatal nicotine challenge to release norepinephrine from rat brain regions. *Brain Research. Developmental Brain Research*, *69*(2), 288–291.

Shaywitz, B. A., Cohen, D. J., & Bowers, M. B., Jr. (1977). CSF monoamine metabolites in children with minimal brain dysfunction: evidence for alteration of brain dopamine. A preliminary report. *The Journal of Pediatrics*, *90*(1), 67–71.

Shekim, W. O., Javaid, J., Davis, J. M., & Bylund, D. B. (1983). Urinary MHPG and HVA excretion in boys with attention deficit disorder and hyperactivity treated with d-amphetamine. *Biological Psychiatry*, *18*(6), 707–714.

Shekim, W. O., Sinclair, E., Glaser, R., Horwitz, E., Javaid, J., & Bylund, D. B. (1987). Norepinephrine and dopamine metabolites and educational variables in boys with attention deficit disorder and hyperactivity. *Journal of Child Neurology*, *2*(1), 50–56.

Slotkin, T. A. (1998). Fetal nicotine or cocaine exposure: which one is worse? *The Journal of Pharmacology and Experimental Therapeutics*, *285*(3), 931–945.

Slotkin, T. A., MacKillop, E. A., Rudder, C. L., Ryde, I. T., Tate, C. A., & Seidler, F. J. (2007). Permanent, sex-selective effects of prenatal or adolescent nicotine exposure, separately or sequentially, in rat brain regions: indices of cholinergic and serotonergic synaptic function, cell signaling, and neural cell number and size at 6 months of age. *Neuropsychopharmacology*, *32*(5), 1082–1097.

Suzuki, M., Matsuda, T., Asano, S., Somboonthum, P., Takuma, K., & Baba, A. (1995). Increase of noradrenaline release in the hypothalamus of freely moving rat by postsynaptic 5-hydroxytryptamine1A receptor activation. *British Journal of Pharmacology*, *115*(4), 703–711.

Thapar, A., Fowler, T., Rice, F., Scourfield, J., van den Bree, M., Thomas, H., et al. (2003). Maternal smoking during pregnancy and attention deficit hyperactivity disorder symptoms in offspring. *The American Journal of Psychiatry*, *160*(11), 1985–1989.

Xu, Z., Seidler, F. J., Ali, S. F., Slikker, W., Jr., & Slotkin, T. A. (2001). Fetal and adolescent nicotine administration: effects on CNS serotonergic systems. *Brain Research*, *914*(1–2), 166–178.

Zhu, J., Fan, F., McCarthy, D. M., Zhang, L., Cannon, E. N., Spencer, T. J., et al. (2017). A prenatal nicotine exposure mouse model of methylphenidate responsive ADHD-associated cognitive phenotypes. *International Journal of Developmental Neuroscience*, *58*, 26–34.

Zhu, J., Lee, K. P., Spencer, T. J., Biederman, J., & Bhide, P. G. (2014). Transgenerational transmission of hyperactivity in a mouse model of ADHD. *The Journal of Neuroscience*, *34*(8), 2768–2773.

Zhu, J., Zhang, X., Xu, Y., Spencer, T. J., Biederman, J., & Bhide, P. G. (2012). Prenatal nicotine exposure mouse model showing hyperactivity, reduced cingulate cortex volume, reduced dopamine turnover, and responsiveness to oral methylphenidate treatment. *The Journal of Neuroscience*, 32(27), 9410–9418.

CHAPTER

25

Craving in Substance Use Disorders With a Focus on Cigarette Smoking

Stephen J. Wilson, Michael A. Sayette†*

*Department of Psychology, The Pennsylvania State University, University Park, PA, United States

†Department of Psychology, University of Pittsburgh, Pittsburgh, PA, United States

Abbreviation

fMRI functional magnetic resonance imaging

25.1 INTRODUCTION

Craving has been one of the most intensely studied—and most contentious—constructs in the field of addiction. This emphasis reflects widespread agreement that craving is a clinically significant feature of drug addiction (Tiffany & Wray, 2012). In particular, the links between craving and substance use have been documented extensively in research on cigarette smoking (Ferguson & Shiffman, 2009), which remains one of the leading preventable causes of death in the world (Reitsma et al., 2017). Although the precise nature of the relationship between craving and cigarette use remains an active area of research and ongoing debate (e.g., see Sayette & Tiffany, 2013; Wray, Gass, & Tiffany, 2013), results from several studies suggest that craving undermines attempts to quit smoking (e.g., Allen, Bade, Hatsukami, & Center, 2008; Bagot, Heishman, & Moolchan, 2007; Killen & Fortmann, 1997; Shiffman, Paty, Gnys, Kassel, & Hickcox, 1996). Acute craving episodes—rapid and intense increases in urge that are triggered by particular situations or stimuli (e.g., by exposure to cigarette cues)—may have an especially pronounced impact on behavior (Ferguson & Shiffman, 2009).

This chapter provides a brief overview of select issues and observations pertaining to the role of craving in substance use, with a focus on craving for cigarettes and smoking behavior. We first consider the association between craving and desire and question whether craving manifests as a linear or nonlinear phenomenon—a fundamental issue that remains unresolved within the field of addiction. We then briefly address the effects of craving on self-regulation, including research demonstrating that craving leads to shifts in cognitive processing that increase the likelihood of drug use. Finally, we discuss the increasingly common approach of studying craving using brain imaging methods and the complementary role that neuroimaging and behavioral methods can serve in advancing the science of craving.

25.2 THE CONSTRUCT OF CRAVING

Drug craving is a hypothetical construct that is invoked to explain observations and advance research on smoking and other forms of drug use (Sayette, 2016). Accordingly, it is essential that craving be carefully defined and meaningfully tied to observable variables within a coherent theoretical framework (Cronbach & Meehl, 1955). This has proved to be a challenge, however, as addiction theorists have struggled to define and conceptualize craving. While there is general agreement that craving is best conceptualized as a state of heightened motivation to seek, obtain, and use drugs (although see Tiffany, 1990), there are numerous definitional and conceptual issues that remain unresolved (for a more detailed review, see Sayette et al., 2000).

One unsettled issue with significant implications concerns whether craving should be conceptualized as a continuous phenomenon that can be measured along a continuum of desire or, alternatively, whether craving should instead be thought of as an extreme state of yearning that is qualitatively distinct from less intense desires (Abrams, 2000; Kozlowski & Wilkinson, 1987; Sayette et al., 2000). In the empirical literature, it is common for addiction researchers to assess craving as a continuous

Neuroscience of Nicotine
https://doi.org/10.1016/B978-0-12-813035-3.00025-3

variable that ranges in intensity from nonexistent/low to high, with the terms craving and desire often used interchangeably. From a conceptual and clinical standpoint, however, it may be best to view craving as a construct that, at its core, represents not only any desire to use drugs but also specifically an intense or extreme desire to do so. As we have noted elsewhere (Wilson & Sayette, 2015), leading addiction researchers such as Nora Volkow and George Koob have recognized the "overwhelming" and "intense" nature of craving, considering this acute experience to reflect an "unstoppable force" leading to drug intake and to represent a core feature of addiction (George & Koob, 2013; Volkow et al., 2010).

The emphasis on the penetrating nature of craving and the notion that a marked desire to use drugs is clinically significant is also reflected in the latest edition of the *Diagnostic and Statistical Manual of Mental Disorders* (DSM-5; American Psychiatric Association, 2013; p. 483). Craving was added to the DSM-5 as one of the criteria for substance use disorders, with the manual defining it as "an intense desire or urge" that should be assessed by "asking [individuals] if there has ever been a time when they had such strong urges to use the drug that they could not think of anything else." Thus, there appears to be conceptual, clinical, and diagnostic utility to distinguishing between craving and weak-to-moderate states of desire, even though this distinction is seldom made in studies that involve the manipulation and assessment of craving.

25.3 CRAVING, COGNITION, AND SELF-REGULATION

Despite continued uncertainty regarding the conceptualization and boundaries of craving, there is compelling evidence that it is a clinically meaningful construct (Sayette, 2016; Tiffany & Wray, 2012). Among other things, craving has been found to change cognitive processing in ways that are thought to contribute to drug use and relapse (for elaboration, see Sayette & Creswell, 2016). For instance, results from studies using a variety of methods indicate that craving demands nonautomatic (limited capacity) cognitive processing resources (Tiffany, 1990; Waters & Sayette, 2006). In other words, craving may "tie up" mental resources, such as those required for the active maintenance of information in working memory. Furthermore, the maintenance of craving-related representations may lead to an increasing amount of similar information entering the contents of working memory via selective attention to relevant stimuli in the environment (e.g., cigarette butts on the pavement) and the active elaboration of thoughts related to drug use desires (Kavanagh, Andrade, & May, 2005). Recent findings suggest that the impact of these effects on cognitive processing may be exacerbated because craving also interferes with the ability to recognize these effects while they are occurring (e.g., Sayette, Schooler, & Reichle, 2010).

One approach that has proved to be particularly useful for studying the effects of craving on cognitive processing involves producing robust ("peak-provoked") cravings using a combination of acute drug deprivation and the presentation of highly salient drug cues (Sayette & Tiffany, 2013). For example, one study used this strategy to examine the way in which craving influences how smokers generate and evaluate smoking-related information by comparing a peak-craving condition to a low-craving control condition in which participants had recently smoked and were exposed to a neutral cue (Sayette & Hufford, 1997). In the peak-craving condition, participants generated a list of smoking characteristics that was positively biased, relative to when they were not craving (i.e., peak craving was associated with a significant increase in the number of positive, but not negative, characteristics that were generated). In addition to affecting the generation of drug-related information, craving also appears to alter the evaluation of such material. In two studies, smokers in a peak-craving condition tended to judge positive smoking consequences to be more probable, relative to negative ones, than did those in a low-craving condition (Sayette, Loewenstein, Kirchner, & Travis, 2005; Sayette, Martin, Wertz, Shiffman, & Perrott, 2001). These findings indicate that craving may cause a shift in expectations regarding the anticipated outcomes of drug use, such that positive outcomes appear more salient and likely than negative ones. This may be compounded by the blunting effect that craving appears to have on the valuation of nondrug incentives (Piper, 2015; Wilson, Delgado, et al., 2014). It is not hard to imagine how these shifts in cognitive processing might contribute to drug use. For example, a smoker who is craving and about to relapse might focus on the perceived benefits of smoking (e.g., how a cigarette would taste) and underestimate the potential negative consequences (e.g., likelihood of dying from a smoking-related disease), such that the decision to smoke may appear rational in that moment.

Craving also appears to affect temporal cognition in ways that are likely to contribute to the maintenance of drug use and self-regulation failure in those trying to quit. Namely, smokers in a state of craving describe time as passing more slowly than those who are not craving (Klein, Corwin, & Stine, 2003; Sayette et al., 2005). Additionally, smokers in a peak-craving state anticipate that their urges will steadily climb over time if they go without smoking, increasing the perceived utility (perhaps even necessity) of cigarette use as a way to alleviate this unpleasant state (Sayette et al., 2005). Importantly, this expectation is inaccurate, as cravings tend to dissipate naturally within a relatively short period of time regardless of whether smoking occurs (Marlatt, 1985; Niaura et al., 1999).

In summary, there is converging evidence that craving affects cognitive processing in ways that may increase the likelihood of drug use and relapse (see also Baker, Morse, & Sherman, 1986). During cravings states, individuals who use drugs may be more likely to notice drug-related cues in their environment and, in turn, may consider the act of consuming to be more attractive, whereas cues associated with nondrug activities become less attractive than they would in a neutral state. Moreover, altered time perception may hinder attempts to resist this appeal. These shifts in cognitive change processes are unlikely to be an exhaustive set of the potential links between craving and drug uses. Whether these changes are manifestations of craving or merely effects of craving depends on one's conceptualization (Sayette et al., 2000). Regardless, these changes highlight the importance of carefully considering the assessment context when measuring attitudes about drug use. A treatment provider who learns, for example, that a smoker holds a negative view of smoking and is highly motivated to quit may be surprised to learn that the individual has relapsed days later. However, it may have become clear that the smoker was prone to ambivalence about quitting had they been evaluated while in a state of craving. Such motivational shifts can undermine a quit attempt, as it takes only a single moment of weakness during a high-risk situation for an individual who is otherwise committed to quitting to reconsider and resume smoking (Sayette & Creswell, 2016). An important objective for future research is to characterize the various ways in which drug craving affects cognitive processing and subsequent decision-making."INTEGRATING NEUROBIOLOGICALAND BEHAVIORAL INVESTIGATIONS OFCRAVING"6,1

25.4 INTEGRATING NEUROBIOLOGICAL AND BEHAVIORAL INVESTIGATIONS OF CRAVING

The use of functional brain imaging techniques (in particular, functional magnetic resonance imaging or fMRI) has become an increasingly common approach to examining craving, resulting in a dramatic increase in the number of such studies in recent years (Wilson, 2015). Functional brain imaging methods allow researchers to measure brain activity under conditions designed to stimulate drug cravings. Over a relatively short period of time, the field has advanced beyond a focus on localizing brain responses to drug-related cues and increasingly addresses more nuanced questions pertaining to neurobiology and craving (Ekhtiari, Nasseri, Yavari, Mokri, & Monterosso, 2016, chap. 7). As reviewed elsewhere, this work presents both challenges and opportunities (Sayette & Wilson, 2015).

Arguably, many studies have been too quick to adopt methods for inducing craving from the behavioral literature that may be less than ideal for neuroimaging research. Due to the low signal-to-noise ratio associated with most brain imaging methods (Culham, 2006), studies have often utilized designs in which cues designed to provoke craving are presented repeatedly, typically interspersed with ostensibly neutral cues designed to evoke minimal changes in desire. In line with basic concepts from measurement and statistics (Nunnally, 1978), the rationale of this approach is that the reliability of the estimated brain response to craving-related versus neutral cues increases as more stimuli of each type are presented and responses are averaged across trials or blocks within each condition. While well reasoned from a psychometric perspective, the rapid multiple presentation design used by many studies may not be an optimal match to the typical time course of craving responses, which persist for several minutes (Heishman, Lee, Taylor, & Singleton, 2010). Another potential drawback of the interleaved designs often used in neuroimaging research is the brain activity associated with craving-related cues may carry over and become confounded with the signal measured during subsequent control trials/blocks, reducing the ability to detect differences between conditions (Franklin et al., 2011; Wilson, Sayette, Fiez, & Brough, 2007). A more general concern is that several neuroimaging studies seeking to evaluate craving instead have examined mild states of desire, which may have important implications for the nature and interpretation of associated results (Wilson & Sayette, 2015).

Notwithstanding these challenges, neuroimaging research on patterns of brain activity related to the experience and regulation of craving has the potential to lead to important scientific advances and inform a new generation of behavioral research. Indeed, neuroimaging methods can serve as a valuable compliment to approaches that involve measuring subjective responses and behavior (Berkman, Falk, & Lieberman, 2011; Wilson, Smyth, & MacLean, 2014). In this manner, insights derived using both behavioral and neuroimaging methods can feed into a cycle of research in which new ideas are generated, tested, and refined over time. For example, one line of behavioral research suggested that perceived drug use opportunity can affect craving in significant ways (Wertz & Sayette, 2001). This behavioral research was used to help organize results from neuroimaging studies of craving that previously were inconsistent (see Wilson, Sayette, & Fiez, 2004) and subsequently led to empirical studies identifying distinct neural patterns associated with contexts that varied in terms of perceived drug use opportunity (e.g., Wilson, Sayette, & Fiez, 2012). Future research integrating brain imaging with other methods for assessing behavior, such as ecological momentary assessment and the coding of affective facial expressions, may lead to further advances in understanding craving.

25.5 CONCLUSIONS

It is clear that, despite unresolved conceptual and methodological issues, craving continues to be a central construct in the study and treatment of addiction. In clinical settings, the craving experienced by individuals who use drugs can be assessed easily and reliably via self-report. Similarly, methods exist for inducing robust craving states under controlled laboratory conditions so that craving-related responses and subsequent effects on decision-making and behavior can be examined. Nonetheless, although there is widespread agreement that craving is a clinically and conceptually meaningful concept, more research is still needed to better understand this complex, multimodal experience (Abrams, 2000; Sayette et al., 2000).

Previous studies have underscored the potential importance of craving intensity, and additional research is required to determine if distinctions between mild and powerful desires to use represent two points on a craving continuum or fundamentally distinct experiences. More work is also needed to characterize the mechanisms that link craving and behavior and information that would be particularly useful for improving the treatment of addiction. This chapter outlined some explanations for why craving may promote relapse, including the effect that it has on the processing of drug- and nondrug-related information and temporal cognition. These are likely to represent only a subset of the processes that underlie the association between craving and drug use, and additional mechanisms are likely to be discovered with continued experimental research.

Given the ease with which they are implemented, it is likely that self-report measures will continue to play a major role in research on craving, though it is also a priority for the field to continue developing and refining nonverbal approaches to the assessment of craving. Toward this end, neuroimaging methods have proved to be particularly valuable for measuring nonverbal craving responses. As illustrated by examples offered in this chapter, brain imaging methods can be combined with cutting-edge behavioral methods and findings in creative ways that hold tremendous promise for advancing our understanding of the link between craving and substance use. The results from such integrative work will also help to refine interventions aimed at reducing craving to facilitate behavior change.

MINI-DICTIONARY OF TERMS

Craving Generally conceptualized as a state of heightened motivation to seek, obtain, and use drugs, with debate regarding whether craving should be thought of as an extreme state of yearning that is qualitatively distinct from less intense desires.

Cue reactivity paradigm Exposing substance users to drug cues in order to elicit and measure concomitant changes in one or more response systems (e.g., self-reported urge and cognitive task performance).

Functional magnetic resonance imaging A type of functional neuroimaging that involves measuring a signal related to changes in blood flow that occur in conjunction with changes in neural activity.

Functional neuroimaging A group of methods that noninvasively measure direct or indirect signals associated with neural activity.

Self-regulation The capacity to alter or override one's own responses.

Key Facts of Functional Brain Imaging

- Functional brain imaging methods allow researchers to study how the brain works as humans perform various tasks.
- Functional magnetic resonance imaging, or fMRI, is currently the most widely used functional brain imaging technique.
- The availability of fMRI and other functional brain imaging methods has helped to stimulate new interdisciplinary areas of science, such as cognitive neuroscience and social neuroscience.

Summary Points

- Craving is a central construct in the field of addiction research.
- While craving is generally conceptualized as a state of heightened motivation to seek, obtain, and use drugs, many definitional and conceptual issues about the construct remain unresolved.
- One important unsettled issue concerns the association between desire and craving and the extent to which mild and intense desires are quantitatively versus qualitatively different.
- Despite continued uncertainty regarding the conceptualization and boundaries of craving, there is compelling evidence that it is a clinically meaningful construct.
- Research suggests that craving changes cognitive processing in various ways that may contribute to drug use and relapse.
- The use of functional brain imaging methods has become an increasingly common approach to examining the effects of craving on cognition and behavior.
- Additional research, particularly interdisciplinary work that integrates multiple methods, is needed to advance our understanding of the link between craving and substance use.

References

Abrams, D. B. (2000). Transdisciplinary concepts and measures of craving: commentary and future directions. *Addiction, 95*(8s2), 237–246.

Allen, S. S., Bade, T., Hatsukami, D., & Center, B. (2008). Craving, withdrawal, and smoking urges on days immediately prior to smoking relapse. *Nicotine & Tobacco Research, 10*(1), 35–45. https://doi.org/10.1080/14622200701705076.

American Psychiatric Association (2013). *Diagnostic and statistical manual of mental disorders (DSM-5®)*. Arlington, VA: American Psychiatric Publishing.

Bagot, K. S., Heishman, S. J., & Moolchan, E. T. (2007). Tobacco craving predicts lapse to smoking among adolescent smokers in cessation treatment. *Nicotine & Tobacco Research, 9*(6), 647–652. https://doi.org/10.1080/14622200701365178.

Baker, T. B., Morse, E., & Sherman, J. E. (1986). The motivation to use drugs: a psychobiological analysis of urges. *Nebraska Symposium on Motivation, 34*, 257–323.

Berkman, E. T., Falk, E. B., & Lieberman, M. D. (2011). In the trenches of real-world self-control: neural correlates of breaking the link between craving and smoking. *Psychological Science, 22*(4), 498–506. https://doi.org/10.1177/0956797611400918.

Cronbach, L. J., & Meehl, P. E. (1955). Construct validity in psychological tests. *Psychological Bulletin, 52*(4), 281.

Culham, J. C. (2006). *Functional neuroimaging: experimental design and analysis*. Cambridge, MA: MIT Press.

Ekhtiari, H., Nasseri, P., Yavari, F., Mokri, A., & Monterosso, J. (2016). Neuroscience of drug craving for addiction medicine: from circuits to therapies. In: H. Ekhtiari, & M. Paulus (Eds.), *Progress in brain research*, Vol. 223 (pp. 115–141). Cambridge. MA: Elsevier.

Ferguson, S. G., & Shiffman, S. (2009). The relevance and treatment of cue-induced cravings in tobacco dependence. *Journal of Substance Abuse Treatment, 36*(3), 235–243.

Franklin, T. R., Wang, Z., Suh, J. J., Hazan, R., Cruz, J., Li, Y., et al. (2011). Effects of varenicline on smoking cue-triggered neural and craving responses. *Archives of General Psychiatry, 68*(5), 516–526. https://doi.org/10.1001/archgenpsychiatry.2010.190.

George, O., & Koob, G. F. (2013). Control of craving by the prefrontal cortex. *Proceedings of the National Academy of Sciences, 110*(11), 4165–4166.

Heishman, S. J., Lee, D. C., Taylor, R. C., & Singleton, E. G. (2010). Prolonged duration of craving, mood, and autonomic responses elicited by cues and imagery in smokers: effects of tobacco deprivation and sex. *Experimental and Clinical Psychopharmacology, 18*(3), 245–256. https://doi.org/10.1037/a0019401 pii: 2010-11933-006.

Kavanagh, D. J., Andrade, J., & May, J. (2005). Imaginary relish and exquisite torture: the elaborated intrusion theory of desire. *Psychological Review, 112*(2), 446.

Killen, J. D., & Fortmann, S. P. (1997). Craving is associated with smoking relapse: findings from three prospective studies. *Experimental and Clinical Psychopharmacology, 5*(2), 137–142.

Klein, L., Corwin, E., & Stine, M. (2003). Smoking abstinence impairs time estimation accuracy in cigarette smokers. *Psychopharmacology Bulletin, 37*(1), 90–95.

Kozlowski, L. T., & Wilkinson, D. A. (1987). Use and misuse of the concept of craving by alcohol, tobacco, and drug researchers. *Addiction, 82*(1), 31–36.

Marlatt, G. A. (1985). Cognitive factors in the relapse process. In G. A. Marlatt, & J. R. Gordon (Eds.), *Relapse prevention: Maintenance strategies in the treatment of addictive behaviors* (pp. 128–200). New York: Guilford.

Niaura, R., Abrams, D. B., Shadel, W. G., Rohsenow, D. J., Monti, P. M., & Sirota, A. D. (1999). Cue exposure treatment for smoking relapse prevention: a controlled clinical trial. *Addiction, 94*(5), 685–695.

Nunnally, J. C. (1978). *Psychometric theory* (2nd ed.). New York: McGraw-Hill.

Piper, M. E. (2015). Withdrawal: expanding a key addiction construct. *Nicotine & Tobacco Research, 17*(12), 1405–1415.

Reitsma, M. B., Fullman, N., Ng, M., Salama, J. S., Abajobir, A., Abate, K. H., et al. (2017). Smoking prevalence and attributable disease burden in 195 countries and territories, 1990–2015: a systematic analysis from the Global Burden of Disease Study 2015. *The Lancet, 389*(10082), 1885–1906. https://doi.org/10.1016/S0140-6736(17)30819-X.

Sayette, M. A. (2016). The role of craving in substance use disorders: theoretical and methodological issues. *Annual Review of Clinical Psychology, 12*(1), 407–433. https://doi.org/10.1146/annurev-clinpsy-021815-093351.

Sayette, M. A., & Creswell, K. G. (2016). Self-regulatory failure and addiction. *Handbook of self-regulation: Research, theory, and applications*, Vol. 3 (pp. 571–590). New York: Guilford Press.

Sayette, M. A., & Hufford, M. R. (1997). Effects of smoking urge on generation of smoking-related information. *Journal of Applied Social Psychology, 27*(16), 1395–1405.

Sayette, M. A., Loewenstein, G., Kirchner, T. R., & Travis, T. (2005). Effects of smoking urge on temporal cognition. *Psychology of Addictive Behaviors, 19*(1), 88.

Sayette, M. A., Martin, C. S., Wertz, J. M., Shiffman, S., & Perrott, M. A. (2001). A multi-dimensional analysis of cue-elicited craving in heavy smokers and tobacco chippers. *Addiction, 96*(10), 1419–1432.

Sayette, M. A., Schooler, J. W., & Reichle, E. D. (2010). Out for a smoke: the impact of cigarette craving on zoning out during reading. *Psychological Science, 21*(1), 26–30.

Sayette, M. A., Shiffman, S., Tiffany, S. T., Niaura, R. S., Martin, C. S., & Schadel, W. G. (2000). The measurement of drug craving. *Addiction, 95*(8s2), 189–210. https://doi.org/10.1046/j.1360-0443.95.8s2.8.x.

Sayette, M. A., & Tiffany, S. T. (2013). Peak provoked craving: an alternative to smoking cue-reactivity. *Addiction, 108*(6), 1019–1025.

Sayette, M. A., & Wilson, S. J. (2015). The measurement of craving and desires. In W. Hofmann, & L. Nordgren (Eds.), *The psychology of desire* (pp. 104–126). New York: Guilford.

Shiffman, S., Paty, J. A., Gnys, M., Kassel, J. A., & Hickcox, M. (1996). First lapses to smoking: within-subjects analysis of real-time reports. *Journal of Consulting and Clinical Psychology, 64*(2), 366–379.

Tiffany, S. T. (1990). A cognitive model of drug urges and drug-use behavior: role of automatic and nonautomatic processes. *Psychological Review, 97*(2), 147.

Tiffany, S. T., & Wray, J. M. (2012). The clinical significance of drug craving. *Annals of the New York Academy of Sciences, 1248*(1), 1–17. https://doi.org/10.1111/j.1749-6632.2011.06298.x.

Volkow, N. D., Wang, G. J., Fowler, J. S., Tomasi, D., Telang, F., & Baler, R. (2010). Addiction: decreased reward sensitivity and increased expectation sensitivity conspire to overwhelm the brain's control circuit. *BioEssays, 32*(9), 748–755.

Waters, A. J., & Sayette, M. A. (2006). Implicit cognition and tobacco addiction. In *Handbook of implicit cognition and addiction* (pp. 309–338). London: SAGE.

Wertz, J. M., & Sayette, M. A. (2001). A review of the effects of perceived drug use opportunity on self-reported urge. *Experimental and Clinical Psychopharmacology, 9*(1), 3.

Wilson, S. J. (Ed.). (2015). *The Wiley handbook on the cognitive neuroscience of addiction*. Chichester, UK: John Wiley & Sons, Ltd.

Wilson, S. J., Delgado, M. R., McKee, S. A., Grigson, P. S., MacLean, R. R., Nichols, T. T., et al. (2014). Weak ventral striatal responses to monetary outcomes predict an unwillingness to resist cigarette smoking. *Cognitive, Affective, & Behavioral Neuroscience, 14*(4), 1196–1207.

Wilson, S. J., & Sayette, M. A. (2015). Neuroimaging craving: urge intensity matters. *Addiction, 110*(2), 195–203. https://doi.org/10.1111/add.12676.

Wilson, S. J., Sayette, M. A., & Fiez, J. A. (2004). Prefrontal responses to drug cues: a neurocognitive analysis. *Nature Neuroscience, 7*(3), 211–214.

Wilson, S. J., Sayette, M. A., & Fiez, J. A. (2012). Quitting-unmotivated and quitting-motivated cigarette smokers exhibit different patterns of cue-elicited brain activation when anticipating an opportunity to smoke. *Journal of Abnormal Psychology, 121*(1), 198–211. https://doi.org/10.1037/a0025112.

Wilson, S. J., Sayette, M. A., Fiez, J. A., & Brough, E. (2007). Carry-over effects of smoking cue exposure on working memory performance. *Nicotine & Tobacco Research, 9*(5), 613–619. https://doi.org/10.1080/14622200701243144.

Wilson, S. J., Smyth, J. M., & MacLean, R. R. (2014). Integrating ecological momentary assessment and functional brain imaging methods: new avenues for studying and treating tobacco dependence. *Nicotine & Tobacco Research, 16*(Suppl. 2), S102–S110. https://doi.org/10.1093/ntr/ntt129.

Wray, J. M., Gass, J. C., & Tiffany, S. T. (2013). A systematic review of the relationships between craving and smoking cessation. *Nicotine & Tobacco Research, 15*(7), 1167–1182.

C H A P T E R

26

The Acute Effect of Exercise on Cravings and Withdrawal Symptoms

Wuyou Sui, Scott Rollo, Harry Prapavessis

Faculty of Health Sciences, School of Kinesiology, The University of Western Ontario, London, ON, Canada

Abbreviations

DtS	desire to smoke
HR_{max}	heart rate max
MPSS	Mood and Physical Symptoms Scale
NRT	nicotine replacement therapy
PA	physical activity
RPE	rating of perceived exertion
SJWS	Shiffman-Jarvik Withdrawal Scale
SoD	strength of desire to smoke
TWS	tobacco withdrawal symptoms

This chapter will foremost summarize the effects of an acute bout of exercise on smoking, specifically the acute effect of exercise on cravings and tobacco withdrawal symptoms (TWS). This chapter will also delve into proposed mechanisms for these effects and present clinical recommendations for implementing acute bouts of exercise as a therapy/intervention.

26.1 CRAVINGS

All smokers, whether they attempt to quit or not, experience some form of cravings or desire to smoke. These cravings may be episodic in nature, strong, intermittent or habitual, and often triggered by cues or the environment (Shiffman, 2000). Cravings may also manifest as background cravings, which persist throughout the day as a chronic desire to smoke. Both types of cravings tend to present more symptomatically in those with higher nicotine dependence (Dunbar, Shiffman, Kirchner, Tindle, & Scholl, 2014). Craving strength can be assessed by a number of ways: self-report questionnaires, cue reactivity, attentional bias, or through fMRI. Of these, self-report questionnaires are primarily used, measuring either desire to smoke (DtS; Tiffany & Drobes, 1991) or strength of desire to smoke (SoD; West & Hajek, 2004).

Pharmacological therapy (i.e., nicotine replacement therapy (NRT)), is the standard treatment prescribed to alleviate cravings. NRT binds to the nicotinic acetylcholine receptors in the brain, replacing nicotine from cigarettes. NRT can be administered through a variety of modes: gum, lozenge, oral spray, or dissolved through a patch. Nicotine gum, lozenges, and spray offer a relatively quick dose of nicotine to the body—hence, are more effective at curbing episodic cravings. By contrast, the patch is more effective at curbing background cravings; nicotine release occurs over a longer period of time. Overall, NRT is effective for alleviating acute cravings (Kotlyar et al., 2007).

26.1.1 Effect of Acute Exercise on Cravings

A single bout of exercise significantly and immediately decreases craving levels in both temporary abstinent smokers and smokers in a quit attempt. The beneficial effects of exercise are most established during and immediately following an acute bout and can last up to 30 min post exercise (Ussher, Cropley, Playle, Mohidin, & West, 2009). The craving reduction effect of acute exercise is comparable with or greater than NRT aids (Haasova et al., 2013). Additionally, the speed at which acute exercise can provide craving relief may be faster than NRT, which can take up to 30 min to attain maximal relief (Hansson, Hajek, Perfekt, & Kraiczi, 2012). The versatility and cost-effectiveness of exercise means barriers like financial burden or availability of NRT are removed.

The strength of craving reduction does vary based upon a number of factors: the intensity, duration, and

Neuroscience of Nicotine
https://doi.org/10.1016/B978-0-12-813035-3.00026-5

type of exercise; the presence of stressors; and concurrent therapy must be considered when determining the effectiveness of acute exercise on cravings (Figs. 26.1–26.3).

Intensity of exercise. The role of exercise intensity has been examined in light-, moderate-, and vigorous-intensity protocols. While all three intensities are effective in reducing cravings, moderate and vigorous levels of intensity are significantly more effective; vigorous bouts do not seem to reduce cravings significantly more than moderate bouts (Haasova et al., 2014). Some research has suggested that vigorous bouts can have a negative impact on affect (Everson, Daley, & Ussher, 2008). From a clinical perspective, moderate intensity is easier for the exerciser—which may increase adherence and enjoyment from exercise.

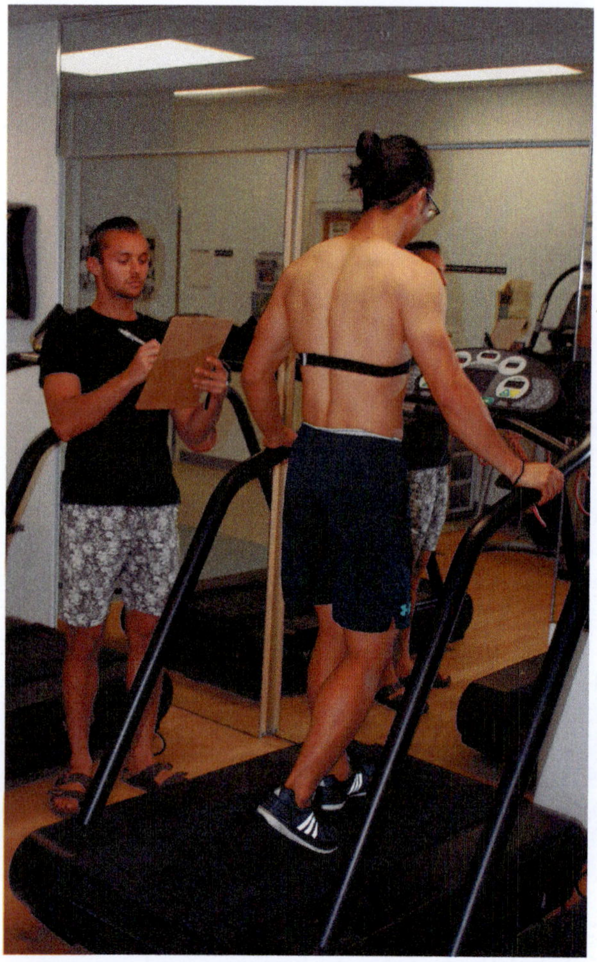

FIG. 26.1 Typical aerobic exercise intervention setup.
The researcher is monitoring the subject's rating of perceived exertion (RPE) and heart rate in order to maintain a specific intensity. *Figure is unpublished.*

Duration of exercise. The duration of exercise has an effect on craving reduction. Medium (10 min) and long bouts (15+ min) of acute exercise have elicited significant craving relief in temporary abstinent smokers (Haasova et al., 2014). Longer durations show slightly larger reduction in measures of cravings; however, from a clinical standpoint, medium duration bouts may be adequate for sufficient craving relief (see Fig. 26.4).

Type of exercise. Some forms of activity have a more pronounced effect on cravings. Forms of aerobic exercise attenuate the strength of cravings similarly. Isometric exercises have also been shown to reduce cravings in temporarily abstinent smokers, although to a much lesser degree than aerobic exercise (Haasova et al., 2014). Yoga can also significantly decrease cravings (Elibero, Janse Van Rensburg, & Drobes, 2011). Resistance training, albeit briefly examined, has shown no significant craving reduction (Ho et al., 2014); future research into resistance training and cravings is needed. The sheer variety in types of exercise speaks to its application as a therapy: convenient, immediate relief, and with similar or better results as NRT.

Presence of stressors. The presence of other smokers, smoking paraphernalia, cognitive stress, or triggers for smoking behavior can all spike a smoker's (episodic) craving levels. Given how common these environmental triggers are, "cue-elicited cravings" (Taylor & Katomeri, 2007) pose a serious threat for those looking to avoid relapse.

Exercise can still attenuate craving levels (by as much as 8%) when presented with a stressor 30 min postexercise

FIG. 26.2 Classification of exercise intensities.
The figure describes the hallmark measures of light-, moderate-, and vigorous-intensity activities through examples of walking, cycling, and sprinting, respectively. *Figure is unpublished; photos used through CC0 license via stocksnap.io.*

FIG. 26.3 Classification of types of exercise and associated effect on cravings/withdrawal symptoms.
The figure summarizes the effect of specific types/modes of exercise on cravings and tobacco withdrawal symptoms (TWS). *Figure is unpublished; photos used through CC0 license via stocksnap.io.*

(Taylor & Katomeri, 2007). However, this sequential presentation of a stressor is less ecologically valid than concurrent stressors. Fong, De Jesus, Bray, and Prapavessis (2014) found that when stressors are met with acute exercise concurrently, exercise still significantly reduced SoD by 37% at postexercise, which attests to the pervasive effect of exercise on cravings. However, researchers did not report any differences in time to first puff between groups, suggesting concurrent stressors may limit exercise's effect on cravings post exercise.

Concurrent therapies. Because acute exercise can produce similar reductions in cravings as NRT, the potential combined effect of acute exercise and NRT, or other therapies, together has been examined in detail.

Acute exercise, in conjunction with NRT, can significantly reduce cravings, as compared to NRT alone. In temporarily abstinent smokers, combining NRT and a 15 min bout of acute exercise exhibits a greater, consistent reduction in cravings for at least 40 min post exercise (Tritter, Fitzgeorge, & Prapavessis, 2015). Even among those in a cessation attempt who currently use NRT, acute exercise maintains its additive reduction in cravings. Interestingly, acute exercise is able to maintain a consistent reductive effect, even as the dose of NRT

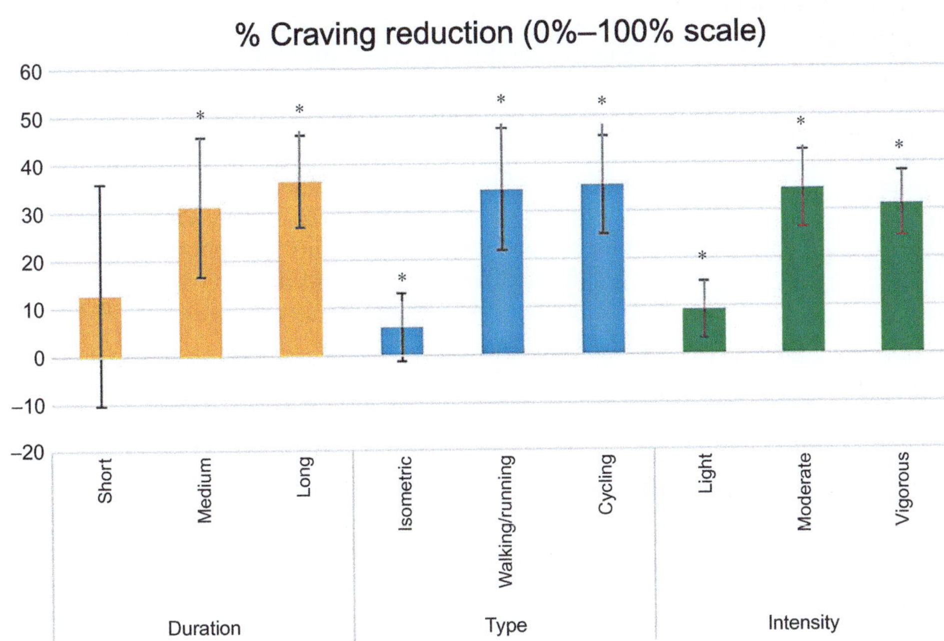

FIG. 26.4 Percent craving reduction associated with parameters of exercise.
Bars denote the average reduction in cravings associated with a particular parameter of exercise. Asterisks indicate significance ($P < .05$). *Figure adapted from data by Haasova et al. (2014); figure is unpublished.*

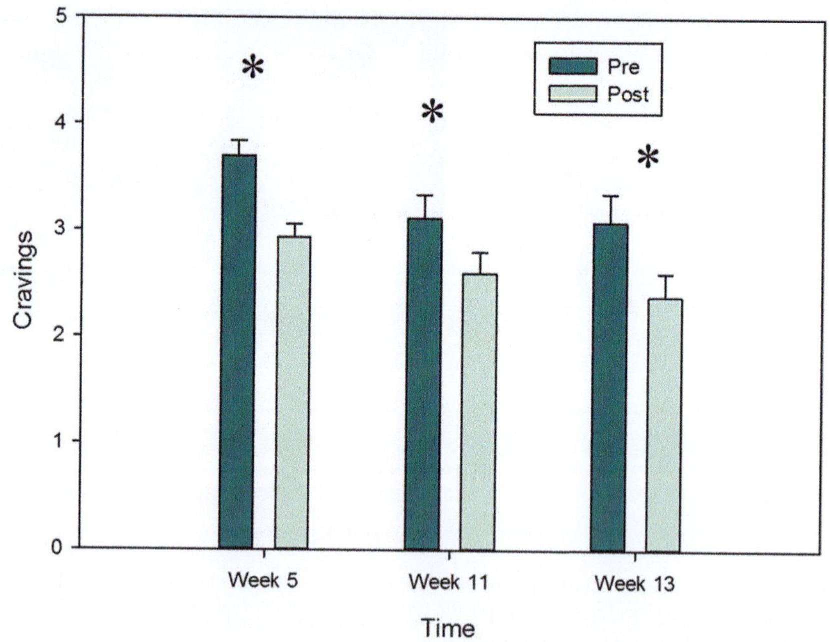

FIG. 26.5 Craving levels following a bout of acute exercise during cessation program. The graph indicates level of cravings before and after a bout of acute exercise during a cessation program. Week 5, week 7, and week 13 represent 1-, 7-, and 9-week post quit date, respectively. Asterisk indicates significance ($P < .05$). *Adapted from data by Harper et al. (2012); figure is unpublished.*

decreased (21 mg → 14 mg → 7 mg) over the cessation program (see Fig. 26.5; Harper, Fitzgeorge, Tritter, & Prapavessis, 2012). The additive effect of exercise has been theorized to occur due to physical activity and nicotine providing craving relief through separate and distinct mechanisms.

Even in an NRT cessation program, where background cravings may play a more significant role in an individual's cravings, acute exercise significantly moderates the level of cravings (see Fig. 26.5; Harper et al., 2012). Acute exercise is an effective means to immediately reduce cravings in smokers experiencing elevated craving levels in the presence of concurrent or imminent stressors.

Behavioral therapy and acute exercise have also been briefly examined as a concurrent model for craving reduction; combining self-regulatory strategies with acute exercise does not appear to impact the postexercise DtS (Hatzigeorgiadis et al., 2016). However, further exploration into differing behavioral models and delivery of a combined intervention are a ripe avenue for research.

Special populations. Research with temporarily abstinent pregnant smokers has revealed that pregnant women respond similarly to a bout of acute exercise, as the normal population. A 20 min bout of mild-moderate intensity elicits a significant reduction in cravings immediately post exercise, with a significant sustained lowered craving level 10 min post exercise (Prapavessis et al., 2014).

Researchers examining the effect of a 10 min moderate-intensity cycling on cravings in temporarily abstinent adolescents found no reduction in DtS during or after exercise (Everson, Daley, & Ussher, 2006). These results allude to a differing response to exercise in adolescents, compared to an adult population, and inspire future research in the area.

The absence of evidence for other special populations (e.g., older adults, obese adults, and diseased populations) indicates a need for more specialized acute exercise paradigms to be investigated.

26.1.2 Clinical Importance

An acute bout of exercise offers immediate and sustained craving relief for smokers, for up to 30 min post exercise. Specifically, bouts of 10+ min of moderate-intensity aerobic exercise offer the greatest reduction in cravings and may be more sustainable for novice exercisers. Versatility in types of exercise allows for convenient craving relief. Acute exercise, regardless of smoking status, offers similar relief as NRT and an additive reduction in cravings when combined. There is limited evidence that these effects can be replicated in special populations. Overall, an acute bout of exercise should be prescribed as a monotherapy or concurrently with NRT to suppress cravings.

26.2 TOBACCO WITHDRAWAL SYMPTOMS

Cessation of smoking leads to feelings of anxiety, stress, irritability, poor concentration, and sleep disturbance. These tobacco withdrawal symptoms (TWS) pose a challenge to smokers who may want to quit.

Additionally, relief from these symptoms occurs shortly after smoking a cigarette, encouraging further smoking behavior. Effective management of TWS can reduce harmful smoking behavior in the smoker and may improve the odds of a successful quit attempt through preventing relapse.

Reliable representation of the diverse TWS is important. The mood and physical symptoms scale (MPSS; West & Hajek, 2004) is a validated five-point self-report questionnaire that assesses feelings of depression, anxiety, irritability, restlessness, hunger, concentration, and sleep at night and physical symptoms such as mouth sores, constipation, and cough/sore throat in single-item-scales. Another commonly used questionnaire is the Shiffman-Jarvik withdrawal scale (SJWS; Shiffman & Jarvik, 1976), which assesses cravings (five items), psychological symptoms (five items), physical symptoms (three items), sedation (single item), and appetite (single item). Despite being validated measures for TWS, the MPSS and SJWS are self-reported questionnaires, unable to capture objective differences in cognitive variables, such as attention, memory, or executive function. Hence, specific cognitive tests are used in order to test cognition; common tests include the Stroop task (attention; Homack & Riccio, 2004) and the n-back test (working memory; Kirchner, 1958).

26.2.1 Effect of Acute Exercise on TWS

Acute exercise has shown positive effects for a number of TWS. Irritability, depression, tension, restlessness, difficulty concentrating, and stress (Roberts, Maddison, Simpson, Bullen, & Prapavessis, 2012; Taylor, Ussher, & Faulkner, 2007), as well as psychological symptoms and sedation (Harper et al., 2012), have all shown significant reductions after acute exercise. Similar to cravings, the potency of exercise for reducing TWS is greatest during, and immediately after, a bout of exercise. These effects have been elicited during walking, cycling, and isometric exercises, suggesting different types of exercise can elicit similar responses. Adherence to exercise may be more feasible if smokers can pick from a variety of exercises, although further research is needed into the effects of resistance training, yoga, and other alternative exercises (Fig. 26.6).

Intensity of exercise also seems to play a role for TWS. Exercise's effect on TWS has been primarily examined using moderate-intensity protocols. There is evidence to suggest TWS may increase during vigorous-intensity exercise but then ameliorate postexercise (Everson et al., 2008). Interestingly, the negative effects of vigorous-intensity exercise are seemingly absent when engaged in a cessation program; Harper et al. (2012) note that vigorous-intensity exercise improved psychological and sedation withdrawal symptoms at week 11 of an exercise-aided NRT cessation program (i.e., 7 weeks post quit date).

An acute bout of moderate-intensity exercise may also attenuate TWS in the presence of concurrent stressors. Fong et al. (2014) found a significant decrease in psychological withdrawal symptoms following concurrent stressors, in the exercise arm only. These results are encouraging, as being exposed to cues and triggers for smoking while experiencing TWS presents as a more externally valid scenario.

Early evidence also alludes to features of cognition that may be positively affected an acute bout of exercise in temporarily abstinent smokers. Guirguis, Sui, and Prapavessis (2016) examined working memory via n-back task, by comparing an acute exercise condition with a nicotine (NRT) condition. Results suggest that exercise is just as effective, if not better, than nicotine for improving working memory. Future work applying this paradigm in smokers is currently being researched in our lab.

26.2.2 Clinical Importance

Taken together, acute bouts of moderate-intensity exercise seem effective in attenuating a number of TWS

FIG. 26.6 Representation of positive and negative affect.
Affect is the experiencing of emotions. Positive affect is how positive emotions are experienced; negative affect is how negative emotions are experienced. *Figure is unpublished; photos used through CC0 license via stocksnap.io.*

(e.g., irritability, stress, and depression), even in the presence of concurrent stressors. Vigorous-intensity exercise may have adverse effects on TWS in the early stages of cessation.

26.3 MECHANISMS

The exact mechanism by which exercise acts upon cravings and TWS is not well understood. From a physiological/psychobiological perspective, cortisol, heart-rate variability, catecholamines, and activation shifts of the reward and visuospatial areas of the brain have received some research attention, but results have been equivocal and/or yet to be reproduced (e.g., Janse Van Rensburg, Taylor, Benattayallah, & Hodgson, 2012; Roberts et al., 2015). From a psychological perspective, expectancy and distraction hypotheses have not been supported (Daniel, Cropley, & Fife-Schaw, 2007).

Currently, affect remains one of the most widely studied psychological mechanisms for mediating the relationship between exercise and cravings (Haasova et al., 2014; Roberts et al., 2012). Unfortunately, only a handful of studies have found support for mediation (Janse Van Rensburg, Elibero, Kilpatrick, & Drobes, 2013; Tart et al., 2010; Taylor, Katomeri, & Ussher, 2006). Janse Van Rensburg et al. (2013) showed that exercise reduced cravings through positive affect, whereas Tart et al. (2010) and Taylor et al. (2006) found exercise reduced cravings through negative affect. Reasons for the low number of studies supporting the affect mediation hypothesis include inconsistent measurable dimensions of affect, insufficiently powered sample, nonrobust statistical mediation methodology, and inferior design (i.e., nonrandomized controlled trial). Addressing these issues, De Jesus and Prapavessis (2017) explored the role of positive and negative affect as a mechanism for the exercise-craving relationship. Results were encouraging, as both positive and negative affect were found to be potent mediators. The authors hypothesize numerous reasons for this mediation effect: Exercise may counteract the negative affect derived from cravings and TWS; dopamine release via positive affect from exercise may imitate a reward pathway associated with nicotine cravings; and/or improved affect may ameliorate the self-regulatory strength normally lost through psychological distress/stressors (De Jesus & Prapavessis, 2017). This finding draws to attention the importance of designing acute exercise interventions as being enjoyable and versatile and minimizing discomfort and negative affect. Notably, all of the proposed mechanisms for the relationship of exercise on nicotine rely on pathways exclusive from the mechanism of nicotine (i.e., nicotinic acetylcholine receptors), which supports the additive effect of exercise as a therapy, in conjunction with NRT, for smoking cessation. Implementing enjoyable and feasible bouts of acute exercise in combination with NRT will likely increase a smoker's odds of quitting than just NRT. Future work into the exact method and proportion to which exercise is influenced by affect, as well as the potential role of other mechanisms, requires further exploration.

26.4 SUMMARY OF THE CLINICAL IMPORTANCE OF ACUTE EXERCISE

The positive effect of acute exercise on cravings and TWS is clear. Exercise is and should be prescribed as an effective therapy for acutely reducing nicotine cravings. Medium bouts or longer (10 + min) of moderate physical activity (11–13 RPE or 40%–60% heart rate max) have been shown to be the most effective for reducing cravings; higher intensities do not appear to offer further benefits for cravings or TWS. Current evidence suggests that affect is a likely mechanism for exercise in an acute paradigm, and as such, exercise interventions should be tailored to improve positive affect and reduce negative affect. While the exact mechanism for exercise is still debated, it likely operates under a separate and distinct pathway from nicotine and thus offers an additive craving reduction when paired with NRT. Overall, acute exercise is an effective means to reduce cravings in smokers engaged in a temporary or real quit attempt.

MINI-DICTIONARY OF TERMS

Acute exercise A single bout of exercise; ranges in degree of intensity, duration, and/or type of exercise.

Aerobic exercise A category of exercises that primarily challenge the cardiorespiratory system. Includes activities such as walking, cycling, running, and swimming.

Affect The experiencing of emotions. Can be further divided into positive or negative affect—how an individual experiences positive or negative mood(s), respectively.

Duration of exercise The length of a bout of exercise. For this review, the duration of a bout of exercise will be categorized as short (5 min), medium (10 min), and long (>15 min).

Exercise A subcategory of physical activity; structured and intentional activity aimed at improving measures of physical fitness and/or mental health. Purposeful exercise is typically performed at a moderate-vigorous intensity (MVPA) and accumulated in bouts of >10 min.

Frequency of exercise How often an exercise is performed—can range from a single acute bout of exercise to more regular exercise performed multiple times a week.

Heart rate max (HR_{max}) The predicted maximum heart rate an individual can attain through physical activity. Typically calculated as 220 − age.

Intensity of exercise The level of exertion required during an exercise. Can be measured objectively (e.g., heart-rate monitor) or subjectively (self-reported exertion scale). Typically categorized as light (<50% HR_{max}), moderate (50%–70% HR_{max}), and vigorous (70%–85% HR_{max}) intensity.

Isometric exercise A category of exercises that involves static, sustained contraction of a muscle/muscle group. Can be performed using a handgrip, resistance band, or body weight (e.g., plank).
Physical activity (PA) Movement of the body through skeletal muscle contraction that results in energy expenditure.
Physical inactivity A term describing those individuals who are unable to attain recommended levels of physical activity.
Rating of perceived exertion (RPE) The subjectively perceived effort that an individual feels in the moment during a bout of exercise; a self-reported measure of momentary exercise intensity.
Resistance exercise A category of exercises that primarily challenge the musculoskeletal system. Includes activities such as weightlifting, conditioning, and powerlifting.
Type of exercise The mode in which exercise is performed. Encompasses a variety of exercise forms, such as aerobic, resistance, isometric, and sports.

Key Facts of Physical Activity and Exercise

- Physical inactivity is the fourth leading cause of death worldwide.
- Researchers have appraised the risk from physical inactivity to be similar to that of smoking or obesity (Lee et al., 2012).
- Long-term benefits of physical activity include reduced weight gain and risk of obesity, reduced risk of coronary heart disease, reduced risk of developing type 2 diabetes, and reduced incidence of Alzheimer's disease and dementia.
- A single bout of moderate-to-vigorous intensity exercise can increase circulating immune and inflammatory markers (Brown, Davison, McClean, & Murphy, 2015), enhance numerous cognitive processes (Tomporowski, 2003), diminish stress-related blood pressure responses (Hamer, Taylor, & Steptoe, 2006), and improve metabolic markers (Dunstan et al., 2012).
- There are four parameters used to classify and prescribe exercise: frequency, intensity, time (duration), and type (mode) of activity. Collectively, these terms comprise the FITT principle often used among researchers and health-care professionals who prescribe exercise (Bushman, 2017).
- To be considered physically active in most countries (e.g., Canada and the United States), individuals must accumulate 150 min of moderate-vigorous intensity exercise a week, in bouts of >10 min.

Summary Points

- This chapter focuses on the acute effect of exercise on cravings and tobacco withdrawal symptoms.
- An acute bout of exercise is a single session of exercise, ranging from light to vigorous intensity, performed at durations of less than 5 min to greater than 15 min, and completed through any of a variety of modes (e.g., aerobic, resistance, and stretching).
- An acute bout of exercise offers immediate and sustained craving relief for smokers, for up to 30 min post exercise—specifically, medium bouts or longer (10 + min) of moderate physical activity (11–13 rating of perceived exertion (RPE) or 40%–60% heart rate max (HR_{max})) have been shown to be the most effective for reducing cravings.
- Exercise also offers an additive effect when used in combination with nicotine replacement therapy (NRT).
- An acute bout of moderate-intensity exercise seems effective in attenuating a number of tobacco withdrawal symptoms (TWS; e.g., irritability, stress, and depression), even in the presence of concurrent stressors.
- Vigorous-intensity exercise may have adverse effects on TWS in the early stages of cessation.
- Currently, affect appears to be the best supported mechanism for the exercise-craving relationship.
- Exercise is an effective therapy for acutely reducing nicotine cravings and should be prescribed as such.

References

Brown, W. M. C., Davison, G. W., McClean, C. M., & Murphy, M. H. (2015). A systematic review of the acute effects of exercise on immune and inflammatory indices in untrained adults. *Sports Medicine: Open*, *1*(1), 35. https://doi.org/10.1186/s40798-015-0032-x.

Bushman, B. (2017). In B. Bushman (Ed.), *ACSM's complete guide to fitness & health* (2nd ed.). Champaign, IL: Human Kinetics.

Daniel, J. Z., Cropley, M., & Fife-Schaw, C. (2007). Acute exercise effects on smoking withdrawal symptoms and desire to smoke are not related to expectation. *Psychopharmacology*, *195*(1), 125–129. https://doi.org/10.1007/s00213-007-0889-6.

De Jesus, S., & Prapavessis, H. (2017). *Cortisol and affect mechanisms through which acute exercise attenuates cigarette cravings during a temporary quit attempt.* Unpublished work.

Dunbar, M. S., Shiffman, S., Kirchner, T. R., Tindle, H. A., & Scholl, S. M. (2014). Nicotine dependence, "background" and cue-induced craving and smoking in the laboratory. *Drug and Alcohol Dependence*, *142*, 197–203. https://doi.org/10.1016/j.drugalcdep.2014.06.018.

Dunstan, D. W., Kingwell, B. A., Larsen, R., Healy, G. N., Cerin, E., Hamilton, M. T., et al. (2012). Breaking up prolonged sitting reduces postprandial glucose and insulin responses. *Diabetes Care*. https://doi.org/10.2337/dc11-1931.

Elibero, A., Janse Van Rensburg, K., & Drobes, D. J. (2011). Acute effects of aerobic exercise and hatha yoga on craving to smoke. *Nicotine & Tobacco Research*, *13*(11), 1140–1148. https://doi.org/10.1093/ntr/ntr163.

Everson, E. S., Daley, A. J., & Ussher, M. (2006). Does exercise have an acute effect on desire to smoke, mood and withdrawal symptoms in abstaining adolescent smokers? *Addictive Behaviors*, *31*(9), 1547–1558. https://doi.org/10.1016/j.addbeh.2005.11.007.

Everson, E. S., Daley, A. J., & Ussher, M. (2008). The effects of moderate and vigorous exercise on desire to smoke, withdrawal symptoms and mood in abstaining young adult smokers. *Mental Health and Physical Activity*, *1*(1), 26–31. https://doi.org/10.1016/j.mhpa.2008.06.001.

Fong, A. J., De Jesus, S., Bray, S. R., & Prapavessis, H. (2014). Effect of exercise on cigarette cravings and ad libitum smoking following concurrent stressors. *Addictive Behaviors*, *39*(10), 1516–1521. https://doi.org/10.1016/j.addbeh.2014.05.027.

Guirguis, S., Sui, W., & Prapavessis, H. (2016). *The acute effects of nicotine and exercise on working memory in non-smokers.* Unpublished work.

Haasova, M., Warren, F. C., Ussher, M., Janse Van Rensburg, K., Faulkner, G., Cropley, M., et al. (2013). The acute effects of physical activity on cigarette cravings: systematic review and meta-analysis with individual participant data. *Addiction, 108*(1), 26–37. https://doi.org/10.1111/j.1360-0443.2012.04034.x.

Haasova, M., Warren, F. C., Ussher, M., Janse Van Rensburg, K., Faulkner, G., Cropley, M., et al. (2014). The acute effects of physical activity on cigarette cravings: exploration of potential moderators, mediators and physical activity attributes using individual participant data (IPD) meta-analyses. *Psychopharmacology, 231*(7), 1267–1275. https://doi.org/10.1007/s00213-014-3450-4.

Hamer, M., Taylor, A., & Steptoe, A. (2006). The effect of acute aerobic exercise on stress related blood pressure responses: a systematic review and meta-analysis. *Biological Psychology, 71*(2), 183–190. https://doi.org/10.1016/j.biopsycho.2005.04.004.

Hansson, A., Hajek, P., Perfekt, R., & Kraiczi, H. (2012). Effects of nicotine mouth spray on urges to smoke, a randomised clinical trial. *BMJ Open, 2*(5). e001618. https://doi.org/10.1136/bmjopen-2012-001618.

Harper, T., Fitzgeorge, L., Tritter, A., & Prapavessis, H. (2012). Acute exercise effects on craving and withdrawal symptoms among women attempting to quit smoking using nicotine replacement therapy. *Journal of Smoking Cessation*, 1–8. https://doi.org/10.1017/jsc.2012.15.

Hatzigeorgiadis, A., Pappa, V., Tsiami, A., Tzatzaki, T., Georgakouli, K., Zourbanos, N., et al. (2016). Self-regulation strategies may enhance the acute effect of exercise on smoking delay. *Addictive Behaviors, 57*, 35–37. https://doi.org/10.1016/j.addbeh.2016.01.012.

Ho, J. Y., Kraemer, W. J., Volek, J. S., Vingren, J. L., Fragala, M. S., Flanagan, S. D., et al. (2014). Effects of resistance exercise on the HPA axis response to psychological stress during short-term smoking abstinence in men. *Addictive Behaviors, 39*(3), 695–698. https://doi.org/10.1016/j.addbeh.2013.10.027.

Homack, S., & Riccio, C. A. (2004). A meta-analysis of the sensitivity and specificity of the Stroop Color and Word Test with children. *Archives of Clinical Neuropsychology, 19*(6), 725–743. https://doi.org/10.1016/j.acn.2003.09.003.

Janse Van Rensburg, K., Elibero, A., Kilpatrick, M., & Drobes, D. J. (2013). Impact of aerobic exercise intensity on craving and reactivity to smoking cues. *Experimental and Clinical Psychopharmacology, 21* (3), 196–203. https://doi.org/10.1037/a0032768.

Janse Van Rensburg, K., Taylor, A., Benattayallah, A., & Hodgson, T. (2012). The effects of exercise on cigarette cravings and brain activation in response to smoking-related images. *Psychopharmacology, 221*(4), 659–666. https://doi.org/10.1007/s00213-011-2610-z.

Kirchner, W. K. (1958). Age differences in short-term retention of rapidly changing information. *Journal of Experimental Psychology, 55*(4), 352–358.

Kotlyar, M., Mendoza-Baumgart, M. I., Li, Z.-Z., Pentel, P. R., Barnett, B. C., Feuer, R. M., et al. (2007). Nicotine pharmacokinetics and subjective effects of three potential reduced exposure products, moist snuff and nicotine lozenge. *Tobacco Control, 16*(2), 138–142. https://doi.org/10.1136/tc.2006.018440.

Lee, I.-M., Shiroma, E. J., Lobelo, F., Puska, P., Blair, S. N., Katzmarzyk, P. T., et al. (2012). Impact of physical inactivity on the World's Major Non-Communicable Diseases. *The Lancet, 380* (9838), 219–229. https://doi.org/10.1016/S0140-6736(12)61031-9.

Prapavessis, H., De Jesus, S., Harper, T., Cramp, A., Fitzgeorge, L., Mottola, M. F., et al. (2014). The effects of acute exercise on tobacco cravings and withdrawal symptoms in temporary abstinent pregnant smokers. *Addictive Behaviors, 39*(3), 703–708. https://doi.org/10.1016/j.addbeh.2013.10.034.

Roberts, V., Gant, N., Sollers, J. J., Bullen, C., Jiang, Y., & Maddison, R. (2015). Effects of exercise on the desire to smoke and physiological responses to temporary smoking abstinence: a crossover trial. *Psychopharmacology, 232*(6), 1071–1081. https://doi.org/10.1007/s00213-014-3742-8.

Roberts, V., Maddison, R., Simpson, C., Bullen, C., & Prapavessis, H. (2012). The acute effects of exercise on cigarette cravings, withdrawal symptoms, affect, and smoking behaviour: systematic review update and meta-analysis. *Psychopharmacology, 222*(1), 1–15. https://doi.org/10.1007/s00213-012-2731-z.

Shiffman, S. (2000). Comments on craving. *Addiction, 95*(8s2), 171–175. https://doi.org/10.1046/j.1360-0443.95.8s2.6.x.

Shiffman, S. M., & Jarvik, M. E. (1976). Smoking withdrawal symptoms in two weeks of abstinence. *Psychopharmacology, 50*(1), 35–39. https://doi.org/10.1007/BF00634151.

Tart, C. D., Leyro, T. M., Richter, A., Zvolensky, M. J., Rosenfield, D., & Smits, J. A. J. (2010). Negative affect as a mediator of the relationship between vigorous-intensity exercise and smoking. *Addictive Behaviors, 35*(6), 580–585. https://doi.org/10.1016/j.addbeh.2010.01.009.

Taylor, A., & Katomeri, M. (2007). Walking reduces cue-elicited cigarette cravings and withdrawal symptoms, and delays ad libitum smoking. *Nicotine & Tobacco Research, 9*(11), 1183–1190. https://doi.org/10.1080/14622200701648896.

Taylor, A., Katomeri, M., & Ussher, M. (2006). Effects of walking on cigarette cravings and affect in the context of Nesbitt's paradox and the Circumplex model. *Journal of Sport & Exercise Psychology, 28*. https://doi.org/10.1123/jsep.28.1.18.

Taylor, A. H., Ussher, M. H., & Faulkner, G. (2007). The acute effects of exercise on cigarette cravings, withdrawal symptoms, affect, and smoking behaviour: a systematic review. *Addiction, 102*, 534–543. https://doi.org/10.1007/s00213-012-2731-z.

Tiffany, S. T., & Drobes, D. J. (1991). The development and initial validation of a questionnaire on smoking urges. *British Journal of Addiction, 86*(11), 1467–1476. https://doi.org/10.1111/j.1360-0443.1991.tb01732.x.

Tomporowski, P. D. (2003). Effects of acute bouts of exercise on cognition. *Acta Psychologica, 112*(3), 297–324. https://doi.org/10.1016/S0001-6918(02)00134-8.

Tritter, A., Fitzgeorge, L., & Prapavessis, H. (2015). The effect of acute exercise on cigarette cravings while using a nicotine lozenge. *Psychopharmacology, 232*(14), 2531–2539. https://doi.org/10.1007/s00213-015-3887-0.

Ussher, M., Cropley, M., Playle, S., Mohidin, R., & West, R. (2009). Effect of isometric exercise and body scanning on cigarette cravings and withdrawal symptoms. *Addiction, 104*(7), 1251–1257. https://doi.org/10.1111/j.1360-0443.2009.02605.x.

West, R., & Hajek, P. (2004). Evaluation of the mood and physical symptoms scale (MPSS) to assess cigarette withdrawal. *Psychopharmacology, 177*(1–2), 195–199. https://doi.org/10.1007/s00213-004-1923-6.

CHAPTER

27

CRF2 Receptor Agonists and Nicotine Withdrawal

Zsolt Bagosi

Department of Pathophysiology, Faculty of Medicine, University of Szeged, Szeged, Hungary

Abbreviations

ACTH	adrenocorticotropic hormone
AVP	arginine vasopressin
CFLP mice	Carworth Farm Lane-Petter mice
CNS	central nervous system
CRF	corticotropin-releasing factor
CRF1	corticotropin-releasing factor receptor type 1
CRF2	corticotropin-releasing factor receptor type 2
CRF-BP	corticotropin-releasing factor-binding protein
HPA axis	hypothalamic-pituitary-adrenal axis
icv	intracerebroventricularly
ip	intraperitoneally
SNS	sympathetic nervous system
Ucn1	urocortin 1
Ucn2	urocortin 2
Ucn3	urocortin 3

Corticotropin-releasing factor (CRF) and the urocortins are members of the mammalian CRF family of peptides, having similar biochemical structures but different anatomical distributions, physiological functions, and pharmacological profiles (Fekete & Zorrilla, 2007; Suda, Kageyama, Sakihara, & Nigawara, 2004).

CRF is a 41-amino acid mammalian neuropeptide that shows 54% similarity with its fish homolog urotensin and 48% with its frog homolog sauvagine (Vale, Spiess, Rivier, & Rivier, 1981). CRF is synthesized in the paraventricular nucleus of the hypothalamus and the central nucleus of the amygdala from where it regulates the neuroendocrine, autonomic, and behavioral responses to stress (Bale, Lee, & Vale, 2002; Bale & Vale, 2004). The neuroendocrine response is represented by the activation of the hypothalamic-pituitary-adrenal (HPA) axis and mediated by paraventricular CRF that, along with the synergistic arginine vasopressin (AVP), stimulates the release of adrenocorticotropic hormone (ACTH) from the anterior pituitary and the subsequent release of glucocorticoids from the adrenal cortex (Bale et al., 2002; Bale & Vale, 2004). The autonomic response is represented by the activation of the sympathetic nervous system (SNS) and mediated by amygdalar CRF that stimulates the release of catecholamines from the locus coeruleus that, in turn, stimulates the release of catecholamines from the adrenal medulla (Bale et al., 2002; Bale & Vale, 2004). Thus, CRF acts not only as a hypothalamic neurohormone but also as an extrahypothalamic neurotransmitter that modulates the behavioral responses to stress, manifested by increased locomotor activity, decreased food and water intake, etc. (Bale et al., 2002; Bale & Vale, 2004).

Urocortin 1 (Ucn1) is a 40-amino acid neuropeptide that shares 63% similarity with the fish urotensin and 45% with the human CRF, from which the name urocortin was derived (Vaughan et al., 1995). Ucn1 is synthesized predominantly in the Edinger-Westphal nucleus and the lateral superior olive that project caudally to the spinal cord and the lateral septum (Reul & Holsboer, 2002). Despite of the oculomotor, pupillary, and auditory functions attributed to these brain regions, Ucn1 rather modulates the neuroendocrine and behavioral responses to stress, being a less potent activator of locomotion and a more potent suppressor of food and water ingestion, when compared to CRF (Skelton, Owens, & Nemeroff, 2000).

The actions of CRF and Ucn1 are mediated by two distinct G-protein-coupled receptors, CRF1 and CRF2 (Chang, Pearse II, O'Connell, & Rosenfeld, 1993; Lovenberg et al., 1995), and inhibited by CRF-binding protein (CRF-BP) (Behan, Cepoi, et al., 1996; Behan, De Souza, et al., 1996). CRF acts preferentially through CRF1, binding with 15-fold higher affinity to CRF1 than to CRF2, whereas Ucn1 acts equipotently through both CRF receptors, binding with sevenfold higher affinity

Neuroscience of Nicotine
https://doi.org/10.1016/B978-0-12-813035-3.00027-7

to CRF1 than CRF itself (Reul & Holsboer, 2002). In the central nervous system (CNS), CRF1 is distributed abundantly in the cerebral cortex, the cerebellum, the amygdala, and the anterior pituitary (Dautzenberg & Hauger, 2002; Van Pett et al., 2000).

Urocortin 2 (Ucn2), also known as stresscopin-related peptide in humans, is a 38-amino acid neuropeptide that presents 34% identity with CRF (Reyes et al., 2001). Ucn2 expression was shown in the paraventricular, the supraoptic, and the arcuate nuclei of the hypothalamus and the locus coeruleus and brain regions involved in neuroendocrine control, food intake, water intake, and autonomic control, but Ucn2 projections are unknown yet. Ucn2 was demonstrated to have mild locomotor suppressive and delayed anxiolytic-like effects (Valdez et al., 2002).

Urocortin 3 (Ucn3), also known as stresscopin in humans, is another 38-amino acid neuropeptide that exhibits 36% identity with CRF (Lewis et al., 2001). Ucn3 expression was found in the perifornical area of the hypothalamus, the medial nucleus of the amygdala, and the bed nucleus of the stria terminalis, which are adjacent or project to the paraventricular nucleus of the hypothalamus and the lateral septum, brain regions involved in stress coping (Reul & Holsboer, 2002). When compared to Ucn2, Ucn3 was demonstrated to have more acute locomotor suppressive and anxiolytic-like effects (Valdez, Zorrilla, Rivier, Vale, & Koob, 2003).

The actions of Ucn2 and Ucn3 are mediated exclusively by CRF2, both having more than 1000-fold higher affinity for CRF2 than for CRF1 (Reul & Holsboer, 2002). Therefore, they are considered selective agonists of CRF2 (Hsu & Hsueh, 2001). CRF2 is limited centrally to the subcortical regions of the brain: the lateral septum, the hypothalamus, the amygdala, and the hippocampus (Dautzenberg & Hauger, 2002; Van Pett et al., 2000).

Central administration of CRF and Ucn1 induces activation of the HPA axis reflected by the elevation of the plasma corticosterone concentration and anxiety- and depression-like behavior in mice and rats (Bagosi et al., 2014; Spina et al., 1996, 2002; Tanaka & Telegdy, 2008). In contrast, central administration of Ucn2 and Ucn3 produces anxiolytic and antidepressant actions in rodents (Inoue et al., 2003; Tanaka & Telegdy, 2008; Valdez et al., 2002, 2003). It was hereby hypothesized that in physiological conditions, CRF would initiate the responses to stress activating CRF1 in the anterior pituitary, whereas the urocortins would terminate these responses activating CRF2 in the paraventricular nucleus of the hypothalamus (Bale et al., 2002; Bale & Vale, 2004). Overwhelming stress may induce a pathological stimulation of CRF/CRF1 system in the cerebral cortex and the amygdala over the urocortin/CRF2 system in the lateral septum and the hippocampus that may result in hyperactivity of the HPA axis, anxiety, and depression (Bale et al., 2002; Bale & Vale, 2004). However, the physiological role of urocortin/CRF2 system in the regulation HPA axis is still under debate, since studies in mice and rats led to contradictory results (Bagosi et al., 2013; Jamieson, Li, Kukura, Vaughan, & Vale, 2006; Maruyama, Makino, Noguchi, Nishioka, & Hashimoto, 2007; Pelleymounter, Joppa, Ling, & Foster, 2004). Nevertheless, a recent hypothesis questions such dualistic and complementary actions of CRF1 and CRF2 and suggests that stress would recruit CRF systems in a brain region and neuron population-specific manner as conditions dictate (Janssen & Kozicz, 2013).

Besides the regulation of the stress responses, CRF and the urocortins have been implicated in nicotine addiction (Bruijnzeel & Gold, 2005; Sarnyai, Shaham, & Heinrichs, 2001). Inspired by the original hypothesis according to which CRF1 and CRF2 exert (mainly) antagonistic effects in the CNS, our recent study aimed to investigate whether the activation of CRF2 by central administration of Ucn2 and Ucn3 would attenuate the anxiety- and the depression-like state developed during chronic nicotine treatment and consequent acute withdrawal in mice (Bagosi et al., 2016).

In this purpose, 72 male CFLP mice were exposed to intraperitoneal treatment with nicotine or saline solution for 7 days, 4 times/day, and then to 1 day of withdrawal. A single intracerebroventricular injection with Ucn2, Ucn3, or saline solution was performed at 12 h or at 24 h following the the last intraperitoneal treatment. After 30 min, the mice were evaluated in an elevated plus-maze test (see "Key Facts of the Elevated Plus-Maze Test") and a forced swim test (see "Key Facts of the Forced Swim Test") for signs of anxiety and depression, respectively. States of anxiety and depression are usually associated with the hyperactivity of the HPA axis; hence, after 5 min, the plasma corticosterone concentration was also determined by a chemofluorescent assay (see "Key Facts of the Chemofluorescent Assay"). Half of the mice were tested on the 8th day (12 h after the last intraperitoneal treatment) and half of them on the 9th day (24 h after the last intraperitoneal treatment) (Bagosi et al., 2016).

On the 8th day, nicotine-treated mice presented signs of anxiolysis and depression and no significant elevation of the plasma corticosterone concentration (Bagosi et al., 2016). Actually, in the elevated plus-maze test, the time spent in the open arms increased significantly after nicotine treatment, and this parameter increased even more after urocortin treatment, but the number of entries into the open arms and the total number of entries did not change considerably in nicotine-treated mice, compared to the saline-treated ones (Fig. 27.1). In parallel, in the forced swim test, the time spent with climbing and swimming decreased remarkably after nicotine treatment, and these parameters decreased even more after urocortin treatment, but the time spent immobile did not change considerably in nicotine-treated mice, compared to the

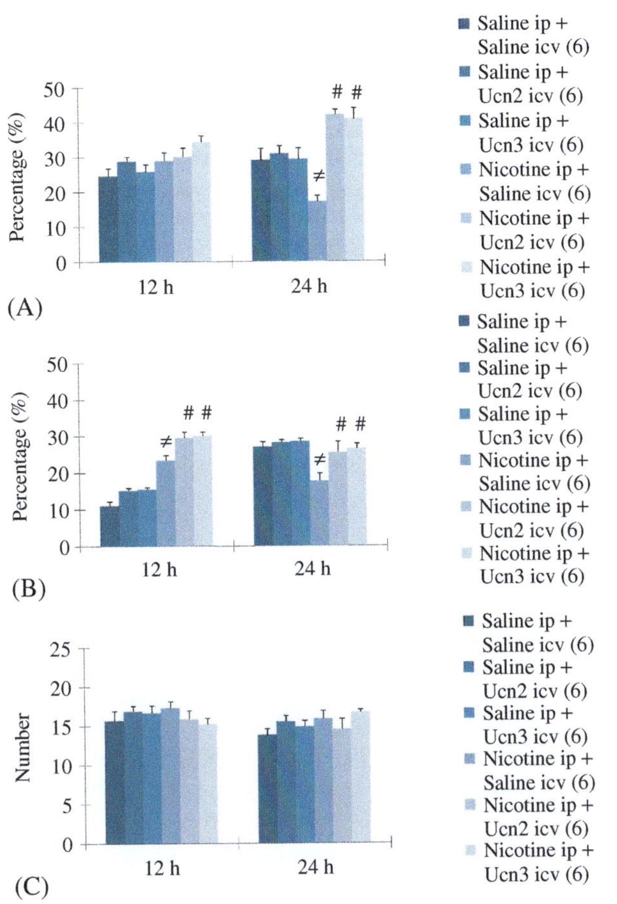

FIG. 27.1 The effects of Ucn2 and Ucn3 in mice exposed to chronic nicotine treatment and acute nicotine withdrawal and investigated in an elevated plus-maze test.
(A) Number of entries into the open arms/total number of entries, (B) time spent in the open arms/total time, and (C) total number of entries. Values are presented as means ± SEM; statistically significant difference was accepted for $P < .05$ and indicated with ≠ for nicotine ip + saline icv vs saline ip + saline icv and with # for nicotine ip + Ucn2 or Ucn3 icv vs nicotine ip + saline icv. Abbreviations: *icv*, intracerebroventricularly; *ip*, intraperitoneally; *Ucn2*, urocortin 2; *Ucn3*, urocortin 3. *Adapted from Bagosi et al. (2016), with permission from Elsevier.*

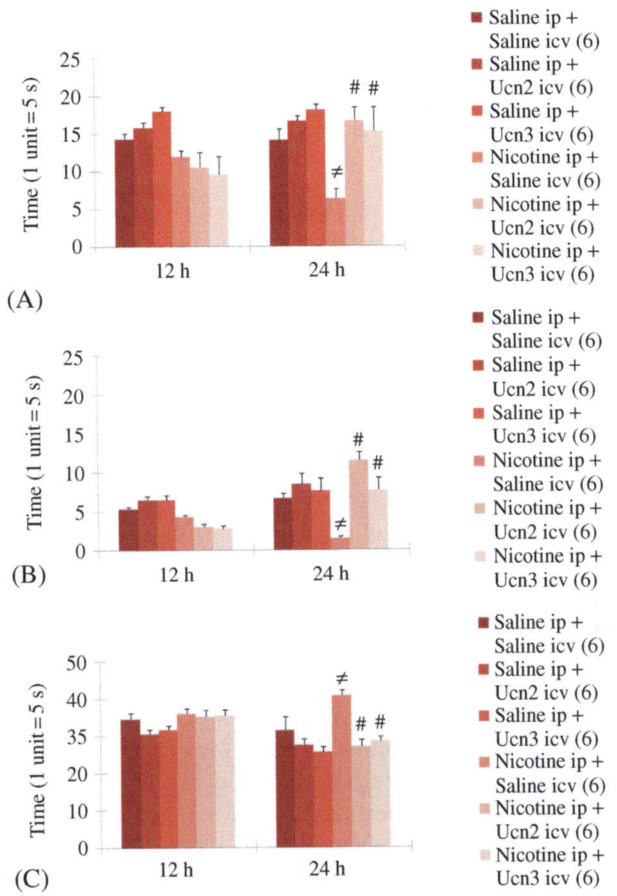

FIG. 27.2 The effects of Ucn2 and Ucn3 in mice exposed to chronic nicotine treatment and acute nicotine withdrawal and investigated in a forced swim test.
(A) Climbing activity, (B) swimming activity, and (C) immobility. Values are presented as means ± SEM; statistically significant difference was accepted for $P < .05$ and indicated with ≠ for nicotine ip + saline icv vs saline ip + saline icv and with # for nicotine ip + Ucn2 or Ucn3 icv vs nicotine ip + saline icv. Abbreviations: *icv*, intracerebroventricularly; *ip*, intraperitoneally; *Ucn2*, urocortin 2; *Ucn3*, urocortin 3. *Adapted from Bagosi et al. (2016), with permission from Elsevier.*

saline-treated ones (Fig. 27.2). Most of the behavioral changes described after 12 h of nicotine withdrawal were nonsignificant and accompanied with a slight but nonsignificant elevation of the plasma corticosterone concentration that was reduced significantly after urocortin treatment (Fig. 27.3).

On the 9th day, nicotine-treated mice exhibited signs of anxiety and depression and a significant increase of the plasma corticosterone concentration (Bagosi et al., 2016). Accordingly, in the elevated plus-maze test, the number of entries into the open arms and the time spent in the open arms decreased significantly, while the total number of entries did not change significantly in nicotine-treated mice, compared to the saline-treated ones; after treating these mice with urocortins, the first two parameters were increased or normalized (Fig. 27.1). In addition, in the forced swim test, the time spent with climbing and swimming decreased

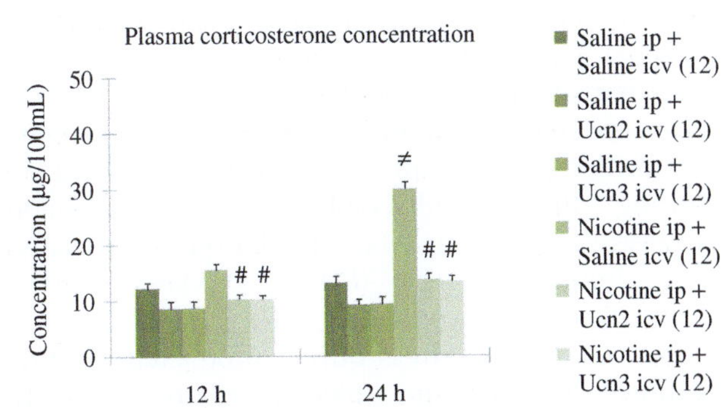

FIG. 27.3 The effects of Ucn2 and Ucn3 on the plasma corticosterone concentration determined by a chemofluorescent assay in mice exposed to chronic nicotine treatment and acute nicotine withdrawal.
Values are presented as means ± SEM; statistically significant difference was accepted for $P < .05$ and indicated with ≠ for nicotine ip + saline icv vs saline ip + saline icv and with # for nicotine ip + Ucn2 or Ucn3 icv vs nicotine ip + saline icv. Abbreviations: *icv*, intracerebroventricularly; *ip*, intraperitoneally; *Ucn2*, urocortin 2; *Ucn3*, urocortin 3. *Adapted from Bagosi et al. (2016), with permission from Elsevier.*

significantly, while the time spent immobile increased significantly in nicotine-treated mice, compared to the saline-treated ones; after treating these mice with urocortins, all these parameters were returned to normal (Fig. 27.2). The behavioral response described after 24h of nicotine withdrawal was associated with a significant neuroendocrine response, both of which were reversed completely after urocortin treatment (Fig. 27.3).

Previous studies have already suggested that CRF and CRF-like peptides contribute to the acute, chronic, and withdrawal actions of nicotine, based on three observations (Bruijnzeel & Gold, 2005; Sarnyai et al., 2001).

First, acute administration of nicotine evokes a dose-dependent activation of the HPA axis that is reflected by the accumulation of salivary cortisol in humans and plasma corticosterone in mice and rats and seems to be initiated by hypothalamic CRF (Bruijnzeel & Gold, 2005; Sarnyai et al., 2001). Despite of the rapid desensitization of this effect that may occur even after a single intake and the tolerance that may develop after repeated intake, chronic administration of nicotine often leads to elevation of the glucocorticoid concentration (Bruijnzeel & Gold, 2005; Sarnyai et al., 2001).

Second, acute nicotine withdrawal provokes a group of symptoms in both humans and rodents that resembles stress responses and seems to be mediated by extrahypothalamic CRF (Bruijnzeel & Gold, 2005; Sarnyai et al., 2001). This nicotine withdrawal syndrome is composed of a somatic (physical) and an affective (emotional) component that appear in few hours and disappear in few days or weeks after cessation of chronic administration of nicotine (Kenny & Markou, 2001; Markou, 2008; Wonnacott, Sidhpura, & Balfour, 2005). The somatic symptoms in humans include bradycardia, gastrointestinal discomfort, and increased appetite; in rodents, they correspond to abdominal constrictions, facial fasciculations, eyeblinks, ptosis and gasps, escape attempts, foot licks, genital grooming shakes, scratches, and yawns (Kenny & Markou, 2001; Markou, 2008; Wonnacott et al., 2005). The affective symptoms in humans incorporate craving, anxiety, depression, dysphoria, irritability, and difficulty concentrating; in rodents, they correlate with anhedonia (a diminished interest or pleasure to rewarding stimuli) and conditioned place aversion (an aversive motivational state that becomes associated with environmental cues) (Kenny & Markou, 2001; Markou, 2008; Wonnacott et al., 2005).

Third, exposure to stress during protracted nicotine withdrawal increases the vulnerability to relapse to smoking in humans and nicotine self-administration in rodents (Bruijnzeel & Gold, 2005; Sarnyai et al., 2001). The negative affective state that emerges during acute withdrawal may persist during protracted withdrawal, and its avoidance plays an essential role in this relapse and thereby in the maintenance of nicotine addiction (Kenny & Markou, 2001; Markou, 2008; Wonnacott et al., 2005).

Moreover, recent studies have suggested that the affective and the somatic components of nicotine withdrawal syndrome are mediated by CRF1 and CRF2, respectively (Bruijnzeel, 2012; Bruijnzeel et al., 2012; Bruijnzeel & Gold, 2005; Bruijnzeel, Prado, & Isaac, 2009; Bruijnzeel, Zislis, Wilson, & Gold, 2007; George et al., 2007; Kamdi, Nakhate, Dandekar, Kokare, & Subhedar, 2009; Marcinkiewcz et al., 2009). Several studies reported that the administration of selective CRF1 antagonists prevents the dysphoria and the reward deficit observed during nicotine withdrawal (Bruijnzeel et al., 2012; Marcinkiewcz et al., 2009). Another study referred that the administration of nonselective CRF2 agonists, such as CRF and Ucn1, prevents the hyperphagia and the weight gain assessed during nicotine withdrawal (Kamdi et al., 2009). However, our study was the first to demonstrate that intracerebroventricular administration of Ucn2 and Ucn3 ameliorates the anxiety- and depression-like state developed during chronic nicotine treatment and consequent acute withdrawal, suggesting that the selective CRF2 agonists can be potential candidates in the therapy of nicotine withdrawal (Bagosi et al., 2016). For a better understanding of the physiological and pharmacological actions, future studies should perform the infusion of these peptides in specific brain regions, such as the paraventricular nucleus of the hypothalamus and the lateral septum, where CRF2 is colocalized with Ucn2 neurons and axonal projections of Ucn3 neurons (Reul & Holsboer, 2002; Van Pett et al., 2000).

MINI-DICTIONARY OF TERMS

Acute nicotine withdrawal Sudden cessation of nicotine following repeated exposure; in rodents, it can be spontaneous or precipitated by the administration of a nicotinic acetylcholine or an opioid receptor antagonist and may last between 2 and 3h and 2–3 days following the cessation of nicotine.

Anxiety A pathological condition characterized by the presence of excessive worry about a variety of topics, events, or activities; in rodents, it can be investigated in open-field exploration test, elevated plus-maze test, light-dark exploration test, social interaction test, etc.

CFLP mice An outbred strain of mice, which are commonly used in genetic, toxicological, and pharmacological research; the term CFLP comes from Carworth Farms and Lane-Petter companies.

Chronic nicotine administration Repeated exposure to nicotine; in rodents, it can be performed by water drinking; vapor inhalation; osmotic minipumps; and intravenous, subcutaneous, or intraperitoneal injection that may last from 7 to 270 days.

Corticotropin-releasing factor (CRF) Also known as corticotropin-releasing hormone. Not only a hypothalamic neurohormone but also an extrahypothalamic neurotransmitter that mediates the neuroendocrine, autonomic, and behavioral responses to stress.

Depression A pathological condition characterized by depressed mood and diminished pleasure or interest in daily activities; in rodents, it can be investigated in forced swim test, tail-suspension test, anhedonia test, conditioned place preference test, etc.

Hypothalamic-pituitary-adrenal (HPA) axis A neuroendocrine system that is represented by the release of CRF from the paraventricular nucleus of the hypothalamus that, along with the synergistic AVP, stimulates the release of ACTH from the anterior pituitary and the subsequent release of glucocorticoids from the adrenal cortex.

Urocortins (Ucn1, Ucn2, and Ucn3) CRF-like neuropeptides that have similar amino acid structure but different pharmacological profiles compared to CRF.

Key Facts of the Elevated Plus-Maze Test

- The elevated plus-maze test is a method validated by Pellow, Chopin, File, and Briley (1985) to investigate the anxiety-like behavior and the efficacy of anxiolytic drugs in rodents.
- The apparatus consists of a plus-shaped wooden platform elevated at 40 cm from the floor, made up by four opposing arms of 30 cm x 5 cm. Two of the opposing arms are enclosed by 15 cm-high side and end walls (closed arms), whereas the other two arms have no walls (open arms).
- Each mouse is placed in the central area (5 cm x 5 cm) of the maze, facing one of the open arms, and their behavior is assessed by an observer sitting at 1 m distance from the center of the plus maze.
- For 5 min period, the following parameters are recorded: (a) the ratio between the number of entries into the open arms and the total number of entries, (b) the ratio between the time spent in the open arms and the total time, and (c) the total number of entries (an entry into an arm is defined as the entry of all four feet of the animal into that arm).
- The principle of the test is that open arms are more fear-provoking, and the ratio of the time spent in open versus closed arms or the ratio of the entries into open versus closed arms reflects the relative safety of closed arms as compared with the relative danger of open arms.

The above points list the key facts of the elevated plus--maze test used to evaluate anxiety-like signs in rodents.

Key Facts of the Forced Swim Test

- The forced swim test is a method invented by Porsolt, Bertin, and Jalfre (1977) to investigate the depression-like behavior and the antidepressant properties of drugs in rodents.
- The apparatus consists of a plexiglass cylinder of 40 cm height and 12 cm diameter containing 1.5 L of water maintained at 25 ± 1°C temperature that is positioned on a table.
- Each mouse is dropped individually into the water, and their behavior is assessed by an observer sitting at 1 m distance from the table.
- For 5 min period, the following parameters are recorded and expressed in time units (1 time unit = 5 s): (a) the climbing activity (the time that mice spent with climbing the walls, in their attempt to escape the cylinder), (b) the swimming activity (the time that mice spent with swimming in the water, in their attempt to remain at the surface), and (c) the immobility (the time that mice spent in an upright position on the surface with its front paws together).
- The principle of the test is that in such a situation, from which they cannot escape, animals rapidly became immobile, floating in an upright position and making only small movements to keep their heads above water. In parallel, their attempt to escape the cylinder by climbing or swimming may decrease or cease eventually.

The above points list the key facts of the forced swim test used to investigate depression-like signs in rodents.

Key Facts of the Chemofluorescent Assay

- The chemofluorescent assay is a method described by Zenker and Bernstein (1958) and modified by Purves and Sirett (1965), used to determine the plasma corticosterone concentration.
- The assay is based on the principle that the hydrophobic corticosterone can be extracted from the plasma of rodents with methylene chloride and determined with a fluorescent mixture of sulfuric acid and ethyl alcohol.
- The chemical substances used are heparin, distilled water, corticosterone standards, methylene chloride, sulfuric acid, and ethyl alcohol.
- The apparatus used is a Hitachi 204-A fluorescent spectrophotometer set at 456 nm extinction and 515 emission wavelengths.
- The plasma corticosterone concentration is calculated from the values of the standards and expressed in μg/100 mL.

The above points list the key facts of the chemofluorescent assay used to determine plasma corticosterone concentration.

Summary Points

- Corticotropin-releasing factor (CRF) and the urocortins (Ucn1, Ucn2, and Ucn3) belong to the mammalian CRF family of peptides.
- CRF and Ucn1 bind to both CRF receptors (CRF1 and CRF2), whereas Ucn2 and Ucn3 bind selectively to CRF2.
- Besides the regulation of the stress responses, CRF and the urocortins have been implicated in nicotine addiction.

- Ucn2 and Ucn3 ameliorate the anxiety- and the depression-like state developed during chronic nicotine treatment and consequent acute withdrawal.
- Therefore, selective CRF2 agonists can be potential candidates in the therapy of nicotine withdrawal.

References

Bagosi, Z., Csabafi, K., Palotai, M., Jaszberenyi, M., Foldesi, I., Gardi, J., et al. (2013). The interaction of urocortin II and urocortin III with amygdalar and hypothalamic cotricotropin-releasing factor (CRF)– reflections on the regulation of the hypothalamic-pituitary-adrenal (HPA) axis. *Neuropeptides*, *47*(5), 333–338.

Bagosi, Z., Csabafi, K., Palotai, M., Jaszberenyi, M., Foldesi, I., Gardi, J., et al. (2014). The effect of urocortin I on the hypothalamic ACTH secretagogues and its impact on the hypothalamic-pituitary-adrenal axis. *Neuropeptides*, *48*(1), 15–20.

Bagosi, Z., Palotai, M., Simon, B., Bokor, P., Buzas, A., Balango, B., et al. (2016). Selective CRF2 receptor agonists ameliorate the anxiety- and depression-like state developed during chronic nicotine treatment and consequent acute withdrawal in mice. *Brain Research*, *1652*, 21–29.

Bale, T. L., Lee, K. F., & Vale, W. W. (2002). The role of corticotropin-releasing factor receptors in stress and anxiety. *Integrative and Comparative Biology*, *42*(3), 552–555.

Bale, T. L., & Vale, W. W. (2004). CRF and CRF receptors: role in stress responsivity and other behaviors. *Annual Review of Pharmacology and Toxicology*, *44*, 525–557.

Behan, D. P., Cepoi, D., Fischer, W. H., Park, M., Sutton, S., Lowry, P. J., et al. (1996). Characterization of a sheep brain corticotropin releasing factor binding protein. *Brain Research*, *709*(2), 265–274.

Behan, D. P., De Souza, E. B., Potter, E., Sawchenko, P., Lowry, P. J., & Vale, W. W. (1996). Modulatory actions of corticotropin-releasing factor-binding protein. *Annals of the New York Academy of Sciences*, *780*, 81–95.

Bruijnzeel, A. W. (2012). Tobacco addiction and the dysregulation of brain stress systems. *Neuroscience and Biobehavioral Reviews*, *36*(5), 1418–1441.

Bruijnzeel, A. W., Ford, J., Rogers, J. A., Scheick, S., Ji, Y., Bishnoi, M., et al. (2012). Blockade of CRF1 receptors in the central nucleus of the amygdala attenuates the dysphoria associated with nicotine withdrawal in rats. *Pharmacology, Biochemistry, and Behavior*, *101*(1), 62–68.

Bruijnzeel, A. W., & Gold, M. S. (2005). The role of corticotropin-releasing factor-like peptides in cannabis, nicotine, and alcohol dependence. *Brain Research. Brain Research Reviews*. *49*(3), 505–528.

Bruijnzeel, A. W., Prado, M., & Isaac, S. (2009). Corticotropin-releasing factor-1 receptor activation mediates nicotine withdrawal-induced deficit in brain reward function and stress-induced relapse. *Biological Psychiatry*, *66*(2), 110–117.

Bruijnzeel, A. W., Zislis, G., Wilson, C., & Gold, M. S. (2007). Antagonism of CRF receptors prevents the deficit in brain reward function associated with precipitated nicotine withdrawal in rats. *Neuropsychopharmacology*, *32*(4), 955–963.

Chang, C. P., Pearse, R. V., II, O'Connell, S., & Rosenfeld, M. G. (1993). Identification of a seven transmembrane helix receptor for corticotropin-releasing factor and sauvagine in mammalian brain. *Neuron*, *11*(6), 1187–1195.

Dautzenberg, F. M., & Hauger, R. L. (2002). The CRF peptide family and their receptors: yet more partners discovered. *Trends in Pharmacological Sciences*, *23*(2), 71–77.

Fekete, E. M., & Zorrilla, E. P. (2007). Physiology, pharmacology, and therapeutic relevance of urocortins in mammals: ancient CRF paralogs. *Frontiers in Neuroendocrinology*, *28*(1), 1–27.

George, O., Ghozland, S., Azar, M. R., Cottone, P., Zorrilla, E. P., Parsons, L. H., et al. (2007). CRF-CRF1 system activation mediates withdrawal-induced increases in nicotine self-administration in nicotine-dependent rats. *Proceedings of the National Academy of Sciences of the United States of America*, *104*(43), 17198–17203.

Hsu, S. Y., & Hsueh, A. J. (2001). Human stresscopin and stresscopin-related peptide are selective ligands for the type 2 corticotropin-releasing hormone receptor. *Nature Medicine*, *7*(5), 605–611.

Inoue, K., Valdez, G. R., Reyes, T. M., Reinhardt, L. E., Tabarin, A., Rivier, J., et al. (2003). Human urocortin II, a selective agonist for the type 2 corticotropin-releasing factor receptor, decreases feeding and drinking in the rat. *The Journal of Pharmacology and Experimental Therapeutics*, *305*(1), 385–393.

Jamieson, P. M., Li, C., Kukura, C., Vaughan, J., & Vale, W. (2006). Urocortin 3 modulates the neuroendocrine stress response and is regulated in rat amygdala and hypothalamus by stress and glucocorticoids. *Endocrinology*, *147*(10), 4578–4588.

Janssen, D., & Kozicz, T. (2013). Is it really a matter of simple dualism? Corticotropin-releasing factor receptors in body and mental health. *Front Endocrinol (Lausanne)*, *4*, 28.

Kamdi, S. P., Nakhate, K. T., Dandekar, M. P., Kokare, D. M., & Subhedar, N. K. (2009). Participation of corticotropin-releasing factor type 2 receptors in the acute, chronic and withdrawal actions of nicotine associated with feeding behavior in rats. *Appetite*, *53*(3), 354–362.

Kenny, P. J., & Markou, A. (2001). Neurobiology of the nicotine withdrawal syndrome. *Pharmacology, Biochemistry, and Behavior*, *70*(4), 531–549.

Lewis, K., Li, C., Perrin, M. H., Blount, A., Kunitake, K., Donaldson, C., et al. (2001). Identification of urocortin III, an additional member of the corticotropin-releasing factor (CRF) family with high affinity for the CRF2 receptor. *Proceedings of the National Academy of Sciences of the United States of America*, *98*(13), 7570–7575.

Lovenberg, T. W., Liaw, C. W., Grigoriadis, D. E., Clevenger, W., Chalmers, D. T., De Souza, E. B., et al. (1995). Cloning and characterization of a functionally distinct corticotropin-releasing factor receptor subtype from rat brain. *Proceedings of the National Academy of Sciences of the United States of America*, *92*(3), 836–840.

Marcinkiewcz, C. A., Prado, M. M., Isaac, S. K., Marshall, A., Rylkova, D., & Bruijnzeel, A. W. (2009). Corticotropin-releasing factor within the central nucleus of the amygdala and the nucleus accumbens shell mediates the negative affective state of nicotine withdrawal in rats. *Neuropsychopharmacology*, *34*(7), 1743–1752.

Markou, A. (2008). Review. Neurobiology of nicotine dependence. *Philosophical Transactions of the Royal Society of London. Series B, Biological Sciences*, *363*(1507), 3159–3168.

Maruyama, H., Makino, S., Noguchi, T., Nishioka, T., & Hashimoto, K. (2007). Central type 2 corticotropin-releasing hormone receptor mediates hypothalamic-pituitary-adrenocortical axis activation in the rat. *Neuroendocrinology*, *86*(1), 1–16.

Pelleymounter, M. A., Joppa, M., Ling, N., & Foster, A. C. (2004). Behavioral and neuroendocrine effects of the selective CRF2 receptor agonists urocortin II and urocortin III. *Peptides*, *25*(4), 659–666.

Pellow, S., Chopin, P., File, S. E., & Briley, M. (1985). Validation of open: closed arm entries in an elevated plus-maze as a measure of anxiety in the rat. *Journal of Neuroscience Methods*, *14*(3), 149–167.

Porsolt, R. D., Bertin, A., & Jalfre, M. (1977). Behavioral despair in mice: a primary screening test for antidepressants. *Archives Internationales de Pharmacodynamie et de Thérapie*, *229*(2), 327–336.

Purves, H. D., & Sirett, N. E. (1965). Assay of corticotrophin in dexamethasone-treated rats. *Endocrinology*, *77*(2), 366–374.

Reul, J. M., & Holsboer, F. (2002). Corticotropin-releasing factor receptors 1 and 2 in anxiety and depression. *Current Opinion in Pharmacology*, *2*(1), 23–33.

Reyes, T. M., Lewis, K., Perrin, M. H., Kunitake, K. S., Vaughan, J., Arias, C. A., et al. (2001). Urocortin II: a member of the corticotropin-releasing factor (CRF) neuropeptide family that is selectively bound by type 2 CRF receptors. *Proceedings of the National Academy of Sciences of the United States of America*, *98*(5), 2843–2848.

Sarnyai, Z., Shaham, Y., & Heinrichs, S. C. (2001). The role of corticotropin-releasing factor in drug addiction. *Pharmacological Reviews*, *53*(2), 209–243.

Skelton, K. H., Owens, M. J., & Nemeroff, C. B. (2000). The neurobiology of urocortin. *Regulatory Peptides*, *93*(1–3), 85–92.

Spina, M. G., Merlo-Pich, E., Akwa, Y., Balducci, C., Basso, A. M., Zorrilla, E. P., et al. (2002). Time-dependent induction of anxiogenic-like effects after central infusion of urocortin or corticotropin-releasing factor in the rat. *Psychopharmacology*, *160*(2), 113–121.

Spina, M., Merlo-Pich, E., Chan, R. K., Basso, A. M., Rivier, J., Vale, W., et al. (1996). Appetite-suppressing effects of urocortin, a CRF-related neuropeptide. *Science*, *273*(5281), 1561–1564.

Suda, T., Kageyama, K., Sakihara, S., & Nigawara, T. (2004). Physiological roles of urocortins, human homologues of fish urotensin I, and their receptors. *Peptides*, *25*(10), 1689–1701.

Tanaka, M., & Telegdy, G. (2008). Antidepressant-like effects of the CRF family peptides, urocortin 1, urocortin 2 and urocortin 3 in a modified forced swimming test in mice. *Brain Research Bulletin*, *75*(5), 509–512.

Valdez, G. R., Inoue, K., Koob, G. F., Rivier, J., Vale, W., & Zorrilla, E. P. (2002). Human urocortin II: mild locomotor suppressive and delayed anxiolytic-like effects of a novel corticotropin-releasing factor related peptide. *Brain Research*, *943*(1), 142–150.

Valdez, G. R., Zorrilla, E. P., Rivier, J., Vale, W. W., & Koob, G. F. (2003). Locomotor suppressive and anxiolytic-like effects of urocortin 3, a highly selective type 2 corticotropin-releasing factor agonist. *Brain Research*, *980*(2), 206–212.

Vale, W., Spiess, J., Rivier, C., & Rivier, J. (1981). Characterization of a 41-residue ovine hypothalamic peptide that stimulates secretion of corticotropin and beta-endorphin. *Science*, *213*(4514), 1394–1397.

Van Pett, K., Viau, V., Bittencourt, J. C., Chan, R. K., Li, H. Y., Arias, C., et al. (2000). Distribution of mRNAs encoding CRF receptors in brain and pituitary of rat and mouse. *The Journal of Comparative Neurology*, *428*(2), 191–212.

Vaughan, J., Donaldson, C., Bittencourt, J., Perrin, M. H., Lewis, K., Sutton, S., et al. (1995). Urocortin, a mammalian neuropeptide related to fish urotensin I and to corticotropin-releasing factor. *Nature*, *378*(6554), 287–292.

Wonnacott, S., Sidhpura, N., & Balfour, D. J. (2005). Nicotine: from molecular mechanisms to behaviour. *Current Opinion in Pharmacology*, *5*(1), 53–59.

Zenker, N., & Bernstein, D. E. (1958). The estimation of small amounts of corticosterone in rat plasma. *The Journal of Biological Chemistry*, *231* (2), 695–701.

C H A P T E R

28

Delirium and Nicotine Withdrawal

Kataria Dinesh, Goel Ankit, Tiwari Sucheta, Kukreti Prerna

Department of Psychiatry, Lady Hardinge Medical College, New Delhi, India

Abbreviations

CRF	corticotropin-releasing factor
DSM-5	Diagnostic and Statistical Manual of Mental Disorders, Fifth Edition
EEG	electroencephalogram
GABA	gamma-aminobutyric acid
HPA axis	hypothalamic-pituitary-adrenal axis
ICU	intensive care unit
NMDA	*N*-methyl-D-aspartate
NRT	nicotine replacement therapy
REM	rapid eye movement

28.1 INTRODUCTION

Tobacco use disorders constitute a major health hazard worldwide. Tobacco smoking is the most common addiction of the world, and nicotine is the chief ingredient of tobacco, which leads to the development of dependence (Benowitz, 2009; Hatsukami, Stead, & Gupta, 2008). Cessation of tobacco use may cause an unpleasant nicotine withdrawal syndrome. Symptoms of nicotine withdrawal include frustration, irritability, anger, anxiety, depressed mood, insomnia, and restlessness. These symptoms peak within the 7 days of cessation of smoking and may last to about 2–4 weeks (Awissi, Lebrun, Fagnan, Skrobik, & Regroupement de Soins Critiques, Réseau de Soins Respiratoires, Québec, 2013).

Delirium is a clinical syndrome characterized by an acute alteration in attention and cognition (American Psychiatric Association, 2013; World Health Organisation, 1992). The clinical picture of delirium is variable, with core symptoms including disorientation in time, place, or person, impairment of working memory; psychomotor disturbances; and a disturbed sleep-wake cycle (World Health Organisation, 1992). It is commonly found in the critically ill with a reported incidence of 15%–80% (Ely et al., 2004; Jacobi et al., 2002). Delirium may be classified into three subtypes based on arousal and psychomotor behavior. Hyperactive delirium is characterized by agitation, disorientation, and hallucinations, while hypoactive delirium is predominated by lethargy, sedation, and confusion. Third subtype is the mixed type with combined features (Lipowski, 1983).

Delirium resulting from withdrawal of some psychoactive substances is a well-known entity (World Health Organisation, 1992). However, evidences for nicotine withdrawal delirium have started to emerge in recent past decade only (Brody, 2006; Kukreti, Garg, & Gautam, 2015). Substance withdrawal delirium is associated with poor outcomes, including increased morbidity, mortality, longer hospital stay, and more financial burden (Dubos et al., 1996; Inouye, Rushing, Foreman, Palmer, & Pompei, 1998).

28.2 NICOTINE USE AS A RISK FACTOR FOR DELIRIUM: A THEORETICAL CONCEPTUALIZATION

28.2.1 Clinical Features of Delirium and Nicotine Withdrawal

Delirium has been described by DSM-5 as a disturbance in attention, awareness, and cognition, including memory deficits (American Psychiatric Association, 2013). Nicotine withdrawal delirium is characterized by confusion, agitation, increased psychomotor activity, and irritability mimicking the features of hyperactive delirium (Park, Kim, & Yoon, 2016). Few authors have also reported inattention and working memory impairment during nicotine withdrawal (Brody, 2006; Heishman, 1998).

Functional brain imaging in relation to tobacco use shows that nicotine enhances neurotransmission through the prefrontal and paralimbic cortico-basal ganglia-thalamic circuit (Brody, 2006). This has been shown to improve performance on tasks testing arousal and sustenance of attention.

https://doi.org/10.1016/B978-0-12-813035-3.00028-9

Total duration of nicotine withdrawal appears to be between 2 and 4 weeks, with a peak under 2 weeks (Hughes, 2007). Anger/irritation peaks within the first week and may last up to 4 weeks. Anxiety symptoms increase in the first 3 days of abstinence and total duration of 2 weeks. Features of delirium in nicotine withdrawal peak within the first several days after the beginning of nicotine abstinence especially hyperactive delirium and may last up to 2–4 weeks (Papaioannou, Fraidakis, Michaloudis, Balalis, & Askitopoulou, 2005; Vandrey, Budney, Hughes, & Liguori, 2008).

There is evidence to suggest that the effect of nicotine on memory and attention might vary between smokers and nonsmokers. Difficulty in concentrating peaks at 2–3 days, lasting for 3–4 weeks. On nicotine administration, an improvement in recognition memory was seen in both abstinent smokers and nonsmokers, but improvement in attention was significant only among abstinent smokers (Heishman, 1998). This neuronal adaptation may also contribute to attention deficits in abstinent state.

Both delirium and nicotine withdrawal are characterized by disturbed sleep. In delirium, the disturbance is severe, with sleep-wake cycles occasionally getting reversed and worsening of symptoms in the evening. While sleep disturbances are not as severe in nicotine withdrawal, there is increased fragmentation of sleep in nicotine withdrawal (Hughes, 2007). This may lead to a greater frequency of REM sleep and increased remembrance of dreams, suggesting that there might be common neurological factors influencing sleep disturbances in both situations. Vivid, disturbing dreams may persist in the awake state as hallucinations or illusions (American Psychiatric Association, 2013; Webster & Holroyd, 2000). Thus, phenomenologically, delirium and nicotine withdrawal share few features common to both states (Table 28.1).

TABLE 28.1 Features Common to Nicotine Withdrawal and Delirium

- Agitation
- Irritability
- Confusion
- Restlessness
- Impaired attention
- Sleep disturbances

28.3 NEUROBIOLOGICAL LINKS BETWEEN NICOTINE WITHDRAWAL AND DELIRIUM

Various neurobiological mechanisms that can mediate nicotine withdrawal delirium are discussed below (Fig. 28.1 and Table 28.2).

28.3.1 Dysregulation of the Cholinergic System

Acetylcholine mediates cortical arousal, attentional processes, learning, memory, REM sleep induction, motor components of behavior, mood thought, perception, and

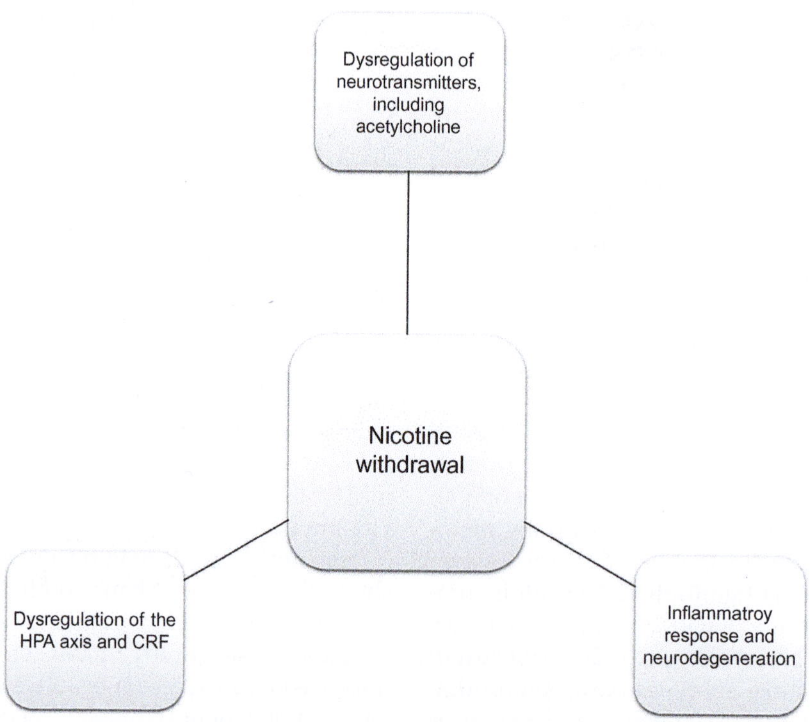

FIG. 28.1 Possible mechanisms of nicotine withdrawal delirium.

TABLE 28.2 Role of Various Neurotransmitters in Nicotine Withdrawal Delirium

Neurotransmitter	Role in nicotine withdrawal delirium
Acetylcholine (inhibition)	Interference with memory and attention, hyperactivity, slowing of EEG rhythm
Dopamine (release/loss of inhibition)	Excessive neuronal excitation
GABA (inhibition/desensitization)	Neuronal excitation due to loss of inhibition, the normal GABAergic function
Glutamate (release/desensitization of receptors)	Startle response, neuronal excitation, acute symptoms of delirium
Serotonin (inhibition)	Startle response, hypoactive delirium

orientation. Withdrawal of cholinergic stimulation can thus lead to interference with attention and memory, hyperactivity, and slowing of EEG rhythm, which is also the classical findings of delirium (Trzepacz, 2000). Acetylcholine-deficient states are known to predispose for the development of delirium (Hshieh, Fong, Marcantonio, & Inouye, 2008).

Nicotine produces its central and peripheral actions by binding to the nicotinic acetylcholine receptor (nAChR) complex. Various nAChR alpha and beta subunit combinations, especially the alpha4beta2 subtype, are present throughout the mesolimbic pathway including the ventral tegmental area (VTA), prefrontal cortex, amygdala, septal area, and nucleus accumbens. These receptors are potential binding sites through which nicotine may activate neurons within these structures to stimulate the release of several neurotransmitters (Marks et al., 1992; Picciotto, Caldarone, King, & Zachariou, 2000; Sargent, 1993). Animal studies show role of both central and peripheral cholinergic nervous systems in mediating the signs of nicotine withdrawal (Malin et al., 1997). Abrupt withdrawal of nicotine leads to decrease in cholinergic inputs for thalamocortical pathways that are responsible for attention, alertness, and vigilance (Dani & Heinemann, 1996). This hypocholinergic state induced by withdrawal may be responsible for the development of delirium.

28.3.2 Dysregulation of Other Neurotransmitters

Nicotine also mediates its effects via multiple other neurotransmitters that are released after the initial nAChR stimulation, including dopamine, glutamate, and serotonin (Watkins, Koob, & Markou, 2000). Effect of nicotine on attention may be mediated through direct stimulation of nicotinic acetylcholine receptors or via increase of dopamine and suppression of monoamine oxidase in the basal ganglia (Vandrey et al., 2008).

Chronic nicotine exposure leads to decline in extracellular dopamine levels in the nucleus accumbens and central nucleus of amygdala, and sudden withdrawal is associated with GABAergic neuronal desensitization (Fung, Schmid, Anderson, & Lau, 1996; Hildebrand, Nomikos, Bondjers, Nisell, & Svensson, 1997). Thus, loss of inhibitory effect on the dopamine release during abstinent states can clinically present as hyperactive delirium (Zhu & Chiappinelli, 1999).

A subset of glutamatergic receptors increase the acoustic startle response, which is a measure of reactivity to environmental stimuli, associated with nicotine withdrawal (Wiley, 1998). Nicotine administration activates nAChRs located presynaptically on glutamatergic terminals and leads to increased evoked glutamate release (Gray, Rajan, Radcliffe, Yakehiro, & Dani, 1996). In turn, glutamate increases the burst firing of the ventral tegmental dopaminergic neurons through excitatory actions at the NMDA receptors, leading to dopamine release in the nucleus accumbens. They lead to acute state of delirium. It is seen that these effects can be reversed by the administration of NMDA receptor antagonists (Shoaib, Benwell, Akbar, Stolerman, & Balfour, 1994; Watkins et al., 2000). How this may be linked to nicotine withdrawals remains to be elucidated upon, but possibly desensitization of glutaminergic receptors may be responsible for the same.

28.3.3 Decreased Serotonergic Function in Midbrain

Nicotine stimulates nAChRs located in the somatodendritic region in the median raphe nucleus, and the terminal fields in the forebrain facilitate serotonin release. It is suggested that the increased startle response during nicotine withdrawal is due to a decrease in the availability of synaptic serotonin, which has an inhibitory influence on the startle response (Geyer, Petersen, & Rose, 1980). In nicotine withdrawal state, the absence of nicotine combined with upregulated postsynaptic 5-HT1A serotonergic receptors leads to hypofunctioning serotonergic system, which can be postulated to precipitate a hypoactive delirious state (Hughes, Gust, Skoog, Keenan, & Fenwick, 1991).

28.3.4 Hypothalamic-Pituitary Axis (HPA) and Corticotrophin Releasing Factor Dysregulation

Administration of nicotine suppresses corticotropin-releasing factor (CRF) function, and withdrawal states are associated with elevated function (George et al., 2007; Watkins et al., 2000), thus explaining the immunosuppressive action of nicotine. CRF is known to have pro-inflammatory actions (Agelaki, Tsatsanis, Gravanis, & Margioris, 2002). Inflammatory markers are elevated in both delirium and dementia, suggesting them to lie on the two ends of same continuum possibly then being discrete entities (Simone & Tan, 2011).

Thus, acute nicotine withdrawal may also be associated with increase in circulating corticosterone due to CRF activation, which in turn may affect HPA axis, contributing to the development of delirium in the same way as seen in withdrawal states of other major drugs of abuse like ethanol and cannabinoids (Epping-Jordan, Watkins, Koob, & Markou, 1998).

28.3.5 Neuronal Inflammation and Increased Susceptibility to Delirium

Development of delirium is linked even to minor neuronal insults. Van Gool et al. suggest that long-term nicotine use is associated with neurodegeneration of cholinergic neurons in the basal ganglia due to inflammatory cytokines (van Gool, van de Beek, & Eikelenboom, 2010). Another review proposes that nicotine is likely to have antiinflammatory effects, possibly mediated via CRF suppression, as previously discussed (George et al., 2007; Piao et al., 2009; Watkins et al., 2000). This is the putative reason as to why smoking appears to be protective against neurodegenerative diseases such as Alzheimer's disease. Thus, dysregulation of inflammatory markers may also contribute to delirium (Table 28.3).

TABLE 28.3 Role of Inflammatory Markers in Nicotine Withdrawal Delirium

Role of CRF and inflammatory mediators	Role in nicotine withdrawal delirium
Increased CRF levels during nicotine withdrawal	• Predispose to delirium by facilitating a pro-inflammatory state • Increase in circulating corticosterone
Inflammatory cytokines	Neurodegeneration of cholinergic neurons in the basal ganglia

28.4 CLINICAL EVIDENCE

There have been published case reports of nicotine withdrawal delirium in ICU setting that were not explained by other comorbidities (Kukreti et al., 2015; Mayer et al., 2001) and resolved completely after administration of nicotine replacement. Although the cases reported had other significant medical and psychiatric comorbidities, they did not respond to usual interventions, and the response of delirium to nicotine replacement indicates that nicotine may have a role to play in the development of delirium in these specific subsets of patients.

A prospective study analyzing risk factors for developing delirium in ICU patients found that history of heavy smoking (more than 10 cigarettes a day) and previous history of any level of exposure to smoke are important risk factors for the development of delirium, independent of other risk factors (Van Rompaey et al., 2009).

A retrospective case-control study comparing smokers with nonsmokers in the ICU setting found significantly greater odds of developing hyperactive delirium among those with abrupt cessation of nicotine intake (Park et al., 2016). Another study found a statistically significant increase in agitation, but not delirium, in relation to nicotine withdrawal (Lucidarme et al., 2010). In light of these clinical experiences, the use of objective scales for assessment of nicotine withdrawal is recommended for patients with nicotine dependence admitted in ICU settings.

Though the individual studies suggest possibility of a general association between nicotine withdrawal and delirium, a comprehensive systematic review reported inconclusive results (Hsieh, Shum, Lee, Hasselmark, & Gong, 2013). This review recommended future studies with uniform definitions for exposure and outcome variables, larger sample sizes, and methodological rigor that will be needed to establish conclusive link. In addition, most studies use smoking as the only form of nicotine exposure with no comment on smokeless tobacco (Table 28.4).

TABLE 28.4 Clinical Evidence for Nicotine Withdrawal Delirium

	Author	Type of study	Conclusion
1.	Van Rompaey et al. (2009)	Prospective study	Smoking ≥10 cigarettes per day was significantly associated with the development of delirium in ICU patients
2.	Park et al. (2016)	Retrospective case-control study	Found significantly greater odds of developing hyperactive delirium among those with abrupt cessation of nicotine intake in ICU setting
3.	Lucidarme et al. (2010)	Cross-sectional observation study	Found a statistically significant increase in agitation, but not delirium, in relation to nicotine withdrawal in critically ill patients
4.	Kukreti et al. (2015)	Case report	If promptly diagnosed, nicotine withdrawal can be a treatable cause of delirium
5.	Mayer et al. (2001)	Case report	Nicotine withdrawal may be an underrecognized cause of delirium in patients with acute brain injury
6.	Hsieh et al. (2013)	Systematic review	Evidence linking nicotine withdrawal to delirium was inconclusive

28.5 NICOTINE REPLACEMENT THERAPY IN DELIRIUM

Individual case reports have mentioned promising results with nicotine replacement in chronic smokers presenting with delirium, but clinical trials have shown mixed results (Table 28.5).

In a review of six studies (Kowalski, Udy, McRobbie, & Dooley, 2016), one study did not find any difference in outcome related to the use of NRT. Three of these studies actually found an increase in agitation and delirium with NRT use, and two found a decrease in symptoms of nicotine withdrawal. However, only one of the studies included in the review was a randomized double-blinded case-control study (Pathak et al., 2013). This study reported reduction in ventilator days in ICU patients given NRT, but the results were not statistically significant.

Similarly, another study reported of hospitalized smokers who developed agitated delirium within 2–10 days after smoking cessation. And it either completely resolved or showed significant improvement following the use of NRT (Mayer et al., 2001).

Contrary to the above findings, a retrospective case-control study with 90 cases (smokers who received NRT in the first 24 h of their medical ICU admission) and 90 controls (smokers who did not receive NRT) concluded that NRT was associated with increased hospital mortality. This is a remarkable finding, but being a retrospective study performed in a single center makes it difficult to establish any conclusive evidence (Lee & Afessa, 2007).

Another study raise concerns over the use of NRT as they reported that patients undergoing cardiac surgery who received NRT had a significantly increased mortality (Paciullo, Short, Steinke, & Jennings, 2009). However, since it included only patients of cardiac disease, the deleterious cardiovascular effects of nicotine may be contributory to these findings. Thus, the factors that lead to improvement and worsening of symptoms with NRT need to be explored in a rigorous manner before NRT can be considered a legitimate method of treatment of delirium.

Not only in the patient profile but also in various studies have heterogeneous mode of NRT used. Results are mixed; however, it is imperative to consider that in ICU settings, transdermal nicotine patch appears to be the most feasible and efficient mode or NRT, especially in the background of altered sensorium and reduced alertness in such patients. Secondly, other existing comorbid medical conditions should also be taken into account, for example, cardiovascular conditions where the use of nicotine can worsen the general medical condition. Role of NRT in delirium management in heavy smokers is interesting but as of now an unexplored area of research. Rigorously designed studies are needed before recommending NRT in delirium management, but clinical studies warrant for cautious consideration of nicotine withdrawal as a possible etiologic agent for the development of delirium in heavy smokers.

TABLE 28.5 Clinical Evidence for Use of Nicotine Replacement Therapy in Delirium

	Author	Type of study	Conclusion
1.	Kowalski et al. (2016)	Systematic review	Inconclusive evidence of NRT in delirium management in the ICU
2.	Pathak et al. (2013)	Randomized double-blinded case-control study	Numerical decrease in ventilator days and ICU stay among those who required NRT (but the results were not statistically significant)
3.	Mayer et al. (2001)	Case report	Nicotine withdrawal delirium among brain-injured patients either completely resolved or showed significant improvement following the use of NRT
4.	Lee and Afessa (2007)	Retrospective case-control study	NRT was associated with increased hospital mortality
5.	Paciullo et al. (2009)	Retrospective matched cohort pilot study	Patients undergoing cardiac surgery who received NRT had significantly increased mortality

28.6 CONCLUSION

Nicotine addiction is a burgeoning epidemic with panaromic health hazards. It is often associated with significant comorbid psychiatric and medical ailments. Clinicians are utmost sensitive in diagnosing and managing nicotine dependence and associated health hazards, but sensitivity toward nicotine withdrawal assessment and treatment is still lacking.

Theoretical basis for nicotine withdrawal delirium does exist; however, clinical studies still have methodological limitations to establish a conclusive link. However, evidence is emerging to suggest that nicotine withdrawal can have variable presentation ranging from mild manifestations as irritability and dysphoria to severe manifestations as delirium, specifically in heavy smokers with medical illnesses.

Nicotine withdrawal often goes unnoticed, and nicotine withdrawal is not even considered as a possible differential of delirium in routine care. Thus, there is a need to sensitize all health professionals for actively screening for nicotine use and withdrawal. Especially in critically ill patients with history of heavy nicotine use, possibility of agitation and delirium from nicotine withdrawal should always be kept in mind.

Future research is needed to elucidate conclusive neurobiological basis of nicotine withdrawal delirium and framing systematic guidelines for the use of NRT in such situations.

MINI-DICTIONARY OF TERMS

Delirium An acute syndrome of altered mental status characterized by rapid onset fluctuating disturbance in attention, orientation, and other higher mental functions. Depending on the psychomotor activity, it can be (a) hyperactive delirium: characterized by increased psychomotor activity as seen in form of restlessness and agitation and "positive symptoms" of delusions and hallucinations may also be present, and (b) hypoactive delirium: characterized by reduced psychomotor activity as characterized by lethargy, slowness of movement, and reduced responsiveness.

Dependence A chronic often relapsing and remitting disease characterized by repeated indulgence in a behavior despite adverse consequences.

Dysregulation of neurotransmitters Upregulation or downregulation of a neurotransmitter in synaptic cleft as a result of abnormal synthesis, release, reuptake, or receptor activity.

Hypothalamic-pituitary-adrenal axis Neuroendocrine axis between the hypothalamus, pituitary, and the adrenal glands that mediates, among other things, the body's response to stress, including the stress of substance withdrawal states.

Neuroadaptation The process by which the body adapts to the presence of a chemical in the body by bringing necessary changes so that it can function normally.

Neurodegeneration The process of physiological or structural damage to neurons.

Neurotransmitters Chemicals within the nervous system that help in conduction of nerve impulses. These are released within synapses (space between neurons or between neurons and another structure) as a result of electric impulse carried by presynaptic neurons and act on specific postsynaptic receptors, thereby leading to successful transmission of impulse from one neuron to another.

Nicotine replacement therapy Medical treatment that involves delivery of nicotine via various routes in a tobacco-free form. Available as patches, chewing gums, lozenges, and intranasal spray.

Reward circuit A dopaminergic pathway within the brain that connects the ventral tegmental area to the ventral striatum. Also known as the mesolimbic pathway. Responsible for pleasure, motivation, and desire mediated by such as food, water, sex, and psychoactive substances.

Withdrawal A cluster of symptoms that occurs when there is a reduction or cessation of intake of a substance for which individual has developed dependence.

Key Facts of Nicotine Withdrawal Delirium

- Nicotine withdrawal includes frustration, irritability, anxiety, depressed mood, insomnia, and restlessness.
- These symptoms peak within the 7 days of cessation of smoking and may last 2–4 weeks.
- Abrupt cessation of nicotine intake especially in ICU settings may lead also to delirium especially hyperactive delirium.
- Replacement with nicotine in such patient may prove useful for management of delirium.

Summary Points

- This chapter focuses on delirium caused by nicotine withdrawal.
- Clinical evidence exists in terms of case-control studies and case reports supporting the association of nicotine withdrawal and delirium.
- The use of nicotine replacement therapy (NRT) in delirium has shown mixed results.
- We argue that nicotine use should always be assessed in patients with delirium as a possible cause.
- Therefore, we strongly suggest more research in the future for better understanding of association between nicotine withdrawal and delirium.

References

Agelaki, S., Tsatsanis, C., Gravanis, A., & Margioris, A. N. (2002). Corticotropin-releasing hormone augments proinflammatory cytokine production from macrophages in vitro and in lipopolysaccharide-induced endotoxin shock in mice. *Infection and Immunity, 70*(11), 6068–6074. https://doi.org/10.1128/IAI.70.11.6068-6074.2002.

American Psychiatric Association & DSM-5 Task Force (2013). *Diagnostic and statistical manual of mental disorders: DSM-5*. Washington, DC: American Psychiatric Association.

Awissi, D.-K., Lebrun, G., Fagnan, M., Skrobik, Y., & Regroupement de Soins Critiques, Réseau de Soins Respiratoires, Québec (2013). Alcohol, nicotine, and iatrogenic withdrawals in the ICU. *Critical Care Medicine, 41*(9 Suppl. 1), S57–S68. https://doi.org/10.1097/CCM.0b013e3182a16919.

Benowitz, N. L. (2009). Pharmacology of nicotine: addiction, smoking-induced disease, and therapeutics. *Annual Review of Pharmacology and Toxicology, 49*, 57–71. https://doi.org/10.1146/annurev.pharmtox.48.113006.094742.

Brody, A. L. (2006). Functional brain imaging of tobacco use and dependence. *Journal of Psychiatric Research, 40*(5), 404–418. https://doi.org/10.1016/j.jpsychires.2005.04.012.

Dani, J. A., & Heinemann, S. (1996). Molecular and cellular aspects of nicotine abuse. *Neuron, 16*(5), 905–908.

Dubos, G., Gonthier, R., Simeone, I., Camus, V., Schwed, P., Cadec, B., et al. (1996). Confusion syndromes in hospitalized aged patients: polymorphism of symptoms and course. Prospective study of 183 patients. *La Revue De Medecine Interne, 17*(12), 979–986.

Ely, E. W., Shintani, A., Truman, B., Speroff, T., Gordon, S. M., Harrell, F. E., Jr., et al. (2004). Delirium as a predictor of mortality in mechanically ventilated patients in the intensive care unit. *JAMA, 291*(14), 1753–1762. https://doi.org/10.1001/jama.291.14.1753.

Epping-Jordan, M. P., Watkins, S. S., Koob, G. F., & Markou, A. (1998). Dramatic decreases in brain reward function during nicotine withdrawal. *Nature, 393*(6680), 76–79. https://doi.org/10.1038/30001.

Fung, Y. K., Schmid, M. J., Anderson, T. M., & Lau, Y.-S. (1996). Effects of nicotine withdrawal on central dopaminergic systems. *Pharmacology Biochemistry and Behavior, 53*(3), 635–640. https://doi.org/10.1016/0091-3057(95)02063-2.

George, O., Ghozland, S., Azar, M. R., Cottone, P., Zorrilla, E. P., Parsons, L. H., et al. (2007). CRF–CRF1 system activation mediates withdrawal-induced increases in nicotine self-administration in nicotine-dependent rats. *Proceedings of the National Academy of*

Sciences of the United States of America, *104*(43), 17198–17203. https://doi.org/10.1073/pnas.0707585104.

Geyer, M. A., Petersen, L. R., & Rose, G. J. (1980). Effects of serotonergic lesions on investigatory responding by rats in a holeboard. *Behavioral and Neural Biology*, *30*(2), 160–177.

Gray, R., Rajan, A. S., Radcliffe, K. A., Yakehiro, M., & Dani, J. A. (1996). Hippocampal synaptic transmission enhanced by low concentrations of nicotine. *Nature*, *383*(6602), 713–716. https://doi.org/10.1038/383713a0.

Hatsukami, D. K., Stead, L. F., & Gupta, P. C. (2008). Tobacco addiction. *Lancet (London, England)*, *371*(9629), 2027–2038. https://doi.org/10.1016/S0140-6736(08)60871-5.

Heishman, S. J. (1998). What aspects of human performance are truly enhanced by nicotine? *Addiction (Abingdon, England)*, *93*(3), 317–320.

Hildebrand, B. E., Nomikos, G. G., Bondjers, C., Nisell, M., & Svensson, T. H. (1997). Behavioral manifestations of the nicotine abstinence syndrome in the rat: peripheral versus central mechanisms. *Psychopharmacology*, *129*(4), 348–356.

Hshieh, T. T., Fong, T. G., Marcantonio, E. R., & Inouye, S. K. (2008). Cholinergic deficiency hypothesis in delirium: a synthesis of current evidence. *The Journals of Gerontology Series A, Biological Sciences and Medical Sciences*, *63*(7), 764–772.

Hsieh, S. J., Shum, M., Lee, A. N., Hasselmark, F., & Gong, M. N. (2013). Cigarette smoking as a risk factor for delirium in hospitalized and intensive care unit patients. A systematic review. *Annals of the American Thoracic Society*, *10*(5), 496–503. https://doi.org/10.1513/AnnalsATS.201301-001OC.

Hughes, J. R. (2007). Effects of abstinence from tobacco: etiology, animal models, epidemiology, and significance: a subjective review. *Nicotine & Tobacco Research*, *9*(3), 329–339. https://doi.org/10.1080/14622200701188927.

Hughes, J. R., Gust, S. W., Skoog, K., Keenan, R. M., & Fenwick, J. W. (1991). Symptoms of tobacco withdrawal. A replication and extension. *Archives of General Psychiatry*, *48*(1), 52–59.

Inouye, S. K., Rushing, J. T., Foreman, M. D., Palmer, R. M., & Pompei, P. (1998). Does delirium contribute to poor hospital outcomes? A three-site epidemiologic study. *Journal of General Internal Medicine*, *13*(4), 234–242.

Jacobi, J., Fraser, G. L., Coursin, D. B., Riker, R. R., Fontaine, D., Wittbrodt, E. T., et al. (2002). Clinical practice guidelines for the sustained use of sedatives and analgesics in the critically ill adult. *Critical Care Medicine*, *30*(1), 119–141.

Kowalski, M., Udy, A. A., McRobbie, H. J., & Dooley, M. J. (2016). Nicotine replacement therapy for agitation and delirium management in the intensive care unit: a systematic review of the literature. *Journal of Intensive Care*, *4*, 69. https://doi.org/10.1186/s40560-016-0184-x.

Kukreti, P., Garg, A., & Gautam, P. (2015). Delirium: nicotine withdrawal as a rare etiology. *Delhi Psychiatry Journal*, *18*(1), 210–211.

Lee, A. H., & Afessa, B. (2007). The association of nicotine replacement therapy with mortality in a medical intensive care unit. *Critical Care Medicine*, *35*(6), 1517–1521. https://doi.org/10.1097/01.CCM.0000266537.86437.38.

Lipowski, Z. J. (1983). Transient cognitive disorders (delirium, acute confusional states) in the elderly. In *Psychosomatic medicine and liaison psychiatry* (pp. 289–306). Boston, MA: Springer. https://doi.org/10.1007/978-1-4613-2509-3_16.

Lucidarme, O., Seguin, A., Daubin, C., Ramakers, M., Terzi, N., Beck, P., et al. (2010). Nicotine withdrawal and agitation in ventilated critically ill patients. *Critical Care*, *14*(2), R58. https://doi.org/10.1186/cc8954.

Malin, D. H., Lake, J. R., Schopen, C. K., Kirk, J. W., Sailer, E. E., Lawless, B. A., et al. (1997). Nicotine abstinence syndrome precipitated by central but not peripheral hexamethonium. *Pharmacology Biochemistry and Behavior*, *58*(3), 695–699. https://doi.org/10.1016/S0091-3057(97)90006-X.

Marks, M. J., Pauly, J. R., Gross, S. D., Deneris, E. S., Hermans-Borgmeyer, I., Heinemann, S. F., et al. (1992). Nicotine binding and nicotinic receptor subunit RNA after chronic nicotine treatment. *The Journal of Neuroscience*, *12*(7), 2765–2784.

Mayer, S. A., Chong, J. Y., Ridgway, E., Min, K. C., Commichau, C., & Bernardini, G. L. (2001). Delirium from nicotine withdrawal in neuro-ICU patients. *Neurology*, *57*(3), 551–553.

Paciullo, C. A., Short, M. R., Steinke, D. T., & Jennings, H. R. (2009). Impact of nicotine replacement therapy on postoperative mortality following coronary artery bypass graft surgery. *The Annals of Pharmacotherapy*, *43*(7), 1197–1202. https://doi.org/10.1345/aph.1L423.

Papaioannou, A., Fraidakis, O., Michaloudis, D., Balalis, C., & Askitopoulou, H. (2005). The impact of the type of anaesthesia on cognitive status and delirium during the first postoperative days in elderly patients. *European Journal of Anaesthesiology*, *22*(7), 492–499.

Park, H., Kim, K. W., & Yoon, I.-Y. (2016). Smoking cessation and the risk of hyperactive delirium in hospitalized patients: a retrospective study. *Canadian Journal of Psychiatry Revue Canadienne de Psychiatrie*, *61*(10), 643–651. https://doi.org/10.1177/0706743716652401.

Pathak, V., Rendon, I. S. H., Lupu, R., Tactuk, N., Olutade, T., Durham, C., et al. (2013). Outcome of nicotine replacement therapy in patients admitted to ICU: a randomized controlled double-blind prospective pilot study. *Respiratory Care*, *58*(10), 1625–1629. https://doi.org/10.4187/respcare.01791.

Piao, W.-H., Campagnolo, D., Dayao, C., Lukas, R. J., Wu, J., & Shi, F.-D. (2009). Nicotine and inflammatory neurological disorders. *Acta Pharmacologica Sinica*, *30*(6), 715–722. https://doi.org/10.1038/aps.2009.67.

Picciotto, M. R., Caldarone, B. J., King, S. L., & Zachariou, V. (2000). Nicotinic receptors in the brain. Links between molecular biology and behavior. *Neuropsychopharmacology*, *22*(5), 451–465. https://doi.org/10.1016/S0893-133X(99)00146-3.

Sargent, P. B. (1993). The diversity of neuronal nicotinic acetylcholine receptors. *Annual Review of Neuroscience*, *16*, 403–443. https://doi.org/10.1146/annurev.ne.16.030193.002155.

Shoaib, M., Benwell, M. E. M., Akbar, M. T., Stolerman, I. P., & Balfour, D. J. K. (1994). Behavioural and neurochemical adaptations to nicotine in rats: influence of NMDA antagonists. *British Journal of Pharmacology*, *111*(4), 1073–1080. https://doi.org/10.1111/j.1476-5381.1994.tb14854.x.

Simone, M. J., & Tan, Z. S. (2011). The role of inflammation in the pathogenesis of delirium and dementia in older adults: a review. *CNS Neuroscience & Therapeutics*, *17*(5), 506–513. https://doi.org/10.1111/j.1755-5949.2010.00173.x.

Trzepacz, P. T. (2000). Is there a final common neural pathway in delirium? Focus on acetylcholine and dopamine. *Seminars in Clinical Neuropsychiatry*, *5*(2), 132–148. (2000). https://doi.org/10.153/SCNP00500132.

van Gool, W. A., van de Beek, D., & Eikelenboom, P. (2010). Systemic infection and delirium: when cytokines and acetylcholine collide. *Lancet (London, England)*, *375*(9716), 773–775. https://doi.org/10.1016/S0140-6736(09)61158-2.

Van Rompaey, B., Elseviers, M. M., Schuurmans, M. J., Shortridge-Baggett, L. M., Truijen, S., & Bossaert, L. (2009). Risk factors for delirium in intensive care patients: a prospective cohort study. *Critical Care*, *13*(3), R77. https://doi.org/10.1186/cc7892.

Vandrey, R. G., Budney, A. J., Hughes, J. R., & Liguori, A. (2008). A within-subject comparison of withdrawal symptoms during abstinence from cannabis, tobacco, and both substances. *Drug and Alcohol Dependence*, *92*(1–3), 48–54. https://doi.org/10.1016/j.drugalcdep.2007.06.010.

Watkins, S. S., Koob, G. F., & Markou, A. (2000). Neural mechanisms underlying nicotine addiction: acute positive reinforcement and withdrawal. *Nicotine & Tobacco Research*, *2*(1), 19–37.

Webster, R., & Holroyd, S. (2000). Prevalence of psychotic symptoms in delirium. *Psychosomatics*, *41*(6), 519–522. https://doi.org/10.1176/appi.psy.41.6.519.

Wiley, J. L. (1998). Nitric oxide synthase inhibitors attenuate phencyclidine-induced disruption of prepulse inhibition. *Neuropsychopharmacology*, *19*(1), 86–94. https://doi.org/10.1016/S0893-133X(98)00008-6.

World Health Organisation (1992). *The ICD-10 classification of mental and behavioural disorders: Clinical descriptions and diagnostic guidelines.*

Zhu, P. J., & Chiappinelli, V. A. (1999). Nicotine modulates evoked GABAergic transmission in the brain. *Journal of Neurophysiology*, *82*(6), 3041–3045.

CHAPTER

29

Postoperative Nicotine Withdrawal

Paul Zammit

Department of Geriatrics, Karen Grech Hospital, Pieta, Malta

Abbreviations

CAM	confusion assessment method
CAM-ICU	confusion assessment method for the intensive care unit
DSI	delirium symptom interview
ICDSC	intensive care delirium screening checklist
ITU	intensive treatment unit
NuDesc	Nursing Delirium Screening Scale
POD	postoperative delirium

29.1 INTRODUCTION

The word 'delirium' was first described as a medical term as early as the first century AD and was used to describe mental disorders occurring during fever or head trauma. Delirium is a common clinical syndrome especially in older people characterized by inattention and acute cognitive dysfunction. A varied range of terms have since emerged to describe delirium. These terms include "acute confusional state," "acute brain syndrome," "acute cerebral insufficiency," and "toxic-metabolic encephalopathy," but "delirium" should still be used as the standard term for this syndrome. Over time, the term delirium has evolved to describe a transient, reversible syndrome that is both acute and fluctuating and that would occur in the setting of a medical condition (Fong, Tulebaev, & Inouye, 2009).

Clinical experience and evidenced-based research have shown that delirium can become chronic or result in permanent sequelae. In older age persons, delirium can initiate or otherwise be a key component in a cascade of events that lead to a downward spiral in functional decline, loss of independence, institutionalization, and finally death. Delirium affects an estimated 14%–56% of all hospitalized older age patients. At least 20% of the 12.5 million patients over 65 years of age hospitalized each year in the United States experience complications during hospitalization because of delirium (Inouye, 1998).

29.2 POSTOPERATIVE DELIRIUM

The number of people over 60 years is rising exponentially and will continue to do so over the next 10–20 years. Those requiring surgery will thus also increase in number. Surgical studies have quoted postoperative delirium (POD) in as many as 47% of patients, more commonly present in older persons. More significantly, POD is associated with increased morbidity, mortality, length of stay, and care home placement (Noimark, 2009). There is also a burden on the health service due to the substantial increase in the health-care cost per patient episode.

POD is not time-related to emergence from anesthesia. By definition, patients with POD do not have an identifiable cause, although there can be other contributing factors. These patients often emerge well immediately postoperatively and may be lucid in the postanesthesia care unit. However, after this initial time interval, the patients develop the classic fluctuating mental status. This most commonly happens between postoperative days 1 and 3. Some postoperative patients may reside in the intensive treatment unit (ITU). However, the term ITU delirium (previously known as ITU psychosis) may include both medical and surgical patients (Brauer, Morrison, Silberzweig, & Siu, 2000). POD may differ from delirium in medical patients because the admission characteristics of the two groups can be different. By definition, patients hospitalized for acute medical indications

Neuroscience of Nicotine
https://doi.org/10.1016/B978-0-12-813035-3.00029-0

are usually acutely ill and may have exacerbations of chronic diseases. Most surgical operations are elective, and patients have been managed to ensure optimal physical status before entering the hospital. Surgery and the associated anesthetics and analgesics are generally absent in medical patients but can contribute to POD (Deiner & Silverstein, 2009).

29.3 PATHOGENESIS

There have been a number of hypotheses that have been proposed to explain the pathogenesis of postoperative delirium (Lipowski, 1987). One suggests that when the oxidative metabolism of the brain decreases, the levels of neurotransmitters within the brain, such as acetylcholine, decline that then cause mental dysfunction. Studies have shown that cerebral acetylcholine synthesis is sensitive to hypoxia. Moreover, an association between postoperative confusion and anticholinergic drug activity has also been observed (Hshieh, Fong, Marcantonio, & Inouye, 2008).

The second hypothesis suggests that an increase of serum cortisol from the stress of surgery or anesthesia may be responsible for postoperative confusion (MacLullich, Ferguson, Miller, de Rooij, & Cunningham, 2008). Reduced availability of tryptophan after cardiopulmonary bypass has also been incriminated in the pathogenesis of postcardiotomy psychosis (Parikh & Chung, 1995).

TABLE 29.1 Risk Factors for Postoperative Delirium

Age greater than 60 years
Dementia
Multiple comorbidities
Hearing or vision problems
Admitted with a hip fracture
The presence of an acute infection
Inadequately controlled pain
Depression
Alcohol abuse
Sleep deprivation or disturbance
Renal failure
Anemia
Hypoxia
Inadequate nutrition
Dehydration
Electrolyte abnormalities such as hyper- or hyponatremia
Poor functional status
Immobilization or limited mobility
Polypharmacy especially the use of psychotropic medications (benzodiazepines, anticholinergics, antihistamines, and antipsychotics)
Risk of urinary retention or constipation
Insertion of a urinary catheter

29.4 CAUSES OF POSTOPERATIVE DELIRIUM

Delirium is usually the result of a physiological stressor (e.g., an operation) and predisposing patient risk factors. Postoperative precipitants may include constipation, urinary retention medications (see below), infection, electrolyte abnormalities, and environmental causes. Refer to Table 29.1 for more causes of delirium. When a diagnosis of delirium is suspected, a standardized workup to exclude organic or identifiable causes of delirium is needed. The first step of the evaluation is a complete history and physical exam with cognitive assessment for delirium. As mentioned previously, patient assessment for delirium can be performed with the help of screening tools.

29.5 CLINICAL PRESENTATION

Presentation of POD is varied, and there are various signs with which this condition can be associated with (MacLullich et al., 2008). The *Diagnostic and Statistical Manual of Mental Disorders, Fifth Edition* (American Psychiatric Association, 2013) diagnostic criteria for delirium are as follows:

- Disturbance in attention (i.e., reduced ability to direct, focus, sustain, and shift attention) and awareness.
- Change in cognition (e.g., memory deficit, disorientation, language disturbance, and perceptual disturbance) that is not better accounted for by a preexisting, established, or evolving dementia.
- The disturbance develops over a short period (usually hours to days) and tends to fluctuate during the course of the day.
- There is evidence from the history, physical examination, or laboratory findings that the disturbance is caused by a direct physiological consequence of a general medical condition, an intoxicating substance, medication use, or more than one cause.

The various clinical presentations of delirium postoperatively can be viewed in Table 29.2.

TABLE 29.2 Symptoms Associated With Delirium

Change in the level of alertness: drowsiness or decreased alertness (hypoactive delirium) or increased alertness with hypervigilance (hyperactive delirium)
Delayed awakening from anesthesia
Sudden change in cognitive function (worsening confusion over hours or days). These may include problems with attention, difficulty concentrating, new memory problems, and new disorientation
Difficulty tracking conversations and following simple instructions
Thinking and speech that is more disorganized, difficult to follow, slow, or rapid
Quick-changing emotions, becoming easily irritable, tearfulness, uncharacteristic refusals to engage with postoperative care
Development of new paranoid thoughts, ideas, or delusions (i.e., fixed false beliefs)
New perceptual disturbances (e.g., illusions and hallucinations)
Motor changes that may include slowed or decreased movements, purposeless fidgeting or restlessness, and new difficulties in maintaining posture such as sitting or standing
Sleep/wake cycle disturbances such as sleeping during the day and/or being awake and active at night
Decreased appetite
New incontinence of urine or stool
Fluctuating symptoms and/or level of arousal over the course of minutes to hours

29.6 DIAGNOSIS

A number of studies have demonstrated that nurses and medical doctors do not accurately diagnose delirium on the basis of their bedside evaluation, including in the ITU, medical, and surgical wards (Spronk, Riekerk, Hofhuis, & Rommes, 2009). Screening tools or brief instruments for use by nonspecialist bedside personnel, for the detection of delirium, are numerous. These include the confusion assessment method (CAM), delirium symptom interview (DSI), and the Nursing Delirium Screening Scale (NuDesc). In an ITU setting, the confusion assessment method for the intensive care unit (CAM-ICU) or intensive care delirium screening checklist (ICDSC) is more appropriate. These diagnostic tests vary significantly and depend on the patient population in which they are used. Accordingly, evidence finds that when a patient is screened for delirium, multidisciplinary health-care professionals should be trained in and use a screening instrument that has been validated against a reference standard (American Geriatrics Society, 2015).

Review of treatment is important as drug-induced delirium is common. Sedative-hypnotics, anticholinergic medications, and pethidine contribute significantly to the risk of postoperative delirium in older adults (Marcantonio et al., 1994). The medication itself or medications within these classes have been shown to more than double the odds of an older patient developing delirium. Diphenhydramine increases the odds ratio of developing delirium to 2.3 (95% CI 1.4–3.6) in older adults (Agostini, Leo-Summers, & Inouye, 2001). Pethidine was associated with delirium in adults older than 50 years with an odds ratio of 2.7 (95% CI 1.3–5.5), and benzodiazepines had an increased odds of 3.0 (95% CI 1.3–6.8). Clinical guidelines to improve the safety of medication use in older adults recommend the avoidance of agents prone to increasing the risk or severity of delirium. The role of anesthesia in the development of delirium remains unclear. Recent metaanalyses have concluded that general anesthesia had an increased risk of developing postoperative cognitive dysfunction as compared to regional anesthesia (Vijayakumar, Elango, & Ganessan, 2014).

A careful review of the possibility of the patient having a substance issue such as alcohol abuse is needed as withdrawal of this needs specific treatment. Routine blood tests include a complete blood count, electrolytes, glucose, and a septic screen if an infective source is suspected. Neuroimaging such as a computerized tomography of the brain is limited to patients who have had recent falls or head trauma, use of anticoagulation, focal neurological signs, or fever without other explanation (Inouye, Westendorp, & Saczynski, 2014).

29.7 NICOTINE WITHDRAWAL IN POSTOPERATIVE PATIENTS

Nicotine is the principal alkaloid of snuff, responsible for its addictive power. It is rapidly absorbed and reaches the brain, causing the release of various neurotransmitters. Stimulation of the dopaminergic system mediated by the nucleus accumbens in the limbic system explains the addictiveness of nicotine (Adinoff, 2004).

Nicotine ingestion results in pleasure, relief of anxiety, increased ability to perform tasks, increased memory, modulation of mood, and muscle relaxation (Knott et al., 2011). Low-level nicotine in dependent smokers causes withdrawal symptoms that include anger, irritability, anxiety, insomnia, difficulty concentrating, drowsiness, fatigue, hunger, weight gain, restlessness, agitation, depressed mood, and more rarely delirium (Hughes, 2006). These symptoms begin within the first 24 h, reaching a maximum at 1–2 weeks, and usually dissipate after 1 month (Talwar, Jain, & Vijayan, 2004).

The frequency of nicotine withdrawal syndrome is difficult to predict but probably underestimated. This syndrome has already been described in the literature in several case reports (Kallel, Ellouze, Triki, & Karoui, 2012; Miranda, Slachevsky, & Venegas, 2005; Zammit, Cordina, Vassallo, & Dalli, 2015). Acute nicotine withdrawal was incriminated in arrays of acute delirium in an ITU setting postbrain injury. In intensive care, a series of five cases were published by an American neurological intensive care team (Mayer et al., 2001). These included patients with stroke and meningeal hemorrhage who experienced agitation and confusion 2–10 days after quitting. In all five cases, the establishment of a nicotine patch delivering 21 mg/24 h resulted in a rapid improvement in the situation. For smoking patients admitted to ITU, nicotine withdrawal syndrome is likely to occur because the cessation of smoking is complete and sudden.

There are also studies done on delirium in smoking patients in a surgical and ITU setting that have conflicting results. A systematic review of 14 studies in this subject was done by Hsieh, Andrew, Lee, Hasselmark, and Gong (2013). It identified several deficiencies in the existing literature that would account for the inconclusive nature of the findings. First and foremost, the existing literature was limited by suboptimal assessment of active smoking status. Some studies did not use a validated scale. There were also studies that had an incomplete adjustment for potential confounding factors of delirium such as preexisting cognitive impairment and depression. This systematic review also had several limitations. It concluded that since there is high prevalence of cigarette smokers and delirium in hospitalized and critically ill patients, a study is needed to carefully and decisively determine whether this association exists. This would have major potential preventative and therapeutic implications. It also concluded that future studies on delirium in smokers should be designed to specifically investigate this association and should use biochemical measures of cigarette smoking to objectively quantify smoking behavior.

29.8 DIAGNOSIS OF NICOTINE WITHDRAWAL

Diagnosis of nicotine withdrawal is on clinical suspicion. It is important to exclude other causes of delirium such as infection, metabolic disturbances, neurological pathology, and other possible pathologies as mentioned previously. Preventive measures include sensory enhancement, increasing mobility, cognitive orientation, pain control, adequate fluid intake, optimizing sleep quality, medication review, and daily multidisciplinary review that include a geriatrician for older age patients. Delirium education is an essential part of the prevention and treatment of postoperative delirium in older adults. Educational content should be based on the recognition of delirium, screening tools, outcomes, risk factors, and nonpharmacological and pharmacological approaches for prevention and management. Education is most effective when combined with reinforcement, booster sessions, peer support, one-to-one interactions, and feedback (American Geriatrics Society, 2015).

29.9 TREATMENT

Specific treatment is by using a nicotine patch of 20 mg and if no response to increase to 30 mg daily (Miranda et al., 2005). The diagnosis of this syndrome could be facilitated by the use of a transdermal patch of nicotine that constitutes a diagnostic test and a therapeutic means if rapid resolution of delirium occurs by using this medication. Nicotine patches have been linked with severe cardiovascular complications such as myocardial infarction and cerebral hemorrhage (Ottervanger, Festen, de Vries, & Stricker, 1995). Though there are no large randomized control studies to disprove, this caution should be used in patients who have undergone coronary surgery.

In agitated patients, titrated doses of neuroleptic agents, particularly haloperidol, may help (Robinson & Eiseman, 2008). The potential benefit of antipsychotics is decreased delirium severity, although the results of clinical trials have not been consistent. There are various potential harms associated with antipsychotic medication such as lethargy, retention of urine, constipation, and neuroleptic malignant syndrome to mention just a few. There is no evidence of benefit from the treatment of antipsychotics in patients without agitation. The use of antipsychotics should be reserved for short-term management of acute agitation when there is the possibility of possible substantial harm, that is, for the treatment of postoperative delirium in older surgical patients with behavior such as agitation that substantially threatens the patient's safety or the safety of others (American Geriatrics Society, 2015). No current evidence supports the routine use of benzodiazepines in the treatment of delirium. There is substantial evidence that benzodiazepines promote delirium (Zaal et al., 2015).

Nonpharmacological intervention in the management of nicotine withdrawal with subsequent delirium is important especially in older people. These interventions include environment (not having excessive, inadequate, or ambiguous sensory input; medication not interrupting sleep; and presenting one stimulus or task at a time), orientation (room should have a clock, calendar, and chart of the day's schedule; evaluate need for glasses, hearing aid, and interpreter), therapeutic activities (avoid physical restraint, allow movement, and encourage self-care and personal activities), vision/hearing optimization, communication (clear, slow, simple, repetitive, facing patient, warm, firm kindness, address patient by name, identify self, and encourage verbal expression), oral volume repletion, sleep enhancement, and familiarization of the hospital setting (objects from home, same staff, family members staying with patient, and discussion of familiar areas of interest) (Hipp & Ely, 2012). These interventional programs that recognize and address the multifactorial nature of delirium have shown success in reducing delirium.

29.10 CONCLUSION

POD is an important and often unrecognized complication that has important consequences for both the individual and health-care organizations. Nicotine withdrawal is an uncommon cause of delirium postoperatively. Diagnosis is clinical on exclusion of other more common causes. Treatment is by nicotine replacement. Optimal treatment of postoperative delirium reduces the incidence, duration, and side effects of this complication in geriatric postoperative patients.

MINI-DICTIONARY OF TERMS

Delirium Delirium is a condition of severe confusion and rapid changes in brain function. Delirium is in itself is not a disease, but rather a cluster of symptoms that may result from a disease or other clinical processes. Delirium may also be referred to as "acute confusional state" or "acute brain syndrome."

Multidisciplinary care Multidisciplinary care occurs when professionals from a range of disciplines with different but complementary skills, knowledge, and experience work together to deliver comprehensive health care aimed at providing the best possible outcome for the physical and psychosocial needs of a patient and their careers.

Older people This cohort is generally defined according to a range of characteristics including chronological age, change in social role, and changes in functional abilities. In Western countries, older age is generally defined in relation to retirement from paid employment and receipt of a pension, at 60 or 65 years.

Postoperative period This refers to the period of time after surgery. It begins with the patient's emergence from anesthesia and continues through the time required for the acute effects of the anesthesia and surgical procedures to resolve.

Withdrawal state A group of symptoms of variable clustering and degree of severity that occur on cessation or reduction of use of a psychoactive substance that has been taken repeatedly, usually for a prolonged period and/or in high doses.

Key Facts of Postoperative Delirium

- This may occur in the immediate period postsurgery.
- It is more common in older people and people with multiple comorbidities.
- It is associated with significant morbidity and mortality.
- Causes may be multifactorial including infection, cardiac and drug administration, or withdrawal.
- Management is through a multidisciplinary approach and involves preventative measures, treatment of the causes, and nonpharmacological interventions.

Summary Points

- POD is a common complication of postsurgical intervention and is associated with significant morbidity and mortality.
- Nicotine withdrawal is one of the many causes of POD and should be suspected in patients who stopped smoking suddenly in the immediate postoperative period.
- Presentation may be mild with symptoms such as anxiety or more severe including drowsiness, agitation, and confusion with fluctuation during the course of the day.
- Diagnosis of nicotine withdrawal is by exclusion for other metabolic causes of delirium such as infection, retention of urine, and electrolyte disturbances.
- Investigation for nicotine withdrawal is similar to the one for delirium and involves routine bloods and rarely imaging to exclude other causes for the symptoms.
- Management after diagnosis is by a multidisciplinary approach and involves nicotine replacement through a nicotine patch and nonpharmacological interventions.

References

Adinoff, B. (2004). Neurobiologic processes in drug reward and addiction. *Harvard Review of Psychiatry*, *12*(6), 305–320.

Agostini, J. V., Leo-Summers, L. S., & Inouye, S. K. (2001). Cognitive and other adverse effects of diphenhydramine use in hospitalized older patients. *Archives of Internal Medicine*, *161*, 2091–2097.

American Geriatrics Society. (2015). Postoperative delirium in older adults: best practice statement from the American Geriatrics Society. *Journal of the American College of Surgeons*, *220*(2), 136–148.e1.

American Psychiatric Association. (2013). *Diagnostic and statistical manual of mental disorders* (5th ed.). Washington, DC: American Psychiatric Association.

Brauer, C., Morrison, R. S., Silberzweig, S. B., & Siu, A. L. (2000). The cause of delirium in patients with hip fracture. *Archives of Internal Medicine*, *160*(12), 1856–1860.

Deiner, S., & Silverstein, J. H. (2009). Postoperative delirium and cognitive dysfunction. *British Journal of Anaesthesia*, *103*(Suppl. 1), i41–i46.

Fong, T. G., Tulebaev, S., & Inouye, S. (2009). Delirium in elderly adults: diagnosis, prevention and treatment. *Nature Reviews. Neurology*, *5*(4), 210–220.

Hipp, D. M., & Ely, E. W. (2012). Pharmacological and nonpharmacological management of delirium in critically ill patients. *Neurotherapeutics*, *9*(1), 158–175.

Hshieh, T. T., Fong, T. G., Marcantonio, E. R., & Inouye, K. S. (2008). Cholinergic deficiency hypothesis in delirium: A synthesis of current evidence. *The Journals of Gerontology. Series A, Biological Sciences and Medical Sciences*, *63*(7), 764–772.

Hsieh, S. J., Andrew, M. S., Lee, N., Hasselmark, F., & Gong, M. N. (2013). Cigarette smoking as a risk factor for delirium in hospitalized and intensive care unit patients. A systematic review. *Annals of the American Thoracic Society*, *10*(5), 496–503.

Hughes, J. R. (2006). Clinical significance of tobacco withdrawal. *Nicotine & Tobacco Research*, *8*, 153–156.

Inouye, S. K. (1998). Delirium in hospitalized older patients: recognition and risk factors. *Journal of Geriatric Psychiatry and Neurology*, *11*, 118–125.

Inouye, S. K., Westendorp, R. G., & Saczynski, J. S. (2014). Delirium in elderly people. *Lancet*, *383*, 911–922.

Kallel, S., Ellouze, M., Triki, Z., & Karoui, A. (2012). Le syndrome de sevrage nicotinique après chirurgie cardiaque: à propos d'un cas. *The Pan African Medical Journal*, *13*, 10 [in French].

Knott, V., Heenan, A., Shah, D., Bolton, K., Fisher, D., & Villeneuve, C. (2011). Electrophysiological evidence of nicotine's distracter-filtering properties in non-smokers. *Journal of Psychopharmacology, 25*(2), 239–248.

Lipowski, Z. J. (1987). Delirium (acute confusional states). *JAMA, 258*, 1789–1792.

MacLullich, A. M. J., Ferguson, K. J., Miller, T., de Rooij, S. E. J. A., & Cunningham, C. (2008). Unravelling the pathophysiology of delirium: a focus on the role of aberrant stress responses. *Journal of Psychosomatic Research, 65*, 229–238.

Marcantonio, E. R., Juarez, G., Goldman, L., Mangione, C. M., Ludwig, L. E., Lind, L., et al. (1994). The relationship of postoperative delirium with psychoactive medications. *JAMA, 272*(19), 1518–1522.

Mayer, S. A., Chong, J. Y., Ridgway, E., Min, K. C., Commichau, C., & Bernardini, G. L. (2001). Delirium from nicotine withdrawal in neuro-ICU patients. *Neurology, 57*(3), 551–553.

Miranda, M., Slachevsky, A., & Venegas, P. (2005). Delirium from nicotine withdrawal in a post-operative adult patient. *Revista Médica de Chile, 133*, 385–386 [in Spanish].

Noimark, D. (2009). Predicting the onset of delirium in the post-operative patient. *Age Ageing, 38*(4), 368–373.

Ottervanger, J. P., Festen, J. M., de Vries, A. G., & Stricker, B. H. (1995). Acute myocardial infarction while using the nicotine patch. *Chest, 107*(6), 1765–1766.

Parikh, S. S., & Chung, F. (1995). Postoperative delirium in the elderly. *Anesthesia and Analgesia, 80*(6), 1223–1232.

Robinson, T., & Eiseman, B. (2008). Postoperative delirium in the elderly: diagnosis and management. *Clinical Interventions in Aging, 3*(2), 351–355.

Spronk, P. E., Riekerk, B., Hofhuis, J., & Rommes, J. H. (2009). Occurrence of delirium is severely underestimated in the ICU during daily care. *Intensive Care Medicine, 35*(7), 1276–1280. https://doi.org/10.1007/s00134-009-1466-8.

Talwar, A., Jain, M., & Vijayan, V. (2004). Pharmacotherapy of tobacco dependence. *Medical Clinics of North America, 88*, 1517–1534.

Vijayakumar, B., Elango, P., & Ganessan, R. (2014). Post-operative delirium in elderly patients. *Indian Journal of Anaesthesia, 58*(3), 251–256.

Zaal, I. J˙, Devlin, J. W., Hazelbag, M˙, Klein Klouwenberg, P. M., van der Kooi, A. W., Ong, D. S., et al. (2015). Benzodiazepine-associated delirium in critically ill adults. *Intensive Care Medicine, 41*(12), 2130–2137.

Zammit, P., Cordina, J., Vassallo, M., & Dalli, S. (2015). Post-operative nicotine withdrawal in an elderly patient. *European Geriatric Medicine, 6*(6), 607–608.

C H A P T E R

30

Nicotine and Alpha3beta2 Neuronal Nicotinic Acetylcholine Receptors

Doris Clark Jackson, Sterling N. Sudweeks

Brigham Young University, Physiology and Developmental Biology, Provo, UT, United States

Abbreviations

ACh	acetylcholine
EC_{50}	effective concentration for 50% activation
HEK-293	human embryonic kidney cells
M1–M4	transmembrane region
nAChR	nicotinic acetylcholine receptor
SNP	single-nucleotide polymorphism
VTA	ventral tegmental area

30.1 INTRODUCTION

Nicotinic acetylcholine receptors (nAChRs) are pentameric, ligand-gated ion channels found both neuronally and at the neuromuscular junction. Each subunit consists of four transmembrane regions (M1–M4) with the intracellular M3–M4 loop being the least conserved regions between subunits. The M2 of each subunit forms the receptor's pore (Fig. 30.1). nAChRs are cation-selective with some subtypes permeable to Ca^{2+} (α3β2 are somewhat permeable; inclusion of α5 significantly increases Ca^{2+} permeability). Ca^{2+}-permeable nAChRs play the unique role of facilitating and even independently stimulating the release of other neurotransmitters like dopamine when these receptors are found presynaptically. The nAChR found at the neuromuscular junction is made of five different subunits (α1, β1, γ, δ, and ε). Neuronal nAChRs can be both homomeric and heteromeric. However, only the α7 and α9 subunits can form homomeric receptors. Most neuronal nAChR are a combination of α and β subunits (α2, α3, α4, α5 (must have an additional α subunit), α6, α7, α8 (chicken), α9, α10, β2, β3, and β4). The most common heteromeric nAChRs are those with two α2 and three βs ($\alpha^*{}_2\beta^*{}_3$) or three αs and two β2 ($\alpha^*{}_3\beta^*{}_2$). The most characterized nAChR subtypes are the homomeric α7 and the α4β2. We have characterized at least two subtypes of the α3β2 nAChR (Fig. 30.2) and will detail the possible role the α3β2 nAChRs play in nicotine addiction.

30.2 NICOTINE SENSITIVITY BY SUBTYPE

The α3β2 nAChRs are shown to be one of the least nicotine sensitive nAChRs subtypes and only a partial agonist of nicotine (Gerzanich, Wang, Kuryatov, & Lindstrom, 1998; Olale, Gerzanich, Kuryatov, Wang, & Lindstrom, 1997; Wang et al., 1996, 1998). Traditionally, only the α4β2 nAChRs were considered high-affinity nicotine binding sites. However, the α4β2 nAChRs only represent 90% of the high-affinity nicotine binding sites. Both the α and β subunits contribute to the nicotine affinity and efficacy. The β2 subunit has been shown to have greater nicotine binding affinity and more easily desensitized when compared to the β4 subunit (Fenster, Rains, Noerager, Quick, & Lester, 1997; Parker, Beck, & Luetje, 1998), whereas α3* receptors are less desensitized than α4* receptors (Fenster et al., 1997). Likewise, inclusion of the α5 subunit in the α3β2 increases nicotine efficacy significantly (Wang et al., 1996, 1998). However, this is unlikely due to a change in nicotine affinity but in channel gating (Wang et al., 1998). The most likely location of the α5 subunit would not change the binding site but instead impact the channel lining (Wang et al., 1996).

Differences in amino acids of M1 (224 and 226) account for the higher nicotine affinity of the β2 subunit as compared to the β4 (Rush, Kuryatov, Nelson, & Lindstrom, 2002). Differences in the amino acids at position 226 of

Neuroscience of Nicotine
https://doi.org/10.1016/B978-0-12-813035-3.00030-7

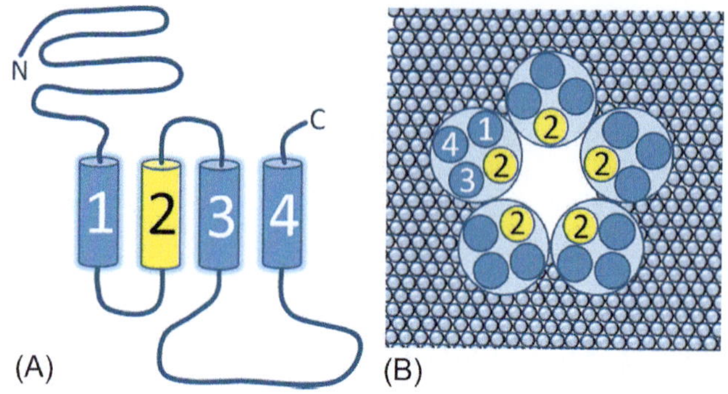

FIG. 30.1 nAChR subunit and embedded assembly.
(A) Each nicotinic acetylcholine receptor (nAChR) subunit contains four transmembrane regions. The intracellular loop between transmembrane regions 3 and 4 (M3–M4) is the most variable between nAChR subunits. (B) Five subunits (usually a combination of αs and βs) form a functional receptor with M2 forming the channel pore.

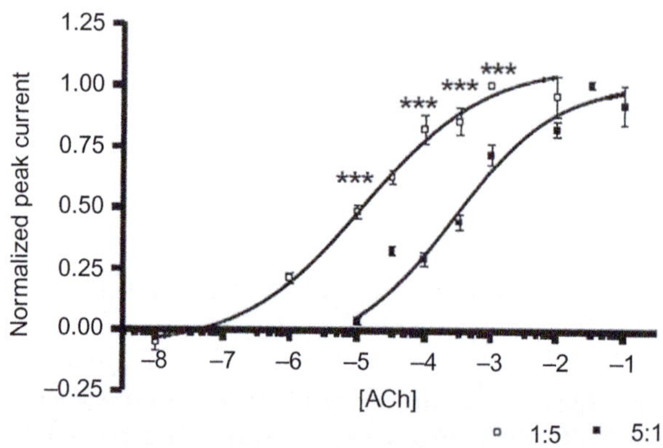

FIG. 30.2 ACh dose-response curves differentiate α3β2 subtypes. Upon injection of α3 and β2 human mRNA into *X. laevis* oocytes in 1:5 and 5:1 ratios, two distinct populations of nicotinic acetylcholine receptors (nAChRs) arise. More β and less α expression resulted in more acetylcholine (ACh)-sensitive α3β2 subtypes ($EC_{50}=12.2\pm1.7\,\mu M$; $n_H=0.49\pm0.13$ ($R^2=0.74$)) ($n=14$, replicates of 4, 1 outlier removed), whereas more β and less α expression resulted in less ACh-sensitive α3β2 subtypes ($EC_{50}=263.8\pm1.6\,\mu M$; $n_H=0.55\pm0.15$ ($R^2=0.77$)) ($n=12$, replicates of 4, 2 outliers removed). One-way ANOVA analysis resulted in significant differences ($F_{[15,\ 258]}=54.644$; $^{***}P<.001$). Data represented are mean ± SEM.

M1 (Fig. 30.1) account for the difference in α3* and α4* nAChR subtype nicotine binding affinity (Rush et al., 2002). Even though the α3 has greater binding affinity than α4, the α3β2 nAChR only has partial efficacy. Rush et al. (2002) suggest several mechanisms for the partial agonist activity of nicotine on α3β2 nAChRs. One mechanism may be that nicotine binding does not effectively shift the channel to the open state. Another mechanism for the partial agonist activity of nicotine may be that nicotine only weakly binds to the binding site. Kuryatov, Olale, Cooper, Choi, and Lindstrom (2000) showed that when using a chimera that consisted of the α4β2 binding site with the α3β2 channel, nicotine acts as a partial agonist; whereas, when using the α3β2 binding site with the α4β2 channel nicotine, had full efficacy. Their conclusion was that the α3β2 nAChRs have channel block when activated with nicotine. A study by Rush et al. (2002) supports the claim that nicotine may cause a channel block through binding to a low-affinity binding site that occludes the channel. However, the concentration of nicotine required for channel blocking is larger than would be expected by smokers (Benowitz, Porchet, & Jacob, 1990). Additionally, when an α6 subunit is incorporated into the α3β2 nAChR, nicotine has 100% efficacy (Kuryatov et al., 2000). Therefore, multiple α3β2 interfaces may be required for the channel block (Rush et al., 2002).

30.3 NICOTINE-INDUCED UPREGULATION

Like other nAChR subtypes, the α3β2 nAChR is upregulated upon nicotine exposure. General nAChR upregulation is observed in rat, mouse, and postmortem human brains (Breese et al., 1997; Perry, Davila-Garcia, Stockmeier, & Kellar, 1999; Yates, Bencherif, Fluhler, & Lippiello, 1995). Like nicotine binding affinity, subunit composition affects different phases of nicotine-induced upregulation. For example, β2* nAChRs have a much larger rate of nicotine-induced upregulation than β4* containing nAChRs (Gahring, Osborne-Hereford, Vasquez-Opazo, & Rogers, 2008; Wang et al., 1998). However, Meyer, Xiao, and Kellar (2001) did not show a change in upregulation with the addition of β4. When comparing three subtypes, the α3β2, α4β2, and α6β2, the α3β2 receptor required the highest nicotine concentration for upregulation and was about 10 times faster than the α4β2 upregulation but slightly slower than the α6β2 nAChR upregulation (Walsh et al., 2008). Therefore, the α3β2 nAChR may do little during initial nicotine exposure but play a more significant role in chronic exposure if nicotine levels are high.

Nicotine upregulation of α3β2 receptors occurs in human embryonic kidney cells (HEK-293) and human neuroblastoma SH-SY5Y, as well as neurons as evidenced by postmortem brains, indicating that upregulation is dependent on the receptor not the response of the neuron cellular processes (Wang et al., 1998). However, Wang et al. (1998) did not show upregulation in *Xenopus laevis* oocytes. The *X. laevis* oocytes may not have the cellular machinery required, like Rapsyn, to allow for upregulation. Nonetheless, the human cell lines and neurons are more directly related to human physiology than *X. laevis* oocytes. This should be considered when interpreting data regarding nAChR upregulation in oocytes.

Upregulation is likely both pre- and posttranscriptional. In HEK-293 cells (tsA201), chronic nicotine exposure (for upregulation, $EC_{50}=2\pm0.3\,\mu M$; for nicotine activation, $EC_{50} = 70 \pm 6\,\mu M$) resulted in a 24-fold increase in α3β2 nAChRs (as well as the α3β2α5). Upregulation was observed as early as 15 min following initial nicotine exposure but continued to increase with prolonged exposure (Wang et al., 1998). It is important to note that the nicotine concentration required for upregulation is much lower and physiologically more similar to serum nicotine levels in smokers than simply the nicotine level required for α3β2 receptor activation. The change was noted as an increase in cell surface expression and an increase in the total number of nAChRs. Upregulation that began almost immediately would likely indicate increased trafficking and increased assembly.

In addition, to an increase in receptor membrane integration, the endoplasmic reticulum slows the degradation of the α3β2 nAChR resulting in an increase in the overall receptor expression. Specifically, Rezvani et al. (2009) showed that nicotine can prevent protein degradation by partially inhibiting the ubiquitin proteasome system. As a note, chronic nicotine exposure to the human neuroblastoma SH-SY5Y cell line resulted in only a sixfold increase in α3β2 nAChRs. Yet, this was simply an increase in receptor number, not membrane integration. Interestingly, chronic nicotine exposure in the HEK-293 and the human neuroblastoma SH-SY5Y did not result in upregulation of either the α3β4 or the α3β4α5 (Wang et al., 1998).

In addition, temperature and pro-inflammatory cytokines can influence nicotine-induced upregulation complicating the analysis and interpretation of results. TNF-α increases α3β2 upregulation during transcription and translational phases although to a smaller degree than α4β2 and α4β4 (Gahring et al., 2008). Inflammation caused by smoking may further compound the effects of smoking. In general, lowering the temperature decreases receptor turnover that may influence results for upregulation giving a falsely high result (Devreotes & Fambrough, 1975; Paulson & Claudio, 1990). For the α4β2 nAChR, lowering the temperature to 29°C or 30°C significantly increases the cell surface expression (Cooper, Harkness, Baker, & Millar, 1999). This is an interesting effect considering that cigarette smoking causes a 1.5°C drop in skin temperature (Benowitz, Jacob III, & Herrera, 2006).

Upregulation can also be impacted through changes in phosphorylation. For example, the muscle nAChR is upregulation via cAMP and protein kinase A. This mechanism increases assembly efficiency and prevents degradation (Green, Ross, & Claudio, 1991). Wang et al. (1998) show that H-7, a protein kinase inhibitor, blocks at least 50% of nicotine-induced upregulation of α3β2 nAChRs in HEK-293 cells (tsA201). Therefore, there are many likely factors in the cellular process for nicotine upregulation including subunit composition, cell expression system, temperature, cytokines, and phosphorylation.

30.4 LOCATION

When considering the location of α3β2 nAChRs, the limiting subunit is the α3. The β2 subunit is quite ubiquitous, at least in the central nervous system, but is also readily found in the peripheral nervous system (Fig. 30.3, Hill Jr, Zoli, Bourgeois, & Changeux, 1993). In the rat and mouse, the α3 subunit is found in the thalamus, medial habenula, superior colliculus, and pineal body (Fig. 30.4, Cimino, Marini, Fornasari, Cattabeni, & Clementi, 1992). With a more limited expression pattern, the α3 subunit is more likely to be the limiting subunit in α3β2 expression (Fig. 30.5). The β2 knockout mice do not self-administer nicotine, and the mesolimbic neurons do not release dopamine in response to nicotine virtually abolishing addiction (Picciotto et al., 1998), whereas the α3 knockout mice results in multiple autonomic system problems (Xu et al., 1999).

In addition to these locations, the *Macaca fascicularis* (cynomolgus monkey) identifies the α3 subunit in the hippocampus (Cimino et al., 1992). This difference may be important in translating research to humans considering the much higher conservation between humans and nonhuman primates (Shorey-Kendrick,

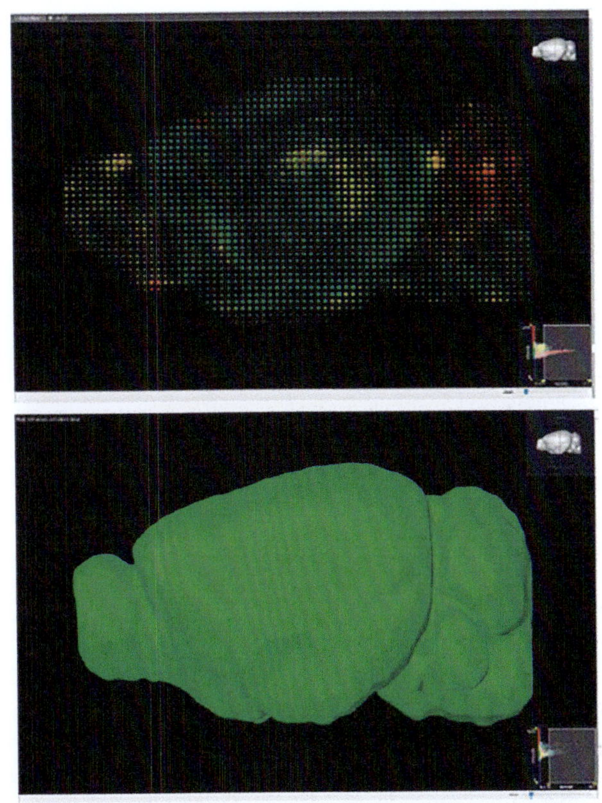

FIG. 30.3 β2 nAChR expression in the mouse brain. Heat map *(red high, blue low)* detailing the ubiquitous β2 subunit expression in the mouse brain. The expression profile is quantified in Fig. 30.5. *Image: With permission: Allen Brain Atlas, http://mouse.brain-map.org/experiment/show/2098.*

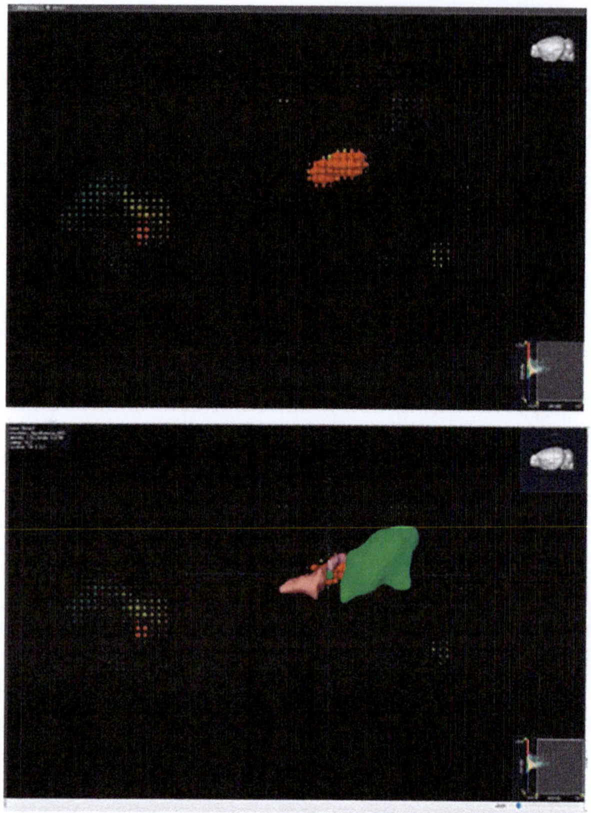

FIG. 30.4 α3 nAChR expression in the mouse brain. Heat map *(red high, blue low)* detailing the limited α3 subunit expression in the mouse brain. Lateral habenula *(pink)*, olivary pretectal nucleus *(purple)*, and superior colliculus *(green)*. *Image: With permission: Allen Brain Atlas, http://mouse.brain-map.org/experiment/show/69734723.*

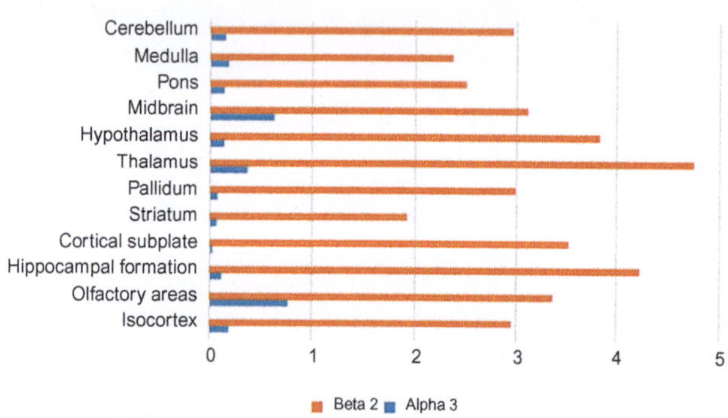

Mouse raw expression values from Allen Brain Atlas

	Alpha 3	Beta 2
Isocortex	0.19	2.96
Olfactory areas	0.19	3.37
Hippocampal formation	0.12	4.23
Cortical subplate	0.03	3.53
Striatum	0.07	1.93
Pallidum	0.08	3
Thalamus	0.37	4.76
Hypothalamus	0.14	3.84
Midbrain	0.63	3.12
Pons	0.14	2.51
Medulla	0.18	2.38
Cerebellum	0.15	2.97

FIG. 30.5 Mouse raw expression values from Allen Brain Atlas. The relative expression profile of mouse (C57BL/6J) brain regions containing both α3 and β2 subunits. Results were obtained using antisense in situ hybridization of an adult (56 days) male mouse. As indicated, the β2 subunit has higher expression than the α3 in all regions analyzed. *Data from Allen Brain Atlas with permission.*

2015). Additionally, both the *M. fascicularis* and in human tissue identify the α3 mRNA subunit at high levels in sympathetic and parasympathetic ganglia (Cimino et al., 1992). The location of the α3 subunit will likely dictate the presence of α3β2 nAChRs in the central nervous system, and the β2 subunit will likely dictate α3β2 expression in the peripheral nervous system.

In the central nervous system, α3β2 nAChRs influence striatal dopamine release and eventual addiction (Picciotto et al., 1998). However, the α3β4 nAChRs are shown to contribute more significantly than the α3β2 nAChRs in the peripheral nervous system (Covernton, Kojima, Sivilotti, Gibb, & Colquhoun, 1994). Yet, the α3 subunit remains the predominant subunit in the peripheral nervous system. Therefore, the α3β2 nAChRs may play a larger role in the autonomic physiological nicotine response.

The dopamine neurons of the midbrain (superior colliculus) likely contribute to the physical locomotion required for nicotine self-administration. Blocking dopamine receptors prevents nicotine self-administration (Corrigall & Coen, 1991). With a moderately high expression of the α3 subunit in the superior colliculus, the α3β2 nAChR may impact dopamine release and indirectly effect nicotine self-administration.

Additionally, the α3 and β2 mRNA subunits have been identified in the ventral tegmental area (VTA) and the nucleus accumbens using high-affinity nicotine and n-bungarotoxin binding (Azam, Winzer-Serhan, Chen, & Leslie, 2002). The α3β2 nAChR has not been identified as the most important contributor to nicotine addiction in the VTA, but it may contribute to a different stage in addiction considering its upregulation and increased nicotine affinity following nicotine exposure as compared to α4β2 (Azam et al., 2002).

Autoradiography and experiments using 125I-nBgt or 125I-α-conotoxin MII label nAChRs at the presynaptic dopaminergic neurons of the nucleus accumbens. Although α-conotoxin MII was originally thought to be α3β2 specific, it has since been shown to bind α6* nAChRs as well (Kuryatov et al., 2000; Parker et al., 2004). Therefore, either/both α3 and α6 containing nAChRs may contribute to dopamine release in the nucleus accumbens. Furthermore, the α3 subunit has relatively high levels in the habenula (Fig. 30.4) and therefore may additionally mediate dopamine release in this pathway and contribute to the nicotine-addicted state.

30.5 PHYSIOLOGICAL EFFECT OF NICOTINE

The nicotine dose-response curve for the α3β2 nAChRs expressed in HEK 293 cells with a 1:1 α3-β2 ratio by Wang et al. (1998) has a nicotine $EC_{50} = 70 \pm 26\,\mu M$, which is higher than would be expected for blood nicotine levels for smokers. Nicotine serum concentration is 0.2 μM for a typical smoker (Benowitz et al., 1990). Therefore, the α3β2 nAChRs may not play a role in initial nicotine addiction. However, considering the high upregulation of α3β2 nAChRs following nicotine exposure, α3β2 nAChRs may be important in sustaining addiction. The α4β2 nAChR is highly sensitive to nicotine and may play an important role in initial addiction. At typical nicotine serum levels, the α4β2 nAChRs would be upregulated like the α3β2 nAChRs, but they would also be almost fully desensitized. However, α3β2 nAChRs (not specific stoichiometries) are not as easily desensitized and would likely maintain normal functioning in smokers as they remain 80% active at normal serum nicotine level in smokers (Fenster et al., 1997; Olale et al., 1997; Picciotto et al., 1998). The α3β2 nAChRs may contribute more significantly to acetylcholine (ACh)-mediated signaling in a nicotine-addicted brain than the α4β2 nAChRs (Olale et al., 1997; Wang et al., 1998). Furthermore, the upregulation or lack thereof may vary between brain regions just as there was variability between cell lines (HEK-293, neuroblastoma, and oocytes) (Wang et al., 1998). This possible and likely variability in upregulation would likely influence behavior considering that brain regions are affected more than others. However, none of these upregulation studies were done on specific α3β2 nAChR subtypes. Our preliminary data, like that of other nAChR subtypes, show a difference in nicotine affinity and efficacy between different stoichiometries of the α3β2 nAChRs.

The α3 nAChR is by far the most populous subunit in the autonomic nervous system. Although the predominant nAChR in the autonomic ganglia is the α3β4 (Vernallis, Conroy, & Berg, 1993), the α3β2 likely also plays a significant role in signaling and nicotine response. Nicotine has many effects on the autonomic nervous system including increased heart rate, increased myocardial contractility, increased respiration, constriction of arteries, sweating, nausea, diarrhea, and increased blood pressure (Haass & Kübler, 1997).

30.6 VARIANTS OF THE α3 SUBUNIT

The α3 nAChR subunit has single-nucleotide variants (rs3743078, rs6495308, and rs1051730) associated with increased nicotine cravings, increased smoking, or increased nicotine-related behaviors (Shmulewitz et al., 2016; Wu et al., 2015). Nees et al. (2013) conclude that the single-nucleotide polymorphism (SNP) rs578776 found on the α3 gene is associated with increased risk of smoking because it dampens the reward response of the anterior cingulate cortex. Several of these polymorphisms are included in a gene cluster between the α5-α 3-β4 subunits that is associated with nicotine dependence. A study by Polina et al. (2014) shows that two polymorphisms (rs578776 and rs3743078) in the α3 gene are associated with an increased risk of smoking in patients with ADHD but are protective for smoking in non-ADHD populations.

Additionally, there multiple polymorphisms (rs578776, rs938682, rs6495309, and rs3743073) identified in the α3 nAChR subunit that increase susceptibility to lung cancer (Qu et al., 2016; Shen et al., 2012; Xiao, Chen, Wu, & Wen, 2014). Two other polymorphisms (rs8042059 and rs7177514) in the α3 nAChR subunit likely increase susceptibility to lung cancer only indirectly through smoking behavior (Zhou et al., 2015). The SNPs rs12910984, rs6495309, and rs1051730 have been shown to be associated with an increased risk of chronic obstructive pulmonary disease (Kaur-Knudsen, Nordestgaard, & Bojesen, 2012; Kim et al., 2013; Yang et al., 2012). The SNP rs6495308 has been found to be associated with an increased risk of hypertension (Wu et al., 2015). The rs8042374 a SNP on the α3 subunit is associated with increased risk of adenocarcinoma (He et al., 2014).

30.7 IMPLICATIONS FOR TREATMENT

As highlighted, variants in the α3 subunit may affect the likelihood of addiction and likelihood of nicotine-related diseases and even help predict response to treatment. By identifying and characterizing the effects of these variants in various populations, nicotine addiction treatment may prove to be more personalized and more effective. We are still in the early stages of characterizing the α3β2 nAChR and its relation to nicotine addiction, but the evidence suggests that by targeting the α3β2, we may be able to counteract the autonomic effects of nicotine considering the location of the α3 subunit. Additionally, considering the location and function of the α3β2 nAChR, pharmacological approaches to treatment targeting the α3β2 nAChR may alter the dopamine release associated with addiction and therefore make nicotine less pleasurable. Lastly, since pro-inflammatory cytokines result in the upregulation of the α3β2 nAChR and other subtypes, the use of antiinflammatory drugs or diet should be considered when developing combinatory treatment options.

30.8 CONCLUSION

In summary, the α3β2 nAChR may play a role in both central and peripheral nervous system nicotine responses. The α3β2 nAChR is upregulated upon nicotine exposure and has many variants that can influence the addictive and disease effects of nicotine.

MINI-DICTIONARY OF TERMS

Affinity Affinity refers to the concentration required for receptor activation. A nicotine higher affinity means a smaller concentration of nicotine is required for activation, whereas a low nicotine affinity means a larger concentration of nicotine is required for activation.

Efficacy Efficacy refers to the maximum activation that can be reached. Efficacy can only be compared across subtypes or agonists.

Nicotine-induced upregulation In this context, nicotine-induced upregulation refers to an increase in the number of nAChRs expressed following nicotine exposure. Upregulation may alter the cell response to chronic nicotine exposure.

Nicotinic acetylcholine receptor (nAChR) A neurotransmitter receptor that conducts positive ions resulting in a change in electric potential. This change in electric potential may induce cell-to-cell communication between neurons or other cell types.

Transmembrane region The region of the nAChR protein that spans the membrane. The transmembrane region is mostly hydrophobic and helps form the channel pore. nAChR subunits have four transmembrane regions.

Variants Variants are small differences in subunit proteins. These differences may result from changes in the gene sequence or protein production.

Key Facts of Neuronal Nicotinic Acetylcholine Receptors

- Neuronal nicotinic acetylcholine receptors (nAChRs) are pentameric, ligand-gated ion channels that are activated using the endogenous neurotransmitter acetylcholine or the exogenous chemical nicotine.
- Most neuronal nAChRs are made of α and β subunits.
- Neuronal nAChRs are cation-selective with specific subtypes being calcium (Ca^{2+}) permeable.
- Neuronal nAChRs can be located presynaptically, postsynaptically, periterminally, and even on nonneuron cells.
- Depending on location and Ca^{2+} permeability, specific subtypes can induce neurotransmitter release.
- Nicotine upregulates and desensitizes many subtypes of neuronal nAChRs.
- nAChR subunit variants may alter nicotine sensitivity or physiological response to nicotine.

Summary Points

- The α3β2 neuronal nicotinic acetylcholine receptor (nAChR) and its relationship to nicotine and nicotine addiction.
- The α3β2 neuronal nAChR is only a partial nicotine agonist and requires a higher nicotine concentration for activation as compared to other subtypes.
- The α3β2 neuronal nAChR is significantly upregulated more than the α4β2 nAChR.
- The α3β2 neuronal nAChR remains 80% active at nicotine concentrations that desensitize other nAChR subtypes.
- The α3 subunit is the most prominent nAChR subtype in the autonomic ganglia.
- Nicotine affects many aspects of the autonomic nervous system including heart rate, blood pressure, and digestion.
- The location of the α3 subunit likely dictates the presence of α3β2 nAChRs in the central nervous system, whereas the location of the β2 subunit likely dictates the presence of the α3β2 nAChRs in the autonomic ganglia.
- Several variants of the α3 subunit increase the likelihood of nicotine addiction and/or the susceptibility to nicotine-related diseases.

References

Azam, L., Winzer-Serhan, U. H., Chen, Y., & Leslie, F. M. (2002). Expression of neuronal nicotinic acetylcholine receptor subunit mRNAs within midbrain dopamine neurons. *The Journal of Comparative Neurology*, *444*, 260–274.

Benowitz, N. L., Jacob, P., III, & Herrera, B. (2006). Nicotine intake and dose response when smoking reduced nicotine content cigarettes. *Clinical Pharmacology and Therapeutics*, *80*, 703–714.

Benowitz, N., Porchet, H., & Jacob, P. (1990). In S. Wonnacott, M. Russell, & I. Stolerman (Eds.), *Nicotine psychopharmacology* (pp. 112–157). Oxford, England: Oxford University.

Breese, C. R., Marks, M. J., Logel, J., Adams, C. E., Sullivan, B., Collins, A. C., et al. (1997). Effect of smoking history on [3H]nicotine binding in human postmortem brain. *The Journal of Pharmacology and Experimental Therapeutics*, *282*, 7–13.

Cimino, M., Marini, P., Fornasari, D., Cattabeni, F., & Clementi, F. (1992). Distribution of nicotinic receptors in cynomolgus monkey brain. *Neuroscience*, *51*, 77–86.

Cooper, S. T., Harkness, P. C., Baker, E. R., & Millar, N. S. (1999). Up-regulation of cell-surface alpha4beta2 neuronal nicotinic receptors by lower temperature and expression of chimeric subunits. *The Journal of Biological Chemistry*, *274*, 27145–27152.

Corrigall, W. A., & Coen, K. M. (1991). Selective dopamine antagonists reduce nicotine self-administration. *Psychopharmacology*, *104*, 171–176.

Covernton, P. J., Kojima, H., Sivilotti, L. G., Gibb, A. J., & Colquhoun, D. (1994). Comparison of neuronal nicotinic receptors in rat sympathetic neurones with subunit pairs expressed in Xenopus oocytes. *The Journal of Physiology*, *481*(Pt1), 27–34.

Devreotes, P. N., & Fambrough, D. M. (1975). Acetylcholine receptor turnover in membranes of developing muscle fibers. *The Journal of Cell Biology*, *65*, 335–358.

Fenster, C. P., Rains, M. F., Noerager, B., Quick, M. W., & Lester, R. A. (1997). Influence of subunit composition on desensitization of neuronal acetylcholine receptors at low concentrations of nicotine. *The Journal of Neuroscience*, *17*, 5747–5759.

Gahring, L. C., Osborne-Hereford, A. V., Vasquez-Opazo, G. A., & Rogers, S. W. (2008). Tumor necrosis factor alpha enhances nicotinic receptor up-regulation via a p38MAPK-dependent pathway. *The Journal of Biological Chemistry*, *283*, 693–699.

Gerzanich, V., Wang, F., Kuryatov, A., & Lindstrom, J. (1998). Alpha 5 subunit alters desensitization pharmacology Ca++ permeability and Ca++ modulation of human neuronal alpha 3 nicotinic receptors. *The Journal of Pharmacology and Experimental Therapeutics*, *286*, 311–320.

Green, W. N., Ross, A. F., & Claudio, T. (1991). Acetylcholine receptor assembly is stimulated by phosphorylation of its gamma subunit. *Neuron, 7*, 659–666.

Haass, M., & Kübler, W. (1997). Nicotine and sympathetic neurotransmission. *Cardiovascular Drugs and Therapy, 10*, 657–665.

He, P., Yang, X. X., He, X. Q., Chen, J., Li, F. X., Gu, X., et al. (2014). CHRNA3 polymorphism modifies lung adenocarcinoma risk in the Chinese Han population. *International Journal of Molecular Sciences, 15*, 5446–5457.

Hill, J. A., Jr., Zoli, M., Bourgeois, J. P., & Changeux, J. P. (1993). Immunocytochemical localization of a neuronal nicotinic receptor: the beta 2-subunit. *The Journal of Neuroscience, 13*(4), 1551–1568.

Kaur-Knudsen, D., Nordestgaard, B. G., & Bojesen, S. E. (2012). CHRNA3 genotype, nicotine dependence, lung function and disease in the general population. *The European Respiratory Journal, 40*, 1538–1544.

Kim, W. J., Oh, Y. M., Kim, T. H., Lee, J. H., Kim, E. K., Lee, J. H., et al. (2013). CHRNA3 variant for lung cancer is associated with chronic obstructive pulmonary disease in Korea. *Respiration, 86*, 117–122.

Kuryatov, A., Olale, F., Cooper, J., Choi, C., & Lindstrom, J. (2000). Human alpha6 AChR subtypes: subunit composition, assembly and pharmacological responses. *Neuropharmacology, 39*, 2570–2590.

Meyer, E. L., Xiao, Y., & Kellar, K. J. (2001). Agonist regulation of rat alpha 3 beta 4 nicotinic acetylcholine receptors stably expressed in human embryonic kidney 293 cells. *Molecular Pharmacology, 60*, 568–576.

Nees, F., Witt, S. H., Lourdusamy, A., Vollstädt-Klein, S., Steiner, S., Poustka, L., et al. (2013). Genetic risk for nicotine dependence in the cholinergic system and activation of the brain reward system in healthy adolescents. *Neuropsychopharmacology, 38*(11), 2081–2089.

Olale, F., Gerzanich, V., Kuryatov, A., Wang, F., & Lindstrom, J. (1997). Chronic nicotine exposure differentially affects the function of human alpha3, alpha4, and alpha7 neuronal nicotinic receptor subtypes. *The Journal of Pharmacology and Experimental Therapeutics, 283*, 675–683.

Parker, M. J., Beck, A., & Luetje, C. W. (1998). Neuronal nicotinic receptor beta2 and beta4 subunits confer large differences in agonist binding affinity. *Molecular Pharmacology, 54*(6), 1132–1139.

Parker, S. L., Fu, Y., McAllen, K., Luo, J., McIntosh, J. M., Lindstrom, J. M., et al. (2004). Upregulation of brain nicotinic acetylcholine receptors in the rat during long-term self-administration of nicotine: disproportionate increase of the alpha6 subunit. *Molecular Pharmacology, 65*, 611–622.

Paulson, H. L., & Claudio, T. (1990). Temperature-sensitive expression of all-Torpedo and Torpedorat hybrid AChR in mammalian muscle cells. *The Journal of Cell Biology, 110*, 1705–1717.

Perry, D. C., Davila-Garcia, M. I., Stockmeier, C. A., & Kellar, K. J. (1999). Increased nicotinic receptors in brains from smokers: membrane binding and autoradiography studies. *The Journal of Pharmacology and Experimental Therapeutics, 289*, 1545–1552.

Picciotto, M. R., Zoli, M., Rimondini, R., Léna, C., Marubio, L. M., Pich, E. M., et al. (1998). Acetylcholine receptors containing the beta2 subunit are involved in the reinforcing properties of nicotine. *Nature, 391*, 173–177.

Polina, E. R., Rovaris, D. L., de Azeredo, L. A., Mota, N. R., Vitola, E. S., Silva, K. L., et al. (2014). ADHD diagnosis may influence the association between polymorphisms in nicotinic acetylcholine receptor genes and tobacco smoking. *Neuro Molecular Medicine, 16*, 389–397.

Qu, X., Wang, K., Dong, W., Shen, H., Wang, Y., Liu, Q., et al. (2016). Association between two CHRNA3 variants and susceptibility of lung cancer: a meta-analysis. *Scientific Reports, 6*, 20149.

Rezvani, K., Teng, Y., Pan, Y., Dani, J. A., Lindstrom, J., García Gras, E. A., et al. (2009). UBXD4, a UBX-containing protein, regulates the cell surface number and stability of alpha3-containing nicotinic acetylcholine receptors. *The Journal of Neuroscience, 29*(21), 6883–6896.

Rush, R., Kuryatov, A., Nelson, M. E., & Lindstrom, J. (2002). First and second transmembrane segments of alpha3, alpha4, beta2, and beta4 nicotinic acetylcholine receptor subunits influence the efficacy and potency of nicotine. *Molecular Pharmacology, 61*, 1416–1422.

Shen, B., Shi, M. Q., Zheng, M. Q., Hu, S. N., Chen, J., & Feng, J. F. (2012). Correlation between polymorphisms of nicotine acetylcholine acceptor subunit CHRNA3 and lung cancer susceptibility. *Molecular Medicine Reports, 6*, 1389–1392.

Shmulewitz, D., Meyers, J. L., Wall, M. M., Aharonovich, E., Frisch, A., Spivak, B., et al. (2016). CHRNA5/A3/B4 variant rs3743078 and nicotine-related phenotypes: indirect effects through nicotine craving. *Journal of Studies on Alcohol and Drugs, 77*(2), 227–237.

Shorey-Kendrick, L. E., Ford, M. M., Allen, D. C., Kuryatov, A., Lindstrom, J., Wilhelm, L., et al. (2015). Nicotinic receptors in non-human primates: Analysis of genetic and functional conservation with humans. *Neuropharmacology, 96*(Pt B), 163–173.

Vernallis, A. B., Conroy, W. G., & Berg, D. K. (1993). Neurons assemble acetylcholine receptors with as many as three kinds of subunits while maintaining subunit segregation among receptor subtypes. *Neuron, 10*(3), 451–464.

Walsh, H., Govind, A. P., Mastro, R., Hoda, J. C., Bertrand, D., Vallejo, Y., et al. (2008). Upregulation of nicotinic receptors by nicotine varies with receptor subtype. *The Journal of Biological Chemistry, 283*, 6022–6032.

Wang, F., Gerzanich, V., Wells, G. B., Anand, R., Peng, X., Keyser, K., et al. (1996). Assembly of human neuronal nicotinic receptor alpha5 subunits with alpha3, beta2 and beta4 subunits. *The Journal of Biological Chemistry, 271*, 17656–17665.

Wang, F., Nelson, M. E., Kuryatov, A., Olale, F., Cooper, J., Keyser, K., et al. (1998). Chronic nicotine treatment up-regulates human alpha3 beta2 but not alpha3 beta4 acetylcholine receptors stably transfected in human embryonic kidney cells. *The Journal of Biological Chemistry, 273*, 28721–28732.

Wu, X. Y., Zhou, S. Y., Niu, Z. Z., Liu, T., Xie, C. B., & Chen, W. Q. (2015). CHRNA3 rs6495308 genotype as an effect modifier of the association between daily cigarette consumption and hypertension in Chinese male smokers. *International Journal of Environmental Research and Public Health, 12*, 4156–4169.

Xiao, M., Chen, L., Wu, X., & Wen, F. (2014). The association between the rs6495309 polymorphism in CHRNA3 gene and lung cancer risk in Chinese: a meta-analysis. *Scientific Reports, 4*, 6372.

Xu, W., Orr-Urtreger, A., Nigro, F., Gelber, S., Sutcliffe, C. B., Armstrong, D., et al. (1999). Multiorgan autonomic dysfunction in mice lacking the beta2 and the beta4 subunits of neuronal nicotinic acetylcholine receptors. *The Journal of Neuroscience, 19*, 9298–9305.

Yang, L., Qiu, F., Lu, X., Huang, D., Ma, G., & Guo, Y. (2012). Functional polymorphisms of CHRNA3 predict risks of chronic obstructive pulmonary disease and lung cancer in Chinese. *PLoS ONE, 7*(10). e46071.

Yates, S. L., Bencherif, M., Fluhler, E. N., & Lippiello, P. M. (1995). Up-regulation of nicotinic acetylcholine receptors following chronic exposure of rats to mainstream cigarette smoke or alpha 4 beta 2 receptors to nicotine. *Biochemical Pharmacology, 50*, 2001–2008.

Zhou, W., Geng, T., Wang, H., Xun, X., Feng, T., Zou, H., et al. (2015). CHRNA3 genetic polymorphism and the risk of lung cancer in the Chinese Han smoking population. *Tumour Biology, 36*, 4987–4992.

CHAPTER

31

Nicotine Addiction and Alpha4beta2* Nicotinic Acetylcholine Receptors

John J. Maurer,†, Heath D. Schmidt†,‡*

*Pharmacology Graduate Group, Perelman School of Medicine, University of Pennsylvania, Philadelphia, PA, United States

†Department of Biobehavioral Health Sciences, School of Nursing, University of Pennsylvania, Philadelphia, PA, United States

‡Department of Psychiatry, Perelman School of Medicine, University of Pennsylvania, Philadelphia, PA, United States

Abbreviations

dFBr desformylflustrabromine
DHβE dihydro-beta-erythroidine
NAc nucleus accumbens
nAChR(s) nicotinic acetylcholine receptor(s)
PAMs positive allosteric modulators
VTA ventral tegmental area

31.1 INTRODUCTION

The primary psychoactive compound in tobacco and most e-cigarettes is nicotine, which functions as a nonselective agonist at nicotinic acetylcholine receptors (nAChRs). Neuronal nAChRs are pentameric ligand-gated ion channels that can be heteromeric protein complexes consisting of various combinations of α (α2–6) and β (β2–4) subunits or homomeric protein complexes consisting of α7 subunits. The stoichiometry of individual nAChRs imparts distinct pharmacokinetic properties of each receptor subtype (Le Novere, Corringer, & Changeux, 2002).

The most abundant subtypes of neuronal nAChRs are comprised of α4 and β2 subunits (α4β2* nAChRs, where the asterisks indicates a nAChR that contains the indicated subunits but the exact stoichiometry remains unknown) (Flores, Rogers, Pabreza, Wolfe, & Kellar, 1992). α4β2* nAChRs are expressed throughout the brain including the mesolimbic dopamine system, a neuronal network that regulates the reinforcing and rewarding effects of all major drugs of abuse (Tuesta, Fowler, & Kenny, 2011). The stimulatory actions of nicotine on ventral tegmental area (VTA) dopamine neurons that project to the nucleus accumbens (NAc) mediate the positive reinforcing effects of nicotine (Corrigall, Franklin, Coen, & Clarke, 1992). Indeed, direct infusions of nicotine into the VTA increase dopamine release in the NAc (Ferrari, Le Novere, Picciotto, Changeux, & Zoli, 2002). The ability of nicotine to activate mesoaccumbens dopamine projections is complex and relies on α4β2* nAChRs expressed on dopamine and GABA neurons in the VTA (Mameli-Engvall et al., 2006) (Fig. 31.1). It should be noted that glutamate transmission in the brain also plays a critical role in nicotine addiction (Li, Semenova, D'Souza, Stoker, & Markou, 2014). Specifically, nicotine activates presynaptic α7 nAChRs, which increases glutamate release in the VTA and excites midbrain dopamine neurons (Mansvelder & McGehee, 2000) (Fig. 31.1).

The reinforcing efficacy of nicotine is measured in laboratory animals and humans using the drug self-administration paradigm (O'Dell & Khroyan, 2009). Nicotine self-administration has the highest degree of face validity of all animal models of nicotine addiction, primarily because it mimics voluntary tobacco consumption in humans (O'Dell & Khroyan, 2009). Therefore, the drug self-administration model is a critical component of key studies examining the neural mechanisms underlying voluntary nicotine taking and seeking in rodents (Rose & Corrigall, 1997; Tuesta et al., 2011). While nicotine self-administration studies in human subjects

Neuroscience of Nicotine
https://doi.org/10.1016/B978-0-12-813035-3.00031-9

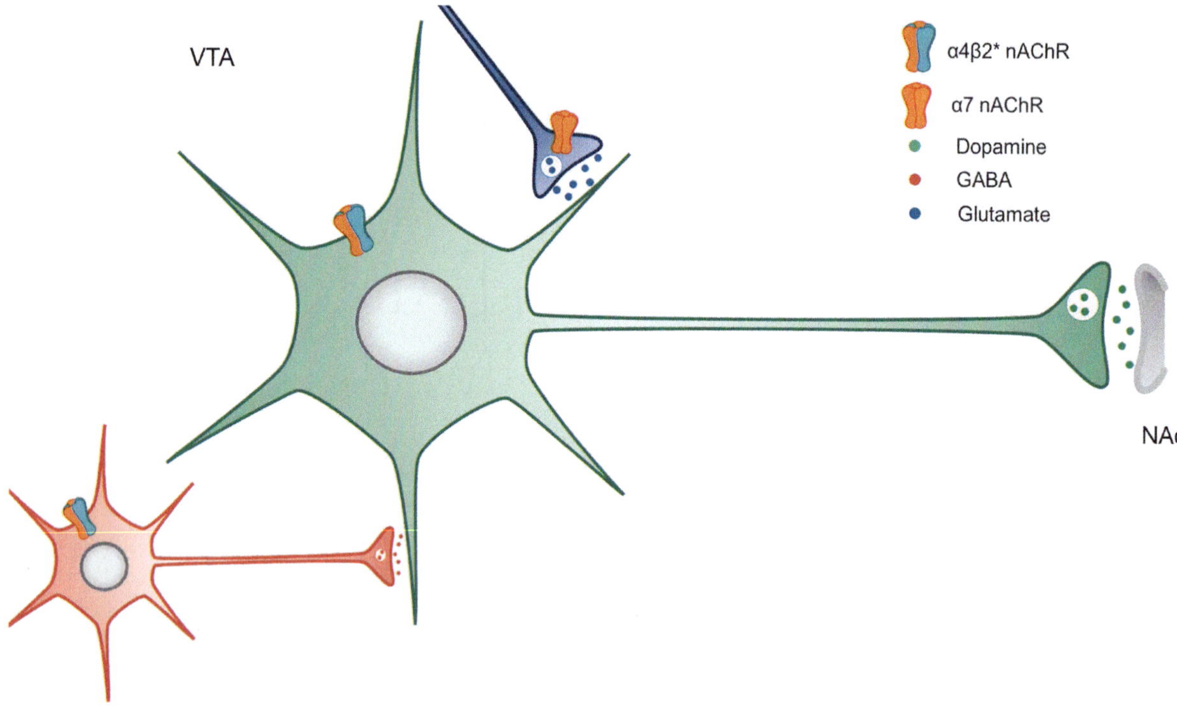

FIG. 31.1 Nicotine activates α4β2* and α7 nAChRs.
Nicotine also activates α4β2* nAChRs expressed on GABAergic interneurons that synapse on dopamine neurons in the VTA. Furthermore, nicotine activates presynaptic α7 nAChRs on glutamate terminals in the VTA. Since α4β2* nAChRs on GABA interneurons desensitize quickly, the effects of nicotine on VTA dopamine cell firing are due mainly to the activation of α4β2* nAChRs on dopamine neurons and enhanced glutamate signaling in the VTA.

typically investigate cigarette-smoking behavior, the majority of nicotine self-administration studies in rodents involve intravenous infusion immediately following an operant response (i.e., lever press or nose poke). Intravenous self-administration is used primarily because this route simulates the rapid rise in arterial nicotine and rapid distribution of nicotine to the brain that occurs via the typical pulmonary route of exposure in humans (Gourlay & Benowitz, 1997). Plasma nicotine levels are similar in animals self-administering nicotine and human tobacco smokers, further validating this animal model of nicotine addiction (Shoaib & Stolerman, 1999). Moreover, rodent nicotine self-administration recapitulates some of the neuronal adaptations observed in human smokers. For example, the expression of α4 and β2 nAChR subunits is increased in the brains of rodents self-administering nicotine (Gaimarri et al., 2007) and human smokers (Staley et al., 2006). Furthermore, the predictive validity associated with altered nicotine self-administration and the effectiveness of smoking cessation medications appears to be relatively high (Lerman et al., 2007).

Smoking relapse is typically modeled in animals using the self-administration/extinction/reinstatement paradigm. The same stimuli (i.e., stress, re-exposure to drug-associated cues, and re-exposure to nicotine itself) that precipitate smoking relapse during abstinence in humans can be used to reinstate nicotine-seeking behavior in rodents (Stoker & Markou, 2015). Following extinction of nicotine self-administration, systemic injections of nicotine and/or cues previously paired with nicotine taking reinstate operant responding in the absence of nicotine reinforcement in rodents (Stoker & Markou, 2015). The validity of the nicotine reinstatement model as an in vivo medication screen appears promising for smoking relapse (Epstein, Preston, Stewart, & Shaham, 2006).

Here, we discuss the role of α4β2* nAChRs in the reinforcing effects of nicotine. Specifically, we review the literature supporting a critical role for α4β2* nAChRs in nicotine self-administration and the reinstatement of nicotine-seeking behavior.

31.2 α4β2* nAChRs IN NICOTINE SELF-ADMINISTRATION AND REINSTATEMENT

Antagonists of α4β2* nAChRs have been shown to attenuate nicotine taking and seeking in rodents. Corrigall and Coen were the first to show that systemic infusions of the noncompetitive and nonselective nAChR antagonist mecamylamine reduced voluntary nicotine taking in rats (Corrigall & Coen, 1989). These results have been replicated by many labs and clearly indicate that the reinforcing effects of nicotine are mediated by the activation of nAChRs. Consistent with these results, mecamylamine reduces nicotine reinstatement suggesting that pharmacological inhibition of nAChRs may reduce smoking relapse in abstinent smokers (Liu, Palmatier,

Caggiula, Donny, & Sved, 2007). Based on these studies, mecamylamine was investigated as an antismoking medication. Clinical trials showed that mecamylamine had modest efficacy in reducing smoking behavior and promoting smoking abstinence (Tennant Jr., Tarver, & Rawson, 1984). However, adverse effects limit the clinical utility of mecamylamine in humans (Tennant Jr. et al., 1984).

Development of subtype-selective nAChR antagonists afforded the opportunity to investigate the role of individual nAChR subtypes in nicotine reinforcement. Systemic administration of dihydro-beta-erythroidine (DHβE), a competitive nAChR antagonist with moderate selectivity for α4β2* nAChRs (Williams & Robinson, 1984), reduced nicotine self-administration in rats (Watkins, Epping-Jordan, Koob, & Markou, 1999). Consistent with these effects, direct infusions of DHβE into the VTA reduced voluntary nicotine taking indicating that α4β2* nAChRs in the VTA play a critical role in nicotine reinforcement (Corrigall, Coen, & Adamson, 1994). Taken together, these findings provide strong pharmacological evidence supporting a role for α4β2* nAChRs in nicotine-taking and nicotine-seeking behaviors.

31.3 β2-CONTAINING nAChRs IN NICOTINE SELF-ADMINISTRATION

The development of genetically engineered mice permitted in vivo studies of the functional role of individual nAChR subunits in animal models of nicotine addiction. Basal firing rates of VTA dopamine neurons, nicotine-evoked dopamine cell firing, and release of extracellular dopamine in the NAc following nicotine administration are reduced in mutant mice lacking β2 nAChR subunits compared to wild-type controls (Mameli-Engvall et al., 2006; Picciotto et al., 1998; Zhou, Liang, & Dani, 2001). With regard to voluntary drug taking, β2 knockout mice do not acquire nicotine self-administration (Epping-Jordan, Picciotto, Changeux, & Pich, 1999; Orejarena et al., 2012; Picciotto et al., 1998). Moreover, in contrast to wild-type mice, β2 knockout mice do not self-administer nicotine directly into the VTA suggesting that β2-containing nAChRs in the VTA are required for the positive reinforcing effects of nicotine (Besson et al., 2006). Importantly, virus-mediated re-expression of β2 subunits in the VTA restores nicotine-evoked dopamine release in the NAc and nicotine self-administration (Maskos et al., 2005; Orejarena et al., 2012; Pons et al., 2008). Collectively, these findings support a critical role for β2-containing nAChRs in nicotine reinforcement and indicate that this receptor population is necessary for nicotine self-administration.

31.4 α4-CONTAINING nAChRs IN NICOTINE SELF-ADMINISTRATION

Similar to β2 knockout mice, mutant mice lacking α4 nAChR subunits also fail to display increases in nicotine-evoked dopamine release in the NAc (Marubio et al., 2003). Moreover, α4-containing nAChRs are necessary for nicotine-induced burst firing of VTA dopamine neurons (Exley et al., 2011). Since α4β2* nAChRs are abundantly expressed in the VTA (Klink, de Kerchove d'Exaerde, Zoli, & Changeux, 2001), these results, together with findings from β2 knockout mice, suggest that α4β2* nAChRs are critically important for nicotine addiction. However, studies examining α4-containing nAChRs in nicotine self-administration are mixed. An initial report indicated that mutant mice lacking α4 subunits failed to acquire nicotine self-administration during a single operant test session (Pons et al., 2008). Virus-mediated re-expression of α4 subunits in the VTA of these knockout mice restored nicotine self-administration indicating that α4-containing nAChRs play an important role in nicotine reinforcement (Pons et al., 2008). In contrast, a more recent study demonstrated no effect of constitutive knockdown of α4 subunits on repeated nicotine self-administration, indicating that α4-containing nAChRs are not involved in nicotine taking (Cahir, Pillidge, Drago, & Lawrence, 2011). These discordant findings are likely due to methodological differences between studies. In the original study by Pons et al. (2008), mice were restrained throughout a single nicotine self-administration test session to preserve temporary catheters implanted in their tail veins. This is a significant confound as restraint forcibly orients mice toward the manipulandum and permits only one self-administration test session. The study by Cahir et al. (2011) uses a more traditional drug self-administration paradigm in which mice self-administer nicotine through indwelling jugular catheters, which allow for repeated testing. Transgenic mice expressing mutated α4 subunits have also been developed to study the role of α4-containing nAChRs in nicotine addiction. The first α4 transgenic mouse was created with a single point mutation, leucine replaced at amino acid residue 9 with alanine (L9A), within the putative pore-forming domain of the α4 subunit that renders α4-containing nAChRs hypersensitive to nicotine (Tapper et al., 2004). These transgenic mice have increased sensitivity to the rewarding effects of nicotine at doses subthreshold for producing reinforcing effects in wild-type mice (Tapper et al., 2004). A second line of hypersensitive α4 transgenic mice was developed in which serine at amino acid residue 248 was replaced with phenylalanine (S248F) in the α4 subunit (Cahir et al., 2011). These transgenic mice robustly self-administer more nicotine at lower nicotine doses compared to wild-type mice (Cahir et al., 2011). Finally, a recent study showed that conditional deletion of α4 subunits in the VTA of adult mice increased self-administration of relatively high doses of nicotine, suggesting that α4-containing nAChRs may mediate the aversive effects of higher unit doses of nicotine (Peng et al., 2017). Collectively, these studies indicate a complex modulatory role for of α4-containing nAChRs in voluntary nicotine taking.

31.5 α4β2* nAChR PARTIAL AGONISTS IN NICOTINE SELF-ADMINISTRATION AND REINSTATEMENT

Based on their role in regulating nicotine reinforcement, α4β2* nAChRs were identified as molecular targets for development of smoking cessation pharmacotherapies. nAChR agonists substitute for the reinforcing effects of nicotine, alleviate some of the adverse symptoms associated with nicotine withdrawal, and are generally well tolerated (Dwoskin et al., 2009). While nicotine replacement therapies have been the mainstay of smoking cessation pharmacotherapies, partial agonists of α4β2* nAChRs have gained increasing attention as potential treatments for smoking relapse. Compared to full agonists, partial agonists produce less than maximal stimulation of nAChRs, thereby substituting for the reinforcing effects of nicotine with less abuse liability. Partial agonists of nAChRs also function as antagonists in that they reduce nicotine reinforcement and nicotine-evoked neurotransmitter release in the brain (Dwoskin et al., 2009).

Varenicline was rationally designed as a smoking cessation medication based on its ability to function as a partial agonist at α4β2* nAChRs (Coe et al., 2005). As a partial agonist, varenicline increases dopamine release in the NAc (~60% of the maximal nicotine response) on its own and reduces nicotine-evoked dopamine release in the NAc (Coe et al., 2005). These effects are absent in mutant mice lacking β2 subunits suggesting that varenicline's effects on nicotine reinforcement are mediated by the activation of β2-containing nAChRs in the brain (Reperant et al., 2010). Varenicline fully substitutes for nicotine in drug discrimination studies and attenuates nicotine self-administration in rodents (Coe et al., 2005; Le Foll et al., 2012; O'Connor, Parker, Rollema, & Mead, 2010). Furthermore, varenicline attenuates the reinstatement of nicotine-seeking behavior (Le Foll et al., 2012; O'Connor et al., 2010), which suggests that varenicline may reduce smoking relapse in human smokers. Consistent with these preclinical findings, clinical trials have shown that abstinence rates are higher in smokers treated with varenicline compared to placebo-treated controls (Gonzales et al., 2006). Despite these results, the efficacy of varenicline in promoting long-term abstinence is modest (Cahill, Stevens, & Lancaster, 2014). Moreover, adverse effects limit patient compliance (Gonzales et al., 2006). Varenicline also functions as a full agonist at α7 nAChRs and partial agonist at α3β4* and α6* nAChRs, which may account for its side-effect profile (Mihalak, Carroll, & Luetje, 2006). Recently, the efficacy of varenicline was compared to nicotine replacement therapies. No differences in smoking abstinence were found between treatments, which raises questions about the relative effectiveness of varenicline (Baker et al., 2016).

Varenicline and nicotine increase expression of α4β2* nAChRs in the brain, effects that are thought to facilitate persistent smoking behavior and smoking relapse (Hussmann et al., 2012). Therefore, recent efforts have focused on developing α4β2* nAChR partial agonists that do not increase expression of nAChRs in the brain. Sazetidine-A was designed as a partial agonist with higher affinity and selectivity for α4β2* nAChRs compared to varenicline (Xiao et al., 2006). Sazetidine-A has been shown to reduce nicotine self-administration (Levin et al., 2010). Importantly, sazetidine-A does not increase nAChR expression in the brain, in contrast to nicotine and varenicline (Hussmann et al., 2012). If increased expression of α4β2* nAChRs is a neuroadaptation that promotes smoking relapse and limits the efficacy of varenicline for smoking cessation, then nAChR partial agonists such as sazetidine-A that do not increase α4β2* nAChR expression may be more efficacious smoking cessation medications. Unfortunately, rats self-administer sazetidine-A, which suggests abuse liability in humans (Paterson et al., 2010). Thus, sazetidine-A has not been tested in clinical trials.

More potent analogs of sazetidine-A, including VMY-2-95, have recently been developed for smoking cessation (Yenugonda et al., 2013). Repeated administration of VMY-2-95 decreases nicotine self-administration in rats (Rezvani et al., 2017; Yenugonda et al., 2013). VMY-2-95 crosses the blood-brain barrier, has a long half-life, and can be administered orally, making it a good candidate for clinical trials (Kong et al., 2015). However, it remains unclear if VMY-2-95 has abuse liability or increases α4β2* nAChR expression in the brain. Investigators continue to expand these studies and identify novel partial agonists of α4β2* nAChRs that may prevent smoking craving and relapse (Lee, Arreola, Kimmey, & Schmidt, 2014).

31.6 POSITIVE ALLOSTERIC MODULATORS OF α4β2* nAChRs IN NICOTINE ADDICTION AND REINSTATEMENT

An additional approach to developing novel smoking cessation medications has focused on positive allosteric modulators (PAMs) of α4β2* nAChRs. PAMs bind nAChRs at allosteric sites that are distinct from the orthosteric binding site for nicotine. Therefore, PAMs have low intrinsic activity in the absence of acetylcholine or nicotine (Uteshev, 2014). By enhancing receptor activity and the probability of nicotine-induced channel opening, PAMs can substantially increase and prolong α4β2* nAChR responses to nicotine from tobacco smoke (Williams, Wang, & Papke, 2011). Based on this unique pharmacological mechanism of action, PAMs of α4β2* nAChRs may reduce the amount of nicotine consumed by enhancing the reinforcing effects of lower unit doses of nicotine—analogous to higher nicotine doses (Maurer, Sandager-Nielsen, & Schmidt, 2017). Indeed, an emerging literature supports this hypothesis. PAMs of α4β2* nAChRs including desformylflustrabromine (dFBr) and galantamine (which also acts as a cholinesterase inhibitor) reduced nicotine self-administration and nicotine reinstatement (Hopkins, Rupprecht, Hayes, Blendy, & Schmidt, 2012; Liu, 2013). Consistent with their mechanism of action, PAMs of α4β2* nAChRs are not reinforcing and do not produce nicotine-like discriminative stimulus effects on their own, suggesting low potential for abuse liability in humans (Liu, 2013; Mohler, Franklin, & Rueter, 2014). These preclinical studies were the basis for a recent translational study that showed reduced nicotine intake in human smokers treated with galantamine (Ashare et al., 2016). This clinical trial is provocative as it suggests that PAMs of α4β2* nAChRs may promote smoking abstinence in humans.

While dFBr and galantamine selectively bind to α4β2* nAChRs, they do not distinguish between α4β2* nAChR subtypes consisting of different ratios of α4β2 subunits (Williams et al., 2011). α4β2* nAChRs assemble into two stoichiometrically and functionally different combinations characterized by the α4:β2 subunit ratio (Fig. 31.2). The 2(α4)3(β2) and 3(α4)2(β2) nAChR subtypes represent channels with high or low sensitivity, respectively, to acetylcholine (Moroni, Zwart, Sher, Cassels, & Bermudez, 2006). Nicotine and varenicline bind to and activate both high- and low-sensitivity populations of α4β2 nAChRs, albeit with different potencies and binding affinities (Moroni et al., 2006). While the relevance of these findings to the therapeutic effects of nicotine and varenicline is unknown, it raises the intriguing possibility that compounds targeting high- or low-sensitivity α4β2 nAChRs alone may be more efficacious antismoking medications. Recently, PAMs that bind selectively to low-sensitivity 3(α4)2(β2) nAChRs have been identified and allow, for the first time, an opportunity to identify the precise role of 3(α4)2(β2) nAChR subtypes in nicotine addiction (Grupe, Jensen, Ahring, Christensen, & Grunnet, 2013; Timmermann et al., 2012). NS9283, a stoichiometry-selective PAM of low-sensitivity 3(α4)2(β2) nAChRs, attenuates nicotine self-administration and reinstatement in rats (Maurer et al., 2017). NS9283 alone does not support self-administration behavior (Maurer et al., 2017) or produce subjective effects similar to nicotine (Mohler et al., 2014) indicating no potential abuse liability in humans. These results raise the intriguing possibility that PAMs targeting low-sensitivity 3(α4)2(β2) nAChRs may reduce chronic smoking behavior and relapse in humans.

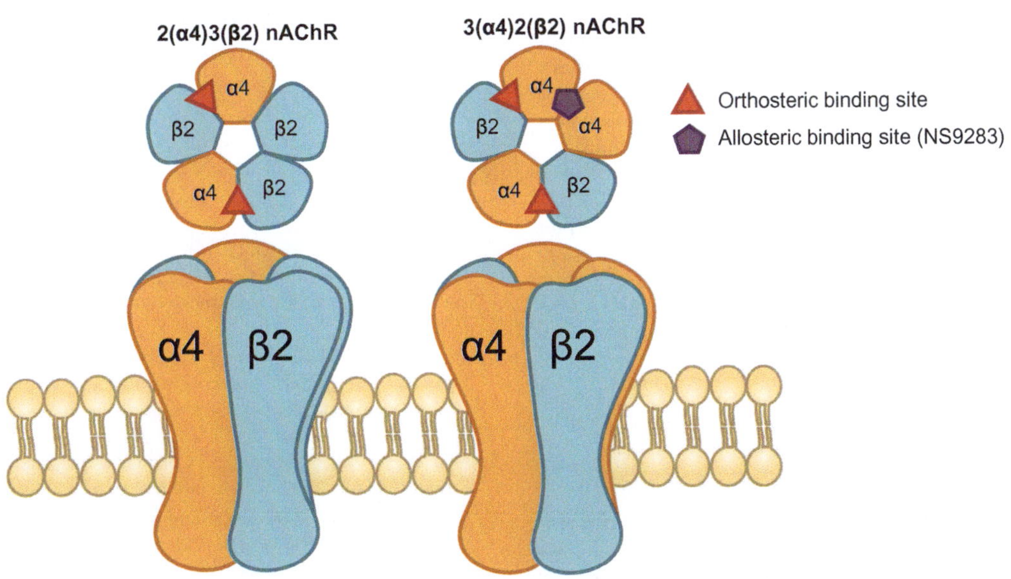

FIG. 31.2 α4β2* nAChRs assemble into two stoichiometrically distinct protein complexes characterized by the α4:β2 subunit ratio. It is possible that targeting one population of α4β2 nAChRs may be more efficacious for smoking cessation therapeutics.

31.7 IMPLICATIONS FOR TREATMENT AND CONCLUSIONS

A large body of evidence clearly indicates that α4β2* nAChRs play an important role in nicotine reinforcement and supports drug discovery approaches aimed at identifying novel pharmacotherapies targeting α4β2* nAChRs for smoking cessation. Since partial agonists of α4β2* nAChRs have been shown to have modest efficacy for smoking relapse, more recent approaches have focused on PAMs of α4β2* nAChRs and stoichiometry-selective α4β2 nAChR compounds. An improved understanding of the roles of specific α4β2 nAChR populations in nicotine addiction will aid in developing more efficacious medications. Defining the exact stoichiometries of α4β2 nAChRs expressed in neural circuits contributing to nicotine-mediated behaviors will be critical to improving upon existing drugs targeting α4β2* nAChRs.

MINI-DICTIONARY OF TERMS

Drug self-administration Laboratory animals will complete an operant response (i.e., lever press) when paired with contingent infusion of a drug that is positively reinforcing. The reinforcing efficacy of a drug can be determined using this model of voluntary drug taking.

Mesolimbic dopamine system The mesolimbic dopamine system is the prime target of drugs of abuse. This system originates in the VTA with dopamine-producing neurons that project to forebrain areas including the nucleus accumbens. As a general rule, all drugs of abuse activate the mesolimbic dopamine system, thereby increasing dopamine signaling in the nucleus accumbens.

Receptor subunits and subtypes nAChRs are protein complexes consisting of five different subunits surrounding a central pore. The subunit composition of nAChRs varies and results in a number of diverse receptor subtypes. As each nAChR subtype has distinct properties, there is great interest in determining the role of individual nAChR subtypes in nicotine addiction.

Reinstatement of drug-seeking behavior Smoking relapse can be modeled in laboratory animals using the reinstatement paradigm. After subjects acquire stable nicotine self-administration, nicotine taking is extinguished. During this abstinence period, reinstatement of nicotine-seeking behavior can be elicited by the same stimuli that precipitate smoking relapse in humans (i.e., re-exposure to nicotine and/or conditioned cues associated with voluntary nicotine taking).

Stoichiometry The exact number (or ratio) of subunits in a receptor complex is referred to as the stoichiometry of a receptor. For example, α4β2* nAChRs are composed of five different α and β subunits in different combinations. The exact combination, whether it is a ratio of 3α-2β versus 2α-3β imparts distinct pharmacokinetic properties and functionality.

Key Facts of Nicotine Self-Administration

- Nicotine is self-administered by laboratory animals including nonhuman primates, dogs, and rodents.
- Intravenous nicotine is the most common route of self-administration. However, some studies investigate self-administration of nicotine in the drinking water, inhaling nicotine/tobacco vapor, or direct infusions into the brain.
- Motivation to self-administer nicotine is studied using a progressive ratio schedule of reinforcement. Briefly, the response requirement for each subsequent infusion of nicotine increases exponentially. Eventually, the animal gives up, and this break point is used to measure motivation to consume a drug or how hard an animal is willing to work for the drug reward.
- When given access to a range of nicotine doses, rodents will self-administer nicotine according to an inverted "U"-shaped dose-response curve. The shape of the dose-response curve reflects competing rewarding (ascending limb) and aversive (descending limb) properties of nicotine.
- Nicotine self-administration is considered a reliable and homologous model of voluntary nicotine taking because it mimics both behavioral (voluntary taking) and physiological (maintenance of stable nicotine plasma levels) aspects of nicotine consumption in humans.

Summary Points

- α4β2* nAChRs play a critical role in nicotine addiction.
- The reinforcing effects of nicotine are mediated primarily by α4β2* nAChRs in the brain.
- α4β2* nAChRs are expressed throughout the brain including the mesolimbic dopamine system.
- Partial agonists of α4β2* nAChRs attenuate voluntary nicotine taking in rodents and humans but have modest efficacy in promoting long-term smoking abstinence.
- New approaches to developing novel smoking cessation medications focus on positive allosteric modulators (PAMs) of α4β2* nAChRs and stoichiometry-selective compounds that target α4β2 nAChR subtypes.

References

Ashare, R. L., Kimmey, B. A., Rupprecht, L. E., Bowers, M. E., Hayes, M. R., & Schmidt, H. D. (2016). Repeated administration of an acetylcholinesterase inhibitor attenuates nicotine taking in rats and smoking behavior in human smokers. *Translational Psychiatry*, *6*, e713.

Baker, T. B., Piper, M. E., Stein, J. H., Smith, S. S., Bolt, D. M., Fraser, D. L., et al. (2016). Effects of nicotine patch vs varenicline vs combination nicotine replacement therapy on smoking cessation at 26 weeks: a randomized clinical trial. *JAMA*, *315*(4), 371–379.

Besson, M., David, V., Suarez, S., Cormier, A., Cazala, P., Changeux, J. P., et al. (2006). Genetic dissociation of two behaviors associated with nicotine addiction: beta-2 containing nicotinic receptors are involved in nicotine reinforcement but not in withdrawal syndrome. *Psychopharmacology*, *187*(2), 189–199.

Cahill, K., Stevens, S., & Lancaster, T. (2014). Pharmacological treatments for smoking cessation. *JAMA*, *311*(2), 193–194.

Cahir, E., Pillidge, K., Drago, J., & Lawrence, A. J. (2011). The necessity of alpha4* nicotinic receptors in nicotine-driven behaviors: dissociation between reinforcing and motor effects of nicotine. *Neuropsychopharmacology, 36*(7), 1505–1517.

Coe, J. W., Brooks, P. R., Vetelino, M. G., Wirtz, M. C., Arnold, E. P., Huang, J., et al. (2005). Varenicline: an alpha4beta2 nicotinic receptor partial agonist for smoking cessation. *Journal of Medicinal Chemistry, 48*(10), 3474–3477.

Corrigall, W. A., & Coen, K. M. (1989). Nicotine maintains robust self-administration in rats on a limited-access schedule. *Psychopharmacology, 99*(4), 473–478.

Corrigall, W. A., Coen, K. M., & Adamson, K. L. (1994). Self-administered nicotine activates the mesolimbic dopamine system through the ventral tegmental area. *Brain Research, 653*(1–2), 278–284.

Corrigall, W. A., Franklin, K. B., Coen, K. M., & Clarke, P. B. (1992). The mesolimbic dopaminergic system is implicated in the reinforcing effects of nicotine. *Psychopharmacology, 107*(2–3), 285–289.

Dwoskin, L. P., Pivavarchyk, M., Joyce, B. M., Neugebauer, N. M., Zheng, G., Zhang, Z., et al. (2009). Targeting reward-relevant nicotinic receptors in the discovery of novel pharmacotherapeutic agents to treat tobacco dependence. *Nebraska Symposium on Motivation, 55*, 31–63.

Epping-Jordan, M. P., Picciotto, M. R., Changeux, J. P., & Pich, E. M. (1999). Assessment of nicotinic acetylcholine receptor subunit contributions to nicotine self-administration in mutant mice. *Psychopharmacology, 147*(1), 25–26.

Epstein, D. H., Preston, K. L., Stewart, J., & Shaham, Y. (2006). Toward a model of drug relapse: an assessment of the validity of the reinstatement procedure. *Psychopharmacology, 189*(1), 1–16.

Exley, R., Maubourguet, N., David, V., Eddine, R., Evrard, A., Pons, S., et al. (2011). Distinct contributions of nicotinic acetylcholine receptor subunit alpha4 and subunit alpha6 to the reinforcing effects of nicotine. *Proceedings of the National Academy of Sciences of the United States of America, 108*(18), 7577–7582.

Ferrari, R., Le Novere, N., Picciotto, M. R., Changeux, J. P., & Zoli, M. (2002). Acute and long-term changes in the mesolimbic dopamine pathway after systemic or local single nicotine injections. *The European Journal of Neuroscience, 15*(11), 1810–1818.

Flores, C. M., Rogers, S. W., Pabreza, L. A., Wolfe, B. B., & Kellar, K. J. (1992). A subtype of nicotinic cholinergic receptor in rat brain is composed of alpha 4 and beta 2 subunits and is up-regulated by chronic nicotine treatment. *Molecular Pharmacology, 41*(1), 31–37.

Gaimarri, A., Moretti, M., Riganti, L., Zanardi, A., Clementi, F., & Gotti, C. (2007). Regulation of neuronal nicotinic receptor traffic and expression. *Brain Research Reviews, 55*(1), 134–143.

Gonzales, D., Rennard, S. I., Nides, M., Oncken, C., Azoulay, S., Billing, C. B., et al. (2006). Varenicline, an alpha4beta2 nicotinic acetylcholine receptor partial agonist, vs sustained-release bupropion and placebo for smoking cessation: a randomized controlled trial. *JAMA, 296*(1), 47–55.

Gourlay, S. G., & Benowitz, N. L. (1997). Arteriovenous differences in plasma concentration of nicotine and catecholamines and related cardiovascular effects after smoking, nicotine nasal spray, and intravenous nicotine. *Clinical Pharmacology and Therapeutics, 62*(4), 453–463.

Grupe, M., Jensen, A. A., Ahring, P. K., Christensen, J. K., & Grunnet, M. (2013). Unravelling the mechanism of action of NS9283, a positive allosteric modulator of (alpha4)3(beta2)2 nicotinic ACh receptors. *British Journal of Pharmacology, 168*(8), 2000–2010.

Hopkins, T. J., Rupprecht, L. E., Hayes, M. R., Blendy, J. A., & Schmidt, H. D. (2012). Galantamine, an acetylcholinesterase inhibitor and positive allosteric modulator of nicotinic acetylcholine receptors, attenuates nicotine taking and seeking in rats. *Neuropsychopharmacology, 37*(10), 2310–2321.

Hussmann, G. P., Turner, J. R., Lomazzo, E., Venkatesh, R., Cousins, V., Xiao, Y., et al. (2012). Chronic sazetidine-A at behaviorally active doses does not increase nicotinic cholinergic receptors in rodent brain. *The Journal of Pharmacology and Experimental Therapeutics 343* (2), 441–450.

Klink, R., de Kerchove d'Exaerde, A., Zoli, M., & Changeux, J. P. (2001). Molecular and physiological diversity of nicotinic acetylcholine receptors in the midbrain dopaminergic nuclei. *The Journal of Neuroscience, 21*(5), 1452–1463.

Kong, H., Song, J. K., Yenugonda, V. M., Zhang, L., Shuo, T., Cheema, A. K., et al. (2015). Preclinical studies of the potent and selective nicotinic alpha4beta2 receptor ligand VMY-2-95. *Molecular Pharmaceutics, 12*(2), 393–402.

Le Foll, B., Chakraborty-Chatterjee, M., Lev-Ran, S., Barnes, C., Pushparaj, A., Gamaleddin, I., et al. (2012). Varenicline decreases nicotine self-administration and cue-induced reinstatement of nicotine-seeking behaviour in rats when a long pretreatment time is used. *The International Journal of Neuropsychopharmacology, 15*(9), 1265–1274.

Le Novere, N., Corringer, P. J., & Changeux, J. P. (2002). The diversity of subunit composition in nAChRs: evolutionary origins, physiologic and pharmacologic consequences. *Journal of Neurobiology, 53*(4), 447–456.

Lee, A. M., Arreola, A. C., Kimmey, B. A., & Schmidt, H. D. (2014). Administration of the nicotinic acetylcholine receptor agonists ABT-089 and ABT-107 attenuates the reinstatement of nicotine-seeking behavior in rats. *Behavioural Brain Research, 274*, 168–175.

Lerman, C., LeSage, M. G., Perkins, K. A., O'Malley, S. S., Siegel, S. J., Benowitz, N. L., et al. (2007). Translational research in medication development for nicotine dependence. *Nature Reviews. Drug Discovery, 6*(9), 746–762.

Levin, E. D., Rezvani, A. H., Xiao, Y., Slade, S., Cauley, M., Wells, C., et al. (2010). Sazetidine-A, a selective alpha4beta2 nicotinic receptor desensitizing agent and partial agonist, reduces nicotine self-administration in rats. *The Journal of Pharmacology and Experimental Therapeutics, 332*(3), 933–939.

Li, X., Semenova, S., D'Souza, M. S., Stoker, A. K., & Markou, A. (2014). Involvement of glutamatergic and GABAergic systems in nicotine dependence: Implications for novel pharmacotherapies for smoking cessation. *Neuropharmacology, 76*(Pt B), 554–565.

Liu, X. (2013). Positive allosteric modulation of alpha4beta2 nicotinic acetylcholine receptors as a new approach to smoking reduction: evidence from a rat model of nicotine self-administration. *Psychopharmacology, 230*(2), 203–213.

Liu, X., Palmatier, M. I., Caggiula, A. R., Donny, E. C., & Sved, A. F. (2007). Reinforcement enhancing effect of nicotine and its attenuation by nicotinic antagonists in rats. *Psychopharmacology, 194*(4), 463–473.

Mameli-Engvall, M., Evrard, A., Pons, S., Maskos, U., Svensson, T. H., Changeux, J. P., et al. (2006). Hierarchical control of dopamine neuron-firing patterns by nicotinic receptors. *Neuron, 50*(6), 911–921.

Mansvelder, H. D., & McGehee, D. S. (2000). Long-term potentiation of excitatory inputs to brain reward areas by nicotine. *Neuron, 27*(2), 349–357.

Marubio, L. M., Gardier, A. M., Durier, S., David, D., Klink, R., Arroyo-Jimenez, M. M., et al. (2003). Effects of nicotine in the dopaminergic system of mice lacking the alpha4 subunit of neuronal nicotinic acetylcholine receptors. *The European Journal of Neuroscience, 17*(7), 1329–1337.

Maskos, U., Molles, B. E., Pons, S., Besson, M., Guiard, B. P., Guilloux, J. P., et al. (2005). Nicotine reinforcement and cognition restored by targeted expression of nicotinic receptors. *Nature, 436*(7047), 103–107.

Maurer, J. J., Sandager-Nielsen, K., & Schmidt, H. D. (2017). Attenuation of nicotine taking and seeking in rats by the stoichiometry-selective alpha4beta2 nicotinic acetylcholine receptor positive allosteric modulator NS9283. *Psychopharmacology, 234*(3), 475–484.

Mihalak, K. B., Carroll, F. I., & Luetje, C. W. (2006). Varenicline is a partial agonist at alpha4beta2 and a full agonist at alpha7 neuronal nicotinic receptors. *Molecular Pharmacology, 70*(3), 801–805.

Mohler, E. G., Franklin, S. R., & Rueter, L. E. (2014). Discriminative-stimulus effects of NS9283, a nicotinic alpha4beta2* positive allosteric modulator, in nicotine-discriminating rats. *Psychopharmacology, 231*(1), 67–74.

Moroni, M., Zwart, R., Sher, E., Cassels, B. K., & Bermudez, I. (2006). Alpha4beta2 nicotinic receptors with high and low acetylcholine sensitivity: pharmacology, stoichiometry, and sensitivity to long-term exposure to nicotine. *Molecular Pharmacology, 70*(2), 755–768.

O'Connor, E. C., Parker, D., Rollema, H., & Mead, A. N. (2010). The alpha4beta2 nicotinic acetylcholine-receptor partial agonist varenicline inhibits both nicotine self-administration following repeated dosing and reinstatement of nicotine seeking in rats. *Psychopharmacology, 208*(3), 365–376.

O'Dell, L. E., & Khroyan, T. V. (2009). Rodent models of nicotine reward: what do they tell us about tobacco abuse in humans? *Pharmacology, Biochemistry, and Behavior, 91*(4), 481–488.

Orejarena, M. J., Herrera-Solis, A., Pons, S., Maskos, U., Maldonado, R., & Robledo, P. (2012). Selective re-expression of beta2 nicotinic acetylcholine receptor subunits in the ventral tegmental area of the mouse restores intravenous nicotine self-administration. *Neuropharmacology, 63*(2), 235–241.

Paterson, N. E., Min, W., Hackett, A., Lowe, D., Hanania, T., Caldarone, B., et al. (2010). The high-affinity nAChR partial agonists varenicline and sazetidine-A exhibit reinforcing properties in rats. *Progress in Neuro-Psychopharmacology & Biological Psychiatry, 34*(8), 1455–1464.

Peng, C., Engle, S. E., Yan, Y., Weera, M. M., Berry, J. N., Arvin, M. C., et al. (2017). Altered nicotine reward-associated behavior following alpha4 nAChR subunit deletion in ventral midbrain. *PLoS ONE, 12*(7). e0182142.

Picciotto, M. R., Zoli, M., Rimondini, R., Lena, C., Marubio, L. M., Pich, E. M., et al. (1998). Acetylcholine receptors containing the beta2 subunit are involved in the reinforcing properties of nicotine. *Nature, 391*(6663), 173–177.

Pons, S., Fattore, L., Cossu, G., Tolu, S., Porcu, E., McIntosh, J. M., et al. (2008). Crucial role of alpha4 and alpha6 nicotinic acetylcholine receptor subunits from ventral tegmental area in systemic nicotine self-administration. *The Journal of Neuroscience, 28*(47), 12318–12327.

Reperant, C., Pons, S., Dufour, E., Rollema, H., Gardier, A. M., & Maskos, U. (2010). Effect of the alpha4beta2* nicotinic acetylcholine receptor partial agonist varenicline on dopamine release in beta2 knock-out mice with selective re-expression of the beta2 subunit in the ventral tegmental area. *Neuropharmacology, 58*(2), 346–350.

Rezvani, A. H., Slade, S., Wells, C., Yenugonda, V. M., Liu, Y., Brown, M. L., et al. (2017). Differential efficacies of the nicotinic alpha4beta2 desensitizing agents in reducing nicotine self-administration in female rats. *Psychopharmacology, 234*(17), 2517–2523.

Rose, J. E., & Corrigall, W. A. (1997). Nicotine self-administration in animals and humans: similarities and differences. *Psychopharmacology, 130*(1), 28–40.

Shoaib, M., & Stolerman, I. P. (1999). Plasma nicotine and cotinine levels following intravenous nicotine self-administration in rats. *Psychopharmacology, 143*(3), 318–321.

Staley, J. K., Krishnan-Sarin, S., Cosgrove, K. P., Krantzler, E., Frohlich, E., Perry, E., et al. (2006). Human tobacco smokers in early abstinence have higher levels of beta2* nicotinic acetylcholine receptors than nonsmokers. *The Journal of Neuroscience, 26*(34), 8707–8714.

Stoker, A. K., & Markou, A. (2015). Neurobiological bases of cue- and nicotine-induced reinstatement of nicotine seeking: implications for the development of smoking cessation medications. *Current Topics in Behavioral Neurosciences, 24*, 125–154.

Tapper, A. R., McKinney, S. L., Nashmi, R., Schwarz, J., Deshpande, P., Labarca, C., et al. (2004). Nicotine activation of alpha4* receptors: sufficient for reward, tolerance, and sensitization. *Science, 306*(5698), 1029–1032.

Tennant, F. S., Jr., Tarver, A. L., & Rawson, R. A. (1984). Clinical evaluation of mecamylamine for withdrawal from nicotine dependence. *NIDA Research Monograph, 49*, 239–246.

Timmermann, D. B., Sandager-Nielsen, K., Dyhring, T., Smith, M., Jacobsen, A. M., Nielsen, E. O., et al. (2012). Augmentation of cognitive function by NS9283, a stoichiometry-dependent positive allosteric modulator of alpha2- and alpha4-containing nicotinic acetylcholine receptors. *British Journal of Pharmacology, 167*(1), 164–182.

Tuesta, L. M., Fowler, C. D., & Kenny, P. J. (2011). Recent advances in understanding nicotinic receptor signaling mechanisms that regulate drug self-administration behavior. *Biochemical Pharmacology, 82*(8), 984–995.

Uteshev, V. V. (2014). The therapeutic promise of positive allosteric modulation of nicotinic receptors. *European Journal of Pharmacology, 727*, 181–185.

Watkins, S. S., Epping-Jordan, M. P., Koob, G. F., & Markou, A. (1999). Blockade of nicotine self-administration with nicotinic antagonists in rats. *Pharmacology, Biochemistry, and Behavior, 62*(4), 743–751.

Williams, M., & Robinson, J. L. (1984). Binding of the nicotinic cholinergic antagonist, dihydro-beta-erythroidine, to rat brain tissue. *The Journal of Neuroscience, 4*(12), 2906–2911.

Williams, D. K., Wang, J., & Papke, R. L. (2011). Positive allosteric modulators as an approach to nicotinic acetylcholine receptor-targeted therapeutics: advantages and limitations. *Biochemical Pharmacology, 82*(8), 915–930.

Xiao, Y., Fan, H., Musachio, J. L., Wei, Z. L., Chellappan, S. K., Kozikowski, A. P., et al. (2006). Sazetidine-A, a novel ligand that desensitizes alpha4beta2 nicotinic acetylcholine receptors without activating them. *Molecular Pharmacology, 70*(4), 1454–1460.

Yenugonda, V. M., Xiao, Y., Levin, E. D., Rezvani, A. H., Tran, T., Al-Muhtasib, N., et al. (2013). Design, synthesis and discovery of picomolar selective alpha4beta2 nicotinic acetylcholine receptor ligands. *Journal of Medicinal Chemistry, 56*(21), 8404–8421.

Zhou, F. M., Liang, Y., & Dani, J. A. (2001). Endogenous nicotinic cholinergic activity regulates dopamine release in the striatum. *Nature Neuroscience, 4*(12), 1224–1229.

CHAPTER

32

The Medial Habenula-Interpeduncular Nucleus Pathway in Nicotine Sensitization: The Role of α3β4 Nicotinic Acetylcholine Receptors and Substance P

Branden Eggan, Sarah McCallum*[†]

*Department of Liberal Arts and Science, Maria College, Albany, NY, United States

[†]Department of Neuroscience and Experimental Therapeutics, Albany Medical College, Albany, NY, United States

Abbreviations

ACh acetylcholine
ChAT choline acetyltransferase
IPN interpeduncular nucleus
MHb medial habenula
NAc nucleus accumbens
nAChR nicotinic acetylcholine receptor
NK1R neurokinin-1 receptor
VTA ventral tegmental area

32.1 INTRODUCTION

The α3β4 nicotinic acetylcholine receptor (nAChR) is a subtype of nicotinic receptors that is densely expressed in the sensory and autonomic ganglia as well as in the brain's medial habenula (MHb) and interpeduncular nucleus (IPN). While α4β2 nAChRs are densely expressed in the brain's dopamine reward pathway and have been shown to directly regulate nicotine addiction and dependence (Epping-Jordan, Picciotto, Changeux, & Pich, 1999; Tapper et al., 2004), the role of other nicotinic receptors including the α3β4 nAChR has been far less studied.

Human genome studies have revealed that individuals with a mutation in the gene coding for the α3β4 nAChR have a greater likelihood to not only use nicotine but also start using nicotine early in life (Liu et al., 2010; Schlaepfer et al., 2008); receptor subtypes are not expressed in the mesolimbic pathway; rather, they are densely expressed in an "alternate reward pathway known" as the MHb-IPN pathway (Sutherland & Nakajima, 1981). Recent animal studies have largely focused on a role for habenular nAChRs in mediating nicotine aversion and withdrawal with less emphasis on their contribution to the rewarding effects of nicotine (Antolin-Fontes, Ables, Gorlich, & Ibanez-Tallon, 2015).

32.2 EXPRESSION OF THE α3β4 nAChR

In the MHb, the majority of α3β4 nAChRs are expressed in the ventral two-thirds, which correlates with the expression of choline acetyltransferase (ChAT; Contestabile et al., 1987; Quick, Ceballos, Kasten, McIntosh, & Lester, 1999) suggesting the likely location of these receptors to be postsynaptic on acetylcholine (ACh)-containing neurons of the MHb. Studies show α3β4 nAChRs on ACh neurons dominate MHb function (Quick et al., 1999). In the IPN, α3β4 nAChRs are located presynaptically on ACh neurons projecting from the MHb (Grady et al., 2009). Synaptosome studies in the IPN have shown the attenuation of Ca^{2+}-mediated release of ACh by the addition of an α3β4 nAChR antagonist (Grady et al., 2001). These receptors can also be expressed on GABAergic interneurons in the IPN (Lena, Changeux, & Mulle, 1993) suggesting another possible contribution to the regulation of ACh release in MHb-IPN communication.

Neuroscience of Nicotine
https://doi.org/10.1016/B978-0-12-813035-3.00032-0

32.3 THE α3β4 nAChR IN NICOTINE ADDICTION

Animal models of nicotine addiction have also demonstrated an important role for MHb and IPN α3β4 nAChRs. While MHb and IPN neurons are rich in a variety of nAChRs containing α2–α6 and β2–β4 subunits, only stimulation of α3β4* and α3β3β4* subtypes causes subsequent ACh release in the IPN (Grady et al., 2009). The α5 subunit forms functioning heteromers with α3 and β4 (Groot-Kormelink, Boorman, & Sivilotti, 2001) and when activated causes glutamate release in the IPN (Fowler, Tuesta, & Kenny, 2011). This glutamate is likely coreleased with ACh from MHb neurons (Ren et al., 2011) and results in activation of neurons in the IPN.

Recently, increasing evidence supports the theory that MHb-IPN α3β4* (possibly with α5) nAChRs mediate the negative reward components of nicotine addiction. Deletion of α5 nAChR subunits decreases aversion to high concentrations of nicotine, and this effect is reversed by reexpression of the α5 nAChR subunit in the MHb (Fowler, Lu, Johnson, Marks, & Kenny, 2011). In transgenic mouse model with overexpressed β4 receptors, increasing α3β4* nAChR levels enhances aversion to nicotine, which is reversed by the expression of the α5 nAChR subunit to the MHb (Frahm et al., 2011). Additionally, β4 and α5 nAChR knockout mice show decreased signs of nicotine withdrawal (Salas, Pieri, & De Biasi, 2004; Salas, Sturm, Boutler, & De Biasi, 2009).

Transgenic overexpression of the CHRNA4/A3/B4 genome cluster causes a significant increase in β4* receptor binding and increased nicotine self-administration (Gallego et al., 2012). In parallel, intra-MHb administration of the α3β4* nAChR antagonist 18-methoxycoronaridine decreases nicotine self-administration and acute nicotine-induced increases in accumbal dopamine (Glick, Sell, McCallum, & Maisonneuve, 2011; McCallum, Cowe, Lewis, & Glick, 2012). Nicotine self-administration is also blocked by systemic administration of the α3β4* nAChR antagonist AT-1001 (Toll et al., 2012) suggesting that these receptors are unique in the fact that they play a role in both the positive and negative rewarding components of nicotine addiction.

32.4 NICOTINE SENSITIZATION: AN IMPORTANT MODEL IN ADDICTION RESEARCH

Sensitization occurs as an initial component of drug addiction, when casual drug use starts to escalate to drug-seeking and compulsive drug-taking behavior (Koob & LeMoal, 1997). Sensitized responses occur with any psychostimulant drug, including nicotine. This phenomenon can be broken into two primary components: **induction** and **expression** (DiFranza & Wellman, 2007). Induction begins to develop with the first use of a drug, but because it strongly depends on the frequency of drug use and the dose. Expression manifests at different rates across animal models and humans.

Sensitization is unique in that it is a persistent effect of drug use. Sensitized responding to nicotine can be seen for months after the last drug exposure and, as a result, may be a strong contributor to relapse (Miller, Wilkins, Bardo, Crooks, & Dwoskin, 2001). Models of nicotine sensitization are critical to addiction research because they best reflect the neurological changes during the initial development of an addicted behavior rather than effects traditionally measured once the disease has fully manifested. Changes in dopamine transmission in the mesolimbic reward pathway seem to underlie sensitized behavior, specifically neuroadaptations in the ventral tegmental area (Nisell, Nomikos, Hertel, Panagis, & Svensson, 1996; Saal, Dong, Bonci, & Malenka, 2003). In addition to relapse, brain changes that occur during sensitization could underlie drug craving that is traditionally seen in many smokers, especially after periods of abstinence (Steketee & Kalivas, 2011).

32.5 ANIMAL MODELS OF NICOTINE SENSITIZATION

Experimentally, nicotine sensitization can be measured in two separate models: **behavioral** and **neurochemical sensitization**. Early models of nicotine sensitization in rats revealed that 5 days of nicotine administration is sufficient to induce both behavioral and neurochemical sensitizations to nicotine (Benwell & Balfour, 1992). Fig. 32.1A illustrates a traditional sensitization model used experimentally. In this model, animals in the sensitized treatment group are given a once daily challenge of nicotine at a specific dose. Control groups are run in parallel to confirm that the effects seen of a treatment drug on sensitized responding are indeed due to an effect on the sensitized response itself rather than just overall locomotion or dopamine release. Control groups include a vehicle group, receiving once daily saline injections and an acute group, receiving four saline injections followed by a 5-day single nicotine injection.

Fig. 32.1B illustrates the induction of behavioral sensitization, with the sensitized animal having a significantly higher activity count (as measured in a **photocell activity chamber** for 30 min post nicotine challenge) after only three daily nicotine challenge injections. Animals given an acute nicotine challenge may cause an increase, a decrease, or no effect on locomotor activity (Benwell & Balfour, 1992; Dwoskin, Crooks, Teng, Green, & Bardo, 1999; Eggan & McCallum, 2016) making acute locomotor

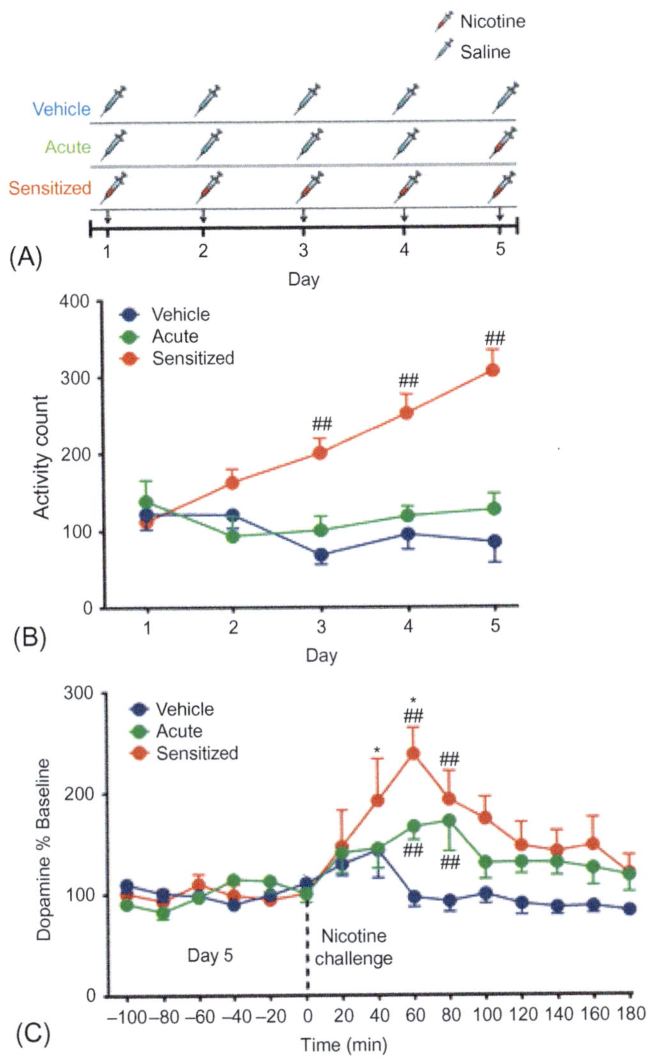

FIG. 32.1 Nicotine sensitization administration regimen and behavioral and neurochemical results from dosing.
(A) Traditional experimental dosing regimen of nicotine to cause sensitization and appropriate controls. Injections are administered once daily over the course of 5days. (B) Locomotor sensitization as reflected in an increased daily activity count in freely moving animals as measured by a photocell activity chamber. (C) Neurochemical sensitization as reflected in an increased accumbal dopamine release following 5days of nicotine treatment as measured by in vivo microdialysis. *Data from Eggan & McCallum, unpublished.*

responses to nicotine a difficult effect to study. The effect of treatment can be measured both on the induction (over the course of 5days of nicotine challenges) or on the expression (5-day locomotor activity counts).

Neurochemical sensitization can be measured via **in vivo microdialysis, by determining** dopamine release in the nucleus accumbens following nicotine administration. Daily monitoring of dopamine release is difficult due to the technicality of the experimental set up, so expression (5-day nicotine challenge) data are traditionally used for analysis of the effects of a treatment drug on neurochemical sensitization. Fig. 32.1C illustrates that after 5days of nicotine challenges, the sensitized accumbal dopamine release is significantly higher than that of the acute and vehicle control groups.

32.6 THE α3β4 nAChR IN NICOTINE SENSITIZATION

In a set of experiments from our laboratory, effects of MHb α3β4 nAChR antagonism on the induction (development) and expression (manifestation) of locomotor sensitization to nicotine were studied (Eggan & McCallum, 2016). These studies were conducted using 18-methoxycoronaridine, a congener of the antiaddiction medication ibogaine and a noncompetitive α3β4 nAChR antagonist (IC50 = 0.75 μM; Kuehne et al., 2003). Results showed that with daily administration of this α3β4 nAChR antagonist prior to nicotine administration, the increased activity count or locomotor sensitization was significantly reduced, suggesting that α3β4 nAChR antagonism attenuates the induction of sensitization. Because in the CNS, α3β4 nAChRs are predominantly centralized in the MHb and are not densely expressed in any other brain region, it is speculated that this effect is centralized to the drug's action in this region.

In regard to the expression of sensitization, the effects of α3β4 nAChR antagonism in the MHb was measured both behaviorally and neurochemically. In animals sensitized to nicotine, when MHb α3β4 nAChRs are blocked via either the systemically or intra-MHb, sensitized locomotor effects are significantly reduced. Similarly, when the noncompetitive α3β4 nAChR antagonist α-conotoxin AuIB (IC50 = 0.75 μM; Luo et al., 1998) is administered in the MHb, locomotor sensitization is also attenuated. These data support the localization and action of α3β4 nAChRs primarily in the MHb and not in other brain regions. In a neurochemical sensitization paradigm, MHb α3β4 nAChR antagonism has a similar effect on the expression of sensitization in that direct administration of an α3β4 nAChR antagonist significantly reduces increased accumbal dopamine following nicotine sensitization (Eggan & McCallum, 2017). Together, these data suggest an important role for MHb α3β4 nAChRs receptors in mediating sensitized responding to nicotine. Moreover, when other drugs were tested in the MHb including antagonists of other nicotinic receptors, glutamate modulators, and neurokinin-1 antagonists, there was no result on sensitized responding (Eggan & McCallum, 2017), suggesting this is a selective effect of α3β4 nAChR antagonism in the MHb and not related to other signaling in the nucleus.

32.7 DOWNSTREAM TARGETS OF THE MHb

The MHb sends efferent fibers exclusively to the interpeduncular nucleus (IPN) via the fasciculus retroflexus (Herkenham & Nauta, 1977). While less well characterized than the MHb, the IPN contains a rich population of cholinergic fibers from the MHb and presynaptic nAChRs, including α3β4 nAChRs (Grady et al., 2009). In addition to the cholinergic populations of the IPN, the IPN is also known to be rich in substance P-positive fibers (Nakaya, Kaneko, Shigemoto, Shigetada, & Mizuno, 1994) and the receptor for substance P, the neurokinin-1 receptor (NK1R; Yip & Chahl, 2001). This population could play a role in MHb-IPN communication, for ACh and substance P are the two primary transmitter systems found in the fasciculus retroflexus (Cuello, Emson, Paxinos, & Jessell, 1978).

The NK1R is located in many brain regions that have been implicated in the regulation of affective behaviors. These regions include the amygdala, striatum, raphe nuclei, hippocampus, MHb-IPN pathway, and several other monoaminergic brainstem areas (Quirion et al., 1983; Yip & Chahl, 2001). Other brain regions, including the locus coeruleus, hypothalamus, midbrain, basal ganglia, and dorsal tegmental areas are also rich in NK1Rs (Maeno, Kiyama, & Tohyama, 1993). While there are NK1Rs in the mesolimbic reward pathway, there is little expression of substance P in cell bodies in the VTA, suggesting that substance P signaling in the mesolimbic dopamine pathway is from outside of the nucleus accumbens (NAc) and ventral tegmental area (VTA; Lessard, Savard, Gobeil, Pierce, & Pickel, 2009). Brain regions rich in substance P-containing cell bodies include the neocortex, hippocampus, caudate-putamen, hypothalamus, MHb, IPN, superior colliculus, central gray, and raphe nuclei (Warden & Young, 1988).

32.8 SUBSTANCE P AS A CO-NEUROTRANSMITTER

Traditional neuropeptides, including substance P, are often packaged within the neuron and released in a vesicle along with common neurotransmitters. Double immunofluorescence studies show that in several brain regions, substance P immunoreactivity near completely overlaps with ChAT expression, supporting the idea that these transmitters are coreleased. Near all ChAT positive neurons in the basal ganglia and neighboring basal forebrain regions show substance P-positive immunoreactivity (Chen, Wei, Liu, & Chan, 2001) suggesting that substance P directly modulates the activity of ACh-containing neurons. Additionally, immunohistochemical experiments have also shown that substance P colocalizes with 30%–70% of serotonin-containing neurons in several raphe nuclei (Thor & Helke, 1989), 15%–30% of dorsal root ganglia glutamate neurons (Battaglia & Rustioni, 1988), and 3% of GABAergic retinal ganglion cells (Caruso, Owczarzak, & Pourcho, 1990) again supporting the corelease of substance P with other transmitter systems.

32.9 THE NK1R IN ADDICTION AND NICOTINE SENSITIZATION

The precise mechanism by which substance P influences addictive phenotypes is not yet known. In regard to the role of substance P in the mesolimbic reward pathway, a direct infusion of substance P to the VTA causes a subsequent activation of VTA dopamine neurons (Stinus, Kelley, & Iversen, 1978) suggesting the ability of substance P to directly modulate the mesolimbic reward pathway. Additionally, NK1R agonism in the VTA causes subsequent dopamine release in the NAc, while NK1R antagonism in the VTA causes reduced NAc dopaminergic cell firing (Minabe, Emori, Toor, Stutzman, & Ashby, 1996). The precise input or existence of VTA substance P-containing afferent fibers is yet to be identified.

Limited studies have investigated the role of substance P transmission in addiction. Morphine reward was shown to be significantly reduced with NK1R global deletion and systemic NK1R blockade (Murtra, Sheasby, Hunt, & De Felipe, 2000; Ripley, Gadd, De Felipe, Hunt, & Stephens, 2002). Similarly, rewarding effects of alcohol were decreased both in a NK1R global knockout (George et al., 2008) and when an NK1R antagonist was administered intracerebroventricular, to target CNS NK1Rs. In regard to cocaine, systemic NK1R antagonism resulted in a blunted responding in accumbal dopamine upon receiving an acute cocaine challenge (Loonam, Noailles, Yu, Zhu, & Angulo, 2003).

32.10 THE NK1 RECEPTOR IN NICOTINE SENSITIZATION

The role of substance P in nicotine addiction has received no attention outside of a single set of sensitization studies. Results show that when the competitive NK1R antagonist ezlopitant (Ki = 1.6–34 nm; Margolis & Obach, 2003) is administered systemically, there is no effect on locomotor sensitization (Eggan & McCallum, unpublished data). Interestingly, while there is no effect of NK1R blockade in the MHb on sensitization, if ezlopitant is administered directly to the IPN, both behavioral and neurochemical sensitizations are significantly attenuated (Eggan & McCallum, 2017). These data suggest an important role for IPN NK1Rs in mediating this reward-based component of addiction and are the first to show a critical role for these receptors in mediating any responding to nicotine.

32.11 PUTATIVE NEUROCIRCUITRY REGULATING THE MHb-IPN EFFECTS ON NICOTINE SENSITIZATION

It is not known how the MHb-IPN pathway regulates sensitized responding, but further investigation to this question could potentially provide new insight to addiction signaling in the brain. If the fasciculus retroflexus is severed, a near complete loss of ACh in the IPN but only a fractional loss of substance P occurs (Artymyshyn & Murray, 1985). These data suggest that substance P-containing afferents to the IPN likely originate in nuclei additional to the MHb.

Of all of the previously identified IPN afferents, the septum and laterodorsal tegmental nucleus (LDTg) are the only two to contain substance P- and ACh-containing cell bodies (Baker et al., 1991; Brownstein, Mroz, Kizer, Palkovits, & Leeman, 1976; Contestabile & Flumerfelt, 1981; Perry & Kellar, 1995). Because substance P has been shown to cotransmit with ACh in other brain regions and because of the neurochemistry of the MHb-IPN pathway, it is likely that ACh could be IPN afferent cotransmitter for the substance P arriving outside of the MHb (Chen et al., 2001).

We investigated this possibility more closely via immunohistochemical analysis. In a set of experiments (Eggan & McCallum, unpublished data), Fluoro-Gold retrograde dye was injected into the IPN and allowed 10 days to be trafficked to afferent cell bodies. The brain was sectioned to determine afferent connectivity, and these regions were colabeled with a marker for substance P or ACh neuron marker, ChAT. The septum and laterodorsal tegmental nucleus (LDTg) were promising targets for analysis due to their previously identified ACh and substance P-containing cell bodies, but we found that these cell bodies only colocalized with Fluoro-Gold in the LDTg (Fig. 32.2). These data are preliminary though they support the possibility of LDTg substance P playing an important role in mediating nicotine sensitized responding. Further research is necessary to determine the role of this potential target and to further understand MHb-IPN modulation of the mesolimbic reward pathway and the role these receptors and brain regions play in sensitized responding to nicotine (Fig. 32.3).

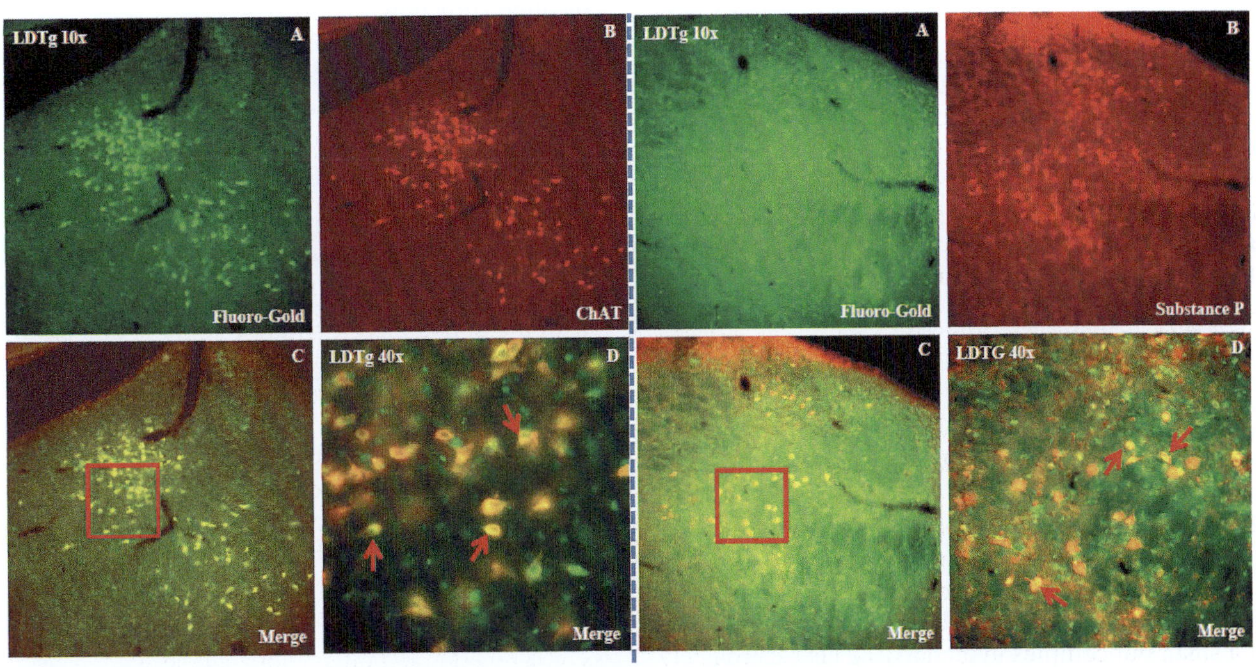

FIG. 32.2 Fluoro-Gold colabeling with an ACh and substance P marker in the LDTg.
Left: (A) Fluoro-Gold-labeled cell bodies in the LDTg following a single IPN Fluoro-Gold injection 10 days prior. (B) ChAT-labeled cell bodies in the LDTg. (C) Fluoro-Gold and ChAT merge at 10 × magnification. (D) Fluoro-Gold and ChAT merge at 40 × magnification showing colabeling of LDTg cell bodies.
Right: (A) Fluoro-Gold-labeled cell bodies in the LDTg following a single IPN Fluoro-Gold injection 10 days prior. (B) Substance P-labeled cell bodies in the LDTg. (C) Fluoro-Gold and substance P merge at 10 × magnification. (D) Fluoro-Gold and substance P merge at 40 × magnification showing colabeling of LDTg cell bodies. *Images from Eggan & McCallum, unpublished.*

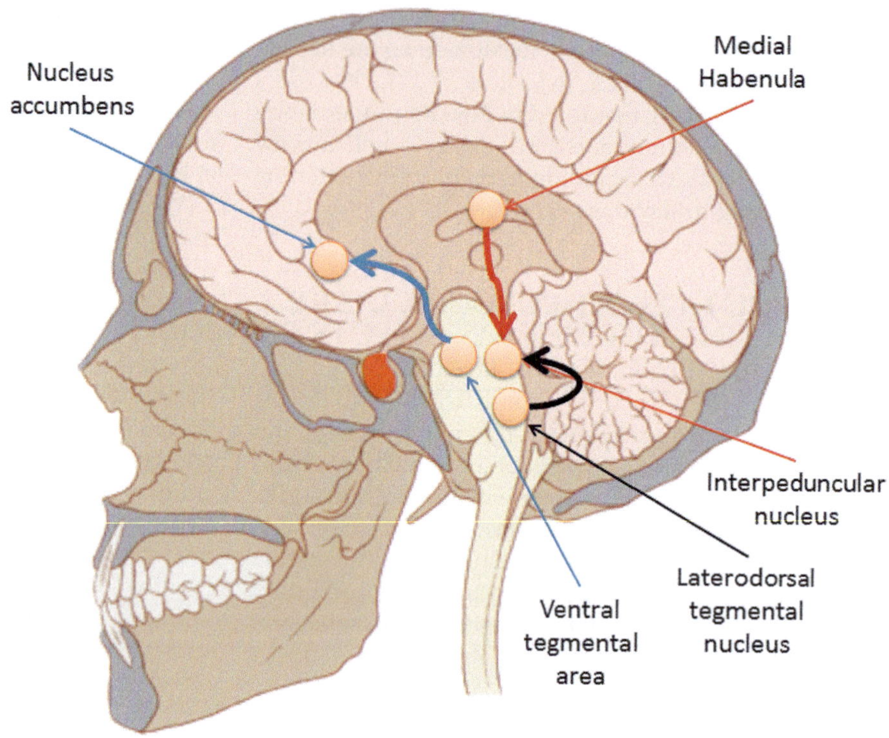

FIG. 32.3 Neuroanatomical location of relevant structures.
Neuroanatomical location of the mesolimbic reward pathway (*blue*; dopaminergic), MHb-IPN pathway (*red*; cholinergic and substance P-ergic), and LDTg-IPN connectivity (*black*; unknown).

MINI-DICTIONARY OF TERMS

Behavioral sensitization Subsequent increases in locomotor activity that result from repeated exposure to a psychostimulant drug.

Expression The manifestation of a sensitized response that is significantly higher than that of a single psychostimulant challenge.

Fluoro-Gold A neuronal retrograde tracer used experimentally to visualize afferent fibers and cell bodies connecting to the targeted injection site.

In vivo microdialysis An experimental technique that samples and collects the contents of the extracellular space in a target brain region while the animal is freely moving and allows for analysis of the specific components in the collected sample.

Induction The developmental or acquisition phase of sensitization.

Neurochemical sensitization Subsequent increases in accumbal dopamine release that result from repeated exposure to a psychostimulant drug.

Photocell activity chamber An experimental apparatus used to test open-field locomotor activity by measuring the amount of beam breaks of intersecting photobeams by a freely moving animal.

Key Facts of the Habenula-Interpeduncular Nucleus Pathway

- The habenula, Latin for "little rein" based on its shape, is a small but complex evolutionarily conserved epithalamic structure located on the lateral portions of the third ventricle.
- The habenula is divided into two sections, the medial and lateral habenula, which are both neuroanatomically and neurochemically distinct from one another.
- The medial habenula has been implicated in regulating stress, depression, memory, and more recently nicotine addiction.
- The interpeduncular nucleus is the primary output of the medial habenula and is located, as implicated by its name, between the cerebral peduncles.
- The interpeduncular nucleus has a tonic inhibitory tone over many brain regions and has been implicated in playing an important role in the regulation of rapid eye movement sleep.
- The medial habenula-interpeduncular nucleus pathway has unique afferent and efferent connectivity with the dopaminergic striatum and the limbic forebrain, allowing the pathway to act as an important modulator of the cross talk between these brain regions.
- The medial habenula-interpeduncular nucleus pathway has been implicated in aversive mood states, motivation, and most importantly addiction.
- The medial habenula-interpeduncular nucleus pathway has earned itself the title of an "alternate reward pathway" in the brain.

Summary Points

- The medial habenula-interpeduncular nucleus pathway has been implicated in mediating multiple components of addiction, including nicotine addiction.

- Connectivity between the medial habenula and interpeduncular is primarily via the neurotransmitters acetylcholine and substance P.
- The medial habenula and interpeduncular nucleus are rich in α3β4 nicotinic acetylcholine receptors. Outside of the medial habenula-interpeduncular nucleus pathway, these receptors are not densely expressed in any brain region.
- Pharmacologically blocking medial habenula α3β4 nicotinic acetylcholine receptors significantly reduces both behavioral and neurochemical sensitizations to nicotine. Blockade of α3β4 nicotinic acetylcholine receptors in the interpeduncular nucleus has no effect on sensitized responding.
- The medial habenula and interpeduncular nucleus are rich in neurokinin-1 receptors and substance P-containing cell bodies.
- Pharmacologically blocking interpeduncular nucleus neurokinin-1 receptors significantly reduces both behavioral and neurochemical sensitizations to nicotine. Blockade of neurokinin-1 receptors in the medial habenula has no effect on sensitized responding.
- The medial habenula contributes near all acetylcholine input to the interpeduncular nucleus but only a small fraction of the substance P input.
- Other brain regions such as the laterodorsal tegmental nucleus contain both substance P- and acetylcholine-containing cell bodies, making them a likely contributor to the interpeduncular nucleus substance P.

References

Antolin-Fontes, B., Ables, J. L., Gorlich, A., & Ibanez-Tallon, I. (2015). The habenulo-interpeduncular pathway in nicotine aversion and withdrawal. *Neuropharmacology, 96*, 213–222.

Artymyshyn, R., & Murray, M. (1985). Substance P in the interpeduncular nucleus of the rat: Normal distribution and the effects of deafferentation. *Journal of Comparative Neurology, 231*, 78–90.

Baker, K. G., Halliday, G. M., Hornung, J. P., Geffen, L. B., Cotton, R. G., & Tork, I. (1991). Distribution, morphology and number of monoamine-synthesizing and substance-P containing neurons in the human dorsal raphe nucleus. *Neuroscience, 42*, 757–775.

Battaglia, G., & Rustioni, A. (1988). Coexistence of glutamate and substance P in dorsal root ganglion neurons of rat and monkey. *Journal of Comparative Neurology, 277*, 302–312.

Benwell, M. E., & Balfour, D. J. (1992). The effects of acute and repeated nicotine treatment on nucleus accumbens dopamine and locomotor activity. *British Journal of Pharmacology, 105*, 849–856.

Brownstein, M. J., Mroz, E. A., Kizer, J. S., Palkovits, M., & Leeman, S. E. (1976). Regional distribution of substance P in the brain of the rat. *Brain Research, 116*, 299–305.

Caruso, D. M., Owczarzak, M. T., & Pourcho, R. G. (1990). Colocalization of substance P and GABA in retinal ganglion cells: a computer-assisted visualization. *Visual Neuroscience, 5*, 389–394.

Chen, L. W., Wei, L. C., Liu, H. L., & Chan, Y. S. (2001). Cholinergic neurons expressing substance P receptor (NK1) in the basal forebrain of the rat: a double immunocytochemical study. *Brain Research, 904*, 161–166.

Contestabile, A., & Flumerfelt, B. A. (1981). Afferent connections of the interpeduncular nucleus and the topographical organization of the habenulo-interpeduncular pathway: an HRP study in the rat. *Journal of Comparative Neurology, 169*, 253–270.

Contestabile, A., Villani, L., Fasolo, A., Franzoni, M. F., Gribaudo, L., Oktedalen, O., et al. (1987). Topography of cholinergic and substance P pathways in the habenulo-interpeduncular system of the rat. An immunocytochemical and microchemical approach. *Neuroscience, 21*, 253–270.

Cuello, A. C., Emson, P. C., Paxinos, G., & Jessell, T. (1978). Substance P containing and cholinergic projections from the habenula. *Brain Research, 149*, 413–429.

DiFranza, J. R., & Wellman, R. J. (2007). Sensitization to nicotine: how the animal literature might inform future human research. *Nicotine & Tobacco Research, 9*, 9–20.

Dwoskin, L. P., Crooks, P. A., Teng, L., Green, T. A., & Bardo, M. T. (1999). Acute and chronic effects of nornicotine on locomotor activity in rats: altered response to nicotine. *Psychopharmacology, 145*, 442–451.

Eggan, B. L., & McCallum, S. E. (2016). 18-Methoxycoronaridine acts in the medial habenula to attenuate behavioral and neurochemical sensitization to nicotine. *Behavioural Brain Research, 307*, 186–193.

Eggan, B. L., & McCallum, S. E. (2017). α3β4 nicotinic acetylcholine receptors in the medial habenula and substance P transmission in the interpeduncular nucleus modulate nicotine sensitization. *Behavioural Brain Research, 316*, 94–103.

Epping-Jordan, M. P., Picciotto, M. R., Changeux, J. P., & Pich, E. M. (1999). Assessment of nicotinic acetylcholine receptor subunit contributions to nicotine self-administration in mutant mice. *Psychopharmacology (Berlin), 147*, 25–26.

Fowler, C. D., Lu, Q., Johnson, P. M., Marks, M. J., & Kenny, P. J. (2011). Habenular α5 nicotinic receptor signaling controls nicotine intake. *Nature, 471*, 597–601.

Fowler, C. D., Tuesta, L., & Kenny, P. J. (2011). Role of α5* nicotinic acetylcholine receptors in the effects of acute and chronic nicotine treatment on brain reward function in mice. *Psychopharmacology, 229*, 503–513.

Frahm, S., Slimak, M. A., Ferrarese, L., Santos-Torres, J., Antolin-f, B., Auer, S., et al. (2011). Aversion to nicotine is regulated by the balanced activity of β4 and α5 nicotinic receptor subunits in the medial habenula. *Neuron, 70*, 522–535.

Gallego, X., Molas, S., Amador-Arjona, A., Marks, M. J., Robles, N., Murta, P., et al. (2012). Overexpression of the CHRNA5/A3/B4 genomic cluster in mice increases the sensitivity to nicotine and modifies its reinforcing effects. *Amino Acids, 44*, 897–909.

George, D. T., Gilman, J., Hersh, J., Thorsell, A., Herion, D., Geyer, C., et al. (2008). Neurokinin 1 receptor antagonism as a possible therapy for alcoholism. *Science, 319*, 1535–1539.

Glick, S. D., Sell, E. M., McCallum, S. E., & Maisonneuve, I. M. (2011). Brain regions mediating α3β4 nicotinic antagonist effects of 18-MC on nicotine self-administration. *European Journal of Pharmacology, 699*, 71–75.

Grady, S. R., Meinerz, N. M., Cao, J., Reynolds, A. M., Piccioto, M. R., Changeux, J. P., et al. (2001). Nicotinic agonists stimulate acetylcholine release from mouse interpeduncular nucleus: a function mediated by a different nAChR than dopamine release from striatum. *Journal of Neurochemistry, 76*, 258–268.

Grady, S. R., Moretti, M., Zoli, M., Marks, M. J., Zanardi, A., Pucci, L., et al. (2009). Rodent habenulo-interpeduncular pathway expresses a large variety of uncommon nAChR subtypes, but only the α3β4* and α3β3β4* subtypes mediate acetylcholine release. *Journal of Neuroscience, 29*, 2272–2282.

Groot-Kormelink, P. J., Boorman, J. P., & Sivilotti, L. G. (2001). Formation of functional α3β4α5 human neuronal nicotinic receptors in *Xenopus* oocytes: a reporter mutation approach. *British Journal of Pharmacology, 134*, 789–796.

Herkenham, M., & Nauta, W. J. H. (1977). Afferent connections of the habenular nuclei in the rat. A horseradish peroxidase study, with a note on the fiber-of-passage problem. *Journal of Comparative Neurology, 173*, 277–299.

Koob, G. F., & LeMoal, M. L. (1997). Drug abuse: hedonic homeostatic dysregulation. *Science, 278*, 52–58.

Kuehne, M. E., He, L., Joikel, P. A., Pace, C. J., Fleck, M. W., Maisonneuve, I. M., et al. (2003). Synthesis and biological evaluation of 18-methoxycoronaridine congeners. Potential antiaddiction agents. *Journal of Medicinal Chemistry, 46*, 2716–2730.

Lena, C., Changeux, J. P., & Mulle, C. (1993). Evidence for "preterminal" nicotinic receptors on GABAergic axons in the rat interpeduncular nucleus. *Journal of Neuroscience, 13*, 2680–2688.

Lessard, A., Savard, M., Gobeil, F., Pierce, J. P., & Pickel, V. M. (2009). The neurokinin-3 (NK3) and the neurokinin-1 (NK1) receptors are differentially targeted to mesocortical and mesolimbic projection neurons, and to neuronal nuclei in the rat ventral tegmental area. *Synapse, 63*, 484–501.

Liu, J. Z., Tozzi, F., Waterworth, D. M., et al. (2010). Meta-analysis and imputation refines the association of 15q25 with smoking quantity. *Nature Genetics, 42*, 436–440.

Loonam, T. M., Noailles, P. A., Yu, J., Zhu, J. P., & Angulo, J. A. (2003). Substance P and cholecystokinin regulate neurochemical responses to cocaine and methamphetamine in the striatum. *Life Sciences, 73*, 727–739.

Luo, S., Kulak, J. M., Cartier, G. E., Jacobsen, R. B., Yoshikami, D., Olivera, B. M., et al. (1998). α-Conotoxin AuIB selectively blocks α3β4 nicotinic acetylcholine receptors and nicotine-evoked norepinephrine release. *Journal of Neuroscience, 18*, 8571–8579.

Maeno, H., Kiyama, H., & Tohyama, M. (1993). Distribution of the substance P receptor (NK-1 receptor) in the central nervous system. *Molecular Brain Research, 18*, 43–58.

Margolis, J. M., & Obach, R. S. (2003). Impact of nonspecific binding to microsomes and phospholipid on the inhibition of cytochrome P4502D6: implications for relating in vitro inhibition data to in vivo drug interactions. *Drug Metabolism and Disposition, 31*, 606–611.

McCallum, S. E., Cowe, M. A., Lewis, S. W., & Glick, S. D. (2012). α3β4 nicotinic acetylcholine receptors in the medial habenula modulate the mesolimbic dopaminergic response to acute nicotine *in vivo*. *Neuropharmacology, 63*, 434–440.

Miller, D. K., Wilkins, L. H., Bardo, M. T., Crooks, P. A., & Dwoskin, L. B. (2001). Once weekly administration of nicotine produces long-lasting locomotor sensitization in rats via a nicotinic receptor-mediated mechanism. *Journal of Psychopharmacology, 156*, 469–476.

Minabe, Y., Emori, K., Toor, A., Stutzman, G. E., & Ashby, C. R. (1996). The effect of acute and chronic administration of CP 96,345, a selective neurokinin$_1$ antagonist, on midbrain dopamine neurons in the rat. *Synapse, 22*, 35–45.

Murtra, P., Sheasby, A. M., Hunt, S. P., & De Felipe, C. (2000). Rewarding effects of opiates are absent in mice lacking the receptor for substance P. *Nature, 405*, 180–183.

Nakaya, Y., Kaneko, T., Shigemoto, R., Shigetada, N., & Mizuno, N. (1994). Immunohisotchemical localization of substance P receptor in the central nervous system of the adult rat. *Journal of Comparative Neurology, 347*, 249–274.

Nisell, M., Nomikos, G. G., Hertel, P., Panagis, G., & Svensson, T. H. (1996). Condition-independent sensitization of locomotor stimulation and mesocortical dopamine release following chronic nicotine treatment in the rat. *Synapse, 22*, 369–381.

Perry, D. C., & Kellar, K. J. (1995). [3H] Epibatidine labels nicotinic receptors in rat brain: an autoradiographic study. *Journal of Pharmacology and Experimental Therapeutics, 276*, 1030–1034.

Quick, M. W., Ceballos, M. R., Kasten, M., McIntosh, M. J., & Lester, R. A. (1999). α3β4 subunit-containing nicotinic receptors dominate function in the medial habenula neurons. *Neuropharmacology, 38*, 769–783.

Quirion, R., Shults, C. W., Moody, T. W., Pert, C. B., Chase, T. N., & O'Donohue, T. L. (1983). Autoradiographic distribution of substance P receptors in the rat central nervous system. *Nature, 303*, 714–716.

Ren, J., Qin, C., Hu, F., Tan, J., Qui, L., Zhao, S., et al. (2011). Habenula "cholinergic" neurons corelease glutamate and acetylcholine and activate postsynaptic neurons via distinct transmission modes. *Neuron, 69*, 445–452.

Ripley, T. L., Gadd, C. A., De Felipe, C., Hunt, S. P., & Stephens, D. N. (2002). Lack of self administration and behavioral sensitization to morphine, but not cocaine, in mice lacking NK1 receptors. *Neuropharmacology, 43*, 1258–1268.

Saal, D., Dong, Y., Bonci, A., & Malenka, R. C. (2003). Drugs of abuse and stress trigger a common synaptic adaptation in dopamine neurons. *Neuron, 37*, 577–582.

Salas, R., Pieri, F., & De Biasi, M. (2004). Decreased signs of nicotine withdrawal in mice null for the beta4 nicotinic acetylcholine receptor subunit. *Journal of Neuroscience, 24*, 10035–10039.

Salas, R., Sturm, R., Boutler, J., & De Biasi, M. (2009). Nicotinic receptors in the habenulo-interpeduncular system are necessary for nicotine withdrawal in mice. *Journal of Neuroscience, 29*, 4934–4938.

Schlaepfer, I. R., Hoft, N. R., Collins, A. C., Corley, R. P., Hewitt, J. K., Hopfer, C. J., et al. (2008). The CHRNA5/A3/B4 gene cluster variability as an important determinant of early alcohol and tobacco initiation in young adults. *Biological Psychiatry, 63*, 1039–1046.

Steketee, J. D., & Kalivas, P. W. (2011). Drug wanting: behavioral sensitization and relapse to drug-seeking behavior. *Pharmacology Review, 63*, 348–365.

Stinus, L., Kelley, A. E., & Iversen, S. D. (1978). Increased spontaneous activity following substance P infusion into A10 dopaminergic area. *Nature, 276*, 616–618.

Sutherland, R. J., & Nakajima, S. (1981). Self-stimulation of the habenular complex in the rat. *Journal of Comparative and Physiological Psychology, 95*, 781–791.

Tapper, A. R., McKinney, S. L., Dashmi, R., Schwarz, J., Deshpande, P., Labarca, C., et al. (2004). Nicotine activation of α4* receptors: sufficient for reward, tolerance, and sensitization. *Science, 306*, 1029–1032.

Thor, K. B., & Helke, C. J. (1989). Serotonin and substance P colocalization in the medullary projections to the nucleus tractus solitaries: dual-colour immunohistochemistry combined with retrograde tracing. *Journal of Chemical Neuroanatomy, 2*, 139–148.

Toll, L., Zaveri, N. T., Polgar, W. E., Jiang, F., Khroyan, T. V., Zhou, W., et al. (2012). AT-1001: a high affinity and selective α3β4 nicotinic acetylcholine receptor antagonist blocks nicotine self-administration in rats. *Neuropsychopharmacology, 37*, 1367–1376.

Warden, M. K., & Young, W. S. (1988). Distribution of cells containing mRNAs encoding substance P and neurokinin B in the rat central nervous system. *Journal of Comparative Neurology, 272*, 90–113.

Yip, J., & Chahl, L. A. (2001). Localization of NK1 and NK3 receptors in Guinea-pig brain. *Regulatory Peptides, 98*, 55–62.

C H A P T E R

33

Targeting Nicotinic Acetylcholine Receptors for the Treatment of Pain

Deniz Bagdas, S. Lauren Kyte*, Wisam Toma*, M. Sibel Gurun*[†], *M. Imad Damaj**

*Department of Pharmacology & Toxicology, Virginia Commonwealth University, Richmond, VA, United States
[†]Department of Pharmacology, Faculty of Medicine, Uludag University, Bursa, Turkey

Abbreviations

ACh	acetylcholine
ago-PAM	allosteric agonist-positive allosteric modulator
nAChRs	nicotinic acetylcholine receptors
PAMs	positive allosteric modulators

33.1 INTRODUCTION TO NICOTINIC ACETYLCHOLINE RECEPTORS

Currently, nonsteroidal antiinflammatory drugs and opioids remain the most common forms of pain treatment, followed by antidepressants and antiepileptics. However, these drugs either provide insufficient efficacy or cause adverse effects. These medications are also not very effective for several types of chronic pain, such as neuropathic pain (Woolf, 2010). Consequently, there is a critical need for novel pharmacotherapies for pain management. One of the new targets proposed for alleviating pain is nicotinic acetylcholine receptors (nAChRs).

These receptors are members of the Cys-loop superfamily of pentameric ligand-gated ion channels consisting of five subunits surrounding an ion pore. They are involved in the signal transduction of acetylcholine (ACh)-mediated signals throughout the peripheral and central nervous systems. In addition, nAChRs play an important role in the mediation of pain and several nicotine-evoked responses (Decker, Rueter, & Bitner, 2004).

To date, 12 neuronal subunits (α2–α10 and β2–β4) of nAChRs have been identified. nAChRs may come in heteromeric forms, encompassing various combinations of α and β subunits, or homomeric forms, in which they express only α subunits. The most prevalent nAChR subtypes are the heteromeric $\alpha4\beta2^*$ and homomeric $\alpha7$ receptors (*the * denotes that these nAChRs can contain other α and β subunits as well*, reviewed in Gotti, Zoli, & Clementi, 2006). In addition to ACh, nAChRs also respond to several types of ligands, such as agonists, antagonists, and allosteric modulators. When an agonist binds to its orthosteric binding site, it stabilizes the open and desensitized states. When a channel opens by an agonist, it allows sodium and calcium ions to enter the cell and potassium ions to exit (Fig. 33.1).

Positive allosteric modulators (PAMs), a new class of nAChR ligands, bind to nAChRs at sites distinct from the orthosteric binding sites without activating the receptor in the absence of endogenous or exogenous agonists (Fig. 33.2). However, they potentiate agonist-induced responses by increasing its potency, efficacy, and/or nAChR opening (Williams, Wang, & Papke, 2011). While the potentiating effect of PAMs is caused by their ability to prevent receptor desensitization and prolong synaptic transmission, orthosteric agonists do the exact opposite, making them less effective overall. The main classification of PAMs has been introduced as type I and type II. While type I PAMs increase agonist response with little or no effect on desensitization of nAChRs, type II PAMs increase agonist response and delay the desensitization rate of the agonist, as seen with PAMs for $\alpha7$ nAChRs (Williams et al., 2011).

Another new class of selective nAChR ligands consists of silent agonists, which possess the unique ability to bind to the receptors with very low efficacy, but instead of inducing a current, they stabilize a nonconducting state associated with desensitization (Fig. 33.1). Therefore, these ligands alone can modulate signal transduction (Chojnacka, Papke, & Horenstein, 2013).

Neuroscience of Nicotine
https://doi.org/10.1016/B978-0-12-813035-3.00033-2

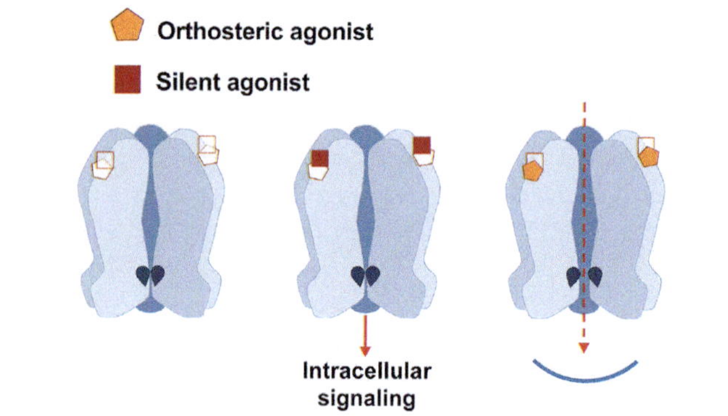

FIG. 33.1 Orthosteric agonists bind to the orthosteric binding sites ( ), open the ion channel, and activate the receptor, whereas silent agonists bind to the silent agonist binding sites ( ), but do not activate the receptor. Silent agonists elicit their effects through an unknown intracellular mechanism.

33.2 THE α4β2* nAChRs IN PAIN MODULATION

In addition to the reinforcing effects of nicotine in tobacco, it has significant pain-relieving effects in human and rodents, particularly through the α4β2* nAChRs, which are expressed on pain transmission pathways (see review Damaj, Freitas, Bagdas, & Flood, 2014). It has been shown in genetically modified mice that the absence of the α4 or β2 nAChR subunit reduces sensitivity to the antinociceptive effects of nicotinic compounds in acute pain tests (Marubio et al., 1999). However, nicotine's short half-life and harmful side effects limit its use. Yet, antinociception provided by nicotine still provides insight into the fundamental mechanisms of α4β2* nAChRs in pain and also facilitates the development of therapeutic α4β2* ligands.

Various α4β2* agonists have been reported for their antinociceptive effects in a range of rodent pain models (Table 33.1). One of the important compounds is epibatidine, a potent agonist at a variety of nAChRs. Although epibatidine has strong antinociceptive properties, it has a rather short duration of action (Bannon, Decker, Kim, Campbell, & Arneric, 1998; Curzon, Nikkel, Bannon, Arneric, & Decker, 1998). Another important compound is ABT-594, which was the first of its kind to demonstrate efficacy in treating painful diabetic neuropathy in humans. However, ABT-594 induces side effects, such as nausea, dizziness, vomiting, abnormal dreams, and asthenia (Rowbotham, Rachel Duan, Thomas, Nothaft, & Backonja, 2009).

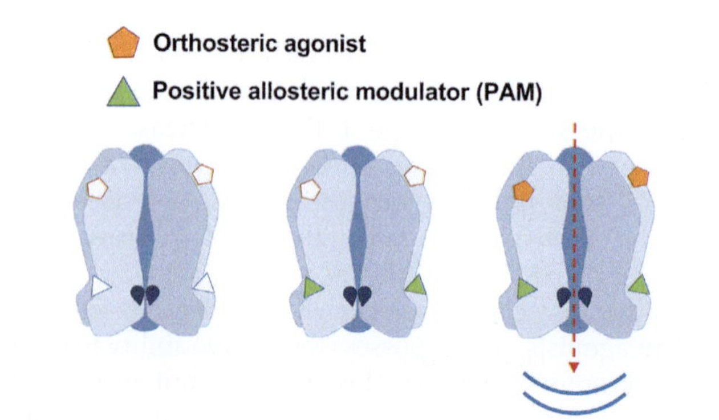

FIG. 33.2 PAMs bind to positive allosteric modulatory sites ( ), but do not activate the receptor. In the presence of an orthosteric agonist, which binds to the orthosteric binding site ( ), PAMs enhance the effect of the agonist.

α5 nicotinic subunits are accessory subunits that can only form functional receptors when coexpressed with a principal subunit (such as α3 or α4) and one complementary subunit (β2 or β4; e.g., as α4β2α5, α3β2α5, or α3β4α5 receptors) (Brown, Collins, Lindstrom, & Whiteaker, 2007). The α5 subunit greatly influences nicotine's modulation of nAChR function in animal models of pain. For example, a possible role for the α5 nAChR subunit in the processing of nociceptive information in neuropathic pain has been suggested (Vincler & Eisenach, 2004). We have also reported that α5 nAChR subunit deletion in the mouse impacts the development and intensity of nociceptive behavior in chronic pain models. Furthermore, nicotine's antinociceptive properties are reduced or absent in α5 subunit knockout mice (Bagdas, AlSharari, Freitas, Tracy, & Damaj, 2015). Both sazetidine-A and varenicline, two α4β2 partial agonists, have antinociceptive properties in the formalin test (AlSharari, Carroll, McIntosh, & Damaj, 2012), yet the effect of sazetidine-A is α5-dependent, which is not the case with varenicline (Bagdas, AlSharari, et al., 2015).

While α4β2* agonists are limited in their use due to their relatively low functional selectivity among the different nAChR subtypes (Zhang et al., 2012), α4β2* PAMs can enhance properties of nicotinic agonists. For example, NS9283 potentiated the antinociceptive effects of ABT-594 and NS3956 in rodent models of pain (Table 33.1). Also, the combination of desformylflustrabromine, an α4β2* PAM, with nicotine enhanced reversal of neuropathic pain behavior without motor impairment, confirming the dissociation of analgesic activity and adverse effects (Bagdas et al., 2017). Because NS9283 and desformylflustrabromine lacked activity alone in the abovementioned studies, it suggests the absence of a β2* nAChR-mediated endogenous antinociceptive cholinergic tone in neuropathic pain. This is in contrast to the α7 nAChR PAMs, which have been shown to be active when administered alone in preclinical models (see below).

33.3 THE α7 nAChRs IN PAIN MODULATION

A great number of studies suggest that activation of the α7 nAChR subtype has a potential role in pain management (Table 33.2). The α7 nAChR has been shown to

TABLE 33.1 Efficacy of α4β2* nAChR Ligands in Various Animal Models of Pain

Compound	Ligand type	Animal model/assay	Animal	Response	Reference
Nicotine	Agonist	Tail flick	Mice	Antinociceptive	Damaj et al. (1998)
		Hot plate	Rats	Antinociceptive	Boyce et al. (2000)
		CCI	Mice	Reversed mechanical allodynia	Bagdas et al. (2017) and Bagdas, AlSharari, et al. (2015)
Epibatidine	Agonist	Hot plate	Mice	Antinociceptive	Badio and Daly (1994)
			Rats	Antinociceptive	Boyce et al. (2000)
		Tail flick	Mice	Antinociceptive	Damaj et al. (1998)
		Formalin	Rats	Antinociceptive	Curzon et al. (1998)
		CIIP	Rats	↓ Thermal hyperalgesia	Lawand, Lu, and Westlund (1999)
		CFA	Rats	Reversed inflammatory and neuropathic hyperalgesia	Kesingland et al. (2000)
		PSNL			
ABT-594	Agonist	Hot plate	Rats	Antinociceptive	Boyce et al. (2000)
		Tail flick		Antinociceptive	Kesingland et al. (2000)
		CFA		Reversed inflammatory and neuropathic hyperalgesia	
		PSNL			
		CIIP		↓ Thermal hyperalgesia, mechanical allodynia, and osteoarthritic pain	Zhu et al. (2011)
		Paw skin incision			
		Monoiodoacetate-induced arthritis			
		SNL		↓ Mechanical allodynia and hyperalgesia	Bannon et al. (1998)
		Diabetic neuropathy			
		Vincristine-induced CIPN		↓ Mechanical allodynia	Lynch, Wade, Mikusa, Decker, and Honore (2005)
NS3956	Agonist	Formalin	Rats	Antinociceptive	Rode et al. (2012)
5-Iodo-A-85380	Agonist	Hargreaves	Rats	Antinociceptive	Rueter, Meyer, and Decker (2000)
Sazetidine-A	Partial agonist	Tail flick, hot plate	Mice	No effect	AlSharari et al. (2012)
		Formalin		Antinociceptive	
		Formalin		Antinociceptive	Bagdas, AlSharari, et al. (2015)
Varenicline	Partial agonist	Tail flick, hot plate	Mice	No effect	AlSharari et al. (2012)
		Formalin		Antinociceptive	Bagdas, AlSharari, et al. (2015)
NS9283	Type II PAM	Formalin	Rats	Potentiated antinociceptive effect of NS3956	Rode et al. (2012)
		CIIP		Potentiated antinociceptive effects of ABT-594	Zhu et al. (2011)
		Paw skin incision			
		Monoiodoacetate-induced arthritis			
		SNL		Reversed mechanical allodynia in combination with ABT-594	Lee et al. (2011)
Desformylflustrabromine	Type II PAM	CCI	Mice	Potentiated antiallodynic effect of nicotine	Bagdas et al. (2017)

CCI, chronic constrictive nerve injury; *CFA*, complete Freund's adjuvant; *CIIP*, carrageenan-induced inflammatory pain; *CIPN*, chemotherapy-induced peripheral neuropathy; *PAM*, positive allosteric modulator; *PSNL*, partial sciatic nerve ligation; *SNL*, spinal nerve ligation. (↓) decreased.

TABLE 33.2 Efficacy of α7 nAChR Ligands in Various Animal Models of Pain

Compound	Ligand type	Animal model/assay	Animal	Response	Reference
Choline	Agonist	Tail flick	Mice	Antinociceptive	Damaj, Meyer, and Martin (2000)
		Formalin		Antinociceptive	Freitas, Negus, et al. (2013)
		Incisional postoperative pain		↓ Thermal hyperalgesia and mechanical allodynia	Rowley, McKinstry, Greenidge, Smith, and Flood (2010)
PNU-282987	Agonist	Formalin	Mice	Antinociceptive	Donvito et al. (2017)
PHA-543613	Agonist	Formalin	Rats	Antinociceptive via centrally mediated mechanism	Umana, Daniele, Miller, Gallagher, and Brown (2017)
NS6740	Silent agonist	Formalin	Mice	Antinociceptive	Papke et al. (2015)
		CCI		↓ Mechanical allodynia	
		CPA		reversed CPA	
PNU-120596	Type II PAM	Tail flick, hot plate	Mice	No effect	Freitas, Negus, et al. (2013), Freitas, Carroll, and Damaj (2013), and Freitas, Ghosh, et al. (2013)
		Formalin		Antinociceptive alone and potentiated effect of PHA-543613	
		CIIP		↓ Thermal hyperalgesia	
		CCI		↓ Thermal hyperalgesia and mechanical allodynia	
3-Furan-2-yl-N-p-tolyl-acrylamide	Type II PAM	CIIP	Mice	↓ Mechanical allodynia and potentiated antiallodynic effect of choline	Bagdas, Targowska-Duda, et al. (2015)
		CFA		↓ Thermal hyperalgesia	
		CCI		↓ Mechanical allodynia	
		CPA		Reversed CPA	
GAT107	Ago-PAM (allosteric agonist and type II PAM)	Tail flick, hot plate	Mice	No effect	Bagdas et al. (2016)
		Formalin		Antinociceptive	
		CFA		↓ Thermal hyperalgesia	
		LPS		↓ Mechanical allodynia	
		Acid-induced stretching		↓ Stretching	
		CCI		↓ Mechanical allodynia	
		CPA		Reversed CPA	

CCI, chronic constrictive nerve injury; *CFA*, complete Freund's adjuvant; *CIIP*, carrageenan-induced inflammatory pain; *CPA*, conditioned place aversion; *LPS*, lipopolysaccharide; *PAM*, positive allosteric modulator. (↓) decreased.

be distributed in the pain transmission pathway, including neuronal and nonneuronal cells and specifically immune cells (Corradi & Bouzat, 2016; Khan et al., 2003). The α7 nAChRs downregulate cytokine production and the associated inflammatory response, making them an essential component of the cholinergic antiinflammatory pathway, which is defined as neural signals transmitted via the vagus nerve that inhibit cytokine release (Pavlov et al., 2009). For example, α7 nAChR knockout mice showed a significant increase in the incidence and severity of arthritis (van Maanen, Stoof, LaRosa, Vervoordeldonk, & Tak, 2010). In addition, a marked increase in edema, hyperalgesia, and allodynia associated with intraplantar complete Freund's adjuvant injection was observed in α7 knockout mice compared with wild-type littermates (AlSharari, Freitas, & Damaj, 2013). Furthermore, studies have shown that the selective α7 nAChR agonists choline and PHA-543613 significantly reduce paw-licking time in the formalin test (Freitas, Carroll, & Damaj, 2013; Freitas, Negus, Carroll, & Damaj,

2013). It has been reported that the type II PAM PNU-120596 reverses mechanical allodynia in a neuropathic pain model (Freitas, Ghosh, Ivy Carroll, Lichtman, & Imad Damaj, 2013). Another type II PAM, 3-furan-2-yl-N-p-tolyl-acrylamide, reversed mechanical allodynia and thermal hyperalgesia in chronic pain models (Bagdas, Targowska-Duda, et al., 2015). GAT107, a new α7 selective dual allosteric agonist-PAM (ago-PAM; Fig. 33.3), showed antiinflammatory and antinociceptive properties (Bagdas et al., 2016). Also, the silent agonist NS6740 reduced time spent paw licking in the formalin test and attenuated mechanical allodynia in a neuropathic pain model via an unknown signaling mechanism (Papke et al., 2015). Furthermore, these compounds effectively reduced acetic acid-induced conditioned place aversion (Bagdas et al., 2016; Bagdas, Targowska-Duda, et al., 2015; Papke et al., 2015). Finally, we recently demonstrated an in vivo cross talk between α7 nAChRs and the nuclear peroxisome proliferator-activated receptor type-α; the activation of α7 nAChRs by a selective full agonist (i.e., PNU-282987) in the formalin test may lead to an increase of an endogenous peroxisome proliferator-activated receptor type-α tone, mediating the observed antinociceptive effects (Donvito et al., 2017).

Although α7 nAChRs are classically known to be homomeric, emerging evidence demonstrates the presence of heteromeric α7 nAChRs, which coassemble with β2 subunits (see review Wu et al., 2016). However, the role of α7β2 nAChRs in pain modulation is unknown.

33.4 THE α9/α9α10 nAChRs IN PAIN MODULATION

The α9 nAChR subunit can be assembled into a homopentamer with other α9 subunits and a heteropentamer with α10 nAChR subunits, to form functional nAChRs. Both the α9 and α10 subunits are expressed in the peripheral nervous system, but not in the brain (see review Gotti et al., 2006). Like other nAChRs, α9α10 receptors are activated by ACh; however, nicotine acts as an antagonist by inhibiting ACh-evoked currents (Elgoyhen et al., 2001; Elgoyhen, Johnson, Boulter, Vetter, & Heinemann, 1994).

A drug class that is showing promise for nAChR-mediated pain management is the α-conotoxin family (Table 33.3), which consists of short peptides isolated from the venom of *Conus* marine snails, some of which are potent antagonists of the α9α10 nAChRs (McIntosh,

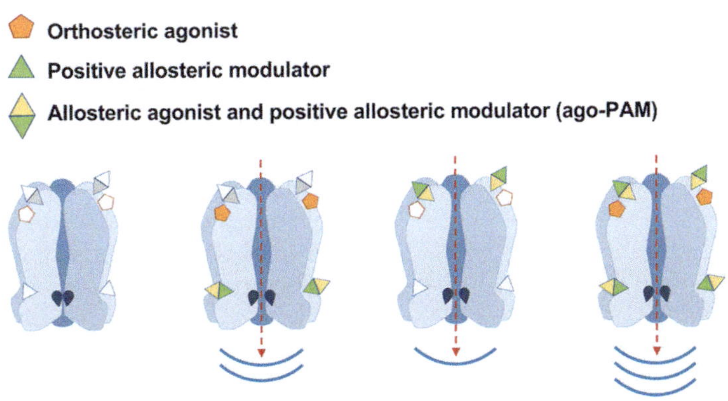

FIG. 33.3 An ago-PAM can activate the receptor on its own by binding to the direct allosteric activation binding site (). An ago-PAM can bind to positive allosteric modulatory sites () and enhance the effect of an orthosteric agonist, which binds to the orthosteric binding site ().

TABLE 33.3 Efficacy of α9α10 nAChR Ligands in Various Animal Models of Pain

Compound	Ligand type	Animal model/assay	Animal	Response	Reference
RgIA	Antagonist	CCI	Rats	Prevented and reversed mechanical allodynia	Di Cesare Mannelli et al. (2014)
		CCI		Reversed mechanical hyperalgesia	Vincler et al. (2006)
RgIA4	Antagonist	Oxaliplatin-induced CIPN	Rats	Prevented cold allodynia and mechanical hyperalgesia	Romero et al. (2017)
Vc1.1	Antagonist	CCI, PSNL	Rats	Reversed mechanical hyperalgesia	Satkunanathan et al. (2005)
ZZ1-61c	Antagonist	Vincristine-induced CIPN	Rats	Prevented/reversed mechanical allodynia	Wala et al. (2012)

CCI, chronic constrictive nerve injury; *CIPN*, chemotherapy-induced peripheral neuropathy; *PSNL*, partial sciatic nerve ligation.

Absalom, Chebib, Elgoyhen, & Vincler, 2009; Napier et al., 2012). The α-conotoxin RgIA, the most selective mammalian α9α10 antagonist (McIntosh et al., 2009), has been shown to reverse mechanical allodynia and hyperalgesia in a neuropathic pain model (Di Cesare Mannelli et al., 2014; Vincler et al., 2006); however, the drug is 300-fold less potent at human α9α10 nAChRs (Azam & McIntosh, 2012). A similar compound, RgIA4, has a high affinity for both the rat and human α9α10 nAChR and has been shown to prevent oxaliplatin-induced neuropathic pain (Romero et al., 2017). Another α-conotoxin of interest is Vc1.1, which successfully passed phase I of clinical trials, but did not continue past phase 2A after in vitro data suggested that Vc1.1 is less selective for the human α9α10 nAChR than that of the rat (Azam & McIntosh, 2012). In addition, studies have shown that the selective α9α10 antagonist ZZ1-61c, a *bis*-azaaromatic quaternary ammonium analogue of nicotine, prevents and reverses chemotherapy-induced mechanical allodynia when administered in combination with or after vincristine (Wala, Crooks, McIntosh, & Holtman, 2012). In conjunction with other α9α10 antagonists, this effect of ZZ1-61c was not transient, indicating that the acute administration induces lasting physiological changes.

33.5 CONCLUSIONS

The ongoing discovery and characterization of nAChR agonists, silent agonists, PAMs, ago-PAMs, and antagonists in animal models of pain and inflammation has created new opportunities for targeting these receptors for the development of alternative analgesics. To facilitate this process, future drug development will be focused on identifying the nAChR-mediated mechanisms responsible for antinociception and neuroprotection, with the goal of not only relieving pain but also preventing the initiation and/or progression of chronic pain states.

MINI-DICTIONARY OF TERMS

Agonist A drug that binds to the orthosteric site of a receptor, thereby changing the receptor state and producing a physiological response (Neubig, Spedding, Kenakin, & Christopoulos, 2003).

Allodynia Pain induced by a stimulus that usually does not provoke pain (Merskey & Bogduk, 1994).

Allosteric agonist-positive allosteric modulator A ligand that binds to two allosteric sites, one on the extracellular domain and one on the transmembrane domain, to induce and enhance receptor activation (Papke et al., 2017).

Analgesia The lack of pain following stimulation that is normally painful (Merskey & Bogduk, 1994).

Antagonist A drug that blocks the action of an agonist (Neubig et al., 2003).

Hyperalgesia Heightened pain caused by a stimulus that usually provokes pain (Merskey & Bogduk, 1994).

Nociception The encoding and transmission of noxious stimuli (Merskey & Bogduk, 1994).

Pain An unpleasant sensory and emotional reaction to an actual or possible physical injury (Bonica, 1979).

Positive allosteric modulator A drug that binds to the allosteric site of a receptor to enhance activation in the presence of an orthosteric agonist (Papke et al., 2017).

Silent agonist A drug that binds to the orthosteric site of a receptor to induce a conformational change associated with a desensitized, non-conducting state (Papke et al., 2015).

Key Facts of nAChRs and Their Ligands

- Nicotine acts as an agonist on α7 and α4β2* nAChRs but as an antagonist on α9α10 nAChRs.
- Upon agonist binding, α7 nAChRs rapidly desensitize after opening.
- nAChR agonists induce pore opening and positive ion flux.
- Although silent agonists do not allow for nAChR opening, they induce a conformational change that mediates downstream signaling.
- PAMs bind to the transmembrane modulatory site of the nAChR, while ago-PAMs can bind to both this site and an extracellular domain site.

Summary Points

- This chapter focuses on investigating nAChR ligands for the treatment of pain.
- Agonists for the α4β2* nAChRs alleviate pain in various nociception assays, such as acute and chronic pain.
- Silent agonists, PAMs, and ago-PAMs for the α7 nAChR reduce chronic pain behaviors but have no effect in acute pain sensitivity.
- The α4β2* PAMs lack activity when administered alone in preclinical pain models, but the α7 PAMs are capable of producing antinociceptive effects alone.
- Antagonists for the α9α10 nAChR attenuate neuropathic pain behavior.

Acknowledgments

Given the reference limitations, we apologize that some original works are not cited.

References

AlSharari, S. D., Carroll, F. I., McIntosh, J. M., & Damaj, M. I. (2012). The Antinociceptive effects of nicotinic partial agonists varenicline and sazetidine-A in murine acute and tonic pain models. *Journal of Pharmacology and Experimental Therapeutics*, *342*(3), 742–749. https://doi.org/10.1124/jpet.112.194506.

AlSharari, S. D., Freitas, K., & Damaj, M. I. (2013). Functional role of alpha7 nicotinic receptor in chronic neuropathic and inflammatory pain: studies in transgenic mice. *Biochemical Pharmacology*, *86*(8), 1201–1207. https://doi.org/10.1016/j.bcp.2013.06.018.

Azam, L., & McIntosh, J. M. (2012). Molecular basis for the differential sensitivity of rat and human α9α10 nAChRs to α-conotoxin RgIA. *Journal of Neurochemistry, 122*(6), 1137–1144. https://doi.org/10.1111/j.1471-4159.2012.07867.x.

Badio, B., & Daly, J. W. (1994). Epibatidine, a potent analgetic and nicotinic agonist. *Molecular Pharmacology, 45*(4), 563–569.

Bagdas, D., AlSharari, S. D., Freitas, K., Tracy, M., & Damaj, M. I. (2015). The role of alpha5 nicotinic acetylcholine receptors in mouse models of chronic inflammatory and neuropathic pain. *Biochemical Pharmacology, 97*(4), 590–600. https://doi.org/10.1016/j.bcp.2015.04.013.

Bagdas, D., Ergun, D., Jackson, A., Toma, W., Schulte, M. K., & Damaj, M. I. (2017). Allosteric modulation of α4β2* nicotinic acetylcholine receptors: desformylflustrabromine potentiates antiallodynic response of nicotine in a mouse model of neuropathic pain. *European Journal of Pain*, 1–10. https://doi.org/10.1002/ejp.1092.

Bagdas, D., Targowska-Duda, K. M., López, J. J., Perez, E. G., Arias, H. R., & Damaj, M. I. (2015). The antinociceptive and antiinflammatory properties of 3-furan-2-yl-N-p-tolyl-acrylamide, a positive allosteric modulator of α7 nicotinic acetylcholine receptors in mice. *Anesthesia & Analgesia, 121*(5), 1369–1377. https://doi.org/10.1213/ANE.0000000000000902.

Bagdas, D., Wilkerson, J. L., Kulkarni, A., Toma, W., AlSharari, S., Gul, Z., et al. (2016). The α7 nicotinic receptor dual allosteric agonist and positive allosteric modulator GAT107 reverses nociception in mouse models of inflammatory and neuropathic pain. *British Journal of Pharmacology*. https://doi.org/10.1111/bph.13528.

Bannon, A. W., Decker, M. W., Kim, D. J. B., Campbell, J. E., & Arneric, S. P. (1998). ABT-594, a novel cholinergic channel modulator, is efficacious in nerve ligation and diabetic neuropathy models of neuropathic pain. *Brain Research, 801*(1–2), 158–163. https://doi.org/10.1016/S0006-8993(98)00596-4.

Bonica, J. J. (1979). The need of a taxonomy. *Pain 6*(3), 247–248.

Boyce, S., Webb, J. K., Shepheard, S. L., Russell, M. G. N., Hill, R. G., & Rupniak, N. M. J. (2000). Analgesic and toxic effects of ABT-594 resemble epibatidine and nicotine in rats. *Pain, 85*(3), 443–450. https://doi.org/10.1016/S0304-3959(99)00303-6.

Brown, R. W. B., Collins, A. C., Lindstrom, J. M., & Whiteaker, P. (2007). Nicotinic alpha5 subunit deletion locally reduces high-affinity agonist activation without altering nicotinic receptor numbers. *Journal of Neurochemistry*. https://doi.org/10.1111/j.1471-4159.2007.04700.x.

Chojnacka, K., Papke, R. L., & Horenstein, N. a. (2013). Synthesis and evaluation of a conditionally-silent agonist for the α7 nicotinic acetylcholine receptor. *Bioorganic & Medicinal Chemistry Letters, 23*(14), 4145–4149. https://doi.org/10.1016/j.bmcl.2013.05.039.

Corradi, J., & Bouzat, C. (2016). Understanding the bases of function and modulation of α7 nicotinic receptors: implications for drug discovery. *Molecular Pharmacology*, 288–299. https://doi.org/10.1124/mol.116.104240.

Curzon, P., Nikkel, A. L., Bannon, A. W., Arneric, S. P., & Decker, M. W. (1998). Differences between the antinociceptive effects of the cholinergic channel activators A-85380 and (+/-)-epibatidine in rats. *The Journal of Pharmacology and Experimental Therapeutics, 287*(3), 847–853. Retrieved from: http://www.ncbi.nlm.nih.gov/pubmed/9864263.

Damaj, M. I., Fei-Yin, M., Dukat, M., Glassco, W., Glennon, R. A., & Martin, B. R. (1998). Antinociceptive responses to nicotinic acetylcholine receptor ligands after systemic and intrathecal administration in mice. *The Journal of Pharmacology and Experimental Therapeutics, 284*(3), 1058–1065.

Damaj, M. I., Freitas, K., Bagdas, D., & Flood, P. (2014). Nicotinic receptors as targets for novel analgesics and anti-inflammatory drugs. In: R. A. J. Lester (Ed.), *Nicotinic receptors* (pp. 239–254). Vol. 11(pp. 239–254). New York, NY: Springer New York. https://doi.org/10.1007/978-1-4939-1167-7_12.

Damaj, M. I., Meyer, E. M., & Martin, B. R. (2000). The antinociceptive effects of alpha7 nicotinic agonists in an acute pain model. *Neuropharmacology, 39*(13), 2785–2791. pii: S0028390800001398.

Decker, M. W., Rueter, L. E., & Bitner, R. S. (2004). Nicotinic acetylcholine receptor agonists: a potential new class of analgesics. *Current Topics in Medicinal Chemistry, 4*(3), 369–384.

Di Cesare Mannelli, L., Cinci, L., Micheli, L., Zanardelli, M., Pacini, A., McIntosh, J. M., et al. (2014). α-Conotoxin RgIA protects against the development of nerve injury-induced chronic pain and prevents both neuronal and glial derangement. *Pain, 155*(10), 1986–1995. https://doi.org/10.1016/j.pain.2014.06.023.

Donvito, G., Bagdas, D., Toma, W., Rahimpour, E., Jackson, A., Meade, J. A., et al. (2017). The interaction between alpha 7 nicotinic acetylcholine receptor and nuclear peroxisome proliferator-activated receptor-α represents a new antinociceptive signaling pathway in mice. *Experimental Neurology, 295*, 194–201. https://doi.org/10.1016/j.expneurol.2017.06.014.

Elgoyhen, A. B., Johnson, D. S., Boulter, J., Vetter, D. E., & Heinemann, S. (1994). α9: an acetylcholine receptor with novel pharmacological properties expressed in rat cochlear hair cells. *Cell, 79*(4), 705–715. https://doi.org/10.1016/0092-8674(94)90555-X.

Elgoyhen, A. B., Vetter, D. E., Katz, E., Rothlin, C. V., Heinemann, S. F., & Boulter, J. (2001). Alpha10: a determinant of nicotinic cholinergic receptor function in mammalian vestibular and cochlear mechanosensory hair cells. *Proceedings of the National Academy of Sciences of the United States of America, 98*(6), 3501–3506. https://doi.org/10.1073/pnas.051622798.

Freitas, K., Carroll, F. I., & Damaj, M. I. (2013). The antinociceptive effects of nicotinic receptors α7-positive allosteric modulators in murine acute and tonic pain models. *The Journal of Pharmacology and Experimental Therapeutics, 344*(1), 264–275. https://doi.org/10.1124/jpet.112.197871.

Freitas, K., Ghosh, S., Ivy Carroll, F., Lichtman, A. H., & Imad Damaj, M. (2013). Effects of alpha 7 positive allosteric modulators in murine inflammatory and chronic neuropathic pain models. *Neuropharmacology, 65*, 156–164. https://doi.org/10.1016/j.neuropharm.2012.08.022.

Freitas, K., Negus, S., Carroll, F. I., & Damaj, M. I. (2013). In vivo pharmacological interactions between a type II positive allosteric modulator of α7 nicotinic ACh receptors and nicotinic agonists in a murine tonic pain model. *British Journal of Pharmacology, 169*(3), 567–579. https://doi.org/10.1111/j.1476-5381.2012.02226.x.

Gotti, C., Zoli, M., & Clementi, F. (2006). Brain nicotinic acetylcholine receptors: native subtypes and their relevance. *Trends in Pharmacological Sciences, 27*(9), 482–491. https://doi.org/10.1016/j.tips.2006.07.004.

Kesingland, A. C., Gentry, C. T., Panesar, M. S., Bowes, M. A., Vernier, J. M., Cube, R., et al. (2000). Analgesic profile of the nicotinic acetylcholine receptor agonists, (+)-epibatidine and ABT-594 in models of persistent inflammatory and neuropathic pain. *Pain, 86*(1–2), 113–118. https://doi.org/10.1016/S0304-3959(00)00233-5.

Khan, I., Osaka, H., Stanislaus, S., Calvo, R. M., Deerinck, T., Yaksh, T. L., et al. (2003). Nicotinic acetylcholine receptor distribution in relation to spinal neurotransmission pathways. *Journal of Comparative Neurology, 467*(1), 44–59. https://doi.org/10.1002/cne.10913.

Lawand, N. B., Lu, Y., & Westlund, K. N. (1999). Nicotinic cholinergic receptors: potential targets for inflammatory pain relief. *Pain, 80* (1–2), 291–299. https://doi.org/10.1016/s0304-3959(98)00221-8.

Lee, C. H., Zhu, C., Malysz, J., Campbell, T., Shaughnessy, T., Honore, P., et al. (2011). α4β2 neuronal nicotinic receptor positive allosteric modulation: an approach for improving the therapeutic index of α4β2 nAChR agonists in pain. *Biochemical Pharmacology, 82*(8), 959–966. https://doi.org/10.1016/j.bcp.2011.06.044.

Lynch, J. J., Wade, C. L., Mikusa, J. P., Decker, M. W., & Honore, P. (2005). ABT-594 (a nicotinic acetylcholine agonist): anti-allodynia in a rat chemotherapy-induced pain model. *European Journal of Pharmacology, 509*(1), 43–48. https://doi.org/10.1016/j.ejphar.2004.12.034.

Marubio, L. M., del Mar Arroyo-Jimenez, M., Cordero-Erausquin, M., Léna, C., Le Novère, N., de Kerchove d'Exaerde, M., et al. (1999).

Reduced antinociception in mice lacking neuronal nicotinic receptor subunits. *Nature, 398*(6730), 805–810. https://doi.org/10.1038/19756.

McIntosh, J. M., Absalom, N., Chebib, M., Elgoyhen, A. B., & Vincler, M. (2009). Alpha9 nicotinic acetylcholine receptors and the treatment of pain. *Biochemical Pharmacology, 78*(7), 693–702. https://doi.org/10.1016/j.bcp.2009.05.020.

Merskey, H., & Bogduk, N. (1994). Classification of chronic pain. *IASP Pain Terminology*. https://doi.org/10.1002/ana.20394.

Napier, I. A., Klimis, H., Rycroft, B. K., Jin, A. H., Alewood, P. F., Motin, L., et al. (2012). Intrathecal α-conotoxins Vc1.1, AuIB and MII acting on distinct nicotinic receptor subtypes reverse signs of neuropathic pain. *Neuropharmacology, 62*(7), 2201–2206. https://doi.org/10.1016/j.neuropharm.2012.01.016.

Neubig, R. R., Spedding, M., Kenakin, T., & Christopoulos, A. (2003). International Union of Pharmacology Committee on Receptor Nomenclature and Drug Classification. XXXVIII. Update on terms and symbols in quantitative pharmacology. *Pharmacological Reviews, 55*(4), 597–606. https://doi.org/10.1124/pr.55.4.4.

Papke, R. L., Bagdas, D., Kulkarni, A. R., Gould, T., AlSharari, S. D., Thakur, G. a., et al. (2015). The analgesic-like properties of the alpha7 nAChR silent agonist NS6740 is associated with non-conducting conformations of the receptor. *Neuropharmacology, 91*, 34–42. https://doi.org/10.1016/j.neuropharm.2014.12.002.

Papke, R. L., Stokes, C., Damaj, M. I., Thakur, G. A., Manther, K., Treinin, M., et al. (2017). Persistent activation of α7 nicotinic ACh receptors associated with stable induction of different desensitized states. *British Journal of Pharmacology*, 1–17. https://doi.org/10.1111/bph.13851.

Pavlov, V. A., Parrish, W. R., Rosas-Ballina, M., Ochani, M., Puerta, M., Ochani, K., et al. (2009). Brain acetylcholinesterase activity controls systemic cytokine levels through the cholinergic anti-inflammatory pathway. *Brain, Behavior, and Immunity, 23*(1), 41–45. https://doi.org/10.1016/j.bbi.2008.06.011.

Rode, F., Munro, G., Holst, D., Nielsen, E., Troelsen, K. B., Timmermann, D. B., et al. (2012). Positive allosteric modulation of α4β2 nAChR agonist induced behaviour. *Brain Research, 1458*, 67–75. https://doi.org/10.1016/j.brainres.2012.03.064.

Romero, H. K., Christensen, S. B., Di Cesare Mannelli, L., Gajewiak, J., Ramachandra, R., Elmslie, K. S., et al. (2017). Inhibition of α9α10 nicotinic acetylcholine receptors prevents chemotherapy-induced neuropathic pain. *Proceedings of the National Academy of Sciences*. https://doi.org/10.1073/pnas.1621433114.

Rowbotham, M. C., Rachel Duan, W., Thomas, J., Nothaft, W., & Backonja, M. M. (2009). A randomized, double-blind, placebo-controlled trial evaluating the efficacy and safety of ABT-594 in patients with diabetic peripheral neuropathic pain. *Pain, 146*(3), 245–252. https://doi.org/10.1016/j.pain.2009.06.013.

Rowley, T. J., McKinstry, A., Greenidge, E., Smith, W., & Flood, P. (2010). Antinociceptive and anti-inflammatory effects of choline in a mouse model of postoperative pain. *British Journal of Anaesthesia, 105*(2), 201–207. https://doi.org/10.1093/bja/aeq113.

Rueter, L. E., Meyer, M. D., & Decker, M. W. (2000). Spinal mechanisms underlying A-85380-induced effects on acute thermal pain. *Brain Research, 872*(1–2), 93–101. Retrieved from: http://www.ncbi.nlm.nih.gov/pubmed/10924680.

Satkunanathan, N., Livett, B., Gayler, K., Sandall, D., Down, J., & Khalil, Z. (2005). Alpha-conotoxin Vc1.1 alleviates neuropathic pain and accelerates functional recovery of injured neurones. *Brain Research, 1059*(2), 149–158. https://doi.org/10.1016/j.brainres.2005.08.009.

Umana, I. C., Daniele, C. A., Miller, B. A., Gallagher, K., & Brown, M. A. (2017). Nicotinic modulation of descending pain control circuitry. *Pain, 158*, 1938–1950.

van Maanen, M. A., Stoof, S. P., LaRosa, G. J., Vervoordeldonk, M. J., & Tak, P. P. (2010). Role of the cholinergic nervous system in rheumatoid arthritis: aggravation of arthritis in nicotinic acetylcholine receptor α7 subunit gene knockout mice. *Annals of the Rheumatic Diseases, 69*(9), 1717–1723. https://doi.org/10.1136/ard.2009.118554.

Vincler, M., & Eisenach, J. C. (2004). Plasticity of spinal nicotinic acetylcholine receptors following spinal nerve ligation. *Neuroscience Research, 48*(2), 139–145. Retrieved from: http://www.ncbi.nlm.nih.gov/pubmed/14741388.

Vincler, M., Wittenauer, S., Parker, R., Ellison, M., Olivera, B. M., & McIntosh, J. M. (2006). Molecular mechanism for analgesia involving specific antagonism of α9α10 nicotinic acetylcholine receptors. *Proceedings of the National Academy of Sciences, 103*(47), 17880–17884. https://doi.org/10.1073/pnas.0608715103.

Wala, E. P., Crooks, P. A., McIntosh, J. M., & Holtman, J. R. (2012). Novel small molecule α9α10 nicotinic receptor antagonist prevents and reverses chemotherapy-evoked neuropathic pain in rats. *Anesthesia and Analgesia, 115*(3), 713–720. https://doi.org/10.1213/ANE.0b013e31825a3c72.

Williams, D. K., Wang, J., & Papke, R. L. (2011). Positive allosteric modulators as an approach to nicotinic acetylcholine receptor-targeted therapeutics: advantages and limitations. *Biochemical Pharmacology, 82*(8), 915–930. https://doi.org/10.1016/j.bcp.2011.05.001.

Woolf, C. J. (2010). Overcoming obstacles to developing new analgesics. *Nature Medicine, 16*(11), 1241–1247. https://doi.org/10.1038/nm.2230.

Wu, J., Liu, Q., Tang, P., Mikkelsen, J. D., Shen, J., Whiteaker, P., et al. (2016). Heteromeric α7β2 nicotinic acetylcholine receptors in the brain. *Trends in Pharmacological Sciences, 37*(7), 562–574. https://doi.org/10.1016/j.tips.2016.03.005.

Zhang, J., Xiao, Y. D., Jordan, K. G., Hammond, P. S., Van Dyke, K. M., Mazurov, A. A., et al. (2012). Analgesic effects mediated by neuronal nicotinic acetylcholine receptor agonists: correlation with desensitization of α4β2* receptors. *European Journal of Pharmaceutical Sciences, 47*(5), 813–823. https://doi.org/10.1016/j.ejps.2012.09.014.

Zhu, C. Z., Chin, C.-L., Rustay, N. R., Zhong, C., Mikusa, J., Chandran, P., et al. (2011). Potentiation of analgesic efficacy but not side effects: co-administration of an α4β2 neuronal nicotinic acetylcholine receptor agonist and its positive allosteric modulator in experimental models of pain in rats. *Biochemical Pharmacology, 82*(8), 967–976. https://doi.org/10.1016/j.bcp.2011.05.007.

CHAPTER

34

Pharmacology of Muscle-Type Nicotinic Receptors

Armando Alberola-Die, Raúl Cobo, Isabel Ivorra, Andrés Morales

Division of Physiology, Department of Physiology, Genetics and Microbiology, Universidad de Alicante (Spain), Alicante, Spain

Abbreviations

A	agonist
ACh	acetylcholine
AM	allosteric modulator
CCh	carbamylcholine
CI	competitive inhibitor
ECD	extracellular domain
epp	end-plate potential
FCS	fast-channel syndrome
I_{ACh}	membrane current elicited by ACh
ICD	intracellular domain
LGIC	ligand-gated ion channel
nAChR$_m$	muscle-type nicotinic acetylcholine receptor
NAM	negative allosteric modulator
NCI	noncompetitive inhibitor
NMJ	neuromuscular junction
PAM	positive allosteric modulator
SCS	slow-channel syndrome
TMD	transmembrane domain

34.1 INTRODUCTION

For over a century, some of the most brilliant neuroscientists (many of them are Nobel Prize laureates) have been involved in deciphering the mechanisms underlying synaptic transmission. For this purpose, the synapsis between a motor nerve and a striated muscle fiber, the neuromuscular junction (NMJ), became one of the most useful models. At the NMJ, muscle-type nicotinic acetylcholine (ACh) receptors (nAChR$_m$s) are key elements localized in the own muscle fiber. Likely, the first pharmacological study involving nAChR$_m$s was carried out in the 19th century by C. Bernard, who discovered the lack of skeletal muscle contraction after nerve stimulation in curarized frogs. Later on, Langley (1905) found that nicotine, an alkaloid present in *Nicotiana tabacum* leaves, induces potent contractions of fowl muscle, which are suppressed by curare; he concluded that "muscles have some accessory substance that combines with nicotine and curare." Then, Dale, Feldberg, and Vogt (1936) showed that stimulation of motoneuron fibers in perfused voluntary muscles induces the appearance of ACh in the venous fluid, even when preventing the muscle contraction by curare. Soon after, Eccles, Katz, and Kuffler (1941) explored the neuromuscular transmission by extracellularly recording end-plate potentials (epps) and tested the effects of eserine, an anticholinesterase, on them. Afterward, Katz's group studied in detail the characteristics of epps by recording muscle fibers with intracellular microelectrodes, establishing the basis for synaptic transmission (Fig. 34.1). Besides, they accurately released ACh by iontophoresis at the NMJ using ACh-filled micropipettes. By this localized ACh application, they could mimic the effect of nerve stimulation on muscle (Del Castillo & Katz, 1955) and mapped the distribution of nAChR$_m$s on the muscle fiber, which lead them to detect junctional and extrajunctional receptors (Miledi, 1960).

A deeper knowledge of structural and functional properties of nAChR$_m$s, at a molecular level, has been possible thanks to several landmarks: (i) Introduction of the patch-clamp technique by Neher and Sakmann (1976), which enables recording the activity of a single nAChR$_m$. (ii) Biochemical purification of the nAChR$_m$ from *Torpedo marmorata* electroplaques by Miledi, Molinoff, and Potter (1971), using radioactive labeled α-bungarotoxin, a component of *Bungarus multicinctus* snake venom that specifically and irreversibly blocks the depolarizing effect of ACh on the NMJ (Lee & Tseng, 1966). Afterward, purified nAChR$_m$s were functionally reconstituted in artificial lipid matrices, allowing detailed studies of its activity either in cell-free models, as lipid bilayers (Nelson, Anholt, Lindstrom, & Montal, 1980) or in host cells, as

Neuroscience of Nicotine
https://doi.org/10.1016/B978-0-12-813035-3.00034-4

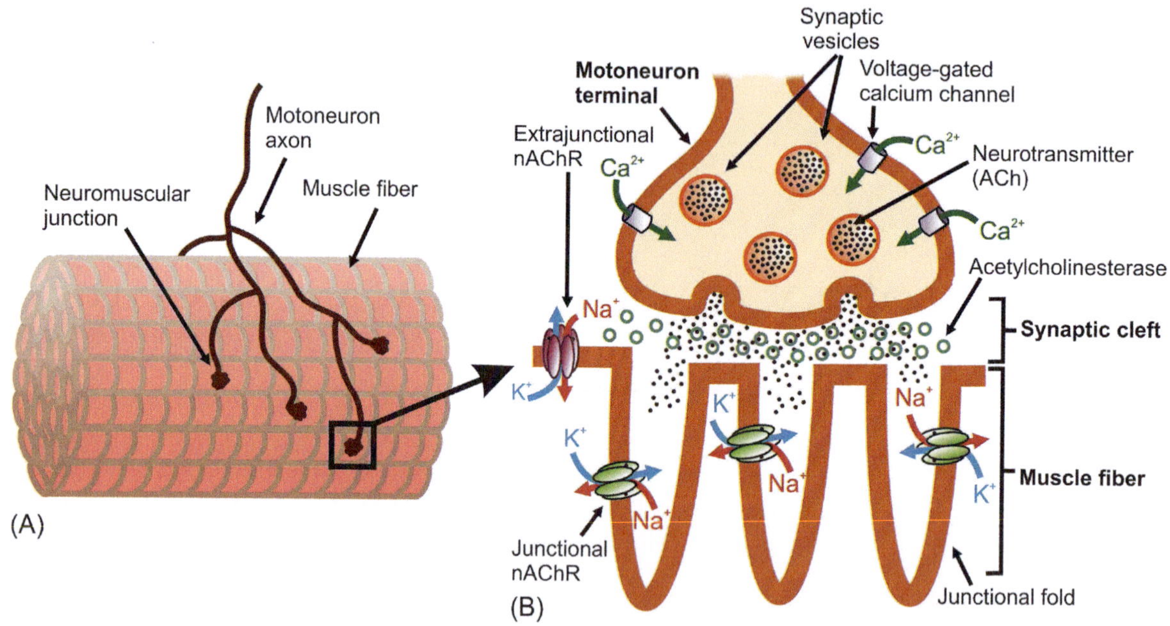

FIG. 34.1 Scheme of synaptic transmission at the neuromuscular junction (NMJ).
(A) A motoneuron axon branches off and ends on the surface of each muscle fiber that it controls, at the NMJ. (B) Expanded view of the NMJ region indicated in (A), showing the presynaptic nerve terminal containing ACh vesicles and voltage-dependent Ca^{2+} channels involved in transmitter release. ACh is released to the synaptic cleft, rich in acetylcholinesterase, and eventually reaches junctional folds containing numerous (up to 20.000/μm^2) junctional $nAChR_m$s. Scarce extrajunctional (fetal subtype) $nAChR_m$s are beyond this synaptic region.

Xenopus oocytes, after their microtransplantation (Morales et al., 1995). (iii) With the progress of molecular biology techniques, it was possible to solve the α-subunit sequence (Noda et al., 1982) and later to clone the cDNA coding for other $nAChR_m$ subunits (reviewed by Changeux & Edelstein, 2005), which paved the way for mutagenesis experiments addressed to unravel the functional role of key structural $nAChR_m$ residues. (iv) High-resolution electron microscopy of *Torpedo* electroplax membranes, in which $nAChR_m$s are densely packed, enabling to develop atomic-scale models of the $nAChR_m$ structure (Unwin, 2005), which have become crucial tools for in silico studies of nAChR function and modulation.

34.2 OVERALL STRUCTURE AND FUNCTION OF $nAChR_m$

The $nAChR_m$ is a ligand-gated ion channel (LGIC) belonging to the Cys-loop subfamily of receptors, which are involved in fast synaptic transmission. All of them share a pentameric structure, four transmembrane (M1–M4) hydrophobic regions (TMD) in each subunit, and a disulfide bond between two cysteine residues, forming a distinctive loop of 13 amino acids at the extracellular side. $nAChR_m$s from adult vertebrates are constituted by $2\alpha_1$, $1\beta_1$, 1δ, and 1ε subunits, being the ε subunit (junctional type) equivalent to the γ subunit from *Torpedo* nAChRs. In embryonic and extrajunctional (Fig. 34.1) $nAChR_m$s, the ε subunit is substituted by a different γ subunit. This heterogeneity of $nAChR_m$ subunit combinations is relevant, because it determines, at least in part, both their functional and pharmacological properties (see below).

Structural models from high-resolution electron microscopy images of *Torpedo* $nAChR_m$ show its five subunits arranged around a central axis, constituting the channel pore, and both extracellular and intracellular domains (ECD and ICD, respectively) protruding from the membrane bilayer (Unwin, 2005; Fig. 34.2A_1 and A_2). The channel pore is lined by the M2 segments of each subunit (channel inner ring), whereas M4 segments constitute the outermost ring, interacting directly with bilayer lipids (see Fig. 34.2A_2). In the ECD, the $nAChR_m$ has two ligand-binding (orthosteric) sites located at α-γ (or α-ε) and α-δ interfaces (Fig. 34.2A_1).

$nAChR_m$s are allosteric proteins that may adopt, at least, three different interconvertible conformational states (reviewed by Albuquerque, Pereira, Alkondon, & Rogers, 2009; Bouzat & Sine, 2017; Changeux, 2012; Fig. 34.2B). In the absence of agonist, the hydrophobic channel gate, located deep in the membrane, is closed, and the $nAChR_m$ is in the resting state. When ACh or other agonists, as carbamylcholine (CCh), bind to the orthosteric sites, the $nAChR_m$ quickly activates (in the microsecond range), and the channel opens, allowing the flux of, mainly, Na^+ and K^+ through the membrane, generating ionic currents (see Fig. 34.2C_1 and C_2). A prolonged exposure to ACh elicits a conformational

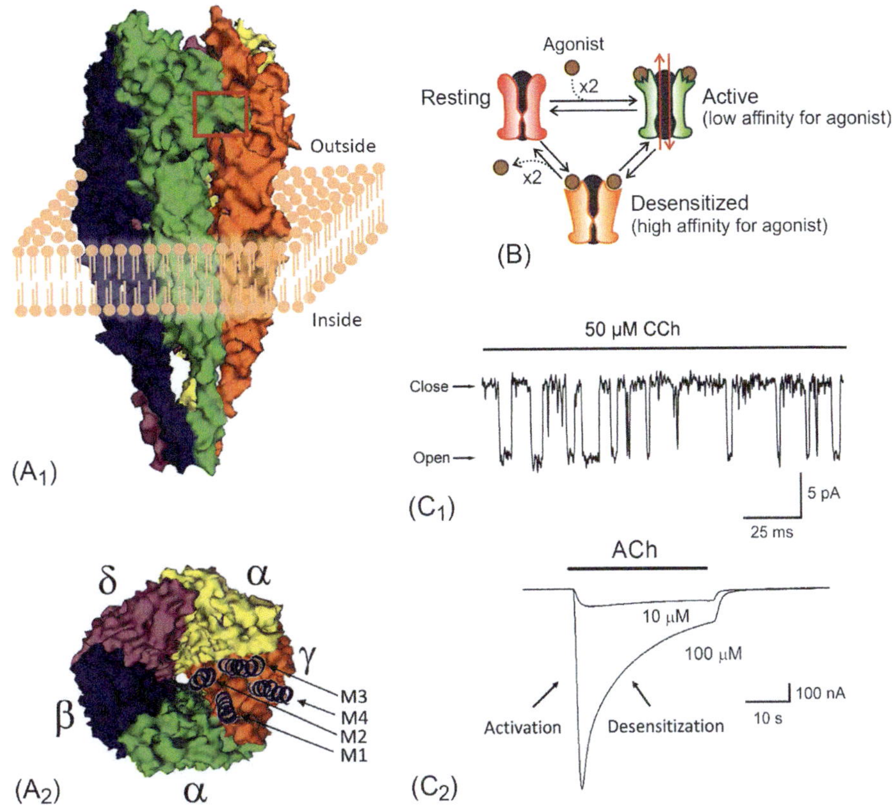

FIG. 34.2 Structure and function of nAChR$_m$s.
(A) Lateral (A_1), at the membrane plane, and top (A_2), from the synaptic cleft, views of the nAChR$_m$, showing its pentameric structure. *Red square* in A_1 shows α-γ orthosteric binding site. Each subunit contains four transmembrane segments (M1–M4, shown as spirals in γ subunit of panel A_2) arranged around the channel pore. (B) nAChR$_m$ can exchange between three functional states: (i) resting, without bound agonist and channel closed; (ii) active, after binding of two agonist molecules the channel opens; and (iii) desensitized, channel closes though agonist remains bound. (C) Functional activity of nAChR$_m$s monitored by recording single-channel (C_1) or whole-cell membrane (C_2) currents. Carbamylcholine (CCh) binding activates the nAChR$_m$, eliciting channel openings (C_1, discrete downward deflections at −80 mV). nAChR$_m$ desensitization is better shown in slower, whole-cell recordings; ACh-elicited current (I_{ACh}) decreases during steady presence of agonist because of desensitization, which increases with agonist concentration (C_2). Bars above recordings indicate the time of agonist application, and downward deflections denote inward currents, here and thereafter. Panels A_1 and A_2 derived from *Torpedo* nAChR$_m$ structure (code 2BG9; Unwin, 2005). Recordings in this and subsequent figures are authors' unpublished results from *Torpedo* nAChR$_m$s.

shift in the nAChR$_m$ called desensitized state (Thesleff, 1955), which is characterized by a nonconducting conformation with an enhanced affinity for the agonist. Most likely, there are several intermediate desensitized states, each one with its own kinetics, but the overall desensitization rate is markedly dependent on agonist concentration, increasing when ACh concentration rises (Fig. 34.2C_2).

34.3 MECHANISMS OF nAChR$_m$ MODULATION

nAChR$_m$s are relevant therapeutic targets, since their dysfunction is related to the genesis of several pathophysiological processes that lead to impaired motor activity, as occurs in diverse congenital myasthenic syndromes (reviewed by Kalamida et al., 2007). Hence, it is important to understand the mechanisms by which these receptors are modulated by different molecules and to unravel their specific binding sites (reviewed by Chatzidaki & Millar, 2015), in order to develop new therapeutic agents.

In the last few decades, it has been shown that a broad number of molecules, with heterogeneous chemical structures, modulate nAChR$_m$ function by their binding to different regions of this receptor and, therefore, acting through distinct mechanisms:

(i) *Competitive*: mediated by molecules that bind into the orthosteric binding site (Fig. 34.3; site 1), interfering with the agonist-receptor interaction (see Table 34.1 for examples). Many of these molecules can also act as full or partial agonists. The pharmacological profile of all these molecules is characterized by the rightward shift of the dose-response curve (Fig. 34.4A_2 and B).

(ii) *Steric:* elicited by molecules that interact with (a) residues at positions close to the orthosteric

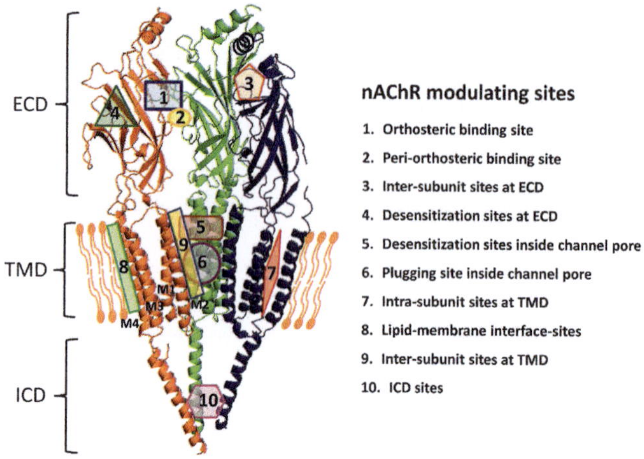

FIG. 34.3 Main modulating sites at the $nAChR_m$. All three $nAChR_m$ domains, extracellular (ECD), transmembrane (TMD), and intracellular (ICD), have relevant loci for its modulation. The main modulating sites are indicated on the $nAChR_m$ structure (same template as in Fig. 34.2A) and summarized on the right.

binding sites in such a way that they hinder/restrict agonist binding to the orthosteric site (Fig. 34.3, site 2; Table 34.1), and (b) residues located into the channel pore when the receptor is in the active state, acting as open-channel blockers (Fig. 34.3, site 6). Steric blockers elicit a noncompetitive $nAChR_m$ inhibition (NCI; Fig. 34.4A_2 and B); nevertheless, open-channel blockers are continuously associating and dissociating from their interacting site, causing a dynamic blockade, which is observed as "flickering" when recording single-channel currents (Neher & Steinbach, 1978). Usually, the open-channel blockade is exerted by charged molecules at physiological pH, as tacrine, edrophonium, or fluoxetine (Table 34.1), and their extent of blockade is largely dependent on the cell membrane potential (Fig. 34.4A_1 and A_2).

(iii) *Allosteric*: molecules that interact at loci distinct from orthosteric sites and modify the functional activity of $nAChR_m$s. They are classified as either negative

TABLE 34.1 Selected Molecules That Modulate $nAChR_m$ Function

Category	Molecules	Action(s)/binding site(s)	Potency	Reference(s)
Muscle relaxants	Succinylcholine	A; (NCI)/1, 5, 6	EC_{50} = 10.8 μM; IC_{50} = 126 μM	Jonsson et al. (2006)
	Pancuronium	CI/1	IC_{50} = 15 nM	Liu and Dilger (2009)
Acetylcholinesterase inhibitors	Tacrine	NAM (NCI)/3, 6	IC_{50} = 1.6–4.6 μM	Prince, Pennington, and Sine (2002)
	Edrophonium	NAM (NCI)/6	IC_{50} = 10 μM	Olivera-Bravo, Ivorra, and Morales (2007)
	Physostigmine	PAM; NAM/2, 3	IC_{50} = 10 mM	Hamouda, Kimm, and Cohen (2013)
	Galanthamine	PAM; NAM/2, 3	IC_{50} = 2.8 mM	Hamouda et al. (2013)
Cations	Ca^{2+}	NAM/4	0.1–1 mM range	Ochoa et al. (1989)
	Zn^{2+}	PAM/3	Tested concentration = 200 μM	García-Colunga, Vázquez-Gómez, and Miledi (2004)
Endogenous molecules	Progesterone	NAM (NCI)/3, 7, 8, 9	IC_{50} = 1.0–6.1 μM	Ke and Lukas (1996)
	Estradiol	NAM (NCI)/3, 7, 8, 9	IC_{50} = 20–56 μM	Ke and Lukas (1996)
	Corticosterone	NAM (NCI)/3, 7, 8, 9	IC_{50} = 30–92 μM	Ke and Lukas (1996)
	Cholesterol	AM/7, 8, 9		Barrantes (2004)
	Substance P	NAM (NCI)/5		Arias (1997)
	Serotonin	NAM (NCI)/6		Arias (1997)
	Fatty acids	NAM (NCI)/8, 9		Barrantes (2004)
	Protein kinase A	NAM/10		Hoffman, Ravindran, and Huganir (1994)
	Protein kinase C	NAM/10		Ochoa et al. (1989)
Antimalarials	Quinacrine	NAM (NCI)/8, 9	10–100 μM	Arias (1997) and Kaldany and Karlin (1983)
	Quinine	NAM (NCI)/3, 6	50 μM	Sieb, Milone, and Engel (1996)

TABLE 34.1 Selected Molecules That Modulate $nAChR_m$ Function—cont'd

Category	Molecules	Action(s)/binding site(s)	Potency	Reference(s)
Antibiotics	Gentamicin	NAM (NCI, CI)/1, 4	$IC_{50}=25$ μM	Amici, Eusebi, and Miledi (2005)
	Penicillins	NAM (NCI, CI)/1, 6	$IC_{50}=0.71$ mM	Schlesinger, Krampfl, Haeseler, Dengler, and Bufler (2004)
Antipsychotics	Chlorpromazine	NAM (NCI)/4, 6	$IC_{50}>300$ nM	Changeux and Edelstein (2005)
	Fluoxetine	NAM (NCI)/3, 5	Tested concentration = 2 μM	García-Colunga, Awad, and Miledi (1997)
	Bupropion	NAM (NCI)/3, 4, 5, 6	$IC_{50}=0.40$–40.1 μM	Arias et al. (2009)
	Imipramine	NAM (NCI)/5	$K_i=0.85$–3.8 μM	Sanghvi et al. (2008)
General anesthetics	Propofol	NAM (NCI)/5, 6, 7, 9	$IC_{50}=40$–125 μM	Jayakar, Dailey, Eckenhoff, and Cohen (2013)
	Isoflurane	NAM (NCI)/5, 6	$K_d=0.36$ mM	Arias and Bhumireddy (2005)
	Ketamine	NAM (NCI, CI)/2, 5, 6	$K_d=2$ μM	Scheller et al. (1996)
	Pentobarbital	NAM (NCI, CI)/1, 2, 6	$K_d=15$–30 μM	Krampfl, Schlesinger, Dengler, and Bufler (2000)
Local anesthetics	Procaine	NAM (NCI)/6, 8	$K_d=690$–790 μM	Arias and Bhumireddy (2005)
	Lidocaine	NAM (NCI)/3, 4, 7, 9, 5, 6	$IC_{50}=70$ μM	Alberola-Die et al. (2011)
	Proadifen	NAM (NCI, CI)/1, 3, 4	$IC_{50}=19$ μM	Spitzmaul, Gumilar, Dilger, and Bouzat (2009)
	Adiphenine	NAM (NCI)/4, 5	$IC_{50}=15$ μM	Spitzmaul et al. (2009)
Toxins	Anatoxin A	A/1	$EC_{50}=50$ nM	Wonnacott and Barik (2007)
	Histrionicotoxins	NAM (NCI)/4, 6	K_i ~0.1–1 μM	Changeux and Edelstein (2005)
	D-Tubocurarine	CI/1	$IC_{50}=50$–100 nM	Arias (1997) and Wonnacott and Barik (2007)
	α-Bungarotoxin	CI/1	Tested concentrations = 0.01–10 nM	Wonnacott and Barik (2007)
	α-Conotoxin	CI/1	$K_d=0.1$–1 nM	Wonnacott and Barik (2007)
	Nicotine	A/1	$ED_{20}=20$ μM	Wonnacott and Barik (2007)
	Epibatidine	A/1	Tested concentrations 2–300 μM	Prince and Sine (1998)
	Strychnine	NCI/3	$IC_{50}=7.3$ μM	García-Colunga and Miledi (1999)
Other amine/ammonium compounds	Mecamylamine	NCI/5, 6	1–10 μM up to 100 μM	Varanda et al. (1985)
	Tetraethylammonium	A; NAM (CI)/1, 3, 6	$IC_{50}=2$–3 mM	Akk and Steinbach (2003)
	Diethylamine	NAM/2, 3, 6, 7	$IC_{50}=70$ μM	Alberola-Die, Fernández-Ballester, González-Ros, Ivorra, and Morales (2016a)
	2,6-Dimethylaniline	NAM/4, 9, 5, 6	$IC_{50}=2.1$ mM	Alberola-Die, Fernández-Ballester, González-Ros, Ivorra, and Morales (2016b)

Some of the most representative modulators of $nAChR_m$ (grouped by their therapeutic action or structure) are listed. Their main effects, putative acting site(s), and potencies, according to the listed references, are indicated.

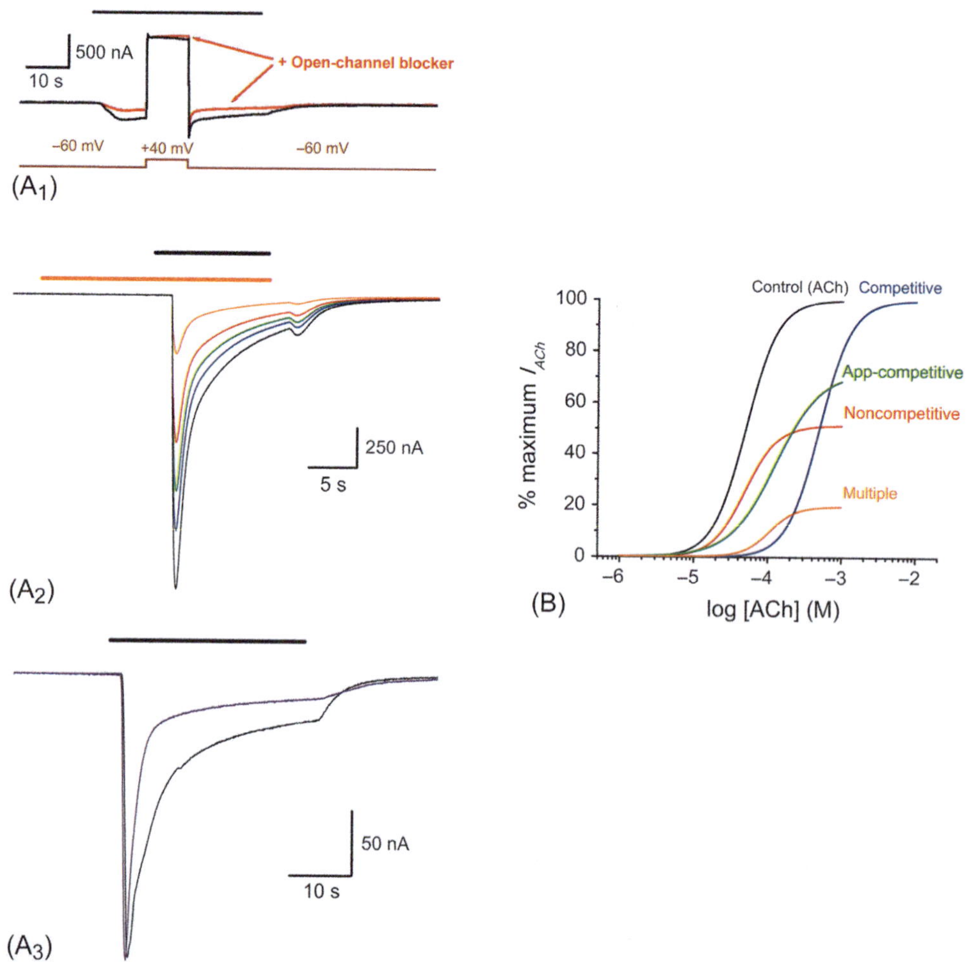

FIG. 34.4 Mechanisms of $nAChR_m$ blockade.
(A) Recordings illustrate different mechanisms of $nAChR_m$ blockade: (A_1) Voltage-dependent (open-channel) blockade, usually elicited by charged molecules bound inside the channel pore. At −60 mV (*brown trace* indicates the imposed voltage potential), the blocker diminishes I_{ACh} (compare *black* and *red recordings*), but blockade vanishes at +40 mV. (A_2) Noncompetitive blockers, at their IC_{50}, reduce control I_{ACh} *(black recording)* to one-half *(red recording)*; competitive *(blue recording)* and apparent-competitive *(green recording)* blockers decrease I_{ACh} depending on ACh concentration; a larger and complex I_{ACh} inhibition is elicited by molecules acting both as closed- and open-channel blockers when they are preapplied and then coapplied with the agonist *(orange recording)*. (A_3) Modifying the rate of $nAChR_m$ desensitization, I_{ACh} desensitization elicited by ACh *(black recording)* is markedly enhanced when ACh is coapplied with the modulator *(purple recording)*. (B) Models of dose-response curves elicited by ACh either alone *(black)* or together with modulating compounds acting by different mechanisms: a noncompetitive way *(red curve)*, for instance, when drugs bind into the channel pore; a competitive manner *(blue curve)* when molecules interact at the orthosteric site; an apparent-competitive mechanism *(green curve)* elicited by molecules acting by open- and closed-channel blockade (see text for details); finally, certain molecules interact with $nAChR_m$s at different sites, causing a large and complex inhibition *(orange curve)*.

allosteric modulators (NAMs) or positive allosteric modulators (PAMs) whether they decrease or enhance $nAChR_m$ function, respectively. These modulators can act by binding to different $nAChR_m$ locations:

(a) *ECD acting modulators*: these molecules bind to residues located at the ECD, and hence, they may act on resting-state $nAChR_m$, triggering a conformational change that can either hinder or promote $nAChR_m$ opening (Fig. 34.3; site 3) or modify its desensitization (Fig. 34.3, site 4; Fig. 34.4A_3; Table 34.1). NAMs acting at these sites do not significantly shift the dose-response curves and typically behave as NCI, eliciting a closed-channel blockade (Fig. 34.4A_2 and B). However, some of them show an "apparent-competitive" pharmacological profile when coapplied with the agonist, slightly rightward shifting the dose-response curves (Fig. 34.4A_2 and B). This is so when the NAM mainly evokes closed-channel blockade, but it also induces open-channel blockade. Thus, the higher the ACh concentration, the more $nAChR_m$s get open, and hence, the fewer remain in the resting state and are prone to the blocker action outside the pore. Accordingly, an increase in ACh concentration will decrease the extent of inhibition, resembling a competitive mechanism of inhibition, as it happens for lidocaine (Alberola-Die, Martinez-Pinna, González-Ros, Ivorra, & Morales, 2011; Table 34.1).

(b) *TMD acting modulators*: they are mostly hydrophobic molecules, which bind to residues located either at intrasubunit cavities among M1–M4 segments (Fig. 34.3, site 7) or at intersubunit crevices (Fig. 34.3, site 9; Table 34.1). Besides, these hydrophobic molecules can diffuse through the membrane and interact with amino acids located at the lipid-protein interface (Barrantes, 2004; Fig. 34.3, site 8). Furthermore, some TMD modulators can modify $nAChR_m$ desensitization by interacting with residues located into the ion channel, at specific positions near the extracellular side (Arias, 2010; Fig. 34.3, site 5).

(c) *ICD acting modulators*: these molecules modify $nAChR_m$ function by binding to intracellular residues (Fig. 34.3, site 10). At least, some of them change the desensitization by acting through an indirect pathway, involving the generation of intracellular signaling molecules that phosphorylate ICD residues at the $nAChR_m$ (Ochoa, Chattopadhyay, & McNamee, 1989).

Owing to the presence of different functional groups in the chemical structure of many ligands, not all the above-mentioned mechanisms of action are mutually exclusive. Actually, a single molecule can bind to the $nAChR_m$ at distinct and even distant sites, giving rise to complex pharmacological profiles, which depend on both the way of application and the concentration of modulator used. Moreover, $nAChR_m$ modulators can attain their binding sites through two different and nonexclusive routes (Hille, 1977; Fig. 34.5): first, a hydrophilic pathway, used by polar ligands to interact with residues located at either the ECD or the channel pore (Fig. 34.5, red arrows), and second, a hydrophobic way, followed by nonpolar and lipophilic molecules to bind $nAChR_m$ residues at different TMD regions, including lipid-protein interface, and the ICD (Fig. 34.5, blue arrows). The path followed by each molecule to act on $nAChR_m$s mainly depends on the pK_a of the modulator and the pH at the receptor environment, since both determine the molecule protonation. Many amphipathic molecules, as the local anesthetic lidocaine, are partially protonated at physiological pH, and so, they can follow both pathways to reach their binding sites.

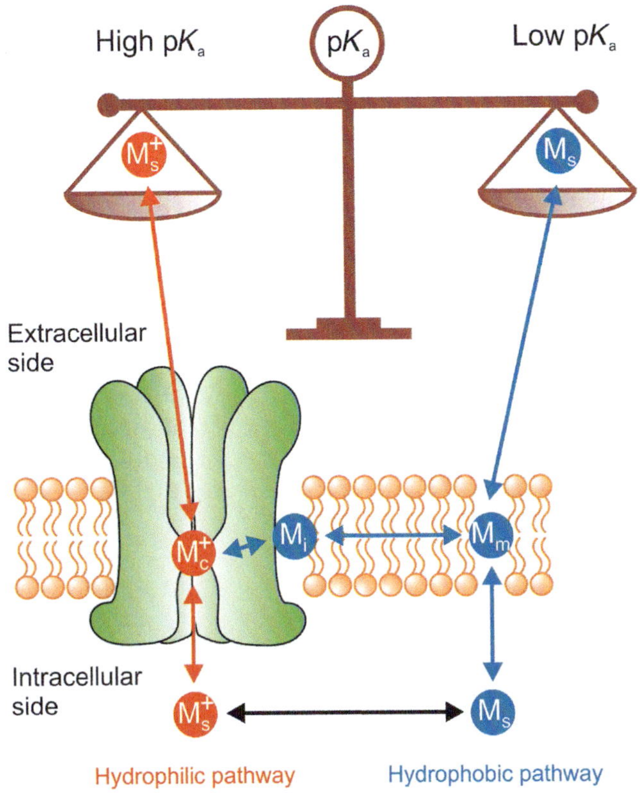

FIG. 34.5 Scheme of hydrophilic and hydrophobic pathways of drug action.
Many molecules modulating $nAChR_m$s are amphipathic; thus, charged and uncharged forms are present in the medium. Charged (protonated) molecules mostly reach their binding sites by the hydrophilic pathway *(red arrows)*, whereas uncharged forms mainly follow the hydrophobic pathway *(blue arrows)*. The proportion of each form is dependent on the molecule pK_a and the medium pH; in the scheme, the balance "senses" the molecule pK_a, rising the percentage of charged molecules as pK_a increases.

34.4 DIFFERENT THERAPEUTIC DRUGS MODULATE $nAChR_m$ FUNCTION

Regarding functional modulation of $nAChR_m$s, three characteristics of the NMJ should be considered: (i) The presence of two types of $nAChR_m$s differentially expressed in fetal and adult life. Junctional and extrajunctional $nAChR_m$s differ in both functional and pharmacological properties. Thus, the junctional subtype shows larger channel conductance and shorter mean-open time (Katz & Miledi, 1972; Neher & Sakmann, 1976) and, furthermore, faster desensitization (Morales & Sumikawa, 1992) than extrajunctional ones. Moreover, extrajunctional $nAChR_m$s are selectively blocked by αA-conotoxin OIVB, purified from *Conus obscurus* venom, since its affinity for the junctional subtype is almost two thousand times lower (Teichert et al., 2005). (ii) The large amplitude of epps, which confers a very high safety factor for signal transmission and makes the NMJ a singular synapsis. (iii) $nAChR_m$s are directly exposed to circulating compounds (including toxins), in contrast to neuronal nAChRs, which are partially protected by either blood-brain or blood-nerve barriers. Notably, a large number of heterogeneous molecules, many amphipathic, interact with $nAChR_m$ modulating its function. Among them (Table 34.1), there are endogenous molecules (including hormones) and numerous compounds commonly used in clinical practice, as muscle relaxants, acetylcholinesterase inhibitors, antibiotics, antimalarials, antipsychotics, and local and general anesthetics, many of them acting in the micromolar range (Table 34.1). Besides, plenty of toxins, amine/ammonium compounds, and other molecules have powerful effects as either NAM or PAM on $nAChR_m$s. Even more, the action of certain molecules on $nAChR_m$s is markedly concentration-dependent, most likely because they act upon several modulating sites with different affinities and, therefore, they can elicit diverse, even antagonistic, effects.

34.5 FUTURE PERSPECTIVES

Further studies on allosteric modulation of $nAChR_m$s are still required to better understand the precise mechanisms by which modulators act on $nAChR_m$s. This knowledge is relevant to reduce/prevent certain side effects triggered by different therapeutic molecules. For instance, dexamethasone is widely used to ameliorate patients suffering myasthenia gravis because of its immunosuppressive action; nevertheless, at first, this synthetic glucocorticoid might inhibit nAChRs, leading to worsening patient symptoms. Furthermore, these studies are fundamental to establish the bases to develop new therapeutic molecules, more proficient and with fewer side effects, to treat pathophysiological processes related with $nAChR_m$s dysfunctions.

MINI-DICTIONARY OF TERMS

Acetylcholine activated current (I_{ACh}) Ionic membrane current elicited by ACh, acting on $nAChR_m$s.

Acetylcholinesterase Esterase located at the synaptic cleft that cleaves ACh.

Allosteric modulator Molecule that binds outside the $nAChR_m$ orthosteric sites and modifies its function. They are classified as negative or positive allosteric modulators, depending if they decrease or enhance $nAChR_m$ performance, respectively.

Closed-channel blocker Prevents the receptor from being activated by the agonist.

Desensitization Reduction of $nAChR_m$ response when steadily exposed to agonist.

Ligand-gated ion channel (LGIC) receptor Membrane ion channel activated by specific ligands, which is involved in fast synaptic transmission.

Muscular-type nicotinic receptor ($nAChR_m$) Heterologous pentameric ($2\alpha_1$, $1\beta_1$, 1ε, and 1δ subunits) membrane protein activated by ACh, present in adult neuromuscular junction (synaptic-type) or in fetal, extrasynaptic, or denervated muscle (composed by γ instead of ε subunit).

Myasthenia Muscle weakness that can be elicited by different $nAChR_m$ dysfunctions.

Open-channel blocker Molecule that binds into the pore, plugging the open channel.

Orthosteric binding site Loci where the agonist specifically binds to activate the receptor.

Key Facts of Desensitization

- First described by Thesleff (1955) for $nAChR_m$.
- When steadily exposed to the agonist, $nAChR_m$ switches to a high-affinity ligand-binding state, which is nonconducting (closed state).
- ACh and some partial agonists evoke $nAChR_m$ desensitization, this increasing with agonist concentration.
- Not all LGICs desensitize. The rate of nAChR desensitization is largely dependent on the subunit composition.
- There are, at least, two desensitization states of the $nAChR_m$, with their own kinetic constants.
- $nAChR_m$ recovery from desensitization might last seconds or even minutes after agonist withdrawal.
- $nAChR_m$ modulators affect desensitization rates by acting at extracellular, transmembrane, or intracellular loci.
- Some negative allosteric modulators increase $nAChR_m$ desensitization, whereas certain positive allosteric modulators decrease or even prevent it.
- Some congenital myasthenic syndromes modify $nAChR_m$ desensitization.

Key Facts of $nAChR_m$ Disorders

- Most $nAChR_m$ dysfunctions cause myasthenia, that is, muscle weakness.
- Some toxins (α-bungarotoxin and α-conotoxin) or alkaloids (D-tubocurarine) act on $nAChR_m$ leading to severe myasthenia or even paralysis.
- Myasthenia might be triggered by genetic alterations in one or several $nAChR_m$ subunits (often causing either slow- or fast-channel syndrome) or by autoimmune reduction in either the number of postsynaptic $nAChR_m$s (myasthenia gravis) or the amount of ACh released (Lambert-Eaton syndrome).
- Muscle weakness associated with slow-channel syndrome (SCS) is caused by prolonged postsynaptic depolarization, elicited by delayed channel closure, decreased desensitization, or enhanced affinity for ACh.
- Certain open-channel blockers of $nAChR_m$ (fluoxetine and quinidine) ameliorate myasthenia due to SCS, whereas cholinesterase inhibitors exacerbate it (Rodríguez Cruz, Palace, & Beeson, 2014).

- Patients suffering myasthenia by fast-channel syndrome (FCS) have reduced end-plate potentials because of subunit mutations eliciting a slow channel opening, reduced open-channel probability, short open dwell-time, enhanced desensitization, or decreased ACh-binding affinity.
- Anticholinesterase drugs (pyridostigmine) attenuate both FCS and myasthenia gravis symptoms.
- The K^+-channel blocker 3,4-diaminopyridine is useful for the treatment of myasthenia elicited by both FCS and Lambert-Eaton syndrome.

Summary Points

- Two subtypes of muscle nicotinic acetylcholine receptors ($nAChR_m$s) are differentially expressed throughout the life.
- $nAChR_m$s are key elements for synaptic transmission at the neuromuscular junction.
- A broad number of compounds, including endogenous molecules and widely used therapeutic drugs, interact with $nAChR_m$s.
- $nAChR_m$s can be modulated by competitive, steric, and allosteric mechanisms.
- A single molecule might modulate $nAChR_m$s by different mechanisms and by its binding to different loci.
- Allosteric modulators can decrease or enhance $nAChR_m$ activity.
- A better knowledge of $nAChR_m$ modulation will help to develop new therapeutic molecules addressed to treat diseases by $nAChR_m$ dysfunctions.

Acknowledgment

R.C. held a predoctoral fellowship from Universidad de Alicante (FPUUA36).

References

Akk, G., & Steinbach, J. H. (2003). Activation and block of mouse muscle-type nicotinic receptors by tetraethylammonium. *The Journal of Physiology*, *551*(1), 155–168. https://doi.org/10.1113/jphysiol.2003.043885.

Alberola-Die, A., Fernández-Ballester, G., González-Ros, J. M., Ivorra, I., & Morales, A. (2016a). Muscle-type nicotinic receptor blockade by diethylamine, the hydrophilic moiety of lidocaine. *Frontiers in Molecular Neuroscience*, *9*, 12. https://doi.org/10.3389/fnmol.2016.00012.

Alberola-Die, A., Fernández-Ballester, G., González-Ros, J. M., Ivorra, I., & Morales, A. (2016b). Muscle-type nicotinic receptor modulation by 2,6-dimethylaniline, a molecule resembling the hydrophobic moiety of lidocaine. *Frontiers in Molecular Neuroscience*, *9*, 127. https://doi.org/10.3389/fnmol.2016.00127.

Alberola-Die, A., Martinez-Pinna, J., González-Ros, J. M., Ivorra, I., & Morales, A. (2011). Multiple inhibitory actions of lidocaine on Torpedo nicotinic acetylcholine receptors transplanted to *Xenopus* oocytes. *Journal of Neurochemistry*, *117*(6), 1009–1019. https://doi.org/10.1111/j.1471-4159.2011.07271.x.

Albuquerque, E. X., Pereira, E. F., Alkondon, M., & Rogers, S. W. (2009). Mammalian nicotinic acetylcholine receptors: from structure to function. *Physiological Reviews*, *89*(1), 73–120. https://doi.org/10.1152/physrev.00015.2008.

Amici, M., Eusebi, F., & Miledi, R. (2005). Effects of the antibiotic gentamicin on nicotinic acetylcholine receptors. *Neuropharmacology*, *49*(5), 627–637. https://doi.org/10.1016/j.neuropharm.2005.04.015.

Arias, H. R. (1997). Topology of ligand binding sites on the nicotinic acetylcholine receptor. *Brain Research Reviews*, *25*(2), 133–191. https://doi.org/10.1016/S0165-0173(97)00020-9.

Arias, H. R. (2010). Positive and negative modulation of nicotinic receptors. *Advances in Protein Chemistry and Structural Biology*, *80*, 153–203. https://doi.org/10.1016/B978-0-12-381264-3.00005-9.

Arias, H. R., & Bhumireddy, P. (2005). Anesthetics as chemical tools to study the structure and function of nicotinic acetylcholine receptors. *Current Protein & Peptide Science*, *6*(5), 451–472. https://doi.org/10.2174/138920305774329331.

Arias, H. R., Gumilar, F., Rosenberg, A., Targowska-Duda, K. M., Feuerbach, D., Jozwiak, K., et al. (2009). Interaction of bupropion with muscle-type nicotinic acetylcholine receptors in different conformational states. *Biochemistry*, *48*(21), 4506–4518. https://doi.org/10.1021/bi802206k.

Barrantes, F. J. (2004). Structural basis for lipid modulation of nicotinic acetylcholine receptor function. *Brain Research Reviews*, *47*(1–3), 71–95. https://doi.org/10.1016/j.brainresrev.2004.06.008.

Bouzat, C., & Sine, S. M. (2017). Nicotinic acetylcholine receptors at the single-channel level. *British Journal of Pharmacology*. https://doi.org/10.1111/bph.13770.

Changeux, J. P. (2012). The nicotinic acetylcholine receptor: the founding father of the pentameric ligand-gated ion channel superfamily. *The Journal of Biological Chemistry*, *287*(48), 40207–40215. https://doi.org/10.1074/jbc.R112.407668.

Changeux, J. P., & Edelstein, S. J. (2005). *Nicotinic acetylcholine receptors: From molecular biology to cognition*. New York: Odile Jacob Publishing Corporation.

Chatzidaki, A., & Millar, N. S. (2015). Allosteric modulation of nicotinic acetylcholine receptors. *Biochemical Pharmacology*, *97*(4), 408–417. https://doi.org/10.1016/j.bcp.2015.07.028.

Dale, H. H., Feldberg, W., & Vogt, M. (1936). Release of acetylcholine at voluntary motor nerve endings. *The Journal of Physiology*. *86*, https://doi.org/10.1113/jphysiol.1936.sp003371.

Del Castillo, J., & Katz, B. (1955). On the localization of acetylcholine receptors. *The Journal of Physiology*, *128*(1), 157–181. https://doi.org/10.1113/jphysiol.1955.sp005297.

Eccles, J. C., Katz, B., & Kuffler, S. W. (1941). Nature of the "end-plate potential" in curarized muscle. *Journal of Neurophysiology*, *4*(5), 362–387.

García-Colunga, J., Awad, J. N., & Miledi, R. (1997). Blockage of muscle and neuronal nicotinic acetylcholine receptors by fluoxetine (Prozac). *Proceedings of the National Academy of Sciences of the United States of America*, *94*(5), 2041–2044.

García-Colunga, J., & Miledi, R. (1999). Modulation of nicotinic acetylcholine receptors by strychnine. *Proceedings of the National Academy of Sciences of the United States of America*, *96*(7), 4113–4118. https://doi.org/10.1073/pnas.96.7.4113.

García-Colunga, J., Vázquez-Gómez, E., & Miledi, R. (2004). Combined actions of zinc and fluoxetine on nicotinic acetylcholine receptors. *The Pharmacogenomics Journal*, *4*(6), 388–393. https://doi.org/10.1038/sj.tpj.6500275.

Hamouda, A. K., Kimm, T., & Cohen, J. B. (2013). Physostigmine and galanthamine bind in the presence of agonist at the canonical and non-canonical subunit interfaces of a nicotinic acetylcholine receptor. *The Journal of Neuroscience*, *33*(2), 485–494. https://doi.org/10.1523/JNEUROSCI.3483-12.2013.

Hille, H. R. (1977). Local anesthetics: hydrophilic and hydrophobic pathways for the drug-receptor reaction. *The Journal of General Physiology*, *69*(4), 497–515. https://doi.org/10.1085/jgp.69.4.497.

Hoffman, P. W., Ravindran, A., & Huganir, R. L. (1994). Role of phosphorylation in desensitization of acetylcholine receptors expressed in *Xenopus* oocytes. *The Journal of Neuroscience*, *14*(7), 4185–4195.

Jayakar, S. S., Dailey, W. P., Eckenhoff, R. G., & Cohen, J. B. (2013). Identification of propofol binding sites in a nicotinic acetylcholine receptor with a photoreactive propofol analog. *The Journal of Biological Chemistry*, *288*(9), 6178–6189. https://doi.org/10.1074/jbc.M112.435909.

Jonsson, M., Dabrowski, M., Gurley, D. A., Larsson, O., Johnson, E. C., Fredholm, B. B., et al. (2006). Activation and inhibition of human muscular and neuronal nicotinic acetylcholine receptors by succinylcholine. *Anesthesiology*, *104*, 724–733.

Kalamida, D., Poulas, K., Avramopoulou, V., Fostieri, E., Lagoumintzis, G., Lazaridis, K., et al. (2007). Muscle and neuronal nicotinic acetylcholine receptors. Structure, function and pathogenicity. *The FEBS Journal*, *274*(15), 3799–3845. https://doi.org/10.1111/j.1742-4658.2007.05935.x.

Kaldany, R. R., & Karlin, A. (1983). Reaction of quinacrine mustard with the acetylcholine receptor from *Torpedo californica*. *The Journal of Biological Chemistry*, *258*(10), 6232–6242.

Katz, B., & Miledi, R. (1972). The statistical nature of the acetylcholine potential and its molecular components. *The Journal of Physiology*, *224*(3), 665–699. https://doi.org/10.1113/jphysiol.1972.sp009918.

Ke, L., & Lukas, R. J. (1996). Effects of steroid exposure on ligand binding and functional activities of diverse nicotinic acetylcholine receptor subtypes. *Journal of Neurochemistry*, *67*(3), 1100–1112. https://doi.org/10.1046/j.1471-4159.1996.67031100.x.

Krampfl, K., Schlesinger, F., Dengler, R., & Bufler, J. (2000). Pentobarbital has curare-like effects on adult-type nicotinic acetylcholine receptor channel currents. *Anesthesia and Analgesia*, *90*(4), 970–974.

Langley, J. N. (1905). On the reaction of cells and of nerve-endings to certain poisons, chiefly as regards the reaction of striated muscle to nicotine and to curari. *The Journal of Physiology*, *33*(4–5), 374–413. https://doi.org/10.1113/jphysiol.1905.sp001128.

Lee, C. Y., & Tseng, L. F. (1966). Distribution of *Bungarus multicinctus* venom following envenomation. *Toxicon*, *3*(4), 281–290. https://doi.org/10.1016/0041-0101(66)90076-6.

Liu, M., & Dilger, J. P. (2009). Site selectivity of competitive antagonists for the mouse adult muscle nicotinic acetylcholine receptor. *Molecular Pharmacology*, *75*(1), 166–173. https://doi.org/10.1124/mol.108.051060.

Miledi, R. (1960). Junctional and extra-junctional acetylcholine receptors in skeletal muscle fibres. *The Journal of Physiology*, *151*(1), 24–30. https://doi.org/10.1113/jphysiol.1960.sp006417.

Miledi, R., Molinoff, P., & Potter, L. T. (1971). Isolation of the cholinergic receptor protein of *Torpedo* electric tissue. *Nature*, *229*(5286), 554–557. https://doi.org/10.1038/229554a0.

Morales, A., Aleu, J., Ivorra, I., Ferragut, J. A., González-Ros, J. M., & Miledi, R. (1995). Incorporation of reconstituted acetylcholine receptors from *Torpedo* into the *Xenopus* oocyte membrane. *Proceedings of the National Academy of Sciences of the United States of America*, *92*(18), 8468–8472.

Morales, A., & Sumikawa, K. (1992). Desensitization of junctional and extrajunctional nicotinic ACh receptors expressed in *Xenopus* oocytes. *Brain Research Molecular Brain Research*, *16*(3–4), 323–329. https://doi.org/10.1016/0169-328X(92)90242-4.

Neher, E., & Sakmann, B. (1976). Noise analysis of drug induced voltage clamp currents in denervated frog muscle fibres. *The Journal of Physiology*, *258*(3), 705–729. https://doi.org/10.1113/jphysiol.1976.sp011442.

Neher, E., & Steinbach, J. H. (1978). Local anaesthetics transiently block currents through single acetylcholine-receptor channels. *The Journal of Physiology*, *277*, 153–176. https://doi.org/10.1113/jphysiol.1978.sp012267.

Nelson, N., Anholt, R., Lindstrom, J., & Montal, M. (1980). Reconstitution of purified acetylcholine receptors with functional ion channels in planar lipid bilayers. *Proceedings of the National Academy of Sciences of the United States of America*, *77*(5), 3057–3061.

Noda, M., Takahasi, H., Tanabe, T., Toyosato, M., Furutani, Y., Hirose, T., et al. (1982). Primary structure of α-subunit precursor of *Torpedo californica* acetylcholine receptor deduced from cDNA sequence. *Nature*, *299*(5886), 793–797. https://doi.org/10.1038/299793a0.

Ochoa, E. L., Chattopadhyay, A., & McNamee, M. G. (1989). Desensitization of the nicotinic acetylcholine receptor. Molecular mechanisms and effect of modulators. *Cellular and Molecular Neurobiology*, *9*(2), 141–178. https://doi.org/10.1007/BF00713026.

Olivera-Bravo, S., Ivorra, I., & Morales, A. (2007). Diverse inhibitory actions of quaternary ammonium cholinesterase inhibitors on *Torpedo* nicotinic ACh receptors transplanted to *Xenopus* oocytes. *British Journal of Pharmacology*, *151*(8), 1280–1292. https://doi.org/10.1038/sj.bjp.0707329.

Prince, R. J., Pennington, R. A., & Sine, S. M. (2002). Mechanism of tacrine block at adult human muscle nicotinic acetylcholine receptors. *The Journal of General Physiology*, *120*(3), 369–393. https://doi.org/10.1085/jgp.20028583.

Prince, R. J., & Sine, S. M. (1998). Epibatidine activates muscle acetylcholine receptors with unique site selectivity. *Biophysical Journal*, *75*(4), 1817–1827. https://doi.org/10.1016/S0006-3495(98)77623-4.

Rodríguez Cruz, P. M., Palace, J., & Beeson, D. J. (2014). Inherited disorders of the neuromuscular junction: an update. *Journal of Neurology*, *261*(11), 2234–2243. https://doi.org/10.1007/s00415-014-7520-7.

Sanghvi, M., Hamouda, A. K., Jozwiak, K., Blanton, M. P., Trudell, J. R., & Arias, H. R. (2008). Identifying the binding site(s) for antidepressants on the Torpedo nicotinic acetylcholine receptor: [3H]2-azidoimipramine photolabeling and molecular dynamics studies. *Biochimica et Biophysica Acta*, *1778*(12), 2690–2699. https://doi.org/10.1016/j.bbamem.2008.08.019.

Scheller, M., Bufler, J., Hertle, I., Schneck, H. J., Franke, C., & Kochs, E. (1996). Ketamine blocks currents through mammalian nicotinic acetylcholine receptor channels by interaction with both the open and the closed state. *Anesthesia and Analgesia*, *83*(4), 830–836.

Schlesinger, F., Krampfl, K., Haeseler, G., Dengler, R., & Bufler, J. (2004). Competitive and open channel block of recombinant nAChR channels by different antibiotics. *Neuromuscular Disorders*, *14*(5), 307–312. https://doi.org/10.1016/j.nmd.2004.01.005.

Sieb, J. P., Milone, M., & Engel, A. G. (1996). Effects of the quinoline derivatives quinine, quinidine, and chloroquine on neuromuscular transmission. *Brain Research*, *712*(2), 179–189. https://doi.org/10.1016/0006-8993(95)01349-0.

Spitzmaul, G., Gumilar, F., Dilger, J. P., & Bouzat, C. (2009). The local anaesthetics proadifen and adiphenine inhibit nicotinic receptors by different molecular mechanisms. *British Journal of Pharmacology*, *157*(5), 804–817. https://doi.org/10.1111/j.1476-5381.2009.00214.x.

Teichert, R. W., Rivier, J., Torres, J., Dykert, J., Miller, C., & Olivera, B. M. (2005). A uniquely selective inhibitor of the mammalian fetal neuromuscular nicotinic acetylcholine receptor. *The Journal of Neuroscience*, *25*(3), 732–736. https://doi.org/10.1523/JNEUROSCI.4065-04.2005.

Thesleff, S. (1955). The mode of neuromuscular block caused by acetylcholine, nicotine, decamethonium and succinylcholine. *Acta Physiologica Scandinavica*, *34*(2–3), 218–231. https://doi.org/10.1111/j.1748-1716.1955.tb01242.x.

Unwin, N. (2005). Refined structure of the nicotinic acetylcholine receptor at 4 Å resolution. *Journal of Molecular Biology*, *346*(4), 967–989. https://doi.org/10.1016/j.jmb.2004.12.031.

Varanda, W. A., Aracava, Y., Sherby, S. M., VanMeter, W. G., Eldefrawi, M. E., & Albuquerque, E. X. (1985). The acetylcholine receptor of the neuromuscular junction recognizes mecamylamine as a noncompetitive antagonist. *Molecular Pharmacology*, *28*(2), 128–137.

Wonnacott, S., & Barik, J. (2007). Nicotinic ACh receptors. *Tocris Reviews*, *28*, Tocris Cookson.

C H A P T E R

35

Involvement of Opioid Receptors in Nicotine-Related Reinforcement and Pleasure

Ari P. Kirshenbaum

Department of Psychology, Neuroscience Program, Saint Michael's College, Colchester, VT, United States

Abbreviations

ACCx	anterior cingulate cortex
AL	anterior lobe, pituitary
AON	anterior olfactory nucleus
ARC	arcuate nucleus, hypothalamus
BLA	basolateral nucleus, amygdala
BNST	bed nucleus of the stria terminalis
CeA	central nucleus, amygdala
CPP	conditioned place preference
CPu	caudate putamen
DMH	dorsal-medial hypothalamus
DMR	dorsal-medial raphe nucleus
DOR	δ-opioid receptor
FR	fixed ratio
Fr, Pr, T, and OCx	frontal, parietal, temporal, and occipital cortex
ICSS	intracranial self-stimulation
KOR	κ-opioid receptor
MOR	μ-opioid receptor
NAc	nucleus accumbens
nAChR	nicotinic acetylcholine receptor
NRT	nicotine replacement treatment
OR	opioid receptor
pCREB	phosphorylated response element-binding protein
PET	positron emission tomography
PFC	prefrontal cortex
PiR	piriform cortex
POA	preoptic area
PR	progressive ratio
PVN	paraventricular hypothalamus
RM	raphe magnus
RN	red nucleus
S_D	discriminative stimulus
SON	supraoptic tubercle
VL, VM	ventral lateral and medial thalamus
VTA	ventral tegmentum

35.1 INTRODUCTION

Drug experimentation and subsequent self-administration can be assumed to be reinforced by the effects of those substances on basic satisfaction that is induced by reward. Opioid systems lie at the foundation of hedonic sensation (Castro & Berridge, 2014), so assuming a ubiquitous role of opioid systems in all drug dependencies might stand to reason, and this hypothesis has received considerable support (Le Merrer, Becker, Befort, & Kieffer, 2009; Shippenberg, LeFevour, & Chefer, 2008; Trigo, Martin-García, Berrendero, Robledo, & Maldonado, 2010; for reviews). The principal subject addressed in this chapter is the degree to which ORs participate in the experience of pleasure produced by nicotine. Related to this topic is how the ORs, engaged by nicotine, alter motivational processes that are implicated in reinforcement.

The exploration of the initial effects of nicotine on reward processes must be included as a means toward understanding the neurobiology of tobacco dependence. Positive reinforcement and satisfaction are related but not synonymous, and this distinction is elegantly evidenced by the data on nicotine-related OR involvement. In short, the pleasure incited by nicotine via OR activation falls short of characterizing the abuse liability of tobacco products. The manner in which nicotine-activated ORs alter general reward-related learning and how habitual tobacco use causes a dysregulation of the opioid system to create withdrawal (Berrendero, Robledo, Trigo, Martín-García, & Maldonado, 2010; Pomerleau, 1998; Watkins, Koob, & Markou, 2000) bring us closer to a fuller characterization. For this short review, the neuroadaptive modifications resulting from chronic nicotine exposure (i.e., tobacco dependence) are not examined, nor is OR involvement in withdrawal; these are comprehensively reviewed elsewhere (e.g., Norman & D'Souza, 2017).

Important reviews (Berrendero et al., 2010; Hadjiconstantinou & Neff, 2011) indicate that nicotine and ORs are integrally connected in the cellular pathways

Neuroscience of Nicotine
https://doi.org/10.1016/B978-0-12-813035-3.00035-6

TABLE 35.1 Endogenous Opioid Systems Related to Nicotine

Receptors	MOR	DOR	KOR
Ligands/ peptides with highest affinity	β-Endorphins	Enkephalins	Dynorphins
Precursors	Opiomelanocortin	Proenkephalin	Prodynorphin
Locations	Striatum, NAc, hypothalamus, PFC	Striatum, CPu, hippocampus, NAc	Striatum, CPu, NAc
Intervening systems	Dopamine, corticotropin-releasing factor	Dopamine and glutamate	Dopamine and glutamate

The information contained in Table 35.1 is a synopsis of Berrendero et al. (2010, 2012), Hadjiconstantinou and Neff (2011), Norman and D'Souza (2017), Pomerleau (1998), and Trigo et al. (2010).

that are implicated in reinforcement; many of the pathways rich in ORs have connections to the mesocortical dopaminergic pathway (e.g., Tanda & Di Chiara, 1998) implicated in motivation and affect (Shippenberg et al., 2008). Evidence suggests that μ-opioid receptors (MORs) are principal actors in the response to nicotine; the precise role of δ-opioid receptor (DOR) activation in the reward-related response to nicotine varies depending upon the experimental paradigm. The involvement of κ-opioid receptors (KORs) is peripheral to the reward-related effects of nicotine, but involvement of KORs may be most pertinent to withdrawal. The activation of ORs is related to the release of peptides resulting from nicotine exposure; see Table 35.1 and Fig. 35.1.

35.2 THE HEART OF THE MATTER: NICOTINE INTEROCEPTION

Interoceptive stimuli (Bevins & Besheer, 2014) produced by nicotine may be related to OR activation, and hedonic interoception may be described by drug *liking* rather than drug *wanting* (Castro & Berridge, 2014). In terms of the feel of nicotine, Duke, Johnson, Reissig, and Griffiths (2015) found that nonsmokers given oral nicotine in a double-blind fashion reported positive experiences of pleasure. Additionally, smoking increased the appraisal of happiness and enjoyment of sensory experiences and enhanced responding for various nonnicotine rewards (Perkins, Karelitz, & Michael, 2015; Rukstalis et al., 2005).

Nontreatment, neuropharmacological investigations with smokers show that, on many occasions, antagonism of ORs diminishes the desire to smoke, the pleasure obtained from smoking, and the choice for nicotine-containing versus denicotinized cigarettes; see Table 35.2. King and Meyer (2000) found that naltrexone reduced the desire to smoke and the pleasure of smoking, albeit OR blockade also produced mild unpleasant sensations. Rukstalis et al. (2005) found that naltrexone reduced the choice for nicotine-containing versus denicotinized cigarettes, attenuated the relief from craving that was attributable to nicotine delivery, and reduced the degree of pleasure induced by a nicotine-containing cigarette, though this latter finding was beneath the standard threshold for statistical significance. OR blockade has been shown to evoke withdrawal symptoms and tobacco craving (Krishnan-Sarin, Rosen, & O'Malley, 1999) and interact with external stressors to increase both cigarette desire and negative affect created by stress (Hutchison, Collins, Tassey, & Rosenberg, 1996). These latter findings suggest that ORs participate when nicotine serves as a negative reinforcer (i.e., escape from stress). Important to note is that the results of OR antagonists are not universally affirmative; some studies demonstrate no effect (Table 35.2). Also, a caveat to any findings of nonselective OR antagonists (naltrexone or naloxone) is that they have been shown to affect nAChRs in vitro (Almeida et al., 2000; Tomé, Izaguirre, Rosário, Ceña, & González-García, 2001), so it is plausible that their influence on hedonic responses to nicotine is accomplished by receptor interactions peripheral to ORs.

Tomographic analysis in humans has identified structural locality of nicotine-related MORs; see Table 35.3. For instance, Kuwabara et al. (2014) exposed smokers and nonsmokers to nicotine-containing versus denicotinized cigarettes and examined regional activation of MORs using a radioligand and PET imaging. Nicotine-related pleasure was correlated with left rostral FrCx activation but only for smokers; possibly, this may have been due to lower nicotine absorption in nonsmokers. Negative affect following overnight abstinence has been associated with greater MOR activation in the amygdala and ACCx; interestingly though, smoking does not result in negative affect relief (Falcone et al., 2012). Ray et al. (2011) found that in smokers, G-allele carriers for *OPRM1* differed with regard to MOR binding compared to A-allele homozygous participants; nicotine delivery resulted in greater MOR binding in the G-carriers only, and this change was correlated with reward-related pleasure. Differences in MOR binding were not apparent in nonsmokers regardless of their genetic profile, suggesting that chronic exposure to nicotine may interact with *OPRM1* genotype to produce differential expression of MORs.

35.3 PRECLINICAL EVIDENCE: NICOTINE INTEROCEPTION

Nonhuman investigations regarding nicotine interoception cluster into two paradigms, these being conditioned place preference (CPP) and drug discrimination.

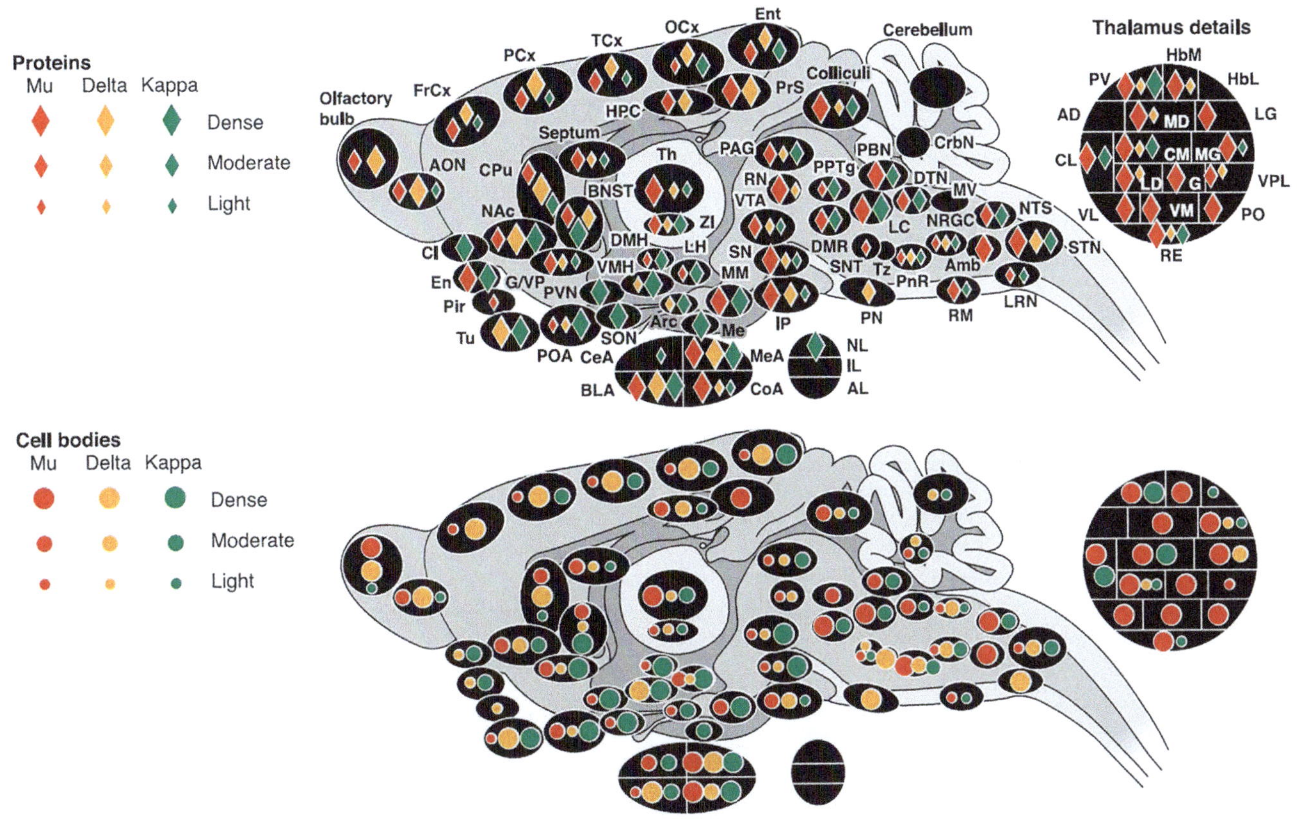

FIG. 35.1 Distribution of opioid receptors in the rat brain.
The distribution of opioid receptors and opioid-receptor-containing cell bodies identified in the rat brain. *This copyrighted figure is presented with permission from Le Merrer et al. (2009).*

In CPP, nicotine is associated with an environmental context, which acts as a conditioned reinforcer in the absence of nicotine, and movement into an area previously associated with nicotine signifies CPP. MOR blockade interfered with nicotine-induced CPP in mice and disrupted nicotine-evoked changes in pCREB-positive cells in brain reward areas (VTA & NAc; Walters, Cleck, Kuo, & Blendy, 2005). Furthermore, mice lacking β-endorphin genes (Trigo et al., 2010) and MOR and DOR knockouts (Berrendero, Kieffer, & Maldonado, 2002; Berrendero et al., 2012, respectively) were less sensitive to nicotine-induced CPP, which suggests a high involvement of these receptors in interoception.

Evidence is inconclusive regarding ORs and the ability of nicotine to serve as a discriminative stimulus (S_D). In drug discrimination, a subject is administered nicotine and then trained to "go left" (e.g., in either a T-maze or two-option lever press), to experience a reinforcing event. Intermixed with drug training are saline-training sessions in which a subject is reinforced to "go right." As such, the nicotine's interoceptive stimuli act as an S_D indicating the direction of reinforcement. During testing, a different drug is delivered, and then, the degree to which the novel drug substitutes for nicotine is evaluated. In rats and monkeys, morphine either fails to substitute for nicotine (Romano, Goldstein, & Jewell, 1981; Takada, Hagen, Cook, Goldberg, & Katz, 1988) or does so imperfectly (Moerke, Zhu, Tyndale, Javors, & McMahon, 2017). There is no reason to expect that nicotine activates ORs to the same degree as pure opioid agonists, so morphine may generate subjective experiences that are dissimilar to nicotine, and rats trained with a morphine S_D readily discriminate nicotine (Romano et al., 1981). Therefore, there are inherent problems in interpretation using OR agonist substitutions, but one can surmise that dose matters, as does the extent of nicotine experienced because drug-related interoception changes with subsequent exposures (Paulus, Tapert, & Schulteis, 2009). Furthermore, general rate-decreasing effects of opioid agonists make this work fraught with confounds. The influence of the coadministration of OR *antagonists* on the nicotine S_D is limited. Romano et al. (1981) found that the coadministration of 2.0 mg/kg naloxone failed to disrupt the S_D qualities of 0.4 mg/kg nicotine, but a single dose is inconclusive. In a more recent work using a related Pavlovian goal-tracking procedure, Palmatier, Peterson, Wilkinson, and Bevins (2004) found that naloxone (0.5–6.0 mg/kg) dose-dependently attenuated the ability of nicotine to provoke "approach" responses in rats that were trained to associate food delivery with nicotine.

TABLE 35.2 Laboratory Human Behavioral Pharmacology of the Subjective Effects of Nicotine, Nontreatment Studies

Citation	Opioid antagonist	Result
Brauer, Behm, Westman, Patel, and Rose (1999)	Naltrexone with NRT	Altered the wakefulness produced by a nicotine-containing versus denicotinized cigarette, and it disrupted the craving relief produced by a nicotine patch, slight reduction in smoking outside of the laboratory
Epstein and King (2004)	Naltrexone	Reduced the number of cigarettes smoked in the lab and decrease positive affect, but increased negative affect and sedation; increased withdrawal in women but not men
Gorelick, Rose, and Jarvik (1989)	10mg naloxone	Reduced the number of cigarettes smoked, no effect on subjective ratings produced by smoking
Hutchison et al. (1999)	Naltrexone with NRT	Reduced reactivity to smoking-related cues in relation to a group that received NRT with placebo
Hutchison et al. (1996)*	Naltrexone	Stress, in combination with opioid blockade, increased key presses for cigarettes
Karras and Kane (1980)	10mg naloxone	Decreased smoking and desirability of cigarettes
King and Meyer (2000)	Naltrexone	Decreased smoking, the desirability and the pleasure of cigarettes
Knott and Fisher (2007)	Naltrexone	Moderated alertness and euphoria ratings provoked by nicotine gum; mediated theta and alpha2, but not delta-wave changes induced by nicotine; no effect nicotine relief of withdrawal symptoms
Krishnan-Sarin et al. (1999)	0.8–3.2mg/70kg naloxone	Produced symptoms of withdrawal and increased cortisol in smokers and nonsmokers
Lee et al. (2005)*	25–50mg naltrexone	Decreased cue-elicited craving but elevated prolactin, cortisol, and adrenocorticotropic hormone; no change in β-endorphin or dynorphin A
Nemeth-Coslett and Griffiths (1986)	0.06–4.0mg naloxone	No effect on smoking behavior; increased sedation
Roche et al. (2010)	Naltrexone	Smoking a nicotine cigarette potentiated naltrexone-induced increases in adrenocorticotropic hormone and cortisol
Rohsenow et al. (2007)	Naltrexone with NRT	Decreased cue-elicited responses to smoke, no effect on noncued responses
Rukstalis et al. (2005)*	Naltrexone	Decreased choice for nicotine-containing vs denicotinized cigarettes; reduced smoking cravings, but no significant effect on satisfaction from either version of a cigarette
Sutherland, Stapelton, Russell, and Feyerabend (1995)	50 and 100mg naltrexone	No effect on subjective measures of satisfaction or smoking behavior but reduced feelings of alertness, perceived difficulty in abstaining and craving, and increased negative affect

Studies listed in the table are double-blind and required overnight abstinence; exception indicated by *. Naltrexone was administered at 50mg unless otherwise indicated.

TABLE 35.3 Opioid Response to Nicotine Administration, Imaging Studies Using Positron Emission Tomography (PET) and MOR Radioligand [^{11}C] Carfentanil

Citation	Result
Kuwabara et al. (2014)	Nicotine-induced changes in left rostral frontal lobe were correlated with subjective enjoyment
Falcone et al. (2012)	Amygdalar differences resulting from overnight abstinence, no effect of smoking on the amygdala, VTA, insula, caudate, thalamus, or ACCx
Nuechterlein, Ni, Domino, and Zubeita (2016)	Basal ganglia and thalamus activation to nicotine was greater in nonsmokers than abstaining smokers; among smokers, bilateral activation of the NAc and amygdala was associated with identification of *OPRM1* genotype
Ray et al. (2011)	Lateralization differences between carriers of the *OPRM1* A118G-allele genotype; amygdala, caudate, ACCx, and thalamus activation
Scott, Domino, Heitzeg, Koeppe, and Ni (2007)	Decreased activity in ACCx, increased VTA activity, and localized hemispheric activation of the amygdala and thalamus; these findings were associated with nicotine-reduction of tobacco craving
Weerts et al. (2013)	*OPRM1* A118G-allele genotype corresponded with global MOR differences; nicotine not administered in this study

The imaging studies listed in Table 35.3 used positron emission tomography (PET) and MOR radioligand [^{11}C]carfentanil.

35.4 NICOTINE AS A REINFORCER

Nicotine self-administration is related to OR activation (Table 35.4). Ismayilova and Shoaib (2010) found that naloxone dose-dependently decreased nicotine self-administration, but did not affect food reinforcement. MOR-selective antagonists reduce nicotine self-administration in rats on most occasions, but DOR- and KOR-selective antagonists fail (Table 35.4). However, disruption of nicotine self-administration by DOR-selective antagonism or genetic knockout has been found in mice (Berrendero et al., 2012). KORs may act in a manner different to their OR counterparts; for instance, KOR-selective *agonism* disrupts nicotine reinforcement at certain doses, but the lowest dose of the agonist marginally increased nicotine self-administration (Ismayilova & Shoaib, 2010). Others (Liu & Jernigan, 2011) found no evidence that KOR *antagonism* alters nicotine self-administration, so at this time, the contribution of KORs on nicotine self-administration appears negligible.

The influence of OR antagonists on nicotine self-administration may be related to alterations of the hedonic response or a disruption in the incentive value of nicotine. Therefore, responses in extinction and recovery after extinction are important in parsing the influences of opioid antagonists on nicotine reinforcement. Liu et al. (2009) found that neither acutely nor repeatedly administered naltrexone reduced nicotine self-administration; however, naltrexone significantly suppressed cue-maintained lever responding in extinction and decreased cue-induced reinstatement. Liu et al. (2009) concluded that naltrexone attenuated nicotine-associated motivational cues and thus may have relevance to the treatment of nicotine dependence because it may mitigate cue-provoked relapse. A range of human behavioral pharmacological research (Table 35.2, e.g., Hutchison et al., 1999) shows that OR blockade can impact the efficacy of nicotine-related cues to precipitate nicotine self-administration; therefore, ORs participate in the learning about reinforcement in the context of nicotine.

TABLE 35.4 Nicotine Self-Administration Studies With Rodents

Citation	Opioid antagonist	Nicotine per infusion	Result
Berrendero et al. (2012)	DOR selective; 2.5 and 5.0 naltrindole	0.015 and 0.03; FR1, PR	FR dose-dependent decrement; dose-independent decrement on PR breakpoints
	DOR knockout		Decreased compared to unmodified, wild-type mice
Corrigall and Coen (1991)	0.1–10 mg/kg naloxone	0.03; FR5	No effect
Corrigall, Coen, Adamson, Chow, and Zhang (2000)	MOR-selective *agonist**; VTA infusion 0.005–0.05 μg DAMGO	0.01–0.03; FR5	Dose-dependent decrement on low nicotine dose; no effect at higher dose
DeNoble and Mele (2006)	0.7, 1.5, and 3.0 naloxone	0.032; FR1	No effect
Liu and Jernigan (2011)	MOR selective; 5.0 and 15.0 naloxanazine	0.03; FR5	Dose-dependent decrement; no effect on food reinforcement
	DOR selective; 0.5 and 5.0 naltrindole		No effect
	KOR selective; 0.25 and 1.0 5-guanidinonaltrindole		No effect
Liu et al. (2009)	0.25–2.0 naltrexone	0.03; FR5	No effect, but increased responses in extinction and reduced cue-induced reinstatement
Ismayilova and Shoaib (2010)	0.3, 1.0, and 3.0 naloxone	0.03; FR3	All doses decreased; no effect on food reinforcement
	DOR selective; 0.3, 1.0, and 3.0 naltrindole		No effect
	KOR-selective *agonism**; 0.3, 1.0, and 3.0 U50,488		Only the highest dose showed modest decrease

Continued

TABLE 35.4 Nicotine Self-Administration Studies With Rodents—cont'd

The studies listed in the table involved nicotine self-administration, and the fixed ratio (FR) requirements per nicotine infusion are listed; doses are in mg/kg, and all drugs are antagonists (see exceptions*). On one occasion, a progressive ratio (PR) reinforcement schedule was used.

35.5 NICOTINE-INDUCED REWARD SENSITIVITY

More than a decade of research shows that nicotine enhances the efficacy of reinforcers experienced in its presence (Donny, Caggiula, Weaver, Levin, & Sved, 2011, for a review). When a progressive ratio (PR) schedule was used to assess the influence of nicotine on sucrose-reinforcement in rats, the administration of nicotine dose-dependently increased the effort expended working for sucrose (Kirshenbaum et al., 2015; Palmatier et al., 2008). Reinforcers are not stimuli, but rather opportunities to engage in a response (Allison, 1993), and nicotine enhances these actions; importantly, the enhancement effect encapsulates a range of reinforcers, not just eating (Kirshenbaum et al., 2015; Perkins et al., 2015). When coadministered with nicotine, naloxone interferes with nicotine enhancement of sucrose reinforcement, but does not influence sucrose reinforcement on its own in the absence of nicotine (Kirshenbaum, Suhaka, Phillips, & Voltolini de Souza Pinto, 2016). This result and others (Ismayilova & Shoaib, 2010; Liu & Jernigan, 2011) show that the effects of opioid antagonism at certain doses are specific to nicotine-related reinforcement in rats. The clinical significance of nicotine's reinforcement-enhancement effect remains unclear, but some evidence exists in human smokers (Perkins et al., 2015); whether OR antagonism negates reinforcement enhancement in humans remains to be tested.

Nicotine enhancement may be related to alterations in brain reward sensitivity. Acute nicotine lowers intracranial self-stimulation (ICSS) thresholds in rats (Huston-Lyons & Kornetsky, 1992), and this effect has become a gold standard in understanding the neurobiology of how drugs become preeminent as reinforcers. OR antagonism with naloxone does not prevent the threshold-lowering influence of nicotine at doses among those that have blocked the threshold-lowering effects of other abuse drugs, such as D-amphetamine (Esposito, Perry, & Kornetsky, 1980). Although naloxone disrupts nicotine-reinforcement enhancement, it does not alter nicotine-induced reductions in ICSS thresholds.

A second and equally reliable finding in the ICSS literature is that, when chronic nicotine exposure is terminated or arrested with a nicotine antagonist (mecamylamine), thresholds skyrocket (Watkins, Stinus, Koob, & Markou, 2000). Brain reward desensitization by nicotine discontinuation is evidenced in other paradigms as well. For instance, Pergadia et al. (2014) presented rats with a probabilistic concurrent choice test for food reinforcement and then chronically exposed them to nicotine. Upon discontinuation of nicotine delivery, the rats became insensitive to reinforcement probabilities. When human smokers were given a similar task (for money), they too became insensitive to differential reinforcer probabilities during withdrawal (Pergadia et al., 2014). Taken together, the ICSS threshold increases and the results of Pergadia et al. (2014) and others (Kirshenbaum et al., 2015, 2016) suggest that nicotine discontinuation produces reinforcer insensitivities; however, changes in reward responsiveness are inconsistent in humans (Hughes et al., 2017) and may or may not be linked to anhedonia.

Naloxone dose-dependently (>1.0 mg/kg) precipitated significant increases in ICSS thresholds in rats chronically treated with nicotine (Watkins, Stinus, et al., 2000). An important qualification is that although there were subtle differences between nicotine and saline-pretreated rats, naloxone also elevated ICSS thresholds in the saline-pretreated animals at doses above 2.0 mg/kg. Because of the nonsignificant between-group differences, the authors (Watkins, Koob, et al., 2000; Watkins, Stinus, et al., 2000) could not affirm that the opioid system governs nicotine-dependent brain reward sensitivity. Other researchers have found that naloxone (up to 16 mg/kg), alone, fails to alter ICSS (Esposito et al., 1980), so there exist important inconsistencies regarding OR blockade on its own. Also, naloxone prevents reinforcement enhancement by nicotine, but it does not prevent the reward desensitizing effects of nicotine discontinuation (Kirshenbaum et al., 2016). These results uniformly suggest that brain reward insensitivity is unrelated to antagonism of ORs.

35.6 INTO THE CLINIC: OPIOID ANTAGONISTS AND TOBACCO DEPENDENCE

Chronic exposure to nicotine reliably results in OR modifications that have important ramifications for tobacco dependence. For instance, pain sensitivity governed by ORs is altered by chronic nicotine exposure (see Yoon, Lane, & Weaver, 2015, for a review). These OR modifications are beyond the scope of this chapter, but briefly, the clinical findings suggest modest efficacy of opioid antagonism when used in combination with NRT. Naltrexone given to patients prior to target quit date can result in better outcomes than a placebo (King, Cao, Zhang, & Rueger, 2013) and in combination with NRT can reduce relapse (Krishnan-Sarin, Meandzija, & O'Malley, 2003), but there are enough negative findings or small effects in the clinical literature to warrant skepticism regarding opioid antagonism as an effective treatment (see David et al., 2014). However, a number of variables that have theoretical associations to OR systems may help to resolve the inconsistent findings; negative affect (Walsh, Epstein, Munisamy, & King, 2008), gender

(King et al., 2012; Roche, Childs, Epstein, & King, 2010), and the *OPRM1* polymorphism (Ray et al., 2006) may be important variables that can be used a priori to enhance opioid antagonist efficacy (Norman & D'Souza, 2017).

MINI-DICTIONARY OF TERMS

Conditioned reinforcer An event that serves as a reinforcer due to its association with a primary reinforcer.
Discriminative stimulus An event that directs behavior; in the case of drug discrimination, these are internal events produced by drug administration.
Hedonic Subjective sensations or pleasure or satisfaction.
Interoception Internal events that influence behavior, in this case stimuli that are provoked or occasioned by nicotine administration.
Mesocortical dopaminergic pathway Brain reward circuitry that involves neurons in the VTA, amygdala, hippocampus, NAc, and PFC.
Opioid receptors Receptors that are activated by endogenous ligands typically involved in nociception and reward.
Pavlovian approach behavior Nicotine administration can provoke food seeking in rodents when nicotine has been previously associated with food.

Key Facts of ORs

- OR activation by pure opioid agonists (e.g., heroin) produces hedonic sensations, and OR antagonists such as naloxone reverses this effect.
- *OPMR1* polymorphisms, or genetic predispositions to OR expression, are associated with a variety of drug dependencies.
- Self-administration, CPP, and drug discrimination are behavioral assays traditionally used in the preclinical identification of opioid abuse liability.
- ICSS threshold alterations produced by drugs of abuse lend to the theory that OR-acting chemicals change brain reward sensitivity.
- Chronic pain is aggravated by nicotine use.
- Opioid abusers are more likely to smoke.

Summary Points

- The evidence of opioid involvement in the hedonic response to nicotine in humans is well established, but the magnitude of the effect is inconsistent and less robust than one might anticipate.
- A combination of research tools and techniques, such as opioid antagonism, [^{11}C]carfentanil, and *OPMR1* genotyping may resolve the inconsistencies regarding the hedonic response to nicotine.
- Opioid receptor blockade reliably interferes with nicotine self-administration in rats and has some effect on cigarette choice, cue reactivity, and smoking behavior in nicotine-dependent humans.
- Studies on conditioned place preference and Pavlovian approach behavior reliably show a disruption of the nicotine's interoceptive stimulus by opioid antagonists, but when nicotine serves as an S_D in drug discrimination, evidence for opioid involvement is negative or lacking.
- Clinical results of opioid antagonism in the treatment of tobacco dependence are variable, and heterogeneity of several patient factors may contribute to the inconsistency of findings.

References

Allison, J. (1993). Response deprivation, reinforcement, and economics. *Journal of the Experimental Analysis of Behavior*, *60*(1), 129–140.

Almeida, L. E., Pereira, E. F., Alkondon, M., Fawcett, W. P., Randall, W. R., & Albuquerque, E. X. (2000). The opioid antagonist naltrexone inhibits activity and alters expression of alpha7 and alpha4beta2 nicotinic receptors in hippocampal neurons: implications for smoking cessation programs. *Neuropharmacology*, *39*(13), 2740–2755.

Berrendero, F., Kieffer, B. L., & Maldonado, R. (2002). Attenuation of nicotine-induced antinociception, rewarding effects, and dependence in mu-opioid receptor knock-out mice. *The Journal of Neuroscience*, *22*(24), 10935–10940.

Berrendero, F., Plaza-Zabala, A., Galeote, L., Flores, Á., Bura, S. A., Kieffer, B. L., et al. (2012). Influence of δ-opioid receptors in the behavioral effects of nicotine. *Neuropsychopharmacology*, *37*(10), 2332–2344.

Berrendero, F., Robledo, P., Trigo, J. M., Martín-García, E., & Maldonado, R. (2010). Neurobiological mechanisms involved in nicotine dependence and reward: participation of the endogenous opioid system. *Neuroscience and Biobehavioral Reviews*, *35*, 220–231.

Bevins, R. A., & Besheer, J. (2014). Interoception and learning: import to understanding and treating diseases and psychopathologies. *ACS Chemical Neuroscience*, *5*(8), 624–631.

Brauer, L. J., Behm, F. M., Westman, E. C., Patel, P., & Rose, J. E. (1999). Naltrexone blockade of nicotine effects in cigarette smokers. *Psychopharmacology*, *143*, 339–346.

Castro, D. C., & Berridge, K. C. (2014). Advances in the neurobiological bases for food 'liking' versus 'wanting'. *Physiology & Behavior*, *136*, 22–30.

Corrigall, W. A., & Coen, K. M. (1991). Opiate antagonists reduce cocaine but not nicotine self-administration. *Psychopharmacology*, *104*, 167–170.

Corrigall, W. A., Coen, K. M., Adamson, K. L., Chow, B. L. C., & Zhang, J. (2000). Response of nicotine self-administration in the rat to manipulations of mu-opioid and γ-aminobutyric acid receptors in the ventral tegmental area. *Psychopharmacology*, *149*, 107–114.

David, S. P., Chu, I. M., Lancaster, T., Stead, L. F., Evins, A. E., & Prochaska, J. J. (2014). Systematic review and meta-analysis of opioid antagonists for smoking cessation. *BMJ Open*, *4*(3)e004393.

DeNoble, V. J., & Mele, P. C. (2006). Intravenous nicotine self-administration in rats: Effects of mecamylamine, hexamethonium and nalaxone. *Psychopharmacology*, *136*, 83–90.

Donny, E. C., Caggiula, A. R., Weaver, M. T., Levin, M. E., & Sved, A. F. (2011). The reinforcement-enhancing effects of nicotine: implications for the relationship between smoking, eating and weight. *Physiology & Behavior*, *104*(1), 143–148.

Duke, A. N., Johnson, M. W., Reissig, C. J., & Griffiths, R. R. (2015). Nicotine reinforcement in never-smokers. *Psychopharmacology*, *232*(23), 4243–4252.

Epstein, A. M., & King, A. C. (2004). Naltrexone attenuates acute acute cigarette smoking behavior. *Pharmacology Biochemistry and Behavior*, *77*, 29–37.

Esposito, R. U., Perry, W., & Kornetsky, C. (1980). Effects of d-amphetamine and naloxone on brain stimulation reward. *Psychopharmacology*, *69*(2), 187–191.

Falcone, M., Gold, A. B., Wileyto, E. P., Ray, R., Ruparel, K., Newberg, A., et al. (2012). μ-Opioid receptor availability in the amygdala is associated with smoking for negative affect relief. *Psychopharmacology*, *222*(4), 701–708.

Gorelick, D. A., Rose, J., & Jarvik, M. E. (1989). Effect of naloxone on cigarette smoking. *Journal of Substance Abuse*, *1*, 153–159.

Hadjiconstantinou, M., & Neff, N. (2011). Nicotine and endogenous opioids: neurochemical and pharmacological evidence. *Neuropharmacology*, *60*, 1209–1220.

Hughes, J. R., Budney, A. J., Muellers, S. R., Lee, D. C., Callas, P. W., Sigmon, S. C., et al. (2017). Does tobacco abstinence decrease reward sensitivity? A human laboratory test. *Nicotine & Tobacco Research*, *19* (6), 677–685.

Huston-Lyons, D., & Kornetsky, C. (1992). Effects of nicotine on the threshold for rewarding brain stimulation in rats. *Pharmacology, Biochemistry, and Behavior*, *41*(4), 755–759.

Hutchison, K. E., Collins, F. R., Tassey, J., & Rosenberg, E. (1996). Stress, naltrexone, and the reinforcement value of nicotine. *Experimental and Clinical Psychopharmacology*, *4*(4), 431–437.

Hutchison, K. E., Monti, P. M., Rohsenow, D. J., Swift, R. M., Colby, S. M., Gnys, M., et al. (1999). Effects of naltrexone with nicotine replacement on smoking cue reactivity: preliminary results. *Psychopharmacology*, *142*(2), 139–143.

Ismayilova, N., & Shoaib, M. (2010). Alteration of intravenous nicotine self-administration by opioid receptor agonist and antagonists in rats. *Psychopharmacology*, *210*(2), 211–220.

Karras, A., & Kane, J. M. (1980). Naloxone reduces cigarette smoking. *Life Sciences*, *27*, 1541–1545.

King, A. C., Cao, D., O'Malley, S. S., Kranzler, H. R., Cai, X., deWit, H., et al. (2012). Effects of naltrexone on smoking cessation outcomes and weight gain in nicotine-dependent men and women. *Journal of Clinical Psychopharmacology*, *32*(5), 630–636.

King, A., Cao, D., Zhang, L., & Rueger, S. Y. (2013). Effects of the opioid receptor antagonist naltrexone on smoking and related behaviors in smokers preparing to quit: a randomized controlled trial. *Addiction*, *108*(10), 1836–1844.

King, A. C., & Meyer, P. J. (2000). Naltrexone alteration of acute smoking response in nicotine-dependent subjects. *Pharmacology, Biochemistry, and Behavior*, *66*(3), 563–572.

Kirshenbaum, A., Green, J., Fay, M., Parks, A., Phillips, J., Stone, J., et al. (2015). Reinforcer devaluation as a consequence of acute nicotine exposure and withdrawal. *Psychopharmacology*, *232*(9), 1583–1594.

Kirshenbaum, A. P., Suhaka, J. A., Phillips, J. L., & Voltolini de Souza Pinto, M. (2016). Nicotine enhancement and reinforcer devaluation: interaction with opioid receptors. *Pharmacology, Biochemistry, and Behavior*, *150–151*, 1–7.

Knott, V. J., & Fisher, D. J. (2007). Naltrexone alteration of nicotine-induced EEG and mood activation response in tobacco-deprived cigarette smokers. *Experimental & Clinical Psychopharmacology*, *15*, 368–381.

Krishnan-Sarin, S., Meandzija, B., & O'Malley, S. (2003). Naltrexone and nicotine patch smoking cessation: a preliminary study. *Nicotine & Tobacco Research*, *5*(6), 851–857.

Krishnan-Sarin, S., Rosen, M. I., & O'Malley, S. S. (1999). Naloxone challenge in smokers. Preliminary evidence of an opioid component in nicotine dependence. *Archives of General Psychiatry*, *56*(7), 663–668.

Kuwabara, H., Heishman, S. J., Brasic, J. R., Contoreggi, C., Cascella, N., Mackowick, K. M., et al. (2014). Mu opioid receptor binding correlates with nicotine dependence and reward in smokers. *PLoS ONE*, *9*(12), e113694.

Lee, Y. S., Joe, K. H., Sohn, I. K., Na, C., Kee, B. S., & Chae, S. L. (2005). Changes of smoking behavior and serum adrenocorticotropic hormone, cortisol, prolactin, and endogenous opioid levels in nicotine dependence after naltrexone treatment. *Progress in Neuro-Psychopharmacology & Biological Psychiatry*, *29*, 639–647.

Le Merrer, J., Becker, J. J., Befort, K., & Kieffer, B. L. (2009). Reward processing by the opioid system in the brain. *Physiological Reviews*, *89*(4), 1379–1412.

Liu, X., & Jernigan, C. (2011). Activation of the opioid μ1, but not δ or κ, receptors is required for nicotine reinforcement in a rat model of drug self-administration. *Progress in Neuro-Psychopharmacology & Biological Psychiatry*, *35*(1), 146–153.

Liu, X., Palmatier, M., Caggiula, A., Sved, A., Donny, E., Gharib, M., et al. (2009). Naltrexone attenuation of conditioned but not primary reinforcement of nicotine in rats. *Psychopharmacology*, *202*(4), 589–598.

Moerke, M. J., Zhu, A. X., Tyndale, R. F., Javors, M. A., & McMahon, L. R. (2017). The discriminative stimulus effects of i.v. nicotine in rhesus monkeys: pharmacokinetics and apparent pA2 analysis with dihydro-β-erythroidine. *Neuropharmacology*, *116*, 9–17.

Nemeth-Coslett, R., & Griffiths, R. R. (1986). Naloxone does not affect cigarette smoking. *Psychopharmacology*, *89*, 261–264.

Norman, H., & D'Souza, M. S. (2017). Endogenous opioid system: a promising target for future smoking cessation medications. *Psychopharmacology*, *234*(9–10), 1371–1394.

Nuechterlein, E. B., Ni, L., Domino, E., & Zubeita, J.-K. (2016). Nicotine-specific and non-specific effects of cigarette smoking on endogenous opioid mechanisms. *Progress in Neuro-Psychopharmacology & Biological Psychiatry*, *69*, 69–77.

Palmatier, M. I., Coddington, S. B., Liu, X., Donny, E. C., Caggiula, A. R., & Sved, A. F. (2008). The motivation to obtain nicotine-conditioned reinforcers depends on nicotine dose. *Neuropharmacology*, *55*, 1425–1430.

Palmatier, M. I., Peterson, J. L., Wilkinson, J. L., & Bevins, R. A. (2004). Nicotine serves as a feature-positive modulator of Pavlovian appetitive conditioning in rats. *Behavioural Pharmacology*, *15*(3), 183–194.

Paulus, M. P., Tapert, S. F., & Schulteis, G. (2009). The role of interoception and alliesthesia in addiction. *Pharmacology, Biochemistry, and Behavior*, *94*(1), 1–7.

Pergadia, M. L., Der-Avakian, A., D'Souza, M. S., Madden, P. F., Heath, A. C., Shiffman, S., et al. (2014). Association between nicotine withdrawal and reward responsiveness in humans and rats. *JAMA Psychiatry*, *71*(11), 1238–1245.

Perkins, K. A., Karelitz, J. L., & Michael, V. C. (2015). Reinforcement enhancing effects of acute nicotine via electronic cigarettes. *Drug and Alcohol Dependence*, *153*, 104–108.

Pomerleau, O. F. (1998). Endogenous opioids and smoking: a review of progress and problems. *Psychoneuroendocrinology*, *23*(2), 115–130.

Ray, R., Jepson, C., Patterson, F., Strasser, A., Rukstalis, M., Perkins, K., et al. (2006). Association of OPRM1 A118G variant with relative reinforcing value of nicotine. *Psychopharmacology*, *188*, 355–363.

Ray, R., Ruparel, K., Newberg, A., Wileyto, E. P., Loughead, J. W., Divgi, C., et al. (2011). Human mu opioid receptor (OPRM1 A118G) polymorphism is associated with brain mu-opioid receptor binding potential in smokers. *Proceedings of the National Academy of Sciences of the United States of America*, *108*(22), 9268–9273.

Roche, D. O., Childs, E., Epstein, A. M., & King, A. C. (2010). Acute HPA axis response to naltrexone differs in female vs. male smokers. *Psychoneuroendocrinology*, *35*(4), 596–606.

Rohsenow, D. J., Monti, P. M., Hutchison, K. E., Swift, R. M., MacKinnon, S. V., Sirota, A. D., et al. (2007). High-dose trandermal nicotine with naltrexone: Effect on nicotine withdrawal, urges, smoking, and effects of smoking. *Experimental & Clinical Psychopharmacology*, *15*, 81–92.

Romano, C., Goldstein, A., & Jewell, N. P. (1981). Characterization of the receptor mediating the nicotine discriminative stimulus. *Psychopharmacology*, *74*(4), 310–315.

Rukstalis, M., Jepson, C., Strasser, A., Lynch, K. G., Perkins, K., Patterson, F., et al. (2005). Naltrexone reduces the relative reinforcing value of nicotine in a cigarette smoking choice paradigm. *Psychopharmacology*, *180*(1), 41–48.

Scott, D. J., Domino, E. F., Heitzeg, M. M., Koeppe, R. A., & Ni, L. (2007). Smoking modulation of mu-opioid and dopamine receptor D2 receptor-mediated neurotransmission in humans. *Neuropsychopharmacology, 32*, 450–457.

Shippenberg, T. S., LeFevour, A., & Chefer, V. I. (2008). Targeting endogenous mu- and delta-opioid receptor systems for the treatment of drug addiction. *CNS & Neurological Disorders Drug Targets, 7*(5), 442–453.

Sutherland, G., Stapelton, J. A., Russell, M. A., & Feyerabend, C. (1995). Naltrexone, smoking behavior, and cigarette withdrawal. *Psychopharmacology, 120*, 418–425.

Takada, K., Hagen, T. J., Cook, J. M., Goldberg, S. R., & Katz, J. L. (1988). Discriminative stimulus effects of intravenous nicotine in squirrel monkeys. *Pharmacology, Biochemistry, and Behavior, 30*(1), 243–247.

Tanda, G., & Di Chiara, G. (1998). A dopamine-mu1 opioid link in the rat ventral tegmentum shared by palatable food (Fonzies) and non-psychostimulant drugs of abuse. *The European Journal of Neuroscience, 10*(3), 1179–1187.

Tomé, A. R., Izaguirre, V., Rosário, L. M., Ceña, V., & González-García, C. (2001). Naloxone inhibits nicotine-induced receptor current and catecholamine secretion in bovine chromaffin cells. *Brain Research, 903*(1–2), 62–65.

Trigo, J. M., Martin-García, E., Berrendero, F., Robledo, P., & Maldonado, R. (2010). The endogenous opioid system: a common substrate in drug addiction. *Drug and Alcohol Dependence, 108*(3), 183–194.

Walsh, Z., Epstein, A., Munisamy, G., & King, A. (2008). The impact of depressive symptoms on the efficacy of naltrexone in smoking cessation. *Journal of Addictive Diseases, 27*(1), 65–72.

Walters, C. L., Cleck, J. N., Kuo, Y., & Blendy, J. A. (2005). Mu-opioid receptor and CREB activation are required for nicotine reward. *Neuron, 46*(6), 933–943.

Watkins, S. S., Koob, G. F., & Markou, A. (2000). Neural mechanisms underlying nicotine addiction: acute positive reinforcement and withdrawal. *Nicotine & Tobacco Research, 2*(1), 19–37.

Watkins, S. S., Stinus, L., Koob, G. F., & Markou, A. (2000). Reward and somatic changes during precipitated nicotine withdrawal in rats: centrally and peripherally mediated effects. *The Journal of Pharmacology and Experimental Therapeutics, 292*(3), 1053–1064.

Weerts, E. M., McCaul, M. E., Kuwabara, H., Yang, X., Xu, X., Dannals, R. F., et al. (2013). Influence of OPMR1 Asn40Asp variant (A118G) on [11C]carfentanil binding potential: preliminary findings in human subjects. *International Journal of Neuropsychopharmacology, 16*, 47–53.

Yoon, J. H., Lane, S. D., & Weaver, M. F. (2015). Opioid analgesics and nicotine: more than blowing smoke. *Journal of Pain & Palliative Care Pharmacotherapy, 29*(3), 281–289.

CHAPTER

36

Nicotine-Induced Kindling: Influences of Age, Sex, and Prevention by Antioxidants

Danielle Macedo, Adriano José Maia Chaves Filho, Patrícia Xavier Lima Gomes, Lia Lira Olivier Sanders, David Freitas de Lucena

Department of Physiology and Pharmacology, Drug Research and Development Centre, Faculty of Medicine, Federal University of Ceara, Fortaleza, Brazil

Abbreviations

BDNF	brain-derived neurotrophic factor
CREB	cAMP-responsive element-binding protein
ERKs	extracellular signal-regulated kinases
GABA	gamma-aminobutyric acid
GSH	reduced glutathione
NAC	*N*-acetylcysteine
nAChR	nicotinic acetylcholine receptors
NMDAR	*N*-methyl-D-aspartate receptors
PTZ	pentylenetetrazol
SOD	superoxide dismutase

36.1 INTRODUCTION

Kindling is a model of synaptic plasticity related to progressive epilepsy (Bertram, 2007; Pinel & Rovner, 1978) and to the spontaneous manifestation of symptoms observed in stress-related disorders such as anxiety, depression, substance use, and bipolar disorders (Post, 2007). It has long been recognized that the repeated administration of central nervous system stimulants, such as cocaine, induces kindling (Post & Kopanda, 1976), but only in the last decade, nicotine-induced kindling was characterized (Bastlund, Berry, & Watson, 2005).

Nicotine is a psychoactive drug known to induce either stimulant (hyperactivity) or depressant (hypoactivity) behavioral effects depending on the specific experimental protocol (Clarke & Kumar, 1983). Apart from nicotine, the main constituent of tobacco, tobacco smoking presents a lot of agents with both proconvulsant (e.g., nicotine and carbon monoxide) or anticonvulsant (e.g., selenium, zinc, and acetone) properties (Rong, Frontera, & Benbadis, 2014). High doses of nicotine induce seizures, being one of the symptoms of this drug poisoning (overdose) in humans. A previous study showed that smokers had double the risk of seizures when compared with nonsmoker controls and that past tobacco smokers had a modestly increased risk of epilepsy, but not reaching statistical significance (Dworetzky, Bromfield, Townsend, & Kang, 2010). As abovementioned, tobacco smoking presents several agents related to increased risk of seizures. Therefore, studies on tobacco smoking cannot directly imply nicotine as the constituent of tobacco related to seizure risk. Some comorbidities associated with tobacco smoking increase the risk of seizures, being other confounding factors. Thus, it is very difficult to determine the influence of nicotine in the magnitude of seizures related to tobacco smoking. One of the most recent reviews dealing with the theme tobacco smoking and seizures/epilepsy concluded that the investigation of the relationship between tobacco smoking and seizures/epilepsy has rather raised more questions than provided answers (Rong et al., 2014).

Hence, for a better evaluation of the role of nicotine in seizures and/or epileptogenesis, the evidence must come from preclinical studies and/or from clinical studies using patients treated with nicotine patches or using e-cigarettes. To date, evidence regarding the induction of seizures by the treatment with nicotine patches is scarce. On the contrary, nicotine patches have been used for the treatment of some types of epilepsy, such as epilepsy associated with a mutation in the nicotinic acetylcholine receptor (Sieciechowicz & Kohrman, 2015). Regarding e-cigarette, reported was a case of a child poisoned by a refill liquid

Neuroscience of Nicotine
https://doi.org/10.1016/B978-0-12-813035-3.00036-8

that contained a nicotine concentration of 1.8% (18 mg/mL). The child presented a cholinergic crisis with signs of central nervous toxicity, which may include ataxia and seizures. The authors called attention to the potentially fatal liquid nicotine poisoning of infants and young children due to the growing use of e-cigarettes (Bassett, Osterhoudt, & Brabazon, 2014).

Evidence from preclinical studies has demonstrated that nicotine can induce opposing effects in relation to seizures ranging from a protective to a proconvulsant effect. This effect is dose-dependent, that is, low doses of nicotine are nonconvulsant, while high doses are convulsant. The proconvulsant effect of nicotine is mediated by the activation of putative nicotinic acetylcholine receptors (nAChR), such as α7 receptors, and by glutamate release and stimulation of *N*-methyl-D-aspartate receptors (NMDAR), leading to nitric oxide formation and seizure production (Damaj, Glassco, Dukat, & Martin, 1999).

Despite the importance of the study of nicotine-induced kindling for a better comprehension of mechanisms of neuroplasticity involved in nicotine long-term effects, to date, there is little information on this subject.

This chapter provides a review of the existing literature on nicotine-induced kindling, the influences of age and sex in its development, and the role of antioxidants in the prevention of kindling development.

36.2 KINDLING PHENOMENON AND ITS RELATION TO EPILEPTOGENESIS AND TO NEUROPSYCHIATRIC DISORDERS

Small electric stimulus, if repeatedly applied, will eventually generate seizures that can lead to fully generalized convulsions, a phenomenon called kindling. The discovery of kindling as a model of focal epilepsy was accidental. By examining the effects of electric stimulation of the amygdaloid complex (the integrative center for emotions, emotional behavior, and motivation) on learning, Graham Goddard noticed that a number of his rats developed seizures after repeated stimulation (Goddard, McIntyre, & Leech, 1969). Goddard recognized that the brain was changing in response to a constant stimulus and that this change constituted a form of plasticity, which could provide a useful neural model of epilepsy. Later on, Ron Racine described the kindling progression in detail, delineating the progression into five distinct behavioral stages from motor arrest accompanied by facial automatisms (stage 1) to fully kindled seizures accompanied by forelimb clonus and hindlimb tonus identified by rearing and bipedal instability (stage 5) (Racine, 1972).

Besides electric stimulation, kindling may also be induced by chemical compounds, such as cocaine (Itzhak, 1996) and pentylenetetrazol (PTZ) (Zhu et al., 2015) and by environmental factors, such as emotional stress (Table 36.1).

TABLE 36.1 Main Stimuli Related to the Development of Experimental Kindling

Type of stimuli	Current intensity or dose	Time of exposures	Reference
ELECTRIC			
• Amygdala	60 Hz, one set train of 1 ms biphasic square-wave pulses, delivered at 500 μA	At least seven stimulations	Hosford et al. (1995)
• Corneal	3 mA (duration of 3 s)	10–12 days	Potschka and Löscher (1999)
CHEMICAL			
• Pentylenetetrazol	35 mg/kg	13 days (on average)	Dhir (2012)
• Cocaine	35 mg/kg	10 days	Itzhak (1996)
• Bicuculline	3.5 mg/kg	28 days	Nutt, Cowen, Batts, Grahame-Smith, and Green (1982)
• Picrotoxin	5 mg/kg	5 days	Nutt et al. (1982)
ENVIRONMENTAL			
• Psychological stress	–	Early life stress from postnatal days 2–14	Kumar et al. (2011)

Spontaneous manifestations of symptoms observed in stress-related disorders such as anxiety, depression, substance use, and bipolar disorders seem to be related to kindling mechanisms. Importantly, these psychiatric conditions present structural and functional changes within the amygdala (Trevor, 2012). In these disorders, the exposure to major life stress changes from an initial episode over subsequent recurrences; for example, manic patients in their initial hospitalization were significantly more likely to have experienced a preadmission event (with mean scaled stress scores significantly higher) in the month prior to admission, as compared to manic patients in a repeat hospitalization (Ambelas, 1987). Therefore, one proposition that has been continuously studied in humans, but not completely elucidated due to methodological issues, is that major life stress is required to induce initial onsets and recurrences of affective episodes, although successive episodes seem to be less tied to stressors or may eventually occur autonomously (Bender & Alloy, 2011).

The changes caused by kindling persist over long periods of time and are generally viewed as being semipermanent (Dennison, Campbell Teskey, & Cain, 1995). Based on this observation, as abovementioned, this model can conceptualize the general mechanisms involved in a wide range of neuropsychiatric disorders that present recurrent and cyclic symptoms (Post, 2007). In line with the evidence of the participation of kindling mechanisms in neuropsychiatric disorders, anticonvulsant drugs have been successfully used for the treatment of some of these disorders based on its antikindling effects, for example, valproate for the treatment of bipolar disorder.

Kindling development involves activity-dependent functional plasticity, and along with this process, a spatiotemporal induction of immediate genes, such as c-Fos, and late effector genes including neuropeptides, neurotrophic factors, and other genes involved in synthesis and release of neurotransmitters (for more details, please read Post, 2007). The most studied mechanisms underlying kindling phenomenon are alterations in NMDARs, neurotrophic factors such as brain-derived neurotrophic factor (BDNF), and oxidative imbalance (Kaminski et al., 2011; Zhu et al., 2015) (Fig. 36.1).

36.3 NICOTINE-INDUCED KINDLING

The first evidence of the nicotine ability to induce kindling came from the study of Bastlund et al. (2005). These authors performed an elegant study in which they compared the well-established model of kindling induced by the gamma-aminobutyric acid (GABA) antagonist, PTZ (a model of temporal lobe epilepsy), with the new proposed model of kindling induced by nicotine. Furthermore, they verified the protective effect of anticonvulsant drugs in these models and evaluated neuronal hyperactivity in several brain areas by the determination of c-Fos. For the induction of nicotine kindling, an intraperitoneal (i.p.) dose of 2.3 mg/kg every weekday for 2 weeks was administered to male adult mice, while PTZ kindling was induced with a dose of 37 mg/kg every other weekday (Mon, Wed, and Fri) for 3 weeks. Additionally, they conducted a protocol with acute seizures induced by nicotine 3.3 mg/kg, i.p., and PTZ 70 mg/kg, i.p.

The results obtained by Bastlund et al. (2005) showed that from the anticonvulsant drugs tested, namely, levetiracetam, tiagabine, and phenytoin, levetiracetam was about 10-fold more potent, while tiagabine was similarly effective against nicotine-kindled seizures when compared to PTZ-kindled seizures. When analyzing the immediate early gene c-Fos, which is induced in neurons following increases in intracellular calcium levels, these authors observed a c-Fos immunoreactivity pattern in nicotine-kindled animals distinct from the one obtained with PTZ. c-Fos immunoreactivity was increased in the substantia nigra pars compacta and in both the medial and lateral part of the medial habenula in nicotine-kindled mice. On the other hand, PTZ-kindled animals presented increased c-Fos immunoreactivity in the cortical amygdala and piriform cortex, that is, limbic structures. It is worth mentioning that the medial habenula shows high expression of $\alpha 6$, $\alpha 4\beta 2$, $\beta 2$, and $\beta 4$ subunits of the nAChR, which means that these subunits could be altered by nicotine kindling, although further confirmation is needed.

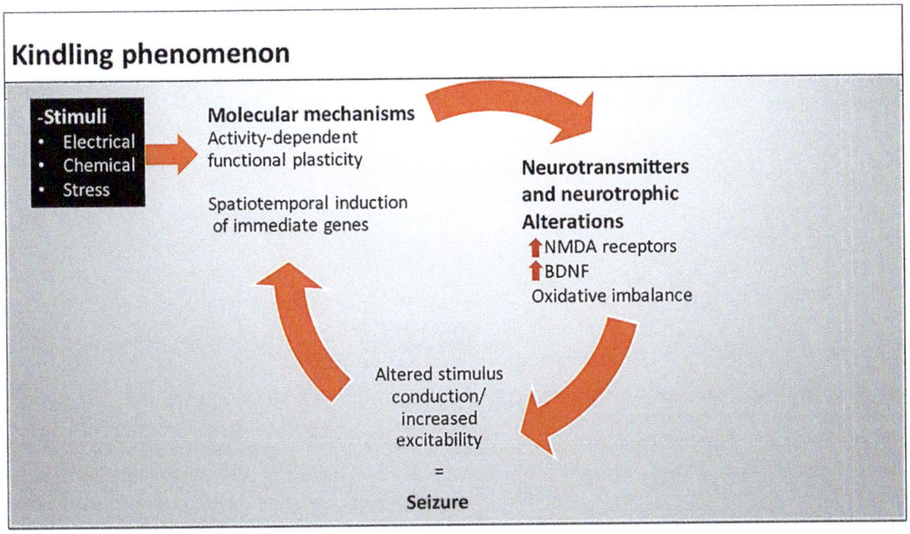

FIG. 36.1 Sequence of events related to the development of kindling. Different kinds of stimuli such as electric, chemical, and stress can induce kindling. Each stimulus triggers molecular mechanisms related to the induction of genes that in turn may alter the expression of receptors and neurotrophic factors such as brain-derived neurotrophic factor (BDNF) and cause oxidative imbalance. Together, these alterations increase neuronal excitability leading to the occurrence of seizures. Once kindling is developed, the cyclic process takes place leading to the spontaneous recurrence of seizures, even in the absence of the initial stimulus. *By D. Macedo, 2017.*

Some other mechanisms, although not studied yet, may be related to the development of nicotine-induced kindling. In fact, alterations in glutamatergic and dopaminergic neurotransmissions that are present in other models of kindling should be studied for a better understanding of mechanisms underlying nicotine kindling development and progression (Fig. 36.2).

In conclusion, the results obtained by Bastlund et al. (2005) revealed that distinct from PTZ-induced kindling, nicotine kindling model shows less limbic activation and is more sensitive in detecting the anticonvulsant effect of levetiracetam, a drug that present unique mechanism of action. To date, preliminary findings show that levetiracetam may have beneficial effects on alcohol use and anxiety symptoms in patients with alcohol dependence and co-occurring anxiety disorders (Mariani & Levin, 2008). Thus, the determination of the effects of this drug in the treatment of nicotine dependence needs further studies.

36.3.1 Sex and Age Influences

More recently, our research group demonstrated a distinct susceptibility to the development of nicotine kindling in female periadolescent rats when compared to their male counterparts. In our experimental conditions, female periadolescent rats administered nicotine 2mg/kg, i.p., during weekdays developed fully kindled seizures at a median time of 19 days, while in periadolescent males, kindled seizures were observed by a median time of 24 days (Fig. 36.3) (Gomes et al., 2013). On the other hand, during adulthood, the time for kindling development in males and females was the same, 19 days (unpublished data).

In this previous report (Gomes et al., 2013), we showed that oxidative alterations were underpinning the sex difference in the development of nicotine kindling. In this regard, female periadolescent nicotine-kindled rats presented low levels of reduced glutathione (GSH), the main endogenous antioxidant, in the brain areas, hippocampus and striatum, accompanied by decreased levels of the enzyme superoxide dismutase (SOD) in the prefrontal cortex and striatum and by increased lipid peroxidation in the prefrontal cortex, hippocampus, and striatum. Contrarily, male rats presented only low levels of GSH in the hippocampus (Fig. 36.3). In this study, the main brain region related to sex-biased alterations was the striatum.

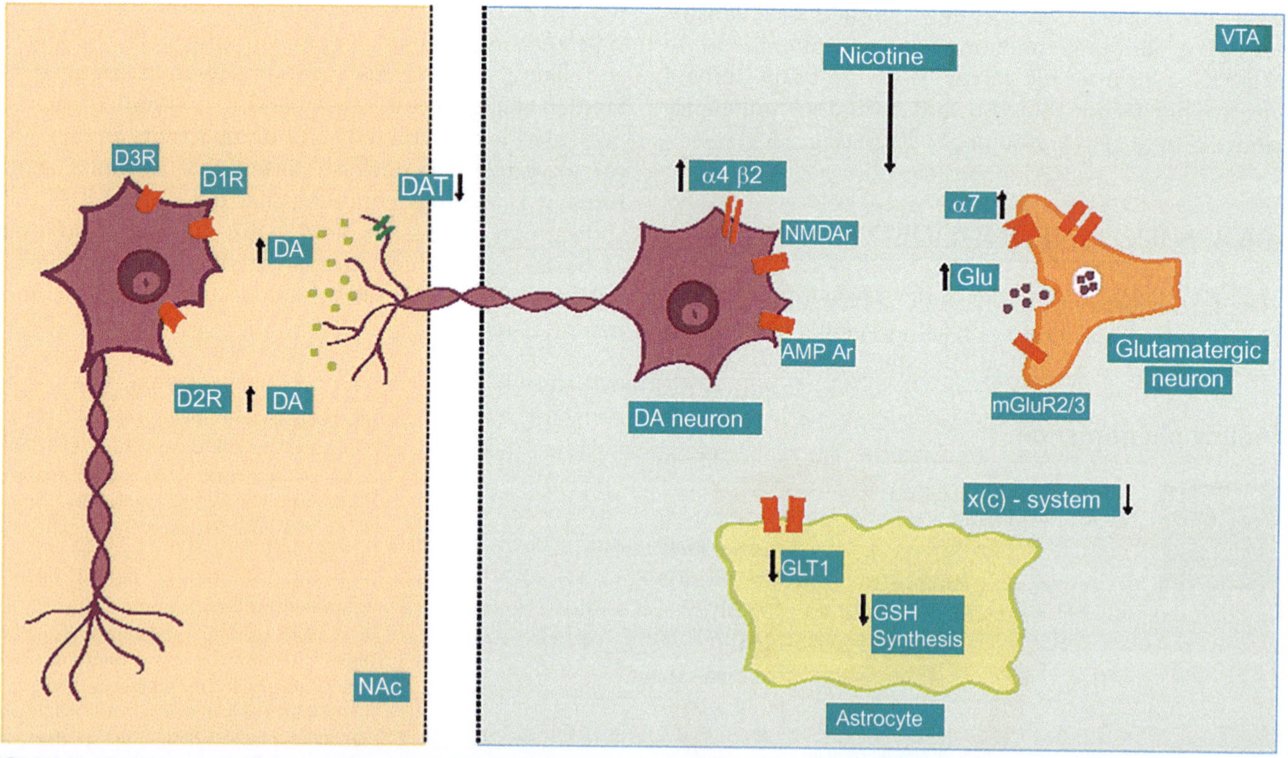

FIG. 36.2 Putative mechanisms involved in nicotine kindling.
In limbic brain areas (the ventral tegmental area), nicotine directly stimulates α4β2 nAChR in dopaminergic neurons, inducing their depolarization and the vesicular release of dopamine from their projections to nucleus accumbens. Nicotine also acts at α7 nAChR expressed in glutamatergic neurons, leading to the synaptic release of glutamate (GLU). This can activate postsynaptic ionotropic glutamate receptors AMPA and NMDA, which promotes the depolarization and the maintenance of an increased dopaminergic tone of dopaminergic neurons. Nicotine could also impair the uptake of glutamate mediated by glial transporters, such as glutamate transporter 1 (GLT1), which also contributes to altered glutamatergic transmission and overactivation of ionotropic receptors caused by nicotine. *By A.J.M. Chaves Filho, 2017.*

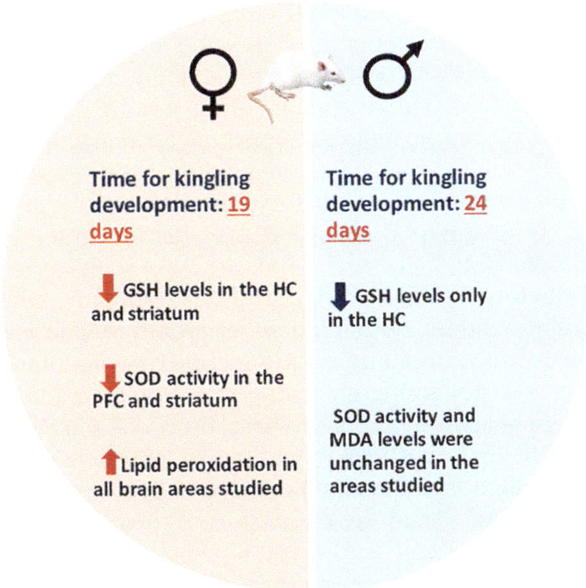

FIG. 36.3 Summary of the differences in nicotine-induced kindling observed in male and female animals.
In the study conducted by Gomes et al. (2013), the median time for kindling development in female periadolescent animals was 19 days, while in male rats, the time extended to 24 days, showing a greater predisposition of female animals for the development of kindling. A prooxidant status in brain areas of female rats when compared to males was observed revealing sex-biased oxidative alterations. Abbreviations: *GSH*, reduced glutathione; *HC*, hippocampus; *MDA*, malondialdehyde; *SOD*, superoxide dismutase. *By P.X.L. Gomes, 2017.*

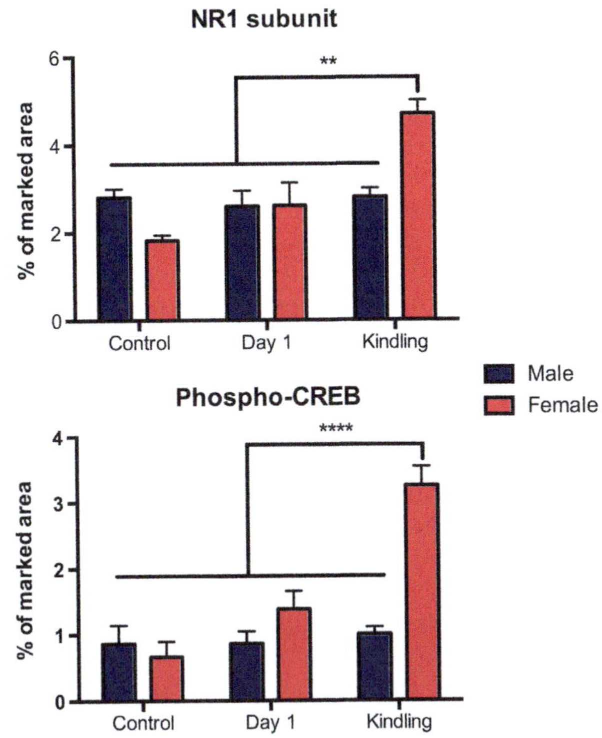

FIG. 36.4 Immunoexpression of striatal NMDAR subunit NR1 and phospho-CREB in periadolescent male and female rats along with the development of nicotine kindling.
Rats were administered during weekdays with nicotine 2 mg/kg, i.p., and dissected for striatum removal on the first day of nicotine administration and on the day of kindling development (day 19 for females and day 24 for males). Control group received saline. Immunofluorescence technique was used for the determination of the immunoexpression of NMDAR subunit NR1 and phospho-CREB. Bars represent mean ± standard error of the mean (SEM). Data were analyzed by two-way ANOVA followed by Tukey post-hoc test. We observed that female nicotine-kindled animals presented significant high levels of NR1 subunit (**$P < .01$) and of phospho-CREB (****$P < .0001$). *Source: unpublished data from author's databases.*

In agreement with our previous results of striatal sex-biased alterations in nicotine-kindled animals, unpublished data from our research group revealed increased expression of the phosphorylated NR1 subunit of NMDARs and phosphorylated CREB transcription factor (cAMP-responsive element-binding protein) in the striatum of female periadolescent nicotine-kindled rats (Fig. 36.4).

In fact, behavioral alterations induced by drugs of abuse relate to striatal abnormalities in transcription and neurotrophic factors. Changes in the activity of the transcription factor CREB are involved in psychomotor sensitization, memory functions, and drug-seeking behavior induced by psychostimulants. Thus, we hypothesize that in female nicotine-kindled animals, the activation of NMDARs could mediate a strong calcium influx to striatal cells, leading to the activation of kinases, such as extracellular signal-regulated kinases (ERKs) 1 and 2, which phosphorylate CREB factor. Of note, CREB participates in long-term forms of synaptic plasticity and modulates the intrinsic excitability of neurons (Benito & Barco, 2010) being thus related to the development of sensitization to drugs of abuse (Guerriero et al., 2005). It is important to mention that we also observed in periadolescent animals decreased levels of BDNF in male nicotine-kindled rats while females presented high levels (data not shown). Taken together, these results point toward brain oxidative imbalance and activation of learning-like mechanisms in nicotine-kindled female periadolescent animals. This mechanism has already been implicated in processes of stress and cocaine-induced behavioral sensitization (Kalivas & Stewart, 1991, for review).

36.3.2 Prevention by the Use of Antioxidants

In our first study, we demonstrated that the oral administration of vitamin E 200 and 400 mg/kg prevented the development of nicotine kindling in both male and female periadolescent rats. The decision to use this antioxidant came from the results showing increased lipid peroxidation in brain areas of nicotine-kindled rats (Gomes et al., 2013). Interception of lipid peroxyl radicals

(LOO•) and termination of lipid peroxidation chain reactions are the main antioxidant mechanisms of vitamin E (Nimse & Pal, 2015).

Since we also observed decreased levels of GSH in brain areas of female nicotine-kindled rats (Gomes et al., 2013), we decided to test *N*-acetylcysteine (NAC) as a preventive antikindling strategy (Okamura et al., 2016). Of note, NAC is a precursor of the amino acid cysteine, which is one of the key determinants for GSH synthesis (Lu, 2013). NAC at dose of 270 mg/kg prevented the development and progression of nicotine-kindled seizures in female periadolescent rats (Okamura et al., 2016).

In recent years, an increasing body of evidence point to the potential use of NAC for the treatment of neuropsychiatric disorders. Some studies report promising effects of NAC in affective (Berk et al., 2011) and substance use disorders, including nicotine dependence (Schmaal et al., 2011). The universality of NAC effects for the treatment of mental disorders is intriguing and possibly related to its action in several pathways, such as oxidative stress, inflammation, mitochondrial dysfunction, and neurotransmitter imbalance (Dean, Giorlando, & Berk, 2011).

N-Acetylcysteine attenuates drug-seeking behavior induced by psychostimulants, such as cocaine and nicotine. These effects have been related to the reduction of vesicular traffic of the excitatory neurotransmitter glutamate (Moussawi et al., 2009), restoring glutamate homeostasis.

Therefore, NAC directly and/or indirectly could modulate glutamatergic and subsequently dopaminergic neurotransmission. Considering the role of glutamatergic and dopaminergic pathways and of oxidative stress in several neuropsychiatric disorders, one could speculate that the broad therapeutic effects of NAC are mediated by its action on these targets (Fig. 36.5). Finally, NAC could represent a new promising and safe strategy as add-on therapy to substance use disorders, whose putative neurobiological mechanisms include a "kindling" process.

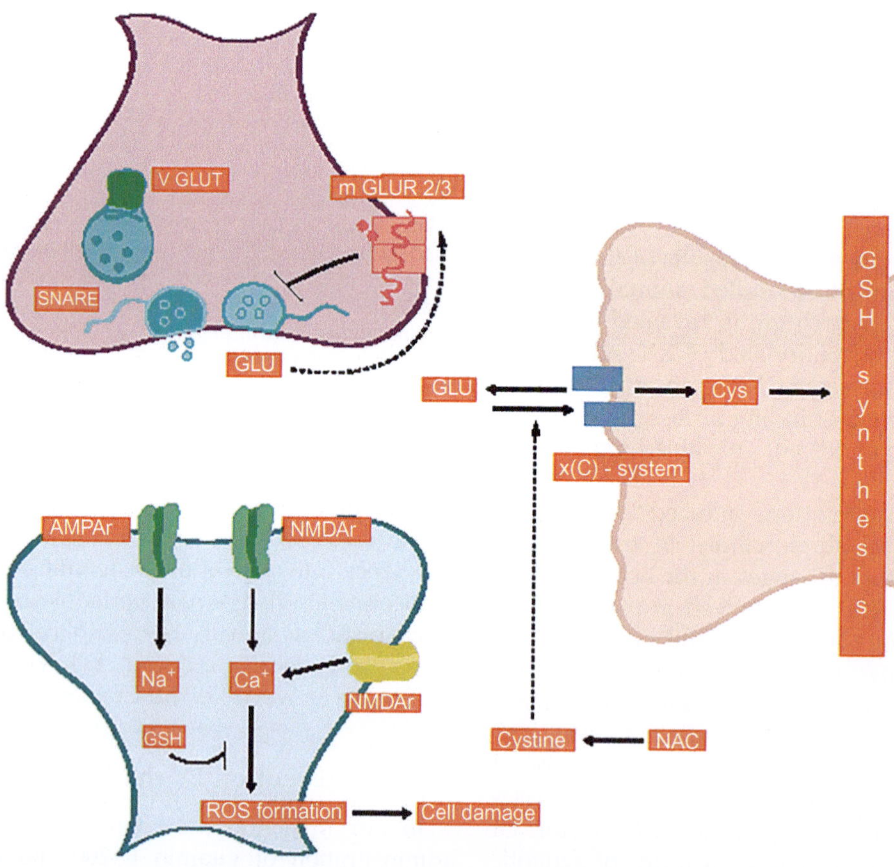

FIG. 36.5 Mechanisms related to the preventive effect of *N*-acetylcysteine against nicotine-induced kindling. *N*-acetylcysteine is metabolized to cystine. Cystine is transported to the intracellular medium through cystine-glutamate antiporter (x(c)-system). Inside the cell, cystine is reduced to cysteine for the synthesis of GSH. While x(c)-system internalizes cystine, it moves glutamate to extracellular medium through a nonvesicular pathway. This release of glutamate preferentially activates the group of presynaptic metabotropic glutamate receptors 2/3 (mGluR2/3) that negatively regulates the vesicular release of this neurotransmitter. This mechanism prevents the activation of glutamate ionotropic receptors AMPA and NMDA and their related excitotoxicity processes as well as reduces the levels of reactive oxygen species (ROS) and increases cell viability. *By A.J.M. Chaves Filho, 2017.*

36.4 CONCLUSION AND FUTURE PERSPECTIVES

The present review indicates that nicotine-induced kindling has emerged as a useful tool for the study of long-lasting neuroplastic changes associated with nicotine repeated exposure. Despite its importance, nicotine kindling currently is poorly studied. To date, there are evidences for the increased susceptibility of female periadolescent rats to the development of kindling when compared to male counterparts. Brain oxidative, glutamatergic, and neurotropic mechanisms underpin the sex-biased results. During adulthood, the susceptibility seems to be the same (Gomes et al., 2013).

Based on the oxidative alterations present in nicotine-kindled animals, the use of antioxidant strategies such as vitamin E and NAC was effective in the prevention of kindling development.

Furthermore, given the involvement of kindling mechanisms in the progressive course of several neuropsychiatric disorders, this model is relevant to the screening of new antikindling agents to improve the therapeutics of these disorders, especially those associated with nicotine use.

MINI-DICTIONARY OF TERMS

Cholinergic crisis Overstimulation of cholinergic receptors, leading to symptoms, such as excessive salivation, cramps, diarrhea, and blurred vision, accompanied or not by central nervous system manifestations such as seizures.

Comorbidities associated with tobacco smoking In medicine, the term comorbidity means the coexistence of two or more diseases. The presence of the first disease, or the exposure to a risk factor, increases the likelihood of occurrence of the second disease. Tobacco smoking is a risk factor that increases the chances of developing stroke, hypoxia, infections, and inflammation, which are known as comorbidities associated with tobacco smoking.

E-cigarettes Handheld electronic devices used to simulate the feeling of smoking tobacco. Different from tobacco cigarettes that deliver nicotine and high amounts of harmful chemical compounds, e-cigarettes present in their composition mainly nicotine ranging from low (6 mg) to extra high (24 mg) quantities.

Epileptogenesis A gradual process by which the previously normal brain develops a functional alteration toward the generation of aberrant electric activity that underpins chronic seizures.

Kindling The name kindling refers to material that can be readily ignited just as a small spark will ignite a flame that eventually can grow into a roaring bonfire. This term in biology refers to a behavioral modification (e.g., seizures) triggered by the repeated exposure to a stimulus of low intensity (e.g., amygdala electric stimulation).

Neuroplasticity This term relates to the brain's ability to form and reorganize synaptic connections. Neuroplasticity is involved in physiological process, such as the formation of memories, and in brain reorganization following injury.

Neuropsychiatric disorders Mental disorders related to alterations of cerebral functions compromising affect, cognition, and behavior. Examples of neuropsychiatric disorders are mood disorders (e.g., bipolar disorder and major depression), schizophrenia, and autism spectrum disorder.

Oxidative stress A state of increased production of reactive oxygen species (free radicals) or/and decreased antioxidant defenses that impairs cell functioning.

Periadolescence Periadolescence in rodents is the developmental period that relates to the adolescence in humans. It occurs between the earliest detection of diurnal gonadotropin cycling (approximately postnatal day 28) and the reproductive maturity (approximately postnatal days 38–42).

Racine scale for kindling progression A scale developed by Racine to assess the intensity of seizures in rodent models of epilepsy. The scale comprises five categories characterized by "mouth and facial movements" (stage 1), "head nodding" (stage 2), "forelimb clonus" (stage 3), seizures characterized by rearing (stage 4), and seizures characterized by "full-blown" seizures (stage 5).

Key Facts of Kindling Model

- Stimuli, such as electric, chemical, and psychological stress, induce kindling.
- Kindling models conceptualize the neuroadaptive mechanisms involved in progressive epilepsy.
- Kindling mechanisms are involved in the recurrence and cyclic symptoms of neuropsychiatric disorders.
- Changes in NMDARs, neurotrophic factors, and oxidative imbalance are involved in kindling development.
- Antikindling drugs are effective for the treatment of affective disorders.

Key Facts of Nicotine-Induced Kindling

- Nicotine kindling was demonstrated in the last decade.
- Nicotine kindling causes hyperactivity of neurons in the substantia nigra pars compacta and in both the medial and lateral part of the medial habenula.
- The anticonvulsant drug levetiracetam was effective in the prevention of nicotine kindling.
- Sex influences the development of kindling in periadolescent animals, being females more susceptible.
- Oxidative, glutamatergic, and neurotropic alterations underpin sex differences in kindling development.
- In adult animals, no influence of sex in the development of nicotine kindling was observed.
- Antioxidants, such as vitamin E and *N*-acetylcysteine, inhibit nicotine kindling development.

Summary Points

- Kindling is a model for the study of long-lasting neuroplasticity.
- Kindling mechanisms are related to epileptogenesis and to the spontaneous manifestation of symptoms of neuropsychiatric disorders, such as affective and substance use disorders.
- The kindling model seems to be an important approach for the study of neuroplasticity mechanism triggered by nicotine repeated administration.

- Kindling helps understanding nicotine-induced neuropsychiatric symptoms and brain structural alterations.
- The study of nicotine kindling has the perspective of developing appropriate targets for therapeutic drug development.

References

Ambelas, A. (1987). Life events and mania. A special relationship? *British Journal of Psychiatry*, *150*, 235–240. https://doi.org/10.1192/bjp.150.2.235.

Bassett, R. A., Osterhoudt, K., & Brabazon, T. (2014). Nicotine poisoning in an infant. *New England Journal of Medicine*, *370*(23), 2249–2250. https://doi.org/10.1056/NEJMc1403843.

Bastlund, J. F., Berry, D., & Watson, W. P. (2005). Pharmacological and histological characterisation of nicotine-kindled seizures in mice. *Neuropharmacology*, *48*(7), 975–983. https://doi.org/10.1016/j.neuropharm.2005.01.015.

Bender, R. E., & Alloy, L. B. (2011). Life stress and kindling in bipolar disorder: review of the evidence and integration with emerging biopsychosocial theories. *Clinical Psychology Review*, *31*(3), 383–398. https://doi.org/10.1016/j.cpr.2011.01.004.

Benito, E., & Barco, A. (2010). CREB's control of intrinsic and synaptic plasticity: implications for CREB-dependent memory models. *Trends in Neurosciences*, *33*(5), 230–240. https://doi.org/10.1016/j.tins.2010.02.001.

Berk, M., Munib, A., Dean, O., Malhi, G. S., Kohlmann, K., Schapkaitz, I., et al. (2011). Qualitative methods in early-phase drug trials: broadening the scope of data and methods from an RCT of N-acetylcysteine in schizophrenia. *The Journal of Clinical Psychiatry*, *72*(7), 909–913. https://doi.org/10.4088/JCP.09m05741yel.

Bertram, E. (2007). The relevance of kindling for human epilepsy. *Epilepsia*, *48*(s2), 65–74. https://doi.org/10.1111/j.1528-1167.2007.01068.x.

Clarke, P. B. S., & Kumar, R. (1983). The effects of nicotine on locomotor activity in non-tolerant and tolerant rats. *British Journal of Pharmacology*, *78*(2), 329–337. https://doi.org/10.1111/j.1476-5381.1983.tb09398.x.

Damaj, M. I., Glassco, W., Dukat, M., & Martin, B. R. (1999). Pharmacological characterization of nicotine-induced seizures in mice. *The Journal of Pharmacology and Experimental Therapeutics*, *291*(3), 1284–1291.

Dean, O., Giorlando, F., & Berk, M. (2011). N-acetylcysteine in psychiatry: current therapeutic evidence and potential mechanisms of action. *Journal of Psychiatry & Neuroscience: JPN*, *36*(2), 78–86. https://doi.org/10.1503/jpn.100057.

Dennison, Z., Campbell Teskey, G., & Cain, D. P. (1995). Persistence of kindling: effect of partial kindling, retention interval, kindling site, and stimulation parameters. *Epilepsy Research*, *21*(3), 171–182. https://doi.org/10.1016/0920-1211(95)00025-6.

Dhir, A. (2012). Pentylenetetrazol (PTZ) kindling model of epilepsy. *Current Protocols in Neuroscience*, *58*, 9.37.1–9.37.12. (2012). http://doi.wiley.com/10.1002/0471142301.ns0937s58.

Dworetzky, B. A., Bromfield, E. B., Townsend, M. K., & Kang, J. H. (2010). A prospective study of smoking, caffeine, and alcohol as risk factors for seizures or epilepsy in young adult women: data from the Nurses' Health Study II. *Epilepsia*, *51*(2), 198–205. https://doi.org/10.1111/j.1528-1167.2009.02268.x.

Goddard, G. V., McIntyre, D. C., & Leech, C. K. (1969). A permanent change in brain function resulting from daily electrical stimulation. *Experimental Neurology*, *25*(3), 295–330.

Gomes, P. X. L., De Oliveira, G. V., De Araújo, F. Y. R., De Barros Viana, G. S., De Sousa, F. C. F., Hyphantis, T. N., et al. (2013). Differences in vulnerability to nicotine-induced kindling between female and male periadolescent rats. *Psychopharmacology*, *225*(1)https://doi.org/10.1007/s00213-012-2799-5.

Guerriero, R. M., Rajadhyaksha, A., Crozatier, C., Giros, B., Nosten-Bertrand, M., & Kosofsky, B. E. (2005). Augmented constitutive CREB expression in the nucleus accumbens and striatum may contribute to the altered behavioral response to cocaine of adult mice exposed to cocaine in utero. *Developmental Neuroscience*, *27*(2–4), 235–248. https://doi.org/10.1159/000085997.

Hosford, D. A., Simonato, M., Cao, Z., Garcia-Cairasco, N., Silver, J. M., Butler, L., et al. (1995). Differences in the anatomic distribution of immediate-early gene expression in amygdala and angular bundle kindling development. *The Journal of Neuroscience_ The Official Journal of the Society for Neuroscience*, *15*(3 Pt 2), 2513–2523. (1995). Available from: http://www.ncbi.nlm.nih.gov/pubmed/7891185.

Itzhak, Y. (1996). Attenuation of cocaine kindling by 7-nitroindazole, an inhibitor of brain nitric oxide synthase. *Neuropharmacology*, *35*(8), 1065–1073.

Kalivas, P. W., & Stewart, J. (1991). Dopamine transmission in the initiation and expression of drug-induced and stress-induced sensitization of motor-activity. *Brain Research Reviews*, *16*(3), 223–244. https://doi.org/10.1016/0165-0173(91)90007-U.

Kaminski, R. M., Núñez-Taltavull, J. F., Budziszewska, B., Lasoń, W., Gasior, M., Zapata, A., et al. (2011). Effects of cocaine-kindling on the expression of NMDA receptors and glutamate levels in mouse brain. *Neurochemical Research*, *36*(1), 146–152. https://doi.org/10.1007/s11064-010-0284-2.

Kumar, G., Jones, N. C., Morris, M. J., Rees, S., O'Brien, T. J., & Salzberg, M. R. (2011). Early life stress enhancement of limbic epileptogenesis in adult rats: Mechanistic insights. M. Avoli (Ed.), *PLoS One*, *6*(9), e24033. Available from: http://www.ncbi.nlm.nih.gov/pmc/articles/PMC3177819/.

Lu, S. C. (2013). Glutathione synthesis. *Biochimica et Biophysica Acta*, *1830*(5), 3143–3153. https://doi.org/10.1016/j.bbagen.2012.09.008.

Mariani, J. J., & Levin, F. R. (2008). Levetiracetam for the treatment of co-occurring alcohol dependence and anxiety: case series and review. *The American Journal of Drug and Alcohol Abuse*, *34*(6), 683–691. https://doi.org/10.1080/00952990802308213.

Moussawi, K., Pacchioni, A., Moran, M., Olive, M. F., Gass, J. T., Lavin, A., et al. (2009). N-acetylcysteine reverses cocaine-induced metaplasticity. *Nature Neuroscience*, *12*(2), 182–189. https://doi.org/10.1038/nn.2250.

Nimse, S. B., & Pal, D. (2015). Free radicals, natural antioxidants, and their reaction mechanisms. *RSC Advances*, *5*(35), 27986–28006. https://doi.org/10.1039/C4RA13315C.

Nutt, D. J., Cowen, P. J., Batts, C. C., Grahame-Smith, D. G., & Green, A. R. (1982). Repeated administration of subconvulsant doses of GABA antagonist drugs— I. Effect on seizure threshold (kindling). *Psychopharmacology (Berlin)*, *76*(1), 84–87.

Okamura, A. M. N. C., Gomes, P. X. L., de Oliveira, G. V., Araújo, F. Y. R. D., Tomaz, V. S., Chaves Filho, A. J. M., et al. (2016). N-acetylcysteine attenuates nicotine-induced kindling in female periadolescent rats. *Progress in Neuro-Psychopharmacology and Biological Psychiatry*, *67*, https://doi.org/10.1016/j.pnpbp.2016.01.010.

Pinel, J. P. J., & Rovner, L. I. (1978). Experimental epileptogenesis: kindling-induced epilepsy in rats. *Experimental Neurology*, *58*(2), 190–202. https://doi.org/10.1016/0014-4886(78)90133-4.

Post, R. M. (2007). Kindling and sensitization as models for affective episode recurrence, cyclicity, and tolerance phenomena. *Neuroscience and Biobehavioral Reviews*, *31*, 858–873. https://doi.org/10.1016/j.neubiorev.2007.04.003.

Post, R. M., & Kopanda, R. T. (1976). Cocaine, kindling, and psychosis. *American Journal of Psychiatry*, *133*(6), 627–634. https://doi.org/10.1176/ajp.133.6.627.

Potschka, H., & Löscher, W. (1999). Corneal kindling in mice: Behavioral and pharmacological differences to conventional kindling. *Epilepsy Research*, *37*(2), 109–120. Available from: http://www.sciencedirect.com/science/article/pii/S0920121199000625.

Racine, R. J. (1972). Modification of seizure activity by electrical stimulation. II. Motor seizure. *Electroencephalography and Clinical Neurophysiology*, *32*(3), 281–294.

Rong, L., Frontera, A. T., & Benbadis, S. R. (2014). Tobacco smoking, epilepsy, and seizures. *Epilepsy & Behavior*, *31*, 210–218. https://doi.org/10.1016/j.yebeh.2013.11.022.

Schmaal, L., Berk, L., Hulstijn, K. P., Cousijn, J., Wiers, R. W., & van den Brink, W. (2011). Efficacy of N-acetylcysteine in the treatment of nicotine dependence: a double-blind placebo-controlled pilot study. *European Addiction Research*, *17*(4), 211–216. https://doi.org/10.1159/000327682.

Sieciechowicz, D., & Kohrman, M. (2015). Transdermal nicotine patch as a novel treatment for epilepsy associated with a mutation in the nicotinic acetylcholine receptor (S35.002). *Neurology*, *84*(14 Suppl).

Trevor, H. (2012). Amygdalar models of neurological and neuropsychiatric disorders. In *The amygdala—A discrete multitasking manager*. InTechhttps://doi.org/10.5772/50244.

Zhu, X., Dong, J., Shen, K., Bai, Y., Zhang, Y., Lv, X., et al. (2015). NMDA receptor NR2B subunits contribute to PTZ-kindling-induced hippocampal astrocytosis and oxidative stress. *Brain Research Bulletin*, *114*, 70–78. https://doi.org/10.1016/j.brainresbull.2015.04.002.

CHAPTER

37

Nicotine Reward and Abstinence: Role of the CB1 Receptors

S. Tannous, S. Caille

Aquitaine Institute for Cognitive and Integrative Neuroscience, University of Bordeaux, CNRS UMR5287, Bordeaux, France

Abbreviations

2-AG	2-arachidonoyl glycerol
AEA	anandamide
BLA	basolateral amygdala
BNST	bed nucleus of the stria terminalis
CB1	cannabinoid type 1 receptor
CPA	conditioned place aversion
CPP	conditioned place preference
D9-THC	delta-9-tetrahydrocannabinol
DA	dopamine
FAAH	fatty acid amide hydrolase
IM	intramuscular injection
IP	intraperitoneal injection
IV	intravenous injection
IVSA	intravenous self-administration
LTP	long-term potentiation
MAGL	monoacylglycerol lipase
NAC	nucleus accumbens
PFC	prefrontal cortex
VTA	ventral tegmental area

37.1 INTRODUCTION

37.1.1 Nicotine Addiction

Cigarette smoking is the most preventable cause of death with regard to global health. It is well accepted now that nicotine is the actual addictive compound of tobacco, its repeated use leading to compulsive drug seeking and use despite negative consequences, withdrawal syndrome upon drug cessation and relapse in about 80% of those who attempt to quit. Moreover, smoking behavior is not even a declining addiction but actually increasing since there is now consumption of vaporized nicotine through the e-cigarette device. Indeed, the number of worldwide e-cigarette users has been continuously increasing since 2010, most of them being ex-smokers (*World Health Organization*). Whatever the route of nicotine consumption, altogether, these data emphasize how crucial is the understanding of nicotine addiction mechanisms if we want to resolve this public health problem. Even though large amount of research has been done to develop therapeutic tools, clinical treatments against nicotine addiction that are currently available have very low success rate after 1-year therapy (Cahill, Stevens, Perera, & Lancaster, 2013).

Nicotine, like drugs of abuse, promotes reward and learning through an increase in neuronal activity of the dopaminergic neurons in the ventral tegmental area ("reward system"). By targeting and stimulating the nicotinic acetylcholine receptors, which are ligand-gated ion channels, and more specifically the heteromeric $\alpha4\beta2$ and homomeric $\alpha7$ subtypes located within the VTA, nicotine increases dopamine (DA) release in the nucleus accumbens (NAC) and in the prefrontal cortex (PFC). On the opposite, nicotine withdrawal decreases the brain reward system sensitivity (Kenny & Markou, 2005) and decreases extracellular DA levels within the accumbens (Zhang, Dong, Doyon, & Dani, 2012), leading to the dysphoric state associated with nicotine withdrawal and depressive-like behaviors.

37.1.2 How to Probe Nicotine Addictive Behaviors?

One of the pitfalls for drug development could be that animal models can't mimic all aspects of nicotine dependence and abstinence, but still, available models are efficient at measuring several behavioral parameters of the addictive process. Moreover, using those paradigms contributed to the thorough investigation of several neuronal

Neuroscience of Nicotine
https://doi.org/10.1016/B978-0-12-813035-3.00037-X

mechanisms involved in the appetitive and rewarding effects of nicotine. The first of the two main paradigms is the intravenous nicotine self-administration where the animal "works" specifically by targeting at an active operandum for an intravenous delivery of nicotine associated with the delivery of a cue light to facilitate the associative learning. It allows the quantification of drug-taking and drug-seeking behaviors. The second paradigm is nicotine-induced conditioned place preference where the animal is conditioned to associate the administration of nicotine to a certain environment, and the test consists in allowing the drug-free animal to return to or avoid the environment previously associated with nicotine. The appetitive effects of nicotine will promote the rat to "prefer" the drug-associated rather than the saline-associated environment. Moreover, withdrawal syndrome and abstinence neural substrates have been widely studied in rats following a period of chronic nicotine administration via repeated injections or subcutaneous osmotic minipumps. Both spontaneous and antagonist-precipitated drug discontinuation induce (i) a negative motivational state that contributes to the maintenance of drug-seeking behavior and (ii) a withdrawal syndrome that includes several somatic symptoms of variable intensity. The somatic signs that are observed during nicotine withdrawal syndrome are the following: body shakes, cheek tremors, escape attempts, eye blinks, gasps, genital licks, head shakes, ptosis, teeth chattering, writhes, and yawns (for review, see Kenny & Markou, 2001). Altogether, these animal models have validated the use of several "classical" therapeutic tools such as nicotinic receptor agonist pharmacology with the nicotine replacement therapy (NRT), nicotinic receptor partial agonist pharmacology with varenicline (trade name Champix), or noradrenergic/dopaminergic receptor indirect agonist pharmacotherapy such as bupropion (trade name Zyban) (Anthenelli et al., 2016; Cahill et al., 2013). It also brought in recent endeavors in medication development for nicotine addiction.

37.2 THE CB1 RECEPTOR DIRECT PHARMACOTHERAPY

One of the most promising pharmacotherapies involves several key players of the endocannabinoid system. Identifying the central Δ9-tetrahydrocannabinol (the active component of *Cannabis sativa indica*) and cloning of its target led to the discovery of the first endogenous receptor: the cannabinoid type 1 (CB1) receptor. Soon after, anandamide (AEA) and 2-arachidonoyl glycerol (2-AG), the two main endogenous ligands for the CB1 receptor, were isolated as well as their associated degradation enzymes, fatty amide acid hydrolase (FAAH) and monoacylglycerol lipase (MAGL), respectively. These CB1 receptors are the most abundant G-protein-coupled receptors in the brain of animals and humans, and they participate to a wide range of physiological functions. Their localization has indicated high densities in brain structures such as the basal ganglia, cerebellum, and hippocampus and to less extent in the brain reward structures as well. There is strong evidence that CB1 receptors are mostly presynaptic receptors from where they can control neurotransmitter release. Rapidly after these discoveries, strong evidence acknowledged that CB1 receptor is a key player in the modulation of the positive reinforcing effects of natural and pharmacological rewards. Subsequently, a selective CB1 receptor antagonist has been suggested as a pharmacotherapy for the treatment of obesity. At the same time, rimonabant (SR141716A, trade name Acomplia) has also been tested against nicotine addiction because a pioneer preclinical study had revealed that CB1 receptor knockout could block the appetitive properties of nicotine. This compound was a promising candidate since it was also likely to assist smokers in regulating the weight gain associated with their quit attempt. Thus, acute injection of rimonabant has been tested on several nicotine addiction-related behaviors, and it has demonstrated that CB1 receptors controlled nicotine-induced conditioned place preference, nicotine-taking and nicotine-seeking behaviors in the intravenous self-administration paradigm (Cohen, Perrault, Griebel, & Soubrie, 2005; Forget, Hamon, & Thiebot, 2005; Simonnet, Cador, & Caille, 2013). Surprisingly, acute rimonabant did not contribute to the attenuation of mecamylamine-induced withdrawal somatic signs in nicotine-dependent animals. Finally, CB1 receptors seemed to be involved in nicotine-induced DA release in the NAC (Cohen, Perrault, Voltz, Steinberg, & Soubrie, 2002). Altogether, these results were very promising in terms of potential use of the selective CB1 receptor antagonists in the treatment of nicotine addiction. Unfortunately, long-term use of CB1 receptor antagonists was accompanied by anxiogenic and depressive side effects (Cahill & Ussher, 2011). Thus, development of both rimonabant and taranabant (another CB1 receptor antagonist) was discontinued in 2008 due to their links with mental disorders (Rigotti et al., 2009) and unacceptable side effects. Here, we review the most recent evidence on the functional role of CB1 receptor signaling in nicotine-induced reward, seeking, and abstinence including the presentation of the latest endocannabinoid pharmacological tools (see Tables 37.1 and 37.2). A more extensive list of the early findings regarding CB1 receptors and endocannabinoid involvement in nicotine addictive processes has already been reviewed elsewhere (Gamaleddin et al., 2015).

TABLE 37.1 Effects of Drugs Targeting Directly or Indirectly CB1 Receptors on Nicotine-Induced Reward and Nicotine-Seeking Behavior

CB1r-dependent pharmacology	Route of administration	Behavioral effects	References
SR141716A	IP IM	↘ IVSA ↘ drug seeking ↘ CPP	De Vries, de Vries, Janssen, and Schoffelmeer (2005), Le Foll and Goldberg (2004), Cohen et al. (2005), Merritt et al. (2008), and Schindler et al. (2016)
	NACshell BLA PFC	↘ drug seeking = spontaneous locomotor activity	Kodas, Cohen, Louis, and Griebel (2007)
AM 251	VTA NAC BLA	↘ IVSA = IVSA ↘ CPP	Simonnet et al. (2013) and Hashemizadeh, Sardari, and Rezayof (2014)
	IV BNST	↘ IVSA-induced LTP in the BNST	Reisiger et al. (2014)
	IP IV	↘ IVSA ↘ seeking	Shoaib (2008) and Simonnet et al. (2013)
AM4113	IP IM	↘ IVSA and motivation ↘ drug- and stress-induced seeking ↘ activity of VTA dopamine neurons	Gueye et al. (2016) and Schindler et al. (2016)
URB597	IP IV	↘ or ↗ CPP ↘ IVSA ↘ drug seeking	Merritt et al. (2008), Scherma et al. (2008), and Justinova et al. (2015)
URB694	IV	↘ IVSA ↘ drug seeking	Justinova et al. (2015)
JZL184	IP	= IVSA ↗ cue-induced seeking	Trigo and Le Foll (2016)
AM404	IP	↘ CPP and seeking ↘ nicotine-induced DA in the NACshell	Scherma et al. (2012)
WIN55, 212-2	IP	↗ IVSA and motivation ↗ drug seeking	Gamaleddin et al. (2012)

Effect of the endocannabinoid pharmacology on appetitive nicotine-related behaviors. Effects either ↗ increase, ↘ decrease, or stay = unchanged (no modification). Compounds are ranked based on the ability to inhibit/activate CB1 receptors. First are the inverse agonists (SR141716A and AM251) and then the neutral antagonist (AM4113); following are for FAAH inhibitors (URB597 and URB694), MAGL inhibitor (JZL184), or AEA transporter inhibitors (AM404) and finally the CB1 receptor agonist WIN55, 212-2.

TABLE 37.2 Behavioral Effects of Drugs Targeting CB1 Receptor on Abstinence and Withdrawal

CB1 r-dependent pharmacology	Route of administration	Behavioral effects	References
SR141716A	IP	↘ somatic withdrawal signs ↗ withdrawal memory impairment	Merritt et al. (2008) and Saravia et al. (2017)
URB597	IP	↘ withdrawal-induced anxiety = or ↗withdrawal	Cippitelli et al. (2011) and Merritt et al. (2008)
O7460	IP	↘ abstinence memory impairment ↗ somatic withdrawal severity	Saravia et al. (2017)
JZL184	IP	= cognitive deficits ↘ withdrawal severity	Saravia et al. (2017) and Muldoon et al. (2015)
D9-THC	IP	↘ withdrawal signs ↘ c-Fos in the caudate putamen and the dentate gyrus	Balerio, Aso, Berrendero, Murtra, and Maldonado (2004)

Effect of the endocannabinoid pharmacology on nicotine withdrawal-induced behaviors. Effects either ↗ increase, ↘ decrease, or stay = unchanged (no modification). Compounds are ranked based on the ability to inhibit/activate CB1 receptors. First are the inverse agonists (SR141716A) and then the FAAH inhibitor (URB597); following layers are for 2-AG biosynthesis enzyme inhibitor (O7460) and MAGL inhibitor (JZL184) and finally the CB1 receptor agonist D9-THC.

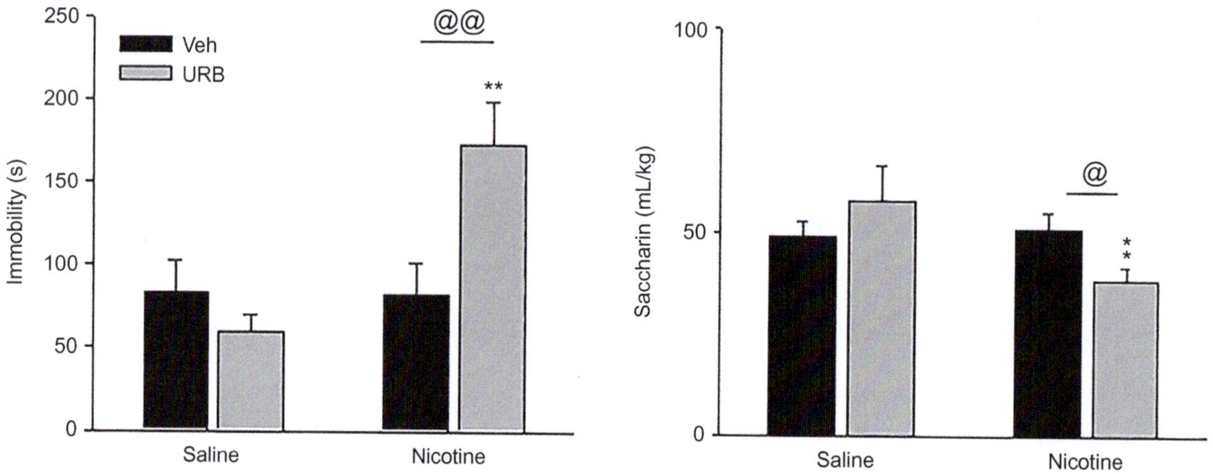

FIG. 37.1 FAAH inhibition and depressive-like behavioral symptoms. Effect of chronic URB597 on depressive-like symptoms during protracted abstinence. *Left panel*, time spent in immobility time (sec) is measured in the forced swimming test measured on day 38 of abstinence. *Right panel*, relative saccharin intake (mL/kg) is measured on day 41 of abstinence. Saline vs nicotine self-administration exposed rats, vehicle (Veh, *black bars*)- vs URB597 (URB, *gray bars*)-treated rats. Values are expressed as mean ± SEM. Fisher LSD post hoc @$P < .05$ or @@$P < .01$ NIC-V vs NIC-URB; **$P < .01$ SAL-URB vs NIC-URB. *From Simonnet et al. (2017); with permission from the publishers.*

37.3 THE CB1 RECEPTOR INDIRECT PHARMACOTHERAPY

Another strategy to assist with smoking cessation has been to restore the balance of the endocannabinoid system with the use of the inhibitors of the endocannabinoid degradation enzymes. Indeed, the hypothesis was that if targeting at CB1 receptors could prevent nicotine reward, it suggested that the endogenous cannabinoids were tonically involved with the induction or expression of nicotine addiction. Moreover, the endogenous cannabinoid system had been implicated in the stress response as it was shown that endocannabinoids are recruited during stress to attenuate the corticosterone rise and to alleviate anxiogenic and depressive-like behaviors (for review, see Patel, Hill, Cheer, Wotjak, & Holmes, 2017). Thus, these additional therapeutic properties further supported the beneficial use of FAAH/MAGL inhibitors considering that nicotine abstinence promoted negative emotional states. There were few studies done aiming at the MAGL inhibition with JZL184. While Muldoon et al. (2015) have shown that genetic deletion of MAGL or administration with JZL184 reduced nicotine withdrawal syndrome, another study has indicated that JZL184 enhanced cue-induced nicotine relapse (Trigo & Le Foll, 2016). Altogether, these data have suggested a complex involvement of the 2-AG tonic tone. On the other hand, several studies were done with the FAAH inhibitor URB597. It has been shown that acute URB597 decreased nicotine rewarding properties in rodent using the conditioned place preference paradigm (Scherma et al., 2008), reduced acquisition of nicotine intravenous self-administration and nicotine seeking (Forget, Coen, & Le Foll, 2009; Forget, Guranda, Gamaleddin, Goldberg, & Le Foll, 2016), and improved nicotine withdrawal-induced affective but not somatic symptoms (Cippitelli et al., 2011; Merritt, Martin, Walters, Lichtman, & Damaj, 2008). In addition, FAAH inhibition seemed to prevent nicotine-induced extracellular DA elevations in the NAC (Scherma et al., 2008). Thus, FAAH inhibition seemed to be the best pharmacotherapy against nicotine addiction.

37.4 ACUTE VS CHRONIC TREATMENT WITH THE CB1-TARGETING PHARMACOTHERAPY

Importantly, most studies have examined the acute use of these pharmacotherapies, often neglecting the fact that cessation medications are usually given over several weeks or months until the outcome of prolonged abstinence has been reached (for at least 6 months) (see Table 37.3 key facts). However, that's under such conditions of chronic treatment with rimonabant in clinical and preclinical settings that the occurrence of serious adverse effects has been demonstrated (Beyer et al., 2010; Cahill & Ussher, 2011; O'Brien et al., 2013) and that decision of rimonabant discontinuation has been made. Recently developed, a neutral CB1 receptor

TABLE 37.3 Key Facts About the Acute vs Chronic Use of CB1 Direct and Indirect Pharmacology

- Nicotine passive administration and intravenous self-administration increase extracellular endocannabinoid levels in several brain structures, including areas from the reward pathway (VTA, NAC, PFC, etc.)
- Acute administration with CB1 receptor indirect agonists such as degradation enzyme inhibitor compounds promotes similar inhibitory control over nicotine addictive processes than acute treatment with CB1 receptor antagonists. (See Tables 37.1 and 37.2.)
- Preclinical studies have suggested that chronic administration of rimonabant (Beyer et al., 2010) or FAAH inhibitor in rodents induces cognitive deficits and depressive-like behaviors (Basavarajappa, Nagre, Xie, & Subbanna, 2014; Ceci et al., 2014; Wu et al., 2014).
- However, endocannabinoid pharmacology had never been tested with chronic schedule of treatment in drug abstinent animals until recent studies were reported. (Gueye et al., 2016; Simonnet et al., 2017).
- It has been demonstrated that prolonged FAAH inhibition with a low dose of URB597 (0.3 mg/kg/day) in nicotine-abstinent rats worsen rather than improved withdrawal symptoms. It increased depressive-like symptoms and stress-induced plasma corticosterone rise (Figs. 37.1 and 37.2).
- Both behavioral and physiological depression biomarkers were correlated with alterations of the habenular CB1 receptor density and activity (Fig. 37.3).

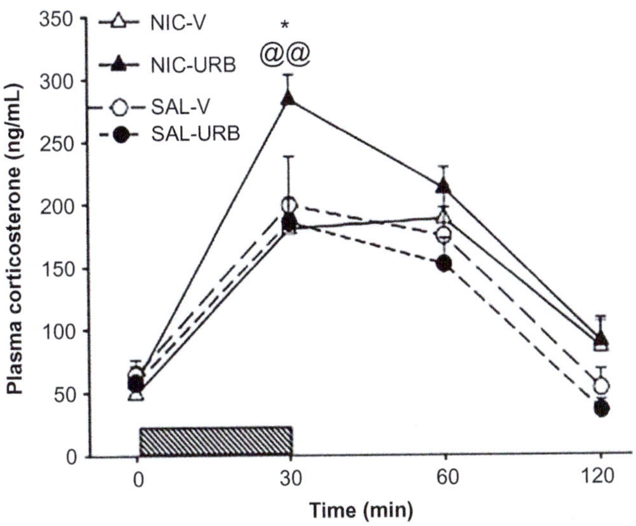

FIG. 37.2 FAAH inhibition on stress-induced plasma corticosterone rise, a biomarker of depression. Effect of chronic URB597 on plasmatic corticosterone levels in response to restraint stress during nicotine abstinence. Plasma concentrations of corticosterone at 0, 30, 60 and 120 min post stress (30 min restraint indicated by the striped bar) are measured on day 44. Values are expressed as mean ± SEM. Saline vs nicotine self-administration exposed rats, vehicle (V, *white symbols*)- vs URB597 (URB, *black symbols*)-treated rats. Fisher LSD post hoc @@$P<.01$ NIC-V vs NIC-URB; *$P<.01$ SAL-URB vs NIC-URB. *From Simonnet et al. (2017); with permission from the publishers.*

antagonist AM4113 (Gueye et al., 2016) has opened new prospects for therapeutic intervention helping smoking cessation. Indeed, it attenuates motivation in rats self-administering intravenous nicotine under a progressive-ratio schedule of reinforcement; it also decreases nicotine-seeking behavior induced by cue, stress, or drug priming. Moreover, it seems to show a better psychiatric tolerability following chronic exposure. No study had been done on the long-term use of JZL184 or URB597 in the treatment of nicotine addiction until we recently investigated the effect of chronic FAAH inhibition during prolonged abstinence in nicotine-dependent rats (Simonnet, Zamberletti, Cador, Rubino, & Caille, 2017). In our conditions, rats were first exposed to a prolonged period of nicotine intravenous self-administration, which made the animals nicotine-dependent, as shown by the withdrawal-induced anhedonia upon nicotine cessation. Then, we sought to assess the effects of chronic treatment with URB597 in nicotine-abstinent and control rats. Our hypothesis was that by restoring the AEA tone, URB597 would prevent the emergence of negative emotion-related symptoms during nicotine abstinence and be a good candidate for smoking cessation therapy. In contrast to what was expected, our results demonstrated that URB597 induced persistent anhedonia and severe depressive-like behavior and biomarker (Simonnet et al., 2017). Together, these new data suggested that maybe long-term use of URB597 had more of a CB1 receptor antagonist-like action, promoting the same kind of affective side effects. Thus, to test our new hypothesis, we examined CB1 receptor binding and functionality in several brain areas, including the habenula based on its implication in both nicotine addiction and depressive symptoms (Baldwin, Alanis, & Salas, 2011; Fowler & Kenny, 2014). Instead of restoring the endocannabinoid balance, we actually confirmed that chronic URB597 specifically decreased both CB1 receptor density and activity in the habenula of nicotine-abstinent rats. Removing the CB1-receptor-mediated inhibition might have contributed to the emergence of depressive-like symptoms (Lecca et al., 2016; Simonnet et al., 2017).

The fact that we still don't understand the neuroadaptations that take place following prolonged treatment targeting the endocannabinoid is a serious "brake" to its therapeutic use. Thus, we provide here a brief synopsis of endocannabinoid pharmacology relevance to chronic treatment of nicotine addiction highlighting the fact that current animal and human laboratory models have limited predictive clinical validity, often neglecting the long-term period of use of these medications (see Table 37.3).

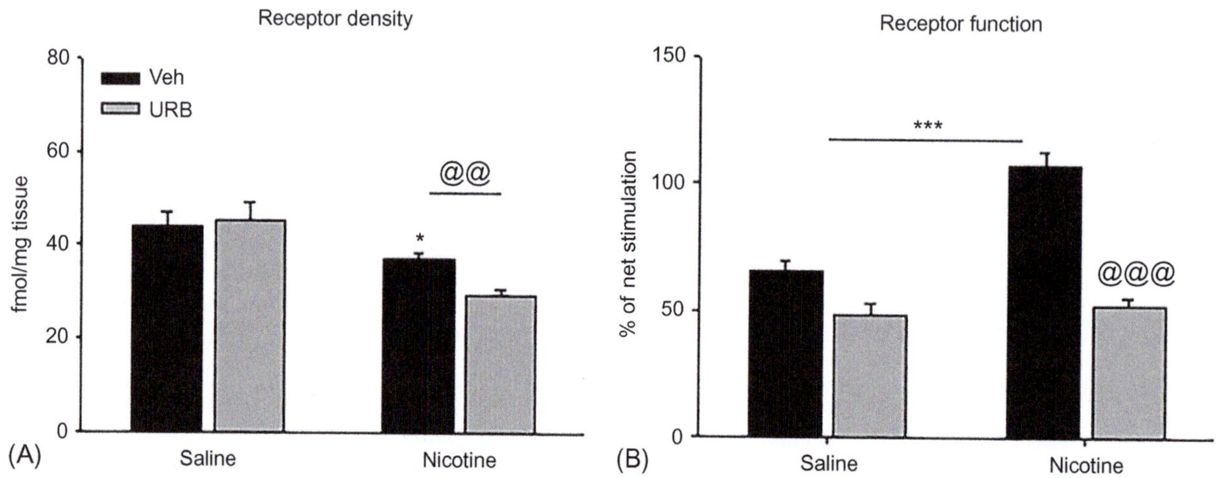

FIG. 37.3 Effects of FAAH inhibition during nicotine abstinence on CB1 receptor density and function. CB1 receptor binding and CP-55,940-stimulated [35S]GTPgS binding in autoradiography in the habenula. (A) CB1 receptor density expressed in "fmol/mg tissue" and (B) CB1 receptor function expressed in "% of net stimulation." Values are expressed as mean ± SEM. Saline (SAL) vs nicotine (NIC) self-administration exposed rats. Vehicle (Veh, *black bars*)- vs URB597 (URB, *gray bars*)-treated rats. Fisher LSD post hoc test $*P < .05$ or $***P < .001$ SAL-V vs NIC-V; $@@P > .01$ or $@@@P < .001$ NIC-V vs NIC-URB. *From Simonnet et al. (2017); with permission from the publishers.*

MINI-DICTIONARY OF TERMS

Abstinence This part of the addiction process describes a drug cessation episode without extinction of drug-taking behavior.

Cannabinoid type 1 receptor Cannabinoid type 1 (CB1) receptors are Gi/o-protein-coupled receptors that are primarily localized on axon terminals. Thus, activation of CB1 receptors by endogenous ligands or exogenous ligands such as delta-9-tetrahydrocannabinol results in a robust suppression of neurotransmitter release into the synapse.

Conditioned place preference/aversion Paradigm used as a behavioral preclinical and nonoperant procedure to explore the positive/negative properties of nicotine/nicotine withdrawal. At first, animals are allowed to explore all environments. Then, nicotine or nicotine withdrawal effects are associated with a specific environment. During the test, animals will spend more time in drug-associated compartment in case of positive conditioned memories or will avoid the withdrawal-associated compartment in case of aversive conditioned memories.

Dependence Pathological homeostatic state in which neuroadaptations develop to compensate for chronic drug use, usually referred as physical dependence. Psychological dependence refers to the mental condition that leads to compulsive drug use.

Endocannabinoid system The endogenous cannabinoid system (ECS) is a neuromodulatory lipid system necessary for a variety of physiological processes, including reward control. This system includes cannabinoid receptors (CB1 and CB2); endogenous ligands such as anandamide (AEA) and 2-arachidonoyl glycerol (2-AG); and their degradation enzymes, fatty acid amide hydrolase (FAAH) and monoacylglycerol lipase (MAGL), respectively.

Intravenous drug self-administration Operant behavior that allows an animal to self-administer a drug, and it is used in laboratories to assess its addiction potential. Nicotine self-administration is usually associated with a cue-light delivery that signals the drug delivery. Robust self-administration indicates the rewarding effects of nicotine. It allows the measure of nicotine consumption under a fixed-ratio schedule of reinforcement, and it allows to examine the motivation for nicotine (ratio workload/reward—breaking point) under a progressive-ratio schedule of reinforcement.

Reinstatement/drug seeking It's the animal model of drug relapse in human. Term used to designate the reinstatement of drug self-administration behavior after extinction or abstinence, also called "drug-seeking" behavior. Reinstatement can be initiated by a dose of the drug itself, a cue, or environment associated with the drug availability or a stressor.

Withdrawal syndrome Negative physical and affective state upon drug withdrawal that results from the unmasking of the "opponent" processes occurring during drug dependence. The symptoms are specific to the drug. Nicotine withdrawal can be precipitated in animals either by drug cessation (spontaneous syndrome) or by the administration of a selective nicotinic antagonist such as mecamylamine. In rodents, nicotine withdrawal involves a physical component that included paw and body tremors, head shakes, backing, jumps, curls, and ptosis and a negative affective component that included anxiety- and depressive-like behaviors.

Key Facts of Depressive-Like Behaviors

- Up to now, there is no agreement on a specific rodent model for depression. Thus, behavioral studies dealing with this question are talking about "depressive-like" behaviors.
- It gathers several behavioral tests among which we have decreased interest or pleasure in most activities (anhedonia), significant weight change, change in sleep, despair behavior, and high stress response/irritable mood.
- Decreased consumption of sweet solution reflects anhedonia.
- Increased immobility in a forced swimming test reflects behavioral despair.
- Higher response to stress correlates with higher rise in plasmatic corticosterone levels (a depression biomarker).

Summary Points

- Among the human population, 30% of those who try smoking develop a nicotine addiction.
- Clinically effective smoking cessation treatments are few in number, mainly varenicline, bupropion, and nicotine replacement therapy being prescribed by health organizations.
- Challenges to develop new pharmacotherapies for the treatment of nicotine dependence persist.
- The endocannabinoid system, more specifically molecules targeting at CB1 receptors, has opened new therapeutic opportunities.
- However, chronic use of endocannabinoid pharmacotherapy raises several issues and needs further investigation to refine its therapeutic use.

References

Anthenelli, R. M., Benowitz, N. L., West, R., St Aubin, L., McRae, T., Lawrence, D., et al. (2016). Neuropsychiatric safety and efficacy of varenicline, bupropion, and nicotine patch in smokers with and without psychiatric disorders (EAGLES): a double-blind, randomised, placebo-controlled clinical trial. *Lancet*, *387*(10037), 2507–2520.

Baldwin, P. R., Alanis, R., & Salas, R. (2011). The role of the habenula in nicotine addiction. *Journal of Addiction Research & Therapy*, *S1*(2).

Balerio, G. N., Aso, E., Berrendero, F., Murtra, P., & Maldonado, R. (2004). Delta9-tetrahydrocannabinol decreases somatic and motivational manifestations of nicotine withdrawal in mice. *The European Journal of Neuroscience*, *20*(10), 2737–2748.

Basavarajappa, B. S., Nagre, N. N., Xie, S., & Subbanna, S. (2014). Elevation of endogenous anandamide impairs LTP, learning, and memory through CB1 receptor signaling in mice. *Hippocampus*, *24*(7), 808–818.

Beyer, C. E., Dwyer, J. M., Piesla, M. J., Platt, B. J., Shen, R., Rahman, Z., et al. (2010). Depression-like phenotype following chronic CB1 receptor antagonism. *Neurobiology of Disease*, *39*(2), 148–155.

Cahill, K., Stevens, S., Perera, R., & Lancaster, T. (2013). Pharmacological interventions for smoking cessation: an overview and network meta-analysis. *Cochrane Database of Systematic Reviews*, *5*, CD009329.

Cahill, K., & Ussher, M. H. (2011). Cannabinoid type 1 receptor antagonists for smoking cessation. *Cochrane Database of Systematic Reviews*, *3*, CD005353.

Ceci, C., Mela, V., Macri, S., Marco, E. M., Viveros, M. P., & Laviola, G. (2014). Prenatal corticosterone and adolescent URB597 administration modulate emotionality and CB1 receptor expression in mice. *Psychopharmacology*, *231*(10), 2131–2144.

Cippitelli, A., Astarita, G., Duranti, A., Caprioli, G., Ubaldi, M., Stopponi, S., et al. (2011). Endocannabinoid regulation of acute and protracted nicotine withdrawal: effect of FAAH inhibition. *PLoS ONE*, *6*(11) e28142.

Cohen, C., Perrault, G., Griebel, G., & Soubrie, P. (2005). Nicotine-associated cues maintain nicotine-seeking behavior in rats several weeks after nicotine withdrawal: reversal by the cannabinoid (CB1) receptor antagonist, rimonabant (SR141716). *Neuropsychopharmacology*, *30*(1), 145–155.

Cohen, C., Perrault, G., Voltz, C., Steinberg, R., & Soubrie, P. (2002). SR141716, a central cannabinoid (CB(1)) receptor antagonist, blocks the motivational and dopamine-releasing effects of nicotine in rats. *Behavioural Pharmacology*, *13*(5–6), 451–463.

De Vries, T. J., de Vries, W., Janssen, M. C., & Schoffelmeer, A. N. (2005). Suppression of conditioned nicotine and sucrose seeking by the cannabinoid-1 receptor antagonist SR141716A. *Behavioural Brain Research*, *161*(1), 164–168.

Forget, B., Coen, K. M., & Le Foll, B. (2009). Inhibition of fatty acid amide hydrolase reduces reinstatement of nicotine seeking but not break point for nicotine self-administration–comparison with CB(1) receptor blockade. *Psychopharmacology*. https://doi.org/10.1007/s00213-009-1569-5.

Forget, B., Guranda, M., Gamaleddin, I., Goldberg, S. R., & Le Foll, B. (2016). Attenuation of cue-induced reinstatement of nicotine seeking by URB597 through cannabinoid CB1 receptor in rats. *Psychopharmacology*, *233*(10), 1823–1828. https://doi.org/10.1007/s00213-016-4232-y.

Forget, B., Hamon, M., & Thiebot, M. H. (2005). Cannabinoid CB1 receptors are involved in motivational effects of nicotine in rats. *Psychopharmacology*, *181*(4), 722–734.

Fowler, C. D., & Kenny, P. J. (2014). Nicotine aversion: neurobiological mechanisms and relevance to tobacco dependence vulnerability. *Neuropharmacology*, *76*(Pt B), 533–544.

Gamaleddin, I. H., Trigo, J. M., Gueye, A. B., Zvonok, A., Makriyannis, A., Goldberg, S. R., et al. (2015). Role of the endogenous cannabinoid system in nicotine addiction: novel insights. *Frontiers in Psychiatry*, *6*, 41.

Gamaleddin, I., Wertheim, C., Zhu, A. Z., Coen, K. M., Vemuri, K., Makryannis, A., et al. (2012). Cannabinoid receptor stimulation increases motivation for nicotine and nicotine seeking. *Addiction Biology*, *17*(1), 47–61.

Gueye, A. B., Pryslawsky, Y., Trigo, J. M., Poulia, N., Delis, F., Antoniou, K., et al. (2016). The CB1 neutral antagonist AM4113 retains the therapeutic efficacy of the inverse agonist rimonabant for nicotine dependence and weight loss with better psychiatric tolerability. *The International Journal of Neuropsychopharmacology*, *19*(12).

Hashemizadeh, S., Sardari, M., & Rezayof, A. (2014). Basolateral amygdala CB1 cannabinoid receptors mediate nicotine-induced place preference. *Progress in Neuro-Psychopharmacology & Biological Psychiatry*, *51*, 65–71.

Justinova, Z., Panlilio, L. V., Moreno-Sanz, G., Redhi, G. H., Auber, A., Secci, M. E., et al. (2015). Effects of fatty acid amide hydrolase (FAAH) inhibitors in non-human primate models of nicotine reward and relapse. *Neuropsychopharmacology*, *40*(9), 2185–2197.

Kenny, P. J., & Markou, A. (2001). Neurobiology of the nicotine withdrawal syndrome. *Pharmacology, Biochemistry, and Behavior*, *70*(4), 531–549.

Kenny, P. J., & Markou, A. (2005). Conditioned nicotine withdrawal profoundly decreases the activity of brain reward systems. *The Journal of Neuroscience*, *25*(26), 6208–6212.

Kodas, E., Cohen, C., Louis, C., & Griebel, G. (2007). Cortico-limbic circuitry for conditioned nicotine-seeking behavior in rats involves endocannabinoid signaling. *Psychopharmacology*, *194*(2), 161–171.

Le Foll, B., & Goldberg, S. R. (2004). Rimonabant, a CB1 antagonist, blocks nicotine-conditioned place preferences. *NeuroReport*, *15*(13), 2139–2143.

Lecca, S., Pelosi, A., Tchenio, A., Moutkine, I., Lujan, R., Herve, D., et al. (2016). Rescue of GABAB and GIRK function in the lateral habenula by protein phosphatase 2A inhibition ameliorates depression-like phenotypes in mice. *Nature Medicine*, *22*(3), 254–261.

Merritt, L. L., Martin, B. R., Walters, C., Lichtman, A. H., & Damaj, M. I. (2008). The endogenous cannabinoid system modulates nicotine reward and dependence. *The Journal of Pharmacology and Experimental Therapeutics*, *326*(2), 483–492.

Muldoon, P. P., Chen, J., Harenza, J. L., Abdullah, R. A., Sim-Selley, L. J., Cravatt, B. F., et al. (2015). Inhibition of monoacylglycerol lipase reduces nicotine withdrawal. *British Journal of Pharmacology*, *172*(3), 869–882.

O'Brien, L. D., Wills, K. L., Segsworth, B., Dashney, B., Rock, E. M., Limebeer, C. L., et al. (2013). Effect of chronic exposure to rimonabant and phytocannabinoids on anxiety-like behavior and saccharin palatability. *Pharmacology, Biochemistry, and Behavior, 103*(3), 597–602.

Patel, S., Hill, M. N., Cheer, J. F., Wotjak, C. T., & Holmes, A. (2017). The endocannabinoid system as a target for novel anxiolytic drugs. *Neuroscience & Biobehavioral Reviews, 76*(Pt A), 56–66.

Reisiger, A. R., Kaufling, J., Manzoni, O., Cador, M., Georges, F., & Caille, S. (2014). Nicotine self-administration induces CB1-dependent LTP in the bed nucleus of the stria terminalis. *The Journal of Neuroscience, 34*(12), 4285–4292.

Rigotti, N. A., Gonzales, D., Dale, L. C., Lawrence, D., Chang, Y., & CIRRUS Study Group. (2009). A randomized controlled trial of adding the nicotine patch to rimonabant for smoking cessation: efficacy, safety and weight gain. *Addiction, 104*(2), 266–276.

Saravia, R., Flores, A., Plaza-Zabala, A., Busquets-Garcia, A., Pastor, A., de la Torre, R., et al. (2017). CB1 cannabinoid receptors mediate cognitive deficits and structural plasticity changes during nicotine withdrawal. *Biological Psychiatry, 81*(7), 625–634.

Scherma, M., Justinova, Z., Zanettini, C., Panlilio, L. V., Mascia, P., Fadda, P., et al. (2012). The anandamide transport inhibitor AM404 reduces the rewarding effects of nicotine and nicotine-induced dopamine elevations in the nucleus accumbens shell in rats. *British Journal of Pharmacology, 165*(8), 2539–2548.

Scherma, M., Panlilio, L. V., Fadda, P., Fattore, L., Gamaleddin, I., Le Foll, B., et al. (2008). Inhibition of anandamide hydrolysis by cyclohexyl carbamic acid 3′-carbamoyl-3-yl ester (URB597) reverses abuse-related behavioral and neurochemical effects of nicotine in rats. *The Journal of Pharmacology and Experimental Therapeutics, 327*(2), 482–490.

Schindler, C. W., Redhi, G. H., Vemuri, K., Makriyannis, A., Le Foll, B., Bergman, J., et al. (2016). Blockade of nicotine and cannabinoid reinforcement and relapse by a cannabinoid CB1-receptor neutral antagonist AM4113 and inverse agonist rimonabant in squirrel monkeys. *Neuropsychopharmacology, 41*(9), 2283–2293.

Shoaib, M. (2008). The cannabinoid antagonist AM251 attenuates nicotine self-administration and nicotine-seeking behaviour in rats. *Neuropharmacology, 54*(2), 438–444.

Simonnet, A., Cador, M., & Caille, S. (2013). Nicotine reinforcement is reduced by cannabinoid CB1 receptor blockade in the ventral tegmental area. *Addiction Biology, 18*(6), 930–936.

Simonnet, A., Zamberletti, E., Cador, M., Rubino, T., & Caille, S. (2017). Chronic FAAH inhibition during nicotine abstinence alters habenular CB1 receptor activity and precipitates depressive-like behaviors. *Neuropharmacology, 113*(Pt A), 252–259.

Trigo, J. M., & Le Foll, B. (2016). Inhibition of monoacylglycerol lipase (MAGL) enhances cue-induced reinstatement of nicotine-seeking behavior in mice. *Psychopharmacology, 233*(10), 1815–1822.

Wu, C. S., Morgan, D., Jew, C. P., Haskins, C., Andrews, M. J., Leishman, E., et al. (2014). Long-term consequences of perinatal fatty acid amino hydrolase inhibition. *British Journal of Pharmacology, 171*(6), 1420–1434.

Zhang, L., Dong, Y., Doyon, W. M., & Dani, J. A. (2012). Withdrawal from chronic nicotine exposure alters dopamine signaling dynamics in the nucleus accumbens. *Biological Psychiatry, 71*(3), 184–191.

CHAPTER

38

The Therapeutic Potential of the Cognitive-Enhancing Effects of Nicotine and Other Nicotinic Acetylcholine Receptor Agonists

Britta Hahn

University of Maryland School of Medicine, Maryland Psychiatric Research Center, Baltimore, MD, United States

Abbreviations

b.i.d. twice a day
nAChR nicotinic acetylcholine receptor

Laboratory evidence that the prototypical nicotinic acetylcholine receptor (nAChR) agonist nicotine can facilitate cognitive performance, in particular attentional processes, is plentiful (Hahn, 2015; Heishman, Kleykamp, & Singleton, 2010). Not all reports have been positive, and observed effects tended to be of small to moderate size (Heishman et al., 2010); however, more profound benefits were expected in disorders marked by cognitive deficits and nAChR hypofunction, such as mild cognitive impairment, Alzheimer's disease (AD), and schizophrenia (Adams & Stevens, 2007; Kendziorra et al., 2011). To date, a range of novel agonists with greater selectivity for subtypes of the nAChR have been developed and advanced to clinical trial stage for the treatment of cognitive deficits in the above disorders (Haydar & Dunlop, 2010). However, despite over two decades of drug development efforts, cognitive benefits of the compounds tested have been small and of uncertain clinical significance. No nAChR ligand has achieved FDA approval for treating cognitive deficits to date. The present chapter discusses possible reasons for this disappointing trend and potential future strategies.

38.1 WHICH nAChR SUBTYPES TO TARGET?

The development of novel nAChR agonists as cognitive enhancers has targeted the $\alpha4\beta2^*$ and the $\alpha7$ subtypes of the nAChR (Gundisch & Eibl, 2011; Wallace, Ballard, Pouzet, Riedel, & Wettstein, 2011). These are the two most abundantly and widely expressed nAChR subtypes in the brain, leaving few systems unaffected (Gotti et al., 2007; Seguela, Wadiche, Dineley-Miller, Dani, & Patrick, 1993). Thus, while reduced peripheral side effects can be expected (and have been observed) with these compounds relative to the nonselective nAChR agonist nicotine, the potential of preserved cognitive benefits with lower centrally mediated unwanted effects, as for example, related to reward processing or dependence, is low.

The basic problem is that the brain systems critical for mediating the cognitive enhancing effects of nAChR agonists are largely unknown. Evidence from preclinical and human fMRI studies suggests that modulation of noradrenergic and glutamatergic neurons and of fronto-parietal and default network regions may be important (e.g., Hahn et al., 2007; Hahn & Stolerman, 2005; Lawrence, Ross, & Stein, 2002; Quarta et al., 2007; Sutherland et al., 2015), but more research is warranted. As a result of the uncertainty regarding the target system(s), the nAChR subtypes to be targeted for the purpose of selectively

Neuroscience of Nicotine
https://doi.org/10.1016/B978-0-12-813035-3.00038-1

modulating the relevant systems are also unknown. The subtypes modulating the mesolimbic dopamine system have been studied in depth due to the evidence that this system plays a central role in the positive reinforcing effects of nicotine (Exley, Clements, Hartung, McIntosh, & Cragg, 2008; Gotti et al., 2010). Knowledge about the target system(s) central to the cognitive enhancing effects of nicotine would lead to similarly detailed analysis of the nAChR subtypes modulating these systems and channel drug development toward more selectively expressed subtypes, allowing for a narrower effect profile. Thus, the focus of drug development efforts on the two most widely expressed nAChR subtypes may be a reflection of a profound lack of knowledge regarding the relevant target systems.

The fact that small benefits have been observed both with α4β2*-selective and with α7-selective nAChR agonists (Haydar & Dunlop, 2010) suggests that greater efficacy may be expected with compounds acting on multiple nAChR subtypes. For example, nAChR agonist development for the cognitive deficits associated with schizophrenia has focused on the α7 subtype, based on findings that sensory gating deficits in people with schizophrenia and their relatives and the diagnosis of schizophrenia itself are associated with polymorphisms near the α7 subunit gene (Freedman et al., 1997; Leonard et al., 2002; Martin, Kem, & Freedman, 2004). However, both α7 and α4β2 nAChR binding sites are decreased in schizophrenia (Martin et al., 2004), and polymorphisms in the α5 subunit gene have also been linked to the diagnosis of schizophrenia (Hong et al., 2011). The critical challenge will be to find the right mix of nAChR subunit activity to maximize desirable relative to undesirable brain system modulations. This can only be achieved with a combination of selective nAChR ligands that leave large parts of the brain unaffected.

38.2 OVERDOSED?

Preclinical studies suggest an inverted U-shaped dose-response function of the cognitive enhancing effects of nicotine and other nAChR agonists (Bushnell, Oshiro, & Padnos, 1997; Hahn, Sharples, Wonnacott, Shoaib, & Stolerman, 2003; Hahn, Shoaib, & Stolerman, 2002; Levin, 1992; McGaughy, Decker, & Sarter, 1999). Reasons for this may involve not only aversive effects at larger doses (Perkins et al., 1994; Stolerman, 1999) but also pharmacological properties of the nAChR, whose predominant signaling mechanism appears to involve responses to low and sustained rises in acetylcholine and choline concentration from nonsynaptic diffuse volume transmission (Dani & Bertrand, 2007). Because nAChR desensitization becomes increasingly fast and persistent with higher agonist concentrations, the maximum receptor response can occur at concentrations below its EC_{50} (Mike, Castro, & Albuquerque, 2000). Indeed, under sustained equilibrium conditions, low but not high agonist concentrations can evoke steady-state nAChR activation (Papke et al., 2011).

In an early human phase I trial testing two doses of the partial α7 nAChR agonist DMXB-A in people with schizophrenia, the smaller dose tended to have more pronounced acute benefits in a cognitive battery (Olincy et al., 2006), consistent with the preclinical literature. The follow-up phase II trial of this compound, testing the same doses b.i.d., found comparable small cognitive benefits with both doses but more adverse effects with the larger dose (Freedman et al., 2008). Fig. 38.1 sketches the hypothetical dose-response relationship of cognitive benefits relative to adverse side effects. In the development of nAChR agonists for cognitive enhancement, decision-making processes characteristic of early-stage drug development, where the dose of choice tends to be close to the largest dose meeting safety standards with the aim of demonstrating efficacy, may be a maladaptive strategy. Choosing lower than maximal possible doses may result in larger benefits with fewer side effects and less neuroadaptation.

The development of nAChR positive allosteric modulators (PAMs) is in line with nAChR modulation strategies that are more subtle and take the receptor's endogenous physiological parameters into account. PAMs do not activate the nAChR on their own but facilitate agonist-induced responses by binding to a second, modulatory site (Schrattenholz et al., 1996; Williams, Wang, & Papke, 2011). By amplifying a preexisting signal but leaving dormant receptors unaffected, PAMs may augment nicotinic tone while being sparing of native circuit dynamics. The combination of very small doses of nAChR agonists with nAChR PAMs may be another fruitful avenue. Given that nAChR subtype selectivity can

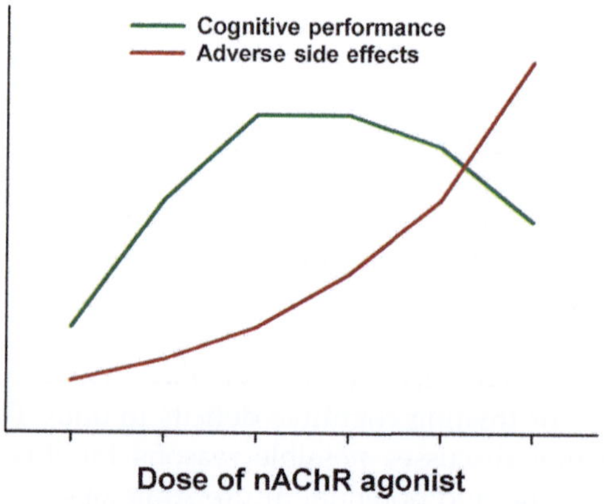

FIG. 38.1 A hypothetical dose-response relationship of the cognitive enhancing effects of a nAChR agonist relative to its adverse side effects. Increasing the dosage beyond an optimal concentration is likely to increase side effects without any further cognitive benefits.

be achieved with both compound classes, a combination strategy may also achieve greater flexibility and fine-tuning when it comes to targeting the right set of nAChR subtypes. For example, selective modulation of a specific nAChR subtype may be enabled by combining subthreshold doses of a nAChR agonist selective for a group of nAChR subtypes with subthreshold doses of a PAM selective for a different group of subtypes but overlapping with the first group on the targeted subtype (Fig. 38.2).

38.3 NO SUCH THING AS A FREE LUNCH?

When facing the question of whether long-lasting cognitive enhancement is achievable with a nAChR-based pharmacological strategy, perhaps the biggest uncertainty is whether neuroadaptive changes will neutralize benefits relative to the predrug baseline. The nicotine literature suggests that caution is warranted in this regard. Performance-enhancing effects of nicotine can be shown in nonsmokers but have been seen more reliably in studies assessing deprived smokers (Foulds et al., 1996; Heishman, Taylor, & Henningfield, 1994). This suggests that the reversal of withdrawal-induced deficits substantially contributes to benefits seen in chronically nicotine-exposed individuals. Indeed, difficulty concentrating is a main complaint of over 65% of individuals undergoing smoking cessation (Ward, Swan, & Jack, 2001) and is one of the diagnostic criteria for nicotine withdrawal (American Psychiatric Association, 2013).

From a drug development perspective, the neuronal mechanisms underlying cognitive impairment associated with nAChR agonist withdrawal are no less important than the mechanisms underlying drug-induced cognitive benefits. Neuroadaptations with chronic nAChR agonist exposure can occur far downstream from nAChR activation (or desensitization) and may manifest in secondary neurotransmitter systems (e.g., Avila-Ruiz et al., 2014; Rademacher et al., 2016). However, the perhaps most studied neuroadaptation seen with cigarette smoking, or chronic nicotine exposure in an experimental context, is nAChR upregulation (e.g., Colombo, Mazzo, Pistillo, & Gotti, 2013; Cosgrove et al., 2009), likely a consequence of chronic nAChR desensitization (Fenster, Whitworth, Sheffield, Quick, & Lester, 1999). A study in mice reports that persistent hippocampal nAChR upregulation after chronic nicotine exposure parallels the time course of withdrawal-induced mnemonic deficits (Gould et al., 2012). However, aside from this, little is known about the neural substrates underlying withdrawal-induced cognitive deficits.

For compounds and administration regimens to effect long-lasting benefits relative to a predrug baseline, they would have to be devoid of neuroadaptations responsible for shifting drug-free cognitive performance in a negative direction. Thus, understanding these mechanisms would benefit drug development because novel compounds could be screened for these properties long before clinical trial stage. With regard to novel nAChR agonists that have made it to clinical trial stage to date, the author is not aware that such potential baseline shifts have been addressed by, for example, obtaining a follow-up outcome measurement shortly after washout and comparing the pre- to posttreatment difference with a placebo-treated time control. It is conceivable that a lower baseline in later parts of the treatment phase accounts for the lack of efficacy relative to a placebo group (see Fig. 38.3). In other words, while the tested compound may still exert acute improvement, these benefits may occur against an impaired baseline and do not represent net benefits relative to before the treatment started.

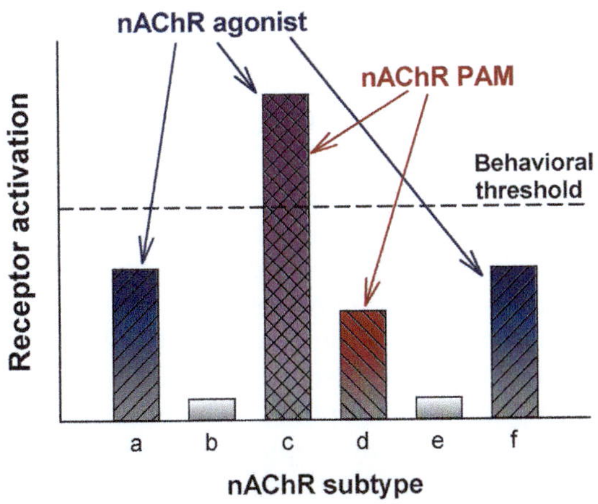

FIG. 38.2 nAChR agonist and PAM combination strategy.
The combination of low doses of an agonist acting on a subset of nAChR subtypes with low doses of a PAM acting on another, partially overlapping, subset may help create a narrower profile of desirable behavioral effects.

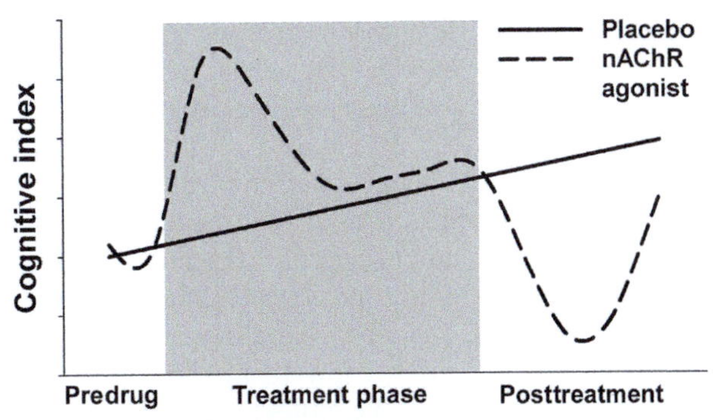

FIG. 38.3 A hypothetical time course of nAChR agonist effects over a chronic treatment and washout period.
Counteradaptive baseline shifts may annihilate cognitive benefits seen upon initial drug exposure. Such shifts would manifest as net deficits upon drug washout. Outcome measurements are typically performed prior to and at the end of the treatment phase prior to washout. Thus, without a follow-up assessment, no or minimal efficacy is indistinguishable from the scenario depicted here.

38.4 A NAIVE PREMISE?

Cognitive performance depends on the complex and correctly timed interplay of heterogeneous neuronal populations, circuits, and structures. Accordingly, cognitive processing deficits may derive from a large number of potential sources. Even in disorders with evidence for nAChR hypofunction such as AD and schizophrenia, there is broad underlying neuropathology (e.g., Bakhshi & Chance, 2015; Elahi & Miller, 2017; Lodge & Grace, 2011), suggesting that correcting a low nAChR tone is unlikely to remedy cognitive deficits in a substantial manner. Other compromised components will likely continue to prevent smooth and efficient neuronal clockwork.

Another line of thought questioning the therapeutic potential of pharmacotherapies more generally is that cognitive deficits associated with CNS disorders are a chronic and likely self-perpetuating phenomenon. Cognitive deficits can be expected to reduce cognitive engagement, and this reduction in use is likely to result in further cognitive decline, accompanied at the physiological level by a loss of neuronal connections and decay in circuit tuning, analogous to the primary sensory cortices after sensory deprivation (e.g., Kral, Yusuf, & Land, 2017). Against this backdrop, too, an immediate and significant amelioration of cognitive deficits from the acute optimization of nAChR (or any other system's) tone may seem like a naive expectation.

However, it is in the context of self-perpetuation that the potential of nAChR agonists in the treatment of cognitive deficits may lie. The slight gain of function brought about by the presence of a nAChR agonist, be it due to enhanced sensory processing, alertness, or higher attention or mnemonic functions, may subtly augment the engagement of cognitive processing resources over months or years, and if sustained, this may contribute to preventing further decline or even result in incremental improvement. As discussed above, this would require sustained drug effects, relative to the predrug baseline. Clinical trials conducted to date likely have been too short in duration to test such a phenomenon.

Another way nAChR agonists may benefit the treatment of cognitive deficits is as an augmentation tool for cognitive training interventions (Hahn, Gold, & Buchanan, 2013). Training-induced cognitive improvements in conditions such as Alzheimer's disease and schizophrenia tend to be of low to medium effect size (Hill et al., 2017; McGurk, Twamley, Sitzer, McHugo, & Mueser, 2007). The time demands for such interventions to be effective may limit their clinical applicability; however, a means of augmenting and speeding the training effects could enhance their feasibility and clinical significance. A pharmacological agent administered at the time of training may optimize the neuronal environment for training effects. With this approach, treatment success would manifest itself in enhanced performance long after the drug has cleared the system. In addition to potentially more long-lasting benefits, another advantage is that periods of drug exposure are fundamentally shortened, thus minimizing the potential for counteradaptive neuroadaptations. In this context, a short half-life would even be a preferred property, so long as the window during which drug effects unfold is repeatedly coupled with the information processing challenges of the cognitive training.

nAChR agonists may be candidates for promoting experience-dependent changes. Nicotine facilitates early sensory processes (Fisher et al., 2010; Knott, Fisher, & Millar, 2010; Phillips, Ehrlichman, & Siegel, 2007), and the same appears true for α7-selective nAChR agonists (Martin & Freedman, 2007). More accurate sensory representations may form better building blocks for higher functions, and more efficient sensory processing may free up resources, allowing training challenges to be met at a higher level of difficulty. Furthermore, an acute facilitation of alertness and attention during the training sessions by a nAChR agonist (Newhouse, Potter, & Singh, 2004) may enable deeper engagement in the training exercises and may enable participants to stay on task longer. Finally, nAChR activation induces long-lasting plastic changes in the brain via diverse cellular mechanisms (Buccafusco, Letchworth, Bencherif, & Lippiello, 2005; Castner et al., 2011; Hasselmo, 2006). Notably, it has been shown that nicotine can promote the induction of long-term potentiation (LTP) (Kenney & Gould, 2008; Matsuyama, Matsumoto, Enomoto, & Nishizaki, 2000). Through enhanced neural plasticity, nAChR agonists may facilitate learning, memory, and skill development by the training exercises. Indeed, there is evidence that the cholinergic system is critical for cortical plasticity associated with motor skill learning (Conner, Culberson, Packowski, Chiba, & Tuszynski, 2003).

38.5 CONCLUSIONS

In many ways, the development of nAChR agonists for the treatment of cognitive deficits appears to have gotten ahead of itself. A lack of knowledge regarding the critical target systems may have promoted the focus on the two nAChR subtypes most widely expressed in the brain. The choice of dose tended to disregard preclinical and in vitro data on nAChR dose-response dynamics. The mechanisms underlying neuroadaptations and withdrawal-induced deficits, both well-known phenomena with chronic nicotine administration, are largely unknown and have received little consideration in drug development and clinical trial design. Finally, cognitive functions, as a target outcome, may require a longer-term approach, perhaps integrating additional interventions such as training programs. Disappointing results from clinical trials testing nAChR agonists as cognitive enhancers may indicate that, rather than throwing the baby out with the bathwater, it is time to take the strategy back to the drawing board.

MINI-DICTIONARY OF TERMS

Experience-dependent changes Long-lasting changes in neuronal function brought about by the organism's interaction with the environment.

Inverted U-shaped dose-response function The phenomenon whereby an ascending dose is associated with more drug effect only up to a specific dose, beyond which further dose increases would lead to less effect.

nAChR desensitization Diminished nAChR response after exposure to a nAChR agonist.

nAChR subtypes Each nAChR is composed of five subunits. Different nAChR subunits can assemble in different combinations to form nAChRs with different pharmacological properties and receptor dynamics.

Neural plasticity The ability of neurons to change their responsivity or connectivity based on the degree or type of their engagement.

Neuroadaptation With frequent use of a drug, the action of neurons will mechanistically adjust to preserve homeostasis. This commonly leads to tolerance to the drug's effects, and a state opposite to these effects when the drug is not present.

Nonsynaptic diffuse volume transmission The signaling by neurotransmitters released at points remote from the target and diffusing broadly through the extracellular fluid.

Receptor tone Level of activation on a broader timescale.

Secondary neurotransmitter systems Systems on which a drug does not act directly, but that are modulated by systems directly altered by the drug.

Sustained equilibrium A state in which processes opposite to each other, such as absorption versus elimination or receptor activation versus desensitization, are balanced, maintaining the current state.

Key Facts on Nicotinic Agonist Therapy for Cognitive Deficits

- The idea of using nicotinic agonists to treat cognitive deficits dates back over 40 years.
- The development of nicotinic agonists as cognitive enhancers began in the early 1990s.
- Aside from nicotine and varenicline, over 20 different nicotinic agonists have been tested for cognitive enhancing potential in clinical populations.
- Over 100 clinical trials testing novel nicotinic agonists as cognitive enhancers have been posted on https://ClinicalTrials.gov.
- No nicotinic compound has received FDA approval as a treatment for cognitive deficits to date.

Summary Points

- The development of nAChR agonists has not resulted in any new treatments for cognitive deficits.
- Critical target systems, and hence the most promising nAChR subtypes to be targeted, are largely unknown.
- Cognitive enhancing effects of nAChR agonists follow an inverted U-shaped dose-response function; thus, the choice of large doses may have obscured beneficial effects.
- Counteradaptive neuronal changes with chronic exposure may account for a lack of net efficacy, but these neuroadaptations are largely unstudied and thus not considered in drug development.
- The broad underlying neuropathology and chronic nature of cognitive deficits associated with disease states may call for prolonged treatment periods and integrated cognitive training interventions.

References

Adams, C. E., & Stevens, K. E. (2007). Evidence for a role of nicotinic acetylcholine receptors in schizophrenia. *Frontiers in Bioscience*, *12*, 4755–4772.

American Psychiatric Association. (2013). *Diagnostic and statistical manual of mental disorders* (5th ed.). Washington, DC/London, England: American Psychiatric Publishing.

Avila-Ruiz, T., Carranza, V., Gustavo, L. L., Limon, D. I., Martinez, I., Flores, G., et al. (2014). Chronic administration of nicotine enhances NMDA-activated currents in the prefrontal cortex and core part of the nucleus accumbens of rats. *Synapse*, *68*(6), 248–256. https://doi.org/10.1002/syn.21726.

Bakhshi, K., & Chance, S. A. (2015). The neuropathology of schizophrenia: a selective review of past studies and emerging themes in brain structure and cytoarchitecture. *Neuroscience*, *303*, 82–102. https://doi.org/10.1016/j.neuroscience.2015.06.028.

Buccafusco, J. J., Letchworth, S. R., Bencherif, M., & Lippiello, P. M. (2005). Long-lasting cognitive improvement with nicotinic receptor agonists: mechanisms of pharmacokinetic-pharmacodynamic discordance. *Trends in Pharmacological Sciences*, *26*(7), 352–360. https://doi.org/10.1016/j.tips.2005.05.007.

Bushnell, P. J., Oshiro, W. M., & Padnos, B. K. (1997). Detection of visual signals by rats: effects of chlordiazepoxide and cholinergic and adrenergic drugs on sustained attention. *Psychopharmacology*, *134*(3), 230–241.

Castner, S. A., Smagin, G. N., Piser, T. M., Wang, Y., Smith, J. S., Christian, E. P., et al. (2011). Immediate and sustained improvements in working memory after selective stimulation of alpha 7 nicotinic acetylcholine receptors. *Biological Psychiatry*, *69*(1), 12–18.

Colombo, S. F., Mazzo, F., Pistillo, F., & Gotti, C. (2013). Biogenesis, trafficking and up-regulation of nicotinic ACh receptors. *Biochemical Pharmacology*, *86*(8), 1063–1073. https://doi.org/10.1016/j.bcp.2013.06.023.

Conner, J. M., Culberson, A., Packowski, C., Chiba, A. A., & Tuszynski, M. H. (2003). Lesions of the basal forebrain cholinergic system impair task acquisition and abolish cortical plasticity associated with motor skill learning. *Neuron*, *38*(5), 819–829.

Cosgrove, K. P., Batis, J., Bois, F., Maciejewski, P. K., Esterlis, I., Kloczynski, T., et al. (2009). Beta2-nicotinic acetylcholine receptor availability during acute and prolonged abstinence from tobacco smoking. *Archives of General Psychiatry*, *66*(6), 666–676. https://doi.org/10.1001/archgenpsychiatry.2009.41.

Dani, J. A., & Bertrand, D. (2007). Nicotinic acetylcholine receptors and nicotinic cholinergic mechanisms of the central nervous system. *Annual Review of Pharmacology and Toxicology*, *47*, 699–729. https://doi.org/10.1146/annurev.pharmtox.47.120505.105214.

Elahi, F. M., & Miller, B. L. (2017). A clinicopathological approach to the diagnosis of dementia. *Nature Reviews. Neurology*, *13*(8), 457–476. https://doi.org/10.1038/nrneurol.2017.96.

Exley, R., Clements, M. A., Hartung, H., McIntosh, J. M., & Cragg, S. J. (2008). Alpha 6-containing nicotinic acetylcholine receptors dominate the nicotine control of dopamine neurotransmission in nucleus accumbens. *Neuropsychopharmacology, 33*(9), 2158–2166.

Fenster, C. P., Whitworth, T. L., Sheffield, E. B., Quick, M. W., & Lester, R. A. (1999). Upregulation of surface alpha4beta2 nicotinic receptors is initiated by receptor desensitization after chronic exposure to nicotine. *The Journal of Neuroscience, 19*(12), 4804–4814.

Fisher, D. J., Scott, T. L., Shah, D. K., Prise, S., Thompson, M., & Knott, V. J. (2010). Light up and see: enhancement of the visual mismatch negativity (vMMN) by nicotine. *Brain Research, 1313*, 162–171.

Foulds, J., Stapleton, J., Swettenham, J., Bell, N., McSorley, K., & Russell, M. A. (1996). Cognitive performance effects of subcutaneous nicotine in smokers and never-smokers. *Psychopharmacology, 127*(1), 31–38.

Freedman, R., Coon, H., Myles Worsley, M., Orr Urtreger, A., Olincy, A., Davis, A., et al. (1997). Linkage of a neurophysiological deficit in schizophrenia to a chromosome 15 locus. *Proceedings of the National Academy of Sciences of the United States of America, 94*(2), 587–592.

Freedman, R., Olincy, A., Buchanan, R. W., Harris, J. G., Gold, J. M., Johnson, L., et al. (2008). Initial phase 2 trial of a nicotinic agonist in schizophrenia. *The American Journal of Psychiatry, 165*(8), 1040–1047.

Gotti, C., Guiducci, S., Tedesco, V., Corbioli, S., Zanetti, L., Moretti, M., et al. (2010). Nicotinic acetylcholine receptors in the mesolimbic pathway: primary role of ventral tegmental area alpha 6 beta 2*receptors in mediating systemic nicotine effects on dopamine release, locomotion, and reinforcement. *Journal of Neuroscience, 30* (15), 5311–5325.

Gotti, C., Moretti, M., Gaimarri, A., Zanardi, A., Clementi, F., & Zoli, M. (2007). Heterogeneity and complexity of native brain nicotinic receptors. *Biochemical Pharmacology, 74*(8), 1102–1111.

Gould, T. J., Portugal, G. S., Andre, J. M., Tadman, M. P., Marks, M. J., Kenney, J. W., et al. (2012). The duration of nicotine withdrawal-associated deficits in contextual fear conditioning parallels changes in hippocampal high affinity nicotinic acetylcholine receptor upregulation. *Neuropharmacology, 62*(5–6), 2118–2125. https://doi.org/10.1016/j.neuropharm.2012.01.003.

Gundisch, D., & Eibl, C. (2011). Nicotinic acetylcholine receptor ligands, a patent review (2006-2011). *Expert Opinion on Therapeutic Patents, 21* (12), 1867–1896.

Hahn, B. (2015). Nicotinic receptors and attention. *Current Topics in Behavioral Neurosciences, 23*, 103–135. https://doi.org/10.1007/978-3-319-13665-3_5.

Hahn, B., Gold, J. M., & Buchanan, R. W. (2013). The potential of nicotinic enhancement of cognitive remediation training in schizophrenia. *Neuropharmacology, 64*, 185–190. https://doi.org/10.1016/j.neuropharm.2012.05.050.

Hahn, B., Ross, T. J., Yang, Y., Kim, I., Huestis, M. A., & Stein, E. A. (2007). Nicotine enhances visuospatial attention by deactivating areas of the resting brain default network. *The Journal of Neuroscience, 27*(13), 3477–3489.

Hahn, B., Sharples, C. G., Wonnacott, S., Shoaib, M., & Stolerman, I. P. (2003). Attentional effects of nicotinic agonists in rats. *Neuropharmacology, 44*(8), 1054–1067.

Hahn, B., Shoaib, M., & Stolerman, I. P. (2002). Nicotine-induced enhancement of attention in the five-choice serial reaction time task: the influence of task-demands. *Psychopharmacology, 162*(2), 129–137.

Hahn, B., & Stolerman, I. P. (2005). Modulation of nicotine-induced attentional enhancement in rats by adrenoceptor antagonists. *Psychopharmacology, 177*(4), 438–447.

Hasselmo, M. E. (2006). The role of acetylcholine in learning and memory. *Current Opinion in Neurobiology, 16*(6), 710–715. https://doi.org/10.1016/j.conb.2006.09.002.

Haydar, S. N., & Dunlop, J. (2010). Neuronal nicotinic acetylcholine receptors - targets for the development of drugs to treat cognitive impairment associated with schizophrenia and Alzheimer's disease. *Current Topics in Medicinal Chemistry, 10*(2), 144–152.

Heishman, S. J., Kleykamp, B. A., & Singleton, E. G. (2010). Meta-analysis of the acute effects of nicotine and smoking on human performance. *Psychopharmacology, 210*(4), 453–469.

Heishman, S. J., Taylor, R. C., & Henningfield, J. E. (1994). Nicotine and smoking: a review of effects on human performance. *Experimental and Clinical Psychopharmacology, 2*(4), 345–395.

Hill, N. T., Mowszowski, L., Naismith, S. L., Chadwick, V. L., Valenzuela, M., & Lampit, A. (2017). Computerized cognitive training in older adults with mild cognitive impairment or dementia: a systematic review and meta-analysis. *The American Journal of Psychiatry, 174*(4), 329–340. https://doi.org/10.1176/appi.ajp.2016.16030360.

Hong, L. E., Yang, X., Wonodi, I., Hodgkinson, C. A., Goldman, D., Stine, O. C., et al. (2011). A CHRNA5 allele related to nicotine addiction and schizophrenia. *Genes, Brain, and Behavior, 10*(5), 530–535. https://doi.org/10.1111/j.1601-183X.2011.00689.x.

Kendziorra, K., Wolf, H., Meyer, P. M., Barthel, H., Hesse, S., Becker, G. A., et al. (2011). Decreased cerebral a4ß2* nicotinic acetylcholine receptor availability in patients with mild cognitive impairment and Alzheimer's disease assessed with positron emission tomography. *European Journal of Nuclear Medicine and Molecular Imaging, 38*, 515–525.

Kenney, J. W., & Gould, T. J. (2008). Modulation of hippocampus-dependent learning and synaptic plasticity by nicotine. *Molecular Neurobiology, 38*(1), 101–121. https://doi.org/10.1007/s12035-008-8037-9.

Knott, V. J., Fisher, D. J., & Millar, A. M. (2010). Differential effects of nicotine on P50 amplitude, its gating, and their neural sources in low and high suppressors. *Neuroscience, 170*(3), 816–826.

Kral, A., Yusuf, P. A., & Land, R. (2017). Higher-order auditory areas in congenital deafness: top-down interactions and corticocortical decoupling. *Hearing Research, 343*, 50–63. https://doi.org/10.1016/j.heares.2016.08.017.

Lawrence, N. S., Ross, T. J., & Stein, E. A. (2002). Cognitive mechanisms of nicotine on visual attention. *Neuron, 36*(3), 539–548.

Leonard, S., Gault, J., Hopkins, J., Logel, J., Vianzon, R., Short, M., et al. (2002). Association of promoter variants in the alpha 7 nicotinic acetylcholine receptor subunit gene with an inhibitory deficit found in schizophrenia. *Archives of General Psychiatry, 59*(12), 1085–1096.

Levin, E. D. (1992). Nicotinic systems and cognitive function. *Psychopharmacology, 108*(4), 417–431.

Lodge, D. J., & Grace, A. A. (2011). Hippocampal dysregulation of dopamine system function and the pathophysiology of schizophrenia. *Trends in Pharmacological Sciences, 32*(9), 507–513. https://doi.org/10.1016/j.tips.2011.05.001.

Martin, L. F., & Freedman, R. (2007). Schizophrenia and the alpha 7 nicotinic acetylcholine receptor. *Integrating the Neurobiology of Schizophrenia, 78*, 225–246.

Martin, L. F., Kem, W. R., & Freedman, R. (2004). Alpha-7 nicotinic receptor agonists: potential new candidates for the treatment of schizophrenia. *Psychopharmacology, 174*(1), 54–64.

Matsuyama, S., Matsumoto, A., Enomoto, T., & Nishizaki, T. (2000). Activation of nicotinic acetylcholine receptors induces long-term potentiation in vivo in the intact mouse dentate gyrus. *The European Journal of Neuroscience, 12*(10), 3741–3747.

McGaughy, J., Decker, M. W., & Sarter, M. (1999). Enhancement of sustained attention performance by the nicotinic acetylcholine receptor agonist ABT-418 in intact but not basal forebrain-lesioned rats. *Psychopharmacology, 144*(2), 175–182.

McGurk, S. R., Twamley, E. W., Sitzer, D. I., McHugo, G. J., & Mueser, K. T. (2007). A meta-analysis of cognitive remediation in schizophrenia.

The American Journal of Psychiatry, *164*(12), 1791–1802. https://doi.org/10.1176/appi.ajp.2007.07060906.

Mike, A., Castro, N. G., & Albuquerque, E. X. (2000). Choline and acetylcholine have similar kinetic properties of activation and desensitization on the alpha7 nicotinic receptors in rat hippocampal neurons. *Brain Research*, *882*(1–2), 155–168.

Newhouse, P. A., Potter, A., & Singh, A. (2004). Effects of nicotinic stimulation on cognitive performance. *Current Opinion in Pharmacology*, *4*(1), 36–46.

Olincy, A., Harris, J. G., Johnson, L. L., Pender, V., Kongs, S., Allensworth, D., et al. (2006). Proof-of-concept trial of an alpha7 nicotinic agonist in schizophrenia. *Archives of General Psychiatry*, *63*(6), 630–638.

Papke, R. L., Trocme-Thibierge, C., Guendisch, D., Abdullah, S., Rubaiy, A. A., & Bloom, S. A. (2011). Electrophysiological perspectives on the therapeutic use of nicotinic acetylcholine receptor partial agonists. *Journal of Pharmacology and Experimental Therapeutics*, *337* (2), 367–379.

Perkins, K. A., Grobe, J. E., Fonte, C., Goettler, J., Caggiula, A. R., Reynolds, W. A., et al. (1994). Chronic and acute tolerance to subjective, behavioral and cardiovascular effects of nicotine in humans. *The Journal of Pharmacology and Experimental Therapeutics*, *270*(2), 628–638.

Phillips, J. M., Ehrlichman, R. S., & Siegel, S. J. (2007). Mecamylamine blocks nicotine-induced enhancement of the P20 auditory event-related potential and evoked gamma. *Neuroscience*, *144*(4), 1314–1323. https://doi.org/10.1016/j.neuroscience.2006.11.003.

Quarta, D., Naylor, C. G., Morris, H. V., Patel, S., Genn, R. F., & Stolerman, I. P. (2007). Different effects of ionotropic and metabotropic glutamate receptor antagonists on attention and the attentional properties of nicotine. *Neuropharmacology*, *53*(3), 421–430.

Rademacher, L., Prinz, S., Winz, O., Henkel, K., Dietrich, C. A., Schmaljohann, J., et al. (2016). Effects of smoking cessation on presynaptic dopamine function of addicted male smokers. *Biological Psychiatry*, *80*(3), 198–206. https://doi.org/10.1016/j.biopsych.2015.11.009.

Schrattenholz, A., Pereira, E. F., Roth, U., Weber, K. H., Albuquerque, E. X., & Maelicke, A. (1996). Agonist responses of neuronal nicotinic acetylcholine receptors are potentiated by a novel class of allosterically acting ligands. *Molecular Pharmacology*, *49*(1), 1–6.

Seguela, P., Wadiche, J., Dineley-Miller, K., Dani, J. A., & Patrick, J. W. (1993). Molecular cloning, functional properties, and distribution of rat brain alpha 7: a nicotinic cation channel highly permeable to calcium. *The Journal of Neuroscience*, *13*(2), 596–604.

Stolerman, I. P. (1999). Inter-species consistency in the behavioural pharmacology of nicotine dependence. *Behavioural Pharmacology*, *10*(6–7), 559–580.

Sutherland, M. T., Ray, K. L., Riedel, M. C., Yanes, J. A., Stein, E. A., & Laird, A. R. (2015). Neurobiological impact of nicotinic acetylcholine receptor agonists: an activation likelihood estimation meta-analysis of pharmacologic neuroimaging studies. *Biological Psychiatry*, *78*(10), 711–720. https://doi.org/10.1016/j.biopsych.2014.12.021.

Wallace, T. L., Ballard, T. M., Pouzet, B., Riedel, W. J., & Wettstein, J. G. (2011). Drug targets for cognitive enhancement in neuropsychiatric disorders. *Pharmacology Biochemistry and Behavior*, *99*(2), 130–145.

Ward, M. M., Swan, G. E., & Jack, L. M. (2001). Self-reported abstinence effects in the first month after smoking cessation. *Addictive Behaviors*, *26*(3), 311–327.

Williams, D. K., Wang, J. Y., & Papke, R. L. (2011). Positive allosteric modulators as an approach to nicotinic acetylcholine receptor-targeted therapeutics: advantages and limitations. *Biochemical Pharmacology*, *82*(8), 915–930.

C H A P T E R

39

Nicotine and Dopamine DA1 Receptor Pharmacology

Agnieszka Michalak, Barbara Budzyńska

Department of Pharmacology and Pharmacodynamics, Medical University of Lublin, Lublin, Poland

Abbreviations

AC	adenyl cyclase
CAMKII	Ca^{2+}/calmodulin-dependent protein kinase II
cAMP	cyclic adenosine monophosphate
CaN	calcineurin
CNS	central nervous system
CPA	conditioned place aversion
CPP	conditioned place preference
CREB	cAMP response element-binding protein
DA	dopamine
DA1	dopaminergic type 1 receptors
DAG	diacylglycerol
DARPP-32	dopamine and cAMP-regulated phosphoprotein, 32 kDa
DARPP-32-P	phosphorylated dopamine and cAMP-regulated phosphoprotein
DRD1	dopamine DA1 receptor gene
ERK	extracellular signal-regulated kinases
GP	globus pallidus
IA	inhibitory avoidance
ip	intraperitoneally
IP3	inositol triphosphate
LTP	long-term potentiation
MEK	MAPK/ERK kinase
mRNA	messenger ribonucleic acid
NAc	nucleus accumbens
nAChRs	nicotinic acetylcholine receptors
NK1	neurokinin receptor 1
NMDA	*N*-methyl-D-aspartate receptor
PKA	protein kinase A
PKC	protein kinase C
PLC	phospholipase C
PP1	protein phosphatase-1
sc	subcutaneously
SN	substantia nigra
SP	substance P
SPLI	substance P-like immunoreactivity
VTA	ventral tegmental area

39.1 INTRODUCTION

39.1.1 Dopamine DA1 Receptors, Localization and Signaling

Dopamine (DA) receptors are divided into two classes of G protein-coupled receptors, DA1- and DA2-like receptors. The subfamily of DA1-like receptors includes DA1 and DA5 subtypes, while the DA2-like subfamily comprises DA2, DA3, and DA4 receptors (Fig. 39.1). DA1 receptors are the most prevalent DA receptor subtype in the central nervous system (CNS), with different density levels depending on the brain structure. The highest level of DA1 expression is found in the nigrostriatal, mesolimbic, and mesocortical pathways, especially in areas such as the dorsal striatum, amygdala, nucleus accumbens (NAc), substantia nigra (SN), frontal cortex, and olfactory bulb, while lower DA1 levels are observed in the cerebellum, hippocampus, thalamus, and hypothalamus (Beaulieu & Gainetdinov, 2011). Distribution of DA1 receptors corresponds to their functions. DA1 receptors are involved in learning and memory, they play an important role in motor control, and their activation is critical for reward mechanisms (Figs. 39.2 and 39.3). Furthermore, abnormalities in DA1 receptors are contributed to mental disorders such as Parkinson's disease (PD), schizophrenia, or Huntington's disease (Beaulieu & Gainetdinov, 2011; Komatsu, 2015).

DA1 receptors are predominantly coupled to $G\alpha_s$ and $G\alpha_{olf}$ subtypes of G protein and activate adenylyl cyclase (AC) stimulating the production of cyclic adenosine

Neuroscience of Nicotine
https://doi.org/10.1016/B978-0-12-813035-3.00039-3

monophosphate (cAMP) and activating protein kinase A (PKA). PKA phosphorylates dopamine and cAMP-regulated phosphoprotein, 32 kDa (DARPP-32) leading to the inactivation of protein phosphatase-1 (PP1), therefore preventing its inhibitory function against MAPK/ERK kinase (MEK). It allows for the activation of extracellular signal-regulated kinases (ERK) and ERK-mediated cAMP response element-binding protein (CREB) phosphorylation (Beaulieu & Gainetdinov, 2011; Hisahara & Shimohama, 2011). DARPP-32 activation can be countered inter alia by calcineurin (CaN) activated due to the stimulation of NMDA receptors (Svenningsson et al., 2004). DA1 receptors may also be coupled to $G\alpha_q$ that is linked with phospholipase C (PLC). Activated PLC results in an increase in diacylglycerol (DAG) and inositol triphosphate (IP3). DAG activates protein kinase C (PKC), whereas IP3 increases the mobilization of calcium (Ca^{2+}) from the endoplasmic reticulum activating CaMKII (Ca^{2+}/calmodulin-dependent protein kinase II). Both PKC and CAMKII lead to CREB-related gene transcription (Beaulieu & Gainetdinov, 2011; Hisahara & Shimohama, 2011) (Fig. 39.4).

39.1.2 Nicotine and DA1 Receptors—Molecular Insight

Strong involvement of DA1 receptors in the effects of nicotine was evaluated in many reports; nicotine indirectly stimulates DA1 receptors increasing the release of DA via nicotinic acetylcholine receptors (nAChRs). Most likely, two subtypes of nAChRs, α4β2 and α7, are critical for nicotine/DA1 interaction. It is assumed that indirect activation of DA1 receptors by nicotine is related to stimulation of α4β2 nAChRs at dopaminergic terminals, as well as α7 nAChRs at glutamatergic terminals responsible for modulation of dopaminergic transmission in the nigrostriatal pathway (Hamada et al., 2005). Interestingly, nicotine modulates DA release in a frequency-dependent manner and, at least partly, is conditional on DA1 receptors. It was revealed that nicotine attenuates the striatal release of DA evoked by low-frequency electric field stimulation (one or five pulses of 10 Hz) but intensifies DA efflux at high-frequency (five pulses of 100 Hz) stimulation. To evaluate the mechanism involved, both a DA1 receptor agonist (SKF-38393) and antagonist (SCH-23390) were used. Although neither of ligands influenced nicotine effects on DA release evoked by five pulses of 10 and 100 Hz, both blocked its inhibitory effect on DA release triggered by one pulse of 10 Hz stimulation (Goutier, O'Connor, Lowry, & McCreary, 2016).

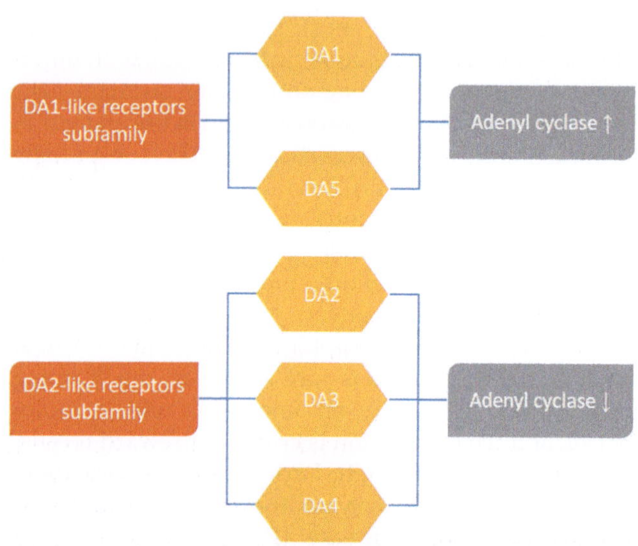

FIG. 39.1 DA1 receptor subtypes and their effects on adenyl cyclase.

39.1.2.1 *PKA/DARPP-32/PP1 Signaling Cascade*

Nicotine mediates the release of DA via activation of α4β2 and α7 nAChRs, which leads to the subsequent activation of DA1 receptors and activation of $G\alpha_{s/olf}$/PKA/DARPP-32 signaling cascade. Following DARPP-32, phosphorylation may occur at its various sites with different modulatory effects on PP1 activity. DARPP-32 phosphorylation at Thr34 is essential for the inactivation of PP1, while phosphorylation of DARPP-32 at Thr75 leads to the inactivation of PKA and thus inhibits the signaling cascade. Moreover, DARPP-32 phosphorylation at Thr34 by PKA can be increased when it is simultaneously phosphorylated at Ser97 and phosphorylation of DARPP-32 at Ser130 decreases dephosphorylation of Thr34 by CaN. It has been found that higher concentrations of nicotine (100 μM) enhance PKA/DARPP-32/PP1 cascade via increasing phosphorylation of DARPP-32 at Ser97 and Ser130 and decreasing phosphorylation state at Thr75-DARPP-32 (Hamada et al., 2005). Furthermore, Abdolahi, Acosta, Breslin, Hemby, and Lynch (2010) revealed that increasing over an abstinence period drug-seeking to nicotine is associated with enhanced PKA/DARPP-32 signaling, as a result of increased

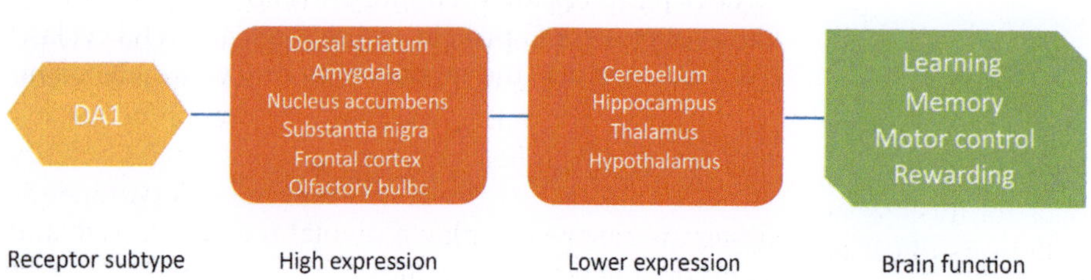

FIG. 39.2 The localization, expression, and brain function of DA1 receptors.

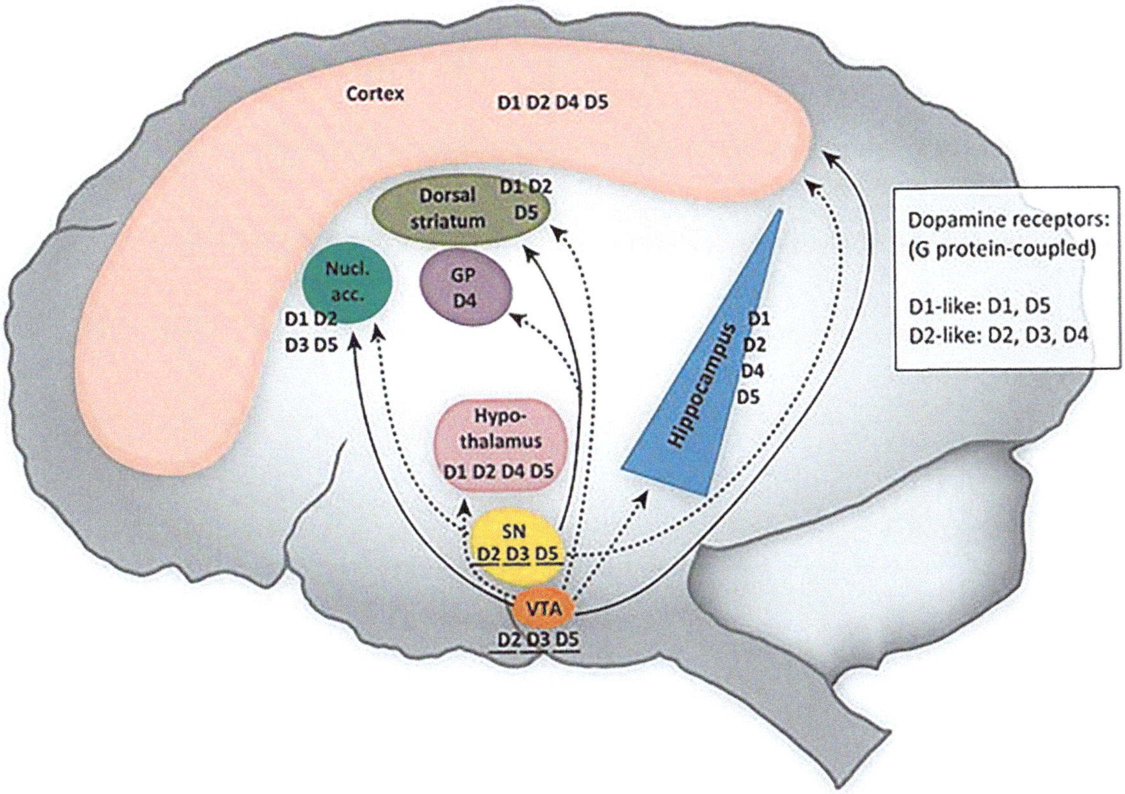

FIG. 39.3 Dopaminergic receptor expression and DA projections in the human brain.
DA neurons from SN projects principally not only to the dorsal striatum but also to the cortex, GP, and nucleus accumbens. Axons of DA neurons in the VTA project mainly to the nucleus accumbens and cortex, as well as the dorsal striatum, hippocampus, and hypothalamus. Location of DA1 receptors in the brain is described in the text. Autoreceptors are underlined. Abbreviations: *DA*, dopamine; *DA1*, dopaminergic type 1 receptors; *GP*, globus pallidus; *Nucl. acc.*, nucleus accumbens; *SN*, substantia nigra; *VTA*, ventral tegmental area. *With permission from Brichta, Greengard, and Flajolet (2013).*

phosphorylation of Thr34-DARPP-32 in the insular cortex and decreased levels of phosphorylated Thr75-DARPP-32 in NAc at 7 days of drug abstinence.

39.1.2.2 *Genetic Background*

Robust data indicate that there is a significant linkage between nicotine dependency and DA1 receptor gene (DRD1). It has been found that subchronic nicotine exposure (0.4 mg/kg/day, subcutaneously (sc), 15 days) increases DRD1 mRNA expression in rat prefrontal cortex, which can be connected with increased histone H4 acetylation level (Gozen, Balkan, Yildirim, Koylu, & Pogun, 2013). Furthermore, it was shown that a single nicotine injection enhances DRD1 mRNA level in NAc shell in rats preexposed in adolescence to nicotine (Wheeler et al., 2013). Significant association of DRD1 and nicotine dependence has been confirmed in human smokers. Moreover, genetic variation of rs686 (a single-nucleotide polymorphism located in the promoter region of DRD1) was found as a causative locus for this association, and microRNA plays a great role in its effect on DRD1 expression (Huang & Li, 2009).

39.2 DA1 RECEPTORS AND BEHAVIORAL EFFECTS OF NICOTINE

39.2.1 Motivational Effects

There is ample evidence of the role of DA1 receptors in the aversive and reinforcing effects of nicotine. It was revealed that DA1 receptors are localized in reward-related brain regions, which receive cholinergic input. Because recent developments in neurobiology have been focused on DA efflux in the brain, interesting information has emerged. It was revealed that DA is released from dopaminergic neurons in two manners: phasic and tonic. Large and fast phasic release of DA activates primarily DA1 receptors, whereas tonic release of DA activates primarily DA2 receptors (Floresco, West, Ash, Moore, & Grace, 2003). Additionally, it is thought that cue-reward association and acquisition of incentive salience is mediated by phasic activity, whereas response inhibition and behavioral flexibility is facilitated by tonic activity (Grieder et al., 2012). Thus, activation of a dopaminergic system is believed to play a pivotal role in aversive and rewarding effects of nicotine.

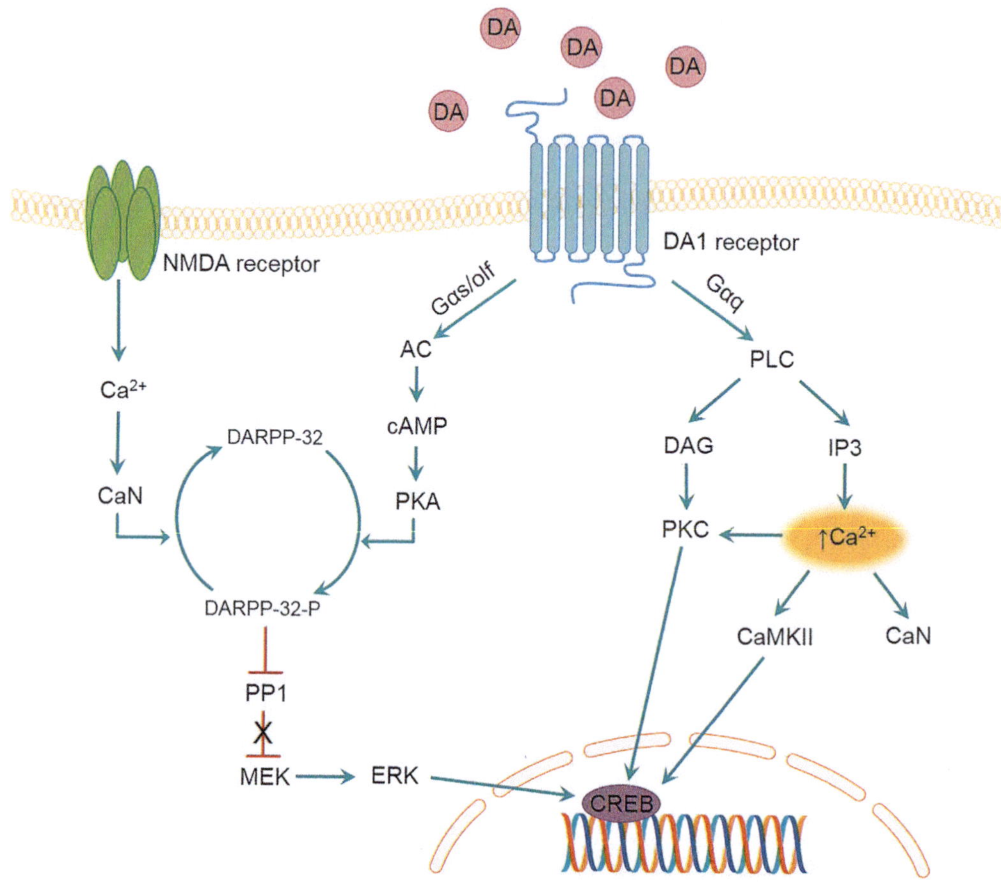

FIG. 39.4 D1-mediated signaling pathways.
Binding of DA to DA1 receptors leads to activation of $G\alpha_{s/olf}$/PKA/DARPP-32 and $G\alpha_q$/PLC signaling cascades. *Blue arrows* indicate activation, while *red T-arrows* show inhibition. Inactivation of PP1 by DARPP-32-P eradicates the PP1 inhibitory effect on MEK (unpublished, author: Agnieszka Michalak). Abbreviations: *AC*, adenyl cyclase; *CAMKII*, Ca^{2+}/calmodulin-dependent protein kinase II; *CaN*, calcineurin; *CREB*, cAMP response element-binding protein; *DA*, dopamine; *DA1*, dopaminergic type 1 receptors; *DAG*, diacylglycerol; *DARPP-32*, dopamine and cAMP-regulated phosphoprotein, 32kDa; *DARPP-32-P*, phosphorylated dopamine and cAMP-regulated phosphoprotein; *ERK*, extracellular signal-regulated kinases; *MEK*, MAPK/ERK kinase; *PKA*, protein kinase A; *PKC*, protein kinase C; *PP1*, protein phosphatase-1.

Self-administration paradigm. The role of DA1 receptors in the reinstatement of nicotine-seeking behavior induced by cues after extinction of nicotine self-administration was demonstrated in many studies. First of all, it should be mentioned that nicotine evokes seeking behavior; for example, observed in self-administration paradigm, this behavior can be extinguished after few days without exposure to nicotine. However, nicotine-associated cues can maintain responding after extensive testing in the absence of the drug, even after few months. It was shown that environmental signals play a pivotal role in the acquisition of nicotine self-administration (Caggiula et al., 2001), and a dopaminergic system is crucial for this effect (Bossert, Poles, Wihbey, Koya, & Shaham, 2007).

The important role of DA1 receptors in the reinstatement of conditioned incentive properties of nicotine-associated cues was evaluated by Liu et al. (2010). In this study, the rats were trained for self-administration with nicotine (30days, 0.03mg/kg/infusion) with a presence of light/tone cues. After extinction, the response was reinstated by the introduction of two levers and light. To examine the role of DA1 receptors in cue-induced reinstatement of extinguished nicotine-seeking behavior, DA1 antagonist—SCH-23390—was administered (5, 10, and 30μg/kg, intraperitoneally, ip). The study revealed that SCH-23390 administered at the highest dose reduced the nicotine-seeking behaviors evoked by introduction of an illumination cue previously associated with nicotine delivery. In summary, the results suggest that stimulation of DA1 receptors mediate the conditioned motivation by nicotine cues. Similarly, Guy and Fletcher (2014) showed that DA1 receptor blockade by SCH-23390 reduced lever-pressing response induced by presentations of the conditioned stimuli and impaired capacity of acute administration of nicotine to increase this response.

The role of DA1 receptors in nicotine reinforcement has been confirmed by Hall, Slade, Allenby, Kutlu, and Levin (2015) in self-administration paradigm. Administration of SCH-23390 at the doses of 1–4μg per side into the NAc, parietal association cortex, anterior cingulate cortex, anterior cingulate cortex, and anterior cingulate cortex reduced nicotine self-administration. These results confirm a crucial role of the DA1 receptor in nicotine addiction and outline a neuroanatomical map of this association. Moreover, blockade of the DA1 receptor by SCH-23390 reduces the reinforcing effects of nicotine in rats in a sex-independent manner (Barrett, Geary, Steiner, & Bevins, 2017).

Conditioned place preference/aversion paradigm. The involvement of DA1 receptors located in different regions of the NAc in the reinforcing and aversive effects of nicotine was assessed by Laviolette, Lauzon, Bishop, Sun, and Tan (2008) in conditioned place preference (CPP)/conditioned place aversion (CPA) paradigms. It is worth to emphasize that one of the regions critical to the motivational effects of drugs of abuse is the NAc. The NAc comprises two regions, the core and the shell (Zahm, 1999). Several lines of evidence show that the NAc shell is involved in the control of the rewarding effects of psychoactive substance and drug-seeking behaviors by spatial/contextual cues, whereas the NAc core is important for discrete cues (Bossert et al., 2007; Chaudhri, Sahuque, Schairer, & Janak, 2010). In the CPP test, the DA1 antagonist administered into the shell of the NAc attenuated the acquisition of nicotine-induced CPP (Spina, Fenu, Longoni, Rivas, & Di Chiara, 2006). Furthermore, it was revealed that nicotine administered at the low dose (0.008 nmol/0.5 μL) in the intraventral tegmental area (VTA) or sc at the dose of 0.8 mg/kg induces CPA and the microinfusion of SCH-23390 into the NAc shell (1 μg/0.5 μL) do not change this aversion. Meanwhile, animals injected with SCH-23390 into the NAc core developed strong place preference when administrated with the same dose of nicotine, thus switching CPA into CPP. Intra-VTA nicotine injection in a neutral dose (0.8 nmol/0.5 μL) coadministered with SCH-23390 (1 μg/0.5 μL) into the NAc shell did not evoke either aversion or preference of nicotine-coupled environments. However, animals injected with SCH-23390 into the NAc core revealed strong place preference when coadministrated with a subreward dose of nicotine. Furthermore, withdrawal of chronic exposure to nicotine has been shown to reduce the level of DA in the striatum of rodents (Rada, Jensen, & Hoebel, 2001). It was revealed that inhibition of DA1 receptors in the NAc shell blocked nicotine withdrawal aversion, while this effect was not observed after SCH-23390 administration into the NAc core (Laviolette et al., 2008). Thus it can be concluded that DA1 receptors localized in the NAc core are involved in switching nicotine aversion to reward and increase the reinforcing effects of nicotine, while DA1 receptors localized in NAc shell are important for nicotine withdrawal.

Additionally, Grieder et al. (2012) showed that both the activation and inhibition of DA1 receptors attenuated the aversive motivational response observed after acute administration of nicotine in nondependent mice. There was no effect on nicotine-dependent and nicotine-withdrawn rodents. These results suggest an important role of DA1 receptors only in the acute aversive effects of nicotine. Interestingly, experiments engaging discriminative stimulus tests showed that the DA1 antagonist SCH-23390 diminished the discriminative stimulus effects of nicotine (Corrigall & Coen, 1994), whereas agonist of DA1 receptors, SKF82958, substituted nicotine in this paradigm (Gasior, Shoaib, Yasar, Jaszyna, & Goldberg, 1999). The aforementioned results unambiguously confirm the role of DA1 receptors in rewarding effects of nicotine.

Withdrawal. To evaluate the role of DA1 receptors in the alleviation of spontaneous withdrawal symptoms 1 day after nicotine cessation (9 mg/kg/day, 7 days via osmotic minipumps), DA1/5 agonist (SKF81297) was administered. It was revealed that activation of DA1-like receptors with SKF81297 at the dose of 0.32 mg/kg diminished teeth chattering/chews; there was no influence on wet dog shakes. Because DA2 receptor activation reduced both mentioned somatic abstinence signs, the authors suggested that DA2 rather than DA1 agonists should be considered for the treatment of physical symptoms of nicotine dependence (Ohmura, Jutkiewicz, Zhang, & Domino, 2011). These studies together with the aforementioned experiments conducted by Grieder et al. (2012) and Laviolette et al. (2008) failed to clearly indicate the involvement of DA1 receptors in nicotine withdrawal. Hence, further investigation is needed into this matter.

Sensitization on locomotor effects. Behavioral sensitization concerns the enhancement of activity of a drug after repeated, intermittent administration (Vanderschuren & Kalivas, 2000). Nicotine injections (5 days, 0.4 mg/kg, sc) induce an increase in locomotor activity in rodents, which suggests the occurrence of synaptic neuroplasticity. A noticeable increase of this effect was observed after the last dose was injected after a 5 days' break (*challenge dose*), which indicates the development of nicotine-induced locomotor sensitization. A significant decrease of the locomotor activity in mice with developed locomotor sensitization to nicotine was demonstrated after administration of DA1 antagonist SCH-23390 (0.03 mg/kg, ip). Consequently, nicotine-induced behavioral sensitization is connected with the activation of cAMP-dependent signaling cascade resulting from activation of DA1 receptors (Goutier, Lowry, McCreary, & O'Connor, 2015).

39.2.2 Other Nicotine-DA1 Receptors Associations

Memory, anxiety and neuronal remodeling. Dopaminergic transmission involving DA1 receptors in the NAc takes also part in memory retrieval improvement induced by nicotine in the inhibitory avoidance (IA) discrimination task. Nicotine itself improves retrieval of IA memory impaired by morphine. While DA1 receptor antagonist (SCH-23390)-administered intra-NAc by itself has no effect on memory retrieval performance, when injected before systemic nicotine, it prevents reinstatement of memory retrieval induced by nicotine. It suggests that nicotine shows its effects on retrieval of IA memory by interacting with DA1 receptors (Azizbeigi, Ahmadi, Babapour, Rezayof, & Zarrindast, 2011). It is well established that acute exposure to nicotine exert an anxiogenic effect, which was also confirmed by Zarrindast, Eslahi, Rezayof, Rostami, and Zahmatkesh (2013). They showed that injection of nicotine (1 µg/rat) into the central amygdala exerts the anxiogenic effect observed as a decrease of time spent on open arms and entries into open arms in the elevated plus maze test. Interestingly, administration of SCH-23390 at the dose of 0.25 µg per rat intra-VTA attenuated the anxiogenic effect of nicotine. It suggests that DA1 receptors play also an important role in the mechanisms underlying anxiogenic-like effects of nicotine. Finally, nicotine-induced remodeling of neurons in adolescence is DA1 receptor-dependent. Most smokers start smoking during adolescence (NIDA, 2012). Therefore, evaluation of the influence of nicotine on developing brain is of urgent need. Nicotine-induced remodeling of dendrites is highly expressed in the NAc shell of adolescent (McDonald et al., 2007). Formation of new dendritic branches and spines was observed after 28–42 days postnatal and persist for 21 days. Coadministration of nicotine with DA1 receptor antagonist SCH-23390 blocked this remodeling effect on dendrites. Thus, it can be suggested that dopaminergic system is critical for neuroadaptive changes induced by nicotine in adolescence (Ehlinger et al., 2016).

Nicotine, DA1 receptors and neuropeptides. Neuropeptides are related to the pathology of drugs of abuse, mainly through the influence on the dopaminergic neurotransmission. Substance P (SP) is a neuropeptide, which influences the limbic and extrapyramidal dopaminergic neurotransmission that may play a role in the development and maintenance of drug dependence. It was revealed that SP exerts its effects by activation of the neurokinin receptor 1 (NK1), as opioids do not produce rewarding effects in NK1 knockout mice, in either CPP or self-administration paradigms. Additionally, SP is involved in abstinence syndrome after opioids withdrawal (Murtra, Sheasby, Hunt, & De Felipe, 2000). Nicotine influence on SP level and interaction of SP/DA1 receptors have been studied by Alburges, Frankel, Hoonakker, and Hanson (2009). They revealed that repeated administration of nicotine 4.0 mg/kg/day (five injections every 2 h) diminished the tissue level of SP-like immunoreactivity (SPLI) in the limbic system and basal ganglia, which reflects an increased release and turnover of SP. The alteration in level of SPLI persisted 12–18 h after the last injection and was no longer observed after 48 h. Interestingly, it was shown that pretreatment with DA1 receptor antagonist (SCH-23390) abolished nicotine-induced changes in SPLI level in the VTA, but not in the SN. Thus, it can be suggested that DA1 receptor blockade, among other mechanisms, mediates the nicotine-induced SP changes in the mesolimbic pathway.

Nicotine in pharmacotherapy—The role of DA1 receptors. It has long been recognized that the loss of dopaminergic neurons in the SN underlies PD pathogenesis. Additionally, it was found that tobacco smokers are at lower risk of PD incidents and, additionally, that motor dysfunctions in PD are alleviated by nicotine. In rodent model of PD induced by administration of 6-hydroxydopamine, chronic intake of nicotine (increasing doses 15–30 mg/L, 27 weeks) accelerated the activity of DA1 receptors localized in the striatum. Thus, it can be suggested that exposure to nicotine by modification of dopaminergic balance may be useful in the alleviation of PD symptoms (García-Montes et al., 2012).

Also the mechanisms involved in the motivational effects of nicotine in rats with diabetes induced by streptozotocin, a substance toxic to pancreatic beta cell producing insulin, have been evaluated. There is ample evidence that activity of the mesolimbic system remains under the influence of insulin. The insulin receptors are localized in the NAc and in the VTA. It has been shown that the activation of insulin receptors exerts negative modulatory effect on the reward response (Figlewicz, Evans, Murphy, Hoen, & Baskin, 2003). Epidemiological studies indicate that diabetic patients show an increased tendency to smoke cigarettes. The desire for nicotine has been found to be enhanced in rats with streptozotocin-induced diabetes in a self-administration paradigm. Increased DA transporter level and reduction of DA1 receptor were observed in the NAc of investigated rats. Thus, it can be concluded that a decrease in dopaminergic neurotransmission in the NAc of streptozotocin-treated rodents increase nicotine self-administration (O'Dell et al., 2014).

MINI-DICTIONARY OF TERMS

Behavioral sensitization A phenomenon involving the enhanced behavioral responses to repeated intermittent exposure to a stimulus such as a drug of abuse.

Conditioned place aversion A simple, noninvasive technique, based on Pavlovian conditioning, where an animal learns to avoid a compartment that was previously paired with an aversive stimulus.

Conditioned place preference A form of classic Pavlovian conditioning. Drugs of abuse, which possess reinforcing properties, provide an unconditioned stimulus. Their administration combined with a definite environment makes this environment acquire the rewarding properties, thus becoming a conditioned stimulus.

Self-administration A form of operant conditioning, where an animal makes an effort, for example, presses a lever, in order to deliver a dose of a drug.

Key Facts of Link Between Nicotine and D1 Receptors

- Nicotine mediates the release of dopamine via activation of $\alpha4\beta2$ and $\alpha7$ nicotinic acetylcholine receptors, which leads to the subsequent activation of dopaminergic type 1 receptors and activation of $G\alpha_{s/olf}$/PKA/DARPP-32 signaling cascade.
- rs686 is a causative locus for the association of dopaminergic type 1 receptor gene and nicotine dependence.
- Dopaminergic type 1 receptors are involved in rewarding and aversive effects of nicotine. However, evidence concerning involvement of dopaminergic type 1 receptors in nicotine withdrawal is contradictory.
- Nicotine by increasing the activity of dopaminergic type 1 receptors alleviates Parkinson's disease symptoms.
- Nicotine induces remodeling on dendrites in adolescence in dopaminergic type 1 receptor-dependent mechanism.
- Dopaminergic type 1 receptors play an important role in anxiety and memory processes induced by nicotine.
- Decrease of dopaminergic neurotransmission in nucleus accumbens in rodents with diabetes elevates nicotine self-administration.

Summary Points

- The aim of this chapter is to summarize the findings concerning the role of dopaminergic type 1 receptors on the effects of nicotine.
- Dopaminergic type 1 receptors are localized in the reward-related brain regions, for example, mesolimbic and mesocortical pathways, especially in areas such as the striatum, amygdala, nucleus accumbens, substantia nigra, frontal cortex, and olfactory bulb, which also receive cholinergic input. Mesolimbic system remains under the influence of peptides, that is, insulin.
- There is a significant link between nicotine dependence and dopaminergic type 1.
- Dopaminergic type 1 receptors play a key role in memory and anxiety processes and contribute to mental disorders such as Parkinson's disease, schizophrenia, or Huntington's disease.
- Experimental data have shown that substance P has an influence on the limbic and extrapyramidal dopaminergic neurotransmission and may play a role in the development and maintenance of drug dependence.
- Nicotine induces remodeling of dendrites in developing brain.

References

Abdolahi, A., Acosta, G., Breslin, F. J., Hemby, S. E., & Lynch, W. J. (2010). Incubation of nicotine seeking is associated with enhanced protein kinase A-regulated signaling of dopamine- and cAMP-regulated phosphoprotein of 32 kDa in the insular cortex. *European Journal of Neuroscience*, *31*(4), 733–741.

Alburges, M. E., Frankel, P. S., Hoonakker, A. J., & Hanson, G. R. (2009). Responses of limbic and extrapyramidal substance P systems to nicotine treatment. *Psychopharmacology*, *201*(4), 517–527.

Azizbeigi, R., Ahmadi, S., Babapour, V., Rezayof, A., & Zarrindast, M. R. (2011). Nicotine restores morphine-induced memory deficit through the D1 and D2 dopamine receptor mechanisms in the nucleus accumbens. *Journal of Psychopharmacology*, *25*(8), 1126–1133.

Barrett, S. T., Geary, T. N., Steiner, A. N., & Bevins, R. A. (2017). Sex differences and the role of dopamine receptors in the reward-enhancing effects of nicotine and bupropion. *Psychopharmacology*, *234*(2), 187–198.

Beaulieu, J. M., & Gainetdinov, R. R. (2011). The physiology, signaling, and pharmacology of dopamine receptors. *Pharmacological Reviews*, *63*(1), 182–217.

Bossert, J. M., Poles, G. C., Wihbey, K. A., Koya, E., & Shaham, Y. (2007). Differential effects of blockade of dopamine D1-family receptors in nucleus accumbens core or shell on reinstatement of heroin seeking induced by contextual and discrete cues. *Journal of Neuroscience*, *27*, 12655–12663.

Brichta, L., Greengard, P., & Flajolet, M. (2013). Advances in the pharmacological treatment of Parkinson's disease: targeting neurotransmitter systems. *Trends in Neurosciences*, *36*(9), 543–554.

Caggiula, A. R., Donny, E. C., White, A. R., Chaudhri, N., Booth, S., Gharib, M. A., et al. (2001). Cue dependency of nicotine self-administration and smoking. *Pharmacology, Biochemistry and Behavior*, *70*(4), 515–530.

Chaudhri, N., Sahuque, L. L., Schairer, W. W., & Janak, P. H. (2010). Separable roles of the nucleus accumbens core and shell in context- and cue-induced alcohol-seeking. *Neuropsychopharmacology*, *35*, 783–791.

Corrigall, W. A., & Coen, K. M. (1994). Dopamine mechanisms play at best a small role in the nicotine discriminative stimulus. *Pharmacology, Biochemistry and Behavior*, *48*(3), 817–820.

Ehlinger, D. G., Bergstrom, H. C., Burke, J. C., Fernandez, G. M., McDonald, C. G., & Smith, R. F. (2016). Adolescent nicotine-induced dendrite remodeling in the nucleus accumbens is rapid, persistent, and D1-dopamine receptor dependent. *Brain Structure and Function*, *221*(1), 133–145.

Figlewicz, D. P., Evans, S. B., Murphy, J., Hoen, M., & Baskin, D. G. (2003). Expression of receptors for insulin and leptin in the ventral tegmental area/substantia nigra (VTA/SN) of the rat. *Brain Research*, *964*, 107–115.

Floresco, S. B., West, A. R., Ash, B., Moore, H., & Grace, A. A. (2003). Afferent modulation of dopamine neuron firing differentially regulates tonic and phasic dopamine transmission. *Nature Neuroscience*, *6*, 968–973.

García-Montes, J. R., Boronat-García, A., López-Colomé, A. M., Bargas, J., Guerra-Crespo, M., & Drucker-Colín, R. (2012). Is nicotine protective against Parkinson's disease? An experimental analysis. *CNS & Neurological Disorders: Drug Targets*, *11*(7), 897–906.

Gasior, M., Shoaib, M., Yasar, S., Jaszyna, M., & Goldberg, S. R. (1999). Acquisition of nicotine discrimination and discriminative stimulus effects of nicotine in rats chronically exposed to caffeine. *Journal of Pharmacology and Experimental Therapeutics*, *288*(3), 1053–1073.

Goutier, W., Lowry, J. P., McCreary, A. C., & O'Connor, J. J. (2015). Frequency-dependent modulation of dopamine release by nicotine and dopamine D1 receptor ligands: an in vitro fast cyclic voltammetry study in rat striatum. *Neurochemical Research*, *41*(5), 945–950.

Goutier, W., O'Connor, J. J., Lowry, J. P., & McCreary, A. C. (2016). The effect of nicotine induced behavioral sensitization on dopamine D1 receptor pharmacology: an in vivo and ex vivo study in the rat. *European Neuropsychopharmacology*, *25*(6), 933–943.

Gozen, O., Balkan, B., Yildirim, E., Koylu, E. O., & Pogun, S. (2013). The epigenetic effect of nicotine on dopamine D1 receptor expression in rat prefrontal cortex. *Synapse*, *67*(9), 545–552.

Grieder, T. E., George, O., Tan, H., George, S. R., Le Foll, B., Laviolette, S. R., et al. (2012). Phasic D1 and tonic D2 dopamine receptor signaling double dissociate the motivational effects of acute nicotine and chronic nicotine withdrawal. *Proceedings of the National Academy of Sciences of the United States of America*, *109*(8), 3101–3106.

Guy, E. G., & Fletcher, P. J. (2014). Responding for a conditioned reinforcer, and its enhancement by nicotine, is blocked by dopamine receptor antagonists and a 5-HT(2C) receptor agonist but not by a 5-HT(2A) receptor antagonist. *Pharmacology, Biochemistry, and Behavior*, *125*, 40–47.

Hall, B. J., Slade, S., Allenby, C., Kutlu, M. G., & Levin, E. D. (2015). Neuro-anatomic mapping of dopamine D1 receptor involvement in nicotine self-administration in rats. *Neuropharmacology*, *99*, 689–695.

Hamada, M., Hendrick, J. P., Ryan, G. R., Kuroiwa, M., Higashi, H., Tanaka, M., et al. (2005). Nicotine regulates DARPP-32 (dopamine- and cAMP-regulated phosphoprotein of 32 kDa) phosphorylation at multiple sites in neostriatal neurons. *Journal of Pharmacology and Experimental Therapeutics*, *315*(2), 872–878.

Hisahara, S., & Shimohama, S. (2011). Dopamine receptors and Parkinson's disease. *International Journal of Medicinal Chemistry*. https://doi.org/10.1155/2011/403039.

Huang, W., & Li, M. D. (2009). Differential allelic expression of dopamine D1 receptor gene (DRD1) is modulated by microRNA miR-504. *Biological Psychiatry*, *65*(8), 702–705.

Komatsu, H. (2015). Novel therapeutic GPCRs for psychiatric disorders. *International Journal of Molecular Sciences*, *16*(6), 14109–141021.

Laviolette, S. R., Lauzon, N. M., Bishop, S. F., Sun, N., & Tan, H. (2008). Dopamine signaling through D1-like versus D2-like receptors in the nucleus accumbens core versus shell differentially modulates nicotine reward sensitivity. *Journal of Neuroscience*, *28*(32), 8025–8033.

Liu, X., Jernigen, C., Gharib, M., Booth, S., Caggiula, A. R., & Sved, A. F. (2010). Effects of dopamine antagonists on drug cue-induced reinstatement of nicotine-seeking behavior in rats. *Behavioral Pharmacology*, *21*(2), 153–160.

McDonald, C. G., Eppolito, A. K., Brielmaier, J. M., Smith, L. N., Bergstrom, H. C., Lawhead, M. R., et al. (2007). Evidence for elevated nicotine-induced structural plasticity in nucleus accumbens of adolescent rats. *Brain Research*, *1151*, 211–218.

Murtra, P., Sheasby, A. M., Hunt, S. P., & De Felipe, C. (2000). Rewarding effects of opiates are absent in mice lacking the receptor for substance P. *Nature*, *405*, 180–183.

National Institute on Drug Abuse (2012). *Research report series: Tobacco addiction.* Washington: Department of Health and Human Services (US) [NIH Pub No. 12-4342].

O'Dell, L. E., Natividad, L. A., Pipkin, J. A., Roman, F., Torres, I., Jurado, J., et al. (2014). Enhanced nicotine self-administration and suppressed dopaminergic systems in a rat model of diabetes. *Addiction Biology*, *19*(6), 1006–1019.

Ohmura, Y., Jutkiewicz, E. M., Zhang, A., & Domino, E. F. (2011). Dopamine D1/5 and D2/3 agonists differentially attenuate somatic signs of nicotine withdrawal in rats. *Pharmacology Biochemistry and Behavior*, *99*(4), 552–556.

Rada, P., Jensen, K., & Hoebel, B. G. (2001). Effects of nicotine and mecamylamine induced withdrawal on extracellular dopamine and acetylcholine in the rat nucleus accumbens. *Psychopharmacology*, *157*, 105–110.

Spina, L., Fenu, S., Longoni, R., Rivas, E., & Di Chiara, G. (2006). Nicotine-conditioned single-trial place preference: selective role of nucleus accumbens shell dopamine D1 receptors in acquisition. *Psychopharmacology*, *184*(3–4), 447–455.

Svenningsson, P., Nishi, A., Fisone, G., Girault, J. A., Nairn, A. C., & Greengard, P. (2004). DARPP-32: an integrator of neurotransmission. *Annual Review of Pharmacology and Toxicology*, *44*, 269–296.

Vanderschuren, L. J., & Kalivas, P. W. (2000). Alterations in dopaminergic and glutamatergic transmission in the induction and expression of behavioral sensitization: a critical review of preclinical studies. *Psychopharmacology*, *151*(2–3), 99–120.

Wheeler, T. L., Smith, L. N., Bachus, S. E., McDonald, C. G., Fryxell, K. J., & Smith, R. F. (2013). Low-dose adolescent nicotine and methylphenidate have additive effects on adult behavior and neurochemistry. *Pharmacology Biochemistry and Behavior*, *103*(4), 723–734.

Zahm, D. S. (1999). Functional-anatomical implications of the nucleus accumbens core and shell subterritories. *Annals of the New York Academy of Sciences*, *877*, 113–128.

Zarrindast, M. R., Eslahi, N., Rezayof, A., Rostami, P., & Zahmatkesh, M. (2013). Modulation of ventral tegmental area dopamine receptors inhibit nicotine-induced anxiogenic-like behavior in the central amygdala. *Progress in Neuro-Psychopharmacology and Biological Psychiatry*, *41*, 11–17.

CHAPTER

40

Brain Gene Expression in the Context of Nicotine Rewards: A Focus on Cholinergic Genes

Mark D. Namba, Gregory L. Powell*, Armani P. Del Franco†, Julianna G. Goenaga*, Cassandra D. Gipson**

*Department of Psychology, Arizona State University, Tempe, AZ, United States

†Department of Neuroscience, University of Minnesota, Minneapolis, MN, United States

Abbreviations

GWAS Genome-Wide Association Study
KO knockout
nAChR nicotinic acetylcholine receptor
ND nicotine dependence
SNP single-nucleotide polymorphism

40.1 INTRODUCTION

Advances in the study of genetics have greatly improved our understanding of the etiology of smoking behaviors. Several studies have found that the heritability of nicotine dependence (ND) is roughly 50% (Rose, Broms, Korhonen, Dick, & Kaprio, 2009) and that these genetic contributors are subject to socioeconomic, cultural, and other environmental influences (Mackillop, Obasi, Amlung, McGeary, & Knopik, 2010). Importantly, the genetic contributions underlying the initiation of tobacco use appear to be distinct from those underlying its persistence (Morley et al., 2007), which highlights the dynamic role that genes play in the acquisition and maintenance of smoking habits. Relapsing use of drugs is generally thought to be the consequence of multiple gene interactions. Many clinical and preclinical studies examining genetic contributors to nicotine addiction vulnerability have revealed several variants in multiple neurotransmitter, neuromodulator, and metabolic systems that are associated with the acquisition, maintenance, and relapse of nicotine use (Mackillop et al., 2010). Particularly, genetic variation in several cholinergic genes is associated with heavy smoking, whereas light smokers retain a lower genetic risk for ND (Belsky et al., 2013). In this chapter, we will briefly discuss both clinical and preclinical studies examining cholinergic gene targets underlying tobacco use, identifying gaps in the literature wherever necessary. In addition, we will discuss other potential gene targets that warrant future study. Finally, we will address potential challenges and important considerations of applying this knowledge in a clinical setting.

40.2 CLINICAL EXAMINATION OF CHOLINERGIC GENE VARIATION IN NICOTINE ADDICTION

Clinical data have highlighted the importance of a number of gene clusters as mediators of nicotine dependence. Among these, cholinergic genes stand out as the primary driver of the rewarding properties of nicotine and may be beneficial targets for treatment to promote cessation of nicotine use. As reviewed by Bühler and colleagues, a majority of single-nucleotide polymorphisms (SNPs) associated with nicotine-related phenotypes found in the literature involve nicotinic acetylcholine receptor (nAChR) genes (Bühler et al., 2015). Nicotine is known to exert its rewarding and reinforcing effects primarily through the activation of nAChRs (including α4β2 and α7 containing) in regions such as the ventral tegmental area (VTA) and nucleus accumbens (NA) (Laviolette & van der Kooy, 2004). Genetic

Neuroscience of Nicotine
https://doi.org/10.1016/B978-0-12-813035-3.00040-X

variations in nAChRs are the most consistent translational findings between clinical and preclinical studies and are significantly associated with dependence to nicotine and comorbidities such as lung cancer vulnerability (Li & Burmeister, 2009). Furthermore, different nAChR subtypes have been shown to possess different desensitization profiles, with c-based receptors undergoing rapid desensitization, while α4β2-based receptors have increased affinity for nicotine and bind nicotine for longer durations (Quick & Lester, 2002; see Fig. 40.1 for α4β2 and α7 structure). Potential genetic variations in these nAChR subtypes or differential expression of subtype by population would impact receptor functionality and response to long-term nicotine exposure. It is therefore crucial to examine the genetic heterogeneity of nAChR expression as it relates to ND. *CHRNA4*, which encodes the α4 subunit, has been shown to share a significant association with measures of ND in multiple ethnicities (primarily European-American and African-American samples reported) and in both men and women (Feng et al., 2004; Hutchison et al., 2007; Li et al., 2005). Recently, a metaanalysis of two separate longitudinal family studies found a moderate association between *CHRNA4* SNPs and ND (Kamens et al., 2013). Other studies have reported no association between *CHRNB2* (which encodes β2 subunits) and ND (Feng et al., 2004; Li et al., 2005; Silverman et al., 2000); however, the effects of *CHRNB2* variants may depend on variation in *CHRNA4* (Li & Burmeister, 2009; Li, Lou, Chen, Ma, & Elston, 2008).

Genome-Wide Association Studies (GWAS) (Fig. 40.2) have revealed that variants in the *CHRNA5-CHRNA3-CHRNB4* gene locus, located on chromosome 15 (15q25.1), are associated with both ND and disease susceptibility (Fig. 40.3). For instance, Saccone and colleagues discovered that SNPs in CHRNB3 and CHRNA5 were associated with ND, revealing potential genetic risk loci underlying ND susceptibility (Saccone et al., 2006). Other studies also support such findings, where genetic variation among this gene cluster is associated with not only ND (Berrettini et al., 2008; Bierut et al., 2008) but also lung cancer and other cardiopulmonary diseases (Thorgeirsson et al., 2008).

One recent study examining the *CHRNA5-CHRNA3-CHRNB4* gene cluster found that the *CHRNA3* SNP rs578776 in nonsmoking adolescents is associated with blunted activation of the anterior cingulate cortex (ACC) during a reward-based task, potentially underlying altered reward processing and increased ND susceptibility (Nees et al., 2013). The ACC is involved in

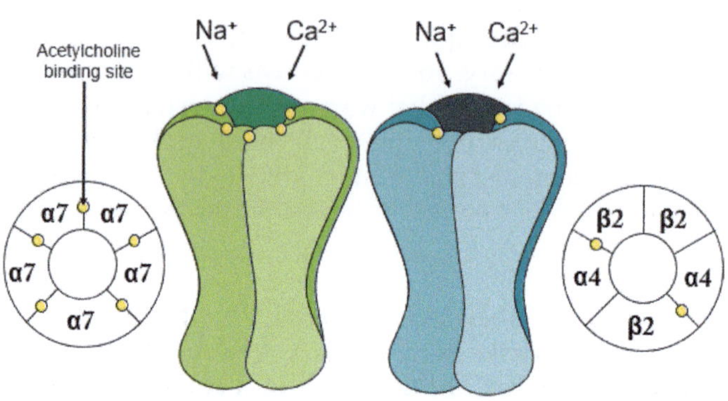

FIG. 40.1 nAChR structures.
Genetic variation in homomeric α7 *(green)* and heteromeric α4β2 *(blue)* nAChRs is associated with ND. Both α7 and α4β2 gate calcium, which is a likely mechanism through which these receptors alter cell signaling in various brain regions and cell types.

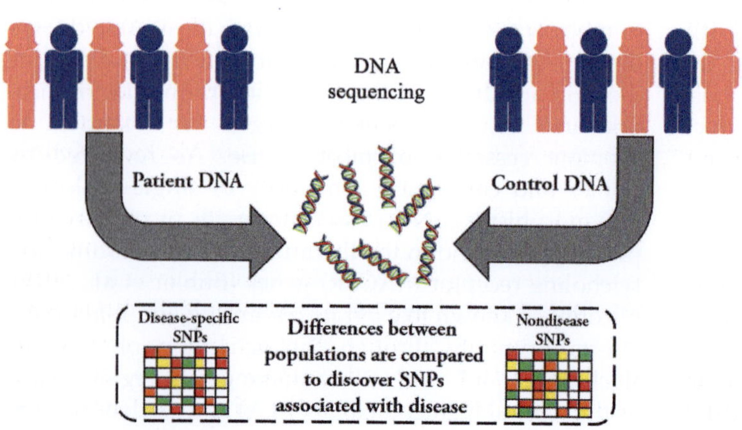

FIG. 40.2 Summary of GWAS.
DNA from disease-susceptible individuals and healthy controls is sequenced and compared across the entire genome for population differences in specific SNPs that are associated with the disease of interest.

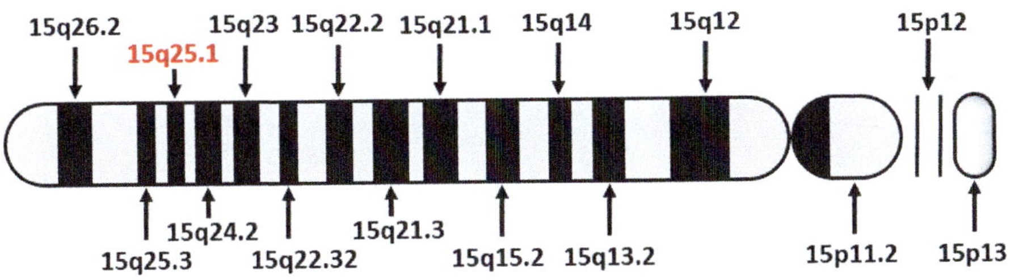

FIG. 40.3 Schematic of chromosome 15. The *CHRNA5-CHRNA3-CHRNB4* gene locus is located on 15q25.1 (highlighted in *red*) and is associated with ND and disease vulnerability.

several higher-order executive functions, such as reward processing, impulse control, and decision-making. Furthermore, the influence of peers on ND has been found to be less influential among individuals containing a functional SNP at *CHRNA5* in adolescents, while overall, four specific SNPs in *CHRNA5*, *CHRNA3*, *CHRNB3*, and *CHRND* were associated with ND (Johnson et al., 2010). Recent pathway analysis of two substance dependence GWAS datasets identified cholinergic receptors among the gene ontology (GO) terms, in addition to sensory perception, ribosome, and retinoid terms (Harari et al., 2012), strengthening the association of nAChR gene expression variation with ND. Investigation of genetic variability for the enzyme choline acetyltransferase (ChAT) has also revealed significant effects on smoking cessation and ND. In particular, a subset of ChAT SNPs are significantly associated with cessation success following an 8-week replacement therapy and counseling treatment regimen (Ray et al., 2010), whereas other SNPs are associated with ND in European-American and African-American samples (Wei et al., 2010).

These genetic studies have provided compelling evidence implicating cholinergic genes in ND and disease vulnerability. However, it is important to note that nAChRs also mediate physiological mechanisms such as inflammation (Bencherif, 2009) and pain (Decker, Meyer, & Sullivan, 2001) and are involved in other neuropsychiatric conditions such as anxiety and depression (Picciotto, Brunzell, & Caldarone, 2002). These neuropsychiatric conditions are often comorbid and may exacerbate ND. As such, the clinical relevance of cholinergic gene variants should be examined holistically within the context of a broader disease state that extends beyond ND.

40.3 PRECLINICAL MODELS EXAMINING CHOLINERGIC GENE VARIATION IN NICOTINE ADDICTION

Pharmacological and genetic approaches have revealed important information regarding the role of these subunits in mediating the rewarding and reinforcing properties of nicotine. Early studies have demonstrated that the nicotinic system is a critical regulator of cognitive function (Levin, 2002) and that genetic deletion of the β2 nAChR subunit mediates associative learning processes and nicotine-induced dopamine release from the VTA (Picciotto et al., 1995, 1998). Several nAChR subunits (α3, α5, α6, and β4) are expressed in dopaminergic neurons in the VTA, where mRNA signals for these subunits are absent following chemical ablation of the VTA (Charpantier, Barnéoud, Moser, Besnard, & Sgard, 1998). Klink, de Kerchove d'Exaerde, Zoli, and Changeux (2001) found that nearly all cells in the VTA contained mRNA for α4 and β2 subunits. This study also observed differential expression of α5, α6, α7, and β3 mRNA between dopaminergic and GABAergic cells in the VTA. Until recently, evidence elucidating the role of nAChR subunits in regulating the reinforcing effects of nicotine has been limited to the β2 subunit (Epping-Jordan, Picciotto, Changeux, & Pich, 1999; Picciotto et al., 1998). However, one study using a virus to reexpress (or "rescue") β2 in knockout (KO) mice found that rescued β2 KO mice exhibited increased dopamine release in the NA similar to wild-type mice when exposed to nicotine. Rescued β2 KO mice preferred the nicotine-paired arm in a Y-maze similar to wild types, whereas β2 KO mice did not exhibit such neurobehavioral traits (Maskos et al., 2005). Currently, newly innovative genetic techniques to study nAChRs in the brain are being implemented, such as Cre-Lox (or FLP-FRT) recombinase technology (Fig. 40.4) (Fowler & Kenny, 2012; Hernandez et al., 2014; Ngolab et al., 2015).

As previously mentioned, genetic variation in *CHRNA5*, which encodes the α5 nAChR subunit, is associated with ND. Recent studies have demonstrated that α5-containing nAChRs may be involved in the regulation of nicotine intake and may also mediate nicotine aversion. In one study, mice containing a null mutation in *CHRNA5* demonstrated increased nicotine intake, an effect that was rescued by reexpressing α5 in the medial habenula (MHb) (Fowler, Lu, Johnson, Marks, & Kenny, 2011). This study also showed that knocking down α5 in the MHb did not alter nicotine reward but did reduce intracranial self-stimulation thresholds at higher nicotine doses, suggesting that MHb α5 regulates inhibitory control over nicotine consumption (Fowler et al., 2011). Similarly, a more recent study observed that the deletion of α5 in nonnicotine-dependent mice (α5 −/−) enhances nicotine reward at lower, nonmotivational doses and eliminates conditioned taste avoidance to higher nicotine

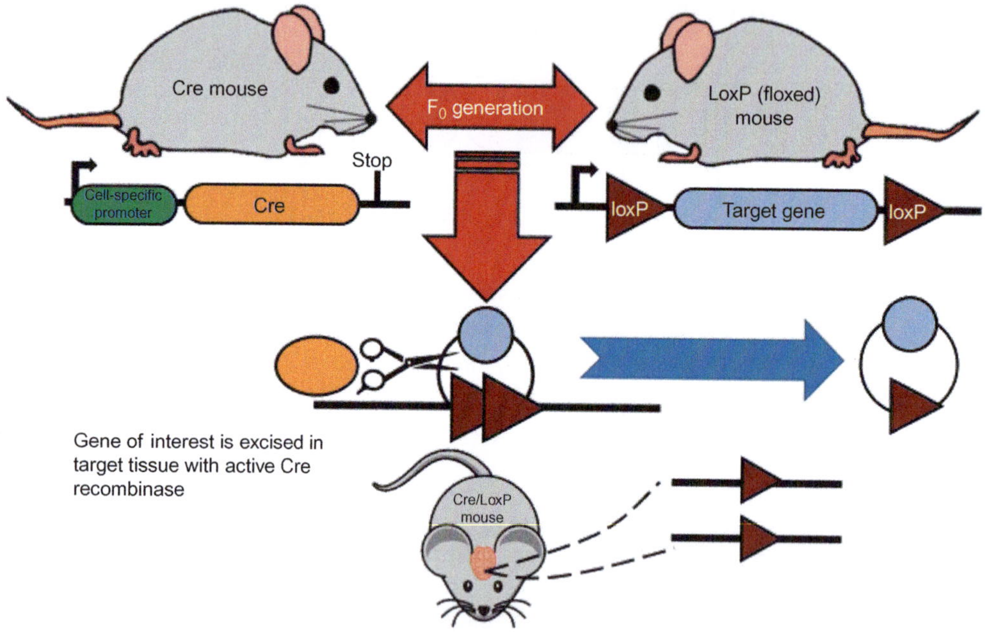

FIG. 40.4 Summary of Cre-Lox conditional knockout.
Cre mice are bred with loxP ("floxed") mice, and a portion of the offspring expresses active Cre recombinase, where the gene of interest is excised by active Cre in target tissue.

doses (Grieder et al., 2017). As well, this study showed that nicotine-dependent α5 −/− mice did not display an aversive conditioned response to nicotine withdrawal and that these effects mimic those observed with dopamine receptor antagonism (Grieder et al., 2017). Nicotine withdrawal symptoms contribute to the high rate of relapse among smokers (Benowitz, 2009), and projections from the MHb to the interpeduncular nucleus (IPN) may mediate these symptoms (Gorlich et al., 2013; Zhao-Shea et al., 2015). Recently, it has been demonstrated that viral upregulation of α4 function in MHb cholinergic neurons increases anxiety-like behaviors in nicotine-withdrawn mice (Pang et al., 2016). Thus, genetic variation in this particular nAChR subunit may contribute to nicotine dependence and relapse vulnerability (McClure-Begley et al., 2014). Fig. 40.5 provides a nonexhaustive schematic of habenular circuitry and associated nAChR subunits implicated in ND.

40.4 OTHER POTENTIAL GENES UNDERLYING ND

Current smoking cessation treatments primarily function to replace nicotinic stimulation of nAChRs to attenuate drug craving without producing rewarding effects (Nides, 2008). For example, varenicline (Chantix®) is a partial agonist of α4β2 and a full agonist of α7 nAChRs and has shown some clinical efficacy in helping individuals quit smoking (Kasza et al., 2015; McClure, Vandrey, Johnson, & Stitzer, 2013). Regardless, high rates of relapse persist even for individuals receiving replacement therapy (Leshner & Stapleton, 1997), which highlights the need for a more holistic understanding of the genetic predispositions underlying ND vulnerability. Table 40.1 summarizes noncholinergic gene targets that may be involved in ND susceptibility and relapse vulnerability.

40.5 CONCLUSION

Over the past several decades, scientists have made great progress in elucidating the genetic etiology of both the development and maintenance of ND. Genetic variants in cholinergic substrates associated with ND are the most consistent findings in the literature, although other noncholinergic gene targets may also be associated with ND. Undoubtedly, these genetic studies have the potential to guide the development of treatment strategies that promote smoking cessation and long-term abstinence. However, it is important to note that ND is highly comorbid with other neuropsychiatric conditions (Grant, Hasin, Chou, Stinson, & Dawson, 2004) and many gene variants are not specifically predictive of ND vulnerability alone. The cholinergic system is ubiquitously expressed throughout both central and peripheral nervous tissue and thus serves a modulatory role over other neurotransmitter systems. As such, genetic variation of cholinergic substrates is likely not specific to ND. Genetic risk for ND is also subject to developmental and environmental influence where phenotypic expression of ND for a given haplotype can depend on factors such as age of

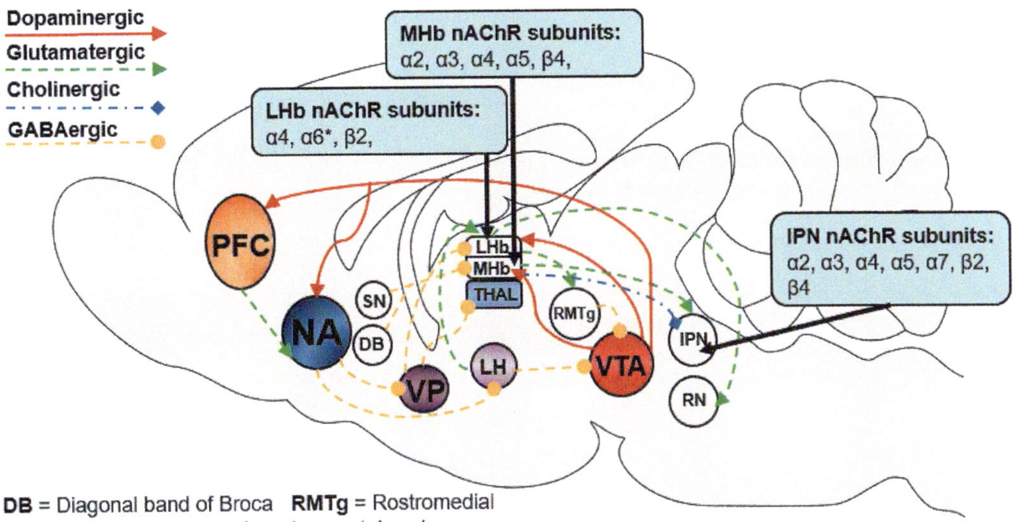

FIG. 40.5 Habenular circuitry and nAChR subunit distribution in ND.
Genetic studies have revealed several nAChR subunits within the habenula and IPN that are involved in ND. The lateral habenula (LHb) receives inhibitory and excitatory input from the ventral pallidum (VP) and lateral hypothalamus (LH), respectively, as well as excitatory input from other cortical and subcortical structures (not shown). The MHb primarily receives inhibitory input from the SN (septal nuclei) and DB (diagonal band of broca). The LHb is activated by aversive stimuli and regulates nicotine intake by indirectly exerting inhibitory control over the VTA through excitatory projections through the rostromedial tegmental nucleus (RMTg). In addition to the α4 and β2 subunits, α6* nAChRs also enhance LHb activity and may modulate nicotine aversion or reward. The IPN receives excitatory projections from the MHb and regulates nicotine consumption through the α5* nAChRs. The β4 subunit along the MHb-IPN tract also regulates nicotine aversion and intake.

TABLE 40.1 Other Potential Gene Targets Underlying ND

Gene	Function	References
DBH	Dopamine β-hydroxylase—catalyzes the conversion of dopamine to norepinephrine	McKinney et al. (2000)
COMT	Catechol-O-methyltransferase—catalyzes the degradation of catecholamines and other compounds with a catechol structure	Beuten, Payne, Ma, and Li (2006)
DRD2/3/4	Dopamine $D_{2/3/4}$ (G_i-coupled) receptors—inhibit adenylate cyclase (AC) and cyclic AMP (cAMP)	Vandenbergh et al. (2007) and David and Munafò (2008)
OPRM1	μ_1-Opioid receptor (G_i-coupled)—inhibits AC and cAMP	Kuwabara et al. (2014)
SLC6A4	Serotonin transporter (5-HTT)—mediates presynaptic reuptake of serotonin	Herman and Balogh (2012)
GABBR1/2	$GABA_{B1}$ and $GABA_{B2}$ ($G_{i/o}$-coupled) receptors—inhibit AC and cAMP and activate K^+ channels	Cui, Seneviratne, Gu, and Li (2012)
CNR1	Cannabinoid receptor type 1 ($G_{i/o}$-coupled)—inhibits AC and cAMP and increases mitogen-activated protein kinase	Chen et al. (2008)
BDNF	Brain-derived neurotrophic factor—neurotrophic growth factor highly expressed in the hippocampus, cortex, and basal forebrain	Jamal, Van der Does, Elzinga, Molendijk, and Penninx (2015)

onset, stress, and emotional state (Baker et al., 2009; Mackillop et al., 2010). Moreover, genetic variants associated with ND may also be subject to epigenetic modification caused by stress, developmental nicotine exposure, previous drug use, and many other factors (Renthal & Nestler, 2008; Volkow, 2011). It is likely that these complications have contributed to inconsistencies between human genetic studies. Nevertheless, understanding the genetic mechanisms underlying ND and relapse vulnerability will help guide future development of novel treatment strategies that better promote smoking cessation and long-term abstinence.

MINI-DICTIONARY OF TERMS

Choline acetyltransferase A transferase enzyme that acts to synthesize acetylcholine.
Epigenetics Study of the environmental modifications to gene expression that occur without alterations to the genetic code.
Etiology The cause, set of causes, or manner of causation of a disease or condition.
Gene cluster A set of nearby genes that code for similar peptides and functions.
Genotype A gene or set of genes carried by an individual.
Haplotype A set of genetic variations that tend to be inherited together.
Knockout An animal (usually a mouse) that does not express the gene of interest.
Phenotype Expression of a specific characteristics or traits.
Single-nucleotide polymorphisms The variation in a single base pair in a DNA sequence.
Wild type An animal that expresses the gene of interest.

Key Facts of Genetic Techniques in Preclinical/ Clinical Models

- A GWAS examines genetic variants across a whole genome to link genetic traits with certain diseases in humans:
  - Typically, a GWAS focuses on associations between specific SNPs and physical disease traits.
  - The case-control design is the most common type of GWAS (Fig. 40.2). Allele frequency is used as a measure between the control and disease groups.
- The Cre-Lox recombinase system allows researchers to have temporal and cell-type-specific control over gene expression in animal models (Nagy, 2000), which preserves endogenous activity of the target gene(s) throughout development unlike traditional knockout methods:
  - Cre-Lox recombination involves the deletion, inversion, or translocation of DNA sequences that are "flanked" by loxP, depending on the directionality and orientation of the lox sites.

Summary Points

- Genetic polymorphisms among cholinergic genes underlie potential risk factors for the development and maintenance of nicotine dependence (ND).
- Cholinergic genes are potential biomarkers for ND susceptibility and treatment.
- Polymorphisms in *CHRNA4* (encoding for the α4 nAChR subunit) have significant associations with ND.
- *CHRNA5-CHRNA3-CHRNB4* gene locus is associated with ND susceptibility.
- Single-nucleotide polymorphisms (SNPs) clustered around choline acetyltransferase (ChAT) have a significant association with successful sustained cessation of nicotine use.
- Cholinergic genes are associated with other psychiatric disorders, illustrating important neurobiological substrates of comorbidity with ND.
- Genetic variants of noncholinergic genes may also be associated with ND.
- Genetic risk for ND is subject to developmental and environmental factors.
- Epigenetic modification of cholinergic genes may explain inconsistencies between human genetic studies.

References

Baker, T. B., Weiss, R. B., Bolt, D., von Niederhausern, A., Fiore, M. C., Dunn, D. M., et al. (2009). Human neuronal acetylcholine receptor A5-A3-B4 haplotypes are associated with multiple nicotine dependence phenotypes. *Nicotine & Tobacco Research, 11*(7), 785–796.

Belsky, D. W., Moffitt, T. E., Baker, T. B., Biddle, A. K., Evans, J. P., Harrington, H., et al. (2013). Polygenic risk and the developmental progression to heavy, persistent smoking and nicotine dependence: evidence from a 4-decade longitudinal study. *JAMA Psychiatry, 70*(5), 534–542.

Bencherif, M. (2009). Neuronal nicotinic receptors as novel targets for inflammation and neuroprotection: mechanistic considerations and clinical relevance. *Acta Pharmacologica Sinica, 30*(6), 702–714.

Benowitz, N. L. (2009). Pharmacology of nicotine: addiction, smoking-induced disease, and therapeutics. *Annual Review of Pharmacology and Toxicology, 49*(1), 57–71.

Berrettini, W., Yuan, X., Tozzi, F., Song, K., Francks, C., Chilcoat, H., et al. (2008). α-5/α-3 nicotinic receptor subunit alleles increase risk for heavy smoking. *Molecular Psychiatry, 13*(4), 368–373.

Beuten, J., Payne, T. J., Ma, J. Z., & Li, M. D. (2006). Significant association of catechol-O-methyltransferase (COMT) haplotypes with nicotine dependence in male and female smokers of two ethnic populations. *Neuropsychopharmacology, 31*(3), 675–684.

Bierut, L. J., Stitzel, J. A., Wang, J. C., Hinrichs, A. L., Grucza, R. A., Xuei, X., et al. (2008). Variants in nicotinic receptors and risk for nicotine dependence. *The American Journal of Psychiatry, 165*(9), 1163–1171.

Bühler, K. M., Giné, E., Echeverry-Alzate, V., Calleja-Conde, J., De Fonseca, F. R., & Lõpez-Moreno, J. A. (2015). Common single nucleotide variants underlying drug addiction: more than a decade of research. *Addiction Biology 20*(5), 845–871.

Charpantier, E., Barnéoud, P., Moser, P., Besnard, F., & Sgard, F. (1998). Nicotinic acetylcholine subunit mRNA expression in dopaminergic neurons of the rat substantia nigra and ventral tegmental area. *NeuroReport, 9*(13), 3097–3101.

Chen, X., Williamson, V. S., An, S., Hettema, J. M., Aggen, S. H., Neale, M. C., et al. (2008). Cannabinoid receptor 1 gene association with nicotine dependence. *Archives of General Psychiatry, 65*(7), 816–824.

Cui, W.-Y., Seneviratne, C., Gu, J., & Li, M. D. (2012). Genetics of GABAergic signaling in nicotine and alcohol dependence. *Human Genetics, 131*(6), 843–855.

David, S. P., & Munafò, M. R. (2008). Genetic variation in the dopamine pathway and smoking cessation. *Pharmacogenomics, 9*(9), 1307–1321.

Decker, M. W., Meyer, M. D., & Sullivan, J. P. (2001). The therapeutic potential of nicotinic acetylcholine receptor agonists for pain control. *Expert Opinion on Investigational Drugs, 10*(10), 1819–1830.

Epping-Jordan, M. P., Picciotto, M. R., Changeux, J. P., & Pich, E. M. (1999). Assessment of nicotinic acetylcholine receptor subunit contributions to nicotine self-administration in mutant mice. *Psychopharmacology*, *147*(1), 25–26.

Feng, Y., Niu, T., Xing, H., Xu, X., Chen, C., Peng, S., et al. (2004). A common haplotype of the nicotine acetylcholine receptor alpha 4 subunit gene is associated with vulnerability to nicotine addiction in men. *The American Journal of Human Genetics*, *75*(1), 112–121.

Fowler, C. D., & Kenny, P. J. (2012). Utility of genetically modified mice for understanding the neurobiology of substance use disorders. *Human Genetics*, *131*(6), 941–957.

Fowler, C. D., Lu, Q., Johnson, P. M., Marks, M. J., & Kenny, P. J. (2011). Habenular α5 nicotinic receptor subunit signalling controls nicotine intake. *Nature*, *471*(7340), 597–601.

Gorlich, A., Antolin-Fontes, B., Ables, J. L., Frahm, S., Slimak, M. A., Dougherty, J. D., et al. (2013). Reexposure to nicotine during withdrawal increases the pacemaking activity of cholinergic habenular neurons. *Proceedings of the National Academy of Sciences*, *110*(42), 17077–17082.

Grant, B. F., Hasin, D. S., Chou, S. P., Stinson, F. S., & Dawson, D. A. (2004). Nicotine dependence and psychiatric disorders in the United States. *Archives of General Psychiatry*, *61*(11), 1107.

Grieder, T. E., George, O., Yee, M., Bergamini, M. A., Chwalek, M., Maal-Bared, G., et al. (2017). Deletion of α5 nicotine receptor subunits abolishes nicotinic aversive motivational effects in a manner that phenocopies dopamine receptor antagonism. *European Journal of Neuroscience*, *46*(1), 1673–1681.

Harari, O., Wang, J.-C., Bucholz, K., Edenberg, H. J., Heath, A., Martin, N. G., et al. (2012). Pathway analysis of smoking quantity in multiple GWAS identifies cholinergic and sensory pathways. *PLoS ONE*, *7*(12), e50913.

Herman, A. I., & Balogh, K. N. (2012). Polymorphisms of the serotonin transporter and receptor genes: Susceptibility to substance abuse. *Substance Abuse and Rehabilitation*, *3*(1), 49–57.

Hernandez, C. M., Cortez, I., Gu, Z., Colón-Sáez, J. O., Lamb, P. W., Wakamiya, M., et al. (2014). Research tool: validation of floxed α7 nicotinic acetylcholine receptor conditional knockout mice using *in vitro* and *in vivo* approaches. *The Journal of Physiology*, *592*(15), 3201–3214.

Hutchison, K. E., Allen, D. L., Filbey, F. M., Jepson, C., Lerman, C., Benowitz, N. L., et al. (2007). CHRNA4 and tobacco dependence: from gene regulation to treatment outcome. *Archives of General Psychiatry*, *64*(9), 1078.

Jamal, M., Van der Does, W., Elzinga, B. M., Molendijk, M. L., & Penninx, B. W. J. H. (2015). Association between smoking, nicotine dependence, and BDNF Val66Met polymorphism with BDNF concentrations in serum. *Nicotine & Tobacco Research*, *17* (3), 323–329.

Johnson, E. O., Chen, L.-S., Breslau, N., Hatsukami, D., Robbins, T., Saccone, N. L., et al. (2010). Peer smoking and the nicotinic receptor genes: an examination of genetic and environmental risks for nicotine dependence. *Addiction*, *105*(11), 2014–2022.

Kamens, H. M., Corley, R. P., McQueen, M. B., Stallings, M. C., Hopfer, C. J., Crowley, T. J., et al. (2013). Nominal association with CHRNA4 variants and nicotine dependence. *Genes, Brain and Behavior*, *12*(3), 297–304.

Kasza, K. A., Cummings, K. M., Carpenter, M. J., Cornelius, M. E., Hyland, A. J., & Fong, G. T. (2015). Use of stop-smoking medications in the United States before and after the introduction of varenicline. *Addiction (Abingdon, England)*, *110*(2), 346–355.

Klink, R., de Kerchove d'Exaerde, A., Zoli, M., & Changeux, J. P. (2001). Molecular and physiological diversity of nicotinic acetylcholine receptors in the midbrain dopaminergic nuclei. *The Journal of Neuroscience*, *21*(5), 1452–1463.

Kuwabara, H., Heishman, S. J., Brasic, J. R., Contoreggi, C., Cascella, N., Mackowick, K. M., et al. (2014). Mu opioid receptor binding correlates with nicotine dependence and reward in smokers. *PLoS ONE*, *9*(12), e113694.

Laviolette, S. R., & van der Kooy, D. (2004). The neurobiology of nicotine addiction: bridging the gap from molecules to behaviour. *Nature Reviews Neuroscience*, *5*(1), 55–65.

Leshner, A. I., & Stapleton, J. A. (1997). Addiction is a brain disease, and it matters. *Science (New York, NY)*, *278*(5335), 45–47.

Levin, E. D. (2002). Nicotinic receptor subtypes and cognitive function. *Journal of Neurobiology*, *53*(4), 633–640.

Li, M. D., Beuten, J., Ma, J. Z., Payne, T. J., Lou, X.-Y., Garcia, V., et al. (2005). Ethnic- and gender-specific association of the nicotinic acetylcholine receptor alpha4 subunit gene (CHRNA4) with nicotine dependence. *Human Molecular Genetics*, *14*(9), 1211–1219.

Li, M. D., & Burmeister, M. (2009). New insights into the genetics of addiction. *Nature Reviews Genetics*, *10*(4), 225–231.

Li, M. D., Lou, X. Y., Chen, G., Ma, J. Z., & Elston, R. C. (2008). Gene-gene interactions among CHRNA4, CHRNB2, BDNF, and NTRK2 in nicotine dependence. *Biological Psychiatry*, *64*(11), 951–957.

Mackillop, J., Obasi, E., Amlung, M. T., McGeary, J. E., & Knopik, V. S. (2010). The role of genetics in nicotine dependence: mapping the pathways from genome to syndrome. *Current Cardiovascular Risk Reports*, *4*(6), 446–453.

Maskos, U., Molles, B. E., Pons, S., Besson, M., Guiard, B. P., Guilloux, J.-P., et al. (2005). Nicotine reinforcement and cognition restored by targeted expression of nicotinic receptors. *Nature*, *436*(7047), 103–107.

McClure, E. A., Vandrey, R. G., Johnson, M. W., & Stitzer, M. L. (2013). Effects of varenicline on abstinence and smoking reward following a programmed lapse. *Nicotine & Tobacco Research*, *15*(1), 139–148.

McClure-Begley, T. D., Papke, R. L., Stone, K. L., Stokes, C., Levy, A. D., Gelernter, J., et al. (2014). Rare human nicotinic acetylcholine receptor 4 subunit (CHRNA4) variants affect expression and function of high-affinity nicotinic acetylcholine receptors. *Journal of Pharmacology and Experimental Therapeutics*, *348*(3), 410–420.

McKinney, E. F., Walton, R. T., Yudkin, P., Fuller, A., Haldar, N. A., Mant, D., et al. (2000). Association between polymorphisms in dopamine metabolic enzymes and tobacco consumption in smokers. *Pharmacogenetics*, *10*(6), 483–491.

Morley, K. I., Lynskey, M. T., Madden, P. A., Treloar, S. A., Heath, A. C., & Martin, N. G. (2007). Exploring the inter-relationship of smoking age-at-onset, cigarette consumption and smoking persistence: genes or environment? *Psychological Medicine*, *37*(9), 1357.

Nagy, A. (2000). Cre recombinase: the universal reagent for genome tailoring. *Genesis*, *26*(2), 99–109.

Nees, F., Witt, S. H., Lourdusamy, A., Vollstädt-Klein, S., Steiner, S., Poustka, L., et al. (2013). Genetic risk for nicotine dependence in the cholinergic system and activation of the brain reward system in healthy adolescents. *Neuropsychopharmacology*, *38*(11), 2081–2089.

Ngolab, J., Liu, L., Zhao-Shea, R., Gao, G., Gardner, P. D., & Tapper, A. R. (2015). Functional upregulation of α4* nicotinic acetylcholine receptors in VTA GABAergic neurons increases sensitivity to nicotine reward. *The Journal of Neuroscience*, *35*(22), 8570–8578.

Nides, M. (2008). Update on pharmacologic options for smoking cessation treatment. *The American Journal of Medicine*, *121*(4), S20–S31.

Pang, X., Liu, L., Ngolab, J., Zhao-Shea, R., McIntosh, J. M., Gardner, P. D., et al. (2016). Habenula cholinergic neurons regulate anxiety during nicotine withdrawal via nicotinic acetylcholine receptors. *Neuropharmacology*, *107*, 294–304.

Picciotto, M. R., Brunzell, D. H., & Caldarone, B. J. (2002). Effect of nicotine and nicotinic receptors on anxiety and depression. *NeuroReport*, *13*(9), 1097–1106.

Picciotto, M. R., Zoli, M., Léna, C., Bessis, A., Lallemand, Y., LeNovère, N., et al. (1995). Abnormal avoidance learning in mice lacking functional high-affinity nicotine receptor in the brain. *Nature, 374*(6517), 65–67.

Picciotto, M. R., Zoli, M., Rimondini, R., Léna, C., Marubio, L. M., Pich, E. M., et al. (1998). Acetylcholine receptors containing the beta2 subunit are involved in the reinforcing properties of nicotine. *Nature, 391*(6663), 173–177.

Quick, M. W., & Lester, R. A. (2002). Desensitization of neuronal nicotinic receptors. *Journal of Neurobiology, 53*(4), 457–478.

Ray, R., Mitra, N., Baldwin, D., Guo, M., Patterson, F., Heitjan, D. F., et al. (2010). Convergent evidence that choline acetyltransferase gene variation is associated with prospective smoking cessation and nicotine dependence. *Neuropsychopharmacology, 35*(6), 1374–1382.

Renthal, W., & Nestler, E. J. (2008). Epigenetic mechanisms in drug addiction. *Trends in Molecular Medicine, 14*(8), 341–350.

Rose, R. J., Broms, U., Korhonen, T., Dick, D. M., & Kaprio, J. (2009). Genetics of smoking behavior. In *Handbook of behavior genetics* (pp. 411–432). New York, NY: Springer New York.

Saccone, S. F., Hinrichs, A. L., Saccone, N. L., Chase, G. A., Konvicka, K., Madden, P. A. F., et al. (2006). Cholinergic nicotinic receptor genes implicated in a nicotine dependence association study targeting 348 candidate genes with 3713 SNPs. *Human Molecular Genetics, 16* (1), 36–49.

Silverman, M. A., Neale, M. C., Sullivan, P. F., Harris-Kerr, C., Wormley, B., Sadek, H., et al. (2000). Haplotypes of four novel single nucleotide polymorphisms in the nicotinic acetylcholine receptor β2-subunit (CHRNB2) gene show no association with smoking initiation or nicotine dependence. *American Journal of Medical Genetics, 96*(5), 646–653.

Thorgeirsson, T. E., Geller, F., Sulem, P., Rafnar, T., Wiste, A., Magnusson, K. P., et al. (2008). A variant associated with nicotine dependence, lung cancer and peripheral arterial disease. *Nature, 452*(7187), 638–642.

Vandenbergh, D. J., O'Connor, R. J., Grant, M. D., Jefferson, A. L., Vogler, G. P., Strasser, A. A., et al. (2007). Dopamine receptor genes (DRD2, DRD3 and DRD4) and gene-gene interactions associated with smoking-related behaviors. *Addiction Biology, 12*(1), 106–116.

Volkow, N. D. (2011). Epigenetics of nicotine: another nail in the coughing. *Science Translational Medicine, 3*(107), 107ps43.

Wei, J., Ma, J. Z., Payne, T. J., Cui, W., Ray, R., Mitra, N., et al. (2010). Replication and extension of association of choline acetyltransferase with nicotine dependence in European and African American smokers. *Human Genetics, 127*(6), 691–698.

Zhao-Shea, R., DeGroot, S. R., Liu, L., Vallaster, M., Pang, X., Su, Q., et al. (2015). Increased CRF signalling in a ventral tegmental area-interpeduncular nucleus-medial habenula circuit induces anxiety during nicotine withdrawal. *Nature Communications, 6*, 6770.

CHAPTER

41

HIV-Infected Subjects and Tobacco Smoking: A Focus on Nicotine Effects in the Brain

Manuel Delgado-Vélez, José A. Lasalde-Dominicci

Department of Biology, University of Puerto Rico and the Clinical Bioreagent Center, Molecular Sciences Research Center, San Juan, Puerto Rico

Abbreviations

AIDS	acquired immunodeficiency syndrome
ANI	asymptomatic neurocognitive impairment
BBB	blood-brain barrier
CCR5	C—C chemokine receptor type 5
CDC	centers for disease control and prevention
CNS	central nervous system
CXCR4	C-X-C chemokine receptor type 4
gp120	envelope glycoprotein GP120
HAD	HIV-associated dementia
HAND	HIV-associated neurocognitive disorders
HIV	human immunodeficiency virus
MND	mild neurocognitive disorder
nAChRs	nicotinic acetylcholine receptors
PAMs	positive allosteric modulators
α3β2-nAChR	alpha3 beta2 nicotinic acetylcholine receptor
α4β2-nAChR	alpha4 beta2 nicotinic acetylcholine receptor
α6β2-nAChR	alpha6 beta2 nicotinic acetylcholine receptor
α7-nAChR	alpha7 nicotinic acetylcholine receptor

41.1 INTRODUCTION

According to World Health Organization (WHO) data, in 2015, approximately 1.1 billion people in the world smoked tobacco. Estimates suggest that the smoking habit is responsible for about 6 million deaths per year, including about 890,000 people who die as a result of secondhand smoke exposure (CDC, 2017; WHO, 2017). Moreover, it is hypothesized that by 2030, tobacco smoke will be the cause of approximately 8 million deaths worldwide (CDC, 2017). Although the most well-known component of cigarettes is nicotine, they contain a myriad of harmful compounds, including carbon monoxide, toluene, tar, nickel, and many other constituents that represent a serious health threat. Even in people who do not suffer from compromising illnesses or infections, smoking tobacco is harmful. However, in people suffering from immunocompromising diseases, their ability to effectively combat pathogens is significantly reduced by tobacco smoking. HIV infection offers a classic example of the detrimental effects of tobacco smoking during an immunocompromised disease state.

In HIV-infected individuals, smoking prevalence is higher than in the normal population. Nicotine obtained from smoke interacts with nicotinic acetylcholine receptors (nAChRs) in the central nervous system (CNS), thus triggering dopamine release that generates pleasurable effects and reinforces the pernicious smoking habit. Nicotine is the addictive component of tobacco smoke and represents a major health burden for HIV-infected subjects because it not only promotes dependence and addiction but also accelerates the appearance of HIV-associated neurological disorders (HAND). Moreover, smoking is associated with progression toward acquired immunodeficiency syndrome (AIDS). Nicotine consumption has mixed cognitive effects, but overall, the balance of these effects is tipped toward its being more pernicious than beneficial.

HIV infection is characterized by productive infection of CD4+ expressing cells. However, although neurons do not express CD4, some neuronal subsets do express HIV coreceptors (CXCR4 and CCR5) that interact with HIV soluble virotoxins (e.g., gp120), thus triggering signal transduction pathways. Over time, the persistent and chronic viral infection destroys immune cells, particularly CD4+ T-lymphocytes, promoting the appearance of an immunodeficient immune system unable to counteract foreign invaders effectively. To make matters worse, the addition of smoking/nicotine to this immunocompromised condition further diminishes and complicates the immune system's ability to combat foreign invaders. Smoking and nicotine have a range of effects on HIV

Neuroscience of Nicotine
https://doi.org/10.1016/B978-0-12-813035-3.00041-1

infection, such as increased viral replication (Abbud, Finegan, Guay, & Rich, 1995; Rock et al., 2008), impaired macrophage phagocytosis, and modulated function of T-lymphocytes.

41.2 NICOTINE FROM SMOKING AND BRAIN NICOTINIC ACETYLCHOLINE RECEPTORS

Typically, nicotine is obtained directly from smoking cigarettes. Nicotine, 3-(1-methyl-2-pyrrolidinyl)pyridine, is a volatile alkaloid with a molecular weight of 162.23 g/mol. Also, it is the major psychoactive ingredient in tobacco, and its absorption and renal excretion are highly dependent on pH. At a high (alkaline) pH, nicotine is in the nonionized state, which is associated with the ability to more easily pass through lipoprotein membranes (CDC, 2010). When cigarette smoke is inhaled, 25% of the nicotine reaches the brain in about 7 s, about twice as fast as when the drug is administered intravenously. Though some inhaled nicotine is metabolized in the lungs, about 80%–90% of it is metabolized in the liver with small amounts metabolized in the kidneys. The eliminating half-life of nicotine varies among individuals but is typically close to 2 h (Benowitz, 1988), and the average nicotine concentration in smoker's blood is 203 nM, ranging from 25 to 444 nM (Russell, Jarvis, Iyer, & Feyerabend, 1980). In the CNS, cerebrospinal fluid studies show nicotine concentrations of 37 nM to 1.3 μM, while cotinine levels range from 155 nM to 2.6 μM (Malkawi, Al-Ghananeem, de Leon, & Crooks, 2009). Cotinine is the principal oxidative metabolite of nicotine and represents a marker of nicotine consumption or exposure. Actually, cotinine quantification in serum samples recovered from HIV-infected light/moderate and heavy smokers shows median values of 2.1 and 2.8 μM, respectively. Moreover, in HIV-infected intravenous drug users with light/moderate (2.5 μM) and heavy smoking (3.6 μM) habits, even higher mean cotinine values have been observed (Marshall et al., 2011), suggesting higher smoking rates. Consequently, HIV-infected drug abusers are less likely to have undetectable viral loads and frequently have psychosocial comorbidities (e.g., depression and mood disorders) that compromise antiretroviral therapy and smoking cessation medication adherence (Fig. 41.1).

As noted, nicotine reaches the CNS rapidly. The primary sites of action of nicotine are nAChRs in the brain. These transmembrane proteins are pentameric ligand-gated ion-conducting channels that open in response to endogenous agonists such as acetylcholine and choline or exogenous agents such as nicotine. In the brain, the presence and regular intake of nicotine induce the upregulation of several nAChRs, including heteromeric α4β2-nAChRs, α3β2-nAChRs, and α6β2-nAChRs and homomeric alpha7 nicotinic acetylcholine receptors (α7-nAChRs), phenomena that have been linked to nicotine addiction (Govind, Vezina, & Green, 2009). In fact, with regard to α4β2, chronic nicotine exposure causes receptor desensitization and upregulation, which are thought to play critical roles in nicotine reinforcement leading to addiction. Hence, nicotine exposure enhances the synaptic strength and produces long-term potentiation of excitatory synapses onto ventral tegmental area dopaminergic neurons (Mansvelder & McGehee, 2000), thus strengthening synaptic connections and perpetuating addiction.

41.3 SMOKING CESSATION EFFORTS IN HIV-INFECTED SMOKERS

The addictive properties of nicotine from tobacco smoke make it difficult to quit. To date, several smoking cessation drugs have been tested in HIV-infected subjects showing limited smoking abstinence rates (Calvo-Sánchez & Martinez, 2015) (Table 41.1). In this population, low adherence to smoking cessation pharmacotherapy and low abstinence rates are typical (Table 41.1) and are linked to serious health-threatening risk behaviors (Fig. 41.1). The principal pharmacological strategies tested include nicotine replacement therapy, bupropion, and varenicline, supported by nonpharmacological intervention approaches such as motivational counseling, motivational intervention, self-help, and behavioral support (Table 41.1). Currently, six clinical trials are underway focused on smoking cessation in HIV-infected smokers. These trials involve the use of pharmacological and nonpharmacological intervention approaches to increase smoking cessation abstinence (Table 41.2). Notably, three of the trials are using the same drugs previously tested, bupropion and varenicline (Table 41.2). Therefore, additional drugs are needed to treat this population that wants to quit smoking and desperately needs to increase their adherence to antiretroviral therapy.

41.4 DEPENDENCE AND PREVALENCE OF SMOKING IN HIV-INFECTED INDIVIDUALS

HIV-1-infected individuals are more likely to become dependent on nicotine and less likely to quit smoking than noninfected individuals. A recent smoking cessation study found that 83% of patients receiving smoking cessation medications (nicotine replacement therapy, varenicline, bupropion, and/or combination treatment) were not adherent to cessation therapy on at least one or more follow-up visits (Chew, Steinberg, Thomas, Swaminathan, & Hodder, 2014). Importantly, smoking dependence and risk behaviors (e.g., impulsivity) are

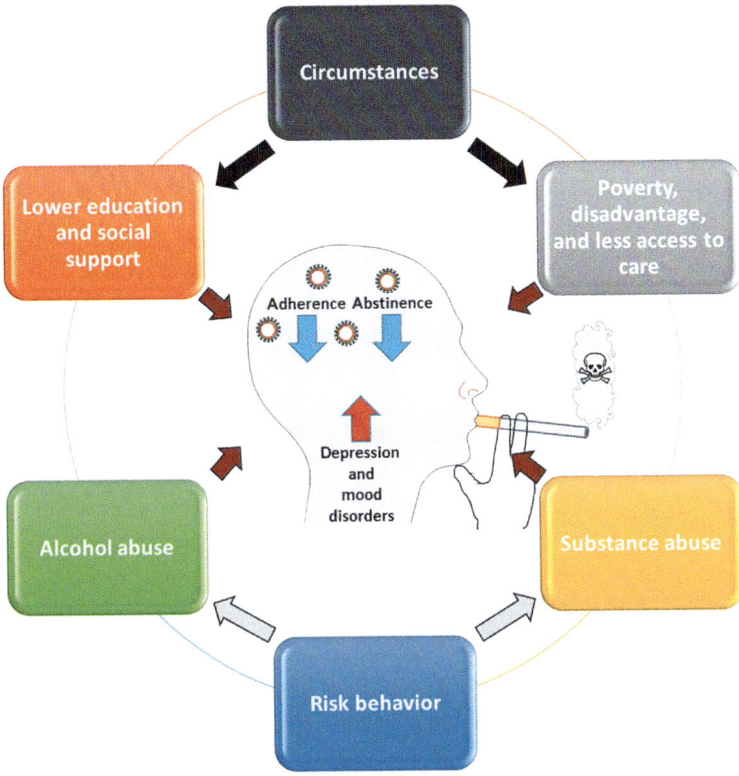

FIG. 41.1 Risk behaviors compromising adherence to smoking pharmacotherapy and abstinence in HIV-infected smokers.
HIV-infected smokers exhibit low adherence to smoking cessation therapies and low abstinence rates, both of which are linked to risk behaviors.

TABLE 41.1 Smoking Cessation Studies in HIV-Infected Smokers

Design	Interventions	Smoking abstinence rate (%)
NR	NRT, counseling, skills training	50
NR	NRT, motivational counseling, diary	22
NR	Counseling, NRT	38
NR	Bupropion	38
NR	Counseling, self-help material, NRT, cell phone intervention, usual care	10.3
NR	Counseling, bupropion, varenicline, and/or NRT	25
NR	Varenicline	24
R	NRT, self-help	22
R	NRT, motivational intervention	9
NR	Varenicline	42
R	Counseling, NRT computer-based intervention, NRT self-help, NRT	25.6 20.4 19.7
NR	Tailored group counseling, NRT	10

Continued

TABLE 41.1 Smoking Cessation Studies in HIV-Infected Smokers—cont'd

Design	Interventions	Smoking abstinence rate (%)
RS	NRT, varenicline, behavioral support	13
R	Varenicline, face-to-face counseling	15[a] 18[b]

NR, nonrandomized; *R*, randomized; *NRT*, nicotine replacement therapy; *RS*, retrospective study.
Description of studies performed using pharmacotherapy.
[a] *As determined by intention-to-treat analysis.*
[b] *As determined by modified intention-to-treat analysis.*
Adapted from Calvo-Sánchez and Martinez (2015).

significantly correlated. Accordingly, HIV-infected crack-smoking injectors have higher nicotine dependence than exclusive injectors or exclusive crack smokers (Hershberger, Fisher, Reynolds, Klahn, & Wood, 2004). Of note, in a cross-sectional study performed in Vietnam to determine whether smoking influences antiretroviral treatment adherence, a high prevalence of suboptimal adherence among examined patients was detected, and those patients with greater nicotine dependence were more likely to be nonadherent to antiretroviral therapy (Nguyen et al., 2016).

TABLE 41.2 Ongoing Smoking Cessation Clinical Trials Targeting HIV-Infected Smokers

Study title	Interventions	Study type	Phase	Enrollment
Smoking cessation intervention for women living with HIV	*Behavioral*: cognitive, behavioral therapy	Interventional	N/A	50
Storytelling narrative communication intervention for smoking cessation in women living with HIV	*Combination product*: storytelling narrative intervention	Interventional	Phase 1	50
Smoking cessation for people living with HIV/AIDS	*Drug*: bupropion *Behavioral*: brief counseling, high-magnitude prize contingency management, monitored support to quit smoking, prize contingency management for abstinence, and low intensity prize contingency management	Interventional	N/A	400
Patch study intervention for HIV-positive smokers	*Other*: combination therapy and standard care intervention	Interventional	Phase 4	600
Optimizing smoking cessation for people with HIV/AIDS who smoke	*Drug*: varenicline and placebo *Behavioral*: positively smoke-free and standard of care	Interventional	Phase 3	300
Impact of smoking and its cessation on systemic and airway immune activation	*Behavioral*: counseling *Drugs*: smoking cessation drugs	Interventional	N/A	120

N/A, not available.
Current clinical trials testing pharmacological and/or behavioral approaches in HIV-infected smokers.
Data compiled from the National Institute of Health, National Library of Medicine (ClinicalTrials.gov) on March 24, 2018.

Several studies have established that the prevalence of tobacco smoking in HIV-1-positive individuals is three- to fourfold higher than in the HIV-1-negative population (Manda, Mittapalli, Geldenhuys, & Lockman, 2010), with smokers representing 40%–84% of infected subjects (Browning, Wewers, Ferketich, & Diaz, 2013). The consequences of smoking in HIV-infected individuals are devastating. In a Danish HIV-infected cohort, it was demonstrated that smoking versus never smoking significantly reduced life years by 12.3; in HIV nonsmokers versus never-smoking controls, the number of life year lost was 5.1; and in smoking vs never-smoking controls, the number was 3.6 (Helleberg et al., 2013) (Fig. 41.2). Therefore, HIV-infected individuals lost more life years due to tobacco smoking than to HIV infection. Consistent with this, the HIV-1 positive population is known to be at higher risk for tobacco-associated morbidity and mortality, and tobacco smoking has emerged as a leading killer among persons living with HIV (Stanton et al., 2015). Demographic studies have also indicated a positive correlation between tobacco use and HIV acquisition. Cross-sectional surveys of pregnant women in Rwanda (Chao et al., 1994) and Haiti (Boulos et al., 1990), which controlled for numerous risk factors, found tobacco use increased the rate of HIV seroconversion. Moreover, a cohort study of homosexual men adjusted for risk in the United States also found a positive association between tobacco smoke and HIV acquisition (Burns et al., 1991). In addition to these studies, an increasing body of evidence shows that tobacco smoking is associated with a more rapid progression to AIDS (Zhao et al., 2010) and HIV-associated dementia (HAD) (Manda et al., 2010). Therefore, HIV infection and smoking are a lethal combination for this population because it promotes risky behaviors that can lead to illicit drug use, failed adherence to antiretroviral therapy, and HAD development.

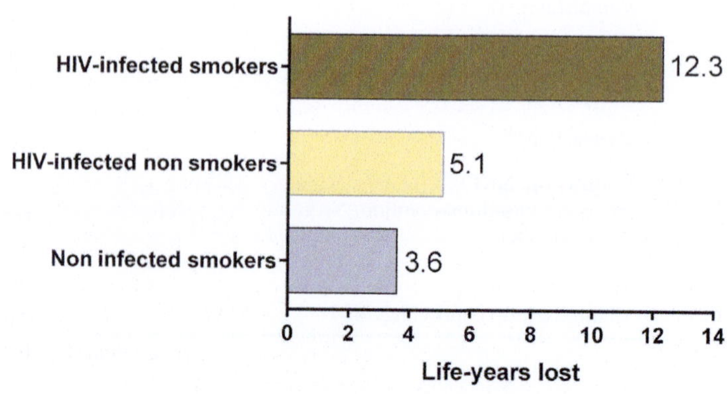

FIG. 41.2 Number of life years lost by HIV-infected smokers. HIV-infected smokers lose more life years to smoking than to HIV. HIV-infected smokers lost more than twice as many years of life than never-smoking controls. Controls were obtained from the Copenhagen General Population Study. *Adapted from Helleberg et al. (2013).*

41.5 SMOKING AND NICOTINE CONSEQUENCES IN HIV-INFECTED INDIVIDUALS

HIV-infected subjects suffer from neurological complications that persist despite effective pharmacological viral suppression, acceptable CD4+ cell counts, and the absence of symptoms. Detailed and comprehensive neurological studies have characterized and classified patients' neurological symptoms as HAND (Table 41.3). HAND prevalence in this population ranges from 20% to 69% (Carroll & Brew, 2017) and comprises asymptomatic neurocognitive impairment (ANI), HIV-associated mild neurocognitive disorder (MND), and HIV-associated dementia (HAD), with HAD being the most severe form of neurocognitive deterioration (Antinori et al., 2007) (Table 41.3). As mentioned, smoking accelerates HAND's appearance as evidenced by worse performance on working memory, processing speed, and intraindividual variability as compared to HIV-uninfected smokers (Harrison et al., 2017). Importantly, a recent study established a unique association by which nicotine and HIV synergize to negatively regulate the synaptic plasticity gene expression and spine density that may contribute to the increased risk of HAND in smokers (Atluri et al., 2014). Furthermore, it is also known that nicotine compromises the blood-brain barrier (BBB) integrity as determined by *in vitro* and *in vivo* studies (Hawkins et al., 2004; Manda et al., 2010), thus facilitating the recruitment of HIV-infected macrophages into the CNS and accelerating HAND appearance (Fig. 41.3).

TABLE 41.3 HAND Classification

HAND stage	Description and characteristics
ANI[a]	Acquired impairment in cognitive functioning, involving at least two ability domains. This impairment does not interfere with everyday functioning and does not meet the criteria for delirium or dementia
MND[a]	This cognitive impairment produces marked interference with day-to-day functioning (work, home life, and social activities) The pattern of cognitive impairment does not meet criteria for delirium (e.g., clouding of consciousness is not a prominent feature)
HAD	Marked acquired impairment in cognitive functioning, involving at least two ability domains; typically, the impairment is in multiple domains, especially in learning of new information, slowed information processing, and defective attention/concentration. HAD produces marked interference with day-to-day functioning

Criteria to classify and diagnose HAND in HIV-infected individuals.

[a] *The neuropsychological assessment must survey at least the following abilities: verbal/language, attention/working memory, abstraction/executive, memory (learning; recall); speed of information processing, and sensory-perceptual-motor skills.*

Adapted from Ramachandran, V.S. (Eds.). (2002). Grant I: The neurocognitive complications of HIV infection. In Encyclopedia of the human brain (pp. 475–489). San Diego: Academic Press.

Once in the brain, HIV infects microglia and astrocytes. Astrocytes are critical for the formation, establishment, and maintenance of the BBB. However, once astrocytes are infected, the presence of viral constituents (e.g., gp120) and cytokines in the CNS, together with systemic circulating of virotoxins (e.g., Tat), compromises the BBB's integrity, thus facilitating infiltration of newly infected macrophages that ultimately exerts higher neuronal damage (Fig. 41.3). Moreover, in the presence of nicotine, the neuronal damage is exacerbated because nicotine also permeates the BBB, which also contributes to HAND development (Fig. 41.3). In the case of neurons, they are also exposed to HIV, viral proteins, and nicotine in HIV-infected smokers. Available studies in rats demonstrate that viral proteins (e.g., Tat) affect the mesocorticolimbic dopaminergic system altering cAMP response element-binding protein and extracellular signal-regulated kinases signaling (Midde, Gomez, Harrod, & Zhu, 2011) in this neural circuit populated by nicotine-responsive dopaminergic neurons expressing α4β2-nAChRs. Notably, activation of α4β2-nAChRs by nicotine increases the firing rate phasic burst of dopaminergic neurons, elevating dopamine levels in the prefrontal cortex and nucleus accumbens, thus providing pleasure. Moreover, activation of α4β2-nAChRs mediates nicotine reward and anxiety relief (McGranahan, Patzlaff, Grady, Heinemann, & Booker, 2011), which reinforce the smoking habit and perpetuate addiction.

Smoking affects not only the CNS of HIV-infected smokers but also their immune system. Tobacco's nicotine has been reported to enhance the production of HIV-1 in *in vitro*-infected alveolar macrophages from otherwise healthy cigarette smokers (Abbud et al., 1995) and in microglia (brain macrophages) as it potentiates HIV-1 levels in a TGF-β1-dependent manner (Rock et al., 2008). Interestingly, in T-lymphocytes, extracts of tobacco smoke increase HIV infectivity and virus production that is nicotine- and carbon monoxide-independent. It does so by upregulating genes known to be capable of enhancing HIV infection or protecting the HIV itself and downregulating several genes involved in cellular defense and antigen presentation (Zhao et al., 2010). This suggests that the enhancing proviral effects of nicotine are cell-dependent, and other constituents of tobacco smoking can also facilitate HIV proliferation in these patients. Overall, smoking and nicotine affect the CNS and exacerbates HIV infection in HIV-infected individuals.

On the other hand, nicotine and smoking have been associated with α7-nAChR upregulation either *in vitro* or *in vivo*. In fact, in HIV-infected smokers, their monocytes are upregulated for α7-nAChR. Moreover, monocytes, T-lymphocytes, and monocyte-derived macrophages are upregulated for α7-nAChR despite smoke status.

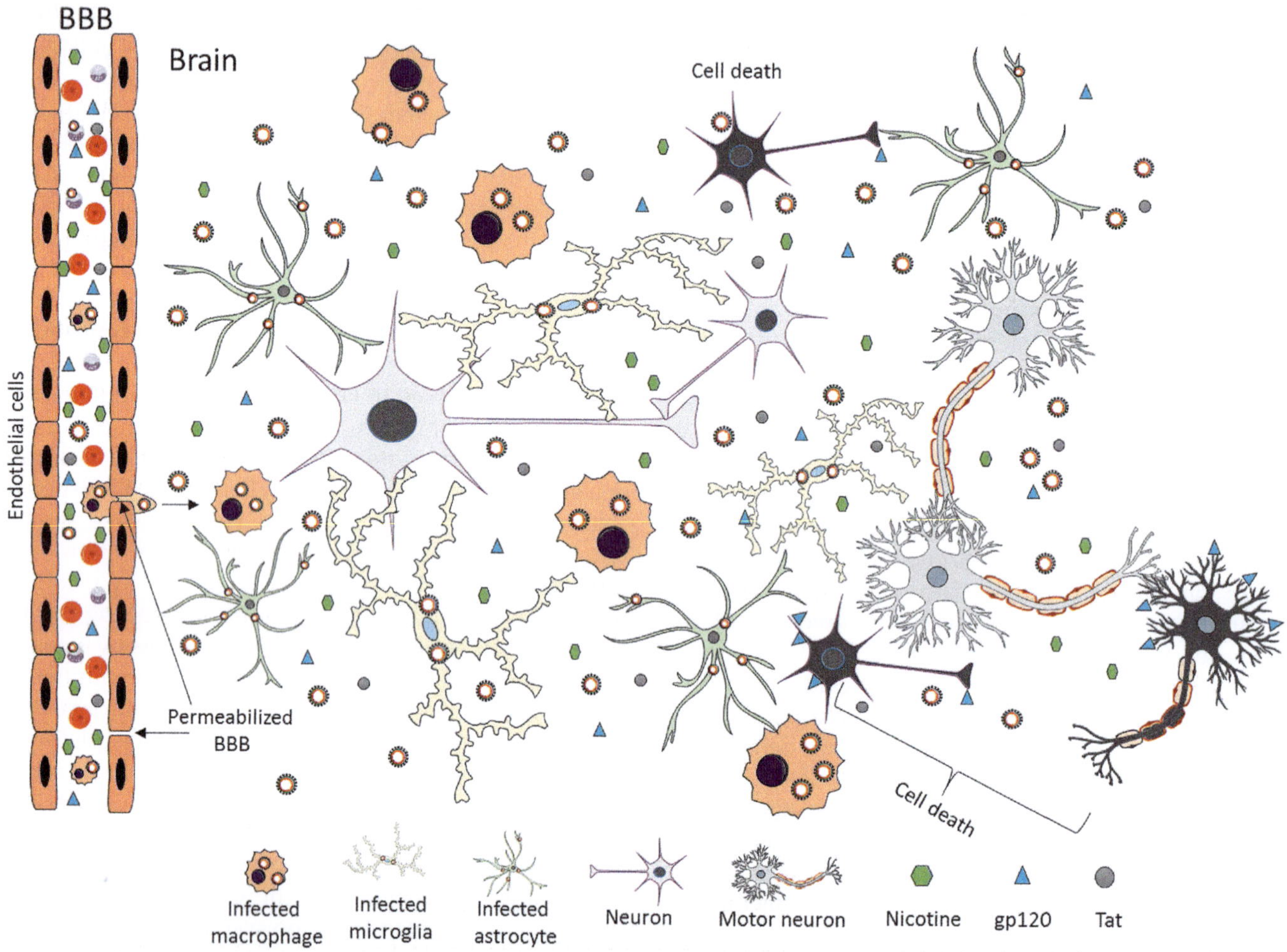

FIG. 41.3 HIV, its constituents, and nicotine facilitate appearance of HAND.
Circulating HIV virotoxins and nicotine permeabilizes the BBB, facilitating entrance of infected macrophages into the brain. Once there, HIV virions infect astrocytes and microglia, and soluble gp120 and Tat promote neuronal death, thus advancing HAND appearance.

41.6 COGNITION IN HIV-INFECTED SMOKERS

Smoking influences cognition in HIV-infected subjects. The few existing studies examining the effects of tobacco smoking in HIV-infected smokers have had contradictory findings. The first such study performed in HIV-infected heavy drinkers who smoked found that cigarette smoking created an additional burden to the neurobiology and neurocognition of the examined patients, resulting in poorer cognitive performance in those who smoked than those who did not smoke (Durazzo et al., 2007). The second study was an observational cross-sectional study showing that history of smoking correlated with a better frontal/executive cognitive domain performance in HIV-seropositive women and with a worse viral immune profile (Wojna et al., 2007). Similarly, evaluation of neural structures using an HIV-1 transgenic rat showed that nicotine "corrected" several pathways altered by HIV-1 infection in the prefrontal cortex, dorsal striatum, and hippocampus (Cao et al., 2013). By contrast, results from a third study of patients being treated for HIV infection suggested that current smoking is negatively associated with learning, memory, and global cognitive functioning. Those authors concluded that smoking may merely reflect a general tendency to more widespread deficits and comorbidities rather than directly impacting cognitive function (Bryant, Kahler, Devlin, Monti, & Cohen, 2013). Finally, a recent study reported that HIV-infected chronic smokers exhibited impulsivity, depression, and cognitive dysfunction (Chang, Lim, Lau, & Alicata, 2017) (Fig. 41.1). The differences between these studies may be accounted for by gender and ethnic differences in the patient groups, the difficulties inherent in the determination of nicotine concentration, and smoking frequency and the type of cigarette smoked. Taken together, the general conclusions about smoking in HIV settings are in contrast to nicotine's cognition-enhancing effects (Evans & Drobes, 2009; Wojna et al., 2007) reported in people with schizophrenia, attention-deficit hyperactivity disorder, and Alzheimer's disease (Jasinska, Zorick, Brody, & Stein, 2014). Nicotine, but not smoking, opens the possibility of using cholinergic agonists, positive allosteric modulators (PAMs), or antagonists to treat HIV-induced neurologic deficits.

MINI DICTIONARY OF TERMS

Acquired immunodeficiency syndrome (AIDS) A disease of the immune system caused by HIV.
Blood-brain barrier Protects neural tissues from circulating toxins and permit the passage of water, gases, and nutrients.
Central nervous system System comprising the brain and spinal cord.
Cholinergic agonist Drugs that stimulate or activate nicotinic acetylcholine receptors.
Cholinergic antagonist Drugs that inhibit or block nicotinic acetylcholine receptors.
Desensitization Occurs when a ligand-gated receptor experiences repeated or chronic exposure to ligands that promotes its decreased responsiveness and results in a nonconductive state.
HAND Term used to describe the myriad of neurocognitive dysfunctions associated with HIV infection.
HIV A retrovirus that preferentially infects and destroys cells expressing the surface receptor CD4, including T-lymphocytes.
HIV-associated dementia (HAD) The most severe form of cognitive impairment suffered by HIV-infected individuals that emerges from HIV establishment in the CNS and inflammation.
Long-term potentiation Refers to a persistent increase in excitatory postsynaptic potential amplitude at a synapse that is typically induced by brief, high-frequency afferent stimulation.
Positive allosteric modulator Cholinergic enhancers that rescue desensitized nicotinic acetylcholine receptors.
Upregulation An increase in the number of nAChRs in cells.
Ventral tegmental area A structure populated by dopaminergic neurons, within the midbrain that extends neural tracts to the limbic and cortical areas.
α4β2-nAChR A heteromeric nicotinic acetylcholine receptor composed of five subunits mainly expressed in the mammalian brain that exhibit high affinity for nicotine.
α7-nAChR A homomeric nicotinic acetylcholine receptor composed of five subunits expressed by neural and immune cells.

Key Facts About HAND

- Almost 30 years has passed since the first report showing HIV-related neurological complications.
- The occurrence of the most severe form of HAND and HAD is relatively low (2%–4%), but the milder forms vary between 30% in HIV-infected subjects without AIDS and 50% in HIV-infected with AIDS.
- In patients receiving antiretroviral therapy or those receiving early treatment (virally suppressed), the HAND prevalence ranges between 20% and 69%.
- Clinical researchers agree that identification of moderate to severe forms of HAND are accurate using neuropsychological approaches; however, distinguishing between MND and ANI is challenging, and they prefer to report MND and HAD.
- Exclusion of ANI diagnosis is a concern because this mild level of HAND is predictive of more cognitive decline.
- Recent evidence suggests that the antiretroviral drug, maraviroc, improved neurocognitive performance in virologically suppressed HIV-infected subjects.

Key Facts About Nicotine-Induced Upregulation of α4β2-nAChRs

- Upon consumption, nicotine reaches the brain and interacts with a basal pool of α4β2 nAChRs, promoting changes. Three mechanisms have been proposed to explain the nicotine-induced upregulation.
- Mechanism 1: increase in the number of α4β2-nAChRs. This mechanism involves decreased surface turnover, increased assembly and maturation, increased trafficking to cell's surface, and slower subunit degradation.
- Mechanism 2: nicotine-induced alterations in stoichiometry that involves changes in subunits without an increase in the number of α4β2-nAChRs.
- Mechanism 3: the nicotine-induced upregulation can occur via conformational changes without alterations in the number of α4β2-nAChRs.
- Upregulation of α4β2-nAChRs is initiated by receptor desensitization after nicotine exposure.

Summary Points

- This chapter focuses on effects of smoking and nicotine consumption in HIV-infected individuals.
- The prevalence of smoking in HIV-infected individuals is three- to fourfold higher than in noninfected persons.
- Nicotine diffuses rapidly to the brain activating, desensitizing, and/or upregulating nAChRs such as α4β2-nAChR and α7-nAChR.
- HIV-infected individuals suffer from neurological impairments termed as HAND.
- HAND prevalence is relatively high (20%–69%) and includes ANI, MND, and HAD.
- Although some cognitive benefits have been detected, smoking and nicotine accelerate the appearance of HAND.
- Smoking and nicotine exacerbate HIV.
- Cholinergic agonists, PAMs, and antagonist can be exploited as therapeutic options to treat smoking and HAND.

References

Abbud, R. A., Finegan, C. K., Guay, L. A., & Rich, E. A. (1995). Enhanced production of human immunodeficiency virus type 1 by in vitro-infected alveolar macrophages from otherwise healthy cigarette smokers. *The Journal of Infectious Diseases*, 172, 859–863.

Antinori, A., Arendt, G., Becker, J. T., Brew, B. J., Byrd, D. A., Cherner, M., et al. (2007). Updated research nosology for HIV-associated neurocognitive disorders. *Neurology, 69*, 1789–1799.

Atluri, V. S. R., Pilakka-Kanthikeel, S., Samikkannu, T., Sagar, V., Kurapati, K. R. V., Saxena, S. K., et al. (2014). Vorinostat positively regulates synaptic plasticity genes expression and spine density in HIV infected neurons: role of nicotine in progression of HIV-associated neurocognitive disorder. *Molecular Brain, 7*, 37.

Benowitz, N. L. (1988). Nicotine and smokeless tobacco. *CA: A Cancer Journal for Clinicians, 38*, 244–247.

Boulos, R., Halsey, N. A., Holt, E., Ruff, A., Brutus, J. R., Quinn, T. C., et al. (1990). HIV-1 in Haitian women 1982-1988. The Cite Soleil/JHU AIDS Project Team. *Journal of Acquired Immune Deficiency Syndrome, 3*, 721–728.

Browning, K. K., Wewers, M. E., Ferketich, A. K., & Diaz, P. (2013). Tobacco use and cessation in HIV-infected individuals. *Clinics in Chest Medicine, 34*, 181–190.

Bryant, V. E., Kahler, C. W., Devlin, K. N., Monti, P. M., & Cohen, R. A. (2013). The effects of cigarette smoking on learning and memory performance among people living with HIV/AIDS. *AIDS Care, 25*, 1308–1316.

Burns, D. N., Kramer, A., Yellin, F., Fuchs, D., Wachter, H., DiGioia, R. A., et al. (1991). Cigarette smoking: a modifier of human immunodeficiency virus type 1 infection? *Journal of Acquired Immune Deficiency Syndromes, 4*, 76–83.

Calvo-Sánchez, M., & Martinez, E. (2015). How to address smoking cessation in HIV patients. *HIV Medicine, 16*, 201–210.

Cao, J., Wang, S., Wang, J., Cui, W., Nesil, T., Vigorito, M., et al. (2013). RNA deep sequencing analysis reveals that nicotine restores impaired gene expression by viral proteins in the brains of HIV-1 transgenic rats. *PLoS One, 8*, e68517.

Carroll, A., & Brew, B. (2017). HIV-associated neurocognitive disorders: recent advances in pathogenesis, biomarkers, and treatment. *F1000Research, 6*.

CDC (2017) Announcement: World No Tobacco Day. (2017). MMWR Morb. Mortal. Wkly. Rep. *66*, 545.

CDC, National Center for Chronic Disease Prevention and Health Promotion (US), & Office on Smoking and Health (US) (2010) Nicotine addiction: past and present. Centers for Disease Control and Prevention (US).

Chang, L., Lim, A., Lau, E., & Alicata, D. (2017). Chronic tobacco-smoking on psychopathological symptoms, impulsivity and cognitive deficits in HIV-infected individuals. *Journal of Neuroimmune Pharmacology: The Official Journal of Society on NeuroImmune Pharmacology, 12*, 389–401.

Chao, A., Bulterys, M., Musanganire, F., Habimana, P., Nawrocki, P., Taylor, E., et al. (1994). Risk factors associated with prevalent HIV-1 infection among pregnant women in Rwanda. National University of Rwanda-Johns Hopkins University AIDS Research Team. *International of Journal of Epidemiology, 23*, 371–380.

Chew, D., Steinberg, M. B., Thomas, P., Swaminathan, S., & Hodder, S. L. (2014). Evaluation of a smoking cessation program for HIV infected individuals in an urban HIV clinic: challenges and lessons learned. *AIDS Research and Treatment, 2014*.

Durazzo, T. C., Rothlind, J. C., Cardenas, V. A., Studholme, C., Weiner, M. W., & Meyerhoff, D. J. (2007). Chronic cigarette smoking and heavy drinking in human immunodeficiency virus: consequences for neurocognition and brain morphology. *Alcohol Fayetteville N, 41*, 489–501.

Evans, D. E., & Drobes, D. J. (2009). Nicotine self-medication of cognitive-attentional processing. *Addiction Biology, 14*, 32–42.

Govind, A. P., Vezina, P., & Green, W. N. (2009). Nicotine-induced upregulation of nicotinic receptors: underlying mechanisms and relevance to nicotine addiction. *Biochemical Pharmacology, 78*, 756–765.

Harrison, J. D., Dochney, J. A., Blazekovic, S., Leone, F., Metzger, D., Frank, I., et al. (2017). The nature and consequences of cognitive deficits among tobacco smokers with HIV: a comparison to tobacco smokers without HIV. *Journal of Neurovirology, 23*, 550–557.

Hawkins, B. T., Abbruscato, T. J., Egleton, R. D., Brown, R. C., Huber, J. D., Campos, C. R., et al. (2004). Nicotine increases in vivo blood-brain barrier permeability and alters cerebral microvascular tight junction protein distribution. *Brain Research, 1027*, 48–58.

Helleberg, M., Afzal, S., Kronborg, G., Larsen, C. S., Pedersen, G., Pedersen, C., et al. (2013). Mortality attributable to smoking among HIV-1-infected individuals: a nationwide, population-based cohort study. *Clinical Infectious Diseases, 56*, 727–734.

Hershberger, S. L., Fisher, D. G., Reynolds, G. L., Klahn, J. A., & Wood, M. M. (2004). Nicotine dependence and HIV risk behaviors among illicit drug users. *Addictive Behaviors, 29*, 623–625.

Jasinska, A. J., Zorick, T., Brody, A. L., & Stein, E. A. (2014). Dual role of nicotine in addiction and cognition: a review of neuroimaging studies in humans. *Neuropharmacology, 84*, 111–122.

Malkawi, A. H., Al-Ghananeem, A. M., de Leon, J., & Crooks, P. A. (2009). Nicotine exposure can be detected in cerebrospinal fluid of active and passive smokers. *Journal of Pharmaceutical and Biomedical Analysis, 49*, 129–132.

Manda, V. K., Mittapalli, R. K., Geldenhuys, W. J., & Lockman, P. R. (2010). Chronic exposure to nicotine and saquinavir decreases endothelial Notch-4 expression and disrupts blood-brain barrier integrity. *Journal of Neurochemistry, 115*, 515–525.

Mansvelder, H. D., & McGehee, D. S. (2000). Long-term potentiation of excitatory inputs to brain reward areas by nicotine. *Neuron, 27*, 349–357.

Marshall, M. M., Kirk, G. D., Caporaso, N. E., McCormack, M. C., Merlo, C. A., Hague, J. C., et al. (2011). Tobacco use and nicotine dependence among HIV-infected and uninfected injection drug users. *Addictive Behaviors, 36*, 61–67.

McGranahan, T. M., Patzlaff, N. E., Grady, S. R., Heinemann, S. F., & Booker, T. K. (2011). alpha4beta2 nicotinic acetylcholine receptors on dopaminergic neurons mediate nicotine reward and anxiety relief. *Journal of Neuroscience: The Official Journal of the Society for Neuroscience, 31*, 10891–10902.

Midde, N. M., Gomez, A. M., Harrod, S. B., & Zhu, J. (2011). Genetically expressed HIV-1 viral proteins attenuate nicotine-induced behavioral sensitization and alter mesocorticolimbic ERK and CREB signaling in rats. *Pharmacology, Biochemistry, and Behavior, 98*, 587–597.

Nguyen, N. T. P., Tran, B. X., Hwang, L. Y., Markham, C. M., Swartz, M. D., Vidrine, J. I., et al. (2016). Effects of cigarette smoking and nicotine dependence on adherence to antiretroviral therapy among HIV-positive patients in Vietnam. *AIDS Care, 28*, 359–364.

Rock, R. B., Gekker, G., Aravalli, R. N., Hu, S., Sheng, W. S., & Peterson, P. K. (2008). Potentiation of HIV-1 expression in microglial cells by nicotine: Involvement of transforming growth factor-beta 1. *Journal of Neuroimmune Pharmacology: An Official Journal of Society Neuroimmune Pharmacology, 3*, 143–149.

Russell, M. A., Jarvis, M., Iyer, R., & Feyerabend, C. (1980). Relation of nicotine yield of cigarettes to blood nicotine concentrations in smokers. *British Medical Journal, 280*, 972–976.

Stanton, C. A., Papandonatos, G. D., Shuter, J., Bicki, A., Lloyd-Richardson, E. E., de Dios, M. A., et al. (2015). Outcomes of a tailored intervention for cigarette smoking cessation among Latinos living with HIV/AIDS. *Nicotine & Tobacco Research: The Official Journal of the Society for Research on Nicotine and Tobacco, 17*, 975–982.

WHO. (2017). *WHO | tobacco [WWW document]. WHO, URL(2017). http://www.who.int/mediacentre/factsheets/fs339/en/.*

Wojna, V., Robles, L., Skolasky, R. L., Mayo, R., Selnes, O., de la Torre, T., et al. (2007). Associations of cigarette smoking with viral immune and cognitive function in human immunodeficiency virus-seropositive women. *Journal of Neurovirology, 13*, 561–568.

Zhao, L., Li, F., Zhang, Y., Elbourkadi, N., Wang, Z., Yu, C., et al. (2010). Mechanisms and genes involved in enhancement of HIV infectivity by tobacco smoke. *Toxicology, 278*, 242–248.

CHAPTER

42

Renin-Angiotensin System Genes and Nicotine Dependence

Sergej Nadalin, Hrvoje Jakovac*[†]

*Department of Biology and Medical Genetics, School of Medicine, University of Rijeka, Rijeka, Croatia

[†]Department of Physiology, Immunology and Pathophysiology, School of Medicine, University of Rijeka, Rijeka, Croatia

Abbreviations

ACE	angiotensin-converting enzyme
AT_1	angiotensin II receptor type 1
AT_2	angiotensin II receptor type 2
AT_4	angiotensin II receptor type 4
I/D polymorphism	insertion/deletion polymorphism
MS	multiple sclerosis
RAS	renin-angiotensin system

42.1 INTRODUCTION

In the brain, astroglia synthesize angiotensinogen, which is classically cleaved by renin; renin is present in brain tissue at very low concentrations. This cleavage generates the inactive decapeptide, angiotensin I. Angiotensin I is converted by angiotensin-converting enzyme (ACE) to the active angiotensin II, which is metabolized to angiotensin III and then angiotensin IV; angiotensin II and angiotensin IV can be converted to angiotensin (1–7) and angiotensin (3–7), respectively (Guimond & Gallo-Payet, 2012; McKinley et al., 2003; Wright & Harding, 2013) (Fig. 42.1). As the main effector of the brain RAS, angiotensin II is both a circulating hormone, which influences cardiovascular and electrolyte homeostasis, and a neuropeptide, which acts as a neurotransmitter. Its biological actions are predominantly mediated by angiotensin II receptors, type 1 (AT_1) and type 2 (AT_2). AT_1 receptors are present throughout the brain, but they are particularly abundant in dopaminergic areas, such as the basal ganglia and hypothalamus. AT_2 receptors are expressed at low density in the brain, and they are limited to specific areas (Guimond & Gallo-Payet, 2012; Marchese et al., 2016; Wright & Harding, 2013). Furthermore, activation of the AT_2 receptor has been shown to lead to actions that oppose those classically mediated by AT_1 receptors; thus, AT_2 antagonizes many of the effects of AT_1 (Fig. 42.1) (Guimond & Gallo-Payet, 2012; Labandeira-García et al., 2014). Finally, in addition, other RAS-related receptors have been identified in the brain, including the (pro)renin receptor, the Mas receptor, and angiotensin II receptor type 4 (AT_4) (Guimond & Gallo-Payet, 2012; Wright & Harding, 2013) (Fig. 42.2).

Scientists have been continually interested in the ability of brain RAS to influence dopaminergic signaling in the mesolimbic pathway, which is known to play a critical role in the addictive properties of substance abuse. In the rat striatum, angiotensin II stimulates dopamine neurotransmission, and several ACE inhibitors (captopril, enalaprilat, and perindopril) have been shown to modulate dopamine turnover (Jenkins, Chai, & Mendelsohn, 1997; Jenkins, Mendelsohn, & Chai, 1997; Obata, Takahashi, Kashiwagi, & Kubota, 2008). Behavioral studies have suggested that brain RAS functions are mediated by striatal dopaminergic systems. For example, the dopamine antagonist, haloperidol, influences locomotor and stereotypic behaviors, and several ACE inhibitors abolish apomorphine-induced stereotypy in rats and mice; the ACE inhibitor, captopril, decreases morphine-induced conditional place preference and morphine self-administration (Banks, Mozley, & Dourish, 1994; Georgiev, György, Getova, & Markovska, 1985; Hosseini, Sharifi, Alaei, Shafei, & Karimooy, 2007). Novel findings have suggested that RAS components change significantly in response to drugs of abuse. In rats, ethanol consumption decreases AT_1 receptor density and increases angiotensinogen expression in the medial prefrontal cortex (Sommer et al., 2007); the effects of cocaine on the central nervous system are proposed to be caused by increases in ACE activity and expression in the frontal cortex and striatum (Visniauskas et al., 2012). In addition,

Neuroscience of Nicotine
https://doi.org/10.1016/B978-0-12-813035-3.00042-3

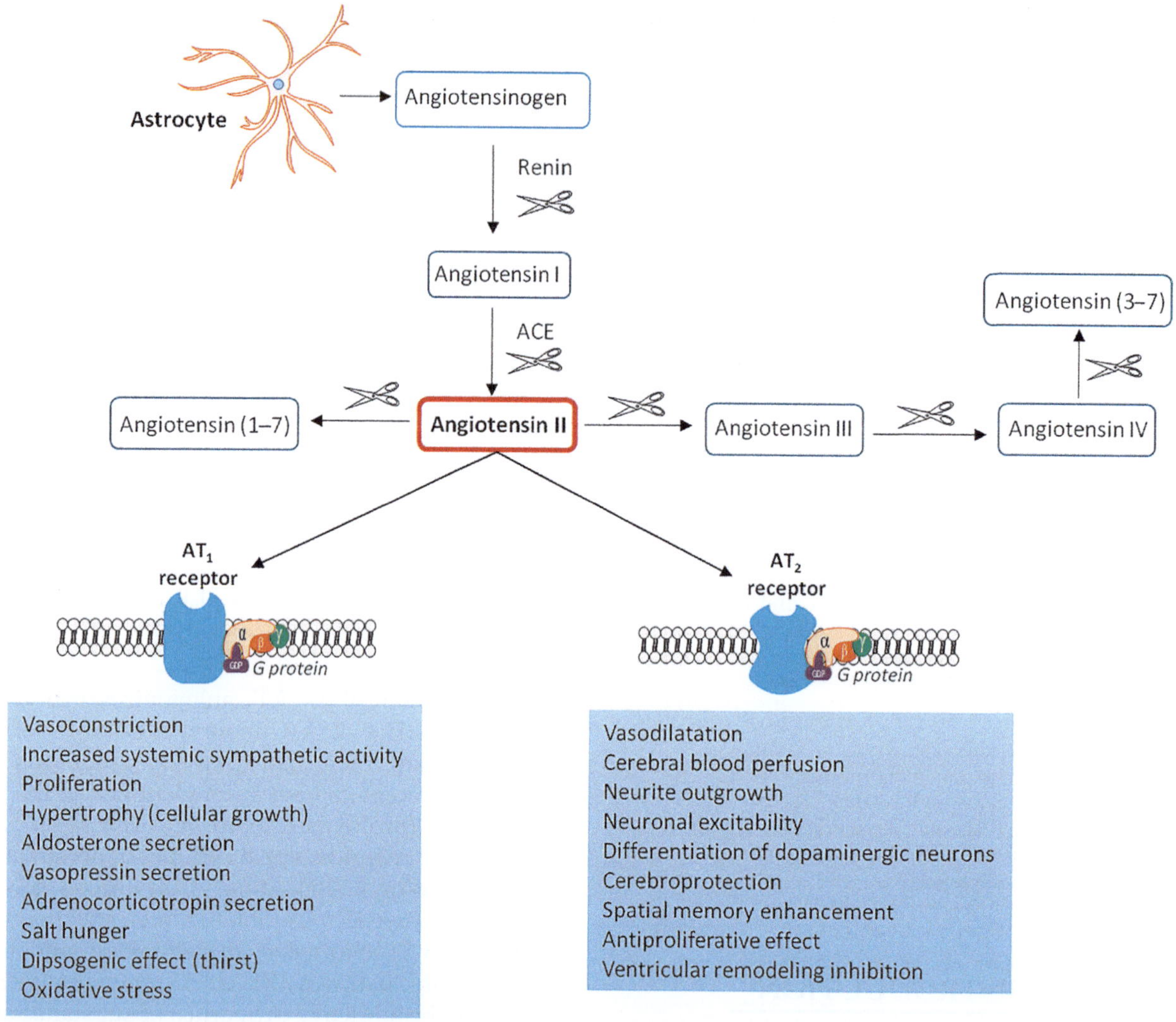

FIG. 42.1 Interrelationships among brain renin-angiotensin system (RAS) components.
Inactive angiotensinogen is secreted by astroglia and then successively enzymatically cleaved (scissors) by angiotensin-converting enzyme (ACE). This results in the formation of angiotensin II, the main effector of the RAS. Angiotensin II effects are predominantly mediated through angiotensin II receptors, type 1 (AT_1) and type 2 (AT_2).

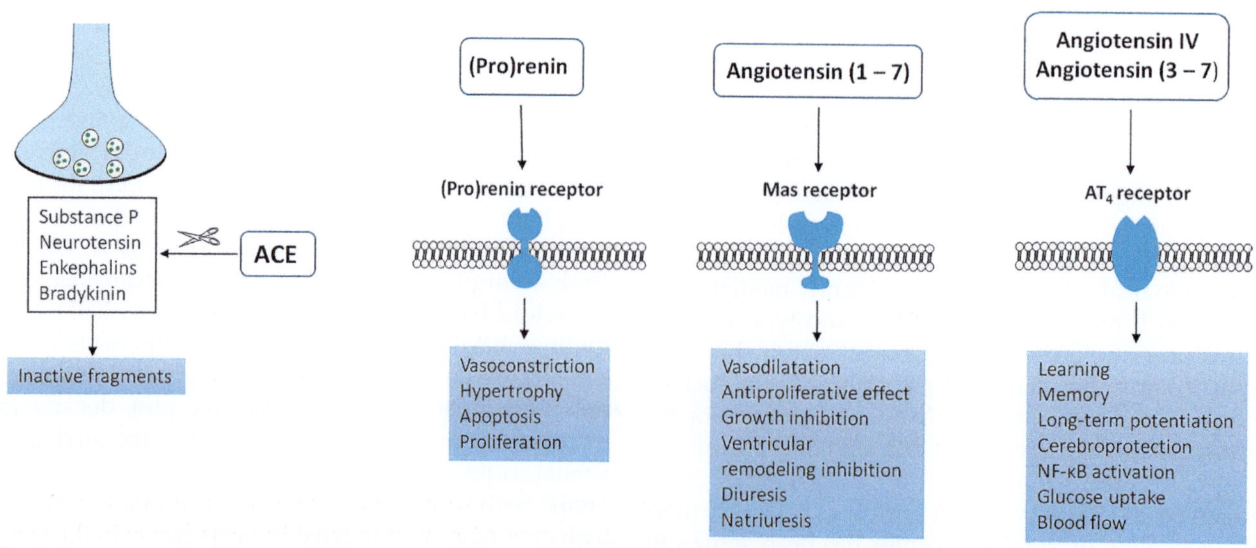

FIG. 42.2 Nonclassical effects of the brain renin-angiotensin system (RAS).
(*Left*) In addition to its primary substrate (angiotensinogen), angiotensin-converting enzyme (ACE) can cleave (scissors) and inactivate peptide neurotransmitters, like substance P, neurotensin, enkephalins, and bradykinin. (*Right*) The indicated RAS components bind and activate (pro)renin receptors, Mas receptors, and angiotensin II receptors type 4 (AT_4), which result in the indicated effects.

amphetamine-induced locomotor sensitization and neurocognitive alterations in rats are associated with changes in AT_1 receptor density in the nucleus accumbens and caudate putamen (Marchese et al., 2016; Paz et al., 2014).

RAS components are genetically determined, and genetics has been proposed to play a substantial role in substance abuse vulnerability. However, data are sparse that support a role for brain RAS in the etiology of addiction in human populations; only a few genetic studies have investigated polymorphic variations in some RAS-related genes (Baghai et al., 2008; Garrib & Peters, 1998; Hubacek, Adamkova, Skodova, Lanska, & Poledne, 2004; Hubacek, Pitha, Skodova, & Poledne, 2001; Nadalin, Buretić-Tomljanović, et al., 2017; Nadalin, Ristić, et al., 2017; Serý, Vojtová, & Zvolský, 2001). These studies primarily focused on nicotine dependence; they included healthy subjects and individuals with several diseases/conditions, such as depression, multiple sclerosis (MS), and schizophrenia. The etiology of nicotine dependence in the context of these diseases seems to be complex; a plausible underlying mechanism might be that some genetic variations might confer dual susceptabilities to a disease and nicotine dependence (de Leon & Diaz, 2005). Thus, studying the genetics of nicotine dependence in the context of disease may provide new insights into nicotine dependence and, in addition, improve our understanding of etiopathogenetic mechanisms involved in those diseases. In this chapter, we discuss studies on RAS-related genes and nicotine dependence. Finally, we will shift our perspective to future genetic and/or environmental studies that might reveal links between RAS components and nicotine dependence.

42.2 BRAIN RAS GENES MIGHT CONTRIBUTE TO NICOTINE DEPENDENCE

A functional insertion/deletion (I/D) polymorphism (rs1799752) in intron 16 of the *ACE* gene (17q23), characterized by the presence or absence of a 287 bp Alu repetitive sequence, has been correlated with the circulating, intracellular, and brain tissue levels of ACE. It has been established that the ACE-I/D polymorphism accounts for ~50% of ACE levels. Individuals homozygous for the D allele have the highest ACE levels, individuals homozygous for the ACE-I allele have the lowest ACE levels, and those with heterozygous for the I and D alleles exhibit intermediate levels (Rigat et al., 1990). The ACE-I/D polymorphism is the most studied RAS-related polymorphic variant. To date, it is the only RAS-related polymorphism investigated in relation to the etiology of smoking behavior. The majority of studies that investigated the relevance of ACE-I/D polymorphisms in smoking behavior studied healthy individuals (Tables 42.1 and 42.2). Hubacek et al. (2001) first speculated that the ACE-I/D polymorphism might play a role in smoking behavior by influencing dopaminergic neurotransmission. They conducted two studies with healthy individuals from the Czech Republic; one small pilot study included only male subjects, and one larger study comprised individuals of both genders. Baghai et al. (2008) hypothesized that ACE-I/D polymorphisms might genetically modulate smoking behavior, and thus, this polymorphism might underlie the high correlation between cardiovascular disease and depression. They investigated whether this polymorphic variant

TABLE 42.1 Risk of Nicotine Dependence Related to ACE-I/D Polymorphisms

Study	Population	Smokers/ nonsmokers	Studied group	Main outcome
Hubacek et al. (2001)	Czech Republic	Males: 189/113	Healthy subjects	ACE-I/D polymorphism is not associated with smoking risk ($P>0.05$)
Hubacek et al. (2004)	Czech Republic	Males: 707/497 Females: 478/897	Healthy subjects	ACE-I/D polymorphism is not associated with smoking risk ($P>0.05$)
Baghai et al. (2008)	German	Total: 17/93	Healthy subjects	ACE-I/D polymorphism is not associated with smoking risk ($P>0.05$)
Baghai et al. (2008)	German	Total: 53/430	Major depression	ACE-DD homozygous genotype contributes to increased risk of smoking in total patient group ($\chi^2=7.0$; $P=0.03$)
Nadalin, Buretić-Tomljanović, et al. (2017), Nadalin, Ristić, et al. (2017)	Croatian and Slovenian	Males: 66/73 Females: 150/232	Multiple sclerosis	ACE-I/D polymorphism is not associated with smoking risk ($P>0.05$)
Nadalin, Buretić-Tomljanović, et al. (2017), Nadalin, Ristić, et al. (2017)	Croatian	Males: 99/41 Females: 77/50	Schizophrenia	ACE-ID homozygous genotype contributes to increased risk of smoking among female patients (OR=2.3; 95% CI=1.1–4.7; $P=0.03$)

TABLE 42.2 Smoking Severity Related to ACE-I/D Polymorphisms

Study	Studied group	Main outcome
Hubacek et al. (2001)	Healthy subjects (males only)	ACE-II homozygous genotype contributes to a high number of cigarettes smoked per week ($P=0.005$)
Hubacek et al. (2004)	Healthy subjects	ACE-I/D polymorphism is not associated with the number of cigarettes smoked per week ($P>0.05$)
Baghai et al. (2008)	Healthy subjects	ACE-DD homozygous genotype contributes to a longer smoking history, measured in pack-years ($F=3.3$; $P=0.04$)
Baghai et al. (2008)	Major depression	ACE-DD homozygous genotype contributes to high daily cigarette consumption ($F=3.2$; $P=0.04$)
Nadalin, Buretić-Tomljanović, et al. (2017), Nadalin, Ristić, et al. (2017)	Multiple sclerosis	ACE-I/D polymorphism is not associated with either the daily cigarette consumption or the number of cigarettes smoked per week ($P>0.05$)

Assessment of smoking severity was not performed for patients with schizophrenia.

influenced smoking habits in patients with depression in the German population. To control for the potential effect of the depressed state on smoking behavior, they also included a small sample of psychiatrically and cardiovascularly healthy controls in the study. In previous studies by our group, conducted in Croatian and Slovenian populations, we addressed the potential relevance of ACE-I/D variants in smoking behavior among patients with MS and schizophrenia (Nadalin, Buretić-Tomljanović, et al., 2017). Evidence has suggested that smoking has detrimental effects in MS: components of tobacco smoke have been associated with demyelinization and axonal degeneration; moreover, nicotine compromises the blood-brain barrier and has immunomodulatory effects on T-lymphocytes (Ramanujam et al., 2015; Zhang et al., 2016). We hypothesized that the ACE-I/D polymorphism might be indirectly associated with an elevated MS risk, primarily by causing habitual smoking. We also proposed that the ACE-I/D polymorphism, by influencing dopaminergic signaling in the mesolimbic pathway (which mediates positive and negative schizophrenia symptoms), may be implicated in the correlation between schizophrenia and nicotine dependence. Smoking prevalence was estimated to be 70%–80% in schizophrenia, and approximately two-thirds of patients that smoked were classified as heavy smokers (smoking ≥30 cigarettes [1.5 packs] daily) (de Leon & Diaz, 2005). Several studies have reported interesting correlations between the ACE-I/D polymorphism, MS, and schizophrenia, which prompted us to investigate whether this polymorphic variant might be linked to smoking behavior in patients with these diseases. As shown in Table 42.3, ACE-I/D polymorphisms contributed to elevated MS and schizophrenia risks in some population studies (Crescenti et al., 2009; Kucukali et al., 2010; Lovrečić et al., 2006; Mazaheri & Saadat, 2015; Živković et al., 2016) and influenced the severity of schizophrenia, based on Positive and Negative Syndrome Scale psychopathology evaluations (Hui et al., 2014; Hui et al., 2015; Nadalin et al., 2012). Our previous studies, conducted in Croatian and Slovenian populations, also found some of these positive associations; moreover, our data suggested that the effects of the polymorphisms might be gender-specific. Positive associations were found between the ACE-I/D polymorphism and depression; indeed, the ACE-I/D variant contributed to disease onset (Stewart et al., 2009). The dexamethasone/corticotropin-releasing hormone test results indicated an association between the ACE-I/D variant and alterations in the hypothalamic-pituitary-adrenocortical system, a major neuroendocrine abnormality in depression (Baghai et al., 2002). Finally, a meta-analysis suggested profound effects of the ACE-I/D polymorphic variant in depression; in fact, it is considered a risk factor for depression in some populations (Wu, Wang, Shen, Tan, & Yuan, 2012) (Table 42.4).

Based on the link between the ACE-I/D polymorphic variant and the risk of nicotine dependence (Table 42.1), we might speculate that the ACE-I/D polymorphism might be more relevant in psychiatric disorders (e.g., depression and schizophrenia) than in other diseases and/or conditions. Similar to schizophrenia, depression is associated with changes in the mesolimbic pathway, and there is a high correlation between cigarette smoking and depression (Dichter, Damiano, & Allen, 2012; Munafò & Araya, 2010). Nevertheless, neuroimaging studies have indicated that mesolimbic signaling is different between these two diseases; in depression, mesolimbic activity is blunted, and in schizophrenia, mesolimbic dopamine transmission is hyperactive (Brisch et al., 2014; Dichter et al., 2012). Consistent with these differences, it is plausible that elevated risks of nicotine dependence are associated with the ACE-DD genotype (implicating high ACE activity) in

TABLE 42.3 Potential Relevance of ACE-I/D Polymorphisms in Multiple Sclerosis and Schizophrenia

Study	Population	Patients/ controls	Studied group	Main outcome
Lovrečić et al. (2006)	Croatian and Slovenian	313/376	Multiple sclerosis	ACE-DD homozygous genotype contributes to elevated multiple sclerosis risk in male patients
Živković et al. (2016)	Serbian	384/395	Multiple sclerosis	Both ACE homozygous genotypes (ACE-DD and ACE-II) contribute to elevated multiple sclerosis risk
Crescenti et al. (2009)	Spanish	243/291	Schizophrenia and related disorders	ACE-D allele is identified as protective for schizophrenia, schizoaffective disorder, acute psychotic disorder, and delusional disorder
Kucukali et al. (2010)	Turkish	239/210	Schizophrenia	ACE-I allele is identified as protective for schizophrenia
Nadalin et al. (2012)	Croatian	211/187	Schizophrenia	ACE-D allele contributes to increased negative and total PANSS scores in total patient group and higher general PANSS scores among male patients
Hui et al. (2014)	Chinese	212/538	Patients with first-episode schizophrenia	ACE-DD homozygous genotype and ACE-D allele are associated with higher negative PANSS scores
Hui et al. (2015)	Chinese	382/538	Schizophrenia	ACE-D allele increases schizophrenia risk in total patient group and in separate groups of male and female patients; PANSS depressive scores were lower in subjects with ACE-DD than in those with ACE-II homozygous genotypes subjects
Mazaheri and Saadat (2015)	Iranian	363/363	Schizophrenia	ACE-II homozygous genotype decreases schizophrenia risk among female patients

PANSS, positive and negative syndrome scale.

TABLE 42.4 Potential Relevance of ACE-I/D Polymorphisms in Major Depression

Study	Population	Study sample size	Main outcome
Baghai et al. (2002)	German	115 patients	ACE-D allele is associated with higher cortisol stimulation during the dexamethasone/corticotropin-releasing hormone test
Stewart et al. (2009)	Finish	119 treatment-resistant patients	ACE-I allele contributes to earlier onset of depression among males, but in females, earlier onset is associated with the ACE-ID heterozygous genotype
Wu et al. (2012)	Asian and European	2479 patients and 7744 controls	ACE-DD homozygous genotype is associated with an elevated risk of depression in a Caucasian population and in a mixed ethnic group, which consisted of several European populations (German, British, Belgian, Finish, and Israeli) and Asian populations (Japanese and Chinese)

Some recent pharmacogenetic studies also indicated that the therapy responses to particular antidepressants were associated with ACE-I/D polymorphisms (not shown).

patients with depression and with the ACE-ID genotype (implicating intermediate ACE activity) in females with schizophrenia. The ACE-DD genotype also contributes to a longer smoking history (pack-years) among healthy German individuals; these findings suggested that the mechanism underlying the influence of ACE-I/D polymorphic variants on smoking severity is similar in healthy and depressed states. However, other reports on the potential relevance of the ACE-I/D polymorphism in smoking severity among healthy subjects do not support this speculation (Table 42.2). In fact, it is difficult to evaluate the role of ACE-I/D polymorphisms in smoking severity in healthy populations, because studies on healthy subjects have assessed smoking severity with different measures (i.e., numbers of cigarettes smoked per day, per week, or duration (pack-years) of smoking history, etc.). Indeed, the Czech Republic pilot study on the potential role of ACE-I/D polymorphisms in smoking severity was not replicated in a larger sample, probably due to different inclusion criteria. The pilot study included current and past smokers, and the larger study only included current smokers.

Several findings might explain the gender-specific differences regarding the association between ACE-I/D polymorphisms and nicotine dependence. One factor could be that estrogen replacement therapy contributes to decreases in serum or plasma ACE activity in postmenopausal women with particular ACE-I/D genotypes. One study showed that plasma ACE activity was significantly reduced in women with the ACE-ID and ACE-II genotypes (Sanada et al., 2001), and another study showed reductions in women with the ACE-ID and ACE-DD genotypes (Sumino et al., 2003). Another factor might be that estrogen reduces dopaminergic neurotransmission. In rats, estrogen treatment reduced dopamine receptor D2 levels in several brain regions (Chavez et al., 2010).

We should be aware of numerous limitations in studies that investigated the roles of ACE-I/D variants in nicotine dependence. The most important limitation was that only one RAS-related polymorphism was investigated. Furthermore, none of the studies assessed nicotine dependence with more specific methods, for example, the Fagerstrom test for nicotine dependence. In addition, most studies included a small number of participants, and the studies on patients with schizophrenia and MS lacked control groups. Finally, an imbalance in the male/female ratio in MS or in the smoker/nonsmoker ratio in schizophrenia might have led to biases in the statistical analyses.

42.3 CONCLUSION AND FUTURE REMARKS

Although results of genetic studies may be disputable, they argue that ACE-I/D polymorphic variants play weak modulatory roles in smoking habits. Furthermore, the relevance of ACE-I/D polymorphisms in nicotine dependence differs in the contexts of particular diseases and/or conditions; indeed, this polymorphism might be most relevant in smoking risk for patients with psychiatric disorders.

Future studies should examine other functional *ACE* polymorphisms, investigate the relevance of other RAS-related polymorphic variants (e.g., variants of angiotensinogen and AT_1 or AT_2 receptor genes), and include populations from different ethnic backgrounds. Another important issue to clarify is that, in addition to angiotensin II, ACE has other potential substrates, such as substance P, neurotensin, and enkephalins (Fig. 42.2), which also influence dopaminergic neurotransmission in the mesolimbic system (Binder, Kinkead, Owens, & Nemeroff, 2001; Gerfen, McGinty, & Young 3rd., 1991). Furthermore, it would be interesting to investigate the ACE-I/D polymorphism association with nicotine dependence in the context of other diseases, particularly diseases that might be caused by smoking (e.g., lung cancer). Genome-wide association studies have shown that numerous polymorphic variants of genes in the mesolimbic system (e.g., particularly nicotinic acetylcholine receptor subunit genes) are linked to lung cancer (Amos et al., 2008); other studies have reported that ACE-I/D polymorphisms were associated with lung cancer susceptibility (Wang, Yang, Ji, & Li, 2015).

Several environmental factors may interact with RAS-related genes and, consequently, influence smoking behavior in the contexts of particular diseases and/or conditions. For instance, the administration of monoamine uptake inhibitors diminished substance P levels in the striatum (Porcelli et al., 2011), and antipsychotic medications could reduce or increase ACE expression/activity (Segman et al., 2002). A novel finding was that interferon-β, which is widely used in MS therapy, interfered with some RAS components. For example, patients with MS, in relapse and remission, were treated with interferon-β and concomitant AT_1 receptor antagonists and ACE inhibitors; these patients trended toward a higher relapse rate compared to patients treated solely with interferon-β (Doerner et al., 2014). Finally, smoking may also influence ACE expression/activity; nicotine and its metabolites increased ACE expression and activity in human endothelial cells. Moreover, chronic smokers exhibited enhanced RAS activation and elevated circulating ACE levels (Ljungberg et al., 2011).

The conditions of depression, schizophrenia, and MS may also modify smoking habits. The self-medication hypothesis holds that the high rate of nicotine dependence and/or high frequency of heavy smoking in the psychiatric population is due to the beneficial effects of nicotine. Smoking restores dopaminergic transmission through the central effects of nicotine on the dopaminergic system; consequently, smoking has been proposed to alleviate symptoms of depression and anxiety, negative and positive schizophrenia symptoms, the extrapyramidal side effects of antipsychotic medications, and cognitive deficits (Fluharty, Taylor, Grabski, & Munafò, 2017; Sagud et al., 2009). Studies have also suggested that patients with MS that smoked experienced a more severe disease course and more rapid disability progression, measured with the Expanded Disability Status Scale (Healy et al., 2009). Therefore, due to the illness, over the long term, patients with MS might reduce or quit a smoking habit. Finally, in determining whether smoking increased the risk of depression, animal experiments have indicated that chronic exposure to cigarette smoke substantially contributed to hypothalamic-pituitary-adrenocortical dysregulation in animals with depression (Fluharty et al., 2017).

42.4 IMPLICATIONS FOR TREATMENTS

In our opinion, it would be interesting to address whether long-term use of medications that interfere with RAS components (i.e., ACE inhibitors or AT_1 receptor blockers), which are safely used for treating hypertension and cardioprotection, might be an effective treatment approach for nicotine dependence. This treatment might be particularly beneficial for patients with psychiatric conditions, because their elevated smoking habits put them at substantially elevated cardiovascular risk. Consistent with the functional properties of the ACE-I/D polymorphism, it would be also worthwhile to elucidate whether responses to nicotine dependence treatments might be related to particular ACE genotype(s). Additionally, cigarette smoking has been shown to increase the metabolism of many drugs, including antidepressants and antipsychotics, by increasing the activity of enzymes in the cytochrome family (Oliveira, Ribeiro, Donato, & Madeira, 2017; Sagud et al., 2009). Thus, based on the evidence that ACE-I/D polymorphisms contribute to smoking behavior in patients with depression and schizophrenia, genotyping these patients may also provide useful information in tailoring the medications appropriately.

MINI-DICTIONARY OF TERMS

Angiotensin-converting enzyme (ACE) It is the key enzyme in RAS; it converts the inactive decapeptide, angiotensin I, into the active octapeptide, angiotensin II.

Functional polymorphisms They are gene variations that have been shown to alter gene expression and/or function.

Gene polymorphisms They are common variations in a single gene found in a population.

Insertion/deletion polymorphism It is a gene alteration that comprises either the inclusion of an extra sequence (insertion) or the deletion of a sequence (deletion).

Major depression It is a mental disorder characterized by a depressed mood or loss of interest and pleasure in daily activities, lasting for more than 2 weeks.

Mesolimbic pathway It is a circuit in the brain, in which dopamine transmission is conducted from the ventral tegmental area to the nucleus accumbens, amygdala, hippocampus, and prefrontal cortex. In addition to generating reward stimuli, it is implicated in addictive behaviors, depression, and schizophrenia.

Multiple sclerosis It is a chronic autoimmune disease of the central nervous system, characterized by inflammation, myelin sheath destruction, and neuronal degeneration.

Renin-angiotensin system (RAS) It is a peptidergic system classically described as a hormone system. Its principal functions are to regulate systemic blood pressure and maintain sodium and fluid homeostasis. In addition, many tissues have local RAS components, regulated independently of the classical RAS.

Schizophrenia It is a chronic mental illness characterized by positive symptoms (delusions and hallucinations), negative symptoms (blunted emotions and social isolation), and cognitive deficits, including impairments in attention, memory, and executive functions.

Substance abuse It is the excessive use of a potentially addictive substance; addiction is characterized by an inability to reduce consumption and by impairments in social or occupational functions.

Key Facts on the Renin-Angiotensin System (RAS)

- The classical functions of RAS were to regulate systemic blood pressure and maintain sodium and fluid homeostasis.
- Interest in RAS was renewed when RAS effectors were identified in different human tissues, particularly in brain.
- All classical RAS components (angiotensinogen, peptidases, angiotensins, and receptor proteins) were identified in different brain areas within the blood-brain barrier.
- The brain RAS controls many central functions, including cerebral blood flow regulation, neuronal regeneration, memory consolidation, alcohol consumption, and stress; moreover, RAS was proposed to play various roles in the etiology of numerous disorders.
- An accumulating body of evidence has suggested that the brain RAS may be an important determinant in substance abuse etiology, by influencing dopaminergic signaling in mesolimbic pathways.

Summary Points

- The brain renin-angiotensin system (RAS) may play an important role in the etiology of substance abuse by influencing dopaminergic signaling.
- Data that link the RAS to addictions in human populations are sparse; they are mostly from a few genetic studies that investigated whether the functional insertion/deletion (I/D) polymorphism in the angiotensin-converting enzyme (ACE) gene affected smoking behavior.
- In addition to healthy subjects, these studies included individuals with depression, multiple sclerosis, and schizophrenia.
- Results argued that ACE-I/D polymorphisms played a weak modulator role in smoking behavior and that the relevance of this polymorphism in nicotine dependence may differ in particular diseases and/or conditions.
- Findings suggested that ACE-I/D polymorphisms might be more relevant in predicting smoking risk in psychiatric disorders compared to other diseases and/or conditions.

References

Amos, C. I., Wu, X., Broderick, P., Gorlov, I. P., Gu, J., Eisen, T., et al. (2008). Genome-wide association scan of tag SNPs identifies a susceptibility locus for lung cancer at 15q25.1. *Nature Genetics*, *40*, 616–622.

Baghai, T. C., Schule, C., Zwanzger, P., Minov, C., Zill, P., Ella, R., et al. (2002). Hypothalamic-pituitary-adrenocortical axis dysregulation in patients with major depression is influenced by the insertion/deletion polymorphism in the angiotensin I-converting enzyme gene. *Neuroscience Letters*, *328*, 299–303.

Baghai, T. C., Varallo-Bedarida, G., Born, C., Häfner, S., Schule, C., Eser, D., et al. (2008). A polymorphism in the angiotensin-converting enzyme gene is associated with smoking behavior. *Journal of Clinical Psychiatry*, *69*, 1983–1985.

Banks, R. J., Mozley, L., & Dourish, C. T. (1994). The angiotensin converting enzyme inhibitors captopril and enalapril inhibit apomorphine-induced oral stereotypy in the rat. *Neuroscience, 58*, 799–805.

Binder, E. B., Kinkead, B., Owens, M. J., & Nemeroff, C. B. (2001). Neurotensin and dopamine interactions. *Pharmacological Reviews, 53*, 453–486.

Brisch, R., Saniotis, A., Wolf, R., Bielau, H., Bernstein, H. G., Steiner, J., et al. (2014). The role of dopamine in schizophrenia from a neurobiological and evolutionary perspective: old fashioned, but still in vogue. *Frontiers in Psychiatry. 5*, https://doi.org/10.3389/fpsyt.2014.00047.

Chavez, C., Hollaus, M., Scarr, E., Pavey, G., Gogos, A., & van den Buuse, M. (2010). The effect of estrogen on dopamine and serotonin receptor and transporter levels in the brain: an autoradiography study. *Brain Research, 1321*, 51–59.

Crescenti, A., Gassó, P., Mas, S., Abellana, R., Deulofeu, R., Parellada, E., et al. (2009). Insertion/deletion polymorphism of the angiotensin-converting enzyme gene is associated with schizophrenia in a Spanish population. *Psychiatry Research, 165*, 175–180.

de Leon, J., & Diaz, F. J. (2005). A meta-analysis of worldwide studies demonstrates an association between schizophrenia and tobacco smoking behaviors. *Schizophrenia Research, 76*, 135–157.

Dichter, G. S., Damiano, C. A., & Allen, J. A. (2012). Reward circuitry dysfunction in psychiatric and neurodevelopmental disorders and genetic syndromes: animal models and clinical findings. *Journal of Neurodevelopmental Disorders, 4*, 19. https://doi.org/10.1186/1866-1955-4-19.

Doerner, M., Beckmann, K., Knappertz, V., Kappos, L., Hartung, H. P., Filippi, M., et al. (2014). Effects of inhibitors of the renin-angiotensin system on the efficacy of interferon beta-1b: a post hoc analysis of the BEYOND study. *European Neurology, 71*, 173–179.

Fluharty, M., Taylor, A. E., Grabski, M., & Munafò, M. R. (2017). The association of cigarette smoking with depression and anxiety: a systematic review. *Nicotine & Tobacco Research, 19*, 3–13.

Garrib, A., & Peters, T. (1998). Angiotensin-converting enzyme (ACE) gene polymorphism and alcoholism. *Biochemical Society Transactions, 26*, S136.

Georgiev, V., György, L., Getova, D., & Markovska, V. (1985). Some central effects of angiotensin II. Interactions with dopaminergic transmission. *Acta physiologica et pharmacologica Bulgarica, 11*, 19–26.

Gerfen, C. R., McGinty, J. F., & Young, W. S., 3rd. (1991). Dopamine differentially regulates dynorphin, substance P, and enkephalin expression in striatal neurons: in situ hybridization histochemical analysis. *Journal of Neuroscience, 11*, 1016–1031.

Guimond, M. O., & Gallo-Payet, N. (2012). The angiotensin II type 2 receptor in brain functions: an update. *International Journal of Hypertension, 2012*, 351758. https://doi.org/10.1155/2012/351758.

Healy, B. C., Ali, E. N., Guttmann, C. R., Chitnis, T., Glanz, B. I., Buckle, G., et al. (2009). Smoking and disease progression in multiple sclerosis. *Archives of Neurology, 66*, 858–864.

Hosseini, M., Sharifi, M. R., Alaei, H., Shafei, M. N., & Karimooy, H. A. (2007). Effects of angiotensin II and captopril on rewarding properties of morphine. *Indian Journal of Experimental Biology, 45*, 770–777.

Hubacek, J. A., Adamkova, V., Skodova, Z., Lanska, V., & Poledne, R. (2004). No relation between angiotensin-converting enzyme gene polymorphism and smoking dependence. *Scandinavian Journal of Clinical and Laboratory Investigation, 64*, 575–578.

Hubacek, J. A., Pitha, J., Skodova, Z., & Poledne, R. (2001). Angiotensin converting enzyme gene—a candidate gene for addiction to smoking? *Atherosclerosis, 159*, 237–238.

Hui, L., Wu, J. Q., Ye, M. J., Zheng, K., He, J. C., Zhang, X., et al. (2015). Association of angiotensin-converting enzyme gene polymorphism with schizophrenia and depressive symptom severity in a Chinese population. *Human Psychopharmacology, 30*, 100–107.

Hui, L., Wu, J. Q., Zhang, X., Lv, J., Du, W. L., Kou, C. G., et al. (2014). Association between the angiotensin-converting enzyme gene insertion/deletion polymorphism and first-episode patients with schizophrenia in a Chinese Han population. *Human Psychopharmacology, 29*, 274–279.

Jenkins, T. A., Chai, S. Y., & Mendelsohn, F. A. (1997). Effect of angiotensin II on striatal dopamine release in the spontaneous hypertensive rat. *Clinical and Experimental Hypertension, 19*, 645–658.

Jenkins, T. A., Mendelsohn, F. A., & Chai, S. Y. (1997). Angiotensin-converting enzyme modulates dopamine turnover in the striatum. *Journal of Neurochemistry, 68*, 1304–1311.

Kucukali, C. I., Aydin, M., Ozkok, E., Bilge, E., Zengin, A., Cakir, U., et al. (2010). Angiotensin-converting enzyme polymorphism in schizophrenia, bipolar disorders, and their first-degree relatives. *Psychiatric Genetics, 20*, 14–19.

Labandeira-García, J. L., Garrido-Gil, P., Rodriguez-Pallares, J., Valenzuela, R., Borrajo, A., & Rodríguez-Perez, A. I. (2014). Brain renin-angiotensin system and dopaminergic cell vulnerability. *Frontiers in Neuroanatomy, 8*, 67. https://doi.org/10.3389/fnana.2014.00067.

Ljungberg, L., Alehagen, U., Länne, T., Björck, H., De Basso, R., Dahlström, U., et al. (2011). The association between circulating angiotensin-converting enzyme and cardiovascular risk in the elderly: a cross-sectional study. *Journal of the Renin-Angiotensin-Aldosterone System, 12*, 281–289.

Lovrečić, L., Ristić, S., Starčević-Čizmarević, N., Jazbec, S. S., Sepčić, J., Kapović, M., et al. (2006). Angiotensin-converting enzyme I/D gene polymorphism and risk of multiple sclerosis. *Acta Neurologica Scandinavica, 114*, 374–377.

Marchese, N. A., Artur de la Villarmois, E., Basmadjian, O. M., Perez, M. F., Baiardi, G., & Bregonzio, C. (2016). Brain angiotensin II AT1 receptors are involved in the acute and long-term amphetamine-induced neurocognitive alterations. *Psychopharmacology (Berlin), 233*, 795–807.

Mazaheri, H., & Saadat, M. (2015). Association between insertion/deletion polymorphism in angiotension converting enzyme and susceptibility to schizophrenia. *Iranian Journal of Public Health, 44*, 369–373.

McKinley, M. J., Albiston, A. L., Allen, A. M., Mathai, M. L., May, C. N., McAllen, R. M., et al. (2003). The brain renin-angiotensin system: location and physiological roles. *The International Journal of Biochemistry & Cell Biology, 35*, 901–918.

Munafò, M. R., & Araya, R. (2010). Cigarette smoking and depression: a question of causation. *British Journal of Psychiatry, 196*, 425–426.

Nadalin, S., Buretić-Tomljanović, A., Lavtar, P., Starčević Čizmarević, N., Hodžić, A., Sepčić, J., et al. (2017). The lack of association between angiotensin-converting enzyme gene insertion/deletion polymorphism and nicotine dependence in multiple sclerosis. *Brain and Behavior. 7*, e00600https://doi.org/10.1002/brb3.600.

Nadalin, S., Buretić-Tomljanović, A., Rubeša, G., Jonovska, S., Tomljanović, D., & Ristić, S. (2012). Angiotensin-converting enzyme gene insertion/deletion polymorphism is not associated with schizophrenia in a Croatian population. *Psychiatric Genetics, 22*, 267–268.

Nadalin, S., Ristić, S., Rebić, J., Šendula Jengić, V., Kapović, M., & Buretić-Tomljanović, A. (2017). The insertion/deletion polymorphism in the angiotensin-converting enzyme gene and nicotine dependence in schizophrenia patients. *Journal of Neural Transmission (Vienna), 124*, 511–518.

Obata, T., Takahashi, S., Kashiwagi, Y., & Kubota, S. (2008). Protective effect of captopril and enalaprilat, angiotensin-converting enzyme inhibitors, on para-nonylphenol-induced *OH generation and dopamine efflux in rat striatum. *Toxicology, 250*, 96–99.

Oliveira, P., Ribeiro, J., Donato, H., & Madeira, N. (2017). Smoking and antidepressants pharmacokinetics: a systematic review. *Annals of General Psychiatry, 16*, 17. https://doi.org/10.1186/s12991-017-0140-8.

Paz, M. C., Marchese, N. A., Stroppa, M. M., Gerez de Burgos, N. M., Imboden, H., Baiardi, G., et al. (2014). Involvement of the brain renin-angiotensin system (RAS) in the neuroadaptive responses induced by amphetamine in a two-injection protocol. *Behavioural Brain Research, 272*, 314–323.

Porcelli, S., Drago, A., Fabbri, C., Gibiino, S., Calati, R., & Serretti, A. (2011). Pharmacogenetics of antidepressant response. *Journal of Psychiatry and Neuroscience*, *36*, 87–113.

Ramanujam, R., Hedström, A. K., Manouchehrinia, A., Alfredsson, L., Olsson, T., Bottai, M., et al. (2015). Effect of smoking cessation on multiple sclerosis prognosis. *JAMA Neurology*, *72*, 1117–1123.

Rigat, B., Hubert, C., Alhenc-Gelas, F., Cambien, F., Corvol, P., & Soubrier, F. (1990). An insertion/deletion polymorphism in the angiotensin I-converting enzyme gene accounting for half the variance of serum enzyme levels. *Journal of Clinical Investigation*, *86*, 1343–1346.

Sagud, M., Mihaljević-Peles, A., Mück-Seler, D., Pivac, N., Vuksan-Cusa, B., Brataljenović, T., et al. (2009). Smoking and schizophrenia. *Psychiatria Danubina*, *21*, 371–375.

Sanada, M., Higashi, Y., Nakagawa, K., Sasaki, S., Kodama, I., Tsuda, M., et al. (2001). Relationship between the angiotensin-converting enzyme genotype and the forearm vasodilator response to estrogen replacement therapy in postmenopausal women. *Journal of the American College of Cardiology*, *37*, 1529–1535.

Segman, R. H., Shapira, Y., Modai, I., Hamdan, A., Zislin, J., Heresco-Levy, U., et al. (2002). Angiotensin converting enzyme gene insertion/deletion polymorphism: case-control association studies in schizophrenia, major affective disorder, and tardive dyskinesia and a family-based association study in schizophrenia. *American Journal of Medical Genetics*, *114*, 310–314.

Serý, O., Vojtová, V., & Zvolský, P. (2001). The association study of DRD2, ACE and AGT gene polymorphisms and metamphetamine dependence. *Physiological Research*, *50*, 43–50.

Sommer, W. H., Rimondini, R., Marquitz, M., Lidström, J., Siems, W. E., Bader, M., et al. (2007). Plasticity and impact of the central renin-angiotensin system during development of ethanol dependence. *Journal of Molecular Medicine (Berlin)*, *85*, 1089–1097.

Stewart, J. A., Kampman, O., Huuhka, M., Anttila, S., Huuhka, K., Lehtimäki, T., et al. (2009). ACE polymorphism and response to electroconvulsive therapy in major depression. *Neuroscience Letters*, *458*, 122–125.

Sumino, H., Ichikawa, S., Ohyama, Y., Nakamura, T., Kanda, T., Sakamoto, H., et al. (2003). Effects of hormone replacement therapy on serum angiotensin-converting enzyme activity and plasma bradykinin in postmenopausal women according to angiotensin-converting enzyme-genotype. *Hypertension Research*, *26*, 53–58.

Visniauskas, B., Perry, J. C., Oliveira, V., Dalio, F. M., Andersen, M. L., Tufik, S., et al. (2012). Cocaine administration increases angiotensin I-converting enzyme (ACE) expression and activity in the rat striatum and frontal cortex. *Neuroscience Letters*, *506*, 84–88.

Wang, N., Yang, D., Ji, B., & Li, J. (2015). Angiotensin-converting enzyme insertion/deletion gene polymorphism and lung cancer risk: a meta-analysis. *Journal of the Renin-Angiotensin-Aldosterone System*, *16*, 189–194.

Wright, J. W., & Harding, J. W. (2013). The brain renin-angiotensin system: a diversity of functions and implications for CNS diseases. *Pflügers Archiv—European Journal of Physiology*, *465*, 133–151.

Wu, Y., Wang, X., Shen, X., Tan, Z., & Yuan, Y. (2012). The I/D polymorphism of angiotensin-converting enzyme gene in major depressive disorder and therapeutic outcome: a case-control study and meta-analysis. *Journal of Affective Disorders*, *136*, 971–978.

Zhang, P., Wang, R., Li, Z., Wang, Y., Gao, C., Lv, X., et al. (2016). The risk of smoking on multiple sclerosis: a meta-analysis based on 20,626 cases from case-control and cohort studies. *PeerJ*. *4*, e1797 https://doi.org/10.7717/peerj.1797.

Živković, M., Kolaković, A., Stojković, L., Dinčić, E., Kostić, S., Alavantić, D., et al. (2016). Renin-angiotensin system gene polymorphisms as risk factors for multiple sclerosis. *Journal of the Neurological Sciences*, *15*(363), 29–32.

CHAPTER

43

Nicotine Dependence and the *CHRNA5/CHRNA3/CHRNB4* Nicotinic Receptor Regulome

Sung-Ha Lee, Elizabeth S. Barrie†, Wolfgang Sadee‡, Ryan M. Smith§*

*Center for Pharmacogenomics, College of Medicine, The Ohio State University, Columbus, OH, United States
†Institute for Genomic Medicine, Nationwide Children's Hospital, Columbus, OH, United States
‡Center for Pharmacogenomics, Department of Cancer Biology and Genetics, College of Medicine, The Ohio State University, Columbus, OH, United States
§Division of Pharmaceutics and Translational Therapeutics, Department of Pharmaceutical Sciences and Experimental Therapeutics, College of Pharmacy, University of Iowa, Iowa City, IA, United States

Abbreviations

CNS	central nervous system
CpG	cytosine phosphate guanine
DNA	deoxyribonucleic acid
GTEx	genotype-tissue expression
GWAS	genome-wide association study
LD	linkage disequilibrium
MAF	minor allele frequency
meQTL	methylation quantitative trait locus
mRNA	messenger ribonucleic acid
ND	nicotine dependence
RNA	ribonucleic acid
SNP	single-nucleotide polymorphism
UTR	untranslated region

43.1 INTRODUCTION: TISSUE EXPRESSION AND FUNCTION OF NICOTINIC α5, α3, β4 SUBUNITS

The genes encoding the α5, α3, and β4 subunits (*CHRNA5, CHRNA3,* and *CHRNB4*) are located in a single gene cluster on chromosome 15q25.1. *CHRNA5* is encoded in the 5′ to 3′ orientation, while *CHRNA3, CHRNB4,* and a noncoding ribonucleic acid (RNA) *RP11-650L12.2* (ENST00000567141) are in the opposite orientation (3′–5′) (Fig. 43.1). This region spans ~76kb, containing blocks of tight linkage disequilibrium (LD; Fig. 43.1). *CHRNA5* and *CHRNA3* share a 466-base-pair overlap in the 3′ ends of their annotated genomic sequence, resulting in complementary base pairs between their extended mRNA transcripts.

Pairwise correlation shows that *CHRNA5, CHRNA3, CHRNB4,* and *RP11-650L12.2* mRNA coexpress in most tissues, suggesting common regulatory mechanisms (Barrie et al., 2017). The biological significance of high correlation between *CHRNA5* and its antisense RNA *RP11-650L12.2* remains unknown. Most protein-coding genes adjacent to the cluster are not coexpressed, with the exception of *PSMA4* and *MORF4L1* (Barrie et al., 2017), indicating the influence of the local "regulome" may extend to these two genes as well.

Following translation, the α3 and β4 subunits coassemble to form functional α3β4* channels, which sometimes incorporate the α5 subunit (α3β4α5) (Vernallis, Conroy, & Berg, 1993; Wang et al., 1996). Varying receptor stoichiometries are observed in adrenal glands (Hone et al., 2015); in peripheral ganglia (Mao, Yasuda, Fan, Wolfe, & Kellar, 2006; Vernallis et al., 1993; Yeh et al., 2001); and in epithalamic structures, notably the habenula and interpeduncular nucleus (Frahm et al., 2011; Yeh et al., 2001; Zoli, Le Novere, Hill Jr., & Changeux, 1995) (Fig. 43.2). In general, α3β4* channels regulate the excitability of ganglionic cells to modulate activity-dependent neurotransmitter release, i.e., noradrenaline from autonomic ganglia (Yokotani, Wang, Okada, Murakami, & Hirata, 2000) and during sensory neuron excitation (Genzen & McGehee, 2003; Mao et al., 2006; Zhang, Albers, & Gold, 2015).

Neuroscience of Nicotine
https://doi.org/10.1016/B978-0-12-813035-3.00043-5

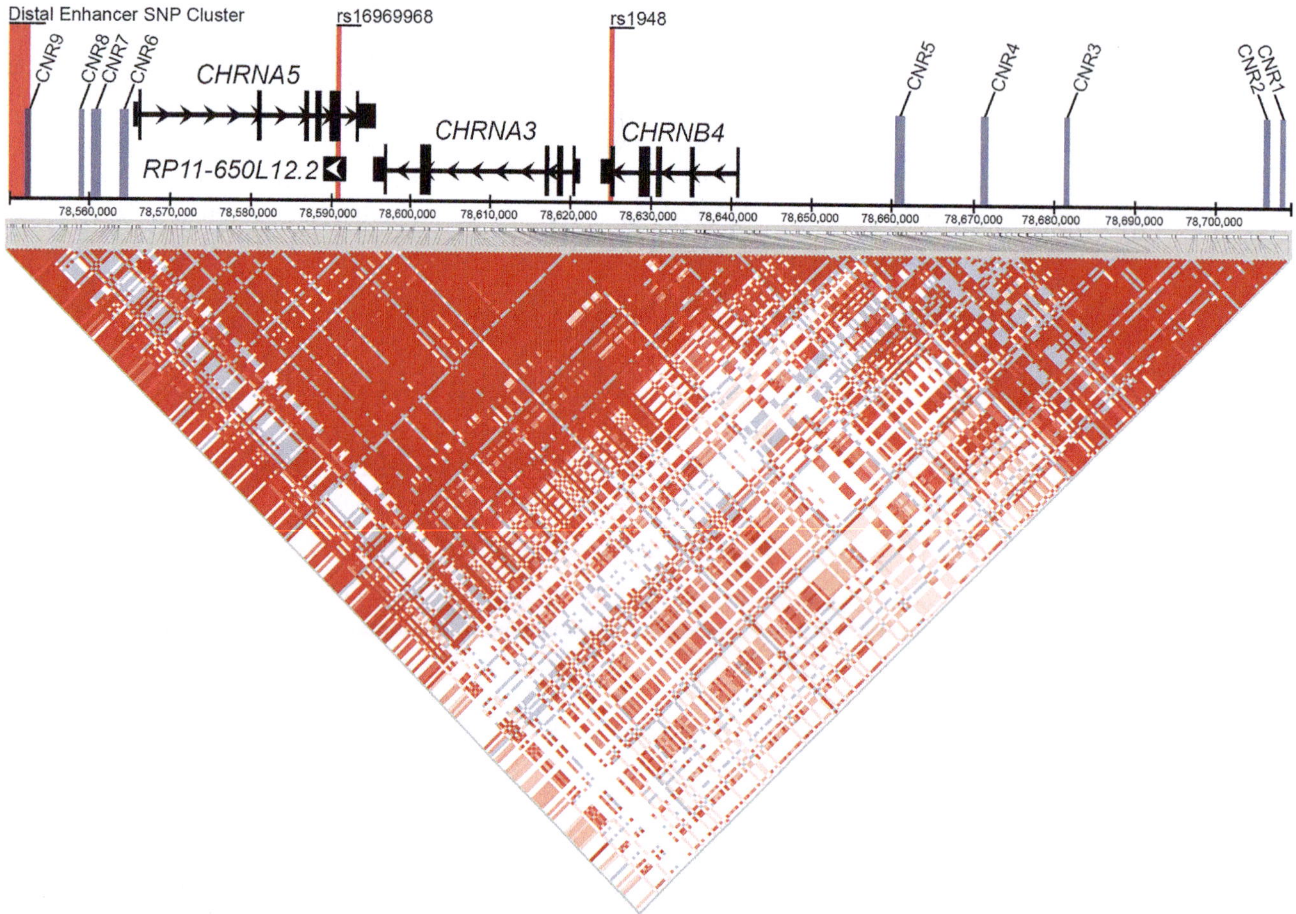

FIG. 43.1 *CHRNA5/CHRNA3/CHRNB4* gene locus.
The *CHRNA5/CHRNA3/CHRNB4* gene locus is located on chromosome 15q25.1. Depicted in black are the protein-coding and noncoding genes. The thickest gene regions indicate protein-coding potential, moderate thickness indicates untranslated regions, thin regions indicate introns, and chevrons indicate transcriptional direction. Red-shaded regions indicate functional polymorphisms with clinical associations. Blue-shaded regions indicate conserved nicotinic regulatory regions (CNR1–9), according to Xu et al. (2006). Below the gene diagram is a linkage disequilibrium plot showing strong linkage disequilibrium, indicated by red shading, extending across the entire gene cluster.

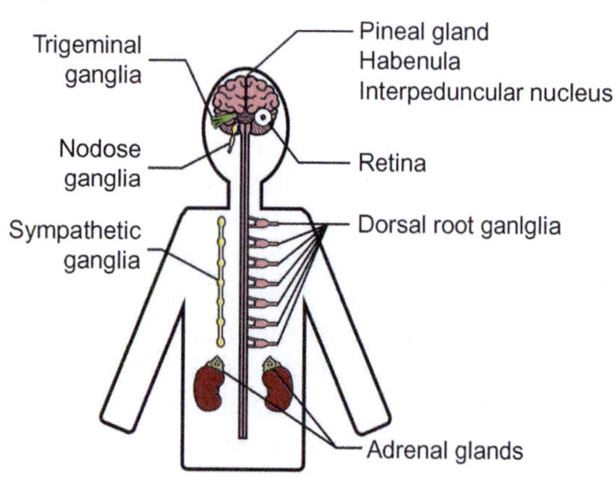

FIG. 43.2 Expression of α3β4* receptors in the body.
α3β4* receptors are expressed throughout the central and peripheral nervous system and in peripheral tissues. Within the brain, α3β4* receptors are expressed in epithalamic structures mediating reward. In the peripheral nervous system, α3β4* receptors are located on multiple ganglia types, which innervate most organs, regulating excitability and neurotransmitter release.

Relevant to addiction, α3β4* channels are well expressed in epithalamic structures (Shih et al., 2014), including the habenula, pituitary, fasciculus retroflexus, and interpeduncular nucleus. The habenula and associated structures modulate negative reward, activated in response to negative or aversive stimuli. As a result, α3β4* channels play an important role in reward behaviors underlying addiction. Transgenic mice with greater α3β4* channel expression in the medial habenula decrease nicotine consumption (Frahm et al., 2011), as excessive nicotine is aversive. Many of the epithalamic α3β4* channels also contain the α5 receptor subunit. While this subunit is not required for the manifestation of nicotine-related reward or aversion behaviors (Jackson, Sanjakdar, Muldoon, McIntosh, & Damaj, 2013), the significance of this subunit is apparent in the context of the α5/α3/β4 regulome.

43.2 PROTEINS REGULATING *CHRNA5/CHRNA3/CHRNB4* TRANSCRIPTION

Context is critical for determining the proteins regulating the expression of *CHRNA5*, *CHRNA3*, or *CHRNB4*, since each gene has its own promoter and interacting enhancers that can act in a tissue-dependent manner, across developmental time lines, or in response to external perturbations. The specificity protein (SP) family of transcription factors regulates the expression of all three genes, with each promoter having multiple SP binding sites (Bigger, Casanova, & Gardner, 1996, Bigger, Melnikova, & Gardner, 1997; Campos-Caro et al., 1999; Fornasari, Battaglioli, Flora, Terzano, & Clementi, 1997). The near-ubiquitous expression of SP transcription factors, in contrast to tissue-specific expression of the three subunits, strongly suggests additional proteins are involved in regulating subunit expression. Human and rodent expressions demonstrate tissue-specific regulation, mediated by *Pou3f1* and *PHOX2A* for α3 in neuronal cells (Benfante et al., 2007; Yang et al., 1994), *Sox10* for α3 and β4 in neuronal cells (Liu, Melnikova, Hu, & Gardner, 1999), and *menin* for α5 in hippocampal cells (Getz, Xu, Visser, Persson, & Syed, 2017). Artemin induces the expression of α3 and β4 in the trigeminal ganglia and neurons innervating the skin following inflammation (Albers et al., 2014).

Proximity of nicotinic genes to form a phylogenetically conserved cluster, along with distinct tissue-specific coexpression patterns, suggests coregulation of their expression by shared regulatory elements. Xu, Scott, and Deneris (2006) mapped conserved regions around the *A5/A3/B4* cluster to identify conserved regulatory elements, finding nine regions >200bp in length sharing >70% sequence identity (Fig. 43.1). α3 and β4 expression is coregulated in the pineal and superior cervical ganglia by an element upstream of β4 (Xu et al., 2006). In summary, strong evidence supports multiple regulatory elements coordinating the expression of all three nicotinic genes in a tissue-dependent manner.

43.3 POLYMORPHISMS REGULATING CHRNA5/CHRNA3/CHRNB4 mRNA EXPRESSION

CHRNA5 mRNA expression is modulated by genetic variation (Smith et al., 2011; Wang et al., 2009; Wang et al., 2009), but the tight LD across the *A5/A3/B4* locus (see Fig. 43.1) makes it challenging to disentangle causative functional polymorphisms from surrogate markers. Using allelic expression to mitigate confounding LD, Smith et al. (2011) uncovered an enhancer region approximately 15kb upstream of the *CHRNA5* gene locus, coinciding with the conserved region identified by Xu et al. (2006) (Fig. 43.1) and replicated in subcortical structures by Doyle et al. (2011) and Ramsay, Rhodes, Thirtamara-Rajamani, and Smith (2015). This enhancer region, marked by six SNPs in high LD (rs1979905, rs1979906, rs1979907, rs880395, rs905740, and rs7164030), increases *CHRNA5* expression in the prefrontal cortex up to fourfold. The enhancer also affects the expression in many peripheral tissues (GTEx Consortium, 2013; Ramsay et al., 2015).

The effect of this enhancer region on *CHRNA3* and *CHRNB4* expression was unknown or regarded as insignificant, as it was not as dramatic as seen for *CHRNA5*. However, using a multiomics approach, Barrie et al. (2017) demonstrated that the distal enhancer region also affected the expression of *CHRNA3* and the antisense, *RP11-650L12.2*, in the CNS and periphery. This suggests coregulated mRNA expression for these nicotinic subunits by a clinically significant functional polymorphism and a distal enhancer region that serves as a master regulator of mRNA expression for the cluster. The low expression of *CHRNB4* in CNS tissues available for study precludes analyses in the Genotype-Tissue Expression (GTEx) data to determine whether the enhancer modulates *CHRNB4* expression, but surveying *CHRNB4* expression in epithalamic tissues, where expression is higher (Shih et al., 2014), could resolve this question.

While polymorphisms in the distal enhancer constitutively modulate the expression of *CHRNA5* and *CHRNA3* across central and peripheral tissues, there is one notable exception with strong implications for addiction. In the striatum, including the caudate, putamen, and nucleus accumbens, rs1948 most strongly correlates with *CHRNA3* expression in the GTEx data (Barrie et al., 2017), where the minor allele associated with increased *CHRNA3* mRNA expression. This brain region is strongly implicated in dopamine-dependent reward behaviors underlying addiction. The α3 subunit, in conjunction with α6 and β2 subunits, is responsible for the majority of the activity- or nicotine-evoked dopamine released in the putamen measured in nonhuman primates (Perez, O'Leary, Parameswaran, McIntosh, & Quik, 2009). An effect of rs1948 on *CHRNB4* expression is evident for constructs modeling different 3′UTR lengths of *CHRNB4* in vitro (Flora et al., 2013; Gallego, Cox, Laughlin, Stitzel, & Ehringer, 2013).

43.4 EPIGENETICS AND ENVIRONMENTAL FACTORS REGULATING CHRNA5/CHRNA3/CHRNB4

Environmental factors can dynamically regulate the methylation of cytosine nucleotides adjacent to guanine nucleotides (*CpGs*) in deoxyribonucleic acid (DNA). *CpG* islands at the *CHRNA5* promoter are hypermethylated in children exposed to adversity (Zhang, Wang, Kranzler, Zhao, & Gelernter, 2013). In an example of sex-specific gene–environment interaction, male carriers of the rs16969968 minor allele exposed to childhood adversity had increased risk for ND (Xie et al., 2012). Furthermore, significant methylation quantitative trait loci (meQTL) in the *A5/A3/B4* cluster are evident for specific *CpG* sites in this region across multiple tissues (Hancock et al., 2015; Ramsay et al., 2015; Volkov et al., 2016). These findings establish a nexus between genetic and epigenetic (including environmental) factors. Consequently, one might expect childhood adversity and the presence of meQTL polymorphisms to converge, leading to hypermethylation of the *CHRNA5* gene promoter, reduced *CHRNA5* mRNA expression, and increased ND risk. However, follow-up analyses that included expression, methylation, and genotypes from the same samples find no direct relationship between methylation and expression (Hancock et al., 2015; Ramsay et al., 2015). Beyond *CpG* methylation, epigenetic modulation of the *A5/A3/B4* gene cluster can occur through chemical modifications to histones and structural changes to chromatin. Research along these lines for the *A5/A3/B4* locus is sparse, but findings of a master regulatory element should intensify further efforts.

43.5 CLINICAL ASSOCIATIONS WITH THE *CHRNA5/CHRNA3/CHRNB4* REGULOME

Candidate gene and single-SNP studies identify the *A5/A3/B4* cluster as susceptibility locus for ND (Bierut et al., 2007; Saccone et al., 2007), later confirmed with genome-wide association studies (GWAS) on ND and related clinical phenotypes (Table 43.1) (Amos et al., 2008; Saccone et al., 2010; Tobacco and Genetics Consortium, 2010). A nonsynonymous SNP in *CHRNA5*, rs16969968 (D398N), strongly relates to ND. The variant reduces the agonist-evoked activity of channels in which it incorporates, without altering orthosteric binding of agonists or antagonists (George et al., 2012; Tammimäki et al., 2012). rs16969968 dampens the signaling of α5-containing receptors in the habenula, thereby decreasing aversion to addictive substances (Fowler, Lu, Johnson, Marks, & Kenny, 2011; Frahm et al., 2011; Jensen et al., 2015).

It is necessary to consider how the *A5/A3/B4* regulome associates with smoking behaviors in different ethnic populations with distinct LD structure and variations. In African Americans, rs16969968 has a similar effect on smoking behavior (Olfson et al., 2016); however, the minor allele is uncommon (1%–5%) in Africa and Asian populations. Other polymorphisms, such as rs1317286 and rs8040806 in *CHRNA3*, show associations with ND in African Americans (Li et al., 2010), but in Koreans, the nicotinic receptor cluster is only weakly associated with smoking behavior (Li et al., 2010). Additionally, regulatory polymorphisms are not well studied in these populations, making knowledge-driven haplotype analyses more difficult to construct and interpret.

Clusters of homologous genes with long haplotypes created by frequent SNPs in high LD suggest positive evolutionary selection, creating a local "regulome" (Barrie et al., 2017; Sadler et al., 2015). Since rs16969968 provides the most tangible evidence for ND risk, researchers have measured the combined effect of additional candidate SNPs in the context of rs16969968. Saccone et al. (2010) reported a reduced but significant protective effect of rs578776 on smoking after controlling for rs16969968 genotype. SNPs in the distal enhancer for *CHRNA5* and rs1948 convey ND risk only when considered with rs16969968 (Barrie et al., 2017; Smith et al., 2011). In short, four major haplotypes constructed from rs16969968, rs880395, and rs1948 convey risk in an allele-dose-dependent fashion (Barrie et al., 2017; Table 43.2), consistent with their known impacts on gene expression and ligand-related signaling properties (Table 43.3). Guided by functional effects of common SNPs in this locus, it is now possible to build haplotypes that reflect divergent nicotinic receptor biology conferred by the regulome.

TABLE 43.1 Reported Association of rs1051730, rs8034191, rs16969968, rs578776, and rs588765 With Nicotine Dependence and/or Related Phenotypes in European American Population

SNP	Location[a] (associated gene symbol)	Allele(MAF)	Functional annotation	Clinical association
rs8034191	15: 78513681 (*HYKK*)	T>C (0.37)	Intronic variant	Lung cancer
rs588765	15: 78573083 (*CHRNA5*)	C>T (0.4)	Intronic variant	Smoking quantity
rs16969968	15: 78590583 (*CHRNA5*)	G>A (0.36)	Nonsynonymous variant (D(Asp)->N(Asn))	Smoking quantity Nicotine dependence Lung cancer
rs578776	15: 78596058 (*CHRNA3*)	G>A (0.28)	3′-UTR	Smoking quantity
rs1051730	15:78601997 (*CHRNA3*)	G>A (0.36)	Synonymous variant	Nicotine dependence Lung cancer Smoking quantity

SNP, single-nucleotide polymorphism; *MAF*, minor allele frequency; *UTR*, untranslated region.
[a] *Based on GRCh39/hg38 assembly.*

TABLE 43.2 Haplotypes of Key SNPs, Effect on Smoking Behavior, and Distribution in Population

rs16969968/ rs880395/ rs1948	Effect on smoking behavior	Average % of population with haplotype
AGG	Increased risk	35
GAA	Moderate increased risk	31
GGG	Baseline risk	20
GAG	Decreased risk	11

TABLE 43.3 Effect of Key SNPs on mRNA Expression and Smoking Behavior

	rs16969968	rs880395	rs1948
Smoking risk allele	A	G	A
Functional effect	Decreased *CHRNA5* signaling	Decreased *CHRNA5* mRNA	Increased *CHRNA3* mRNA
Major allele	G	G	G

MINI-DICTIONARY OF TERMS

Enhancer A DNA region that binds proteins that facilitate interaction with a gene promoter to increase messenger RNA expression and can act in a tissue-specific manner.

Epigenetic Heritable and dynamic modifications to DNA affecting gene expression and phenotypic presentation, including but not limited to histone modifications and DNA methylation.

Epithalamic Brain regions adjacent to the thalamus, including the habenula and pineal gland.

Functional polymorphism A DNA variation that demonstrably affects biology. Examples include changes in gene expression or alterations in the encoded protein that affect ligand binding.

Haplotype A group of polymorphisms inherited on the same allele.

Linkage disequilibrium The nonrandom *co*-occurrence of two or more polymorphisms on the same allele across a population; e.g., polymorphisms that reliably predict the presence of other polymorphisms of the same allele frequency.

Nonsynonymous A polymorphism that changes an amino acid in the resulting protein.

Phenotype An observable trait that differs among individuals in a population due to genetics or environment.

Promoter A DNA region, typically immediately upstream of a gene, encoding protein binding sites necessary for initiating gene transcription.

Quantitative trait locus A polymorphic region of DNA that correlates with a quantitative phenotype, such as levels of gene expression or methylation.

Regulatory element A DNA region encoding protein binding sites that modulate RNA expression. They include enhancers and promoters.

Regulome The collection of regulatory factors that modulate the coordinated expression of a gene or genes. The regulome can vary based on context, such as developmental stage or tissue.

Key Facts of Gene Expression

- Genes express in a time- and tissue-dependent manner, guided by regulatory elements.
- Genetic polymorphisms can have dramatic effects on all aspects of gene expression, including transcription, alternative splicing, trafficking, and translation.
- Nearly all multiexon genes, including noncoding RNAs, undergo alternative splicing in humans to produce unique RNA isoforms that vary across tissues.
- Although genes are transcribed in the nucleus, they can be transported to specific subcellular locations where they perform regulatory functions or are locally translated.
- Protein-coding genes account for only a small proportion of the total number of transcribed genes, which also include noncoding RNAs such as small RNAs (microRNAs and small nucleolar RNA), long noncoding RNAs, and regulatory RNAs (transfer RNA and ribosomal RNA).

Summary Points

- The *CHRNA5/CHRNA3/CHRNB4* gene cluster is located in a region of tight linkage disequilibrium on chromosome 15q25.1.
- *CHRNA5*, *CHRNA3*, and *CHRNB4* are often coexpressed and form functional ligand-gated ion channels in the periphery and in brain regions directly related to reward processing.
- This nicotinic gene cluster harbors numerous regulatory elements controlling coordinated expression of these subunits, some of which also harbor genetic variants that further modulate expression.
- Polymorphisms in a distal enhancer significantly increase the expression of *CHRNA5* and *CHRNA3* in the brain and periphery, with the strongest effects noted for *CHRNA5*.
- A polymorphism in the 3′ untranslated region of *CHRNB4*, rs1948, modulates *CHRNA3* expression in a tissue-specific manner, significantly increasing expression in the striatum.
- Epigenetic factors, such as *CpG* methylation, are significantly associated with single-nucleotide polymorphisms in this gene cluster, but their significance for addiction is yet to be determined.
- The nonsynonymous polymorphism, rs16969968, drives the strongest and most reproducible single signal in clinical association studies on nicotine addiction.
- Understanding the functional effects of polymorphisms in the *CHRNA5/CHRNA3/CHRNB4* locus allows the creation of haplotypes with

rs16969968 for clinical analyses that reveal associations not observed by measuring associations with single variants.

References

Albers, K. M., Zhang, X. L., Diges, C. M., Schwartz, E. S., Yang, C. I., Davis, B. M., et al. (2014). Artemin growth factor increases nicotinic cholinergic receptor subunit expression and activity in nociceptive sensory neurons. *Molecular Pain, 10,* 31. https://doi.org/10.1186/1744-8069-10-31.

Amos, C. I., Wu, X., Broderick, P., Gorlov, I. P., Gu, J., Eisen, T., et al. (2008). Genome-wide association scan of tag SNPs identifies a susceptibility locus for lung cancer at 15q25.1. *Nature Genetics, 40* (5), 616–622. https://doi.org/10.1038/ng.109.

Barrie, E. S., Hartmann, K., Lee, S. H., Frater, J. T., Seweryn, M., Wang, D., et al. (2017). The CHRNA5/CHRNA3/CHRNB4 nicotinic receptor regulome: genomic architecture, regulatory polymorphisms, and clinical associations. *Human Mutation, 38*(1), 112–119. https://doi.org/10.1002/humu.23135.

Benfante, R., Flora, A., Di Lascio, S., Cargnin, F., Longhi, R., Colombo, S., et al. (2007). Transcription factor PHOX2A regulates the human alpha3 nicotinic receptor subunit gene promoter. *The Journal of Biological Chemistry, 282*(18), 13290–13302. https://doi.org/10.1074/jbc.M608616200.

Bierut, L. J., Madden, P. A., Breslau, N., Johnson, E. O., Hatsukami, D., Pomerleau, O. F., et al. (2007). Novel genes identified in a high-density genome wide association study for nicotine dependence. *Human Molecular Genetics, 16*(1), 24–35. https://doi.org/10.1093/hmg/ddl441.

Bigger, C. B., Casanova, E. A., & Gardner, P. D. (1996). Transcriptional regulation of neuronal nicotinic acetylcholine receptor genes. Functional interactions between Sp1 and the rat beta4 subunit gene promoter. *The Journal of Biological Chemistry, 272*(51), 32842–32848.

Bigger, C. B., Melnikova, I. N., & Gardner, P. D. (1997). Sp1 and Sp3 regulate expression of the neuronal nicotinic acetylcholine receptor beta4 subunit gene. *The Journal of Biological Chemistry, 272*(41), 25976–25982.

Campos-Caro, A., Carrasco-Serrano, C., Valor, L. M., Viniegra, S., Ballesta, J. J., & Criado, M. (1999). Multiple functional Sp1 domains in the minimal promoter region of the neuronal nicotinic receptor alpha5 subunit gene. *The Journal of Biological Chemistry, 274*(8), 4693–4701.

Doyle, G. A., Wang, M. J., Chou, A. D., Oleynick, J. U., Arnold, S. E., Buono, R. J., et al. (2011). In vitro and ex vivo analysis of CHRNA3 and CHRNA5 haplotype expression. *PLoS One. 6*(8), e23373, https://doi.org/10.1371/journal.pone.0023373.

Flora, A. V., Zambrano, C. A., Gallego, X., Miyamoto, J. H., Johnson, K. A., Cowan, K. A., et al. (2013). Functional characterization of SNPs in CHRNA3/B4 intergenic region associated with drug behaviors. *Brain Research, 1529,* 1–15. https://doi.org/10.1016/j.brainres.2013.07.017.

Fornasari, D., Battaglioli, E., Flora, A., Terzano, S., & Clementi, F. (1997). Structural and functional characterization of the human alpha3 nicotinic subunit gene promoter. *Molecular Pharmacology, 51*(2), 250–261.

Fowler, C. D., Lu, Q., Johnson, P. M., Marks, M. J., & Kenny, P. J. (2011). Habenular alpha5 nicotinic receptor subunit signalling controls nicotine intake. *Nature, 471*(7340), 597–601. https://doi.org/10.1038/nature09797.

Frahm, S., Slimak, M. A., Ferrarese, L., Santos-Torres, J., Antolin-Fontes, B., Auer, S., et al. (2011). Aversion to nicotine is regulated by the balanced activity of beta4 and alpha5 nicotinic receptor subunits in the medial habenula. *Neuron, 70*(3), 522–535. https://doi.org/10.1016/j.neuron.2011.04.013.

Gallego, X., Cox, R. J., Laughlin, J. R., Stitzel, J. A., & Ehringer, M. A. (2013). Alternative CHRNB4 3′-UTRs mediate the allelic effects of SNP rs1948 on gene expression. *PLoS One. 8*(5)e63699https://doi.org/10.1371/journal.pone.0063699.

Genzen, J. R., & McGehee, D. S. (2003). Short- and long-term enhancement of excitatory transmission in the spinal cord dorsal horn by nicotinic acetylcholine receptors. *Proceedings of the National Academy of Sciences of the United States of America, 100*(11), 6807–6812. https://doi.org/10.1073/pnas.1131709100.

George, A. A., Lucero, L. M., Damaj, M. I., Lukas, R. J., Chen, X., & Whiteaker, P. (2012). Function of human alpha3beta4alpha5 nicotinic acetylcholine receptors is reduced by the alpha5(D398N) polymorphism. *The Journal of Biological Chemistry, 287*(30), 25151–25162. https://doi.org/10.1074/jbc.M112.379339.

Getz, A. M., Xu, F., Visser, F., Persson, R., & Syed, N. I. (2017). Tumor suppressor menin is required for subunit-specific nAChR alpha5 transcription and nAChR-dependent presynaptic facilitation in cultured mouse hippocampal neurons. *Scientific Reports, 7*(1), 1768. https://doi.org/10.1038/s41598-017-01825-x.

GTEx Consortium. (2013). The genotype-tissue expression (GTEx) project. *Nature Genetics, 45*(6), 580–585. https://doi.org/10.1038/ng.2653.

Hancock, D. B., Wang, J. C., Gaddis, N. C., Levy, J. L., Saccone, N. L., Stitzel, J. A., et al. (2015). A multiancestry study identifies novel genetic associations with CHRNA5 methylation in human brain and risk of nicotine dependence. *Human Molecular Genetics, 24*(20), 5940–5954. https://doi.org/10.1093/hmg/ddv303.

Hone, A. J., McIntosh, J. M., Azam, L., Lindstrom, J., Lucero, L., Whiteaker, P., et al. (2015). Alpha-conotoxins identify the alpha3-beta4* subtype as the predominant nicotinic acetylcholine receptor expressed in human adrenal chromaffin cells. *Molecular Pharmacology, 88*(5), 881–893. https://doi.org/10.1124/mol.115.100982.

Jackson, K. J., Sanjakdar, S. S., Muldoon, P. P., McIntosh, J. M., & Damaj, M. I. (2013). The alpha3beta4* nicotinic acetylcholine receptor subtype mediates nicotine reward and physical nicotine withdrawal signs independently of the alpha5 subunit in the mouse. *Neuropharmacology, 70,* 228–235. https://doi.org/10.1016/j.neuropharm.2013.01.017.

Jensen, K. P., DeVito, E. E., Herman, A. I., Valentine, G. W., Gelernter, J., & Sofuoglu, M. (2015). A CHRNA5 smoking risk polymorphism decreases the aversive effects of nicotine in humans. *Neuropsychopharmacology, 40*(12), 2813–2821. https://doi.org/10.1038/npp.2015.131.

Li, M. D., Xu, Q., Lou, X. Y., Payne, T. J., Niu, T., & Ma, J. Z. (2010). Association and interaction analysis of polymorphisms in CHRNA5/CHRNA3/CHRNB4 gene cluster with nicotine dependence in African and European Americans. *American Journal of Medical Genetics Part B, Neuropsychiatric Genetics, 153B*(3), 745–756. https://doi.org/10.1002/ajmg.b.31043.

Li, M. D., Yoon, D., Lee, J. Y., Han, B. G., Niu, T., Payne, T. J., et al. (2010). Associations of polymorphisms in CHRNA5/A3/B4 gene cluster with smoking behaviors in a Korean population. *PLoS One. 5*(8) e12183, https://doi.org/10.1371/journal.pone.0012183.

Liu, Q., Melnikova, I. N., Hu, M., & Gardner, P. D. (1999). Cell type-specific activation of neuronal nicotinic acetylcholine receptor subunit genes by Sox10. *The Journal of Neuroscience, 19*(22), 9747–9755.

Mao, D., Yasuda, R. P., Fan, H., Wolfe, B. B., & Kellar, K. J. (2006). Heterogeneity of nicotinic cholinergic receptors in rat superior cervical and nodose Ganglia. *Molecular Pharmacology, 70*(5), 1693–1699. https://doi.org/10.1124/mol.106.027458.

Olfson, E., Saccone, N. L., Johnson, E. O., Chen, L. S., Culverhouse, R., Doheny, K., et al. (2016). Rare, low frequency and common coding polymorphisms in CHRNA5 and their contribution to nicotine dependence in European and African Americans. *Molecular Psychiatry, 21*(5), 601–607. https://doi.org/10.1038/mp.2015.105.

Perez, X. A., O'Leary, K. T., Parameswaran, N., McIntosh, J. M., & Quik, M. (2009). Prominent role of alpha3/alpha6beta2* nAChRs in regulating evoked dopamine release in primate putamen: effect of long-term nicotine treatment. *Molecular Pharmacology, 75*(4), 938–946. https://doi.org/10.1124/mol.108.053801.

Ramsay, J. E., Rhodes, C. H., Thirtamara-Rajamani, K., & Smith, R. M. (2015). Genetic influences on nicotinic alpha5 receptor (CHRNA5) CpG methylation and mRNA expression in brain and adipose tissue. *Genes Environment, 37,* 14. https://doi.org/10.1186/s41021-015-0020-x.

Saccone, N. L., Culverhouse, R. C., Schwantes-An, T. H., Cannon, D. S., Chen, X., Cichon, S., et al. (2010). Multiple independent loci at chromosome 15q25.1 affect smoking quantity: a meta-analysis and comparison with lung cancer and COPD. *PLoS Genet. 6*(8), https://doi.org/10.1371/journal.pgen.1001053.

Saccone, S. F., Hinrichs, A. L., Saccone, N. L., Chase, G. A., Konvicka, K., Madden, P. A., et al. (2007). Cholinergic nicotinic receptor genes implicated in a nicotine dependence association study targeting 348 candidate genes with 3713 SNPs. *Human Molecular Genetics, 16* (1), 36–49. https://doi.org/10.1093/hmg/ddl438.

Sadler, B., Haller, G., Edenberg, H., Tischfield, J., Brooks, A., Kramer, J., et al. (2015). Positive selection on loci associated with drug and alcohol dependence. *PLoS One. 10*(8)e0134393https://doi.org/10.1371/journal.pone.0134393.

Shih, P. Y., Engle, S. E., Oh, G., Deshpande, P., Puskar, N. L., Lester, H. A., et al. (2014). Differential expression and function of nicotinic acetylcholine receptors in subdivisions of medial habenula. *The Journal of Neuroscience, 34*(29), 9789–9802. https://doi.org/10.1523/JNEUROSCI.0476-14.2014.

Smith, R. M., Alachkar, H., Papp, A. C., Wang, D., Mash, D. C., Wang, J. C., et al. (2011). Nicotinic alpha5 receptor subunit mRNA expression is associated with distant 5′ upstream polymorphisms. *European Journal of Human Genetics, 19*(1), 76–83. https://doi.org/10.1038/ejhg.2010.120.

Tammimäki, A., Herder, P., Li, P., Esch, C., Laughlin, J. R., Akk, G., et al. (2012). Impact of human D398N single nucleotide polymorphism on intracellular calcium response mediated by alpha3beta4alpha5 nicotinic acetylcholine receptors. *Neuropharmacology, 63*(6), 1002–1011. https://doi.org/10.1016/j.neuropharm.2012.07.022.

Tobacco & Genetics Consortium (2010). Genome-wide meta-analyses identify multiple loci associated with smoking behavior. *Nature Genetics, 42*(5), 441–447. https://doi.org/10.1038/ng.571.

Vernallis, A. B., Conroy, W. G., & Berg, D. K. (1993). Neurons assemble acetylcholine receptors with as many as three kinds of subunits while maintaining subunit segregation among receptor subtypes. *Neuron, 10*(3), 451–464.

Volkov, P., Olsson, A. H., Gillberg, L., Jorgensen, S. W., Brons, C., Eriksson, K. F., et al. (2016). A genome-wide mQTL analysis in human adipose tissue identifies genetic polymorphisms associated with DNA methylation, gene expression and metabolic traits. *PLoS One. 11*(6)e0157776, https://doi.org/10.1371/journal.pone.0157776.

Wang, F., Gerzanich, V., Wells, G. B., Anand, R., Peng, X., Keyser, K., et al. (1996). Assembly of human neuronal nicotinic receptor alpha5 subunits with alpha3, beta2, and beta4 subunits. *The Journal of Biological Chemistry, 271*(30), 17656–17665.

Wang, J. C., Cruchaga, C., Saccone, N. L., Bertelsen, S., Liu, P., Budde, J. P., et al. (2009). Risk for nicotine dependence and lung cancer is conferred by mRNA expression levels and amino acid change in CHRNA5. *Human Molecular Genetics, 18*(16), 3125–3135. https://doi.org/10.1093/hmg/ddp231.

Wang, J. C., Grucza, R., Cruchaga, C., Hinrichs, A. L., Bertelsen, S., Budde, J. P., et al. (2009). Genetic variation in the CHRNA5 gene affects mRNA levels and is associated with risk for alcohol dependence. *Molecular Psychiatry, 14*(5), 501–510. https://doi.org/10.1038/mp.2008.42.

Xie, P., Kranzler, H. R., Zhang, H., Oslin, D., Anton, R. F., Farrer, L. A., et al. (2012). Childhood adversity increases risk for nicotine dependence and interacts with alpha5 nicotinic acetylcholine receptor genotype specifically in males. *Neuropsychopharmacology, 37*(3), 669–676. https://doi.org/10.1038/npp.2011.240.

Xu, X., Scott, M. M., & Deneris, E. S. (2006). Shared long-range regulatory elements coordinate expression of a gene cluster encoding nicotinic receptor heteromeric subtypes. *Molecular and Cellular Biology, 26*(15), 5636–5649. https://doi.org/10.1128/MCB.00456-06.

Yang, X., McDonough, J., Fyodorov, D., Morris, M., Wang, F., & Deneris, E. S. (1994). Characterization of an acetylcholine receptor alpha 3 gene promoter and its activation by the POU domain factor SCIP/Tst-1. *The Journal of Biological Chemistry, 269*(14), 10252–10264.

Yeh, J. J., Yasuda, R. P., Davila-Garcia, M. I., Xiao, Y., Ebert, S., Gupta, T., et al. (2001). Neuronal nicotinic acetylcholine receptor alpha3 subunit protein in rat brain and sympathetic ganglion measured using a subunit-specific antibody: regional and ontogenic expression. *Journal of Neurochemistry, 77*(1), 336–346.

Yokotani, K., Wang, M., Okada, S., Murakami, Y., & Hirata, M. (2000). Characterization of nicotinic acetylcholine receptor-mediated noradrenaline release from the isolated rat stomach. *European Journal of Pharmacology, 402*(3), 223–229.

Zhang, H., Wang, F., Kranzler, H. R., Zhao, H., & Gelernter, J. (2013). Profiling of childhood adversity-associated DNA methylation changes in alcoholic patients and healthy controls. *PLoS One. 8*(6) e65648, https://doi.org/10.1371/journal.pone.0065648.

Zhang, X. L., Albers, K. M., & Gold, M. S. (2015). Inflammation-induced increase in nicotinic acetylcholine receptor current in cutaneous nociceptive DRG neurons from the adult rat. *Neuroscience, 284*, 483–499. https://doi.org/10.1016/j.neuroscience.2014.10.018.

Zoli, M., Le Novere, N., Hill, J. A., Jr., & Changeux, J. P. (1995). Developmental regulation of nicotinic ACh receptor subunit mRNAs in the rat central and peripheral nervous systems. *The Journal of Neuroscience, 15*(3 Pt 1), 1912–1939.

CHAPTER

44

Brain, Nrf2, and Tobacco: Mechanisms and Countermechanisms Underlying Oxidative-Stress-Mediated Cerebrovascular Effects of Cigarette Smoking

Shikha Prasad, Taylor Liles†, Luca Cucullo†*

*Department of Neurology, Northwestern University Feinberg School of Medicine, Chicago, IL, United States

†Department of Pharmaceutical Sciences, TTUHSC, School of Pharmacy, Amarillo, TX, United States

Abbreviations

ARE antioxidant response element
BBB blood-brain barrier
COPD chronic obstructive pulmonary disease
CS cigarette smoke
CSE cigarette smoke extract
GSH glutathione
MF metformin
NQO-1 NAD(P)H quinone reductase I
NRF2 nuclear factor erythroid 2-related factor
Nrf2 nuclear factor erythroid 2-related factor
OS oxidative stress
ROS reactive oxygen species
SOD superoxide dismutase
TJ tight junction
TS tobacco smoke
ZO-1 zonulae occludentes-1

44.1 INTRODUCTION

Globally, tobacco use causes approximately 6 million deaths per year, and predictions report that with current trends, more than 8 million deaths are expected annually by 2030. Tobacco smoking (TS) is currently accountable for more than 480,000 deaths each year in the United States (US) and is the leading cause of preventable death in the United States. On average, smokers die 10 years earlier than nonsmokers, and if smoking continues at its current proportion among adolescents, 1 in every 13 Americans aged 17 years or younger is expected to die prematurely from a smoking-related illness. Even though there has been a marginal smoking decline of around 5% in recent years (2005 vs 2015), smokers still account for 15% of the US adult population. What is also concerning is that 41,000 out of 480,000 deaths result from secondhand smoke exposure. In addition, the past decade has witnessed a rise in the consumption of alternative vaping products branded as safer alternative to tobacco smoking although these claims are unsubstantiated by scientific evidences (CDC, 2016b).

Addiction to TS is primarily caused by nicotine. However, recent studies have also shown that nonnicotine components in tobacco such as anabasine, anatabine, and norharmane have addictive properties on their own and can further reinforce that of nicotine. Tobacco smoke contains more than 7000 compounds including 63 different carcinogens and a number of oxidative elements that can severely impact cell and tissue function and are prodromal to the onset of major health disorders (How Tobacco Smoke Causes Disease, 2010).

44.2 HEALTH STATISTICS

Annual deaths in the United States due to cigarette smoking can be statistically categorized as follows: lung cancer ~29%, other cancers ~8%, ischemic heart disease ~28%, chronic obstructive pulmonary disease (COPD) ~21%, stroke ~4%, and other diagnoses ~10% (Surgeon General's Report, 2015). Smoking causes about 90% of all lung cancer deaths in patients (both genders).

https://doi.org/10.1016/B978-0-12-813035-3.00044-7

For every individual who dies due to smoking, at least 30 people are affected and live with a severe smoking-related illness (CDC, 2016a).

Smokers in comparison to nonsmokers are 2–4 times more likely to suffer from coronary heart disease and stroke and approximately 25 times more likely to develop lung cancer. Further, smoking has been associated with the onset of diabetes mellitus, rheumatoid arthritis, pneumonia, asthma, blindness, hardening of the arteries, reduced fertility, and impairment of the immune system leading to enhanced risk and progression of infections of all kinds. The risk of developing diabetes is 30%–40% higher for smokers in comparison to nonsmokers, and the impact is dependent upon the number of cigarettes smoked. Smoking during pregnancy increases the risk of ectopic pregnancy, preterm delivery, stillbirth, low birth weight, orofacial clefts in infants, and sudden infant death syndrome (CDC, 2016b; The Health Consequences, 2014).

Furthermore, cigarette smoking is a prodromal risk factor for numerous cerebrovascular and neurological disorders including stroke (Surgeon General's Report, 2015), Alzheimer's disease (Cataldo, Prochaska, & Glantz, 2010), depression (Martini, Wagner, & Anthony, 2002), cognitive impairment, and vascular dementia (Anstey, von, Salim, & O'Kearney, 2007). The negative cerebrovascular and neurological impact of smoking is largely due to reactive oxygen species (ROS) of which tobacco smoke is highly enriched (Naik et al., 2014; Sobczak, Golka, & Szoltysek-Boldys, 2004) (see also Fig. 44.1), consequent inflammation (Arnson, Shoenfeld, & Amital, 2010), and blood-brain barrier (BBB) impairment (Rosenberg, 2012). As a matter of fact, smoking during pregnancy impacts the cerebrovascular development in the fetus (How Tobacco Smoke Causes Disease, 2010; The Health Consequences, 2014).

44.3 SMOKERS: HEALTH MANAGEMENT AND CURRENT CHALLENGES

Tobacco addiction and dependence makes it important to ensure that both effective behavioral and pharmacological cessation treatments are made available to smokers who want to quit (Galanti, 2008). Smoking cessation is the only effective way of restoring the well-being of smokers. According to the Centers for Disease Control and Prevention (CDC), 1 year of smoking cessation significantly decreases the risk of heart attack and stroke. The risk for lung cancer drops by half, 10 years after quitting smoking (Larzelere & Williams, 2012).

The Fagerström test for nicotine dependence helps in understanding the severity of addiction. The higher the total Fagerström score, the more intense is the patient's physical dependence on nicotine. Clinically, treatments targeting different aspects of nicotine addiction, such as reinforcement, withdrawal, and cue-associated learning, ranging from complete cessation to substitution with less harmful products are enforced accordingly. Pharmacological interventions such as nicotine replacement therapy and varenicline, a partial $\alpha4\beta2$ agonist (nicotinic acetylcholine receptor ligands), are used to reduce the cravings (Wilkes, 2008). In many cases, bupropion is used to dull the negative emotions associated with quitting (Wilkes, 2008). Unfortunately, there is no guideline to clinically treat these patients (ex-smokers) or people exposed to secondhand smoke (SHS) to alleviate the health impact of smoking. The Food and Nutrition Board of the National Academy of Sciences suggests a higher recommended dietary allowance (RDA) of vitamin C for smokers (35 mg/day more compared to nonsmokers) (ODS, 2016); however, the health benefit and overall impact of this regimen on chronic smokers is still uncertain. As for the prophylactic treatment with other antioxidants, the results are quite controversial and highly dependent upon the experimental settings, purity of the agent/s, and regimen of administration (Frei, 2004; Kaisar, Prasad, & Cucullo, 2015). Studies have reported the beneficial effects of a number of popular antioxidants and health supplements emphasizing their ROS scavenging and/or anti-inflammatory properties (Liu et al., 2014) such as vitamins, resveratrol, melatonin, and lipoic acid, without taking into consideration their impact on the growth and proliferation of the cancerous cells (Muller & Hengstermann, 2012).

Factors such as the length of smoking cessation for ex-smokers, the level of smoke exposure in the case of SHS, preestablished health conditions, and genetics (and epigenetics modification caused by chronic smoking) are few of the criteria that need to be evaluated to begin assessing the prophylactic and/or therapeutic impact of treatments aimed at chronic and former smokers (especially early stage ex-smokers) including those frequently subjected to secondhand tobacco smoke exposure.

Further alarming is that in the past decade, a number of alternative vaping products have hit the market, rapidly gaining consumers among adults and, especially, adolescents (FDA, 2016). Electronic nicotine delivery systems or e-cigarettes (e-cigs) have become the sought-after product partly due to the belief that they are much safer than traditional cigarettes. Moreover, it is well established that the association between TS and vascular endothelial dysfunctions (Chen, Chien, Chaung, Lii, & Wang, 2004; Naik et al., 2014; Raij, Demaster, & Jaimes, 2001) in a causative and dose-dependent manner (Gill et al., 1989) is largely related to the TS content of reactive oxygen species (ROS) (Naik et al., 2014; Panda, Chattopadhyay, Ghosh, Chattopadhyay, & Chatterjee, 1999), tissue oxidative stress (OS)-driven/tissue oxidative stress (OS)-driven inflammation (Arnson et al., 2010; Naik et al., 2014), and nicotine (Das, Gautam, Dey, Maiti, & Roy, 2009; Paulson et al., 2010). Current scientific opinion considers

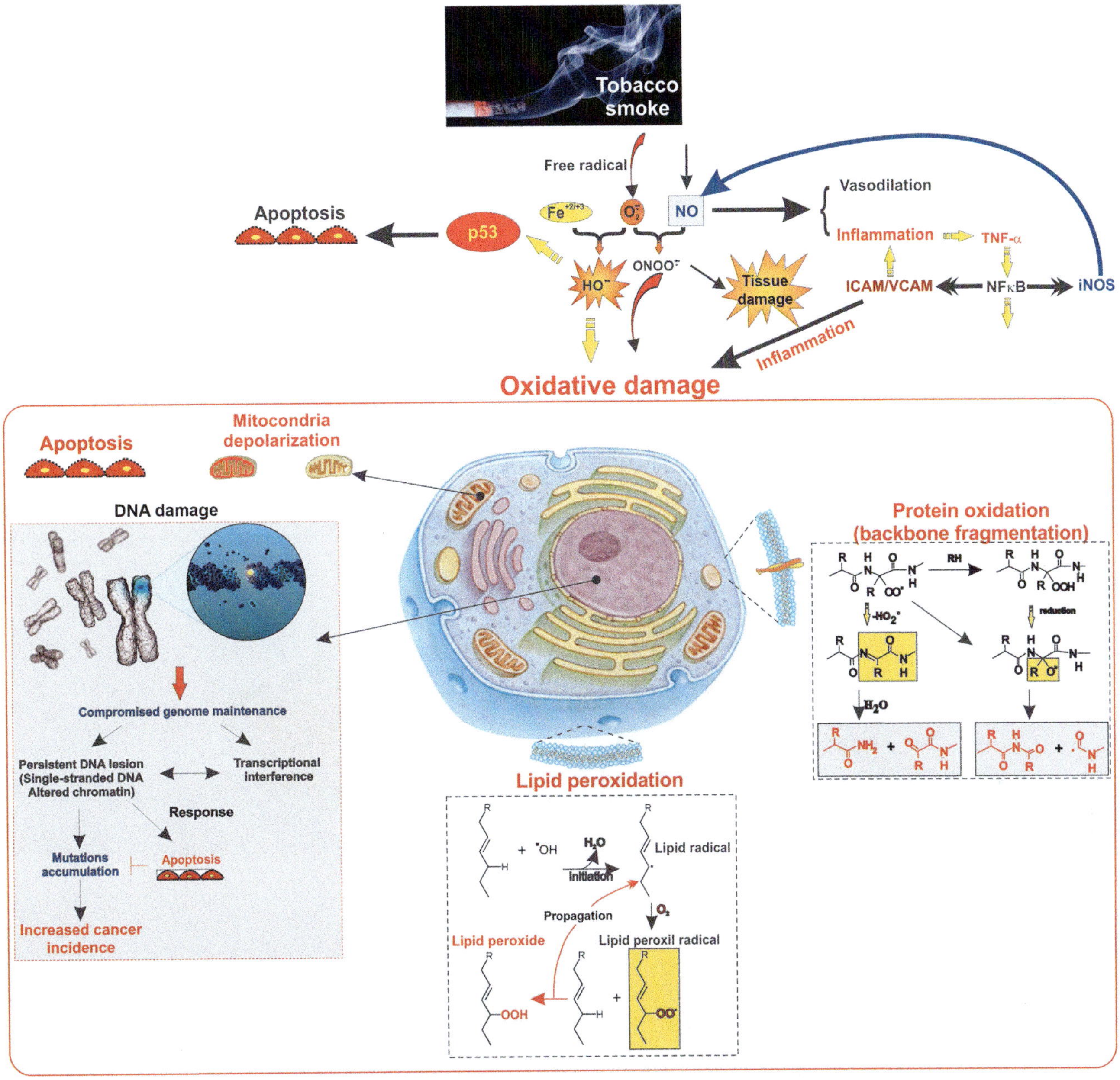

FIG. 44.1 TS-derived ROS promotes cellular inflammatory response and oxidative damage.
Schematic illustration depicting major OS-activated pathways through which exposure to ROS originated from tobacco combustion can induce lipid peroxidation, protein degradation, and DNA damage, thus leading to cellular damage and inflammation.

OS-mediated pathways to play a major role in the pathogenesis of these disorders, especially stroke (Cojocaru, Cojocaru, Sapira, & Ionescu, 2013). Preclinical studies have also shown that nicotine (the principal e-liquid's ingredient used in e-cigarettes) can also cause OS, exacerbation of cerebral ischemia, and secondary brain injury (Bradford, Stamatovic, Dondeti, Keep, & Andjelkovic, 2011; Li, Sun, Arrick, & Mayhan, 2016). Likewise, very recent studies have shown that chronic e-cig vaping could be prodromal to cerebrovascular impairment and promote cerebrovascular conditions favoring the onset of stroke and worsening postischemic brain injury (Kaisar et al., 2017; Prasad et al., 2017). The health impact of e-cig vaping is now a major public and regulatory concern as clearly demonstrated by the recent report from the US Surgeon General (E-Cigarette Use Among Youth and Young Adults, 2016) and the Food & Drug Administration (FDA).

44.4 CEREBROVASCULAR PERSPECTIVE: SMOKING, NICOTINE, AND BBB

At the cellular level, the BBB is constituted by vascular endothelium lining, the cerebral microvessels with the closely apposed astrocytic end-foot processes (Fig. 44.2) (Abbott, Patabendige, Dolman, Yusof, & Begley, 2010). The BBB endothelium is characterized by distinctive expression patterns (cellular polarization) of transmembrane transport systems to regulate traffic of substances in and out of the brain parenchyma. In addition, the expression of interendothelial tight junctions, lack of fenestrations, and minimal pinocytotic transport (which concur in the regulation and maintenance of the homeostasis of brain microenvironment) are unique features of the BBB endothelium. Tight junctions (TJs) between adjacent endothelial cells form a diffusion barrier, which selectively excludes most blood-borne (including electrolytes and other water-soluble compounds) and xenobiotic polar substances from entering the brain through paracellular routes (Abbott et al., 2010).

Specialized efflux transport mechanisms (e.g., P-glycoprotein, breast cancer resistance protein (BCRP), and multidrug resistance protein-4 (MRP-4)) are also in place to regulate the passage of amphipathic and hydrophobic molecules and protect the brain from potentially harmful substances (Abbott et al., 2010).

The BBB plays a vital role in protecting the brain against cigarette-smoke-generated oxidative and inflammatory stress. However, tobacco smoke contains various ROS species such as hydrogen peroxide, epoxides, nitrogen dioxide, and peroxynitrite ($ONOO^-$), all having oxidative damage and capable of causing BBB breakdown via TJ modification and activation of proinflammatory/antiinflammatory pathways (Naik et al., 2014; Pun, Lu, & Moochhala, 2009). Under normal conditions, ROS is scavenged or converted into less reactive molecules by the intracellular action of superoxide dismutase (SOD), catalase, glutathione (GSH) peroxidase (Hayes & Strange, 1995), or (extracellular) antioxidant vitamins such as ascorbic acid (vitamin C) and α-tocopherol (vitamin E) (Davitashvili, Museridze, Svanidze, Pavliashvili, & Sanikidze, 2010; Gallo et al., 2010). However, both active and passive smoking can generate ROS beyond the levels that the human body can effectively

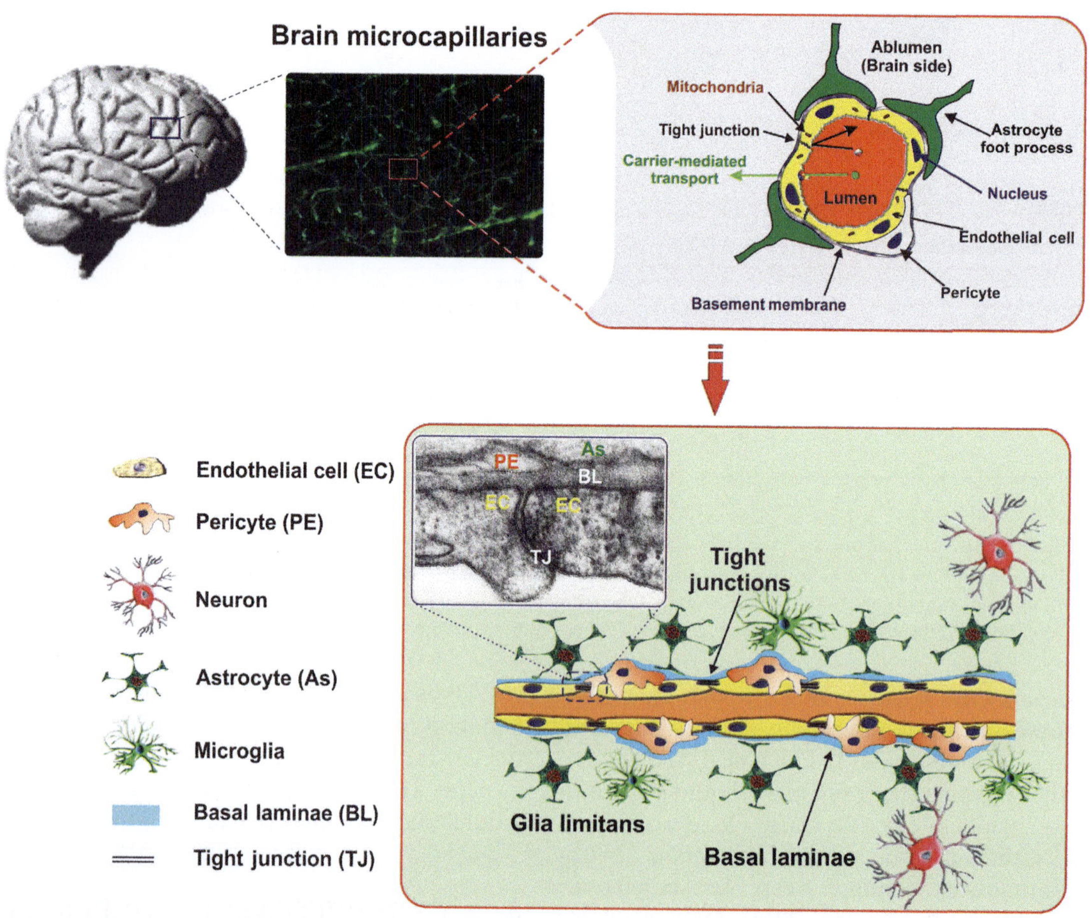

FIG. 44.2 Schematic illustration of the BBB anatomy.
A cross section of brain microcapillary revealing the innermost luminal compartment of endothelial cell (EC) on a support of basal lamina (BL) surrounded by an additional continuous envelop of pericyte (Pe) and astrocyte (As) foot processes that separate the blood vessels from the brain parenchyma. Tight junctions (TJs), present between the cerebral endothelial cells, selectively block paracellular trafficking of polar molecules into the brain, while active transport systems allow for the passage of nutrients and other essential factors.

eliminate. Supporting this fact, several studies have shown that chronic smokers suffer from antioxidant shortage due to its increased mobilization to combat systemic oxidative stress evoked by ROS-enriched TS (Dietrich et al., 2003; Tsuchiya et al., 2002). Further, the oxidation and inflammation induced by TS in animals and cells appear to get reduced (to different degrees) when supplementing certain antioxidants (Hossain, Mazzone, Tierney, & Cucullo, 2011; Kaisar et al., 2015). Recent study demonstrated that cigarette smoke extract (CSE) contains high concentrations of nitric oxide and hydrogen peroxide that corresponded with significant increase in cellular oxidative stress (measured using CellROX® Green Reagent) and induced BBB endothelial dysfunction and a strong inflammatory response. This included the (1) downregulation and redistribution of the TJ protein expression ZO-1 and occludin, (2) upregulation of VE-cadherin and claudin-5 (perhaps a cytoprotective counterregulatory activity), (3) enhanced expression of adhesion molecules, and (4) release of proinflammatory cytokines including interleukin-6 (IL-6) and matrix metalloproteinase-2 (MMP-2) (Naik et al., 2014). These events can lead to the oxidative damage to the brain vascular system and the BBB over a period of sustained (chronic) CS exposure (e.g., chronic smokers) and facilitate the pathogenesis and progression of cerebrovascular disorders (Hossain et al., 2009). Recently published data strongly suggest that a host of cerebrovascular and BBB impairments by chronic TS and e-cig exposure develop largely in response to alterations of the endogenous antioxidant response element (ARE) regulated by the nuclear factor erythroid 2-related factor (Nrf2), an ubiquitously expressed redox-sensitive transcription factor and a master regulator of constitutive or inducible expression of an elaborate network of molecular systems implicated in redox homeostasis via antioxidant, drug metabolism, antiinflammatory, detoxification, and radical scavenging functions (Hayes & Dinkova-Kostova, 2014a). Conversely, deletion or downregulation of Nrf2 has been shown to potentiate cell susceptibility to the toxic effects of prooxidant and inflammatory stimuli, including the collapse of cellular bioenergetics, thus leading to cellular and tissue damage (Holmstrom et al., 2013; Mitsuishi et al., 2012) (see Fig. 44.3).

44.5 THE Nrf2 PATHWAY AND ITS IMPLICATIONS IN TS-INDUCED CEREBROVASCULAR DYSFUNCTIONS

Nrf2 is rapidly polyubiquitinated through the kelch-like ECH-associated protein 1(keap1)/cullin-3/Rbx-1 and degraded by the 26S proteasome in cellular cytoplasm under normal conditions. However, cellular homeostasis is maintained by basal accumulation of the Nrf2 in the nucleus that mediates normal expression of ARE-dependent genes. Under stress conditions or in the presence of ARE inducers, polyubiquitination is inhibited resulting in the activation of Nrf2. An increase in endogenous p62 (a selective substrate of autophagy) causes the inhibition of KEAP1-mediated Nrf2 ubiquitination and degradation leading to the activation of Nrf2 and induction of ARE-dependent genes. Nrf2 is also regulated by phosphorylation (by various protein kinases (e.g., PKC and AMPK)) and acetylation/deacetylation processes. Other than these, histone phosphorylation and acetylation, methylation of CpG islands (a cytosine (C) base followed immediately by a guanine (G) base), and synthesis of specific miRNAs (epigenetic

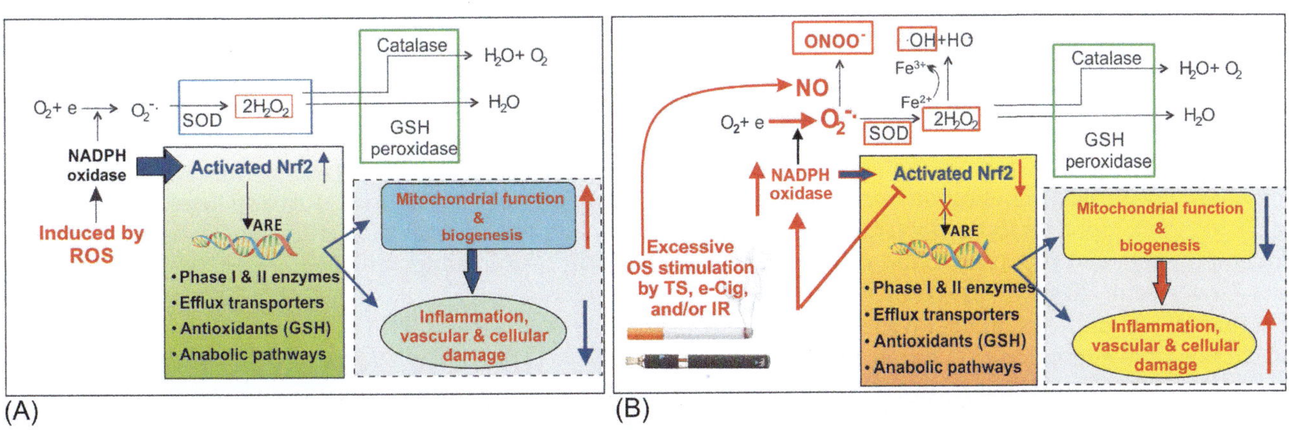

FIG. 44.3 Activation of the cellular antioxidant response element under normal and stress condition.
Panel A: Under normal conditions, the response to injury is adaptive, designed to restore homoeostasis and to protect the cell from further injury.
Panel B: In response to TS-promoted oxidative stress, NADPH oxidase is activated, producing an excess of O2—which in the presence of nitric oxide (NO also, abundant in TS) results in the formation of peroxynitrite ($ONOO^-$). Furthermore, the excess of H_2O_2 leads to the formation of hydroxyl radicals (OH; Fenton's reaction). Dysregulation of Nrf2 activity prevents the cells from responding to these OS stimuli leading to the impairments of mitochondrial function and biogenesis and cellular damage.

mechanisms) are additional means by which cells can regulate the levels of Nrf2 (Hayes & Dinkova-Kostova, 2014b; Sandberg, Patil, D'Angelo, Weber, & Mallard, 2014).

Exogenous Nrf2-ARE inducers are derived from a variety of sources ranging from natural phytochemical derivatives (e.g., oleanolic acid, sulforaphane, genistein, and curcumin to name some) to synthetic products such as acetaminophen and nimesulide. In certain cases, these agents have been found to be more potent than endogenous inducers such as prostaglandins. To date, many ARE inducers are being heavily examined for preventive and therapeutic applications in animals and humans (Ma & He, 2012).

44.6 TARGETING Nrf2 FOR PREVENTING CIGARETTE SMOKE-INDUCED CEREBROVASCULAR DYSFUNCTIONS

Proof-of-concept experiments have demonstrated that Nrf2-deficient mice developed early and extensive emphysema upon chronic CS exposure for 6 months (Rangasamy et al., 2004). Further, the synthetic triterpenoid 1[2-Cyano-3,12-dioxooleana-1,9(11)-dien-28-oyl]imidazole (CDDO-Im), a known Nrf2 inducer, significantly reduced lung oxidative stress, alveolar destruction, and emphysema caused by chronic CS exposure (Sussan et al., 2009). TS per se contains countless monofunctional and bifunctional Nrf2 inducers, which have shown to activate the expression of numerous cytoprotective genes in acute in vitro experimental settings. However, chronic TS exposure both in vitro and in vivo has been shown to negatively impact Nrf2 levels, thus increasing the oxidative burden of cells and tissues. For example, Goven et al. showed that TS condensate activated nuclear localization of Nrf2 and caused subsequent increase in HO-1 expression after 6 h of exposure in human macrophage THP-1 cells followed by decreased Nrf2 nuclear localization and HO-1 expression after 72 h (Goven, Boutten, Lecon-Malas, Boczkowski, & Bonay, 2009). His group later also reported that the Nrf2 protein level decreased significantly in whole lung tissue and alveolar macrophages (cytosol and nucleus) in patients with emphysema in comparison to those without emphysema (Goven et al., 2008).

In this context of the brain microvascular system, several studies have attempted to determine the role of Nrf2 in preserving the functional integrity of the BBB and prevent the onset of cerebrovascular dysfunction associated with various CNS pathologies (Li et al., 2014; Zhao, Moore, Redell, & Dash, 2007). For example, previous findings from Zhao et al. (Zhao et al., 2007) have suggested that pharmacological activation of Nrf2 signaling post brain injury can restore the loss of tight junctions (TJs) and prevent BBB disruption. Similarly, activation of the Nrf2-ARE system has been suggested to prevent BBB breakdown and neurological dysfunction following ischemic stroke (Alfieri et al., 2013). Nrf2 importance with respect to BBB integrity and viability has also been recently demonstrated both in vitro and in vivo, whereas induction of Nrf2 expression levels and activation (nuclear translocation) by metformin promoted the upregulation of TJ proteins, while silencing Nrf2 expression abrogated their expression (Sajja, Green, & Cucullo, 2015).

Overall, these findings suggest that targeting the Nrf2 system could be a viable (although still poorly explored) therapeutic option to develop novel and more effective strategies to prevent neurovascular dysfunction and secondary CNS injury in response not only to TS and similar products but also to several other OS-related pathogenic stimuli. Thus, it is not surprising that Nrf2 activators (e.g., sulforaphane, curcumin, and acetaminophen) are now receiving substantial interest from the pharmaceutical industry for the development of preventive and therapeutic applications in animals and humans (Crunkhorn, 2012). Very recently, metformin (a well-established antidiabetic drug) has been shown to effectively prevent TS from impairing BBB function and promote procoagulant conditions that enhance the risk of stroke. The effect appears to be mediated by positive modulation (activation) of Nrf2-ARE functions (Ashabi, Khalaj, Khodagholi, Goudarzvand, & Sarkaki, 2014; Kaisar et al., 2017; Prasad et al., 2017).

In conclusion, further in-depth study of the mechanisms involved in TS-induced vascular and cerebrovascular impairments, especially in view of the host of alternative products released on the market, will be necessary. Identification of vascular markers in smokers, ex-smokers, and secondhand smokers indicating the level of oxidative stress and the associated cerebrovascular damage will be quite useful for risk assessment of the development of CNS disorders and begin clinical regimen to reduce risk and quit smoking. This also includes the identification of conditions such as epigenetic mutations in smokers (Nrf2 mutation has been linked to progressive growth of cancerous cells upon further stimulation) that indicate further susceptibility of OS damage and/or exclusion from clinical treatments based on Nrf2-ARE induction.

MINI-DICTIONARY OF TERMS

Agonist A chemical that activates a receptor upon binding.
Bifunctional A chemical structure having two highly reactive sites.
Blood-brain barrier A highly selective vascular barrier present at the brain microcapillary level.
Ectopic pregnancy Fetus develops outside the uterus, mostly in the fallopian tube.
Efflux Pumping out of a substance.
Epigenetic Changes in gene expression that are heritable.

Monofunctional A chemical structure having a single highly reactive site.
Mutation Refers to changes in genetic structure.
Polyubiquitinated A process to mark a protein for degradation.
Proinflammatory Agents that lead to inflammation.
Prophylactic Preventative.
Reactive oxygen species Chemically reactive molecules containing oxygen.
Stillbirth Dead fetus after 24 weeks of pregnancy.
Xenobiotic A substance that is foreign to the body.

Key Facts of Tobacco Smoke

- *Nicotiana tabacum* is a species of herbaceous plants (Solanaceae) indigenous to the Americas, Australia, South West Africa, and South Pacific used for the production of tobacco leaf to be used in the manufacturing of cigarettes.
- Tobacco smoke contains about 7000 chemicals of which 60 components have established a human inhalation risk value for cancer and 38 noncancer inhalation risk values.
- Noncancer risks involve direct nasal lesions; respiratory effects; chronic active inflammation and lung fibrosis; neurological disorders; and effects on digestion, the liver/kidney, and reproduction.
- The main addictive component of tobacco is nicotine that can reach the brain in 10–20 s.

Key Facts of Nicotine

- Nicotine is the principal tobacco alkaloid, occurring to the extent of about 1.5% by weight in commercial cigarette tobacco and comprising about 95% of the total alkaloid content.
- Nicotine per se is well absorbed through the small airways of the respiratory tract and alveoli of the lung and skin (not absorbed from the stomach but well absorbed in the small intestine).
- After absorption, nicotine enters the bloodstream where, at pH 7.4, it is about 69% ionized and 31% unionized.
- Smoking a cigarette delivers nicotine rapidly to the pulmonary venous circulation, from which it moves quickly to the systemic arterial circulation and to the brain.
- Nicotine is primarily metabolized in the liver, and the main metabolite is cotinine, which is commonly used as a biomarker for exposure to tobacco smoke.

Key Facts of Nuclear Factor Erythroid 2-Related Factor (Nrf2)

- Nrf2 is a transcription factor that regulates the expression of antioxidant proteins that protect against oxidative damage triggered by injury and inflammations.
- Nrf2 is also a master regulator of detoxification and cellular defense gene expression.
- Nrf2 is ubiquitously expressed with the highest concentrations in the brain, kidney, muscle, heart, lung, and liver.

Summary Points

- This chapter focuses on the impact of tobacco smoke on the cerebrovascular system and the role of nuclear factor erythroid 2-related factor (Nrf2) as a coping mechanism against TS-induced oxidative stress.
- Current scientific opinion considers OS to play a major role in the pathogenesis of major cerebrovascular and neuroinflammatory disorders associated with TS including stroke, vascular dementia, small vessel ischemic disease, and Alzheimer's disease (AD).
- Chronic TS exposure compromises the antioxidant response system at the cerebrovascular level, thus becoming a strong prodromal factor for the onset on neurological disorders.
- Nrf2 is a redox-sensitive transcription factor that acts as a master regulator of cellular antioxidant defense.
- Nrf2 activity is critical for cellular redox homeostasis, and Nrf2 signaling deficits such as that observed in response to chronic TS exposure have been implicated in various CNS disorders.

References

Abbott, N. J., Patabendige, A. A., Dolman, D. E., Yusof, S. R., & Begley, D. J. (2010). Structure and function of the blood-brain barrier. *Neurobiology of Disease*, *37*, 13–25.

Alfieri, A., Srivastava, S., Siow, R. C., Cash, D., Modo, M., Duchen, M. R., et al. (2013). Sulforaphane preconditioning of the Nrf2/HO-1 defense pathway protects the cerebral vasculature against blood-brain barrier disruption and neurological deficits in stroke. *Free Radical Biology & Medicine*, *65*, 1012–1022.

Anstey, K. J., von, S. C., Salim, A., & O'Kearney, R. (2007). Smoking as a risk factor for dementia and cognitive decline: a meta-analysis of prospective studies. *American Journal of Epidemiology*, *166*, 367–378.

Arnson, Y., Shoenfeld, Y., & Amital, H. (2010). Effects of tobacco smoke on immunity, inflammation and autoimmunity. *Journal of Autoimmunity*, *34*, J258–J265.

Ashabi, G., Khalaj, L., Khodagholi, F., Goudarzvand, M., & Sarkaki, A. (2014). Pre-treatment with metformin activates Nrf2 antioxidant pathways and inhibits inflammatory responses through induction of AMPK after transient global cerebral ischemia. *Metabolic Brain Disease 30*(3), 747–754. (2014). https://doi.org/10.1007/s11011-014-9632-2.

Bradford, S. T., Stamatovic, S. M., Dondeti, R. S., Keep, R. F., & Andjelkovic, A. V. (2011). Nicotine aggravates the brain postischemic inflammatory response. *American Journal of Physiology—Heart and Circulatory Physiology*, *300*, H1518–H1529.

Cataldo, J. K., Prochaska, J. J., & Glantz, S. A. (2010). Cigarette smoking is a risk factor for Alzheimer's disease: an analysis controlling for tobacco industry affiliation. *Journal of Alzheimer's Disease*, *19*, 465–480.

CDC (Ed.). (2016). *Fast facts* (12-20-2016). https://www.cdc.gov/tobacco/data_statistics/fact_sheets/fast_facts/index.htm.

CDC (Ed.). (2016). *Health effects of cigarette smoking* (12-1-2016). https://www.cdc.gov/tobacco/data_statistics/fact_sheets/health_effects/effects_cig_smoking/.

Chen, H. W., Chien, M. L., Chaung, Y. H., Lii, C. K., & Wang, T. S. (2004). Extracts from cigarette smoke induce DNA damage and cell adhesion molecule expression through different pathways. *Chemico-Biological Interactions, 150*, 233–241.

Cojocaru, I. M., Cojocaru, M., Sapira, V., & Ionescu, A. (2013). Evaluation of oxidative stress in patients with acute ischemic stroke. *Romanian Journal of Internal Medicine, 51*, 97–106.

Crunkhorn, S. (2012). Deal watch: abbott boosts investment in NRF2 activators for reducing oxidative stress. *Nature Reviews Drug Discovery, 11*, 96.

Das, S., Gautam, N., Dey, S. K., Maiti, T., & Roy, S. (2009). Oxidative stress in the brain of nicotine-induced toxicity: protective role of Andrographis paniculata Nees and vitamin E. *Applied Physiology, Nutrition, and Metabolism, 34*, 124–135.

Davitashvili, D. T., Museridze, D. P., Svanidze, I. K., Pavliashvili, N. S., & Sanikidze, T. V. (2010). Correction of oxidative stress in the rat brain cortical cellular culture with vitamines E and C. *Georgian Medical News*, (180), 56–60.

Dietrich, M., Block, G., Norkus, E. P., Hudes, M., Traber, M. G., Cross, C. E., et al. (2003). Smoking and exposure to environmental tobacco smoke decrease some plasma antioxidants and increase gamma-tocopherol in vivo after adjustment for dietary antioxidant intakes. *The American Journal of Clinical Nutrition, 77*, 160–166.

E-Cigarette Use Among Youth and Young Adults. (2016). *A report of the surgeon general.* U.S.Department of health and human services.

FDA (2016). *Youth tobacco use: results from the 2014 national youth tobacco survey.*

Frei, B. (2004). Efficacy of dietary antioxidants to prevent oxidative damage and inhibit chronic disease. *Journal of Nutrition, 134*, 3196S–3198S.

Galanti, L. M. (2008). Tobacco smoking cessation management: integrating varenicline in current practice. *Vascular Health and Risk Management, 4*, 837–845.

Gallo, C., Renzi, P., Loizzo, S., Loizzo, A., Piacente, S., Festa, M., et al. (2010). Potential therapeutic effects of vitamin e and C on placental oxidative stress induced by nicotine: an in vitro evidence. *Open Biochemistry Journal, 4*, 77–82.

Gill, J. S., Shipley, M. J., Tsementzis, S. A., Hornby, R., Gill, S. K., Hitchcock, E. R., et al. (1989). Cigarette smoking. A risk factor for hemorrhagic and nonhemorrhagic stroke. *Archives of Internal Medicine, 149*, 2053–2057.

Goven, D., Boutten, A., Lecon-Malas, V., Boczkowski, J., & Bonay, M. (2009). Prolonged cigarette smoke exposure decreases heme oxygenase-1 and alters Nrf2 and Bach1 expression in human macrophages: roles of the MAP kinases ERK(1/2) and JNK. *FEBS Letters, 583*, 3508–3518.

Goven, D., Boutten, A., Lecon-Malas, V., Marchal-Somme, J., Amara, N., Crestani, B., et al. (2008). Altered Nrf2/Keap1-Bach1 equilibrium in pulmonary emphysema. *Thorax, 63*, 916–924.

Hayes, J. D., & Dinkova-Kostova, A. T. (2014a). The Nrf2 regulatory network provides an interface between redox and intermediary metabolism. *Trends in Biochemical Sciences, 39*, 199–218.

Hayes, J. D., & Strange, R. C. (1995). Potential contribution of the glutathione S-transferase supergene family to resistance to oxidative stress. *Free Radical Research, 22*, 193–207.

Holmstrom, K. M., Baird, L., Zhang, Y., Hargreaves, I., Chalasani, A., Land, J. M., et al. (2013). Nrf2 impacts cellular bioenergetics by controlling substrate availability for mitochondrial respiration. *Biology Open, 2*, 761–770.

Hossain, M., Mazzone, P., Tierney, W., & Cucullo, L. (2011). In vitro assessment of tobacco smoke toxicity at the BBB: do antioxidant supplements have a protective role? *BMC Neuroscience, 12*, 92.

Hossain, M., Sathe, T., Fazio, V., Mazzone, P., Weksler, B., Janigro, D., et al. (2009). Tobacco smoke: a critical etiological factor for vascular impairment at the blood-brain barrier. *Brain-Research, 1287*, 192–205.

How Tobacco Smoke Causes Disease. (2010). *The biology and behavioral basis for smoking-attributable disease: a report of the surgeon general. Chapter 3.* Publications and Reports of the Surgeon General.

Kaisar, M. A., Prasad, S., & Cucullo, L. (2015). Protecting the BBB endothelium against cigarette smoke-induced oxidative stress using popular antioxidants: are they really beneficial? *Brain-Research, 1627*, 90–100.

Kaisar, M. A., Villalba, H., Prasad, S., Liles, T., Sifat, A. E., Sajja, R. K., et al. (2017). Offsetting the impact of smoking and e-cigarette vaping on the cerebrovascular system and stroke injury: is metformin a viable countermeasure? *Redox Biology, 13*, 353–362.

Larzelere, M. M., & Williams, D. E. (2012). Promoting smoking cessation. *American Family Physician, 85*, 591–598.

Li, C., Sun, H., Arrick, D. M., & Mayhan, W. G. (2016). Chronic nicotine exposure exacerbates transient focal cerebral ischemia-induced brain injury. *Journal of Applied Physiology, 120*, 328–333.

Li, T., Wang, H., Ding, Y., Zhou, M., Zhou, X., Zhang, X., et al. (2014). Genetic elimination of Nrf2 aggravates secondary complications except for vasospasm after experimental subarachnoid hemorrhage in mice. *Brain Research, 1558*, 90–99.

Liu, H., Ren, J., Chen, H., Huang, Y., Li, H., Zhang, Z., et al. (2014). Resveratrol protects against cigarette smoke-induced oxidative damage and pulmonary inflammation. *Journal of Biochemical and Molecular Toxicology, 28*, 465–471.

Ma, Q., & He, X. (2012). Molecular basis of electrophilic and oxidative defense: promises and perils of Nrf2. *Pharmacological Reviews, 64*, 1055–1081.

Martini, S., Wagner, F. A., & Anthony, J. C. (2002). The association of tobacco smoking and depression in adolescence: evidence from the United States. *Substance Use & Misuse, 37*, 1853–1867.

Mitsuishi, Y., Taguchi, K., Kawatani, Y., Shibata, T., Nukiwa, T., Aburatani, H., et al. (2012). Nrf2 redirects glucose and glutamine into anabolic pathways in metabolic reprogramming. *Cancer Cell, 22*, 66–79.

Muller, T., & Hengstermann, A. (2012). Nrf2: friend and foe in preventing cigarette smoking-dependent lung disease. *Chemical Research in Toxicology, 25*, 1805–1824.

Naik, P., Fofaria, N., Prasad, S., Sajja, R. K., Weksler, B., Couraud, P. O., et al. (2014). Oxidative and pro-inflammatory impact of regular and denicotinized cigarettes on blood brain barrier endothelial cells: is smoking reduced or nicotine-free products really safe? *BMC Neuroscience, 15*, 51.

ODS. (Ed.), (2016). *Vitamin C* [2-11-2016]. https://ods.od.nih.gov/factsheets/VitaminC-HealthProfessional/#en8.

Panda, K., Chattopadhyay, R., Ghosh, M. K., Chattopadhyay, D. J., & Chatterjee, I. B. (1999). Vitamin C prevents cigarette smoke induced oxidative damage of proteins and increased proteolysis. *Free Radical Biology & Medicine, 27*, 1064–1079.

Paulson, J. R., Yang, T., Selvaraj, P. K., Mdzinarishvili, A., Van der Schyf, C. J., Klein, J., et al. (2010). Nicotine exacerbates brain edema during in vitro and in vivo focal ischemic conditions. *Journal of Pharmacology and Experimental Therapeutics, 332*, 371–379.

Prasad, S., Sajja, R. K., Kaisar, M. A., Park, J. H., Villalba, H., Liles, T., et al. (2017). Role of Nrf2 and protective effects of metformin against tobacco smoke-induced cerebrovascular toxicity. *Redox Biology, 12*, 58–69.

Pun, P. B., Lu, J., & Moochhala, S. (2009). Involvement of ROS in BBB dysfunction. *Free Radical Research, 43*, 348–364.

Raij, L., Demaster, E. G., & Jaimes, E. A. (2001). Cigarette smoke-induced endothelium dysfunction: role of superoxide anion. *Journal of Hypertension, 19*, 891–897.

Rangasamy, T., Cho, C. Y., Thimmulappa, R. K., Zhen, L., Srisuma, S. S., Kensler, T. W., et al. (2004). Genetic ablation of Nrf2 enhances susceptibility to cigarette smoke-induced emphysema in mice. *Journal of Clinical Investigation, 114*, 1248–1259.

Rosenberg, G. A. (2012). Neurological diseases in relation to the blood-brain barrier. *Journal of Cerebral Blood Flow & Metabolism, 32*, 1139–1151.

Sajja, R. K., Green, K. N., & Cucullo, L. (2015). Altered Nrf2 signaling mediates hypoglycemia-induced blood-brain barrier endothelial dysfunction in vitro. *PLoS One, 10*, e0122358.

Sandberg, M., Patil, J., D'Angelo, B., Weber, S. G., & Mallard, C. (2014). NRF2-regulation in brain health and disease: implication of cerebral inflammation. *Neuropharmacology, 79*, 298–306.

Sobczak, A., Golka, D., & Szoltysek-Boldys, I. (2004). The effects of tobacco smoke on plasma alpha- and gamma-tocopherol levels in passive and active cigarette smokers. *Toxicology Letters, 151*, 429–437.

Surgeon General's Report. (2015). *Surgeon general's report: the health consequences of smoking—50 years of progress.* .

Sussan, T. E., Rangasamy, T., Blake, D. J., Malhotra, D., El-Haddad, H., Bedja, D., et al. (2009). Targeting Nrf2 with the triterpenoid CDDO-imidazolide attenuates cigarette smoke-induced emphysema and cardiac dysfunction in mice. *Proceedings of the National Academy of Sciences of the United States of America, 106*, 250–255.

The Health Consequences. (2014). *The health consequences of smoking-50 years of progress: a report of the surgeon general.* .

Tsuchiya, M., Asada, A., Kasahara, E., Sato, E. F., Shindo, M., & Inoue, M. (2002). Smoking a single cigarette rapidly reduces combined concentrations of nitrate and nitrite and concentrations of antioxidants in plasma. *Circulation, 105*, 1155–1157.

Wilkes, S. (2008). The use of bupropion SR in cigarette smoking cessation. *International Journal of Chronic Obstructive Pulmonary Disease, 3*, 45–53.

Zhao, J., Moore, A. N., Redell, J. B., & Dash, P. K. (2007). Enhancing expression of Nrf2-driven genes protects the blood brain barrier after brain injury. *The Journal of Neuroscience, 27*, 10240–10248.

CHAPTER

45

Effects of Nicotine and Histone Deacetylase Inhibitors on the Brain

Maria Paula Faillace, Ramon O. Bernabeu

Departamento de Fisiología e Instituto de Fisiología y Biofísica (IFIBIO-Houssay, UBA-CONICET), Universidad de Buenos Aires (UBA), Paraguay 2155, Buenos Aires, Argentina

Abbreviations

CPA conditioning place aversion
CpG cytosine-guanine island
CPP conditioning place preference
CREB cyclic-AMP-responsive element-binding protein
DNMT DNA methyltransferase
dStr dorsal striatum
H3 histone 3
H4 histone 4
HAT histone acetyltransferase
HDAC histone deacetylase
HDACi HDAC inhibitor
NaB sodium butyrate
Nacc nucleus accumbens
nAChR nicotinic acetylcholine receptor
NOR novel object recognition
OR object recognition
pCREB phosphorylated CREB
PFC prefrontal cortex
PhB phenylbutyrate
PrL prelimbic frontal area
TSA trichostatin A
VTA ventral tegmental area

45.1 INTRODUCTION

Genetic information is stored in extremely long linear sequences of nucleotides, so the genetic material in eukaryotes is arranged into complex structures, which efficiently pack DNA within the cell nucleus (Horn & Peterson, 2002). DNA chains together with nuclear proteins form the chromatin. The histones (H) are the chief structural and constitutive regulatory proteins in chromatin. One hundred and forty-six base pairs of DNA are wrapped twice around an octamer of histones. Each octamer consists of two subunits of each type of histone: H2A, H2B, H3, and H4. This basic structural unit of chromatin is termed nucleosome.

Epigenesis involves effects determined by a huge array of chromatin changes, which are principally induced by DNA methylation and histone postranslational modifications (Jiang et al., 2008). Other processes can also control gene transcription by chromatin structural remodeling, and all epigenetic changes are heritable and reversible. However, no changes in DNA or protein sequence occur during epigenetic regulation. DNA methylation is the core of long-term epigenetic changes, and for several years, DNA methylation was conceived only as cytosine methylation. However, a few years ago, it was found that guanine and adenine can be methylated (O'Brown & Greer, 2016). Furthermore, histone enzymatic modifications are the most common and studied epigenetic changes that can induce long- and short-term changes in the chromatin structure. Histones have a tail that extends out of the octamer core of the nucleosome. The amino-terminal tails undergo postranslational modifications including acetylation, methylation, phosphorylation, ADP-ribosylation, farnesylation, proline isomerization, and ubiquitination (Tollefsbol, 2011). These chemical groups modify histone-DNA and histone-protein interactions remodeling chromatin with consequences for transcription and DNA replication and repair (Jenuwein & Allis, 2001). Hence, gene expression can be driven by the acetylation and methylation of histones, which modifies DNA-protein associations and chromatin structure (Fig. 45.1). Chromatin relaxed or condensed forms favor or reduce transcription, respectively. The acetylation and demethylation of residues in histones induce DNA relaxation and promote transcription factor and remodeling protein interaction (Kouzarides, 2007; Marmorstein & Zhou, 2014).

In this chapter, we analyzed epigenetic mechanisms that involve histone acetylation. Moreover, we described published results about histone deacetylase inhibitor

Neuroscience of Nicotine
https://doi.org/10.1016/B978-0-12-813035-3.00045-9

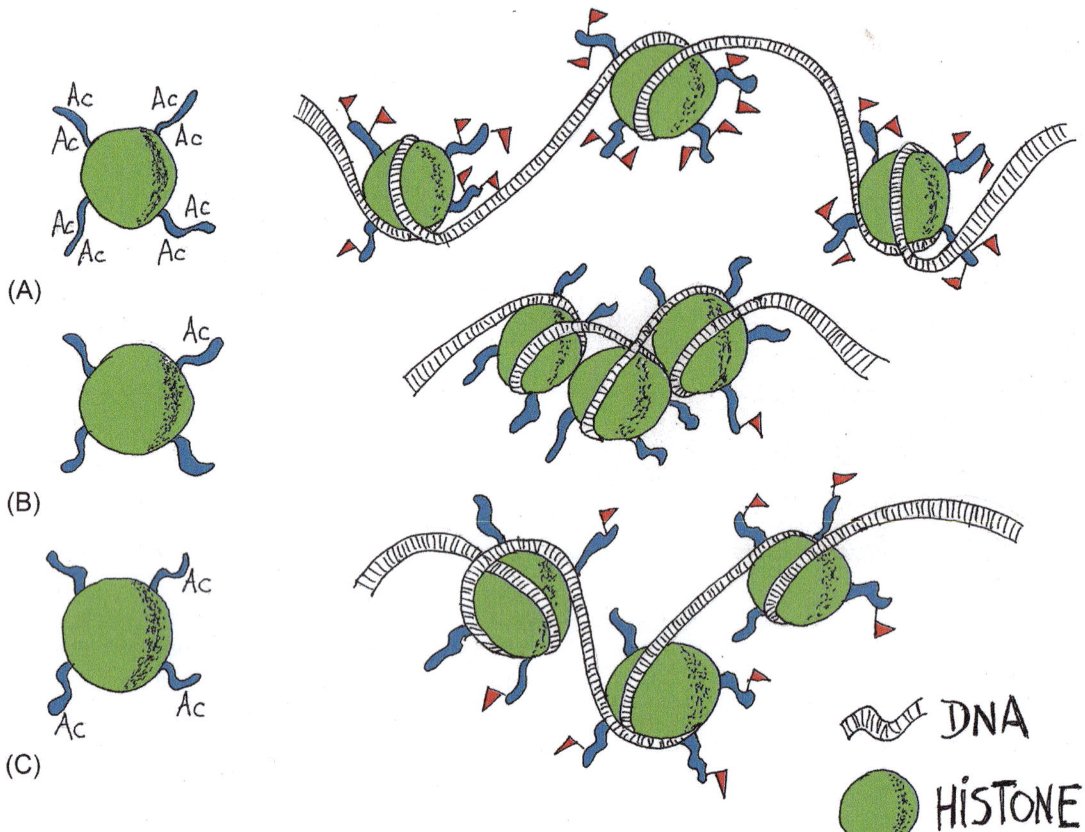

FIG. 45.1 Histone acetylation and deacetylation are epigenetic mechanisms, which provoke opposite effects on chromatin structure and gene expression.
Cartoon representing the epigenetic mechanism of histone acetylation of lysine residues located at histone 3 and 4 (H3 and H4) tails. Acetylation and deacetylation of histones mediate conversion between relaxed and condensed chromatin forms, respectively. (A) Histone tails are fully acetylated that induces an open state of the chromatin, which is more accessible to transcription factor interaction promoting transcription. The acetylated state is promoted by histone acetyltransferase activity. (B) Histone tail deacetylation condenses chromatin, which reduces transcription. The removal of acetyl groups from lysine residues is achieved by histone deacetylase activity. (C) Intermediate acetylated states promote a heterogeneous chromatin structure that differentially regulates transcription. Histone deacetylase inhibitors (HDACi) induce higher degrees of acetylation generally favoring gene expression. Epigenetic regulators such as nicotine can induce acetylation of specific portions of the chromatin differentially regulating gene expression. "Unpublished and original art work by R. O. Bernabeu."

(HDACi) effects on animal behavior and protein expression related to nicotine rewarding properties.

45.2 HISTONE ACETYLATION AND DEACETYLATION BY ACETYLTRANSFERASES AND DEACETYLASES

Many factors from the environment can activate the enzymes that regulate histone modifications. Histone acetylation plays a prominent role affecting chromatin structure. Acetylation of histones is a dynamic and reversible process controlled by a family of enzymes called histone acetyltransferases (HATs) and histone deacetylases (HDACs). HATs transfer acetyl groups to lysine residues, and HDACs remove acetyl groups from these amino acids. The first HAT identified called Hatl was found in yeast in 1995, and 1 year later, the first HDAC was purified and termed HDAC-l (Yang & Seto, 2007). HAT and HDAC activation provides a mechanism to understand how histone acetylation and transcription are regulated. For instance, studies dealing with type A HATs (GCN5, p300/CBP, TAFII250, p55, and PCAF) provided information about acetylation of nucleosome histones, whereas studies dealing with type B HATs informed about acetylation mechanisms of newly synthesized histones in the cytoplasm (Jeanteur, 2005).

45.2.1 Histone Deacetylase Inhibitors

HDAC inhibitors maintain histone acetylation and can tip the balance to an "open" configuration state of chromatin leading to a transcriptional permissive state. Different drugs characterized as HDAC inhibitors have been used for the treatment of neuropsychiatric and drug addiction diseases. For instance, valproate is an HDAC inhibitor that has been extensively used for treating neuropsychiatric disorders. Valproate induces DNA demethylation in the adult mouse brain (Dong, Chen, Gavin, Grayson, & Guidotti, 2010). Likewise, trichostatin A (TSA) increases histone acetylation and decreases DNA methylation. However, treatments with an inhibitor of DNA methyltransferases did not affect histone acetylation suggesting that acetylation can modulate methylation but not the opposite (Vaissière, Sawan, & Herceg, 2008). SIRT1 (class III HDAC) deacetylates DNA methyltransferase 1 (DNMT1), so it is likely that HDAC activity inhibitors regulate DNA methylation levels by modifying a key methyltransferase activity (Peng et al., 2011). Furthermore, the administration of HDAC inhibitors at the appropriate time point alleviates the negative dysphoric effects observed in alcohol withdrawal (Sakharkar et al., 2014).

At present, more than 40 drugs that work as HDAC inhibitors have been described. Some of them were approved by the *Food and Drug Administration* (FDA) and are principally used to treat different cancer types, whereas other compounds are waiting for approval for clinical use. In this chapter, we focused in two HDAC inhibitors: sodium butyrate (NaB) and phenylbutyrate (PhB), which have demonstrated significant biological and behavioral effects on animal models for the study of nicotine dependence.

45.2.2 Histone Deacetylase and Nicotine Effects on the Nervous System

45.2.2.1 Nicotine Mimics the Action of HDAC Inhibitors

Consumption of abused drugs, including nicotine, provokes acute increments and accumulation of FosB in neurons of the reward pathway, chiefly in the nucleus accumbens (Nacc) (Soderstrom, Qin, Williams, Taylor, & McMillen, 2007). The FosB protein exhibits a truncated and stable splicing variant denominated delta-FosB, whose accumulation is critical for the establishment of addictive behaviors (Kowiański et al., 2018). Cocaine induces *fosb* gene expression through inhibition of HDACs (Nestler, 2008). Likewise, it has been found that nicotine inhibits HDACs and produces robust H3 and H4 acetylation at the FosB promoter in the mouse striatum inducing FosB expression (Levine et al., 2011).

Nevertheless, it remains to be examined whether nicotine inhibition of HDAC activity leads to a more widespread acetylation of histones at other gene's promoters. The ability of nicotine to hyperacetylate chromatin by inhibiting HDACs could cause both transient and stable long-term acetylation states.

45.2.2.2 Effect of HDACi and Nicotine on Memory

The widespread distribution throughout the nervous system of nicotinic acetylcholine receptors (nAChRs) may account for nicotine-induced multiple physiological effects. Several studies have demonstrated that nicotine enhances cognitive performance in smokers. It has also been repeatedly reported that nicotine is a positive regulator of learning, memory, and attention. Genetic studies have demonstrated that both dopamine receptors (DRs) and nAChRs participate in nicotine-induced cognitive improvement (Herman & Sofuoglu, 2010). Furthermore, it has been reported that nicotine regulates the synaptic mechanisms of learning that contribute to the addiction process (Subramaniyan & Dani, 2015).

In an interesting work, a preexposure to nicotine effect on cocaine reward was analyzed in rodents. They found that application of the histone deacetylase inhibitor suberoylanilide hydroxamic acid mimicked the potentiating effect of nicotine on the cocaine-induced long-term potentiation in the dentate gyrus (Huang et al., 2014). These findings agreed with previous studies indicating that nicotine could have a role as an HDAC inhibitor favoring specific gene transcription (Levine et al., 2011).

A few studies have demonstrated the effect of sodium butyrate (an HDAC inhibitor) on novel object recognition (NOR) memory. Long-term memory for NOR is especially sensitive to modifications in histone acetylation. The findings showed that favoring a histone hyperacetylated state via HDAC inhibition, a cognitive event that would normally be forgotten was converted into a long-term memory (Stefanko, Barrett, Ly, Reolon, & Wood, 2009). Thus, long-term memory requires structural chromatin modifications caused by histone acetylation, which provides epigenetic cues that retain gene expression associated with consolidated memories.

It has been found that reducing histone acetylation, by mutating histone acetyltransferases, impaired long-term memory for object location, whereas enhancing histone acetylation, by inhibiting histone deacetylases, improved long-term object location memory. Moreover, intrahippocampal administration of the nonspecific HDAC inhibitor TSA or the class I HDAC-selective inhibitor MS275 enhanced long-term object location memory indicating the importance of inhibiting class I HDAC activity for consolidating hippocampus-dependent spatial memory (Hawk, Florian, & Abel, 2011).

On the other hand, using zebra fish as an experimental model system, we evaluated memory for objects in the presence of drugs that affect attention and memory retention in rodents, such as nicotine and the HDACi PhB. An originally modified object recognition task was used to distinguish between familiar and novel objects (Faillace, Pisera-Fuster, Medrano, Bejarano, & Bernabeu, 2017). We have demonstrated for the first time that zebra fish have an innate preference for exploring some colored objects rather than others. Moreover, zebra fish were better at discriminating color changes than modest changes in shape or size. These findings also indicated that objects must be previously evaluated to understand zebra fish innate preference or aversion toward the object (Fig. 45.2). Moreover, nicotine significantly enhanced short-term innate novel object preference or recognition, whereas PhB showed similar effects on long-term innate object preference or recognition. The physiological consequences of HDAC inhibition by the PhB treatment can only be observed several hours after exposure to PhB since its effects are mediated by gene transcription. In contrast, acute nicotine effects can be observed more rapidly and transiently. The findings suggested that nicotine and PhB can facilitate object recognition in zebra fish, which may involve memory. Interestingly, when one of the paired objects in the testing session was a naturally preferred object whereas the other was a nonpreferred object, nicotine and PhB generally potentiated but in some cases inverted naive preferences.

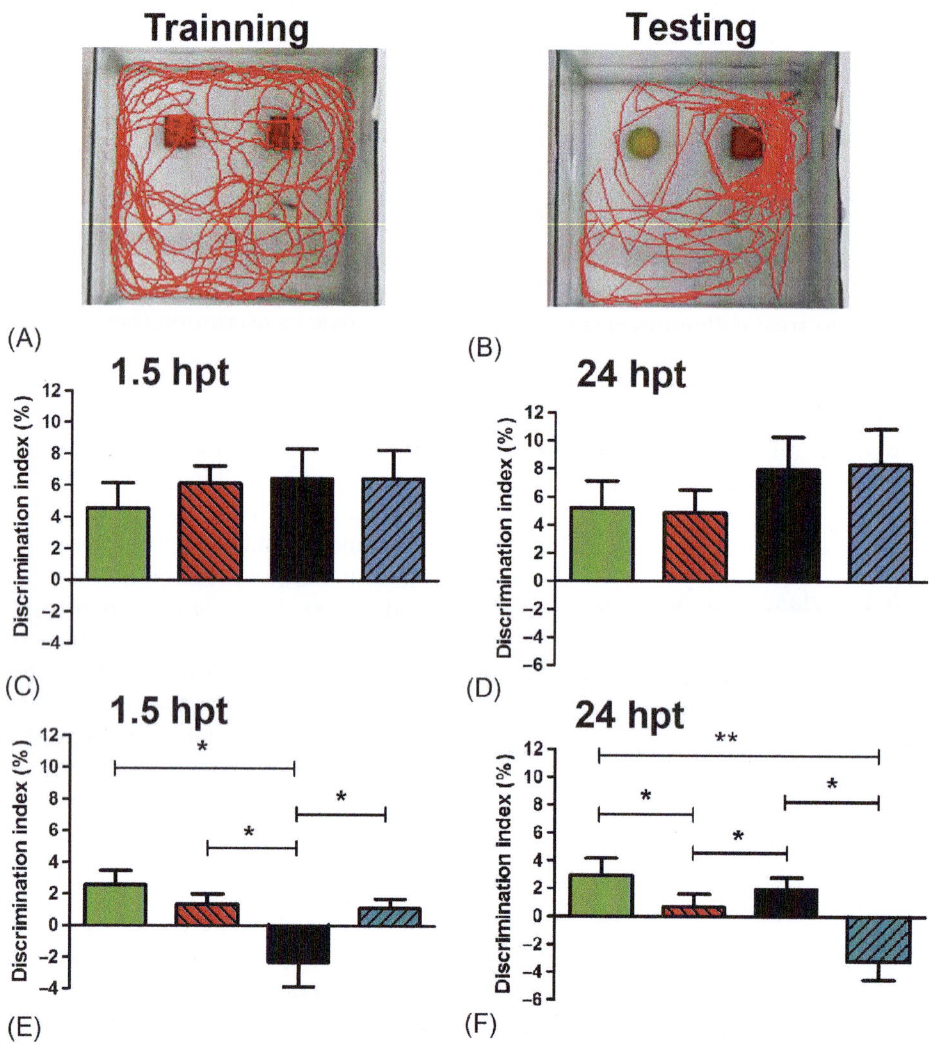

FIG. 45.2 Nicotine and phenylbutyrate effect on novel object preference for zebra fish.
Zebra fish were exposed to two identical red cubes (in this case) during the training session and then either to the same red cubes (A) or to one red cube (familiar object) and one yellow ball (novel object) (B) during the testing session. Zebra fish were exposed to tank water, nicotine, phenylbutyrate (PhB, a histone deacetylase inhibitor), or phenylbutyrate + nicotine immediately after training for 10 min. Drugs were dissolved in the water tank immediately before introducing the fish. Testing sessions were performed 1.5 h (C and E) or 24 h after training (D and F). Object preference in testing sessions is depicted in the bar graphs and was assessed contrasting two red cubes (C and D) or a red cube with a yellow sphere (E and F) in the absence of the drugs. Zebra fish showed naive preference for red and aversion for yellow objects. Zebra fish can discriminate color changes and modest changes in shape of objects of the same size. Nicotine significantly enhanced short-term novel object preference or aversion (depending on the color of the object), whereas phenylbutyrate showed a similar effect at 24 h after training (hpt). Discrimination index percentage (DI) was calculated as follows: ((time of exploration of the novel object) − (time of exploration of the familiar object)/(time of exploration of the novel object + time of exploration of the familiar object)) × 100 (Stefanko et al., 2009). Using the DI, positive values indicate the animal spent more time exploring the novel object, whereas negative values indicate the animal preferred exploring the familiar object. The figure shows that zebra fish explored for longer periods the familiar red object than the novel yellow object when they were exposed to nicotine and tested after an interval of 1.5 h or to PhB and tested with a delay of 24 h. Data are depicted as mean ± SEM; $n = 10$–12 animals per group. $^*P < 0.05$; $^{**}P < 0.01$; by Newman-Keuls test after ANOVA. "Unpublished figure and data."

These effects do not necessarily imply memory but may involve perception and attention improvement. So, nicotine and PhB changed naive perceptual acuity and modified short- and long-term memory, respectively.

45.2.2.3 *Effects of HDACi on the Reinforcing Properties of Nicotine*

The ability of nicotine to alter firing of dopamine neurons of the ventral tegmental area (VTA) that release DA in the NAcc is the first step leading to nicotine reward, but activation of intracellular signaling pathways downstream of nAChR is likely to be critical for the long-term consequences of nicotine exposure including conditioned reward (Walters, Cleck, Kuo, & Blendy, 2005). These studies also identified the VTA and the NAcc as the brain regions where cyclic-AMP-responsive element-binding protein (CREB) activity is essential for the establishment of nicotine-induced conditioning place preference (CPP). Our studies demonstrated that nicotine during CPP induced the phosphorylation of CREB (pCREB) and Fos expression in mesolimbic brain areas (Pascual, Pastor, & Bernabeu, 2009).

The acquisition and maintenance of nicotine-induced CPP but not of nicotine-induced conditioning place aversion (CPA) increased both pCREB and Fos protein in the VTA, as well as in the NAcc, prefrontal cortex (PFC), and dorsal striatum (dStr) suggesting that nicotine associated with specific environmental cues induces changes in the neuronal activity of the different structures forming the mesolimbic pathway (Pascual et al., 2009; Walters et al., 2005). Moreover, the increase in pCREB was abolished by a treatment with mecamylamine (a nonselective nAChR antagonist) suggesting a specific activation of CREB induced by the nicotine-environment association.

Histone deacetylases differentially regulate target genes of pCREB by contributing to either the activation or cessation of transcription (Fass, Butler, & Goodman, 2003). It has been described that HDAC1 can form a complex with pCREB and protein phosphatase 1 (PP1), which can cause pCREB dephosphorylation and acetyl group removal inhibiting gene transcription (Canettieri et al., 2003). In fact, the HDAC inhibitor TSA enhances activation of CRE reporter genes by cAMP. So, pCREB activity promoting specific gene transcription is at least in part inhibited by the recruitment of HDACs.

The effect of HDACs on drug addiction was principally examined in experiments performed with self-administered cocaine (Kumar et al., 2005; Romieu et al., 2008) or conditioning place preference in rats (Raybuck, McCleery, Cunningham, Wood, & Lattal, 2013). There are only two studies that assessed the effect of HDAC inhibition on the rewarding properties of nicotine in rats (Castino, Cornish, & Clemens, 2015; Pastor, Host, Zwiller, & Bernabeu, 2011). Nicotine provokes long-term modifications in brain structures, but little is known about the mechanism involved in nicotine-preference long-lasting neuroplasticity (Barik & Wonnacott, 2009).

The epigenetic machinery is one of the main candidates to be considered for understanding nicotine biological action on the brain. As aforementioned, histone acetylation is the main modification studied so far not only for its crucial role in controlling gene expression but also because HDAC inhibitors are currently used for the treatment of cancer and neurological disorders. Regarding addiction, the effect of NaB on the extinction and reinstatement of nicotine self-administration was carefully examined. The authors found that HDAC inhibition facilitated the extinction of intravenous nicotine self-administration by enhancing extinction memory consolidation in rats (Castino et al., 2015). However, the epigenetic mechanisms involved in nicotine self-administration and the extinction process remain unclear.

In our laboratory, we used a conditioning place preference task as the alternative test to self-administration to study the rewarding properties of nicotine. By studying nicotine-induced CPP, extinction, and relapse in rats, we have demonstrated that HDAC2 protein levels increased in neurons of the reward pathway during CPP expression and relapse. In contrast, no significant change in the level of H3 acetylated at lysine 9 (H3-K9Ac) was found (Figs. 45.3 and 45.4).

Furthermore, we have demonstrated that PhB dramatically reduced the preference for nicotine in a CPP task without altering the aversive properties of the drug (Pastor et al., 2011). In fact, high doses of nicotine induced place aversion (Le Foll & Goldberg, 2005; Pascual et al., 2009), which was not affected by the PhB treatment. These findings suggested that the mechanism underlying nicotine preference differs from that underlying aversion for this drug. Behavioral data obtained from experiments that assessed HDAC inhibitor effects on nicotine self-administration and nicotine-induced CPP indicated that HDAC activity participates in appetitive drug-seeking behaviors without apparent effects on aversive behaviors. HDAC activity was relevant in at least two time windows: during the consolidation of unconditioned stimulus-conditioned stimulus association (observed by using the pavlovian CPP) and during the extinction of the conditioned memory, which was observed by using nicotine self-administration. Development and extinction of drug-cue association behaviors represent appropriate time windows for reducing the reinforcing properties of abused drugs using HDAC inhibitors.

In order to characterize some mechanisms likely involved in the inhibitory effect of PhB on nicotine-induced CPP, the level of histone acetylation, the expression of HDAC2 and methyl-CpG-binding protein 2 (MeCP2), and the phosphorylation of CREB were analyzed in our laboratory. To evaluate the effect of PhB, the acetylation level of H3-K9Ac was assessed. The

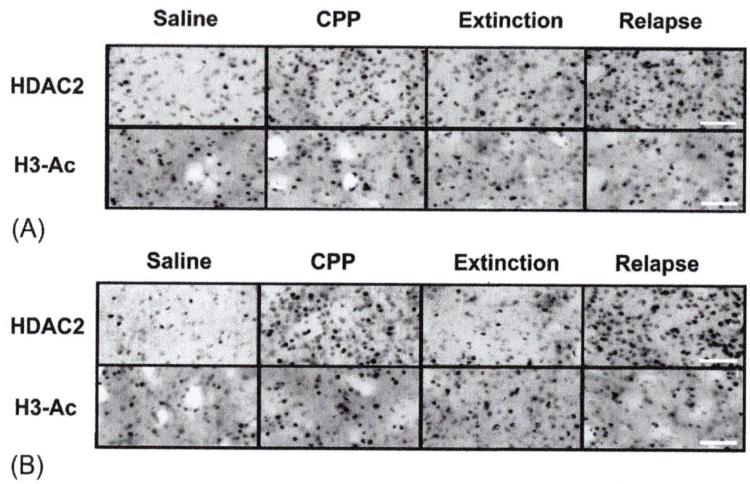

FIG. 45.3 HDAC2 and histone 3 acetylated at lysine 9 (H3-K9Ac) immunoreactivity in the nucleus accumbens (Nacc) of rat.
The photomicrographs show representative images of HDAC2- and H3-Ac-positive immunostaining in the Nacc, core (A) and shell (B), of different experimental groups of rats: saline (control group), CPP (nicotine-conditioned place preference group), extinction (extinction of the nicotine-induced CPP group), and relapse (reinstatement of the nicotine-induced CPP group). See Pascual et al. (2009) for details about the behavioral groups of rats. CPP, conditioning place preference. Scale bar, 50 μm. "Unpublished images."

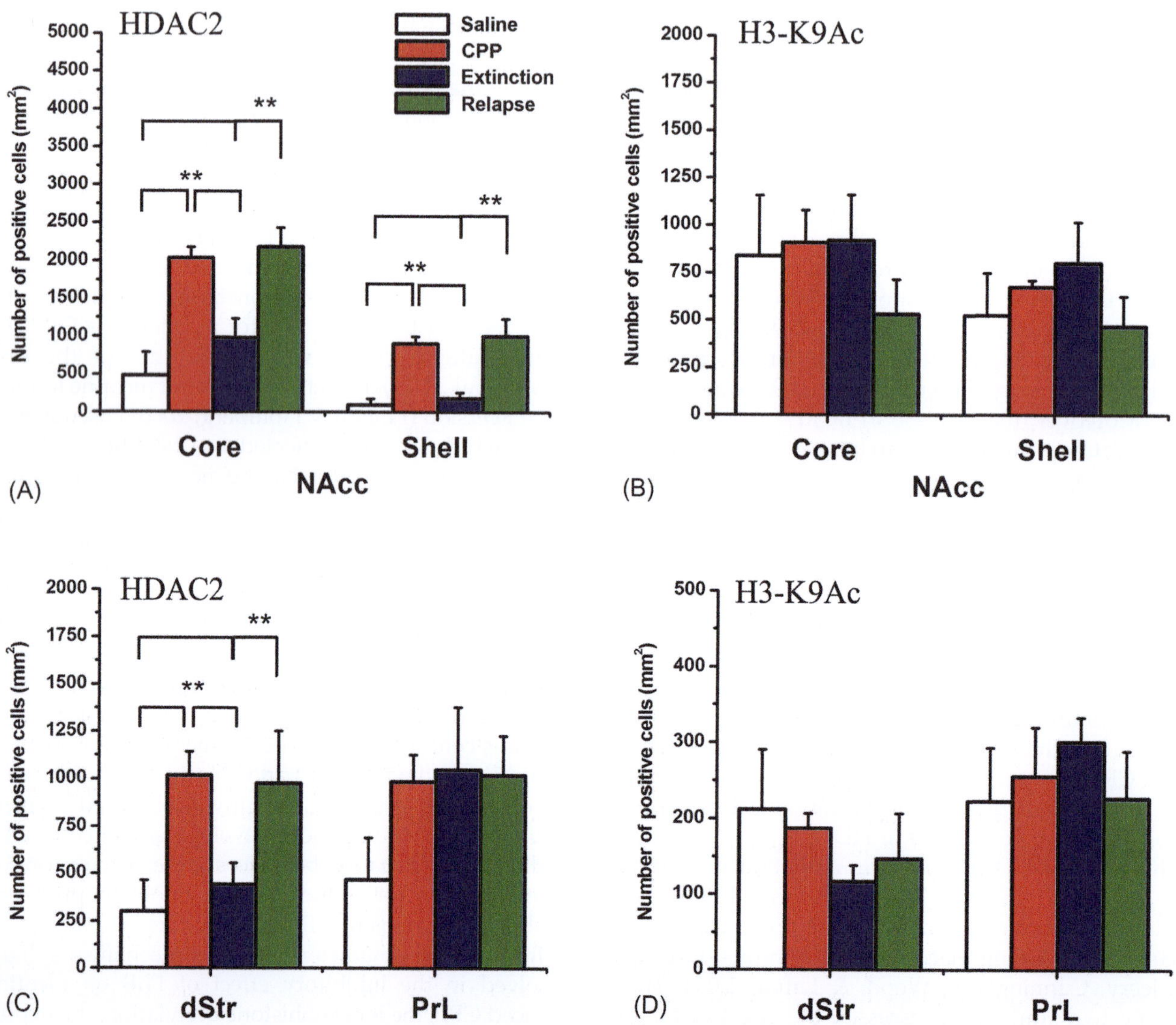

FIG. 45.4 Quantitative analysis of HDAC2- and H3-Ac-positive cells in different structures of the mesolimbic pathway of rat.
HDAC2- (A, C) and H3-K9Ac-immunopositive cells (B, D) were found in the nucleus accumbens (NAcc, core and shell), dorsal striatum (dStr), and prelimbic frontal area (PrL) in different experimental groups of rats (see Fig. 45.3). A stringent criterion was applied to count as immunopositive cells only densities up of the mean values (darker) since every cell constitutively expresses both markers. Bars indicate mean values ± SEM; $n = 7$–9 animals per group. $^{**}P < 0.01$ between behavioral groups using Newman-Keuls test after ANOVA. HDAC2, histone deacetylase 2; H3-K9Ac, histone 3 acetylated at lysine 9. "Unpublished data."

animals treated with PhB showed an enhancement in the acetylation level of two- to threefolds. Moreover, the number of cells expressing pCREB in different structures of the mesolimbic pathway in nicotine-conditioned rats exhibited a slight decrease caused by the HDAC inhibitor treatment. These findings suggested that whereas CREB phosphorylation is necessary to establish CPP (Pascual et al., 2009), it apparently plays no essential role in the reduction of CPP induced by the HDAC inhibitor during conditioning (Pastor et al., 2011). Interestingly, the treatment with PhB significantly reduced the number of HDAC2-immunopositive cells in the striatum of rats of the nicotine-induced CPP group, whereas this parameter remained unchanged in the striatum of animals of the nicotine-induced CPA group. Therefore, HDAC2 was required for nicotine-conditioned place preference, and it was downregulated when the drug-seeking behavior was reduced. These findings strengthened the role of HDAC2 in learning (Franklin & Mansuy, 2010).

Finally, MeCP2 builds a complex with mSin3A (an essential component of the HDAC1/2 repressor complex), and HDAC2 binds to methylated DNA and strongly inhibits gene transcription (Yang & Seto, 2008). Similarly to the experiments performed with cocaine (Cassel et al., 2006), nicotine-induced CPP increases the number of cells expressing MeCP2 in the dStr without affecting the NAcc or PFC in rats (Pastor et al., 2011).

Taken together, our findings indicated that HDAC inhibition was able to modulate the drug-seeking behavior associated with nicotine reward. Furthermore, HDAC2 plays a prominent role in developing the synaptic plasticity in the NAcc and dStr of the mesolimbic pathway necessary for establishing place-conditioned nicotine preference.

45.3 APPLICATIONS TO TREATMENTS

Widely use and abuse of tobacco cigarettes and nicotine within them by humans are considered the principal factors causing addiction. Pharmaceutical companies and public health programs have developed and assess compounds to inhibit histone deacetylase activity (HDAC). These compounds are therapeutic tools for effectively treating different types of cancer and neuropsychiatric disorders. Furthermore, scientific reports indicate that HDAC inhibitors can control addictive behaviors such as alcohol and cocaine consumption and could be very effective for avoiding nicotine dependence and relapse. Moreover, pCREB and deltaFosB activation is responsible for nicotine reward, and these molecules may also be considered as therapeutic targets to help finishing with human tobacco use and abuse.

MINI-DICTIONARY OF TERMS

Addiction A life-lasting modification of the synaptic plasticity in the mesolimbic dopamine system in the animal's brain, which once established induces a seeking behavior to obtain a particular reward or to avoid displeasure. The animal's brain rewarding circuit (mesolimbic pathway) has been potentiated and controls so much the animal's behavior that is almost exclusively oriented to seek and obtain the reward.

Conditioning place preference Pavlovian or classical conditioning task that assays an animal's free will to visit more frequently or spend more time in a place that has been previously associated with a rewarding drug. The unconditioned stimulus is a particular drug, whereas the conditioned stimulus is represented by a specific environment where the animal receives the drug. The drug-environment association can be established in a neutral environment for the animal or in a naively nonpreferred environment, which can further evidence the induction of a drug-seeking behavior provided the conditioning was established.

DeltaFosB and FosB DeltaFosB and FosB are transcription factors and members of the Fos family. The *fosb* gene can express FosB and deltaFosB as splicing variants, which are involved in motivation and reward for abused drugs and induce the neuroadaptations that establish addictive behaviors. Addictive behaviors can be generated in mice by overexpressing deltaFosB in the striatum without exposure to drugs (McClung et al., 2004).

Epigenetic Permanent or long-lasting chemical enzymatic modifications, such as methylation or acetylation of the DNA or the histones in the nucleosome (or sometimes in the cytoplasm in the case of histones) that modify chromatin structure (relaxation vs condensation) and the likelihood of specific gene expression. These modifications can be inherited but do not affect DNA nucleotide sequence or protein identity. Epigenetic mechanisms favor or repress the transcription of specific genes involved in different biological processes.

Histone acetyltransferase Enzyme with a catalytic activity that deals with transferring acetyl groups mainly to lysine residues of the histones from other molecules in the cytoplasm or nucleus of the cell or in an appropriate artificial reaction media.

Histone deacetylases (HDACs) Enzymes localized in the nucleus and cytoplasm of cells that regulate the degree of acetylation of histones forming the nucleosome. They catalyze the deacetylation of mainly lysine residues in histones. A high histone deacetylase activity conducts to a low degree of acetylation (a low number of acetylated lysine residues), which drives to DNA condensation inhibiting gene transcription (Fig. 45.1). Histone deacetylase 2 is the most prominent HDAC expressed in the mammalian brain.

Methyl-CpG-binding protein 2 (MeCP2) This protein can repress gene transcription by specifically binding to methylated promoters on genomic regions with a high frequency of cytosines and guanines (CpG islands). MeCP2 binding to 5-methylcytosines facilitates the recruitment of chromatin remodeling and transcriptional repressor complexes, which results in a condensed chromatin state. The protein has a critical role in breast cancer progression and embryonic development. Mutations of the gene encoding this protein (MBD2) may cause Rett syndrome. It may also function as a demethylase to promote transcription.

Phenylbutyrate (PhB) A chemical compound that inhibits histone deacetylase activity. PhB may increase the likelihood of gene transcription by tilting the balance toward a hyperacetylated state of histones forming the nucleosomes (Fig. 45.1).

Phosphorylated cyclic AMP response element-binding transcription factor (pCREB) CREB protein is activated by phosphorylation regulated by kinases in the cytoplasm of cells that respond to different cell signaling pathways. pCREB translocates to the nucleus where it regulates the transcription of a battery of genes by binding to CRE sequences in gene promoter regions. pCREB enhancement suggests functional activation of a cell, such as depolarization of an excitable cell.

Synaptic plasticity or neuroadaptations Any long-lasting morphological or molecular change that significantly modifies synaptic efficacy or activity in a neural circuit. Sprouting, dendritic or synaptic terminal retraction, spine formation, pruning, membrane receptor up- and downregulation, neurotransmitter release or uptake up- and downregulation, pre- and/or postsynaptic efficacy modifications such as long-term potentiation, long-term depression, and synaptic facilitation.

Key Facts

- Key fact of epigenetics: The epigenetic machinery is one of the main candidates to be considered for understanding nicotine-induced long-lasting changes in synaptic plasticity in the brain.
- Key fact of nicotine-environment associations and the mesolimbic pathway: Nicotine associated with specific environmental cues in a conditioning task induces changes in the neuronal activity of different structures of the mesolimbic pathway.
- Key fact of histone deacetylases (HDAC): HDAC inhibition facilitates the extinction of intravenous nicotine self-administration in rats.
- Key fact of histone deacetylase 2 (HDAC2): HDAC2 plays a prominent role in triggering changes in the neuronal activity of the nucleus accumbens and dorsal striatum of the mesolimbic pathway necessary for establishing nicotine-conditioned place preference.
- Key fact of histone deacetylases and phosphorylated CREB: Histone deacetylases differentially regulate the target genes of phosphorylated CREB by contributing to either the activation or cessation of transcription.
- Key fact of phosphorylated CREB: CREB phosphorylation is necessary to establish nicotine-conditioned place preference but apparently plays no essential role in the reduction of nicotine-conditioned place preference caused by the histone deacetylase inhibitor phenylbutyrate.
- Key fact of histone deacetylase inhibitors: These inhibitors are currently used for medical treatment of life-threatening diseases, and they could be used at appropriate doses to treat nicotine addiction and help smokers to quit.

Summary Points

- We have described herein published results about histone deacetylase inhibitor effects on animal behavior and protein expression involved in the rewarding properties of nicotine.
- Histone deacetylase activity inhibition enhanced consolidation of the extinction memory and hence attenuated nicotine reinstatement.
- Histone deacetylase 2 was required for nicotine-induced place preference conditioning and was downregulated when the drug-seeking behavior was reduced.
- Empirical evidences indicate that nicotine could have a role as a histone deacetylase inhibitor regulating synaptic plasticity in particular brain areas by favoring specific gene transcription.
- Histone deacetylase activity inhibitors increased aversive long-term memory but reduced nicotine place preference.
- Nicotine induced short-term improvement, while histone deacetylase inhibition induced long-term enhancement of object perception and memory for object recognition.

References

Barik, J., & Wonnacott, S. (2009). Molecular and cellular mechanisms of action of nicotine in the CNS. *Handbook of Experimental Pharmacology*, *192*, 173–207.

Canettieri, G., Morantte, I., Guzmán, E., Asahara, H., Herzig, S., Anderson, S. D., et al. (2003). Attenuation of a phosphorylation-dependent activator by an HDAC-PP1 complex. *Nature Structural Biology*, *10*(3), 175–181.

Cassel, S., Carouge, D., Gensburger, C., Anglard, P., Burgun, C., Dietrich, J. B., et al. (2006). Fluoxetine and cocaine induce the epigenetic factors MeCP2 and MBD1 in adult rat brain. *Molecular Pharmacology*, *70*(2), 487–492.

Castino, M. R., Cornish, J. L., & Clemens, K. J. (2015). Inhibition of histone deacetylases facilitates extinction and attenuates reinstatement of nicotine self-administration in rats. *PLoS One*, *10*(4)e0124796.

Dong, E., Chen, Y., Gavin, D. P., Grayson, D. R., & Guidotti, A. (2010). Valproate induces DNA demethylation in nuclear extracts from adult mouse brain. *Epigenetics*, *5*(8), 730–735.

Faillace, M. P., Pisera-Fuster, A., Medrano, M. P., Bejarano, A. C., & Bernabeu, R. O. (2017). Short- and long-term effects of nicotine and the histone deacetylase inhibitor phenylbutyrate on novel object recognition in zebrafish. *Psychopharmacology*, *234*(6), 943–955.

Fass, D. M., Butler, J. E., & Goodman, R. H. (2003). Deacetylase activity is required for cAMP activation of a subset of CREB target genes. *The Journal of Biological Chemistry*, *278*(44), 43014–43019.

Franklin, T. B., & Mansuy, I. M. (2010). The prevalence of epigenetic mechanisms in the regulation of cognitive functions and behaviour. *Current Opinion in Neurobiology*, *20*(4), 441–449.

Hawk, J. D., Florian, C., & Abel, T. (2011). Post-training intrahippocampal inhibition of class I histone deacetylases enhances long-term object-location memory. *Learning & Memory*, *18*(6), 367–370.

Herman, A. I., & Sofuoglu, M. (2010). Cognitive effects of nicotine: genetic moderators. *Addiction Biology*, *15*(3), 250–265.

Horn, P. J., & Peterson, C. L. (2002). Chromatin higher order folding-wrapping up transcription. *Science*, *297*(5588), 1824–1827.

Huang, Y. Y., Levine, A., Kandel, D. B., Yin, D., Colnaghi, L., Drisaldi, B., et al. (2014). D1/D5 receptors and histone deacetylation mediate the Gateway. Effect of LTP in hippocampal dentate gyrus. *Learning & Memory*, *21*(3), 153–160.

Jeanteur, P. (2005). *Epigenetics and chromatin* (1st ed.). Germany: Springer-Verlag.

Jenuwein, T., & Allis, C. D. (2001). Translating the histone code. *Science*, *293*(5532), 1074–1080.

Jiang, Y., Langley, B., Lubin, F. D., Renthal, W., Wood, M. A., Yasui, D. H., et al. (2008). Epigenetics in the nervous system. *The Journal of Neuroscience*, *28*(46), 11753–11759.

Kouzarides, T. (2007). Chromatin modifications and their function. *Cell*, *128*(4), 693–705.

Kowiański, P., Lietzau, G., Steliga, A., Czuba, E., Ludkiewicz, B., Waśkow, M., et al. (2018). Nicotine-induced CREB and DeltaFosB activity is modified by caffeine in the brain reward system of the rat. *Journal of Chemical Neuroanatomy*, *88*, 1–12.

Kumar, A., Choi, K. H., Renthal, W., Tsankova, N. M., Theobald, D. E., Truong, H. T., et al. (2005). Chromatin remodeling is a key mechanism underlying cocaine-induced plasticity in striatum. *Neuron*, *48*(2), 303–314.

Le Foll, B., & Goldberg, S. R. (2005). Nicotine induces conditioned place preferences over a large range of doses in rats. *Psychopharmacology*, *178*(4), 481–492.

Levine, A., Huang, Y., Drisaldi, B., Griffin, E.A Jr., Pollak, D.D., Xu, S., Yin, D., Schaffran, C., Kandel, D.B., & Kandel, E.R. (2011). Molecular mechanism for a gateway drug: epigenetic changes initiated by nicotine prime gene expression by cocaine. Sci Transl med, 3 (107):107ra109.

Marmorstein, R., & Zhou, M. M. (2014). Writers and readers of histone acetylation: structure, mechanism, and inhibition. *Cold Spring Harbor Perspectives in Biology*, *6*(7), a018762.

McClung, C. A., Ulery, P. G., Perrotti, L. I., Zachariou, V., Berton, O., & Nestler, E. J. (2004). DeltaFosB: a molecular switch for long-term adaptation in the brain. *Brain Research. Molecular Brain Research*. *132*, 146–154.

Nestler, E. J. (2008). Transcriptional mechanisms of addiction: role of DeltaFosB. Review. *Philosophical Transactions of the Royal Society of London. Series B, Biological Sciences*, *363*(1507), 3245–3255.

O'Brown, Z. K., & Greer, E. L. (2016). N6-methyladenine: a conserved and dynamic DNA mark. *Advances in Experimental Medicine and Biology*, *945*, 213–246.

Pascual, M. M., Pastor, V., & Bernabeu, R. O. (2009). Nicotine-conditioned place preference induced CREB phosphorylation and Fos expression in the adult rat brain. *Psychopharmacology*, *207*(1), 57–71.

Pastor, V., Host, L., Zwiller, J., & Bernabeu, R. O. (2011). Histone deacetylase inhibition decreases preference without affecting aversion for nicotine. *Journal of Neurochemistry*, *116*(4), 636–645.

Peng, L., Yuan, Z., Ling, H., Fukasawa, K., Robertson, K., Olashaw, N., et al. (2011). SIRT1 deacetylates the DNA methyltransferase 1 (DNMT1) protein and alters its activities. *Molecular and Cellular Biology*, *31*(23), 4720–4734.

Raybuck, J. D., McCleery, E. J., Cunningham, C. L., Wood, M. A., & Lattal, K. M. (2013). The histone deacetylase inhibitor sodium butyrate modulates acquisition and extinction of cocaine-induced conditioned place preference. *Pharmacology, Biochemistry, and Behavior*, *106*, 109–116.

Romieu, P., Host, L., Gobaille, S., Sandner, G., Aunis, D., & Zwiller, J. (2008). Histone deacetylase inhibitors decrease cocaine but not sucrose self-administration in rats. *The Journal of Neuroscience*, *28*(38), 9342–9348.

Sakharkar, A. J., Zhang, H., Tang, L., Baxstrom, K., Shi, G., Moonat, S., et al. (2014). Effects of histone deacetylase inhibitors on amygdaloid histone acetylation and neuropeptide Y expression: a role in anxiety-like and alcohol-drinking behaviours. *The International Journal of Neuropsychopharmacology*, *17*(8), 1207–1220.

Soderstrom, K., Qin, W., Williams, H., Taylor, D. A., & McMillen, B. A. (2007). Nicotine increases FosB expression within a subset of reward- and memory-related brain regions during both peri- and post-adolescence. *Psychopharmacology*, *191*, 891–897.

Stefanko, D. P., Barrett, R. M., Ly, A. R., Reolon, G. K., & Wood, M. A. (2009). Modulation of long-term memory for object recognition via HDAC inhibition. *Proceedings of the National Academy of Sciences of the United States of America*, *106*(23), 9447–9452.

Subramaniyan, M., & Dani, J. A. (2015). Dopaminergic and cholinergic learning mechanisms in nicotine addiction. *Annals of the New York Academy of Sciences*, *1349*, 46–63.

Tollefsbol, T. (2011). *Handbook of epigenetics, the new molecular and medical genetics* (1st ed.). London: Academic Press.

Vaissière, T., Sawan, C., & Herceg, Z. (2008). Epigenetic interplay between histone modifications and DNA methylation in gene silencing. *Mutation Research*, *659*(1–2), 40–48.

Walters, C. L., Cleck, J. N., Kuo, Y. C., & Blendy, J. A. (2005). Mu-opioid receptor and CREB activation are required for nicotine reward. *Neuron*, *46*(6), 933–943.

Yang, X. J., & Seto, E. (2007). HATs and HDACs: from structure, function and regulation to novel strategies for therapy and prevention. *Oncogene*, *26*(37), 5310–5318.

Yang, X. J., & Seto, E. (2008). Lysine acetylation: codified crosstalk with other posttranslational modifications. *Molecular Cell*, *31*(4), 449–461.

CHAPTER

46

L-Type Calcium Channels and Nicotine

Yudan Liu, Meghan Harding*†

*Department of Neuroendocrine Pharmacology, School of Pharmacy, China Medical University, Shenyang, Liaoning, China

†Family Medicine Physician Candidate, UCONN School of Medicine, Hartford, CT, United States

Abbreviations

AChR	acetylcholine receptor
CaMKII	calcium/calmodulin-dependent protein kinase II
CPA	conditioned place aversion
CPP	conditioned place preference
DA	dopamine
DBI	diazepam binding inhibitor
DHP	dihydropyridine
EPM	elevated plus maze
LTCC	L-type calcium channel
NAc	nucleus accumbens
VTA	ventral tegmental area

46.1 WHAT ARE L-TYPE CALCIUM CHANNELS?

In 1985, three scientists in Yale University defined L-type calcium channels (LTCCs) as "L" because they have long-lasting currents during depolarization compared to T-type calcium channels with transient inward currents (Nowycky, Fox, & Tsien, 1985). LTCCs are also called Ca_v1 Ca^{2+} channels, named by the main subunit of Ca^{2+} channels. LTCCs play influential roles in physiological and pathological processes, such as excitation-contraction coupling in skeletal muscles. LTCCs were the first purified voltage-gated calcium channels, initially from rabbit skeletal muscle, and found to have five subunits (Fig. 46.1): α1 (170 kDa), α2 (150 kDa), β (52 kDa), δ (17–25 kDa), and γ (32 kDa) subunits (Takahashi, Seagar, Jones, Reber, & Catterall, 1987; Tanabe et al., 1987), in which α2 and δ are normally integrated into one as α2δ subunit. Among these, the α1 subunit is the most momentous because it is the pore-forming subunit and the binding sites for most vital regulatory modulators and drugs (such as 1,4-dihydropyridine (DHP) chemicals). α2δ, β, and γ subunits are accessory subunits involved in regulatory functions (Hofmann, Flockerzi, Kahl, & Wegener, 2014). Actually, the feature that distinguishes LTCCs from other voltage calcium channels (such as N-, P/Q-, and T-type calcium channels) is their high sensitivity to DHP chemicals (e.g., nifedipine), which are not only the essential pharmacological tools used to isolate LTCC but also the major key medication to treat cardiovascular diseases, e.g., hypertension. The LTCC family has four isoforms regarding their α1 subunits (Table 46.1): $Ca_v1.1$, $Ca_v1.2$, $Ca_v1.3$, and $Ca_v1.4$. All these four isoforms share the similar physiological properties, e.g., long-lasting inward currents, but differ in subunit composition; e.g., $Ca_v1.2$ channels are composed of only three subunits, α1, α2δ, and β without γ subunits (Hofmann et al., 2014), and tissue distributions along with some electrophysiological properties. For example, the $Ca_v1.1$ isoform is predominantly distributed in the skeletal muscles, while the $Ca_v1.4$ isoform is primarily restricted to the retina. However, $Ca_v1.2$ and $Ca_v1.3$ isoforms are abundantly expressed in many tissues, such as the heart and brain. In particular occasions, $Ca_v1.2$ and $Ca_v1.3$ isoforms share the same location, e.g., both in the postsynaptic dendrites of the neurons in the brain. As for their electrophysiological properties, $Ca_v1.2$ possesses the typical LTCC properties: a higher activation threshold (e.g., −40 to −20 mV), with long-lasting currents during depolarization and a high sensitivity to the DHP drugs. Nonetheless, $Ca_v1.3$, compared to $Ca_v1.2$, are activated by a lower voltage (−60 to −40 mV) and have a reduced sensitivity to DHP drugs. The molecular difference between subtypes implies that they might couple to different signaling pathways and consequently mediate different physiological roles in neurons. For example, $Ca_v1.3$ channels are questionably vital in

Neuroscience of Nicotine
https://doi.org/10.1016/B978-0-12-813035-3.00046-0

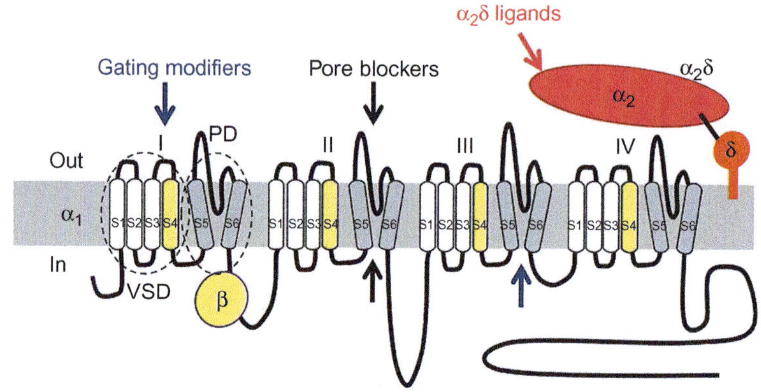

FIG. 46.1 Schematic representation of an L-type calcium channel structure model.
This figure shows the major drug binding sites in the L-type calcium channels. The drugs block channels by modifying channel gating (*blue arrows*, gating modifiers, e.g., nifedipine), by blocking the pore directly (*black arrows*, pore blockers), or through both mechanisms (e.g., phenylalkylamine blockers). The *α2δ* ligands (*magenta arrow*) can modify channel trafficking. *From Zamponi, G. W., Striessnig, J., Koschak, A., & Dolphin, A. C. (2015). The physiology, pathology, and pharmacology of voltage-gated calcium channels and their future therapeutic potential. Pharmacological Reviews, 67(4), 821–870.*

TABLE 46.1 The Main Properties of Different L-Type Calcium Channel Isoforms

Isoforms	α1 subunit	Encoding genes	Major distribution
$Ca_v1.1$	α1S	CACNA1S	Skeletal muscles
$Ca_v1.2$	α1C	CACNA1C	Brain, heart, vascular \intestinal\bladder smooth muscles, pancreas, adrenal medulla
$Ca_v1.3$	α1D	CACNA1D	Brain, auditory hair cells, vestibular hair cells, heart, pancreas, adrenal gland
$Ca_v1.4$	α1F	CACNA1F	Retina

This table introduces the four isoforms of L-type calcium channels with their main subunit, encoding genes, and major distributions.

TABLE 46.2 The Main Subgroups of L-Type Calcium Channel (LTCC) Activators and Inhibitors

Chemical classes	Actions on LTCCs	Typical drugs
1,4-Dihydropyridine	Inhibitor	Nifedipine, nimodipine, nicardipine, amlodipine, isradipine
Phenylalkylamine	Inhibitor	Verapamil
Benzothiazepines	Inhibitor	Diltiazem
Diphenylalkylamine	Inhibitor	Flunarizine
Benzoylpyrrole	Activator	FPL 64176
1,4-Dihydropyridine	Activator	(*S*)-(−)-Bay K8644

This table lists the main L-type calcium channel drugs used in animal studies and in clinics.

mediating Ca^{2+} influx by responding to small membrane depolarization, which is essential for maintaining spontaneous rhythmic firing in neurons. Consistent with this, the results from our group displayed that $Ca_v1.3$ set the basic tone to maintain the basic single spike firing, while both $Ca_v1.2$ and $Ca_v1.3$ mediated the specific firing pattern "burst firing" in the dopamine (DA) cell of the ventral tegmental area (VTA) (Liu et al., 2014). In the VTA, single spike firing keeps the basic DA levels in the terminal areas and signals expected environmental stimuli; however, "burst firing" dramatically augments DA synaptic transmission and behaviorally encodes unexpected stimuli. For disease treatment and research purposes, various classes of LTCC antagonists (Table 46.2) have been developed, including DHP (e.g., nifedipine), phenylalkylamines (e.g., verapamil), benzothiazepines (e.g., diltiazem), and diphenylalkylamines (e.g., flunarizine).

46.2 THE RELATIONSHIP BETWEEN LTCCS AND NICOTINE TREATMENTS

Nicotine abuse is a complexity of multiple brain area dysfunctions (Fig. 46.2). Classical theory proposes that

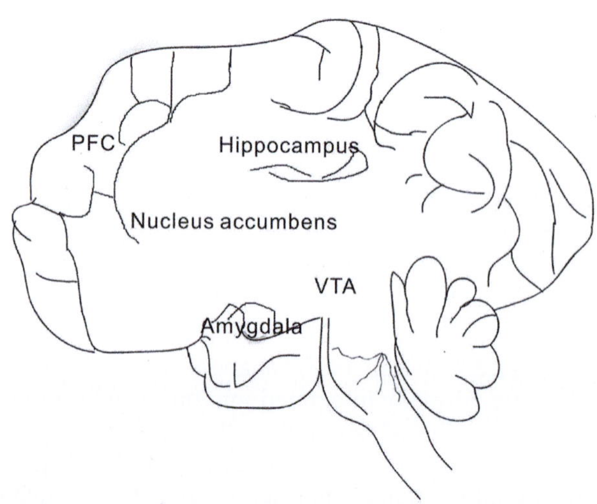

FIG. 46.2 A schematic figure showing the main brain areas involving nicotine addiction.
The main brain areas involving nicotine addiction: *PFC*, prefrontal cortex; *VTA*, ventral tegmental area.

nicotine activates DA neurons in the mesolimbic system originating in the VTA to the target areas mainly in the nucleus accumbens (NAc); disrupts the brain areas including those involved in reward and learning, such as the hippocampus and the frontal cortex; and then recruits interconnected brain circuits involved in stress and anxiety, e.g., the amygdala. In all these brain areas, only $Ca_v1.2$ and $Ca_v1.3$ LTCC isoforms are highly distributed (Chan et al., 2007; Dragicevic, Schiemann, & Liss, 2015; Olson et al., 2005). Of these brain areas, 89% of the LTCCs comprise the $Ca_v1.2$ isoform, whereas $Ca_v1.3$ isoform only accounts for 11% (Hell et al., 1993; Sinnegger-Brauns et al., 2009). In certain areas, their expression ratio could be reversed; for example, Takada et al. (Takada, Kang, & Imanishi, 2001) spoke of $Ca_v1.3$ LTCCs that are immunohistochemically localized in most of the midbrain neurons and distributed throughout the entire neuron, whereas fewer $Ca_v1.2$ LTCCs are displayed in the midbrain. The dense expression of different LTCC isotypes in nicotine-related brain areas implies a connection between LTCCs and nicotine dependence. In fact, the impact of LTCCs has been analyzed in nicotine use and abuse at both the behavioral and molecular levels in animal studies (Table 46.3).

46.3 ANTINOCICEPTION

At the early stage, many researchers focused on the effects of LTCC drugs on nicotine-induced antinociception. The tail-flick technique is a major measurement of nociception, in which the animal's body except the tail is restrained and a beam of light is projected on the animal's tail. Damaj's group demonstrates that the LTCC activator (±)-Bay K8644 potentiates acute or chronic nicotine-induced antinociception, while the LTCC inhibitors nimodipine and nifedipine block the antinociceptive effects by nicotine alone or the combination of (±)-Bay K8644 and nicotine in mice (Damaj, 1997, 2005, 2007; Damaj & Martin, 1993; Damaj, Welch, & Martin, 1993). Damaj (2007) also indicates that the effects of LTCCs on nicotine-induced antinociception are mediated by β_2-containing acetylcholine receptors (AchR) in the spinal cord, which require calcium influx through LTCCs to activate calcium/calmodulin-dependent protein kinase II (CaMKII) to produce nicotinic analgesia. In addition, LTCCs are involved in expression and development of nicotine tolerance to nicotine-induced antinociception in mice (Damaj, 2005). However, the results from rat studies are different from that of Damaj's group, which show that nifedipine (15 mg/kg i.p.) potentiates nicotine-induced antinociception (Zbuzek, Cohen, & Wu, 1997).

46.4 LOCOMOTION

The effects of LTCCs on nicotine-induced locomotive changes are investigated as well. Damaj et al. indicates that pretreatment of Bay K8644 (0.75 mg/kg i.p.) produces 5- to 10-fold potentiation in decreased spontaneous activity induced by nicotine administration (Damaj & Martin, 1993), which is blocked by nifedipine. Hart et al. note that LTCC inhibitor nimodipine (10 and 20 mg/kg) could reverse the increase of locomotion following acute nicotine administration (Hart, Kisro, Robinson, & Ksir, 1996). A characteristic in the addictive behavior is a phenomenon called sensitization: the progressive enhancement of certain drug-induced effects, e.g., locomotion, which develops following repeated, intermittent treatment with addictive drugs. Biala et al. (Biala, 2003; Biala & Weglinska, 2004a, 2004b) and Bernardi et al. (Bernardi, Uhrig, Spanagel, & Hansson, 2014) show that four LTCC blockers nifedipine, nimodipine, verapamil, and diltiazem all could block the acquisition and expression of nicotine-induced locomotor sensitization; verapamil and diltiazem could attenuate cross sensitization to the locomotor effects of nicotine and ethanol; but nimodipine and verapamil could decrease the cross sensitization of nicotine and morphine or MK-801.

46.5 DEPENDENCE

Conditioned place preference (CPP) paradigm is a popular and effective method to demonstrate nicotine reinforcing property. Biala et al. (Biala, 2003) find that pretreatment with nimodipine, verapamil, or diltiazem at 10 μM i.p. blocks the acquisition of nicotine-induced CPP (nicotine at 0.5 mg/kg i.p.). In addition, in the cross sensitization tests, morphine (5 mg/kg) only induces CPP in the animals with nicotine injections previously, and the administration of nimodipine, verapamil, or diltiazem at 10 and 20 mg/kg i.p. dose-dependently prevents this sensitization (Biala & Weglinska, 2004b). Our group further studies the subtypes of LTCCs involved, indicating $Ca_v1.2$ but not $Ca_v1.3$ subtype mediates the acquisition of CPP using transgenic mouse models that do not express $Ca_v1.3$ ($Ca_v1.3^{-/-}$) or contain a mutation in the DHP site of the $Ca_v1.2$ ($Ca_v1.2DHP^{-/-}$) (Liu, Harding, Dore, & Chen, 2017).

CPP is also an effective paradigm to investigate the reinstatement of drugs. Drug addiction is a relapsing brain dysfunction with the neurobiological changes resulting in compulsive drug-seeking behavior. The high relapse rate after a long period of abstinence is one of the main problems in the treatment of addiction. Biala et al.

TABLE 46.3 A Summary of Animal Studies of L-Type Calcium Channels (LTCCs) and Nicotine

Index	Nicotine model	Animal	LTCC drugs	References
Amphetamine-induced dopamine release in prefrontal cortex	5 μM	Sprague Dawley rats	Nitrendipine	J Neurochem 77 (2001) 839–848
Antinociception	1.5 mg/kg s.c.	Institute for Cancer Research (ICR) mice	Nimodipine, nifedipine	J Pharmacol Exp Ther 266 (1993) 1330–8
Antinociception	6 mg/kg per day for 28 days	Sprague Dawley rats	Nifedipine, verapamil	Life Sci 60 (1997) 1651–1658
Antinociception	24 mg/kg per day for 14 days	ICR mice	Nifedipine, verapamil	J Pharmacol Exp Ther 315 (2005) 959–964
Antinociception	20 μg per mouse intrathecal injections	C57BL/6 mice	Nimodipine and verapamil	J Pharmacol Exp Ther 320 (2007) 244–249
Antinociception Locomotion	1.75 mg/kg s.c. or 1.5 mg/kg s.c.	ICR mice	Nifedipine, nimodipine, verapamil Bay K8644	Drug Alcohol Depend 32 (1993) 73–79
Antinociception Locomotion	2 mg/kg, 2 times per day for 10 days	ICR mice	Bay K8644	Eur J Pharmacol 322 (1997)129–135
Anxiety	Acute, 0.1 or 0.5 mg/kg Chronic, 0.1 mg/kg for 6 days	Swiss mice	Nimodipine, flunarizine, verapamil, and diltiazem	Life Sci 79 (2006) 81–88
Anxiety	0.1 mg/kg s.c., 1 day or 6 days	Swiss mice	Nimodipine, flunarizine, verapamil, diltiazem	Prog Neuropsychopharmacol Biol Psychiatry 32 (2008) 54–61
Discrimination	0.4 mg/kg s.c.	Rat	Isradipine	Pharmacol Biochem Behav 41 (1992) 807–12
Expression of cerebral diazepam binding inhibitor mRNA in cerebral cortical neurons	0.1 μM 24 h	ddY strain mice	Cerebral diazepam binding inhibitor mRNA	Brain Res Mol Brain Res 80 (2000) 132–141
Expression of LTCC subtypes Sensitization	0.175 mg/kg, 1 or 14 days	C57Bl/6N mice	Cav1.2 Cav1.3 mRNA expression Nifedipine	Nicotine Tob Res 16 (2014) 774–85
Expression of LTCCs and ERK phosphorylation in primary cortical neurons	100 μM	C57Bl/6J mice	Nifedipine, diltiazem	J Neurochem 103(2007) 666–678
Expression of LTCCs in cerebral cortical neurons	0.1 μM 72 h 1 mg/kg s.c., three times per day for 7 days	ddY strain mice	[3H]Verapamil binding assay Protein electrophoresis and immunoblots	J Biol Chem 277 (2002) 7979–7988
Expression of LTCCs in the mouse brain	1 mg/kg, three times per day for 7 days	ddY strain mice	α1C, α1D, α1F, α2/δ1, β4 subunits	Brain Res Mol Brain Res 135 (2005) 280–284
Locomotion	0.4 mg/kg i.p.	Sprague Dawley rats	Nimodipine	Psychopharmacology 128 (1996) 359–361
Locomotion	0.5 mg/kg i.p. for 5 days	Swiss mice	Verapamil, diltiazem, nimodipine	Pol J Pharmacol 56 (2004) 391–397
Locomotion Place preference	0.5 mg/kg, 5 days for locomotor tests, 0.5 mg/kg, 4 days for CPP tests	Wistar rats	Nimodipine, verapamil, diltiazem	Pol J Pharmacol 55 (2003) 327–335
Locomotion Place preference	0.5 mg/kg	Wistar rats Swiss mice	Nimodipine, verapamil, diltiazem	J Pharm Pharmacol 56 (2004) 1021–1028

TABLE 46.3 A Summary of Animal Studies of L-Type Calcium Channels (LTCCs) and Nicotine—cont'd

Index	Nicotine model	Animal	LTCC drugs	References
Memory	0.0175, 0.035, 0.175, or 0.35 mg/kg s.c.	Swiss mice	Flunarizine, verapamil, amlodipine, nimodipine, nifedipine, and nicardipine	Naunyn Schmiedebergs Arch Pharmacol 386 (2013) 651–664
Memory (short and long term)	0.05, 0.1, and 0.5 mg/kg s.c.	Swiss mice	Amlodipine, nicardipine, verapamil	Behav Brain Res 317 (2017) 27–36
Memory-related response	0.035, 0.175, and 0.35 mg/kg	Swiss mice	Nimodipine, flunarizine, verapamil, diltiazem	Pharmacol Rep 61 (2009) 236–244
Motivational effect	1.168 mg/kg s.c., three times per day for 11 days	Wistar rats	Nimodipine, verapamil and flunarizine	Behav Brain Res 228 (2012) 144–150
Motivational effect	0.1–1 mg/kg, 0.5 mg/kg optimum s.c. for 3 days	C57BL/6J mice	Nifedipine Cav1.2 Cav1.3 transgenic mice	Prog Neuropsychopharmacol Biol Psychiatry 75 (2017) 176–182
Nerve damage	Cigarette smoke exposure, 20 min per day for 14 days	Albino Wistar rats	Nifedipine, verapamil	Ann Plast Surg 70 (2013) 222–226
Relapse	0.5 mg/kg i.p. for 3 days	Wistar rats	Nimodipine, flunarizine	Pharmacol Biochem Behav 89 (2008) 116–125
Skin flap necrosis	Cigarette smoke exposure, 20 min per day for 21 days	Albino Wistar rats	Nifedipine, verapamil	Plast Reconstr Surg 125 (2010) 866–871
Somatic signs	2.5 mg/kg s.c., four times per day for 7 days	Swiss mice	Nimodipine, verapamil, diltiazem, flunarizine	Pharmacol Res 51 (2005) 483–488
Somatic signs Motivational signs	36 mg/kg per day for 14- or 28-day minipumps	C57BL/6J mice	Nimodipine or verapamil (±)Bay K8644	J Pharmacol Exp Ther 330 (2009) 152–161
Electrophysiology	Carbachol Anatoxin A	Sprague Dawley rats	Nifedipine	J Physiol 568 (2005) 469–81
Electrophysiology		Sprague Dawley rats	(*S*)-(−)-Bay K8644 FPL 64176 Nifedipine	J Biol Chem 282 (2007) 8594–603
Electrophysiology		C57BL/6J mice	Nifedipine Subtypes of LTCCs	J Neurophysiol 112 (2014) 1119–30
Electrophysiology	Carbachol	Sprague Dawley rats	Nifedipine	Brain Res 1245 (2008) 41–51

This table lists key researches about L-type calcium channels and nicotine in the animals with detailed reference information.

indicate LTCC antagonists nimodipine and flunarizine at 5 and 10 mg/kg i.p. attenuate the reinstatement of nicotine CPP induced by a cannabinoid receptor agonist WIN55,212-2 and ethanol (Biala & Budzynska, 2008).

Furthermore, nicotine withdrawal symptoms promote continued tobacco use and relapse after smoking cessation. Withdrawal from nicotine after chronic administration causes an abstinence syndrome characterized by physical and affective signs (Biala & Weglinska, 2005). Both Biala et al. (Biala & Weglinska, 2005) and Damaj et al. (Jackson & Damaj, 2009) find that the application of LTCC antagonists such as nimodipine at 1 or 10 mg/kg i.p., verapamil at 1 or 10 mg/kg i.p., or diltiazem at 10 mg/kg i.p. significantly diminishes physical signs after mecamylamine-precipitated nicotine withdrawal, and LTCC activator (±)-Bay K8644 at 0.5 mg/kg exhibits significantly more somatic signs, all of which indicating the important function of LTCCs in the somatic signs of nicotine withdrawal. Conditioned place aversion (CPA) is used to investigate the aversive motivational state of drug withdrawal or a conditioned avoidance for the environmental stimuli paired with withdrawal. The results about LTCCs and nicotine-induced CPA are opposite and complex. Biala et al. (Budzynska, Polak, & Biala, 2012) show that LTCC antagonists nimodipine, verapamil, and flunarizine at 5 and 10 mg/kg i.p. injected before the administration of mecamylamine attenuate nicotine CPA, but Damaj et al. (Jackson & Damaj, 2009) find that pretreatment of nimodipine at 1 mg/kg i.p. before mecamylamine administration even strengthens the CPA response. Therefore, the role of LTCCs in nicotine withdrawal remains less clear and more complicated.

46.6 ANXIETY

Anxiety is a common symptom in both physical dependence and withdrawal syndrome of nicotine. Anxiety level increases during nicotine withdrawal, which is the main reason of nicotine withdrawal failure, whereas anxiety level decreases with nicotine dependence, which is why smokers may continue. Elevated plus maze (EPM) is used for the measurement of anxiety, in which the increase of the entry to open arm and the time inside the open arm means the increase of anxiety. Biala et al. (Biala & Budzynska, 2006; Biala & Kruk, 2008) find that four LTCC inhibitors nimodipine, flunarizine, verapamil, and diltiazem attenuate anxiogenic effects at dose of 10 mg/kg after acute low dose (0.1 mg/kg) of nicotine and anxiolytic effects at dose of 5 mg/kg (nimodipine or flunarizine) or 20 mg/kg (verapamil or diltiazem) after 6 days of 0.1 mg/kg nicotine administration. But in withdrawal tests, Damaj et al. (Jackson & Damaj, 2009) do not see any effects of LTCC blocker nimodipine and verapamil at 1 mg/kg on the anxiogenic effects after nicotine withdrawal.

46.7 COGNITIVE FUNCTIONS

Nicotine has been shown to have neuroprotective properties, and in humans, many studies describe deficits or improved efficiency of cognitive function after nicotine administration. Therefore, Biala et al. study the effects of LTCCs on the nicotine-induced memory changes though their results themselves contradict. They reveal that the nicotine-induced memory improvement in the short- and long-term memory is enhanced by the pretreatment with LTCC inhibitor amlodipine, nicardipine, and verapamil, which is possibly mediated by ERK1/2 activation for both terms or CaN inhibition for long-term memory (Michalak & Biala, 2017). However, their previous findings are in contrary showing that acute amlodipine, nicardipine, and verapamil pretreatment reverses nicotine-induced improvement of memory acquisition and in the case of verapamil also memory consolidation; chronic pretreatment with amlodipine and verapamil reverses both nicotine-improved acquisition and consolidation (Biala & Kruk, 2009; Biala, Kruk-Slomka, & Jozwiak, 2013). The author himself thinks "factors that determine differences in results concerning influence of LTCC blockers in memory effects of nicotine may include route of administration (central vs peripheral), dosage schedule (acute vs chronic), time of treatment (pretraining vs post-training), and other conditions related to behavioral test, e.g., used stimulus" (Michalak & Biala, 2017). This comment also can explain the contrary results that exist in anxiety, CPA, and locomotion experiments from different labs.

Along with these major behavior tests, LTCCs also have been shown to be involved in blocking nicotine discrimination (isradipine 15 mg/kg) (Schechter & Meehan, 1992), reducing the effects of cigarette smoking on peripheral nerve ischemia/reperfusion injury (nifedipine 10 mg/kg/d or verapamil 20 mg/kg/d) (Rinker, Fink, Stoker, Milan, & Nelson, 2013) and improving the smoking-induced skin flap necrosis in flap survival (nifedipine 10 mg/kg/d or verapamil 20 mg/kg/d) (Rinker, Fink, Barry, Fife, & Milan, 2010).

46.8 LTCCS' EXPRESSION

LTCCs definitely perform significant roles on nicotine-induced behaviors, but the underlying brain areas and molecular mechanisms are still not well understood. Ohkuma's group has discussed the relationship between LTCCs and nicotine exposure in the mice cerebral cortex in primary cultures and nicotine-treated mice (Hayashida, Katsura, Torigoe, Tsujimura, & Ohkuma, 2005; Katsura et al., 2000; Katsura et al., 2002). They found that the 24 h nicotine exposure increases the mRNA expression of diazepam binding inhibitor (DBI), which is mediated by CaMKII activation resulting from the increase in intracellular Ca^{2+} through LTCCs subsequent to the nAChR activation-induced membrane depolarization (Katsura et al., 2000). They also find that nicotine exposure after 72 h in cell culture (Katsura et al., 2002) or 7 days in the nicotine administration of mice (Hayashida et al., 2005) significantly increases the expression of α1C, α1D, α1F, and $\alpha 2/\delta 1$ subunits of LTCCs and $[^3H]$diltiazem binding sites in the cerebral cortex. They also investigate (Katsura et al., 2002) how LTCC inhibitor nifedipine (1 μM) suppresses 30 mM KCl-induced Ca^{2+} influx, how nicotine exposure enhances Bay K8644-induced Ca^{2+} influx that is inhibited by nifedipine, and how nicardipine (1 μM) reduces the increase of peak amplitude of Ba^{2+} current after 72 h exposure of 0.1 μM nicotine. All of these studies imply that nicotine exposure in nAChR can activate LTCC to induce Ca^{2+} influx to modify the intracellular processes in the cerebral cortex, which might be the underlying mechanism on nicotine dependence and withdrawal syndrome. Recently, Bernardi et al. (Bernardi et al., 2014) investigate the individual contribution of LTCC subtypes to the long-term nicotine exposure in wider brain areas and find $Ca_v1.2$ mRNA is not changed following a single nicotine (0.175 mg/kg) exposure, whereas a strong upregulation of $Ca_v1.3$ mRNA is observed in several brain regions including the prefrontal cortex, caudate putamen, and NAc shell. Following 14 days of nicotine treatment and 24 h of abstinence, $Ca_v1.2$ mRNA is downregulated throughout the brain areas including prefrontal cortex, caudate putamen, VTA, and ventral hippocampal subregions (CA3), whereas $Ca_v1.3$ mRNA is unchanged. Following 7 days of abstinence, $Ca_v1.2$ transcripts are upregulated including prefrontal cortex, caudate putamen, NAc shell, and ventral hippocampal subregions (CA1 and CA3), whereas $Ca_v1.3$ mRNA is largely unaffected. The author suggestes that $Ca_v1.2$ may be involved in the mediation of abstinence-related symptoms, whereas $Ca_v1.3$ appears mainly to be involved in the early stages of nicotine exposure.

46.9 FIRING PATTERNS

Our group tells this story from the electrophysiological side (Fig. 46.3). In the VTA, DA neurons have two different firing modes, single spike firing or burst firing, which encode different behaviors and molecular signals. Nicotine exposure is supposed to increase the percentage of burst firing in the midbrain. Therefore, the firing modes could be linked to the theoretical molecular pathways of nicotine dependence. Our group finds that the general cholinergic activator carbachol or $\alpha_4\beta_2$ nAChR agonist anatoxin A induces burst firings in the VTA DA neurons, which could be blocked or changed back to single spike firings after LTCC inhibitor nifedipine administration (Zhang, Liu, & Chen, 2005). This bursting also can be induced by direct activation of LTCCs by (*S*)-(−)-Bay K8644 or FPL 64176, which is mediated by protein kinase M produced by the cleavage of protein kinase C (Liu, Dore, & Chen, 2007). It is also noted that different subtypes of LTCCs mediate different firing modes, in that only Cav1.3 regulates basal single spike firing but the activation of both Cav1.2 and Cav1.3 supports burst firing (Liu et al., 2014). All our results imply that different subtypes of LTCCs and different firing modes in the VTA might be the underlying reason of nicotine dependence and take different responsibilities.

In summary, much of the animal research about nicotine and LTCCs suggest that LTCCs do play a key role in both nicotine exposure and addiction in many aspects

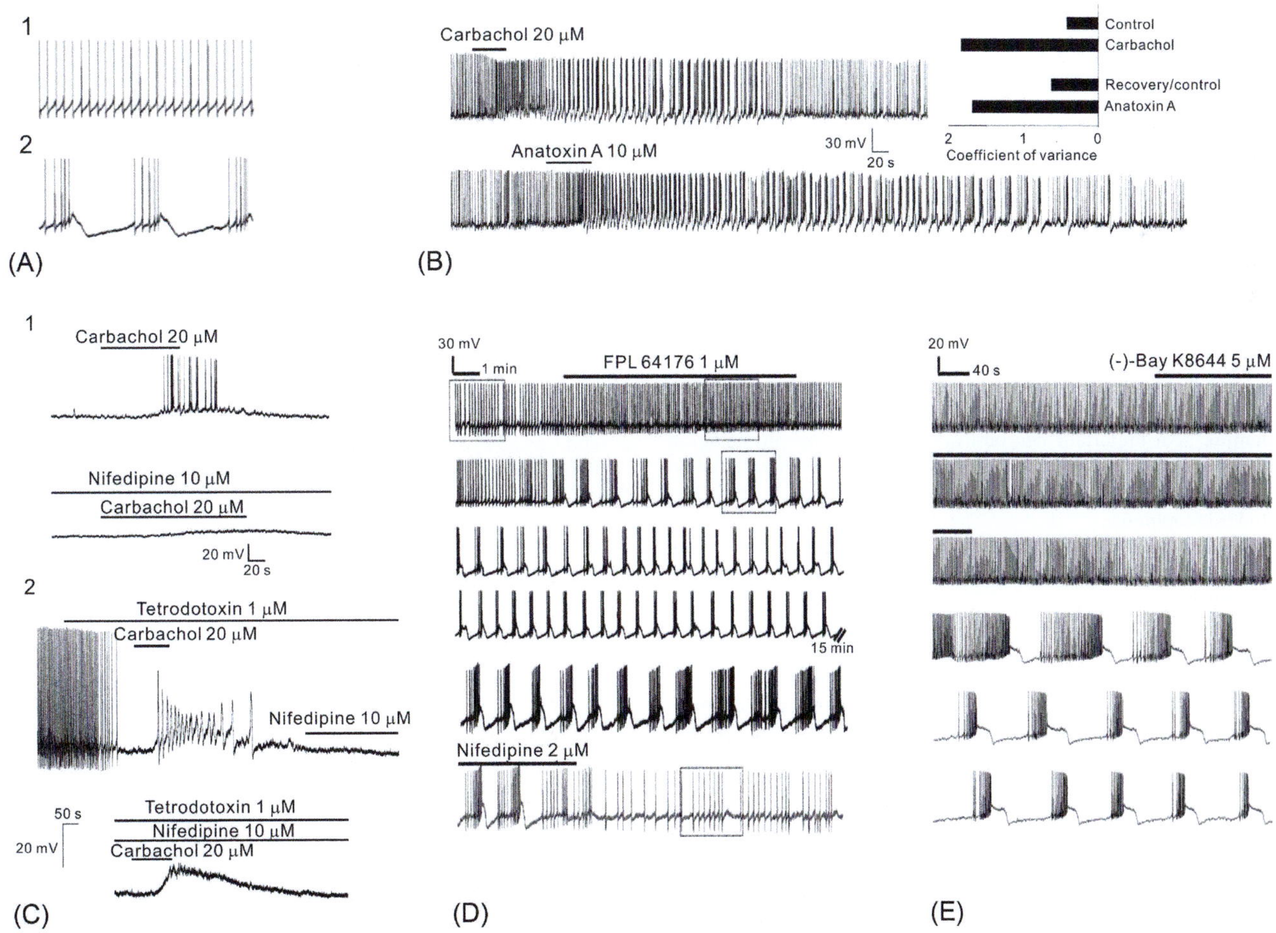

FIG. 46.3 L-type calcium channels mediate nicotine-related burst firing in the ventral tegmental area.
(A) Two firing modes: (1) single spike firing and (2) burst firing. (B) Both carbachol and $\alpha_4\beta_2$ agonist anatoxin A induced similar burst firing at the same neuron. (C) Carbachol-induced burst firing (1) and membrane oscillation (2) were abolished by L-type calcium channel blocker nifedipine. (D and E). Both L-type calcium channel activator FPL 64176 (D) and (S)-(−)-Bay K8644 (E) converted single firing to burst firing. *B and C from Zhang et al. (2005) published in J Physiol; A, D, and E from Liu et al. (2007) published in J Biol Chem.*

including nicotine dependence and withdrawal syndromes, due to the effects seen after LTCC inhibitors are administered in animals, which reduce these symptoms. Therefore, further evaluation of LTCC inhibitors in either treating nicotine dependence or withdrawal may be the promising area. However, some important issues must be kept in mind First, these treatment effects have only been shown in lab animals and therefore cannot be directly correlated to human treatment or experiences yet. Second, LTCC blockers at clinical doses do not affect brain function in humans, although some changes in corticospinal metaplasticity have been detected (Wankerl, Weise, Gentner, Rumpf, & Classen, 2010). Third, systemic administration of LTCC drugs may cause pronounced cardiovascular effects with the dosage most studies have used in the experimentation, although effects on spontaneous locomotion have not yet been reported. Fourth, the effects of LTCC drugs on nicotine-related behaviors might involve multiple nicotine-related areas and signaling pathways; therefore, one theory alone is not yet capable of explaining the entire process.

MINI-DICTIONARY OF TERMS

Agonist/activator A chemical with the ability to promote the receptor activation or channel opening.

Antagonist/inhibitor A chemical with the ability to block the receptor activation or channel opening.

Voltage-gated calcium channel An ion channel sensitive to voltage changes. When it opens, calcium ions flow into the cell.

Conditioned place preference (CPP) An effective method to test the reinforcing property of the drug.

Cross sensitization One drug's sensitization is induced by another drug.

Depolarization The membrane potential changes from lower value to higher value, for example, from −60 to −40 mV.

Dopamine An important neurotransmitter, involving physiological states such as reward processing and movement fine regulation and disease states such as Parkinson's disease, drug addiction, and schizophrenia.

Firing It means action potentials in the cells. Firing modes mean the firing pattern with different time intervals between two action potentials.

L-type calcium channel One type of calcium channels. When it opens, calcium influx time is relatively longer, and channel inactivation is relatively slower compared to other calcium channels.

Reinstatement Restore the drug intake behaviors after drug withdrawal.

Sensitization The progressive enhancement of some drug-related effects such as locomotion.

Ventral tegmental area A small midbrain area with the major source of dopaminergic neurons in the brain. Its functions are related to reward and disease states such as drug addiction and schizophrenia.

Withdrawal syndrome The complex and multiple symptoms appeared after addictive drug withdrawal, such as increased anxiety and tremor.

Key Facts of L-Type Calcium Channel (LTCC)

- LTCC is one type of calcium channels with long-lasting current during depolarization.
- LTCC has four isotypes: $Ca_v1.1$, $Ca_v1.2$, $Ca_v1.3$, and $Ca_v1.4$.
- LTCC is the first purified calcium channel.
- LTCC inhibitors are popular and effective clinical drugs in treating cardiovascular diseases.
- LTCCs are distributed in the brain and involved in neuronal functions such as transmitter release, firing regulation, and brain processing such as reward and memory.

Summary Points

- This chapter focuses on the L-type calcium channels (LTCCs) and nicotine.
- LTCCs are distributed in the brain, mainly $Ca_v1.2$ and $Ca_v1.3$ subtypes, which can be on the same neuron with the same location such as in the dendrites.
- LTCC activators potentiate, while LTCC inhibitors block some nicotine-related behaviors, such as locomotion sensitization and conditioned place preference, in animal studies.
- Nicotine administration increases LTCC expression in related brain areas, e.g., the cortex, nucleus accumbens, and ventral tegmental area in animals.
- Nicotine receptor activation induces firing mode transition that is involved in LTCC activation in animals.
- Bear in mind that the drug dosages used in animal studies might be higher than that in clinics.

References

Bernardi, R. E., Uhrig, S., Spanagel, R., & Hansson, A. C. (2014). Transcriptional regulation of L-type calcium channel subtypes Cav1.2 and Cav1.3 by nicotine and their potential role in nicotine sensitization. *Nicotine & Tobacco Research*, *16*(6), 774–785.

Biala, G. (2003). Calcium channel antagonists suppress nicotine-induced place preference and locomotor sensitization in rodents. *Polish Journal of Pharmacology*, *55*(3), 327–335.

Biala, G., & Budzynska, B. (2006). Effects of acute and chronic nicotine on elevated plus maze in mice: involvement of calcium channels. *Life Science*, *79*(1), 81–88.

Biala, G., & Budzynska, B. (2008). Calcium-dependent mechanisms of the reinstatement of nicotine-conditioned place preference by drug priming in rats. *Pharmacology Biochemistry and Behavior*, *89*(1), 116–125.

Biala, G., & Kruk, M. (2008). Calcium channel antagonists suppress cross-tolerance to the anxiogenic effects of D-amphetamine and nicotine in the mouse elevated plus maze test. *Progress in Neuropsychopharmacology & Biological Psychiatry*, *32*(1), 54–61.

Biala, G., & Kruk, M. (2009). Influence of bupropion and calcium channel antagonists on the nicotine-induced memory-related response of mice in the elevated plus maze. *Pharmacological Reports*, *61*(2), 236–244.

Biala, G., Kruk-Slomka, M., & Jozwiak, K. (2013). Influence of acute or chronic calcium channel antagonists on the acquisition and consolidation of memory and nicotine-induced cognitive effects in mice. *Naunyn-Schmiedeberg's Archives of Pharmacology*, *386*(7), 651–664.

Biala, G., & Weglinska, B. (2004a). Calcium channel antagonists attenuate cross-sensitization to the locomotor effects of nicotine and ethanol in mice. *Polish Journal of Pharmacology, 56*(4), 391–397.

Biala, G., & Weglinska, B. (2004b). Calcium channel antagonists attenuate cross-sensitization to the rewarding and/or locomotor effects of nicotine, morphine and MK-801. *Journal of Pharmacy and Pharmacology, 56*(8), 1021–1028.

Biala, G., & Weglinska, B. (2005). Blockade of the expression of mecamylamine-precipitated nicotine withdrawal by calcium channel antagonists. *Pharmacological Research, 51*(5), 483–488.

Budzynska, B., Polak, P., & Biala, G. (2012). Effects of calcium channel antagonists on the motivational effects of nicotine and morphine in conditioned place aversion paradigm. *Behavioural Brain Research, 228*(1), 144–150.

Chan, C. S., Guzman, J. N., Ilijic, E., Mercer, J. N., Rick, C., Tkatch, T., … Surmeier, D. J. (2007). 'Rejuvenation' protects neurons in mouse models of Parkinson's disease. Nature, 447(7148), 1081–1086.

Damaj, M. I. (1997). Altered behavioral sensitivity of Ca(2+)-modulating drugs after chronic nicotine administration in mice. *European Journal of Pharmacology, 322*(2–3), 129–135.

Damaj, M. I. (2005). Calcium-acting drugs modulate expression and development of chronic tolerance to nicotine-induced antinociception in mice. *Journal of Pharmacology and Experimental Therapeutics, 315*(2), 959–964.

Damaj, M. I. (2007). Nicotinic regulation of calcium/calmodulin-dependent protein kinase II activation in the spinal cord. *Journal of Pharmacology and Experimental Therapeutics, 320*(1), 244–249.

Damaj, M. I., & Martin, B. R. (1993). Calcium agonists and antagonists of the dihydropyridine type: effect on nicotine-induced antinociception and hypomotility. *Drug and Alcohol Dependence, 32*(1), 73–79.

Damaj, M. I., Welch, S. P., & Martin, B. R. (1993). Involvement of calcium and L-type channels in nicotine-induced antinociception. *Journal of Pharmacology and Experimental Therapeutics, 266*(3), 1330–1338.

Dragicevic, E., Schiemann, J., & Liss, B. (2015). Dopamine midbrain neurons in health and Parkinson's disease: emerging roles of voltage-gated calcium channels and ATP-sensitive potassium channels. *Neuroscience, 284*, 798–814.

Hart, C., Kisro, N. A., Robinson, S. L., & Ksir, C. (1996). Effects of the calcium channel blocker nimodipine on nicotine-induced locomotion in rats. *Psychopharmacology (Berlin), 128*(4), 359–361.

Hayashida, S., Katsura, M., Torigoe, F., Tsujimura, A., & Ohkuma, S. (2005). Increased expression of L-type high voltage-gated calcium channel alpha1 and alpha2/delta subunits in mouse brain after chronic nicotine administration. *Molecular Brain Research, 135*(1–2), 280–284.

Hell, J. W., Westenbroek, R. E., Warner, C., Ahlijanian, M. K., Prystay, W., Gilbert, M. M., et al. (1993). Identification and differential subcellular localization of the neuronal class C and class D L-type calcium channel alpha 1 subunits. *Journal of Cell Biology, 123*(4), 949–962.

Hofmann, F., Flockerzi, V., Kahl, S., & Wegener, J. W. (2014). L-type CaV1.2 calcium channels: from in vitro findings to in vivo function. *Physiological Reviews, 94*(1), 303–326.

Jackson, K. J., & Damaj, M. I. (2009). L-type calcium channels and calcium/calmodulin-dependent kinase II differentially mediate behaviors associated with nicotine withdrawal in mice. *Journal of Pharmacology and Experimental Therapeutics, 330*(1), 152–161.

Katsura, M., Higo, A., Tarumi, C., Tsujimura, A., Takesue, M., Mohri, Y., et al. (2000). Mechanism for increase in expression of cerebral diazepam binding inhibitor mRNA by nicotine: involvement of L-type voltage-dependent calcium channels. *Molecular Brain Research, 80* (2), 132–141.

Katsura, M., Mohri, Y., Shuto, K., Hai-Du, Y., Amano, T., Tsujimura, A., et al. (2002). Up-regulation of L-type voltage-dependent calcium channels after long term exposure to nicotine in cerebral cortical neurons. *The Journal of Biological Chemistry, 277*(10), 7979–7988.

Liu, Y., Dore, J., & Chen, X. (2007). Calcium influx through L-type channels generates protein kinase M to induce burst firing of dopamine cells in the rat ventral tegmental area. *The Journal of Biological Chemistry, 282*(12), 8594–8603.

Liu, Y., Harding, M., Dore, J., & Chen, X. (2017). Cav1.2, but not Cav1.3, L-type calcium channel subtype mediates nicotine-induced conditioned place preference in miceo. *Progress in Neuropsychopharmacology & Biological Psychiatry, 75*, 176–182.

Liu, Y., Harding, M., Pittman, A., Dore, J., Striessnig, J., Rajadhyaksha, A., et al. (2014). Cav1.2 and Cav1.3L-type calcium channels regulate dopaminergic firing activity in the mouse ventral tegmental area. *Journal of Neurophysiology, 112*(5), 1119–1130.

Michalak, A., & Biala, G. (2017). Calcium homeostasis and protein kinase/phosphatase balance participate in nicotine-induced memory improvement in passive avoidance task in mice. *Behavioural Brain Research, 317*, 27–36.

Nowycky, M. C., Fox, A. P., & Tsien, R. W. (1985). Three types of neuronal calcium channel with different calcium agonist sensitivity. *Nature, 316*(6027), 440–443.

Olson, P. A., Tkatch, T., Hernandez-Lopez, S., Ulrich, S., Ilijic, E., Mugnaini, E., et al. (2005). G-protein-coupled receptor modulation of striatal CaV1.3L-type Ca2+ channels is dependent on a Shank-binding domain. *Journal of Neuroscience, 25*(5), 1050–1062.

Rinker, B., Fink, B. F., Barry, N. G., Fife, J. A., & Milan, M. E. (2010). The effect of calcium channel blockers on smoking-induced skin flap necrosis. *Plastic and Reconstructive Surgery, 125*(3), 866–871.

Rinker, B., Fink, B. F., Stoker, A. R., Milan, M. E., & Nelson, P. T. (2013). Calcium channel blockers reduce the effects of cigarette smoking on peripheral nerve ischemia/reperfusion injury. *Annals of Plastic Surgery, 70*(2), 222–226.

Schechter, M. D., & Meehan, S. M. (1992). Further evidence for the mechanisms that may mediate nicotine discrimination. *Pharmacology Biochemistry and Behavior, 41*(4), 807–812.

Sinnegger-Brauns, M. J., Huber, I. G., Koschak, A., Wild, C., Obermair, G. J., Einzinger, U., et al. (2009). Expression and 1,4-dihydropyridine-binding properties of brain L-type calcium channel isoforms. *Molecular Pharmacology, 75*(2), 407–414.

Takada, M., Kang, Y., & Imanishi, M. (2001). Immunohistochemical localization of voltage-gated calcium channels in substantia nigra dopamine neurons. *European Journal of Neuroscience, 13*(4), 757–762.

Takahashi, M., Seagar, M. J., Jones, J. F., Reber, B. F., & Catterall, W. A. (1987). Subunit structure of dihydropyridine-sensitive calcium channels from skeletal muscle. *Proceedings of the National Academy of Sciences of the United States of America, 84*(15), 5478–5482.

Tanabe, T., Takeshima, H., Mikami, A., Flockerzi, V., Takahashi, H., Kangawa, K., et al. (1987). Primary structure of the receptor for calcium channel blockers from skeletal muscle. *Nature, 328*(6128), 313–318.

Wankerl, K., Weise, D., Gentner, R., Rumpf, J. J., & Classen, J. (2010). L-type voltage-gated Ca2+ channels: a single molecular switch for long-term potentiation/long-term depression-like plasticity and activity-dependent metaplasticity in humans. *Journal of Neuroscience, 30*(18), 6197–6204.

Zbuzek, V. K., Cohen, B., & Wu, W. (1997). Antinociceptive effect of nifedipine and verapamil tested on rats chronically exposed to nicotine and after its withdrawal. *Life Science, 60*(19), 1651–1658.

Zhang, L., Liu, Y., & Chen, X. (2005). Carbachol induces burst firing of dopamine cells in the ventral tegmental area by promoting calcium entry through L-type channels in the rat. *The Journal of Physiology, 568* (Pt 2), 469–481.

CHAPTER

47

The Co-occurrence of Nicotine With Other Substance Use and Addiction: Risks, Mechanisms, Consequences, and Implications for Practice, With a Focus on Youth

Linda Richter

Director of Policy Research and Analysis, Center on Addiction, New York, NY, United States

47.1 INTRODUCTION

National statistics from the United States demonstrate that nicotine product use remains a significant public health problem, despite recent declines in cigarette smoking. This is because many of those who continue to smoke combustible cigarettes are more socioeconomically disadvantaged and more vulnerable to tobacco's devastating health effects relative to those who no longer smoke or who have not taken up smoking (Jamal et al., 2016). It also reflects the fact that while rates of cigarette use have declined overall, there has been a general increase in the use of other nicotine products, such as electronic cigarettes and water pipe (hookah), especially among youth (Agaku et al., 2014; Jamal et al., 2017). While the decline in combustible cigarette use certainly represents commendable progress from a public health perspective due to the tremendous toll smoking takes on morbidity and mortality, the growth in the use of other nicotine products (even those that do not contain tobacco) remains a significant cause for concern (The National Center on Addiction and Substance Abuse, 2015; The National Center on Addiction and Substance Abuse, 2016).

Regardless of the device through which it is delivered, nicotine is not a harmless drug, especially when ingested by young people (Klein, 2015). Nicotine increases blood pressure, respiration, and heart rate and adversely affects the nervous, cardiovascular, respiratory, and reproductive systems (Benowitz, 2009). It may contribute to cancerous tumor development (Grando, 2014) and can be lethal if orally ingested (Mayer, 2014). Most importantly for young people, all nicotine products, including electronic cigarettes, deliver nicotine, a highly addictive drug. In fact, nicotine is the most addictive drug (Palmer et al., 2009), with the greatest proportion of weekly users developing addiction relative to those who use other addictive substances (67.3% vs 25.0% for cannabis and 15.6% for alcohol) (Cougle, Hakes, Macatee, Zvolensky, & Chavarria, 2016).

Today's cigarettes are more addictive than in the past due to design changes that have increased the efficiency of nicotine delivery and the nicotine yield in tobacco products (Land et al., 2014). Additives and flavorings found in traditional cigarettes and in noncombustible nicotine products, which facilitate nicotine delivery and increase the appeal of the products, also may exacerbate their health consequences and the risk of nicotine addiction (Alpert, Agaku, & Connolly, 2016).

47.1.1 The Vulnerability of Youth

The addictive potential of nicotine is of particular concern when it comes to young people. Numerous studies have shown that the risk of addiction increases with earlier age of first exposure to a substance (Lanza & Vasilenko, 2015). The vast majority of people who are addicted to nicotine began smoking in adolescence or early adulthood: 84% before age 18% and 95% before age 21 (The National Center on Addiction and Substance Abuse, 2011). Nearly half the adolescents who smoke cigarettes develop at least one symptom of

Neuroscience of Nicotine
https://doi.org/10.1016/B978-0-12-813035-3.00047-2

nicotine addiction before age 18 (Apelberg et al., 2014), and one in five meet the full diagnostic criteria (Dierker, Swendsen, Rose, He, & Merikangas, 2012). For those who start young, symptoms of addiction can develop rapidly, even with only "experimental" or occasional use (Dierker et al., 2012; DiFranza, 2015; McQuown, Belluzzi, & Leslie, 2007). Youth who use multiple nicotine products are at especially high risk of developing addiction, and twice as many youth in the United States report using two or more of these products than using cigarettes alone (Apelberg et al., 2014; Lee, Hebert, Nonnemaker, & Kim, 2015).

47.1.2 Increased Risk of Other Substance Use and Addiction

Nicotine use is associated not only with nicotine addiction but also with an increased risk of other forms of substance use and addiction. Richter, Pugh, Smith, and Ball (2017) found that the odds of alcohol use, marijuana use, other drug use, polysubstance use, and substance use disorders all are significantly higher among youth and adults who report any type of nicotine product use or who have nicotine addiction relative to those who never used nicotine; the odds of co-occurrence generally are highest among those who report current use of both cigarettes and other nicotine products. The relationship between nicotine use and other substance use was especially strong in younger respondents, suggesting a potentially greater risk of co-occurrence among younger than older youth.

Additional analysis of data from the 2014 National Survey on Drug Use and Health (Substance Abuse and Mental Health Services Administration, 2015) demonstrates the association between nicotine addiction and alcohol and other drug use and addiction. Respondents who met criteria for nicotine addiction were more likely to report past 30-day alcohol use (63.3% vs 46.2%) and a past-year alcohol use disorder (16.5% vs 6.3%), as well as past 30-day use of other drugs (32.1% vs 11.2%) and a past-year drug use disorder (12.7% vs 3.0%) (see Fig. 47.1).

More specifically, those with nicotine addiction were more likely than those without to report past 30-day marijuana use (28.2% vs 9.4%) and a past-year marijuana use disorder (6.8% vs 2.1%), past 30-day use of any drug other than marijuana (11.2% vs 3.4%) and a past-year drug use disorder (7.9% vs 1.1%), and past 30-day misuse of opioid medications (5.9% vs 1.6%) and a past-year opioid use disorder (4.2% vs 0.6%). The differences in the rates of past-year alcohol and other drug use disorders among those with versus without nicotine addiction were more pronounced among youth, aged 12–20. More than six times as many young people with versus without nicotine addiction had an alcohol use disorder (25.1% vs 4.2%), whereas among adult respondents aged 21 and older, the discrepancy was smaller (15.4% vs 7.4%). Similarly, nearly seven times as many young people with vs without nicotine addiction had a drug use disorder (27.6% vs 4.0%), whereas the discrepancy was smaller among adults (10.8% vs 2.3%) (see Fig. 47.2).

Other research confirms the strong link between youth nicotine use and the risk of alcohol and other drug use (Camenga et al., 2014; Cohen et al., 2015; Miech, O'Malley, Johnston, & Patrick, 2016) and alcohol and other drug use disorders (Richter, Pugh, & Ball, 2017; Richter, Pugh, Peters, Vaughan, & Foster, 2016).

47.2 THE NEUROBIOLOGICAL UNDERPINNINGS OF CO-OCCURRING NICOTINE AND OTHER SUBSTANCE USE AND ADDICTION

Nicotine use, especially during adolescence, not only perpetuates the use of harmful tobacco products and increases the risk of nicotine addiction but also increases the risk of other substance use and primes the brain to be more susceptible to the addicting effects of alcohol and other drugs. This is one reason why the popularity of electronic cigarettes and other noncigarette nicotine products among young people is so troubling. In 2016, electronic cigarettes were the most commonly used tobacco product among middle and high school students in the United States (Jamal et al., 2017). We know that the use of these products among young people not only exposes them to nicotine at a time when their brains are still developing and especially susceptible to its addicting effects but also leads to or perpetuates cigarette smoking and all its associated harms, even among those who are not otherwise inclined to smoke (Soneji et al., 2017).

The increased vulnerability to addiction among young people who engage in any type of substance use, including nicotine, appears to be due to biological, psychological, and environmental risk factors to which this age group is especially sensitive (Smith, McDonald, Bergstrom, Ehlinger, & Brielmaier, 2015; Yuan, Cross, Loughlin, & Leslie, 2015). The adolescent brain is very impressionable and undergoes essential developmental transformations on the structural, functional, and neurochemical levels that allow young people to learn quickly and adapt rapidly. However, this also means that the adolescent brain is more responsive to and affected by addictive substances, including nicotine, and that these effects, particularly those related to neural connectivity and behavioral regulation, can persist into adulthood (Yuan et al., 2015).

Nicotinic acetylcholine receptors (nAChRs), receptor proteins in the brain, play a critical role in the

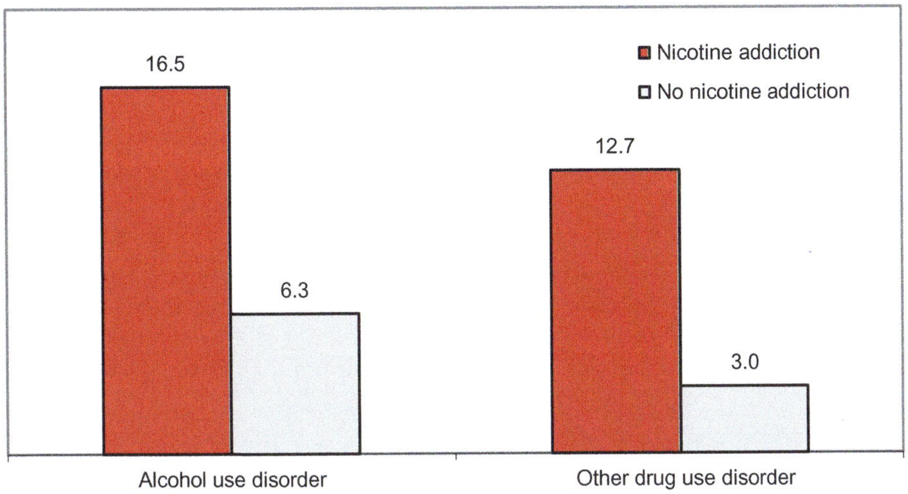

FIG. 47.1 Alcohol and other drug use disorders by nicotine addiction. Analysis of data from the 2014 National Survey on Drug Use and Health.
Percentage of respondents with and without nicotine addiction who met diagnostic criteria for an alcohol or other drug use disorder. Based on an analysis of data from the 2014 National Survey on Drug Use and Health (Substance Abuse and Mental Health Services Administration, 2015).

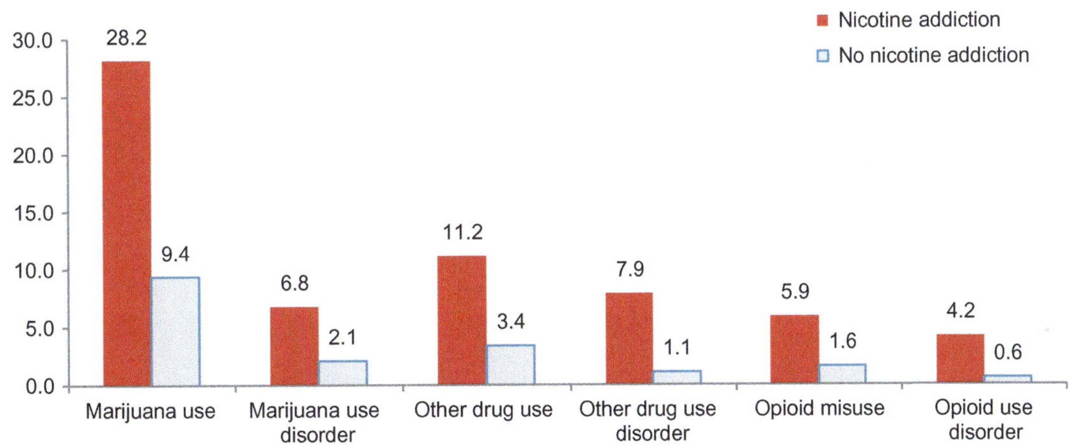

FIG. 47.2 Specific types of substance use and substance use disorders by nicotine addiction. Analysis of data from the 2014 National Survey on Drug Use and Health.
Percentage of respondents with and without nicotine addiction who engaged in past 30-day use of marijuana, a drug other than marijuana, or opioid misuse or who had a past-year marijuana use disorder, other drug use disorder, or opioid use disorder. Based on analysis of data from the 2014 National Survey on Drug Use and Health (Substance Abuse and Mental Health Services Administration, 2015).

neurophysiology of nicotine addiction. They are involved in the regulation of neurotransmitter systems, particularly dopamine, that are strongly implicated in reward processing and the reinforcing effects of addictive substances. They also play a significant role in brain maturation throughout adolescence (Yuan et al., 2015).

Nicotine affects the adolescent brain more intensely than the adult brain, producing more powerful reward and less aversive sensations. Animal research indicates that exposure to nicotine during adolescence has a more profound, adverse, and long-term effect on the brain's limbic system relative to exposure during adulthood, which in turn affects reward processing, cognitive functioning (e.g., reduced attention and increased impulsivity), and emotional regulation (in relation to anxiety and depression) (Yuan et al., 2015). Nicotine's effects on the limbic system also make young people more susceptible to using and becoming addicted to other substances (Dao, McQuown, Loughlin, Belluzzi, & Leslie, 2011). Even brief exposure to nicotine during adolescence has been associated in studies of rats with greater self-administration of cocaine, methamphetamine, and alcohol, suggesting that nicotine acts on the brain to sensitize it to other addictive substances (Dao et al., 2011; Kandel & Kandel, 2014) (see Table 47.1).

The co-occurrence of multiple forms of substance use and substance use disorders is common, especially among adolescents, possibly due to shared genetic risks or common vulnerabilities or to neurobiological priming or sensitization of the brain to addictive substances through upregulation of receptors (Falk, Yi, & Hiller-Sturmhöfel, 2006; Melroy-Greif, Stitzel, & Ehringer, 2016; Palmer et al., 2009) (see Table 47.2).

TABLE 47.1 Factors Influencing Nicotine Exposure

Smoke pH level	The more basic (alkaline) the smoke/tobacco pH, the more readily it is absorbed through the oral mucosa—regardless of inhalation—and the greater the exposure
Inhalation behavior	More inhalation relates to greater absorption of nicotine through both the oral mucosa and the lungs, increasing exposure
Length of time to finish the product/amount of product consumed	The longer the time spent using, the greater the nicotine exposure

Nicotine content and concentrations vary depending on the size and type of product being used, the smoke pH levels, inhalation behavior, and time spent with the product/the amount of product consumed.

TABLE 47.2 Alcohol and Other Drug Use Disorders by Nicotine Addiction, by Age Categories

	Past-year substance use disorder				
	Alcohol (%)	**Marijuana (%)**	**Other illicit drugs (%)**	**Prescription drugs (%)**	**Polysubstance (%)**
Total (n=48,750)	5.7	1.4	0.5	0.8	0.9
Never used nicotine					
12–17	0.6	0.5	<0.1	0.4	0.2
18–25	3.7	0.6	<0.1	0.1	0.3
26 and older	1.4	<0.1	<0.1	0.2	<0.1
Past 30-day nicotine dependence					
12–17	25.7	25.3	7.4	9.4	16.7
18–25	23.6	12.6	6.3	7.5	12.7
26 and older	10.9	2.8	2.2	2.9	2.9

An analysis of data from the 2014 National Survey on Drug Use and Health shows that alcohol and other drug use disorders are much more prevalent among individuals with nicotine dependence relative to those who have never used nicotine, especially among adolescents (Richter, Pugh, Smith, & Ball, 2017).

47.2.1 Nicotine and Alcohol

The two addictive substances most commonly used together are nicotine and alcohol (Falk et al., 2006). Smokers are more likely to consume alcohol and develop an alcohol use disorder relative to nonsmokers (Richter, Pugh, Smith, & Ball, 2017). The adverse health effects of nicotine use and of alcohol use, such as cancer, are exacerbated in those who use both addictive substances (Adams, 2017; Falk et al., 2006). The prevalent concurrent use of nicotine and alcohol may be due to several factors. Each of the substances has reward enhancing or potentiating effects on the other, known as cross reinforcement, in which there are additive effects of both substances acting on nAChRs and the mesolimbic dopamine pathways. Cross reinforcement effects are evidenced by increased craving, greater experience of reward, and an enhanced motivation among smokers to consume alcohol and the potential for smoking to undermine attempts to reduce drinking and for drinking to undermine attempts at smoking cessation. Each of the substances also serves to diminish or buffer the negative effects of the other, known as cross-tolerance, such as when alcohol mitigates some of the arousing or stimulating effects of nicotine or nicotine attenuates some of the sedating, intoxicating, or depressant effects of alcohol that otherwise might lead to reduced alcohol consumption (Adams, 2017; Tolu et al., 2017).

47.2.2 Nicotine and Cannabis

As is true of alcohol, nicotine and cannabis use and addiction have high rates of co-occurrence, and the use of one of the substances both increases the risk of becoming dependent on the other and interferes with attempts to quit or reduce the use of the other (Kristman-Valente et al., 2017; Rabin & George, 2015; Richter, Pugh, & Ball, 2017).

Frequently, cannabis is rolled into cigarettes to make a joint, tobacco is frequently added to joints to facilitate use and reduce costs, and cigars are hollowed out and filled with cannabis to make blunts. Individuals who engage in the simultaneous use of both tobacco and cannabis experience more harmful and severe consequences than those who use just one of the two substances alone (Agrawal et al., 2009).

Certain neurobiological mechanisms can help account for the appeal of co-occurring use of these two substances. Similar to nicotine and alcohol, there is some evidence of a synergistic or reinforcing effect of cannabis on nicotine, in which the priming of the brain's endocannabinoid system helps to enhance the rewarding effects of nicotine (Rabin & George, 2015). Likewise, chronic nicotine administration affects the density of cannabinoid receptors and the levels of endogenous cannabinoids in the brain, such that nicotine enhances the response to cannabis (Kristman-Valente et al., 2017). Nicotine increases arousal and attentiveness, whereas cannabis has the opposite effect. Nicotine use may help to mask the cognitive deficits associated with cannabis use, especially among youth (Jacobsen, Pugh, Constable, Westerveld, & Mencl, 2007). Unlike nicotine and alcohol, nicotine and cannabis use have a common route of administration— inhalation—which is the most efficient path to addiction; as such, the use of one substance may cue or activate cravings for the other (Rabin & George, 2015). A common genetic vulnerability might also account for the high rate of nicotine addiction among those who use cannabis (Agrawal, Silberg, Lynskey, Maes, & Eaves, 2010).

47.2.3 Nicotine and Cocaine

The most common substance for which nicotine is considered a "gateway" drug is cocaine. Rates of cocaine addiction are highest among those who began to use cocaine after having smoked cigarettes, relative to those who used cocaine before smoking or who were not regular smokers. Chronic exposure to nicotine in mice has been found to sensitize or prime the brain to be more receptive and responsive to cocaine (Kandel & Kandel, 2014; Yuan et al., 2015). Consistent with the notion that the adolescent brain is especially sensitive to the reinforcing effects of addictive substances, the potential for nicotine exposure to prime the brain to be more inclined toward cocaine use is pronounced in adolescents. Adolescent rats exposed to nicotine (vs saline) demonstrate stronger responses to cocaine (self-administration of cocaine) relative to adult rats who were similarly exposed (McQuown et al., 2007).

47.2.4 Multiple Nicotine Product Use

The co-occurrence of nicotine and other substance use and addiction is most evident in those who use multiple types of nicotine products. Richter, Pugh, and Ball (2017) and Richter, Pugh, Smith, and Ball (2017) found that respondents to a US national survey who reported using nicotine products had significantly higher rates of co-occurring substance use and substance use disorders relative to those who had not used nicotine products, and this was especially true of those who reported using both cigarettes and noncigarette nicotine products. The study also found that adolescents who reported any type of nicotine use had higher rates of alcohol use, drug use, and polysubstance use, but this association was strongest among adolescents who used multiple types of nicotine products.

47.3 WHAT IS NEEDED TO REDUCE NICOTINE USE AND ITS CONSEQUENCES

Reducing the use of nicotine products requires the sustained implementation of population-wide evidence-based prevention programs, early interventions for those at risk, and timely and effective treatments for those with addiction. It also requires a careful look at the burgeoning popularity of noncigarette nicotine products like electronic cigarettes. Although likely safer than tobacco products, electronic cigarettes and other vaping devices are inaccurately perceived as safe when, in fact, most contain nicotine and a host of other toxins. Their use has been linked repeatedly to adverse health effects, the initiation of cigarette smoking, and alcohol and other drug use and addiction, especially among youth (U.S. Department of Health and Human Services, 2016).

Despite the considerable co-occurrence with other forms of substance use, nicotine typically is not addressed within the same prevention initiatives that address alcohol and other drugs. The artificial separation of nicotine from other addictive substances in prevention programming may perpetuate an inaccurate perception of nicotine use as unrelated to, and perhaps less problematic than, other substance use and addiction. To be effective, prevention programs should address the use of all nicotine products in conjunction with other addictive substances in order to prevent nicotine product use and its health consequences and decrease the risk of other substance use and addiction (Richter & Foster, 2013; Richter, Pugh, Smith, & Ball, 2017).

Given the high prevalence of co-occurring alcohol and other drug use among those who use nicotine, failing to screen for nicotine use and addiction in educational and health-care settings might contribute to missed opportunities to address addiction among youth and adults. All forms of nicotine use should be included in substance-related screening and assessment instruments; currently, most such assessment instruments exclude nicotine products or only address cigarette use (Richter & Foster, 2013). Attending to those who screen positive for any nicotine use (and particularly polynicotine use), and not just addiction, seems merited given the high levels of co-occurring substance use and addiction found among nicotine users, even those who do not meet criteria for nicotine addiction. Providing or connecting people who use nicotine products with evidence-based interventions should be a routine part of health-care practice.

47.3.1 Implications for Treatment

Despite the obvious link between nicotine use and addiction and other substance use and addiction, tobacco cessation services are not available in the majority of mental health and addiction treatment programs (Knudsen, Studts, Boyd, & Roman, 2010). There is a prevailing assumption that encouraging tobacco cessation might compromise successful recovery from alcohol or other drug addiction, when in fact the opposite is true. Addressing nicotine addiction as a standard part of treatment for alcohol and other drug addiction has been found to improve treatment outcomes (Lemon, Friedmann, & Stein, 2003; Prochaska, Delucchi, & Hall, 2004). Nicotine product use interferes with recovery from substance use disorders and increases the risk of relapse. Treating nicotine and co-occurring substance use disorders simultaneously capitalizes on both the psychological and neurobiological mechanisms of successful recovery.

MINI-DICTIONARY OF TERMS

Co-occurrence The presence or comorbidity of more than one type of addictive substance use or substance use disorder.

Endogenous cannabinoids Receptors found in the brain and throughout the body that function to maintain homeostasis, or stability, and allow for communication and coordination across different types of cells within the brain and body.

Nicotine dependence/addiction Meeting diagnostic criteria for past-month "nicotine dependence," as measured by the nicotine dependence syndrome scale (NDSS), which is the measure of addiction used in the National Survey on Drug Use and Health. NDSS criteria generally parallel those for "tobacco use disorder" in the *Diagnostic and Statistical Manual of Mental Disorders, Fourth Edition (DSM-IV)*.

Nicotinic acetylcholine receptors (nAChRs) Brain receptors for the neurotransmitter acetylcholine, which also respond to nicotine.

Upregulation An increase in response, on the cellular level, to a stimulus by increasing the number of postsynaptic receptors.

Key Facts about Nicotine's Relationship With Other Addictive Substances

- Nicotine is the most addictive of all drugs. Two-thirds of weekly users of nicotine products become addicted to it.
- People with nicotine addiction are more than three times as likely to use marijuana, opioids, and other drugs and to have a marijuana, opioid, or other drug use disorder as people who are not addicted to nicotine.
- Young people with nicotine addiction are at higher risk than adults of engaging in co-occurring substance use or having a co-occurring substance use disorder.
- The high rates of co-occurrence may be due to shared risk factors or to the neurobiological effects of nicotine on the brain.
- Relapse to alcohol and other drug use disorders is more likely for people who continue to use nicotine than for people who engage in successful cessation.

Summary Points

- This chapter focuses on the co-occurrence of nicotine with other substance use and addiction.
- While rates of cigarette use have been declining, overall rates of nicotine product use in the United States have remained relatively steady, largely due to the increase in the use of noncigarette nicotine products like electronic cigarettes.
- Although noncombustible nicotine products are less hazardous overall, they are not risk-free.
- Aside from the risk that nicotine use will lead to addiction, it is strongly associated with the use of and addiction to alcohol and other drugs and with relapse for those in recovery from substance use disorders.
- Therefore, prevention and treatment practices should address nicotine as part of overall addiction care and not avoid addressing it for fear that it will compromise the prevention or treatment of alcohol and other drug use disorders.

References

Adams, S. (2017). Psychopharmacology of tobacco and alcohol comorbidity: a review of current evidence. *Current Addiction Reports*, *4*(1), 25–34. https://doi.org/10.1007/s40429-017-0129-z.

Agaku, I. T., King, B. A., Husten, C. G., Bunnell, R., Ambrose, B. K., Hu, S. S., et al. (2014). Tobacco product use among adults—United States, 2012-2013. *Morbidity and Mortality Weekly Report*, *63*(25), 542–547.

Agrawal, A., Lynskey, M. T., Madden, P. A., Pergadia, M. L., Bucholz, K. K., & Heath, A. C. (2009). Simultaneous cannabis and tobacco use and cannabis-related outcomes in young women. *Drug and Alcohol Dependence*, *101*(1–2), 8–12. https://doi.org/10.1016/j.drugalcdep.2008.10.019.

Agrawal, A., Silberg, J. L., Lynskey, M. T., Maes, H. H., & Eaves, L. J. (2010). Mechanisms underlying the lifetime co-occurrence of tobacco and cannabis use in adolescent and young adult twins. *Drug and Alcohol Dependence*, *108*(1–2), 49–55. https://doi.org/10.1016/j.drugalcdep.2009.11.016.

Alpert, H. R., Agaku, I. T., & Connolly, G. N. (2016). A study of pyrazines in cigarettes and how additives might be used to enhance tobacco addiction. *Tobacco Control*, *25*(4), 444–450. https://doi.org/10.1136/tobaccocontrol-2014-051943.

Apelberg, B. J., Corey, C. G., Hoffman, A. C., Schroeder, M. J., Husten, C. G., Caraballo, R. S., et al. (2014). Symptoms of tobacco dependence among middle and high school tobacco users: results from the 2012 National Youth Tobacco Survey. *American Journal of Preventive Medicine*, *47*(2 Suppl 1), S4–S14. https://doi.org/10.1016/j.amepre.2014.04.013.

Benowitz, N. L. (2009). Pharmacology of nicotine: addiction, smoking-induced disease, and therapeutics. *Annual Review of Pharmacology and Toxicology*, *49*, 57–71. https://doi.org/10.1146/annurev.pharmtox.48.113006.094742.

Camenga, D. R., Kong, G., Cavallo, D. A., Liss, A., Hyland, A., Delmerico, J., et al. (2014). Alternate tobacco product and drug use among adolescents who use electronic cigarettes, cigarettes only, and never smokers. *Journal of Adolescent Health*, *55*(4), 588–591. https://doi.org/10.1016/j.jadohealth.2014.06.016.

Cohen, A., Villanti, A., Richardson, A., Rath, J. M., Williams, V., Stanton, C., et al. (2015). The association between alcohol, marijuana use, and new and emerging tobacco products in a young adult population. *Addictive Behaviors*, *48*, 79–88. https://doi.org/10.1016/j.addbeh.2015.02.005.

Cougle, J. R., Hakes, J. K., Macatee, R. J., Zvolensky, M. J., & Chavarria, J. (2016). Probability and correlates of dependence among regular users of alcohol, nicotine, cannabis, and cocaine: concurrent and prospective analyses of the National Epidemiologic Survey on alcohol and related conditions. *Journal of Clinical Psychiatry*, *77*(4), 444–450. https://doi.org/10.4088/JCP.14m09469.

Dao, J. M., McQuown, S. C., Loughlin, S. E., Belluzzi, J. D., & Leslie, F. M. (2011). Nicotine alters limbic function in adolescent rat by a 5-HT1A receptor mechanism. *Neuropsychopharmacology*, *36*(7), 1319–1331. https://doi.org/10.1038/npp.2011.8.

Dierker, L., Swendsen, J., Rose, J., He, J., & Merikangas, K. (2012). Transitions to regular smoking and nicotine dependence in the Adolescent National Comorbidity Survey (NCS-A). *Annals of Behavioral Medicine*, *43*(3), 394–401. https://doi.org/10.1007/s12160-011-9330-9.

DiFranza, J. R. (2015). A 2015 update on the natural history and diagnosis of nicotine addiction. *Current Pediatric Reviews*, *11*(1), 43–55.

Falk, D. E., Yi, H. Y., & Hiller-Sturmhöfel, S. (2006). An epidemiologic analysis of co-occurring alcohol and tobacco use and disorders: findings from the National Epidemiologic Survey on Alcohol and Related Conditions. *Alcohol Research & Health*, *29*(3), 162–171.

Grando, S. A. (2014). Connections of nicotine to cancer. *Nature Reviews Cancer*, *14*(6), 419–429. https://doi.org/10.1038/nrc3725.

Jacobsen, L. K., Pugh, K. R., Constable, R. T., Westerveld, M., & Mencl, W. E. (2007). Functional correlates of verbal memory deficits emerging during nicotine withdrawal in abstinent adolescent cannabis users. *Biological Psychiatry*, *61*(1), 31–40.

Jamal, A., Gentzke, A., Hu, S. S., Cullen, K. A., Apelberg, B. J., Homa, D. M., et al. (2017). Tobacco use among middle and high school students—United States, 2011-2016. *Morbidity and Mortality Weekly Report*, *66*(23), 597–603. https://doi.org/10.15585/mmwr.mm6623a1.

Jamal, A., King, B. A., Neff, L. J., Whitmill, J., Babb, S. D., & Graffunder, C. M. (2016). Current cigarette smoking among adults—United States, 2005–2015. *Morbidity and Mortality Weekly Report*, *65*, 1205–1211. (2016). https://doi.org/10.15585/mmwr.mm6544a2.

Kandel, E. R., & Kandel, D. B. (2014). A molecular basis for nicotine as a gateway drug. *New England Journal of Medicine*, *371*(10), 932–943. https://doi.org/10.1056/NEJMsa1405092.

Klein, J. D. (2015). Electronic cigarettes are another route to nicotine addiction for youth. *JAMA Pediatrics*, *169*(11), 993–994. https://doi.org/10.1001/jamapediatrics.2015.1929.

Knudsen, H. K., Studts, J. L., Boyd, S., & Roman, P. M. (2010). Structural and cultural barriers to the adoption of smoking cessation services in addiction treatment organizations. *Journal of Addictive Diseases*, *29*(3), 294–305. https://doi.org/10.1080/10550887.2010.489446.

Kristman-Valente, A. N., Hill, K. G., Epstein, M., Kosterman, R., Bailey, J. A., Steeger, C. M., et al. (2017). The relationship between marijuana and conventional cigarette smoking behavior from early adolescence to adulthood. *Prevention Science*, *18*(4), 428–438. https://doi.org/10.1007/s11121-017-0774-4.

Land, T., Keithly, L., Kane, K., Chen, L., Paskowsky, M., Cullen, D., et al. (2014). Recent increases in efficiency in cigarette nicotine delivery: implications for tobacco control. *Nicotine & Tobacco Research*, *16*(6), 753–758. https://doi.org/10.1093/ntr/ntt219.

Lanza, S. T., & Vasilenko, S. A. (2015). New methods shed light on age of onset as a risk factor for nicotine dependence. *Addictive Behaviors*, *50*, 161–164. https://doi.org/10.1016/j.addbeh.2015.06.024.

Lee, Y. O., Hebert, C. J., Nonnemaker, J. M., & Kim, A. E. (2015). Youth tobacco product use in the United States. *Pediatrics*, *135*(3), 409–415. https://doi.org/10.1542/peds.2014-3202.

Lemon, S. C., Friedmann, P. D., & Stein, M. D. (2003). The impact of smoking cessation on drug abuse treatment outcome. *Addictive Behaviors*, *28*(7), 1323–1331.

Mayer, B. (2014). How much nicotine kills a human? Tracing back the generally accepted lethal dose to dubious self-experiments in the nineteenth century. *Archives of Toxicology*, *88*(1), 5–7. https://doi.org/10.1007/s00204-013-1127-0.

McQuown, S. C., Belluzzi, J. D., & Leslie, F. M. (2007). Low dose nicotine treatment during early adolescence increases subsequent cocaine reward. *Neurotoxicology and Teratology*, *29*(1), 66–73.

Melroy-Greif, W. E., Stitzel, J. A., & Ehringer, M. A. (2016). Nicotinic acetylcholine receptors: upregulation, age-related effects and associations with drug use. *Genes, Brain, and Behavior*, *15*(1), 89–107. https://doi.org/10.1111/gbb.12251.

Miech, R. A., O'Malley, P. M., Johnston, L. D., & Patrick, M. E. (2016). E-cigarettes and the drug use patterns of adolescents. *Nicotine and Tobacco Research*, *18*(5), 654–659. https://doi.org/10.1093/ntr/ntv217.

Palmer, R. H., Young, S. E., Hopfer, C. J., Corley, R. P., Stallings, M. C., Crowley, T. J., et al. (2009). Developmental epidemiology of drug use and abuse in adolescence and young adulthood: evidence of generalized risk. *Drug and Alcohol Dependence*, *102*(1–3), 78–87. https://doi.org/10.1016/j.drugalcdep.2009.01.012.

Prochaska, J. J., Delucchi, K., & Hall, S. M. (2004). A meta-analysis of smoking cessation interventions with individuals in substance abuse treatment or recovery. *Journal of Consulting and Clinical Psychology*, *72*(6), 1144–1156.

Rabin, R. A., & George, T. P. (2015). A review of co-morbid tobacco and cannabis use disorders: possible mechanisms to explain high rates of co-use. *American Journal on Addictions*, *24*(2), 105–116. https://doi.org/10.1111/ajad.12186.

Richter, L., & Foster, S. E. (2013). The exclusion of nicotine: closing the gap in addiction policy and practice. *American Journal of Public Health*, *103*, e14–e16. https://doi.org/10.2105/AJPH.2013.301448.

Richter, L., Pugh, B. S., & Ball, S. A. (2017). Assessing the risk of marijuana use disorder among adolescents and adults who use marijuana. *American Journal of Drug and Alcohol Abuse*, *43*(3), 247–260. https://doi.org/10.3109/00952990.2016.

Richter, L., Pugh, B. S., Peters, E., Vaughan, R. D., & Foster, S. E. (2016). Underage drinking: prevalence and correlates of risky drinking measures among youth aged 12-20. *American Journal of Drug and Alcohol Abuse*, *42*(4), 385–394. https://doi.org/10.3109/00952990.2015.1102923.

Richter, L., Pugh, B. S., Smith, P. H., & Ball, S. A. (2017). The co-occurrence of nicotine and other substance use and addiction among youth and adults in the United States: implications for research, practice, and policy. *American Journal of Drug and Alcohol Abuse*, *43*(2), 132–145. https://doi.org/10.1080/00952990.2016.1193511.

Smith, R. F., McDonald, C. G., Bergstrom, H. C., Ehlinger, D. G., & Brielmaier, J. M. (2015). Adolescent nicotine induces persisting changes in development of neural connectivity. *Neuroscience and Biobehavioral Reviews*, *55*, 432–443. https://doi.org/10.1016/j.neubiorev.2015.05.019.

Soneji, S., Barrington-Trimis, J. L., Wills, T. A., Leventhal, A. M., Unger, J. B., Gibson, L. A., et al. (2017). Association between initial use of e-cigarettes and subsequent cigarette smoking among adolescents and young adults: a systematic review and meta-analysis. *JAMA Pediatrics*. https://doi.org/10.1001/jamapediatrics.2017.1488.

Substance Abuse and Mental Health Services Administration. (2015). *Analysis of the National Survey on Drug Use and Health (NSDUH), 2014 [Public Use Files]*. Retrieved from (2015). https://www.samhsa.gov/samhsa-data-outcomes-quality/major-data-collections/public-use-files-2014-nsduh/.

The National Center on Addiction and Substance Abuse. (2011). *Adolescent substance use: America's #1 public health problem.* New York: Author.

The National Center on Addiction and Substance Abuse. (2015). *Understanding and addressing nicotine addiction: A science-based approach to policy and practice.* New York: Author.

The National Center on Addiction and Substance Abuse. (2016). *Beyond cigarettes: The risks of non-cigarette nicotine products and implications for tobacco control.* New York: Author.

Tolu, S., Marti, F., Morel, C., Perrier, C., Torquet, N., Pons, S., et al. (2017). Nicotine enhances alcohol intake and dopaminergic responses through β2* and β4* nicotinic acetylcholine receptors. *Scientific Reports*, *7*, 45116. https://doi.org/10.1038/srep45116.

U.S. Department of Health and Human Services. (2016). *E-cigarette use among youth and young adults: A report of the surgeon general.* Atlanta, GA: Office on Smoking and Health.

Yuan, M., Cross, S. J., Loughlin, S. E., & Leslie, F. M. (2015). Nicotine and the adolescent brain. *Journal of Physiology*, *593*(16), 3397–3412. https://doi.org/10.1113/JP270492.

CHAPTER

48

Comorbid Smoking and Gambling Disorder: Potential Underlying Mechanisms and Future Explorations

Emma V. Ritchie, David C. Hodgins, Daniel S. McGrath

Department of Psychology, University of Calgary, Calgary, AB, Canada

Abbreviations

APA American Psychiatric Association
AUD alcohol use disorder
DGs disordered gambling/disordered gamblers
DSM-5 the Diagnostic and Statistical Manual of Mental Disorders, 5th ed.
GD gambling disorder
MAO monoamine oxidase
NRT nicotine replacement therapy
SUD substance use disorder
TUD tobacco use disorder
VLT video lottery terminal

48.1 INTRODUCTION

Problematic use of tobacco typically includes repeated failed efforts to quit; development of tolerance; withdrawal symptoms; strong physiological cravings; and continued use of tobacco products despite negative effects on one's health, relationships, or personal safety (American Psychiatric Association, 2013). These are some of the criteria for tobacco use disorder (TUD), which is classified as a substance use disorder (SUD) in the latest edition of the *Diagnostic and Statistical Manual for Mental Disorders* (DSM-5; APA, 2013). Nicotine is an alkaloid chemical contained in tobacco products and has been identified as the putative psychoactive constituent of tobacco (Benowitz, 2008). However, based on the accumulation of evidence identifying the contribution of individual and synergistic effects of other tobacco constituents in TUD (Baker, Breslau, Covey, & Shiffman, 2012; Fagerström, 2012), the term "tobacco dependence" has now principally supplanted "nicotine dependence" in the literature. Tobacco dependence usually does not occur in isolation, but rather is highly comorbid with other addictive behaviors and mental health conditions (Lasser, Boyd, & Woolhandler, 2000). While there is a strong relationship between alcohol use disorder (AUD) and TUD in particular (Falk, Yi, & Hiller-Sturmhöfel, 2006), other addictive behaviors are also frequently comorbid with smoking, including gambling disorder (GD) (Lorains, Cowlishaw, & Thomas, 2011). For instance, a recent study found elevated rates of GD in individuals who are tobacco dependent (~14%) compared to the general population (~5%) (Barnes, Welte, Tidwell, & Hoffman, 2015). Indeed, recent studies have begun to focus more intently on the contribution of nicotine and tobacco in the maintenance of GD and subclinical disordered gambling (DG).

Recently reclassified as a behavioral addiction in the DSM-5 (APA, 2013), gambling is frequently concurrent with other addictive behaviors, particularly tobacco use (McGrath & Barrett, 2009). In their systematic review of comorbid disorders commonly found among DGs, Lorains et al. (2011) identified numerous psychological ailments such as mood disorders and SUDs. However, tobacco use displayed the highest prevalence rate with approximately 60% of DGs also meeting the criteria for TUD. There is also evidence that DGs who smoke may represent a more severe form of gambling pathology. One study comparing DGs who smoke to those who did not found that the smokers displayed more severe gambling symptoms, were more likely to have another psychological disorder, and were more likely to have sought previous treatment (Odlaug, Stinchfield, Golberstein, & Grant, 2013) (Figs. 48.1 and 48.2).

Neuroscience of Nicotine
https://doi.org/10.1016/B978-0-12-813035-3.00048-4

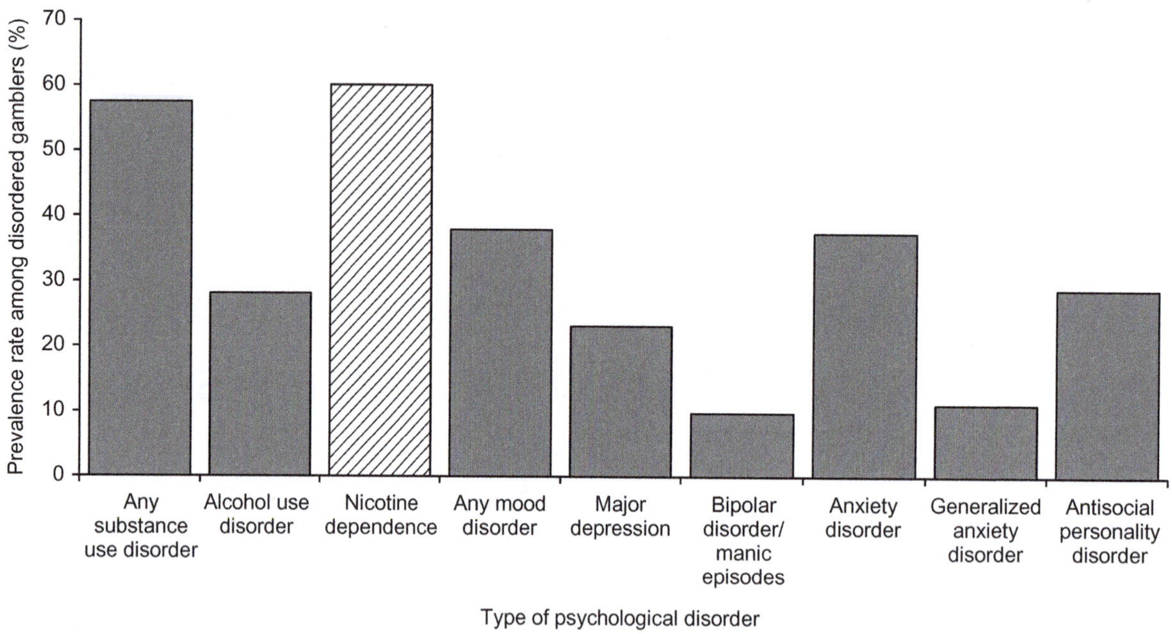

FIG. 48.1 The reported prevalence rates of various psychological disorders among disordered gamblers.
The prevalence rates (mean effect sizes) of various psychological disorders in individuals with gambling disorder. Data are from a meta-analysis conducted by Lorains et al. (2011).

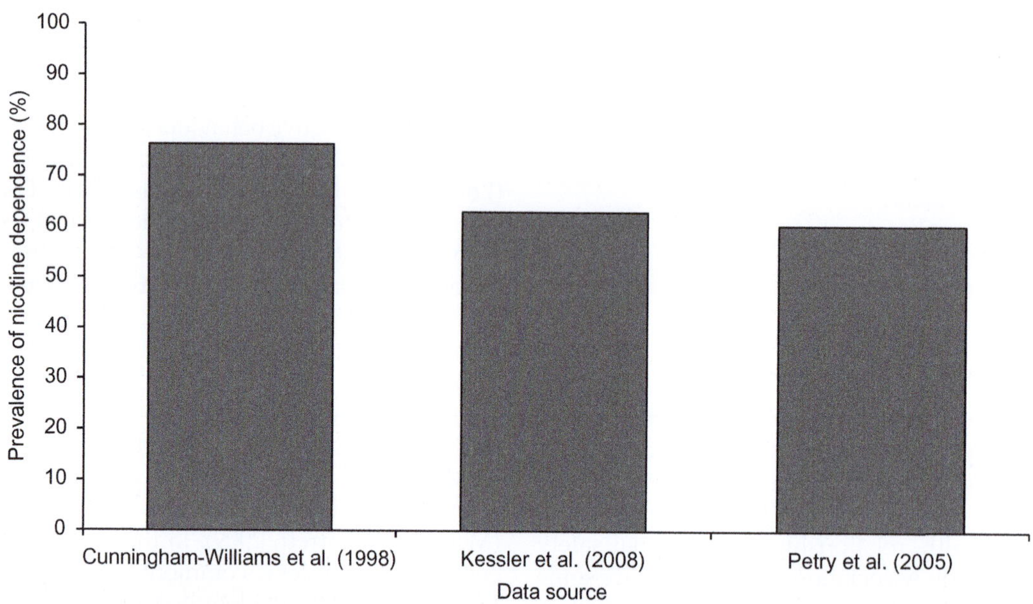

FIG. 48.2 The reported prevalence rates of nicotine dependence in epidemiological studies of disordered gamblers.
Reported prevalence rate of nicotine dependence, among samples of disordered gamblers, according to three different studies.

48.2 UNDERLYING MECHANISMS INVOLVED IN CO-MORBID SMOKING AND GAMBLING

Numerous potential underlying contributors to the smoking-gambling relationship have been proposed. For example, conditioning effects may play a role. Classical conditioning occurs when an individual learns to associate one type of stimulus with a different stimulus following repeated pairings. Gamblers may come to associate smoking with gambling if they smoke while gambling (McGrath & Barrett, 2009). Essentially, both activities may serve as a conditioned cue for the other, where the act of smoking may trigger a craving to gamble or vice versa. This phenomenon, known as cross cue reactivity, has been empirically shown to at least partially explain the TUD and AUD relationship (e.g., Piasecki et al., 2011), but it has only recently been studied in relation to gambling and smoking (Wulfert, Harris, & Broussard, 2016).

Another body of research centers on the possibility that there are dispositional differences between gamblers who smoke and gamblers who do not smoke. Studies measuring gambling severity have found that DGs who smoke score higher on these indices (Grant, Kim, Odlaug, & Potenza, 2008; McGrath, Barrett, Stewart, & McGrath, 2012). DGs who smoke also tend to display elevated psychiatric symptoms (anxiety, depression, and increased suicidality) and are more likely to have another SUD (Grant et al., 2008; Petry & Oncken, 2002; Potenza et al., 2004). Additionally, a recent study reported that DGs who smoke scored higher on a facet of impulsivity called negative urgency, whereby one reacts impulsively when faced with negative feelings or situations (Boothby, Kim, Romanow, Hodgins, & McGrath, 2017). Others have proposed that DGs who smoke may do so to dampen their impulsivity due to nicotine's cognitive-enhancing effects (Mooney, Odlaug, Kim, & Grant, 2011). Lastly, smokers and DGs have also both been shown to have elevated neuroticism scores and lower conscientiousness scores (Bagby et al., 2007; Terracciano & Costa Jr., 2004), although personality dimensions in DGs who smoke have received comparatively less empirical focus.

In addition, certain components of tobacco smoke have been shown to inhibit monoamine oxidase (MAO), an enzyme that has been linked to depression (Lewis, Miller, & Lea, 2007). It has been posited that some individuals smoke as a way of self-medicating their depression (Benowitz, 2008). Evidence indicates that MAO platelet activity is also depleted in DGs, and this is related to lower impulse control due to the effect of low MAO activity on the serotonergic system (Carrasco, Sáiz-Ruiz, Hollander, César, & López-Ibor, 1994). MAO inhibitors have also been used to ameliorate self-control issues in individuals with borderline personality disorder; it is possible that a similar process occurs with DGs (i.e., they smoke to deal with their impulsivity) (Carrasco et al., 1994).

48.3 INFLUENCE OF NICOTINE/ TOBACCO ON REINFORCED BEHAVIOUR

Findings from both animal and human studies indicate that in addition to primary reinforcing effects and conditioned reinforcing properties (e.g., smoking cues), nicotine also has reinforcement-enhancing effects for other behavior. That is, acute administration of nicotine can augment reinforcement for rewards that are *not* directly related to smoking itself (see Perkins, Karelitz, & Boldry, 2017 for a review). Importantly, these reinforcement-enhancing effects occur through nonassociative mechanisms and therefore do not require prior pairing of the drug directly with stimuli that are predictive of reward. Empirical investigations of these effects were first conducted in rodent models, which revealed that both contingent and noncontingent nicotine administration can potentiate responding for primary rewards such as food; however, these effects appear to be especially robust for sensory rewards such as visual lights and auditory tones (Rupprecht et al., 2015). The exact neurobiological mechanisms responsible for the reinforcement-enhancing effects of nicotine have yet to be established. However, nicotine's modulation of dopamine via the mesocorticolimbic pathway has long been known to play a critical role in tobacco dependence (Balfour, 2015). There is some evidence to suggest that administration of selective D3 receptor antagonists potentiate reinforcement-enhancing effects specifically but do not promote further nicotine self-administration (Paterson, 2009) (Table 48.1).

A smaller body of research has focused on human subjects, yet this work does reveal parallel nicotine reinforcement-enhancing effects in smokers. Studies have found that acute nicotine administration via cigarettes can augment ratings of rewarding visual stimuli such as pictures of attractive faces (Attwood et al., 2009; Attwood et al., 2012). A separate line of research has focused on both the form of reward and route of administration. In a series of studies, Perkins and Karelitz assessed nicotine's influence on reinforced performance on a laboratory operant task. Generally, they found that nicotine enhances the reinforcing value of sensory rewards such as music (Perkins & Karelitz, 2013a; Perkins & Karelitz, 2013b; Perkins, Karelitz, Jao, & Stratton, 2013) and video (Perkins et al., 2015; Perkins & Karelitz, 2016), but not monetary reinforcement. These effects were more consistently witnessed when nicotine was delivered via cigarette compared to pure forms such as nicotine replacement therapy (NRT). Yet, one study from another research group found that acute NRT amplified performance on a card-sorting task reinforced by money (Dawkins et al., 2009). The aggregate of these findings provides support for nicotine's reinforcement-enhancing effects for sensory rewards in humans and possibly similar influences on monetary rewards.

48.4 INFLUENCE OF NICOTINE/ TOBACCO ON RISK-TAKING

A modest line of research has explored nicotine's effects on laboratory risk-taking tasks. For instance, Pilarski, Skeel, and Reilly (2014) examined the influence

TABLE 48.1 Summary of the Reinforcement-Enhancing Effects of Nicotine on Humans in Laboratory Studies

Study	Method of nicotine administration	Reward type	Outcome
Attwood, Penton-Voak, and Munafò (2009)	Nicotine cigarette (0.6 mg nicotine) vs denicotinized cigarette (0.005 mg nicotine)	Photographs of faces	Ratings of attractiveness were significantly higher among participants who smoked the nicotine cigarette
Attwood, Penton-Voak, Goodwin, and Munafò (2012)	Nicotine cigarette (0.6 mg nicotine) vs denicotinized cigarette (0.005 mg nicotine), which were administered after ingesting an alcoholic (0.4 g/kg) or placebo beverage	Photographs of faces and of natural scenes	Ratings of attractiveness for both types of photographs were higher among participants who smoked the nicotine cigarette, regardless of alcohol; highest ratings of attractiveness were in participants who received alcohol and nicotine cigarette
Dawkins, Powell, Pickering, Powell, and West (2009)	Lozenge (4 mg nicotine) vs placebo lozenge	Money rewards in a card-sorting task	Reinforced responding for task higher in lozenge condition in heavy smokers only
Perkins and Karelitz (2013a)	Half nicotine cigarette (eight puffs) vs full nicotine cigarette (16 puffs) after overnight abstinence	Music (varying in levels of preference) and no reward (control)	Reinforced responding for high preference music rewards were enhanced following smoking of full nicotine cigarette
Perkins and Karelitz (2013b)	Nicotine cigarette (0.6 mg nicotine) vs denicotinized cigarette (<0.05 mg nicotine)	Money, music, ceasing of a loud noise, and no reward (control)	Reinforced responding for music rewards only were enhanced following smoking of nicotine cigarette
Perkins and Karelitz (2014)	Nicotine cigarette (0.6 mg) vs denicotinized cigarette (<0.05 mg nicotine) vs no smoking after overnight abstinence	Music, video, money, and no reward (control)	Reinforced responding for music and video rewards were enhancing following smoking of nicotine cigarette; no difference between denicotinized cigarette and no smoking
Perkins, Karelitz, and Michael (2015)	*E*-cigarettes containing nicotine (36 mg nicotine) or placebo (0 mg nicotine) after overnight abstinence	Music, video, money, and no reward (control)	Reinforced responding for video rewards only were enhanced following use of the nicotine-containing e-cigarette
Perkins and Karelitz (2016)	Overnight abstinence vs nicotine cigarette of their choosing (six puffs before each trial)	Music, video, and no reward (control)	Reinforced responding for music and video rewards were enhancing following smoking of nicotine cigarette; effect for video rewards stronger in males, while effect of music rewards stronger in females

This table provides a summary of research that examines how nicotine enhances responding to different sensory and nonsensory rewards in humans.

of nicotine on risk-taking in nonsmokers. Risk-taking was measured with experimental tasks including finger tapping, the Balloon Analogue Risk Task and the Stoplight Task. Varying doses of nicotine gum (2 and 4 mg) were compared to placebo. Contrary to predictions, nicotine (4 mg) actually reduced risk-taking on the Balloon Analogue Risk Task compared to placebo. Similarly, males who received nicotine also experienced a decline in risk-taking on the Stoplight Task. Their initial hypotheses were based on research suggesting that nicotine's influence on the dopaminergic system should increase risk-taking behaviors (Sevy et al., 2006). The researchers theorized that these contrary findings may have been due to a combination of the specific demands of the tasks (e.g., task relevance) and individual differences (e.g., personality) of participants in their sample.

48.5 INFLUENCE OF NICOTINE/ TOBACCO ON GAMBLING

Very few laboratory studies have been devoted to studying how nicotine may impact gambling specifically. This is somewhat surprising as gambling provides a real-world illustration of reinforced behavior that co-occurs with tobacco use. Gambling contains elements of sensory

TABLE 48.2 Effects of Nicotine on Gambling and Risk-Taking Tasks in Laboratory Studies

Study	Method of nicotine administration	Task	Outcome
Barrett, Collins, and Stewart (2015)	Nicotine cigarette (0.6 mg nicotine) vs denicotinized cigarette (0.05 mg nicotine), which were administered after ingesting an alcoholic beverage or a placebo	$40 worth of play on a single spinning reel VLT; played for as long as participants wished (up to 15 min) or until no credits remained	Smoking the nicotine cigarette resulted in higher average bets; ingesting alcohol increased desire to gamble
McGrath, Barrett, Stewart, and Schmid (2012)	Nicotine inhalers (10 mg nicotine) vs placebo inhalers after overnight abstinence	$60 worth of play on a single spinning reel VLT; played for as long as participants wished (up to 30 min) or until no credits remained	There were no significant differences between nicotine and placebo conditions on any measure of gambling behavior (e.g., money spent, average bet size, and bets per minute) or desire to gamble
McGrath et al. (2013)	Nicotine lozenges (4 mg nicotine) vs placebo (breath mints) after overnight abstinence	Option to gamble $10 on a VLT or to add the money to their compensation for the session	There was no effect of nicotine on the decision to gamble or keep the money; also no significant differences between nicotine and placebo conditions on gambling cravings
Pilarski et al. (2014)	Nicotine gum (2 and 4 mg) vs placebo gum	Balloon Analogue Risk Task	Risk-taking scores were higher in placebo group compared to 4 mg nicotine group

This table summarizes research that has been conducted on the effect of nicotine on gambling behaviors and cravings, as well as risk-taking tasks, in controlled laboratory settings.

stimulation, monetary reinforcement, and risk-taking, each of which has been investigated in isolation in previous nicotine drug-challenge experiments. In one study, McGrath, Barrett, Stewart, and Schmid (2012) used nicotine and placebo inhalers to administer nicotine to regular video lottery terminal (VLT) gamblers who were daily smokers. They did not find any differences in gambling cravings, enjoyment of gambling, or gambling behavior (e.g., number of bets and amount per bet) between the nicotine and placebo conditions. In a follow-up study, McGrath, Dorbeck, and Barrett (2013) provided nicotine and placebo lozenges to assess the influence of nicotine on gambling cravings and behaviors on at-risk/disordered gamblers who smoke. Again, no differences between nicotine and placebo conditions on gambling were found (Table 48.2).

Importantly, the studies described above administered nicotine via NRTs (inhalers and lozenges) and not tobacco cigarettes. To address the possibility that nicotine via smoking may have unique effects on gambling, Barrett et al. (2015) provided nicotine-containing and denicotinized cigarettes, as well as alcoholic and placebo beverages, to gamblers in a VLT study. When participants smoked a nicotinized cigarette, they spent, on average, more per bet compared to when they had a denicotinized cigarette. This finding suggests that nicotine administered via tobacco smoke did have a measurable effect on at least one aspect of gambling behavior.

Thus, it is possible that nicotine via tobacco smoke has unique and distinguishable effects and that the frequent co-occurrence of smoking and gambling is related to tobacco smoking rather than nicotine per se. The largely null findings from research in this area may be explained by a special relationship between *smoking* and gambling, not just nicotine in isolation. While nicotine is the main putative chemical in tobacco, there are other distinctive features of smoking that contribute to its addictive qualities. For instance, smoking is presumed to deliver a nicotine bolus to the brain and does so more efficiently than other methods of administration (Benowitz, Hukkanen, & Jacob III, 2009). Drugs that result in a rapid onset of physiological effects shortly after administration tend to be more addictive (Samaha & Robinson, 2005). Similarly, when a rewarding stimulus (e.g., the pleasurable effects of smoking) is presented quickly after inhalation, it is more likely that the behavior will be reinforced (Perkins & Karelitz, 2013b). Since nicotine administration via smoking is more reinforcing, this may be directly related to how it influences other rewarding behavior such as gambling.

48.6 THE POTENTIAL INFLUENCE OF GAMBLING BEHAVIOUR ON TOBACCO USE

While tobacco dependence and GD are clearly associated, the temporal order of these disorders remains unclear. Very little attention has been given to the possible contributing role of gambling on smoking itself. Recently, Bussu and Detotto (2015) examined whether a bidirectional relationship exists between gambling and several SUDs including tobacco. They surveyed over 700 gamblers at gambling venues on whether they consumed alcohol, tobacco, or other drugs while they gambled and how much they spent gambling. Interestingly, they did not find evidence of a bidirectional relationship between gambling and tobacco use; however, greater within-session gambling involvement did appear to increase an individual's propensity to smoke.

A recent laboratory study investigated cross cue reactivity between smoking and gambling (Wulfert et al., 2016). Cross cue reactivity describes how a conditioned cue for one addiction-related stimulus may elicit reactions (e.g., cravings) for a second addictive behavior with which it is frequently paired. The researchers recruited poker players who smoked, poker players who did not smoke, and smokers who did not gamble. Participants were exposed to a tray containing a deck of cards (gambling cue), a tray with cigarettes and a lighter (smoking cue), and a pencil and an eraser (neutral cue). Subjective craving ratings and skin conductance were measured several times. Exposure to smoking and gambling cues increased urges for tobacco in the poker players who smoked, but not pure smokers or gamblers. Interestingly, cross cue reactivity was not found for desire to gamble.

Unfortunately, no known studies have directly assessed the influence of gambling on actual nicotine self-administration. However, studies from the drinking literature have shown that alcohol enhances the pleasurable effects of smoking, more so than smoking enhances the pleasurable effects of alcohol consumption (e.g., Piasecki et al., 2011). In the only known laboratory-based experiment on this topic in the gambling field, Stewart, McWilliams, Blackburn, and Klein (2002) randomly assigned community-recruited VLT players (regular gamblers without an AUD) to either 90 min of VLT play or watching a movie (control task); each was followed by the opportunity to purchase alcohol vs nonalcoholic beverages with their own money. A ceiling was placed on the number of alcoholic drinks participants could purchase to ensure a blood alcohol concentration (BAC) < 0.08 g%. As predicted, significantly more people in the VLT condition (11 of 15) purchased at least one alcoholic beverage than in the movie condition (6 of 15). Eight participants reached the ceiling BAC and were unable to purchase any additional alcoholic beverages. These findings suggest that exposure to gambling increased the likelihood to consume alcohol. It is conceivable that a similar pattern would be observed with tobacco administration; however, this possibility has not yet been empirically explored.

48.7 AREAS IN NEED OF FUTURE EXPLORATION

In sum, the co-occurrence and maintenance of smoking and gambling remains a complex phenomenon, and a number of potential underlying mechanisms are likely involved in this relationship. While there is some evidence in support of nicotine's reinforcement-enhancing effects on gambling behavior (e.g., average bet size on a VLT), the magnitude of these effects are uncertain. It's plausible that tobacco smoke and not nicotine per se contributes to gambling cravings and behaviors, but more research on how smoking impacts gambling needs to be conducted. The possible impact of gambling outcomes on smoking behavior is just now receiving modest empirical focus. The directionality of this association is far from clear, and the common transdiagnostic contributions to this relationship have yet to be concretely identified.

Going forward, we can envision several laboratory studies that would help to further elucidate the smoking-gambling association. For instance, the majority of research has largely ignored recreational gamblers who smoke. Drawing on the literature examining the co-use of alcohol and tobacco, smoking increases while under the influence of alcohol and many people who "socially smoke" only do so under the influence of alcohol (McKee, Hinson, Rounsaville, & Petrelli, 2004). It is possible that a similar relationship between smoking and recreational gambling exists. Another area that would be valuable to study relates to how smoking cessation impacts gambling behaviors, especially longitudinally. Most NRT products are designed for prolonged use (e.g., weeks), whereas most drug-challenge studies have used these products for acute administration. In addition, future research could also further examine motivations underlying smoking in DGs, as there has been little research devoted to this specifically. Previous research has indicated that some DGs find that smoking encourages them to gamble, while some find that it soothes them when they are trying to abstain from gambling. Understanding more about what motivates smoking in this group may be informative. Finally, while Bussu and Detotto (2015) suggested that gambling directly impacts smoking, no controlled laboratory studies have assessed whether different gambling outcomes (e.g., a big win vs a win disguised as a loss) directly influence smoking behavior (Table 48.3).

MINI-DICTIONARY OF TERMS

Bidirectional relationship In this chapter, this refers to the potential relationship between smoking and gambling whereby each exerts some kind of influence on the other; i.e., gambling encourages further use of tobacco or vice versa.

TABLE 48.3 Key Facts of Comorbid Smoking and Gambling

- Tobacco dependence is one of the most frequent comorbid disorders in individuals with another psychological disorder.
- Rates of tobacco dependence among individuals with a current or past psychological disorder are almost twice those of individuals who have never been diagnosed with one.
- The rate of tobacco dependence among individuals with gambling disorder (GD) has been estimated to be between 40% and 60%.
- Studies cite tobacco dependence as the most frequent or second most frequent addiction comorbid with GD.
- GD was originally classified as an impulse control disorder, but this was recently changed due to research showing that GD more closely resembled substance addictions.
- Smoking in casinos used to be a common practice until concerns over the effects of second-hand smoke became more prominent. Many casinos are now banning smoking, and significant interest has been paid to the effects of this ban on gambling revenues and the behaviors of their patrons.

This table lists some key facts of tobacco dependence and gambling disorder, including comorbidity rates and facts, as well as key facts about each. *GD*, gambling disorder.

Classical conditioning A type of unconscious learning where an individual learns to associate one type of stimulus with a different stimulus following repeated pairings of both stimuli together.

Comorbidity In psychology, this refers to when an individual experiences at least two different psychological disorders concurrently; in this chapter, it is mostly used to refer to individuals with both gambling disorder and tobacco use disorder.

Conditioned reinforcement Also known as secondary reinforcement, this refers to a *learned* response to stimuli that is presented following a behavior in order to encourage future instances of the behavior.

Cross cue reactivity A psychological phenomenon in which stimuli related to one type of addictive behavior (e.g., smoking) elicits a measurable response (i.e., cravings) for a different addiction (e.g., gambling) due to the two frequently co-occurring together (i.e., smoking when gambling).

Gambling disorder A behavioral addiction where an individual has difficulty controlling their gambling behaviors to the extent that they experience withdrawal symptoms when they cannot gamble, cannot quit gambling despite repeated attempts to, and continue to gamble despite financial or personal harms.

Impulsivity A psychological construct that has been linked to disordered gambling and smoking. Impulsivity is characterized by acting without thinking about the consequences of one's actions and the inability to control one's actions, resulting in reckless or inappropriate behavior.

Primary reinforcement This refers to a *natural* (not learned) process whereby future instances of a behavior are encouraged by a stimulus that is presented following a behavior. Nicotine is considered a primary reinforcer because the physiological effects of nicotine can be pleasurable and rewarding and encourage future use.

Reinforcement A type of learning where a stimulus is presented after a behavior in order to encourage that behavior to occur again in the future.

Reinforcement-enhancing effects This refers to a unique property of nicotine where it can strengthen the reinforcement of rewarding stimuli that are not related to smoking.

Summary Points

- This chapter focuses on comorbid smoking and gambling, particularly in people who have a gambling disorder.
- While numerous studies have noted that smoking and gambling are highly comorbid, less is known about why this occurs.
- A number of underlying mechanisms have been proposed to explain this relationship, including the influence of classical conditioning, underlying psychological or dispositional characteristics, and the effects of components of tobacco smoke on undesirable psychological states such as depression and impulsivity.
- Nicotine influences reinforcement in three distinct ways. It acts as a primary and conditioned reinforcer, and it also enhances reinforcement for rewarding stimuli that are unrelated to smoking and nicotine itself.
- Little research has been done on how nicotine specifically affects gambling, although the modest body of research currently suggests that it is smoking specifically and not nicotine per se that has an impact of gambling behaviors.
- The effect of gambling on smoking behaviors must also be considered. Two recent studies have suggested that gambling may encourage smoking to take place, either through eliciting cravings or because it enhances the pleasurable effects associated with smoking.
- This remains a topic that would be fruitful for future research, and we have outlined several ideas that would contribute to the literature in this area.

References

American Psychiatric Association. (2013). *Diagnostic and statistical manual of mental disorders* (5th ed.). Washington, DC: Author.

Attwood, A. S., Penton-Voak, I. S., Goodwin, C., & Munafò, M. R. (2012). Effects of acute nicotine and alcohol on the rating of attractiveness in social smokers and alcohol drinkers. *Drug and Alcohol Dependence, 125*(1–2), 43–48. https://doi.org/10.1016/j. drugalcdep. 2012.03.014.

Attwood, A. S., Penton-Voak, I. S., & Munafò, M. R. (2009). Effects of acute nicotine administration on ratings of attractiveness of facial cues. *Nicotine and Tobacco Research, 11*(1), 44–48. https://doi.org/10.1093/ntr/ntn006.

Bagby, R. M., Vachon, D. D., Bulmash, E. L., Toneatto, T., Quilty, L. C., & Costa, P. T. (2007). Pathological gambling and the five-factor model of personality. *Personality and Individual Differences, 43*(4), 873–880. https://doi.org/10.1016/j.paid.2007.02.011.

Baker, T. B., Breslau, N., Covey, L., & Shiffman, S. (2012). DSM criteria for tobacco use disorder and tobacco withdrawal: a critique and proposed revisions for DSM-5. *Addiction, 107*(2), 263–275. https://doi.org/10.1111/j.1360-0443.2011.03657.x.

Balfour, D. J. K. (2015). The role of mesoaccumbens dopamine in nicotine dependence. D. Balfour, & M. Munafó (Eds.), *Current topics in behavioural neurosciences* (pp. 55–98). *The neuropharmacology of nicotine dependence: Vol. 24*(pp. 55–98). Basel, Switzerland: Springer International Publishing. https://doi.org/10.1007/978-3-319-13482-6_3.

Barnes, G. M., Welte, J. W., Tidwell, M. O., & Hoffman, J. H. (2015). Gambling and substance use: co-occurrence among adults in a recent general population study in the United States. *International Gambling Studies, 15*(1), 55–71. https://doi.org/10.1080/14459795.2014.990396.

Barrett, S. P., Collins, P., & Stewart, S. H. (2015). The acute effects of tobacco smoking and alcohol consumption on video-lottery terminal gambling. *Pharmacology, Biochemistry, and Behaviour*, *130*, 34–39. https://doi.org/10.1016/j.pbb.2014.12.015.

Benowitz, N. L. (2008). Clinical pharmacology of nicotine: Implications for understanding, preventing, and treating tobacco addiction. *Clinical Pharmacology and Therapeutics*, *83*(4), 531–541. https://doi.org/10.1038/clpt.2008.3.

Benowitz, N. L., Hukkanen, J., & Jacob, P., III (2009). Nicotine chemistry, metabolism, kinetics, and biomarkers. In J. E. Henningfieldet al. (Ed.) *Handbook of experimental pharmacology* (pp. 29–60). Berlin: Springer-Verlag. https://doi.org/10.1007/978-3-540-69248-5_2.

Boothby, C. A., Kim, H. S., Romanow, N. K., Hodgins, D. C., & McGrath, D. S. (2017). Assessing the role of impulsivity in smoking and non-smoking disordered gamblers. *Addictive Behaviours*, *70*, 35–41. https://doi.org/10.1016/j.addbeh.2017.02.002.

Bussu, A., & Detotto, C. (2015). The bidirectional relationship between gambling and addictive substances. *International Gambling Studies*, *15*(2), 285–308. https://doi.org/10.1080/14459795.2015.1042493.

Carrasco, J. L., Sáiz-Ruiz, J., Hollander, E., César, J., & López-Ibor, J. J. (1994). Low platelet monoamine oxidase activity in pathological gambling. *Acta Psychiatrica Scandinavica*, *90*(6), 427–431. https://doi.org/10.1111/j.1600-0447.1994.tb01619.x.

Dawkins, L., Powell, J. H., Pickering, A., Powell, J., & West, R. (2009). Patterns of change in withdrawal symptoms, desire to smoke, reward motivation and response inhibition across 3 months of smoking abstinence. *Addiction*, *104*(5), 850–858. https://doi.org/10.1111/j.1360-0443.2009.02522.x.

Fagerström, K. (2012). Determinants of tobacco use and renamed the FTND to the Fagerström test for cigarette dependence. *Nicotine and Tobacco Research*, *14*(1), 75–78. https://doi.org/10.1093/ntr/ntr137.

Falk, D. E., Yi, H., & Hiller-Sturmhöfel, S. (2006). An epidemiologic analysis of co-occurring alcohol and tobacco use and disorders: findings from the National Epidemiologic Survey on alcohol and related conditions. *Alcohol Research and Health*, *29*(3), 162–171.

Grant, J. E., Kim, S. W., Odlaug, B. L., & Potenza, M. N. (2008). Daily tobacco smoking in treatment-seeking pathological gamblers: clinical correlates and co-occurring psychiatric disorders. *Journal of Addiction Medicine*, *2*(4), 178–184. https://doi.org/10.1097/ADM.0b013e3181878673.

Kessler, R. C., Hwang, I., LaBrie, R., Petukhova, M., Sampson, N. A., Winters, K. C., et al. (2008). DSM-IV pathological gambling in the National Comorbidity Survey Replication. *Psychological Medicine*, *38*, 1351–1360. https://doi.org/10.1017/S0033291708002900.

Lasser, K., Boyd, J. W., & Woolhandler, S. (2000). Smoking and mental illness: a population based prevalence study. *Journal of the American Medical Association*, *284*(20), 2606–2610. https://doi.org/10.1001/jama.284.20.2606.

Lewis, A., Miller, J. H., & Lea, R. A. (2007). Monoamine oxidase and tobacco dependence. *NeuroToxciology*, *28*(1), 182–195. https://doi.org/10.1016/j.neuro.2006.05.019.

Lorains, F. K., Cowlishaw, S., & Thomas, S. A. (2011). Prevalence of comorbid disorders in problem and pathological gambling: systematic review and meta-analysis of population surveys. *Addiction*, *106*, 490–498. https://doi.org/10.1111/j.1360-0443.2010.03300.x.

McGrath, D. S., & Barrett, S. P. (2009). The comorbidity of tobacco smoking and gambling: a review of the literature. *Drug and Alcohol Review*, *28*(6), 676–681. https://doi.org/10.1111/j.1465-3362.2009.00097.x.

McGrath, D. S., Barrett, S. P., Stewart, S. H., & McGrath, P. R. (2012). A comparison of gambling behaviour, problem gambling indices, and reasons for gambling among smokers and nonsmokers who gamble: evidence from a provincial gambling prevalence study. *Nicotine and Tobacco Research*, *14*(7), 833–839. https://doi.org/10.1093/ntr/ntr294.

McGrath, D. S., Barrett, S. P., Stewart, S. H., & Schmid, E. A. (2012). The effects of acute doses of nicotine on video lottery terminal gambling in daily smokers. *Psychopharmacology*, *220*, 155–161. https://doi.org/10.1007/s00213-011-2465-3.

McGrath, D. S., Dorbeck, A., & Barrett, S. P. (2013). The influence of acutely administered nicotine on cue-induced cravings for gambling in at-risk video lottery terminal gamblers who smoke. *Behavioural Pharamcology*, *24*(2), 124–132. https://doi.org/10.1097/FBP.0b013e32835f3cff.

McKee, S. A., Hinson, R., Rounsaville, D., & Petrelli, P. (2004). Survey of subjective effects of smoking while drinking among college students. *Nicotine and Tobacco Research*, *6*(1), 111–117. https://doi.org/10.1080/14622200310001656939.

Mooney, M. E., Odlaug, B. L., Kim, S. W., & Grant, J. E. (2011). Cigarette smoking status in pathological gamblers: association with impulsivity and cognitive flexibility. *Drug and Alcohol Dependence*, *117*(1), 74–77. https://doi.org/10.1016/j.drugalcdep.2010.12.017.

Odlaug, B. L., Stinchfield, R., Golberstein, E., & Grant, J. E. (2013). The relationship of tobacco use with gambling problem severity and gambling treatment outcome. *Psychology of Addictive Behaviours*, *27* (3). https://doi.org/10.1037/a0029812.

Paterson, N. E. (2009). The neuropharmacological substrates of nicotine reward: reinforcing versus reinforcement-enhancing effects of nicotine. *Behavioral Pharmacology*, *20*(3), 211–225. https://doi.org/10.1097/FBP.0b013e32832c7083.

Perkins, K. A., & Karelitz, J. L. (2013a). Influence of reinforcer magnitude and nicotine amount on smoking's acute reinforcement enhancing effects. *Drug and Alcohol Dependence*, *133*(1), 167–171. https://doi.org/10.1016/j.drugalcdep.2013.05.016.

Perkins, K. A., & Karelitz, J. L. (2013b). Reinforcement enhancing effects of nicotine via smoking. *Psychopharmacology*, *228*(3), 479–486. https://doi.org/10.1007/s00213-013-3054-4.

Perkins, K. A., & Karelitz, J. L. (2014). Sensory reinforcement-enhancing effects of nicotine via smoking. Experimental and Clinical Psychopharmacology, 22(6), 511-516. Doi: 10.137/a0037823

Perkins, K. A., & Karelitz, J. L. (2016). Potential sex differences in the pattern of sensory reinforcers enhanced by nicotine. *Experimental and Clinical Psychopharmacology*, *24*(3), 156–161. https://doi.org/10.1037/pha0000073.

Perkins, K. A., Karelitz, J. L., & Boldry, M. C. (2017). Nicotine acutely enhances reinforcement from non-drug rewards in humans. *Frontiers in Psychiatry*. *8*(65). https://doi.org/10.3389/fpsyt.2017.00066.

Perkins, K. A., Karelitz, J. L., Jao, N. C., & Stratton, E. (2013). Possible reinforcement enhancing effects of burpropion during initial smoking abstinence. *Nicotine and Tobacco Research*, *15*, 1141–1145. https://doi.org/10.1093/ntr/nts224.

Perkins, K. A., Karelitz, J. L., & Michael, V. C. (2015). Reinforcement enhancing effects of acute nicotine via electronic cigarettes. *Drug and Alcohol Dependence*, *153*, 104–108. https://doi.org/10.1016/j.drugalcdep.2015.05.041.

Petry, N. M., & Oncken, C. (2002). Cigarette smoking is associated with increased severity of gambling problems in treatment-seeking gamblers. *Addiction*, *97*(6), 745–753. https://doi.org/10.1046/j.1360-0443.2002.00163.x.

Petry, N. M., Stinson, F. S., & Grant, B. F. (2005). Comorbidity of DSM-IV pathological gambling and other psychiatric disorders: results from the National Epidemiologic Survey on alcohol and related conditions. *The Journal of Clinical Psychiatry*, *66*(5), 564–574. https://doi.org/10.4088/JCP.v66n0504.

Piasecki, T. M., Jahng, S., Wood, P. K., Robertson, B. M., Epler, A. J., Cronk, N. J., et al. (2011). The subjective effects of alcohol-tobacco co-use: an ecological momentary assessment investigation. *Journal of Abnormal Psychology*, *120*(3), 557–571. https://doi.org/10.1037/a0023033.

Pilarski, C. R., Skeel, R. L., & Reilly, M. P. (2014). Acute effects of nicotine on risky choice among non-smokers. *The Psychological Record*, *64*, 151–159. https://doi.org/10.1007/s40732-014-0057-4.

Potenza, M. N., Steinberg, M. A., McLaughlin, S. D., Wu, R., Rounsaville, B. J., Krishnan-Sarin, S., et al. (2004). Characteristics of tobacco-smoking problem gamblers calling a gambling helpline. *The American Journal on Addictions*, *13*(5), 471–493. https://doi.org/10.1080/10550490490483044.

Rupprecht, L. E., Smith, T. T., Schassburger, R. L., Buffalari, D. M., Sved, A. F., & Donny, E. C. (2015). Behavioural mechanisms underlying nicotine reinforcement. In: D. Balfour, & M. Munafó (Eds.), *Current topics in behavioural neurosciences* (pp. 19–53). *The neuropharmacology of nicotine dependence: Vol. 24*(pp. 19–53). Basel, Switzerland: Springer International Publishing.

Samaha, A., & Robinson, T. E. (2005). Why does the rapid delivery of drugs to the brain promote addiction? *TREND in Pharmacological Sciences*, *26*(2), 82–87. https://doi.org/10.1016/j.tips.2004.12.007.

Sevy, S., Hassoun, Y., Bechara, A., Yechiam, E., Napolitano, B., Burdick, K., et al. (2006). Emotion-based decision-making in healthy subjects: short-term effects of reducing dopamine levels. *Psychopharmacology*, *188*(2), 228–235. https://doi.org/10.1007/s00213-006-0450-z.

Stewart, S. H., McWilliams, L. A., Blackburn, J. R., & Klein, R. M. (2002). A laboratory-based investigation of relations among video lottery terminal (VLT) play, negative mood, and alcohol consumption in regular VLT players. *Addictive Behaviours*, *27*(5), 819–835. https://doi.org/10.1016/S0306-4603(01)00213-1.

Terracciano, A., & Costa, P. T., Jr. (2004). Smoking and the five-factor model of personality. *Addiction*, *99*(4), 472–481. https://doi.org/10.1111/j.1360-0443.2004.00687.x.

Wulfert, E., Harris, K., & Broussard, J. (2016). The role of cross-cue reactivity in coexisting smoking and gambling habits. *Journal of Gambling Issues*, *32*, 28–43. https://doi.org/10.4309/jgi.2016.32.3.

CHAPTER

49

Neuroscience of Tobacco and Crack Cocaine Use: Metabolism, Effects, and Symptomatology

*Antonio Gomes de Castro-Neto**[,a], Rossana Carla Rameh-de-Albuquerque*[†,a], *Pollyanna Fausta Pimentel de Medeiros**[,a], *Roberta Uchôa*[‡,a], *Beate Saegesser Santos**[,a]

***Study Group on Alcohol and other Drugs, Research Group on Biomedical Nanotechnology, Department of Pharmaceutical Sciences, Federal University of Pernambuco, Cidade Universitária, Recife, Pernambuco, Brazil**

[†]Study Group on Alcohol and other Drugs, Federal Institute of Education, Science and Technology of Pernambuco, Cidade Universitária, Recife, Pernambuco, Brazil

[‡]Department of Social Work, Federal University of Pernambuco, Economistas Avenue, Cidade Universitária, Recife, Pernambuco, Brazil

Abbreviations

ACTH	adrenocorticotropic hormone
ADHD	attention deficit hyperactivity disorder
AEME	anhydroecgonine methyl ester
CHT	choline high-affinity presynaptic transporter
CPP	conditioned place preference
DHEA	dehydroepiandrosterone
HDAC	histone deacetylase
KO	knockout rats
LH	luteinizing hormone
LTSP	long-term synaptic potentiation
MDU	multiple drug use
mGluR2/3	Group II metabotropic glutamate receptors
nAChR	nicotinic acetylcholine receptor
SAHA	suberoylanilide hydroxamic acid
US	United States
VTA	ventral tegmental area
WT	Wild-type rats

49.1 INTRODUCTION

Although scientific reports from the late 1960s pointed out the beneficial use of tobacco (or cigarette), it is very well known nowadays that this habit is extremely harmful. Despite what has been established about nicotine, tobacco remains one of the most consumed drugs in the world, having important reinforcing factors: a valuable market (and its profits) and many cultural aspects related to its using patterns (Pinto et al., 2017; Reitsma et al., 2017).

On the other hand, crack cocaine use is now an established form of drug abuse around the world. Brazil is today the greatest user of this form of cocaine, and it has been observed that most crack cocaine users combine this drug with tobacco.

The combination of tobacco and cocaine (in different forms) has been in fact performed in different ways by users around the world. Some South American studies reported the smoking of *basuco*, a drug made of coca paste and tobacco or marijuana, in the form of a cigarette known as *pitillos* or *papilloes* (Jeri, 1980; Siegel, 1982). Although that habit is a major health issue concern, there are still only a few quantitative and qualitative studies that describe the users experience when they combine tobacco and crack cocaine and that try to understand the reasons for that (Gonçalves & Nappo, 2015).

In Recife, located at the Northeast coast of Brazil, crack cocaine combined with tobacco is called *capeta* and with marijuana is called *mesclado* or *melado* (Fig. 49.1).

Users report that these combinations reduce the craving, especially with marijuana. *Capeta* users state that they have more control over crack cocaine use than without, although the intensity of this control varies

[a] All authors approve the submission of the manuscript and adhere to the ethics relating to copying and plagiarism.

Neuroscience of Nicotine
https://doi.org/10.1016/B978-0-12-813035-3.00049-6

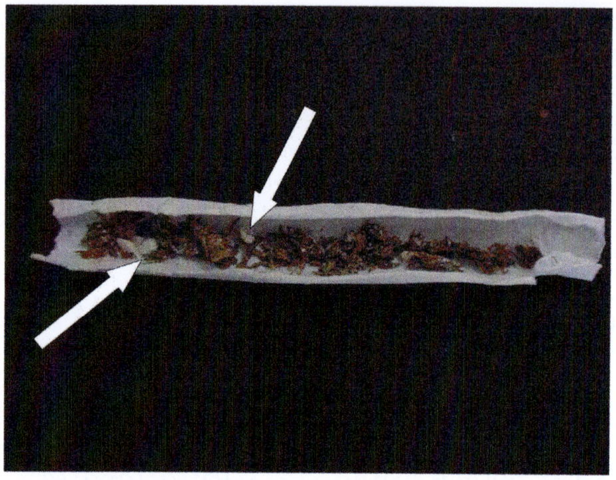

FIG. 49.1 Capeta cigarette.
Handmade tobacco cigarette with crack cocaine. White arrows indicate some crack cocaine rocks.

among users. The following descriptions exemplify some data reported by users:

> In addition to its biomechanical similarity, to use the "rock," according to patients, it is necessary to insert the cigarette ashes at the bottom of the pipe or tin can to perform the combustion. In other words, [...] it is imperative to use tobacco to use crack cocaine (Zeni & Araujo, 2011, p. 32).
>
> There is one thing that all the crack cocaine users agree: the ashes remaining in the pipe after burning the "rock" are highly valuable and considered as having greater power than the "rock" itself (Alves, 2016, p.5).

It has been observed that the association of crack cocaine with tobacco is a preferential use pattern. However, many users say that they only use tobacco ashes to ease the combustion process:

> [...] The cigarette, this way, is inevitable among "rock circles," and new tobacco ash supplies are always welcome. The tobacco cigarette itself is not always smoked, as many crack cocaine users are tobacco-free. For that, the cigarette is set alight and left to burn in a vertical position to obtain the ashes without inhaling the smoke [...] (Alves, 2016, p.5).

Other important characteristics that have to be mentioned on the simultaneous use of tobacco and crack cocaine are the similarity related to their stimulant effects and the way they are consumed. The type of pipe to smoke tobacco—"*boris*"—is very similar to the one to smoke crack cocaine, especially its curvilinear form (Jorge, Quindere, Yasui, & Albuquerque, 2013). In other words, its behavioral aspect brings synergic correspondences that facilitate simultaneous use:

> Cigarette is essential to crack cocaine consumption. This is socially observed through the "cigaretter," person responsible at the drug use scene to supply crack cocaine users with tobacco cigarettes, sold in packs or by unit or even by smoked cigarette ashes [...] (Alves, 2016, p. 9).

Moreover, when tobacco is used alone, it acts as a trigger for the stimulation of crack cocaine use. However, the craving is attenuated when tobacco is used simultaneously with crack cocaine—"*capeta.*" Losing control is one of the most common problems among crack cocaine users; therefore, further studies related to the use of tobacco or marijuana combined with crack cocaine as a control of harm reduction strategies should be carried out (Acioli Neto, 2014).

49.2 TOBACCO AS A GATEWAY

Some authors consider tobacco use as a gateway to other drugs such as marijuana, cocaine, and crack cocaine. However, it is still unclear how nicotine, the active substance of tobacco, promotes the gateway mechanism. The gateway theory was developed by Kandel (1975), when observing that young people engage in drug use in phases and sequence. She showed that, among US and other west society's general population, there is a clear developmental drug use sequence that starts with legal drugs and proceeds to illegal drugs (Fig. 49.2). Specifically, tobacco and alcohol use leads to marijuana, next to cocaine, and then to other illicit drugs (Degenhardt et al., 2010; Kandel, 2002; Kandel, Yamaguchi, & Chen, 1992; Wagner & Anthony, 2002).

Alternatively to the gateway theory, MDU is based on the idea of common drug use susceptibility. That means, the use of more than one drug increases the risk of progressing to MDU (Palmer et al., 2009; Vanyukov et al., 2012).

Epidemiological and biological studies combined suggest a model in which nicotine plays its initial effect over cocaine by HDAC inhibition in the striatum. This inhibition induces general changes in histone acetylation in the nucleus accumbens—changes that are known for transcription alteration of different FosB genes when cocaine is administered—suggesting that nicotine increases FosB transcription in response to cocaine (Kandel & Kandel, 2014; Levine et al., 2011). These factors create an environment to induce gene expression. Long-term potentiation in the nucleus accumbens is blocked when long-term exposure to nicotine is followed by cocaine use, which

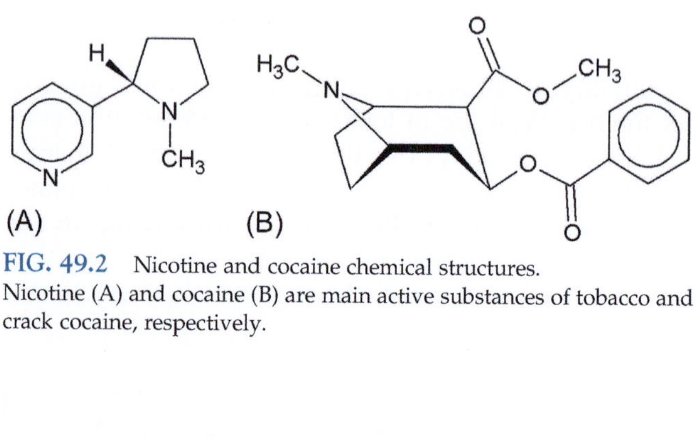

FIG. 49.2 Nicotine and cocaine chemical structures.
Nicotine (A) and cocaine (B) are main active substances of tobacco and crack cocaine, respectively.

presumably decreases restrictions at dopaminergic neurons in VTA, that leads to a better dopamine release.

A metabonomic model of ^{1}H nuclear magnetic resonance was used to investigate metabolic profiles of the nucleus accumbens and striatum in rats. This model showed that preadministrated nicotine drastically increased the CPP induced by 2 mg/kg of cocaine, being similar to CPP induced by high dose (20 mg/kg) of cocaine alone. Curiously, metabolic profiles showed considerable overlap between these two models. These overlapped metabolites included mainly neurotransmitters and molecules that participated on energetic homeostasis and cell metabolism. These results show that nicotine reinforcement effect in behavioral responses to cocaine can be related to changes of some specific metabolites in the nucleus accumbens and striatum, hence creating a favorable metabolic environment to increase cocaine conditioning and reward effects (Li et al., 2014).

49.3 SMOKE AS DRUG ABUSE POTENTIATION

Cocaine pyrolysis product, AEME (Fig. 49.3), is a partial agonist of M1 and M3 muscarinic receptors of acetylcholine and an antagonist of other subtypes. Cocaine and AEME combined substantially potentiate cocaine effects in adult Wistar rats than cocaine alone. An increase in dopamine levels in the caudate-putamen and a decrease of protein levels in D1 receptors in this brain region were observed. Within the nucleus accumbens occurs the increase of dopamine level without changes in protein levels in D1 and D2 receptors. These behavioral and neurochemical data demonstrate that AEME alone does not provoke behavioral sensitivity. On the other hand, it significantly potentiates cocaine effects when coadministrated, resulting in dopamine increase in the caudate-putamen and nucleus accumbens, brain regions where dopamine release is mediated by cholinergic activity (Garcia et al., 2016).

49.4 NICOTINE MECHANISM OF ACTION ASSOCIATED TO COCAINE

Nicotine and cocaine produce their positive reinforcement effects through extracellular dopamine increase. The combination of marginal reinforcement doses can produce an addictive dopamine increase, which may explain that tobacco and crack cocaine simultaneous use is more intense than the intake of these two drugs alone. In fact, nicotine increase induces levels of extracellular dopamine by cocaine in the nucleus accumbens in an addictive pharmacological manner (Mello & Newman, 2011). Moreover, pharmacological studies demonstrated that some nicotine and cocaine effects can be mimicked by dopamine agonists or blocked by dopamine antagonists (Corrigall & Coen, 1991; Desai, Barber, & Terry, 2003).

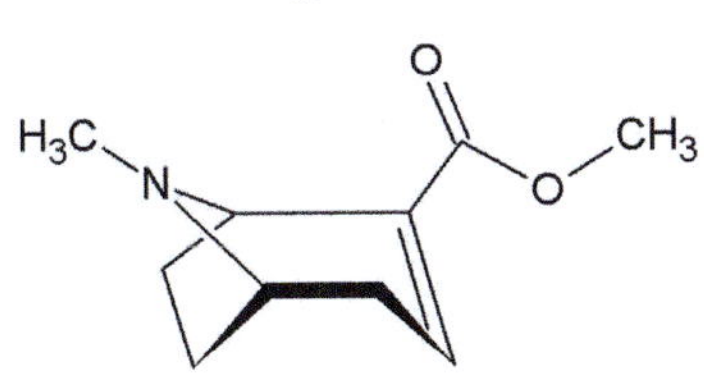

FIG. 49.3 AEME chemical structure.
AEME is the main crack cocaine pyrolysis product.

Mecamylamine, a noncompetitive antagonist nicotine receptor, can avoid cocaine self-administration increase in rats. It has been reported in clinical studies that mecamylamine reduced cocaine craving in nicotine and cocaine codependent individuals (Reid, Mickalian, Delucchi, & Berger, 1999).

Nicotine significantly improved changes in synaptic plasticity in the striatum induced by cocaine. Short-term (1 day) and long-term (7 days) preexposure to nicotine facilitates LTSP induction by cocaine in amygdala—the brain region involved in emotion orchestrations and in drug dependence. This effect in LTSP is unidirectional, and the nicotine exposure after cocaine use is ineffective. This nicotine metaplastic effect over cocaine is longstanding, though reversible. The LTSP facilitation can be reached for 24 days, but not for 40 days after nicotine cessation. Like the striatum, pretreatment with SAHA, a HDAC inhibitor, simulates nicotine initiation effect. These results give additional evidence that nicotine initiation effect can be reached, at least partially, by histone acetylation inhibition, and indicate that the amygdala seems to be an important structural part of the brain to process metaplastic nicotine effect over cocaine (Huang, Kandel, Kandel, & Levine, 2013).

Nicotine and cocaine stimulate hormone in hypothalamus-hypophysis-adrenal-gonadal axis. There is increased evidence that hormonal environment can modulate effects related to the abuse of these drugs. Nicotine and cocaine stimulate LH and ACTH quick increase, followed by gradual cortisol and DHEA increase. Estimates of positive subjective effects increased immediately after nicotine or cocaine use (Mello, 2010).

49.5 NICOTINE MOLECULAR MECHANISM IN COMBINATION WITH COCAINE

Cholinergic signaling through nAChR in the mesolimbic pathway is involved with drug abuse reward effects like nicotine and cocaine. Studies in rats with mecamylamine, a nAChR nonselective antagonist, inhibit CPP and behavioral sensitization induced by cocaine. When nAChR subtype selective antagonists, such as dihydro-beta-erythroidine, β2 selective antagonist, and methyllycaconitine, α7 selective antagonist, are simultaneously administered, they avoid cocaine behavioral sensitization (Khroyan, Yasuda, Toll, Polgar, & Zaveri, 2015).

Two strong α3β4 nAChR selective ligands, AT-1001 and AT-1012, produce robust CPP in cocaine doses of 5–30 mg/kg in rats, while locomotor activity behavioral sensitization is only observed in higher doses (20–30 mg/kg). Pretreatment with AT-1001 (1–10 mg/kg) or AT-1012 (3–10 mg/kg) blocked CPP induced by 5 mg/kg, but not with 30 mg/kg of cocaine. Lower doses of AT-1001 (0.3 mg/kg) and AT-1012 (1–3 mg/kg) do not increase locomotor activity by 5 or 30 mg/kg of cocaine. However, AT-1001, in these doses, blocks induced locomotor sensitization by 30 mg/kg of cocaine. This indicates that α3β4 nAChR play a role over cocaine reward and behavioral effects and its selective ligands can attenuate cocaine-induced behavioral expressions. Functional α3β4 nAChR antagonist AT-1001 also blocked nicotine self-administration in rats (Khroyan et al., 2015).

Other nicotine receptors such as α6 nAChR (and its many subtypes) have an important role in nicotine behavioral effects. Yet, little is known about the role of α6 nAChR many subtypes on nicotine reward effects. In an animal model, α6 KO exhibited a direct change in nicotine dose-answer curve compared with WT; but α4 KO did not have preference to nicotine, suggesting that α6α4β2 nAChR (and its subtypes) are involved with positive effects of the drug. Besides, α6β2 nAChR found in the nucleus accumbens has an important role in conditioned nicotine reward, and intraaccumbal injection of α-conotoxin MII (α6β2 selective) blocks nicotine CPP. In contrast with nicotine, α6 KO was not conditioned to cocaine, but cocaine CPP in α4 KO was preserved. Intriguingly, α-conotoxin MII blocks cocaine conditioned in α4 KO, implying in cocaine reward by α6β2 nAChR (and its subtypes). It is important to notice that these effects cannot be generalized, as α6 KO showed climate aversion, conditioned to lithium chloride, and palatable food CPP. In addition, dopamine absorption was not different in α6 KO or WT (Sanjakdar et al., 2015).

Dopamine signaling in the central nervous system mediates the capacity to provoke dependence of multiple drugs, such as nicotine and cocaine. Dopaminergic neuron activation in VTA and dopamine release by neuron projections in the nucleus accumbens are closely controlled by cholinergic signaling mediated by nAChR. Cholinergic signaling capacity is ruled by CHT Slc5a7 availability and activity. Depending on the synaptic activity, choline reuptake promotes acetylcholine synthesis. The loss of CHT constitutive expression, mediated by gene elimination of a Slc5a7 gene copy in rats (CHT +/−), leads to significant reduction of basal extracellular dopamine levels in the nucleus accumbens. In addition, CHT heterozygosity results in attenuated dopamine increase after nicotine and cocaine systemic administration. This reinforces the critical role of acetylcholine capacity in tonic signaling and drug modulation. Yet, they demonstrate that gene reductions imposed by CHT lead to dopamine signaling decrease and to weak answers to reward stimulus. These mechanisms possibly contribute to disorders linked to cholinergic signaling disturbs (Fig. 49.4), including depression and ADHD (Dong, Dani, & Blakely, 2013).

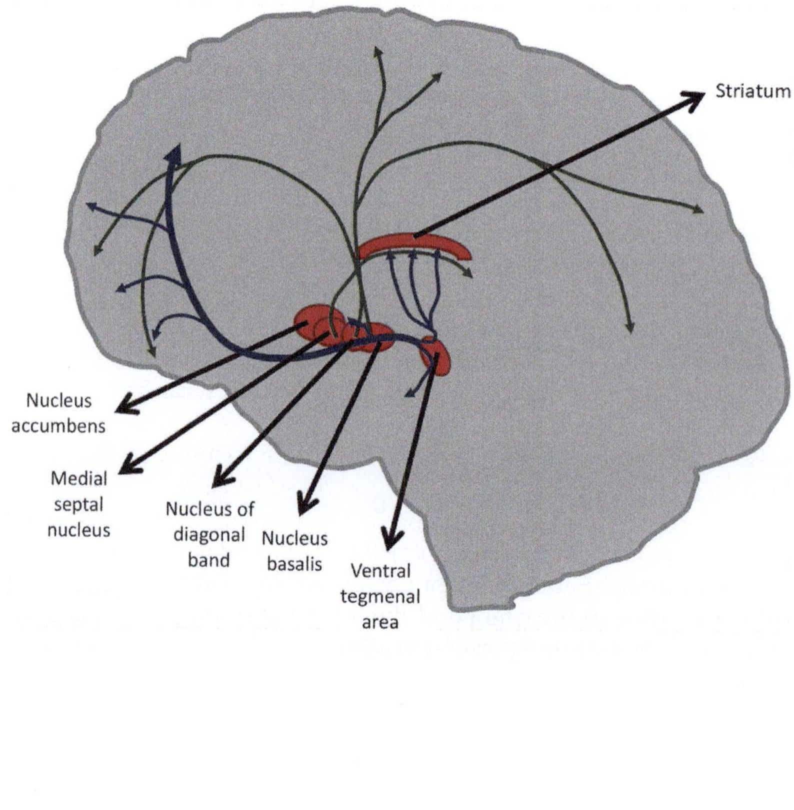

FIG. 49.4 Dopaminergic and cholinergic pathways. Main brain structures affected by nicotine and cocaine effects. Blue arrows show dopaminergic pathways, and green arrows show cholinergic pathways.

Metabotropic glutamate receptors, mainly mGluR2/3, have an important role in behavioral mediation of drug reward and in relapse mechanism. A study applying a prototype agonist of mGluR2/3, LY379268, showed in rat models nicotine reinforcement attenuation and reintegration induced by search suggestion for the drug and, moreover, reintegration induced by stimulus and contexts associated with different drugs of abuse (cocaine, heroin, and methamphetamine). However, in primates, LY379268 showed opposite results related to cocaine abuse effects. This substance blocked nicotine, but not cocaine in self-administration model. There was partial overposition between doses that affected nicotine and food self-administration. In abstinent primates, LY379268 blocked mGluR2/3 depending on the amount of nicotine, but not on the amount of cocaine, inducing drug relapse. For nicotine and cocaine expert animal groups, LY379268 strongly reduced induced readjustment to drug-seeking behavior. These findings give robust support to the potential use of mGluR2/3 agonists for nicotine dependence treatment and suggest their use for relapse prevention induced by environmental factors associated to drug use (Justinova, Foll, Redhi, Markou, & Goldberg, 2016).

49.6 NICOTINE AND COCAINE DEPENDENCE AND CRAVING

In a double-blind cross study, placebo-controlled, 20 patients all tobacco smokers were randomly distributed for nicotine and placebo in a single dose. Desire and anxiety were measured before and after crack cocaine suggestions with visual analogic scales for crack cocaine desire and mood. Skin conductance and temperature were registered before and after crack cocaine suggestions. After exposure to crack cocaine suggestions, all subjects reported an increase of desire and anxiety in relation to initial records and also increased skin conductance and decreased skin temperature. The increased anxiety induced by crack cocaine suggestions was strongly reinforced by tobacco. Responses related to skin measurements were not affected by tobacco. It was then concluded that cocaine use is highly induced by nicotine. This response occurred without tobacco suggestions, indicating that nicotine may have psychopharmacological effects directly linked to cocaine desire (Reid, Mickalian, Delucchi, Hall, & Berger, 1998).

Nicotine and cocaine promote locomotor activity changes. Cocaine-pretreated rats have suffered sensitization to cocaine motor answers and nicotine-crossed locomotor sensitization. In self-administrated experiments, repeated administration of cocaine caused an increase of the nicotine rupture point and the ingestion during a compulsive 24h session (Leão et al., 2013).

Delay discount is an impulsivity measurement that describes how reinforcement loses its value while use delay increases. A higher delay discount is observed in a confident form among those with disorders related to MDU compared with general population. Delay discount was compared in four groups: tobacco alone dependents, crack cocaine alone dependents, tobacco and crack cocaine dependents, and control group. Results showed that delay discount was significantly higher among tobacco and crack cocaine dependents in comparison to tobacco alone and control group. Delay discount was also higher among crack cocaine alone in relation to tobacco alone and control group, but differences between tobacco and crack cocaine and crack cocaine alone groups were not observed. Delay discount differs depending on MDU type, but not in MDU quantities (García-Rodriguez, Secades-Villa, Weidberg, & Yoon, 2013).

Crack cocaine high subjective classification and desire and other subjective effects were changed by tobacco. Smokers responded that crack cocaine use strongly increased their desire to smoke both nicotine and crack cocaine. Data suggest that the act of simultaneous tobacco and crack cocaine smoke increases crack cocaine desire (Brewer III, Mahoney III, Nerumalla, Newton, & Garza II, 2013). Fig. 49.5 depicts the craving mechanism cycle.

Tobacco use shows a linear increase in the frequency of tobacco and crack cocaine craving random evaluations. Tobacco significantly reduced crack cocaine craving during abstinence periods, and there was a tendency (not significant) to reduce morning tobacco use. So, tobacco use is strongly associated to crack cocaine use and its craving and vice versa. Dependence treatment for tobacco should be offered simultaneously with other drug treatment from the very beginning and not isolated (Epstein, Marrone, Heishman, Schmittner, & Preston, 2010).

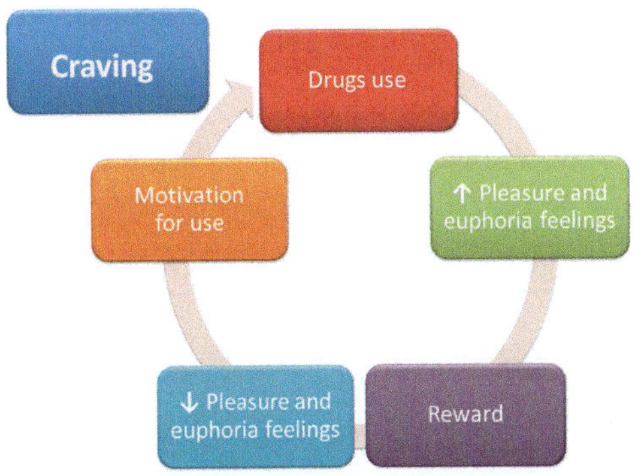

FIG. 49.5 Craving mechanism.
How craving settles in the individual. Drug use raises pleasure and euphoria feelings, which are mediated by reward mechanisms. The decrease of these feelings motivates the individual to look for drugs again. If this quest is incessant, then one has craving.

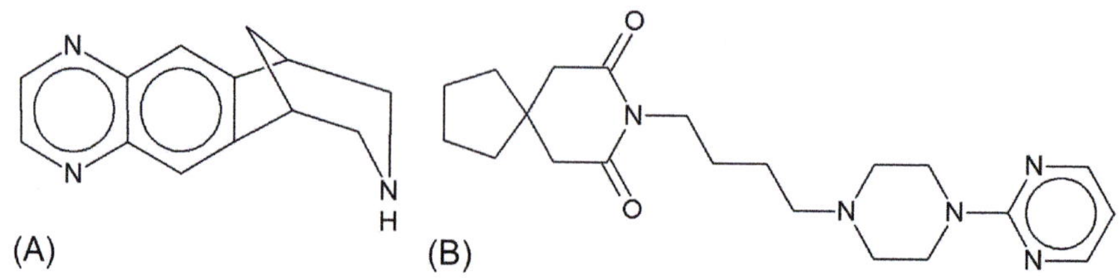

FIG. 49.6 Main drugs used in tobacco and crack cocaine pharmacological treatment. Varenicline (A) and buspirone (B) are the main drugs used in tobacco and crack cocaine pharmacological treatment.

49.7 NICOTINE AND COCAINE DEPENDENCE PHARMACOLOGICAL TREATMENT

The ideal treatment should reduce both tobacco and crack cocaine misuse. Varenicline is a partial agonist clinically used of α4β2 and α6β2 nAChR (and its subtypes) and a complete agonist of α7 nAChR. It facilitates tobacco cessation in clinical studies, reduces nicotine self-administration, and substitutes nicotine discriminatory stimulus in preclinical studies. Varenicline (Fig. 49.6) responds to either isolated nicotine or combined with cocaine, in a dose-dependent way without significant effect on the eating pattern. Though, varenicline did not significantly reduce isolated nicotine, isolated cocaine, and combined nicotine and cocaine in a low-dose self-administration (0.001, 0.0032, and 0.01 mg/kg, respectively). Therefore, varenicline selectively attenuates isolated nicotine reinforcement effects, but not cocaine effects as it depends on cocaine doses on combined nicotine and cocaine (Mello, Fivel, Kohut, & Carroll, 2014).

Buspirone (Fig. 49.6) is a nonbenzodiazepine anxiolytic clinically used that attenuates serotoninergic and dopaminergic systems. In preclinical studies, cocaine self-administration was reduced after acute and chronic treatment in rhesus monkeys. Buspirone presents a dose-dependent response not only to isolated nicotine doses but also to combined nicotine and cocaine doses without significant effects on eating pattern. Buspirone selectively attenuates isolated nicotine and also combined nicotine/cocaine reinforcement effects in primate nonhuman self-administration drug model (Mello, Fivel, & Kohut, 2013).

49.8 FINAL CONSIDERATIONS

MDU enhances a greater drug use among users with this profile. Tobacco use acts as a physiological stimulus to crack cocaine users, but at the same time, tobacco decreases crack cocaine craving, functioning to reduce crack cocaine use frequency. These mechanisms are due to nicotine and cocaine action in dopaminergic and cholinergic receptors in central nervous system regions such as the nucleus accumbens and striatum.

Some implications for treatments of tobacco and crack cocaine misuse showed in this chapter are that the treatment should focus not only on one of these drugs but also on both at the same time, as one serves as reinforcement to the other. Thus, tobacco and crack cocaine treatment should be combined, with the support of pharmacological therapies because of the high degree of chemical dependence of these two drugs. Finally, new approach methods, behavioral therapies, and medication should be developed as result of a better understanding of tobacco and crack cocaine combined neuropharmacological mechanisms.

MINI-DICTIONARY OF TERMS

Basuco Mixture of cocaine base with tobacco or marijuana.

Boris Type of pipe used to smoke crack cocaine.

Capeta Cigarette made of a mixture of crack cocaine and tobacco

Conditioned place preference Animal model test that reflects the capability of environment stimuli, originally neutral, to acquire positive motivational properties after being presented repeatedly to a substance with potential of misuse.

FosB Cell membrane protein decoded by FosB gene involved in drug dependence development.

Knockout rats Genetically modified rats with a deactivated, changed, or blocked existing gene by a DNA artificial component.

Long-term synaptic potentiation Longstanding improvement in signal transmission between two neurons that results in their stimulus in a synchronic way.

Mesclado or melado Cigarette made by a mixture of crack cocaine and marijuana.

Metabonomic Quantitative measurement of dynamic multiparametric metabolic answer of living systems to physiopathology's stimuli or gene modifications.

Metabotropic receptors Postsynaptic receptors made by an extracellular domain that binds to neurotransmitters and to an intracellular domain bound to a G-protein.

Pitillos or papilloes Other denominations for basuco.

Relapse When a person tries to modify or quit a problematic behavior—in this case, problematic drug use—but fails.

Rock Popular term to describe crack cocaine.

Key Facts of Nicotine and Cocaine Misuse

- Crack cocaine users are also, most of the time, tobacco users.
- Tobacco decreases crack cocaine craving effects, although it acts as a trigger to its consumption.
- Crack cocaine pyrolysis products potentiate cocaine effects.

- Nicotine and cocaine acts in a synergic way in the reward pathway.
- Nicotine and cocaine dependence treatment must be performed together due to drug synergic actions.

Summary Points

- Most of crack cocaine users use tobacco to decrease the craving and also because cigarette ashes facilitate crack cocaine burn.
- Nicotine and cocaine acts in the reward pathway mediated by dopamine and acetylcholine in a synergic way.
- Nicotine use promotes changes in brain regions, mainly in the nucleus accumbens and striatum, favoring effects promoted by cocaine.
- Nicotine and cocaine synergic action increase positive reinforcement effects making even more difficult the treatment for dependence.
- Crack cocaine and tobacco chemical dependence treatment must be performed together, being pharmacological drug therapy one of the most important strategies for this condition.

References

Acioli Neto, M. L. (2014). *Os contextos de uso do crack: representações e práticas sociais entre usuários*. Recife: Novas Edições Acadêmicas.

Alves, Y. D. D. (2016). O uso do crack como ele é: o cachimbo, o "bloco" e o usuário. *Etnográfica, 20*, 495–515.

Brewer, A. J., III, Mahoney, J. J., III, Nerumalla, C. S., Newton, T. F., & Garza, R. D. L., II. (2013). The influence of smoking cigarettes on the high and desire for cocaine among active cocaine users. *Pharmacology, Biochemistry and Behavior, 106*, 132–136.

Corrigall, W. A., & Coen, K. M. (1991). Selective dopamine antagonists reduce nicotine self-administration. *Psychopharmacology, 104*, 171–176.

Degenhardt, L., Dierker, L., Chiu, W. T., Medina-Mora, M. E., Neumark, Y., Sampson, N., et al. (2010). Evaluating the drug use "gateway" theory using cross-national data: consistency and associations of the order of initiation of drug use among participants in the WHO World Mental Health Surveys. *Drug and Alcohol Dependence, 108*, 84–97.

Desai, R. I., Barber, D. J., & Terry, P. (2003). Dopaminergic and cholinergic involvement in the discriminative stimulus effects of nicotine and cocaine in rats. *Psychopharmacology, 167*, 335–343.

Dong, Y., Dani, J. A., & Blakely, R. D. (2013). Choline transporter hemizygosity results in diminished basal extracellular dopamine levels in nucleus accumbens and blunts dopamine elevations following cocaine or nicotine. *Biochemical Pharmacology, 86*, 1084–1088.

Epstein, D. H., Marrone, G. F., Heishman, S. J., Schmittner, J., & Preston, K. L. (2010). Tobacco, cocaine, and heroin: craving and use during daily life. *Addictive Behaviors, 35*, 318–324.

Garcia, R. C. T., Torres, L. H., Balestrin, N. T, Andrioli, T. C., Flório, J. C., Oliveira, C. D., Costa, J. L., Yonamine, M., Sandoval, M. R., Camarini, R. & Marcourakis, T. (2016) Anhydroecgonine methyl ester, a cocaine pyrolysis product, may contribute to cocaine behavioral sensitization, Toxicology, 376, 44–50.

García-Rodriguez, O., Secades-Villa, R., Weidberg, S., & Yoon, J. H. (2013). A systematic assessment of delay discounting in relation to cocaine and nicotine dependence. *Behavioural Processes, 99*, 100–105.

Gonçalves, J. R., & Nappo, S. A. (2015). Factors that lead to the use of crack cocaine in combination with marijuana in Brazil: a qualitative study. *BMC Public Health, 15*, 706–713.

Huang, Y. Y., Kandel, D. B., Kandel, E. R., & Levine, A. (2013). Nicotine primes the effect of cocaine on the induction of LTP in the amygdala. *Neuropharmacology, 74*, 126–134.

Jeri, F. R. (1980). Nuevas observaciones sobre los síndromes producidos por fumar pasta de coca. In F. R. Jeri (Ed.), *Cocaína 1980*. Lima: Pacific Press.

Jorge, M. S. B., Quindere, P. H. D., Yasui, S., & Albuquerque, R. A. (2013). Ritual de consumo do crack: aspectos socioantropológicos e repercussões para a saúde dos usuários. *Ciência & Saúde Coletiva, 18*, 2909–2918.

Justinova, Z., Foll, B. L., Redhi, G. H., Markou, A., & Goldberg, S. R. (2016). Differential effects of the metabotropic glutamate 2/3 receptor agonist LY379268 on nicotine versus cocaine self administration and relapse in squirrel monkeys. *Psychopharmacology, 233*, 1791–1800.

Kandel, D. B. (1975). Stages in adolescent involvement in drug use. *Science, 190*, 912–914.

Kandel, D. B. (2002). *Stages and pathways of drug involvement: examining the gateway hypothesis*. Cambridge: Cambridge University Press.

Kandel, D. B., Yamaguchi, K., & Chen, K. (1992). Stages of progression in drug involvement from adolescence to adulthood: further evidence for the gateway theory. *Journal of Studies on Alcohol and Drugs, 53*, 447–457.

Kandel, E. R., & Kandel, D. B. (2014). A molecular basis for nicotine as a gateway drug. *The New England Journal of Medicine, 371*, 932–943.

Khroyan, T. V., Yasuda, D., Toll, L., Polgar, W. E., & Zaveri, N. T. (2015). High affinity α3β4 nicotinic acetylcholine receptor ligands AT-1001 and AT-1012 attenuate cocaine-induced conditioned place preference and behavioral sensitization in mice. *Biochemical Pharmacology, 97*, 531–541.

Leão, R. M., Cruz, F. C., Carneiro-de-Oliveira, P. E., Rossetto, D. B., Valentini, S. R., Zanelli, C. F., et al. (2013). Enhanced nicotine-seeking behavior following pre-exposure to repeated cocaine is accompanied by changes in BDNF in the nucleus accumbens of rats. *Pharmacology, Biochemistry and Behavior, 104*, 169–176.

Levine, A., Huang, Y. Y., Drisaldi, B., Griffin, E. A., Jr., Pollak, D. D., Xu, S., et al. (2011). Molecular mechanism for a gateway drug: epigenetic changes initiated by nicotine prime gene expression by cocaine. *Science Translational Medicine, 3*, 107–109.

Li, H., Bu, Q., Chen, B., Shao, X., Hu, Z., Deng, P., et al. (2014). Mechanisms of metabonomic for a gateway drug: nicotine priming enhances behavioral response to cocaine with modification in energy metabolism and neurotransmitter level. *PLoS One, 9*, e87040.

Mello, N. K. (2010). Hormones, nicotine, and cocaine: clinical studies. *Hormones and Behavior, 58*, 57–71.

Mello, N. K., Fivel, P. A., & Kohut, S. J. (2013). Effects of chronic Buspirone treatment on nicotine and concurrent nicotine + cocaine self-administration. *Neuropsychopharmacology, 38*, 1264–1275.

Mello, N. K., Fivel, P. A., Kohut, S. J., & Carroll, F. I. (2014). Effects of chronic varenicline treatment on nicotine, cocaine, and concurrent nicotine + cocaine self-administration. *Neuropsychopharmacology, 39*, 1222–1231.

Mello, N. K., & Newman, J. L. (2011). Discriminative and reinforcing stimulus effects of nicotine, cocaine, and cocaine + nicotine combinations in rhesus monkeys. *Experimental and Clinical Psychopharmacology, 19*, 203–214.

Palmer, R. H. C., Young, S. E., Hopfer, C. J., Corley, R. P., Stallings, M. C., Crowley, T. J., et al. (2009). Developmental epidemiology of drug use and abuse in adolescence and young adulthood: evidence of generalized risk. *Drug and Alcohol Dependence, 102*, 78–87.

Pinto, M., Bardach, A., Palacios, A., Biz, A. N., Alcaraz, A., Rodríguez, B., et al. (2017). *Carga de doença atribuível ao uso do tabaco no Brasil e potencial impacto do aumento de preços por meio de impostos. Documento técnico IECS N 21*. Buenos Aires: Instituto de Efectividad Clínica y Sanitaria.

Reid, M. S., Mickalian, J. D., Delucchi, K. L., & Berger, S. P. (1999). A nicotine antagonist, mecamylamine, reduces cueinduced cocaine craving in cocaine-dependent subjects. *Neuropsychopharmacology, 20*, 297–307.

Reid, M. S., Mickalian, J. D., Delucchi, K. L., Hall, S. M., & Berger, S. P. (1998). An acute dose of nicotine enhances cue-induced cocaine craving. *Drug and Alcohol Dependence, 49*, 95–104.

Reitsma, M. B., et al. (2017). Smoking prevalence and attributable disease burden in 195 countries and territories, 1990–2015: a systematic analysis from the global burden of disease study 2015. *The Lancet, 389*, 1885–1906.

Sanjakdar, S. S., Maldoon, P. P., Marks, M. J., Brunzell, D. H., Maskos, U., McIntosh, J. M., et al. (2015). Differential roles of α6β2* and α4β2* neuronal nicotinic receptors in nicotine- and cocaine-conditioned reward in mice. *Neuropsychopharmacology, 40*, 350–360.

Siegel, R. K. (1982). Cocaine smoking. *Journal of Psychoactive Drugs, 14*, 277–559.

Vanyukov, M. M., Tarter, R. E., Kirillova, G. P., Kirisci, L., Reynolds, M. D., Kreek, M. J., et al. (2012). Common liability to addiction and "gateway hypothesis": theoretical, empirical and evolutionary perspective. *Drug and Alcohol Dependence, 123*(Suppl 1), S3–S17.

Wagner, F. A., & Anthony, J. C. (2002). Into the world of illegal drug use: exposure opportunity and other mechanisms linking the use of alcohol, tobacco, marijuana, and cocaine. *American Journal of Epidemiology, 155*, 918–925.

Zeni, T. C., & Araujo, R. B. (2011). Relationship between craving for tobacco and craving for crack in patients hospitalized for detoxification. *Brazilian Journal of Psychiatry, 60*, 28–33.

CHAPTER

50

Salivary Cotinine Assays

M. Inês G.S. Almeida, Luisa Barreiros†, Spas D. Kolev*, Marcela A. Segundo†*

*School of Chemistry, The University of Melbourne, Parkville, Melbourne, VIC, Australia
†Department of Chemical Sciences, Faculty of Pharmacy, University of Porto, Porto, Portugal

Abbreviations

ASAP	atmospheric pressure solid analysis probe
ELISA	enzyme-linked immunosorbent assay
ETS	environmental tobacco smoking
GC-FID	gas chromatography coupled to flame ionization detector
GC-MS	gas chromatography coupled to mass spectrometry
IgG	immunoglobulin G
LC-DAD	liquid chromatography coupled to diode array detector
LC-MS	liquid chromatography coupled to mass spectrometry
LC-MS/MS	liquid chromatography coupled to tandem mass spectrometry
LLE	liquid-liquid extraction
MS-qTOF	quadrupole time-of-flight tandem mass spectrometry
POC	point-of-care
SPE	solid-phase extraction

50.1 INTRODUCTION

Cotinine is a major metabolite of the tobacco alkaloid nicotine (Fig. 50.1) (Hukkanen, Jacob, & Benowitz, 2005). As a biological marker, cotinine has been selected for monitoring compliance to smoking cessation programs (Gorber, Schofield-Hurwitz, Hardt, Levasseur, & Tremblay, 2009) and also for the evaluation of environmental tobacco smoking (ETS) exposure (Avila-Tang et al., 2013), because it offers advantages over other compounds (e.g., carbon monoxide and thiocyanate ions) since cotinine levels are not influenced by diet or common pollution exposure (Jarvis, Tunstallpedoe, Feyerabend, Vesey, & Saloojee, 1987).

Regarding cotinine concentration in saliva defined as the cut point to distinguish between smokers and nonsmokers, the Society for Research on Nicotine and Tobacco has recommended a value of 15ng/mL in plasma or saliva (SNRT, 2002). Nevertheless, recent work (Kim, 2016) has pointed out that the bimodal distribution of cotinine values, arising from the two populations (nonsmokers and smokers) may change over time. With increasing effective tobacco control in public places through legislation enforcement, the distribution of cotinine concentrations in nonsmokers will tend to shift to lower values, and more nonsmokers may have nondetectable concentrations with current analytical techniques. Consequently, the cutoff point will likely also shift to lower values. In fact, this was observed within the US National Health and Nutrition Examination Survey (1999–2004), on the basis of which a lower cutoff value of 3ng/mL was proposed (Benowitz, Bernert, Caraballo, Holiday, & Wang, 2009). Therefore, suitable analytical methods, able to detect trace cotinine concentrations, are needed, and increasing challenges in developing such methods have been posed over recent years.

In this context, saliva is more suitable than plasma for biomonitoring schemes as its collection is noninvasive, being particularly fitted for sampling of children in ETS exposure studies. As cotinine is a weak base, it is transported from the blood into the saliva by passive diffusion, requiring no additional binding to other molecules. Moreover, several studies have shown that saliva/plasma cotinine concentration ratio is above unity and is almost constant, despite variations in salivary flow (Michalke, Rossbach, Goen, Schaferhenrich, & Scherer, 2015). For these reasons, salivary cotinine can provide an accurate picture of plasma levels, without requiring blood collection.

From an analytical chemistry point of view, though, saliva is a challenging sample (Table 50.1). Saliva is mainly composed of water (97%–99%), but it also contains lipids, proteins, and inorganic compounds that

Neuroscience of Nicotine
https://doi.org/10.1016/B978-0-12-813035-3.00050-2

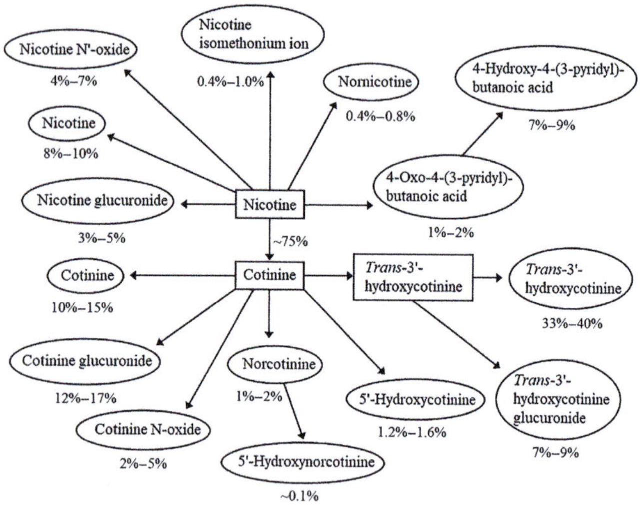

FIG. 50.1 Main metabolites of nicotine.
Quantitative scheme of nicotine metabolism, based on estimates of average excretion of metabolites as percent of total urinary nicotine. *From Pharmacological Reviews, Metabolism and Disposition Kinetics of Nicotine, 57, 2005, 79, J. Hukkanen, P. Jacob, N. L. Benowitz. © 2005 ASPET, reproduced with permission from The American Society for Pharmacology and Experimental Therapeutics.*

TABLE 50.1 Key Facts of Saliva as Analytical Sample

- Saliva is a challenging sample matrix due to its complex, dynamic composition and inherent inter- and intraindividual variability
- Saliva is produced by three paired major salivary glands and by minor salivary glands in the mucosa of the tongue, cheeks, lips, and palate
- Healthy human adults produce 500–2500mL of saliva per day, with salivary flow rate up to 6mL/min (unstimulated) or up to 10mL/min (stimulated)
- Saliva mainly consists of water (97%–99%), lipids, proteins, and inorganic compounds
- The main lipids in saliva are cholesterol and cholesterol esters, triglycerides, free fatty acids, and phospholipids, while proteins are mainly represented by glycoproteins, albumin, metabolic enzymes, and immunoglobulins
- The inorganic fraction of saliva contains a variety of electrolytes, including sodium, potassium, calcium, magnesium, bicarbonate, and phosphates, whose levels may be affected by saliva rate production
- Saliva pH can range from 5.8 to 7.8, depending mainly on bicarbonate ion levels, which are enhanced when salivation is stimulated
- Saliva collection should be done preferentially under unstimulated conditions because pH increase can change the fraction of the ionized form of biomarkers
- Whole saliva can be collected by draining, spitting, suction, rinsing, or using an absorbent media

This table lists the key facts of saliva as an analytical sample, including aspects of human physiology, methods of collection, and chemical composition that affect its utilization in analytical methods for assay of biomarkers.
Information compiled from Michalke, B., Rossbach, B., Goen, T., Schaferhenrich, A., & Scherer, G. (2015).

can significantly interfere when quantifying cotinine by instrumental methods of analysis. These interferences can arise when macromolecules are present in amounts that are several orders of magnitude higher than that of cotinine (e.g., total proteins are present at microgram-per-milliliter levels while cotinine at nanogram-per-milliliter levels). Moreover, several spectrophotometric detection systems suffer from interference from species that absorb radiation at the same wavelengths applied for cotinine detection. Another common cause of interference, particularly in mass spectrometric detection, is the presence of ions that cause enhancement or suppression of the ionization of cotinine, thus hindering its accurate determination. Therefore, sample pretreatment is almost mandatory when targeting low nanogram-per-milliliter levels of cotinine in saliva.

50.2 METHODS OF ANALYSIS

As mentioned above, salivary cotinine assays are used to evaluate compliance to tobacco cessation programs or to assess ETS exposure in a given population. Assay requirements can differ significantly in these situations. In the first case, if saliva is collected from someone that has been smoking, high levels of cotinine (>100ng/mL) are expected (Avila-Tang et al., 2013). Hence, a screening method, with semiquantitative features for distinguishing between background and positive values, is suitable. In the second situation, expected levels are <15ng/mL for nonexposed subjects, and more recent data have suggested a cutoff point of 3ng/mL (Benowitz et al., 2009), thus requiring state-of-the-art analytical techniques.

Point-of-care (POC) techniques (Table 50.2) meet the requirements of screening methods. Several devices are commercially available for direct measurement of cotinine in saliva, including NicAlert strips, SafeCare Biotech strips, iScreen OFD strips, and Salimetrics ELISA kit. All methods rely on immunochemical detection of cotinine

TABLE 50.2 Key Facts of Point-of-Care Methods

- Point-of-care (POC) methods are aimed to provide an analytical result on site, meaning that a sample is collected and analyzed within the physician's office or any other location where the patient/target individual is
- POC methods are designed to work with noninvasive samples (saliva and urine) or minute amounts of blood (one or two drops)
- Current devices are based on membrane-based test strips with immobilized reagents, enclosed in plastic cases or connected to transducers and electronic components
- Results are mainly semiquantitative, indicating a positive or negative result (e.g., pregnancy tests) or a result within a range of concentrations (e.g., urine glucose strips)
- There are also more precise POC methods whenever a quantitative result has relevance on the implementation of therapeutics (e.g., results from blood glucose analyzers used for setting insulin bolus)

This table includes some key facts about point-of-care methods regarding sample collection, scope of application, and type of response as qualitative or semiquantitative.

by visual inspection or spectrophotometry using a microplate reader. This is the case for Salimetrics ELISA kit where cotinine present in samples competes with cotinine-horseradish peroxidase conjugate for the antibody binding sites on a microtiter plate, where bound conjugates are measured using tetramethylbenzidine as the colorimetric reagent. A quantitative result is provided with a detection limit of 0.15 ng/mL, making this test suitable for the screening of ETS exposure. Despite its easy-to-use features as an analytical POC technique, its main disadvantage is that it still requires lab facilities for the analytic readout.

SafeCare Biotech strips and iScreen OFD strips can only provide qualitative results, with a high cutoff value (>30 ng/mL) for a positive response. Therefore, these strips are applicable for the evaluation of smoking status but are unsuitable for screening of ETS. NicAlert strips offer a semiquantitative approach, based on the use of gold-particle-coated monoclonal antibodies and a range of molecular traps with different avidity toward these antibodies, placed sequentially in the analytical strip. Therefore, a cotinine molecule binds to an antibody immobilized on a gold particle, which carries several antibodies (40–100 receptor sites). The number of occupied sites is a function of cotinine concentration present in the sample. Hence, particles with more unbound antibodies (less cotinine in the sample) will be retained first by a low-avidity trap, while particles with less unbound antibodies (more cotinine in the sample) will be retained in a further zone by a high-avidity trap. Using this strategy, evaluation of smoking status is possible with the determination of cotinine in the range between 10 and 1000 ng/mL.

State-of-the-art analytical techniques, reaching subnanogram-per-milliliter levels, can only be performed so far under laboratory conditions, using, for instance, mass spectrometry (Table 50.3) detectors. Besides immunoassays under ELISA format (Langone & Bjercke, 1989), liquid chromatography coupled to tandem mass spectrometry (LC-MS/MS), considered as the gold standard in this area, offers the advantage of simultaneous determination of other nicotinine metabolites (*trans*-3′-hydroxycotinine and norcotinine (Fig. 50.1), for instance, (Bentley, Abrar, Kelk, Cook, & Phillips, 1999; Jacob et al., 2011; Shakleya & Huestis, 2009)) or even of other controlled substances (Concheiro, Gray, Shakleya, & Huestis, 2010). In Fig. 50.2, the mass spectra of cotinine obtained in a triple quadrupole spectrometer are shown (Perez-Ortuno, Martinez-Sanchez, Fernandez, & Pascual, 2015). Two major fragments obtained from the molecular ion (*m/z* 180.2) at electrospray ionization in positive mode can be identified. The more abundant fragment at *m/z* 80.1 corresponds to the pyridine moiety and it has been used for quantification, while the fragment at *m/z* 98.1 is attributed to methylpyrrolidone ring and it has been used for identification purposes. For accurate quantification, deuterated cotinine-d_3 is frequently used as internal standard, generating the same quantification fragment at *m/z* 80.1 and the identification fragment at 101.1, accounting for the three deuterated hydrogens added to the structure.

TABLE 50.3 Key Facts of Mass Spectrometry Coupled to Liquid Interfaces

- Mass spectrometry is based on the measurement of mass-to-charge ratio, indicated as *m/z*, of ionized molecules in gas phase
- Removal of solvent, ionization of target molecules, and transition to gas phase are required for connecting liquid interfaces to the mass analyzers
- The most common interfaces between liquid streams and the mass analyzers are based on electrospray ionization (ESI) and atmospheric pressure chemical ionization (APCI)
- ESI can be performed in positive or negative mode, with the gain or loss of a hydrogen atom, resulting in parental ions with *m/z* values corresponding to 1 Dalton more or less than the molecular mass of the target molecule
- There are several types of mass analyzers (quadrupole, time of flight, ion trap, among others), differing mainly on the accuracy and precision of reported *m/z*
- Tandem mass spectrometry refers to the serial connection of two or more mass analyzers, allowing the identification/quantification of fragment ions produced by the parental ion
- High-resolution mass spectrometry is based on the accurate measurement of *m/z* values, providing values to several decimal places and enabling the determination of elemental composition

This table includes some key facts of mass spectrometry coupled to liquid interfaces, namely, the steps required for the detection of molecules through ionization and conversion to gas phase, the most common interfaces, and how they produce ions. Some types of detectors are introduced as well as features of high-resolution and tandem mass spectrometry.

Besides LC-MS/MS, LC-MS (Kataoka, Inoue, Yagi, & Saito, 2009), LC-DAD (Machacek & Jiang, 1986), GC–MS (Shin, Kim, Shin, & Jee, 2002), and GC-FID (Feyerabend & Russell, 1980) have been proposed for quantification of cotinine in saliva at levels compatible to evaluation of ETS exposure. Those methods cover a working range of 0.5–5000 ng/mL, with detection limits of 0.02–1 ng/mL. To attain such low detection levels, sample pretreatment is required to remove interferences and also to preconcentrate cotinine (Fig. 50.3).

Liquid-liquid extraction (LLE), comprising the addition of a liquid extracting solvent (e.g., dichloromethane), the separation of liquid phases through centrifugation, and the recovery of the extractant, has been regarded as a simple, convenient, and straightforward method for quantitative separation of cotinine, with recoveries >83%. Solid-phase extraction (SPE), regarded as a more environmentally friendly sample pretreatment method, has also been applied for cotinine determination in saliva. Several types of sorbents have been proposed, including conventional silica-C18, hydrophilic-lipophilic polymers, and molecularly imprinted polymers.

Other types of sample pretreatment are required according to the chosen instrumental method of analysis.

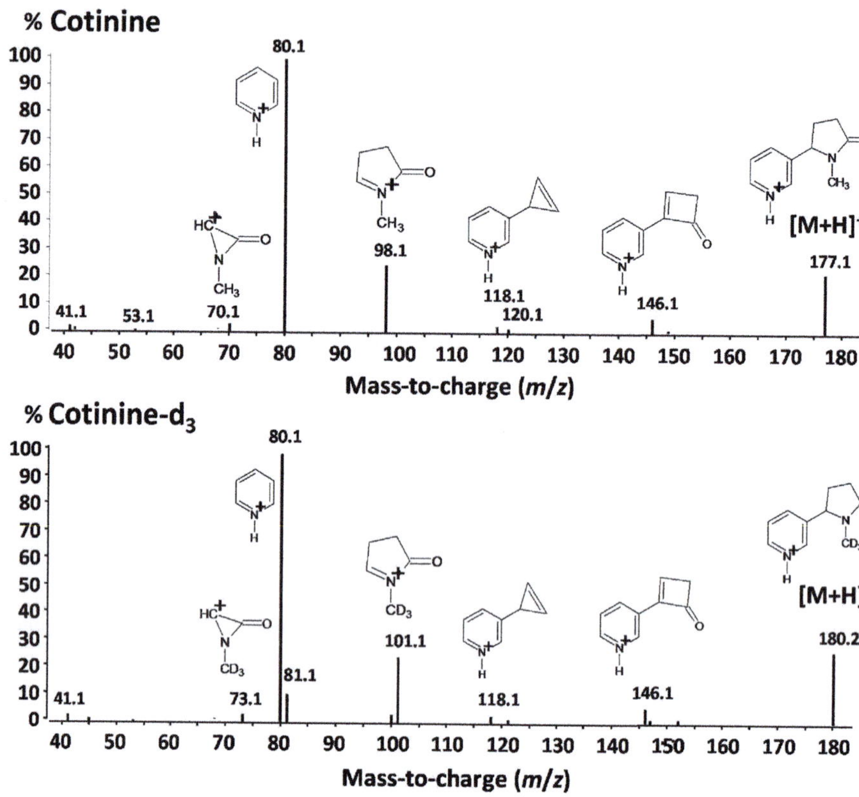

FIG. 50.2 Mass spectra of cotinine and cotinine-d_3.
Mass spectra of cotinine and cotinine-d_3 obtained using a triple quadrupole equipment (tandem mass spectrometry) showing the product ion mass spectra obtained for cotinine based on the fragmentation of the precursor ion $[M+H]^+$ at m/z 177. The product ion mass spectrum of the internal standard cotinine-d_3 (precursor ion $[M+H]^+$ at m/z 180) is also shown. The chemical structure proposed for each observed peak is also presented. *Modified from Analytical and Bioanalytical Chemistry, High-throughput wide dynamic range procedure for the simultaneous quantification of nicotine and cotinine in multiple biological matrices using hydrophilic interaction liquid chromatography-tandem mass spectrometry, 407, 2015, 8467; R. Perez-Ortuno, J. M. Martinez-Sanchez, E. Fernandez & J. A. Pascual. ©Springer-Verlag Berlin Heidelberg 2015, reproduced with permission of Springer.*

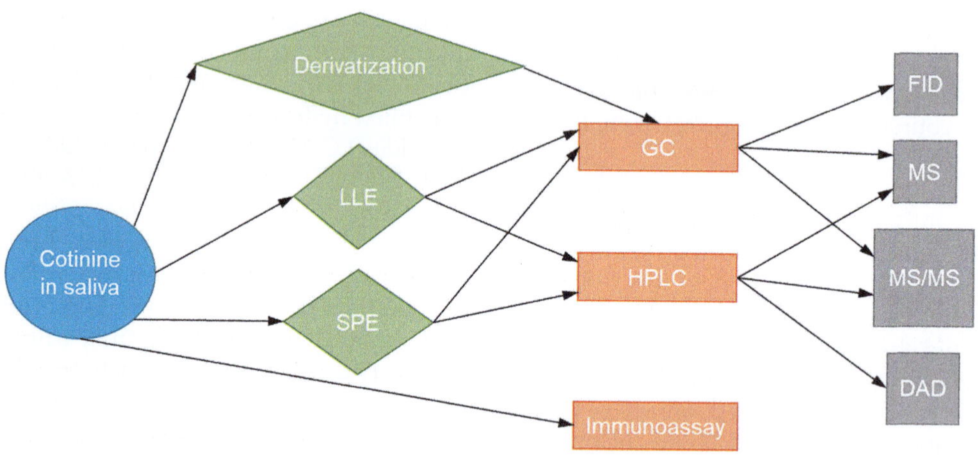

FIG. 50.3 Analytical workflow for salivary cotinine assays.
Schematic representation of procedures adopted for saliva pretreatment (operations inserted in rhombi) prior to the application of analytical techniques (placed in rectangles and squares). The arrows indicate the possible combinations described so far in literature for this determination. *DAD*, diode-array detection; *FID*, flame ionization detection; *GC*, gas chromatography; *HPLC*, high-performance liquid chromatography; *LLE*, liquid-liquid extraction; *MS*, mass spectrometry; *MS/MS*, tandem mass spectrometry; *SPE*, solid-phase extraction.

For instance, sensitive GC analysis requires that the species under separation are volatile. Hence, GC methods proposed for cotinine determination in saliva required previous derivatization using *N*-methyl-*N*-trimethylsilyl-trifluoroacetamide and trimethylchlorosilane (da Fonseca et al., 2012), generating more volatile compounds from cotinine.

50.3 RECENT DEVELOPMENTS IN SAMPLE PRETREATMENT AND COTININE DETERMINATION

In the past few years, novel trends regarding minimization or elimination of sample pretreatment alongside miniaturization have been observed. Regarding

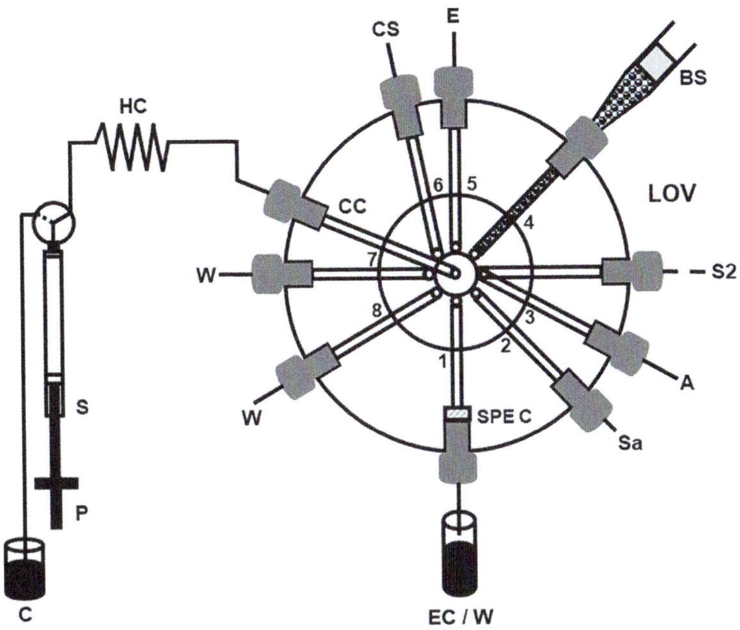

FIG. 50.4 Representation of bead injection lab-on-valve system for micro-SPE of salivary cotinine.
Schematic representation of the lab-on-valve equipment designed for automatic SPE of cotinine from saliva. Reagents and sorbent are placed around a central channel, from which solutions and bead suspension are aspirated or delivered to perform all steps required for SPE, including column packing and conditioning, sample loading and removal of matrix interferences, cotinine elution, and disposal of sorbent. All these operations are performed automatically through computer control. *P*, low-pressure piston pump; *S*, syringe; *HC*, holding coil; *C*, carrier (1.20 mM KH_2PO_4, pH 6.8); *E*, eluent; *SPE C*, SPE column; *Sa*, sample; *A*, air; *BS*, bead suspension; *CS*, conditioning solution (acetonitrile-water, 50:50, *v/v*); *CC*, central channel; *EC*, eluate collection; *W*, waste. *From Journal of Chromatography A, Determination of salivary cotinine through solid phase extraction using a bead-injection lab-on-valve approach hyphenated to hydrophilic interaction liquid chromatography, 1429, 2016, 284, A. N. Ramdzan, L. Barreiros, M. I. G. S. Almeida, S. D. Kolev & M. A. Segundo. © 2015 Elsevier B. V, reproduced with permission from Elsevier.*

automation of saliva pretreatment, an automatic micro solid-phase extraction method based on the bead injection concept was proposed recently (Ramdzan, Barreiros, Almeida, Kolev, & Segundo, 2016). Using the mesofluidic lab-on-valve platform, schematically depicted in Fig. 50.4, all steps of SPE were performed through computer control, fostering the precise metering of solutions and strict control of applied flow rates. Several reversed-phase polymeric sorbents were tested, with better packing and recovery features observed for spherically shaped Oasis HLB sorbent. Cotinine elution was performed using a solution with the same composition as the chromatographic mobile phase (acetonitrile:100 mM ammonium acetate buffer, pH 5.8 in 95:5 *v/v*), fostering the direct introduction of eluate into the liquid chromatograph with separation based on hydrophilic interaction (HILIC). This mode of operation is particularly suited for polar compounds, and as the chromatographic mobile phase is mostly composed of an organic solvent, band-broadening effects are not observed when injecting SPE eluates directly, thus saving time because solvent switching through evaporation and resuspension is not required. The proposed method was suitable for the direct introduction of saliva, despite its viscous composition, because the sample was diluted in phosphate buffer. This dilution did not affect the method sensitivity because all cotinine present in the original sample was retained in the SPE column, fostering limits of detection and quantification of 1.5 and 3 ng/mL, respectively, with a sampling rate of six samples per hour, including sample pretreatment.

Toward further downscaling, cotinine was determined using ELISA performed under microchip format (Cheng, Zhao, Liu, & Sui, 2013). Microfluidic chips, enabling up to eight parallel determinations, were designed and fabricated in polydimethylsiloxane with standard soft lithography (Fig. 50.5). Cotinine determination was based

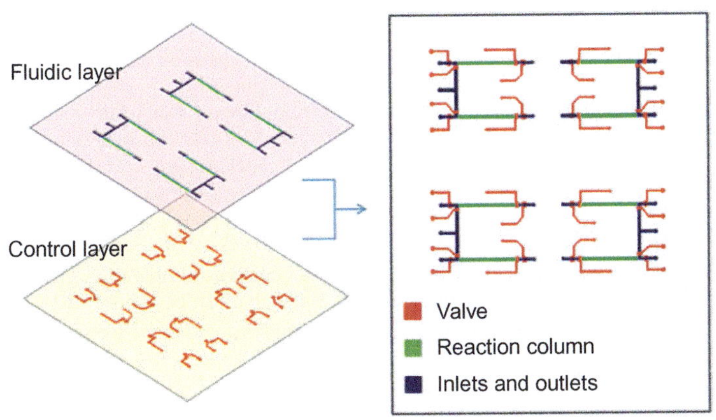

FIG. 50.5 Schematic diagram of microfluidic chips fabricated for the rapid detection of cotinine.
The microfluidic analytical chip was fabricated from two separate pieces built in polydimethylsiloxane using molds with different patterns fabricated by photolithographic processes (left part of the scheme). The pattern designed in the mold and the functions of different parts of the chip are depicted in the right part of the figure, including valves for fluid passage and immunosorbent entrapment by channel restriction, channel for immunosorbent entrapment (reaction column), and points for solution entrance and exit (inlets and outlets). *From Biomedical Microdevices, Microfluidic immunoassay for rapid detection of cotinine in saliva, 15, 2013, 951; K. P. Cheng, W. Zhao, S. X. Liu & G. D. Sui. ©Springer Science+Business Media New York 2013, reproduced with permission of Springer.*

on cotinine-IgG immobilized in protein A-coated polystyrene microspheres. The assay protocol comprised immobilization of cotinine-IgG microspheres inside the microchip, perfusion with sample diluted in tracer (cotinine-horseradish peroxidase) solution, incubation, and fluorescence detection after reaction with Amplex Red, where lower fluorescence values corresponded to higher levels of cotinine. The minimum detectable concentration was 1 ng/mL using 12 μL of saliva with a sample throughput of 12 samples per hour. The method was successfully applied to measuring cotinine in saliva samples from smokers and nonsmokers submitted to ETS.

Epidemiological studies require methods with even higher throughput and with easy and straightforward sample handling. Targeting these features, an HPLC-MS/MS method was proposed for application to saliva and also to urine and hair samples (Perez-Ortuno et al., 2015) with each chromatographic run taking only 2 min for the determination of cotinine and nicotine. Taking advantage of the built-in features of the autosamplers coupled to the LC (or GC) equipment, where the depth of needle penetration in vials can be programmed, LLE was performed directly in the vial, and the bottom organic phase was directly probed. Hence, 0.5 mL of saliva was mixed with alkaline internal standard solution and dichloromethane directly in the vial. As mentioned before, HILIC was implemented for its compatibility with high levels of organic solvents and separation capabilities regarding polar compounds. The calibration range for cotinine in saliva was 0.1–2000 ng/mL.

Finally, recent trends in trace analysis focus on complete elimination of sample pretreatment, for instance, when analytes are transferred directly from the sample into the ionization chamber of mass spectrometers. This approach has been proposed for cotinine screening using atmospheric pressure solid analysis probe (ASAP) coupled to a quadrupole time-of-flight tandem mass spectrometer (MS-qTOF) (Carrizo, Nerin, Domeno, Alfaro, & Nerin, 2016). A glass capillary was dipped twice in saliva (or urine) sample and introduced in the ionization chamber for atmospheric pressure ionization in positive mode. Samples from individuals under the smoking cessation program were analyzed, and relative differences in cotinine values during the day were observed, though the corresponding exact cotinine concentrations were not determined. Hence, despite the convenience of this analytical approach, its application is still semiquantitative, but it shows an enormous potential for multianalyte detection and comprehensive diagnosis as the qTOF detector provides accurate mass determination, enabling the identification of numerous compounds of clinical interest.

Using a more conventional ionization approach based on the application of a high-voltage direct current to the saliva sample placed in a paper support, quantitative analysis was enabled through the application of a deuterated internal standard (Wang et al., 2013). No sample pretreatment was necessary, i.e., 5 μL of saliva were directly impregnated in a paper triangle (base 7 mm and height 12 mm), followed by the addition of 20 μL of acetonitrile. Hence, mass spectra features were acquired after ionization in positive mode by placing the paper support in front of the mass spectrometer inlet and applying high voltage. Though not applied to epidemiological studies, the method showed potential for quantitative and fast determination of cotinine.

In conclusion, despite the efforts made so far, fast methods for quantification of cotinine in saliva are still needed, especially those enabling POC measurements with a better precision and lower detection limits.

MINI-DICTIONARY OF TERMS

Autosampler Equipment currently present in most chromatographic equipment that allows the automatic introduction of samples into the chromatographic column. Recent models allow external programming, enabling automatic sample pretreatment within the autosampler.

Bead injection Bead injection is a flow analysis technique where a suspension of particles is trapped inside the flow conduits (usually the measuring cell) and discarded automatically after the analytical operation is finished.

ELISA ELISA is the abbreviation of enzyme-linked immunosorbent assay. It is based on the immobilization of the target compound (through specific interactions with a surface, via reactive groups or using a specific antibody), followed by its detection using an immunocomplex.

Lab-on-valve Lab-on-valve is a mesofludic platform, built in polymeric material that, coupled to a selection valve and reversible piston pump, allows the development of a flow analysis system for automation of (bio)chemical analysis.

Limit of detection The limit of detection is the lowest concentration that provides an analytical signal that can be distinguished from the signal obtained when no analyte is present. There are several ways of assessing the limit of detection. One of the most prevalent is based on signal-to-noise ratio, where the lowest detectable concentration corresponds to the signal obtained in the absence of analyte plus three times its standard deviation.

Liquid-liquid extraction This is a separation process based on mass transfer between two immiscible liquid phases.

Mass spectrum Representation of current intensity (or relative abundance) as a function of mass-to-charge ratio. It is important to notice that a given substance can provide different spectra upon the type of mass detector applied and the selected ionization process as different ions (positive, negative, or adducts) and different fragments can be formed.

Sampling rate Sampling rate, also designated as sample throughput, corresponds to the number of samples that can be processed by a given analytical technique within a given time period (usually 1 h or 1 working day).

Silica C18 This is the common designation of octadecyl silica, which is the most frequently applied material as stationary phase in reversed-phase liquid chromatography. It is also frequently employed in solid-phase extraction protocols.

Solid-phase extraction This is a separation process based on mass transfer from a liquid phase to a solid phase. For most analytical processes, the analyte is eluted from the solid phase to a new liquid phase prior to quantification using an instrumental method of analysis.

Sorbent Sorbent is the solid phase where analytes are retained in solid-phase extraction.

Working range For analytical methods, the working range is defined by the lowest and the highest concentration that can be determined by the application of the established calibration curve.

Summary Points

- This chapter focuses on the determination of cotinine in saliva as a biomarker of tobacco smoking or environmental exposure to tobacco smoke.
- Saliva is collected by noninvasive procedures, making it suitable to epidemiological studies.
- Point-of-care methods provide qualitative or semiquantitative results concerning tobacco smoking habits.
- Current point-of-care methods are not suitable for assessing environmental tobacco smoke exposure due to inadequate sensitivity.
- Liquid chromatographic methods coupled to tandem mass spectrometry are currently the gold standard for quantification of cotinine in saliva, with the advantage of providing simultaneous results for other biomarkers.
- Analytical methods proposed recently tend to automate, minimize, or eliminate sample pretreatment using paper-based and meso- and microfluidic devices.
- Fast methods for quantification of cotinine in saliva with better precision and lower detection limits are still needed for point-of-care applications.

References

Avila-Tang, E., Al-Delaimy, W. K., Ashley, D. L., Benowitz, N., Bernert, J. T., Kim, S., et al. (2013). Assessing secondhand smoke using biological markers. *Tobacco Control*, *22*, 164–171.

Benowitz, N. L., Bernert, J. T., Caraballo, R. S., Holiday, D. B., & Wang, J. T. (2009). Optimal serum cotinine levels for distinguishing cigarette smokers and nonsmokers within different racial/ethnic groups in the United States between 1999 and 2004. *American Journal of Epidemiology*, *169*, 236–248.

Bentley, M. C., Abrar, M., Kelk, M., Cook, J., & Phillips, K. (1999). Validation of an assay for the determination of cotinine and 3-hydroxycotinine in human saliva using automated solid-phase extraction and liquid chromatography with tandem mass spectrometric detection. *Journal of Chromatography B*, *723*, 185–194.

Carrizo, D., Nerin, I., Domeno, C., Alfaro, P., & Nerin, C. (2016). Direct screening of tobacco indicators in urine and saliva by atmospheric pressure solid analysis probe coupled to quadrupole-time of flight mass spectrometry (ASAP-MS-Q-TOF-). *Journal of Pharmaceutical and Biomedical Analysis*, *124*, 149–156.

Cheng, K. P., Zhao, W., Liu, S. X., & Sui, G. D. (2013). Microfluidic immunoassay for rapid detection of cotinine in saliva. *Biomedical Microdevices*, *15*, 949–957.

Concheiro, M., Gray, T. R., Shakleya, D. M., & Huestis, M. A. (2010). High-throughput simultaneous analysis of buprenorphine, methadone, cocaine, opiates, nicotine, and metabolites in oral fluid by liquid chromatography tandem mass spectrometry. *Analytical and Bioanalytical Chemistry*, *398*, 915–924.

da Fonseca, B. M., Moreno, I. E. D., Magalhaes, A. R., Barroso, M., Queiroz, J. A., Ravara, S., et al. (2012). Determination of biomarkers of tobacco smoke exposure in oral fluid using solid-phase extraction and gas chromatography-tandem mass spectrometry. *Journal of Chromatography B-Analytical Technologies in the Biomedical and Life Sciences*, *889*, 116–122.

Feyerabend, C., & Russell, M. A. H. (1980). Rapid gas-liquid-chromatographic determination of cotinine in biological-fluids. *Analyst*, *105*, 998–1001.

Gorber, S. C., Schofield-Hurwitz, S., Hardt, J., Levasseur, G., & Tremblay, M. (2009). The accuracy of self-reported smoking: a systematic review of the relationship between self-reported and cotinine-assessed smoking status. *Nicotine & Tobacco Research*, *11*, 12–24.

Hukkanen, J., Jacob, P., & Benowitz, N. L. (2005). Metabolism and disposition kinetics of nicotine. *Pharmacological Reviews*, *57*, 79–115.

Jacob, P., Yu, L. S., Duan, M. J., Ramos, L., Yturralde, O., & Benowitz, N. L. (2011). Determination of the nicotine metabolites cotinine and trans-3 ′-hydroxycotinine in biologic fluids of smokers and non-smokers using liquid chromatography-tandem mass spectrometry: biomarkers for tobacco smoke exposure and for phenotyping cytochrome P450 2A6 activity. *Journal of Chromatography B-Analytical Technologies in the Biomedical and Life Sciences*, *879*, 267–276.

Jarvis, M. J., Tunstallpedoe, H., Feyerabend, C., Vesey, C., & Saloojee, Y. (1987). Comparison of tests used to distinguish smokers from nonsmokers. *American Journal of Public Health*, *77*, 1435–1438.

Kataoka, H., Inoue, R., Yagi, K., & Saito, K. (2009). Determination of nicotine, cotinine, and related alkaloids in human urine and saliva by automated in-tube solid-phase microextraction coupled with liquid chromatography-mass spectrometry. *Journal of Pharmaceutical and Biomedical Analysis*, *49*, 108–114.

Kim, S. (2016). Overview of cotinine cutoff values for smoking status classification. *International Journal of Environmental Research and Public Health*, *13*. https://dx.doi.org/10.3390/ijerph13121236.

Langone, J. J., & Bjercke, R. J. (1989). Idiotype anti-idiotype hapten immunoassays—assay for cotinine. *Analytical Biochemistry*, *182*, 187–192.

Machacek, D. A., & Jiang, N. S. (1986). Quantification of cotinine in plasma and saliva by liquid-chromatography. *Clinical Chemistry*, *32*, 979–982.

Michalke, B., Rossbach, B., Goen, T., Schaferhenrich, A., & Scherer, G. (2015). Saliva as a matrix for human biomonitoring in occupational and environmental medicine. *International Archives of Occupational and Environmental Health*, *88*, 1–44.

Perez-Ortuno, R., Martinez-Sanchez, J. M., Fernandez, E., & Pascual, J. A. (2015). High-throughput wide dynamic range procedure for the simultaneous quantification of nicotine and cotinine in multiple biological matrices using hydrophilic interaction liquid chromatography-tandem mass spectrometry. *Analytical and Bioanalytical Chemistry*, *407*, 8463–8473.

Ramdzan, A. N., Barreiros, L., Almeida, M. I. G. S., Kolev, S. D., & Segundo, M. A. (2016). Determination of salivary cotinine through solid phase extraction using a bead-injection lab-on-valve approach hyphenated to hydrophilic interaction liquid chromatography. *Journal of Chromatography A*, *1429*, 284–291.

Shakleya, D. M., & Huestis, M. A. (2009). Optimization and validation of a liquid chromatography-tandem mass spectrometry method for the simultaneous quantification of nicotine, cotinine, trans-3′-hydroxycotinine and norcotinine in human oral fluid. *Analytical and Bioanalytical Chemistry*, *395*, 2349–2357.

Shin, H. S., Kim, J. G., Shin, Y. J., & Jee, S. H. (2002). Sensitive and simple method for the determination of nicotine and cotinine in human urine, plasma and saliva by gas chromatography-mass spectrometry. *Journal of Chromatography B-Analytical Technologies in the Biomedical and Life Sciences*, *769*, 177–183.

Society for Research on Nicotine and Tobacco - Subcommittee on Biochemical Verification. (2002). Biochemical verification of tobacco and cessation. *Nicotine & Tobacco Research*, *4*, 149–159.

Wang, H., Ren, Y., McLuckey, M. N., Manicke, N. E., Park, J., Zheng, L. X., et al. (2013). Direct quantitative analysis of nicotine alkaloids from biofluid samples using paper spray mass spectrometry. *Analytical Chemistry*, *85*, 11540–11544.

CHAPTER

51

Overview of Cotinine Cutoff Values for Smoking Status Classification

Sungroul Kim

Department of Environmental Health Sciences, SoonChunHyang University, Asan, South Korea

Abbreviations

IARC International Agency for Research on Cancer
NRC National Research Council
USDHHS U.S. Department of Health and Human Services

There are many reasons why accurate assessment of smoking status is critical; these include its utility in evaluating the progress of worldwide tobacco control programs, documenting tobacco epidemics, and estimating smoking risk to populations (IARC, 2004; NRC, 1986; USDHHS, 2006; Warren, Jones, Eriksen, & Asma, 2006). When assessing smoking prevalence and exposure to secondhand smoke (SHS), researchers have most commonly used standardized questionnaire.

However, this method is limited in its ability to accurately assess subjects' true smoking status. These limitations include subjects' difficulty recalling the details of past smoking and respondent bias, the extent of which may increase with time as tobacco use becomes less socially acceptable due to changing social norms. Moreover, questionnaires do not evaluate nonsmokers' exposure to SHS in a fully accurate way, and questionnaires are especially inadequate when attempting to determine the smoking status of recent former smokers whose SHS exposure levels may vary (Kawachi & Colditz, 1996; Wells, English, Posner, Wagenknecht, & Perez-Stable, 1998).

For these reasons, researchers conducting prevalence surveys and epidemiological studies have for many years made use of biological markers of tobacco exposure to validate reported smoking status (Benowitz, 1983, 1999; Coultas, Howard, Peake, Skipper, & Samet, 1987; DeLorenze, Kharrazi, Kaufman, Eskenazi, & Bernert, 2002; Eliopoulos, Klein, & Koren, 1996; Etzel, 1990; NRC, 1986, 1989) and to monitor and track exposure to tobacco considering people, place, and time. Such biomarkers can include concentrations or metabolites of smoke-related components and the products of interaction between biological materials and chemicals in smoke.

Nicotine and its metabolites are tobacco-specific biomarkers (Benowitz & Jacob III, 1994; Benowitz, Jacob III, Fong, & Gupta, 1994; Jacob & Byrd, 1999). Nicotine binds to particles in the smoke inhaled by active smokers and is exclusively in the vapor phase in SHS (Benner et al., 1986; Ogden, Maiolo, Nelson, Heavner, & Green, 1993). The liver primarily metabolizes absorbed nicotine to cotinine (Benowitz & Jacob III, 1994; Benowitz et al., 1994); nicotine has a half-life in the body of approximately 2–3 h, but that of cotinine is 12–20 h (Benowitz, 1996; Hammond & Leaderer, 1987; Jaakkola & Jaakkola, 1997). Excreted small amount of cotinine (10%–15%) in urine is further metabolized to trans-3′-hydroxycotinine and other by-products (Benowitz et al., 1994). Cotinine is a commonly used biomarker for tobacco smoke exposure due to its longer half-life and is measured in the saliva, serum, or urine. Benowitz (1999) published a comprehensive review of tobacco smoke exposure biomarkers, levels of which are influenced by uptake, distribution, metabolism, and elimination inside the body after exposure.

The distribution of a biomarker specific to tobacco smoke such as nicotine or cotinine should be bimodal and reflect underlying distributions for both smokers and nonsmokers in a population that includes both. Two key self-report validity indicators can be calculated in a dichotomous template comparing self-reported smoking status and biomarkers: sensitivity (%) (the percentage of self-reported smokers classified as smokers by biomonitoring) and specificity (%) (the percentage of self-reported nonsmokers classified as nonsmokers by biomonitoring). A particularly critical issue here is underreporting, that is, "false negatives," those who report as nonsmokers but have a positive level of biomarkers. Nonsmokers are typically not expected to self-report as

Neuroscience of Nicotine
https://doi.org/10.1016/B978-0-12-813035-3.00051-4

smokers, but may have a positive biomarker level regardless due to exposure to SHS, which makes the specificity >100%. The lower the specificity for exposure status, the more the underestimation of association between the exposure and a disease occurrence (Flegal, Brownie, & Haas, 1986).

As mentioned above, a passive smoking pattern should be reflected in biomarker distribution among nonsmokers, while this distribution among active smokers should reflect their tobacco use. The purpose of setting a cutoff value is to optimize sensitivity and specificity while recognizing that there is inevitably a trade-off between these two. A lower cutoff value in a bimodal distribution generally achieves higher sensitivity, but a higher value will have greater specificity; in other words, more nonsmokers will exhibit biomarker levels that exceed the cutoff. A higher cutoff value has greater specificity but lower sensitivity than a lower cutoff value.

When choosing a cutoff value, the underlying distributions of biomarkers in both smokers and nonsmokers should ideally be taken into consideration; however, these distributions may change over time. Effective control of tobacco would shift the distribution of these values to a lower level in nonsmokers, with fewer nonsmokers having a detectable level of smoking biomarkers. This shift would also decrease the cutoff point. These biomarker distributions can potentially vary between populations due to different active smoking patterns and SHS exposure.

Cotinine is widely used as a validation standard in classifying exposure status (Boyd, Windsor, Perkins, & Lowe, 1998; Etter, Vu, & Perneger, 2000; Etzel, 1990; Hegaard, Kjaergaard, Moller, Wachmann, & Ottesen, 2007; Jarvis, Tunstall-Pedoe, Feyerabend, Vesey, & Saloojee, 1984, 1987; Luepker, Pallonen, Murray, & Pirie, 1989; McNeill, West, Jarvis, Jackson, & Bryant, 1986; Murray, Connett, Lauger, & Voelker, 1993; Pierce et al., 1987; Stookey, Katz, Olson, Drook, & Cohen, 1987). The utility of cotinine, carbon monoxide, and thiocyanate as biomarkers for tobacco use and cessation has been evaluated by a subcommittee of the Society for Research on Nicotine and Tobacco, and their application for cutoff values in clinical trials has been recommended (SRNT, 2002).

Articles reviewed in this study were selected using the keywords "cotinine" and "cutoff" or "self-reported" and "smoking status" and "validation" in the public literature database, PubMed, from 1985 to 2014 ($n=104$). We included articles if (1) it demonstrated information on agreement and discrepancy between self-reported smoking status and cotinine testing using self-reporting or cotinine testing as the gold standard and (2) the results described how a given cutoff value of cotinine concentration was determined. Bibliographies of the identified articles were also searched to determine cited methods for cutoff determination. This overview included a comparison of results from both adolescent and adult study populations. Articles describing a cotinine test strip or cotinine dipstick method were excluded. This overview includes a total of 32 articles.

The author has summarized cutoff values, their determination methods, and study population characteristics. Sensitivity and specificity are also summarized to differentiate smokers and nonsmokers with self-reporting as a gold standard test using three biospecimens: saliva, serum, and urine. The author also evaluated changes in cutoff values over time and their differences in terms of both populations and regions, and issues arising from the process of determining cutoff value were described for the purposes of generalization.

The characteristics of the study population are summarized in Tables 51.1–51.3 along with methods used to determine cutoff values; saliva, serum, and urinary cotinine cutoff values; and the number of self-reported smokers and nonsmokers. Most of the studies reviewed here were conducted in either North America or Europe from the mid-1980s to 2013, and their sample sizes ranged from 24 to 24,332. A number of methods were used to select cutoff values. These included (1) cotinine levels in nonsmokers (Benowitz, 1983), passive smokers (Hoffmann, Haley, Adams, & Brunnemann, 1984), and smokers (Slattery, Hunt, French, Ford, & Williams, 1989); (2) a 2 × 2 table (Jarvis et al., 1987; Luepker et al., 1989; Martinez, Reid, Jiang, Einspahr, & Alberts, 2004; Murray et al., 1993; Pierce et al., 1987); and (3) a separation point in the bimodal distribution of cotinine concentrations in smokers and nonsmokers (Pirkle et al., 1996). A more recent method of selecting a cutoff value is the receiver operating characteristic (ROC) curve, which optimizes sensitivity and specificity while correctly classifying smoking status at the highest rate (Benowitz, Bernert, Caraballo, Holiday, & Wang, 2009; Boyd et al., 1998; Goniewicz et al., 2011; Hegaard et al., 2007; Jarvis, Fidler, Mindell, Feyerabend, & West, 2008). Cutoff value details and determination methods using biospecimens (serum, saliva, and urine) are also presented.

51.1 SALIVARY COTININE

Cutoff values of salivary cotinine used to differentiate smokers from nonsmokers ranged from 10 to 44 ng/mL within the 11 studies reviewed. The overall ranges of sensitivity and specificity were 69%–99% and 74%–99%, respectively (Fig. 51.1). Jarvis et al. (1987) studied 211 adult outpatients in London, United Kingdom, and reported a salivary cotinine cutoff value of 14.2 ng/mL; this was selected using a 2 × 2 table (Table 51.1). This value had a sensitivity of 96.4% and specificity of 99.0%. Pierce et al. (1987) reported 44 ng/mL as the cutoff value in a study of an Australian population ($n=975$)

TABLE 51.1 Salivary Cotinine Cutoff Values and Methods of Cutoff Determination

		Study population characteristics				Cutoff value (ng/mL)	Number of self-reported		
Author	**Year**	**Description**	***N***	**Age**	**Female (%)**		**Smokers**	**Nonsmokers**	**Method for cutoff determination**
Jarvis et al.	1987	Outpatients at St Mary's Hospital, London, United Kingdom	211	Mean age, 55 years	24.6	14.2	111	100	2 × 2 table: The cutoff value providing the highest numbers of correctly classified smokers and nonsmokers
McNeill et al.	1987	Students attending girls' comprehensive school in London, United Kingdom	508	11–16 years	100	14.7	173	335	Cutoff value was adopted from the study results of Jarvis et al. (1987)
Pierce et al.	1987	Residents randomly selected in a community, Australia	975	14 years or older	49.2	44	353	622	2 × 2 table
Stookey et al.	1987	Participants in a clinical trial for evaluating measures to facilitate smoking cessation, United States	236	NA	NA	10	216	20	Cutoff value was adopted from the study results of Benowitz (1983): "there was no nonsmokers with blood cotinine values greater than 10 ng/mL" (page 21)
Luepker et al.	1989	High school students randomly selected for a survey in Minneapolis, MN, United States	263	17–21 years	NA	20	87	176	2 × 2 table
Etzel[a]	1990	Participants in 22 studies published between 1973 and 1989	NA	NA	NA	10	NA	NA	Cutoff value was selected after comparisons of the distributions of salivary cotinine concentrations with respect to smoking status between 22 study papers published between 1973 and 1989
Murray et al.	1993	Participants under routine care in a clinical trial of "Lung Health Study" in United States and Canada with evidence of early-stage chronic obstructive lung disease	1498	35–60 years (mean, 48.5 years)	36	20	1345	153	Cutoff value was selected as the salivary cotinine level that provided the highest percentage of correctly classified smoking status
Boyd et al.	1998	Pregnant women in the Birmingham trial II conducted at four public health maternity clinics in	548	Mean, 24.6 years	100	24	441	107	Receiver operating characteristic (ROC) curve: the cutoff value providing the greatest percentage

Continued

TABLE 51.1 Salivary Cotinine Cutoff Values and Methods of Cutoff Determination—cont'd

		Study population characteristics				Cutoff value (ng/mL)	Number of self-reported		
Author	Year	Description	*N*	Age	Female (%)		Smokers	Nonsmokers	Method for cutoff determination
		Birmingham, AL, United States							of correctly classified smoking status
Jarvis et al.	2008	Participants in the Health Survey for England: 1996–2004	24,332	All adults	NA	12	8808	15,524	ROC
Stragierowicz et al.	2013	Pregnant women in Polish Mother and Child Cohort Study	69	Mean, 26.4 years	100	12.9	19	50	ROC
Smith et al.	2014	Pregnant women in Southcentral Foundation's Anchorage Primary Health Care Center: 2006–2010	376	Mean, 26.9 years	100	1.07	116	260	ROC

[a] *Etzel did not provide gold standard information, but the cutoff value was included in this paper because the value was obtained after reviewing a large number of papers.*

aged at least 14 years. Etzel (1990) reviewed the results of 22 studies from an English-language search of MEDLINE from 1973 to 1989 with the keywords "saliva" and "cotinine" and reported a cutoff value of 10 ng/mL as a result. This cutoff value was determined by comparing the distribution of salivary cotinine in relation to smoking status: Passive smokers had values not exceeding 5 ng/mL, heavy passive smokers exhibited values of 10 ng/mL or slightly higher, infrequent regular smokers were found to have values between 10 and 100 ng/mL, and the values of regular active smokers equaled or surpassed 100 ng/mL. However, the sensitivity and specificity for the 10 ng/mL value determined by Etzel (1990) were unavailable. Jarvis et al. (2008) recently reported a new salivary cotinine cutoff value of 12 ng/mL; this was determined from a study of participants aged 16 and older in the Health Survey for England (HSE) from 1996 to 2003 by means of an ROC curve and had a sensitivity of 96.7% and specificity of 96.9% to distinguish cigarette smokers from those who had never smoked. Fig. 51.1 shows the distributions of sensitivity and specificity values for the various cotinine cutoff levels summarized in Tables 51.1–51.3.

51.2 SERUM COTININE

Serum cotinine cutoff values from 14 studies identified to evaluate them ranged from 3.0 to 20 ng/mL, yielding a sensitivity range of 73.2%–98.9% and specificity range of 78.7%–99.0%. Eight of these studies provided a cutoff value of 14 or 15 ng/mL between 1989 and 2004 (Caraballo, Giovino, Pechacek, & Mowery, 2001; Heller, Scherer, Sennewald, & Adlkofer, 1998; Nafstad et al., 1996; Pichini et al., 2000; Pirkle et al., 1996; Seccareccia et al., 2003; Slattery et al., 1989; Wagenknecht, Burke, Perkins, Haley, & Friedman, 1992) (Table 51.2). Slattery et al. (1989) reported a cutoff value of 15 ng/mL, which was equal to 6% of the mean serum cotinine level of smokers in their study. Pirkle et al. (1996) selected 15 ng/mL as a serum cotinine cutoff value as this value marks the separation point in the bimodal distribution of serum cotinine levels in both users and nonusers of tobacco as reported in the American 3rd National Health and Nutrition Examination Survey (NHANES III, 1988–1991) (n = 10,270, aged ≥4 years). Pirkle's study did not include information on sensitivity and specificity. The NHANES study was based on a relatively large number of healthy participants of varying ages and ethnicities, and its reported value of 15 ng/mL was consistent with the cutoff value from an ROC curve of the Italian population from a cross-sectional study of cardiovascular diseases (MATISS project, n = 3379) (Seccareccia et al., 2003). Caraballo et al. (2001) used the NHANES III dataset but with an extended survey period (1988–94) and adult participants aged ≥17 years (n = 15,357) to validate and suggest a new sensitivity (89.5%) and specificity (98.5%) using the cutoff value indicated by Pirkle et al. (1996). They reported their own new cutoff of 11.4 ng/mL later with corresponding sensitivity (73.2%) and specificity

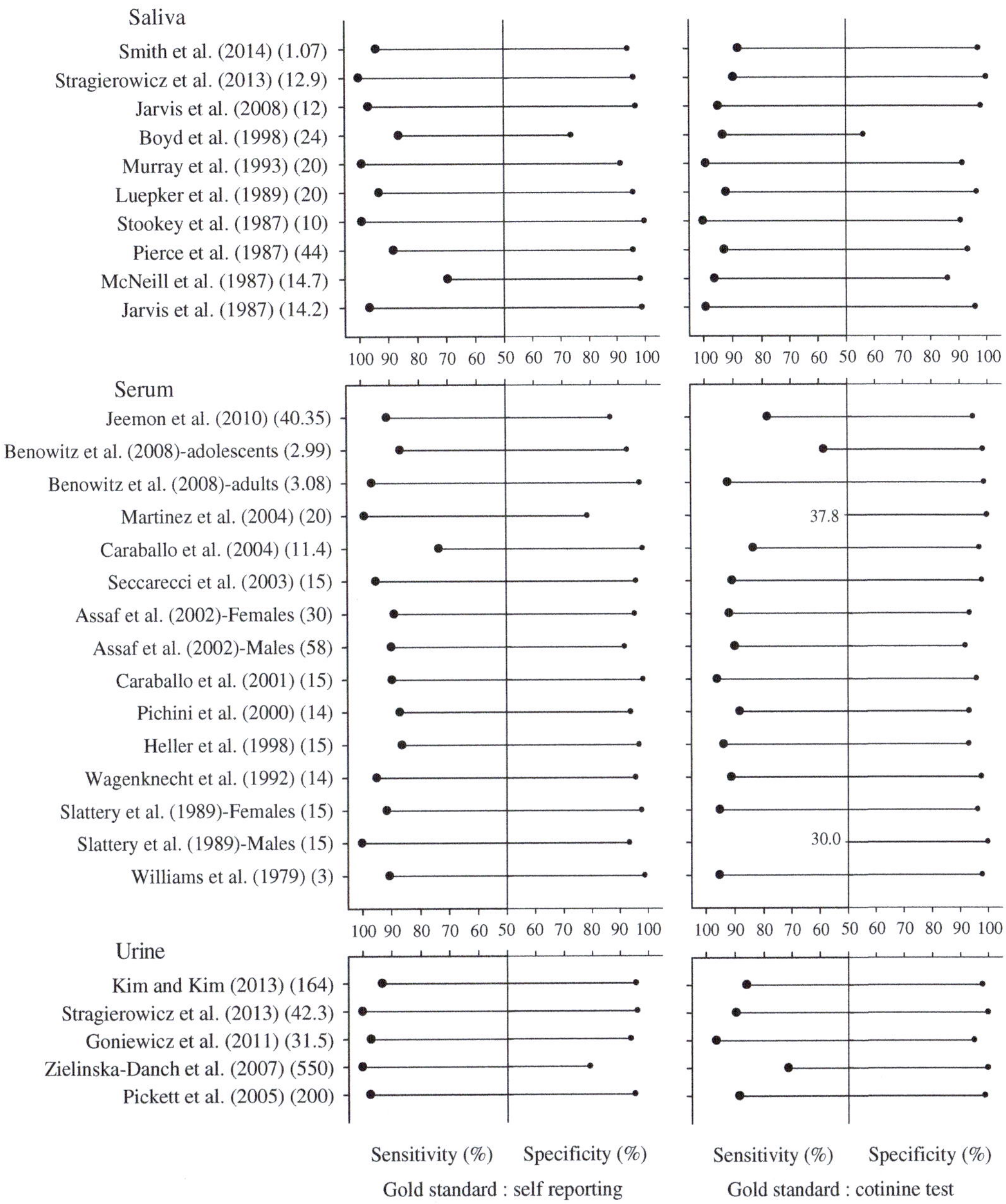

FIG. 51.1 Plots of sensitivities and specificities according to the cutoff values of salivary or serum cotinine. (The numbers in parentheses represent cutoff values selected by the authors.)

(98.4%) for the adolescent population ($n=2107$) from the same dataset (NHANES III, 1988–1994), and this was calculated using an ROC curve. The serum cotinine cutoff values shown in Table 51.3 are largely comparable with only a few exceptions (Assaf, Parker, Lapane, McKenney Jr., & Carleton, 2002; Jeemon et al., 2010; Williams, Eng, Botvin, Hill, & Wynder, 1979) and fell within a range of 8–20 ng/mL. However, much lower cutoff values of 3.08 ng/mL for adults and 2.99 ng/mL for adolescents were recently reported by Benowitz et al. using the ROC analysis of the 1999–2004 NHANES data.

51.3 URINARY COTININE

Our review included only a few studies ($n=5$) of urinary cotinine with cutoff values of between 31.5 and 550 ng/mL (Goniewicz et al., 2011; Hoffmann et al., 1984; Pickett, Rathouz, Kasza, Wakschlag, & Wright, 2005; Riboli et al., 1990; Zielinska-Danch, Wardas, Sobczak, & Szoltysek-Boldys, 2007). Pickett et al. (2005) provided a cutoff value of 200 ng/mL using an ROC curve and calculated based on data from 998 pregnant women who were patients at an East Boston Neighborhood Health Center in the United States. A cutoff value

TABLE 51.2 Serum Cotinine Cutoff Values and Methods of Determination

		Study population characteristics				Cutoff value (ng/mL)	Number of self-reported		
Author	Year	Description	*N*	Age	Female (%)		Smokers	Nonsmokers	Method for cutoff determination
Williams et al.	1979	High school students participating in a school health education program. Blinded for the purpose of blood collection in United States	118	14–17 years	53	3.0	21	97	2×2 table
Benowitz	1983	Participants in a test evaluating an elimination half-life for cotinine. They stopped smoking in a research ward (average, 19.1h; range, 10.9–37.0h)	16	NA	NA	10	NA	NA	Authors selected the cutoff value from a range of concentrations among nonsmokers and reported that "no nonsmoker had blood cotinine values higher than 10ng/mL"
Slattery et al.	1989	Participants in (1) a cross-sectional study on dietary intake and hormone, (2) a case control study of squamous cell cervical cancer in Utah, United States	**(1)** 112 **(2)** 547	17years or older	**(1)** Male only **(2)** Female only	15	**(1)** 3 **(2)** 163	**(1)** 109 **(2)** 379	Authors selected the cutoff value by calculating 6% of the mean serum cotinine levels in smokers in the study
Wagenknecht et al.	1992	Young adults in a cohort of cardiovascular disease study in United States	4984	17–30years	NA	14	1540	3444	ROC
Pirkle et al.[a]	1996	Participants in the 3rd National Health and Nutrition Examination Survey (NHANES), United States	10,270	4years or older	50	15	NA	NA	Authors selected the cutoff value from a separation point in the bimodal distribution of serum cotinine in tobacco users and nonusers
Nafstad et al.[a]	1996	Pregnant women in the Oslo Birth Cohort, Norway	202	Mean, 30years (range, 19–43)	100	14	42 regular +24 occasional	136	The traditionally used cutoff value (14ng/mL) was chosen at the authors' discretion
Heller et al.	1998	Followers among participants in the WHO monitoring trends and determinants in cardiovascular disease (MONICA) project in 1987–88, Germany	3661	TBA	50.9	15	1227	2434	Cutoff value was adopted from the study results of Wagenknecht et al. (1992)
Pichini et al.	2000	Pregnant women attending the Hospital del Mar in Barcelona, Spain	404	TBA	100	14	136	268	Cutoff value was adopted from the study results of Nafstad et al. (1996)
Caraballo et al.	2001	Adults in the 3rd National Health and Nutrition Examination Survey (NHANES) 1988–1994, United States	15,357	17years or older	53.8	15	4274	11,083	Cutoff value was adopted from the study results of Pirkle et al. (1996)

Assaf et al.	2002	Adults those who conducted cotinine tests in Pawtucket Heart Health Program 1985–1986, United States	784	18–65 years	57.5	58 for male (M) 30 for female (F)	131(M) 141(F)	172 (M) 279 (F)	ROC
Seccareccia et al.	2003	Providers of serum samples among participants in the project of MATISS Malattie cardiovascolari Aterosclerotiche, Istituto Superiore di Sanità (MATISS), Italy	3379	20–79 years	39.5	15	977	2402	ROC
Caraballo et al.	2004	Adolescents in the 3rd National Health and Nutrition Examination Survey (NHANES) 1988–1994, United States	2107	12–17 years	53.8	11.4	213	1894	ROC
Martinez et al.	2004	Participants in a dietary trial on adenoma recurrence, Phoenix, AZ, United States	824	40–80 years	31	20	95	729	2 × 2 table
Benowitz et al.	2008	Participants in the National Health and Nutrition Examination Survey (NHANES) for 1999–2004, United States	9901 for adults 5138 for adolescents	20 years or older 12–19 years	50.6 for adults 49.6 for adolescents	3.08 for adults 2.99 for adolescents	2340 515	7561 4623	ROC
Jeemon et al.	2010	Participants in the cardiovascular disease surveillance program at New Delhi, India	426	18 years or older	TBA	40.35	142	284	ROC

[a] *Benowitz (1983), Pirkle et al. (1996), and Nafstad et al. (1996) did not provide sensitivity and specificity values, but the cutoff value was included in this overview because the values were obtained after reviewing a large number of study populations or the cutoff value was referred by other studies.*

TABLE 51.3 Urine Cotinine Cutoff Values and Methods of Determination

		Study population characteristics				Cutoff value (ng/mL)	Number of self-reported		
Author	Year	Description	*N*	Age	Female (%)		Smokers	Nonsmokers	Method for cutoff determination
Hoffmann et al.[a]	1984	Volunteers joined a study on uptake of sidestream smoke	NA	NA	NA	55	NA	NA	The value was obtained from urine samples that were collected when saliva nicotine levels returned to baseline levels (i.e., 5h after study subjects were exposed to SHS in closed chamber (280 mg/m^3 for air nicotine concentration) for 1h)
Riboli et al.[a]	1990	Married nonsmoking women from 10 countries	1369	Age, 42–60 years	NA	50	NA	NA	Cutoff value was chosen as the value that provided 3.4% misclassification. It was also compared with the study results of Hoffmann et al. (1984)
Pickett et al.	2005	Pregnant women attending the East Boston Neighborhood Health Center, United States, between 1986 and 1992 with allowance of multiple visits	998	19 years or more	NA	200	1272	3566	ROC
Zielinska-Danch	2007	Volunteers living in Sosnowiec, Poland	327	19–60	57.2	550	111	216	Authors selected the cutoff value from a separation point in the bimodal distribution of urine cotinine in self-reported smokers and nonsmoker
Goniewicz et al.	2011	Smokers from three different studies conducted in San Francisco, United States; Silesia, Poland; and Pittsburgh, United States Nonsmokers from the other three studies conducted in United States, Poland, and Mexico	601	18 years or older	52.7	31.5	373	228: passive smokers only	ROC
Stragierowicz et al.	2013		69	Mean, 26.4 years	100	53.0	17	52	ROC

TABLE 51.3 Urine Cotinine Cutoff Values and Methods of Determination—cont'd

		Study population characteristics				Cutoff value	Number of self-reported		
Author	Year	Description	*N*	Age	Female (%)	(ng/mL)	Smokers	Nonsmokers	Method for cutoff determination
		Pregnant women in Polish mother and child cohort study							
Kim and Jung	2013	Participants in Korea National Health and Nutrition Examination Survey (KNHANES) for 2008–10, South Korea	11,629	19 years or older	55.5	164	2547	9082	ROC

[a] *Hoffmann et al. (1984) and Riboli et al. (1990) did not provide sensitivity and specificity values, but the cutoff values were included in this overview because the values were one of few available values referred by other studies or obtained after reviewing a large number of study populations, respectively.*

of 550 ng/mL was obtained from 327 subjects of a Polish study (Zielinska-Danch et al., 2007). This cutoff value was selected based on a separation point in the bimodal distribution of urinary cotinine among self-reported smokers and nonsmokers. An ROC analysis of data from the subjects of six different studies was the basis for the cutoff value of 31.5 ng/mL reported by Goniewicz et al. (2011) to distinguish adult active and passive smokers. Kim and Jung (2013) recently found an optimum urinary cotinine cutoff value of 164 ng/mL using the Korea National Health and Nutrition Examination Survey database (2008–10, $n = 11{,}629$). This cutoff value for urinary cotinine had a sensitivity of 93.2% and specificity of 95.7%.

51.4 ISSUES IN DETERMINING CUTOFFS

The results of salivary cotinine studies conducted to determine cutoff values varied widely depending on the characteristics of study populations (Table 51.1). Three studies on salivary cotinine (Boyd et al., 1998; Murray et al., 1993; Stookey et al., 1987) were conducted with subjects from a clinical trial of smoking cessation, while Jarvis et al. (1987) conducted a study of outpatients in a London hospital. Table 51.2 covers the results of studies of serum cotinine (Heller et al., 1998; Martinez et al., 2004; Slattery et al., 1989; Wagenknecht et al., 1992; Williams et al., 1979) with participants from a cardiovascular disease study, health education program, and clinical trial. Participants in smoking cessation clinical trials or other studies related to health are likely to underreport their levels of exposure to smoke. Previous studies have also indicated that social norm pressures can influence smokers to underreport their own smoking (Fendrich, Mackesy-Amiti, Johnson, Hubbell, & Wislar, 2005; Perez-Stable, Marin, Marin, Brody, & Benowitz, 1990).

Pregnant women were the focus of three saliva studies (Boyd et al., 1998; Smith, Robinson, Khan, Sosnoff, & Dillard, 2014; Stragierowicz, Mikołajewska, Zawadzka-Stolarz, Polańska, & Ligocka, 2013), two serum studies (Nafstad et al., 1996; Pichini et al., 2000), and one urine study (Kim & Jung, 2013). It is widely recognized that pregnancy affects both nicotine uptake and metabolism (Rebagliato et al., 1998). The aforementioned studies determined that the ratio of salivary cotinine per cigarette smoked was much lower during pregnancy (median of 3.53 ng/mL per cigarette) than after pregnancy (median of 9.87 ng/mL per cigarette). These findings indicate that cutoff values for pregnant populations cannot be applied to the general population.

Three (Murray et al., 1993; Pierce et al., 1987; Smith et al., 2014) of the 11 studies in Table 51.1 were conducted exclusively on Caucasian populations. A recent study by Wells et al. (1998) reported that the misclassification rate of female occasional or regular smokers who were classified as never having smoked was higher among American minorities (15.3% or 2.8%) than among American non-Hispanic Caucasians (6.0% or 0.8%). This pattern of misclassification was similar among men but slightly higher than that among women. Differences in specificities of given cutoff values were reported to be different among races (Jarvis et al., 2008): 98.4% in Caucasian participants and 90.2% in South Asian immigrants, thereby indicating that it may be necessary to apply a different cutoff value to these Asian populations to achieve a similar specificity to that of Caucasian participants. Oral tobacco products are very frequently consumed in India, Pakistan, and Bangladesh, while smoking among women, once taboo, is now dramatically increasing in some Asian countries such as Japan, Korea, and China (WHO, 2008). Serum cotinine cutoff values were reported to differ with race/ethnicity by Benowitz et al. (2009) based on the 1999–2004 NHANES data (3078 smokers and 13,078 nonsmokers); these values were found to be 5.92ng/mL for non-Hispanic Blacks, 4.85ng/mL for non-Hispanic Whites, and 0.84ng/mL for Mexican-Americans. The estimated cutoff values for urinary cotinine were 15 and 60ng/mL, separately converted from the American serum cotinine cutoff value (3ng/mL) and English salivary cotinine cutoff value (12ng/mL). These values were substantially lower than reported Korean urinary cotinine cutoff values. This suggests that values obtained from Caucasian populations cannot be generalized to other racial/ethnic groups and that country-, ethnic-, and gender-specific cutoff values should be determined along with validation of women's smoking status.

Our review suggests insufficient consistency between study designs to facilitate comparison. Sensitivity and specificity tests were conducted differently in different studies (i.e., using self-reporting or cotinine testing). When the same cutoff value was chosen, the sensitivity dropped to <80% from ≥90% when cotinine tests were used rather than self-reporting in several studies (Benowitz et al., 2009; Jeemon et al., 2010; Martinez et al., 2004; Slattery et al., 1989) (Fig. 51.1); this indicates that self-reporting is not an accurate indicator of true smoking status. Based on this overview, it is recommended to provide sensitivity and specificity for a given cutoff value based on the two gold stands with their corresponding 2×2 tables for the purposes of comparison and validation. Sensitivity should also be reported when specificity is 95%, or vice versa, to compare similar study types conducted at different times and/or among different populations and regions. Such information from a large population-based study would be extremely useful.

51.5 DROP IN CUTOFF VALUES OVER THE LAST 20 YEARS

While it is true that cotinine levels in saliva, serum, or urine can be used to differentiate smokers from nonsmokers within an individual study, it is as yet difficult to generalize any particular study as sensitivity and specificity are prone to vary by ethnicity, population characteristics such a gender and location, tobacco products used and their usage patterns, SHS exposure, and absorption pathway. The optimal cutoff for the classification of smokers and nonsmokers can therefore vary with population characteristics, region, or time. A comparison of the salivary cotinine cutoff values obtained from the 1988–94 and 1999–2004 NHANES data with the two salivary cotinine cutoff values reported in 1987 and 2008 in the United Kingdom by Jarvis et al. (Table 51.1) shows this to be true.

The recent cutoff value of 12ng/mL reported by Jarvis et al. (2008) indicates that the salivary cotinine cutoff value in the UK population has decline in the past 20 years from 14.2 to 12ng/mL; however, this was a relatively smaller difference than that found in the US population, as discussed below.

Cutoff values recently reported by Benowitz et al. (2009) for adults (3.08ng/mL) and adolescents (2.99ng/mL) based on ROC analysis from the 1999–2004 NHANES data (9901 adults and 5138 adolescents) were much lower than the earlier ones reported by Caraballo et al. (2001) and Caraballo, Giovino, and Pechacek (2004) using the 1988–94 NHANES data: 15ng/mL for adults (n=15,357) and 11.4ng/mL for adolescents (n=2107) (Caraballo et al., 2001, 2004). ROC analysis was used by both researchers. These results indicate that the serum cotinine cutoff value in the US population dropped over a period of 20 years. This lower cutoff value may be accounted for by a reduction in the amount of tobacco smoking among smokers and less SHS exposure among nonsmokers in the past two decades due to increased tobacco control policy efforts in the United States, including tobacco excise taxes, smoke-free workplaces, youth access laws, and increased funding provided for tobacco control (Stoner & Foley, 2006).

Jarvis et al. (2008)'s recent salivary cotinine cutoff value based on the HSE data was compared with Benowitz et al.'s recently reported serum cotinine cutoff value from the US NHANES datasets using a conversion factor (1.16) from serum to salivary cotinine. This indicated that Jarvis et al. (2008)'s result was rather higher than that of Benowitz et al. (2009). The difference in smoking prevalence between the two countries may account for this difference in cutoff values. A report by the World Health Organization on the global tobacco epidemic (WHO, 2008) suggested that age-standardized adult smoking prevalence in the United Kingdom and United States was 28.4% and 18.7%, respectively. The distribution of tobacco smoke biomarkers in nonsmokers should reflect a pattern of never smoking or passive smoking, whereas that in active smokers will reflect their tobacco use. The distribution may change over time; with effective tobacco control, the distribution of values in nonsmokers will shift to the left, and more nonsmokers will have undetectable level. This shift to the left may also result in the cutoff point shifting to the left, that is, reaching a lower value.

Limitations of this overview include its use primarily of the PubMed database to extract articles while ignoring other databases that may also contain useful articles. However, the number of articles included in our review was higher as the bibliographies of included articles were also searched. Another limitation is that the calculation of sensitivity and specificity through self-reporting or cotinine testing using the provided 2 × 2 table was done with the assumption that the cotinine tests used in each study were 100% accurate.

The purpose of this study was to examine studies that have reported cotinine cutoff values and potential issues in applying these studies' reported cutoff values for salivary or serum cotinine to the general population. It may be more desirable to provide country-specific or multiple biomarker cutoff values for different exposure categories in large populations, that is, active versus passive smokers and passive versus nonsmokers. This would enable future researchers to compare and apply the most relevant data for their own studies. Furthermore, more studies to evaluate the utility of the new serum cutoff value (3 ng/mL) obtained from the 1999–2004 NHANES dataset would be beneficial.

MINI-DICTIONARY OF TERMS

Sensitivity (%) The percentage of self-reported smokers classified as smokers by biomonitoring

Specificity (%) The percentage of self-reported nonsmokers classified as nonsmokers by biomonitoring

Key Facts of Cutoff Value

- Biomarkers specific to tobacco smoke include nicotine and cotinine; these are expected to be bimodal and to reflect two underlying distributions for smokers and nonsmokers in populations that include both.
- Two key self-report validity indicators can be calculated in a dichotomous template comparing self-reported smoking status and biomarkers: sensitivity (%) (the percentage of self-reported smokers classified as smokers by biomonitoring) and specificity (%) (the percentage of self-reported nonsmokers classified as nonsmokers by biomonitoring).
- The purpose of setting a cutoff value is to optimize both sensitivity and specificity; however, there will inevitably be a trade-off between the two.
- Higher sensitivity can generally be achieved by a lower cutoff value in a bimodal distribution; however, the specificity is then lower than it would be with a higher cutoff value. In other words, more nonsmokers have biomarker levels exceeding the cutoff.
- In contrast, a higher cutoff value means greater specificity but lower sensitivity.

Summary Points

- We describe how cutoff values were derived and explain issues that occur when generalizing a cutoff value.
- We conducted a search of English literature from 1985 to 2014 in PubMed with keywords "cotinine" and "cutoff" or "self-reported," "smoking status," and "validation." Out of a total 104 articles obtained, 32 provided (1) sensitivity and specificity and (2) methods used to determine the cutoff value applied.
- Cotinine cutoff values were found to range from 10 to 25 ng/mL (salivary cotinine), 10 to 20 ng/mL (serum cotinine), and 50 to 200 ng/mL (urinary cotinine). These have been commonly used to validate self-reported smoking status by means of a 2 × 2 table or a receiver operating characteristic curve.
- While traditionally accepted, cotinine cutoff values in serum and saliva were 15 and 14 ng/mL, respectively; recent large population-based studies conducted in the United States and United Kingdom reported lower values of 3 ng/mL in serum and 12 ng/mL in saliva.
- Time, location, and ethnicity can affect the optimal cutoff value for the classification of smoking status.

References

Assaf, A. R., Parker, D., Lapane, K. L., McKenney, L., Jr., & Carleton, R. A. (2002). Are there gender differences in self-reported smoking practices? Correlation with thiocyanate and cotinine levels in smokers and nonsmokers from the Pawtucket Heart Health Program. *Journal of Women's Health, 11*(10), 899–906.

Benner, C. L., Bayona, J. M., Caka, F. M., Tang, H., Kewis, L., Crawford, J., et al. (1986). Chemical composition of environmental tobacco smoke. 2. Particulate-phase compounds. *Environmental Science & Technology, 23*, 688–699.

Benowitz, N. L. (1983). The use of biologic fluid samples in assessing tobacco smoke consumption. In: NIDA research monograph, no. 48, *Measurement in the analysis and treatment of smoking behavior*. Rockville, MD: National Institute on Drug Abuse.

Benowitz, N. L. (1996). Cotinine as a biomarker of environmental tobacco smoke exposure. *Epidemiologic Reviews, 18*(2), 188–204.

Benowitz, N. L. (1999). Biomarkers of environmental tobacco smoke exposure. *Environmental Health Perspectives, 107*(Suppl. 2), 349–355.

Benowitz, N. L., Bernert, J. T., Caraballo, R. S., Holiday, D. B., & Wang, J. (2009). Optimal serum cotinine levels for distinguishing cigarette smokers and nonsmokers within different racial/ethnic groups in the United States between 1999 and 2004. *American Journal of Epidemiology, 169*(2), 236–248.

Benowitz, N. L., & Jacob, P., III (1994). Metabolism of nicotine to cotinine studied by a dual stable isotope method. *Clinical Pharmacology and Therapeutics, 56*(5), 483–493.

Benowitz, N. L., Jacob, P., III, Fong, I., & Gupta, S. (1994). Nicotine metabolic profile in man: comparison of cigarette smoking and transdermal nicotine. *The Journal of Pharmacology and Experimental Therapeutics, 268*(1), 296–303.

Boyd, N. R., Windsor, R. A., Perkins, L. L., & Lowe, J. B. (1998). Quality of measurement of smoking status by self-report and saliva cotinine among pregnant women. *Maternal and Child Health Journal, 2*(2), 77–83.

Caraballo, R. S., Giovino, G. A., & Pechacek, T. F. (2004). Self-reported cigarette smoking vs. serum cotinine among U.S. adolescents. *Nicotine & Tobacco Research, 6*(1), 19–25.

Caraballo, R. S., Giovino, G. A., Pechacek, T. F., & Mowery, P. D. (2001). Factors associated with discrepancies between self-reports on cigarette smoking and measured serum cotinine levels among persons aged 17 years or older, Third National Health and Nutrition Examination Survey, 1988-1994. *American Journal of Epidemiology, 153*(8), 807–814.

Coultas, D. B., Howard, C. A., Peake, G. T., Skipper, B. J., & Samet, J. M. (1987). Salivary cotinine levels and involuntary tobacco smoke exposure in children and adults in New Mexico. *The American Review of Respiratory Disease, 136*(2), 305–309.

DeLorenze, G. N., Kharrazi, M., Kaufman, F. L., Eskenazi, B., & Bernert, J. T. (2002). Exposure to environmental tobacco smoke in pregnant women: the association between self-report and serum cotinine. *Environmental Research, 90*(1), 21–32.

Eliopoulos, C., Klein, J., & Koren, G. (1996). Validation of self-reported smoking by analysis of hair for nicotine and cotinine. *Therapeutic Drug Monitoring, 18*(5), 532–536.

Etter, J. F., Vu, D. T., & Perneger, T. V. (2000). Saliva cotinine levels in smokers and nonsmokers. *American Journal of Epidemiology, 151*(3), 251–258.

Etzel, R. A. (1990). A review of the use of saliva cotinine as a marker of tobacco smoke exposure. *Preventive Medicine, 19*(2), 190–197.

Fendrich, M., Mackesy-Amiti, M. E., Johnson, T. P., Hubbell, A., & Wislar, J. S. (2005). Tobacco-reporting validity in an epidemiological drug-use survey. *Addictive Behaviors, 30*(1), 175–181.

Flegal, K. M., Brownie, C., & Haas, J. D. (1986). The effects of exposure misclassification on estimates of relative risk. *American Journal of Epidemiology, 123*, 736–751.

Goniewicz, M. L., Eisner, M. D., Lazcano-Ponce, E., Zielinska-Danch, W., Koszowski, B., Sobczak, A., et al. (2011). Comparison of urine cotinine and the tobacco-specific nitrosamine metabolite 4-(methylnitrosamino)-1-(3-pyridyl)-1-butanol (NNAL) and their ratio to discriminate active from passive smoking. *Nicotine & Tobacco Research, 13*(3), 202–208.

Hammond, S. K., & Leaderer, B. P. (1987). A diffusion monitor to measure exposure to passive smoking. *Environmental Science & Technology, 21*(5), 494–497.

Hegaard, H. K., Kjaergaard, H., Moller, L. F., Wachmann, H., & Ottesen, B. (2007). Determination of a saliva cotinine cut-off to distinguish pregnant smokers from pregnant non-smokers. *Acta Obstetricia et Gynecologica Scandinavica, 86*(4), 401–406.

Heller, W. D., Scherer, G., Sennewald, E., & Adlkofer, F. (1998). Misclassification of smoking in a follow-up population study in southern Germany. *Journal of Clinical Epidemiology, 51*(3), 211–218.

Hoffmann, D., Haley, N. J., Adams, J. D., & Brunnemann, K. D. (1984). Tobacco sidestream smoke: uptake by nonsmokers. *Preventive Medicine, 13*(6), 608–617.

International Agency for Research on Cancer. (2004). IARC monographs on the evaluation of carcinogenic risks to humans. *Tobacco smoke and involuntary smoking: Vol. 38*. Lyon: International Agency for Research on Cancer.

Jaakkola, M. S., & Jaakkola, J. J. (1997). Assessment of exposure to environmental tobacco smoke. *The European Respiratory Journal, 10*(10), 2384–2397.

Jacob, P. I., & Byrd, G. D. (1999). Use of gas chromatographic and mass spectrometric techniques for the determination of nicotine and its metabolites. In: J. W. Gorrod, & P. I. Jacob (Eds.), *Analytical determination of nicotine and related compounds and their metabolites* (pp. 191–224). *1999*. Amsterdam: Elsevier Science.

Jarvis, M. J., Fidler, J., Mindell, J., Feyerabend, C., & West, R. (2008). Assessing smoking status in children, adolescents and adults: cotinine cut-points revisited. *Addiction, 103*(9), 1553–1561.

Jarvis, M., Tunstall-Pedoe, H., Feyerabend, C., Vesey, C., & Salloojee, Y. (1984). Biochemical markers of smoke absorption and self reported exposure to passive smoking. *Journal of Epidemiology and Community Health, 38*(4), 335–339.

Jarvis, M., Tunstall-Pedoe, H., Feyerabend, C., Vesey, C., & Salloojee, Y. (1987). Comparison of tests used to distinguish smokers from nonsmokers. *American Journal of Public Health, 77* (11), 1435–1438.

Jeemon, P., Agarwal, S., Ramakrishnan, L., Gupta, R., Snehi, U., Chaturvedi, V., et al. (2010). Validation of self-reported smoking status by measuring serum cotinine levels: an Indian perspective. *The National Medical Journal of India, 23*(3), 134–136.

Kawachi, I., & Colditz, G. A. (1996). Invited commentary: confounding, measurement error, and publication bias in studies of passive smoking. *American Journal of Epidemiology, 144*(10), 909–915.

Kim, S., & Jung, A. (2013). Optimum cutoff value of urinary cotinine distinguishing South Korean adult smokers from nonsmokers using data from the KNHANES (2008-2010). *Nicotine & Tobacco Research, 15*(9), 1608–1616.

Luepker, R. V., Pallonen, U. E., Murray, D. M., & Pirie, P. L. (1989). Validity of telephone surveys in assessing cigarette smoking in young adults. *American Journal of Public Health, 79*(2), 202–204.

Martinez, M. E., Reid, M., Jiang, R., Einspahr, J., & Alberts, D. S. (2004). Accuracy of self-reported smoking status among participants in a chemoprevention trial. *Preventive Medicine, 38*(4), 492–497.

McNeill, A. D., West, R. J., Jarvis, M., Jackson, P., & Bryant, A. (1986). Cigarette withdrawal symptoms in adolescent smokers. *Psychopharmacology, 90*(4), 533–536.

Murray, R. P., Connett, J. E., Lauger, G. G., & Voelker, H. T. (1993). Error in smoking measures: effects of intervention on relations of cotinine and carbon monoxide to self-reported smoking. The Lung Health Study Research Group. *American Journal of Public Health, 83*(9), 1251–1257.

Nafstad, P., Kongerud, J., Botten, G., Urdal, P., Silsand, T., Pedersen, B. S., et al. (1996). Fetal exposure to tobacco smoke products: a comparison between self-reported maternal smoking and concentrations of cotinine and thiocyanate in cord serum. *Acta Obstetricia et Gynecologica Scandinavica, 75*(10), 902–907.

National Research Council. (1986). *Environmental tobacco smoke: Measuring exposures and assessing health effects.* Washington, DC: National Academy Press.

National Research Council. (1989). *Biologic markers of pulmonary toxicology.* Washington, DC: National Academy Press.

Ogden, M. W., Maiolo, K. C., Nelson, P. R., Heavner, D. L., & Green, C. R. (1993). Artifacts in determining the vapor-particulate phase distribution of environmental tobacco smoke nicotine. *Environmental Technology, 14*, 779–785.

Perez-Stable, E. J., Marin, B. V., Marin, G., Brody, D. J., & Benowitz, N. L. (1990). Apparent underreporting of cigarette consumption among Mexican American smokers. *American Journal of Public Health, 80*(9), 1057–1061.

Pichini, S., Basagana, X. B., Pacifici, R., Garcia, O., Puig, C., Vall, O., et al. (2000). Cord serum cotinine as a biomarker of fetal exposure to cigarette smoke at the end of pregnancy. *Environmental Health Perspectives, 108*(11), 1079–1083.

Pickett, K. E., Rathouz, P. J., Kasza, K., Wakschlag, L. S., & Wright, R. (2005). Self-reported smoking, cotinine levels, and patterns of smoking in pregnancy. *Paediatric and Perinatal Epidemiology, 19*(5), 368–376.

Pierce, J. P., Dwyer, T., DiGiusto, E., Carpenter, T., Hannam, C., Amin, A., et al. (1987). Cotinine validation of self-reported smoking in commercially run community surveys. *Journal of Chronic Diseases, 40*(7), 689–695.

Pirkle, J. L., Flegal, K. M., Bernert, J. T., Brody, D. J., Etzel, R. A., & Maurer, K. R. (1996). Exposure of the US population to environmental tobacco smoke: the Third National Health and Nutrition Examination Survey, 1988 to 1991. *JAMA, 275*(16), 1233–1240.

Rebagliato, M., Bolumar, F., Florey, C. V., Jarvis, M. J., Perez-Hoyos, S., Hernandez-Aguado, I., et al. (1998). Variations in cotinine levels in smokers during and after pregnancy. *American Journal of Obstetrics and Gynecology, 178*(3), 568–571.

Riboli, E., Preston-Martin, S., Saracci, R., Haley, N. J., Trichopoulos, D., Becher, H., et al. (1990). Exposure of nonsmoking women to environmental tobacco smoke: a 10-country collaborative study. *Cancer Causes & Control, 1*(3), 243–252.

Seccareccia, F., Zuccaro, P., Pacifici, R., Meli, P., Pannozzo, F., Freeman, K. M., et al. (2003). Serum cotinine as a marker of environmental tobacco smoke exposure in epidemiological studies: the experience of the MATISS project. *European Journal of Epidemiology, 18*(6), 487–492.

Slattery, M. L., Hunt, S. C., French, T. K., Ford, M. H., & Williams, R. R. (1989). Validity of cigarette smoking habits in three epidemiologic studies in Utah. *Preventive Medicine, 18*(1), 11–19.

Smith, J. J., Robinson, R. F., Khan, B. A., Sosnoff, C. S., & Dillard, D. A. (2014). Estimating cotinine associations and a saliva cotinine level to identify active cigarette smoking in Alaska native pregnant women. *Maternal and Child Health Journal, 18*(1), 120–128.

SRNT Subcommittee on Biochemical Verification. (2002). Biochemical verification of tobacco use and cessation. *Nicotine & Tobacco Research, 4*(2), 149–159.

Stoner, W. I., & Foley, B. X. (2006). Current tobacco control policy trends in the United States. *Clinics in Occupational and Environmental Medicine, 5*(1), 85–99. ix.

Stookey, G. K., Katz, B. P., Olson, B. L., Drook, C. A., & Cohen, S. J. (1987). Evaluation of biochemical validation measures in determination of smoking status. *Journal of Dental Research, 66*(10), 1597–1601.

Stragierowicz, J., Mikołajewska, K., Zawadzka-Stolarz, M., Polańska, K., & Ligocka, D. (2013). Estimation of cutoff values of cotinine in urine and saliva for pregnant women in Poland. Article ID 386784, 11 pages. *BioMed Research International. 2013*, https://dx.doi.org/10.1155/2013/386784 [Epub 2013 Oct 21].

U.S. Department of Health and Human Services. (2006). *The health consequences of involuntary smoking: A report of the surgeon general.* Rockville, MD: Public Health Service, Centers for Disease Control, Center for Health Promotion and Education, Office on Smoking and Health.

Wagenknecht, L. E., Burke, G. L., Perkins, L. L., Haley, N. J., & Friedman, G. D. (1992). Misclassification of smoking status in the CARDIA study: a comparison of self-report with serum cotinine levels. *American Journal of Public Health, 82*(1), 33–36.

Warren, C. W., Jones, N. R., Eriksen, M. P., & Asma, S. (2006). Patterns of global tobacco use in young people and implications for future chronic disease burden in adults. *Lancet, 367*(9512), 749–753.

Wells, A. J., English, P. B., Posner, S. F., Wagenknecht, L. E., & Perez-Stable, E. J. (1998). Misclassification rates for current smokers misclassified as nonsmokers. *American Journal of Public Health, 88*(10), 1503–1509.

Williams, C. L., Eng, A., Botvin, G. J., Hill, P., & Wynder, E. L. (1979). Validation of students' self-reported cigarette smoking status with plasma cotinine levels. *American Journal of Public Health, 69*(12), 1272–1274.

World Health Organization. (2008). WHO report on the global tobacco epidemic, 2008. In *The MPOWER package, Geneva*. 08.

Zielinska-Danch, W., Wardas, W., Sobczak, A., & Szoltysek-Boldys, I. (2007). Estimation of urinary cotinine cut-off points distinguishing non-smokers, passive and active smokers. *Biomarkers, 12*(5), 484–496.

CHAPTER

52

Smoking Abstinence Expectancies Questionnaire

Lorra Garey, Fiammetta Cosci†, Michael J. Zvolensky**

*Department of Psychology, University of Houston, Houston, TX, United States

†Department of Health Sciences, University of Florence, Florence, Italy

Abbreviation

SAEQ Smoking Abstinence Expectancies Questionnaire

52.1 SMOKING: PREVALENCE AND GLOBAL IMPACT

Tobacco use is the leading cause of preventable death and disability in the world (World Health Organization, 2008) and contributes to nearly 6 million deaths each year (World Health Organization, 2015b). Tobacco use kills more people than tuberculosis, HIV/AIDS, and malaria combined (World Health Organization, 2015b). Tobacco use also has severe social, environmental, and economic consequences (World Health Organization, 2015b). Indeed, the economic burden of tobacco use worldwide is so great that it is estimated to exceed the total annual spending on health across all low- and middle-income countries (Gibson, Loddenkemper, Lundbäck, & Sibille, 2013). Despite the widely known health and financial consequences, roughly 1.1 billion people worldwide still smoke tobacco cigarettes (World Health Organization, 2015a).

52.2 THEORETICAL AND CLINICAL TIES: SMOKING ABSTINENCE EXPECTANCIES AND SMOKING

In recent years, there has been a growing interest to understand cognitive processes that may influence smoking behavior (King, Marcus, Pinto, Emmons, & Abrams, 1996; Schwarzer, 2008). One promising group of cognitive processes related to smoking is expectancies. Expectancies are conceptualized as the consequences that individuals anticipate from their actions (Bandura, 1977b). Within the smoking literature, researchers have primarily focused on smoker's expectancies for the *outcome effects of smoking use* (Brandon, Juliano, & Copeland, 1999). Yet, in more recent years, there has been a growing interest to understand the expectancies for the *effects of smoking abstinence* (Svicher, Zvolensky, & Cosci, 2017). In contrast to expectancies about the outcome effects of smoking, smoking abstinence expectancies are beliefs about the acute effects of nicotine abstinence; thus, they are short-term psychological and physiological consequences from abstaining from smoking (Abrams, Zvolensky, Dorman, Gonzalez, & Mayer, 2011).

Several prominent theories of health behavior—including response expectancy theory (Kirsch & Lynn, 1999), social learning theory (Bandura, 1977a), health belief model (Rosenstock, 1974), and expectancy theory (see Brandon et al., 1999)—emphasize the appraisal of anticipated outcomes. Applied to substance use, these theories jointly posit that substance use expectancies influence the use and effects of drugs. For instance, theoretically, smokers who anticipate more negative consequences than positive consequences upon cessation may be less likely to make a quit attempt (Hendricks, Wood, Baker, Delucchi, & Hall, 2011). Conversely, those who anticipate predominantly positive outcomes with few negative outcomes may exhibit more success in quitting. Consistent with theoretical work, smoking abstinence expectancies are associated with a host of smoking outcomes. For instance, negative smoking abstinence expectancies are related to more severe withdrawal, more severe tobacco dependence symptoms, and lower probability to quit success (Hendricks et al., 2011, 2014; Hendricks & Leventhal, 2013). Indeed, smoking deprivation expectancies are frequently endorsed by

Neuroscience of Nicotine
https://doi.org/10.1016/B978-0-12-813035-3.00052-6

smokers and often reflect concerns about the negative consequences they might experience during smoking abstinence (Svicher et al., 2017), such as withdrawal symptoms. As a result, short-term smoking abstinence expectancies may, to a degree, represent an etiologic factor that maintains smoking (Farris, Langdon, DiBello, & Zvolensky, 2015). There is evidence, however, that smoking abstinence expectancies can be manipulated in the context of brief smoking deprivation (Tate et al., 1994). Theoretically, challenging smoking abstinence expectancies in the context of a smoking cessation attempt may help facilitate quit successfully.

52.3 MEASUREMENT DEVELOPMENT: SMOKING ABSTINENCE EXPECTANCIES QUESTIONNAIRE

The theoretical importance for the role of smoking abstinence expectancies in the dynamic smoking process led to the development of the Smoking Abstinence Expectancies Questionnaire (SAEQ; Abrams et al., 2011). This measure was developed among a sample of primarily Caucasian adult, daily smokers. The SAEQ consists of 28 items that comprise four unique factors: negative mood, somatic symptoms, harmful consequences, and positive consequences (see Table 52.1). The negative mood factor taps into negative mood elements that one expects would be a consequence of smoking abstinence (i.e., "my temper would be short"). The somatic symptoms factor assesses negative physiological responses that one expects as a consequence of smoking abstinence (i.e., "I would have stomach gas"). The harmful consequences factor measures expected negative psychological or physical consequences of smoking abstinence (i.e., "I would feel like I'm losing control"). The positive consequences factor evaluates the anticipated positive consequences of smoking abstinence (i.e., "I would feel energized") (Abrams et al., 2011). The first three subscales represent negative expectancies from abstaining from smoking, while the fourth represents positive expectancies. Notably, the somatic symptoms and harmful consequences factors are often conceptualized as interoceptive threat-related smoking abstinence expectancies (Farris, Langdon, et al., 2015; Farris, Paulus, et al., 2015) because they assess (mis)interpretations of internal bodily symptoms and sensations (Spielberger & Spielberger, 1966). The scoring algorithm for the SAEQ subscales and the total score are presented in Table 52.2.

Abrams et al. (2011) conducted initial psychometric testing of the SAEQ alongside its development. This group found evidence for relations among the factors, convergent and discriminant validity across the factors and constructs theorized to impede or promote smoking cessation, and test-retest reliability. Specifically, results

TABLE 52.1 Smoking Abstinence Expectancies Questionnaire (SAEQ)

The Smoking Abstinence Expectancies Questionnaire (SAEQ)

Instructions: Rate how likely or unlikely you believe each consequences would be for you if, for a day, you were to stop smoking cigarettes and not use any other forms of nicotine

Item	Very unlikely 0	1	2	3	4	5	6 Very likely
(1) My throat would feel dry	0	1	2	3	4	5	6
(2) I would feel grouchy	0	1	2	3	4	5	6
(3) I would feel like I'm going crazy	0	1	2	3	4	5	6
(4) I would feel energized	0	1	2	3	4	5	6
(5) My feet would feel tingly	0	1	2	3	4	5	6
(6) My temper would be short	0	1	2	3	4	5	6
(7) I would feel like I'm losing control	0	1	2	3	4	5	6
(8) I would feel happy	0	1	2	3	4	5	6
(9) I would feel short of breath	0	1	2	3	4	5	6
(10) I would feel frustrated	0	1	2	3	4	5	6
(11) I would feel like I'm dying	0	1	2	3	4	5	6
(12) I would find it easy to remember things	0	1	2	3	4	5	6
(13) I would have stomach gas	0	1	2	3	4	5	6
(14) I would feel tense	0	1	2	3	4	5	6
(15) I would embarrass myself in front of others	0	1	2	3	4	5	6
(16) I would feel calm	0	1	2	3	4	5	6
(17) I would sweat more than usual	0	1	2	3	4	5	6
(18) I would have trouble getting along with difficult people	0	1	2	3	4	5	6

TABLE 52.1 Smoking Abstinence Expectancies Questionnaire (SAEQ)—cont'd

The Smoking Abstinence Expectancies Questionnaire (SAEQ)

Instructions: Rate how likely or unlikely you believe each consequences would be for you if, for a day, you were to stop smoking cigarettes and not use any other forms of nicotine

Item	Very unlikely						Very likely
(19) Things around me would seem unreal	0	1	2	3	4	5	6
(20) I would sleep through the night	0	1	2	3	4	5	6
(21) I would feel dizzy	0	1	2	3	4	5	6
(22) I would feel anxious	0	1	2	3	4	5	6
(23) I would have a panic attack	0	1	2	3	4	5	6
(24) I would find it easy to concentrate	0	1	2	3	4	5	6
(25) My chest would feel tight	0	1	2	3	4	5	6
(26) My nerves would be jittery	0	1	2	3	4	5	6
(27) I would have a heart attack	0	1	2	3	4	5	6
(28) I would feel physically comfortable	0	1	2	3	4	5	6

Reprint from Abrams, K., Zvolensky, M. J., Dorman, L., Gonzalez, A., & Mayer, M. (2011). Development and validation of the smoking abstinence expectancies questionnaire. Nicotine & Tobacco Research, *13(12), 1296–1304, by permission of Oxford University Press.*

TABLE 52.2 SAEQ Scoring Key

Item numbers	SAEQ subscale
2, 6, 10, 14, 18, 22, 26	Negative mood
1, 5, 9, 13, 17, 21, 25	Somatic symptoms
3, 7, 11, 15, 19, 23, 27	Harmful consequences
4, 8, 12, 16, 20, 24, 28	Positive consequences

Note: Items that load onto corresponding SAEQ subscales summarized in the table. To calculate the score of each SAEQ subscales, sum the items that load onto the subscale. The total score is the sum of all responses for all items after reverse scoring items from the positive consequences subscale.

indicated significant, positive correlations among the negative expectancies from smoking abstinence factors (i.e., negative mood, somatic symptoms, and harmful consequences), all of which also negatively correlated with the positive consequences factor. With regard to convergent validity, all negative expectancy factors positively related to withdrawal severity and tobacco dependence, whereas positive consequences negatively correlated with these constructs. Negative mood and harmful consequences positively related to smoking rate. Expectancies for harmful consequences of abstaining from smoking also were negatively associated with past quit behavior, and positive consequences were positively related to this variable. In tests of discriminant validity, all negative expectancy factors positively related to trait-like propensity toward negative affect. Positive consequences were negatively correlated with negative affect and positively correlated with positive affect. All factors demonstrated acceptable test stability over time.

Importantly, although work using the measure has largely focused on the individual factors, Abrams et al. (2011) included the SAEQ total score in their evaluation of this measure. It is important to note that the SAEQ total score represents more negative expectancies from abstaining from smoking, as the score is derived from summing all items after positive consequences items have been reversely coded (Abrams et al., 2011). The total score positively correlated with factors that assessed negative expectancies from smoking depravation and negatively correlated with positive consequences. Trends in observed convergent and discriminant analyses generally mirrored those observed for the negative expectancies associated with smoking abstinence. Namely, the total score positively associated with withdrawal severity, tobacco dependence, smoking rate, and negative affect, while it negatively correlated with quit history. In part, the similarity in associative patterns observed across the total score and negative expectancies associated with smoking abstinence is a direct consequence of the scoring algorithm for the total score. The total score exhibited high test-retest reliability.

Collectively, results from Abrams et al.'s (2011) study provided initial support for the SAEQ as a psychometrically strong measure of positive and negative expectancies for smoking abstinence. Within the context of its evaluation, extensive construct validity evidence for the SAEQ factors and total score was scrutinized and synthesized. Results support unique yet related factors that evince varying strengths of association with core smoking behaviors and processes and stability over time. The synthesis of this information informed the researchers' position that this measure may serve as an important research and clinical tool in regard to its capability to assist in understanding dynamic models of successful quitting and cognitive processes that influence smoking and treatment planning and development. See Fig. 52.1 for heuristic model of the function for the SAEQ in the dynamic process of smoking behavior, including prequit behavior, cessation, and relapse (Baker et al., 2011; Hendershot, Witkiewitz, George, & Marlatt, 2011).

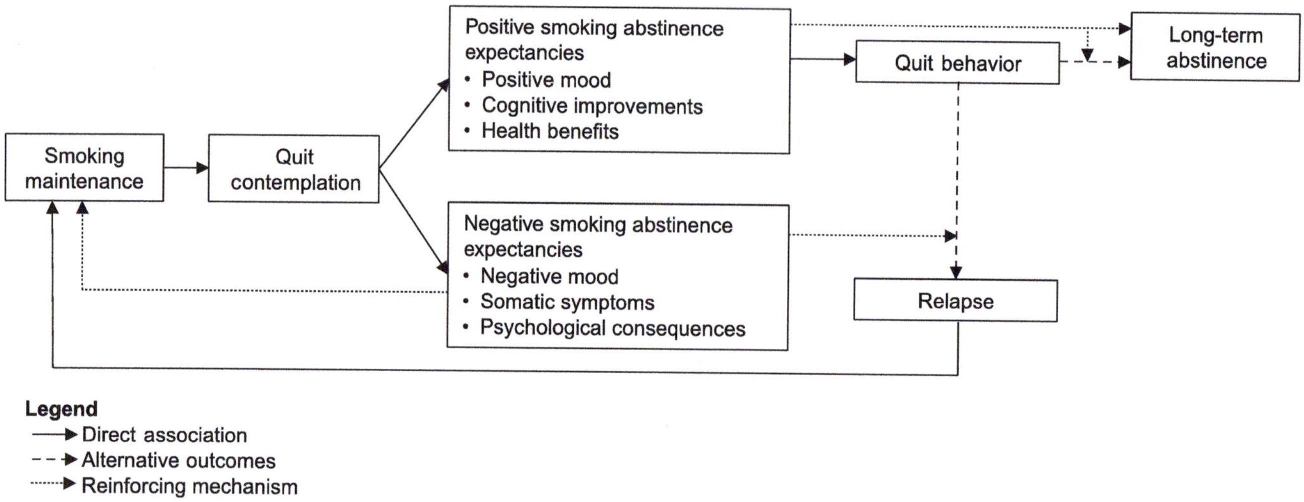

FIG. 52.1 Heuristic model for the role of positive and negative smoking abstinence expectancies in the dynamic quit process. The heuristic model represents the association among smoking behavior, positive and negative smoking abstinence expectancies, and quit behavior. *Solid lines* represent direct associations. *Dashed lines* represent alternative outcomes. *Dotted lines* represent reinforcing mechanisms.

Sex differences. Sex differences for the SAEQ have been reported, yet the robustness of these differences is speculative given their inconsistence across different samples. For example, some work has found that women report high scores on the harmful consequences factor (Abrams et al., 2011), whereas other works have failed to find any association between this factor and sex (Farris, Langdon, et al., 2015; Farris, Paulus, et al., 2015; Kauffman, Farris, Alfano, & Zvolensky, 2017; Zvolensky et al., 2017). Similar mixed findings have been reported for somatic symptoms, with some research indicated an association (Farris, Langdon, et al., 2015) and others reporting no difference or association (Abrams et al., 2011; Farris, Paulus, et al., 2015; Kauffman et al., 2017; Zvolensky et al., 2017). In fact, the negative mood factor is the only facet of smoking abstinence expectancies that has consistently related to sex across studies (Abrams et al., 2011; Robles et al., 2017). Indeed, there is consistent evidence for higher scores on this dimension among women relative to men. Lastly, while Abrams et al. (2011) did not find a difference across sex for positive consequences, this finding has not been investigated in an independent sample. Overall, there is some scientific evidence for possible sex differences in the SAEQ, particularly for negative mood, but additional work is needed.

Cross-validation. Several independent research groups have evaluated associations among the SAEQ factors and smoking behavior and constructs believed to impede or promote quit success. In a concerted effort to contextualize this work within the larger smoking literature, cross validation work with the SAEQ has focused on negative interoceptive threat-related smoking abstinence expectancies (i.e., somatic symptoms and harmful consequences) (Farris, Langdon, et al., 2015;Farris, Paulus, et al., 2015; Kauffman et al., 2017; Zvolensky et al., 2017). In part, the research has primarily focused on these factors to address a critical gap in the growing theoretical and empirical literature that posits cigarettes use as a mean to avoid and escape negative internal stimuli in smokers (Baker, Piper, McCarthy, Majeskie, & Fiore, 2004). Namely, prior to this work, it was unknown how interoceptive threat-related smoking abstinence expectancies fit into these larger conceptual models of tobacco use.

Under this supposition, work has evaluated and found support for interoceptive-threat smoking abstinence expectancy factors as a correlate of smoking behavior (i.e., smoking rate and tobacco dependence) (Farris, Paulus, et al., 2015; Kauffman et al., 2017; Zvolensky et al., 2017) and psychological constructs that maintain and exacerbate tobacco dependence, including the presence of anxiety symptoms, posttraumatic stress disorder symptoms, sleep disturbance symptoms, emotion dysregulation, anxiety sensitivity, and negative affect (Farris, Langdon, et al., 2015; Farris, Paulus, et al., 2015; Kauffman et al., 2017; Zvolensky et al., 2017). Moreover, interoceptive-threat smoking abstinence expectancy factors are strongly related to smoking motivation and expectancies for smoking to alleviate negative affect, perceived barriers to cessation, and prior quit-related problems (Farris, Langdon, et al., 2015; Farris, Paulus, et al., 2015; Kauffman et al., 2017; Zvolensky et al., 2017). Importantly, interoceptive-threat smoking abstinence expectancies serve a mechanistic function for the relation between affective vulnerabilities that position someone for greater problematic use and prequit cognitive processes (Farris, Langdon, et al., 2015; Leventhal & Zvolensky, 2015).

With the foundational understanding for the impact of interoceptive-threat smoking abstinence expectancies on smoking behaviors and processes, work has begun to evaluate other SAEQ factors. In a recent article by

Robles et al. (2017), researchers found support for the indirect effect of financial strain on abstinence expectancies for negative mood consequences through depressive symptoms. In bivariate correlation analyses, negative mood was positively associated with affective processes (i.e., negative affect and depressive symptoms), affect regulation smoking processes (i.e., negative affect reduction smoking motives and perceived barriers to cessation), and education. Interestingly, abstinence expectancies for negative mood consequences and depressive symptoms only shared 11.6% of their variance; thus, each of these constructs taps into unique processes related to smoking maintenance and quit behavior.

Populations. The SAEQ has demonstrated generalizability with regard to its appropriateness across diverse segments of the smoking population. Indeed, the SAEQ has been utilized with treatment-seeking smokers (Farris, Paulus, et al., 2015; Zvolensky et al., 2017) and nontreatment-seeking smokers (Abrams et al., 2011; Kauffman et al., 2017). Comparison of these studies revealed overall higher negative smoking abstinence expectancies among treatment-seeking smokers relative to nontreatment-seeking smokers. This observation, in part, taps into the severity of treatment-seeking smokers. Specifically, treatment-seeking substance users tend to be a more severe substance-using population relative to nontreatment-seeking substance users, as evidenced by great dependence, more negative consequences, less control over their use, and greater use and frequency of use (Ray, Bujarski, Yardley, Roche, & Hartwell, 2017). Thus, treatment-seeking smokers may exhibit characteristics and beliefs unique from nontreatment-seeking smokers. The SAEQ has also been administered to vulnerable populations, including smokers with psychopathology (Farris, Paulus, et al., 2015). Within this work, statistical tests of differences across those with and without psychopathology in harmful consequences and somatic symptoms abstinence expectancies supported significant scores on the harmful consequences factor. Importantly, cross study comparisons support a general trend from higher scores among those with psychopathology across harmful consequences and somatic symptoms abstinence expectancies. The *statistical significance* of these differences, however, needs further investigation. Across different populations, it is important to note that similar patterns of associations between the SAEQ and affective vulnerabilities (i.e., anxiety sensitivity and negative affect) and smoking behaviors and processes have emerged. The magnitude of these associations, however, differs across populations. Thus, there is initial evidence for the SAEQ to differentiate groups of smokers and identify those who may be more afflicted with emotional distress that may impede quit behavior or success.

Clinical and research implications. Clinically, there are several important considerations of the SAEQ. First, the SAEQ may assist in screening smokers for increased quit difficulty and thereby contribute to prediction models of successful quitting. For example, those who expect more negative consequences from abstaining (such as negative affect or unpleasant physiological sensations) may be less likely to successfully quit. The SAEQ offers a brief, standardized method to assess such dimension that could be used to inform treatment options. Second, the SAEQ has the potential to enable clinicians to target specific expectancies during smoking cessation treatment. Indeed, clinical clients with the SAEQ may afford clinicians an opportunity to isolate and target specific abstinence expectancies with greater specificity. For example, clinicians might aim to modify irrational beliefs regarding the likelihood of medical or psychiatric symptoms following smoking abstinence if those particular expectancies were most severe or potentially interfering with the quit process. This approach would allow for a more individualized treatment plan that may promote greater quit success. Finally, continued use of the SAEQ could give researchers and clinicians a broader understanding of cognitive processes that influence smoking, including prequit behavior, cessation, and relapse (see Fig. 52.1).

52.4 ALTERNATIVE MEASURES

Two alternative measures of smoking abstinence expectancies warrant brief comment. First, an Italian version of the SAEQ is in current circulation (Svicher et al., 2017). The translated version of the SAEQ was tested on 366 Italian smokers. Exploratory factor analytic techniques with the translated measure supported a three-factor solution. Thus, cultural consideration may warrant future attention with this measure. As a result, work with the SAEQ should exercise caution when administering it to culturally disparate groups of people. Second, Hendricks et al. (2011) developed the Smoking Abstinence Questionnaire. The measure assesses short-, medium-, and longer-term smoking abstinence expectancies. Although this measure taps into a greater array of smoking abstinence expectancies, it may overburden research participants and clinical clients. Specifically, this measure contains nearly twice as many items as the SAEQ. Moreover, the Smoking Abstinence Questionnaire has considerable overlap with the Smoking Consequences Questionnaire-Adult (Copeland, Brandon, & Quinn, 1995). Thus, this measure may not truly assess the underlying crux of smoking abstinence expectancies as currently conceptualized in theoretical or empirical work.

52.5 FUTURE DIRECTIONS

Construct validity. The four-factor SAEQ structure has not been evaluated using confirmatory factor analysis techniques. Thus, the replicability of this structure remains unknown. Similarly, work has not evaluated the stability of the structure across groups (i.e., sex, ethnicity/race, and age) or over time. This limits the degree to which researchers can conclude that observed differences are the result of true differences and not an artifact of measurement bias (Vandenberg & Lance, 2000). Moreover, theorized prospective relation between the SAEQ and smoking abstinence has not been investigated empirically and remains theoretical in nature. These oversights call for consideration of additional psychometric testing of the SAEQ.

Malleability. An important area for future research will be to evaluate the malleability of the SAEQ in response to treatment. Indeed, despite theoretical work implicating smoking abstinence expectancies as a targetable treatment factor, no work has examined change in the SAEQ after participating in a smoking cessation treatment. As a result, how such changes may impact smoking cessation remain unknown.

Developmental origins. Promising findings have resulted from the scientific inquiry of the SAEQ. Indeed, current understanding of the maintenance of smoking abstinence expectancies is limited. Theoretical work interested in smoking abstinence expectancies posits that learning history or environmental stressors may influence their development and maintenance. Nevertheless, the use and understanding of the SAEQ specifically, including its constituents and total score, are still underdeveloped. Continued investigation is warranted for the unique role of the SAEQ in models that conceptualize risk and protective smoking processes in negative reinforcement models of smoking behavior (see Fig. 52.1).

52.6 CONCLUSION

Smoking abstinence expectancies are positioned to either impede or promote quit behavior and thereby quit success. The Smoking Abstinence Expectancies Questionnaire is a psychometrically sound, clinically relevant tool for assessing smoking abstinence expectancies. Furthermore, work explicating the nature of the Smoking Abstinence Expectancies Questionnaire, its role in the dynamic smoking process, and its relations to psychopathology and processes that may impede quit success continues to mount. Future work is needed to offer further construct validity of this measure, uncover its malleability in response to manipulations, and evaluate its position in larger models of smoking behavior and cognitions.

MINI-DICTIONARY OF TERMS

Affective vulnerabilities Elevated psychopathological symptoms, including anxiety and/or depressive symptoms, that may place a person at increased risk for developing a psychiatric disorder.

Anxiety sensitivity The tendency for a person to fear anxiety-related symptoms and sensations.

Interoceptive Of or relating to the internal state of a person, including cognitions and internal physiological sensations.

Smoking abstinence expectancies Beliefs about the positive and negative consequences of quitting smoking.

Summary Points

- This chapter focuses on smoking abstinence expectancies and a commonly used measure to assess these expectancies, that is, the Smoking Abstinence Expectancies Questionnaire.
- Theoretical and empirical work supports a relation between smoking abstinence expectancies and more problematic smoking and quit outcomes.
- The Smoking Abstinence Expectancies Questionnaire has exhibited strong psychometric characteristics as a valid measure of smoking abstinence expectancies, as demonstrated by strong convergent and discriminant validity and internal consistency across several independent samples.
- Recent advances in statistical modeling provide an area for future work regarding the construct validity argument of the Smoking Abstinence Expectancies Questionnaire.
- We argue that the Smoking Abstinence Expectancies Questionnaire is a psychometrically sound, clinically relevant tool for assessing smoking abstinence expectancies and advocate for its use as a screening tool in clinical and research settings.

References

Abrams, K., Zvolensky, M. J., Dorman, L., Gonzalez, A., & Mayer, M. (2011). Development and validation of the smoking abstinence expectancies questionnaire. *Nicotine & Tobacco Research*. *13*(12), 1296–1304. https://doi.org/10.1093/ntr/ntr184.

Baker, T. B., Mermelstein, R., Collins, L. M., Piper, M. E., Jorenby, D. E., Smith, S. S., et al. (2011). New methods for tobacco dependence treatment research. *Annals of Behavioral Medicine*, *41*(2), 192–207.

Baker, T. B., Piper, M. E., McCarthy, D. E., Majeskie, M. R., & Fiore, M. C. (2004). Addiction motivation reformulated: an affective processing model of negative reinforcement. *Psychological Review*, *111*(1), 33.

Bandura, A. (1977a). Self-efficacy: toward a unifying theory of behavioral change. *Psychological Review*, *84*(2), 191.

Bandura, A. (1977b). *Social learning theory*. Englewood Cliffs, NJ: Prentice-Hall.

Brandon, T. H., Juliano, L. M., & Copeland, A. L. (1999). Expectancies for tobacco smoking. In I. Kirsch (Ed.), *How expectancies shape experience* (pp. 263–299). Washington, DC: American Psychological Association.

Copeland, A. L., Brandon, T. H., & Quinn, E. P. (1995). The Smoking Consequences Questionnaire-Adult: measurement of smoking outcome expectancies of experienced smokers. *Psychological Assessment, 7*(4), 484.

Farris, S. G., Langdon, K. J., DiBello, A. M., & Zvolensky, M. J. (2015). Why do anxiety sensitive smokers perceive quitting as difficult? The role of expecting "interoceptive threat" during acute abstinence. *Cognitive Therapy and Research, 39*(2), 236–244.

Farris, S. G., Paulus, D. J., Gonzalez, A., Mahaffey, B. L., Bromet, E. J., Luft, B. J., et al. (2015). Anxiety sensitivity mediates the association between post-traumatic stress symptom severity and interoceptive threat-related smoking abstinence expectancies among World Trade Center disaster-exposed smokers. *Addictive Behaviors, 51,* 204–210.

Gibson, G. J., Loddenkemper, R., Lundbäck, B., & Sibille, Y. (2013). Respiratory health and disease in Europe: The new European lung white book. *European Respiratory Journal,* **42**(3), 559–563. https://doi.org/10.1183/09031936.00105513.

Hendershot, C. S., Witkiewitz, K., George, W. H., & Marlatt, G. A. (2011). Relapse prevention for addictive behaviors. *Substance Abuse Treatment, Prevention, and Policy, 6*(1), 17.

Hendricks, P. S., & Leventhal, A. M. (2013). Abstinence-related expectancies predict smoking withdrawal effects: Implications for possible causal mechanisms. *Psychopharmacology, 230*(3), 363–373.

Hendricks, P. S., Westmaas, J. L., Ta Park, V. M., Thorne, C. B., Wood, S. B., Baker, M. R., et al. (2014). Smoking abstinence-related expectancies among American Indians, African Americans, and women: potential mechanisms of tobacco-related disparities. *Psychology of Addictive Behaviors, 28*(1), 193.

Hendricks, P. S., Wood, S. B., Baker, M. R., Delucchi, K. L., & Hall, S. M. (2011). The Smoking Abstinence Questionnaire: measurement of smokers' abstinence-related expectancies. *Addiction, 106*(4), 716–728.

Kauffman, B. Y., Farris, S. G., Alfano, C. A., & Zvolensky, M. J. (2017). Emotion dysregulation explains the relation between insomnia symptoms and negative reinforcement smoking cognitions among daily smokers. *Addictive Behaviors, 72,* 33–40.

King, T. K., Marcus, B. H., Pinto, B. M., Emmons, K. M., & Abrams, D. B. (1996). Cognitive–behavioral mediators of changing multiple behaviors: smoking and a sedentary lifestyle. *Preventive Medicine, 25*(6), 684–691.

Kirsch, I., & Lynn, S. J. (1999). Automaticity in clinical psychology. *American Psychologist, 54*(7), 504.

Leventhal, A. M., & Zvolensky, M. J. (2015). Anxiety, depression, and cigarette smoking: a transdiagnostic vulnerability framework to understanding emotion–smoking comorbidity. *Psychological Bulletin, 141*(1), 176.

Ray, L. A., Bujarski, S., Yardley, M. M., Roche, D. J., & Hartwell, E. E. (2017). Differences between treatment-seeking and non-treatment-seeking participants in medication studies for alcoholism: do they matter? *The American Journal of Drug and Alcohol Abuse,* 1–8.

Robles, Z., Anjum, S., Garey, L., Kauffman, B. Y., Rodríguez-Cano, R., Langdon, K. J., et al. (2017). Financial strain and cognitive-based smoking processes: the explanatory role of depressive symptoms among adult daily smokers. *Addictive Behaviors, 70,* 18–22.

Rosenstock, I. M. (1974). The health belief model and preventive health behavior. *Health Education Monographs, 2*(4), 354–386.

Schwarzer, R. (2008). Modeling health behavior change: how to predict and modify the adoption and maintenance of health behaviors. *Applied Psychology, 57*(1), 1–29.

Spielberger, C. D., & Spielberger, C. D. (1966). Theory and research on anxiety. *Anxiety and Behavior, 1,* 3–20.

Svicher, A., Zvolensky, M. J., & Cosci, F. (2017). The Smoking Abstinence Expectancies Questionnaire—Italian version: analysis of psychometric properties. *Journal of Addictive Diseases, 36*(1), 80–87.

Tate, J. C., Stanton, A. L., Green, S. B., Schmitz, J. M., Le, T., & Marshall, B. (1994). Experimental analysis of the role of expectancy in nicotine withdrawal. *Psychology of Addictive Behaviors, 8*(3), 169.

Vandenberg, R. J., & Lance, C. E. (2000). A review and synthesis of the measurement invariance literature: suggestions, practices, and recommendations for organizational research. *Organizational Research Methods, 3*(1), 4–70.

World Health Organization. (2008). *WHO report on the global tobacco epidemic, 2008: The MPOWER package.*

World Health Organization. (2015a). *Global Health Observatory (GHO) data: Prevalence of tobacco use 2015. Available from:(2015a). http://www.who.int/gho/tobacco/use/en/. Accessed 2 November 2015.*

World Health Organization. (2015b). *WHO global report on trends in prevalence of tobacco smoking 2015.* World Health Organization.

Zvolensky, M. J., Paulus, D. J., Langdon, K. J., Robles, Z., Garey, L., Norton, P. J., et al. (2017). Anxiety sensitivity explains associations between anxious arousal symptoms and smoking abstinence expectancies, perceived barriers to cessation, and problems experienced during past quit attempts among low-income smokers. *Journal of Anxiety Disorders 48,* 70–77. https://doi.org/10.1016/j.janxdis.2016.12.003.

CHAPTER

53

Pharmacist-led Smoking Cessation Services: Current and Future Perspectives

*Chee Fai Sui**, *Long Chiau Ming*[†,‡]

*Department of Pharmacy, Hospital Tapah, Perak, Malaysia

[†]Faculty of Pharmacy, Quest International University Perak, Perak, Malaysia

[‡]Unit for Medication Outcomes Research and Education, Pharmacy, University of Tasmania, Hobart, Tasmania, Australia

Abbreviations

CSCSP	Certified Smoking Cessation Service Provider
E-cigarette	electronic cigarette
FDA	Food and Drug Authority
FIP	International Pharmaceutical Federation
mQuit	Malaysia online portal for quitting smoking
Non-NRD	nonnicotine replacement drugs
NRD	nicotine replacement drugs
SCS	smoking cessation services
WHO	World Health Organization

53.1 INTRODUCTION

The historical connection between pharmacies and nicotine began in the early 20th century when tobacco leaves were sold in pharmacies for medicinal purpose. Due to its high demand and its potent effect, the practice evolved to "cigarette-selling pharmacies" within a decade. With the passage of time, the service transformed to "smoking cessation pharmacy" after the hazards and harmful effects of smoking were uncovered and validated (Anderson, 2007). As a result, the community's pharmacy model was converted from tobacco cigarette sellers to smoking cessation advisors in the 1960s. During that time, many over-the-counter preparations were widely sold in pharmacies in countries such as the United Kingdom to aid chronic smokers to quit the habit. The active ingredients were mainly ferrous and lobeline salts that were not tested clinically for either efficacy or safety (Anderson, 2007).

Subsequently, an alternative nicotine delivery system has been introduced as a harm-reduction method by using "pure nicotine," eliminating the various toxic chemicals found in commercialized cigarettes. The first nicotine replacement drug (NRD), nicotine gum 2 mg, was approved by the US Food and Drug Administration (FDA) in 1984 as a prescription drug to aid smoking cessation. The role of pharmacists in smoking cessation became even more significant in counseling, monitoring, and dispensing NRDs. The professional duties and responsibilities doubled when the NRDs could be sold as pharmacist items without prescription in the early 1990s (Callahan-Lyon, 2010). Since then, the range of NRD products have grown due to research and development to cater users with different needs. In the current market, NRDs consist of gum, patches, sprays, inhalers, sublingual, and lozenges. The availability of NRDs depends on the regulation of each country. Apart from NRDs, there are nonnicotine preparations such as varenicline, a nicotine blocker approved in 2006, and bupropion, an antidepressant approved in 1997, both of which may be restricted by prescription.

Pharmacies provide excellent locations in which to provide smoking cessation services (SCS), whether they are in the community, in hospitals, or in primary health clinics, as they have frequent interaction with the public, who are either looking for advice or getting repeat prescriptions. Pharmacists and pharmacy staff have many suitable occasions for smoking cessation activities, such as encouraging and motivating people to initiate attempts at stopping smoking or providing a full SCS, as presented in Fig. 53.1. Globally, support for pharmacist-led SCS has been incorporated into guidelines and protocols. The World Health Organization (World Health Organization, 1998) nominated pharmacists as the key player in helping smokers to quit and preventing future users. In 2009, the International Pharmaceutical Federation (FIP) through

Neuroscience of Nicotine
https://doi.org/10.1016/B978-0-12-813035-3.00053-8

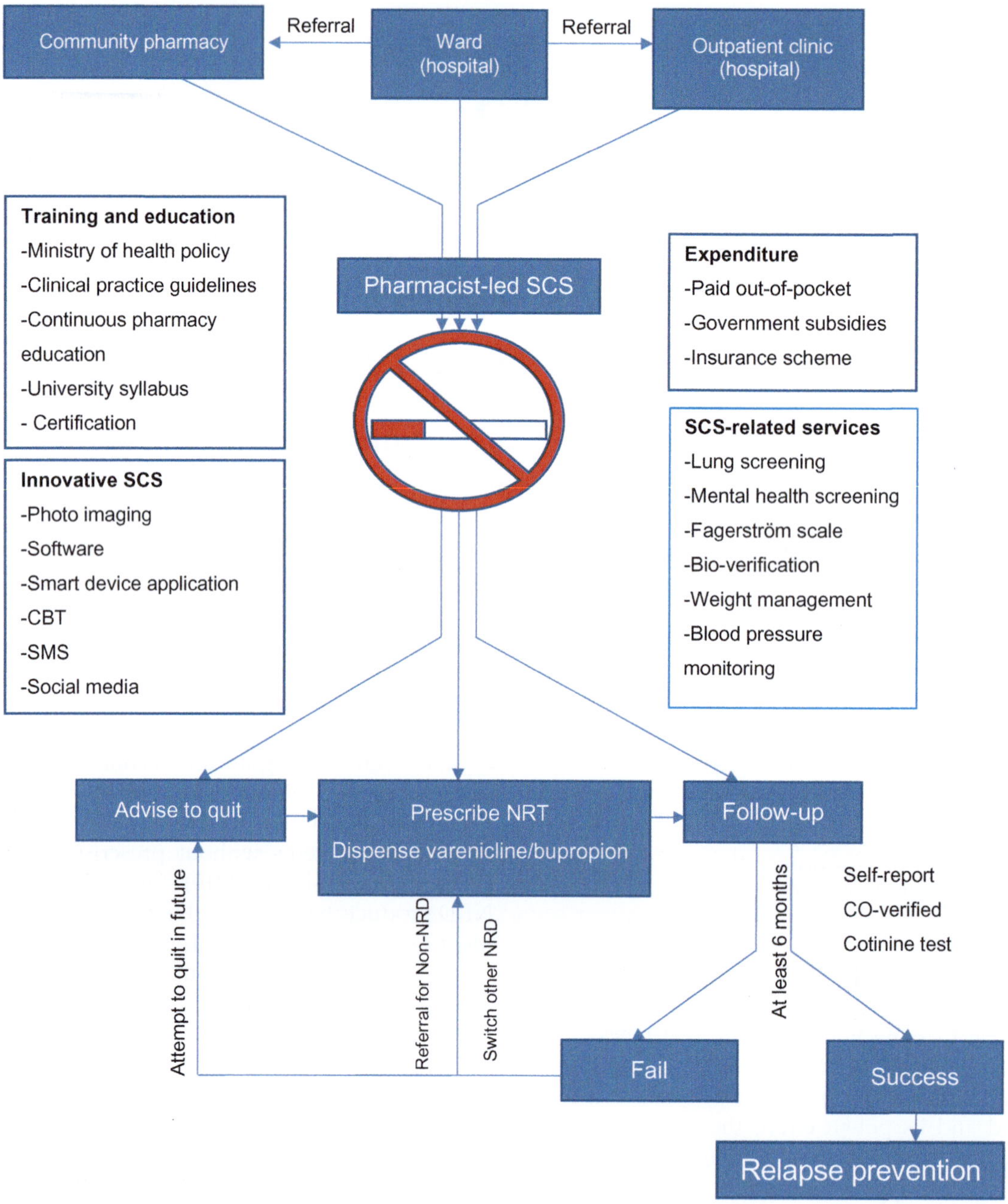

FIG. 53.1 Conceptual framework on pharmacist-led SCS.

its publication *Curbing the Tobacco Pandemic: The Global Role of Pharmacy* (Brock, Taylor, & Wuliji, 2009) and UK National Health Service (NHS) through its guideline "Pharmacy-based stop smoking services: optimising commissioning" (NHS Employers, 2009) promulgated the significant roles of pharmacists in SCS. Both guidelines provide easy-to-follow steps for initiating pharmacist-led SCS and give evidence of the success of pharmacist-led programs. A total of 14 countries across Asia and Europe have implemented and integrated enhanced SCS provided by pharmacists into public-funded health programs.

On top of the guidelines, pharmacy bodies in each country are instrumental in updating and providing evidence-based continuous pharmacy education on SCS to all pharmacy-related staff. For instance, pharmacy chapter bodies in Malaysia and Singapore implemented an online training program to certify pharmacists in SCS (CSCSP, 2017; Pharmaceutical Society of Singapore, 2016). In Finland, the Pharmacist Action on Smoking (PAS) model was used as a guide for providing smoking cessation in pharmacy (Maguire, McElnay, & Drummond, 2001) (refer to Table 53.1).

TABLE 53.1 Education, Training, Guidelines, and Certification on Pharmacist-led SCS

Society/bodies	Year	Title	Type	Author
EuroPharm Forum in collaboration with the WHO Tobacco or Health Unit	1998	Pharmacists and action on tobacco	Guidelines	World Health Organization (1998)
The School of Pharmacy, University of London, International Pharmaceutical Federation (FIP)	2007	Curbing the tobacco pandemic: the global role for pharmacy	Reports	Brock et al. (2009)
PATH is a joint initiative between ASH Scotland, NHS Health Scotland, and the Scottish Government to reduce the prevalence of tobacco use in Scotland	2016	The minimum dataset for Scottish smoking cessation services (including pharmacies offering the national community pharmacy smoking cessation service)	Guidelines	Partnership on Tobacco and Health (2016)
Alabama Pharmacy Association	2014	A pharmacist's guide for smoking cessation	Continuous education	Boutwell et al. (2014)
Health Development Agency (HDA) in collaboration with the Royal Pharmaceutical Society of Great Britain and Pharmacy Health Link	2005	Helping smokers to stop: advice for pharmacists in England	Guidelines	McRobbie and McEwen (2005)
Providence Health and Services, Oregon	NA	Pharmacist-assisted smoking cessation class: a guide to implementation	Guidelines	Bentz, Gray, and Swan (2007)
British Columbia Pharmacy Association	2013	Pharmacist-led smoking cessation services	Clinical service proposal	British Columbia Pharmacy Association (2013)
National Health Service Employers, United Kingdom	2009	Pharmacy-based stop smoking services: optimising commissioning	Guidelines	NHS Employers (2009)
Tobacco Control Network	2017	Access to tobacco cessation medication through pharmacists	Protocol	Tobacco Control Network (2017)
Pharmaceutical Society of Singapore	2016	Practice Guides for Minor Ailments (PGMA) module 5: smoking cessation	Training	Pharmaceutical Society of Singapore (2016)
Ontario Ministry of Health and Long-Term Care	2011	Pharmacy Smoking Cessation Program, the pharmacist's role in a smoking cessation system	Guidelines	Ministry of Health and Long-Term Care (2011)
Malaysia Academy of Pharmacy	2010	Certified Smoking Cessation Service Provider (CSCSP)	Training	CSCSP (2017)
Smoking Cessation Trust Management Services (SCTMS)	2013	Smoking cessation provider guide: US script pharmacy services	Guidelines	Smoking Cessation Trust (2013)

53.2 DIFFERENT SETTINGS OF PHARMACIST-LED SCS

Pharmacists have been voted as the leading trusted profession in the eyes of the American public (Norman, 2016). Similarly, pharmacists in Poland have been ranked (Goniewicz et al., 2010) as the first health professional that smokers would ask about pharmacological treatment for smoking cessation. Pharmacies in community settings have easy access to the public seeking health care and sick care without needing an appointment or having to pay additional consultation charges (Patwardhan & Chewning, 2010). Many potential quitters come to pharmacies with family members, looking for health advice or health supplements. Pharmacists can ask a mixture of closed- and open-ended questions to evoke a patient's own motivations to quit (Boutwell, Cook, Norman, & Lindsey, 2014), for example, by asking "do you ever smoke?" to patients who purchase cough or cold medicines for themselves, their children, their partners, or other family members. In this way, further information can be given to those patients who smoke and are interested to know more about SCS.

In contrast to community pharmacists, hospital pharmacists routinely deal with many inpatients, for example, patients with special medical needs such as respiratory and cardiovascular diseases and complications. So, they are ideally positioned to give advice and provide necessary aids for any smokers to quit during the period of ward stay. Besides that, hospital pharmacists who have direct contact with the public, patients, relatives, and carers have a role to play in getting potential quitters to join SCS. The role of pharmacists in hospital settings in providing SCS is not well documented in literature compared to the equivalent role of community pharmacists. A review of smoking care prevalence in hospitals between 1994 and 2005 by Freund et al. (2008) shows that only 13% of health professionals provided nicotine replacement therapy (NRT) or advised its use and 39% provided referrals or follow-up. Similarly, one Australian study of pharmacist-led smoking cessation in a hospital setting recruited 102 smokers and randomized to three groups: minimal intervention (without NRT patch), hospital-based, and community pharmacy-based. No significant difference was found between hospital-based (38% abstinence) and community pharmacy-based (24% abstinence) after a 6-month follow-up, but the abstinence rates were superior to the group with minimal intervention (4.6% abstinence) (Vial, Jones, Ruffin, & Gilbert, 2002).

Smoking cessation clinics led by pharmacists mostly involve an integrated team such as a medical officer, nurses, and medical assistants and are subsidized by a national health fund. Many countries offer free treatments in selected facilities for patients interested in quitting, for example, the National Health Service in England (Dobbie et al., 2015), US Trust Script (Smoking Cessation Trust, 2013), and Ministry of Health Malaysia (Fai, Yen, & Malik, 2016). Interested patients can choose facilities nearest to them and decide whether to use privately or publicly funded facilities—specialist clinics, hospitals, or community pharmacies—to help them quit smoking. The success of pharmacist-led SCS is not solely dependent on medication. Innovative approaches and add-on health services play a significant role in increasing the chances of someone's successful quitting within a specific period.

53.3 CONTRIBUTING FACTORS LEADING TO SUCCESSFUL PHARMACIST-LED SCS

The development pipeline of medication for smoking cessation is limited. Only a handful of drug companies are willing to spend millions in research and development on new drugs to stop smoking, even though up to 1 billion people worldwide are smokers (World Health Organization, 2017). A recent systemic review has confirmed that baseline willpower alone may generate roughly 3%–5% abstinence, but NRDs increase the chance of success by 50%–70% (Stead et al., 2012). Evidence-based studies evaluating pharmacist-led SCS are well documented in the literature. From Table 53.2, most abstinence rates in pharmacist-led SCS exceeded 10%, except the study by Hoving, Mudde, Dijk, and Vries (2010) with only a 0.8% success rate.

The strategies used by different pharmacist-led SCS (Fig. 53.2) play a vital role in increasing the success rate with the aid of pharmacology therapy. Attempts to quit are not guaranteed to succeed due to the withdrawal symptoms and addictive component of nicotine. Stead et al. (2012) confirmed that short message system (SMS) messages, booklets, telephone calls, and websites provide estimated 52-week quit rates of about 9%, 5%, 8%, and 8%, respectively. Interactive smartphone apps are the newest innovation in SCS, and pharmacist-led SCS should incorporate this strategy to engage more potential quitters. A randomized study by Bricker et al. piloted on a smartphone app for quitting smoking reported a success rate of 13% in smartphone-delivered acceptance and commitment therapy for a smoking cessation app (SmartQuit) and 8% in the National Cancer Institute's app for smoking cessation (QuitGuide) (Bricker et al., 2014). One example of an innovative approach in SCS is a joint public-funded clinic and community pharmacy website called mQuit launched in 2016 in Malaysia, which provides a multiple choice of locations for SCS. Interested users can select an SCS location, decide whether they want private or public funded, and register through the website (mQuit, 2016).

Many smokers may need several attempts to fully quit smoking, while others may relapse occasionally during their quitting period. This chronic relapsing process requires multiple approaches on top of pharmacological and behavioral support. Second-line drugs may not work as effectively as first-line therapy, and they may have more side effects and incur higher costs per treatment. Pharmacists should offer smoking cessation strategies within the value packages of all medical insurance or public health plans, to promote access to these interventions for all potential quitters who are looking for help to quit.

53.4 FUTURE DEVELOPMENTS OF PHARMACIST-LED SCS

Smoking cessation, even though is a proved economical and effective way to prevent noncommunicable diseases, still needs to be reinforced repeatedly by pharmacist to ensure its success. After a coronary event, smoking cessation remains more cost-effective than oral medication or surgical intervention; still, continuous SCS needs to be given to convince a patient to quit smoking. In 2013, Australia initiated a multicenter randomized control trial (GIVE UP FOR GOOD) to evaluate the effectiveness of pharmacist-led smoking cessation in a hospital setting (Thomas et al., 2013). The findings revealed that sustained abstinence rates for the pharmacist-led SCS group and usual care group were not significantly

TABLE 53.2 Evidence on Pharmacist-led SCS

Reference	Setting	Design	Sample size	Intervention/ strategy	Pharmacotherapy	Number of success (%)	Follow-up duration (month)	Verification of success
Sinclair et al. (1998)	Community pharmacies	RCT	492	Pharmacist support program	Yes	26 (12.0)	9	Self-report
Maguire et al. (2001)	Community pharmacies	RCT	484	Pharmacist's action plan	Yes	38 (14.3)	12	Self-report with cotinine validation
Kennedy, Giles, Chang, Small, and Edwards (2002)	Community pharmacies	Single group Unblinded	48	Smoking cessation training manual	Yes	12 (25.0)	12	Questionnaire
Vial et al. (2002)	Hospital to community pharmacies	Cohort	102	Consultation with the research pharmacist	Yes	8 (24.0)	12	Self-report with CO-level
McEwen, West, and McRobbie (2006)	Community primary care	Quasi experimental	1501	One-to-one basic pharmacist/ nurses	Yes	285 (19.0)	4	Not known
Dent, Harris, and Noonan (2009)	Community clinic	RCT	101	Three-session face-to-face group program by pharmacist team	Yes	28 (28.0)	12	Self-report with cotinine validation
Bauld, Chesterman, Ferguson, and Judge (2009)	Community pharmacies	Observational	1785	Pharmacies providing one-to-one support	Yes	256 (18.6)	4	Self-report with CO-level
Philbrick, Newkirk, Farris, McDanel, and Horner (2009)	University clinic	Prospective, single group	21	Structured group counseling	Yes	11 (52.4)	6	Self-report
Hoving et al. (2010)	Community pharmacies	RCT	545	Computer-generated tailored advice	Not mentioned	2 (0.8)	12	Not mentioned
Bock, Hudmon, Christian, Graham, and Bock (2010)	Community pharmacies	Not mentioned	200	Pharmacist counseling with ExperQuit software	Yes	28 (28.0)	6	Not mentioned
Costello et al. (2011)	Not mention	RCT	6987	Pharmacist-led behavioral intervention	Yes	612 (17.5%) 604 (18.0%)	1 weeks 5 weeks	Self-report
Taskila et al. (2012)	Community pharmacies	Randomized factorial trial	160	Behavioral support	Encourage	Not yet published	34 weeks	Self-report with CO-level
Burford, Jiwa, Carter, Parsons, and Hendrie (2013)	Community pharmacies	RCT	160	Digitally photoaged by using the Internet-based APRIL Face Aging software as smoker and nonsmoker	Standard care with advice	11 (13.8)	6	Self-report with CO-level
Fai et al. (2016)	Public health clinic	Cross-sectional	176	Pharmacy-led integrated services	Yes	75 (42.6)	6	Self-report with CO-level
Augustine, Taylor, Pelger, Schiefer, and Warholak (2016)	Medication management center	Retrospective review	238	Pharmacist-provided telephone-based quit counseling services	Yes	24 (55.0)	13	Self-report
El Hajj et al. (2017)	Ambulatory pharmacies	RCT	314	Pharmacist-delivered smoking cessation program	Yes	40 (23.9)	12	Self-report with CO-level

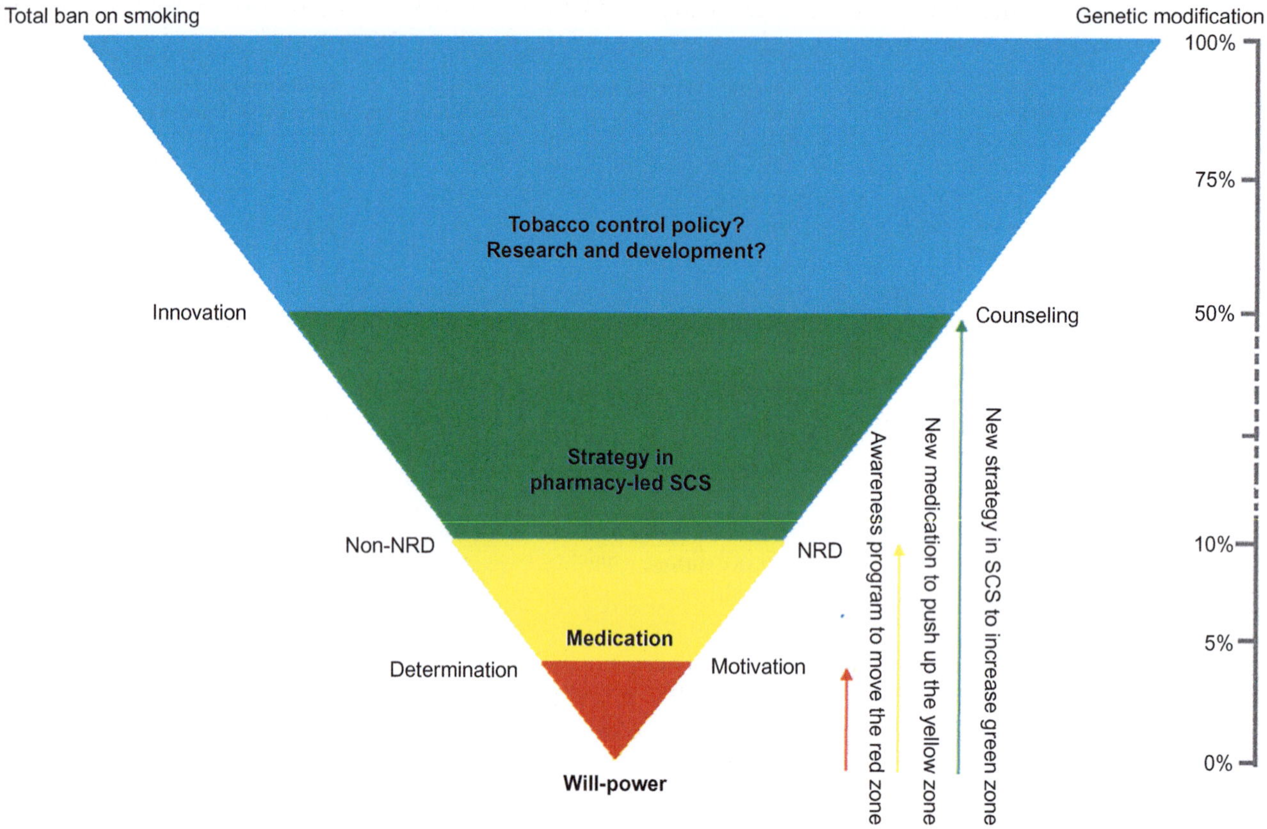

FIG. 53.2 Pyramid of success for SCS (*red*, willpower zone; *yellow*, pharmacotherapy zone; *green*, strategy in SCS zone; and *blue*, tobacco control policy or research and development).

different at both 6 months (11.6% (34 of 294) vs 12.6% (37 of 294); odds ratio (OR) = 0.91, 95% confidence interval (CI) = 0.55–1.50) and 12 months (11.6% (34 of 292) vs 11.2% (33 of 294); OR = 1.04, 95% CI = 0.63–1.73) (Thomas et al., 2016). The findings by Thomas et al. (2016) indicating hospital-based pharmacist-led SCS was not superior to usual care presented the challenges of using clinical trial to corroborate its significance. In their study, the usual care group also received well-organized SCS that was also partially or fully supported by the pharmacy team. For example, patients in the usual care group could also request ward pharmacists and nurses to initiate and counsel the use of NRD and non-NRD medications as part of the usual SCS (Thomas et al., 2016). In this case, the use of cluster randomized trial and crossover study design would be more appropriate to reduce the minimize treatment group contamination and Hawthorne effect. The main lesson learned from this study is pharmacist should intensify smokers' motivation with regular follow-up with their general practitioners and community pharmacists.

From the novel pharmacotherapy viewpoint, there are three new key developments for potentially treating patients with nicotine addiction: (i) electronic cigarettes (e-cigarettes), without the harmful by-products of smoking normal cigarettes (Rahman, Hann, Wilson, Mnatzaganian, & Worrall-Carter, 2015); (ii) vaccines against nicotine receptors; and (iii) cytisine, a natural product, found in plants such as the golden rain acacia and New Zealand kowhai.

NRDs and non-NRDs have been proved to improve the quit rate, with varenicline showing the greatest benefit. However, these products do not have the burst and heat sensation of cigarettes for consumer appeal and product satisfaction. Meanwhile, while e-cigarettes have both appeal and satisfaction, no long-term data are yet available regarding their claim to be safer in terms of ingredients (Mukhtar, Khan, & Long, 2016). E-cigarettes are being misused for recreational purposes rather than as an aid to stop smoking. Many nonsmokers and teenagers have taken up e-cigarettes for glamor purposes, while conventional smokers are becoming dual smokers (Rahman et al., 2015). Further study and regulation are needed to clear up public confusion about the actual meaning of quitting smoking and not merely switch from cigarettes to e-cigarettes. Meanwhile, a vaccine against nicotine receptors is not yet available and currently under clinical trial. Investigators are optimistic the vaccine will prevent future development of nicotine addiction and at the same time treat patients who are already addicted. On a similar note, a naturally occurring substance called cytisine, marketed under the name Tabex, has been used for decades as a traditional remedy for quitting smoking in some European countries. Cytisine has a structure and mechanism of action akin to those of nicotine, but it is much less harmful. However, there are not many available publications about the effectiveness of cytisine (Etter, 2006). A study by Walker et al. (2014) reported that cytisine is superior to NRDs in term of abstinence rates at 1 week, 2 months, and 6 months.

Apart from new drug development, out-of-the-box strategies are needed to make the program more attractive to and accepted by the public as therapeutically important in preventing noncommunicable diseases such as cardiovascular disease and cancer.

53.5 IMPLICATIONS FOR TREATMENT

The involvement of community pharmacies, hospital pharmacies, and health clinic pharmacies in SCS make quitting smoking more effective, with reduced waiting times and multiple locations to choose from. Pharmacist can individualize the treatment plan according to patient need and affordability consideration. SCS pharmacists should provide smoking cessation intervention, whether short or long term, to aid the process of stopping smoking. They should also venture outside community pharmacies or hospitals into schools and universities to help young smokers to kick the habit.

MINI-DICTIONARY OF TERMS

Abstinence A period of not using any forms of cigarettes. Definition may differ from different facility or country.

Addiction A dependency toward the need of continuous dose of cigarettes every day.

Behavioral support A strategic plan in an informed manner to a person using cigarettes for them to change from a smoker to a quitter by reinforcement of motivation.

Biochemical verification A method for confirmation of smoking status by measuring cotinine in the urine or levels of carbon monoxide in exhaled breath or in blood.

Electronic cigarettes A battery-operated device releasing vapors when user inhales from the mouth piece. It mimics the sensation of smoking with less harmful ingredient.

Pharmacotherapy Management of cigarette addiction using pharmaceutical drugs, for example, NRD, bupropion, and varenicline.

Quit guide/quit plan A structural program includes setting quit date, choosing type of pharmacotherapy, signing pledge to quit, technique to refrain from smoking and overcome withdrawal symptom, and managing side effect.

Relapse Smoking during the quit period or restart smoking after successful quit smoking.

Smartphone application A mobile software application for smoking cessation created to run on a mobile device. The term "app" is a shortening of the term "software application."

Smoking cessation service A sophisticated system in recruiting smoker and the process of discontinuing smoking habit in people addicted to cigarettes with the use pharmacotherapy and behavioral support.

Withdrawal symptom A change in human's behavioral and physiological responses during the absence of cigarettes such as intense craving, tiredness, increase appetite, irritability, and difficult concentrating. The symptom can begin within the last dose of cigarettes and up to months.

Key Facts of Pharmacist-led SCS

- The core of pharmacist-led SCS is to guide and counsel smokers to quit and become free from all forms of nicotine addiction.
- The involvement of the pharmacy profession expands the coverage of SCS to communities, primary health settings, and hospital settings.
- Different intensity SCS programs, from low to high intensity, are based on the number of follow-up sessions and time taken for each session.
- Pharmacists are trained to prescribe, dispense, monitor, and counsel the use of NRDs and non-NRDs to any patient who needs the medication.
- Types of NRDs are tailored according to patient suitability, and patients can choose to have a follow-up face-to-face, by phone, by SMS, or via social media.
- The advantages of pharmacist-led SCS in community settings are cost savings, ease of accessibility, and availability of medication.
- Success rates of pharmacist-led SCS are up to 50% with pharmacotherapy and behavioral support.

Summary Points

- Eradication of tobacco use and nicotine addiction has become an important pharmacy service.
- The success of pharmacist-led SCS depends, apart from smokers' own determination, on the modality of the combined pharmacotherapy and behavioral support.
- Training, education, and certification by pharmacy bodies are essential to improve skills and update knowledge of pharmacists and pharmacy-related staff in SCS.
- Pharmacists could help to disseminate the right information in order to prevent smoking habits among the younger generation.
- Support by local government and pharmacy bodies for pharmacist-led SCS has proved to reduce the smoking habit.

References

Anderson, S. (2007). Community pharmacists and tobacco in Great Britain: from selling cigarettes to smoking cessation services. *Addiction*, *102*(5), 704–712.

Augustine, J. M., Taylor, A. M., Pelger, M., Schiefer, D., & Warholak, T. L. (2016). Smoking quit rates among patients receiving pharmacist-provided pharmacotherapy and telephonic smoking cessation counseling. *Journal of the American Pharmaceutical Association, 56*(2), 129–136.

Bauld, L., Chesterman, J., Ferguson, J., & Judge, K. (2009). A comparison of the effectiveness of group-based and pharmacy-led smoking cessation treatment in Glasgow. *Addiction, 104*(2), 308–316.

Bentz, C. J., Gray, M., & Swan, C. (2007). *Pharmacist-assisted smoking cessation class: a guide to implementation. Available from: https://smokingcessationleadership.ucsf.edu/sites/smokingcessationleadership.ucsf.edu/files/Downloads/Toolkits/HospGuide.pdf. Accessed 1 September 2017.*

Bock, B. C., Hudmon, K. S., Christian, J., Graham, A. L., & Bock, F. R. (2010). A tailored intervention to support pharmacy-based counseling for smoking cessation. *Nicotine & Tobacco Research, 12* (3), 217–225.

Boutwell, L., Cook, L., Norman, K., & Lindsey, W. T. (2014). *A pharmacist's guide for smoking cessation. Available from: http://c.ymcdn.com/sites/www.aparx.org/resource/resmgr/imported/CE_Winter%20Smoking%20Cessation.pdf/. Accessed 1 September 2017.*

Bricker, J. B., Mull, K. E., Kientz, J. A., Vilardaga, R., Mercer, L. D., Akioka, K. J., et al. (2014). Randomized, controlled pilot trial of a smartphone app for smoking cessation using acceptance and commitment therapy. *Drug and Alcohol Dependence, 143*, 87–93.

British Columbia Pharmacy Association (2013). *British Columbia Pharmacy Association, clinical service proposal pharmacist-led smoking cessation services. Available from: https://www.bcpharmacy.ca/uploads/Smoking%20Cessation.pdf. Accessed 1 September 2017.*

Brock, T., Taylor, D., & Wuliji, T. (2009). *Curbing the tobacco pandemic: The global role for pharmacy.* London: International Pharmaceutical Federation.

Burford, O., Jiwa, M., Carter, O., Parsons, R., & Hendrie, D. (2013). Internet-based photoaging within Australian pharmacies to promote smoking cessation: randomized controlled trial. *Journal of Medical Internet Research, 15*(3), e64.

Callahan-Lyon, P. (2010). *Nicotine replacement therapy: the CDER experience. Available from: http://www.fda.gov/downloads/AdvisoryCommittees/Commi-tteesMeetingMaterials/TobaccoProductsScientificAdvisoryCommittee/UCM288284.pdf. Accessed 1 June 2017.*

Costello, M. J., Sproule, B., Victor, J. C., Leatherdale, S. T., Zawertailo, L., & Selby, P. (2011). Effectiveness of pharmacist counseling combined with nicotine replacement therapy: a pragmatic randomized trial with 6,987 smokers. *Cancer Causes & Control, 22*(2), 167–180.

CSCSP. (2017). *Certified Smoking Cessation Service Provider, Malaysia Academy of Pharmacy. Available from:(2017). http://www.acadpharm.org.my/index.cfm?&menuid=2. Accessed 1 September 2017.*

Dent, L. A., Harris, K. J., & Noonan, C. W. (2009). Randomized trial assessing the effectiveness of a pharmacist-delivered program for smoking cessation. *The Annals of Pharmacotherapy, 43*(2), 194–201.

Dobbie, F., Hiscock, R., Leonardi-Bee, J., Murray, S., Shahab, L., Aveyard, P., et al. (2015). Evaluating long-term outcomes of NHS stop smoking services (ELONS): a prospective cohort study. *Health Technology Assessment, 19*(95), 1–156.

El Hajj, M. S., Kheir, N., Al Mulla, A. M., Shami, R., Fanous, N., & Mahfoud, Z. R. (2017). Effectiveness of a pharmacist-delivered smoking cessation program in the State of Qatar: a randomized controlled trial. *BMC Public Health, 17*(1), 215.

Etter, J. F. (2006). Cytisine for smoking cessation: a literature review and a meta-analysis. *Archives of Internal Medicine, 166*(15), 1553–1559.

Fai, S. C., Yen, G. K., & Malik, N. (2016). Quit rates at 6 months in a pharmacist-led smoking cessation service in Malaysia. *Canadian Pharmaceutical Journal, 149*(5), 303–312.

Freund, M., Campbell, E., Paul, C., McElduff, P., Walsh, R. A., Sakrouge, R., et al. (2008). Smoking care provision in hospitals: a review of prevalence. *Nicotine & Tobacco Research, 10*(5), 757–774.

Goniewicz, M. L., Lingas, E. O., Czogala, J., Koszowski, B., Zielinska-Danch, W., & Sobczak, A. (2010). The role of pharmacists in smoking cessation in Poland. *Evaluation & the Health Professions, 33*(1), 81–95.

Hoving, C., Mudde, A. N., Dijk, F., & Vries, H. d. (2010). Effectiveness of a smoking cessation intervention in Dutch pharmacies and general practices. *Health Education, 110*(1), 17–29.

Kennedy, D. T., Giles, J. T., Chang, Z. G., Small, R. E., & Edwards, J. H. (2002). Results of a smoking cessation clinic in community pharmacy practice. *Journal of the American Pharmaceutical Association, 42*(1), 51–56.

Maguire, T. A., McElnay, J. C., & Drummond, A. (2001). A randomized controlled trial of a smoking cessation intervention based in community pharmacies. *Addiction, 96*(2), 325–331.

McEwen, A., West, R., & McRobbie, H. (2006). Effectiveness of specialist group treatment for smoking cessation vs. one-to-one treatment in primary care. *Addictive Behaviors, 31*(9), 1650–1660.

McRobbie, H., & McEwen, A. (2005). Helping smokers to stop: advice for pharmacists in England. *National Institute for Health and Clinical Excellence*, 14–15.

Ministry of Health and Long-Term Care (2011). *Pharmacy Smoking Cessation Program, the pharmacist's role in a smoking cessation system. Available from:(2011). http://www.health.gov.on.ca/en/pro/programs/drugs/smoking/. Accessed 1 September 2017.*

mQuit. (2016). *Malaysia Quit Smoking Service (mQuit). Available from: (2016). http://jomquit.moh.gov.my/mquitcenters. Accessed 1 September 2017.*

Mukhtar, M., Khan, T. M., & Long, C. M. (2016). E-cigarette: more harm than good? *Proceedings of Singapore Healthcare, 25*(2), 115–116.

NHS Employers. (2009). *Pharmacy-based stop smoking services: optimising commissioning. Available from:(2009). http://archive.psnc.org.uk/data/files/smoking_cessation_guidance.pdf. Accessed 1 September 2017.*

Norman, J. (2016). Americans rate healthcare providers high on honesty, ethics. *Gallup*.

Partnership on Tobacco and Health. (2016). *The minimum dataset for Scottish smoking cessation services (including pharmacies offering the national community pharmacy smoking cessation service).* .

Patwardhan, P. D., & Chewning, B. A. (2010). Tobacco users' perceptions of a brief tobacco cessation intervention in community pharmacies. *Journal of the American Pharmaceutical Association, 50*(5), 568–574.

Pharmaceutical Society of Singapore. (2016). *Module 5: Smoking cessation, practice guides for minor ailments. Available from:(2016). https://www.pss.org.sg/product/pss-pgma-module-5-smoking-cessation. Accessed 1 September 2017.*

Philbrick, A. M., Newkirk, E. N., Farris, K. B., McDanel, D. L., & Horner, K. E. (2009). Effect of a pharmacist managed smoking cessation clinic on quit rates. *Pharmacy Practice (Granada), 7*(3), 150–156.

Rahman, M. A., Hann, N., Wilson, A., Mnatzaganian, G., & Worrall-Carter, L. (2015). E-cigarettes and smoking cessation: evidence from a systematic review and meta-analysis. *PLoS ONE, 10*(3) e0122544.

Sinclair, H. K., Bond, C. M., Lennox, A. S., Silcock, J., Winfield, A. J., & Donnan, P. T. (1998). Training pharmacists and pharmacy assistants in the stage-of-change model of smoking cessation: a randomised controlled trial in Scotland. *Tobacco Control, 7*(3), 253–261.

Smoking Cessation Trust. (2013). *Smoking cessation provider guide: US script pharmacy services. (2013). https://www.smokingcessationtrust.org/AvailableServices.aspx#.Wdwu2GiCzIU.*

Stead, L. F., Perera, R., Bullen, C., Mant, D., Hartmann-Boyce, J., Cahill, K., et al. (2012). Nicotine replacement therapy for smoking cessation. *Cochrane Database of Systematic Reviews, 11*, CD000146.

Taskila, T., Macaskill, S., Coleman, T., Etter, J. F., Patel, M., Clarke, S., et al. (2012). A randomised trial of nicotine assisted reduction to stop in pharmacies - The RedPharm study. *BMC Public Health, 12*, 182.

Thomas, D., Abramson, M. J., Bonevski, B., Taylor, S., Poole, S. G., Paul, E., et al. (2016). Integrating smoking cessation into routine care in hospitals—a randomized controlled trial. *Addiction*, *111*(4), 714–723.

Thomas, D., Abramson, M. J., Bonevski, B., Taylor, S., Poole, S., Weeks, G. R., et al. (2013). A pharmacist-led system-change smoking cessation intervention for smokers admitted to Australian public hospitals (GIVE UP FOR GOOD): study protocol for a randomised controlled trial. *Trials*, *14*(1), 148.

Tobacco Control Network. (2017). *Access to tobacco cessation medication through pharmacists.* Tobacco Control Network. Available from: *http://www.astho.org/Prevention/Tobacco/Tobacco-Cessation-Via-Pharmacists/* (Accessed 1 September 2017).

Vial, R. J., Jones, T. E., Ruffin, R. E., & Gilbert, A. L. (2002). Smoking cessation program using nicotine patches linking hospital to the community. *Journal of Pharmacy Practice and Research*, *32*(1), 57–62.

Walker, N., Howe, C., Glover, M., McRobbie, H., Barnes, J., Nosa, V., et al. (2014). Cytisine versus nicotine for smoking cessation. *The New England Journal of Medicine*, *371*(25), 2353–2362.

World Health Organization. (1998). *Pharmacists and action on tobacco. Available from:(1998). http://apps.who.int/iris/bitstream/10665/108128/1/E61288.pdf. Accessed 1 September 2017.*

World Health Organization. (2017). *WHO report on the global tobacco epidemic 2017 (Monitoring tobacco use and prevention policies). Available from:(2017). http://www.who.int/tobacco/global_report/2017/en/. Accessed 12 September 2017.*

CHAPTER

54

Nicotine Use and Weight Control in Young People: Implications for Prevention and Early Intervention

Adrian B. Kelly, Rebekah Thomas†, Gary C.K. Chan‡*

*School of Psychology and Counselling, Institute for Health and Biomedical Innovation, Queensland University of Technology, Brisbane, QLD, Australia

†School of Psychology, University of Queensland, Brisbane, QLD, Australia

‡Centre for Youth Substance Abuse Research, University of Queensland, Brisbane, QLD, Australia

Abbreviations

BMI body mass index
NR nationally representative
ATT attrition
prev prevalence

The belief that smoking can facilitate weight control by suppressing appetite is widespread and presents a significant public health challenge. Rates of overweight in Western societies are high, which may increase the risks of smoking in people with weight concerns. In the United Kingdom, 36% of 16–24-year-olds are overweight or obese (Baker, 2017; Public Health England, 2015), and in the United States, around 20.5% of 12–19-year-olds are obese (Ogden, Carroll, Fryar, & Flegal, 2015). In Australia, almost 30% of adolescents aged 14–17 years of age are overweight or obese (AIHW, 2017). Initiation of tobacco use most often occurs in the teenage years, and this is a developmental period when concerns about body weight are high. The likely success of smoking cessation attempts may also be eroded in teenagers where weight concerns are often significant (often the case). In the long term, tobacco use also tends to erode fitness and exercise regimes, which increases the risk of weight gain.

In this chapter, we provide an update on empirical literature on the association of smoking with body weight, overweight, and weight-related beliefs. We begin with an overview of theoretical accounts of the link between tobacco use and body weight and weight control; then, we review empirical literature of the association of these two factors from the past 15 years. We focus on the past 15 years because of changing trends in the prevalence of tobacco use, overweight, and obesity over these years (OECD, 2017; WHO, 2015). Based on data from the last 15 years, the core research questions that we seek to answer are the following:

1. Does being overweight or being concerned about being overweight increases the risk of smoking during adolescence?
2. If weight factors increase the risk of smoking, is this effect stronger in females than males?
3. What implications might these findings have for prevention programs?

54.1 MECHANISMS LINKING WEIGHT CONCERNS AND WEIGHT LOSS WITH TOBACCO USE

Key theoretical accounts of the links between body weight and smoking center around metabolic processes, social developmental mechanisms, and social ecological drivers. Metabolic mechanisms seek to account for the association via research on whether tobacco use has a demonstrable effect on appetite. Social cognitive mechanisms emphasize personal beliefs about smoking and weight control. Social ecological drivers include family and media messages about health risk behavior.

Neuroscience of Nicotine
https://doi.org/10.1016/B978-0-12-813035-3.00054-X

54.1.1 Metabolic Drivers

A long-standing and widespread perception is that smoking may facilitate increases in metabolism by increasing heart rate, thereby expending more calories (Wack & Rodin, 1982). Reviews of the literature indicate that nicotine ingestion increases caloric expenditure and may reduce appetite by decreasing caloric absorption (Chiolero, Faeh, Paccaud, & Cornuz, 2008; Wehby, Murray, Wilcox, & Lie, 2012). For example, smoking has been shown to increase caloric expenditure by 3% in the 30min following tobacco use (Dallosso & James, 1984). While proximal effects have been demonstrated, the findings on the chronic effects of smoking on caloric expenditure are mixed. Based on large-scale studies where effects for the initiation of smoking can be observed and autocorrelational effects can be controlled, the balance of evidence indicates that smoking does not constrain weight gain in the long term (Robertson, McGee, & Hancox, 2014)—weight gain is a general tendency rather than a consequence of smoking cessation. Large longitudinal studies with controls also show that either nonsmokers have lower weight gain than smoking initiators or the effects are inconsistent and vary by cultural background (Chiolero et al., 2008). These findings suggest that proximal effects may be initially rewarding and therefore important drivers of early entrenchment of smoking, but smoking erodes exercise regimes and therefore may contribute to weight gain in the long term (Nagaya, Yoshida, Takahashi, & Kawai, 2007) (Fig. 54.1).

54.1.2 Social Cognitive Drivers

A key tenet of social cognitive approaches to substance use is that individuals develop beliefs about the personal consequences of a certain action (expectancies) that drive health risk behaviors. These expectancies are presumed to arise from the individual's history of learning (e.g., parent/peer/media messages and personal experiences). In the case of smoking, beliefs that smoking can constrain weight gain will drive smoking, particularly for individuals concerned about their weight (see Fig. 54.2). Among adolescents, expectancies that drive smoking mostly relate to control of negative affect and weight management. For example, negative affect relief expectancies prospectively predict smoking outcomes and increases in smoking behavior and nicotine dependence (Heinz, Kassel, Berbaum, & Mermelstein, 2010). For both females and males, expectancies of weight control, boredom relief, and negative affect management are predictive of current smoking and future smoking (Wahl, Turner, Mermelstein, & Flay, 2005). These social drivers of smoking and weight control fit with cognitive research showing beliefs that smoking provides relief from negative affect and evidence that body weight concerns are a key source of distress, depressed mood, and anxiety among adolescents.

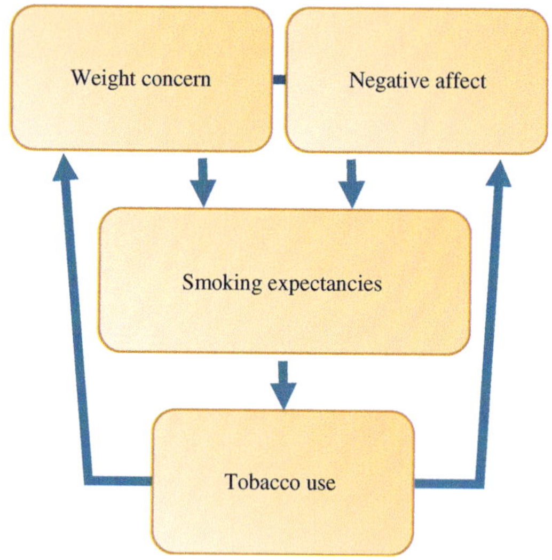

FIG. 54.2 A social cognitive model of smoking and weight control.

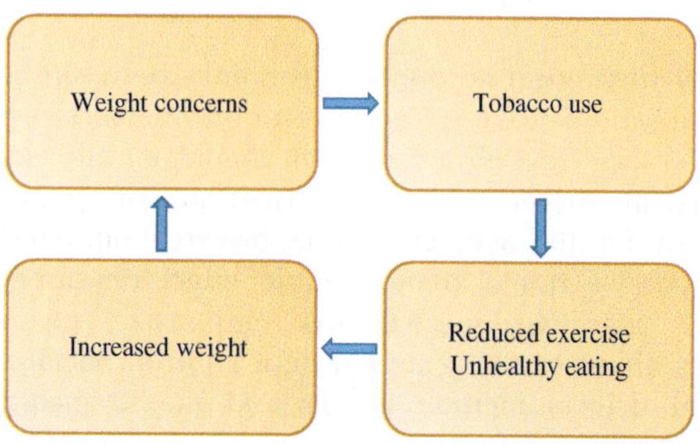

FIG. 54.1 The short- vs long-term metabolic effects of smoking on weight.

54.1.3 Social Ecological Frameworks

Social ecological approaches (Bronfenbrenner, 1989; Ennett et al., 2008) focus on the often-nested influences of families, peers, communities, and the media on adolescent health risk behaviors (Mehus et al., 2018; Smith et al., 2014) (see Fig. 54.3). Tobacco smoking is most likely to begin when susceptibility to peer influence is highest (Monahan, Steinberg, & Cauffman, 2009), and adolescents adjust to major transitions, including puberty and high school. Parent, older sibling, and peer smoking are among the strongest predictors of adolescent tobacco use (Kelly et al., 2011), and family management (conflict, supervision, and monitoring) is predictive of both tobacco use (Kelly et al., 2011; Mehus et al., 2018) and emerging weight control preoccupations (Hinchliff et al., 2016; Thomas et al., 2018). Systemic and persistent exposure to media messages that promote the "thin ideal" (Dittmar, Halliwell, & Stirling, 2009; Harrison, 2000) may increase depression/anxiety and increase the likelihood of

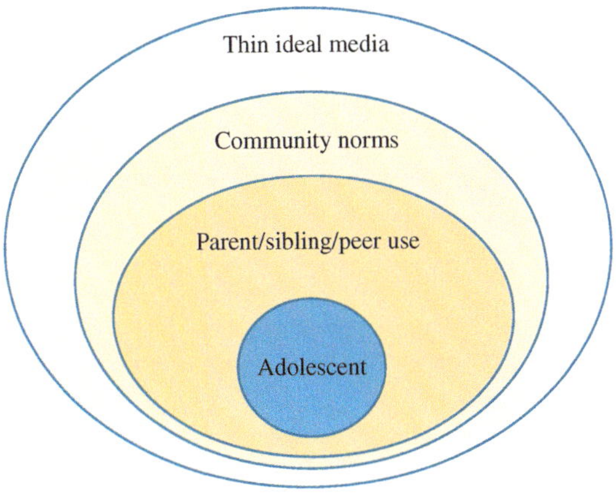

FIG. 54.3 Social ecological systems influencing tobacco use by vulnerable adolescents.

smoking to constrain weight and/or manage negative affect (Dittmar et al., 2009; Thomas et al., 2018). Adolescents living in socioeconomically disadvantaged communities may be at further risk of smoking initiation because smoking and overweight prevalence rates are often higher than in socioeconomically advantaged communities.

54.1.4 Summary and Integration

The association of weight concerns and tobacco use in adolescents may be understood through a focus on metabolic processes, social cognitive processes, and social ecological/learning approaches. Tobacco use may have a proximal effect on appetite, but in the long term, tobacco use may actually contribute to weight problems by increasing the likelihood of a sedentary lifestyle and decreasing regular exercise (Nagaya et al., 2007). Adolescent smoking expectancies are often related to negative affect and weight management, and these may drive early tobacco use. Media messages about the thin ideal may increase the risk of smoking, particularly in girls, and adolescents are at high risk of smoking when exposed to smoking through family and peer networks. Adolescents from low-SES communities are at high risk of tobacco use and overweight.

54.2 WHAT DOES THE RECENT EMPIRICAL LITERATURE TELL US?

To gather empirical research on the association, we searched *Psycinfo*, *PubMed*, *Google Scholar* and reference lists of included papers. Search terms included *adolescent*, *young adult*, *teenager*, *smoking*, *tobacco*, *weight*, and *weight control*. To be included, studies had to focus on adolescents (age <18 years), have at least one measure of smoking and BMI and/or measures of perceived weight and tobacco expectancies, and be published in a peer-reviewed journal in 2004 or later.

54.2.1 Cross-Sectional Research

An overview of study characteristics (country of origin, sample size, measures, and findings) is presented in Table 54.1. Findings on the association of weight concerns/control and smoking were mixed. Six of the 14 studies found that female beliefs about the impact of smoking on weight and/or concerns about weight were related to smoking status. In particular, female smokers report more concerns about gaining weight as a result of quitting (Cavallo, Duhig, McKee, & Krishnan-Sarin, 2006; Seo, Jiang, & Kolbe, 2009); more female smokers than male smokers use tobacco to control weight (Cawley, Dragone, & Von Hinke Kessler Scholder, 2016; Xie et al., 2006).

Among females, BMI is associated with smoking (Pénzes, Czeglédi, Balázs, & Foley, 2012; Seo et al., 2009), beliefs that smoking controls weight mediates this association (Pénzes et al., 2012), and eating and weight loss attitudes are associated with smoking (Thomas et al., 2018). Lange, Thamotharan, Racine, Hirko, and Fields (2014) found that perceived smoking and BMI were associated with smoking in females, but only BMI was associated with smoking in males. Only one study found opposite effects (Delk, Creamer, Perry, & Harrell, 2017), but notably, the significant effect for boys was limited to those in the obese range. The lack of effects for lower weight categories and for females may have arisen because of limited statistical power—the numbers of identified smokers were very small (e.g., $n=42$ 30-day smokers; Delk et al., 2017). The other studies either found no evidence of gender by weight interactions or did not test this interaction (gender often treated as a control/adjustment variable). There were nongender-specific association of weight perceptions/weigh control and smoking (Chung & Joung, 2014; Kendzor, Copeland, Stewart, Businelle, & Williamson, 2007; Leatherdale, Wong, Manske, & Colditz, 2008; Weiss, Merrill, & Gritz, 2007).

54.2.2 Longitudinal Research

Longitudinal studies offer the possibility of assessing the direction of effects linking weight/weight control to smoking. If weight/weight control increases the risk of smoking, we should see a longitudinal association after accounting for potential confounds. Of course, a critical factor in the validity of longitudinal research like this is that attrition rates can be low. Attrition rates in excess of 25% are problematic because adolescents with multiple problems are typically overrepresented in those who drop out of longitudinal research (Gustavson, von Soest, Karevold, & Røysamb, 2012; Western, Braga, Hureau, & Sirois, 2016). To minimize threats to internal and external validity, attrition rates less than 25% are recommended for survey studies of 3 or more years'

TABLE 54.1 Cross-Sectional Studies on the Association of Tobacco Smoking and Weight Concern/Control

First author/year	Sample size	Sample characteristics	Key smoking measures	Key findings
Cavallo (2006)	$N=103$	Smokers, age 14–18 years, United States	Thirty-day smoking; belief that cigs help control weight; concern about weight if quit smoking	Compared to males, females had more concerns about gaining weight
Cawley et al. (2016)	$N=10{,}442$	14–16 years, United States	Present smoking status (nonsmoker, any smoking/week)	46% females, 30% males smoke to control weight. Smoking for weight loss more common when perceived overweight
Chung (2014)	$N=93{,}911$	Korean/American students (grades 9–11)	Number of smoking days in the last 30 days (>1 cig, smoker; 0, nonsmoker)	Weight perception/control → ↑ current smoking
Cook, MacPherson, and Langille (2007)	$N=1133$	Grade 10–12 girls, four schools, Canada	Number of smoking days in the last 30 days (endorsement of all days in month)	Unrelated to perceived weight
Delk et al. (2017)	$N=2733$	Grade 7–11, United States	Lifetime prev, 30-day prev for cigs and e-cigs	Males: compared to healthy weight, obesity → ↑ 30-day smoking. Females: no association
Johnson, Eaton, Pederson, & Lowry (2009)	$N=13{,}917$	Grade 9–12, United States	Number of smoking days in last 30 days (>1 cig, smoker; 0, nonsmoker)	Trying to lose weight associated with smoking. No gender effects
Kendzor (2007)	$N=727$	Grades 2–6, from a prevention trial, five schools, United States	Lifetime prev (yes/no)	Smoking → ↑ dieting, weight concern, compared to nonsmokers. Gender differences unclear
Lange et al. (2015)	$N=15{,}425$	Grades 9–12, United States	Lifetime prev, cigs/day, smoking days/last 30 days (>1 cig, smoker; 0, nonsmoker)	Perceived weight predicted lifetime prev; BMI predicted all smoking measures; perceived weight and BMI predicted female smoking, not males
Leatherdale (2008)	$N=25{,}060$	Grades 9–12, 76 schools, Canada	Smoking susceptibility (three items)	Smoking susceptibility associated with overweight
Pénzes (2012)	$N=2208$	Grades 6–9, 78 schools, Hungary	Thirty-day prev; number of cigs/last 30 days. Perceptions of weight control and smoking	BMI → ↑ smoking; belief that smoking controls weight mediates this association. Associations did not vary by motivation to lose/gain weight. Females: stronger beliefs than boys
Seo (2009)	$N=13$–$15{,}000$/survey	Grades 9–12, United States	Smoking/30 days (>1 cig, smoker; 0, nonsmoker)	Perceived body weight and BMI → ↑ current smoking. Females: perception of overweight → smoker. Males: not significant
Thomas (2018)	$N=10{,}273$	Grades 7–11, Australia	Thirty-day smoking (yes/no)	After controlling for demographics, anxiety/depression, puberty, peer use, and smoking → ↑ weight control for females but not males
Weiss (2007)	$N=3515$	Grade 8–9 students, United States	Smoking/30 days (yes, even a few puffs/no)	Weight concern → ↑ smoking risk increased by 40%. Did not examine gender-weight interactions
Xie (2006)	$N=6863$	Middle/high schoolers, China	Smoking/30 days (yes, even a few puffs/no)	Weight dissatisfaction → ↑ smoking; significant for girls, not boys

Note: This review includes empirical papers from 2004 onward. *ATT*, attrition; *cig*, cigarette.

duration (Hansen, Tobler, & Graham, 1990; Nemes, Wish, Wraight, & Messina, 2002). We therefore focused on longitudinal studies where attrition was less than 25%. Study characteristics are summarized in Table 54.2.

Among the nationally representative studies (American research studies based on the National Longitudinal Survey of Youth, the Add Health study), findings were relatively consistent. The proportion of smoking initiation among 15-year-old girls attributable to feeling overweight could be estimated as close to 20% (Caria, Bellocco, Zambon, Horton, & Galanti, 2009). In the study with one of the lowest attrition rates, females who were overweight or trying to lose weight were more likely to initiate smoking, compared to males (Cawley et al.,

TABLE 54.2 Longitudinal Studies on the Association of Tobacco Smoking and Weight Concern/Control

First author/year	Sample size	Sample characteristics	Method/survey	Key measures	Key findings
Caria (2009)	$N=2922$	11.6 years, Sweden	Seven-year f/u, ATT = 18%	Lifetime prev, lifetime number of cig, regular use	Females: overweight/perceived overweight → smoking
Cawley, Markowitz, and Tauras (2004)	$N=9022$	12–16 years, *NR*	Four waves, ATT = 4%–9%	Smoking initiation	Females: high BMI/report trying to lose weight/report being overweight → initiate smoking, compared to other females
Harakeh (2010)	$N=428$ families	Intact families, Netherlands, *NR*	Two waves, ATT = 3%	Weight control motives, BMI, smoking frequency	Weight control motives → ↑ smoking initiation, general weight measures did not. No gender moderation
Hong, Rice, and Johnson (2011)	$N=921$	Females in 9th grade, United States, *not NR*	F/u 12th grade, ATT = 34.5%	Smoking days/30 days (>1 cig)	Perceptions of overweight and perceived low risks for smoking → ↑ smoking
Kaufman (2008)	$N=6956$	Females only, Grades 7–12	Waves I and II, *Add Health*, ATT = 11.4%	Regular smoking (>1 cig/day for 30 days)	Perceived weight did not predict smoking in multivariate model
Koval (2008)	$N=1598$	Grade 6 students, 107 schools, Canada, *not NR*	Four waves to age 22 years, ATT (last f/u) = 21.5%	Life/30-day prev	Females: perceptions of overweight (grades 8/11) → ↑ recent smokers. Males: BMI (grade 8/11) → ↑ recent smokers
Neumark-Sztainer et al. (2006)	$N=2516$	Junior/seniors, 31 schools, United States, *not NR*	Two waves, 5-year gap, ATT = 22.6%	Smoking frequency in past year, BMI, body satisfaction, dieting/weight control	Females: lower body satisfaction → ↑ dieting health risk behaviors, but not smoking Males: body dissatisfaction → ↑ smoking/month
Rees and Sabia (2010)	$N=$ approx. 15,000	11–23-year-olds, American, *NR*	*Add Health*: 132 schools, 5 year f/u ATT = 11.4%–22.6%	Smoking days/30 days, smoke > ¼ pack, smoke > ½ pack	Being overweight → ↑ smoking initiation
Tanner-Smith (2010)	$N=5591$	Girls 10–15 years, United States	Three waves, ATT = 15.4% (wave 2)	Lifetime prev	BMI moderated association of early puberty and substance use

Note: This review includes empirical papers from 2004 onward. *ATT*, attrition; *NR*, nationally representative.

2004). Other studies have found that smoking is associated with perceived rather than actual weight (Harakeh, Engels, Monshouwer, & Hanssen, 2010; Koval, Pederson, Zhang, Mowery, & McKenna, 2008). Koval et al. (2008) found that female perceptions of being overweight increase the likelihood of smoking, but for males, BMI but not perceptions of weight was related to smoking, suggesting that body image is important and more so for girls than boys. Kaufman and Augustson (2008) also found that in simple models of smoking, perceived weight and trying to lose weight predicted smoking 1 year later, but these effects became nonsignificant once self-esteem was accounted for in multivariate analyses. These findings highlight the importance of self-esteem as a determinant of smoking and point to the possibility that concern about weight is a marker of low self-esteem. Contrasting findings exist for 5-year longitudinal research showing that body dissatisfaction is associated with several unhealthy weight control strategies in females, including frequent dieting and using food substitutes, but there was no significant effect for smoking in females (Neumark-Sztainer, Paxton, Hannon, Haines, & Story, 2006). Rather, this study found an association between body dissatisfaction and smoking in males. It is possible that the long period between assessment waves (5 years) may have weakened the effects for girls (the effect for girls was significant in univariate analyses), and there was differential attrition that may also have weakened effects. The authors note that around 10% of females and 5% of males reported smoking to lose weight.

54.2.3 Summary and Integration

Over the last 15 years, some 23 empirical studies have examined the links between weight concern/control and tobacco use during adolescence. The evidence that smoking is used as a weight control strategy in vulnerable adolescents is reasonably consistent, and there is reasonable evidence that females are at elevated risk of smoking to control weight. Studies with contrary findings are few in number. Longitudinal research studies are mostly conducted in the United States, have relatively low attrition, and have large often nationally representative samples. Overall, the empirical evidence indicates that having weight-related concerns (real or perceived) increases the risk of smoking, and this effect is stronger for females than males. The findings of this review suggest that the evidence around the research questions has got stronger and more consistent over the last 15 years, relative to a 2004 review finding that research support was mixed (Potter, Pederson, Chan, Aubut, & Koval, 2004). In this earlier review that included eight longitudinal studies, three studies found evidence supporting our findings, two found no relationship, and three had conflicting findings. These findings have pressing implications for prevention and early intervention, because although adolescent smoking rates are in decline, the decline in low-SES communities is much slower, and smoking-related policies and programs seem to have limited reach in these communities (AHPRA, 2013). Also, rates of overweight and obesity are much higher in low-SES areas, so adolescents in these communities have the dual exposures to adult smoking and unsatisfactory diet/exercise.

The present review highlights the potential importance of weight concern/control and associated depression and anxiety in the initiation maintenance and resolution of tobacco use. Based on the findings of this review, we make three recommendations for prevention programs (see Table 54.3).

TABLE 54.3 Some Recommendations for Prevention and Early Intervention

Early detection	*Screen* for at-risk adolescents, based on BMI, perceived overweight, and smoking-related beliefs/expectancies: • Weight concerns may be a better predictor of smoking risk than actual weight/BMI • Assess expectancies about tobacco use on weight and negative affect. These may indicate risk for smoking or relapse
Universal prevention	Build the above findings into smoking prevention programs: • Interactive discussions on the links between "thin-ideal" media, smoking expectancies, weight, depression/anxiety, and tobacco use • Challenge smoking expectancies. Expectancy challenge strategies (interactive not information-based) have been shown to be effective in the alcohol literature (Darkes & Goldman, 1993) • Discuss the short-term vs long-term effects of smoking on weight • Provide interactive education on healthy alternatives to weight loss • Discuss the pressures associated with parent/sibling/peer smoking, particularly for adolescents concerned about their weight
Reaching adolescents in vulnerable communities	• Adolescents from low-SES communities are at elevated risk of smoking/overweight yet are the hardest to reach • Community-based coalition approaches are important for building sustainable approaches and imparting prevention skills to community stakeholders (Rowland et al., 2013, 2018) • Community-based coalition approaches result in substantial reductions in adolescent tobacco use (Hawkins et al., 2009)

MINI-DICTIONARY OF TERMS

Body mass index (BMI) Calculated by dividing weight in kilograms by height in meters squared.

Smoking expectancies Beliefs about the personal consequences of tobacco smoking.

Social ecological systems Social systems in which the adolescent is a member that vary in proximity to the adolescent and that are commonly nested (e.g., peers within schools within communities).

Thin-ideal media Portrayals of thinness as desirable and overweight as undesirable.

Weight control preoccupations Persistent and intrusive concerns about weight and weight management via diet and weight management strategies.

Key Facts About Smoking, Eating, and Weight Loss Control

- Smoking most commonly begins during early to middle adolescence.
- Adolescence is a period of high vulnerability to the "thin ideal" (media messages of thinness as the ideal or perfect body shape).
- In vulnerable adolescents, discrepancies between ideal and actual weight are associated with depression and anxiety.
- Tobacco use is widely believed to assist with weight control and management of negative affect.
- Prevention programs for smoking should address beliefs about the utility of weight-related smoking expectancies.

Summary Points

- Adolescents concerned about their weight are at elevated risk of smoking initiation.

- Prevention programs for smoking should incorporate strategies that address the links between weight concerns, depressed mood, and tobacco use.
- Vulnerable adolescents may reside in communities that have low socioeconomic status and high rates of adult smoking.
- Conventional antismoking policies and programs may have less "reach" in high-risk communities.
- Empowering local community coalitions to implement effective prevention programs is likely to enhance prevention programs for at-risk adolescents.

References

AHPRA. (2013). *Smoking and disadvantage: Evidence brief.* Canberra: Commonwealth of Australia.

AIHW. (2017). *Overweight and obesity in Australia: A birth cohort analysis.* Canberra: Commonwealth of Australia.

Baker, C. (2017). *Obesity statistics: Briefing paper number 3336.* London: House of Commons.

Bronfenbrenner, U. (1989). Ecological systems theory. In R. Vasta (Ed.), *Annals of child development. Six theories of child development: Revised formulations and current issues* (pp. 1–103). Greenwich: JAI.

Caria, M. P., Bellocco, R., Zambon, A., Horton, N. J., & Galanti, M. R. (2009).Overweight and perception of overweight as predictors of smokeless tobacco use and of cigarette smoking in a cohort of Swedish adolescents. *Addiction, 104*(4), 661–668.

Cavallo, D. A., Duhig, A. M., McKee, S., & Krishnan-Sarin, S. (2006). Gender and weight concerns in adolescent smokers. *Addictive Behaviors, 31*(11), 2140–2146.

Cawley, J., Dragone, D., & Von Hinke Kessler Scholder, S. (2016). The demand for cigarettes as derived from the demand for weight loss: A theoretical and empirical investigation. *Health Economics, 25*, 8–23.

Cawley, J., Markowitz, S., & Tauras, J. (2004). Lighting up and slimming down: The effects of body weight and cigarette prices on adolescent smoking initiation. *Journal of Health Economics, 23*, 293–311.

Chiolero, A., Faeh, D., Paccaud, F., & Cornuz, J. (2008). Consequences of smoking for body weight, body fat distribution, and insulin resistance. *The American Journal of Clinical Nutrition, 87*(4), 801–809. https://doi.org/10.1093/ajcn/87.4.801.

Chung, S. S., & Joung, K. H. (2014). Risk factors for current smoking among American and south Korean adolescents, 2005-2011. *Journal of Nursing Scholarship, 46*(6), 408–415.

Cook, S. J., MacPherson, K., & Langille, D. B. (2007). Weight perception, weight control, and associated risky behavior of adolescent girls in Nova Scotia. *Canadian Family Physician, 63*, 679–684.

Dallosso, H. M., & James, W. P. (1984). The role of smoking in the regulation of energy balance. *International Journal of Obesity, 8*, 365–375.

Darkes, J., & Goldman, M. S. (1993). Expectancy challenge and drinking reduction: experimental evidence for a mediational process. *Journal of Consulting and Clinical Psychology, 61*(2), 344–353.

Delk, J., Creamer, M. R., Perry, C. L., & Harrell, M. B. (2017). Weight status and cigarette and electronic cigarette use in adolescents. *American Journal of Preventive Medicine, 54*(1), e31–e35.

Dittmar, H., Halliwell, E., & Stirling, E. (2009). Understanding the impact of thin media models on women's body-focused affect: the roles of thin-ideal internalization and weight-related self-discrepancy activation in experimental exposure effects. *Journal of Social and Clinical Psychology, 28*(1), 43–72.

Ennett, S. T., Foshee, V. A., Bauman, K. E., Hussong, A. M., Cai, L., Luz, H., et al. (2008). The social ecology of adolescent alcohol misuse. *Child Development, 79*, 1777–1791.

Gustavson, K., von Soest, T., Karevold, E., & Røysamb, E. (2012). Attrition and generalizability in longitudinal studies: findings from a 15-year population-based study and a Monte Carlo simulation study. *BMC Public Health, 12*, 918.

Hansen, W. B., Tobler, N. S., & Graham, J. W. (1990). Attrition in substance-abuse prevention research - a metaanalysis of 85 longitudinally followed cohorts. *Evaluation Review, 14*(6), 677–685.

Harakeh, Z., Engels, R. C. M. E., Monshouwer, K., & Hanssen, P. F. (2010). Adolescent's weight concerns and the onset of smoking. *Substance Use & Misuse, 45*(12), 1847–1860.

Harrison, K. (2000). The body electric: thin-ideal media and eating disorders in adolescents. *Journal of Communication, 50*(3), 119–143.

Hawkins, J. D., Oesterle, S., Brown, E. C., Arthur, M. W., Abbott, R. D., Fagan, A. A., et al. (2009). Results of a type 2 translational research trial to prevent adolescent drug use and delinquency. *Archives of Pediatrics and Adolescent Medicine, 163*(9), 789–798.

Heinz, A. J., Kassel, J. D., Berbaum, M., & Mermelstein, R. (2010). Adolescents' expectancies for smoking to regulate affect predict smoking behavior and nicotine dependence over time. *Drug and Alcohol Dependence, 111*(1–2), 128–135.

Hinchliff, G. L. M., Kelly, A. B., Chan, G. C. K., Toumbourou, J. W., Patton, G. C., & Williams, J. (2016). Risky dieting behaviors amongst adolescent girls: associations with family relationship problems and depressed mood. *Eating Behaviors, 22*, 222–224. https://doi.org/10.1016/j.eatbeh.2016.06.001.

Hong, T., Rice, J., & Johnson, C. (2011). Social environmental and individual factors associated with smoking among a panel of adolescent girls. *Women and Health, 51*, 187–203.

Johnson, J. L., Eaton, D. K., Pederson, L. L., & Lowry, R. (2009). Associations of trying to lose weight, weight control behaviors, and currentcigarette use among US high schoolstudents. *Journal of School Health, 79*, 355–360.

Kaufman, A. R., & Augustson, E. M. (2008). Predictors of regular cigarette smoking among adolescent females: does body image matter? *Nicotine & Tobacco Research, 10*(8), 1301–1309.

Kelly, A. B., O'Flaherty, M., Connor, J. P., Homel, R., Toumbourou, J. W., Patton, G. C., et al. (2011). The influence of parents, siblings and peers on pre- and early-teen smoking: a multilevel model. *Drug and Alcohol Review, 30*, 381–387.

Kendzor, D. E., Copeland, A. L., Stewart, T. M., Businelle, M. S., & Williamson, D. A. (2007). Weight-related concerns associated with smoking in young children. *Addictive Behaviors, 32*(3), 598–607.

Koval, J. J., Pederson, L. L., Zhang, X., Mowery, P., & McKenna, M. (2008). Can young adult smoking status be predicted from concern about body weight and self-reported BMI among adolescents? Results from a ten-year cohort study. *Nicotine & Tobacco Research, 10*(9), 1449–1455.

Lange, K., Thamotharan, S., Racine, M., Hirko, C., & Fields, S. (2015). The relationship between weight and smoking in a national sample of adolescents: role of gender. *Journal of Health Psychology, 20*(12), 1558–1567.

Leatherdale, S. T., Wong, S. L., Manske, S. R., & Colditz, G. A. (2008). Susceptibility to smoking and its association with physical activity, BMI, and weight concerns among youth. *Nicotine & Tobacco Research, 10*(3), 499–505.

Mehus, C., Doty, J., Chan, G. C. K., Kelly, A. B., Hemphill, S., Toumbourou, J. W., et al. (2018). Testing the social interaction learning model's applicability to adolescent substance misuse in an Australian context. *Substance Use and Misuse.* https://doi.org/10.1080/10826084.2018.1441307.

Monahan, K. C., Steinberg, L., & Cauffman, E. (2009). Affiliation with antisocial peers, susceptibility to peer influence, and antisocial behavior during the transition to adulthood. *Developmental Psychology, 45*(6), 1520–1530.

Nagaya, T., Yoshida, H., Takahashi, H., & Kawai, M. (2007). Cigarette smoking weakens exercise habits in healthy men. *Nicotine & Tobacco Research*, *9*(10), 1027–1032.

Nemes, S., Wish, E., Wraight, B., & Messina, N. (2002). Correlates of treatment follow-up difficulty. *Substance Use & Misuse*, *37*(1), 19–45.

Neumark-Sztainer, D., Paxton, S. J., Hannan, P. J., Haines, J., & Story, M. (2006). Does body satisfaction matter? Five year longitudinal associations between body satisfaction and health behaviors in adolescent females and males. *Journal of Adolescent Health*, *39*, 244–251.

OECD. (2017). *Obesity update 2017*. Organisation for Economic Co-operation and Development. Retrieved from:(2017). *www.oecd.org*. Accessed 21 March 2018.

Ogden, C. L., Carroll, M. D., Fryar, C. D., & Flegal, K. M. (2015). *Prevalence of obesity among adults and youth: United States, 2011–2014*. .

Pénzes, M., Czeglédi, E., Balázs, P., & Foley, K. L. (2012). Factors associated with tobacco smoking and the belief about weight control effects of smoking among Hungarian adolescents. *Central European Journal of Public Health*, *20*(1), 11–17.

Potter, B. K., Pederson, L. L., Chan, S. S. H., Aubut, J.-A. L., & Koval, J. J. (2004). Does a relationship exist between body weight, concerns about weight, and smoking among adolescents? An integration of the literature with an emphasis on gender. *Nicotine & Tobacco Research*, *6*(3), 397–425. https://doi.org/10.1080/14622200410001696529.

Public Health England. Childhood Obesity: Applying All our health, 2015, Retrieved 27/08/2018 from www.gov.uk/government/publications/childhood-obesity-applying-all-our-health.

Rees, D. I., & Sabia, J. J. (2010). Body weight and smoking initiation: Evidence from Add Health. *Journal of Health Economics*, *29*, 774–777.

Robertson, L., McGee, R., & Hancox, R. J. (2014). Smoking cessation and subsequent weight change. *Nicotine & Tobacco Research*, *16*(6), 867–871.

Rowland, B., Toumbourou, J. W., Osborn, A., Smith, R., Hall, J. K., Kremer, P., et al. (2013). A clustered randomised trial examining the effect of social marketing and community mobilisation on the age of uptake and levels of alcohol consumption by Australian adolescents: study protocol. *British Medical Journal Open*. *3*, e002423https://doi.org/10.1136/bmjopen-2012-002423.

Rowland, B. C., Williams, J., Smith, R., Hall, J. K., Osborn, A., Kremer, P., et al. (2018). Social marketing and community mobilisation to reduce underage alcohol consumption in Australia: a cluster randomised community trial. *Preventive Medicine*. https://doi.org/10.1016/j.ypmed.2018.02.032.

Seo, D., Jiang, N., & Kolbe, L. J. (2009). Association of smoking with body weight in US high school students, 1999-2005. *American Journal of Health Behavior*, *33*(2), 202–212.

Smith, D., Kelly, A. B., Chan, G. C. K., Toumbourou, J. W., Patton, G. C., & Williams, J. (2014). Beyond the primary influences of parents and peers on very young adolescent alcohol use: evidence of independent neighbourhood effects. *Journal of Early Adolescence*, *34*(5), 568–583.

Tanner-Smith, E. E. (2010). Negotiating the early developing body: Pubertal timing, body weight, and adolescent girls' substance use. *Journal of Youth and Adolescence*, *39*, 1402–1416.

Thomas, R., Kelly, A. B., Chan, G. C. K., Hides, L. M., Quinn, C. A., Kavanagh, D. J., et al. (2018). An examination of gender differences in the association of eating and weight loss attitudes. *Substance Use & Misuse 53*, 2125–2131. https://doi.org/10.1080/10826084.2018.1455703.

Wack, J. T., & Rodin, J. (1982). Smoking and its effects on body weight and the systems of caloric regulation. *The American Journal of Clinical Nutrition*, *35*(2), 366–380.

Wahl, S. K., Turner, L. R., Mermelstein, R. J., & Flay, B. R. (2005). Adolescents' smoking expectancies: psychometric properties and prediction of behavior change. *Nicotine & Tobacco Research*, *7*(4), 613–623. https://doi.org/10.1080/14622200500185579.

Wehby, G. L., Murray, J. C., Wilcox, A., & Lie, R. T. (2012). Smoking and body weight: evidence using genetic instruments. *Economics and Human Biology*, *10*(2), 113–126.

Weiss, J. W., Merrill, V., & Gritz, E. R. (2007). Ethnic variation in the association between weight concern and adolescent smoking. *Addictive Behaviors*, *32*(10), 2311–2316. https://doi.org/10.1016/j.addbeh.2007.01.020.

Western, B., Braga, A., Hureau, D., & Sirois, C. (2016). Study retention as bias reduction in a hard-to-reach population. *Proceedings of the National Academy of Sciences*, *113*(20), 5477.

WHO. (2015). *WHO global report on trends in prevalence of tobacco smoking 2015*. Geneva: World Health Organization.

Xie, B., Chou, C.-P., Spruijt-Metz, D., Reynolds, K., Clark, F., Palmer, P. H., et al. (2006). Weight perception and weight-related sociocultural and behavioral factors in Chinese adolescents. *Preventive Medicine*, *42* (3), 229–234. https://doi.org/10.1016/j.ypmed.2005.12.013.

CHAPTER

55

Exercise as a Smoking Cessation Aid

Scott Rollo, Wuyou Sui, Harry Prapavessis

Faculty of Health Sciences, School of Kinesiology, The University of Western Ontario, London, ON, Canada

Abbreviations

CBT cognitive behavioral therapy
NRT nicotine replacement therapy
PA physical activity
RCT randomized controlled trial

55.1 INTRODUCTION

Tobacco smoking is the leading cause of preventable deaths worldwide, with over 6 million deaths each year attributable to tobacco use (WHO, 2017). It has been well established that smoking is detrimental to one's health and an important modifiable risk factor for numerous chronic diseases (USDHHS, 2010). Although it is declining worldwide, the prevalence of tobacco smoking is still over 1.1 billion people globally (WHO, 2017). Smoking cessation at any age is associated with numerous health benefits and diminished risk of developing smoking-related diseases, such as lung cancer, heart attack, stroke, and chronic lung disease (USDHHS, 1990).

Unfortunately, smoking cessation is difficult. Despite the known health consequences, many smokers find it challenging to quit smoking, and failure rates are consistently high. Among adult smokers who quit smoking without formal treatment, only 3%–5% are successfully tobacco-free 1 year following their quit attempt (Hughes, Keely, & Naud, 2004). Low quit rates can be explained by a number of factors including the highly addictive properties of nicotine found within cigarettes (Hirschhorn and World Health Organization, 2009), heightened cravings and withdrawal symptoms (Allen, Bade, Hatsukami, & Center, 2008), weight gain associated with smoking cessation (Klesges et al., 1988), and the learned and reinforcing behavior of smoking (CADTH, 2014).

Numerous pharmacological (e.g., nicotine replacement therapies (NRT), nicotine gum, inhaler, nasal spray, transdermal patch, and lozenge; see Figs. 55.1 and 55.2) and behavioral smoking cessation treatment (e.g., cognitive behavioral therapy (CBT)) options exist to aid in the cessation process, promote cessation maintenance, and prevent smoking relapse (Collins, Witkiewitz, Kirouac, & Marlatt, 2010). While the combination of the abovementioned interventions has shown to improve quit success rates, these rates nevertheless remain modest (i.e., 1 year quit rates range between 8% and 23%; CADTH, 2014; Lancaster & Stead, 2017). Hence, there is a need to explore alternative and/or adjunct treatment options that may help to improve existing smoking cessation treatment and maintenance approaches.

55.2 WHY EXERCISE MIGHT BE AN EFFECTIVE SMOKING CESSATION TREATMENT OPTION

Evidence from a number of large cross-sectional surveys indicates that levels of physical activity (PA) are inversely related to smoking rates (Picavet, Wendelvos, Vreeken, Schuit, & Verschuren, 2011). Additionally, greater levels of PA have been positively associated with initiating a quit attempt (Gauthier, Snelling, & King, 2012), confidence in maintaining smoking abstinence (King, Marcus, Pinto, Emmons, & Abrams, 1996), and successful smoking cessation (Abrantes et al., 2009). Evidences from two metaanalyses have shown that an acute bout of exercise reduces the magnitude of urges to smoke following a temporary period of abstinence compared with control conditions (Haasova et al., 2013; Roberts, Maddison, Simpson, Bullen, & Prapavessis, 2012). Numerous nicotine withdrawal symptoms during the cessation process have also been shown to be ameliorated by exercise (Ussher, Taylor, & Faulkner, 2014). In addition, there is research evidence, which indicates exercise may reduce postcessation weight gain in the long term (Farley, Hajek, Lycett, & Aveyard, 2012). Finally, exercise has been shown to have a positive effect on other factors

Neuroscience of Nicotine
https://doi.org/10.1016/B978-0-12-813035-3.00055-1

FIG. 55.1 Nicotine gum. An illustration of a common form of nicotine replacement therapy (NRT). *Figure unpublished.*

that may protect against smoking relapse, including depression (Williams, 2008), general fatigue and sleep disturbances (Hatsukami, Hughes, Pickens, & Suilis, 1984), self-esteem (Spence, McGannon, & Poon, 2005), and perceived coping ability (Tritter, Fitzgeorge, & Prapavessis, 2015). Taken together, there is evidence supporting the benefits of exercise as a component of tobacco dependence treatments. It should be highlighted that being physically active is associated with numerous health benefits. Hence, it is likely that targeting exercise and smoking cessation in combination may maximize health benefits in smokers.

55.3 EXERCISE AS A SMOKING CESSATION AID

A number of studies have examined the influence of exercise, either alone or in combination with traditional smoking cessation approaches (i.e., behavioral therapy and/or pharmacotherapy) on smoking cessation outcomes. When commenting on the trials to date, continuous abstinence will be reported as the primary outcome measure over 7-day point prevalence abstinence when both are reported. This can be justified as continuous abstinence is a more stringent measure of smoking cessation and relates more directly to health outcomes (Marcus et al., 1999).

Cochrane review. Ussher et al. (2014) conducted a systematic review to determine whether exercise-based interventions alone or combined with a smoking cessation program are more effective than a smoking cessation intervention alone. In this review, 20 randomized controlled trials (RCTs; $n = 5870$) examining exercise-aided interventions for smoking cessation with at least a 6-month follow-up period were identified. Only 4 of the 20 trials showed significantly higher abstinence rates in the exercise-aided group versus a control group at the end of treatment (Bock et al., 2012; Marcus et al., 1999; Marcus, Albrecht, Niaura, Abrams, & Thompson, 1991; Martin et al., 1997), while only one study showed a borderline significant benefit of exercise at 1-year follow-up (Marcus et al., 1999—see Fig. 55.3). In addition, only four

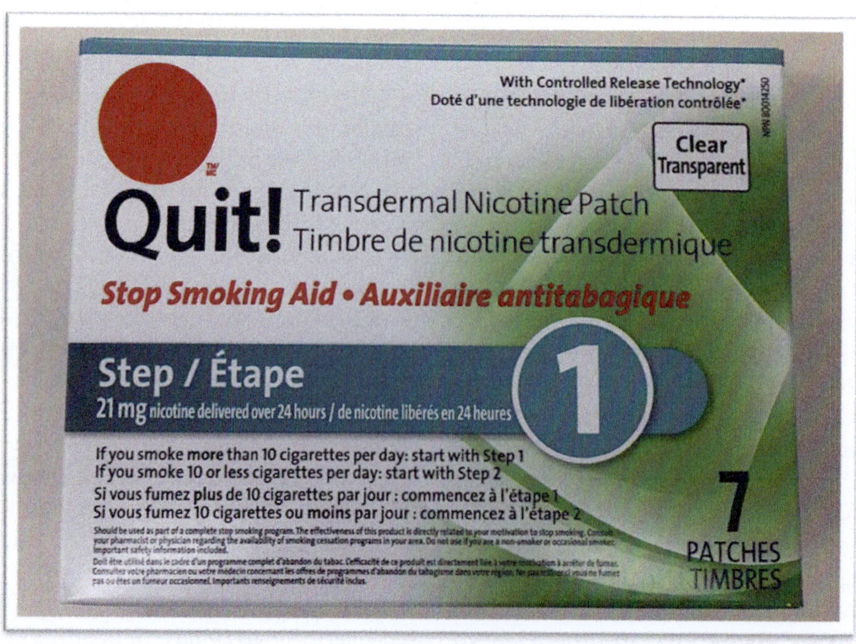

FIG. 55.2 A transdermal nicotine patch.
An illustration of a common form of NRT. *Figure unpublished.*

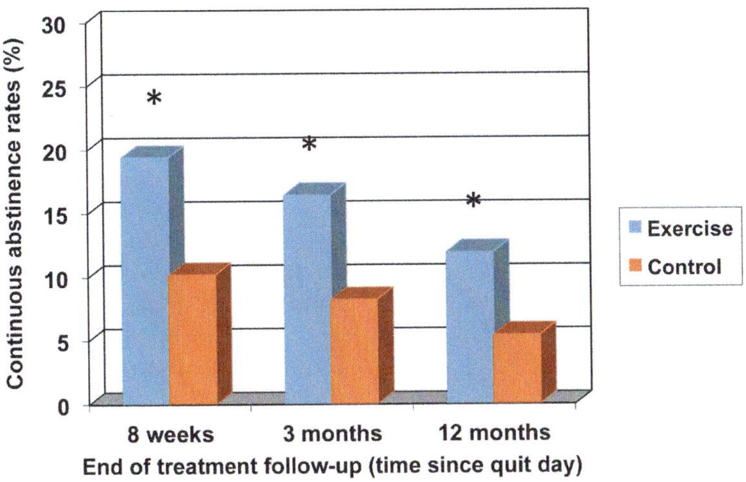

FIG. 55.3 Continuous abstinence rates by treatment condition at the end of treatment and 3 and 12 months following quit day.
Bars denote abstinence rates (%) for the exercise-aided group versus control group at postquit day follow-ups. *Asterisk* indicates significance ($P < .05$). *Adapted from data by Marcus et al. (1999); figure unpublished.*

studies examined the effectiveness of combining exercise with NRT (Hill, Rigdon, & Johnson, 1993; Kinnunen et al., 2008; Martin et al., 1997; Prapavessis et al., 2007). Of these, one study showed significantly higher abstinence rates in the exercise-plus-patch group compared to the exercise-only group at the end of treatment and at 12-month follow-up (Prapavessis et al., 2007). A number of explanations can be offered to explain the lack of findings among the majority of studies. First, only one study was found to be at low risk of bias across all domains, and only seven studies had a sufficiently large sample size to detect a significant difference between treatment and control conditions. Second, studies were found to vary in the timing and intensity of the smoking cessation and exercise programs offered. Third, cessation outcomes varied across studies as continuous abstinence was assessed in eight, prolonged abstinence in two, and point prevalence abstinence in eight; two did not specify. Fourth, many trials included interventions, which may not have been of sufficient intensity to produce the required changes in exercise levels. In conclusion, this review provided evidence warranting further investigation into the role of exercise as a smoking cessation aid. In the following pages, the reader is provided with detailed insight into the "state of affairs" of the exercise-aided smoking cessation literature that has been published since the Ussher et al. (2014) Cochrane review.

Intensive exercise-aided treatment. Abrantes et al. (2014) argued that a number of methodological limitations (e.g., insufficient intervention intensity, lack of efforts to increase adherence, and lack of contact control conditions) inherent in previous exercise intervention trials might have served to diminish its effectiveness as a smoking cessation aid. To address these limitations, Abrantes and colleagues conducted a 12-week RCT to examine the effectiveness of a behavioral exercise intervention for smokers ($n = 61$). Participants in the aerobic exercise (AE) condition received supervised exercise sessions (once per week; see Fig. 55.4), prescriptions to engage in home-based moderate-intensity exercise (2–4 times per week), weekly, cognitive behavioral group counseling sessions for exercise promotion, and a financial incentive component based on treatment compliance. Participants in the health education contact control (HEC) were asked to attend weekly hour-long health education sessions. Independent of condition, all participants received an 8-week (beginning week 5) smoking

FIG. 55.4 Supervised exercise session for smoking cessation.
This figure illustrates an example of an on-site structured aerobic exercise session during an exercise-aided cessation program. *Figure unpublished.*

cessation protocol including 20min weekly telephone counseling sessions and NRT treatment (i.e., transdermal patch). Assessments were administered at baseline, 3- (end of treatment), 6-, and 12-month follow-ups. Treatment adherence in both groups was fairly high (i.e., 75%). Participants in the AE condition demonstrated higher continuous abstinence rates (EOT: 30% vs 25.8%, OR 1.23; 6-month follow-up: 23.3% vs 9.7%, OR 2.83; 12-month follow-up: 13.3% vs 3.2%, OR 4.64) compared to those in the HEC condition; however, these differences did not reach statistical significance (see Fig. 55.5). It was also found that participants in the AE condition reported significantly higher levels of PA following quit day ($b=1.37$, $SE=0.43$, $P<.01$). This intervention included a number of notable features to promote exercise adherence including supervised moderate-intensity aerobic exercise sessions, a sequential approach with exercise initiation preceding the quit date by 1 month, combined supervised plus home-based exercise, a contingency-based financial incentive component, and cognitive behavioral exercise counseling. In conclusion, this preliminary trial provided promising evidence that a behavioral exercise intervention may be a useful adjunct to improve cessation outcomes over standard care alone.

Special populations. Exercise as a smoking cessation aid has also been investigated in particular at-risk populations. For example, Ussher et al. (2015) implemented a large-scale, multicenter RCT to determine the effectiveness of a PA intervention as a smoking cessation aid for pregnant smokers ($n=789$). Participants were randomized to receive 6 weekly sessions of individual behavioral cessation support alone (control) or behavioral cessation support plus a PA intervention, which included PA consultations and 14 sessions of supervised moderate-intensity exercise over 8weeks (2 times per week for weeks 1–6 and once per week for weeks 6–8). Interventions began 1week before the target quit date.

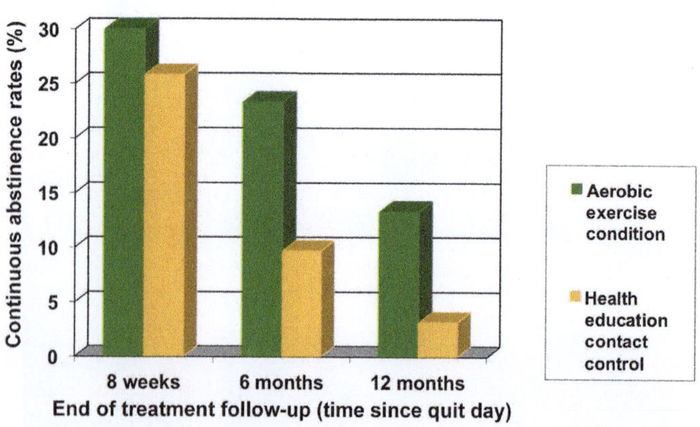

FIG. 55.5 Continuous abstinence rates by treatment condition at the end of treatment and 6 and 12months following quit day.
Bars indicate abstinence rates (%) for the aerobic exercise condition versus health education contact control condition. *Adapted from data by Abrantes et al. (2014); figure unpublished.*

Biochemically verified continuous abstinence was assessed between the quit date and end of pregnancy. Findings indicated there were no significant differences in rates of smoking abstinence between groups at the end of pregnancy. Continuous abstinence rates were 8% and 6% in the PA and control groups, respectively (OR 1.21, 95% CI: 0.70–2.10). However, compared to the control group, increases in self-reported minutes of weekly PA were significantly greater for the PA group by 33% (95% CI: 14%–56%), 28% (95% CI: 7%–52%), and 36% (95% CI: 12%–65%) at 1, 4, and 6weeks postquit day, respectively. These findings indicate that a PA intervention, in conjunction with behavioral cessation support, does not significantly increase cessation rates in pregnant smokers compared to behavioral support alone. Despite low attendance in the PA group (only 29% adherence), the PA consultation component may have still served to increase PA levels. The authors suggested that perhaps being asked to change two health behaviors simultaneously (i.e., smoking and PA) while also dealing with pregnancy was too difficult a task. Future interventions for smoking cessation during pregnancy should initiate exercise well before the anticipated quit date in order to ease the load on those attempting to quit and maximize the potential for exercise to assist smoking cessation.

Economically disadvantaged smokers who are not intending to stop smoking may benefit from treatments aimed at reducing their smoking. Using a pilot RCT, Thompson et al. (2016) assessed the effects of an exercise-assisted reduction then stop-smoking intervention on smoking and PA outcomes at 16weeks compared with usual care. Disadvantaged smokers ($n=99$) who wished to reduce their smoking but not quit were randomized to receive usual care or usual care plus the exercise-assisted reduction then stop-smoking intervention. The intervention consisted of up to 12 weekly client-centered individual motivational support sessions to promote smoking reduction and increased PA. Compared with controls, participants in the intervention group were more likely to make a quit attempt (35.5% vs 9.7%, respectively; OR 5.05, 95% CI: 1.10–23.15), and a greater proportion achieved at least a 50% reduction in cigarettes smoked at 16weeks (63.3% vs 32.3%, respectively; OR 4.21, 95% CI: 1.32–13.39). Furthermore, promising differences between groups were found for postquit abstinence at 4-week follow-up (23% vs 6%; OR 4.91, 95% CI: 0.80–30.24). Unfortunately, no intervention effect on PA was found, and the percentage of support sessions attended (52.5%) could have been greater among intervention smokers. This study was one of the first to examine the effectiveness of a behavioral intervention focusing on PA to promote smoking reduction among smokers who initially did not want to quit. The findings suggest that an exercise-assisted smoking reduction intervention with behavioral counseling appears to be more effective than usual care alone in achieving reduction.

In a recent study, Smits et al. (2016) conducted an RCT to examine the effectiveness of vigorous-intensity exercise as an aid to smoking cessation among sedentary adult daily smokers with high anxiety sensitivity. All participants ($n=136$) received 15 weeks of standard smoking cessation treatment (ST; i.e., 7 weekly cognitive behavioral therapy sessions plus NRT patches starting at week 6) and were asked to make a quit attempt at week 6. In addition, participants were randomly assigned to either a 15-week supervised exercise program (ST+EX) or 15-week wellness education contact control condition (ST+CTRL), each consisting of three 35 min sessions per week. Assessments were administered at 10 (end of treatment), 16 (i.e., 4 months), and 24 weeks (i.e., 6 months) after the target quit date. Results showed adherence to be mediocre with participants attending an average of 24 of the 45 (i.e., 53%) total exercise or wellness education treatment sessions. With regard to smoking abstinence, findings indicated that continuous abstinence rates were significantly higher for those in the ST+EX condition than for those in the ST+CTRL condition at each assessment period among those with high anxiety sensitivity (continuous abstinence: $b=-0.98$, $SE=0.346$, $t(132)=-2.84$, and $P=.005$), but not among those with low anxiety sensitivity. Among those with high anxiety sensitivity, estimated continuous abstinence rates at the end of treatment, 4 months postquit day, and 6 months postquit day were 25.9% versus 11.6%, 24.8% versus 11.0%, and 23.3% versus 10.2% for the ST+EX versus ST+CTRL conditions, respectively. Hence, among adult smokers with high levels of anxiety sensitivity, combining exercise with standard care may increase the odds of quit success compared to standard care alone.

Exercise and cessation maintenance. Prapavessis et al. (2016) conducted the "Getting Physical on Cigarettes" RCT to examine the effectiveness of an exercise-aided NRT smoking cessation intervention program, with built-in exercise and smoking cessation maintenance components, on postintervention 14-, 26-, and 56-week continuous smoking abstinence. Acknowledging the importance of adherence to the success of exercise-aided cessation programs, Prapavessis and colleagues also sought to determine whether adherence throughout the program had an influence on smoking status. Female cigarette smokers ($n=413$) who wished to quit smoking were recruited to participate in a 14-week supervised exercise and NRT (i.e., transdermal patch) smoking cessation program and were subsequently randomized to one of four conditions: exercise maintenance plus smoking cessation maintenance, exercise maintenance plus contact control, smoking cessation maintenance plus contact control, or contact control. A targeted quit date was set for week 4. Those in the exercise maintenance groups received five group-based CBT sessions to promote exercise adherence during weeks 8–14 and telephone counseling to maintain exercise behavior during weeks 26 and 52. Those in the smoking cessation plus contact control and contact control arms received five health education sessions during weeks 8–14 and telephone counseling regarding women's health issues during weeks 26 and 52. At week 14, the smoking cessation maintenance groups received informational booklets targeting cessation and relapse prevention. Adherence rates for the supervised exercise sessions (66%), NRT protocol (68.81%), and CBT exercise maintenance component (54%) were substantially higher than adherence to the postintervention telephone exercise maintenance component (38%). It was found that abstainers had higher treatment adherence rates and received more maintenance than their smoking counterparts. Differences in continuous smoking abstinence rates between the exercise and equal contact nonexercise maintenance groups were found at weeks 14 (57% vs 43%), 26 (27% vs 21%), and 56 (26% vs 23.5%) (see Fig. 55.6).

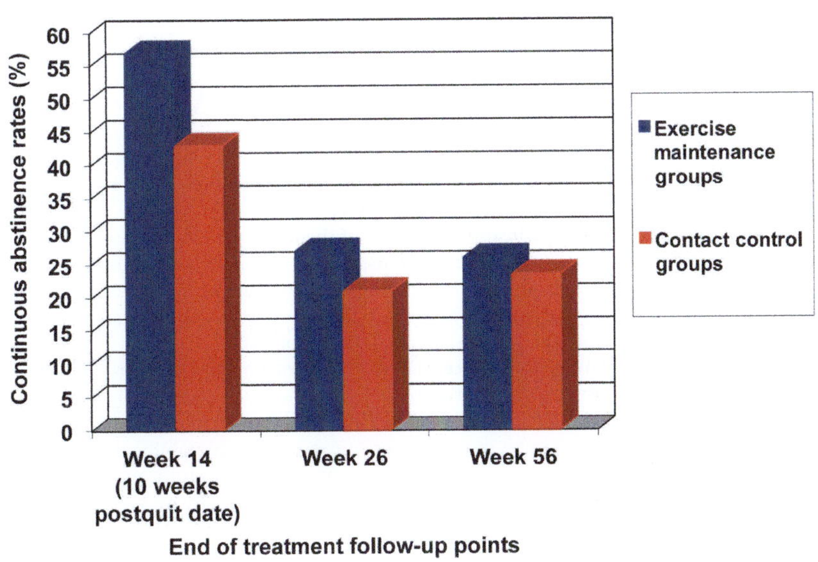

FIG. 55.6 Continuous abstinence rates by treatment condition at the end of treatment and 26- and 56-week follow-ups.
Bars indicate abstinence rates (%) for the exercise maintenance groups versus equal contact nonexercise maintenance groups. *Adapted from data by Prapavessis et al. (2016); figure unpublished.*

Despite these clinically meaningful differences, only the week-14 differences in smoking status approached significance (χ^2 [1, n =409]=2.36, and P =.08). These findings suggest that other methods of delivering exercise maintenance following termination of an exercise-aided NRT smoking cessation program need to be considered.

55.4 FUTURE DIRECTIONS

A number of future recommendations can be made concerning the role of exercise as a smoking cessation aid. Trials with larger sample sizes, structured on-site exercise interventions with at least moderate exercise intensity, equal contact control conditions, biochemically verified continuous smoking abstinence as the primary outcome, and follow-up periods of at least 6 months are needed to ascertain the effects of exercise for smoking cessation. With regard to intervention design, it has been argued that being asked to change two health behaviors simultaneously may be too difficult to achieve (e.g., Prapavessis et al., 2007, 2016; Ussher et al., 2015). Hence, it is recommended that the exercise program begin at a time prior to the smoker initiating a quit attempt, thereby allowing individuals to focus on becoming more active before being asked to significantly change their smoking behavior (Ussher et al., 2014). Based on the findings reported herein, it is also recommended that future trials include pharmacological or behavioral therapies as the foundation in smoking cessation treatment. Evidence suggests that a combined approach incorporating exercise with NRT or behavioral counseling may improve abstinence rates and increase the likelihood of a successful quit attempt over either treatment option alone.

A key focus for future research is to find novel and innovative approaches to maximize adherence to these cessation programs. One possible solution to improve adherence rates to exercise-aided smoking cessation programs that was addressed in the study by Prapavessis et al. (2016) is to focus on the maintenance stage of both exercise and smoking cessation. All smoking cessation programs (including exercise-aided ones) have shown early promise followed by relapse effects at the end of treatment and follow-up. Hence, it is vital that these types of programs integrate both initial cessation and exercise maintenance components into their design to prevent relapse in both behaviors. To date, programs that have weaned smokers off supervised exercise sessions have not succeeded in promoting independent exercise among participants following program termination. Future work is required to investigate more effective ways to sustain both long-term smoking abstinence and exercise. It is possible that face-to-face technology and applications with real-time feedback may provide viable options to improve exercise maintenance and reduce smoking relapse (Free et al., 2011). In order to accurately assess treatment adherence, it is recommended that future research include objective measures of exercise adherence over treatment and follow-up. Further research is also needed to test different forms of exercise prescription and to determine the most effective timing, intensity, and modalities of exercise. The design and implementation of exercise-aided cessation programs in real-world settings are also warranted to allow for future dissemination of this approach as a cessation aid. Finally, additional research is needed to investigate the role of exercise for smoking cessation among more ethnically diverse populations, at-risk populations (e.g., cardiovascular disease patients), and different age groups (e.g., youth smokers).

55.5 CONCLUSION

Exercise and smoking cessation are both independently beneficial for health. Thus, combining exercise with existing smoking cessation treatment options could prove to be the best approach to promote and sustain smoking abstinence. Based on the findings to date, there is some research evidence to support the effectiveness of exercise as an adjunct to behavioral counseling and pharmacotherapy (i.e., NRT) treatment options in smoking cessation. Evidence suggests that exercise has the potential to improve rates of continuous abstinence over and above traditional smoking cessation treatments alone; however, exercise adherence is a consistent problem, particularly after program termination. This is unfortunate as there is a clear association between exercise adherence and cessation outcomes. Future trials need to explore ways to improve adherence to exercise treatment and maintain exercise following the termination of exercise-aided cessation programs in order to generate the greatest effects on smoking cessation and health-related outcomes.

MINI-DICTIONARY OF TERMS

Aerobic exercise A category of exercise that primarily challenges the cardiorespiratory system. It includes activities such as walking, cycling, running, and swimming.

Behavioral counseling A type of therapy that includes motivational support, reinforcement of nonsmoking behavior, relaxation training, coping skills training, contingency management, self-control, and cognitive behavioral therapy. Behavioral therapy approaches vary in both their intensity and delivery format.

Continuous abstinence Refers to having smoked no more than five cigarettes since the quit date.

Exercise A subcategory of physical activity; planned, structured, and repetitive activity focused on improving or maintaining measures of physical fitness and/or mental health. Purposeful exercise is typically performed at a moderate-vigorous intensity and accumulated in bouts >10 min.

Moderate-intensity exercise Exercise that requires a moderate amount of effort and noticeably accelerates the heart rate (i.e., 50%–70% of maximal heart rate). Examples include brisk walking, dancing, and bicycling.

Nicotine replacement therapy *NRT*; for example, gum, transdermal patch, inhaler, nasal spray, and lozenge substitute for tobacco-derived nicotine in the blood and reduce nicotine-related craving and withdrawal symptoms in abstaining smokers.

Physical activity Any body movement produced by skeletal muscles that results in energy expenditure above resting level.

Quit attempt A period of time in which a smoker abstains from smoking cigarettes; defined as a period of at least 24h of abstinence.

Seven-day point prevalence abstinence Defined as not a single puff of a cigarette in the past 7 days.

Smoking cessation The process of discontinuing cigarette smoking; typically refers to the point at which a person attains smoking abstinence.

Smoking relapse The point that the smoker resumes smoking cigarettes following a quit episode.

Vigorous-intensity exercise Exercise that requires a large amount of effort and causes rapid breathing and substantial increase in heart rate (i.e., 70%–85% of maximal heart rate). Examples include running and fast swimming.

Key Facts of Exercise as a Smoking Cessation Aid

- An acute bout of exercise reduces nicotine-related craving and withdrawal symptoms during a quit attempt.
- Exercise may reduce postcessation weight gain.
- Exercise is associated with numerous health benefits.
- In industrial countries, only 15% of the adult population engages in regular physical activity.
- Early physical activity patterns in life are strongly predictive of physical activity patterns later in life.

Summary Points

- This chapter focuses on the effectiveness of exercise, either alone or in combination with traditional smoking cessation treatments, as a smoking cessation aid.
- Despite the availability of existing treatment options, only 8%–23% of individuals remain successfully tobacco-free 1 year following their quit attempt.
- Recently, exercise has been proposed as an aid for smoking cessation because of its demonstrated potential to reduce nicotine-related cravings and withdrawal symptoms.
- There is some evidence suggesting that exercise-aided cessation programs have the potential to improve continuous abstinence rates, compared to traditional smoking cessation treatments alone; however, exercise adherence is a consistent problem, particularly after program termination.
- Novel and innovative approaches to improve adherence to exercise treatment and maintain exercise following the termination of an exercise-aided cessation program are needed.

References

Abrantes, A. M., Bloom, E. L., Strong, D. R., Riebe, D., Marcus, B. H., Desaulniers, J., et al. (2014). A preliminary randomized controlled trial of a behavioral exercise intervention for smoking cessation. *Nicotine & Tobacco Research*, *16*(8), 1094–1103.

Abrantes, A. M., Strong, D. R., Lloyd-Richardson, E. E., Niaura, R., Kahler, C. W., & Brown, R. A. (2009). Regular exercise as a protective factor in relapse following smoking cessation treatment. *American Journal on Addictions*, *18*(1), 100–101.

Allen, S. S., Bade, T., Hatsukami, D., & Center, B. (2008). Craving, withdrawal, and smoking urges on days immediately prior to smoking relapse. *Nicotine & Tobacco Research*, *10*(1), 35–45.

Bock, B. C., Fava, J. L., Gaskins, R., Morrow, K. M., Williams, D. M., Jennings, E., et al. (2012). Yoga as a complementary treatment for smoking cessation in women. *Journal of Women's Health*, *21*(2), 240–248.

Canadian Agency for Drugs and Technologies in Health. (2014). *Nicotine replacement therapy for smoking cessation or reduction: a review of the clinical evidence*. Ottawa, ON: CADTH.

Collins, S. E., Witkiewitz, K., Kirouac, M., & Marlatt, G. A. (2010). Preventing relapse following smoking cessation. *Current Cardiovascular Risk Reports*, *4*(6), 421–428.

Farley, A. C., Hajek, P., Lycett, D., & Aveyard, P. (2012). Interventions for preventing weight gain after smoking cessation. *Cochrane Database of Systematic Reviews*, *1*.

Free, C., Knight, R., Robertson, S., Whittaker, R., Edwards, P., Zhou, W., et al. (2011). Smoking cessation support delivered via mobile phone text messaging (txt2stop): a single-blind, randomised trial. *Lancet*, *378*(9785), 49–55.

Gauthier, A. P., Snelling, S. J., & King, M. (2012). "Thinking outside the pack": examining physically active smokers and implications for practice among Ontario residents. *Health Promotion Practice*, *13*(3), 395–403.

Haasova, M., Warren, F. C., Ussher, M., Janse Van Rensburg, K., Faulkner, G., Cropley, M., et al. (2013). The acute effects of physical activity on cigarette cravings: systematic review and meta-analysis with individual participant data. *Addiction*, *108*(1), 26–37.

Hatsukami, D. K., Hughes, J. R., Pickens, R. W., & Suilis, D. (1984). Tobacco withdrawal symptoms: an experimental analysis. *Psychopharmacology*, *84*, 231–236.

Hill, R. D., Rigdon, M., & Johnson, S. (1993). Behavioral smoking cessation treatment for older chronic smokers. *Behavior Therapy*, *24*(2), 321–329.

Hirschhorn, N., & World Health Organization. (2009). *Evolution of the tobacco industry positions on addiction to nicotine: A report prepared for the Tobacco Free Initiative*. World Health Organization. WHO Tobacco Control Papers (NLM No. HD 9149).

Hughes, J. R., Keely, J., & Naud, S. (2004). Shape of the relapse curve and long-term abstinence among untreated smokers. *Addiction*, *99*, 29–38.

King, T. K., Marcus, B. H., Pinto, B. M., Emmons, K. M., & Abrams, D. B. (1996). Cognitive–behavioral mediators of changing multiple behaviors: smoking and a sedentary lifestyle. *Preventive Medicine*, *25*(6), 684–691.

Kinnunen, T., Leeman, R. F., Korhonen, T., Quiles, Z. N., Terwal, D. M., Garvey, A. J., et al. (2008). Exercise as an adjunct to nicotine gum in treating tobacco dependence among women. *Nicotine & Tobacco Research*, *10*(4), 689–703.

Klesges, R. C., Brown, K., Pascale, R. W., Murphy, M., Williams, E., & Cigrang, J. A. (1988). Factors associated with participation, attrition, and outcome in a smoking cessation program at the workplace. *Health Psychology*, *7*(6), 575–589.

Lancaster, T., & Stead, L. F. (2017). Individual behavioural counselling for smoking cessation. *Cochrane Database of Systematic Reviews*, *3*.

Marcus, B. H., Albrecht, A. E., King, T. K., Parisi, A. F., Pinto, B. M., Roberts, M., et al. (1999). The efficacy of exercise as an aid for smoking cessation in women: a randomized controlled trial. *Archives of Internal Medicine*, *159*(11), 1229–1234.

Marcus, B. H., Albrecht, A. E., Niaura, R. S., Abrams, D. B., & Thompson, P. D. (1991). Usefulness of physical exercise for maintaining smoking cessation in women. *The American Journal of Cardiology*, *68*(4), 406–407.

Martin, J. E., Calfas, K. J., Patten, C. A., Polarek, M., Hofstetter, C. R., Noto, J., et al. (1997). Prospective evaluation of three smoking interventions in 205 recovering alcoholics: one-year results of Project SCRAP-Tobacco. *Journal of Consulting and Clinical Psychology*, *65*(1), 190–194.

Picavet, H. S. J., Wendel-vos, G. W., Vreeken, H. L., Schuit, A. J., & Verschuren, W. M. M. (2011). How stable are physical activity habits among adults? The Doetinchem Cohort Study. *Medicine & Science in Sports & Exercise*, *43*(1), 74–79.

Prapavessis, H., Cameron, L., Baldi, J. C., Robinson, S., Borrie, K., Harper, T., et al. (2007). The effects of exercise and nicotine replacement therapy on smoking rates in women. *Addictive Behaviors*, *32*(7), 1416–1432.

Prapavessis, H., De Jesus, S., Fitzgeorge, L., Faulkner, G., Maddison, R., & Batten, S. (2016). Exercise to enhance smoking cessation: the getting physical on cigarette randomized control trial. *Annals of Behavioral Medicine*, *50*(3), 358–369.

Roberts, V., Maddison, R., Simpson, C., Bullen, C., & Prapavessis, H. (2012).The acute effects of exercise on cigarette cravings, withdrawal symptoms, affect, and smoking behaviour: systematic review update and meta-analysis. *Psychopharmacology*, *222*(1), 1–15.

Smits, J. A., Zvolensky, M. J., Davis, M. L., Rosenfield, D., Marcus, B. H., Church, T. S., et al. (2016). The efficacy of vigorous-intensity exercise as an aid to smoking cessation in adults with high anxiety sensitivity: a randomized controlled trial. *Psychosomatic Medicine*, *78*(3), 354–364.

Spence, J. C., McGannon, K. R., & Poon, P. (2005). The effect of exercise on global self-esteem: a quantitative review. *Journal of Sport and Exercise Psychology*, *27*(3), 311–334.

Thompson, T. P., Greaves, C. J., Ayres, R., Aveyard, P., Warren, F. C., Byng, R., et al. (2016). An exploratory analysis of the smoking and physical activity outcomes from a pilot randomized controlled trial of an exercise assisted reduction to stop smoking intervention in disadvantaged groups. *Nicotine & Tobacco Research*, *18*(3), 289–297.

Tritter, A., Fitzgeorge, L., & Prapavessis, H. (2015). The effect of acute exercise on cigarette cravings while using a nicotine lozenge. *Psychopharmacology*, *232*(14), 2531–2539.

U.S. Department of Health & Human Services. (1990). *The health benefits of smoking cessation: A report of the surgeon general.* Atlanta, GA: U.S. Department of Health and Human Services, Public Health Service, Centers for Disease Control, Center for Chronic Disease Prevention and Health Promotion, Office on Smoking and Health.

U.S. Department of Health & Human Services. (2010). *How tobacco smoke causes disease: The biology and behavioral basis for smoking-attributable disease: A report of the surgeon general.* Atlanta, GA: U.S. Department of Health & Human Services, Centers for Disease Control and Prevention, National Center for Chronic Disease Prevention and Health Promotion, Office on Smoking and Health.

Ussher, M., Lewis, S., Aveyard, P., Manyonda, I., West, R., Lewis, B., et al. (2015). Physical activity for smoking cessation in pregnancy: randomized controlled trial. *BMJ*, *350*, h2145.

Ussher, M. H., Taylor, A. H., & Faulkner, G. E. (2014). Exercise interventions for smoking cessation. *The Cochrane Library*, *8*.

WHO. (2017). *WHO report on the global tobacco epidemic, 2017: Monitoring tobacco use and prevention policies.* Geneva: World Health Organization. Licence: CC BY-NC-SA 3.0 IGO.

Williams, D. M. (2008). Increasing fitness is associated with fewer depressive symptoms during successful smoking abstinence among women. *International Journal of Fitness*, *4*(1), 39–44.

CHAPTER

56

Varenicline: Treating Smoking Addiction and Schizophrenia

Do-Un Jung, Sung-Jin Kim

Department of Psychiatry, Busan Paik Hospital, Inje University College of Medicine, Busan, Republic of Korea

Abbreviations

CBT	cognitive behavioral therapy
CDSS	Calgary Depression Scale for Schizophrenia
EAGLES study	Evaluating Adverse Events in a Global Smoking Cessation study
HIV	human immunodeficiency virus
mCEQ	Modified Cigarette Evaluation Questionnaire
nAChR	nicotinic acetylcholine receptor
NRT	nicotine replacement therapy

56.1 CHRONIC MENTAL ILLNESS: SCHIZOPHRENIA

Schizophrenia is a serious chronic mental illness with a lifetime morbidity rate of approximately 1% in the general population (McGrath, Saha, Chant, & Welham, 2008). Clinical symptoms of schizophrenia can be classified as positive, negative, or cognitive. Positive symptoms, including hallucination, delusion, disorganized speech, and disorganized behavior, are the main reason why patients seek treatment in the acute phase of the disease. Negative symptoms, including affective flattening, alogia, and avolition, cause many patients to experience difficulty in returning to society (Crow, 1985). Cognitive symptoms were once classified as either positive or negative symptoms but recently have been separated into an independent symptom domain (Green & Harvey, 2014). Although various treatment approaches have been attempted to improve cognitive symptoms, these treatments are less effective than those for positive and negative symptoms (Goff, Hill, & Barch, 2011). As presented here, diverse symptoms manifest themselves in schizophrenia, which can affect not only everyday functioning but also social and occupational functioning of the patient (Dickinson, Bellack, & Gold, 2007). Even after treatment, a considerable number of patients with schizophrenia experience residual symptoms and report a low quality of life (Huber, Gross, Schuttler, & Linz, 1980).

56.2 SUBSTANCE RELATED DISORDERS IN THE SCHIZOPHRENIC POPULATION

Schizophrenic patients have a high comorbidity of substance-related disorders. It is thought that schizophrenic patients show higher levels of substance abuse because of developmental alterations in the hippocampus and frontal cortex. These alterations may increase dopamine release and thus facilitate the positive reinforcing effects of the drug reward pathway, which in turn would disinhibit drug-seeking behavior (Chambers, Krystal, & Self, 2001). Approximately 40%–50% of schizophrenic patients experience lifetime substance use disorder, which can negatively influence the treatment and prognosis of schizophrenia (Blanchard, Brown, Horan, & Sherwood, 2000). Indeed, comorbid substance use disorder is associated with lowered treatment adherence and an increase in psychotic symptoms, violent behavior, homelessness, and overall treatment cost (Dixon, 1999). Additionally, substance use disorder can cause other medical problems and negatively affect the treatment of such problems in schizophrenic patients (Dickey, Azeni, Weiss, & Sederer, 2000).

56.3 PREVALENCE OF SMOKING ADDICTION IN SCHIZOPHRENIA

Tobacco is an addictive drug that can lead to substance use disorder and can cause withdrawal symptoms upon cessation of use. Recently, the prevalence of daily tobacco smoking has been declining due to various efforts, such

Neuroscience of Nicotine
https://doi.org/10.1016/B978-0-12-813035-3.00056-3

as antismoking policies by many national governments, but the number of smokers is still gradually increasing as the human population increases (Ng et al., 2014). Smoking is a well-known major global health problem that has been reported to be correlated with many other medical illnesses, including cardiovascular disease and cancer (Ambrose & Barua, 2004; Gandini et al., 2008). Lasser et al. (2000) reported that people suffering from mental illness were approximately twice as likely to smoke compared to the general population. Further, a meta-analysis conducted by de Leon and Diaz (2005) found that the smoking prevalence for schizophrenic patients in particular was approximately 5.9 times (male odds ratio = 7.2 and female odds ratio = 3.3) higher than the general population. Several studies have reported a prevalence ranging from 50% to 90% for patients with schizophrenia. The prevalence of heavy smoking and high nicotine dependence has been reported to be higher among schizophrenic patients than the general population, and the cessation rate has been reported to be lower. Smoking is a preventable cause of death, and the rates of mortality due to medical illness and smoking prevalence are higher in schizophrenic patients than in healthy individuals. Thus, there seems to be much to be gained individually, societally, and economically through smoking cessation in the schizophrenic population (Saha, Chant, & McGrath, 2007).

56.4 HYPOTHESES OF SMOKING ADDICTION AND SCHIZOPHRENIA

A few hypotheses are available that attempt to explain the high smoking addiction rate in the schizophrenic population. First, smoking increases the metabolism of several antipsychotic drugs, and patients may therefore smoke to decrease the side effects of these medications (Desai, Seabolt, & Jann, 2001). Goff, Henderson, and Amico (1992) evaluated the association between medication side effects and smoking in 78 patients with schizophrenia and found that schizophrenic smokers showed a lower level of parkinsonism and took a higher volume of antipsychotic medication than did nonsmokers. Further, Barnes et al. (2006) reported an association between smoking and a low level of akathisia.

A second hypothesis is that schizophrenic patients use nicotine, a psychoactive substance, to self-medicate their symptoms (Potvin, Stip, & Roy, 2003). Nicotine increases mesolimbicocortical dopaminergic activity in the nucleus accumbens and prefrontal cortex, and this increased activity seems to reduce the symptoms of schizophrenia (Lyon, 1999). Sacco and colleagues (Sacco et al., 2005) reported that smokers with schizophrenia showed decreased scores on Visuospatial Working Memory Tests and on Continuous Performance Tests after abstinence. Moreover, transdermal nicotine increased attentional performance and improved impulsive response inhibition in nonsmokers with schizophrenia (Barr et al., 2008). Another study also showed that negative symptoms significantly improved in schizophrenic nonsmokers when they were treated with a nicotinic agonist (Freedman et al., 2008).

A third hypothesis of smoking addiction and schizophrenia is that smoking in schizophrenic patients can be explained by psychosocial factors. Due to the severity of their symptoms, a considerable number of schizophrenic patients experience decreased social and occupational functioning, avoid social relationships, and tend to be generally inactive. In such situation, smoking can become a tool to relieve boredom (Hughes, Hatsukami, Mitchell, & Dahlgren, 1986). In addition, smoking can decrease anxiety, depression, and stress caused by schizophrenia, which can be independent of the psychopathological symptoms of schizophrenia. Thus, an improvement in social relationships can ultimately be expected through smoking (Lyon, 1999). In support of this hypothesis, Smith, Infante, Ali, Nigam, and Kotsaftis (2001) reported that smoking ameliorates negative symptoms in schizophrenic patients regardless of the amount of nicotine contained in the cigarettes. Although cigarettes may exert a self-medicating effect, these results highlight the possible significance of the psychosocial aspects of smoking.

56.5 VARENICLINE, AN ATTRACTIVE TREATMENT OPTION FOR SMOKING ADDICTION IN SCHIZOPHRENIA

Treatments for smoking addiction can be categorized into psychosocial and pharmacological therapies, which can be used alone or in combination depending on the patient's situation. Psychosocial therapy involves diverse approaches, from simple advice or basic education to motivational enhancement therapy, cognitive behavioral therapy (CBT), mindfulness-based therapy, or other methods (Brewer, Elwafi, & Davis, 2013; Grimshaw & Stanton, 2006; Lancaster & Stead, 2004). Pharmacological therapy mainly involves treatment with nicotine replacement therapy (NRT), bupropion, or varenicline (Cahill, Stevens, Perera, & Lancaster, 2013).

Varenicline is currently the most well-known pharmacological therapy for smoking addiction. This drug is a partial agonist for the alpha-4 beta-2 nicotinic acetylcholine receptor (nAChR) and acts as a full agonist for the alpha7 nAChR (Mihalak, Carroll, & Luetje, 2006). It thus helps patients to quit smoking by decreasing nicotine reward and by mitigating the withdrawal symptoms and cravings that occur with smoking cessation by stimulating dopamine secretion at a sustained and moderate level (Coe et al., 2005; Rollema et al., 2007).

The Evaluating Adverse Events in a Global Smoking Cessation (EAGLES) study included 8144 participants (4116 in the psychiatric cohort and 4028 in the nonpsychiatric cohort) and evaluated the safety and efficacy of varenicline, bupropion, and nicotine patches. This study reported that varenicline showed the highest efficacy level and did not significantly increase neuropsychiatric adverse events compared to nicotine patches or a placebo (Anthenelli et al., 2016). Further, Stapleton et al. (2008) compared the effectiveness and safety of NRT and varenicline in tobacco-dependent patients with mental illness who attended weekly group support sessions and found that the short-term cessation rate at 4 weeks after the initiation of smoking cessation was higher in the varenicline group than in the NRT group and the safety results were similar to the results in participants without mental illness. Therefore, varenicline has been considered useful as a treatment for smoking addiction not only in the general population but also in mentally ill patients. From the cost-effectiveness perspective also, varenicline has been found to be more favorable than bupropion, NRT, nortriptyline, and unaided cessation (Hoogendoorn, Welsing, & Rutten-van Molken, 2008).

Postmortem studies on schizophrenic patients have found a noticeable decrease in alpha7 nAChR in several brain areas, including the hippocampus and the cingulate cortex. This decrease may be related not only to schizophrenia symptoms but also to smoking patterns (Brunzell & McIntosh, 2012). Tidey, Colby, and Xavier (2014) compared several symptoms of smokers with schizophrenia and smokers without mental illness after a 72 h smoking abstinence period and found that schizophrenic smokers showed more severe cigarette cravings and withdrawal symptoms, a higher nicotine preference, and an earlier return to smoking. Based on the findings above, varenicline is believed to be an attractive treatment option to treat smoking addiction for schizophrenia patents in particular.

Jeon et al. (2016) conducted a randomized, double-blind, placebo-controlled trial that included 60 schizophrenic patients who smoked and found significant efficacy of varenicline in smoking reduction. The researchers observed that the interaction between time and group was significantly associated with the number of cigarettes, expired CO levels, and results of the Modified Cigarette Evaluation Questionnaire (mCEQ) during the 8-week study period. Further, a study by Smith et al. (2016) that was conducted to examine smoking reduction in schizophrenic smokers also reported significant decreases in the number of cigarettes, expired CO levels, plasma cotinine, and smoking urges in those administered 2 mg/day of varenicline for 8 weeks. Controversy exists as to whether smoking reduction is an appropriate treatment and research goal in schizophrenic patients who do not easily achieve smoking cessation. However, a large smoking reduction can ultimately reduce the likelihood of smoking by decreasing the severity of nicotine dependence and increasing motivation for smoking cessation (Tidey & Miller, 2015).

Several studies have been conducted on the smoking cessation effects of varenicline. Williams et al. (2012) conducted a randomized, double-blind, placebo-controlled trial with 128 schizophrenic and schizoaffective patients and reported that after a 12-week treatment period, the proportion of patients satisfied with smoking cessation was significantly higher in a varenicline treatment group than in a placebo group (19.0% vs 4.7%). Further, the smoking cessation rate was higher in the varenicline group compared to the placebo group at 24 weeks (11.9% vs 2.3%), but this difference was not statistically significant. Additionally, Pachas et al. (2012) conducted a study in which 112 schizophrenic smokers were administered varenicline in combination with weekly group CBT for 12 weeks. After the completion of treatment, the 14-day continuous abstinence rate was 47.3%, and the 28-day continuous abstinence rate was 34%.

Long-term administration of varenicline in schizophrenic patients seems to be associated with a consistent and high tobacco abstinence rate. In a study that included 87 patients with schizophrenia ($n=77$) or bipolar disorder ($n=10$), Evins et al. (2014) administered varenicline in combination with CBT to patients during a 12-week open-label phase, after which they conducted a double-blind, placebo-controlled trial. In the trial, the participants were divided into a varenicline group and a placebo group, administered either a placebo or varenicline and CBT from week 12 to week 52, and followed up without treatment from week 52 to week 76. The varenicline group's point prevalence abstinence rate at week 52 (60%) was significantly higher than that of the placebo group (19%), and the continuous abstinence rates from weeks 12 to 64 and weeks 12 to 76 were also higher in the varenicline group (45% vs 15% and 30% vs 11%, respectively). Studies on the smoking treatment effect of varenicline are summarized in Table 56.1.

Several studies have investigated positive response predictors for varenicline in schizophrenic patients. When schizophrenic patients were administered varenicline with CBT for 12 weeks, Dutra, Stoeckel, Carlini, Pizzagalli, and Evins (2012) reported that the smoking abstinence rate increased and affective flattening decreased. Schuster et al. (2017) conducted a study in patients with schizophrenia ($n=130$) and bipolar disorder ($n=23$) and reported that abstinence was positively associated with fewer withdrawal symptoms, less smoking, better attentional performance, history of alcohol dependence, and anticipation of less support from others. DRD4 variation, known to be involved in nicotine's

TABLE 56.1 Summary of Studies Regarding Smoking Treatment Effects of Varenicline With Schizophrenic Patients

Author	Aim	Method	*N* (MA ± SD; M:F)	Assessments	Results
Jeon et al. (2016)	To investigate smoking reduction effect with varenicline	Double-blind, RCT, outpatients, schizophrenia, 8 weeks	Va = 29 (41.10 ± 8.86; 26:3), Pla = 30 (41.53 ± 8.83; 29:1)	mNWS, QSU-brief, mCEQ, expired CO, cigarette amounts	Va had significant time × group interactions in expired CO ($P = .046$), mCEQ ($P = .002$), and cigarette amounts ($P = .007$)
Smith et al. (2016)	To evaluate effects of varenicline on smoking reduction and other aspects (cognition, psychiatric symptom, and side effect)	Double-blind RCT, outpatients and inpatients, schizophrenia, schizoaffective psychosis, 8 weeks	Va = 42 (46.6 ± 8.9; 35:7), Pla = 45 (43.6 ± 10.6; 39:6)	QSU-brief, expired CO, cigarette amounts, plasma nicotine and cotinine levels, CDS	Va had significant time × group interactions in expired CO ($P = .035$) and cotinine level ($P < .001$)
Williams et al. (2012)	To evaluate smoking cessation effect with varenicline	Double-blind RCT, outpatients, schizophrenia, schizoaffective psychosis, 24 weeks	Va = 84 (40.2 ± 11.9; 65:19), Pla = 43 (43.0 ± 10.2; 33:10)	Abstinence was defined as no smoking 7 days, verified by CO level	At 12 weeks, Va was superior to Pla in smoking cessation effect (Va 19.0% and Pla 4.7%) ($P = .046$). At 24 weeks, Va was 11.9%; Pla was 2.3% ($P = .090$)
Pachas et al. (2012)	To assess effect and safety of varenicline for smoking cessation (combined with weekly group CBT)	Open-label study, outpatients, Schizophrenia, schizoaffective psychosis, 12 weeks	*N* = 112 (47.2 ± 10; 68:44)	Continuous abstinence was defined as consecutive weeks of biochemically verified 7-day point prevalence abstinence	At 12 weeks, 14-day continuous abstinence rate was 47.3%; 28-day rate was 34%
Evins et al. (2014)	To evaluate smoking cessation maintenance effect with varenicline (combined with CBT to week 52 and weeks 52–76 were no treatment follow-up period)	Double-blind RCT, outpatients, schizophrenia, bipolar disorder, schizoaffective psychosis, 76 weeks	Va = 40 (51.4 ± 9.6; 24:16), Pla = 47 (45.7 ± 10.3; 31:16)	Continuous abstinence rate for weeks 12–64 was based on biochemically verified abstinence and weeks 12–76 was based on self-reported smoking behavior	At 52 weeks, point abstinence rate was 60% in Va and 19% in Pla ($P < .001$). In weeks 12–64, continuous abstinence rate was 64.45% in Va and 15% in Pla ($P = .004$). In weeks 12–76, Va was 30%; Pla was 11% ($P = .03$)

Abbreviations: *CDS*, Cigarette Dependence Scale; *CO*, carbon monoxide; *MA ± SD*, participants' mean age and standard deviation; *M:F*, ratio of male to female; *mCEQ*, Modified Cigarette Evaluation Questionnaire; *mNWS*, Minnesota Nicotine Withdrawal Scale; *N*, number of participants; *Pla*, placebo group; *QSU-brief*, brief questionnaire of smoking urge; *RCT*, randomized controlled trial; *Va*, varenicline group.

reward effect, may also be a biological predictor, although no research has been conducted on its direct link to varenicline (Harrell et al., 2016; Jeon et al., 2016).

56.6 OTHER ASPECTS OF VARENICLINE USE IN SCHIZOPHRENIC PATIENTS

During smoking cessation, schizophrenic patients often show additional cognitive impairments, such as reduced attention, due to nicotine withdrawal symptoms. Not only does varenicline show a beneficial effect for smoking addiction in schizophrenia patients, but also it improves cognitive function. These effects are related to the activity of varenicline on nAChR, and they are very significant for schizophrenic patients during smoking cessation.

Shim et al. (2012) examined the relationship between varenicline and cognitive function in a group of 120 schizophrenic patients. A treatment group was administered 2 mg varenicline and was found to perform better on the Digit Symbol Substitution Test and made fewer nonperseverative errors on the Wisconsin Card Sorting Test compared to a control group. Furthermore, hit reaction time on the Continuous Performance Test was shorter in the schizophrenic smokers who were administered varenicline, and interference on the Stroop test was decreased in the nonsmokers. Varenicline has also been shown to be associated with a significant improvement in verbal learning and memory in schizophrenic smokers (Smith et al., 2009) and with a decrease in the severity of cognitive and affective exacerbation after abstinence (Liu et al., 2011). Roh et al. (2014) evaluated cognitive

TABLE 56.2 Summary of Studies Measuring Cognition in Schizophrenic Patients With Varenicline Administration

Author	Aim	Method	*N* (MA ± SD; M:F)	Assessments	Results
Shim et al. (2012)	To evaluate effects of varenicline on cognitive function	Double-blind RCT, outpatients, schizophrenia, 8 weeks	Va = 59 (39.9 ± 8.6; 38:21), Pla = 58 (39.9 ± 9.9; 45:13)	CPT, SCWT, WCST, DSST, DST, VST	Va performed significantly better on DSST (P = .013) and WCST (P = .043) compared to Pla. In smoker, Va is superior to Pla in CPT-HRT (P = .008) and SI (P = .004)
Hong et al. (2011)	To investigate effects of varenicline on neurobiological and cognitive biomarkers	Double-blind RCT, outpatients, schizophrenia, schizoaffective psychosis, 8 weeks	Va = 32 (44.03 ± 1.82; 20:12), Pla = 32 (41.57 ± 1.93; 22:10)	PPI, P50 gating, SPEM, ASER, DSST, CPT	In Va, startle reactivity reduced (P = .015); ASER reduced (P = .034) P50 gating deficit reduced in Va with smokers (P = .006)
Smith et al. (2016)	To evaluate effects of varenicline on cognition and other aspects (smoking, psychiatric symptom, and side effect)	Double-blind RCT, outpatients and inpatients, schizophrenia, schizoaffective psychosis, 8 weeks	Va = 42 (46.6 ± 8.9; 35:7), Pla = 45 (43.6 ± 10.6; 39:6)	MCCB without social cognition module	Significant cognitive function improvement was not found in Va

Abbreviations: *ASER*, antisaccade error rate; *CPT*, Continuous Performance Test; *DSST*, Digital Symbol Substitution Test; *DST*, Digit Span Test; *HRT*, hit reaction time; *MA ± SD*, participants' mean age and standard deviation; *M:F*, ratio of male to female; *MCCB*, MATRICS consensus battery; *N*, number of participants; *Pla*, placebo group; *PPI*, prepulse inhibition; *RCT*, randomized controlled trial; *SCWT*, Stroop Color Word Test; *SI*, Stroop interference; *SPEM*, smooth-pursuit eye movement; *Va*, varenicline group; *VST*, Visual Span Test; *WCST*, Wisconsin Card Sorting Test.

function after administering mecamylamine (nAChR antagonist), varenicline (nAChR agonist), or a placebo to nonsmoking schizophrenic patients and found that attention was decreased in those administered mecamylamine compared to those administered varenicline or a placebo. Hong et al. (2011) conducted a study that included 69 schizophrenic patients in order to examine the correlations between varenicline and neurobiological biomarkers. In their study, several neurobiological biomarkers were found to be correlated with varenicline, including P50 gating, startle response, and antisaccade error rate. Table 56.2 provides a summary of some studies measuring cognition in schizophrenic patients with varenicline administration.

It has also been reported that varenicline improves depressive symptoms in schizophrenic patients. Cather et al. (2017) found a decrease in depressive symptoms (assessed using Calgary Depression Scale for Schizophrenia, CDSS) independent of the effects of tobacco abstinence in schizophrenic patients treated with CBT and varenicline for 12 weeks.

56.7 ADVERSE EVENTS AND VARENICLINE USE IN PATIENTS WITH SCHIZOPHRENIA

It has been reported that patients taking varenicline often complain of nausea and headache, among physical adverse events. However, even the most common adverse event, nausea, is usually only mild and improves over time; therefore, this is not usually a cause for withdrawal of treatment (Cahill, Lindson-Hawley, Thomas, Fanshawe, & Lancaster, 2016). In contrast to these mild side effects, there is much controversy on the relationship between varenicline and serious cardiovascular events. However, a significant correlation between varenicline and serious cardiovascular events was not found in a recent systematic review and meta-analysis (Sterling, Windle, Filion, Touma, & Eisenberg, 2016).

Many studies have reported that varenicline does not significantly increase neuropsychiatric adverse events compared to a placebo or other nicotine addiction treatments, such as NRT (Anthenelli et al., 2016; Molero, Lichtenstein, Zetterqvist, Gumpert, & Fazel, 2015). However, it has continuously been reported that patients taking varenicline experience neuropsychiatric adverse events, including depressed mood, insomnia, suicidal ideation, and aggressive behavior (Ahmed et al., 2013). Williams et al. (2012) reported that schizophrenic and schizoaffective patients taking varenicline experienced diverse neuropsychiatric adverse events, including abnormal dreams (7.1%), anxiety (4.8%), depression (4.8%), insomnia (9.5%), and suicidal ideation (6.0%). However, all patients had a history of smoking, and the varenicline group and the placebo group did not show a significant difference in adverse events. Thus, it is not clear whether the symptoms listed above were due to varenicline or related to smoking withdrawal symptoms.

To summarize, very close monitoring of potential side effects is required when varenicline is used in patients with illnesses such as schizophrenia for whom diverse mental symptoms are present (Yousefi, Folsom, & Fatemi, 2011) (Table 56.3).

TABLE 56.3 Potential Adverse Events to Consider When Administering Varenicline in Patients With Schizophrenia

Neuropsychiatric	Physical
Abnormal dreams	Dizziness
Anxiety	Fatigue
Agitation	Headache
Depression	Nausea
Disturbed sleep	Vomiting
Excitement	
Irritability	
Suicidal idea	

MINI-DICTIONARY OF TERMS

Addiction An obsessive state in which a desire to do certain activities repeatedly while knowing that these activities will affect health and social life.

Bupropion Bupropion acts as a norepinephrine-dopamine reuptake inhibitor and a nicotinic antagonist and is used to treat depression and smoking cessation.

Cognitive behavioral therapy Psychiatric treatment methods affecting emotions and body senses using changes of thought and behavior.

Nicotine replacement therapy The treatment that alleviates withdrawal symptoms without causing a nicotine addiction by supplying a small amount of nicotine through the skin or oral mucosa with low absorption efficiency.

Schizophrenia A psychiatric disorder that causes a wide range of clinical problems to various aspects of thought, emotion, perception, and behavior.

Substance use disorder Substance dependence and substance abuse can both be involved. Substance dependence is characterized by tolerance, withdrawal, and compulsive use, but substance abuse is not.

Key Facts of Smoking Addiction in Schizophrenic Patients

- Smoking prevalence among schizophrenic patients was approximately 5.9 times higher than that among the general population.
- Smoking prevalence among patients with schizophrenia ranges from 50% to 90%.
- Hypotheses related to the association between smoking addiction in schizophrenic patients and drug metabolism, self-medication, and psychosocial aspect exist.
- Psychosocial and pharmacological therapies for smoking addiction may be offered to patients with schizophrenia and the general population.
- In particular, varenicline appears to be an effective pharmacological treatment option for smoking addiction in schizophrenic patients.

Summary Points

- In this chapter, we discuss the smoking addiction of patients with schizophrenia and the application of varenicline as a remedy.
- Schizophrenia is a major mental illness with positive, negative, and cognitive symptoms.
- About 40%–50% of patients with schizophrenia suffer a lifetime substance disorder, which can affect disease prognosis, health, and quality of life.
- Smoking addiction is a commonly observed substance use problem in schizophrenic patients, and its prevalence among schizophrenic patients is higher (approximately 5.9 times) than that among the general population.
- Drug metabolism, self-medication, and psychosocial aspects may be associated with the high prevalence of smoking addiction in patients with schizophrenia.
- Varenicline is a well-known pharmacological agent used for smoking addiction and has excellent efficacy and cost-effectiveness.
- In schizophrenic patients, varenicline was found to be effective in smoking reduction, cessation, and relapse prevention.
- Varenicline may also affect other psychopathologies such as cognitive function in schizophrenic patients.
- Because neuropsychiatric adverse effects associated with varenicline continue to be reported, careful attention should be paid to schizophrenic patients with preexisting mental symptoms.

References

Ahmed, A. I., Ali, A. N., Kramers, C., Harmark, L. V., Burger, D. M., & Verhoeven, W. M. (2013). Neuropsychiatric adverse events of varenicline: a systematic review of published reports. *Journal of Clinical Psychopharmacology*, *33*(1), 55–62.

Ambrose, J. A., & Barua, R. S. (2004). The pathophysiology of cigarette smoking and cardiovascular disease: an update. *Journal of the American College of Cardiology*, *43*(10), 1731–1737.

Anthenelli, R. M., Benowitz, N. L., West, R., St Aubin, L., McRae, T., Lawrence, D., et al. (2016). Neuropsychiatric safety and efficacy of varenicline, bupropion, and nicotine patch in smokers with and without psychiatric disorders (EAGLES): a double-blind, randomised, placebo-controlled clinical trial. *Lancet*, *387*(10037), 2507–2520.

Barnes, M., Lawford, B. R., Burton, S. C., Heslop, K. R., Noble, E. P., Hausdorf, K., et al. (2006). Smoking and schizophrenia: is symptom profile related to smoking and which antipsychotic medication is of benefit in reducing cigarette use? *The Australian and New Zealand Journal of Psychiatry*, *40*(6–7), 575–580.

Barr, R. S., Culhane, M. A., Jubelt, L. E., Mufti, R. S., Dyer, M. A., Weiss, A. P., et al. (2008). The effects of transdermal nicotine on cognition in nonsmokers with schizophrenia and nonpsychiatric controls. *Neuropsychopharmacology, 33*(3), 480–490.

Blanchard, J. J., Brown, S. A., Horan, W. P., & Sherwood, A. R. (2000). Substance use disorders in schizophrenia: review, integration, and a proposed model. *Clinical Psychology Review, 20*(2), 207–234.

Brewer, J. A., Elwafi, H. M., & Davis, J. H. (2013). Craving to quit: psychological models and neurobiological mechanisms of mindfulness training as treatment for addictions. *Psychology of Addictive Behaviors, 27*(2), 366–379.

Brunzell, D. H., & McIntosh, J. M. (2012). Alpha7 nicotinic acetylcholine receptors modulate motivation to self-administer nicotine: implications for smoking and schizophrenia. *Neuropsychopharmacology, 37*(5), 1134–1143.

Cahill, K., Lindson-Hawley, N., Thomas, K. H., Fanshawe, T. R., & Lancaster, T. (2016). Nicotine receptor partial agonists for smoking cessation. *Cochrane Database of Systematic Reviews, 5*, CD006103.

Cahill, K., Stevens, S., Perera, R., & Lancaster, T. (2013). Pharmacological interventions for smoking cessation: an overview and network meta-analysis. *Cochrane Database of Systematic Reviews, 5*, CD009329.

Cather, C., Hoeppner, S., Pachas, G., Pratt, S., Achtyes, E., Cieslak, K. M., et al. (2017). Improved depressive symptoms in adults with schizophrenia during a smoking cessation attempt with Varenicline and behavioral therapy. *Journal of Dual Diagnosis*, 1–11.

Chambers, R. A., Krystal, J. H., & Self, D. W. (2001). A neurobiological basis for substance abuse comorbidity in schizophrenia. *Biological Psychiatry, 50*(2), 71–83.

Coe, J. W., Brooks, P. R., Vetelino, M. G., Wirtz, M. C., Arnold, E. P., Huang, J., et al. (2005). Varenicline: an alpha4beta2 nicotinic receptor partial agonist for smoking cessation. *Journal of Medicinal Chemistry, 48*(10), 3474–3477.

Crow, T. J. (1985). The two-syndrome concept: origins and current status. *Schizophrenia Bulletin, 11*(3), 471–486.

de Leon, J., & Diaz, F. J. (2005). A meta-analysis of worldwide studies demonstrates an association between schizophrenia and tobacco smoking behaviors. *Schizophrenia Research, 76*(2–3), 135–157.

Desai, H. D., Seabolt, J., & Jann, M. W. (2001). Smoking in patients receiving psychotropic medications: a pharmacokinetic perspective. *CNS Drugs, 15*(6), 469–494.

Dickey, B., Azeni, H., Weiss, R., & Sederer, L. (2000). Schizophrenia, substance use disorders and medical co-morbidity. *The Journal of Mental Health Policy and Economics, 3*(1), 27–33.

Dickinson, D., Bellack, A. S., & Gold, J. M. (2007). Social/communication skills, cognition, and vocational functioning in schizophrenia. *Schizophrenia Bulletin, 33*(5), 1213–1220.

Dixon, L. (1999). Dual diagnosis of substance abuse in schizophrenia: prevalence and impact on outcomes. *Schizophrenia Research, 35* (Suppl), S93–100.

Dutra, S. J., Stoeckel, L. E., Carlini, S. V., Pizzagalli, D. A., & Evins, A. E. (2012). Varenicline as a smoking cessation aid in schizophrenia: effects on smoking behavior and reward sensitivity. *Psychopharmacology, 219*(1), 25–34.

Evins, A. E., Cather, C., Pratt, S. A., Pachas, G. N., Hoeppner, S. S., Goff, D. C., et al. (2014). Maintenance treatment with varenicline for smoking cessation in patients with schizophrenia and bipolar disorder: a randomized clinical trial. *JAMA, 311*(2), 145–154.

Freedman, R., Olincy, A., Buchanan, R. W., Harris, J. G., Gold, J. M., Johnson, L., et al. (2008). Initial phase 2 trial of a nicotinic agonist in schizophrenia. *The American Journal of Psychiatry, 165*(8), 1040–1047.

Gandini, S., Botteri, E., Iodice, S., Boniol, M., Lowenfels, A. B., Maisonneuve, P., et al. (2008). Tobacco smoking and cancer: a meta-analysis. *International Journal of Cancer, 122*(1), 155–164.

Goff, D. C., Henderson, D. C., & Amico, E. (1992). Cigarette smoking in schizophrenia: relationship to psychopathology and medication side effects. *The American Journal of Psychiatry, 149*(9), 1189–1194.

Goff, D. C., Hill, M., & Barch, D. (2011). The treatment of cognitive impairment in schizophrenia. *Pharmacology, Biochemistry, and Behavior, 99*(2), 245–253.

Green, M. F., & Harvey, P. D. (2014). Cognition in schizophrenia: past, present, and future. *Schizophrenia Research: Cognition, 1*(1), e1–e9.

Grimshaw, G. M., & Stanton, A. (2006). Tobacco cessation interventions for young people. *Cochrane Database of Systematic Reviews, 4*, CD003289.

Harrell, P. T., Lin, H. Y., Park, J. Y., Blank, M. D., Drobes, D. J., & Evans, D. E. (2016). Dopaminergic genetic variation moderates the effect of nicotine on cigarette reward. *Psychopharmacology, 233*(2), 351–360.

Hong, L. E., Thaker, G. K., McMahon, R. P., Summerfelt, A., Rachbeisel, J., Fuller, R. L., et al. (2011). Effects of moderate-dose treatment with varenicline on neurobiological and cognitive biomarkers in smokers and nonsmokers with schizophrenia or schizoaffective disorder. *Archives of General Psychiatry, 68*(12), 1195–1206.

Hoogendoorn, M., Welsing, P., & Rutten-van Molken, M. P. (2008). Cost-effectiveness of varenicline compared with bupropion, NRT, and nortriptyline for smoking cessation in the Netherlands. *Current Medical Research and Opinion, 24*(1), 51–61.

Huber, G., Gross, G., Schuttler, R., & Linz, M. (1980). Longitudinal studies of schizophrenic patients. *Schizophrenia Bulletin, 6*(4), 592–605.

Hughes, J. R., Hatsukami, D. K., Mitchell, J. E., & Dahlgren, L. A. (1986). Prevalence of smoking among psychiatric outpatients. *The American Journal of Psychiatry, 143*(8), 993–997.

Jeon, D. W., Shim, J. C., Kong, B. G., Moon, J. J., Seo, Y. S., Kim, S. J., et al. (2016). Adjunctive varenicline treatment for smoking reduction in patients with schizophrenia: a randomized double-blind placebo-controlled trial. *Schizophrenia Research, 176*(2–3), 206–211.

Lancaster, T., & Stead, L. (2004). Physician advice for smoking cessation. *Cochrane Database of Systematic Reviews, 4*, CD000165.

Lasser, K., Boyd, J. W., Woolhandler, S., Himmelstein, D. U., McCormick, D., & Bor, D. H. (2000). Smoking and mental illness: a population-based prevalence study. *JAMA, 284*(20), 2606–2610.

Liu, M. E., Tsai, S. J., Jeang, S. Y., Peng, S. L., Wu, S. L., Chen, M. C., et al. (2011). Varenicline prevents affective and cognitive exacerbation during smoking abstinence in male patients with schizophrenia. *Psychiatry Research, 190*(1), 79–84.

Lyon, E. R. (1999). A review of the effects of nicotine on schizophrenia and antipsychotic medications. *Psychiatric Services, 50*(10), 1346–1350.

McGrath, J., Saha, S., Chant, D., & Welham, J. (2008). Schizophrenia: a concise overview of incidence, prevalence, and mortality. *Epidemiologic Reviews, 30*, 67–76.

Mihalak, K. B., Carroll, F. I., & Luetje, C. W. (2006). Varenicline is a partial agonist at alpha4beta2 and a full agonist at alpha7 neuronal nicotinic receptors. *Molecular Pharmacology, 70*(3), 801–805.

Molero, Y., Lichtenstein, P., Zetterqvist, J., Gumpert, C. H., & Fazel, S. (2015). Varenicline and risk of psychiatric conditions, suicidal behaviour, criminal offending, and transport accidents and offences: population based cohort study. *BMJ, 350*, h2388.

Ng, M., Freeman, M. K., Fleming, T. D., Robinson, M., Dwyer-Lindgren, L., Thomson, B., et al. (2014). Smoking prevalence and cigarette consumption in 187 countries, 1980-2012. *JAMA, 311*(2), 183–192.

Pachas, G. N., Cather, C., Pratt, S. A., Hoeppner, B., Nino, J., Carlini, S. V., et al. (2012). Varenicline for smoking cessation in schizophrenia: safety and effectiveness in a 12-week, open-label trial. *Journal of Dual Diagnosis, 8*(2), 117–125.

Potvin, S., Stip, E., & Roy, J. Y. (2003). Schizophrenia and addiction: an evaluation of the self-medication hypothesis. *Encephale, 29*(3 Pt 1), 193–203.

Roh, S., Hoeppner, S. S., Schoenfeld, D., Fullerton, C. A., Stoeckel, L. E., & Evins, A. E. (2014). Acute effects of mecamylamine and varenicline on cognitive performance in non-smokers with and without schizophrenia. *Psychopharmacology*, *231*(4), 765–775.

Rollema, H., Chambers, L. K., Coe, J. W., Glowa, J., Hurst, R. S., Lebel, L. A., et al. (2007). Pharmacological profile of the alpha4beta2 nicotinic acetylcholine receptor partial agonist varenicline, an effective smoking cessation aid. *Neuropharmacology*, *52*(3), 985–994.

Sacco, K. A., Termine, A., Seyal, A., Dudas, M. M., Vessicchio, J. C., Krishnan-Sarin, S., et al. (2005). Effects of cigarette smoking on spatial working memory and attentional deficits in schizophrenia: involvement of nicotinic receptor mechanisms. *Archives of General Psychiatry*, *62*(6), 649–659.

Saha, S., Chant, D., & McGrath, J. (2007). A systematic review of mortality in schizophrenia: is the differential mortality gap worsening over time? *Archives of General Psychiatry*, *64*(10), 1123–1131.

Schuster, R. M., Cather, C., Pachas, G. N., Zhang, H., Cieslak, K. M., Hoeppner, S. S., et al. (2017). Predictors of tobacco abstinence in outpatient smokers with schizophrenia or bipolar disorder treated with varenicline and cognitive behavioral smoking cessation therapy. *Addictive Behaviors*, *71*, 89–95.

Shim, J. C., Jung, D. U., Jung, S. S., Seo, Y. S., Cho, D. M., Lee, J. H., et al. (2012). Adjunctive varenicline treatment with antipsychotic medications for cognitive impairments in people with schizophrenia: a randomized double-blind placebo-controlled trial. *Neuropsychopharmacology*, *37*(3), 660–668.

Smith, R. C., Amiaz, R., Si, T. M., Maayan, L., Jin, H., Boules, S., et al. (2016). Varenicline effects on smoking, cognition, and psychiatric symptoms in schizophrenia: a double-blind randomized trial. *PLoS ONE*, *11*(1):e0143490.

Smith, R. C., Infante, M., Ali, A., Nigam, S., & Kotsaftis, A. (2001). Effects of cigarette smoking on psychopathology scores in patients with schizophrenia: an experimental study. *Substance Abuse*, *22*(3), 175–186.

Smith, R. C., Lindenmayer, J. P., Davis, J. M., Cornwell, J., Noth, K., Gupta, S., et al. (2009). Cognitive and antismoking effects of varenicline in patients with schizophrenia or schizoaffective disorder. *Schizophrenia Research*, *110*(1–3), 149–155.

Stapleton, J. A., Watson, L., Spirling, L. I., Smith, R., Milbrandt, A., Ratcliffe, M., et al. (2008). Varenicline in the routine treatment of tobacco dependence: a pre-post comparison with nicotine replacement therapy and an evaluation in those with mental illness. *Addiction*, *103*(1), 146–154.

Sterling, L. H., Windle, S. B., Filion, K. B., Touma, L., & Eisenberg, M. J. (2016). Varenicline and adverse cardiovascular events: a systematic review and meta-analysis of randomized controlled trials. *Journal of the American Heart Association*, *5*(2).

Tidey, J. W., Colby, S. M., & Xavier, E. M. (2014). Effects of smoking abstinence on cigarette craving, nicotine withdrawal, and nicotine reinforcement in smokers with and without schizophrenia. *Nicotine & Tobacco Research*, *16*(3), 326–334.

Tidey, J. W., & Miller, M. E. (2015). Smoking cessation and reduction in people with chronic mental illness. *BMJ*, *351*, h4065.

Williams, J. M., Anthenelli, R. M., Morris, C. D., Treadow, J., Thompson, J. R., Yunis, C., et al. (2012). A randomized, double-blind, placebo-controlled study evaluating the safety and efficacy of varenicline for smoking cessation in patients with schizophrenia or schizoaffective disorder. *The Journal of Clinical Psychiatry*, *73*(5), 654–660.

Yousefi, M. K., Folsom, T. D., & Fatemi, S. H. (2011). A review of varenicline's efficacy and tolerability in smoking cessation studies in subjects with schizophrenia. *Journal of Addiction Research and Therapy*, *S4*(1).

CHAPTER

57

Nicotine Vaccines: The Past, the Present, and the Future

Yun Hu[a], *Zongmin Zhao*[a], *Kyle Saylor*[a], *Chenming Zhang*

Department of Biological Systems Engineering, Virginia Tech, Blacksburg, VA, United States

Abbreviations

nAChRs	nicotinic acetylcholine receptors
NTR	nicotine replacement therapy
PBPK	physiologically based pharmacokinetic
TLR	toll-like receptor

57.1 INTRODUCTION: THE NEED OF DEVELOPING NICOTINE VACCINES

There are currently over 1.1 billion smokers worldwide (Munoz, Bunge, Barrera, Wickham, & Lee, 2016). In recent years, cigarette smoking has been the leading preventable cause of death, and its devastating impact on health is astonishing (Hu, Zheng, Huang, & Zhang, 2014). Smoking can tremendously increase the risk of numerous diseases, including cardiovascular disease, stroke, and lung cancer (Pope III et al., 2011; Strong, Mathers, & Bonita, 2007). Worldwide, tobacco use causes more than 5 million deaths per year, and studies show that tobacco use will result in more than 10 million deaths annually by 2030 if the pandemic is not controlled (Fagerstrom, 2002). In addition, the World Health Organization (WHO) estimates that smoking causes over US$500 billion in economic damage each year globally (Ekpu & Brown, 2015).

Due to the huge mortality, morbidity, and economic loss, it is critical to stem the prevalence of smoking. However, it was reported that 80% of all smokers who attempt to quit on their own return to smoking within a month, and each year, only 3% of smokers quit successfully (Benowitz, 2010). Because of the challenges in unassisted smoking cessation, numerous pharmacotherapies, including nicotine replacement therapy (NRT), varenicline, and bupropion, have been developed (Raupach & van Schayck, 2011). These therapies work through different mechanisms. For example, smokers can acquire nicotine from NRT and reduce their cravings for tobacco (Schnoll et al., 2015); varenicline, a partial agonist for nicotinic acetylcholine receptors (nAChRs), serves by both partially stimulating receptors and blocking receptor access of nicotine, thereby reducing the rewarding effects of nicotine (Crunelle, Miller, Booij, & van den Brink, 2010); bupropion, an antagonist for nAChRs and an inhibitor of norepinephrine-dopamine reuptake, is thought to act as a smoking cessation aid by inhibiting neuronal access to catecholamines and by preventing the binding of nicotine with nAChRs (Roddy, 2004). These therapies have led to improved success rates in smoking cessation, but they are associated with some limitations. First, the overall abstinence rate achieved by these medications is about 20% (Patel, Feucht, Reid, & Patel, 2010). Second, the long treatment time, which may take up to several months, may cause inconvenience to patients' daily life and work (Jiloha, 2014). Third, under certain conditions, these therapies may cause severe adverse effects (Cahill, Stevens, Perera, & Lancaster, 2013). Fourth, the high cost of these treatments may not be affordable to low-income users (Muilenburg, Laschober, & Eby, 2015). Therefore, it is necessary and urgent to develop new treatments for nicotine addiction.

It is widely accepted that nicotine is the major addictive component in tobacco products that causes long-term addiction despite the apparently harmful outcomes (Dani & De Biasi, 2001). Nicotine initiates its action by binding to nAChRs, which subsequently opens the ion channels and allows the entry of sodium or calcium into nerve cells, leading to the release of neurotransmitters,

[a] Equal contribution.

Neuroscience of Nicotine
https://doi.org/10.1016/B978-0-12-813035-3.00057-5

such as dopamine (Paterson & Nordberg, 2000). Dopamine can induce an experience of euphoria and is responsible for the reinforcing effects of nicotine (Benowitz, 2008; Nestler, 2005). Therefore, smoking addiction is largely a matter of nicotine addiction.

57.2 THE PAST: THE FIRST-GENERATION NICOTINE VACCINES

Nicotine vaccines that can induce the immune system to produce highly specific antibodies against nicotine have been considered a promising therapy against smoking (Hall, 2002; Zhao, Hu, Hoerle, et al., 2017; Zhao, Powers, et al., 2017; Zheng et al., 2015). The antibodies elicited by a nicotine vaccine can retain nicotine in the blood and limit its distribution to the brain, thereby reducing many of nicotine's physiological and behavioral effects (Pentel & Malin, 2002). Nicotine vaccines possess several major advantages over traditional pharmacological therapies. First, frequent administration is not required, and the persistence of antibodies in the blood can ensure long-term efficacy (Hoogsteder, Kotz, van Spiegel, Viechtbauer, & van Schayck, 2014). Second, vaccines have minimal if any adverse side effects (Hu et al., 2016). Third, the low cost of a nicotine vaccine makes it affordable to low-income patients (Goniewicz & Delijewski, 2013).

Due to its small molecular weight and simple structure, nicotine is not able to elicit an immune response on its own (Hu et al., 2016). In traditional nicotine vaccine designs, a nicotine analogue is covalently conjugated to a carrier protein in order to instigate recognition by the immune system (Fig. 57.1). So far, there have been several nicotine-protein conjugate vaccines (Table 57.1),

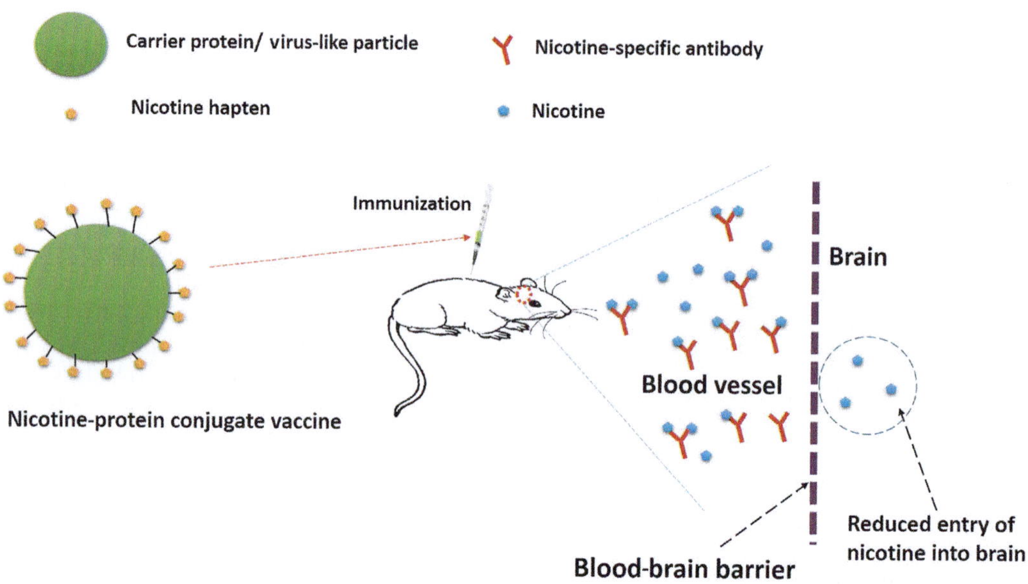

FIG. 57.1 Schematic illustration of antibody response elicited by nicotine-protein conjugate vaccines. Animals or humans, who are immunized with nicotine vaccines, can produce nicotine-specific antibody in the blood. These antibodies can bind with nicotine molecules and limit their entry into the brain, thereby reducing the rewarding effect from nicotine.

TABLE 57.1 First-Generation Nicotine-Protein Conjugate Vaccines

Vaccine	Company	Structure	Status of clinical trials	References
NicVAX	Nabi Biopharmaceuticals	3′-Aminomethylnicotine *Pseudomonas aeruginosa* exoprotein A conjugate vaccine	Phase III failed	Fahim et al. (2013)
TA-NIC	Xenova/Celtic Pharma	A nicotine hapten recombinant B subunit of cholera toxin conjugate vaccine	Phase II halted	Zalewska-Kaszubska (2015)
Niccine	Independent Pharmaceutical	A nicotine hapten tetanus-toxoid conjugate vaccine	Phase II halted	Pentel and LeSage (2014)
NIC002	Cytos Biotechnology	A nicotine hapten virus-like particle conjugate vaccine	Phase II failed	Shen et al. (2012)

including NicVAX, TA-NIC, NIC002 (also known as NicQb), and Niccine, evaluated in clinical trials (Caponnetto, Russo, & Polosa, 2012). Based on the nicotine-protein conjugate design, these vaccines can be categorized as the first-generation nicotine vaccines (Pentel & LeSage, 2014).

Although the first-generation nicotine vaccines showed good safety and the ability to elicit a strong immune response in humans, they ultimately failed in clinical trials due to the less than desired treatment efficacy (Fahim, Kessler, & Kalnik, 2013; Shen, Orson, & Kosten, 2012; Zalewska-Kaszubska, 2015). It was widely believed that the failure of these traditional vaccines might be caused by the insufficient antibody concentration, low antibody affinity, and poor specificity (Hu et al., 2016; Pentel & LeSage, 2014). Therefore, a nicotine vaccine that enhances these parameters is needed to improve treatment outcomes.

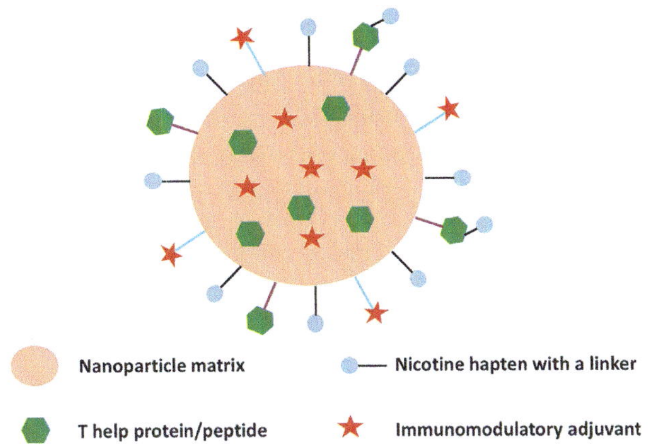

FIG. 57.2 Schematic illustration of structure of the next-generation nanoparticle-based nicotine vaccines.
Nanoparticles are used as particulate vehicles to effectively deliver vaccine components. T-helper proteins/peptides and immunomodulatory adjuvants can be encapsulated within or attached to the surface of nanoparticles. Nicotine haptens can be conjugated to nanoparticle surface or T-helper proteins/peptides via a linker.

57.3 THE PRESENT: NANOPARTICLE-BASED NICOTINE VACCINES

The first-generation conjugate nicotine vaccines suffer from some innate shortcomings that largely limit their immunologic efficacy, such as poor recognition and internalization by immune cells, low bioavailability, difficulty in incorporation with molecular adjuvants, and short immune persistence (Zhao, Hu, Hoerle, et al., 2017). In recent decades, nanoparticles have been widely studied as delivery vehicles for drugs, proteins, and vaccines. Having many superiorities, such as high payload loading capacity, controlled payload release, and tunable physicochemical properties, nanoparticles have the potential to overcome many of the drawbacks of conjugate nicotine vaccines and thus can be a basis for the development of the next-generation nicotine vaccines that can induce a stronger immune response (Ilyinskii & Johnston, 2016).

Promisingly, compared to the first-generation conjugate nicotine vaccines, the next-generation nanoparticle-based nicotine vaccines (Fig. 57.2) hold many advantages: (1) The particulate nature can mimic the geometry of naturally existing pathogens (such as bacteria and viruses), leading to more effective recognition by immune cells. (2) The physicochemical properties (such as size and surface charge) can be easily tuned to achieve enhanced capture and presentation by antigen-presenting cells. (3) Molecular adjuvants can be easily incorporated, effectively delivered, and efficiently released in a controllable manner, to maximize the magnitude of immune response while minimizing systemic toxicity (Ilyinskii & Johnston, 2016; Zhao, Powers, et al., 2017). In the past decade, several nanoparticle-based nicotine vaccines (Table 57.2) have been under development as an effective next-generation immunotherapeutic strategy for smoking cessation.

TABLE 57.2 Examples of the Next-Generation Nanoparticle-Based Nicotine Vaccines

Vaccine	Developer	Nanoparticle platforms	Status of development	References
SEL-068	Selecta Biosciences	Biodegradable polymeric particles	Phase I undergoing	Pittet et al. (2012)
DNA-scaffolded nicotine vaccine	Arizona State University	Self-assembled DNA scaffold	Preclinical	Liu et al. (2016)
Liposome nicotine vaccine	Scripps Institute and Virginia Tech	Liposomes	Preclinical	Hu et al. (2014) and Lockner et al. (2013)
Hybrid nanoparticle-based nicotine vaccine (NanoNicVac)	Virginia Tech	Lipid-polymer hybrid nanoparticles	Preclinical	Hu et al. (2016) and Zhao, Hu, Hoerle, et al. (2017)

All nanoparticle-based nicotine nanovaccines that are currently under development are listed in the table.

SEL-068, a fully synthetic, self-assembling nanoparticle-based nicotine vaccine, is constructed based on a biodegradable and biocompatible (poly(lactic-*co*-glycolic acid) or poly(lactic acid)) matrix. Essential vaccine components are conjugated to or packaged within the nanoparticles. Specifically, nicotine hapten is covalently conjugated to nanoparticle surface, while a novel universal T cell helper peptide and toll-like receptor (TLR) agonist are enclosed within the nanoparticles. Preclinical studies indicated that SEL-068 could result in robust and dose-dependent induction of high levels of nicotine-specific antibodies with high affinity to nicotine in mice and nonhuman primates (Fraser et al., 2014; Ilyinskii & Johnston, 2016; Pittet et al., 2012). SEL-068 has advanced to phase I clinical trials, but the results have not been published yet. Currently, SEL-068 is under reformulation by using a dual-nanoparticle strategy to optimize the human nicotine antibody response.

A DNA-scaffolded nicotine vaccine was developed by using a self-assembled DNA nanostructure as a particulate delivery vehicle of conjugate nicotine vaccines. The DNA-scaffolded nicotine vaccine was constructed by attaching nicotine-streptavidin conjugate to a DNA tetrahedron that was self-assembled from biotinylated DNA strands, CpG-containing oligonucleotides, and other strands. Preclinical studies revealed that using the DNA-scaffolded nicotine vaccine as a priming agent could elicit high titers of nicotine-specific antibodies and significantly block nicotine from entering the brain of mice (Liu et al., 2012; Liu, Hecht, Yan, Pentel, & Chang, 2016). This DNA-scaffolded nicotine vaccine is currently under optimization to further improve the immunologic efficacy.

Liposomes have been studied for the development of the next-generation nanoparticle-based nicotine vaccines (Hu et al., 2014; Lockner et al., 2013). However, liposome-based nicotine nanovaccines have a significant shortfall, instability, which largely limits their further development. By enjoying the superiorities and circumventing the shortfalls of liposomes and polymeric nanoparticles, lipid-polymeric hybrid nanoparticle-based nicotine nanovaccines (NanoNicVac) that can co-deliver nicotine haptens, T-cell epitopes, and molecular adjuvants are under development (Hu et al., 2016; Zhao, Hu, Hoerle, et al., 2017). Preclinical studies revealed that the immunologic efficacy of NanoNicVac could be optimized by multiple approaches, including modulating particle size (Zhao, Hu, Hoerle, et al., 2017), tuning hapten density (Zhao, Powers, et al., 2017), modulating hapten localization (Zhao, Hu, Harmon, et al., 2017), screening carrier proteins (Zhao, Hu, et al., 2018), incorporating molecular adjuvants (Zhao, Harris, et al., 2018), and tuning adjuvant release kinetics (unpublished data). NanoNicVac was demonstrated to exhibit improved recognition and internalization, enhanced immunogenicity, and improved pharmacokinetic efficacy (Fig. 57.3). This hybrid nanoparticle-based nicotine vaccine is under further optimization with hopes of advancing to clinical trials.

Since all the clinically tested conjugate nicotine vaccines have failed, it has been convincingly argued that it is necessary to utilize completely new paradigms to develop novel nicotine vaccines. The use of nanoparticles rather than proteins as carriers is such a new paradigm. Although the next-generation nanoparticle-based nicotine vaccine is still in the early stage of development, it has shown great promise to be further developed for human clinical trials.

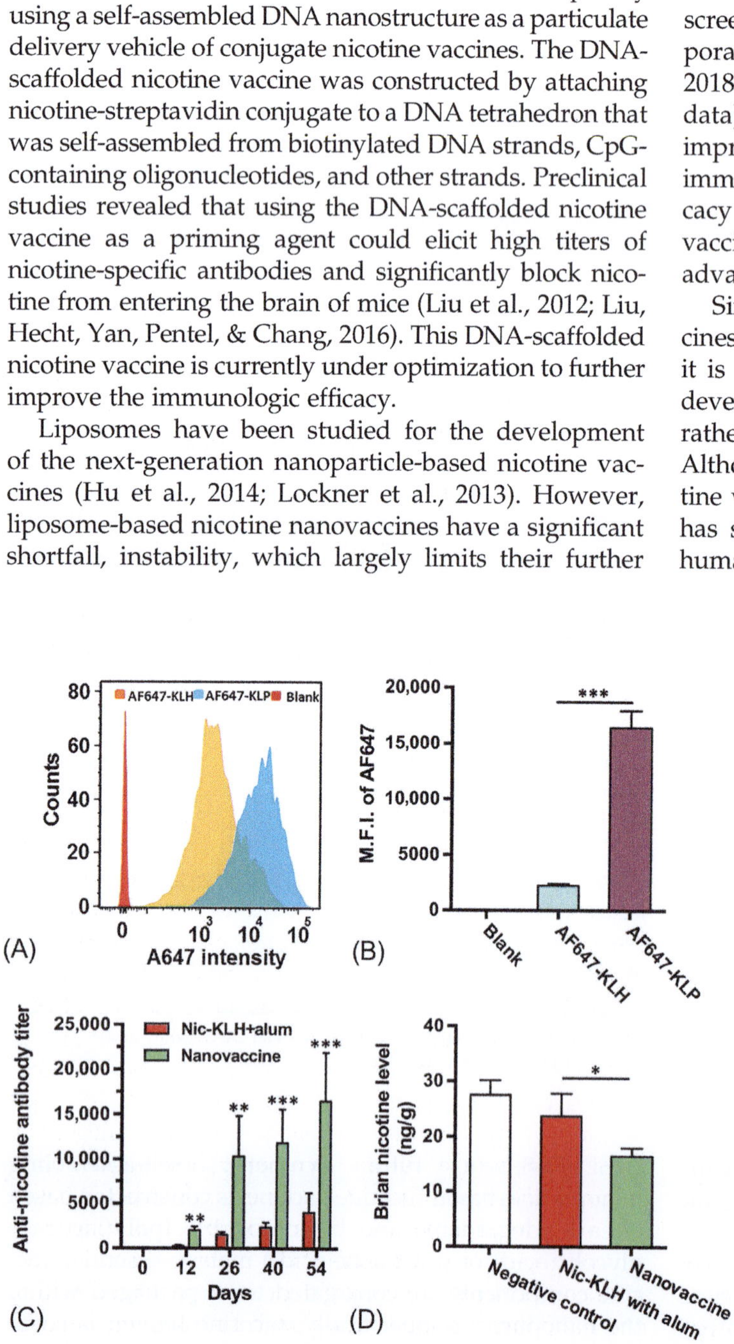

FIG. 57.3 Efficacy of the hybrid nanoparticle-based nicotine nanovaccine (NanoNicVac) (Zhao, Powers, et al., 2017). (A) and (B) Cellular uptake efficiency of free keyhole limpet hemocyanin (KLH) (AF647-KLH) or hybrid nanoparticle-delivered KLH (AF647-KLP) in dendritic cells. KLH was labeled by Alexa Fluor 647 (AF647). MFI represents mean fluorescence intensity. Significantly different: *** $P<.001$. Data were presented as means ± SD ($n=3$). (C) Antinicotine antibody titers induced by Nic-KLH conjugate vaccine with alum or nanovaccine. Significantly different compared to Nic-KLH + alum: ** $P<.05$, *** $P<.001$. Data were presented as means ± SD ($n=8$). (D) Nicotine levels in mouse brains after challenged with 0.03 mg/kg nicotine. Significantly different: * $P<.05$. Data were presented as means ± SD ($n=4$). *Data were reprinted with permission from Elsevier.*

57.4 THE FUTURE: COMBINING PREVIOUS SUCCESSES AND INVESTIGATING NEW CONCEPTS

Past research has established many optima that could be utilized in future vaccine formulations. More specifically, the effects of hapten design, hapten localization, choice of immunogen, immunogen scaffolding, choice of adjuvant, vaccine administration route, and vaccine valency are all well-investigated topics (Pentel & LeSage, 2014). Unfortunately, however, current nicotine vaccine research has generally focused on the investigation of new variables and not on the evaluation of optimal vaccine candidates. This is most likely due to a large number of patents that exist that limit the accessibility of successful technologies and the fact that commercialization is one of the primary end points. If we hope to see a successful human vaccine in the near future, it would be prudent to consider the results from past studies and put forward collaborative vaccination strategies that implement any and all previous, positive findings.

Looking forward, the animal models and benchmarks utilized to evaluate vaccine efficacy should be standardized. In the past, these factors have been inconsistent between test groups. Mice, rats, and rabbits have all been used as animal models, and popular efficacy benchmarks alternate between antibody titers, antibody concentrations, and percent reduction in brain nicotine concentrations (Raupach, Hoogsteder, & Onno van Schayck, 2012). Future studies would benefit by the adoption of an animal model that correlates well with the human immune response to nicotine vaccines. Additionally, standardization of efficacy benchmarks would allow for better comparisons between independent studies.

Nicotine vaccines are very different from traditional vaccines in so much as their efficacy is generally only determined by two factors, antibody binding kinetics and antibody concentration, and the improvement of these parameters has been the central foci of past and present nicotine vaccine research. If one makes the assumption that the optimization of these parameters will continue on into the future, it is possible to speculate on the paths that future nicotine vaccine research might take. A few likely directions for future research include improving upon the next-generation nanoparticle-based nicotine vaccine technology, evaluating the impact of nicotine-specific secretory antibodies on the bioavailability of nicotine, the continued investigation of hapten design as a means to potentiate nicotine vaccine efficacy, and the use of in silico modeling to make predictions on human vaccine efficacy using animal study data.

Further rational design needs to be carried out in order to improve the immunogenicity of the next-generation nanoparticle-based nicotine vaccines, such as rational design of the structure of vaccine nanoparticles, precise control of the physicochemical properties of vaccine particles, and the accurate manipulation of component density and stoichiometry. Overall, developing the next-generation nanoparticle-based nicotine vaccines will potentially provide a novel immunotherapeutic strategy to treat nicotine addiction.

The majority of past and present nicotine vaccine studies have administered vaccine parenterally and have largely failed to target mucosal immune responses. This is disadvantageous because mucosal immunity is characterized by the secretion of IgA, a major antibody isotype that could serve as an additional barrier between inhaled nicotine and the brain (Neutra & Kozlowski, 2006). Promising results have already been seen in one study, where challenge with radiolabeled nicotine via cardiac injection appears to have reduced brain nicotine concentration by 75% when compared with controls (Fraleigh et al., 2016). Additionally, when a prime-boost strategy was implemented in another study, mucosal antibodies were elevated, and systemic antibody titers were comparable to those elicited using parenteral vaccination (Cerny et al., 2002). However, these studies did not investigate the impact secretory IgA has on the bioavailability of nicotine when inhaled. In terms of ease of use, mucosal immunization also offers the advantage of boosting at home without the use of needles. For these reasons, it may be prudent to apply prime-boost mucosal immunization with future nicotine vaccine formulations. Additionally, the use of multivalent nicotine vaccines can improve nicotine-specific antibody titers and decrease the amount of systemic nicotine that can reach the brain (de Villiers, Cornish, Troska, Pravetoni, & Pentel, 2013). As such, the use of a multivalent, multiroute vaccination approach could also prove to be another effective strategy for future nicotine vaccines.

Unfortunately, the efficacy of past nicotine vaccines has not translated well between initial animal studies and later clinical trials. For this reason, a new approach to vaccine design has been proposed in which mathematical modeling is used in concert with animal studies in order to better predict vaccine effectiveness prior to clinical trials. The first model developed for this purpose was a simple, physiologically based pharmacokinetic (PBPK) model that was able to effectively predict the nicotine concentration in rat serum and brain when the effects of antinicotine antibodies were applied (Saylor & Zhang, 2016). A human model was also developed by this same group by superimposing human-specific physiological parameters onto the rat model and calibrating using time-course serum nicotine concentration data found in literature (Fig. 57.4). If correlations established in human studies between brain nicotine reduction and smoking cessation rates are used (Esterlis et al., 2013), mathematical modeling could provide a robust means of predicting vaccine efficacy in humans using animal data prior to costly clinical trials. These models could also

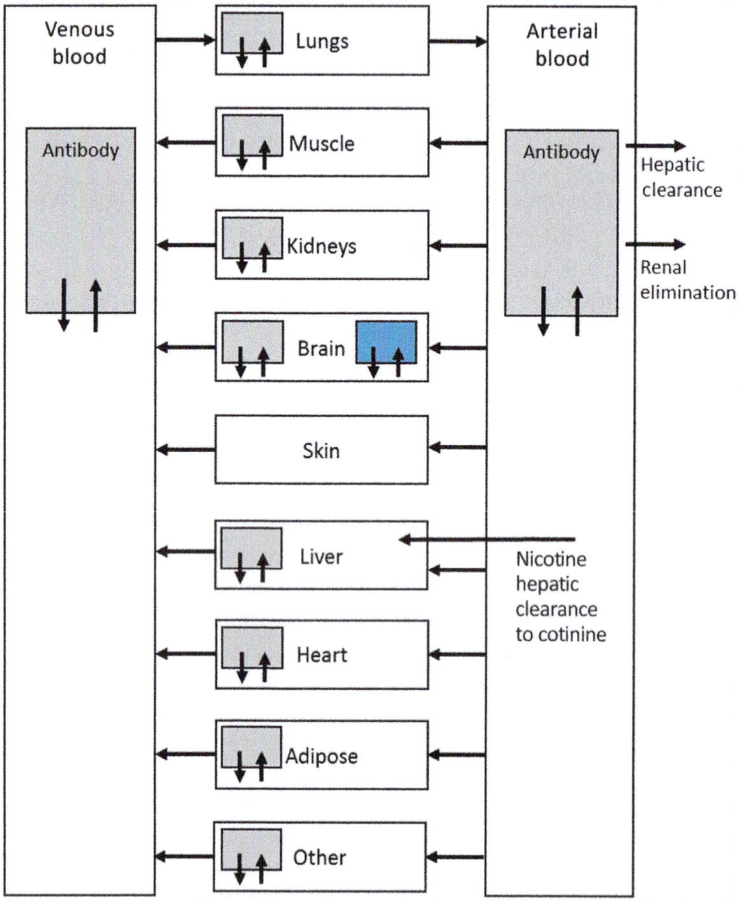

FIG. 57.4 Schematic representation of the PBPK models developed by Saylor and Zhang (2016).
Gray boxes represent antibody binding nicotine and/or cotinine, *blue boxes* represent tissue retention of nicotine, and *arrows* represent nicotine and cotinine mass action kinetics or metabolism. *Figure reused with permission from Elsevier.*

be used to establish vaccine efficacy thresholds (i.e., necessary antibody concentration, antibody affinity, and antibody selectivity), further streamlining the jump from animal studies to clinical trials.

When one considers the steady advancement of nicotine vaccine research over the past decades, it is hard to imagine a future that fails to bring a successful vaccine to market. Every new study that focuses on the implementation of novel ideas and technologies could provide the additional boost in efficacy needed for an effective vaccine in humans. In fact, it is possible that sufficient technology already exists to bring an effective nicotine vaccine to market. However, as was stated earlier, the dissemination of much of this technology is being slowed by licensure issues. To add to this, the specialized nature of most nicotine vaccine projects may be preventing groups from adopting each other's technologies, even when patents are not in play. Ultimately, it seems likely that the success of the nicotine vaccine concept will boil down to time, the degree of cooperation between research groups in the field, and the continued support of funding agencies worldwide.

MINI-DICTIONARY OF TERMS

Adjuvant A substance in a vaccine formulation that can promote the immune response induced by the vaccine.

Antibody affinity The strength of interactions between the binding sites of a given antibody and its epitopes.

Conjugate vaccine A vaccine that is made by attaching haptens to a protein-based carrier via an appropriate linker.

Hapten A small molecule that is not immunogenic by itself but turns to be immunogenic when attached to a large carrier.

Immunogenicity The ability of a vaccine/antigen to induce an immune response.

In silico modeling A computer-simulation-based technique to model physiological or pharmacological processes.

Multivalent vaccination An immunization strategy in which two or more immunogens are simultaneously administered to subjects.

Nanoparticle delivery vehicles Using nanoparticles that can carry pharmaceutical compounds to effectively deliver them to specific places in humans or animals.

PBPK model A mathematical model to predict the absorption, distribution, metabolism, and excretion of a substance in animal models or humans.

Pharmacological therapy for smoking cessation A therapeutic strategy by applying pharmacological principles to aid smoking cessation, including nicotine replacement therapy, bupropion-based therapy, and varenicline-based therapy.

Key Facts of Nicotine Vaccines

- Globally, cigarette smoking has been causing huge mortality, morbidity, and economic loss in the past decades.
- Each year, only 3% of smokers successfully quit smoking on their own.
- Conventional pharmacotherapies, including nicotine replacement therapy, varenicline, and bupropion, prove effective in assisting smoking cessation.
- The pharmacotherapies are facing numerous challenges, including low abstinence rate, long treatment time, and adverse effects.
- Nicotine vaccines are considered promising therapies for smoking addiction.
- The first-generation nicotine vaccines, which are nicotine-protein conjugates, showed limited treatment efficacy in clinical trials due to their low immunogenicity and specificity.
- The next-generation nicotine vaccines, which are nanoparticle-based vaccines, can achieve better treatment outcomes than the traditional nicotine-protein conjugate vaccines.

Summary Points

- This chapter focuses on nicotine vaccines that are potential therapies for nicotine addiction.
- Tobacco addiction is largely a matter of nicotine addiction.
- Nicotine vaccines can induce the production of nicotine-specific antibodies, which can bind with nicotine and prevent it from crossing the blood-brain barrier, thereby reducing the rewarding effect of smoking.
- Traditional nicotine vaccines were constructed by conjugating nicotine analogues to carrier proteins and showed limited efficacy in promoting smoking cessation in clinical trials.
- The next-generation nicotine vaccines are nanoparticle-based nicotine vaccines, which are developed to avoid the shortcomings, such as low immunogenicity, specificity, and short efficacy that are associated with the traditional nicotine-protein conjugate vaccines.
- Nanoparticle-based nicotine vaccine exhibited strong ability in treating nicotine addiction in both preclinical and clinical trials.
- The performance of current nanoparticle-based nicotine vaccines may be further improved by rationalizing the design of the vaccine structure, optimizing the physicochemical properties of the vaccine particles, and incorporating appropriate molecular adjuvants.

References

Benowitz, N. L. (2008). Neurobiology of nicotine addiction: implications for smoking cessation treatment. *The American Journal of Medicine, 121*(4 Suppl. 1), S3–10.

Benowitz, N. L. (2010). Nicotine addiction. *The New England Journal of Medicine, 362*(24), 2295–2303.

Cahill, K., Stevens, S., Perera, R., & Lancaster, T. (2013). Pharmacological interventions for smoking cessation: an overview and network meta-analysis. *Cochrane Database of Systematic Reviews, 5*, CD009329.

Caponnetto, P., Russo, C., & Polosa, R. (2012). Smoking cessation: present status and future perspectives. *Current Opinion in Pharmacology, 12*(3), 229–237.

Cerny, E. H., Levy, R., Mauel, J., Mpandi, M., Mutter, M., Henzelin-Nkubana, C., et al. (2002). Preclinical development of a vaccine 'against smoking'. *Onkologie, 25*(5), 406–411.

Crunelle, C. L., Miller, M. L., Booij, J., & van den Brink, W. (2010). The nicotinic acetylcholine receptor partial agonist varenicline and the treatment of drug dependence: a review. *European Neuropsychopharmacology, 20*(2), 69–79.

Dani, J. A., & De Biasi, M. (2001). Cellular mechanisms of nicotine addiction. *Pharmacology, Biochemistry, and Behavior, 70*(4), 439–446.

de Villiers, S. H., Cornish, K. E., Troska, A. J., Pravetoni, M., & Pentel, P. R. (2013). Increased efficacy of a trivalent nicotine vaccine compared to a dose-matched monovalent vaccine when formulated with alum. *Vaccine, 31*(52), 6185–6193.

Ekpu, V. U., & Brown, A. K. (2015). The economic impact of smoking and of reducing smoking prevalence: review of evidence. *Tobacco Use Insights, 8*, 1–35.

Esterlis, I., Hannestad, J. O., Perkins, E., Bois, F., D'Souza, D. C., Tyndale, R. F., et al. (2013). Effect of a nicotine vaccine on nicotine binding to beta2*-nicotinic acetylcholine receptors in vivo in human tobacco smokers. *The American Journal of Psychiatry, 170*(4), 399–407.

Fagerstrom, K. (2002). The epidemiology of smoking: health consequences and benefits of cessation. *Drugs, 62*(Suppl. 2), 1–9.

Fahim, R. E., Kessler, P. D., & Kalnik, M. W. (2013). Therapeutic vaccines against tobacco addiction. *Expert Review of Vaccines, 12*(3), 333–342.

Fraleigh, N. L., Boudreau, J., Bhardwaj, N., Eng, N. F., Murad, Y., Lafrenie, R., et al. (2016). Evaluating the immunogenicity of an intranasal vaccine against nicotine in mice using the adjuvant Finlay proteoliposome (AFPL1). *Heliyon, 2*(8):e00147.

Fraser, C. C., Altreuter, D. H., Ilyinskii, P., Pittet, L., LaMothe, R. A., Keegan, M., et al. (2014). Generation of a universal CD4 memory T cell recall peptide effective in humans, mice and non-human primates. *Vaccine, 32*(24), 2896–2903.

Goniewicz, M. L., & Delijewski, M. (2013). Nicotine vaccines to treat tobacco dependence. *Human Vaccines & Immunotherapeutics, 9*(1), 13–25.

Hall, W. (2002). The prospects for immunotherapy in smoking cessation. *Lancet, 360*(9339), 1089–1091.

Hoogsteder, P. H., Kotz, D., van Spiegel, P. I., Viechtbauer, W., & van Schayck, O. C. (2014). Efficacy of the nicotine vaccine 3'-AmNic-rEPA (NicVAX) co-administered with varenicline and counselling for smoking cessation: a randomized placebo-controlled trial. *Addiction, 109*(8), 1252–1259.

Hu, Y., Smith, D., Frazier, E., Hoerle, R., Ehrich, M., & Zhang, C. M. (2016). The next-generation nicotine vaccine: a novel and potent hybrid nanoparticle-based nicotine vaccine. *Biomaterials, 106*, 228–239.

Hu, Y., Zheng, H., Huang, W., & Zhang, C. M. (2014). A novel and efficient nicotine vaccine using nano-lipoplex as a delivery vehicle. *Human Vaccines & Immunotherapeutics, 10*(1), 64–72.

Ilyinskii, P. O., & Johnston, L. P. M. (2016). Nanoparticle-based nicotine vaccine. In I. D. Montoya (Ed.), *Biologics to treat substance use disorders: vaccines, monoclonal antibodies, and enzymes* (pp. 249–279): Basel, Switzerland: Springer.

Jiloha, R. C. (2014). Pharmacotherapy of smoking cessation. *Indian Journal of Psychiatry*, *56*(1), 87–95.

Liu, X., Hecht, S. M., Yan, H., Pentel, P. R., & Chang, Y. (2016). Exploration of DNA nanostructures for rational design of vaccines. In I. D. Montoya (Ed.), *Biologics to treat substance use disorders: Vaccines, monoclonal antibodies, and enzymes* (pp. 279–293): Basel, Switzerland: Springer.

Liu, X. W., Xu, Y., Yu, T., Clifford, C., Liu, Y., Yan, H., et al. (2012). A DNA nanostructure platform for directed assembly of synthetic vaccines. *Nano Letters*, *12*(8), 4254–4259.

Lockner, J. W., Ho, S. O., McCague, K. C., Chiang, S. M., Do, T. Q., Fujii, G., et al. (2013). Enhancing nicotine vaccine immunogenicity with liposomes. *Bioorganic & Medicinal Chemistry Letters*, *23*(4), 975–978.

Muilenburg, J. L., Laschober, T. C., & Eby, L. T. (2015). Relationship between low-income patient census and substance use disorder treatment programs' availability of tobacco cessation services. *Journal of Drug Issues*, *45*(1), 69–79.

Munoz, R. F., Bunge, E. L., Barrera, A. Z., Wickham, R. E., & Lee, J. (2016). Using behavioral intervention technologies to help low-income and Latino smokers quit: protocol of a randomized controlled trial. *JMIR Research Protocols*, *5*(2):e127.

Nestler, E. J. (2005). Is there a common molecular pathway for addiction? *Nature Neuroscience*, *8*(11), 1445–1449.

Neutra, M. R., & Kozlowski, P. A. (2006). Mucosal vaccines: the promise and the challenge. *Nature Reviews Immunology*, *6*(2), 148–158. https://doi.org/10.1038/nri1777.

Patel, D. R., Feucht, C., Reid, L., & Patel, N. D. (2010). Pharmacologic agents for smoking cessation: a clinical review. *Clinical Pharmacology*, *2*, 17–29.

Paterson, D., & Nordberg, A. (2000). Neuronal nicotinic receptors in the human brain. *Progress in Neurobiology*, *61*(1), 75–111.

Pentel, P. R., & LeSage, M. G. (2014). New directions in nicotine vaccine design and use. *Advances in Pharmacology*, *69*, 553–580.

Pentel, P., & Malin, D. (2002). A vaccine for nicotine dependence: targeting the drug rather than the brain. *Respiration*, *69*(3), 193–197.

Pittet, L., Altreuter, D., Ilyinskii, P., Fraser, C., Gao, Y., Baldwin, S., et al. (2012). Development and preclinical evaluation of SEL-068, a novel targeted synthetic vaccine particle (tSVP (TM)) for smoking cessation and relapse prevention that generates high titers of antibodies against nicotine. *Journal of Immunology*, *188*.

Pope, C. A., III, Burnett, R. T., Turner, M. C., Cohen, A., Krewski, D., Jerrett, M., et al. (2011). Lung cancer and cardiovascular disease mortality associated with ambient air pollution and cigarette smoke: shape of the exposure-response relationships. *Environmental Health Perspectives*, *119*(11), 1616–1621.

Raupach, T., Hoogsteder, P. H., & Onno van Schayck, C. P. (2012). Nicotine vaccines to assist with smoking cessation: current status of research. *Drugs*, *72*(4), e1–16.

Raupach, T., & van Schayck, C. P. (2011). Pharmacotherapy for smoking cessation: current advances and research topics. *CNS Drugs*, *25*(5), 371–382.

Roddy, E. (2004). Bupropion and other non-nicotine pharmacotherapies. *BMJ*, *328*(7438), 509–511.

Saylor, K., & Zhang, C. (2016). A simple physiologically based pharmacokinetic model evaluating the effect of anti-nicotine antibodies on nicotine disposition in the brains of rats and humans. *Toxicology and Applied Pharmacology*, *307*, 150–164.

Schnoll, R. A., Goelz, P. M., Veluz-Wilkins, A., Blazekovic, S., Powers, L., Leone, F. T., et al. (2015). Long-term nicotine replacement therapy: a randomized clinical trial. *JAMA Internal Medicine*, *175*(4), 504–511.

Shen, X. Y., Orson, F. M., & Kosten, T. R. (2012). Vaccines against drug abuse. *Clinical Pharmacology and Therapeutics*, *91*(1), 60–70.

Strong, K., Mathers, C., & Bonita, R. (2007). Preventing stroke: saving lives around the world. *Lancet Neurology*, *6*(2), 182–187.

Zalewska-Kaszubska, J. (2015). Is immunotherapy an opportunity for effective treatment of drug addiction? *Vaccine*, *33*(48), 6545–6551.

Zhao, Z., Harris, B., Hu, Y., Harmon, T., Pentel, P. R., Ehrich, M., et al. (2018). Rational incorporation of molecular adjuvants into a hybrid nanoparticle-based nicotine vaccine for immunotherapy against nicotine addiction. *Biomaterials*, *155*, 165–175.

Zhao, Z., Hu, Y., Harmon, T., Pentel, P., Ehrich, M., & Zhang, C. (2017). Rationalization of a nanoparticle-based nicotine nanovaccine as an effective next-generation nicotine vaccine: a focus on hapten localization. *Biomaterials*, *138*, 46–56.

Zhao, Z., Hu, Y., Harmon, T., Pentel, P. R., Ehrich, M., & Zhang, C. (2018). Hybrid nanoparticle-based nicotine nanovaccines: Boosting the immunological efficacy by conjugation of potent carrier proteins. *Nanomedicine*, *14*(5), 1655–1665.

Zhao, Z., Hu, Y., Hoerle, R., Devine, M., Raleigh, M., Pentel, P., et al. (2017). A nanoparticle-based nicotine vaccine and the influence of particle size on its immunogenicity and efficacy. *Nanomedicine*, *13*(2), 443–454.

Zhao, Z., Powers, K., Hu, Y., Raleigh, M., Pentel, P., & Zhang, C. M. (2017). Engineering of a hybrid nanoparticle-based nicotine nanovaccine as a next-generation immunotherapeutic strategy against nicotine addiction: a focus on hapten density. *Biomaterials*, *123*, 107–117.

Zheng, H., Hu, Y., Huang, W., de Villiers, S., Pentel, P., Zhang, J., et al. (2015). Negatively charged carbon nanohorn supported cationic liposome nanoparticles: a novel delivery vehicle for anti-nicotine vaccine. *Journal of Biomedical Nanotechnology*, *11*(12), 2197–2210.

CHAPTER

58

Treating Nicotine Dependence in Psychiatric Hospitals

Emily A. Stockings

National Drug and Alcohol Research Centre (NDARC), UNSW Sydney, Randwick, NSW, Australia

Abbreviations

CMS Centers for Medicare & Medicaid Services
NRT nicotine replacement therapy

58.1 INTRODUCTION

Smoking remains a leading global cause of morbidity and mortality. Persons with a mental disorder smoke at higher rates and are more nicotine-dependent, less likely to quit, and more likely to die from smoking-related disease than persons in the general population. Given that rates of smoking and nicotine dependence are highest among persons with the most severe mental disorders, including schizophrenia and psychotic disorders, psychiatric hospitals may prove to be a key setting for the provision of effective nicotine dependence treatment to this vulnerable group.

Treating nicotine dependence in psychiatric hospitals has long been a contentious issue. On one hand, admission to a psychiatric hospital may be viewed as an opportunity to address smoking among a highly dependent and vulnerable group of smokers given the widespread implementation of smoking bans and accompanying provision of pharmacological and behavioral smoking cessation support. On the other hand, smoking bans and nicotine dependence treatment may be viewed as a violation of patients' basic rights and individual freedom, as inpatient psychiatric hospitals act as a temporary residence for patients. There is also a long-standing culture of tobacco smoking within psychiatric hospitals, whereby cigarettes have historically been provided to reward compliant patient behavior and to facilitate social bonds between patients and clinical staff. Despite this entrenched smoking culture, an emerging body of evidence indicates that treating nicotine dependence in psychiatric hospitals is both possible and potentially effective.

This chapter explores psychiatric hospitals as a setting for treating nicotine dependence. It covers the prevalence of smoking among persons with a mental disorder, an overview of the smoking culture within psychiatric hospitals, the introduction of smoke-free policies and barriers to implementation in psychiatric hospitals, the prevalence of treatment for nicotine dependence, the proposed methods to improve care delivery, and the considerations for the continuation of care after discharge from the hospital.

58.2 SMOKING AMONG PERSONS WITH A MENTAL DISORDER

58.2.1 Smoking Prevalence and Health Burden

Population-based studies of smoking prevalence have estimated rates of smoking among persons with a mental disorder to be at least double in the general population (Lawrence, Hafekost, Hull, Mitrou, & Zubrick, 2013), with smoking prevalence increasing as the number and severity of lifetime mental disorders increases (Lasser et al., 2000); see Fig. 58.1.

Prevalence of smoking has been shown to vary depending on diagnosis and setting, with rates found to range between 36% and 39% for persons with anxiety and personality disorders (Lineberry, Allen, Nash, & Galardy, 2009), 36%–49% for persons with depression and mood disorders (Lasser et al., 2000; Lineberry et al., 2009), and between 70% and 88% for persons with psychotic disorders (including schizophrenia and schizoaffective disorders). Some of the highest reported rates

Neuroscience of Nicotine
https://doi.org/10.1016/B978-0-12-813035-3.00058-7

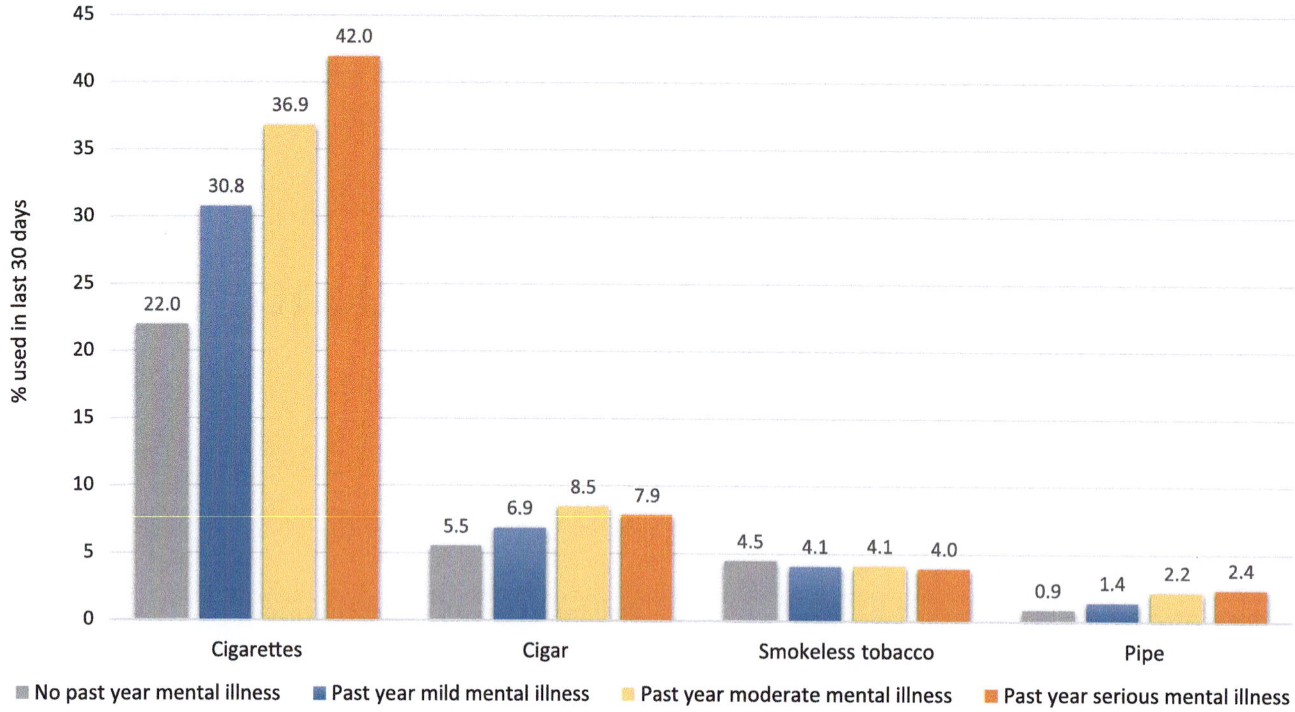

FIG. 58.1 Past 30-day use (%) of specific tobacco products among adults with no past-year mental illness ($n = 34{,}580$), past-year mild mental illness ($n = 4434$), past-year moderate mental illness ($n = 2371$), and past-year serious mental illness ($n = 2176$).
Rates of smoking tend to increase as the severity of mental illness increases. Persons with past-year serious mental illness have the highest rates of smoking, which are approximately double the smoking rate of persons with no past-year mental illness. Graph generated using data from the United States Department of Health and Human Services 2015 National Survey on Drug Use and Health (NSDUH): https://nsduhweb.rti.org/respweb/homepage.cfm. Data refer to adults aged 18 years and older.

of smoking have been identified among individuals attending psychiatric hospitals, with rates of smoking up to 80% identified in this group (McManus, Meltzer, & Campion, 2010); see Fig. 58.2. In addition to increased smoking prevalence, smokers with a mental disorder also smoke more heavily and are more nicotine-dependent than smokers in the general population (Diaz, Rendon, Velasquez, Susce, & de Leon, 2006). As a result of the disproportionate smoking rate among this group, persons with a mental disorder have been estimated to have a reduced life expectancy of 12–15 years compared to the general population, with most excess deaths attributable to cardiovascular disease and cancer (Lawrence, Hancock, & Kisely, 2013).

58.2.2 Factors Associated With Elevated Smoking Rates Among Persons With a Mental Disorder

Biological, social, and psychological factors contribute to the disproportionately high rate of smoking among persons with mental disorder. Some researchers suggest that persons with a mental disorder may have an increased genetic vulnerability to the effects of nicotine and experience greater reward or pleasure than smokers without such disorders (Williams & Ziedonis, 2004). Persons with a mental disorder may also initiate smoking in an attempt to self-medicate (Hall & Prochaska, 2009). There is also some evidence that nicotine may improve cognitive functioning in persons with schizophrenia, including sustained attention and improved filtering of auditory information (Depatie et al., 2002).

Potential social contributors to the disproportionate prevalence of smoking among persons with a mental disorder may include poverty and financial stress, unemployment, homelessness, lower levels of education, access to health services and social support networks, and peer pressure (Australian Institute of Health and Welfare, 2012). Persons with a mental disorder are also more likely to be exposed to cultural norms of smoking at home and in the workplace (Hiscock, Bauld, Amos, Fidler, & Munafò, 2012). Importantly, smokers with a mental disorder have reduced access to smoking cessation resources (Williams, Steinberg, Griffiths, & Cooperman, 2013) and are also less likely to receive support for nicotine dependence when in contact with health-care services, particularly in inpatient psychiatric hospitals (Wye et al., 2010).

58.2.3 A Smoking "Culture"

A long-standing culture of tobacco smoking exists within inpatient psychiatric hospitals. Historically, treating

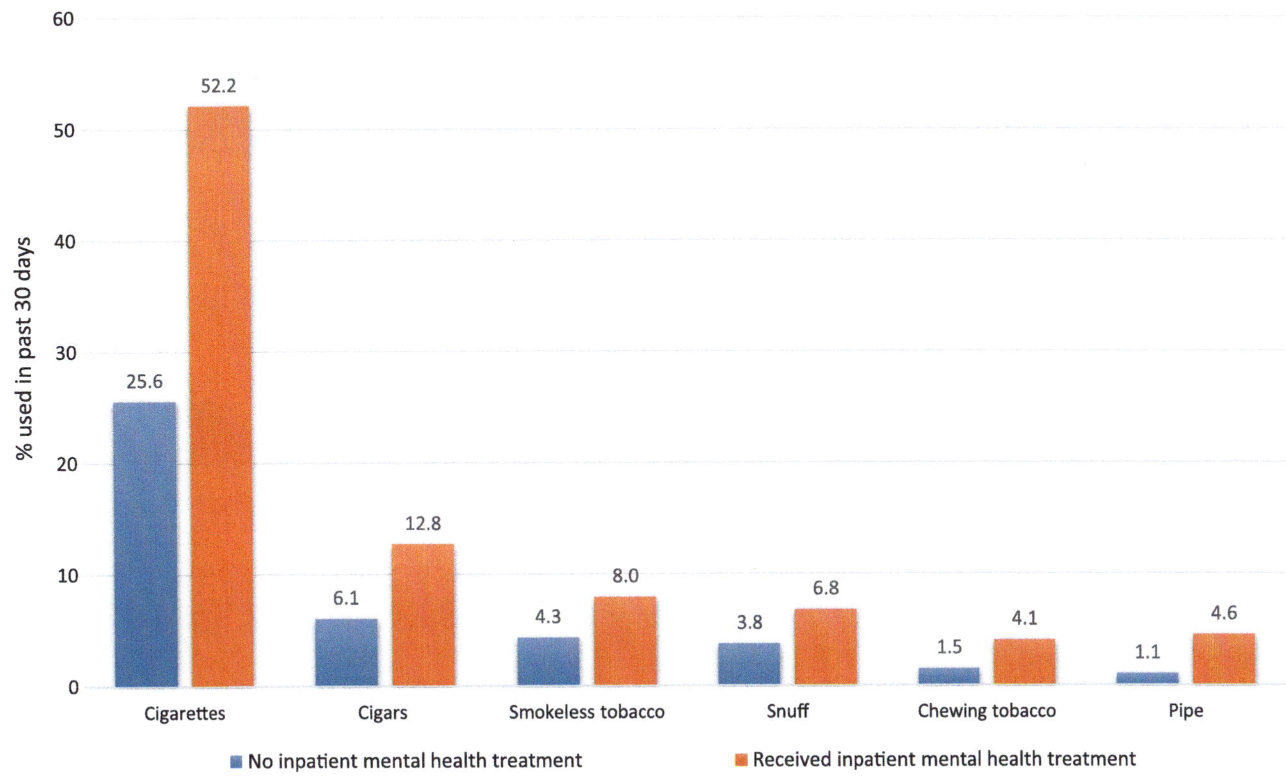

FIG. 58.2 Past 30-day use (%) of specific tobacco products among adults reporting receipt of inpatient mental health treatment in the past year ($n = 429$) compared to adults reporting no use of inpatient mental health treatment in the past year ($n = 41{,}115$).
Smoking rates among persons receiving inpatient psychiatric treatment are substantially higher than persons not receiving mental health treatment, across all forms of tobacco products. Rates of cigarette smoking in particular (52%) are roughly double in the general population estimates (26%). Graph generated using data from the United States Department of Health and Human Services 2014 National Survey on Drug Use and Health (NSDUH): https://nsduhweb.rti.org/respweb/homepage.cfm. Data refer to adults aged 18 years and older.

staff have been known to regularly supply cigarettes to patients and, in some instances, purchase cigarettes for patients who were otherwise unable to do so (Lawn & Campion, 2013). Cigarettes have, even recently, been used as a means of behavior modification, for example, to avoid perceived aggressive or violent behavior or to reward positive behavior such as medication compliance (Lawn, 2010). Smoking has also been found to serve as a common social activity for patients and staff in psychiatric hospitals (Williams & Ziedonis, 2004), thereby reinforcing the "culture" of smoking within psychiatric hospitals.

58.3 SMOKE-FREE POLICIES IN PSYCHIATRIC HOSPITALS

58.3.1 Difficulties Implementing Smoke-Free Policies in Psychiatric Hospitals

Restrictions on smoking in public places and workplaces—including psychiatric hospitals—have been introduced in most nations globally (Fig. 58.3) (Callinan, Clarke, Doherty, & Kelleher, 2010). However, the implementation of such smoking bans has been problematic in inpatient psychiatric facilities. Until recently, many inpatient psychiatric facilities (particularly those that accommodate involuntary patients) have often been made exempt from smoking bans (Ratschen, Britton, & McNeill, 2008). Where smoking bans have been introduced in psychiatric hospitals, they have often been met with resistance from staff (McNally et al., 2006) and patients (Willemsen, Görts, Soelen, Jonkers, & Hilberink, 2004), and as such, implementation of and adherence to smoking bans have been reported to be poor (Stockings et al., 2015).

Barriers to the effective implementation of smoking bans and provision of nicotine dependence treatment in psychiatric hospitals may include the misperception among staff that smoking bans will disrupt the ward's treatment milieu and that negative psychiatric symptoms, particularly aggression and agitation, will be exacerbated with the restriction of smoking (Lawn & Campion, 2013). However, recent evidence indicates that rates of physical violence may actually decrease following the implementation of a comprehensive smoke-free policy (Robson et al., 2017). Another potential barrier to the successful implementation of smoke-free policies in these settings is the commonly held perception among

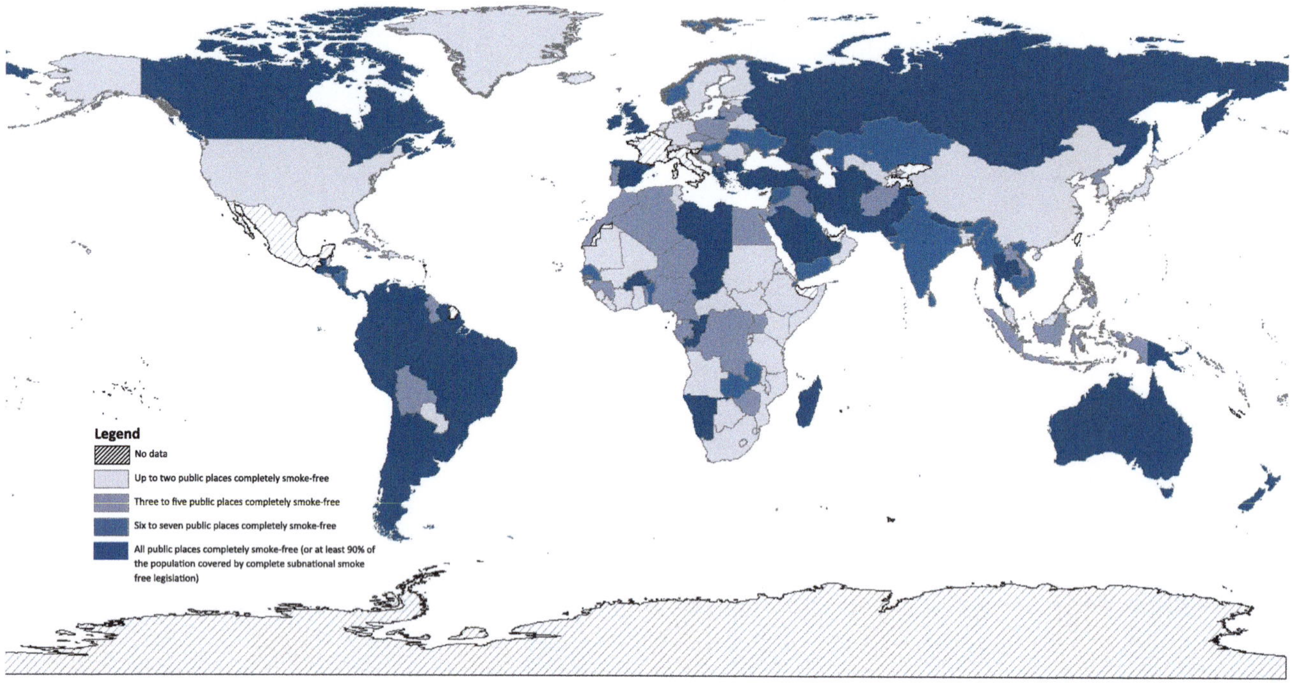

FIG. 58.3 Global coverage of smoke-free policies in public places.
The acknowledgment of the harms of secondhand tobacco smoke has led to the implementation of smoke-free policies in public places, including health-care facilities, across the globe. However, the number of public places with smoke-free policies varies substantially by country. Map generated using data from the WHO Global Health Observatory data repository for tobacco control: http://apps.who.int/gho/data/node.main.1241?lang=en. Data are from 2014 and represent smoke-free policies at either the national or subnational level.

mental health staff that their patients are not motivated or willing or able to quit (Lawn, 2004). Contrary to this view, many smokers in psychiatric hospitals are motivated to quit and have often made several substantial attempts to do so (Stockings et al., 2013).

58.3.2 Prevalence of Treatment for Nicotine Dependence in Psychiatric Hospitals

Clinical practice guidelines recommend the provision of nicotine dependence treatment to patients in psychiatric hospitals in developed nations such as the United States, the United Kingdom, and Australia (Fiore, Jaen, & Baker, 2008; New South Wales Department of Health, 2002; Ratschen, Britton, & McNeill, 2009). These guidelines require treating staff to provide pharmacological and behavioral smoking cessation supports, such as brief advice to quit and nicotine replacement therapy (NRT) to all patients who are identified as smokers on admission. Such cessation supports are provided in health-care settings in varying degrees globally, the costs of which are typically subsidized by governments (Fig. 58.4). Medical record audits of the provision of such nicotine dependence treatment in inpatient psychiatric hospitals in the United States and Australia have indicated that recording of treatment for nicotine dependence occurs rarely, if ever, and provision of nicotine dependence treatment is negligible (Prochaska, Gill, & Hall, 2004; Wye et al., 2010). Studies conducted in Denmark (Willemsen et al., 2004), the United States (Bronaugh & Frances, 1990), the United Kingdom (Ratschen, Britton, Doody, & McNeill, 2010), and Australia (Wye et al., 2014) have also reported that surreptitious smoking continues to occur in "smoke-free" psychiatric hospitals, exposing staff, patients, and visitors to the harmful effects of environmental tobacco smoke. The low rates of treatment for nicotine dependence and violation of smoke-free policies despite the presence of clinical practice guidelines signal the need for strategies to improve the delivery of treatment for nicotine dependence in psychiatric hospitals.

58.4 IMPROVING PROVISION OF TREATMENT FOR NICOTINE DEPENDENCE IN PSYCHIATRIC HOSPITALS

58.4.1 A Systems-Change Approach

A systems-change approach has been recommended as a means to improve the provision of timely and effective treatment for nicotine dependence in health-care settings, including psychiatric hospitals (Bonevski, 2014; Fiore, Keller, & Curry, 2007). The key systems-change

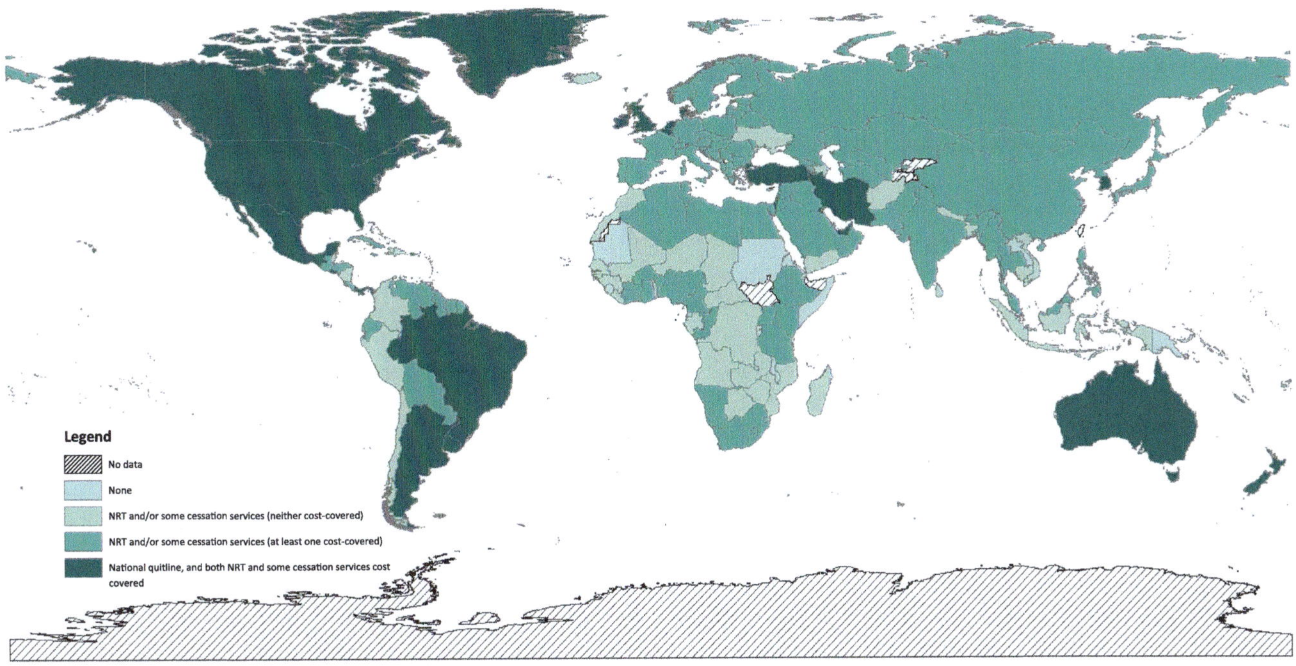

FIG. 58.4 Global availability of smoking cessation support.
Smoking cessation support is available in most countries globally; however, the number of support services available and the degree to which the costs of these products are subsidized by governments vary by country. Smoking cessation support is defined as nicotine replacement therapy (NRT), cessation counseling, or any other cessation support available in any of the following places: health clinics or other primary care facilities, hospitals, office of a health professional, or the community. Map generated using data from the 2014 WHO Global Health Observatory data repository for tobacco control: http://apps.who.int/gho/data/node.main.1241?lang=en.

strategies comprising this approach are summarized in Table 58.1 and include the systematic assessment of smoking status of patients on admission, staff education, promotion of policies that support the provision of treatment for nicotine dependence, subsidies for patients accessing such care, and reimbursement to staff for care provision (Fiore et al., 2007). The implementation of such systems-change strategies has been shown to increase the number of patients who were counseled or assisted to quit in general hospital settings by up to 17% relative to hospitals without such approaches (Freund et al., 2009).

TABLE 58.1 Strategies to Improve the Delivery of Treatment for Nicotine Dependence in Treatment in Health-Care Settings, Including Psychiatric Hospitals

Strategy	
1	Implement systems to identify and record patients' smoking status and nicotine dependence upon admission
2	Provide staff education, resources and develop communication channels to enable feedback and evaluation
3	Dedicate staff to provide treatment for nicotine dependence and evaluate its delivery
4	Promote policies that support the provision of treatment for nicotine dependence
5	Offer treatment for nicotine dependence free of charge or subsidized
6	Reimburse clinicians for delivery of effective treatment for nicotine dependence and include this treatment among the defined duties of clinicians

A number of strategies are available to improve the systematic delivery of treatment for nicotine dependence in psychiatric hospitals. These include both systems-level strategies, such as introducing new admission forms with items to assess tobacco use, and individual-level strategies, such as providing sufficient staff training and feedback (Fiore et al., 2007).

58.4.2 Interventions to Increase the Provision of Treatment for Nicotine Dependence in Psychiatric Hospitals

While there is limited evidence regarding the effectiveness of interventions to increase the treatment of nicotine dependence in psychiatric hospitals, there is emerging evidence that clinical practice and systems-change interventions used in general hospital settings may be similarly effective.

In 2015 in the United States, the Centers for Medicare & Medicaid Services (CMS) introduced new reporting requirements regarding treatment for nicotine dependence in psychiatric hospitals receiving Medicare funds. These reporting requirements included the systematic screening of tobacco use and the offer or provision of tobacco use treatment (including counseling or cessation medication). Facilities that failed to report these measures faced a 2% annual funding penalty. An analysis of one hospital before and after the introduction of such measures found that screening for tobacco use increased to close to 100% of admissions, referral to cessation counseling increased 18-fold, and referral for cessation medication also increased substantially (Carrillo et al., 2017). This study provides good evidence that performance-based penalties may improve the provision of treatment for nicotine dependence in psychiatric hospitals.

Another approach to increasing the treatment of nicotine dependence in psychiatric facilities is to employ multiple strategies that address key barriers to the provision of such care. For example, a multistrategy clinical practice systems-change intervention incorporating leadership and consensus (e.g., consultation with senior staff) and enabling systems and procedures (e.g., modifying admission forms with items to assess tobacco use), staff training and education, provision of resources (e.g., easy-to-access copies of clinical practice guidelines), and audit and feedback found increases in all five measures of nicotine dependence treatment across two inpatient psychiatric hospitals (Wye, Stockings, Bowman, Oldmeadow, & Wiggers, 2017); see Fig. 58.5. Increases in the recording of smoking status and assessment of nicotine dependence, offering of quit advice, prescription of NRT, and provision of treatment for nicotine dependence upon discharge from the facility were identified up to 6 months after the introduction of the intervention.

58.5 CONTINUATION OF NICOTINE DEPENDENCE TREATMENT AFTER DISCHARGE FROM THE HOSPITAL

Admission to hospital is an opportune time to address smoking; however, the effectiveness of inpatient treatment for nicotine dependence is limited if no continuation of care is provided postdischarge (Bowman & Stockings, 2013). It has been suggested that smoke-free policies in inpatient psychiatric settings have had little effect on

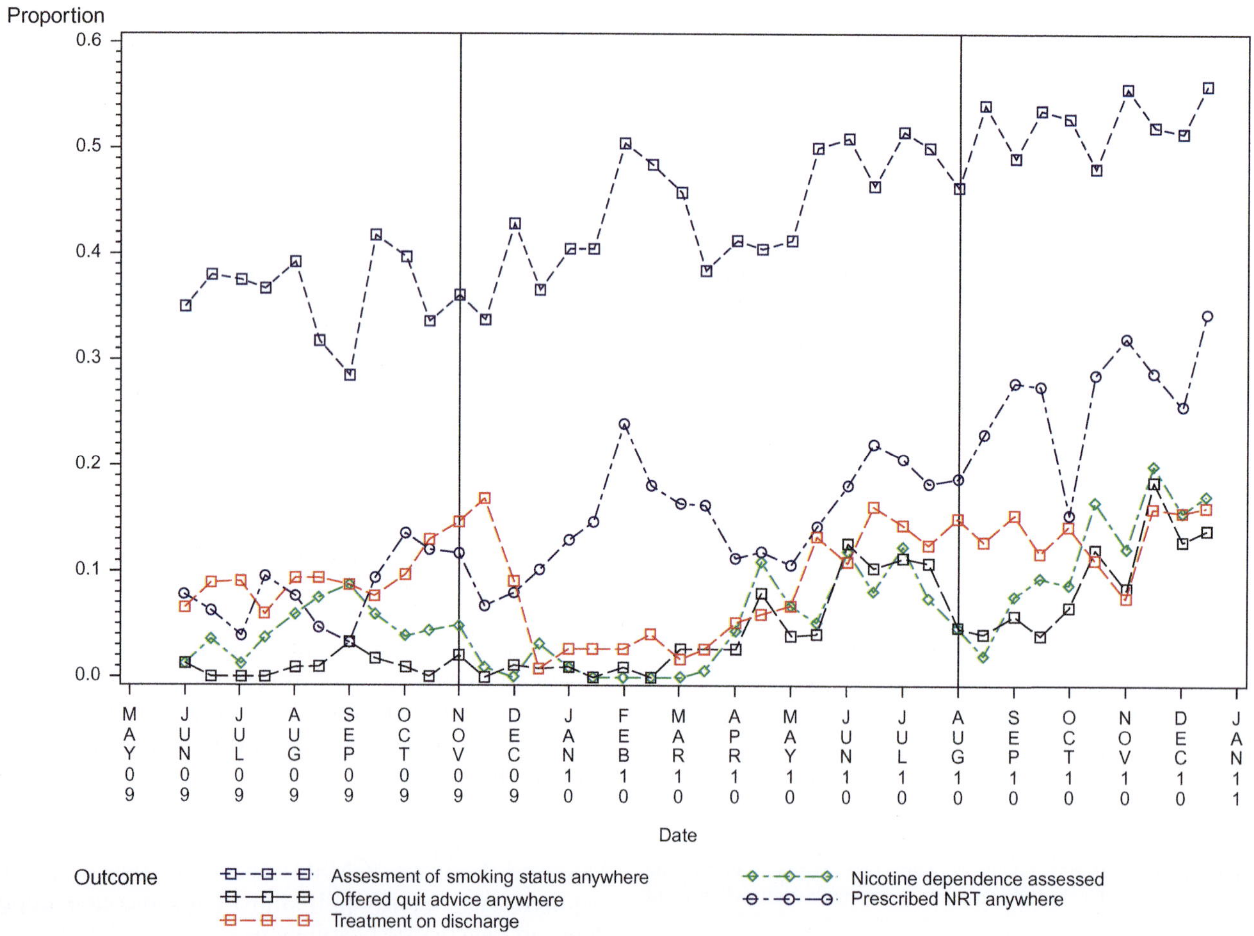

FIG. 58.5 Change in proportion (%) of the recording of five elements of treatment for tobacco dependence in audited medical records of an Australian inpatient psychiatric hospital, pre, during, and postdelivery of a clinical practice change intervention.
Introducing a clinical practice change intervention in an Australian inpatient psychiatric hospital resulted in increases in the recording of five key measures of nicotine dependence treatment to patients, as measured using medical record audit. *The figure originally published by Wye et al. (2017) in BMC Psychiatry is reprinted here under the creative Commons Attribution 4.0 International License: http://creativecommons.org/licenses/by/4.0/.*

long-term smoking cessation due to the lack of coordination between inpatient and community mental health services (Campion, Checinski, & Nurse, 2008). Therefore, in order to increase quit rates, patients should be offered up to a month of ongoing treatment for nicotine dependence upon discharge from a smoke-free hospital facility (Rigotti, Clair, Munafò, & Stead, 2012). Randomized controlled trials conducted in Australia (Metse et al., 2017; Stockings et al., 2014) and the United States (Prochaska, Hall, Delucchi, & Hall, 2014) indicate that continuing treatment for nicotine dependence for up to 4 months following discharge from an inpatient psychiatric hospital can increase quit rates up to 18 months postdischarge.

MINI-DICTIONARY OF TERMS

Behavioral smoking cessation support Any approach to smoking cessation that includes supportive elements to assist in changing smoking behavior, such as supportive counseling, cognitive behavior therapy, and brief advice to quit. Such support can be delivered face-to-face in groups or individually, online, via telephone (e.g., "quitlines"), or via mobile phone applications.

Clinical practice guidelines for nicotine dependence A set of rules requiring clinical staff to provide treatment for nicotine dependence to manage nicotine withdrawal symptoms and promote smoking cessation in settings where a smoke-free policy or smoking ban is in place. This may include **pharmacological smoking cessation support** and **behavioral smoking cessation support**.

Mental disorder Also referred to as "mental illness" or "psychiatric disorder," is any behavior or mental pattern that causes impairment to health and daily life. Common mental disorders include depression and anxiety, and less frequent but more severe mental disorders include bipolar disorder, eating disorders, and schizophrenia. These disorders are diagnosed by trained mental health professionals according to diagnostic criteria; however, persons may suffer the symptoms of a mental disorder without meeting full diagnostic criteria. Mental disorders are typically described as being "current" if they have been present in the past 12 months.

Nicotine dependence A state of dependence upon nicotine that typically includes the presence of frequent, heavy smoking accompanied by cravings or urges to smoke and unpleasant withdrawal symptoms when nicotine is not available.

Nicotine replacement therapy (NRT) Products that allow the delivery of nicotine into the bloodstream in a medically safe manner without the presence of tobacco. Delivery routes include the nose or the mouth (e.g., chewing gum, lozenges, and nasal spray) and the skin (e.g., nicotine patches).

Pharmacological smoking cessation support Any medication used to treat nicotine dependence. This includes medications that replace nicotine using safer delivery systems that do not pose health risks (i.e., **nicotine replacement therapy**), medications that mimic the action of nicotine on the brain's nicotinic receptors (i.e., nicotine receptor partial agonists such as varenicline), and medications that inhibit the action of nicotinic receptors (i.e., nicotinic antagonists such as bupropion).

Psychiatric hospital Hospitals, wards, or dedicated health services that provide treatment to persons suffering symptoms of acute mental disorders. Facilities may provide short- or long-term psychiatric care on an inpatient or outpatient basis. Patients may be admitted on a voluntary or involuntary basis, the latter typically being a legal process for persons who may pose a harm to themselves or others.

Smoke-free policy Also referred to as "smoking bans," are a set of guidelines that require persons to refrain from tobacco smoking in order to protect people from the harms of secondhand tobacco smoke. Such policies can range from comprehensive bans that prohibit smoking in buildings and grounds to partial bans that may prohibit smoking in some areas but permit smoking in other areas (e.g., designated smoking rooms).

Key Facts of Treating Nicotine Dependence in Psychiatric Hospitals

- Rates of smoking range from 36% to 50% for persons with personality disorders, anxiety, and depression and between 70% and 88% for persons with psychotic disorders including schizophrenia and schizoaffective disorders.
- Persons with a mental disorder have a reduced life expectancy of 12–15 years, primarily due to tobacco-related diseases such as cardiovascular disease and cancers.
- Despite common perceptions, smokers with a mental disorder are motivated to quit, and many have often made several substantial attempts to do so.
- There is a long history of a tobacco smoking "culture" in inpatient psychiatric hospitals, and many facilities were made exempt from smoking bans until more recently.
- The introduction of smoking bans has been difficult in these settings, and rates of receipt of nicotine dependence treatment have been low.
- Many common staff perceptions about introducing smoking bans are often not true; for example, introducing comprehensive smoke-free policies has actually been associated with reduced physical violence and aggression among patients.
- Clinical practice change interventions involve multiple strategies across the whole hospital to increase staff support and delivery of nicotine dependence treatment to patients.
- To date, clinical practice change interventions in inpatient psychiatric facilities have resulted in increased screening for tobacco use, delivery of nicotine dependence treatment, and cessation support.
- Staying in a smoke-free psychiatric facility may increase a smoker's chance of quitting after discharge, however, only if smoking cessation treatment is continued.

Summary Points

- This chapter focuses on nicotine dependence treatment in inpatient psychiatric facilities.
- Persons admitted to inpatient psychiatric facilities have among the highest rates of smoking (up to 80%); however, receipt of smoking cessation treatment is low.

- Smoking bans have been introduced in most inpatient psychiatric hospitals in developed countries, providing an opportunity to deliver nicotine dependence treatment to all smokers in a systematic manner.
- However, staff and patients of psychiatric hospitals have historically viewed smoking bans negatively, the bans are often violated, and delivery of nicotine dependence treatment is low.
- Several strategies are available to improve the systematic delivery of nicotine dependence treatment in psychiatric hospitals, including clinical practice change interventions.
- Evidence regarding the effectiveness of clinical practice change interventions to increase the delivery of nicotine dependence treatment in psychiatric hospitals is still emerging, but results to date are promising.
- Receiving nicotine dependence treatment during a smoke-free psychiatric hospitalization may also result in increased quit rates after discharge; however, this only occurs if treatment is continued for at least 1 month postdischarge.
- Linking inpatient nicotine dependence treatment to ongoing community-based smoking cessation supports is needed to reduce the elevated rates of smoking and smoking-related disease among persons with a mental disorder.

References

Australian Institute of Health and Welfare. (2012). *Australia's health 2012*. Canberra: AIHW.

Bonevski, B. (2014). System-centred tobacco management: from 'whole-person' to 'whole-system' change. *Drug and Alcohol Review*, *33*(1), 99–101. https://doi.org/10.1111/dar.12086.

Bowman, J., & Stockings, E. A. (2013). Smoking cessation for hospitalised patients: intensive behavioural counselling started in hospital and continued after discharge increases quit rates; with additional benefit from adding nicotine replacement therapy. *Evidence-Based Nursing*, *16*(1), 21–22. https://doi.org/10.1136/eb-2012-100890.

Bronaugh, T. A., & Frances, R. J. (1990). Establishing a smoke-free inpatient unit: is it feasible? *Hospital & Community Psychiatry*, *41* (12), 1303–1305.

Callinan, J. E., Clarke, A., Doherty, K., & Kelleher, C. (2010). Legislative smoking bans for reducing secondhand smoke exposure, smoking prevalence and tobacco consumption. *Cochrane Database of Systematic Reviews*, *4*: CD005992. https://doi.org/10.1002/14651858.CD005992.pub2.

Campion, J., Checinski, K., & Nurse, J. (2008). Review of smoking cessation treatments for people with mental illness. *Advances in Psychiatric Treatment*, *14*, 208–216.

Carrillo, S., Nazir, N., Howser, E., Shenkman, L., Laxson, M., Scheuermann, T. S., et al. (2017). Impact of the 2015 CMS inpatient psychiatric facility quality reporting rule on tobacco treatment. *Nicotine & Tobacco Research*, *19*(8), 976–982. https://doi.org/10.1093/ntr/ntw386.

Depatie, L., O'Driscoll, G. A., Holahan, A. L., Atkinson, V., Thavundayil, J. X., Kin, N. N., et al. (2002). Nicotine and behavioral markers of risk for schizophrenia: a double-blind, placebo-controlled, cross-over study. *Neuropsychopharmacology*, *27* (6), 1056–1070. https://doi.org/10.1016/s0893-133x(02)00372-x.

Diaz, F., Rendon, D., Velasquez, D., Susce, M., & de Leon, J. (2006). Datapoints: smoking and smoking cessation among persons with severe mental illnesses. *Psychiatric Services*, *57*(4), 462. https://doi.org/10.1176/appi.ps.57.4.462.

Fiore, M., Jaen, C., & Baker, T. (2008). *Treating tobacco use and dependence: 2008 update: Clinical practice guideline*. Rockville, MD: US Department of Health and Human Services, Public Health Service.

Fiore, M. C., Keller, P. A., & Curry, S. J. (2007). Health system changes to facilitate the delivery of tobacco-dependence treatment. *American Journal of Preventive Medicine*, *33*(6 Suppl), S349–S356. https://doi.org/10.1016/j.amepre.2007.09.001.

Freund, M., Campbell, E., Paul, C., Sakrouge, R., McElduff, P., Walsh, R. A., et al. (2009). Increasing smoking cessation care provision in hospitals: a meta-analysis of intervention effect. *Nicotine & Tobacco Research*, *11*(6), 650–662. https://doi.org/10.1093/ntr/ntp056.

Hall, S. M., & Prochaska, J. J. (2009). Treatment of smokers with co-occurring disorders: emphasis on integration in mental health and addiction treatment settings. *Annual Review of Clinical Psychology*, *5*, 409–431.

Hiscock, R., Bauld, L., Amos, A., Fidler, J. A., & Munafò, M. (2012). Socioeconomic status and smoking: a review. *Annals of the New York Academy of Sciences*, (1248), 107–123. https://doi.org/10.1111/j.1749-6632.2011.06202.x.

Lasser, K., Boyd, J. W., Woolhandler, S., Himmelstein, D. U., McCormick, D., & Bor, D. H. (2000). Smoking and mental illness: a population based prevalence study. *Journal of the American Medical Association*, *284*(20), 2606–2610. https://doi.org/10.1001/jama.284.20.2606.

Lawn, S. J. (2004). Systemic barriers to quitting smoking among institutionalised public mental health service populations: a comparison of two Australian sites. *International Journal of Social Psychiatry*, *50*(3), 204–215. https://doi.org/10.1177/0020764004043129.

Lawn, S. (2010). *The culture of smoking in mental health service populations*. Berlin: Lambert Academic Publishing.

Lawn, S., & Campion, J. (2013). Achieving smoke-free mental health services: lessons from the past decade of implementation research. *International Journal of Environmental Research and Public Health*, *10* (9), 4224–4244. https://doi.org/10.3390/ijerph10094224.

Lawrence, D., Hafekost, J., Hull, P., Mitrou, F., & Zubrick, S. R. (2013). Smoking, mental illness and socioeconomic disadvantage: analysis of the Australian National Survey of Mental Health and Wellbeing. *BMC Public Health*, *13*, 462. https://doi.org/10.1186/1471-2458-13-462.

Lawrence, D., Hancock, K. J., & Kisely, S. (2013). The gap in life expectancy from preventable physical illness in psychiatric patients in Western Australia: retrospective analysis of population based registers. *British Medical Journal*, *346*.

Lineberry, T. W., Allen, J. D., Nash, J., & Galardy, C. W. (2009). Population-based prevalence of smoking in psychiatric inpatients: a focus on acute suicide risk and major diagnostic groups. *Comprehensive Psychiatry*, *50*(6), 526–532. https://doi.org/10.1016/j.comppsych.2009.01.004.

McManus, S., Meltzer, H., & Campion, J. (2010). *Cigarette smoking and mental health in England: Data from the Adult Psychiatric Morbidity Survey 2007*. London: National Centre for Social Research.

McNally, L., Oyefeso, A., Annan, J., Perryman, K., Bloor, R., Freeman, S., et al. (2006). A survey of staff attitudes to smoking-related policy and intervention in psychiatric and general health care settings. *Journal of Public Health*, *28*(3), 192–196. https://doi.org/10.1093/pubmed/fdl029.

Metse, A. P., Wiggers, J., Wye, P., Wolfenden, L., Freund, M., Clancy, R., et al. (2017). Efficacy of a universal smoking cessation intervention initiated in inpatient psychiatry and continued post-discharge: a randomised controlled trial. *The Australian and New Zealand Journal of Psychiatry*, *51*(4), 366–381. https://doi.org/10.1177/0004867417692424.

New South Wales Department of Health. (2002). *Guide for the management of nicotine dependent inpatients*. Sydney: State Government of New South Wales.

Prochaska, J. J., Gill, P., & Hall, S. M. (2004). Treatment of tobacco use in an inpatient psychiatric setting. *Psychiatric Services*, *55*(11), 1265–1270. https://doi.org/10.1176/appi.ps.55.11.1265.

Prochaska, J. J., Hall, S. E., Delucchi, K., & Hall, S. M. (2014). Efficacy of initiating tobacco dependence treatment in inpatient psychiatry: a randomized controlled trial. *American Journal of Public Health*, *104* (8), 1557–1565. https://doi.org/10.2105/ajph.2013.301403.

Ratschen, E., Britton, J., Doody, G., & McNeill, A. (2010). Smoking attitudes, behaviour, and nicotine dependence among mental health acute inpatients: an exploratory study. *International Journal of Social Psychiatry*, *56*(2), 107–118. https://doi.org/10.1177/0020764008101855.

Ratschen, E., Britton, J., & McNeill, A. (2008). Smoke-free hospitals - the English experience: results from a survey, interviews and site visits. *BMC Health Services Research*, *8*(1), 41. https://doi.org/10.1186/1472-6963-8-41.

Ratschen, E., Britton, J., & McNeill, A. (2009). Implementation of smoke-free policies in mental health in-patient settings in England. *The British Journal of Psychiatry*, *194*(6), 547–551. https://doi.org/10.1192/bjp.bp.108.051052.

Rigotti, N., Clair, C., Munafò, M., & Stead, L. (2012). Interventions for smoking cessation in hospitalised patients. *Cochrane Database of Systematic Reviews*. (5), CD001837. https://doi.org/10.1002/14651858.CD001837.pub3.

Robson, D., Spaducci, G., McNeill, A., Stewart, D., Craig, T. J. K., Yates, M., et al. (2017). Effect of implementation of a smoke-free policy on physical violence in a psychiatric inpatient setting: an interrupted time series analysis. *Lancet Psychiatry*, *4*(7), 540–546. https://doi.org/10.1016/s2215-0366(17)30209-2.

Stockings, E. A., Bowman, J. A., Baker, A. L., Terry, M., Clancy, R., Wye, P. M., et al. (2014). Impact of a postdischarge smoking cessation intervention for smokers admitted to an inpatient psychiatric facility: a randomized controlled trial. *Nicotine & Tobacco Research*, *16*(11), 1417–1428. https://doi.org/10.1093/ntr/ntu097.

Stockings, E. A., Bowman, J. A., Bartlem, K. M., McElwaine, K. M., Baker, A. L., Terry, M., et al. (2015). Implementation of a smoke-free policy in an inpatient psychiatric facility: patient-reported adherence, support, and receipt of nicotine-dependence treatment. *International Journal of Mental Health Nursing*, *24*(4), 342–349. https://doi.org/10.1111/inm.12128.

Stockings, E., Bowman, J., McElwaine, K., Baker, A., Terry, M., Clancy, R., et al. (2013). Readiness to quit smoking and quit attempts among Australian mental health inpatients. *Nicotine & Tobacco Research*, *15*(5), 942–949. https://doi.org/10.1093/ntr/nts206.

Willemsen, M. C., Görts, C. A., Soelen, P. V., Jonkers, R., & Hilberink, S. R. (2004). Exposure to environmental tobacco smoke (ETS) and determinants of support for complete smoking bans in psychiatric settings. *Tobacco Control*, *13*, 180–185. https://doi.org/10.1136/tc.2003.004804.

Williams, J. M., Steinberg, M. L., Griffiths, K. G., & Cooperman, N. (2013). Smokers with behavioral health comorbidity should be designated a tobacco use disparity group. *American Journal of Public Health*, *103*(9), 1549–1555. https://doi.org/10.2105/ajph.2013.301232.

Williams, J. M., & Ziedonis, D. (2004). Addressing tobacco among individuals with a mental illness or an addiction. *Addictive Behaviors*, *29*, 1067–1083. https://doi.org/10.1016/j.addbeh.2004.03.009.

Wye, P., Bowman, J., Wiggers, J., Baker, A., Carr, V., Terry, M., et al. (2010). An audit of the prevalence of recorded nicotine dependence treatment in an Australian psychiatric hospital. *Australian and New Zealand Journal of Public Health*, *34*(3), 298–303. https://doi.org/10.1111/j.1753-6405.2010.00530.x.

Wye, P., Gow, L. B., Constable, J., Bowman, J., Lawn, S., & Wiggers, J. (2014). Observation of the extent of smoking in a mental health inpatient facility with a smoke-free policy. *BMC Psychiatry*, *14*, 94. https://doi.org/10.1186/1471-244x-14-94.

Wye, P. M., Stockings, E. A., Bowman, J. A., Oldmeadow, C., & Wiggers, J. H. (2017). Effectiveness of a clinical practice change intervention in increasing the provision of nicotine dependence treatment in inpatient psychiatric facilities: an implementation trial. *BMC Psychiatry*, *17*, 56. https://doi.org/10.1186/s12888-017-1220-7.

CHAPTER

59

Oral 18-Methoxycoronaridine (18-MC) Decreases Nicotine Self-Administration in Rats

*Amir H. Rezvani**, Stanley D. Glick*[†], Edward D. Levin***

*Department of Psychiatry and Behavioral Sciences, Duke University Medical Center, Durham, NC, United States
[†]Department of Neuroscience and Experimental Therapeutics, Albany Medical College, Albany, New York, United States

Abbreviations

18-MC	18-methoxycoronaridine
IV	Intravenous
VTA	ventral tegmental area

59.1 INTRODUCTION

Nicotine addiction is a major health problem with severe health and economic consequences throughout the world. More than 430,000 deaths are attributable to smoking each year in the United States alone, and over 3 million smoking-related deaths have been reported annually worldwide. It has been projected that if this disorder is not controlled properly, it will become the largest single health problem, with devastating health and socioeconomic consequences over the next 2–3 decades and resulting in more than 8 million deaths worldwide annually. Several medications, including the nicotine patch, bupropion, and varenicline, have been approved and used for the treatment of nicotine addiction. However, similar to other addictions, nicotine addiction is a heterogeneous and complex disorder; despite the availability of these pharmacological tools, the efficacy of these drugs is relatively small, and they have some unwanted side effects leading to high rates of relapse. Thus, the development of more suitable medications with no or less unwanted side effects to cover more subpopulations remains a challenging goal for the addiction field. Moreover, a greater diversity of effective treatments, at least partially tailored to the need of the individual and based on genetic makeup, may be required. One of these promising therapeutic agents is 18-methoxycoronaridine (18-MC) (Fig. 59.1).

18-Methoxycoronaridine (18-MC) is a nontoxic synthetic iboga alkaloid congener, derived from ibogaine, an indole alkaloid found in the root bark of the West-Central African shrub, *Tabernanthe iboga* (*Apocynaceae* family) (Bandarage, Kuehne, & Glick, 1999). The crude extracts of ibogaine itself in high doses cause a feeling of confusion and sometimes hallucinations. Ibogaine itself and its primary metabolite, noribogaine, have been effective in reducing self-administration of addictive drugs such as nicotine, alcohol, morphine, heroin, and cocaine in rats (Glick, Kuehne, Maisonneuve, Bandarage, & Molinari, 1996; Glick & Maisonneuve, 1998; Glick, Rossman, Steindorf, Maisonneuve, & Carlson, 1991; Mash et al., 1998; Rezvani, Overstreet, & Lee, 1995; Rezvani, Overstreet, Perfumi, & Massi, 2003; Rezvani et al., 1997) by affecting dopaminergic and serotonergic systems in the brain (Glick et al., 1991).

Ibogaine itself, at high doses, can cause severe unwanted side effects (Molinari, Maisonneuve, & Glick, 1996) that may hinder its use; thus, a synthetic ibogaine analog, 18-MC, with no toxicity but with the same suppressing effects on drug self-administration, has been designed.

18-MC has been shown to suppress self-administration of several addictive drugs including alcohol (Rezvani et al., 1997), morphine, cocaine, methamphetamine, and nicotine (Glick et al., 1998; Maisonneuve & Glick, 1999) in rats. Since the central dopaminergic system has been shown to be involved in the reinforcing property of most drugs of abuse, it is believed that 18-MC also exerts its action by modulating the dopaminergic system in the brain.

To help further determine the possible clinical use of 18-MC, it was important to see if it was also effective in reducing drug self-administration following oral

Neuroscience of Nicotine
https://doi.org/10.1016/B978-0-12-813035-3.00059-9

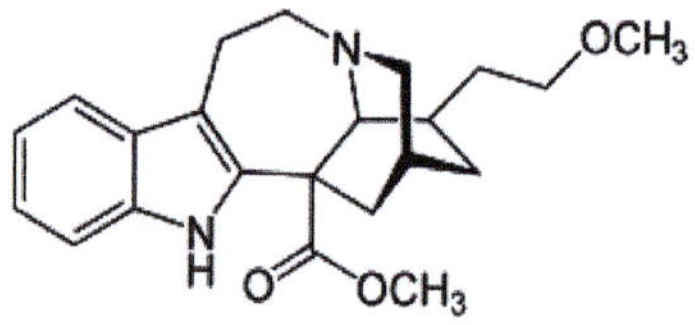

FIG. 59.1 Chemical structure of 18-MC.

administration. Recently, a study was conducted to determine the efficacy of orally administered 18-MC for reducing nicotine self-administration in rats.

59.2 METHODS

Using a standard operant box, rats first were trained for IV nicotine self-administration for several days. Then, at weekly intervals, rats were administered by oral gavage doses of 18-MC (10, 20, and 40 mg/kg) or the control vehicle following a repeated measure counterbalanced design (Levin, Slade, Wells, Petro, & Rose, 2011; Rezvani et al., 2013).

59.3 FINDINGS

It was shown that oral administration of 18-MC resulted in a significant ($P<.05$) decrease in nicotine self-administration. The linear trend of decreasing nicotine self-administration over the dose range was significant ($P<.005$). As illustrated in Fig. 59.2, the 40 mg/kg dose caused a significant ($P<.05$) decrease in nicotine self-administration. When the data were analyzed based on the median level of nicotine self-administration during the baseline performance, it was found that the rats with low response averaged 2.82 ± 0.48 nicotine infusions per session while the rats with higher response averaged 8.20 ± 0.86 nicotine infusions per session. They were significantly ($P<.005$) different from each other. Interestingly, the groups with high and low nicotine self-administration did not have significant differences in food pellet self-administration. As illustrated in Fig. 59.3, the rats with lower baseline performance significantly ($P<.01$) decreased their nicotine self-administration following an oral dose of 18-MC, while those with the higher baseline performance did not show a significant effect of 18-MC.

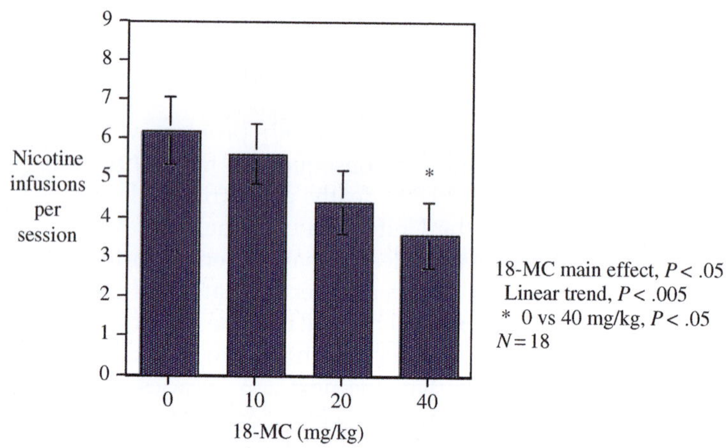

FIG. 59.2 Acute oral 18-MC effects on nicotine self-administration (means ± sem) in female rats.
Data represent means ± sem ($n=18$) (Rezvani et al., 2016).

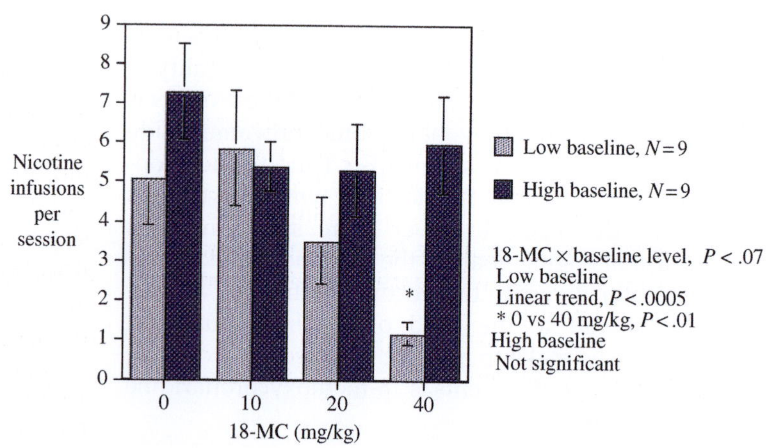

FIG. 59.3 Acute oral 18-MC effects on nicotine self-administration in low- ($n=9$) and high-baseline groups ($n=9$).
Data represent means ± sem (Rezvani et al., 2016).

59.4 DISCUSSION

Systemic administration of 18-MC has been shown to reduce the self-administration of several addictive drugs including nicotine. Regarding the development of 18-MC for possible clinical use, it was important to demonstrate its oral efficacy. Previous studies have shown that oral administration of 18-MC can reduce self-administration of morphine (Maisonneuve & Glick, 1999), cocaine (Glick et al., 1996), methamphetamine (Glick, Maisonneuve, & Dickinson, 2000), and alcohol in rats (Rezvani et al., 1997). The current study was the first study to evaluate the efficacy of oral administration of 18-MC on nicotine self-administration in rats.

The current findings demonstrate that oral administration of 18-MC can also significantly reduce intravenous (IV) nicotine self-administration. These findings replicate and extend earlier findings that acute systemic injection of 18-MC significantly reduced IV nicotine self-administration in rats (Glick et al., 2000). Interestingly, these findings demonstrate that the oral 18-MC was more effective in rats with lower baseline performance compared with the rats with higher baseline performance. The high dose of 18-MC (40 mg/kg) caused a significant effect in reducing nicotine self-administration, while those rats with the higher baseline performance did not show a significant effect of 18-MC. It is noteworthy to mention that the groups with high and low nicotine self-administration did not have significant differences in food pellet self-administration and the number of pellet self-administration did not correlate with the nicotine infusion, indicating that the nicotine self-administration was independent of food reinforcement.

The fact that the oral 18-MC treatment was more effective in rats with the lower baseline performance but not in rats with higher baseline performance may have been due to any of several reasons. The higher avidity of nicotine self-administration in high performers may have presented a tougher target for therapy, requiring a higher dose than was tested in this study. Alternatively, it may be the case that rats with higher levels of nicotine self-administration have a neurobehavioral substrate that is different enough from the low responders to obviate the therapeutic effect of 18-MC. This may have been due to preexisting differences in rats that self-administer greater amounts of nicotine or to the persisting effects of the greater amounts of nicotine self-administered or a combination of these two. These data reinforce our finding that 18-MC is significantly effective in reducing nicotine self-administration, with particular effectiveness in lighter users of nicotine.

The reinforcing properties of most drugs of abuse are partly related to their interactions with the mesolimbic dopaminergic system (Koob, Rassnick, Heinrichs, & Weiss, 1994). Nicotine-induced dopamine release in the nucleus of accumbens of rats has been shown by several investigators (Di Chiara & Imperato, 1988; Grenhoff, Aston-Jones, & Svensson, 1986; Imperato, Mulas, & Di Chiara, 1986; Tizabi, Copeland, Louis, & Taylor, 2002). Interestingly, 18-MC administration has been shown to reduce nicotine-induced dopamine release in the nucleus accumbens of rats (Glick et al., 1998). Thus, it is suggested that 18-MC exerts its effects on the reinforcing effects of addictive drugs including nicotine by blunting their effects on the mesolimbic dopaminergic system and consequently reducing their rewarding effects, which leads to reduced intake.

We previously demonstrated that nicotinic receptors in the brain play a major role in reinforcing effects of nicotine and alcohol (Levin et al., 2010; Rezvani et al., 2010). Indeed, it has been shown that nicotinic $\alpha 3\beta 4$ receptors are blocked by 18-MC (Glick, Maisonneuve, & Kitchen, 2002). Thus, it is possible that, as demonstrated by previous studies, 18-MC exerts its effects by blocking these receptors in the medial habenula (Glick, Sell, & Maisonneuve, 2008; Glick, Sell, McCallum, & Maisonneuve, 2011; Taraschenko, Rubbinaccio, Maisonneuve, & Glick, 2008).

Other possible mechanism is the interaction of 18-MC with endogenous opioid system in the brain. Ibogaine congeners have been reported to have affinity for opioid receptors. It has been demonstrated that ibogaine interacts at the κ-opiate receptor (Deecher et al., 1992). Thus, it is possible that similar to ibogaine, its analog, 18-MC, exerts its suppressant effects on by altering the endogenous opioid system. Other mechanisms also should be considered. These include the interaction of 18-MC with *N*-methyl-D-aspartate (NMDA)-receptor-coupled cation channels (Popik, Layer, & Skolnick, 1994) and with GABAergic systems, which have been implicated in drug addiction.

59.5 CONCLUSIONS

In summary, these findings demonstrated that acute oral administration of 18-MC can significantly reduce nicotine self-administration in rats especially in animals with lower baseline nicotine intake. Although the exact mechanism of action is not fully understood yet, it may exert its suppressant effect on nicotine intake by reducing its reinforcing property as a result of inhibiting or reducing dopaminergic activity in the mesolimbic system and/or interacting with other neuronal systems implicated in nicotine addiction.

MINI-DICTIONARY OF TERMS

18-MC 18-Methoxycoronaridine

IV Intravenous

Nucleus accumbens The nucleus accumbens is a region in the basal forebrain rostral to the preoptic area of the hypothalamus. The nucleus accumbens is a major part of the reward system.

Operant box An operant conditioning chamber, also known as Skinner box, is an enclosed apparatus that contains two levers that an animal can press the active lever in order to obtain food, water, or a particular drug administered into the vein.

Ventral tegmental area The ventral tegmental area (VTA) is a cluster of dopaminergic neurons located close to the midline on the floor of the midbrain. It is implicated in the natural reward circuitry of the brain. Neurons in the VTA project to numerous areas of the brain, including the nucleus acumens and the prefrontal cortex. Stimulation of the VTA leads to dopamine release in these structures.

Key Facts

- Acute oral administration of 18-MC significantly reduces nicotine self-administration in rats.
- 18-MC is more effective in rats with lower nicotine intake.
- 18-MC exerts its action by blunting nicotine-elicited dopamine release in the nucleus accumbens.

Summary Points

- Oral 18-methoxycoronaridine (18-MC) can significantly reduce nicotine self-administration in rats.
- Oral 18-methoxycoronaridine (18-MC) is more effective in animals with lower nicotine intake.
- Nicotine intake is not correlated with food reinforcement.
- It is suggested that 18-MC exerts it effects on nicotine self-administration by blunting nicotine-induced dopamine release in the mesolimbic system.
- With more comprehensive studies, 18-MC should be considered for development as a possible novel therapy for combating smoking addiction.

References

Bandarage, U. K., Kuehne, M. E., & Glick, S. D. (1999). Total syntheses of racemic albifloranine and its anti-addictive congeners, including 18-methoxycoronaridine. *Tetrahedron*, *55*, 9405–9424.

Deecher, D. C., Teitler, M., Soderlund, D. M., Bornmann, W. G., Kuehne, M. E., & Glick, S. D. (1992). Mechanisms of action of ibogaine and harmaline congeners based on radioligand binding studies. *Brain Research*, *571*(2), 242–247.

Di Chiara, G., & Imperato, A. (1988). Drugs abused by humans preferentially increase synaptic dopamine concentrations in the mesolimbic system of freely moving rats. *Proceedings of the National Academy of Sciences of the United States of America*, *85*, 5274–5278.

Glick, S. D., Kuehne, M. E., Maisonneuve, I. M., Bandarage, U. K., & Molinari, H. H. (1996). 18-Methoxycoronaridine, a non-toxic iboga alkaloid congener: effects on morphine and cocaine self-administration and on mesolimbic dopamine release in rats. *Brain Research*, *719*(1–2), 29–35.

Glick, S. D., & Maisonneuve, I. M. (1998). Mechanisms of antiaddictive actions of ibogaine. *Annals of the New York Academy of Sciences*, *844*, 214–226.

Glick, S. D., Maisonneuve, I. M., & Dickinson, H. A. (2000). 18-MC reduces methamphetamine and nicotine self-administration in rats. *NeuroReport*, *11*, 2013–2015.

Glick, S. D., Maisonneuve, I. M., & Kitchen, B. A. (2002). Modulation of nicotine self-administration in rats by combination therapy with agents blocking alpha3beta4 nicotinic receptors. *European Journal of Pharmacology*, *448*(2-3), 185–191.

Glick, S. D., Maisonneuve, I. M., Visker, K. E., Fritz, K. A., Bandarage, U. K., & Kuehne, M. E. (1998). 18-Methoxycoronardine attenuates nicotine-induced dopamine release and nicotine preferences in rats. *Psychopharmacology*, *139*, 274–280.

Glick, S. D., Rossman, K., Steindorf, S., Maisonneuve, I. M., & Carlson, J. N. (1991). Effects and aftereffects of ibogaine on morphine self-administration in rats. *European Journal of Pharmacology*, *195*(3), 341–345.

Glick, S. D., Sell, E. M., & Maisonneuve, I. M. (2008). Brain regions mediating alpha3beta4 nicotinic antagonist effects of 18-MC on methamphetamine and sucrose self-administration. *European Journal of Pharmacology*, *599*(1–3), 91–95.

Glick, S. D., Sell, E. M., McCallum, S. E., & Maisonneuve, I. M. (2011). Brain regions mediating $\alpha 3\beta 4$ nicotinic antagonist effects of 18-MC on nicotine self-administration. *European Journal of Pharmacology*, *669*(1–3), 71–75.

Grenhoff, J., Aston-Jones, G., & Svensson, T. H. (1986). Nicotinic effects on the firing pattern of midbrain dopamine neurons. *Acta Physiologica Scandinavica*, *128*, 151–158.

Imperato, A., Mulas, A., & Di Chiara, G. (1986). Nicotine preferentially stimulates dopamine release in the limbic system of freely moving rats. *European Journal of Pharmacology*, *132*, 337–338.

Koob, G. F., Rassnick, S., Heinrichs, S., & Weiss, F. (1994). Alcohol, the reward system and dependence. *EXS*, *71*, 103–114.

Levin, E. D., Rezvani, A. H., Xiao, Y., Slade, S., Cauley, M., Wells, C., et al. (2010). Sazetidine-A, a selective alpha4beta2 nicotinic receptor desensitizing agent and partial agonist, reduces nicotine self-administration in rats. *The Journal of Pharmacology and Experimental Therapeutics*, *332*(3), 933–939.

Levin, E. D., Slade, S., Wells, C., Petro, A., & Rose, J. E. (2011). D-Cycloserine selectively decreases nicotine self-administration in rats with low baseline levels of response. *Pharmacology, Biochemistry, and Behavior*, *98*, 210–214.

Maisonneuve, I. M., & Glick, S. D. (1999). Attenuation of the reinforcing efficacy of morphine by 18-methoxycoronaridine. *European Journal of Pharmacology*, *383*, 15–21.

Mash, D. C., Kovera, C. A., Buck, B. E., Norenberg, M. D., Shapshak, P., Hearn, W. L., et al. (1998). Medication development of ibogaine as a pharmacotherapy for drug dependence. *Annals of the New York Academy of Sciences*, *844*, 274–292 [Review].

Molinari, H. H., Maisonneuve, I. M., & Glick, S. D. (1996). Ibogaine neurotoxicity: a re-evaluation. *Brain Research*, *737*, 255–262.

Popik, P., Layer, R. T., & Skolnick, P. (1994). The putative anti-addictive drug ibogaine is a competitive inhibitor of [3H]MK-801 binding to the NMDA receptor complex. *Psychopharmacology (Berl)*, *114*(4), 672–674.

Rezvani, A. H., Cauley, M. C., Slade, S., Glick, S., Rose, J. E., & Levin, E. D. (2016). Acute oral 18-methoxycoronaridine (18-MC) decreases both alcohol intake and IV nicotine self-administration in rats. *Pharmacology, Biochemistry, and Behavior*, *150–151*, 153–157.

Rezvani, A. H., Overstreet, D. H., & Lee, Y. W. (1995). Attenuation of alcohol intake by ibogaine in three strains of alcohol-preferring rats. *Pharmacology, Biochemistry, and Behavior, 52* (3), 615–620.

Rezvani, A. H., Overstreet, D. H., Perfumi, M., & Massi, M. (2003). Plant derivatives in the treatment of alcohol dependency. *Pharmacology, Biochemistry, and Behavior, 75*(3), 593–606 [Review].

Rezvani, A. H., Overstreet, D. H., Yang, Y., Maisonneuve, I. M., Bandarage, U. K., Kuehne, M. E., et al. (1997). Attenuation of alcohol consumption by a novel nontoxic ibogaine analogue (18-methoxycoronaridine) in alcohol-preferring rats. *Pharmacology, Biochemistry, and Behavior, 58*(2), 615–619.

Rezvani, A. H., Sexton, H. G., Johnson, J., Wells, C., Gordon, K., & Levin, E. D. (2013). Effects of caffeine on alcohol consumption and nicotine self-administration in rats. *Alcoholism: Clinical and Experimental Research, 37*, 1609–1617.

Rezvani, A. H., Slade, S., Wells, C., Petro, A., Lumeng, L., Li, T. K., et al. (2010). Effects of sazetidine-A, a selective alpha4beta2 nicotinic acetylcholine receptor desensitizing agent on alcohol and nicotine self-administration in selectively bred alcohol-preferring (P) rats. *Psychopharmacology, 211*, 161–174.

Taraschenko, O. D., Rubbinaccio, H. Y., Maisonneuve, I. M., & Glick, S. D. (2008). 18-Methoxycoronaridine: a potential new treatment for obesity in rats? *Psychopharmacology, 201*, 339–350.

Tizabi, Y., Copeland, R. J., Louis, V. A., & Taylor, R. E. (2002). Effects of combined systemic alcohol and central nicotine administration into ventral tegmental area on dopamine release in the nucleus accumbens. *Alcoholism: Clinical and Experimental Research, 26*, 394–399.

CHAPTER

60

Pharmacogenetics and Smoking Cessation

Taraneh Taghavi, Rachel F. Tyndale*†,‡

*Departments of Pharmacology and Toxicology, and Psychiatry, University of Toronto, Toronto, ON, Canada

†Campbell Family Mental Health Research Institute and Addictions Division, Centre for Addiction and Mental Health, Toronto, ON, Canada

‡Department of Psychiatry, Campbell Family Mental Health Research Institute and Addictions Division, University of Toronto, Centre for Addiction and Mental Health, Toronto, ON, Canada

Abbreviations

AGES	additive genetic efficiency score
COMT	catechol-O-methyltransferase
CYP2A6	cytochrome P450 2A6
CYP2B6	cytochrome P450 2B6
CYP2C19	cytochrome P450 2C19
DAT	dopamine transporter
DRD2	dopamine D2 receptor
DRD4	dopamine D4 receptor
nAChR	nicotinic acetylcholine receptor
NMR	nicotine metabolite ratio
OCT2	organic cation transporter 2
UTR	untranslated region
VNTR	variable number of tandem repeats

60.1 INTRODUCTION

Tobacco smoking persists despite the availability of three types of pharmacotherapy, namely, nicotine replacement therapy, bupropion, and varenicline. Approximately 50%–60% of the ability to quit smoking is heritable (Broms, Silventoinen, et al., 2006). Genetic factors that influence smoking cessation outcomes include genetic variation in the pharmacology of nicotine and smoking cessation medications. Single-gene and polygenic influences on smoking cessation, both in the absence and presence of pharmacotherapy, have been investigated, and their results are summarized briefly in this chapter (Table 60.1).

60.2 SMOKING CESSATION USING GENETICS OF DRUG METABOLIZING ENZYMES

60.2.1 The Nicotine Metabolite Ratio

The rate at which nicotine is cleared from the body influences the ability to quit smoking. About 80% of a nicotine dose inhaled from cigarette smoke undergoes metabolic inactivation to cotinine primarily (~90%) by the hepatic enzyme CYP2A6 (Nakajima, Yamamoto, et al., 1996a, 1996b) with minor contributions from CYP2B6 (Al Koudsi & Tyndale, 2010). Cotinine is then further metabolized to 3′-hydroxycotinine, in a reaction mediated entirely by CYP2A6 (Nakajima et al., 1996a, 1996b) (Fig. 60.1). The ratio of these two metabolites (3′-hydroxycotinine and cotinine), known as the nicotine metabolite ratio (NMR), serves as a phenotypic biomarker of CYP2A6 activity in daily smokers such that faster CYP2A6 is reflected by a higher NMR (Dempsey, Tutka, et al., 2004).

60.2.2 CYP2A6 and Smoking Cessation

The *CYP2A6* gene is highly polymorphic, with many variants altering CYP2A6 activity. Individuals can be genotyped for these variants and grouped into CYP2A6 activity groups (e.g., fast and slow metabolizers) based on the predicted impact of their *CYP2A6* genotype on nicotine clearance (Benowitz, Swan, et al., 2006). Likewise,

Neuroscience of Nicotine
https://doi.org/10.1016/B978-0-12-813035-3.00060-5

TABLE 60.1 Summary of Genes Associated With Smoking Cessation

Gene symbol	Gene name	Category	Substrate/ligand
CYP2A6	Cytochrome P450, subfamily IIA, polypeptide 6	Metabolism	NIC/NRT
CYP2B6	Cytochrome P450, subfamily IIB, polypeptide 6	Metabolism	NIC/NRT, BUP
CHRNA5	Cholinergic receptor, neuronal nicotinic, alpha polypeptide 5	Target	NIC/NRT, BUP, Var
DRD2	Dopamine D2 receptor	Target	Neurotransmitters
COMT	Catechol-*O*-methyltransferase	Metabolism	Neurotransmitters
DRD4	Dopamine D4 receptor	Target	Neurotransmitters
SLC6A3 (DAT)	Solute carrier family 6 (dopamine transporter), member 3	Transport	Neurotransmitters

Abbreviations: *BUP*, bupropion; *NIC*, nicotine; *NRT*, nicotine replacement therapy; *Var*, varenicline.

smokers can be grouped as fast or slow metabolizers based on their NMR. No single optimal NMR cut point currently exists to distinguish CYP2A6 slow from fast metabolizers for cessation optimization; slow metabolizers generally represent the lowest 25%–40% of the NMR distribution (Lerman, Jepson, et al., 2010; Lerman, Schnoll, et al., 2015; Lerman, Tyndale, et al., 2006; Schnoll, Patterson, et al., 2009; Schnoll, Wileyto, et al., 2013). Slow nicotine metabolizers, determined by *CYP2A6* genotype or NMR, have lower cigarette consumption (Schoedel, Hoffmann, et al., 2004), nicotine dependence (Wassenaar, Dong, et al., 2011), and brain response to smoking cues (Falcone, Cao, et al., 2016).

Variation in NMR is associated with smoking cessation outcomes on active pharmacotherapy. In smokers randomized to nicotine patch, slow nicotine metabolizers (i.e., those in the lowest NMR quartile) displayed higher quit rates compared with fast metabolizers (Lerman, Tyndale, et al., 2006; Schnoll et al., 2009). Comparing standard (8 weeks) versus extended (6 months) nicotine patch therapy, genotypic (based on *CYP2A6* genotype) and phenotypic (based on the NMR) slow nicotine metabolizers had higher quit rates on extended versus standard nicotine patch therapy. In contrast, fast nicotine metabolizers did not benefit from extended therapy (Lerman et al., 2010). In another small trial of fast nicotine metabolizers (based on the NMR), there was a trend toward higher quit rates with higher dose (42 vs 21 mg nicotine patch) (Schnoll et al., 2013). Thus, nicotine replacement therapy is more effective in slow versus fast nicotine metabolizers; among fast metabolizers, higher doses may potentially be more effective than lower doses.

In smokers randomized to treatment based on NMR, varenicline was more efficacious than nicotine patch in fast metabolizers. In slow metabolizers, quitting did not differ by treatment (Lerman et al., 2015) (Fig. 60.2A). In slow metabolizers, the number needed to treat (NNT) for patch and varenicline was 10 and 8, respectively. In fast metabolizers, the NNT for patch and varenicline was 26 and 5, respectively, further indicating poorer quit rates for fast nicotine metabolizers on nicotine patch therapy relative to placebo (Fig. 60.2B). Additionally, varenicline was associated with greater side effect severity in slow versus fast metabolizers. By contrast, the nicotine patch was well tolerated in both groups. These findings suggest that varenicline is more suitable for fast metabolizers whereas patch is more suitable for slow metabolizers.

Variation in NMR is also associated with smoking cessation outcomes in the absence of pharmacotherapy. Slow nicotine metabolizers, by the NMR, have higher smoking cessation rates in the absence of pharmacotherapy (i.e., in the placebo arm of clinical trials) compared with fast nicotine metabolizers (Patterson, Schnoll, et al., 2008). Similarly, in smokers randomized to placebo, slow nicotine metabolizers by an alternative CYP2A6 activity measure (other than the NMR) had lower relapse risk versus fast metabolizers (Chen, Bloom, et al., 2014).

60.2.3 CYP2B6 and Smoking Cessation

Like *CYP2A6*, *CYP2B6* is highly polymorphic. A common *CYP2B6* haplotype, *CYP2B6*6*, comprises the *CYP2B6*4* and *CYP2B6*9* variants and is associated with lower hepatic CYP2B6 protein expression (Al Koudsi & Tyndale, 2010). CYP2B6 metabolizes nicotine, albeit to a smaller extent compared with CYP2A6 (~10% vs ~90%). In smokers randomized to nicotine patch, CYP2B6 metabolizer status determined by the *CYP2B6*6* genotype did not alter quit rates (Lee, Jepson, et al., 2007a, 2007b), consistent with CYP2B6 enzyme's minor contribution to nicotine metabolism suggesting *CYP2B6* genetics is likely not an ideal candidate for the optimization of nicotine-replacement-therapy-assisted cessation.

CYP2B6 also metabolizes bupropion to hydroxybupropion (Hesse, Venkatakrishnan, et al., 2000; Zanger, Klein, et al., 2007; Zhu, Cox, et al., 2012). Hydroxybupropion demonstrates greater inhibition of nAChR in vitro compared to bupropion (Damaj, Carroll, et al., 2004; Damaj, Grabus, et al., 2010) and has 10 times higher free plasma concentration compared to bupropion (Hsyu,

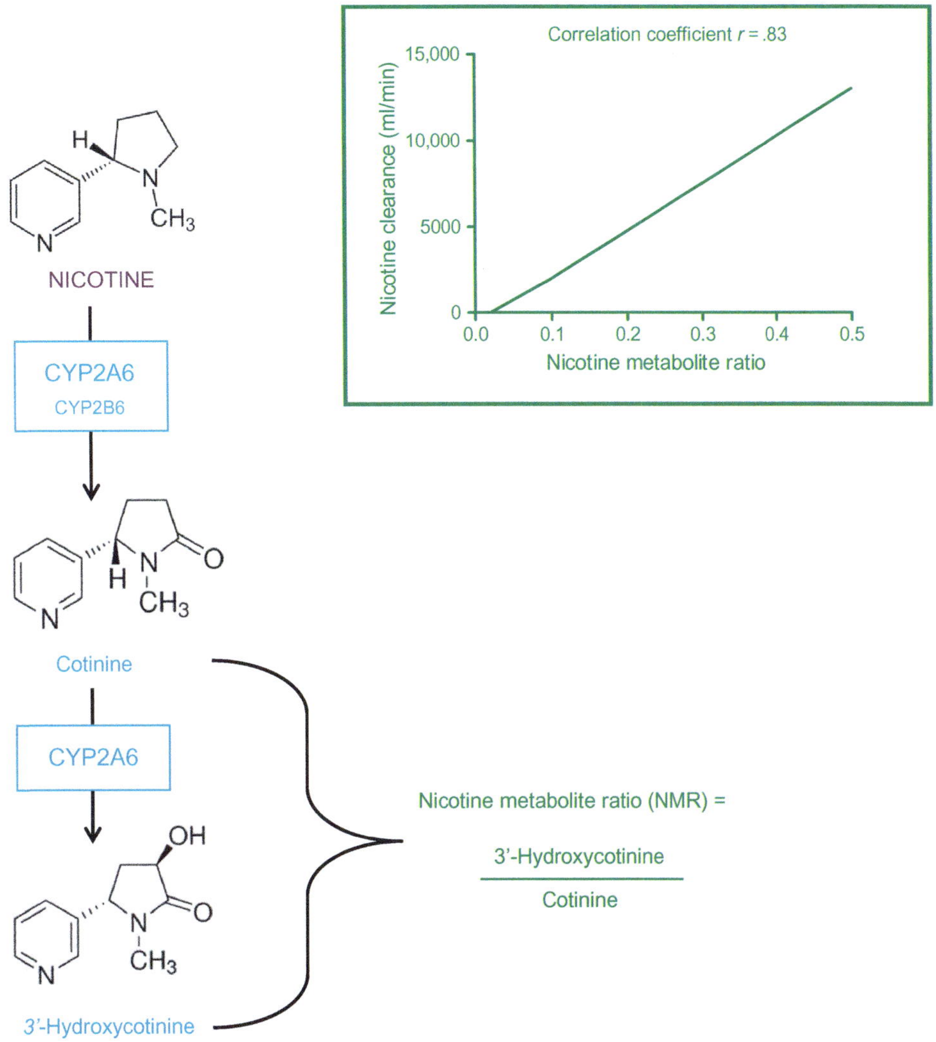

FIG. 60.1 Role of CYP2A6 in the metabolic pathway of nicotine. *This figure was adapted from Dempsey et al. (2004).*

Singh, et al., 1997; Johnston, Ascher, et al., 2002). Variability in the activity of CYP2B6 that alters hydroxybupropion levels could influence bupropion-assisted cessation outcomes.

Among African American smokers treated with bupropion, Zhu et al. demonstrated that slow CYP2B6 metabolizers (including those with the *CYP2B6*6* haplotype) had lower hydroxybupropion levels versus normal metabolizers and that lower hydroxybupropion levels were associated with poorer bupropion-assisted cessation (Zhu et al., 2012). However, *CYP2B6*6* genotype did not directly alter response to bupropion in either Caucasian (Lee et al., 2007a, 2007b) or African American (Zhu et al., 2012) smokers. The lack of a direct association between *CYP2B6* genotype and smoking cessation outcomes may be attributed to lower power or incomplete genotyping, suggesting in a larger study adequately powered for genetic effects that slow CYP2B6 genotypes may be associated with lower bupropion-assisted cessation.

Studies examining potential effects of *CYP2B6* genetics in the placebo arms of clinical trials have found smokers with the *CYP2B6*6* haplotype (relative to wild type) have a higher liability to relapse on placebo (Lee et al., 2007a, 2007b; Lerman, Shields, et al., 2002), suggesting CYP2B6 slow metabolizers are less likely to remain quit compared to CYP2B6 normal metabolizers.

60.3 SMOKING CESSATION USING GENETICS OF CENTRAL NERVOUS SYSTEM TARGETS

Genetic variation in the targets of nicotine within the reward circuitry in the central nervous system including the nicotinic acetylcholine receptors (nAChRs) and components of the dopaminergic pathway (i.e., dopamine receptors and transporters) may assist in the optimization of smoking cessation treatment.

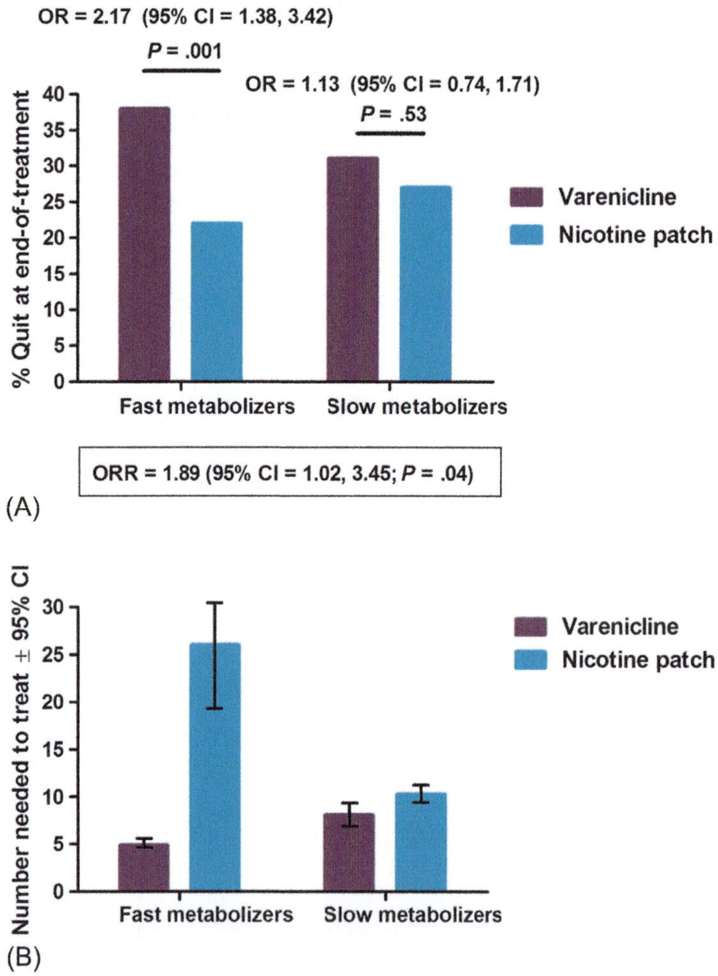

FIG. 60.2 The nicotine metabolite ratio (NMR) influences smoking cessation outcomes on nicotine patch vs varenicline. There was a significant NMR-by-treatment interaction on end-of-treatment quit rates: ratio of OR (ORR) = 1.89 (95% CI = 1.02 and 3.45; $P = .04$) in (A); the number needed to treat was 10 and 8 in slow metabolizers and 26 and 5 in fast metabolizers for patch and varenicline, respectively in (B). *This figure was adapted from Lerman et al. (2015).*

60.3.1 nAChR and Smoking Cessation

Despite the central role of α4β2-containing nAChRs as the target of nicotine and smoking cessation medications, variants in α4β2 subunits are not associated with smoking or cessation outcomes. The α4β2-containing nAChRs display a high degree of evolutionary conservation with little variation (Albuquerque, Pereira, et al., 2009), which may explain the lack of reported association.

Variation in the *CHRNA5-CHRNA3-CHRNB4* cluster located on chromosome 15q25, however, has been extensively examined for associations with smoking cessation success. The *CHRNA5* rs16969968 A allele has been associated with higher cigarette consumption (Munafo, Timofeeva, et al., 2012) and is thought to lead to a lower maximal nAChR response to nicotine (Bierut, Stitzel, et al., 2008). However, a recent metaanalysis in Caucasian smokers receiving nicotine replacement therapy showed no association between rs16969968 and cessation rates (Leung, Bergen, et al., 2015) (Fig. 60.3). Similarly, in smokers receiving varenicline, *CHRNA5* rs16969968 was not associated with cessation outcomes (Chen, Baker, et al., 2015). Among Caucasian smokers, no associations for rs16969968, rs588765, and rs578776 (two other SNPs tagging loci within the *CHRNA5-CHRNA3-CHRNB4* cluster that were previously associated with cigarette consumption) with cessation outcomes for nicotine patch or varenicline were observed (Tyndale, Zhu, et al., 2015). Similarly, among African American smokers, neither rs16969968 nor rs578776 was associated with cessation outcomes on nicotine gum or bupropion treatment (Zhu, Zhou, et al., 2014a, 2014b).

In a metaanalysis of 24 studies in nontreatment-seeking Caucasian smokers, those with *CHRNA5* rs16969968 AA risk genotype quit a median of 4 years later compared to GG smokers (Chen, Hung, et al., 2015), suggesting reduced CHRNA5 activity decreases spontaneous cessation. Similarly, smokers with the *CHRNA5* rs16969968 high-risk genotypes were less likely to quit while receiving placebo medications (Chen, Baker, et al., 2012). However, the rs16969968 effect on cessation on placebo is not consistent. Two recent trials found that *CHRNA5* rs16969968 was not associated with cessation

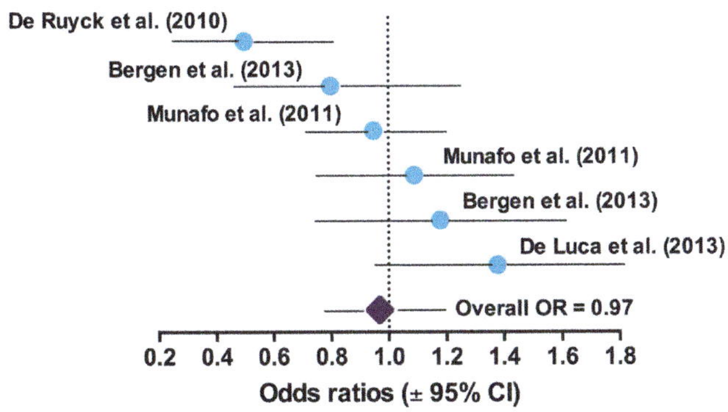

FIG. 60.3 The rs16969968 nicotinic receptor genetic variant does not influence smoking cessation outcomes on nicotine replacement therapy (NRT). *This figure was adapted from Leung et al. (2015).*

among Caucasian or African American smokers receiving placebo medications (Tyndale et al., 2015; Zhu et al., 2014a, 2014b). Together, despite strong evidence for the association of nAChR gene variants with heaviness of smoking and nicotine dependence, the lack of replicated findings for nAChR gene variants and smoking cessation outcomes reduces the likelihood that this region will be useful in the optimization of smoking cessation treatment.

60.3.2 Dopaminergic Pathway and Smoking Cessation

Functional polymorphisms that lead to lower dopaminergic activity are thought to contribute to lower smoking cessation success and have been investigated as potential sources of variability in smoking cessation rates (David, Munafo, et al., 2008).

60.3.2.1 DRD2

The most widely studied genetic variant in dopamine D2 receptor (*DRD2*) with respect to smoking cessation is the Taq1A polymorphism (Neville, Johnstone, et al., 2004). The Taq1A variant may alter the function of the *DRD2* gene, and the A1 allele has been associated with lower striatal D2 receptor density (Blum, Noble, et al., 1990; Jonsson, Nothen, et al., 1999).

In smokers randomized to nicotine patch, *DRD2* A1/A1 or A1/A2 genotype was associated with higher quit rates compared to the A2/A2 genotype (Johnstone, Yudkin, et al., 2004). Contrary to findings with nicotine patch, in smokers randomized to bupropion, *DRD2* A1/A1 or A1/A2 genotype was associated with lower quit rates compared to the A2/A2 genotype (David, Brown, et al., 2007). Of note, the higher bupropion-assisted quit rates in those with the *DRD2* A2/A2 genotype were restricted to those with *CYP2B6* rs3211371 TT or CT genotypes (David et al., 2007), highlighting the potential importance of assessing multiple genes and gene-gene interactions to identify subgroups of smokers who are more likely to benefit from a certain treatment. Similarly, Lerman and colleagues reported that a genetic variant (*DRD2* −141Cdel) leading to lower expression of the *DRD2* gene (similar to Taq1A A1 allele) was associated with greater nicotine replacement therapy efficacy while the high-activity variant (−141Cins) was associated with greater bupropion efficacy (Lerman, Jepson, et al., 2006). A pattern is therefore emerging whereby genotypes associated with reduced D2 receptor density predict better outcomes with nicotine replacement therapy while those associated with normal receptor expression or function predict a better response to bupropion. Together, these data suggest individuals who may not respond to bupropion, in part because of their *DRD2* Taq1A genotype, may derive greater efficacy from the use of nicotine replacement therapy.

60.3.2.2 COMT

Catechol-*O*-methyltransferase (COMT) is an enzyme involved in the metabolic inactivation of dopamine, suggesting that the *COMT* gene is a plausible candidate for pharmacogenetic studies of smoking cessation. The *COMT* rs4680 converts a Val high-activity allele to a Met low-activity allele, resulting in decreased brain enzyme levels and activity (Chen, Lipska, et al., 2004; Shield, Thomae, et al., 2004). Among women randomized to nicotine replacement therapy, the Met allele was associated with higher quit rates (Colilla, Lerman, et al., 2005). A second study by Johnstone and colleagues found higher quit rates with nicotine replacement therapy in the *COMT* Met/Met genotype group in comparison to those in either the Met/Val or Val/Val groups (Johnstone, Elliot, et al., 2007). Similarly, in smokers randomized to bupropion, a *COMT* haplotype of two single nucleotide polymorphisms (including Val/Met) was associated with bupropion-assisted cessation, although no differences in this association were observed between men and women (Breitling, Dahmen, et al., 2009). Thus, *COMT* variation may be useful in optimizing nicotine-replacement-therapy- and bupropion-assisted cessation. Specific attention to the potential moderating effects of sex on *COMT*-optimized cessation outcomes may be needed.

60.3.2.3 DRD4

The most widely studied genetic variant in dopamine D4 receptor (*DRD4*) with respect to smoking cessation is a variable number of tandem repeats (VNTR) polymorphism located in exon 3 of the gene. Most research has focused on the effects of the presence or absence of the 7-repeat (long) allele of the VNTR that is associated with decreased ligand binding and gene expression in vitro compared with 6-repeat or fewer (short) alleles (Van Tol, Wu, et al., 1992). The long allele was associated with overall lower quit rates with nicotine replacement therapy relative to placebo (David et al., 2008). In contrast, the long allele was not associated with overall abstinence with bupropion relative to placebo (Leventhal, David, et al., 2012). However, a genotype by treatment interaction was observed where in smokers with the long allele, bupropion increased cessation relative to placebo; smokers with two copies of the short allele did not benefit from bupropion (Leventhal et al., 2012). Bupropion, may, therefore, be a more suitable treatment for smokers with the long allele. Considering that bupropion diminishes cue-induced cravings (Brody, Mandelkern, et al., 2004), it may buffer those with the long allele and lower D4 density or activity, from relapse following cue exposure.

60.3.2.4 DAT

The most widely studied genetic variant with respect to smoking cessation in *SLC6A3*, the gene encoding the dopamine transporter (*DAT*), is a VNTR polymorphism located in the 3′-untranslated region (3′-UTR) of the gene with a majority of focus on the effect of the 9-repeat allele of the VNTR. The 9-repeat versus 10-repeat allele is associated with increased dopamine transporter availability on the surface of the neurons (Heinz & Goldman, 2000). In a recent metaanalysis of eight studies in Caucasian smokers randomized to bupropion or placebo, the 9-repeat allele is not associated with overall abstinence (Duck & Gyoon, 2016). Thus, *DAT* represents an example of a biologically promising gene that may not assist in the optimization of smoking cessation treatment (Fig. 60.4).

60.4 SMOKING CESSATION USING MULTIPLE GENETIC PREDICTORS

60.4.1 CYP2A6 and nAChR Combined Genetic Risk Scores

Among Caucasian smokers, Chen and colleagues derived genetic risk scores based on *CYP2A6* and *CHRNA5* genotypes (Chen et al., 2014) suggesting on nicotine replacement therapy, but not bupropion, cessation outcomes differed according to the risk score. In the highest-risk group (i.e., *CYP2A6* fast metabolism plus *CHRNA5* high-risk rs16969968-rs680244 diplotype), nicotine replacement therapy had the largest treatment effect versus placebo, while individuals in the lowest-risk group (i.e., *CYP2A6* slow metabolism plus *CHRNA5* low-risk rs16969968-rs680244 diplotype) did not benefit from nicotine replacement therapy. These findings, however, were limited in several ways. When multiple genetic markers and treatment conditions were analyzed, the sample size in certain conditions was very small. As such, studies that examine polygenic (e.g., multiple genetic variants) influences on smoking cessation outcomes, including this one, are often not sufficiently powered and thus should be interpreted with caution.

60.4.2 Additive Genetic Efficiency Score (AGES)

Among Caucasian smokers, David and colleagues calculated an additive genetic efficiency score (AGES) based on the number of alleles hypothesized to promote cessation on bupropion versus placebo (David, Strong, et al., 2013). The four variants selected based on their associations with smoking cessation included *DRD2* Taq1A, *COMT* rs4680, *DRD4* exon 3 VNTR, and *SLC6A3* 3′ VNTR, discussed previously. In smokers randomized to bupropion, AGES was not associated with time to first relapse or abstinence (Fig. 60.5); however, higher AGES was associated with increased relapse risk among smokers randomized to placebo. Current genetic risk score models need further refinement before they can be used to optimize smoking cessation treatment. Future refinements could incorporate additional genes and variants in relevant biological pathways, exclude genes and/or variants that play little role, and determine whether additive versus dominant models are most appropriate for each loci.

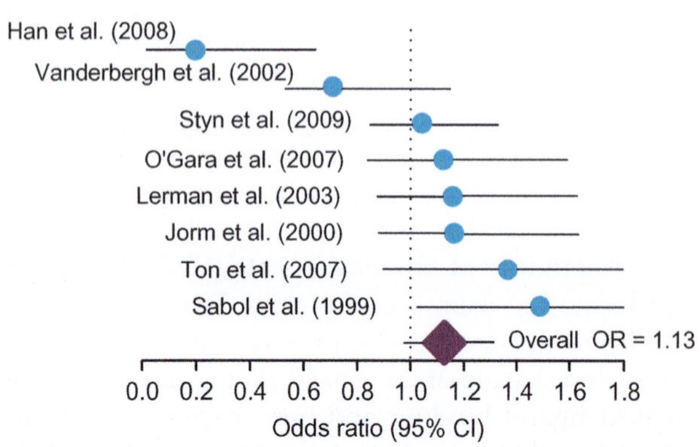

FIG. 60.4 Genetic variation in the *SLC6A3* gene (DAT) does not influence smoking cessation outcomes on bupropion. *This figure was adapted from Duck and Gyoon (2016).*

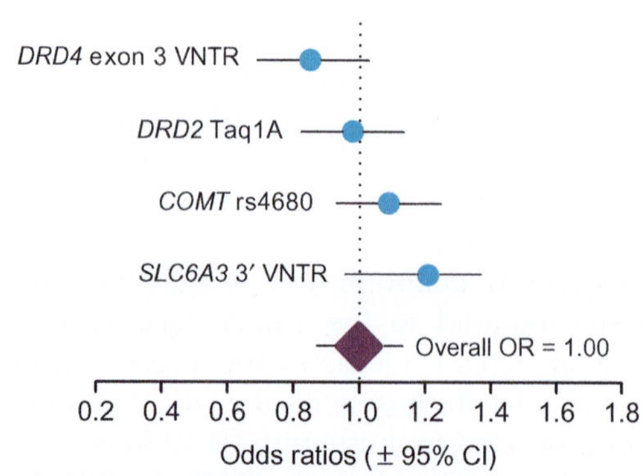

FIG. 60.5 The additive genetic efficiency score (AGES) does not influence smoking cessation outcomes on bupropion. *This figure was adapted from David et al. (2013).*

MINI-DICTIONARY OF TERMS

Bupropion Bupropion (trade names Wellbutrin and Zyban) is a norepinephrine-dopamine reuptake inhibitor and a nicotinic acetylcholine receptor antagonist, used as an antidepressant and as a smoking cessation aid.

Nicotine metabolite ratio The ratio of two of nicotine's primary metabolites, that is, 3′-hydroxycotinine to cotinine; in regular smokers, it is used as a proxy for CYP2A6 activity and the rate of nicotine metabolism.

Number needed to treat (NNT) The number of patients who need to be treated in order for one person to benefit from the medication (e.g., successfully quit) compared with a control in a clinical trial.

Pharmacogenetics The study of how a patient's genetic profile alters their response to a drug, including side effects.

Polygenic influences The additive effect of variation in multiple gene loci, used here to examine response to treatment medication or the development of side effects.

Varenicline Varenicline (trade names Chantix and Champix) is a partial agonist at the nicotinic acetylcholine receptor, used as a smoking cessation aid.

Key Facts of Pharmacogenetics

- How fast or slow a drug is metabolized and how a patient responds to a drug are determined, in part, by their genes.
- Humans share around 99.5% of their genome, but the 0.5% that differs between individuals affects their response to drugs and susceptibility to side effects.
- Understanding how genetic variation affects response to drugs may assist in more accurately determining which drug and which dose is best for individual patients.
- Two types of variation that are common in the human genome include single nucleotide polymorphisms (SNPs) and structural variation.
- SNPs refer to changes in single nucleotide bases (A, C, T, and G).
- Structural variation refers to changes in larger pieces of DNA including those that can alter the structure of the chromosome. Examples include copy number variation, deletion, insertion, and duplication.

Summary Points

- Slow nicotine metabolizers (vs fast), by *CYP2A6* genotype or phenotype (the NMR), have higher smoking cessation rates on nicotine replacement transdermal patch and in the absence of pharmacological treatment.
- Fast nicotine metabolizers (vs slow), by the NMR, have higher smoking cessation rates on varenicline than nicotine patch; slow nicotine metabolizers (vs fast) have greater side effects on varenicline.
- Slow CYP2B6 metabolizers (vs fast), by *CYP2B6* genotype, have lower smoking cessation rates on bupropion and in the absence of treatment.
- Genetic variation in the central nervous system targets of nicotine (e.g., nicotinic acetylcholine receptors and dopaminergic receptors and transporters) has been associated with smoking cessation, albeit inconsistently.
- Polygenic influences and gene-gene interactions need to be taken into account in order to more accurately elucidate genetic effects on smoking cessation outcomes in the presence or absence of pharmacotherapy.

References

Al Koudsi, N., & Tyndale, R. F. (2010). Hepatic CYP2B6 is altered by genetic, physiologic, and environmental factors but plays little role in nicotine metabolism. *Xenobiotica*, *40*(6), 381–392.

Albuquerque, E. X., Pereira, E. F., et al. (2009). Mammalian nicotinic acetylcholine receptors: from structure to function. *Physiological Reviews*, *89*(1), 73–120.

Benowitz, N. L., Swan, G. E., et al. (2006). CYP2A6 genotype and the metabolism and disposition kinetics of nicotine. *Clinical Pharmacology & Therapeutics*, *80*(5), 457–467.

Bergen, A. W., et al. (2013). Nicotinic acetylcholine receptor variation and response to smoking cessation therapies. *Pharmacogenetics and Genomics*, *23*(2), 94–103.

Bierut, L. J., Stitzel, J. A., et al. (2008). Variants in nicotinic receptors and risk for nicotine dependence. *The American Journal of Psychiatry*, *165*(9), 1163–1171.

Blum, K., Noble, E. P., et al. (1990). Allelic association of human dopamine D2 receptor gene in alcoholism. *JAMA*, *263*(15), 2055–2060.

Breitling, L. P., Dahmen, N., et al. (2009). Variants in COMT and spontaneous smoking cessation: retrospective cohort analysis of 925 cessation events. *Pharmacogenetics and Genomics*, *19*(8), 657–659.

Brody, A. L., Mandelkern, M. A., et al. (2004). Attenuation of cue-induced cigarette craving and anterior cingulate cortex activation in bupropion-treated smokers: a preliminary study. *Psychiatry Research*, *130*(3), 269–281.

Broms, U., Silventoinen, K., et al. (2006). Genetic architecture of smoking behavior: a study of Finnish adult twins. *Twin Research and Human Genetics*, *9*(1), 64–72.

Chen, L. S., Baker, T. B., et al. (2012). Interplay of genetic risk factors (CHRNA5-CHRNA3-CHRNB4) and cessation treatments in smoking cessation success. *The American Journal of Psychiatry*, *169*(7), 735–742.

Chen, L. S., Baker, T. B., et al. (2015). Genetic variation (CHRNA5), medication (combination nicotine replacement therapy vs. varenicline), and smoking cessation. *Drug and Alcohol Dependence*, *154*, 278–282.

Chen, L. S., Bloom, A. J., et al. (2014). Pharmacotherapy effects on smoking cessation vary with nicotine metabolism gene (CYP2A6). *Addiction*, *109*(1), 128–137.

Chen, L. S., Hung, R. J., et al. (2015). CHRNA5 risk variant predicts delayed smoking cessation and earlier lung cancer diagnosis–a meta-analysis. *Journal of the National Cancer Institute*, *107*(5).

Chen, J., Lipska, B. K., et al. (2004). Functional analysis of genetic variation in catechol-O-methyltransferase (COMT): effects on mRNA, protein, and enzyme activity in postmortem human brain. *American Journal of Human Genetics*, *75*(5), 807–821.

Colilla, S., Lerman, C., et al. (2005). Association of catechol-O-methyltransferase with smoking cessation in two independent studies of women. *Pharmacogenetics and Genomics*, *15*(6), 393–398.

Damaj, M. I., Carroll, F. I., et al. (2004). Enantioselective effects of hydroxy metabolites of bupropion on behavior and on function of monoamine transporters and nicotinic receptors. *Molecular Pharmacology*, *66*(3), 675–682.

Damaj, M. I., Grabus, S. D., et al. (2010). Effects of hydroxymetabolites of bupropion on nicotine dependence behavior in mice. *The Journal of Pharmacology and Experimental Therapeutics*, *334*(3), 1087–1095.

David, S. P., Brown, R. A., et al. (2007). Pharmacogenetic clinical trial of sustained-release bupropion for smoking cessation. *Nicotine & Tobacco Research*, *9*(8), 821–833.

David, S. P., Munafo, M. R., et al. (2008). Genetic variation in the dopamine D4 receptor (DRD4) gene and smoking cessation: follow-up of a randomised clinical trial of transdermal nicotine patch. *The Pharmacogenomics Journal*, *8*(2), 122–128.

David, S. P., Strong, D. R., et al. (2013). Influence of a dopamine pathway additive genetic efficacy score on smoking cessation: results from two randomized clinical trials of bupropion. *Addiction*, *108*(12), 2202–2211.

De Luca, V., et al. (2013). Analysis of nicotinic receptor genes in nicotine replacement treatment. *European Psychiatry*, *28*(2), 113–118.

Dempsey, D., Tutka, P., et al. (2004). Nicotine metabolite ratio as an index of cytochrome P450 2A6 metabolic activity. *Clinical Pharmacology and Therapeutics*, *76*(1), 64–72.

De Ruyck, K., et al. (2010). Genetic variation in three candidate genes and nicotine dependence, withdrawal and smoking cessation in hospitalized patients. *Pharmacogenomics*, *11*(8), 1053–1063.

Duck, C. H., & Gyoon, S. W. (2016). Meta-analysis update of association between dopamine transporter SLC6A3 gene polymorphism and smoking cessation. *Journal of Health Psychology*, *23*(9), 1250–1257.

Falcone, M., Cao, W., et al. (2016). Brain responses to smoking cues differ based on nicotine metabolism rate. *Biological Psychiatry*, *80*(3), 190–197.

Han, D. H., Joe, K. H., Na, C., et al. (2008). Effect of genetic polymorphisms on smoking cessation: A trial of bupropion in Korean male smokers. *Psychiatric Genetics*, *18*, 11–16.

Heinz, A., & Goldman, D. (2000). Genotype effects on neurodegeneration and neuroadaptation in monoaminergic neurotransmitter systems. *Neurochemistry International*, *37*(5–6), 425–432.

Hesse, L. M., Venkatakrishnan, K., et al. (2000). CYP2B6 mediates the in vitro hydroxylation of bupropion: potential drug interactions with other antidepressants. *Drug Metabolism and Disposition*, *28*(10), 1176–1183.

Hsyu, P. H., Singh, A., et al. (1997). Pharmacokinetics of bupropion and its metabolites in cigarette smokers versus nonsmokers. *Journal of Clinical Pharmacology*, *37*(8), 737–743.

Johnston, A. J., Ascher, J., et al. (2002). Pharmacokinetic optimisation of sustained-release bupropion for smoking cessation. *Drugs*, *2*, 11–24.

Johnstone, E. C., Elliot, K. M., et al. (2007). Association of COMT Val108/158Met genotype with smoking cessation in a nicotine replacement therapy randomized trial. *Cancer Epidemiology, Biomarkers & Prevention*, *16*(6), 1065–1069.

Johnstone, E. C., Yudkin, P. L., et al. (2004). Genetic variation in dopaminergic pathways and short-term effectiveness of the nicotine patch. *Pharmacogenetics*, *14*(2), 83–90.

Jonsson, E. G., Nothen, M. M., et al. (1999). Polymorphisms in the dopamine D2 receptor gene and their relationships to striatal dopamine receptor density of healthy volunteers. *Molecular Psychiatry*, *4*(3), 290–296.

Jorm, A. F., Henderson, A. S., Jacomb, P. A., et al. (2000). Association of smoking and personality with a polymorphism of the dopamine transporter gene: Results from a community survey. *American Journal of Medical Genetics*, *96*, 331–334.

Lee, A. M., Jepson, C., et al. (2007a). CYP2B6 genotype alters abstinence rates in a bupropion smoking cessation trial. *Biological Psychiatry*, *62* (6), 635–641.

Lee, A. M., Jepson, C., et al. (2007b). CYP2B6 genotype does not alter nicotine metabolism, plasma levels, or abstinence with nicotine replacement therapy. *Cancer Epidemiology, Biomarkers & Prevention*, *16*(6), 1312–1314.

Lerman, C., Jepson, C., et al. (2006). Role of functional genetic variation in the dopamine D2 receptor (DRD2) in response to bupropion and nicotine replacement therapy for tobacco dependence: results of two randomized clinical trials. *Neuropsychopharmacology*, *31*(1), 231–242.

Lerman, C., Jepson, C., et al. (2010). Genetic variation in nicotine metabolism predicts the efficacy of extended-duration transdermal nicotine therapy. *Clinical Pharmacology and Therapeutics*, *87*(5), 553–557.

Lerman, C., Schnoll, R. A., et al. (2015). Use of the nicotine metabolite ratio as a genetically informed biomarker of response to nicotine patch or varenicline for smoking cessation: a randomised, double-blind placebo-controlled trial. *The Lancet Respiratory Medicine*, *3*(2), 131–138.

Lerman, C., Shields, P. G., et al. (2002). Pharmacogenetic investigation of smoking cessation treatment. *Pharmacogenetics*, *12*(8), 627–634.

Lerman, C., Shields, P. G., Wileyto, E. P., et al. (2003). Effects of dopamine transporter and receptor polymorphisms on smoking cessation in a bupropion clinical trial. *Health Psychology*, *22*, 541–548.

Lerman, C., Tyndale, R., et al. (2006). Nicotine metabolite ratio predicts efficacy of transdermal nicotine for smoking cessation. *Clinical Pharmacology and Therapeutics*, *79*(6), 600–608.

Leung, T., Bergen, A., et al. (2015). Effect of the rs1051730-rs16969968 variant and smoking cessation treatment: a meta-analysis. *Pharmacogenomics*, *16*(7), 713–720.

Leventhal, A. M., David, S. P., et al. (2012). Dopamine D4 receptor gene variation moderates the efficacy of bupropion for smoking cessation. *The Pharmacogenomics Journal*, *12*(1), 86–92.

Munafo, M. R., et al. (2011). CHRNA3 rs1051730 genotype and short-term smoking cessation. *Nicotine & Tobacco Research*, *13*(10), 982–988.

Munafo, M. R., Timofeeva, M. N., et al. (2012). Association between genetic variants on chromosome 15q25 locus and objective measures of tobacco exposure. *Journal of the National Cancer Institute*, *104*(10), 740–748.

Nakajima, M., Yamamoto, T., et al. (1996a). Characterization of CYP2A6 involved in 3′-hydroxylation of cotinine in human liver microsomes. *The Journal of Pharmacology and Experimental Therapeutics*, *277*(2), 1010–1015.

Nakajima, M., Yamamoto, T., et al. (1996b). Role of human cytochrome P4502A6 in C-oxidation of nicotine. *Drug Metabolism and Disposition*, *24*(11), 1212–1217.

Neville, M. J., Johnstone, E. C., et al. (2004). Identification and characterization of ANKK1: a novel kinase gene closely linked to DRD2 on chromosome band 11q23.1. *Human Mutation*, *23*(6), 540–545.

O'Gara, C., Stapleton, J., Sutherland, G., et al. (2007). Dopamine transporter polymorphisms are associated with short-term response to smoking cessation treatment. *Pharmacogenetics and Genomics*, *17*, 61–67.

Patterson, F., Schnoll, R. A., et al. (2008). Toward personalized therapy for smoking cessation: a randomized placebo-controlled trial of bupropion. *Clinical Pharmacology and Therapeutics*, *84*(3), 320–325.

Sabol, S. Z., Nelson, M. L., Fisher, C., et al. (1999). A genetic association for cigarette smoking behavior. *Health Psychology*, *18*, 7–13.

Schnoll, R. A., Patterson, F., et al. (2009). Nicotine metabolic rate predicts successful smoking cessation with transdermal nicotine: a validation study. *Pharmacology, Biochemistry, and Behavior*, *92*(1), 6–11.

Schnoll, R. A., Wileyto, E. P., et al. (2013). High dose transdermal nicotine for fast metabolizers of nicotine: a proof of concept placebo-controlled trial. *Nicotine & Tobacco Research*, *15*(2), 348–354.

Schoedel, K. A., Hoffmann, E. B., et al. (2004). Ethnic variation in CYP2A6 and association of genetically slow nicotine metabolism and smoking in adult Caucasians. *Pharmacogenetics*, *14*(9), 615–626.

Shield, A. J., Thomae, B. A., et al. (2004). Human catechol O-methyltransferase genetic variation: gene resequencing and functional characterization of variant allozymes. *Molecular Psychiatry*, *9*(2), 151–160.

Styn, M. A., Nukui, T., Romkes, M., et al. (2009). The impact of genetic variation in DRD2 and SLC6A3 on smoking cessation in a cohort of participants 1 year after enrollment in a lung cancer screening study. *American Journal of Medical Genetics Part B: Neuropsychiatric Genetics, 150*, 254–261.

Ton, T. G., Rossing, M. A., Bowen, D. J., et al. (2007). Genetic polymorphisms in dopamine-related genes and smoking cessation in women: A prospective cohort study. *Behavioral and Brain Functions, 23*, 3–22.

Tyndale, R. F., Zhu, A. Z., et al. (2015). Lack of associations of CHRNA5-A3-B4 genetic variants with smoking cessation treatment outcomes in Caucasian smokers despite associations with baseline smoking. *PLoS ONE, 10*(5).

Vandenbergh, D. J., Persico, A. M., Hawkins, A. L., et al. (1992). Human dopamine transporter gene (DAT1) maps to chromosome 5p15.3 and displays a VNTR. *Genomics, 14*, 1104–1106.

Van Tol, H. H., Wu, C. M., et al. (1992). Multiple dopamine D4 receptor variants in the human population. *Nature, 358*(6382), 149–152.

Wassenaar, C. A., Dong, Q., et al. (2011). Relationship between CYP2A6 and CHRNA5-CHRNA3-CHRNB4 variation and smoking behaviors and lung cancer risk. *Journal of the National Cancer Institute, 103*(17), 1342–1346.

Zanger, U. M., Klein, K., et al. (2007). Polymorphic CYP2B6: molecular mechanisms and emerging clinical significance. *Pharmacogenomics, 8*(7), 743–759.

Zhu, A. Z., Cox, L. S., et al. (2012). CYP2B6 and bupropion's smoking-cessation pharmacology: the role of hydroxybupropion. *Clinical Pharmacology and Therapeutics, 92*(6), 771–777.

Zhu, A. Z., Zhou, Q., et al. (2014a). Gene variants in CYP2C19 are associated with altered in vivo bupropion pharmacokinetics but not bupropion-assisted smoking cessation outcomes. *Drug Metabolism and Disposition, 42*(11), 1971–1977.

Zhu, A. Z., Zhou, Q., et al. (2014b). Association of CHRNA5-A3-B4 SNP rs2036527 with smoking cessation therapy response in African-American smokers. *Clinical Pharmacology and Therapeutics, 96*(2), 256–265.

C H A P T E R

61

The Orexin System and Nicotine Addiction: Preclinical Insights

Shaun Yon-Seng Khoo, Gavan P. McNally†, Kelly J. Clemens†*

*Center for Studies in Behavioral Neurobiology, Department of Psychology, Concordia University, Montreal, QC, Canada

†School of Psychology, University of New South Wales, Sydney, NSW, Australia

Abbreviations

Acb nucleus accumbens
DMH dorsomedial hypothalamus
FRn fixed ratio of *n* responses
GPCR G-protein-coupled receptor
LH lateral hypothalamus
nAChR nicotinic acetylcholine receptor
OX-A orexin-A
OX-B orexin-B
PeF perifornical hypothalamus
PFC prefrontal cortex
PR progressive ratio
VTA ventral tegmental area

Current pharmacological therapies for smoking have only modest efficacy with failure rates of up to 90% (Cahill, Lindson-Hawley, Thomas, Fanshawe, & Lancaster, 2016). The orexin/hypocretin system has been suggested as a potential therapeutic target and is in clinical trials for cocaine (The University of Texas Health Science Center, 2016). Here, we review the orexins as a potential therapeutic target for nicotine because there is evidence for both acute and chronic nicotine/orexin interactions and some preliminary evidence that orexin-based medications may be effective in reducing smoking. However, while there is promise, there are some conflicting preclinical results, and only correlational results in human studies are available.

The orexin/hypocretin system is a neuropeptide system involved in arousal, appetite, and reward. Orexin neurons originate exclusively from the lateral hypothalamus (LH), perifornical hypothalamus (PeF), and dorsomedial hypothalamus (DMH; Baldo, Daniel, Berridge, & Kelley, 2003; Elias et al., 1998; Peyron et al., 1998). Orexin fibers project to key mesocorticolimbic reward regions (Fig. 61.1), such as the ventral tegmental area (VTA), nucleus accumbens (Acb), and prefrontal cortex (PFC; Baldo et al., 2003; Peyron et al., 1998), where there are abundant nicotinic acetylcholine receptors (nAChRs; Lerman et al., 2007).

The orexin system is composed of two peptides, orexin-A and orexin-B (OX-A and OX-B, respectively), and two GPCRs, OX_1 and OX_2 (Fig. 61.2; Sakurai et al., 1998). Hypocretin (human, HCRT, and rat/mouse, *Hcrt*) mRNA encodes for the prepro-orexin precursor peptide that is cleaved to form the OX-A and OX-B peptides (Sakurai et al., 1999). OX-A binds nonselectively to both receptors, while OX-B shows some selectivity for the OX_2 receptor. Both OX_1 and OX_2 are excitatory and produce elevated Ca^{2+} suggesting $G\alpha_q$-mediated signaling (de Lecea et al., 1998; Sakurai et al., 1998), but they can also interact with multiple second-messenger pathways (Kukkonen & Leonard, 2014). Orexin receptors also form complexes, which may be homodimers (Xu, Ward, Pediani, & Milligan, 2011), OX_1 and OX_2 heteromers, and CB_1 or CRF receptor complexes (Ellis, Pediani, Canals, Milasta, & Milligan, 2006; Jäntti, Mandrika, & Kukkonen, 2014; Navarro et al., 2015). Thus, orexin signaling is thought to be excitatory with a $G\alpha_q$-mediated signaling cascade but with the potential for interactions with cotransmitters and heteromeric receptor complexes.

61.1 NEUROANATOMICAL AND MOLECULAR INTERACTIONS

61.1.1 Acute Nicotine and Orexin

The orexin system has been suggested as a potential therapeutic target because of the neuroanatomical and

Neuroscience of Nicotine
https://doi.org/10.1016/B978-0-12-813035-3.00061-7

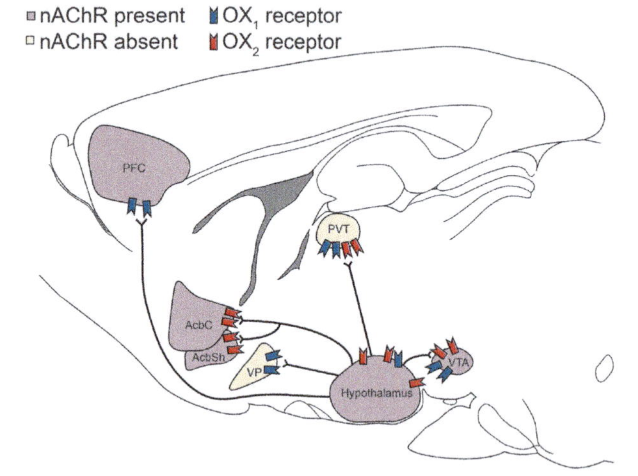

FIG. 61.1 Anatomic overlap between the orexin system and nicotinic acetylcholine receptors.
Orexin receptors are present in varying densities in key brain regions involved in reward. Nicotinic acetylcholine receptors (nAChRs) are present in all of these areas, except for the ventral pallidum (VP) and paraventricular nucleus of the thalamus (PVT).

molecular interactions between the cholinergic and orexin systems. Interactions occur following both acute and chronic nicotine administration. Cholinergic neurons are activated by OX-A application (Fadel, Pasumarthi, & Reznikov, 2005), and orexin neurons are activated by acute systemic nicotine administration (Fig. 61.3; Pasumarthi, Reznikov, & Fadel, 2006). Importantly, this activation, as measured by c-Fos immunohistochemistry, appears to be specific in the hypothalamus for orexin neurons, suggesting that the orexins play a key role in nicotine-mediated effects. While this effect is blocked by both the nonselective nAChR antagonist mecamylamine and the α4β2 nAChR antagonist DHβE, DHβE alone increased medial hypothalamic orexin neuron activation but blocked nicotine-induced activation (Pasumarthi et al., 2006), suggesting that there is endogenous cholinergic regulation of the orexin system. Nicotine-induced activation of orexin neurons appears to be specific to hypothalamic projections to the basal forebrain and paraventricular thalamus (Pasumarthi & Fadel, 2008). In addition to systemic nicotine administration, orexin neurons can also be activated by local nicotine application to the hypothalamus, which produces increased ACh and glutamate efflux in the hypothalamus (Pasumarthi & Fadel, 2010). These results demonstrate interactions between orexin, which activates cholinergic neurons, and acute nicotine administered systemically or directly to the hypothalamus that activates orexin neurons.

61.1.2 Chronic Nicotine and Orexin

Studies of chronic nicotine administration have shown alterations in the regulation of hypocretin gene expression and orexin peptides (Fig. 61.4). Chronic nicotine administration (14 days) increases levels of mRNA encoding the prepro-orexin peptide and orexin receptors (Kane et al., 2000). Blockade of nAChRs in nicotine-exposed rats decreases orexin neuron activation (Simmons et al., 2016). Measurements of protein show that in the DMH, chronic nicotine increases both OX-A and OX-B while OX-B is elevated in the paraventricular nucleus (Kane et al., 2000). However, chronic nicotine also decreases hypothalamic binding sites available for OX-A (Kane, Parker, & Li, 2001). This desensitization appears to occur by obscuring binding sites but without internalization because OX-A continued to bind but with lower affinity (Kane et al., 2001). Although a more direct demonstration of the exact mechanism of nicotine-induced decrease in binding sites has never been reported in the literature, these results demonstrate that chronic nicotine administration has regulatory effects on both hypocretin gene expression and levels of orexin peptide.

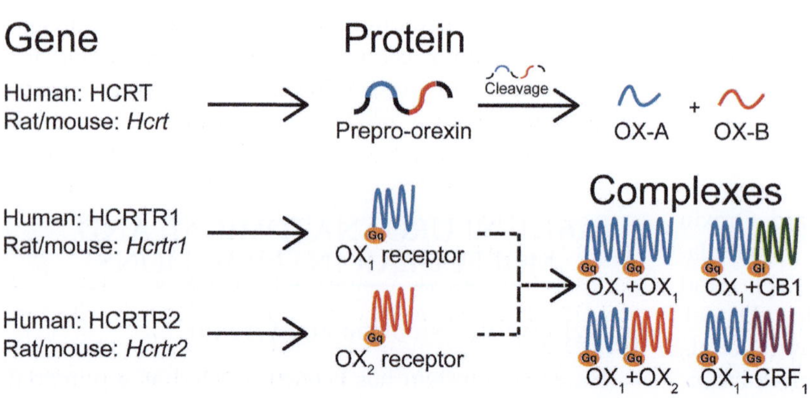

FIG. 61.2 Schematic overview of orexin signaling molecules.
The two orexin peptides are cleaved from a single precursor and bind to the orexin receptors, which may exist as complexes with each other, CB1 or CRF_1 receptors.

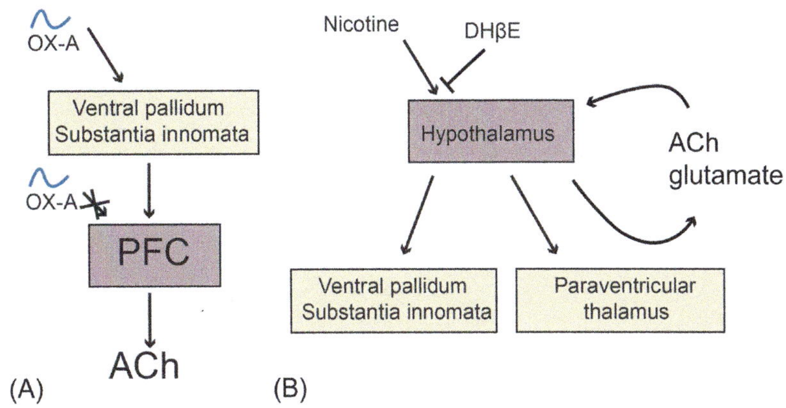

FIG. 61.3 Acute orexin/nicotine interactions. (A) Acute orexin can activate cholinergic neurons, and (B) acute nicotine can activate orexin outputs and result in local glutamate and ACh release.

61.2 OREXIN AS A THERAPEUTIC TARGET FOR NICOTINE DEPENDENCE

61.2.1 Nicotine Self-Administration

Preclinical studies of orexin as a potential therapeutic target for nicotine addiction have found effects of both selective and nonselective (dual) orexin receptor antagonists (Table 61.1). The first direct demonstration of a role for orexins in nicotine-seeking behavior involved the administration of the OX_1 receptor antagonist SB-334867 systemically and into the insular cortex (Hollander, Lu, Cameron, Kamenecka, & Kenny, 2008). OX_1 receptor antagonism selectively suppressed the number of nicotine infusions earned on fixed-ratio 5 (FR5) and progressive-ratio (PR) schedules while leaving the number of food pellets earned unaffected. Similarly, intrainsular administration of SB-334867 reduced FR5 nicotine infusions but not the number of food pellets earned (Hollander et al., 2008). These results have been replicated using SB-334867 and the dual orexin receptor antagonist almorexant (LeSage, Perry, Kotz, Shelley, & Corrigall, 2010). The number of nicotine infusions earned was reduced on an FR5 schedule by both the OX_1 receptor antagonist SB-334867 and the dual orexin receptor antagonist almorexant. However, while SB-334867 in this study also left the number of food pellets earned unaffected, the highest dose of almorexant reduced pellets earned. LeSage et al. (2010) further extended the findings of Hollander et al. (2008) by examining *Hcrtr1* and *Hcrtr2* gene expression and finding decreased *Hcrtr1* in the rostral LH immediately after nicotine self-administration but not 5h after a self-administration session, while arcuate nucleus had increased *Hcrtr1* mRNA 5h but not immediately after self-administration. These consistent results from different groups provided strong indications that the orexins might be a potential target for the treatment of nicotine addiction.

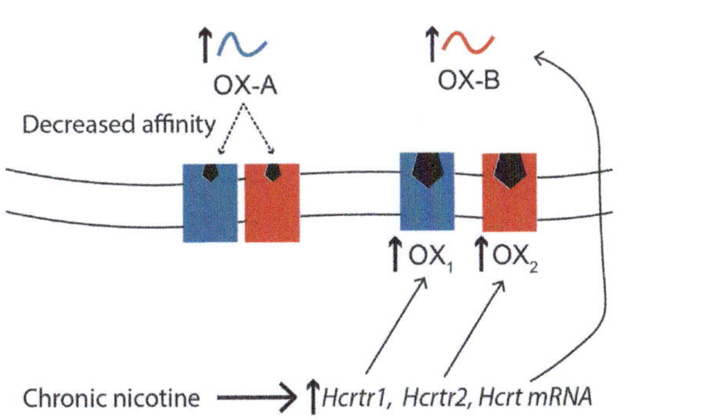

FIG. 61.4 Effects of chronic nicotine on orexin regulation. Chronic nicotine upregulates the expression of orexin peptides and receptors but reduces the availability of binding sites for OX-A without receptor internalization.

61.2.2 Reinstatement of Nicotine Seeking

In addition to regulating the self-administration of nicotine, it has also been shown that orexin signaling plays a role in the reinstatement of nicotine seeking, but results have not been consistent. Intracerebroventricular administration of OX-A can reinstate extinguished nicotine-seeking behavior in mice, an effect that can be blocked by the administration of the OX_1 receptor antagonist SB-334867 (Plaza-Zabala, Martín-García, de Lecea, Maldonado, & Berrendero, 2010). While OX_1 receptor antagonism reduces anxiety-like behavior in the elevated plus maze induced by acute injections of nicotine, suggesting that orexins are involved in the anxiogenic effects of acute nicotine, OX_1 receptor antagonism does not affect footshock-induced reinstatement, indicating that these behaviors are mediated by separate processes (Plaza-Zabala et al., 2010). During cue-induced reinstatement of nicotine seeking in mice, an increased percentage of orexin neurons in the PeF and LH show signs of activation as measured by c-Fos (Plaza-Zabala et al., 2013). Selective OX_1 receptor antagonist using SB-334867 but not OX_2 receptor antagonism using TCS-OX2-29 can attenuate this cue-induced reinstatement (Plaza-Zabala et al., 2013). However, it has been shown that a different selective OX_2 receptor antagonist (SORA-18) was able to

TABLE 61.1 Comparison of Behavioral Studies Examining Nicotine Self-Administration and Reinstatement

Animals	Food restriction	Pretraining	Training	Measure	Drug	Outcome	Reference
Pair-housed male Wistars	To 85% body weight	Food FR5TO20 until stable	Lever, 7–14 days	FR5TO20	SB-334867 (i.p.)	↓	Hollander et al. (2008)
				PR	SB-334867 (i.p.)	↓	
				FR5TO20	SB-334867 (intra-insula)	↓	
Singly housed male Long-Evans rats	18 g/day	Food until 100 pellets/1 h	Lever, 40±4.5 days	FR5TO60	SB-334867 (i.p.)	↓	LeSage et al. (2010)
			66±8.1 days	FR5TO60	Almorexant (i.p.)	↓	
Singly housed male C57BL/6J mice	Ad libitum	None	Daily nicotine prime, nosepoke, 10 days	OX-A-induced reinstatement (i.c.v.)	SB-334867 (i.p.)	↓	Plaza-Zabala et al. (2010)
				Footshock reinstatement	SB-334867 (i.p.)	No effect	
Singly housed male C57BL/6J mice	Ad libitum	None	Daily nicotine prime, nosepoke, 10 days	Cue reinstatement	SB-334867 (i.p.)	↓	Plaza-Zabala et al. (2013)
					TCS-OX2-29 (i.p.)	No effect	
Singly housed male Long-Evans rats	25 g/day	Food FR1	Lever, FR1 × 5 days, FR2 × 2 days, FR5 × 6 days	PR	2-SORA 18 (p.o.)	No effect	Uslaner et al. (2014)
			Animals retrained FR5 × 3 days	Cue reinstatement	2-SORA 18 (p.o.)	↓	
				Nic reinstatement	2-SORA 18 (p.o.)	No effect	
Group-housed Sprague-Dawley rats	20 g/day	None	Nosepoke, 10 days	FR1TO24	TCS 1102 (i.c.v.)	No effect	Khoo et al. (2017)
			Animals retrained (total of 29 days self-admin)	Cue reinstatement	TCS 1102 (i.c.v.)	No effect	
				Nic reinstatement	TCS 1102 (i.c.v.)	No effect	
				Compound reinstatement	TCS 1102 (i.c.v.)	No effect	

Details of key parameters of preclinical studies and the direction of effect.

reduce cue-induced reinstatement in rats, but not nicotine-primed reinstatement or PR self-administration (Uslaner et al., 2014). These studies suggest that the role of orexin signaling in reinstatement behaviors is inconsistent and may depend on the proximal cause of reinstatement or the behavioral model used. In both rats and mice, cue-induced reinstatement has been reduced using orexin receptor antagonists, but it is unclear whether the OX_2 receptor is involved. While these results are suggestive of a role for the orexins in cue-induced reinstatement for nicotine, the relatively small number of studies and the variance in training protocols (Table 61.1) make it difficult to draw strong conclusions.

61.2.3 Withdrawal and Motivation

Preclinical studies have also demonstrated roles for orexin signaling in other nicotine-related behaviors. In mice made dependent by 25 mg/kg/day nicotine administered for 14 days by osmotic minipump, mecamylamine-precipitated withdrawal was attenuated in *Hcrt*$^{-/-}$ mice or mice given the OX_1 receptor antagonist SB-334867 (Plaza-Zabala, Flores, Maldonado, & Berrendero, 2012), suggesting that orexins mediate withdrawal symptoms. Orexin signaling in the paraventricular nucleus of the hypothalamus was particularly implicated because targeted microinjection of SB-334867 attenuated behavioral signs of nicotine withdrawal. Although Plaza-Zabala et al. (2012) did not find effects of the OX_2 receptor antagonist TCS-OX2-29, this may have been because they did not use a sufficiently high dose. Their doses of 5–10 mg/kg TCS-OX2-29 (Plaza-Zabala et al., 2012, 2013) were lower than the 10–30 mg/kg doses used by Smith, See, and Aston-Jones (2009) although these also produced null results for cocaine. A previous study that reported positive effects at OX_2 receptors used a 15 mg/kg dose of SORA-18, which has greater potency than TCS-OX2-29 (Hirose et al., 2003; Uslaner et al., 2014). The dual orexin receptor antagonist, TCS 1102, also attenuated nicotine-potentiated PR responding and nicotine-induced reinstatement for sucrose pellets (Winrow et al., 2010). These results also provide support for a potential use for orexin receptor antagonists in treating nicotine addiction because they might reduce symptoms of withdrawal or reduce other reward-related nicotine responses.

61.2.4 Divergent Findings

However, we recently reported that the dual orexin receptor antagonist TCS 1102 had no effect on nicotine self-administration, cue-induced reinstatement, and nicotine-primed reinstatement and had at most a small transient effect on cue- and prime-induced reinstatement only after chronic nicotine self-administration (Khoo, McNally, & Clemens, 2017). Although TCS 1102 has previously been shown to be effective in reducing behavioral responses to nicotine (Winrow et al., 2010) and we found that intracerebroventricular TCS 1102 could attenuate OX-A-induced increases in feeding behavior, we found no evidence of any effect of dual orexin receptor antagonism that could have clinical relevance. We suspect that this discrepancy between our study and previous findings may be due to several differences in protocols. For example, our rats were trained on FR1, while previous studies have generally used FR5 schedules of reinforcement and food pretraining protocols (Hollander et al., 2008; LeSage et al., 2010; Uslaner et al., 2014). In our study, rats were trained to make nosepokes for nicotine for the entire duration of training with no period of operant responding for food (Khoo et al., 2017). This may have resulted in relatively lower levels of motivation in our animals. It has previously been argued that nicotine has relatively weak primary reinforcing properties but may act to enhance the motivational properties of other reinforcers and conditioned stimuli (Chaudhri et al., 2006). If this is the case, the food pretraining protocols used in previous demonstrations of orexinergic regulation of nicotine seeking or self-administration may have potentiated the motivational properties of the conditioned reinforcers and responses that were ultimately paired with nicotine. It has previously been shown that food pretraining enhances acquisition (Bongiovanni & See, 2008; Clemens, Caillé, & Cador, 2010; Garcia, Lê, & Tyndale, 2014). For example, sucrose pretraining can facilitate later cue-induced reinstatement of nicotine seeking for rats trained to make nosepokes (Clemens et al., 2010), but food pretraining does not appear to affect later reinstatement tests in rats trained to press levers (Garcia et al., 2014). Differences in pretraining protocols could also explain why relatively few rats in our experiments would self-administer under an FR5 schedule of reinforcement (unpublished observations).

These negative results urge caution in seeking to translate orexin antagonists to the clinic, but they do not necessarily exclude orexin-based therapies for nicotine. In humans, nicotine is often used in combination with other drugs (Cross, Lotfipour, & Leslie, 2017). For example, nicotine increases alcohol consumption in men (Acheson, Mahler, Chi, & de Wit, 2006; Barrett, Tichauer, Leyton, & Pihl, 2006) and subjective feelings of drunkenness (Kouri, McCarthy, Faust, & Lukas, 2004), results that have been replicated in animal models (Kalejaiye, Bhatti, Taylor, & Tizabi, 2013; Lê, Funk, Lo, & Coen, 2014). Nicotine use is also associated with using or becoming dependent on cannabis (Taylor et al., 2017), cocaine (Budney, Higgins, Hughes, & Bickel, 1993; Gorelick, Simmons, Carriero, & Tashkin, 1997), and methamphetamine (Grant et al., 2007). Using multiple drugs might increase the motivational salience and render nicotine subject to orexinergic regulation. Other drugs of abuse also have stronger evidence for orexinergic regulation, and there is currently a clinical trial for using the dual orexin receptor antagonist, suvorexant, in the treatment of cocaine dependence (The University of Texas Health Science Center, 2016). However, further preclinical studies in animal models of polydrug use are required to establish whether this is the case.

61.3 HUMAN STUDIES

A few studies in humans have found associations between the orexin system and nicotine (Table 61.2). Human studies have reported a negative correlation between orexin plasma concentration and self-reported nicotine craving (von der Goltz et al., 2010) and reduced *Hcrt* mRNA expression in blood samples in smokers compared to nonsmokers (Rotter et al., 2012). Although these studies are consistent, they are limited by their peripheral measurement of orexin peptide or *Hcrt* mRNA. While the OX-A peptide can diffuse across the blood-brain barrier, OX-B has low lipophilicity and is rapidly metabolized (Kastin & Akerstrom, 1999). Genome-wide association studies have found that in Japanese samples, a single nucleotide polymorphism in the HCRTR2 gene was

TABLE 61.2 Human Correlational Studies

Participants	Measurements	Findings	Reference
60 smokers, 64 nonsmokers	Whole blood—plasma orexin Questionnaire of smoking urges Fagerström test for nicotine dependence	Smoking urges negatively correlated with plasma OX	von der Goltz et al. (2010)
36 cannabis-dependent, 20 smoking, 21 nonsmoking student acquaintances	Whole blood—HCRT and OX-A in peripheral blood lymphocytes	Lower OX-A in smokers No effect on HCRT promoter methylation	Rotter et al. (2012)
Initial GWAS—148 patients	Fagerström test for nicotine dependence	HCRTR2 single nucleotide polymorphism (Val308Ile) correlated with smoking	Nishizawa et al. (2015)
Follow-up—112 abdominal surgery patients, 203 methamphetamine-dependent patients, 311 healthy volunteers	Tobacco dependence screener		
2305 autopsy cases	Autopsy results		

Summary of key findings from human studies to date. *GWAS*, genome-wide association study.

associated with increased risk of smoking (Nishizawa et al., 2015). However, results from these studies should be interpreted with caution because it has been argued that genome-wide association studies produce little useful data when examining complex traits (Boyle, Li, & Pritchard, 2017), although previous studies have yielded associations between targets that are biologically relevant or already targets for current therapeutics (Visscher, Brown, McCarthy, & Yang, 2012). While Nishizawa et al. (2015) found associations between HCRTR2 polymorphism and smoking and methamphetamine and schizotypal trait scores, they also found associations with goiter, aortic aneurysm, and myeloma. The current small array of human studies of the orexin system and nicotine suggests that there may be some linkage, but this is correlational, and further studies in animals and humans are required to better establish whether the orexin system really is involved in nicotine seeking.

61.4 IMPLICATIONS FOR TREATMENTS

There is currently insufficient evidence to recommend orexin-based treatments for smoking. Although there are some promising preclinical results and correlational evidence has been found in humans, there are also some inconsistent findings in the preclinical literature that should be addressed. Further studies are required to examine the relationship between an animal's training history and orexinergic regulation of nicotine seeking.

61.5 CONCLUSION

The orexin system has generated much excitement and interest for its therapeutic potential for a variety of disorders. This potential has recently been realized with the approval of suvorexant for insomnia (Coleman, Gotter, Herring, Winrow, & Renger, 2017) and may be realized for cocaine addiction (The University of Texas Health Science Center, 2016). Animal studies show that acute nicotine stimulates orexin neurons and chronic nicotine can affect the regulation of the orexin system. Preclinical animal models of drug self-administration and seeking have tested orexin antagonists against nicotine less frequently than against other drugs of abuse, but from a relatively small pool of studies, there is a mix of both positive and negative findings regarding the efficacy of orexin antagonists on nicotine self-administration and reinstatement. Research with human participants has found some consistent associations between orexin downregulation and nicotine use and genetic associations between a single nucleotide HCRTR2 polymorphism and smoking, but these results are correlational. For nicotine and orexin, there is currently insufficient evidence that it would be a useful therapeutic target for nicotine addiction, but further studies are required to examine the reasons behind differential outcomes in preclinical research and to follow up on the associations that have been found in human studies.

MINI-DICTIONARY OF TERMS

Almorexant A dual orexin receptor antagonist that progressed to clinical trials.

Dual orexin receptor antagonist A compound that has antagonist effects at both orexin receptors.

Fixed-ratio self-administration Operant reinforcement schedule used in preclinical animal models where a reward is earned for a fixed number of operant responses.

Hypocretin The orexin/hypocretin system named based on its *hypo*thalamic distribution and similarity to the in*cretin* family of hormones. Refers to the genes HCRT/*Hcrt*, which encodes for the two peptides, and HCRTR1/*Hcrtr1* and HCRTR2/*Hcrtr2*, which encode for the two receptors.

Orexin The orexin/hypocretin system named based on its role in appetitive behavior. Refers to the two peptides, orexin-A and orexin-B, which bind to the OX_1 and OX_2 receptors.

Progressive ratio Operant reinforcement schedule used in preclinical animal models where the number of responses required to earn each subsequent reward increases.

Reinstatement A model of relapse-like behavior that involves the return of a previously extinguished behavior. May be precipitated by drug-priming, reward-associated cues, contexts, or stress.

SB-334867 A selective OX_1 receptor antagonist.

SORA-18 A selective OX_2 receptor antagonist.

TCS 110 A dual orexin receptor antagonist.

TCS-OX2-29 A selective OX_2 receptor antagonist.

Key Facts of the Orexin System

- Discovered simultaneously by two groups in 1998.
- Two peptides encoded by the hypocretin gene, orexin-A and orexin-B, bind to excitatory G-protein-coupled receptors, OX_1 and OX_2.
- Loss of orexin neurons involved in human narcolepsy.
- Stimulation of orexin system promotes appetitive behavior.
- Animal studies have demonstrated roles for the orexin in multiple drugs of abuse.
- Suvorexant (Belsomra®) was the first orexin-based pharmacotherapy and was approved in 2014 for the treatment of insomnia.

Summary Points

- This chapter focuses on the orexin system, a neuropeptide system involved in arousal, appetite, and reward, and its potential as a therapeutic target for nicotine addiction.
- The orexin system activates cholinergic neurons and is activated by nicotine administration, with chronic nicotine causing altered expression of the hypocretin genes and orexin peptides.
- A small number of preclinical animal studies have shown effects of various orexin receptor antagonists in self-administration or reinstatement paradigms.
- However, there are substantial differences in methodologies between studies, and some studies have not found effects of orexin antagonism.
- The limited number of studies in animals and humans suggests that there may be some promise in targeting the orexin system, but more research is required to understand its role in nicotine seeking before clinical trials can be proposed.

References

Acheson, A., Mahler, S. V., Chi, H., & de Wit, H. (2006). Differential effects of nicotine on alcohol consumption in men and women. *Psychopharmacology, 186*, 54. https://doi.org/10.1007/s00213-006-0338-y.

Baldo, B. A., Daniel, R. A., Berridge, C. W., & Kelley, A. E. (2003). Overlapping distributions of orexin/hypocretin- and dopamine-β-hydroxylase immunoreactive fibers in rat brain regions mediating arousal, motivation, and stress. *The Journal of Comparative Neurology, 464*, 220–237. https://doi.org/10.1002/cne.10783.

Barrett, S. P., Tichauer, M., Leyton, M., & Pihl, R. O. (2006). Nicotine increases alcohol self-administration in non-dependent male smokers. *Drug and Alcohol Dependence, 81*, 197–204. https://doi.org/10.1016/j.drugalcdep.2005.06.009.

Bongiovanni, M., & See, R. E. (2008). A comparison of the effects of different operant training experiences and dietary restriction on the reinstatement of cocaine-seeking in rats. *Pharmacology, Biochemistry, and Behavior, 89*, 227–233. https://doi.org/10.1016/j.pbb.2007.12.019.

Boyle, E. A., Li, Y. I., & Pritchard, J. K. (2017). An expanded view of complex traits: from polygenic to omnigenic. *Cell, 169*, 1177–1186. https://doi.org/10.1016/j.cell.2017.05.038.

Budney, A. J., Higgins, S. T., Hughes, J. R., & Bickel, W. K. (1993). Nicotine and caffeine use in cocaine-dependent individuals. *Journal of Substance Abuse, 5*, 117–130. https://doi.org/10.1016/0899-3289(93)90056-H.

Cahill, K., Lindson-Hawley, N., Thomas, K. H., Fanshawe, T. R., & Lancaster, T. (2016). Nicotine receptor partial agonists for smoking cessation. *Cochrane Database of Systematic Reviews. 5*, CD006103 https://doi.org/10.1002/14651858.CD006103.pub7.

Chaudhri, N., Caggiula, A. R., Donny, E. C., Booth, S., Gharib, M., Craven, L., et al. (2006). Operant responding for conditioned and unconditioned reinforcers in rats is differentially enhanced by the primary reinforcing and reinforcement-enhancing effects of nicotine. *Psychopharmacology, 189*, 27–36. https://doi.org/10.1007/s00213-006-0522-0.

Clemens, K. J., Caillé, S., & Cador, M. (2010). The effects of response operandum and prior food training on intravenous nicotine self-administration in rats. *Psychopharmacology, 211*, 43–54. https://doi.org/10.1007/s00213-010-1866-z.

Coleman, P. J., Gotter, A. L., Herring, W. J., Winrow, C. J., & Renger, J. J. (2017). The discovery of suvorexant, the first orexin receptor drug for insomnia. *Annual Review of Pharmacology and Toxicology, 57*, 509–533. https://doi.org/10.1146/annurev-pharmtox-010716-104837.

Cross, S. J., Lotfipour, S., & Leslie, F. M. (2017). Mechanisms and genetic factors underlying co-use of nicotine and alcohol or other drugs of abuse. *The American Journal of Drug and Alcohol Abuse, 43*, 171–185. https://doi.org/10.1080/00952990.2016.1209512.

de Lecea, L., Kilduff, T. S., Peyron, C., Gao, X.-B., Foye, P. E., Danielson, P. E., et al. (1998). The hypocretins: hypothalamus-specific peptides with neuroexcitatory activity. *Proceedings of the National Academy of Sciences, 95*, 322–327. https://doi.org/10.1073/pnas.95.1.322.

Elias, C. F., Saper, C. B., Maratos-Flier, E., Tritos, N. A., Lee, C., Kelly, J., et al. (1998). Chemically defined projections linking the mediobasal hypothalamus and the lateral hypothalamic area. *The Journal of Comparative Neurology, 402*, 442–459. https://doi.org/10.1002/(SICI)1096-9861(19981228)402:4<442::AID-CNE2>3.0.CO;2-R.

Ellis, J., Pediani, J. D., Canals, M., Milasta, S., & Milligan, G. (2006). Orexin-1 receptor-cannabinoid CB1 receptor heterodimerization results in both ligand-dependent and -independent coordinated alterations of receptor localization and function. *Journal of Biological Chemistry, 281*, 38812–38824. https://doi.org/10.1074/jbc.M602494200.

Fadel, J., Pasumarthi, R., & Reznikov, L. R. (2005). Stimulation of cortical acetylcholine release by orexin A. *Neuroscience, 130*, 541–547. https://doi.org/10.1016/j.neuroscience.2004.09.050.

Garcia, K. L. P., Lê, A. D., & Tyndale, R. F. (2014). Effect of food training and training dose on nicotine self-administration in rats. *Behavioural Brain Research, 274*, 10–18. https://doi.org/10.1016/j.bbr.2014.07.043.

Gorelick, D. A., Simmons, M. S., Carriero, N., & Tashkin, D. P. (1997). Characteristics of smoked drug use among cocaine smokers. *The American Journal on Addictions, 6*, 237–245. https://doi.org/10.1111/j.1521-0391.1997.tb00403.x.

Grant, K. M., Kelley, S. S., Agrawal, S., Meza, J. L., Meyer, J. R., & Romberger, D. J. (2007). Methamphetamine use in rural midwesterners. *The American Journal on Addictions, 16*, 79–84. https://doi.org/10.1080/10550490601184159.

Hirose, M., Egashira, S.-i., Goto, Y., Hashihayata, T., Ohtake, N., Iwaasa, H., et al. (2003). N-Acyl 6,7-dimethoxy-1,2,3,4-tetrahydroisoquinoline: the first orexin-2 receptor selective non-peptidic antagonist. *Bioorganic & Medicinal Chemistry Letters, 13*, 4497–4499. https://doi.org/10.1016/j.bmcl.2003.08.038.

Hollander, J. A., Lu, Q., Cameron, M. D., Kamenecka, T. M., & Kenny, P. J. (2008). Insular hypocretin transmission regulates nicotine reward. *Proceedings of the National Academy of Sciences, 105*, 19480–19485. https://doi.org/10.1073/pnas.0808023105.

Jäntti, M. H., Mandrika, I., & Kukkonen, J. P. (2014). Human orexin/hypocretin receptors form constitutive homo- and heteromeric complexes with each other and with human CB1 cannabinoid receptors. *Biochemical and Biophysical Research Communications, 445*, 486–490. https://doi.org/10.1016/j.bbrc.2014.02.026.

Kalejaiye, O., Bhatti, B. H., Taylor, R. E., & Tizabi, Y. (2013). Nicotine blocks the depressogenic effects of alcohol: implications for drinking-smoking co-morbidity. *Journal of Drug and Alcohol Research, 2*, 235709. https://doi.org/10.4303/jdar/235709.

Kane, J. K., Parker, S. L., & Li, M. D. (2001). Hypothalamic orexin-A binding sites are downregulated by chronic nicotine treatment in the rat. *Neuroscience Letters, 298*, 1–4. https://doi.org/10.1016/S0304-3940(00)01730-4.

Kane, J. K., Parker, S. L., Matta, S. G., Fu, Y., Sharp, B. M., & Li, M. D. (2000). Nicotine up-regulates expression of orexin and its receptors in rat brain. *Endocrinology, 141*, 3623–3629. https://doi.org/10.1210/endo.141.10.7707.

Kastin, A. J., & Akerstrom, V. (1999). Orexin A but not orexin B rapidly enters brain from blood by simple diffusion. *Journal of Pharmacology and Experimental Therapeutics, 289*, 219–223.

Khoo, S. Y.-S., McNally, G. P., & Clemens, K. J. (2017). The dual orexin receptor antagonist TCS1102 does not affect reinstatement of nicotine-seeking. *PLoS ONE, 12*. e0173967 https://doi.org/10.1371/journal.pone.0173967.

Kouri, E. M., McCarthy, E. M., Faust, A. H., & Lukas, S. E. (2004). Pretreatment with transdermal nicotine enhances some of ethanol's acute effects in men. *Drug and Alcohol Dependence, 75*, 55–65. https://doi.org/10.1016/j.drugalcdep.2004.01.011.

Kukkonen, J. P., & Leonard, C. S. (2014). Orexin/hypocretin receptor signalling cascades. *British Journal of Pharmacology, 171*, 314–331. https://doi.org/10.1111/bph.12324.

Lê, A. D., Funk, D., Lo, S., & Coen, K. (2014). Operant self-administration of alcohol and nicotine in a preclinical model of co-abuse. *Psychopharmacology, 231*, 4019–4029. https://doi.org/10.1007/s00213-014-3541-2.

Lerman, C., LeSage, M. G., Perkins, K. A., O'Malley, S. S., Siegel, S. J., Benowitz, N. L., et al. (2007). Translational research in medication development for nicotine dependence. *Nature Reviews Drug Discovery, 6*, 746–762. https://doi.org/10.1038/nrd2361.

LeSage, M. G., Perry, J. L., Kotz, C. M., Shelley, D., & Corrigall, W. A. (2010). Nicotine self-administration in the rat: effects of hypocretin antagonists and changes in hypocretin mRNA. *Psychopharmacology, 209*, 203–212. https://doi.org/10.1007/s00213-010-1792-0.

Navarro, G., Quiroz, C., Moreno-Delgado, D., Sierakowiak, A., McDowell, K., Moreno, E., et al. (2015). Orexin–corticotropin-releasing factor receptor heteromers in the ventral tegmental area as targets for cocaine. *The Journal of Neuroscience, 35*, 6639–6653. https://doi.org/10.1523/jneurosci.4364-14.2015.

Nishizawa, D., Kasai, S., Hasegawa, J., Sato, N., Yamada, H., Tanioka, F., et al. (2015). Associations between the orexin (hypocretin) receptor 2 gene polymorphism Val308Ile and nicotine dependence in genome-wide and subsequent association studies. *Molecular Brain, 8*, 50. https://doi.org/10.1186/s13041-015-0142-x.

Pasumarthi, R. K., & Fadel, J. (2008). Activation of orexin/hypocretin projections to basal forebrain and paraventricular thalamus by acute nicotine. *Brain Research Bulletin, 77*, 367–373. https://doi.org/10.1016/j.brainresbull.2008.09.014.

Pasumarthi, R. K., & Fadel, J. (2010). Stimulation of lateral hypothalamic glutamate and acetylcholine efflux by nicotine: implications for mechanisms of nicotine-induced activation of orexin neurons. *Journal of Neurochemistry, 113*, 1023–1035. https://doi.org/10.1111/j.1471-4159.2010.06666.x.

Pasumarthi, R. K., Reznikov, L. R., & Fadel, J. (2006). Activation of orexin neurons by acute nicotine. *European Journal of Pharmacology, 535*, 172–176. https://doi.org/10.1016/j.ejphar.2006.02.021.

Peyron, C., Tighe, D. K., van den Pol, A. N., de Lecea, L., Heller, H. C., Sutcliffe, J. G., et al. (1998). Neurons containing hypocretin (orexin) project to multiple neuronal systems. *The Journal of Neuroscience, 18*, 9996–10015.

Plaza-Zabala, A., Flores, Á., Maldonado, R., & Berrendero, F. (2012). Hypocretin/orexin signaling in the hypothalamic paraventricular nucleus is essential for the expression of nicotine withdrawal. *Biological Psychiatry, 71*, 214–223. https://doi.org/10.1016/j.biopsych.2011.06.025.

Plaza-Zabala, A., Flores, A., Martin-Garcia, E., Saravia, R., Maldonado, R., & Berrendero, F. (2013). A role for hypocretin/orexin receptor-1 in cue-induced reinstatement of nicotine-seeking behavior. *Neuropsychopharmacology, 38*, 1724–1736. https://doi.org/10.1038/npp.2013.72.

Plaza-Zabala, A., Martín-García, E., de Lecea, L., Maldonado, R., & Berrendero, F. (2010). Hypocretins regulate the anxiogenic-like effects of nicotine and induce reinstatement of nicotine-seeking behavior. *The Journal of Neuroscience, 30*, 2300–2310. https://doi.org/10.1523/jneurosci.5724-09.2010.

Rotter, A., Bayerlein, K., Hansbauer, M., Weiland, J., Sperling, W., Kornhuber, J., et al. (2012). Orexin A expression and promoter methylation in patients with cannabis dependence in comparison to nicotine-dependent cigarette smokers and nonsmokers. *Neuropsychobiology, 66*, 126–133. https://doi.org/10.1159/000339457.

Sakurai, T., Amemiya, A., Ishii, M., Matsuzaki, I., Chemelli, R. M., Tanaka, H., et al. (1998). Orexins and orexin receptors: a family of hypothalamic neuropeptides and G protein-coupled receptors that regulate feeding behavior. *Cell, 92,* 573–585. https://doi.org/10.1016/s0092-8674(00)80949-6.

Sakurai, T., Moriguchi, T., Furuya, K., Kajiwara, N., Nakamura, T., Yanagisawa, M., et al. (1999). Structure and function of human prepro-orexin gene. *Journal of Biological Chemistry, 274,* 17771–17776. https://doi.org/10.1074/jbc.274.25.17771.

Simmons, S. J., Gentile, T. A., Mo, L., Tran, F. H., Ma, S., & Muschamp, J. W. (2016). Nicotinic receptor blockade decreases fos immunoreactivity within orexin/hypocretin-expressing neurons of nicotine-exposed rats. *Behavioural Brain Research, 314,* 226–233. https://doi.org/10.1016/j.bbr.2016.07.053.

Smith, R. J., See, R. E., & Aston-Jones, G. (2009). Orexin/hypocretin signaling at the orexin 1 receptor regulates cue-elicited cocaine-seeking. *European Journal of Neuroscience, 30,* 493–503. https://doi.org/10.1111/j.1460-9568.2009.06844.x.

Taylor, M., Collin, S. M., Munafò, M. R., MacLeod, J., Hickman, M., & Heron, J. (2017). Patterns of cannabis use during adolescence and their association with harmful substance use behaviour: findings from a UK birth cohort. *Journal of Epidemiology and Community Health,* https://doi.org/10.1136/jech-2016-208503.

The University of Texas Health Science Center. (2016). *Role of the orexin receptor system in stress, sleep and cocaine use. Retrieved from:(2016). clinicaltrials.gov/show/NCT02785406.*

Uslaner, J. M., Winrow, C. J., Gotter, A. L., Roecker, A. J., Coleman, P. J., Hutson, P. H., et al. (2014). Selective orexin 2 receptor antagonism blocks cue-induced reinstatement, but not nicotine self-administration or nicotine-induced reinstatement. *Behavioural Brain Research, 269,* 61–65. https://doi.org/10.1016/j.bbr.2014.04.012.

Visscher, P. M., Brown, M. A., McCarthy, M. I., & Yang, J. (2012). Five years of GWAS discovery. *The American Journal of Human Genetics, 90,* 7–24. https://doi.org/10.1016/j.ajhg.2011.11.029.

von der Goltz, C., Koopmann, A., Dinter, C., Richter, A., Rockenbach, C., Grosshans, M., et al. (2010). Orexin and leptin are associated with nicotine craving: a link between smoking, appetite and reward. *Psychoneuroendocrinology, 35,* 570–577. https://doi.org/10.1016/j.psyneuen.2009.09.005.

Winrow, C. J., Tanis, K. Q., Reiss, D. R., Rigby, A. M., Uslaner, J. M., Uebele, V. N., et al. (2010). Orexin receptor antagonism prevents transcriptional and behavioral plasticity resulting from stimulant exposure. *Neuropharmacology, 58,* 185–194. https://doi.org/10.1016/j.neuropharm.2009.07.008.

Xu, T. R., Ward, R. J., Pediani, J. D., & Milligan, G. (2011). The orexin OX_1 receptor exists predominantly as a homodimer in the basal state: potential regulation of receptor organization by both agonist and antagonist ligands. *Biochemical Journal, 439,* 171–183. https://doi.org/10.1042/bj20110230.

CHAPTER

62

Tobacco Control Policies and Smokers' Responses

Philip DeCicca[*,†], *Erik Nesson*[*]

[*]Department of Economics, Ball State University, Muncie, IN, United States

[†]NBER, Cambridge, MA, United States

Abbreviations

CPS-TUS	Current Population Survey-Tobacco Use Supplement
ETS	environmental tobacco smoke
NHANES	National Health and Nutrition Examination Survey
SFA laws	smoke-free air laws

62.1 INTRODUCTION

Public policies designed to reduce smoking, notably cigarette excise taxes and SFA laws, have been common health policy tools. Cigarette excise taxes are levied at both the state and federal levels in the United States. They increase the price that consumers pay for cigarettes and thus should discourage smoking. In addition, many states have passed SFA laws, legislation restricting where cigarettes may be smoked, affecting smoking in restaurants and bars, private workplaces, government office buildings, and educational facilities, among other venues.

There is substantial variation in smoking prevalence and tobacco control policies across different states, as shown in Figs. 62.1–62.3. Fig. 62.1 shows the state plus federal cigarette excise tax for different states in 2015. In 2015, the lowest state plus federal excise tax was $1.18, while the highest was $5.36. Fig. 62.2 plots state-level SFA laws in private workplaces, restaurants, and bars, and Fig. 62.3 graphs smoking prevalence and shows that most states have adopted SFA laws in private workplaces, restaurants, and bars. However, the restrictiveness of SFA laws across states varies widely. As shown in Fig. 62.3, adult and youth smoking prevalence also varies widely across states, with the highest smoking prevalence generally in the southern and Appalachian states.

There is also temporal variation in tobacco control policies and smoking prevalence. Figs. 62.4 and 62.5 display these trends. Fig. 62.4 plots the average state and federal cigarette excise tax, adjusted for inflation, and the percent of the US population living under a clean indoor air law in restaurants, bars, and private workplaces over the period of 1995–2015. Fig. 62.5 plots the trend in smoking participation for adults and adolescents during the same time periods. The average state plus federal cigarette excise tax has increased from just under $1.00 in 1995 to nearly $2.50 in 2015, and the percentage of Americans living under a SFA law has increased from just over 60% in 1995 to nearly 90% in 2015. Over this same period, smoking participation among adults and adolescents has markedly decreased.

These correlations suggest that tobacco control policies may be responsible for much of the concurrent decrease in smoking, but establishing a causal relationship between tobacco control policies and smoking outcomes requires addressing a number of statistical challenges. Perhaps, the largest challenge is the lack of a counterfactual world where researchers could observe smoking-related outcomes in the absence of tobacco control policies. This leaves open the possibility that while tobacco control policies are correlated with reduced smoking outcomes, the relationship is not causal. Economic research into tobacco control policies has largely been built around a "difference-in-difference" approach, taking advantage of the staggered implementation of these policies across states at different times. Most studies use individual-level data, collected from large surveys, and connect the geographic location of individuals to the tobacco control policies for that location and time. Examining changes in individuals' smoking outcomes in locations that implement tobacco control policies compared to those that

Neuroscience of Nicotine
https://doi.org/10.1016/B978-0-12-813035-3.00062-9

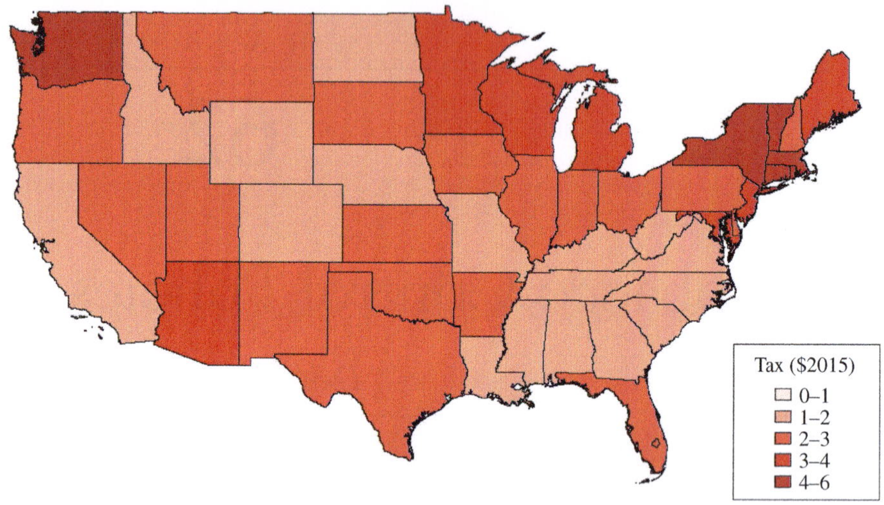

FIG. 62.1 Variation in state plus federal excise taxes in 2015.
Data from the CDC STATE System. Cigarette excise tax represents the state and federal excise tax in 2015 in real $2015 dollars.

did not creates a sort of counterfactual environment or quasi experiment. To the extent that locations that do not implement policies represent what would have happened in the locations that do implement policies, correlations between tobacco control policies and smoking outcomes indicate a causal relationship.

Another challenge is measuring smoking outcomes. Data limitations have until recently limited researchers to measuring smoking outcomes in terms of self-reported smoking participation and the number of cigarettes smoked per day. If tobacco control policies induce smokers to change brands, change the way in which they smoke cigarettes, or misreport their smoking behavior, estimates of the impact of tobacco control policies on the number of cigarettes smoked may not accurately represent changes in smoking.

Below, we summarize the recent advances in research estimating the effects of tobacco control policies. We focus largely on the United States and the most commonly researched tobacco control policies, cigarette excise taxes, and SFA laws. We examine three main smoking-related outcomes: (1) own smoking behavior, (2) ETS exposure, and (3) health effects from smoke exposure. We describe challenges and innovations for each set of outcomes and end with potential areas for future research.

62.2 ADULT SMOKING BEHAVIOR

With respect to taxes and prices, early studies established a negative price elasticity demand for cigarettes of around −0.4, meaning a 10% increase in cigarette prices would decrease smoking by 4% (Chaloupka & Warner, 2000). Early studies generally find that SFA laws also decrease cigarette smoking (Chaloupka, 1992; Chaloupka & Saffer, 1992; Evans, Farrelly, & Montgomery, 1999; Farrelly, Evans, & Sfekas, 1999; Tauras, 2006).

These earliest studies faced statistical and data challenges. First, largely due to data limitations, early studies measured smoking behavior using self-reported smoking participation and the number of cigarettes smoked per day. As noted above, misreported smoking behavior could bias the estimated effects of tobacco control policies on smoking. Relatedly, since cigarette taxes are levied on a per cigarette basis, smokers may respond by changing the way they smoke or changing the types of cigarettes smoked to ingest more nicotine from each cigarette. Thus, reduced cigarette consumption may not translate to reduced nicotine intake or exposure to harmful chemicals (Harris, 1980). Evans and Farrelly (1998) and Farrelly, Nimsch, Hyland, and Cummings (2004) are the first to empirically examine this question, using datasets where cigarette purchases are matched to tar, nicotine, and carbon monoxide contents. Both find that while smokers reduce the number of cigarettes they smoke in response to cigarette excise tax increases, increased tar and nicotine contents of the cigarettes purchased offset these reductions. Adda and Cornaglia (2006) advance this literature by using data from NHANES, which contain both a self-reported cigarette consumption and serum cotinine levels, a biomarker of recent nicotine exposure. Similar to Evans and Farrelly (1998) and Farrelly et al. (2004), Adda and Cornaglia (2006) find that while smokers reduce cigarette consumption due to increased cigarette excise taxes, there are no corresponding decreases in serum cotinine levels, suggesting that smokers either change cigarette brands or change the intensity of how they smoke cigarettes.

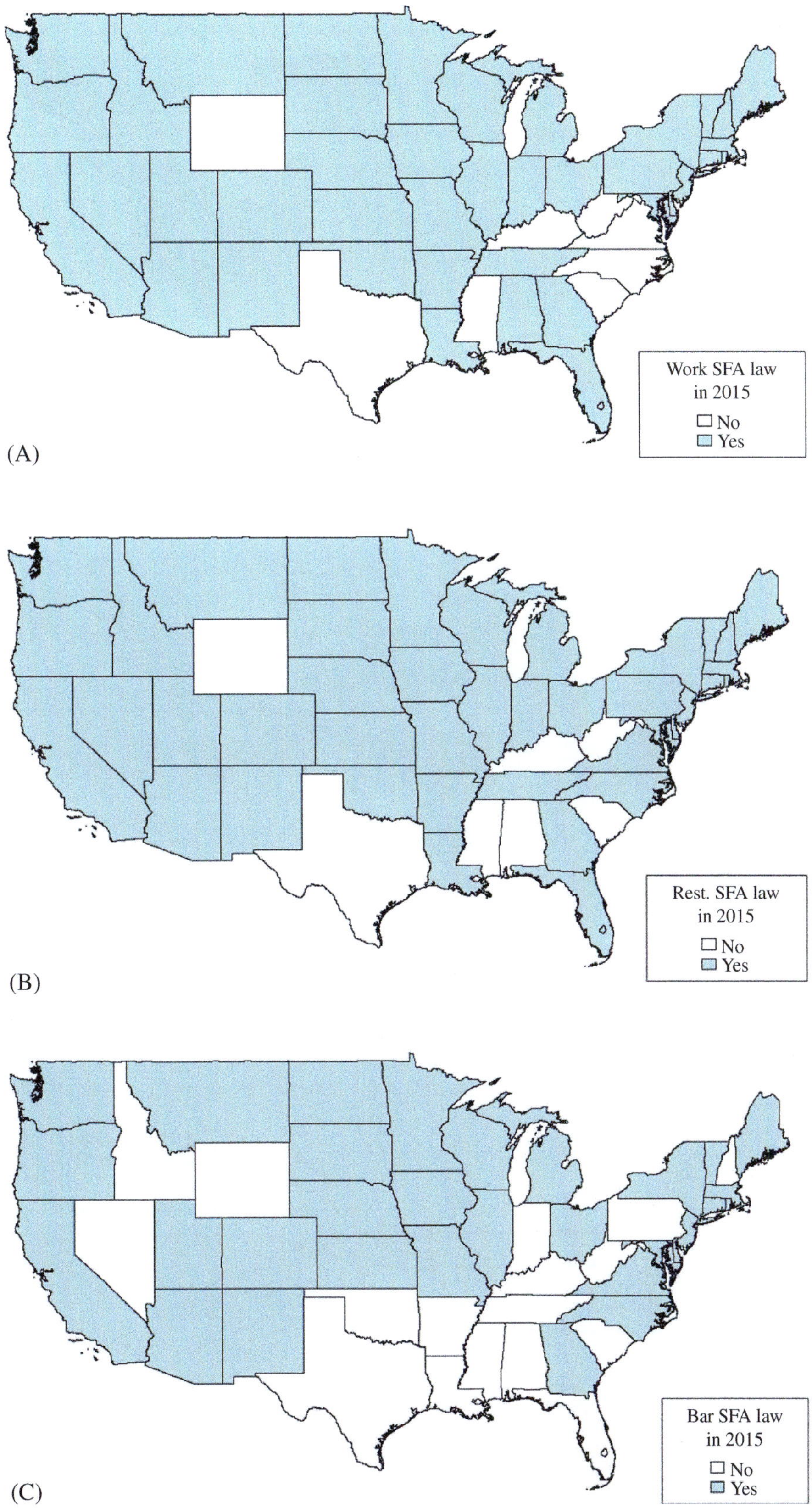

FIG. 62.2 Variation in SFA laws in 2015. (A) Workplace SFA laws, (B) restaurant SFA laws, and (C) bar SFA laws. Data from the CDC STATE System.

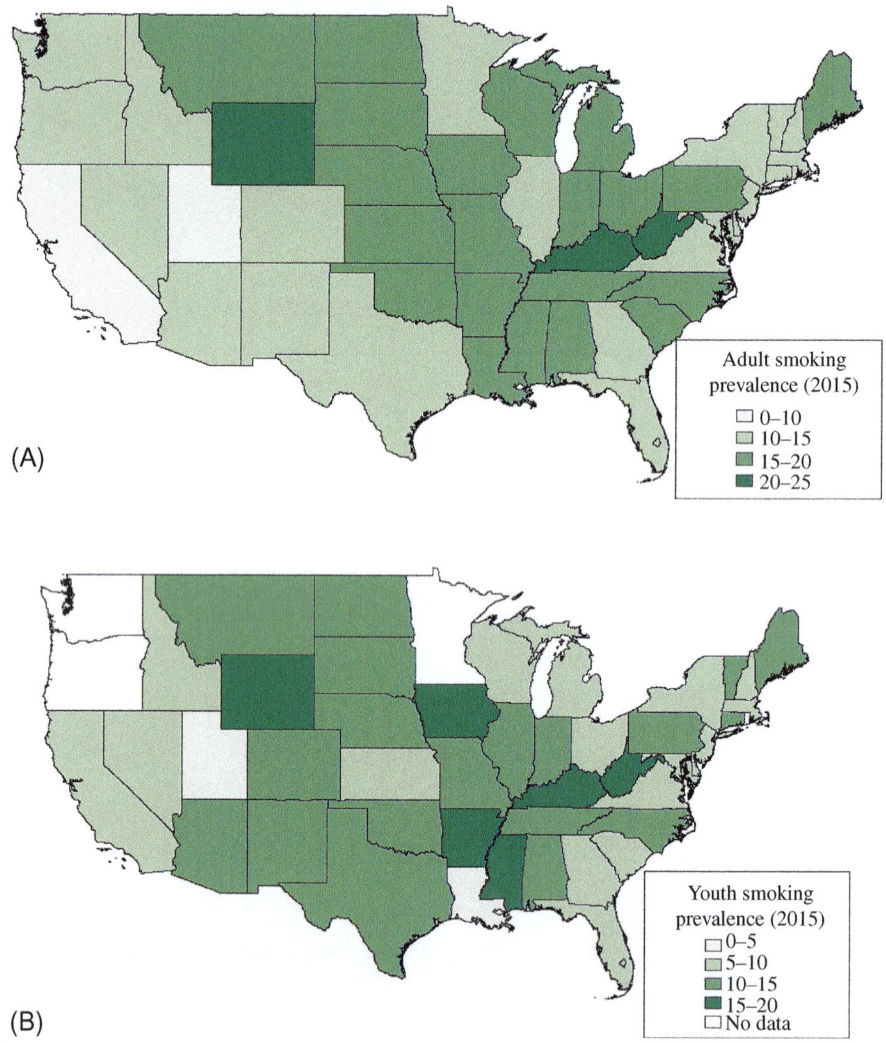

FIG. 62.3 Variation in smoking prevalence in 2015.
(A) Adult smoking prevalence and (B) youth smoking prevalence. Data from the CDC STATE System. Adult smoking prevalence comes from the Current Population Survey-Tobacco Use Supplements from the question, "Do you now smoke cigarettes every day, some days, or not at all?" Adolescent smoking prevalence comes from the Youth Risk Behavioral Surveillance System asking, "During the past 30 days, on how many days did you smoke cigarettes?"

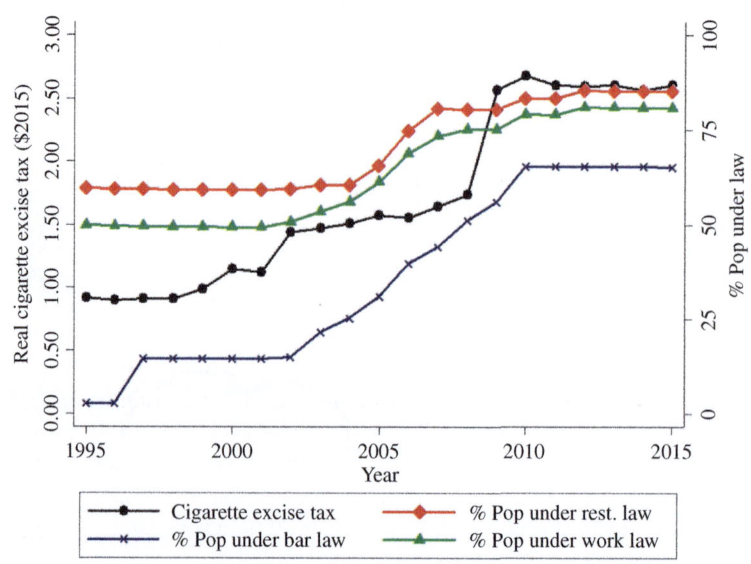

FIG. 62.4 Trends in tobacco control policies 1995–2015. Data from the CDC STATE System and the SEER population data. Cigarette excise tax represents the average state and federal excise tax, weighted by state population in real $2015 dollars. The percentage of the population living under a SFA law is also weighted using state population.

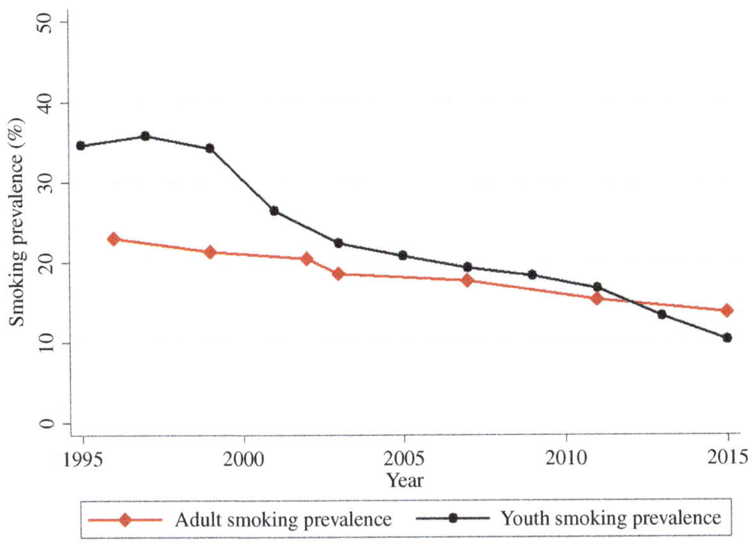

FIG. 62.5 Trends in smoking prevalence 1995–2015. Data from the CDC STATE System and the SEER population data. Adult smoking prevalence comes from the Current Population Survey-Tobacco Use Supplements from the question, "Do you now smoke cigarettes every day, some days, or not at all?" Adolescent smoking prevalence comes from the Youth Risk Behavioral Surveillance System asking, "During the past 30 days, on how many days did you smoke cigarettes?"

However, more recent research finds less evidence of compensating behavior. Abrevaya and Puzzello (2012) find that the Adda and Cornaglia (2006) results are due to using a publicly available subsample of the NHANES data. Nesson (2017a) further expands the NHANES data and also finds little evidence of compensatory behavior. Finally, Cotti, Nesson, and Tefft (2016) use household scanner data that track households' purchases across long periods of time and match cigarette purchases to tar, nicotine, and carbon monoxide contents. They find small increases in tar, nicotine, and carbon monoxide contents that are small compared to the reduction in cigarette purchases.

Another recent innovation regards tax avoidance. Chiou and Muehlegger (2008) use the 2003 wave of the CPS-TUS that contains information on where smokers purchase their cigarettes. They find that heavy smokers are more likely to cross state borders to buy cigarettes in lower-tax venues. DeCicca, Kenkel, and Liu (2013a) and DeCicca, Kenkel, and Liu (2013b) also use the CPS-TUS to further investigate this question. DeCicca et al. (2013a) find taxes are shifted less to consumers who search for lower prices, namely, smokers who buy cartons of cigarettes rather than individual packs and smokers who purchase cigarettes across state borders. DeCicca et al. (2013b) find that the impact of this price search suggests that the optimal cigarette excise tax is about 20% lower than current rates and that federal cigarette taxation may be more efficient. Finally, Harding, Leibtag, and Lovenheim (2012) use household scanner data to examine the pass-through of cigarette excise taxes to prices and also find that households closer to lower-tax states are more likely to purchase cigarettes across that state border.

62.3 YOUTH SMOKING BEHAVIOR

In addition to estimating adults' responses to tobacco control policies, research addresses whether adolescents below legal smoking age respond to tobacco control policies. Early studies suggest that cigarette prices and taxes affect youth smoking but do not come to consistent conclusions regarding SFA laws (Chaloupka, 1991; Chaloupka & Grossman, 1996; Chaloupka & Wechsler, 1997). However, more recent studies come to mixed conclusions regarding excise taxes. Some recent studies make use of panel datasets and often find that higher cigarette taxes lead to weak reductions in smoking (e.g., DeCicca, Kenkel, & Mathios, 2002; Douglas, 1998; Douglas & Hariharan, 1994; Nonnemaker & Farrelly, 2011). On the other hand, Lillard, Molloy, and Sfekas (2013) suggest that these weak tax effects may be due to limited policy variation or the ages at which these studies examine smoking decisions.

Other recent studies have suggested that policy endogeneity, where state-specific antismoking sentiment may drive both the adoption of tobacco control policies and reductions in youth smoking, may bias the results of youth smoking studies. DeCicca, Kenkel, Mathios, Shin, and Lim (2008) derive a measure of antismoking sentiment using respondents' views on various smoking questions from the CPS-TUS. They find that accounting for antismoking sentiment diminishes the response of youth to cigarette prices. Carpenter and Cook (2008) use a much larger repeated cross-sectional dataset and find that cigarette taxes still reduce youth smoking once antismoking is accounted for, although Hansen, Sabia, and Rees (2017) suggest that the effects of cigarette taxes on youth smoking have attenuated in recent years.

Finally, there may be differences between self-reported cigarette consumption and nicotine intake among youth smokers. Like adult smokers, youth smokers may compensate for higher cigarette excise taxes by switching brands or increasing smoking intensity. Moreover, youth may fear reprisals from disclosing smoking behavior and underreport their smoking. If the propensity to underreport is related to tobacco control policies, then reductions in smoking in response to increased tobacco control policies may simply reflect an increase in misreports. Kenkel, Lillard, and Mathios (2004) use longitudinal data containing both contemporaneous measures of smoking and retrospective measures of smoking. They find that accounting for misstated smoking may be important when estimating the effects of tobacco control policies.

Researchers have recently brought smoking biomarkers to bear on this question. Edoka (2017) uses data from the Health Survey for England that contain both self-reported smoking prevalence and serum cotinine levels. She finds that self-reported smoking prevalence understates smoking prevalence as measured using serum cotinine levels and that this bias attenuates marginal effects when estimating the effects of smoking determinants. Nesson (2017b) uses data from NHANES and examines the effects of tobacco control policies on youth smoking as measured by self-reports and serum cotinine levels. While he finds that cigarette excise tax increases decrease smoking participation measured by both self-reports and serum cotinine levels, he notes differences in the intensive margin. Finally, Nesson (2017b) finds evidence that cigarette excise taxes and public policies directed at youth possession, use, or purchase of cigarettes increase misreported smoking status among youth.

62.4 ETS EXPOSURE

Until recently, data limitations confined studies of the effects of tobacco control policies on ETS exposure to geographically specific sites. This study design raises concerns that reductions in ETS exposure correlated with tobacco control policies may not actually be caused by those policies. Examples include Farrelly et al. (2005), who found that a 2003 New York State bar and restaurant SFA law reduced ETS exposure among restaurant and bar employees and a CDC report that found that the law reduced ETS exposure among the general population (CDC, 2007). Other studies examining SFA laws in Massachusetts; Washington, DC; Scotland; Ireland; Sweden; and Norway gather similar results (Ellingsen et al., 2006; Larsson, Boethius, Axelsson, & Montgomery, 2008; Menzies et al., 2006; Mulcahy, Evans, Hammond, Repace, & Byrne, 2005; Pearson, Windsor, El-Mohandes, & Perry, 2009; Siegel, Albers, Cheng, Biener, & Rigotti, 2004).

As large surveys began to ask questions regarding ETS exposure, the economic literature on ETS exposure grew. Bitler, Carpenter, and Zavodny (2010) examine the effects of 12 venue-specific SFA laws on smoking behavior and workplace compliance using the CPS-TUS. They find that although bar SFA laws cause bartenders to report increased smoking restrictions, other SFA laws do not affect workers' reports of smoking restrictions. Using more recent waves of the CPS-TUS, Cheng, Liu, Gonzalez, and Glantz (2017) expand on Bitler et al. (2010) to also examine ETS exposure. They find that clean indoor air laws reduce ETS exposure among nonsmokers by 28%. However, Adda and Cornaglia (2010) use data from NHANES and serum cotinine concentrations and find little evidence that SFA laws reduce individuals' exposure to ETS. In fact, they find evidence that SFA laws increase ETS exposure in children in smoking families, as laws induce the adult smokers to smoke more at home and spend less time at restaurants and bars. That said, they do find that cigarette excise taxes reduce ETS exposure.

Economists have also made use of rich data from Canada. Carpenter (2009) examined local workplace smoking restrictions in Ontario, Canada, from 1997 to 2004 and found that these restrictions reduced ETS exposure for blue-collar workers by 28%–33%. However, these local bans had little effect on ETS exposure for white-collar workers. Carpenter, Postolek, and Warman (2011) find that SFA laws reduce nonsmokers' exposure to ETS in a variety of public venues, with little evidence of displacement inside homes and cars. Similarly, Nguyen (2013) examines Canadian SFA laws in cars and finds they reduce ETS exposure in cars with no evidence of a substitution to smoking at home. Sims et al. (2012) examine whether a 2007 law banning smoking in nearly all indoor public places and workplaces in England affected nonsmokers' ETS exposure. The authors use data from the Health Survey of England, containing serum cotinine levels. They find that the law both reduced the probability that individuals had a measurable level of cotinine in their systems and reduced cotinine levels. The authors find that the effects are most concentrated in higher socioeconomic status households and in households with no smokers.

62.5 SMOKING-RELATED HEALTH OUTCOMES

Another more recent advance concerns the effects of tobacco control policies on smoking-related health outcomes, in particular adult mortality, pregnancy and child health, obesity, and unintended health consequences of tobacco legislation.

In an early study, Moore (1996) examines the relationship between cigarette excise taxes and mortality using data from 1954 to 1988, a period of relatively few and small tax increases. His findings imply a 10% increase in cigarette excise taxes would save about 6000 lives annually. More recent studies utilize natural experiments that may offer more credible research designs, though studies that estimate this effect remain rare. One example is Bedard and Deschênes (2006), who find that access to cheap cigarettes due to conscripted military service increases premature mortality. More recently, Darden, Gilleskie, and Strumpf (2015) show that the causal impact of smoking on mortality is about half of the observed mean difference in years of life between smokers and nonsmokers, after attempting to correct for nonrandom sorting into smoking.

Recent work has also investigated the impact of SFA laws on health outcomes. For example, Pell et al. (2008) find that an SFA law in Scotland reduced coronary-related hospitalizations, although studies by Shetty, DeLeire, White, and Bhattacharya (2011) and Barr, Diez, Wang, Dominici, and Samet (2012) suggest no such health impacts. In sum, estimating the causal impact of smoking on adult mortality and health outcomes is very difficult from a research design perspective.

Estimating the relationship between tobacco control policies and infant and child health offers more credible research designs. Cigarette tax increases, while babies are in utero decrease maternal smoking and improve birth outcomes (Adams et al., 2012; Colman, Grossman, & Joyce, 2003; Evans & Ringel, 1999), although the effect of recent changes in cigarette taxes appears to be smaller and concentrated among lower socioeconomic status women and infants (Markowitz, Kathleen Adams, Dietz, Tong, & Kannan, 2013). Additionally, cigarette tax increases, while infants are in utero can affect children much later, decreasing sick days from school and doctor visits (Simon, 2016). There is also evidence that cigarette tax increases during early childhood decrease instances of sudden infant death syndrome and asthma (King, Markowitz, & Ross, 2015; Markowitz, 2008; Sabia, 2008).

In addition to health effects directly related to smoking, economists have estimated a number of indirect health effects from tobacco control policies, both positive and negative. On the positive side, evidence suggests cigarette taxes decrease obesity, perhaps through indirect effects on exercise and food consumption. Moreover, the decrease in obesity is largest for individuals with high body mass index (Courtemanche, 2009; Wehby & Courtemanche, 2012). Finally, research suggests that the passage of SFA laws in bars induces smokers to switch to smokeless tobacco and drive longer distances to bars where smoking is still allowed, increasing alcohol-related traffic fatalities (Adams & Cotti, 2008; Adams, Cotti, & Fuhrmann, 2013).

62.6 DIRECTIONS FOR FUTURE RESEARCH

Given our constraints, we have omitted several areas within the economics of smoking. We believe that research in the areas discussed will remain robust and that interest in the newer areas we identified will continue to grow. Naturally, the literature will expand in new directions such as the evaluation of novel technologies, like electronic cigarettes, and the evaluation of more recent government regulations on smoking, including recent proposed regulations regarding cigarette nicotine contents by the Food and Drug Administration. Additionally, research that better addresses issues of smoking measurement will be important contributions. As data with more objective biomarker information become more widely available, it will become possible to obtain better estimates of most of the relationships we discussed and gain a broader picture of corresponding behaviors. Together, these should provide better information to policymakers and drive and improve future research in the economics of smoking.

MINI-DICTIONARY OF TERMS

Tobacco control policies Policies enacted by governments, such as taxes on cigarettes or smoke-free air laws, which aim to reduce smoking prevalence or the possible negative external effects of smoking.

Cigarette excise taxes A tax levied on cigarettes when the cigarettes are manufactured rather than when they are sold.

Smoke-free air laws Laws passed at the local or state level which restrict smoking in certain venues, such as private workplaces, restaurants or bars.

Smoking prevalence The percent of a certain population that currently smokes cigarettes.

Environmental Tobacco Smoke Smoke from cigarettes (or other tobacco products) which is not a result of that individual's smoking behavior.

Key Facts

- Both adult and youth smoking prevalence has fallen in the United States over the past 20 years.
- About 14% of adults and 11% of youths report being smokers as of 2015.
- Econometric issues make identifying the relationships between tobacco control policies and smoking-related outcomes difficult.

- Increases in cigarette taxes have reduced adult smoking and improved child and infant health.
- The effects of increases in cigarette taxes on youth smoking are less clear.

Summary Points

- Cigarette excise taxes and SFA laws reduce smoking participation and cigarette consumption among adults.
- Adult smokers compensate for higher cigarette excise taxes by switching cigarette brands and smoking more intensely, although the best evidence suggests this behavior does not completely offset the reduction in cigarette consumption.
- There is an ongoing debate regarding the effectiveness of cigarette excise taxes in reducing youth smoking.
- Estimating the effects of tobacco control policies on adult mortality and morbidity is challenging, although there is convincing evidence that cigarette taxes improve a wide variety of infant health outcomes and long-run child outcomes.

References

Abrevaya, J., & Puzzello, L. (2012). Taxes, cigarette consumption, and smoking intensity: comment. *American Economic Review, 102*(4), 1751–1763.

Adams, S., & Cotti, C. (2008). Drunk driving after the passage of smoking bans in bars. *Journal of Public Economics, 92*(5–6), 1288–1305.

Adams, S., Cotti, C., & Fuhrmann, D. (2013). Smokeless tobacco use following smoking bans in bars. *Southern Economic Journal, 80*(1), 147–161.

Adams, E. K., Markowitz, S., Kannan, V., Dietz, P. M., Tong, V. T., & Malarcher, A. M. (2012). Reducing prenatal smoking: the role of state policies. *American Journal of Preventive Medicine, 43*(1), 34–40.

Adda, J., & Cornaglia, F. (2006). Taxes, cigarette consumption, and smoking intensity. *American Economic Review, 96*(4), 1013.

Adda, J., & Cornaglia, F. (2010). The effect of bans and taxes on passive smoking. *American Economic Journal: Applied Economics, 2*(1), 1–32.

Barr, C. D., Diez, D. M., Wang, Y., Dominici, F., & Samet, J. M. (2012). Comprehensive smoking bans and acute myocardial infarction among Medicare enrollees in 387 US counties: 1999–2008. *American Journal of Epidemiology, 176*(7), 642.

Bedard, K., & Deschênes, O. (2006). The long-term impact of military service on health: evidence from World War II and Korean War veterans. *The American Economic Review, 96*(1), 176–194.

Bitler, M. P., Carpenter, C. S., & Zavodny, M. (2010). Effects of venue-specific state clean indoor air laws on smoking-related outcomes. *Health Economics, 19*(12), 1425–1440.

Carpenter, C. (2009). The effects of local workplace smoking laws on smoking restrictions and exposure to smoke at work. *Journal of Human Resources, 44*(4), 1023–1046.

Carpenter, C., & Cook, P. (2008). Cigarette taxes and youth smoking: new evidence from national, state, and local youth risk behavior surveys. *Journal of Health Economics, 27*(2), 287–299.

Carpenter, C., Postolek, S., & Warman, C. (2011). Public-place smoking laws and exposure to environmental tobacco smoke (ETS) in public places. *American Economic Journal: Economic Policy, 3*(3), 35–61.

Centers for Disease Control and Prevention. (2007). Reduced secondhand smoke exposure after implementation of a comprehensive statewide smoking ban–New York, June 26, 2003-June 30, 2004. *Morbidity and Mortality Weekly Report, 56*(28), 705–708.

Chaloupka, F. (1991). Rational addictive behavior and cigarette smoking. *Journal of Political Economy, 99*(4), 722–742.

Chaloupka, F. J. (1992). Clean indoor air laws, addiction and cigarette smoking. *Applied Economics, 24*(2), 193–205.

Chaloupka, F., & Grossman, M. (1996). *Price, tobacco control policies and youth smoking.* NBER working paper no. 5740.

Chaloupka, F. J., & Saffer, H. (1992). Clean indoor air laws and the demand for cigarettes. *Contemporary Policy Issues, 10*(2), 72–83.

Chaloupka, F., & Warner, K. (2000). The economics of smoking. In: A. J. Culyer, & J. P. Newhouse (Eds.), *Handbook of health economics* (pp. 1539–1627). Vol. 1B(pp. 1539–1627). New York: North-Holland, Elsevier Science B.V.

Chaloupka, F., & Wechsler, H. (1997). Price, tobacco control policies and smoking among young adults. *Journal of Health Economics, 16*(3), 359–373.

Cheng, K.-W., Liu, F., Gonzalez, M. E., & Glantz, S. (2017). The effects of workplace clean indoor air law coverage on workers' smoking-related outcomes. *Health Economics, 2*(26), 226–242.

Chiou, L., & Muehlegger, E. (2008). Crossing the line: direct estimation of cross-border cigarette sales and the effect on tax revenue. *B.E. Journal of Economic Analysis and Policy: Contributions to Economic Analysis and Policy, 8*(1).

Colman, G., Grossman, M., & Joyce, T. (2003). The effect of cigarette excise taxes on smoking before, during and after pregnancy. *Journal of Health Economics, 22*(6), 1053.

Cotti, C., Nesson, E., & Tefft, N. (2016). The effects of tobacco control policies on tobacco products, tar, and nicotine consumption: evidence from household panel data. *American Economic Journal: Economic Policy, 8*(4), 103–123.

Courtemanche, C. (2009). Rising cigarette prices and rising obesity: coincidence or unintended consequence? *Journal of Health Economics, 28* (4), 781–798.

Darden, M., Gilleskie, D. B., & Strumpf, K. (2015). *Smoking and mortality: new evidence from a long panel.* Working paper.

DeCicca, P., Kenkel, D., & Liu, F. (2013a). Who pays cigarette taxes? The impact of consumer price search. *The Review of Economics and Statistics, 95*(2), 516–529. https://doi.org/10.1162/REST_a_00303.

DeCicca, P., Kenkel, D. S., & Liu, F. (2013b). Excise tax avoidance: the case of state cigarette taxes. *Journal of Health Economics 32*(6), 1130–1141.

DeCicca, P., Kenkel, D., & Mathios, A. (2002). Putting out the fires: will higher taxes reduce the onset of youth smoking? *Journal of Political Economy, 110*(1), 144–169.

DeCicca, P., Kenkel, D., Mathios, A., Shin, Y.-J., & Lim, J.-Y. (2008). Youth smoking, cigarette prices, and anti-smoking sentiment. *Health Economics, 17*(6), 733–749.

Douglas, S. (1998). The duration of the smoking habit. *Economic Inquiry, 36*(1), 49–64.

Douglas, S., & Hariharan, G. (1994). The hazard of starting smoking: estimates from a split population duration model. *Journal of Health Economics, 13*(2), 213–230.

Edoka, I. (2017). Implications of misclassification errors in empirical studies of adolescent smoking behaviours. *Health Economics, 26*(4), 486–499.

Ellingsen, D. G., Fladseth, G., Daae, H. L., Gjolstad, M., Kjaerheim, K., Skogstad, M., et al. (2006). Airborne exposure and biological monitoring of bar and restaurant workers before and after the introduction of a smoking ban. *Journal of Environmental Monitoring, 8*(3), 362–368. https://doi.org/10.1039/b600050a.

Evans, W. N., & Farrelly, M. C. (1998). The compensating behavior of smokers: taxes, tar, and nicotine. *RAND Journal of Economics, 29*(3), 578–595.

Evans, W. N., Farrelly, M. C., & Montgomery, E. (1999). Do workplace smoking bans reduce smoking? *American Economic Review, 89*(4), 728–747.

Evans, W. N., & Ringel, J. S. (1999). Can higher cigarette taxes improve birth outcomes? *Journal of Public Economics, 72*, 135–154.

Farrelly, M. C., Evans, W. N., & Sfekas, A. E. (1999). The impact of workplace smoking bans: results from a national survey. *Tobacco Control, 8*(3), 272–277.

Farrelly, M. C., Nimsch, C. T., Hyland, A., & Cummings, M. (2004). The effects of higher cigarette prices on tar and nicotine consumption in a cohort of adult smokers. *Health Economics, 13*(1), 49–58.

Farrelly, M. C., et al. (2005). Changes in hospitality workers' exposure to secondhand smoke following the implementation of New York's smoke-free law. *Tobacco Control, 14*(4), 236–241.

Hansen, B., Sabia, J. J., & Rees, D. I. (2017). Have cigarette taxes lost their bite? New estimates of the relationship between cigarette taxes and youth smoking. *American Journal of Health Economics, 3* (1), 60–75.

Harding, M., Leibtag, E., & Lovenheim, M. F. (2012). The heterogeneous geographic and socioeconomic incidence of cigarette taxes: evidence from Nielsen Homescan data. *American Economic Journal: Economic Policy, 4*(4), 169–198.

Harris, J. E. (1980). Taxing tar and nicotine. *American Economic Review, 70*(3), 300–311.

Kenkel, D., Lillard, D. R., & Mathios, A. (2004). Accounting for misclassification error in retrospective smoking data. *Health Economics, 13* (10), 1031–1044.

King, C., Markowitz, S., & Ross, H. (2015). Tobacco control policies and sudden infant death syndrome in developed nations. *Health Economics, 24*(8), 1042–1048.

Larsson, M., Boethius, G., Axelsson, S., & Montgomery, S. M. (2008). Exposure to environmental tobacco smoke and health effects among hospitality workers in Sweden–before and after the implementation of a smoke-free law. *Scandinavian Journal of Work, Environment & Health, 34*(4), 267–277.

Lillard, D., Molloy, E., & Sfekas, A. (2013). Smoking initiation and the Iron law of demand. *Journal of Health Economics, 32*(1), 114–127.

Markowitz, S. (2008). The effectiveness of cigarette regulations in reducing cases of sudden infant death syndrome. *Journal of Health Economics, 27*(1), 106–133.

Markowitz, S., Kathleen Adams, E., Dietz, P. M., Tong, V. T., & Kannan, V. (2013). Tobacco control policies, birth outcomes, and maternal human capital. *Journal of Human Capital, 7*(2), 130–160.

Menzies, D., Nair, A., Williamson, P. A., Schembri, S., Al-Khairalla, M. Z., Barnes, M., et al. (2006). Respiratory symptoms, pulmonary function, and markers of inflammation among bar workers before and after a legislative ban on smoking in public places. *JAMA, 296*(14), 1742–1748. https://doi.org/10.1001/jama.296.14.1742.

Moore, M. J. (1996). Death and tobacco taxes. *RAND Journal of Economics, 27*(2), 415–428.

Mulcahy, M., Evans, D. S., Hammond, S. K., Repace, J. L., & Byrne, M. (2005). Secondhand smoke exposure and risk following the Irish smoking ban: an assessment of salivary cotinine concentrations in hotel workers and air nicotine levels in bars. *Tobacco Control, 14*(6), 384–388. https://doi.org/10.1136/tc.2005.011635.

Nesson, E. (2017a). Heterogeneity in smokers' responses to tobacco control policies. *Health Economics, 26*(2), 206–225.

Nesson, E. (2017b). The impact of tobacco control policies on adolescent smoking: comparing self-reports and biomarkers. *American Journal of Health Economics, 3*(4), 507–527.

Nguyen, H. V. (2013). Do smoke-free car laws work? Evidence from a quasi-experiment. *Journal of Health Economics, 32*(1), 138–148.

Nonnemaker, J. M., & Farrelly, M. C. (2011). Smoking initiation among youth: the role of cigarette excise taxes and prices by race/ethnicity and gender. *Journal of Health Economics, 30*(3), 560–567.

Pearson, J., Windsor, R., El-Mohandes, A., & Perry, D. C. (2009). Evaluation of the immediate impact of the Washington, D.C., smoke-free indoor air policy on bar employee environmental tobacco smoke exposure. *Public Health Reports, 124*(Suppl. 1), 134–142.

Pell, J. P., Haw, S., Cobbe, S., Newby, D. E., Pell, A. C. H., Fischbacher, C., et al. (2008). Smoke-free legislation and hospitalizations for acute coronary syndrome. *New England Journal of Medicine, 359*(5), 482–491.

Sabia, J. J. (2008). Every breath you take: the effect of postpartum maternal smoking on childhood asthma. *Southern Economic Journal*, 128–158.

Shetty, K. D., DeLeire, T., White, C., & Bhattacharya, J. (2011). Changes in US hospitalization and mortality rates following smoking bans. *Journal of Policy Analysis and Management, 30*(1), 6–28.

Siegel, M., Albers, A. B., Cheng, D. M., Biener, L., & Rigotti, N. A. (2004). Effect of local restaurant smoking regulations on environmental tobacco smoke exposure among youths. *American Journal of Public Health, 94*(2), 321–325.

Simon, D. (2016). Does early life exposure to cigarette smoke permanently harm childhood welfare? Evidence from cigarette tax hikes. *American Economic Journal: Applied Economics, 8*(4), 128–159.

Sims, M., Mindell, J. S., Jarvis, M. J., Feyerabend, C., Wardle, H., & Gilmore, A. (2012). Did smokefree legislation in England reduce exposure to secondhand smoke among nonsmoking adults? Cotinine analysis from the health survey for England. *Environmental Health Perspectives, 120*(3), 425–430. https://doi.org/10.1289/ehp.1103680.

Tauras, J. A. (2006). Smoke-free air Laws, cigarette prices, and adult cigarette demand. *Economic Inquiry, 44*(2), 333–342.

Wehby, G., & Courtemanche, C. (2012). The heterogeneity of the cigarette price effect on body mass index. *Journal of Health Economics, 31*(5), 719–729.

CHAPTER

63

Resources for the Neuroscience of Nicotine

*Rajkumar Rajendram**,†, Victor R. Preedy*†*

*Department of Medicine, King Abdulaziz Medical City, Ministry of National Guard Health Affairs, Riyadh, Saudi Arabia
†Diabetes and Nutritional Sciences Research Division, Faculty of Life Science and Medicine, King's College London, London, United Kingdom

Abbreviations

HIV	human immunodeficiency virus
AIDS	acquired immune deficiency syndrome
UK	United Kingdom
USA	United States of America
WHO	World Health Organisation

63.1 INTRODUCTION

Nicotine is one of the most commonly used recreational drugs, and "smoking" colloquially describes the consumption of nicotine by inhaling the fumes from burning tobacco. Tobacco may also be chewed or sucked. Currently, there are over a billion people who use tobacco-related products (WHO, 2017).

Nicotine has been used as a stimulant for thousands of years (Dani & Balfour, 2011). Nicotine has a plethora of neuroactive effects and is highly addictive (Dani & Balfour, 2011). Use of nicotine has occurred since time immemorial. However, the understanding of the mechanisms relevant to the neurological effects of nicotine and their treatments are relatively recent discoveries in the overall timescale of nicotine research.

By the beginning of the 20th century, nicotine had been synthesized, and John Newport Langley had started studying its actions (Dani & Balfour, 2011; Langley, 1905; Maehle, 2004). In the early 1900s, their studies led to the discovery of what Langley referred to as the "receptive substance" (Langley, 1905). This was eventually found to be nicotinic acetylcholine receptors (Dani & Balfour, 2011).

However, by being addicted to nicotine, there is also exposure to the harmful components in tobacco and related products. The World Health Organization has reported that tobacco use causes the deaths of 7 million people annually (WHO, 2017). This is about 1 in 10 deaths, more than the combined deaths due to HIV/AIDS, malaria, and tuberculosis (WHO, 2017).

Since 1905, when it was recognized that nicotine activates receptors, there has been an explosion in the knowledge and understanding of the neuroscience of nicotine. It is now difficult even for experienced scientists to remain up-to-date. For those new to the field, it is difficult to know which in the myriad of available sources are reliable. To assist colleagues who are interested in understanding more about the neuroscience of nicotine, we have therefore produced tables containing reliable, up-to-date resources in this chapter. Just over 40 experts assisted with the compilation of these tables of resources and are acknowledged below.

Tables 63.1–63.4 list the most up-to-date information on the regulatory bodies and professional societies (Table 63.1), journals on the neuroscience of nicotine (Table 63.2), books (Table 63.3), and online resources (Table 63.4) that are relevant to an evidence-based approach to the neuroscience of nicotine.

MINI-DICTIONARY OF TERMS

Smoking The consumption of nicotine by inhaling the fumes of burning tobacco.

Acetylcholine Acetylcholine is an organic chemical which is a neurotransmitter. There are two main types of cholinergic receptor; nicotinic and muscarinic.

Nicotinic Acetylcholine receptors These transmembrane protein receptors are activated by acetylcholine. They are also activated by nicotine.

Key Facts

- Nicotine has been used as a stimulant for thousands of years.

https://doi.org/10.1016/B978-0-12-813035-3.00063-0

TABLE 63.1 Regulatory Bodies, Professional Societies, and Organizations

Action on Smoking and Health	www.ash.org.uk
Addiction Center	www.addictioncenter.com/nicotine
African American Tobacco Control Leadership Council	www.savingblacklives.org
Agency for Healthcare Research and Quality	www.ahrq.gov
Alcohol and Drug Foundation	adf.org.au
American Association for Cancer Research	www.aacr.org/Pages/Home.aspx
American Cancer Society	www.cancer.org
American College of Neuropsychopharmacology	acnp.org
American Heart Association	www.heart.org/hearorg
American Lung Association	www.lung.org
American Nurses Association	www.ncsbn.org
American Psychiatric Association	www.psychiatry.org
American Society for Health Economics	www.ashecon.org
American Society of Addiction Medicine	www.asam.org
American Thoracic Society	www.thoracic.org
Association for Addiction Professionals	www.naadac.org
Australian National Health and Medical Research Council	www.nhmrc.gov.au
Australian Bureau of Statistics	www.abs.gov.au
Australian Government Quit Now	www.quitnow.gov.au
Campaign for Tobacco-Free Kids	www.tobaccofreekids.org
Canadian Centre on Substance Use and Addiction	www.ccdus.ca
Cancer Council Australia	www.cancer.org.au
Cancer Council Australian Capital Territory	www.actcancer.org
Cancer Institute New South Wales	www.icanquit.com.au
Cancer Council Western Australia	www.cancerwa.asn.au
Centre for Addiction and Mental Health	www.camh.ca
Chinese Association on Tobacco Control	www.catcprc.org.cn
College on Problems of Drug Dependence	www.cpdd.org
Duke University Center for Smoking Cessation	www.dukesmoking.com
The Epilepsy Centre	epilepsycentre.org.au
Epilepsy Foundation	www.epilepsy.com
Epilepsy Society	www.epilepsysociety.org.uk
Food and Drug Administration (United States)	www.fda.gov
Food and Drug Administration (United States), Center for Tobacco Products	www.fda.gov/aboutfda/centersoffices/officeofmedicalproductsandtobacco/aboutthecenterfortobaccoproducts/default.htm
Foundation for a Smokefree America	www.anti-smoking.org
Foundation for a Smoke-Free World	www.smokefreeworld.org
International Health Economics Association	www.healtheconomics.org
Malaysian Pharmaceutical Society, Certified Smoking Cessation Service Provider	www.acadpharm.org.my/index.cfm?&menuid=2
Mayo Clinic	www.mayoclinic.org

TABLE 63.1 Regulatory Bodies, Professional Societies, and Organizations—cont'd

National Tobacco Cessation Collaborative	www.tobacco-cessation.org
National Association for Alcoholism and Drug Abuse Counselors	www.naadac.org
The National Center on Addiction and Substance Abuse	www.centeronaddiction.org
National Health Service	www.nhs.uk
National Institute on Drug Abuse	www.drugabuse.gov
Nicotine Anonymous	nicotine-anonymous.org
Queensland Cancer Council	www.cancerqld.org.au
QuitAssist	www.quitassist.com
QuitCoach	www.quitcoach.org.au
QuitNet	quitnet.meyouhealth.com
Quit South Australia	www.quitsa.org.au
Quit Victoria	www.quit.org.au
Smokefree	www.nhs.uk/smokefree
Smokefree Action Coalition	www.smokefreeaction.org.uk
Society for Neuroscience	www.sfn.org
Society for Research on Nicotine and Tobacco	www.srnt.org
Society for the Study of Addiction	www.addiction-ssa.org
Spanish Society of Stop-Smoking Specialist	www.sedet.org
Substance Abuse and Mental Health Services Administration	www.samhsa.gov
The Loop	theloop.ucsf.edu
Tobacco Centers of Regulatory Science	prevention.nih.gov/tobacco-regulatory-science-program/research-portfolio/centers
Truth Initiative	truthinitiative.org
United States Department of Health & Human Services	betobaccofree.hhs.gov/health-effects/nicotine-health/index.html
Vermont Addiction Professional Association	vapavt.org
World Bank	www.worldbank.org
World Health Organization	www.who.int

This table lists the regulatory bodies, professional societies, and organizations involved with tobacco, nicotine, and addictions. Some of these sites are very comprehensive in that they offer advice, resources, and other information related to nicotine and its use. While some sites are country-specific, some information contained within these sites are useful for other countries. Some resources are also identified in Table 63.4. It is also important to point out that the psychoactive or neurological effects of nicotine should not be considered in isolation as tobacco usage has consequences for a variety of disease states such as cardiovascular disease and cancers. Indeed, some sites relating to cancers have also been identified. Please note, occasionally, the location of the websites or web address changes.

TABLE 63.2 Journals Relevant to the Neuroscience of Nicotine

Nicotine and Tobacco Research
Addictive Behaviors
Psychopharmacology
Drug and Alcohol Dependence
PLOS One
Addiction
Preventive Medicine

Continued

TABLE 63.2 Journals Relevant to the Neuroscience of Nicotine—cont'd

Pharmacology Biochemistry and Behavior
Tobacco Control
Brain Research
BMC Public Health
Neuropharmacology
Journal of Pharmacology and Experimental Therapeutics
European Journal of Pharmacology
Journal of Neuroscience
American Journal of Public Health
Substance Use and Misuse
Journal of Adolescent Health
American Journal of Preventive Medicine
Neuropsychopharmacology

Journals publishing original research and review articles related to nicotine- and tobacco-related products in connection with neuroscience and treatments. Included in this list are the top 20 journals that have published the most number of articles over the past 5 years. *Nature Medicine*, *PLOS One*, and *The New England Journal of Medicine* were also recommended by authors (though only *PLOS One* appears in this list).

TABLE 63.3 Relevant Books Relevant to the Neuroscience of Nicotine

Addiction Biology. Simonnet A, Cador M and Caillé S. Wiley, 2013, UK
Cigarettes, Nicotine and Health. A Biobehavioral Approach. Kozlowski LT, Henningfield JE, Brigham J. SAGE Publications, 2001, UK
Clearing the Smoke: Assessing the Science Base for Tobacco Harm Reduction. Stratton K, Shetty P, Wallace R, et al., (Editors). Institute of Medicine (USA) Committee to Assess the Science Base for Tobacco Harm Reduction, National Academies Press (USA) 2001, USA
Ending the Tobacco Problem: A Blueprint for the Nation. Institute of Medicine. The National Academies Press, 2007, USA
Golden Holocaust: Origins of the Cigarette Catastrophe and the Case for Abolition. Proctor RN. University of California Press, 2012, USA
Handbook of Health Economics, Vol. 1. Culyer A, Newhouse J. Elsevier, 2000, UK
Handbook of Health Economics, Vol. 2. Pauly M, Mcguire T, Parros P. Elsevier, 2011, UK
Helping the Hard-Core Smoker: A Clinician's Guide. Seidman DF, Covey LS. Lawrence Erlbaum Associates, 1999, USA
Individual Decisions for Health. Lindgren B. Routledge, 2002, UK
Methods in Molecular Biology. Caillé S, Clemens K, Stinus L, Cador M. Springer, 2012, Germany
Negative Affective States and Cognitive Impairments in Nicotine Dependence. Hall FS, Young JW, Der-Avakian A. Academic Press, 2017, UK
Neuronal Nicotinic Receptors. Clementi F, Fornasari D, Gotti C. Springer, 2000, Germany
Neuropathology of Drug Addiction and Substance Misuse Volume 1: Foundations of Understanding, Tobacco, Alcohol, Cannabinoids and Opioids. Preedy VR (Editor). Academic Press, 2016, UK
Neuropharmacology. Simonnet A, Zamberletti E, Cador M, Rubino T, Caillé S. Elsevier 2017, The Netherlands
Nicotine. Hens G, Self W, Calleja J. Other Press, 2017, USA
Nicotine Addiction: Prevention, Health Effects and Treatment Options. Di Giovanni G. Nova Science Publishers, 2012, USA
Nicotine Addiction: Principles and Management. Orleans CT, Slade JD. Oxford University Press, UK
Nicotine and Other Tobacco Compounds in Neurodegenerative and Psychiatric Diseases. Xia W, Phillips B, Wong ET, Ho J, Oviedo A, Hoeng J, Peitsch M. Academic Press, 2018, UK
Nicotine in Psychiatry: Psychopathology and Emerging Therapeutics. Piasecki M, Newhouse PA. American Psychiatric Publishing, 2000, USA
Nicotine Psychopharmacology. Henningfield JE, Calvento E, Pogun S. Springer-Verlag, 2009, Germany
Nicotinic Acetylcholine Receptor Technologies. Li M (Editor), Springer, 2016, USA

TABLE 63.3 Relevant Books Relevant to the Neuroscience of Nicotine—cont'd

Nicotinic receptors. Lester RAJ, Humana Press, 2014, USA
Nicotinic Receptors in the Nervous System. Levin ED (Editor). CRC Press, 2001, USA
Preventing Tobacco Use Among Youth and Young Adults: A Report of the Surgeon General. National Center for Chronic Disease Prevention and Health Promotion (USA) Office on Smoking and Health. Centers for Disease Control and Prevention (USA), 2012, USA
Progress in Respiratory Research. Caillé S, Baker AL, Todd J, Turner A, Dayas CV. Karger, 2015, Europe
Psychopharmacology. Clemens KJ, Caillé S, Cador M. European Behavioural Pharmacology Society, 2010, Europe
Public Health Implications of Raising the Minimum Age of Legal Access to Tobacco Products. Committee on the Public Health Implications of Raising the Minimum Age for Purchasing Tobacco Products; Board on Population Health and Public Health Practice; Institute of Medicine; Bonnie RJ, Stratton K, Kwan LY (Editors). National Academies Press (USA), 2015, USA
Research Report Series: Tobacco Addiction. National Institute on Drug Abuse, US Department of Health and Human Services, National Institutes of Health, 2012, USA
Smoking Cessation. Lewis KE. Oxford University Press, 2010, UK
Smoking: Risk, Perception, and Policy. Slovic P. SAGE, 2001, USA
The Addicted Brain: Why We Abuse Drugs, Alcohol, and Nicotine. Kuhar M. Pearson FT Press, 2012, USA
The Cigarette Book: The History and Culture of Smoking. Harrald C, Watkins F. Skyhorse Publishing, 2010, USA
The Cigarette Century: The Rise, Fall, and Deadly Persistence of the Product that Defined America. Brandt A. Basic Books, 2009, USA
The Clinical Management of Nicotine Dependence. Cocores JA. Springer-Verlag, 1991, USA
The Economics of Tobacco and Tobacco Control. US National Cancer Institute (NHU) and WHO. WHO, 2017, Switzerland
The Health Consequences of Smoking—50 Years of Progress: A Report of the Surgeon General. National Center for Chronic Disease Prevention and Health Promotion (USA) Office on Smoking and Health. Centers for Disease Control and Prevention (USA), 2014, USA
The Molecular Basis of Drug Addiction, Volume 137. Rahman S. Elsevier, 2016, USA
The Motivational Impact of Nicotine and its Role in Tobacco Use. Bevins RA, Caggiula AR. Springer-Verlag, 2009, USA
The Neurobiology and Genetics of Nicotine and Tobacco. Balfour D, Munafo M. Springer International Publishing, 2015, Switzerland
The Neuropharmacology of Nicotine Dependence. Balfour DJK, Munafo MR. Springer, 2015, Switzerland
The science of addiction: From neurobiology to treatment. Erickson CK. WW Norton & Company, 2007, USA
This is Nicotine (Addiction). Farrington K. Sanctuary Publishing, 2002, UK
Tobacco: A Cultural History of How an Exotic Plant Seduced Civilization. Gately I. Grove Press, 2002, USA
Treating Nicotine Dependence with Nitrous Oxide/Oxygen (PAN): A manual for Health Professionals. Gillman M. Cerebrum Publishers, 2010, South Africa
Women and Smoking: A Report of the Surgeon General. Office on Smoking and Health (USA). Centers for Disease Control and Prevention (USA), 2001, USA

This table lists books on the neuroscience of nicotine.

TABLE 63.4 Relevant Online Resources and Information on Emerging Techniques

Agency for Healthcare Research and Quality: Pregnancy resources	www.ahrq.gov/professionals/prevention-chronic-care/healthier-pregnancy/preventive/tobacco.html
American Cancer Society: Stopping smoking resources	www.cancer.org/healthy/stay-away-from-tobacco/guide-quitting-smoking.html?from=fast
American Lung Association: Stopping smoking resources	www.lung.org/stop-smoking
American Society of Addiction Medicine: Nicotine and Tobacco resources	www.asam.org/advocacy/find-a-policy-statement/view-policy-statement/public-policy-statements/2011/12/15/nicotine-addiction-and-tobacco
Ashtray Blog	www.ecigarettedirect.co.uk/ashtray-blog/2016/10/nicotine-overdose-vaping.html

Continued

TABLE 63.4 Relevant Online Resources and Information on Emerging Techniques—cont'd

Centers for Disease Control and Prevention (CDC): Fact Sheet—Quitting Smoking—Smoking & Tobacco Use	www.cdc.gov/tobacco/data_statistics/fact_sheets/health_effects/effects_cig_smoking/index.htm
Centers for Disease Control and Prevention (CDC): Tobacco resources	www.cdc.gov/tobacco/campaign/tips/partners/health/hcp/index.html
Food and Drug Administration: Public health resources	www.fda.gov/TobaccoProducts/PublicHealthScienceResearch/default.htm
Lase Ray	www.laserrayli.com
Luyang Health Clinic Android App mQuit	play.google.com/store/apps/details?id=appinventor.ai_h2ocfh2ocf.JomQuitTRIAL&hl=en
Mayo Clinic: Nicotine-dependence resources	www.mayoclinic.org/diseases-conditions/nicotine-dependence/symptoms-causes/syc-20351584
Medscape	www.medscape.com
National Center on Addiction and Substance Abuse: E-cigarette resources	www.centeronaddiction.org/e-cigarettes
National Health Service Inform: Stopping smoking resources	www.nhsinform.scot/healthy-living/stopping-smoking/reasons-to-stop/tobacco
Office champions tobacco cessation behavioral health facts	www.aafp.org/patient-care/public-health/tobacco-nicotine/office-champions/behavioral-health.html
Smoking Cessation Guidelines for Australian General Practice	www.health.gov.au/internet/main/publishing.nsf/Content/health-pubhlth-publicat-document-smoking_cessation-cnt.htm
Tobacconomics	tobacconomics.org
Tobacco-Related Disease Research Program	www.trdrp.org
WebMD	www.webmd.com
World Health Organization: Tobacco	www.who.int/mediacentre/factsheets/fs339/en

This table lists some Internet resources and other relevant materials in relation to the neuroscience of nicotine and treatment for addictions to nicotine. Some of these sites are also listed in Table 63.1. Please note, occasionally, the location of the websites or web address changes.

- The understanding of the mechanisms relevant to the effects of nicotine and their treatments are relatively recent discoveries in nicotine research.
- Since 1905, it was recognized that nicotine activates receptors, there has been an explosion in the knowledge and understanding of the neuroscience of nicotine.
- To assist colleagues who are interested in understanding more about the neuroscience of nicotine, we have therefore produced tables containing reliable, up-to-date resources in this chapter.

Summary Points

- Nicotine is one of the most commonly used recreational drugs.
- There are over 1 billion tobacco users.
- Nicotine is contained in tobacco products that can be either chewed, sucked, or smoked. More recently, vaping has become popular whereby nicotine is obtained via a liquid in an electronic cigarette.
- Nicotine has many neuroactive effects and is highly addictive.
- The mechanisms responsible for the neurological effects of nicotine and their treatment are relatively recent discoveries in the timescale of tobacco usage.
- This chapter lists the resources on the regulatory and professional bodies, journals, books, and websites that are relevant to an evidence-based approach to the neuroscience of nicotine.

Acknowledgements (in alphabetical order)

We would like to thank the following authors for contributing to the development of this resource.

Alkam T, Bagdas D, Beeler J, Budzynska B, Hahn B, Michalak A, Caille S, Chukwueke C, Chylińska-Wrzos P, Cole R, Corsini S, DeCicca P, Nesson E, Ferretti F, Gipson C, Henderson B, Hritcu L, Ibañez-Tallon I, Jodłowska-Jędrych B, Khoo S, Lis-Sochocka M, McNally G, Clemens

K, Kumari V, Liu Y, Macedo D, McCallum S, Nabeshima T, Eggan B, McGrath D, Morales A, Nadalin S, Jakovac H, Nistri A, Richter L, Saad S, Chen H, Schmidt H, Siraj MA, Sudweeks S, Sui CF, and Wawryk-Gawda E.

References

Dani, J. A., & Balfour, D. J. (2011). Historical and current perspective on tobacco use and nicotine addiction. *Trends in Neurosciences*, *34*, 383–392.

Langley, J. N. (1905). On the reaction of cells and of nerve-endings to certain poisons, chiefly as regards the reaction of striated muscle to nicotine and to curari. *The Journal of Physiology*, *33*, 374–413.

Maehle, A. H. (2004). "Receptive substances": John Newport Langley (1852–1925) and his path to a receptor theory of drug action. *Medical History*, *48*, 153–174.

WHO (2017). *Who report on the global tobacco epidemic*. Geneva: World Health Organization.

Index

Note: Page numbers followed by *f* indicate figures and *t* indicate tables.

O

CPI Antony Rowe
Eastbourne, UK
June 18, 2019